SELECTED PAPERS

S. Chandrasekhar

SELECTED PAPERS

S. Chandrasekhar

VOLUME 6

The Mathematical Theory of Black Holes and of Colliding Plane Waves

THE UNIVERSITY OF CHICAGO PRESS

CHICAGO AND LONDON

The University of Chicago Press, Chicago 60637
The University of Chicago Press, Ltd., London

Printed in the United States of America

00 99 98 97 96 95 94 93 92 91 5 4 3 2 1

Acknowledgments appear on page 739.

Library of Congress Cataloging in Publication Data

Chandrasekhar, S. (Subrahmanyan), 1910–
The mathematical theory of black holes and of colliding plane waves / S. Chandrasekhar.
p. cm. — (Selected papers ; v. 6)
Includes bibliographical references.
1. Black holes (Astronomy) 2. Gravity waves. I. Title.
II. Series: Chandrasekhar, S. (Subrahmanyan), 1910– Selections.
1989 ; v. 6.
QB843.B55C483 1991 90-11149
523.8'875—dc20 CIP
ISBN 0-226-10100-2 (vol. 6)
ISBN 0-226-10101-0 (vol. 6, pbk.)

∞The paper used in this publication meets the minimum requirements of the American National Standard for Information Sciences–Permanence of Paper for Printed Library Materials, ANSI Z39.48-1984.

S. Chandrasekhar is the Morton D. Hull Distinguished Service Professor Emeritus in the Department of Astronomy and Astrophysics, the Department of Physics, and the Enrico Fermi Institute at the University of Chicago.

Contents

PART TWO

The Mathematical Theory of Colliding Waves

PART THREE

Cylindrical Waves

PART FOUR

The Two-Center Problem

PART FIVE

Additional Papers

Foreword

S. Chandrasekhar on black holes and gravitational waves: What a story! Forty years after his pioneer investigations, Chandra returned to the study of the dead stars. He had told us back in 1931 that massive stars could not die as ordinary objects, consisting of regular matter, could. Eddington would rebuff the idea as absurd; the scientific community would side mostly with Eddington; and, although the Schwarzschild solution was waiting around for its correct interpretation, the development of black holes would be delayed by a few decades. So, although all the ingredients were there and we did not have to wait too long, it was not until the sixties that we would incorporate black holes into our world view.

Most of the research on the theory of black holes, including the casting of their name, occurred from the early sixties through the middle seventies. There was the discovery of the Kerr solution, the Penrose-Hawking-Geroch singularity theorems, the Penrose extraction of energy, the Teukolsky separation of the perturbation equations, the Christodoulou-Bekenstein-Hawking thermodynamics, and the Israel-Robinson proof of the uniqueness of the Kerr solution. After 1974, the general interest shifted to "quantum black holes" and "everybody" would concentrate on the Hawking radiation and semiclassical methods.

As usual, Chandra was in a phase difference with his contemporaries: a few decades lead in the prediction, with almost a decade's lag in the mathematical analysis of the theory of black holes. Chandra had always been interested in the stability of astrophysical configurations. He had recently completed his investigations of the stability of rotating stars, and was at Caltech when Teukolsky separated the equations for gravitational perturbations. So it seemed natural for Chandra to embark on an investigation of the perturbations of black holes by developing *ab initio* the theory, and eventually putting his benchmark on the field by writing another book. Instead of going through the scientific context of each paper, I will make some general remarks about the entire work.

The initial challenge for Chandra was the numerically established equivalence of the scattering problems for the odd and the even parity perturbations (now referred to as the axial and polar perturbations) of the Schwarzschild solution. So he worked out by himself the metric and the Newman-Penrose perturbations and, through his transformation theory, established analytically the equivalence of the two scattering problems. Next, the Kerr perturbations evolved, step by step, from the "easy" ones (perturbations of external fields, potential barriers for their scatterings, axisymmetric gravitational perturbations) to the more difficult ones, which culminated in two major papers on the Newman-Penrose gravitational perturbations of the Kerr space-time (papers 10 and 11 in this volume). In the next couple of years these papers were supplemented and rectified by three additional contributions (papers 12, 13, and 18), establishing identities among the Teukolsky functions, completing the solution, and analyzing the algebraically special perturbations.

Completeness of the subject was the main motive for Chandra's consideration of the perturbations of the Reissner-Nordström black hole. He considered both the metric and the Newman-Penrose perturbations, established the eqivalence of the axial and the polar scatterings, and investigated issues such as the decoupling of gravitational and electromagnetic waves and the transformation of one kind of wave to the other in the scattering process.

Actually, completion of the theory was never achieved. Chandra wanted—and he tried hard—to decouple and/or separate the perturbations of the Kerr-Newman black hole. It is well known in the scientific community that he did try and did fail. Since Chandra failed, no one seems willing to give this problem a serious try, and the perturbations of the Kerr-Newman solution have remained an unsolved problem for the dozen years since Chandra gave up. Perhaps, for the sake of science, he should have kept his failure secret. On the other hand, it is very likely that the Kerr-Newman perturbations cannot be separated, and his documented failure will have saved many scientist-years of fruitless effort.

The first two papers on the gravitational perturbations of the Kerr black hole (papers 10 and 11) probably represent the peak of his scientific efforts. The circumstances under which Chandra wrote these papers should perhaps be revealed.

In 1976 Chandra was mainly preoccupied with the gravitational perturbations of the Kerr solution. In addition, while teaching a graduate course in early 1977, he came up with his favored derivation of the Kerr solution (paper 4). In the spring of 1977, he worked hard preparing three manuscripts—one on the Kerr solution, and two on gravitational perturbations. (He then thought that these were to be the completion of the Kerr perturbations.) It is worth noting the submission dates for these papers—April 18, May 2, and June 20, all in 1977. During the second week of August, the GRG-8 conference took place in Waterloo, Canada, and there Chandra gave a plenary talk on his recent work. He shared the drive back to Chicago with Gary Horowitz, and disappeared immediately after!

As we discovered later, in the early spring of 1977 Chandra (who had suffered a heart attack in late 1974) learned that he had a serious heart problem that most probably would require open heart surgery. But how? The Kerr perturbations were not completely solved; the manuscripts had not yet been prepared; and GRG-8 was only a few months away. To Chandra the answer was obvious: he was not yet ready to submit to the surgery, and made it clear to his doctors that they must wait! In mid-August he entered the hospital, and, although the operation was successful, every conceivable complication described in the medical books did in fact occur. Eventually the doctors made it clear to him that his recovery depended solely upon his will to recover. Now, twelve years later, we have written records of his amazing will: *The Mathematical Theory of Black Holes,* the work on the colliding waves, and the studies of the two black hole solutions. His lecture at the Gibbs Symposium (paper 37) presents Chandra's current views on relativity.

Of all his work on black holes, the separation of the Dirac equation still remains, I believe, closest to Chandra's heart. For instance, it is very pleasing to him that through this work it would later be learned that the Dirac equation can be separated in Minkowski space-time, in prolate spheroidal coordinates, and also that the development of a new theory on the separability of partial differential equations would be initiated. Moreover, the recollection that he actually started and finished the separation in the same evening provides great satisfaction.

Judging it retrospectively and comparing it with his life-long previous scientific work, Chandra would describe his 1974–1981 work on black holes as his best scientific effort: it required much harder work and a great deal more effort than all the previous fields he had investigated. I was quite surprised to hear him say that his efforts on colliding waves were comparable to the effort required by the study of black holes. However, although he came to be very enchanted with the beauty of general relativity, his work on radiative transfer would always occupy a central place in his heart.

The study of black holes meant a sacrifice for Chandra. Being an astrophysicist of the real stars, he would always consider perfect fluid solutions of the field equations. The vacuum Einstein equations became relevant to him only because they were relevant to black holes. This sacrifice revealed to him the geometrical structure and the richness of Einstein's theory. Then it became clear to him that he should consider charged black holes as well; and the very most his astrophysical conscience would tolerate was an electromagnetic field much weaker than the gravitational field. "Downhill, all the way down" is a favorite expression of his, describing his gradual change in interests: from the stars to black holes, from the fluid configurations to the vacuum space-times and then the electromagnetic ones, and, finally, from the approximate techniques to the exact solutions that were the only means for studying colliding waves. Chandra accepted his downhill slide, and soon became fascinated with the simple exact solutions that describe all sorts of different phenomena in the collisions of plane waves. I can still recall the smile on his face when it was pointed out to him that the real bottom of his slide was the paper on the two black holes attached to strings (paper 33), where the charges of the black holes are actually larger than their masses (but no naked singularities occur). What good fortune to ride with Chandra on his slide downhill!

It could be said that Chandra's work on black holes was one of his scientific objectives. Having predicted them from astrophysical considerations before anybody else, he had been learning relativity all these years. And gradually, from the post-Newtonian approximations to the stability of rotating stars, it was natural for him to move on to the perturbations of black holes. In fact it was his interest (in the sixties) in the ellipsoidal figures of equilibrium, and his determination that this theory of Riemann, Jacobi, MacLaurin, and Dedekind should be completed and presented in a unified treatment, that postponed for several years his entry into the realm of relativistic black holes. The work was done from 1974 to 1980, culminating in the writing of his book in 1980–81.

Contrary to the work on single black holes, there were no blueprints for Chandra's work on colliding waves and on two black holes. These papers were not supposed to have existed or to have been included in this volume. Before explaining how they came into being, let me mention a discourse that took place in Thessaloniki in 1981, while Chandra was finishing his book. According to Chandra,

> When I was eighteen, back in India, I had submitted a paper to the Royal Society in London. The foreign mail would arrive in the central post office every Saturday, but it would not be delivered until the following week. So, every Saturday, I would ride my bike for about six miles to the post office, in order to check whether the answer from the Royal Society had arrived and to learn whether or not the paper had been accepted. You see, a couple of months ago—and some fifty-odd years later—I was expecting, with equal anxiety, the referee's report from the Royal Society for my latest paper! Well, it's ridiculous to feel the same way fifty years later, when you're in your seventies. You should grow old; you should *live* all the stages of your life. I want to grow old! I should retire from science after I finish my book.

Well, Chandra managed to grow a little bit—but not much—older, in the intervening eight years. In 1983, after he finished his book on black holes, the gravitational waves began colliding. They were coupled with electrodynamic and hydrodynamic waves in 1984 and 1985. They developed Cauchy horizons in 1986, and they were almost plane waves in 1987 and 1988. Some cylindrical waves appeared, and the number of black holes increased from one to two. The visits to Chicago by his collaborators also increased, and his one-week visits to Rome (one) and Iraklion (four), where he would usually complete the writing of a paper, became more frequent. Therefore, although Chandra obtained the status of Emeritus Professor in 1986 (by terminating an indefinite appointment with the University of Chicago because "it is better to leave when everybody asks 'why are you leaving?' than to stay while everybody wonders 'when is this guy thinking of retiring?'"), I do not think he managed to grow much older. During these years he became fascinated with Newton and began translating key propositions from the *Principia* into contemporary mathematical language. His "Newton's Selected Works" are not included in this collection; they will be the subject of a forthcoming book. So let us proceed with the work on colliding waves.

Two letters from Europe stimulated Chandra's slide to the exact solutions of colliding waves. First was a communication received in the late seventies, from Yavuz Nutku, a former student of Chandra's, informing him that the solution for colliding plane impulsive gravitational waves with nonparallel polarizations is described by the simplest solution of the (X, Y) equations, recently formulated by Chandra (paper 4). The second was a letter from Roger Penrose, dated February 1984, in which Penrose wondered how to describe collisions of impulsive gravitational waves coupled with electromagnetic waves. Since the energy-momentum tensor of the electromagnetic field is quadratic on the field, impulsive electromagnetic waves would carry infinite energy per unit area. Therefore, for the energy to be finite, one would like to demand that the energy-momentum itself should be impulsive (meaning proportional to Dirac's δ-function). But then the electromagnetic field should behave like the square root of the δ-function, and one would be left wondering how to interpret such fields and how general relativity would cope with all of these difficulties.

After his book appeared in print in the spring of 1983, Chandra was thinking about Nutku's letter. He had always wanted to understand the solutions describing collisions of plane waves, and that meant that he had to work them out *ab initio,* in his own language. By autumn, he had actually done some preliminary calculations. I recall the day I telephoned him—the day he was awarded the Nobel prize. Before I even had the chance to congratulate him, he shifted the discussion to colliding waves, in which he was quite interested and about which I knew nothing.

In the summer of 1983 Chandra had met Valeria Ferrari in Rome, and had invited her to spend some time in Chicago. In October, Valeria unexpectedly appeared in his office, a telex from Rome never having been delivered to Chandra. This was three days after the announcement from Stockholm. He suggested that they reformulate the problem for colliding waves, to take advantage of the experience obtained with the stationary axisymmetric solutions, and rederive the space-time describing the colliding impulsive waves (the Nutku-Halil solution) in this new formalism. In the next two weeks, and simultaneously with the aftershocks of the award of a Nobel prize, the first paper in the series (paper 19) was almost completed. Later, it would always be referred to by Chandra as "the Valeria paper."

Next, Chandra started thinking about the collisions of coupled gravitational and electromagnetic waves. In fact, he actually postponed his departure for Stockholm in order to have a chance to discuss the problem with Penrose, who was then visiting Chicago. This initiated Penrose's letter of February, 1984. The actual work (paper 20) was done in the summer of that year, and it turned out to be more difficult than anticipated.

In addition to the formidable analytical work that was required, it involved, for instance, the conceptual difficulty of accepting a nonflat preinteraction region. But Chandra "knew" that one had to construct the charged version of the Nutku-Halil solution, in complete mathematical analogy with the Kerr-Newman black hole being the charged version of the Kerr black hole. In the end, it was greatly rewarding to realize that general relativity would overcome the difficulty associated with the squared root of the δ-function, by confining the δ-function singularities only in the Weyl part of the curvature.

The first paper on collisions involving fluids (paper 21) was the most difficult conceptually. Here we had to convince ourselves that we must accept a different equation of state for the fluid in the pre- and the afterinteraction regions. It was also difficult to accept and establish that the collisions of plane gravitational waves may actually lead to the formation of Cauchy horizons (paper 23). We—together with everybody else—were expecting a curvature singularity. It was very surprising when computer evaluations of curvature scalars showed them to remain finite on the surface where they were expected to become singular. It was exceedingly difficult for Chandra to be convinced that something novel and interesting might be happening. However, on my next visit to Chicago, in January 1986, I found that Chandra had actually evaluated the curvature scalars in a very elegant way, with pen and paper, so that they could be verified by ordinary humans. In all our joint work, the only equations that Chandra did not work out by himself are equation 30 of paper 23 and equations 37 and 72 of paper 33.

During the years of the colliding waves, Chandra developed a new habit: He would fly to wherever Penrose happened to be to consult with him *for one day.* Upon his return he would always be very enthusiastic about the successful trip. In fact, in one case (see section 7 of paper 23), he worked out Penrose's suggestions on the return flight, and the problem was solved before he landed at Chicago. Chandra undertook these trips three times—once to Houston (1985) and twice to Oxford (1986, 1988). Penrose's suggestions were always very constructive, and are acknowledged in papers 21, 23, and 26.

1987 and 1988 it became clear that the work on the colliding waves would not be Chandra's last contribution to general relativity. This happened when he began considering cylindrical waves and, later, two black holes. My only regret is that I cannot tell "a story" for this period—somehow everything proceeded quite fast.

The cylindrical waves were studied in parallel with the late phases of the colliding plane waves. It came naturally for Chandra to investigate cylindrical waves, since they are also described by space-times possessing two spacelike Killing fields. The first paper (paper 29) exhibits solutions describing stationary monochromatic cylindrical gravitational waves, possibly coupled with electromagnetic waves. The reader can refer to appendix B of that paper to obtain an idea of Chandra's scholarship. In the second paper with Ferrari, they investigated the dispersion of a cylindrical wave packet that was initially impulsive. They provided a very nice solution, expressed in terms of convergent integrals of divergent functions.

The papers on the two black holes were written, I feel, in an effort by Chandra to fulfill an old interest while disassociating himself from the waves that kept colliding with ". . . increasing frequency." He once said, "If I were younger, I would write a book [on colliding waves]," but certainly he did not want to find himself involved in such a project. The study of the Majumdar-Papapetrou solutions turned out to be very difficult, analytically and conceptually. However, he pursued the problem with a youth's persistence, and after an excruciating analysis he succeeded in describing the scattering of gravitational and electromagnetic waves by two black holes as a six-channel scattering problem. On the other hand, when the two black holes were attached to cosmic strings, the work was completed relatively easily, and no visit to Penrose was ever contemplated. In fact, Chandra was so excited about that solution (paper 33) that he even went so far as to ask the Royal Society whether it would be possible to expedite its publication. And they did!

How does one finish the foreword of the last volume of Chandrasekhar's selected works?

Thank you, Chandra! both personally, and on behalf of science.

Basilis C. Xanthopoulos

Author's Note

I am indebted to many who have contributed toward the publication of these six volumes of my *Selected Papers*, and in particular to T. W. Mullikin, Norman Lebovitz, Kip Thorne, and Basilis Xanthopoulos for their selection of the papers to be included in these volumes and for writing the forewords; to the editors and publishers of the various journals, periodicals, and books for allowing the reproduction of the papers originally published in them and waiving the customary royalty charges; to the members of the editorial staff of the University of Chicago Press for their uniform courtesy and cooperation; to Mavis Lozano for assistance with the preparation of the material for the printers and for compiling the bibliography included in this volume; and to my former students and associates whose joint publications with me are represented in these volumes. To all of them I am most beholden.

S. CHANDRASEKHAR

31 July 1990

In remembrance of Basilis Xanthopoulos

Basilis Xanthopoulos was shot to death in an unspeakable act of violence on the evening of November 27, 1990, while he was giving a seminar lecture on advanced computing to his senior colleagues at the Research Center of the University of Crete (Iraklion, Greece). And so ended, without warning, a life of love and joy, rich in promise.

Basilis and I collaborated on a variety of scientific subjects almost continuously from 1978, when he was a graduate student, to the present. Some of the results of that collaboration are represented in this volume. The enduring personal friendship that emerged is amply present in the warm and generous foreword that Basilis wrote for this volume (and for which I am deeply grateful).

My association with Basilis is the most binding in all my sixty years in science.

On the evening of November 28, my wife and I listened on our phonograph to the second and the third movements of the *Eroica* in a renewal of our dedication to a dear departed friend.

S. CHANDRASEKHAR

29 November 1990

PART ONE

The Mathematical Theory of Black Holes

1. *The Schwarzchild Black Hole*

Original page numbers appear in brackets
at the bottoms of pages

1

On the equations governing the perturbations of the Schwarzschild black hole

By S. Chandrasekhar, F.R.S.
University of Chicago, Chicago, Illinois, 60637

(*Received* 7 *October* 1974)

A coherent self-contained account of the equations governing the perturbations of the Schwarzschild black hole is given. In particular, the relations between the equations of Bardeen & Press, of Zerilli and of Regge & Wheeler are explicitly established.

1. Introduction

The equations governing the perturbations of the vacuum Schwarzschild metric – the Schwarzschild black hole – have been the subject of many investigations (Regge & Wheeler 1957; Vishveshwara 1970; Edelstein & Vishveshwara 1970; Zerilli 1970*a*, *b*; Fackerell 1971; Bardeen & Press 1972; Friedman 1973). Nevertheless, there continues to be some elements of mystery shrouding the subject. Thus, Zerilli (1970*a*) showed that the equations governing the perturbation, properly analysed into spherical harmonics (belonging to the different l values) and with a time dependence $e^{i\sigma t}$, can be reduced to a one dimensional Schrödinger equation of the form

$$d^2Z/dr_*^2 + (\sigma^2 - V_Z)\,Z = 0 \quad (+\infty > r_* > -\infty), \tag{1}$$

where

$$V_Z = \frac{2n^2(n+1)\,r^3 + 6n^2mr^2 + 18nm^2r + 18m^3}{r^3(nr+3m)^2}\left(1 - \frac{2m}{r}\right), \tag{2}$$

$$\frac{d}{dr_*} = \left(1 - \frac{2m}{r}\right)\frac{d}{dr} \quad \text{and} \quad n = \tfrac{1}{2}(l-1)\,(l+2). \tag{3}$$

(Here, units are used in which $c = G = 1$.) Accordingly, the reflexion and the transmission coefficients for incident plane waves of various assigned wavenumbers, will clearly suffice to determine the evolution of any initial perturbation of the Schwarzschild black hole. This reduction of the general perturbation problem to the elementary one of determining the reflexion and the transmission coefficients of a one dimensional potential barrier is, of course, a remarkable simplification. But what is the origin of the particular form of the potential V_Z?

Again, considering a particular component of the Riemann tensor – the perturbed Newman–Penrose $\delta\Psi_0$ – Bardeen & Press (1973) showed that *all* the physical results relative to the perturbation of the Schwarzschild black hole can equally be derived from the solutions of the equation

$$\frac{d^2\phi}{dr_*^2} + \left[\sigma^2 - 4i\sigma\frac{r-3m}{r^2} - \frac{l(l+1)\,(r-2m)+2m}{r^3}\right]\phi = 0, \tag{4}$$

with the boundary conditions,

$$\phi \to e^{i\sigma r_*}/(r-2m) \quad \text{as} \quad r_* \to -\infty \quad \text{and} \quad \phi \to e^{-i\sigma r_*} r^{-2} \quad \text{as} \quad r_* \to +\infty, \tag{5}$$

appropriate for ingoing waves at the horizon ($r_* \to -\infty$) and outgoing waves at infinity ($r_* \to +\infty$). If, as claimed, the physical contents of equations (1) and (4) are entirely the same, then there must be a simple relation between the functions Z and ϕ which makes this equivalence manifest. But what is this relation?

In this paper we shall give a coherent, self-contained account of the theory of the perturbations of the Schwarzschild black hole while resolving at the same time some of the unanswered questions, in particular, the relation between the Zerilli and the Bardeen–Press equations.

2. The equations governing the perturbations of the Schwarzschild black hole

The vacuum metric of Schwarzschild written in its standard form is

$$ds^2 = -\left(1 - \frac{2m}{r}\right)(dt)^2 + r^2 \sin^2\theta (d\varphi)^2 + \frac{(dr)^2}{1-2m/r} + r^2 (d\theta)^2. \tag{6}$$

Because of this spherical symmetry, first order perturbations of this metric can be analysed into spherical harmonics belonging to the different l and m values. Perturbations belonging to a particular l and m will be axisymmetric about some axis. Accordingly, there will be no loss of generality in restricting ourselves to axisymmetric perturbations from the outset.

As has been shown (Chandrasekhar & Friedman 1972) a form of the metric adequate for treating time-dependent axisymmetric systems in general relativity is

$$ds^2 = -e^{2\nu}(dt)^2 + e^{2\psi}(d\varphi - \omega\, dt - q_{2,0}\, dx^2 - q_{3,0}\, dx^3)^2 + e^{2\mu_2}(dx^2)^2 + e^{2\mu_3}(dx^3)^2, \tag{7}$$

where ν, ψ, μ_2, μ_3, ω, q_2, and q_3 are all functions of t and the spatial variables x^2 and x^3 but independent of the azimuthal angle φ.

With the identifications

$$x^2 = r, \quad x^3 = \theta, \quad \text{and} \quad \mu_2 = \lambda \quad \text{(say)}, \tag{8}$$

the Schwarzschild metric corresponds to the various metric functions having the values

$$e^{2\nu} = e^{-2\lambda} = 1 - 2m/r, \quad e^{\psi} = r\sin\theta, \quad e^{\mu_3} = r, \tag{9}$$

and

$$\omega = q_2 = q_3 = 0.$$

When the metric is perturbed, we shall write

$$\nu + \delta\nu, \quad \lambda + \delta\lambda, \quad \psi + \delta\psi, \quad \text{and} \quad \mu_3 + \delta\mu_3, \tag{10}$$

for the corresponding metric functions while ω, q_2, and q_3 will be considered as quantities of the first order of smallness like $\delta\nu$, $\delta\lambda$, $\delta\psi$, and $\delta\mu_3$.

The equations governing the various quantities describing the perturbations can be readily written down by transcribing in the present context the various equations given by Chandrasekhar & Friedman (1972; see Part III of this paper). Thus, from the (1, 2)- and the (1, 3)-components of the field equations we find that

$$Q = r^2\left(1-\frac{2m}{r}\right)(q_{2,3}-q_{3,2})\sin^3\theta \tag{11a}$$

satisfies the equation

$$\frac{\partial}{\partial r}\left(\frac{r-2m}{r^3\sin^3\theta}\frac{\partial Q}{\partial r}\right)+\frac{\partial}{\partial\theta}\left(\frac{1}{r^4\sin^3\theta}\frac{\partial Q}{\partial\theta}\right)+\frac{\sigma^2}{r(r-2m)\sin^3\theta}Q = 0. \tag{11b}$$

This equation follows from Chandrasekhar & Friedman (1972; equation (153)) by substituting for ψ, ν, etc., their present values. We observe that (11b) is independent of the other field variables: it governs the 'so-called' odd-parity perturbations. Equation (11b) is considered in the appendix.

Of the remaining field equations, it will suffice to consider the (0, 2)-, (0, 3)-, (2,3)-, and (2, 2)-components. These equations are

$$(\delta\psi+\delta\mu_3)_{,r}+(1/r-\nu_{,r})(\delta\psi+\delta\mu_3)-(2/r)\,\delta\lambda = 0 \quad (\delta R_{02}=0), \tag{12}$$ †

$$(\delta\psi+\delta\lambda)_{,\theta}+(\delta\psi-\delta\mu_3)\cot\theta = 0 \quad (\delta R_{03}=0), \tag{13}$$

$$(\delta\psi+\delta\nu)_{,r\theta}+(\delta\psi-\delta\mu_3)_{,r}\cot\theta+\left(\nu_{,r}-\frac{1}{r}\right)\delta\nu_{,\theta}-\left(\nu_{,r}+\frac{1}{r}\right)\delta\lambda_{,\theta} = 0 \quad (\delta R_{23}=0), \tag{14}$$

and

$$\begin{aligned} &e^{-2\lambda}\left[\frac{2}{r}\delta\nu_{,r}+\left(\frac{1}{r}+\nu_{,r}\right)(\delta\psi+\delta\mu_3)_{,r}-2\delta\lambda\left(\frac{1}{r^2}+2\frac{\nu_{,r}}{r}\right)\right]\\ &+\frac{1}{r^2}[(\delta\psi+\delta\nu)_{,\theta\theta}+(2\delta\psi+\delta\nu-\delta\mu_3)_{,\theta}\cot\theta+2\delta\mu_3]\\ &-e^{-2\nu}(\delta\psi+\delta\mu_3)_{,00} = 0 \quad (\delta G_{22}=0). \end{aligned} \tag{15}$$

In addition, we shall find the following equation, expressing the condition $\delta R_{11}=0$, useful:

$$\begin{aligned} &e^{+2\nu}\left[\delta\psi_{,rr}+2\left(\frac{1}{r}+\nu_{,r}\right)\delta\psi_{,r}+\frac{1}{r}(\delta\psi+\delta\nu+\delta\mu_3-\delta\lambda)_{,r}-2\frac{\delta\lambda}{r}\left(\frac{1}{r}+2\nu_{,r}\right)\right]\\ &+\frac{1}{r^2}[\delta\psi_{,\theta\theta}+\delta\psi_{,\theta}\cot\theta+(\delta\psi+\delta\nu-\delta\mu_3+\delta\lambda)_{,\theta}\cot\theta+2\,\delta\mu_3]-e^{-2\nu}\,\delta\psi_{,00} = 0. \end{aligned} \tag{16}$$

As Friedman (1973) has shown, the variables r and θ in the foregoing equations can be separated by the substitutions

$$\begin{aligned} \delta\nu &= N(r)\,P_l(\cos\theta)\,e^{i\sigma t},\\ \delta\lambda &= \delta\mu_2 = L(r)\,P_l(\cos\theta)\,e^{i\sigma t},\\ \delta\mu_3 &= [T(r)\,P_l+V(r)\,P_{l,\theta\theta}]\,e^{i\sigma t},\\ \text{and}\quad \delta\psi &= [T(r)\,P_l+V(r)\,P_{l,\theta}\cot\theta]\,e^{i\sigma t}, \end{aligned} \tag{17}$$

where we have further assumed a time dependence expressed by the factor $e^{i\sigma t}$.

With the foregoing substitutions, equation (13) at once gives

$$T-V+L = 0. \tag{18}$$

† Commas, in subscripts, signify differentiations with respect to the variable (or variables) that follow.

Accordingly, only three of the four radial functions we have defined are linearly independent; and we shall choose N, L, and V as the independent functions.

The (0, 2)- and the (2,3)-components of the field equations – equations (12) and (14) – give

$$\left[\frac{d}{dr}+\left(\frac{1}{r}-\nu_{,r}\right)\right][2T-l(l+1)\,V]-\frac{2}{r}L=0 \tag{19}$$

and
$$(T-V+N)_{,r}-\left(\frac{1}{r}-\nu_{,r}\right)N-\left(\frac{1}{r}+\nu_{,r}\right)L=0, \tag{20}$$

or, after the elimination of T with the aid of equation (18), we obtain

$$N_{,r}-L_{,r}=\left(\frac{1}{r}-\nu_{,r}\right)N+\left(\frac{1}{r}+\nu_{,r}\right)L \tag{21}$$

and
$$L_{,r}+\left(\frac{2}{r}-\nu_{,r}\right)L=-n\left[V_{,r}+\left(\frac{1}{r}-\nu_{,r}\right)V\right]. \tag{22}$$

Similarly, equations (15) and (16) give

$$\frac{2}{r}N_{,r}+\left(\frac{1}{r}+\nu_{,r}\right)[2T-l(l+1)\,V]_{,r}-\frac{2}{r}\left(\frac{1}{r}+2\nu_{,r}\right)L$$
$$-l(l+1)\frac{e^{-2\nu}}{r^2}N-2n\frac{e^{-2\nu}}{r^2}T+\sigma^2e^{-4\nu}[2T-l(l+1)\,V]=0 \tag{23}$$

and
$$V_{,rr}+2\left(\frac{1}{r}+\nu_{,r}\right)V_{,r}+\frac{e^{-2\nu}}{r^2}(N+L)+\sigma^2e^{-4\nu}V=0. \tag{24}$$

Equation (23), after the elimination of T, takes the form

$$\frac{2}{r}N_{,r}-l(l+1)\frac{e^{-2\nu}}{r^2}N-\frac{2}{r}\left(\frac{1}{r}+2\nu_{,r}\right)L-2\left(\frac{1}{r}+\nu_{,r}\right)(L+nV)_{,r}$$
$$-2n\frac{e^{-2\nu}}{r^2}(V-L)-2\sigma^2e^{-4\nu}(L+nV)=0, \tag{25}$$

where
$$n=\tfrac{1}{2}(l-1)(l+2). \tag{26}$$

It will be observed that equations (21), (22), and (25) provide three linear first order equations for the three radial functions L, N, and V. By suitably combining them, we can express the derivative of each of them as linear combinations of L, N, and V. Thus,

$$N_{,r}=aN+bL+cX, \tag{27}$$

$$L_{,r}=\left(a-\frac{1}{r}+\nu_{,r}\right)N+\left(b-\frac{1}{r}-\nu_{,r}\right)L+cX \tag{28}$$

and
$$X_{,r}=-\left(a-\frac{1}{r}+\nu_{,r}\right)N-\left(b+\frac{1}{r}-2\nu_{,r}\right)L-\left(c+\frac{1}{r}-\nu_{,r}\right)X, \tag{29}$$

where, for the sake of brevity, we have written

$$X = nV = \tfrac{1}{2}(l-1)(l+2)\,V, \tag{30}$$

$$\left.\begin{aligned} a &= \frac{n+1}{r-2m}, \quad \nu_{,r} = \frac{m}{r(r-2m)}, \\ b &= -\frac{1}{r}-\frac{n}{(r-2m)}+\frac{m}{r(r-2m)}+\frac{m^2}{r(r-2m)^2}+\sigma^2\frac{r^3}{(r-2m)^2} \\ \text{and} \qquad c &= -\frac{1}{r}+\frac{1}{(r-2m)}+\frac{m^2}{r(r-2m)^2}+\sigma^2\frac{r^3}{(r-2m)^2}. \end{aligned}\right\} \tag{31}$$

Equations (27), (28), and (29) provide the basic equations of the theory. *All the remaining field equations, including equation* (24), *are verifiable consequences of these equations*†

For later use, we may note here the following equation obtained by adding equations (28) and (29)

$$(L+X)_{,r} = -\left(\frac{2}{r}-\nu_{,r}\right)L-\left(\frac{1}{r}-\nu_{,r}\right)X. \tag{32}$$

3. Zerilli's equation

We shall verify that by virtue of equations (27)–(29), the function,

$$Z = \frac{r^2}{nr+3m}\left(\frac{3m}{r}V-L\right), \tag{33}$$

satisfies Zerilli's equation (1).

First, we observe that an equivalent expression for Z is

$$Z = rV-\frac{r^2}{nr+3m}(L+X). \tag{34}$$

Differentiating this expression with respect to the variable

$$r_* = r+2m\lg\left(\frac{r}{2m}-1\right) \quad (r>2m), \tag{35}$$

† In Regge & Wheeler's discussion (Regge & Wheeler 1957) of this same problem, in a gauge different from the one adopted in this paper, the corresponding situation is somewhat curious. They obtain a set of six equations for three scalar functions. Three of these equations are of the first order while the remaining three are of the second order. They say 'one will expect that the three first order equations will suffice to determine the solution apart from the boundary conditions. Actually, the second order equations contain an additional piece of information...specifically, a rather elaborate investigation shows that [the second order equations] can all be deduced from [the first order equations]' provided a linear combination, F (say), of the three scalar functions vanishes. At the same time Vishveshwara (1970) says that if $F_{,r}$ is evaluated with the aid of the first order equations, $F_{,r} = -\nu_{,r}F$ and that therefore 'if $F = 0$ for some r, it is zero everywhere'. All this is somewhat confusing. In any event, when working in the gauge adopted in this paper, *all* the second order equations are *identically* satisfied by virtue of the first order equations (27)–(29) and no 'additional piece of information' is provided.

and making use of equation (32), we obtain

$$Z_{,r_*} = \left(1-\frac{2m}{r}\right) Z_{,r}$$
$$= (r-2m)\,V_{,r} + \frac{3m(r-2m)}{r(nr+3m)}V + \frac{nr^2-3nmr-3m^2}{(nr+3m)^2}(L+X). \qquad (36)$$

Differentiating this last expression, once again, with respect to r_* and simplifying with the aid of equations (24), (29), and (34), we find after some considerable reductions that, remarkably, we are left with

$$\mathrm{d}^2Z/\mathrm{d}r_*^2 + (\sigma^2 - V_Z)\,Z = 0, \qquad (37)$$

where V_Z is given by equation (2). Thus, Zerilli's equation, while it is a directly verifiable consequence of equations (27)–(29), the manner in which it emerges is shrouded in (apparently) miraculous cancellations.

An interesting integral property of Zerilli's potential V_Z may be noted here:

$$2m\int_{-\infty}^{+\infty} V_Z\,\mathrm{d}r_* = 2n + \tfrac{1}{2}. \qquad (38)$$

4. The expression of $\delta\Psi_0$ in terms of Zerilli's function

In another connexion, Friedman (1973; §3 of this paper) has already considered the perturbed components of the Riemann tensor; and the component $\delta\Psi_0$ to which the Bardeen–Press equation (4) refers is

$$\delta\Psi_0 = = \tfrac{1}{4}\mathrm{e}^{-2\nu}[\delta R_{(t)(\theta)(t)(\theta)} + \delta R_{(r)(\theta)(r)(\theta)} + 2\,\delta R_{(t)(\theta)(r)(\theta)}$$
$$- \delta R_{(t)(\varphi)(t)(\varphi)} - \delta R_{(r)(\varphi)(r)(\varphi)} - 2\,\delta R_{(t)(\varphi)(r)(\varphi)}]. \qquad (39)$$

Expressions for the components of the Riemann tensor (in the tetrad-frame in which they are referred) have been given by Chandrasekhar & Friedman (1972; equations (16)) under general time-dependent axisymmetric conditions. Specializing them to the case on hand, we find that in accordance with equation (39) (cf. Friedman 1973, equation (56))

$$\delta\Psi_0 = -\tfrac{1}{4}\mathrm{e}^{-2\nu}\left\{\frac{1}{r^2}\left(\frac{\partial^2}{\partial\theta^2} - \cot\theta\right)(\delta\nu - \delta\lambda)\right.$$
$$\left. + \left[\mathrm{e}^{-4\nu}\frac{\partial^2}{\partial t^2} + 2\frac{\partial}{\partial t}\left(\frac{\partial}{\partial r} + \frac{1}{r} - \nu_{,r}\right) + \mathrm{e}^{2\nu}\left(\frac{\partial^2}{\partial r^2} + \frac{2}{r}\frac{\partial}{\partial r}\right)\right](\delta\psi - \delta\mu_3)\right\}. \qquad (40)$$

Separating the variables in accordance with equations (17) and making use of equation (24), we obtain

$$\delta\Psi_0 = \tfrac{1}{2}\left\{\mathrm{i}\sigma\mathrm{e}^{-2\nu}\left[V_{,r} + \left(\frac{1}{r} - \nu_{,r}\right)V\right] - \left(\nu_{,r}V_{,r} + \sigma^2\mathrm{e}^{-4\nu}V + \frac{\mathrm{e}^{-2\nu}}{r^2}N\right)\right\}\mathrm{e}^{\mathrm{i}\sigma t}(P_{l,\theta\theta} - P_{l,\theta}\cot\theta). \qquad (41)$$

In the further use of this expression, we shall suppress the angular dependence $(P_{l,\theta\theta} - P_{l,\theta}\cot\theta)$ and the factor $\tfrac{1}{2}\mathrm{e}^{\mathrm{i}\sigma t}$ as well. With this understanding, we can rewrite the expression for $\delta\Psi_0$ in the following form after eliminating $V_{,r}$ in the second term in parenthesis on the right hand side with the aid of equation (29):

$$\delta\Psi_0 = i\sigma\frac{r}{r-2m}\left[V_{,r}+\frac{r-3m}{r(r-2m)}V\right]+\sigma^2\frac{r^2}{(r-2m)^2}\left[\frac{m}{n(r-2m)}(L+X)-V\right]$$
$$-\frac{nr^2-3nmr-3m^2}{nr^2(r-2m)^2}N-\frac{m[nr^2-mr(2n-1)-3m^2]}{nr^2(r-2m)^3}L+\frac{m(r^2-3mr+3m^2)}{r^2(r-2m)^3}V. \tag{42}$$

Making use of equations (34) and (36) giving Z and $Z_{,r_*}$, we find that we can express $\delta\Psi_0$ in terms of these quantities; we find

$$\delta\Psi_0 = \frac{1}{2r(r-2m)}\left\{\frac{2n^2(n+1)r^3+6n^2mr^2+18nm^2r+18m^3}{r^2(nr+3m)^2}Z\right.$$
$$\left.+\left[2i\sigma\frac{r^2}{r-2m}+2\frac{nr^2-3nmr-3m^2}{(r-2m)(nr+3m)}\right](Z_{,r_*}+i\sigma Z)\right\}. \tag{43}$$

We observe that in the first term in braces on the right hand side of equation (43), the coefficient of Z is, apart from a factor $r^2/(r-2m)$ the potential V_Z in Zerilli's equation. In making use of this fact, we shall find it convenient to introduce the operators

$$\Lambda_{\pm} = d/dr_* \pm i\sigma \quad \text{and} \quad \Lambda^2 = \Lambda_+\Lambda_- = \Lambda_-\Lambda_+ = d^2/dr_*^2+\sigma^2 \tag{44}$$

and write Zerilli's equation in the form

$$\Lambda^2 Z = V_Z Z. \tag{45}$$

We now rewrite the expression (43) for $\delta\Psi_0$ in the manner

$$\delta\Psi_0 = \frac{1}{2r(r-2m)}\left[\frac{r^2}{r-2m}(\Lambda^2 Z+2i\sigma\Lambda_+ Z)+2\frac{nr^2-3nmr-3m^2}{(r-2m)(nr+3m)}\Lambda_+ Z\right]$$
$$= \frac{r}{2(r-2m)^2}\left[\Lambda_+(\Lambda_-+2i\sigma)Z+2\frac{nr^2-3nmr-3m^2}{r^2(nr+3m),}\Lambda_+ Z\right], \tag{46}$$

or finally,
$$\delta\Psi_0 = \frac{r}{2(r-2m)^2}\left[\Lambda_+ +2\frac{nr^2-3nmr-3m^2}{r^2(nr+3m)}\right]\Lambda_+ Z. \tag{47}$$

Equation (47) suggests that we define the function

$$Y = 2\frac{(r-2m)^2}{r}\delta\Psi_0; \tag{48}$$

in terms of Y, equation (47) takes the simple form

$$Y = \Lambda_+\Lambda_+ Z + W\Lambda_+ Z, \tag{49}$$

where
$$W = 2\frac{nr^2-3nmr-3m^2}{r^2(nr+3m)}. \tag{50}$$

An equivalent form of equation (49) is

$$Y = V_Z Z+(W+2i\sigma)\Lambda_+ Z. \tag{51}$$

It is noteworthy that in this last equation, expressing Y in terms of Z and $\Lambda_+ Z$, V_Z appears as the coefficient of Z.

5. The derivation of the Bardeen–Press equation from the Zerilli equation and conversely

Applying the operator Λ_- to equation (49) and making use of Zerilli's equation satisfied by Z, we obtain

$$\Lambda_- Y = \left(\frac{dV_Z}{dr_*} + WV_Z\right) Z + \left(V_Z + \frac{dW}{dr_*}\right) \Lambda_+ Z. \tag{52}$$

It can now be directly verified that with the functions V_Z and W as defined

$$\frac{dV_Z}{dr_*} + WV_Z = -6m\frac{(r-2m)^2}{r^6} \tag{53}$$

and

$$\frac{dW}{dr_*} + V_Z = 2\left(1 - \frac{2m}{r}\right)\frac{nr+3m}{r^3}. \tag{54}$$

Accordingly, equation (52) takes the form

$$\Lambda_- Y = -6m\frac{(r-2m)^2}{r^6} Z + 2\left(1-\frac{2m}{r}\right)\frac{nr+3m}{r^3}\Lambda_+ Z. \tag{55}$$

Equations (51) and (55) can now be solved for Z and $\Lambda_+ Z$ to give

$$[\tfrac{2}{3}n(n+1) + 2m\,i\sigma]\, Z = \frac{r^2(nr+3m)}{3(r-2m)} Y - \frac{r^6}{6(r-2m)^2}(W + 2i\sigma)\Lambda_- Y \tag{56}$$

and

$$[\tfrac{2}{3}n(n+1) + 2m i\sigma]\, \Lambda_+ Z = mY + \frac{r^6}{6(r-2m)^2} V_Z \Lambda_- Y. \tag{57}$$

In deriving the foregoing solutions, we have made use of the surprisingly simple relation,

$$\frac{r^2}{r-2m}(nr+3m)V_Z + 3mW = 2n(n+1), \tag{58}$$

which obtains between V_Z and W.

We may parenthetically remark here that the fact that the functions V_Z and W satisfy three simple relations such as equations (53), (54), and (58), to some extent, dispels the 'mystery' about them.

We can now eliminate Z between equations (56) and (57) to obtain an equation for Y. Thus, from equation (57) it follows that

$$\begin{aligned}\Lambda_-\left[\frac{r^6}{6(r-2m)^2} V_Z \Lambda_- Y + mY\right] &= [\tfrac{2}{3}n(n+1) + 2m\,i\sigma]\Lambda^2 Z \\ &= [\tfrac{2}{3}n(n+1) + 2m\,i\sigma] V_Z Z;\end{aligned} \tag{59}$$

while, in accordance with equation (56), we may conclude

$$\Lambda_-\left[\frac{r^6}{6(r-2m)^2} V_Z \Lambda_- Y + mY\right] = V_Z\left[\frac{r^2(nr+3m)}{3(r-2m)} Y - \frac{r^6}{6(r-2m)^2}(W + 2m\,i\sigma)\Lambda_- Y\right]. \tag{60}$$

On simplifying this last equation, we are left with

$$\Lambda^2 Y + 4\frac{r-3m}{r^2}\Lambda_- Y - 2\left(1-\frac{2m}{r}\right)\frac{nr+3m}{r^3}Y = 0. \tag{61}$$

With the substitution

$$Y = \frac{r-2m}{r^3}\phi, \tag{62}$$

it can now be readily verified that we recover equation (4) of Bardeen & Press.

Even as equations (56) and (57), together with Zerilli's equation, enabled us to derive the Bardeen–Press equation, we may, conversely, derive Zerilli's equation from equations (51) and (55) together with the Bardeen–Press equation in the form (61).

6. Concluding remarks

In bringing together in a common framework the different equations which govern the perturbations of the Schwarzschild black hole and clarifying their inner relation, the present paper raises the question whether one may not be able to simplify some, at least, of the aspects of the perturbation problem as it pertains to the Kerr black hole. Thus, Teukolsky's equation (Teukolsky 1973) governing the perturbations of the Kerr black hole reduces, when specialized to the Schwarzschild black hole, to the Bardeen–Press equation (4) and *not* to the Zerilli equation (1). The question naturally arises whether there may not be a transformed version of the Teukolsky equation, which when specialized to the Schwarzschild case becomes the Zerilli equation. Such a transformation, if it were possible, would clearly be advantageous: we shall then need to determine only the reflexion and the transmission coefficients for a one-dimensional potential barrier. In a subsequent paper (Chandrasekhar & Detweiler, in preparation) it will be shown how Teukolsky's equation, for the case of axisymmetric perturbations, can be transformed into a one-dimensional wave equation with, indeed, four possible potentials.

The research reported in this paper has in part been supported by the National Science Foundation under grant GP-34721X1 with the University of Chicago and the Louis Block Fund.

Appendix

(*Communicated* 11 *November* 1974)

In the course of an investigation of Teukolsky's equation (Chandrasekar & Detweiler, in preparation), it became clear that the Regge–Wheeler equation governing the odd-parity perturbations of the Schwarzschild metric must also be equivalent to the Bardeen–Press equation. We give a brief *ab initio* demonstration of this equivalence.

The variables r and θ in equation (11 *b*), governing the odd-parity perturbations, can be separated by the substitution,

$$Q = rX(r)P_{l+2}(\cos\theta/-3),$$

where $P_{l+2}(x/-3)$ is the Gegenbauer, polynomial of order $(l+2)$ and index -3. With the foregoing substitution, equation (11b) reduces to the Regge–Wheeler equation (Regge & Wheeler 1957),

$$\Lambda^2 X = V_0 X, \tag{A 1}$$

where

$$V_0 = 2\left(1-\frac{2m}{r}\right)\frac{(n+1)\,r-3m}{r^3}. \tag{A 2}$$

Defining

$$W_0 = 2\frac{r-3m}{r^2}, \tag{A 3}$$

we can readily verify that the functions V_0 and W_0 satisfy the identities

$$\frac{dV_0}{dr_*}+W_0V_0 = 6m\frac{(r-2m)^2}{r^6}, \tag{A 4}$$

$$\frac{dW_0}{dr_*}+V_0 = 2\left(1-\frac{2m}{r}\right)\frac{nr+3m}{r^3} \tag{A 5}$$

and

$$\frac{r^2}{r-2m}(nr+3m)\,V_0-3mW_0 = 2n(n+1). \tag{A 6}$$

These identities are entirely analogous to those satisfied in the context of Zerilli's equation (cf. equations (53), (54) and (58)).

If we now let

$$Y = \Lambda_+\Lambda_+X+W_0\Lambda_+X, \tag{A 7}$$

or equivalently

$$Y = V_0X+(W_0+2i\sigma)\,\Lambda_+X, \tag{A8}$$

then by virtue of equations (A 4) and (A 5) (cf. equation (55))

$$\Lambda_-Y = 6m\frac{(r-2m)^2}{r^6}X+2\left(1-\frac{2m}{r}\right)\frac{nr+3m}{r^3}\Lambda_+X. \tag{A 9}$$

Solving these equations for X and Λ_+X in terms of Y and Λ_-Y, we find (cf. equations (56) and (57)),

$$[\tfrac{2}{3}n(n+1)-2mi\sigma]\,X = \frac{r^2(nr+3m)}{3(r-2m)}\,Y-\frac{r^6}{6(r-2m)^2}(W_0+2i\sigma)\,\Lambda_-Y \tag{A 10}$$

and

$$[\tfrac{2}{3}n(n+1)-2mi\sigma]\,\Lambda_+X = -mY+\frac{r^6}{6(r-2m)^2}V_0\Lambda_-Y. \tag{A 11}$$

It can now be verified that *equations* (A 8), (A 9), (A 10), and (A 11) *are necessary and sufficient for the Regge–Wheeler equation to imply the Bardeen–Press equation and conversely.*

References

Bardeen, J. M. & Press, W. H. 1973 *J. Math. Phys.* **14**, 7.
Chandrasekhar, S. & Friedman, J. L. 1972 *Astrophys. J.* **175**, 379.
Edelstein, L. & Vishveshwara, C. V. 1970 *Phys. Rev.* D **1**, 3514.
Fackerell, E. D. 1971 *Astrophys. J.* **166**, 197.
Friedman, J. L. 1973 *Proc. R. Soc. Lond.* A **335**, 163.
Regge, T. & Wheeler, J. A. 1957 *Phys. Rev.* **108**, 1063.
Teukolsky, S. A. 1973 *Astrophys. J.* **185**, 635.
Vishveshwara, C. V. 1970 *Phys. Rev.* D **1**, 2870.
Zerilli, F. J. 1970*a* *Phys. Rev. Lett.* **24**, 737.
Zerilli, F. J. 1970*b* *Phys. Rev.* D **2**, 2141.

2

The quasi-normal modes of the Schwarzschild black hole

By S. Chandrasekhar, F.R.S. and S. Detweiler
University of Chicago, Chicago, Illinois, 60637

(*Received* 6 *December* 1974)

The quasi-normal modes of a black hole represent solutions of the relevant perturbation equations which satisfy the boundary conditions appropriate for purely outgoing (gravitational) waves at infinity and purely ingoing waves at the horizon. For the Schwarzschild black hole the problem reduces to one of finding such solutions for a one dimensional wave equation (Zerilli's equation) for a potential which is positive everywhere and is of short-range. The notion of quasi-normal modes of such one-dimensional potential barriers is examined with two illustrative examples; and numerical solutions for Zerilli's potential are obtained by integrating the associated Riccati equation.

1. Introduction

It is known that the evolution of an arbitrary perturbation of the metric coefficients of the Schwarzschild black hole can be fully described in terms of the reflexion (R) and the transmission (T) coefficients of the one-dimensional barrier represented by Zerilli's potential (Zerilli 1970; see also Chandrasekhar 1975). Nevertheless, the notion of *quasi-normal modes* of a black hole has been introduced in the literature in analogy with the normal modes of oscillation of a star. In the context of a black hole these quasi-normal modes are defined as proper solutions of the perturbation equations belonging to certain complex characteristic frequencies which satisfy the boundary conditions appropriate for purely ingoing waves at the horizon and purely outgoing waves at infinity.

It does not appear that the quasi-normal modes of a black hole serve the same purposes as the normal modes of oscillation of a star. Consider, for example, the spherically symmetric perturbations of an initially static configuration either in the Newtonian (Eddington 1918, 1919) or in the relativistic (Chandrasekhar 1964) framework. In either framework, the determination of the characteristic frequencies leads to a two-point boundary-value problem of the classical Sturmian type for a self-adjoint second order differential equation. Consequently, the associated proper solutions (i.e. the normal modes) form a *complete set* in the sense that any arbitrary spherically symmetric perturbation of the star (compatible with the boundary conditions of the problem) can be expressed as a linear superposition of the normal modes. And, therefore, the evolution of any such perturbation can be followed in terms of the normal modes and the characteristic frequencies to which they belong. Also, it follows that if any of the modes should belong to a (purely) imaginary characteristic frequency – this is the only possibility for the spherically symmetric systems under consideration – then the system can be considered as necessarily unstable. It is not at all clear that the enumeration of the quasi-normal modes of a black hole serves the same purposes as the normal modes of radial oscillation of a star. Thus, in the particular case of the Schwarzschild black hole, the evolution of an arbitrary perturbation of its metric coefficients can be followed, as we have already stated, in terms of the reflexion and the transmission coefficients of Zerilli's potential for various frequencies; and a knowledge of its quasi-normal modes is of no relevance for that purpose.

In spite of the adverse comments we have made concerning the usefulness of the notion of quasi-normal modes, their determination in the case of the Schwarzschild black hole has some interest, at least to the extent that it may illuminate their relevance (or, otherwise) for our understanding of the physics of black holes. This paper, then, is devoted to an examination of the general problem of the quasi-normal modes of one-dimensional potential barriers and of Zerilli's potential in particular.

2. The general theory

Consider the simple one-dimensional wave equation,

$$\frac{\mathrm{d}^2\psi}{\mathrm{d}x^2}+[\sigma^2-V(x)]\,\psi=0 \quad (-\infty<x<+\infty), \tag{1}$$

where $V(x)$ is positive everywhere and is of 'short range' in the sense that

$$\int_{-\infty}^{+\infty} V(x)\,\mathrm{d}x \quad \text{is finite.} \tag{2}$$

It is evident that if we have a plane-wave of unit amplitude $\mathrm{e}^{+\mathrm{i}\sigma x}$ incident on the barrier from the right,† then a part of it will be reflected and a part of it will be transmitted; i.e. there will be an admixture of the incident wave with a reflected wave $A\,\mathrm{e}^{-\mathrm{i}\sigma x}$ of amplitude A (say) at $+\infty$ and there will also be a transmitted wave $B\,\mathrm{e}^{+\mathrm{i}\sigma x}$ of amplitude B (say) at $-\infty$. The reflexion and the transmission coefficients are then given by

$$R=|A|^2 \quad \text{and} \quad T=|B|^2; \tag{3}$$

and it must always be true that

$$R+T=1, \tag{4}$$

so long as σ is real. On the other hand, a quasi-normal mode belonging to a complex frequency σ is so defined that there is no wave incident on the barrier from the right and we only have a reflected wave at $+\infty$ and a transmitted wave at $-\infty$.

† The convention that $\mathrm{e}^{+\mathrm{i}\sigma x}$ represents an ingoing wave is the opposite of the one which is normally adopted in the quantum theory; it is a consequence of the assumption, normal in this theory, that the time-dependence of the normal modes is $\mathrm{e}^{\mathrm{i}\sigma t}$.

The problem of determining the reflexion and the transmission coefficients of a rectangular barrier is a standard exercise in elementary quantum mechanics; and from the solution of this problem given in textbooks (cf. Flugge & Marschall 1952, p. 40) one can readily obtain the equations which determine the quasi-normal modes. But for general potential barriers for which explicit solutions cannot be found (as is the case with Zerilli's potential) it is convenient to reformulate the problem explicitly as a standard characteristic-value problem by transforming equation (1) to the form of a Riccati equation by the substitution,

$$\psi = \exp\left(\mathrm{i}\int^{x} \phi \,\mathrm{d}x\right), \tag{5}$$

and obtaining

$$\mathrm{i}\,\mathrm{d}\phi/\mathrm{d}x + \sigma^2 - \phi^2 - V(x) = 0. \tag{6}$$

A quasi-normal mode corresponds to a solution of equation (6) which satisfies the boundary conditions

$$\phi \to -\sigma \quad \text{as} \quad x \to +\infty \quad \text{and} \quad \phi \to +\sigma \quad \text{as} \quad x \to -\infty, \tag{7}$$

with the real part of σ assumed to be positive. Solutions having these properties (generally) exist when σ assumes one of a discrete set of complex values; but the set need not be an enumerable infinity: it can be sometimes, but often it is not.

An identity, which follows from integrating equation (6) over the entire range of x and making use of the boundary conditions (7), is

$$-2\mathrm{i}\sigma + \int_{-\infty}^{+\infty} (\sigma^2 - \phi^2)\,\mathrm{d}x = \int_{-\infty}^{+\infty} V(x)\,\mathrm{d}x. \tag{8}$$

By virtue of the boundary conditions (7) and the assumed short-range character of $V(x)$, both the integrals which appear in equation (8) are finite.

In practice it is useful to separate the real and the imaginary parts of equation (6) by writing

$$\sigma = \sigma_1 + \mathrm{i}\sigma_2 \quad \text{and} \quad \phi = \phi_1 + \mathrm{i}\phi_2 \quad (\sigma_1 \geqslant 0). \tag{9}$$

We then obtain the pair of equations,

$$\mathrm{d}\phi_1/\mathrm{d}x = -2\sigma_1\sigma_2 + 2\phi_1\phi_2 \tag{10}$$

and

$$\mathrm{d}\phi_2/\mathrm{d}x = \sigma_1^2 - \sigma_2^2 - \phi_1^2 + \phi_2^2 - V, \tag{11}$$

together with the boundary conditions

$$\phi_1 \to -\sigma_1 \quad \text{as} \quad x \to +\infty \quad \text{and} \quad \phi_1 \to +\sigma_1 \quad \text{as} \quad x \to -\infty, \tag{12}$$

and

$$\phi_2 \to -\sigma_2 \quad \text{as} \quad x \to +\infty \quad \text{and} \quad \phi_2 \to +\sigma_2 \quad \text{as} \quad x \to -\infty. \tag{13}$$

It can sometimes happen that σ is purely imaginary so that $\sigma_1 = 0$. Since, in this case, ϕ must vanish at $\pm\infty$, it follows from equation (10) that $\phi_1 \equiv 0$; and we are left with asking whether there exist non-trivial solutions of the equation,

$$\mathrm{d}\phi_2/\mathrm{d}x = -\sigma_2^2 + \phi_2^2 - V, \tag{14}$$

which satisfy the boundary conditions (13).

28-2

The problem of the quasi-normal modes as formulated in terms of the phase ϕ appears explicitly as a problem in characteristic values though of a somewhat unconventional kind.

3. Two illustrative examples

Since the notion of quasi-normal modes of a one-dimensional potential barrier is not considered in the standard literature on quantum mechanics, it may be useful to consider them in the context of two elementary situations which admit of explicit solutions.

(a) A rectangular barrier

Let the potential barrier be defined by

$$\begin{aligned} V(x) &= U > 0 \quad \text{for} \quad 0 < x < a \\ &= 0 \quad \text{for} \quad x < 0 \quad \text{and} \quad x > a. \end{aligned} \tag{15}$$

Further, letting

$$\kappa^2 = U - \sigma^2, \tag{16}$$

we can write the required solution of equation (6) in the form

$$\left.\begin{aligned} \phi &= +\sigma & (x \leqslant 0) \\ &= -i\kappa \tanh(\kappa x + \kappa c) & (0 \leqslant x \leqslant a) \\ &= -\sigma & (x \geqslant a), \end{aligned}\right\} \tag{17}$$

where c is a constant. The continuity of ϕ, at $x = 0$ and $x = a$, gives

$$\sigma = -i\kappa \tanh \kappa c$$

and

$$\sigma = +i\kappa \tanh(\kappa a + \kappa c). \tag{18}$$

The characteristic equation for σ will follow from the elimination of κc from the foregoing equations. In carrying out the elimination, it will be convenient to write

$$\sigma = Q \sin \alpha \quad \text{and} \quad \kappa = -Q \cos \alpha \quad \text{where} \quad Q^2 = U; \tag{19}$$

and we shall adopt the convention

$$Q \geqslant 0. \tag{20}$$

With the foregoing definitions, the elimination of κc from the pair of equations (18) gives

$$\cosh \kappa a + i \cot 2\alpha \sinh \kappa a = 0. \tag{21}$$

In considering equation (21), we shall measure σ in the unit a^{-1} and let

$$\alpha = \alpha_1 + i\alpha_2 \quad \text{and} \quad \kappa = \kappa_1 + i\kappa_2. \tag{22}$$

Equation (21) then leads to the pair of equations,

$$\frac{\sin 4\alpha_1}{\cosh 4\alpha_2 - \cos 4\alpha_1} = \frac{\sin 2\kappa_2}{\cosh 2\kappa_1 - \cos 2\kappa_2} \tag{23}$$

and

$$-\frac{\sinh 4\alpha_2}{\cosh 4\alpha_2 - \cos 4\alpha_1} = \frac{\sinh 2\kappa_1}{\cosh 2\kappa_1 - \cos 2\kappa_2}. \tag{24}$$

From these equations it follows that

$$\kappa_1 = -2\alpha_2 \quad \text{and} \quad \kappa_2 = 2\alpha_1 - n\pi, \tag{25}$$

where n is an integer positive, negative, or zero. On the other hand (cf. equation (18))

$$\kappa_1 = -Q\cos\alpha_1\cosh\alpha_2 \quad \text{and} \quad \kappa_2 = +Q\sin\alpha_1\sinh\alpha_2. \tag{26}$$

By combining equations (25) and (26), we obtain the 'characteristic equation'

$$\tan\alpha_1\tanh\alpha_2 = (\alpha_1 - \tfrac{1}{2}n\pi)/\alpha_2. \tag{27}$$

If (α_1, α_2) is a pair of values which satisfies equation (27), then the corresponding values of Q, σ_1, and σ_2 follow from the equations

$$Q = 2[(\alpha_1 - \tfrac{1}{2}n\pi)/\sin\alpha_1)]\operatorname{cosech}\alpha_2, \tag{28}$$

$$\sigma_1 = Q\sin\alpha_1\cosh\alpha_2 = 2\alpha_2\tan\alpha_1 \tag{29}$$

and

$$\sigma_2 = Q\cos\alpha_1\sinh\alpha_2 = 2\alpha_2\tanh\alpha_2. \tag{30}$$

The convention with respect to the outgoing and the ingoing waves requires that

$$\sigma_1 \geqslant 0. \tag{31}$$

From the requirements (20) and (31), it follows from equations (28)–(30) that

$$\text{if} \quad \alpha_2 > 0 \quad \text{then} \quad n \leqslant 0 \quad \text{and} \quad 0 \leqslant \alpha_1 \leqslant \tfrac{1}{2}\pi \tag{32}$$

or

$$\text{if} \quad \alpha_2 < 0 \quad \text{then} \quad n > 0 \quad \text{and} \quad \pi \geqslant \alpha_1 \geqslant \tfrac{1}{2}\pi. \tag{33}$$

We shall adopt the former conditions (32).

We observe that $\sigma_2 > 0$; and this is consistent with the fact that these waves are damped, though the proper solutions themselves diverge exponentially both for $x \to +\infty$ and $x \to -\infty$. These divergences are admissible in the present context.

For the case $n = 0$, corresponding to the lowest mode, the relevant equations are

$$\frac{\tan\alpha_1}{\alpha_1} = \frac{\coth\alpha_2}{\alpha_2}; \quad Q = 2\frac{\alpha_1}{\sin\alpha_1}\operatorname{cosech}\alpha_2, \tag{34}$$

$$\sigma_1 = 2\alpha_2\tan\alpha_1 \quad \text{and} \quad \sigma_2 = 2\alpha_2\tanh\alpha_2.$$

From these equations it follows that $Q \geqslant Q_{\min}$, where $Q_{\min}$ occurs for $\alpha_1 \to 0$ and α_2 is the root of the equation

$$\alpha_2\tanh\alpha_2 = 1. \tag{35}$$

Denoting the root of this equation by α_2^* ($\simeq 1.1997$), we find that

$$Q = 2\operatorname{cosech}\alpha_2^* \,(= 1.3255), \quad \sigma_1 = 0, \quad \text{and} \quad \sigma_2 = 2. \tag{36}$$

For $Q \leqslant Q_{\min}$, the characteristic values of σ are purely imaginary; and the appropriate solution, found in accordance with equation (14), is given by

$$Q = 2\alpha_2\operatorname{sech}\alpha_2 \quad \text{and} \quad \sigma_2 = 2\alpha_2\tanh\alpha_2. \tag{37}$$

Accordingly, along this branch of the solution, $Q = 0$ both when $\alpha_2 \to 0$ and when $\alpha_2 \to \infty$; and Q attains its maximum for precisely the same value of α_2^* (as the solution of equation (35)) when σ_2 has again the value 2.

The higher modes are obtained for $n = -1, -2, -3$, etc., in accordance with equations (28)–(30). It can be readily shown that along these higher modes

$$\alpha_1 \to 0, \quad \alpha_2 \to \infty \quad \text{while} \quad \alpha_1 \alpha_2 \to \tfrac{1}{2}|n|\pi; \tag{38}$$

and simultaneously

$$Q \to 4\alpha_2 e^{-\alpha_2}, \quad \sigma_1 \to |n|\pi, \quad \text{and} \quad \sigma_2 \to 2\alpha_2. \tag{39}$$

The behaviour of the characteristic values belonging to the various modes is illustrated in figure 1.

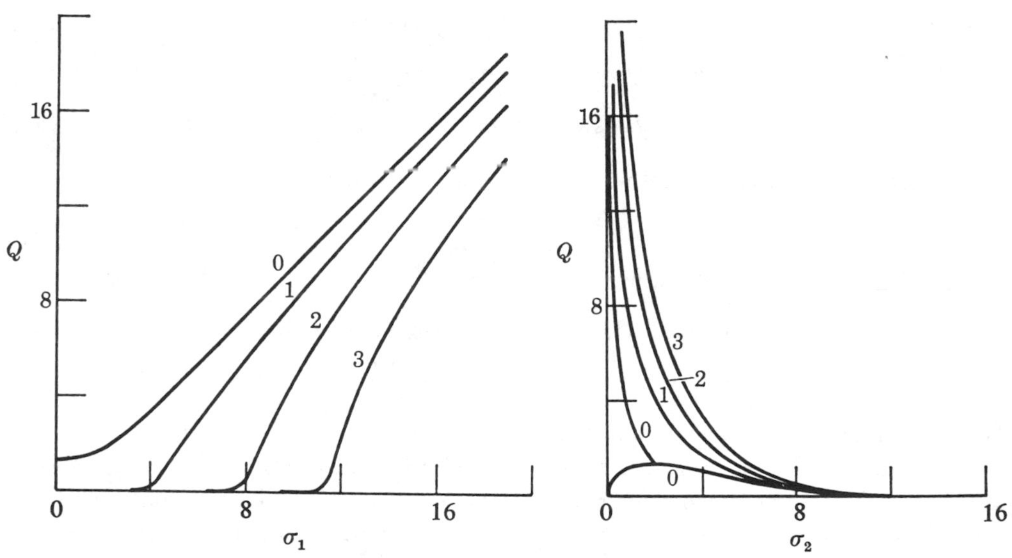

FIGURE 1. The complex frequencies $(\sigma_1 + i\sigma_2)$ belonging to the quasi-normal modes of a rectangular barrier. The frequencies are measured in units of the width of the barrier (a); and Q^2 $(= a^2 U)$ is a measure of the height of the barrier. The curves are labelled by the values of $-n$ to which they belong.

(b) *Price's model potential for the Schwarzschild black hole*

Since we do not expect to solve Zerilli's equation in terms of known functions, Price (1972*a*, *b*) has considered the following potential as having the same general behaviour as Zerilli's potential and for which analytic solutions can be found:

$$\begin{aligned} V(x) &= [l(l+1)]/x^2 \quad \text{for} \quad x \geqslant m \\ &= 0 \quad \text{for} \quad x < m. \end{aligned} \tag{40}$$

In the actual physical problem the variable x is identified with r_* (see §4 below).

The problem of the quasi-normal modes for Price's potential has been solved by Press (1973); but since his solution has not been published, we may briefly outline the solution.

For Price's potential (with the convention regarding outgoing and ingoing waves which we have adopted)

$$\psi(x) = \mathrm{e}^{\mathrm{i}\sigma x} \quad \text{for} \quad x \leqslant m \tag{41}$$

with $\mathrm{Re}(\sigma) \geqslant 0$. For $x \geqslant m$, the general solution of the wave equation is clearly a linear combination of $x^{\frac{1}{2}}J_{l+\frac{1}{2}}(\sigma x)$ and $x^{\frac{1}{2}}J_{-l-\frac{1}{2}}(\sigma x)$; and from the known series expansions for $J_{\pm(l+\frac{1}{2})}(Z)$ (cf. Watson 1922, pp. 53 and 54, equations (1) and (4)) we conclude that

$$\psi(x) = A\,\mathrm{e}^{-\mathrm{i}\sigma x} \sum_{s=0}^{l} \frac{(-\mathrm{i})^s (l+s)!\,(\sigma x)^{-s}}{2^s s!\,(l-s)!} \quad (x \geqslant m), \tag{42}$$

where A is a constant. The continuity of the logarithmic derivative of ψ (i.e. of the phase function ϕ) at $x = m$ leads to the following algebraic equation:

$$2\mathrm{i} \sum_{s=0}^{l} \frac{(-\mathrm{i})^s (l+s)!\,(\sigma m)^{-s}}{2^s s!\,(l-s)!} + \sum_{s=0}^{l} \frac{(-\mathrm{i})^s s(l+s)!\,(\sigma m)^{-s}}{2^s s!\,(l-s)!} = 0. \tag{43}$$

From the character and the order of this equation we infer that if l is odd there are exactly $\frac{1}{2}(l+1)$ complex roots for σ with both its real and imaginary parts positive; while if l is even there is one purely imaginary root and $\frac{1}{2}l$ complex roots with both its real and imaginary parts positive. All these roots correspond to damped waves though they exponentially diverge both for $x \rightarrow +\infty$ and for $x \rightarrow -\infty$.

Press's numerical solutions of equation (43) are listed along with those of Zerilli's equation in table 1 (see §4 below).

We observe that in contrast to the case of the rectangular barrier, there are only a finite number of quasi-normal modes for each l.

4. The quasi-normal modes of the Schwarzschild black hole

We now turn to the determination of the quasi-normal modes of Zerilli's equation

$$\mathrm{d}^2Z/\mathrm{d}r_*^2 + (\sigma^2 - V_Z)\,Z = 0, \tag{44}$$

where

$$V_Z(r) = \frac{2n^2(n+1)\,r^3 + 6n^2mr^2 + 18nm^2r + 18m^3}{r^3(nr+3m)^2}\left(1 - \frac{2m}{r}\right), \tag{45}$$

$$r_* = r + 2m\lg\left(\frac{r}{2m} - 1\right), \quad \text{and} \quad n = \tfrac{1}{2}(l-1)\,(l+2). \tag{46}$$

Zerilli's potential is clearly of short range; indeed, we have (cf. Chandrasekhar 1975, equation (38))

$$2m\int_{-\infty}^{+\infty} V_Z\,\mathrm{d}r_* = 2n + \tfrac{1}{2} = (l-1)\,(l+2) + \tfrac{1}{2}. \tag{47}$$

As $r_* \rightarrow \pm\infty$, equation (44) allows two independent solutions with the asymptotic behaviours

$$Z_\pm \rightarrow \mathrm{e}^{\pm\mathrm{i}\sigma r_*}. \tag{48}$$

As we are assuming a time dependence of the form $e^{i\sigma t}$, Z_- represents an outgoing wave and Z_+ represents an ingoing wave. A quasi-normal mode is one which belongs to a complex σ with $\mathrm{Re}(\sigma) \geqslant 0$, such that it represents a purely outgoing wave at $+\infty$ and a purely ingoing wave at $-\infty$.

A straightforward procedure for finding a quasi-normal mode would appear to be the following. For some chosen complex σ, start from a large positive value of r_*, where $|V(r_*)| \ll |\sigma|^2$, with $Z = Z_- = e^{-i\sigma r_*}$ and integrate backwards to some intermediate value of r_*; and similarly, start from a large negative value of r_*, where again $|V(r_*)| \ll |\sigma|^2$, with $Z = Z_+ = e^{+i\sigma r_*}$ and integrate forwards to the same intermediate value of r_*. The condition that the chosen value of σ belongs to a quasi-normal mode is, clearly, that the Wronskian of the two solutions, at the common point to which we have integrated them, vanishes. The vanishing of the Wronskian provides, then, with a criterion for searching in the positive half of the complex σ-plane for the characteristic values belonging to the quasi-normal modes. But this procedure, simple as it appears, is beset with grave numerical instabilities. They arise from the finite numerical accuracy of all methods of numerical integrations. Thus, for a large positive r_*, the solution for Z, with the asymptotic behaviour $e^{-i\sigma r_*}$, will very soon be contaminated by an admixture with the solution $e^{+i\sigma r_*}$ (which is exponentially small for large r_* when the imaginary part of σ is positive); and by the time we have integrated backwards to some finite r_*, the admixture with the unwanted solution will become appreciable and the solution, we shall be integrating, will no longer be of the kind that we have stipulated. The same thing will happen with the solution integrated forwards from a large negative r_*: the solution started with the asymptotic behaviour $e^{+i\sigma r_*}$ will be contaminated with the solution $e^{-i\sigma r_*}$ $(r_* \to -\infty)$; and, again, the solution will not be of the kind that we have stipulated by the time we have integrated it to some finite r_*.

The numerical instabiities we have described in the foregoing paragraph have prevented, so far, the determination of the quasi-normal modes of the Schwarzschild black hole (see, however, Detweiler 1975). But it appears that the integration of the first-order Riccati equation (6) is not beset with numerical instabilities, at least, to the same extent. The reason for the relative stability of integrating the Riccati equation for the phase function ϕ (as compared with the direct integration of the wave equation) appears to be that we can start the integrations, backwards from a positive r_* and forwards from a negative r_*, without regard to the requirement $|V(r_*)| \ll \sigma^2$, so long as we can ensure that we have convergent series expansions at both ends which are adequate enough to determine ϕ with sufficient accuracy to values of r_* where it differs substantially from its limiting values at $\pm\infty$.

For the particular case of Zerilli's equation the required series expansions are of the forms,

$$Z = e^{-i\sigma r_*} \sum_{j=0}^{\infty} \alpha_j r^{-j} \quad (r_* \to +\infty) \tag{49}$$

and

$$Z = e^{+i\sigma r_*} \sum_{j=0}^{\infty} \beta_j (r-2m)^j \quad (r_* \to -\infty), \tag{50}$$

where the coefficients α_j and β_j can be determined with the aid of the recurrence relations

$$\begin{aligned}2\mathrm{i}\sigma n^2(j+1)\,\alpha_{j+1}+[n^2j(j+1)-2n^2(n+1)+12\mathrm{i}\sigma mjn]\,\alpha_j\\ +m[6nj(j-1)-2n^2(j^2-1)-6n^2+18\mathrm{i}\sigma m(j-1)]\,\alpha_{j-1}\\ +m^2[9(j-1)(j-2)-12nj(j-2)-18n]\,\alpha_{j-2}\\ -18m^3[(j-1)(j-3)+1]\,\alpha_{j-3}=0,\end{aligned} \tag{51}$$

and

$$\begin{aligned}2\mathrm{i}\sigma n^2(j-1)\,\beta_{j-1}+j[n^2(j-1+8\mathrm{i}\sigma m)+12\mathrm{i}\sigma mn(n+1)-(A/j)]\,\beta_j\\ +m(j+1)\,[n^2(2j+2+8\mathrm{i}\sigma m)+6n(n+1)\,(j+8\mathrm{i}\sigma m)+6\mathrm{i}\sigma m(2n+1)\,(2n+3)\\ -(B/m(j+1))]\,\beta_{j+1}\\ +m^2(j+2)\,[6n(n+1)\,(2j+4+8\mathrm{i}\sigma m)+3(2n+1)\,(2n+3)\,(j+1+8\mathrm{i}\sigma m)\\ +4\mathrm{i}\sigma m(2n+3)^2-(C/m^2(j+2))]\,\beta_{j+2}\\ +m^3(j+3)\,[3(2n+1)\,(2n+3)\,(2j+6+8\mathrm{i}\sigma m)+2(2n+3)^2\,(j+2+8\mathrm{i}\sigma m)\\ -(D/m^3(j+3))]\,\beta_{j+3}\\ +4m^4(j+4)\,(2n+3)^2\,(j+4+4\mathrm{i}\sigma m)\,\beta_{j+4}=0,\end{aligned} \tag{52}$$

where
$$A=2n^2(n+1),\quad B=6n^2(2n+3)\,m,$$

$$C=6n(4n^2+8n+3)\,m^2\quad\text{and}\quad D=(16n^3+40n^2+36n+18)\,m^3. \tag{53}$$

With Z determined at both ends by the expansions (49) and (50), the phase function ϕ follows from the equation

$$\phi=\frac{1}{Z}\frac{\mathrm{d}Z}{\mathrm{d}r_*}. \tag{54}$$

The search for the characteristic values of σ belonging to the quasi-normal modes proceeds, then, as follows. We choose a complex value of σ in the positive-half plane ($\mathrm{Re}\,(\sigma)>0$) and determine the expansion coefficients α_j and β_j in accordance with equations (51)–(53) and evaluate ϕ for values of r_* (both positive and negative) for which the series expansions (49) and (50) suffice to determine it accurately enough.† We then continue by numerical integration, backwards from $+\infty$ and forwards from $-\infty$, to a common intermediate value of r_* (generally $3m$ where V_Z is approximately at its maximum). At this common point we find the difference

$$M(\sigma)=\phi_-(r_*)-\phi_+(r_*). \tag{55}$$

The condition that the chosen value of σ belongs to a quasi-normal mode is that $M(\sigma)$ vanishes (in view of the Riccati equation being of the first order).

It was found that the foregoing procedure enables the determination of the quasi-normal modes so long as

$$|\mathrm{Im}\,(\sigma)|\leqslant|\mathrm{Re}\,(\sigma)|. \tag{56}$$

† It is necessary to retain as many terms in the expansions as are necessary to determine ϕ until it is substantially different from its limiting values at $\pm\infty$.

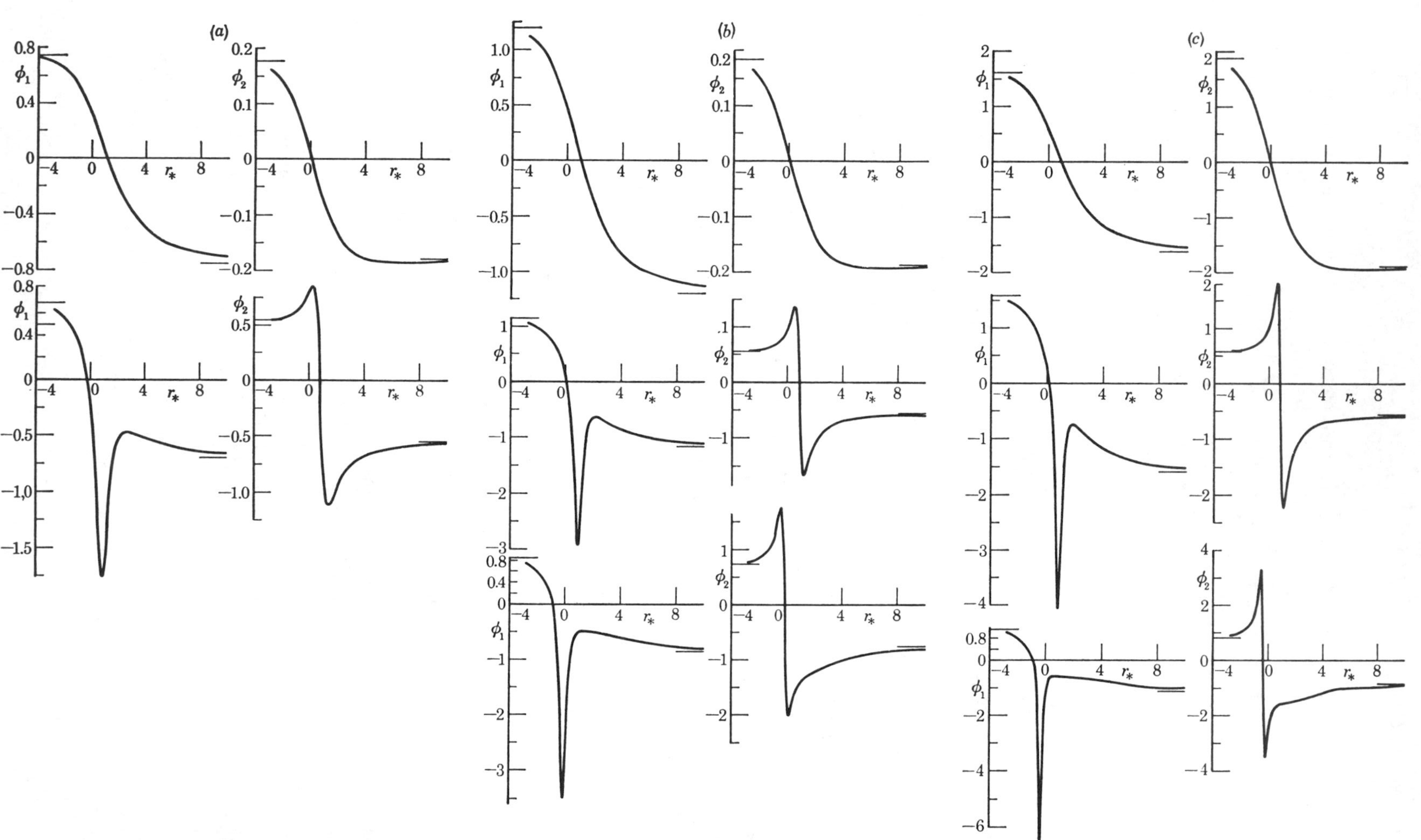

FIGURE 2. The complex phase $(\phi_1 + i\phi_2)$ belonging to the quasi-normal modes of Zerilli's potential for $l = 2$, 3, and 4 (figures (a), (b), and (c), respectively). The asymptotes ($\pm\sigma_1$ and $\pm\sigma_2$) to which the phases ϕ_1 and ϕ_2 tend for $r_* \to +\infty$ and $-\infty$ are indicated. It will be observed that there are two modes for $l = 2$ and three each for $l = 3$ and 4.

When this condition is violated, the numerical integration appears to suffer from instabilities. The underlying cause is probably the same as in the direct integration of the wave equation though it is obscured (and made less unstable!) by the non-linearity of the Riccati equation.

In our search for the characteristic values of σ (for $l \leqslant 4$) in the positive half of the complex σ-plane, we limited ourselves to only those regions in which we were satisfied that the numerical procedure adopted was free from instabilities.

In figure 2 we illustrate the quasi-normal modes determined, in the fashion described, for $l = 2$, 3, and 4; and in table 1 we list the characteristic values of σ to which they belong. In table 1, we also list the characteristic values for Price's potential (as determined by Press). It will be observed that the two sets of values differ very markedly. This difference is perhaps not very surprising in view of the identity (8) and the fact that the integral,

$$2m\int_m^\infty \frac{l(l+1)}{r_*^2}\,\mathrm{d}r_* = 2l(l+1), \tag{57}$$

for Price's potential differs from the corresponding integral for Zerilli's potential (cf. equation (47)) by a factor exceeding 2.

TABLE 1. THE COMPLEX FREQUENCIES BELONGING TO THE QUASI-NORMAL MODES OF ZERILLI'S AND PRICE'S POTENTIALS: σ IS EXPRESSED IN THE UNIT $(2m)^{-1}$

	$2m\sigma$	
l	Zerilli's potential	Price's potential
2	0.74734 + 0.17792i	2.5087 + 1.6871i
	0.69687 + 0.54938i	2.6258i
3	1.19889 + 0.18541i	4.1604 + 2.2210i
	1.16402 + 0.56231i	1.3802 + 3.7790i
	0.85257 + 0.74546i	
4	1.61835 + 0.18832i	5.8840 + 2.6647i
	1.59313 + 0.56877i	2.9090 + 4.6967i
	1.12019 + 0.84658i	5.2771i

5. THE ODD-PARITY MODES

In determining the quasi-normal modes of the Schwarzschild black hole in §4, we restricted ourselves to the even-parity modes described by the Zerilli equation. But it is known that the odd-parity perturbations are described by the Regge–Wheeler equation (Regge & Wheeler 1957; see also Chandrasekhar 1975, appendix):

$$\mathrm{d}^2X/\mathrm{d}r_*^2 + (\sigma^2 - V_0)\,X = 0, \tag{58}$$

where

$$V_0 = 2\left(1-\frac{2m}{r}\right)\frac{(n+1)\,r-3m}{r^3}. \tag{59}$$

It is, however, not necessary to consider this equation separately, since, as we shall now show, it must yield the same complex frequencies (and, indeed, the same reflexion and transmission coefficients) as Zerilli's equation. (It may also be noticed here that the integral of V_0 over the range of r_* has the same value (47), as for V_Z.)

In the equation (Chandrasekhar 1975, equation (A 10)),

$$[\tfrac{2}{3}n(n+1)-2mi\sigma]\,X \doteq \frac{r^2(nr+3m)}{3(r-2m)}\,Y-\frac{r^6}{6(r-2m)^2}(W_0+2i\sigma)\,\Lambda_-Y, \tag{60}$$

substitute for Y and Λ_-Y (in terms of Z) in accordance with equations (51) and (52) of the same paper. On simplifying the resulting equation, we are left with

$$[\tfrac{2}{3}n(n+1)-2mi\sigma]\,X = \left[\tfrac{2}{3}n(n+1)+\frac{6m^2(r-2m)}{r^2(nr+3m)}\right]Z-2m\frac{\mathrm{d}Z}{\mathrm{d}r_*}. \tag{61}$$

Therefore, a solution of Zerilli's equation with the asymptotic behaviour,

$$\begin{aligned} Z &\to \mathrm{e}^{+i\sigma r_*}+A\,\mathrm{e}^{-i\sigma r_*} \quad (r_*\to+\infty)\\ &\to B\,\mathrm{e}^{+i\sigma r_*} \quad (r_*\to-\infty), \end{aligned} \tag{62}$$

will yield a solution of equation (58) with the behaviour

$$\begin{aligned} X &\to \mathrm{e}^{+i\sigma r_*}+\frac{\frac{2}{3}n(n+1)+2mi\sigma}{\frac{2}{3}n(n+1)-2mi\sigma}A\,\mathrm{e}^{-i\sigma r_*} \quad (r_*\to+\infty)\\ &\to B\,\mathrm{e}^{+i\sigma r_*} \quad (r_*\to-\infty). \end{aligned} \tag{63}$$

The equality of the reflexion and transmission coefficients, that are determined by the two equations, is now manifest.† It is also clear that the complex frequencies belonging to the quasi-normal modes of the two parities must be the same; and the modes themselves must be related by equation (61).

We are grateful to Professor W. H. Press for allowing us to quote his unpublished results on the quasi-normal modes for Price's potential.

The research reported in this paper has in part been supported by the National Science Foundation under grant MPS 74-17456 and the Louis Block Fund, The University of Chicago.

References

Chandrasekhar, S. 1964 *Astrophys. J.* **140**, 417.
Chandrasekhar, S. 1975 *Proc. R. Soc. Lond.* A **343**, 289.
Detweiler, S. 1975 *Astrophys. J.* **197**, 203.
Eddington, A. S. 1918 *Mon. Not. R. astr. Soc.* **79**, 2.
Eddington, A. S. 1919 *Mon. Not. R. astr. Soc.* **79**, 177.
Flugge, S. & Marschall, H. 1952 *Rechenmethoden der Quanten Theorie.* Berlin: Springer-Verlag.
Press, W. H. 1973 (private communication).
Price, R. H. 1972*a* *Phys. Rev.* D **5**, 2419.
Price, R. H. 1972*b* *Phys. Rev.* D **5**, 2439.
Regge, T. & Wheeler, J. A. 1957 *Phys. Rev.* **108**, 1063.
Watson, G. N. 1922 *Theory of Bessel functions.* Cambridge University Press.
Zerilli, F. J. 1970 *Phys. Rev.* D **2**, 2141.

† This equality had been suspected by C. V. Vishveshwara (private communication) from the numerical agreement he had found from a direct evaluation of the coefficients. But the reason for the agreement does not appear to have been asked.

On the Potential Barriers Surrounding the Schwarzschild Black Hole

SUBRAHMANYAN CHANDRASEKHAR

1. INTRODUCTION

The remarkable simplicity of the general relativistic theory of black holes derives from the fact that under stationary conditions, Einstein's equations allow for them only a single two-parameter family of solutions. This is the Kerr family of solutions. The two parameters of the solution are the mass M and the angular momentum J of the black hole. If we enlarge Einstein's equations to the Einstein-Maxwell equations (i.e., Einstein's equations with a source represented by the energy-momentum tensor of an electromagnetic field) then the Kerr family is enlarged to the Kerr-Newman family with an additional parameter Q_* which specifies the charge of the black hole.

When we restrict ourselves to the case $J = 0$, the solutions become spherically symmetric and reduce to the ones associated with the names of Schwarzschild and of Reissner and Nordström.

We have now a fairly complete understanding of the Schwarzschild, the Reissner-Nordström, and the Kerr solutions and their perturbations. Indeed, many of the features of the Reissner-Nordström and Kerr solutions are already present in the Schwarzschild solution in simpler contexts; and an understanding of the Schwarzschild solution appears as a prerequisite to an understanding of the other solutions. For this reason, the present account will be restricted to the Schwarzschild solution.

One of the best ways of understanding the attributes of a physical system is to find out how it reacts to external perturbations and, in the first instance, to infinitesimal perturbations. In the case of the black holes this is the only method available to us since there is no way in which an external observer can explore the "other side" of the horizon.

The reaction of an object to an infinitesimal perturbation is determined by the enumeration of the so-called normal modes of oscillation. In the case of the black holes, this enumeration reduces to finding how a black hole reacts to incident waves of different sorts. The solution to this latter problem bears on a number of related questions: the stability of the black hole and the determination of the "quasi-normal modes" (which describe the long-term behavior of a black hole that has been perturbed) (i.e., to its ringing). Further, the analysis, besides suggesting relations to domains which one might not have expected, discloses what appears to be a hidden symmetry of the solutions. Since a complete and self-contained account, of all these different aspects, is not possible within the limits of a single article (even when restricted to the Schwarzschild black hole), we must content ourselves to a bare description of the methods and the results except in cases where no ready reference is available.

2. AN OUTLINE OF THE PROBLEM

It is known that by choosing a coordinate frame consistent with the assumption of spherical symmetry, the metric can be written in the form

$$ds^2 = -e^{2\nu}(dt)^2 + r^2\sin^2\theta\,(d\varphi)^2 + e^{2\mu_2}\,(dr)^2 + r^2(d\theta)^2\,, \tag{1}$$

where ν and μ_2 are, in the first instance, functions of r and t. And when we restrict ourselves to vacuum fields, as we shall, Birkhoff's theorem assures us that ν and μ_2 are functions of r only.

For the Schwarzschild solution

$$e^{2\nu} = e^{-2\mu_2} = \frac{\Delta}{r^2}, \quad \text{where} \quad \Delta = r^2 - 2Mr\,. \tag{2}$$

The horizon of the black hole occurs at

$$r = 2M\,, \quad \text{where} \quad \Delta = 0\,. \tag{3}$$

As is well known, the space interior to $r = 2M$ is incommunicable to the space outside, a consequence of the fact that $r = 2M$ defines a *null surface*, that is, a surface whose tangent at every point is a null vector.

As we have stated, the problem of the perturbation of a black hole is effectively one of determining how waves of different sorts, incident on the black hole, are affected by, and in turn affect, the black hole. In the context of the Schwarzschild black hole, the case of greatest interest is when the incident waves are gravitational in origin.

From general considerations, one may expect that a fraction of the energy in the incident waves will be irreversibly absorbed by the black hole, while the remaining fraction will be scattered (or, reflected) back to infinity. In other words, it would appear that it may be possible to visualize the black hole as presenting a potential barrier to the oncoming waves. We shall see that this expectation is amply fulfilled (not only in the context of the Schwarzschild black hole but in all other contexts as well). Indeed, it will appear that the problem can be reduced to the very simple one of the penetration of one-dimensional potential barriers with which we are familiar in elementary quantum theory. The basic reason why this reduction to a *strict* one-dimensional problem is possible, is the fact that nothing can emerge from the horizon of the black hole and that, in consequence, the region interior to the horizon is of no relevance to our considerations. This latter fact can be expressed in precise mathematical terms by changing to a variable r_* which translates the horizon to minus infinity. Thus, by letting

$$\frac{d}{dr_*} = \frac{\Delta}{r^2}\frac{d}{dr}, \tag{4}$$

or, explicitly,

$$r_* = r + 2M\, ln\left(\frac{r}{2M} - 1\right) \quad (r > 2M), \tag{5}$$

we have the behaviors

$$r_* \rightarrow r \quad \text{as} \quad r \rightarrow \infty \quad \text{and} \quad r_* \rightarrow -\infty \quad \text{as} \quad r \rightarrow 2M + 0. \tag{6}$$

The permitted range of r_* is the entire internal $(-\infty, +\infty)$, and we are debarred from making any reference to $r < 2M$.

In terms of the variable r_*, the boundary conditions are readily stated: for $r_* \rightarrow +\infty$, we can have both incoming and outgoing waves, since, in addition to the postulated incident waves, we can also have waves reflected back to infinity by the black hole; but for $r_* \rightarrow -\infty$, we can, only, have waves going towards $-\infty$ since no waves can emerge from the interior of the black hole.

We conclude this section by stating that the perturbations of the black-hole solutions of general relativity can be studied from two different standpoints and two different bases: *either* directly for the metric perturbations via the Einstein (or the Einstein-Maxwell) equations linearized about the unperturbed solution, *or* for the Weyl (and the Maxwell) scalars via the equations of the Newman-Penrose formalism. It is found that the two methods supplement one another very effectively. For this reason, we shall begin with a brief account of the two methods of treating the problem in the explicit context of the Schwarzschild black hole.

3. THE METHOD OF THE METRIC PERTURBATIONS

Since we are dealing with the perturbations of a system which is initially spherically symmetric, it is clear that, in the equations governing the perturbations, the independent variables r, θ, and φ can be separated. In particular, the dependence on φ can be separated by seeking solutions with a φ-dependence given by $e^{im\varphi}$ where m is an integer positive, negative, or zero. On the other hand, inasmuch as the solutions with a φ-dependence $e^{im\varphi}$ can be obtained by subjecting the solutions independent of φ (i.e., with $m = 0$) to a rotation, it will suffice to linearize the exact equations, governing nonstationary axisymmetric systems, about the given time-independent spherically symmetric solution.

A form of the metric adequate for treating nonstationary axisymmetric systems in general relativity is given by

$$ds^2 = - e^{2\nu}(dt)^2 + e^{2\psi}(d\varphi - \omega dt - q_2 dx^2 - q_3 dx^3)^2 + e^{2\mu_2}(dx^2)^2 + e^{2\mu_3}(dx^3)^2 , \tag{7}$$

where ν, ψ, μ_2, μ_3, ω, q_2, and q_3 are all functions of t and the spatial coordinates x^2 and x^3 (which we may identify with a "radial coordinate" r and an angular coordinate θ). Further, the functions ω, q_2, and q_3 can occur in the field equations only in the combinations

$$q_{2,3} - q_{3,2} , \quad \omega_{,2} - q_{2,0} , \quad \text{and} \quad \omega_{,3} - q_{3,0} , \tag{8}$$

where "0" signifies "t."

We observe that the metric (7) includes the form

$$ds^2 = - e^{2\nu}(dt)^2 + e^{2\psi}(d\varphi - \omega dt)^2 + e^{2\mu_2}(dx^2)^2 + e^{2\mu_3}(dx^3)^2 , \tag{9}$$

which is appropriate for describing stationary axisymmetric systems. And when considering stationary systems, the functions ν, ψ, μ_2, μ_3, and ω are functions only of x^2 and x^3; and, moreover, we have the gauge freedom to impose any coordinate condition on μ_2 and μ_3 which we may choose.

The function ω, which occurs in the metrics (7) and (9), plays a decisive role in distinguishing the character of the perturbations. The physical significance of ω is the following. Let $\Omega(x^2,x^3) = d\varphi/dt$, in the stationary case, denote the angular velocity as measured by an observer at infinity. Then, an observer, who considers himself as locally at rest at (x^2,x^3), will be attributed by an observer at infinity as having an angular velocity $\Omega - \omega$ instead of Ω. This difference in the angular velocities, as perceived by the two observers, is commonly ascribed to the "dragging of the inertial frames" by the angular momentum resident in the field: the angular momentum, J, is in fact determined by the asymptotic behavior, $2J/r^3$, of ω.

Now comparing the metrics (1) and (7), we observe that the Schwarzschild solution ascribes to the various metric functions in (9) the values,

$$e^{2\nu} = e^{-2\mu_2} = \Delta r^{-2} , \quad e^{\psi} = r\sin\theta , \quad e^{\mu_3} = r , \tag{10}$$

and

$$\omega = q_2 = q_3 = 0\,, \tag{11}$$

where, it may be noted that, besides the identifications

$$x^0 = t\,, \quad x^2 = r\,, \quad \text{and} \quad x^3 = \theta\,, \tag{12}$$

the available gauge freedom has been utilized in the choice of the radial coordinate r as defining a "luminosity distance" (which preserves the value $4\pi r^2$ for the area of the two-surface r = constant).

From our earlier remarks, it would appear that we can distinguish among the perturbations of a spherically symmetric solution, two noncombining classes:

(i) Perturbations in which ν, ψ, μ_2, and μ_3 are left unchanged; and perturbation consists only in making ω, q_2, and q_3 nonvanishing (but infinitesimally small so that only terms linear in these quantities need be retained).

(ii) Perturbations in which ν, ψ, μ_2, and μ_3 are subjected to infinitesimal changes while ω, q_2, and q_3 continue to be vanishing.

The perturbations belonging to the two classes are generally referred to as of "odd" and of "even" parities. While the parities of the two classes of perturbations are clearly opposite, the nomenclature "odd" and "even" is unfortunate and misleading. We shall avoid using it.

Equations governing the two classes of perturbations can be obtained by linearizing the Einstein equations appropriate to the metric (7) about the Schwarzschild solution. We shall consider the results of this linearization in § 5 below.

4. THE METHOD BASED ON THE NEWMAN-PENROSE FORMALISM

In the Newman-Penrose formalism, the description of the spacetime is in terms of a local tetrad frame based on four null vectors, $(\mathbf{l}, \mathbf{n}, \mathbf{m}, \overline{\mathbf{m}})$, of which $\mathbf{l}$ and $\mathbf{n}$ are real and $\mathbf{m}$ and $\overline{\mathbf{m}}$ arc complex conjugates. It is further supposed that the chosen vectors satisfy the orthogonality relations,

$$\begin{aligned} &\mathbf{l} \cdot \mathbf{n} = 1\,, \quad \mathbf{m} \cdot \overline{\mathbf{m}} = -1\,, \quad \text{and} \\ &\mathbf{l} \cdot \mathbf{m} = \mathbf{l} \cdot \overline{\mathbf{m}} = \mathbf{n} \cdot \mathbf{m} = \mathbf{n} \cdot \overline{\mathbf{m}} = 0\,. \end{aligned} \tag{13}$$

And any tensor (in spacetime), in which we may be interested, is projected onto the chosen frame by contracting it with the basis vectors in all possible nontrivial ways. Thus, in place of the Weyl tensor C_{pqrs}, we consider the five complex scalars

$$\Psi_0 = -\,C_{pqrs}l^p m^q l^r m^s\,,$$

$$\Psi_1 = -\,C_{pqrs}l^p n^q l^r m^s\,,$$

$$\Psi_2 = - C_{pqrs} l^p m^q \overline{m}^r n^s ,\tag{14}$$

$$\Psi_3 = - C_{pqrs} l^p n^q \overline{m}^r n^s ,$$

$$\Psi_4 = - C_{pqrs} n^p \overline{m}^q n^r \overline{m}^s ;$$

and in place of the Ricci tensor, we consider the scalars

$$\Phi_{00} = -\frac{1}{2} R_{pq} l^p l^q , \qquad \Phi_{01} = -\frac{1}{2} R_{pq} l^p m^q ,$$

$$\Phi_{11} = -\frac{1}{4} R_{pq} (l^p n^q + m^p \overline{m}^q) , \quad \Phi_{12} = -\frac{1}{2} R_{pq} n^p m^q , \tag{15}$$

$$\Phi_{22} = -\frac{1}{2} R_{pq} n^p n^q , \qquad \Phi_{02} = -\frac{1}{2} R_{pq} m^p m^q .$$

The various equations governing the gravitational and other fields in which we may be interested are similarly projected onto the chosen frame and expressed in terms of the various scalars. It is clear that these equations will involve the covariant derivatives of the basis vectors, also, projected onto the null frame. These are the so-called *spin coefficients*; they are equivalent to the Ricci rotation-coefficients when the chosen frame is an orthonormal one consisting of one timelike and three spacelike vectors. In the Newman-Penrose formalism, the spin coefficients are designated by special symbols:

$$\begin{aligned}
\kappa &= m^i l_{i;j} l^j = - l^i m_{i;j} l^j , \\
\sigma &= m^i l_{i;j} m^j = - l^i m_{i;j} m^j , \\
\lambda &= n^i \overline{m}_{i;j} \overline{m}^j = - \overline{m}^i n_{i;j} \overline{m}^j , \\
\nu &= n^i \overline{m}_{i;j} n^j = - \overline{m}^i n_{i;j} n^j , \\
\rho &= m^i l_{i;j} \overline{m}^j = - l^i m_{i;j} \overline{m}^j , \\
\mu &= n^i \overline{m}_{i;j} m^j = - \overline{m}^i n_{i;j} m^j , \\
\tau &= m^i l_{i;j} n^j = - l^i m_{i;j} n^j , \\
\pi &= n^i \overline{m}_{i;j} l^j = - \overline{m}^i n_{i;j} l^j , \\
\varepsilon &= \frac{1}{2} (n^i l_{i;j} l^j + m^i \overline{m}_{i;j} l^j) , \\
\gamma &= \frac{1}{2} (n^i l_{i;j} n^j + m^i \overline{m}_{i;j} n^j) , \\
\alpha &= \frac{1}{2} (n^i l_{i;j} \overline{m}^j + m^i \overline{m}_{i;j} \overline{m}^j) ,
\end{aligned} \tag{16}$$

and

$$\beta = \frac{1}{2}(n^i l_{i;j} m^j + m^i \overline{m}_{i;j} m^j).$$

In the Newman-Penrose formalism (as indeed, in any tetrad formalism) one writes down three sets of equations: the *Bianchi identities*,

$$R_{ij[kl;m]} = 0 \tag{17}$$

expressed in terms of the spin coefficients and the Weyl and the Ricci scalars; the *commutation relations*,

$$[\mathbf{l}^{(a)}, \mathbf{l}^{(b)}] = C^{(a)(b)}{}_{(c)}\mathbf{l}^{(c)}, \tag{18}$$

where we have written $\mathbf{l}^{(1)}$, $\mathbf{l}^{(2)}$, $\mathbf{l}^{(3)}$, and $\mathbf{l}^{(4)}$ in place of $\mathbf{l}$, $\mathbf{n}$, $\mathbf{m}$, and $\overline{\mathbf{m}}$, and the basis vectors are to be interpreted as directional derivatives, and the *structure constants* $C^{(a)(b)}{}_{(c)}$ are expressed in terms of the spin coefficients; and, finally, the components of the Riemann tensor derived from the *Ricci identity*,

$$\mathbf{l}^{(a)}{}_{i;k;l} - \mathbf{l}^{(a)}{}_{i;l;k} = \mathbf{l}^{(a)m} R_{mikl}, \tag{19}$$

and written in terms of the spin coefficients, their directional derivatives, and the Weyl and the Ricci scalars.

The Newman-Penrose formalism has had spectacular successes in treating the perturbations of the black-hole solutions of general relativity. The basic reason for this success can be traced to the circumstance that the geometry of the underlying spacetimes (by virtue of their belonging to the so-called "type-D" class) enables a choice of basis, $\mathbf{l}$, $\mathbf{n}$, $\mathbf{m}$, and $\overline{\mathbf{m}}$, such that the vectors $\mathbf{l}$ and $\mathbf{n}$ form shear-free congruences of null geodesics with the consequence that the Weyl scalars Ψ_0, Ψ_1, Ψ_3, and Ψ_4 and the spin coefficients κ, σ, λ, and ν all vanish. These latter facts entail that we can determine these important quantities, when they are nonvanishing in the perturbed state, without solving for the metric perturbation or, indeed, any other quantity.

The Description of the Schwarzschild Geometry in the Newman-Penrose Formalism

A basic null-tetrad satisfying the requirements

$$\mathbf{l}\cdot\mathbf{n} = 1, \quad \mathbf{m}\cdot\overline{\mathbf{m}} = -1, \quad \text{and} \quad \kappa = \sigma = \lambda = \nu = 0 \tag{20}$$

and appropriate to the Schwarzschild metric is given by

$$l^i = \frac{1}{\Delta}(r^2, +\Delta, 0, 0) \qquad ; \quad l_i = \left(1, -\frac{r^2}{\Delta}, 0, 0\right),$$

$$n^i = \frac{1}{2r^2}(r^2,-\Delta,0,0) \quad ; \quad n_i = \frac{1}{2r^2}(\Delta,+r^2,0,0)\,, \tag{21}$$

$$m^i = \frac{1}{r\sqrt{2}}(0,0,1,i\ \mathrm{cosec}\theta)\,; \quad m_i = \frac{1}{r\sqrt{2}}(0,0,-r^2,-ir^2\sin\theta)\,.$$

The remaining spin coefficients determined by this basis are

$$\rho = -\frac{1}{r}, \quad \beta = \frac{\cot\theta}{r(2\sqrt{2}\,)} = -\alpha, \quad \pi = \tau = \varepsilon = 0\,,$$

$$\mu = -\frac{\Delta}{2r^3}, \quad \text{and} \quad \gamma = \mu + \frac{r-M}{2r^2}. \tag{22}$$

And it can be verified that, with respect to the chosen basis,

$$\Psi_0 = \Psi_1 = \Psi_3 = \Psi_4 = 0 \tag{23}$$

and

$$\Psi_2 = -Mr^{-3}\,. \tag{24}$$

This completes the description of the Schwarzschild geometry in the Newman-Penrose formalism.

5. THE METRIC PERTURBATIONS OF THE SCHWARZSCHILD BLACK HOLE AND THE POTENTIAL BARRIERS SURROUNDING IT

As we have explained in § 3, we can distinguish two classes of perturbations of opposite parity: those of class I in which ν, ψ, μ_2, and μ_3 remain unchanged, while ω, q_2, and q_3 are nonvanishing, and those of class II in which ν, Ψ, μ_2, and μ_3 experience infinitesimal changes, while ω, q_2, and q_3 remain vanishing.

Considering first the perturbations of class I, we find that with the substitution

$$re^{2\nu}(q_{2,3} - q_{3,2})\sin^3\theta = Z^{(-)}(r)P_{l+2}(\cos\theta \mid -3)e^{i\sigma t}\,, \tag{25}$$

where σ is a constant and $P_{l+2}(\cos\theta \mid -3)$ is Gegenbauer's polynomial of order $l+2$ and index -3, $Z^{(-)}(r)$ satisfies the one-dimensional wave equation

$$\left(\frac{d^2}{dr_*^2} + \sigma^2\right) Z^{(-)} = V^{(-)}Z^{(-)}\,, \tag{26}$$

where

$$V^{(-)} = 2\frac{\Delta}{r^5}[(n+1)r - 3M] \tag{27}$$

and

$$n = \frac{1}{2}(l-1)(l+2) = \mu^2 \text{ (say)}. \tag{28}$$

Equation (26) is generally referred to as the Regge-Wheeler equation.

Turning next to the perturbations of class II, we find that the variables can be separated by the substitutions (due to J. Friedman)

$$\begin{aligned} \delta\nu &= N(r)P_l(\cos\theta)e^{i\sigma t}, \\ \delta\mu_2 &= L(r)P_l(\cos\theta)e^{i\sigma t}, \\ \delta\mu_3 &= [T(r)P_l + V(r)P_{l,\theta\theta}]e^{i\sigma t}, \end{aligned} \tag{29}$$

and

$$\delta\psi = [T(r)P_l + V(r)P_{l,\theta}\cot\theta]e^{i\sigma t},$$

where N, L, T, and V are four radial functions. And we find that Einstein's equations, appropriate to the metric (7), linearized about Schwarzschild's solution, lead to the algebraic relation,

$$T - V + L = 0, \tag{30}$$

and three additional equations expressing the radial derivatives of N, L, and V as linear combinations of these same functions. Making use of these equations, we find that the function,

$$Z^{(+)} = \frac{r^2}{nr+3M}\left(\frac{3M}{r}V - L\right), \tag{31}$$

satisfies the one-dimensional wave equation

$$\left(\frac{d^2}{dr_*^2} + \sigma^2\right)Z^{(+)} = V^{(+)}Z^{(+)}, \tag{32}$$

where

$$V^{(+)} = 2\frac{\Delta}{r^5}\frac{n^2(n+1)r^3 + 3n^2Mr^2 + 9nM^2r + 9M^3}{(nr+3M)^2}. \tag{33}$$

Equation (33) is generally referred to as the Zerilli equation.

The potentials $V^{(\pm)}$ are positive for all values of r_* (see Figures 6.1 and 6.2) and have similar asymptotic behaviors:

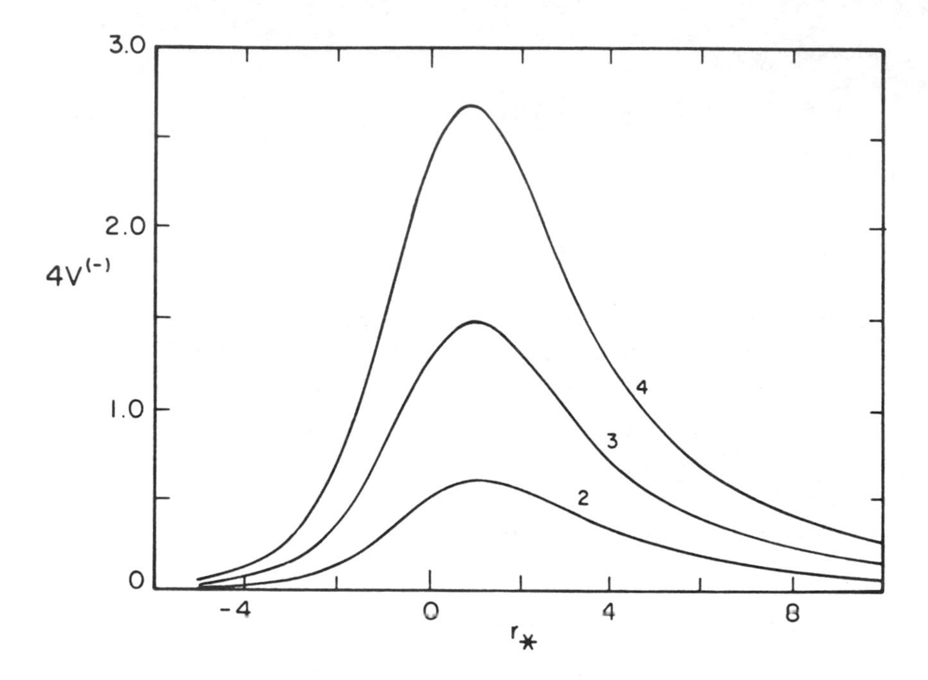

Figure 6.1. The potential barrier $V^{(-)}$ surrounding the Schwarzschild black hole. The curves are labeled by the values of l to which they belong; also r_* is measured in the unit $2M$.

$$\left.\begin{aligned} &V^{(\pm)} \to 2(n+1)r^{-2} && (r \to \infty; r_* \to +\infty) \\ &\text{and} \\ &V^{(\pm)} \to (\text{constant})_{\pm}\, e^{r_*/2M} && (r_* \to -\infty). \end{aligned}\right\} \tag{34}$$

The potentials are, therefore, of *short range* (i.e., their integrals over the range of r_* are finite); indeed, it can be verified that

$$2M \int_{-\infty}^{+\infty} V^{(\pm)}\, dr_* = 2n + \frac{1}{2}. \tag{35}$$

This equality of the integrals of $V^{(+)}$ and $V^{(-)}$ over the range of r_* is the manifestation of an important property of the two classes of perturbations to which we shall return in § 10 below.

The Problem of Reflection and Transmission

From the fact that $Z^{(+)}$ and $Z^{(-)}$ satisfy one-dimensional wave equations, we conclude that the normal modes of the black hole are determined by the

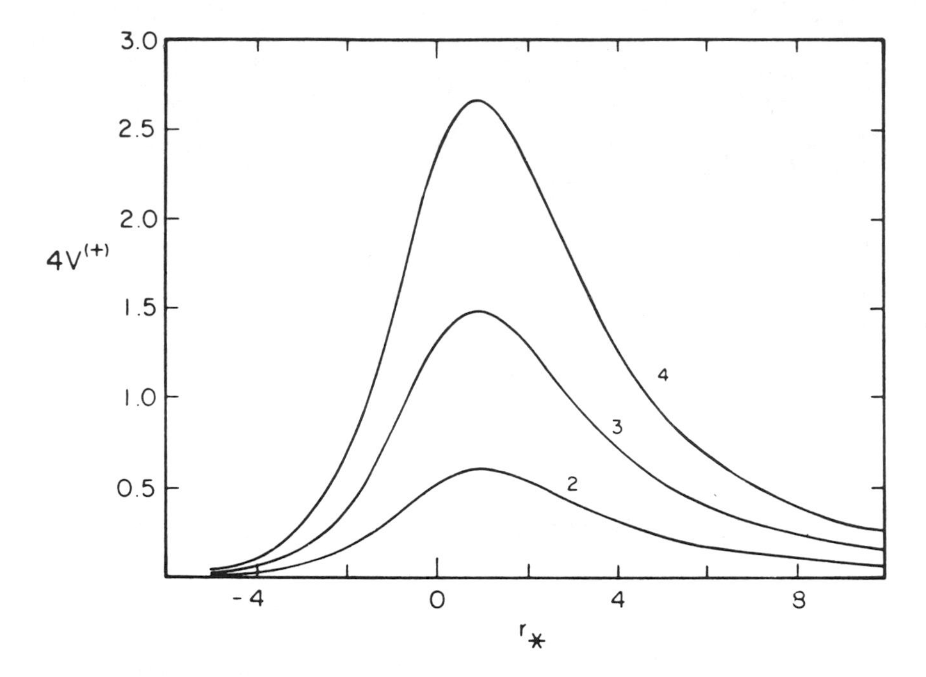

Figure 6.2. The potential barrier $V^{(+)}$ surrounding the Schwarzschild black hole. The curves are labeled by the values of l to which they belong; also r_* is measured in the unit $2M$.

solution to the elementary problem of the penetration of the potential barriers associated with $V^{(+)}$ and $V^{(-)}$. And since the potentials are of short range, they are determined by the solutions of the equations which satisfy the boundary conditions

$$\begin{aligned} Z^{(\pm)} &\rightarrow e^{+i\sigma r_*} + A^{(\pm)} e^{-i\sigma r_*} \quad (r_* \rightarrow +\infty) \\ &\rightarrow \qquad\qquad B^{(\pm)} e^{+i\sigma r_*} \quad (r_* \rightarrow -\infty), \end{aligned} \tag{36}$$

where $A^{(\pm)}$ and $B^{(\pm)}$ are the amplitudes of the waves which are reflected and transmitted by the potential barriers, $V^{(\pm)}$, when a wave of unit amplitude and frequency σ is incident on the black hole from infinity (see Figure 6.3). The fractions of the incident flux of energy that are reflected and transmitted are, therefore, given by

$$\mathbf{R}^{(\pm)} = |A^{(\pm)}|^2 \quad \text{and} \quad \mathbf{T}^{(\pm)} = |B^{(\pm)}|^2. \tag{37}$$

The quantities $\mathbf{R}^{(\pm)}$ and $\mathbf{T}^{(\pm)}$ define the *reflection* and the *transmission* coefficients appropriate to the two classes of perturbations.

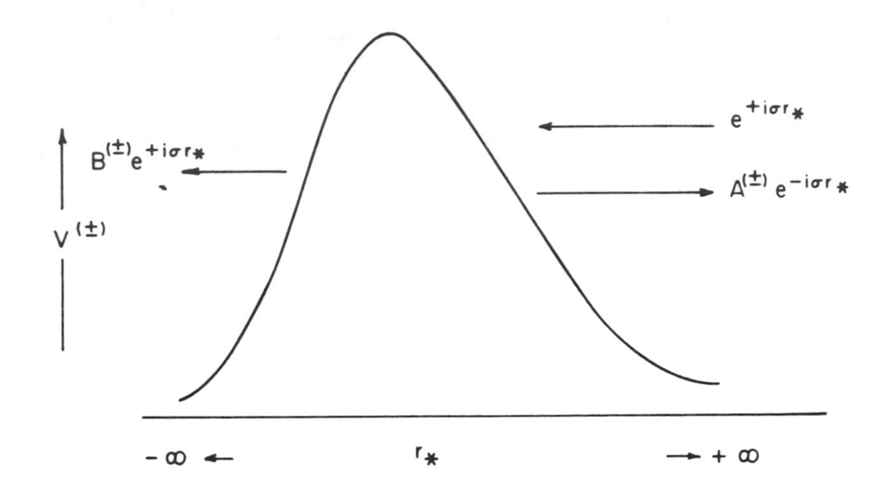

Figure 6.3. Illustrating the problem of reflection and transmission by potential barriers.

From the constancy of the Wronskian of $Z^{(\pm)}$ and their complex conjugates, it follows that, consistent with the meanings of these quantities,

$$\mathrm{R}^{(\pm)} + \mathrm{T}^{(\pm)} = 1 . \tag{38}$$

It is a remarkable fact that the reflection and the transmission coefficients for the two classes of perturbations are *identically equal* (i.e., for all values of σ) even though $V^{(+)}$ and $V^{(-)}$ are such entirely different functions. The meaning and the origin of this equality is what we shall be concerned with, principally, in the rest of this article.

6. A WAVE EQUATION WHICH FOLLOWS FROM THE NEWMAN-PENROSE FORMALISM

The equations of the Newman-Penrose formalism provide equations which are already linearized in the sense that they are linear and homogeneous in the quantities which vanish in the unperturbed state, namely, Ψ_0, Ψ_1, Ψ_3, Ψ_4, κ, σ, λ, and ν. These quantities can, therefore, be determined without any knowledge of how the perturbation affects any of the other quantities. The reduction of these equations is fairly straightforward and leads to wave equations for Ψ_0 and Ψ_4. Thus, writing

$$\Psi_0 = \frac{r^3}{2\Delta^2} Y(r)(\mathrm{P}_{l,\theta\theta} - P_{l,\theta}\cot\theta)e^{i\sigma t} , \tag{39}$$

we find that the radial function $Y(r)$ satisfies the equation

$$\Lambda^2 Y + P\Lambda_- Y - QY = 0\,, \tag{40}$$

where

$$\Lambda_\pm = \frac{d}{dr_*} \pm i\sigma\,, \quad \Lambda^2 = \Lambda_+\Lambda_- = \Lambda_-\Lambda_+ = \frac{d^2}{dr_*^2} + \sigma^2\,, \tag{41}$$

$$P = 4\,\frac{r-3M}{r^2} = \frac{d}{dr_*}\, lg\, \frac{r^8}{\Delta^2}\,, \tag{42}$$

and

$$Q = 2\,\frac{\Delta}{r^5}\,(nr + 3M)\,. \tag{43}$$

A similar consideration of Ψ_4 leads to the complex conjugate of equation (40).

It can be shown that the solutions for the perturbations in all the remaining quantities can be expressed in terms of the solution for Y. Accordingly, it would appear that the functions $Z^{(\pm)}$ must also be related to Y. We consider in the following sections how these relations are to be found.

7. THE TRANSFORMATION THEORY

Our aim is to transform equation (40) to an equation of the form,

$$\Lambda^2 Z = VZ\,, \tag{44}$$

where V is some function of r_*(or r) to be determined. It is clear that Z (on the assumption that one such exists) must be a linear combination of Y and its derivative. There is, therefore, no loss of generality in assuming that Z is related to Y in the manner,

$$Y = f\,VZ + (W + 2i\sigma f)\Lambda_+ Z\,, \tag{45}$$

where f, V, and W are certain functions unspecified for the present. Applying the operator Λ_- to equation (45) and making use of the fact that Z has been assumed to satisfy equation (44), we find that

$$\Lambda_- Y = -\,\beta\,\frac{\Delta^2}{r^8}\,Z + R\Lambda_+ Z\,. \tag{46}$$

where

$$-\,\frac{\Delta^2}{r^8}\,\beta = \frac{d}{dr_*}\,(f\,V) + WV \tag{47}$$

and

$$R = fV + \frac{d}{dr_*}(W + 2i\sigma f). \tag{48}$$

The requirement that equations (45) and (46) are compatible with equation (40), satisfied by Y, leads to the pair of equations,

$$\frac{d}{dr_*}\left(\frac{r^8}{\Delta^2}R\right) = \frac{r^8}{\Delta^2}[Q(W+2i\sigma f) - 2i\sigma R] + \beta \tag{49}$$

and

$$-\frac{\Delta^2}{r^8}\frac{d\beta}{dr_*} = (Qf-R)V. \tag{50}$$

It can be verified that equations (47)–(50) allow the integral

$$\frac{r^8}{\Delta^2}RfV + \beta(W+2i\sigma f) = K = \text{constant}. \tag{51}$$

Accordingly, it will suffice to consider any four of the five equations (47)–(51).

It will be observed that we are provided with only four independent equations for determining the five functions β, f, R, V, and W. We have, therefore, a considerable latitude in seeking solutions of equations (47)–(51).

Dual Transformations and the Conditions for the Existence of Transformations with β = Constant and f = 1

We shall presently verify that for the particular cases in which we are interested, equations (47)–(51) allow solutions consistently with the assumptions,

$$f = 1 \quad \text{and} \quad \beta = \text{constant}. \tag{52}$$

These assumptions are clearly over-restrictive if we consider Q as some arbitrarily given function. We are, therefore, essentially asking how Q should be restricted in order that the transformation equations (47)–(51) may allow solutions compatible with the assumptions of (52).

The assumptions (52) in conjunction with equations (50), (48), (49), and (51), successively require

$$R = Q, \qquad V = Q - \frac{dW}{dr_*}, \tag{53}$$

$$\frac{d}{dr_*}\left(\frac{r^8}{\Delta^2}Q\right) = \frac{r^8}{\Delta^2}QW + \beta, \tag{54}$$

and

$$\frac{r^8}{\Delta^2}QV + \beta W = K - 2i\sigma\beta = \kappa \text{ (say)}. \tag{55}$$

(Note that κ is to be a constant.)

With the definition

$$F = r^8 Q/\Delta^2, \tag{56}$$

equations (54) and (55) give

$$W = \frac{1}{F}\left(\frac{dF}{dr_*} - \beta\right) \tag{57}$$

and

$$FV + \beta W = \kappa. \tag{58}$$

From equations (53) and (58) it now follows that

$$F\left(Q - \frac{dW}{dr_*}\right) + \beta W = \kappa. \tag{59}$$

Finally, eliminating W from equation (59) with the aid of equation (57), we obtain the basic condition,

$$\frac{1}{F}\left(\frac{dF}{dr_*}\right)^2 - \frac{d^2F}{dr_*^2} + \frac{\Delta^2}{r^8}F^2 = \frac{\beta^2}{F} + \kappa. \tag{60}$$

Thus, a necessary and sufficient condition, for the transformation equations (47)–(51) to allow solutions consistently with the assumptions (52), is that equation (60) be satisfied by the given F (i.e., Q) for some suitably chosen constants β^2 and κ. Since β occurs as β^2 in equation (60), it follows that if equation (60) is satisfied for some values of β^2 and κ, then *two* transformations, associated with $+\beta$ and $-\beta$ (but the same κ), are possible. We shall call these *dual transformations*.

8. THE EXPLICIT EXPRESSION OF $Z^{(+)}$ AND $Z^{(-)}$ IN TERMS OF Y

We shall now verify that equations (47)–(51) allow solutions, consistently with the assumptions (52), for the potential

$$V^{(-)} = \frac{\Delta}{r^5}[(\mu^2+2)r - 6M] \quad (\mu^2 = 2n), \tag{61}$$

appropriate for perturbations of class I (cf. equation (27)).

We first observe that the assumptions (52), in conjunction with equation (50), require that (cf. equations (53) and (43))

$$R = Q = \frac{\Delta}{r^5}(\mu^2 r + 6M). \tag{62}$$

From equations (48), (61), and (62) it now follows that

$$\frac{dW^{(-)}}{dr_*} = Q - V^{(-)} = \frac{\Delta}{r^5}(-2r + 12M), \tag{63}$$

or, making use of equation (4),

$$\frac{dW^{(-)}}{dr} = -\frac{2}{r^2} + \frac{12M}{r^3}. \tag{64}$$

The required solution of this equation is

$$W^{(-)} = \frac{2}{r^2}(r - 3M). \tag{65}$$

We next verify that when the expressions (61) and (65) for $V^{(-)}$ and $W^{(-)}$ are substituted in equation (50), we obtain, consistently with the assumptions (52), that

$$\beta^{(-)} = -6M = \text{constant}. \tag{66}$$

And we complete the verification of the entire set of the transformation equations by observing that equation (51) yields the constant value,

$$K^{(-)} = \mu^2(\mu^2+2) - 12i\sigma M. \tag{67}$$

We conclude that equation (40) can be transformed to the one-dimensional wave equation (26) by the substitutions,

$$Y = V^{(-)}Z^{(-)} + (W^{(-)}+2i\sigma)\Lambda_+ Z^{(-)} \tag{68}$$

and

$$\Lambda_- Y = {}_+6M\frac{\Delta^2}{r^8}Z^{(-)} + Q\Lambda_+ Z^{(-)}, \tag{69}$$

where $V^{(-)}$ and $W^{(-)}$ are given by equations (61) and (65).

The inverse of the relations (68) and (69) are

$$\frac{\Delta^2}{r^8}K^{(-)}Z^{(-)} = QY - (W^{(-)}+2i\sigma)\Lambda_- Y \tag{70}$$

and

$$\frac{\Delta^2}{r^8} K^{(-)} \Lambda_+ Z^{(-)} = -6M \frac{\Delta^2}{r^8} Y + V^{(-)} \Lambda_- Y. \tag{71}$$

We have shown that if the transformation equations (47)–(51) allow a solution, consistently with the assumptions (52), for some value of $+\beta$, then they must allow a solution also for $-\beta$. Therefore, in the present context, we shall obtain a further transformation associated with the value,

$$\beta^{(+)} = +6M. \tag{72}$$

We shall now verify that the resulting *dual transformation* leads to the equation for $Z^{(+)}$.

Now, letting (cf. equation (56))

$$F = \frac{r^8}{\Lambda^2} Q = \frac{r^3}{\Lambda} (\mu^2 r + 6M), \tag{73}$$

we deduce from equation (57) that

$$W^{(+)} = \frac{d}{dr_*} lgF - \frac{6M}{F}. \tag{74}$$

Since (as may be readily verified)

$$\begin{aligned} \frac{d}{dr_*} lgF &= \frac{2}{r^2}(r - 3M) - \frac{6M\Delta}{r^3(\mu^2 r+6M)} \\ &= W^{(-)} - \frac{6M\Delta}{r^3(\mu^2 r+6M)}, \end{aligned} \tag{75}$$

it follows that

$$W^{(+)} = W^{(-)} - \frac{12M\Delta}{r^3(\mu^2 r+6M)}. \tag{76}$$

The corresponding expression for $V^{(+)}$ is given by (cf. equation (48))

$$V^{(+)} = V^{(-)} + 12M \frac{d}{dr_*} \frac{\Delta}{r^3(\mu^2 r+6M)}. \tag{77}$$

It can be verified that this expression for $V^{(+)}$ is in agreement with that given in equation (33).

From equation (77) it follows that

$$\int_{-\infty}^{+\infty} V^{(+)} dr_* = \int_{-\infty}^{+\infty} V^{(-)} dr_*, \tag{78}$$

an equality to which we made reference in § 5 (equation (35)).

Finally, we may note that, in agreement with equation (55), we find that

$$\frac{r^8}{\Delta^2} QV^{(+)} + 6M(W^{(+)}+2i\sigma) = K^{(+)} = \mu^2(\mu^2+2) + 12i\sigma M . \quad (79)$$

The transformation equations, appropriate for the perturbations of class II, that are analogous to equations (68)–(71), are

$$Y = V^{(+)}Z^{(+)} + (W^{(+)}+2i\sigma)\Lambda_+ Z^{(+)} , \quad (80)$$

$$\Lambda_- Y = -\,6M \frac{\Delta^2}{r^8} Z^{(+)} + Q\Lambda_+ Z^{(+)} , \quad (81)$$

$$\frac{\Delta^2}{r^8} K^{(+)}Z^{(+)} = QY - (W^{(+)}+2i\sigma)\Lambda_- Y , \quad (82)$$

and

$$\frac{\Delta^2}{r^8} K^{(+)}\Lambda_+ Z^{(+)} = +\,6M\frac{\Delta^2}{r^8} Y + V^{(+)}\Lambda_- Y . \quad (83)$$

9. AN EXPLICIT RELATION BETWEEN THE SOLUTIONS FOR $Z^{(+)}$ AND $Z^{(-)}$ AND THE PROOF OF THE EQUALITY OF THE COEFFICIENTS $R^{(+)}$ AND $R^{(-)}$ AND $T^{(+)}$ AND $T^{(-)}$

We have seen how equation (40) for Y can be transformed to the one-dimensional wave equations governing $Z^{(+)}$ and $Z^{(-)}$ by transformations which we have called dual. We shall now show, with the aid of the transformation equations (68)–(71) and (80)–(83), that there is a simple relation which will enable us to obtain a solution $Z^{(-)}$ appropriate to perturbations of class I, from a solution $Z^{(+)}$ appropriate to perturbations of class II, and conversely.

Let $Z^{(-)}$ denote a solution of the wave equation (26). Then making use of equation (70), relating $Z^{(-)}$ to Y and $\Lambda_- Y$, and then, of equations (80) and (81), relating Y and $\Lambda_- Y$ to $Z^{(+)}$ and $\Lambda_+ Z^{(+)}$, we obtain,

$$\begin{aligned}\frac{\Delta^2}{r^8} K^{(-)}Z^{(-)} &= QY - (W^{(-)}+2i\sigma)\Lambda_- Y \\ &= Q[V^{(+)}Z^{(+)} + (W^{(+)}+2i\sigma)\Lambda_+ Z^{(+)}] \\ &\quad - (W^{(-)}+2i\sigma)\left[-6M \frac{\Delta^2}{r^8} Z^{(+)} + Q\Lambda_+ Z^{(+)} \right] , \end{aligned} \quad (84)$$

or, after some regrouping of the terms, we have

$$\frac{\Delta^2}{r^8} K^{(-)}Z^{(-)} = \frac{\Delta^2}{r^8}\left[\frac{r^8}{\Delta^2} QV^{(+)} + 6M(W^{(+)}+2i\sigma) \right.$$

$$+ 6M(W^{(-)} - W^{(+)})\Big] Z^{(+)}$$

$$+ Q(W^{(+)} - W^{(-)})\Lambda_+ Z^{(+)}. \tag{85}$$

The expression on the right-hand side of this equation can be simplified by making use of equations (67), (76), and (79). We find

$$[\mu^2(\mu^2+2) - 12i\sigma M]Z^{(-)} = \left[\mu^2(\mu^2+2) + \frac{72M^2\Delta}{r^3(\mu^2 r+6M)}\right] Z^{(+)}$$

$$- 12M\frac{dZ^{(+)}}{dr_*}. \tag{86}$$

The analogous equation relating $Z^{(+)}$ to $Z^{(-)}$ is

$$[\mu^2(\mu^2+2) + 12i\sigma M]Z^{(+)} = \left[\mu^2(\mu^2+2) + \frac{72M^2\Delta}{r^3(\mu^2 r+6M)}\right] Z^{(-)}$$

$$+ 12M\frac{dZ^{(-)}}{dr_*}. \tag{87}$$

Equations (86) and (87) clearly enable us to derive a solution (appropriate for describing perturbations of class I) from a solution (appropriate for describing perturbations of class II), and conversely.

Since the second term in square brackets on the right-hand sides of equations (86) and (87) vanishes on the horizon ($r_* \to -\infty$ and $\Delta = 0$) and at infinity ($r_* \to +\infty$), it follows that

$$[\mu^2(\mu^2+2) \pm 12i\sigma M]Z^{(\pm)} \to \mu^2(\mu^2+2)Z^{(\mp)} \pm 12M\frac{dZ^{(\mp)}}{dr_*} \qquad (r_* \to \pm\infty). \tag{88}$$

In particular, solutions for $Z^{(+)}$ derived from solutions for $Z^{(-)}$ (in accordance with equation (87)) having the asymptotic behaviors,

$$Z^{(-)} \to e^{+i\sigma r_*} \quad \text{and} \quad Z^{(-)} \to e^{-i\sigma r_*} \quad (r_* \to \pm\infty), \tag{89}$$

have, respectively, the asymptotic behaviors,

$$Z^{(+)} \to e^{+i\sigma r_*} \text{ and } Z^{(+)} \to \frac{\mu^2(\mu^2+2) - 12i\sigma M}{\mu^2(\mu^2+2) + 12i\sigma M} e^{-i\sigma r_*} \qquad (r_* \to \pm\infty). \tag{90}$$

From these behaviors, it follows that, in equations (36),

$$A^{(+)} = A^{(-)}e^{i\delta} \text{ and } B^{(+)} = B^{(-)}, \tag{91}$$

where

$$e^{i\delta} = \frac{\mu^2(\mu^2+2) - 12i\sigma M}{\mu^2(\mu^2+2) + 12i\sigma M}; \tag{92}$$

and, therefore,

$$|A^{(+)}|^2 = |A^{(-)}|^2 \text{ and } |B^{(+)}|^2 = |B^{(-)}|^2. \tag{93}$$

These relations establish the equality of the reflection and the transmission coefficients for the perturbations belonging to the two classes. It should, however, be noted that, while there is a difference in the relative phases of the reflected amplitudes, there is no such difference in the transmitted amplitudes: they are, in fact, *identically equal*.

10. NECESSARY CONDITIONS FOR DIFFERENT POTENTIALS TO YIELD THE SAME TRANSMISSION AMPLITUDES

The remarkably simple relation between the solutions for $Z^{(+)}$ and $Z^{(-)}$ and the equality of the derived reflection and transmission coefficients suggest the consideration of the following problem in the theory of "inverse scattering."

The one-dimensional wave equation,

$$\frac{d^2\psi}{dx^2} + (\sigma^2 - V)\psi = 0 \quad (-\infty < x < +\infty), \tag{94}$$

where $V(x)$ is a potential whose integral over the range of x is bounded, allows solutions with the asymptotic behaviors,

$$\left.\begin{aligned} \psi &\rightarrow e^{+i\sigma x} + A(\sigma)e^{-i\sigma x} \quad (x \rightarrow +\infty) \\ &\rightarrow \qquad\qquad B(\sigma)e^{+i\sigma x} \quad (x \rightarrow -\infty). \end{aligned}\right\} \tag{95}$$

We now ask for the conditions on V which will yield the same complex amplitude $B(\sigma)$. This problem is more general than the celebrated one of establishing the uniqueness (or, otherwise) of V from a knowledge of the "S-matrix" (i.e., from a knowledge of *both* $A(\sigma)$ and $B(\sigma)$). In considering this latter problem, it is customary to restrict oneself to potentials $V(x)$ which satisfy the requirement,

$$\int_{-\infty}^{+\infty} (1 + |x|)V(x)\, dx \tag{96}$$

is bounded.

When $V(x)$ satisfies this requirement, it has been shown that the amplitude $B(\sigma)$, for complex σ, is an analytic function of σ in the lower half of the complex plane. The potentials in which we are interested have an inverse square behavior for $x \to +\infty$ and, therefore, do not satisfy the requirement (96). We shall, nevertheless, assume that $B(\sigma)$ allows a Laurent expansion for $Im\sigma < 0$—an assumption whose validity remains to be established.

Writing

$$\psi = e^{i\sigma x + w}, \tag{97}$$

we find that w satisfies the differential equation,

$$w_{,xx} + 2i\sigma w_{,x} + (w_{,x})^2 - V = 0. \tag{98}$$

By the further substitutions,

$$w = -\int_x^\infty v(x^1,\sigma)dx^1 \quad \text{and} \quad w_{,x} = -v, \tag{99}$$

we obtain the Riccati equation,

$$v_{,x} + 2i\sigma v + v^2 - V = 0. \tag{100}$$

The boundary conditions (95) imposed on ψ, now require that

$$\left.\begin{aligned} w &\to ln\, B \quad \text{for} \quad x \to -\infty \\ &\to 0 \qquad\ \text{for} \quad x \to +\infty \end{aligned}\right\} (Im\sigma < 0); \tag{101}$$

and, also, that

$$ln\, B = -\int_{-\infty}^{+\infty} v(x,\sigma)dx. \tag{102}$$

And, finally, it follows from equation (100) that

$$v \to \frac{V}{2i\sigma} \quad \text{as} \quad \sigma \to \infty. \tag{103}$$

We now suppose that v can be expanded in a Laurent series of the form,

$$v = \sum_{n=1}^{\infty} \frac{v_n(x)}{(2i\sigma)^n}, \tag{104}$$

where, according to equation (103),

$$v_1 = V. \tag{105}$$

Inserting the series expansion (104) in equation (102), we obtain

$$lnB = -\sum_{n=1}^{\infty} \frac{1}{(2i\sigma)^n} \int_{-\infty}^{+\infty} v_n(x)dx, \tag{106}$$

where we have assumed that integration, term by term, is permissible. Thus, we have an expansion for lnB of the form

$$lnB = -\sum_{n=1}^{\infty} c_n \sigma^{-n}, \tag{107}$$

where

$$(2i)^n c_n = \int_{-\infty}^{+\infty} v_n dx\,. \tag{108}$$

Now inserting the expansion (104) in equation (100) and assuming that we can equate the coefficients of the different inverse powers of σ, we obtain the recurrence relation

$$v_n = -\frac{dv_{n-1}}{dx} - \sum_{l=1}^{n-2} v_l v_{n-1-l}\,. \tag{109}$$

Using this recurrence relation, we can solve for the v_n's, successively, starting with (cf. equation (105))

$$v_1 = V\,. \tag{110}$$

We find

$$\begin{aligned}
v_2 &= -\frac{dv_1}{dx} = -\frac{dV}{dx}\,,\\
v_3 &= -\frac{dv_2}{dx} - v_1^2 = \frac{d^2V}{dx^2} - V^2\,,\\
v_4 &= -\frac{dv_3}{dx} - 2v_1v_2 = -\frac{d^3V}{dx^3} + 2\frac{dV^2}{dx}\,,\\
v_5 &= -\frac{dV_4}{dx} - 2v_1v_3 - v_2^2 = \frac{d^4V}{dx^4} - 3\frac{d^2V^2}{dx^2}\\
&\qquad + \left(\frac{dV}{dx}\right)^2 + 2V^3\,,
\end{aligned} \tag{111}$$

etc.

The coefficients of odd order, c_{2n+1}, in the expansion for $ln\,B$ are, therefore, given by

$$\begin{aligned}
&2ic_1 = \int_{-\infty}^{+\infty} Vdx\,; \quad (2i)^3 c_3 = \int_{-\infty}^{+\infty} V^2 dx\,;\\
&(2i)^5 c_5 = \int_{-\infty}^{+\infty} (2V^3 + V_{,x}^2)dx\,; \quad \text{etc.,}
\end{aligned} \tag{112}$$

while all the coefficients of even order, c_{2n}, vanish.

It would follow from the foregoing, that if different potentials yield the same transmission amplitude B, then the Laurent expansions for $ln\,B$ must coincide and that, therefore, the integrals,

$$\int_{-\infty}^{+\infty} Vdx\,, \quad \int_{-\infty}^{+\infty} V^2 dx\,, \quad \int_{-\infty}^{+\infty} (2V^3 + V_{,x}^2)dx\,, \quad \text{etc.,} \tag{113}$$

must be the same when extended over the different potentials.

We have already verified the equality of the first of the integrals in (113) for $V^{(+)}$ and $V^{(-)}$. The equality of the other two integrals (explicitly listed) in (113) can also be verified directly. But we have, as yet, no general proof that the infinity of the integral relations, which follow from the equality of all the c_{2n+1}'s, are satisfied for $V^{(+)}$ and $V^{(-)}$, unless it be that the present analysis may be considered as having established it!

11. RELATION TO THE THEORY OF THE SOLITONS AND THE THEORY OF THE KORTEWEG-DEVRIES EQUATION

The problem considered in the preceding section is related to the classical theory of one-dimensional solitary waves as described in terms of the Korteweg-Devries equation

$$u_{,t} - 6uu_{,x} + u_{,xxx} = 0. \tag{114}$$

As is well known, this equation allows an infinity of conservation laws. Thus, as may be readily verified, the following identities hold by virtue of equation (114):

$$\begin{aligned} u_{,t} &= (3u^2 - u_{,xx})_{,x}, \\ \left(\frac{1}{2}u^2\right)_{,t} &= \left(2u^3 + \frac{1}{2}u^2_{,x} - uu_{,xx}\right)_{,x}, \\ \left(u^3 + \frac{1}{2}u^2_{,x}\right)_{,t} &= \left(\frac{18}{4}u^4 + 6uu^2_{,x} - 3u^2u_{,xx}\right. \\ &\qquad \left. + \frac{1}{2}u^2_{,xx} - u_{,x}u_{,xxx}\right)_{,x}, \quad \text{etc.} \end{aligned} \tag{115}$$

Therefore,

$$u, \quad u^2, \quad 2u^3 + u^2_{,x}, \quad \text{etc.,} \tag{116}$$

are conserved quantities. In particular, if u vanishes sufficiently rapidly for $x \to \pm\infty$, then

$$\int_{-\infty}^{+\infty} u\,dx, \quad \int_{-\infty}^{+\infty} u^2 dx, \quad \int_{-\infty}^{+\infty} (2u^3 + u^2_{,x})dx, \quad \text{etc.,} \tag{117}$$

are constants of the motion. We recognize that the foregoing integrals over u are the same as the integrals over V which we encountered in § 10 (equation (113)). The appearance of the same integrals, in the problem considered in § 10 and in the present context of the Korteweg-Devries equation, is not surprising since the solutions of the latter equation are often expressed in terms of the theory of inverse scattering.

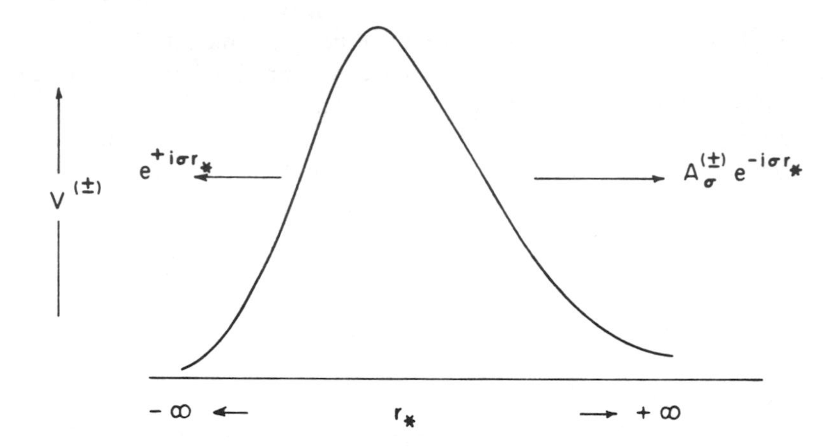

Figure 6.4. Illustrating the nature of the quasi-normal modes.

12. THE QUASI-NORMAL MODES

One may conceive of a black hole being perturbed in a variety of ways: by some object falling into it or by the accretion of matter surrounding it. Or, one may conceive of a black hole being formed by a slightly aspherical collapse of a star and settling toward a final state described by the Schwarzschild solution. In all these cases, the evolution of the initial perturbation—if it can be considered as small—can, in principle, be followed by expressing it as a superposition of the normal modes. However, we may expect that any initial perturbation will, during its last stages, decay in a manner characteristic of the black hole itself and independently of the cause. In other words, we may expect that during these last stages, the black hole emits gravitational waves with frequencies and rates of damping that are characteristic of the black hole itself, in the manner of a bell sounding its last dying notes. These considerations motivate the formulation of the concept of the *quasi-normal modes* of a black hole.

Precisely, quasi-normal modes are defined as proper solutions of the perturbation equations belonging to complex characteristic frequencies which are appropriate for purely ingoing waves at the horizon and purely outgoing waves at infinity (see Figure 6.4).

The problem then is to seek solutions of the equations governing $Z^{\pm}$ which satisfy the boundary conditions,

$$\left.\begin{aligned} Z^{(\pm)} &\rightarrow A_\sigma^{(\pm)} e^{-i\sigma r_*} \quad (r_* \rightarrow +\infty) \\ &\rightarrow \quad e^{+i\sigma r_*} \quad (r_* \rightarrow -\infty)\,. \end{aligned}\right\} \tag{118}$$

This is clearly a characteristic value problem for σ, and the solutions belonging to the different characteristic values define the quasi-normal modes.

First, we may observe that the characteristic frequencies σ are the same for $Z^{(+)}$ and $Z^{(-)}$: for, if σ is a characteristic value and $Z_\sigma^{(-)}$ is the proper solution belonging to it, then the solution $Z_\sigma^{(+)}$, derived from $Z_\sigma^{(-)}$ in accordance with the relation (87), will satisfy the required boundary conditions (120) with

$$A_\sigma^{(+)} = A_\sigma^{(-)} e^{i\delta}, \tag{119}$$

where δ is given by equation (92). It will suffice then to consider the equation governing $Z^{(-)}$ (since the potential $V^{(-)}$ is simpler than $V^{(+)}$).

Letting

$$Z^{(-)} = \exp\left(i \int^{r_*} \phi\, dr_*\right), \tag{120}$$

we have to solve the equation,

$$i\phi_{,r_*} + \sigma^2 - \phi^2 - V^{(-)} = 0\,, \tag{121}$$

which satisfies the boundary conditions

$$\phi \rightarrow -\sigma \text{ as } r_* \rightarrow +\infty \text{ and } \phi \rightarrow +\sigma \text{ as } r_* \rightarrow -\infty\,, \tag{122}$$

assuming (as we shall find) that the real part of σ is positive. Solutions having these properties (generally) exist when σ assumes one of a discrete set of complex values, but the set need not be an enumerable infinity: it often is, but sometimes it is not.

An identity, which follows from integrating equation (121) over the entire range of r_* and making use of the boundary condition (122), is

$$-2i\sigma + \int_{-\infty}^{+\infty} (\sigma^2 - \phi^2)\, dr_* = \int_{-\infty}^{+\infty} V^{(-)} dr_*\,. \tag{123}$$

By virtue of the boundary conditions (122), the integral on the left-hand side of this equation exists, while the integral on the right-hand side has a known value (cf. equation (35)).

In Table 6.1 we list the characteristic values of σ for different values of l. Detailed calculations pertaining to the aspherical collapse of dust clouds and of particles falling into black holes along geodesics do manifest the phenomenon of their ringing with the characteristic frequencies and rates of damping of their quasi-normal modes.

13. CONCLUDING REMARKS

As we have pointed out at the outset, several of the features we encounter in the study of the perturbations of the Schwarzschild black hole reappear in different guises and in more general contexts when we study the perturbations of

TABLE 6.1. The Complex Characteristic Frequencies Belonging to the Quasi-Normal Modes of the Schwarzschild Black Hole

l^a	$2M\sigma^b$	l^a	$2M\sigma^b$
2	$0.74734 + 0.17792i$	4	$1.61835 + 0.18832i$
	$0.69687 + 0.54938i$		$1.59313 + 0.56877i$
3	$1.19889 + 0.18541i$		$1.12019 + 0.84658i$
	$1.16402 + 0.56231i$	5	$2.02458 + 0.18974i$
	$0.85257 + 0.74546i$	6	$2.42402 + 0.19053i$

[a]The entries in the different lines for $l = 2, 3$, and 4 correspond to the characteristic values belonging to different modes.

[b]Here σ is expressed in the unit $(2M)^{-1}$

the other black hole solutions of general relativity. Thus, the treatment of the perturbations of the Reissner-Nordström black hole proceeds along lines which parallel the treatment of the perturbations of the Schwarzschild black hole which we have outlined: two classes of perturbations, of opposite parities, can again be distinguished and relations between them, analogous to those between $Z^{(+)}$ and $Z^{(-)}$, can be established. Similarly, the study of the perturbations of the Kerr metric leads us to consider novel problems in the theory of barrier penetration involving complex and singular potentials and manifesting super-radiance (when $\mathbf{R} + \mathbf{T} = 1$ has to be replaced by $\mathbf{R} - \mathbf{T} = 1$). But in the consideration of all these various problems, a comprehensive understanding of the theory of the perturbations of the Schwarzschild black hole appears as an essential prerequisite.

BIBLIOGRAPHICAL NOTES

We shall not provide a complete bibliography which is extensive even if restricted to the Schwarzschild solution and its perturbations. We shall give, instead, references only to those articles and papers in which the reader will find details of the derivation which are omitted, or supplementary information which bears directly on what is stated in the text. Besides, the present account is a "personal view" which a scholarly reader will doubtless find as grossly unbalanced.

§3. The various components of the Riemann, the Ricci, and the Einstein tensor, which are appropriate to the nonstationary axisymmetric metric (7), are listed in

Chandrasekhar, S., and Friedman, John L., "On the Stability of Axisymmetric Systems to Axisymmetric Perturbations in General Relativity: I. The Equations Concerning Nonstationary, Stationary, and Perturbed Systems," *Astrophys. J.* **175**, 379–405 (1972).

§4. For elementary accounts of the tetrad and the Newman-Penrose formalisms, see

Landau, L. D., and Lifshitz, E. M., *Classical Theory of Fields* (Pergamon Press, Oxford, 4th ed., 1975), pp. 291–294.

Chandrasekhar, S., "An Introduction to the Kerr Metric and Its Perturbations," in *General Relativity—An Einstein Centenary Survey* (edited by W. Israel and S. W. Hawking, Cambridge University Press, Cambridge, 1979*a*), Chapter 7, pp. 370–453.

§§5–9. The account in these sections is a rearrangement of the material contained in

Friedman, John L., "On the Born Approximation for Perturbations of a Spherical Star and the Newman-Penrose Constants," *Proc. R. Soc. London Ser. A* **335**, 163–190 (1973).

Chandrasekhar, S., "On the Equations Governing the Perturbations of the Schwarzschild Black Hole," *Proc. R. Soc. London Ser. A* **343**, 289–298 (1975).

Chandrasekhar, S., and Detweiler, S., "The Quasi-normal Modes of the Schwarzschild Black Hole," *Proc. R. Soc. London Ser. A* **344**, 441–452 (1975*a*).

———, "On the Equations Governing the Axisymmetric Perturbations of the Kerr Black Hole," *Proc. R. Soc. London Ser. A* **345**, 145–167 (1975*b*).

For an account with a different emphasis, see

Thorne, K. S., "General Relativistic Astrophysics," in *Theoretical Principles in Astrophysics and Relativity* (edited by N. R. Lebovitz, W. H. Reid, and P. Vandervoort, University of Chicago Press, 1978), pp. 149–216.

§10. The treatment in this section is derived from

Faddeev, L. D., "On the Relation between the S-matrix and potential for the one-dimensional Schrodinger Equation," *Tr. Mat. Inst. V. A. Steklova* **73**, 314–336 (1964).

§11. For a readable account of the theory of the solitary waves and the Korteweg–Devries equation, see

Whitham, G. B., *Linear and Nonlinear Waves* (J. Wiley & Sons, New York, 1974), §§16.14–16.16 and 17.2–17.4.

§12. The complex characteristic values of σ, belonging to the quasi-normal modes, listed in Table 1 are taken from Chandrasekhar and Detweiler (1975*a*) (for l = 2, 3, and 4) and D. L. Gunter ("A Study of the Coupled Gravitational and Electromagnetic Perturbations to the Reissner-Nordström Black Hole: The Scattering Matrix, Energy Conversion, and Quasi-normal Modes," *Phil. Trans. Roy. Soc. Lond. A* **296**, 497–526 (1980); for l = 5 and 6).

§13. For a treatment of the perturbations of the Reissner-Nordström black hole which parallels the treatment of the perturbations of the Schwarzschild black hole, see

Chandrasekhar, S., "On the Equations Governing the Perturbations of the Reissner-Nordström Black Hole," *Proc. R. Soc. London Ser. A* **365**, 453–465 (1979*b*).

Chandrasekhar, S., and Xanthopoulos, B., "On the Metric Perturbations of the Reissner-Nordström Black Hole," *Proc. R. Soc. London Ser. A* **367**, 1–14 (1979).

And for an account of the more general problems in the theory of the penetration of one-dimensional potential barriers which arise in the context of the Kerr metric, see Chandrasekhar (1979*a*).

2. *The Kerr Black Hole*

4

The Kerr metric and stationary axisymmetric gravitational fields

By S. Chandrasekhar, F.R.S.
University of Chicago, Chicago, Illinois, 60637, U.S.A.

(*Received* 18 *April* 1977)

A treatment of Einstein's equations governing vacuum gravitational fields which are stationary and axisymmetric is shown to divide itself into three parts: a part essentially concerned with a choice of gauge (which can be chosen to ensure the occurrence of an event horizon exactly as in the Kerr metric); a part concerned with two of the basic metric functions which in two combinations satisfy a complex equation (Ernst's equation) and in one combination satisfies a symmetric pair of real equations; and a third part which completes the solution in terms of a single ordinary differential equation of the first order. The treatment along these lines reveals many of the inner relations which characterize the general solutions, provides a derivation of the Kerr metric which is direct and verifiable at all stages, and opens an avenue towards the generation of explicit classes of exact solutions (an example of which is given).

1. Introduction

The remarkable simplicity of the general relativistic theory of black holes derives from the startling fact that the stationary axisymmetric solution of Einstein's equations discovered by Kerr (1963) provides the *unique* basis for their *exact* description (Carter 1971; Robinson 1975). It is natural that, on this account, the various aspects of Kerr's solution have been the subject of the most earnest enquiry. Nevertheless, it is striking that there is no extant derivation of Kerr's solution that is direct and simple. Even the 'outline' of the proof, as set out by Kerr & Schild (1965), occupies some ten pages while the authors find it necessary to warn the reader that 'the calculation giving these results is by no means simple'. Indeed, the following statement by Landau & Lifshitz (1975) in the latest edition of their *Classical Fields* would appear apposite: 'There is no constructive analytic derivation of the metric that is adequate in its physical ideas, and even a check of this solution of Einstein's equations involves cumbersome calculations.' And this fact is the more striking since the Kerr metric, in the most commonly used coordinates of Boyer & Lindquist (1967), is really quite simple in its structure.

It is the object of this paper to provide an *ab initio* derivation of the Kerr metric, which is direct and simply verifiable at all stages, starting from the field equation already in the Boyer–Lindquist coordinates and requiring 'no previous knowledge'.†

† For footnote see following page.

It will, however, appear that the method we shall adopt not only encompasses most of the known results on stationary axisymmetric solutions of Einstein's equations (see Kinnersley (1975) and Meinhardt (1976) for reviews and bibliography) but provides also a basis for deriving explicit classes of exact solutions.

2. The field equations

It is known that a general stationary axisymmetric metric can be written in the form (in units in which $c = 1$ and $G = 1$)

$$ds^2 = -e^{2\nu}(dt)^2 + e^{2\psi}(d\varphi - \omega\, dt)^2 + e^{2\mu_2}(dx^2)^2 + e^{2\mu_3}(dx^3)^2, \tag{1}$$

where φ denotes the azimuthal angle (in the equatorial plane) and $x^2 (= r)$ and x^3 $(= \theta)$ are the two remaining spatial coordinates. In equation (1), ν, ψ, ω, μ_2 and μ_3 are, by the assumptions of stationarity and axisymmetry, functions of x^2 and x^3 only. Besides, we have the freedom of gauge to impose any coordinate condition, we may choose, on μ_2 and μ_3.

Parenthetically, we may note here that, for Kerr's solution, the various metric functions in (1) have the values

$$e^{2\nu} = \rho^2\Delta/\Sigma^2, \quad e^{2\psi} = (\Sigma^2/\rho^2)\sin^2\theta, \quad e^{2\mu_2} = \rho^2/\Delta, \quad e^{2\mu_3} = \rho^2, \quad \omega = 2aMr/\Sigma^2,$$
$$e^{\mu_3-\mu_2} = \Delta^{\frac{1}{2}}, \quad e^{\psi+\nu} = e^{\beta} = \Delta^{\frac{1}{2}}\sin\theta, \quad \text{and} \quad e^{-\psi+\nu} = \chi = \rho^2\Delta^{\frac{1}{2}}/\Sigma^2\sin\theta, \tag{2}$$

where

$$\Delta = r^2 + a^2 - 2Mr, \quad \rho^2 = r^2 + a^2\cos^2\theta \quad \text{and} \quad \Sigma^2 = (r^2+a^2)^2 - a^2\Delta\sin^2\theta. \tag{3}$$

The field equations appropriate to the chosen form for the metric are readily written down, particularly, if one uses Cartan's calculus of exterior forms. They are, as they have been written down by Chandrasekhar & Friedman (1972; the relevant equations are given in part II of this paper):

$$e^{\mu_3-\mu_2}[\nu_{,22} + \nu_{,2}(\psi + \nu + \mu_3 - \mu_2)_{,2}]$$
$$+ e^{\mu_2-\mu_3}[\nu_{,33} + \nu_{,3}(\psi + \nu + \mu_2 - \mu_3)_{,3}] = +\tfrac{1}{2}e^{2\psi-2\nu}X \quad (R^{00}), \tag{4}$$

$$e^{\mu_3-\mu_2}[\psi_{,22} + \psi_{,2}(\psi + \nu + \mu_3 - \mu_2)_{,2}]$$
$$+ e^{\mu_2-\mu_3}[\psi_{,33} + \psi_{,3}(\psi + \nu + \mu_2 - \mu_3)_{,3}] = -\tfrac{1}{2}e^{2\psi-2\nu}X \quad (R^{11}), \tag{5}$$

$$e^{\mu_3-\mu_2}[\nu_{,22} + \nu_{,2}(\nu + \mu_3 - \mu_2)_{,2} + \mu_{3,22} + \mu_{3,2}(\mu_3 - \mu_2)_{,2}]$$
$$+ e^{\mu_2-\mu_3}[\nu_{,33} + \nu_{,3}(\nu + \mu_2 - \mu_3)_{,3} + \mu_{2,33} + \mu_{2,3}(\mu_2 - \mu_3)_{,3}] = \tfrac{3}{4}e^{2\psi-2\nu}X \quad (G^{11}), \tag{6}$$

† The author's attitude in this matter is similar to Dirac's in the following incident described by Condon & Shortley (1935): 'When Dirac visited Princeton in 1928, he gave a seminar report on his paper showing the connection of exchange energy with the spin variables of the electron. In the discussion following the report, Weyl protested that Dirac had said that he would derive the results without the use of group theory, but, as Weyl said, all of Dirac's arguments were really applications of group theory. Dirac replied, 'I said I would obtain the results without *previous knowledge* of group theory.'

$$e^{\mu_3-\mu_2}[\psi_{,22}+\psi_{,2}(\psi+\mu_3-\mu_2)_{,2}+\mu_{3,22}+\mu_{3,2}(\mu_3-\mu_2)_{,2}]$$
$$+e^{\mu_2-\mu_3}[\psi_{,33}+\psi_{,3}(\psi+\mu_2-\mu_3)_{,3}+\mu_{2,33}+\mu_{2,3}(\mu_2-\mu_3)_{,3}] = -\tfrac{1}{4}e^{2\psi-2\nu}X \quad (G^{00}), \quad (7)$$

and
$$(e^{3\psi-\nu+\mu_3-\mu_2}\omega_{,2})_{,2}+(e^{3\psi-\nu+\mu_2-\mu_3}\omega_{,3})_{,3} = 0 \quad (R^{01}), \quad (8)$$

where
$$X = e^{\mu_3-\mu_2}(\omega_{,2})^2+e^{\mu_2-\mu_3}(\omega_{,3})^2, \quad (9)$$

and the 'comma' notation for partial derivatives has been adopted. (We have noted the particular components of the Ricci or the Einstein tensor whose vanishing leads to the different equations.) A further equation which we shall find useful is (Chandrasekhar & Friedman 1972, eq. (76))

$$2e^{-\beta}\{[e^{\mu_3-\mu_2}(e^{\beta})_{,2}]_{,2}-[e^{\mu_2-\mu_3}(e^{\beta})_{,3}]_{,3}\}$$
$$= 4e^{\mu_3-\mu_2}(\beta_{,2}\,\mu_{3,2}+\psi_{,2}\nu_{,2})-4e^{\mu_2-\mu_3}(\beta_{,3}\,\mu_{2,3}+\psi_{,3}\nu_{,3})$$
$$+e^{2\psi-2\nu}[e^{\mu_3-\mu_2}(\omega_{,2})^2-e^{\mu_2-\mu_3}(\omega_{,3})^2] \quad (G^{22}-G^{33}), \quad (10)$$

where
$$\beta = \psi+\nu. \quad (11)$$

Equations (4) and (5) can be rewritten alternatively in the forms

$$(e^{\beta+\mu_3-\mu_2}\nu_{,2})_{,2}+(e^{\beta+\mu_2-\mu_3}\nu_{,3})_{,3} = +\tfrac{1}{2}e^{3\psi-\nu}X \quad (12)$$

and
$$(e^{\beta+\mu_3-\mu_2}\psi_{,2})_{,2}+(e^{\beta+\mu_2-\mu_3}\psi_{,3})_{,3} = -\tfrac{1}{2}e^{3\psi-\nu}X. \quad (13)$$

Adding and subtracting these equations, we obtain

$$[e^{\mu_3-\mu_2}(e^{\beta})_{,2}]_{,2}+[e^{\mu_2-\mu_3}(e^{\beta})_{,3}]_{,3} = 0 \quad (14)$$

and
$$[e^{\beta+\mu_3-\mu_2}(\psi-\nu)_{,2}]_{,2}+[e^{\beta+\mu_2-\mu_3}(\psi-\nu)_{,3}]_{,3} = -e^{3\psi-\nu}X. \quad (15)$$

It may be noted there that equation (15) follows also from equations (6) and (7), while equation (14) is directly a consequence of the vanishing of $G^{22}+G^{33}$ (cf. Chandrasekhar & Friedman 1972, eq. (75)).

An important consequence of equations (8) and (15) is thc following. Since, by virtue of equation (8),

$$(e^{3\psi-\nu+\mu_3-\mu_2}\omega\omega_{,2})_{,2}+(e^{3\psi-\nu+\mu_2-\mu_3}\omega\omega_{,3})_{,3} = e^{3\psi-\nu}X, \quad (16)$$

we can rewrite equation (15) in the form

$$\{e^{3\psi-\nu+\mu_3-\mu_2}[2\,e^{-2\psi+2\nu}(\psi-\nu)_{,2}+(\omega^2)_{,2}]\}_{,2}$$
$$+\{e^{3\psi-\nu+\mu_2-\mu_3}[2\,e^{-2\psi+2\nu}(\psi-\nu)_{,3}+(\omega^2)_{,3}]\}_{,3} = 0, \quad (17)$$

or, equivalently,

$$[e^{3\psi-\nu+\mu_3-\mu_2}(\chi^2-\omega^2)_{,2}]_{,2}+[e^{3\psi-\nu+\mu_2-\mu_3}(\chi^2-\omega^2)_{,3}]_{,3} = 0, \quad (18)$$

where
$$\chi = e^{-\psi+\nu}. \quad (19)$$

A comparison of equations (8) and (18) shows that ω *and* $\chi^2-\omega^2$ *are governed by the same equation.*

3. The choice of gauge and the solution for e^{β}

The particular choice of gauge we shall presently make implies no essential loss of generality; and it is strictly not necessary to make it at this stage. But it provides a meaningful physical motivation for the forms of the metrics we shall seek and, in addition, makes the analysis concrete.

We shall suppose that the metric allows a smooth *event horizon*, i.e. a smooth convex two-dimensional null-surface which is spanned by the two Killing vectors, $\partial/\partial t$ and $\partial/\partial\varphi$, of the assumed space-time geometry (cf. Carter 1973).

In accordance with the assumed stationarity and axisymmetry of the metric, let the equation of the event horizon be

$$N(x^2, x^3) = 0. \tag{20}$$

The condition that it be null is

$$g^{ij}N_{,i}N_{,j} = 0. \tag{21}$$

For the chosen form of the metric, equation (21) gives

$$e^{2(\mu_3-\mu_2)}(N_{,r})^2 + (N_{,\theta})^2 = 0. \tag{22}$$

Since we have the gauge-freedom to impose any coordinate condition, we may choose, on μ_2 and μ_3, we shall suppose that

$$e^{2(\mu_3-\mu_2)} = \Delta(r), \tag{23}$$

where $\Delta(r)$ is some function of r which we shall leave unspecified for the present. From equation (22), it now follows that the equation of the null-surface is, in fact, given by

$$\Delta(r) = 0. \tag{24}$$

The second condition, that the null-surface be spanned by the two Killing vectors, $\partial/\partial t$ and $\partial/\partial\varphi$, requires that (cf. equation (1))

$$e^{2(\psi+\nu)} = e^{2\beta} = 0 \quad \text{on} \quad \Delta(r) = 0. \tag{25}$$

Since we have left $\Delta(r)$ unspecified, we may now suppose that

$$e^{\beta} = \Delta^{\frac{1}{2}} f(r, \theta) \tag{26}$$

where $f(r, \theta)$ is some function of r and θ which is regular on $\Delta(r) = 0$ and on the axis $\theta = 0$. We shall suppose, instead, that e^{β} has the somewhat more restricted form

$$e^{\beta} = \Delta^{\frac{1}{2}} f(\theta), \tag{27}$$

where $f(\theta)$ is a function of θ only; and we shall ask if equation (14) for e^{β} will allow such a separable solution.

With $e^{\mu_3-\mu_2}$ and e^{β} given by equations (23) and (27), equation (14) for e^{β} reduces to

$$[\Delta^{\frac{1}{2}}(\Delta^{\frac{1}{2}})_{,r}]_{,r} + (1/f)f_{,\theta\theta} = 0. \tag{28}$$

A solution of this equation compatible with the requirement of regularity at the poles (and convexity of the horizon) is determined by

$$\Delta_{,rr} = 2 \quad \text{and} \quad f = \sin\theta; \tag{29}$$

and the solution for Δ that is appropriate is

$$\Delta = r^2 + a^2 - 2Mr, \tag{30}$$

where a and M are certain constants.

Thus, with a choice of gauge that is consistent with the existence of an event horizon, we have the solutions

$$e^{\mu_3-\mu_2} = \Delta^{\frac{1}{2}} \quad \text{and} \quad e^{\beta} = \Delta^{\frac{1}{2}} \sin\theta, \tag{31}$$

where Δ is given by equation (30). We observe that the present expressions for $e^{\mu_3-\mu_2}$ and e^{β} are in agreement with those that are given in equation (2) as appropriate for the Kerr metric.†

4. The reduction of the field equations

With the solutions for $e^{\mu_3-\mu_2}$ and e^{β} given in equation (31) and by using

$$\mu = \cos\theta \tag{32}$$

(instead of θ) as the variable indicated by the index '3', we can bring equations (15) and (8), respectively, to the forms

$$[\Delta(\psi-\nu)_{,2}]_{,2} + [\delta(\psi-\nu)_{,3}]_{,3} = -e^{2(\psi-\nu)}[\Delta(\omega_{,2})^2 + \delta(\omega_{,3})^2] \tag{33}$$

and

$$[\Delta\, e^{2\psi-2\nu}\omega_{,2}]_{,2} + [\delta\, e^{2\psi-2\nu}\omega_{,3}]_{,3} = 0, \tag{34}$$

where we have written

$$\delta = 1-\mu^2 = \sin^2\theta. \tag{35}$$

Letting χ have the meaning given to it in equation (19), we can rewrite equations (33) and (34) in the forms

$$\left(\frac{\Delta}{\chi}\chi_{,2}\right)_{,2} + \left(\frac{\delta}{\chi}\chi_{,3}\right)_{,3} = \frac{1}{\chi^2}[\Delta(\omega_{,2})^2 + \delta(\omega_{,3})^2] \tag{36}$$

and

$$\left(\frac{\Delta}{\chi^2}\omega_{,2}\right)_{,2} + \left(\frac{\delta}{\chi^2}\omega_{,3}\right)_{,3} = 0. \tag{37}$$

Alternative forms of these equations are

$$\chi[(\Delta\chi_{,2})_{,2} + (\delta\chi_{,3})_{,3}] = \Delta(\chi_{,2})^2 + \delta(\chi_{,3})^2 + \Delta(\omega_{,2})^2 + \delta(\omega_{,3})^2 \tag{38}$$

and

$$\chi[(\Delta\omega_{,2})_{,2} + (\delta\omega_{,3})_{,3}] = 2\Delta\chi_{,2}\omega_{,2} + 2\delta\chi_{,3}\omega_{,3}. \tag{39}$$

Now, letting

$$X = \chi + \omega \quad \text{and} \quad Y = \chi - \omega, \tag{40‡}$$

we obtain the pair of symmetric equations

$$\tfrac{1}{2}(X+Y)[(\Delta X_{,2})_{,2} + (\delta X_{,3})_{,3}] = \Delta(X_{,2})^2 + \delta(X_{,3})^2, \tag{41}$$

$$\tfrac{1}{2}(X+Y)[(\Delta Y_{,2})_{,2} + (\delta Y_{,3})_{,3}] = \Delta(Y_{,2})^2 + \delta(Y_{,3})^2. \tag{42}$$

† Surprise has sometimes been expressed that the horizon of the Kerr metric is 'spherical' (cf. Adler, Bazin & Schiffer 1975, p. 265). But there is no cause for surprise: it is 'spherical' by design!

‡ The present definition of X will not be confused with the earlier definition in equation (9) as there will be no overlapping of the two usages.

The foregoing are the basic equations of the theory: on them depends the solution of Einstein's equations for stationary axisymmetric fields. Thus, once equations (41) and (42) have been solved for X and Y, the solution for the metric coefficients can be completed by solving equation (10) for $\mu_2+\mu_3$.

Equation (10), after some elementary reductions, becomes

$$2(r-M)\frac{\partial}{\partial r}(\mu_2+\mu_3)+2\mu\frac{\partial}{\partial \mu}(\mu_2+\mu_3)-\frac{1}{\chi^2}(\Delta X_{,2}Y_{,2}-\delta X_{,3}Y_{,3})+3\frac{M^2-a^2}{\Delta}-\frac{1}{\delta}=0. \quad (43)$$

This is a linear first-order partial differential equation for $\mu_2+\mu_3$; its solution can be readily reduced to that of a single ordinary differential equation of the first order.

Three observations concerning the foregoing reductions of the field equations need to be made.

First, by virtue of equations (41) and (42),

$$XY = \chi^2-\omega^2, \quad (44)$$

satisfies the equation

$$\chi\{[\Delta(XY)_{,2}]_{,2}+[\delta(XY)_{,3}]_{,3}\} = 2\Delta\chi_{,2}(XY)_{,2}+2\delta\chi_{,3}(XY)_{,3}, \quad (45)$$

in agreement with the result established in §2 that ω and $\chi^2-\omega^2$ satisfy the same equation.

Second, by the transformation,

$$\xi = (\Delta\delta)^{\frac{1}{2}} \quad \text{and} \quad \zeta = (r-M)\mu, \quad (46)$$

equations (41) and (42) can be brought to the forms

$$\tfrac{1}{2}(X+Y)(X_{,\xi\xi}+(1/\xi)X_{,\xi}+X_{,\zeta\zeta}) = (X_{,\xi})^2+(X_{,\zeta})^2$$

and

$$\tfrac{1}{2}(X+Y)(Y_{,\xi\xi}+(1/\xi)Y_{,\xi}+Y_{,\zeta\zeta}) = (Y_{,\xi})^2+(Y_{,\zeta})^2, \quad (47)$$

in which Δ does not occur explicitly. Accordingly, equations (47) and (48) are *formally* independent of the choice of gauge made in §3; and, indeed, they are independent, since the same equations can be derived, without that choice, from the so-called 'Papapetrou-form' of the metric (cf. Papapetrou 1953).

And *third*, the treatment of the field equations divides, in a very natural way, the consideration of the geometry of the space-time in three distinct parts. A first part in which our concern is with the choice of a suitable gauge and the specification of the metric functions $e^{\mu_3-\mu_2}$ and e^{β} which together enable us to write the metric (1) in the form

$$\begin{aligned} ds^2 &= e^{\beta}\left[-\chi(dt)^2+\frac{1}{\chi}(d\varphi-\omega\,dt)^2\right]+e^{2\mu_2}[(dr)^2+e^{2(\mu_3-\mu_2)}(d\theta)^2] \\ &= (\Delta\delta)^{\frac{1}{2}}\left[-\chi(dt)^2+\frac{1}{\chi}(d\varphi-\omega\,dt)^2\right]+\frac{e^{\mu_2+\mu_3}}{\Delta^{\frac{1}{2}}}[(dr)^2+\Delta(d\theta)^2]; \end{aligned} \quad (48)$$

a central (and a crucial) part in which our attention is focussed on the metric functions χ and ω which are governed by the basic equations (41) and (42) of the theory; and a last part in which our concern is with the solution for $(\mu_2+\mu_3)$ and the geometry of the (r,θ)-subspace.

5. Properties of the equations governing X and Y

A convenient form of equations (41) and (42), which enables one to find by inspection some simple solutions, is obtained, by the transformation

$$X = (1+F)/(1-F) \quad \text{and} \quad Y = (1+G)/(1-G). \tag{49}$$

We find

$$(1-FG)[(\Delta F_{,2})_{,2}+(\delta F_{,3})_{,3}] = -2G[\Delta(F_{,2})^2+\delta(F_{,3})^2] \tag{50}$$

and

$$(1-FG)[(\Delta G_{,2})_{,2}+(\delta G_{,3})_{,3}] = -2F[\Delta(G_{,2})^2+\delta(G_{,3})^2]. \tag{51}$$

The metric functions χ and ω are related to F and G by

$$\chi = \frac{1-FG}{(1-F)(1-G)} \quad \text{and} \quad \omega = \frac{F-G}{(1-F)(1-G)}. \tag{52}$$

From the forms of equations (41) and (42) and (50) and (51) several facts emerge.

Let (X, Y) represent a solution of equations (41) and (42). Then the following are also solutions.

(*a*) (Y, X) also represents a solution. For this solution χ remains unchanged, while ω changes sign – a trivial change which can be compensated by letting $\varphi \to -\varphi$.

(*b*) $(X+c, Y-c)$, where c is an arbitrary constant, also represents a solution. For this solution, χ remains unchanged while $\omega \to \omega+2c$ – again, a trivial change which can be compensated by the coordinate transformation $\varphi \to \varphi - 2ct$.

(*c*) (X^{-1}, Y^{-1}) and (Y^{-1}, X^{-1}) also represent solutions. This fact follows from the invariance of equations (50) and (51) to simultaneous changes in the signs of F and G. The significance of these new solutions is not so immediately apparent.

Let the metric functions derived from the solution (Y^{-1}, X^{-1}) be distinguished from the functions derived from the solution (X, Y) by a tilde. Then

$$2\tilde{\chi} = \left(\frac{1}{Y}+\frac{1}{X}\right) = \frac{X+Y}{XY} = \frac{2\chi}{\chi^2-\omega^2} \quad \text{and} \quad 2\tilde{\omega} = \left(\frac{1}{Y}-\frac{1}{X}\right) = \frac{X-Y}{XY} = \frac{2\omega}{\chi^2-\omega^2}. \tag{53}$$

The transformation $\chi \to \tilde{\chi}$ and $\omega \to \tilde{\omega}$ is equivalent to the 'trivial' transformation

$$t \to i\varphi \quad \text{and} \quad \varphi \to -it, \tag{54†}$$

as can be verified by first effecting this transformation on the metric as written in equation (48) and then recasting it in the form it had. I am indebted to Mr Basilis Xanthopoulos for this clarification of the significance of the transformation $(X, Y) \to (Y^{-1}, X^{-1})$.

(*d*) By combining the results of (*b*) and (*c*), we infer that if (X, Y) represents a solution, then so does $(X^{-1}+c, Y^{-1}-c)$ where c is an arbitrary constant. Next,

† Though we have characterized this transformation as 'trivial' its effect is not simply one of gauge; it must accordingly be considered as 'significant' in the technical sense!

replacing $X^{-1}+c$ and $Y^{-1}-c$ by their reciprocals, we conclude that

$$\text{if } (X, Y) \text{ represents a solution, then so does } [X/(1+cX),\ Y/(1-cY)], \tag{55}$$

where c is an arbitrary constant.

The transformation implied in (55) is equivalent to the so-called 'Ehlers-transformation' for generating new stationary axisymmetric solutions of Einstein's equations (cf. Kinnersley 1975). However, in the present context, it is clear that, by starting with a particular solution (X, Y) we do not obtain any significantly new solution by this transformation (see, however, footnote on page 411). Also, it should be noted that *equation* (43), governing $(\mu_2+\mu_3)$, is *invariant to the transformation*, $(X, Y) \rightarrow [X/(1+cX), Y/(1-cY)]$.

Finally, we note here (for later use) that by changing r to the new variable

$$\eta = (r-M)/(M^2-a^2)^{\frac{1}{2}}, \tag{56}$$

when

$$\Delta = (M^2-a^2)(\eta^2-1), \tag{57}$$

equations (50) and (51) take the forms

$$(1-FG)\{[(\eta^2-1)F_{,\eta}]_{,\eta}+[(1-\mu^2)F_{,\mu}]_{,\mu}\} = -2G[(\eta^2-1)(F_{,\eta})^2+(1-\mu^2)(F_{,\mu})^2] \tag{58}$$

and

$$(1-FG)\{[(\eta^2-1)G_{,\eta}]_{,\eta}+[(1-\mu^2)G_{,\mu}]_{,\mu}\} = -2F[(\eta^2-1)(G_{,\eta})^2+(1-\mu^2)(G_{,\mu})^2]; \tag{59}$$

while equation (43) becomes

$$2\eta\frac{\partial}{\partial\eta}(\mu_2+\mu_3)+2\mu\frac{\partial}{\partial\mu}(\mu_2+\mu_3) - \frac{4}{(1-FG)^2}[(\eta^2-1)F_{,\eta}G_{,\eta}-(1-\mu^2)F_{,\mu}G_{,\mu}]+\frac{3}{\eta^2-1}-\frac{1}{1-\mu^2} = 0. \tag{60}$$

We shall show in §8 how equations (58) and (59) enable us to write down new classes of exact solutions.

6. Combinations of the variables which satisfy Ernst's equation

As is well known, Ernst (1968; also 1974) reduced the equations governing stationary axisymmetric gravitational fields to a single complex equation – equation (71) below – even as they were reduced to the pair of equations (58) and (59) in the present treatment. We shall now show that two distinct combinations of the metric functions we have defined satisfy Ernst's equation.

(a) *The first pair of functions which satisfy Ernst's equation*

First, we observe that equation (37) allows ω to be derived from a potential Φ in the manner

$$\Phi_{2} = (\delta/\chi^2)\,\omega_{,3} \quad\text{and}\quad \Phi_{3} = -(\Delta/\chi^2)\,\omega_{2}. \tag{61}$$

The potential Φ will be governed by the equation

$$[(\chi^2/\delta)\,\Phi_{,2}]_{,2}+[(\chi^2/\Delta)\,\Phi_{,3}]_{,3} = 0; \tag{62}$$

and expressed in terms of Φ, equation (36) becomes

$$[\Delta(\ln\chi)_{,2}]_{,2}+[\delta(\ln\chi)_{,3}]_{,3} = (\chi^2/\Delta)\,(\Phi_{,3})^2+(\chi^2/\delta)\,(\Phi_{,2})^2. \tag{63}$$

Letting

$$\Psi = (\Delta\delta)^{\frac{1}{2}}/\chi, \tag{64}$$

we find that, expressed in terms of Ψ, equations (62) and (63) can be reduced to the forms (cf. equations (38) and (39))

$$\Psi[(\Delta\Psi_{,2})_{,2}+(\delta\Psi_{,3})_{,3}] = \Delta(\Psi_{,2})^2+\delta(\Psi_{,3})^2-\Delta(\Phi_{,2})^2-\delta(\Phi_{,3})^2 \tag{65}$$

and

$$\Psi[\Delta(\Phi_{,2})_{,2}+(\delta\Phi_{,3})_{,3}] = 2\Delta\Psi_{,2}\Phi_{,2}+2\delta\Psi_{,3}\Phi_{,3}. \tag{66}$$

Now expressing Ψ and Φ as the real and the imaginary parts of a complex function

$$Z = \Psi+i\Phi, \tag{67}$$

we can combine equations (65) and (66) into the single equation

$$\mathrm{Re}\,(Z)\,[(\Delta Z_{,2})_{,2}+(\delta Z_{,3})_{,3}] = \Delta(Z_{,2})^2+\delta(Z_{,3})^2. \tag{68}$$

The close similarity of equation (68) with the pair of equations (41) and (42) is apparent.

By the transformation,

$$Z = (1+\mathscr{E})/(1-\mathscr{E}), \tag{69}$$

(analogous to the transformation (49)), equation (68) becomes

$$(1-\mathscr{E}\mathscr{E}^*)\,[(\Delta\mathscr{E}_{,2})_{,2}+(\delta\mathscr{E}_{,3})_{,3}] = -2\mathscr{E}^*[\Delta(\mathscr{E}_{,2})^2+\delta(\mathscr{E}_{,3})^2], \tag{70}$$

or, in terms of the variable η defined in equation (56), we have

$$(1-\mathscr{E}\mathscr{E}^*)\,\{[(\eta^2-1)\,\mathscr{E}_{,\eta}]_{,\eta}+[(1-\mu^2)\,\mathscr{E}_{,\mu}]_{,\mu}\} = -2\mathscr{E}^*[(\eta^2-1)\,(\mathscr{E}_{,\eta})^2+(1-\mu^2)\,(\mathscr{E}_{,\mu})^2]. \tag{71}$$

This is Ernst's equation.

In terms of $\mathscr{E}$, the metric functions Ψ and Φ are given by

$$\Psi = \mathrm{Re}\,(Z) = (1-\mathscr{E}\mathscr{E}^*)/|1-\mathscr{E}|^2 \quad \text{and} \quad \Phi = \mathrm{Im}\,(Z) = -\mathrm{i}(\mathscr{E}-\mathscr{E}^*)/|1-\mathscr{E}|^2. \tag{72}$$

(*b*) *The second pair of functions which satisfy Ernst's equation*

The pair of functions (Ψ, Φ) for which we have derived Ernst's equations is not the pair that was considered by Ernst. In our present notation, his considerations pertain to the pair of functions

$$f = XY\,\mathrm{e}^{2\psi} = (\Delta\delta)^{\frac{1}{2}}\frac{\chi^2-\omega^2}{\chi} \tag{73}$$

and

$$W = \omega/XY = \omega/(\chi^2-\omega^2). \tag{74}$$

Making use of the fact that both ω and XY satisfy the same equation, we readily verify that

$$\left(\frac{f^2}{\delta}W_{,2}\right)_{,2}+\left(\frac{f^2}{\Delta}W_{,3}\right)_{,3} = 0. \tag{75}$$

Accordingly, W may be derived from a potential g in the manner

$$g_{,3} = -(f^2/\delta)\,W_{,2} \quad \text{and} \quad g_{,2} = (f^2/\Delta)\,W_{,3}; \tag{76}$$

and g itself will be governed by the equation

$$\left(\frac{\Delta}{f^2}g_{,2}\right)_{,2} + \left(\frac{\delta}{f^2}g_{,3}\right)_{,3} = 0. \tag{77}$$

Expanding this last equation, we obtain

$$f[(\Delta g_{,2})_{,2} + (\delta g_{,3})_{,3}] = 2\Delta f_{,2}g_{,2} + 2\delta f_{,3}g_{,3}. \tag{78}$$

Again, making use of equation (45) satisfied by XY and the field equation (13) satisfied by ψ, we can directly verify that

$$\left(\frac{\Delta}{f}f_{,2}\right)_{,2} + \left(\frac{\delta}{f}f_{,3}\right)_{,3} = -f^2\left[\frac{1}{\delta}(W_{,2})^2 + \frac{1}{\Delta}(W_{,3})^2\right]; \tag{79}$$

or expressing W in terms of g and expanding, we obtain

$$f[(\Delta f_{,2})_{,2} + (\delta f_{,3})_{,3}] = \Delta(f_{,2})^2 + \delta(f_{,3})^2 - \Delta(g_{,2})^2 - \delta(g_{,3})^2. \tag{80}$$

It is now apparent that, by combining f and g into the single complex function

$$Z = f + \mathrm{i}g, \tag{81}$$

we shall find that Z satisfies the same equation (68).

(c) *The relation between the metric functions derived from the same solution of Ernst's equation*

It is clear from the foregoing reductions that from a given solution of equation (68), we can derive two distinct solutions of Einstein's equations depending on which pair of functions, (f, g) or (Ψ, Φ), we wish to associate with Z. Let the solutions derived from associating Z with (Ψ, Φ) be distinguished by a tilde. Since by assumption $f = \Psi$ and $g = \Phi$, it follows from equation (64) and (73) and, similarly, from equations (61), (73), and (76) that

$$\frac{(\Delta\delta)^{\frac{1}{2}}}{\tilde{\chi}} = (\Delta\delta)^{\frac{1}{2}}\frac{\chi^2-\omega^2}{\chi} \quad \text{or} \quad \tilde{\chi} = \frac{\chi}{\chi^2-\omega^2} \tag{82}$$

and

$$\frac{\delta}{\tilde{\chi}^2}\tilde{\omega}_{,3} = \frac{f^2}{\Delta}W_{,3} = \delta\frac{(\chi^2-\omega^2)^2}{\chi^2}W_{,3} = \frac{\delta}{\tilde{\chi}^2}W_{,3}; \quad \frac{\Delta}{\tilde{\chi}^2}\tilde{\omega}_{,2} = \frac{f^2}{\delta}W_{,2} = \Delta\frac{(\chi^2-\omega^2)^2}{\chi^2}W_{,2} = \frac{\Delta}{\tilde{\chi}^2}W_{,2}. \tag{83}$$

From the last pair of equations, it follows that $\tilde{\omega}$ and W can differ, at most, by a constant. Ignoring the additive constant (which can be absorbed by a simple coordinate-transformation), we conclude that (cf. equation (74))

$$\tilde{\omega} = W = \omega/(\chi^2-\omega^2). \tag{84}$$

From a comparison of the relations (82) and (84) with the relations (53) discussed in §5, it is manifest that two solutions, derived from the same solution of Ernst's equation, can be transformed, one into the other, by the 'trivial' transformation (54). (I am indebted to Mr B. Xanthopoulos for this last observation.)

7. The Kerr metric

As Ernst first showed, the Kerr metric follows from the immediately verifiable solution

$$\mathscr{E} = -p\eta - \mathrm{i}q\mu \quad (p, q \text{ constants}, p^2+q^2 = 1), \tag{85}$$

of his equation. We shall find it convenient to consider for Z the negative of the solution that corresponds to the solution (85); this is permissible since the equation governing Z is invariant to a change in its sign. Thus, writing

$$Z = Z_1 + \mathrm{i}Z_2, \tag{86}$$

we find

$$Z_1 = \frac{p^2\eta^2+q^2\mu^2-1}{1+2p\eta+p^2\eta^2+q^2\mu^2} = \frac{p^2(\eta^2-1)-q^2(1-\mu^2)}{(p\eta+1)^2+q^2\mu^2} \quad \text{and} \quad Z_2 = \frac{2q\mu}{(p\eta+1)^2+q^2\mu^2}. \tag{87}$$

Reverting to the variable r, we have

$$\left.\begin{aligned} Z_1 &= \frac{\Delta - q^2p^{-2}(M^2-a^2)\,\delta}{[r-M+p^{-1}(M^2-a^2)^{\frac{1}{2}}]^2+q^2p^{-2}(M^2-a^2)\,\mu^2} \\ \text{and} \qquad Z_2 &= \frac{2qp^{-2}(M^2-a^2)\,\mu}{[r-M+p^{-1}(M^2-a^2)^{\frac{1}{2}}]^2+q^2p^{-2}(M^2-a^2)\,\mu^2}. \end{aligned}\right\} \tag{88}$$

Since (as may be directly verified) Δ and M^2-a^2 are invariant to the transformation†

$$\left.\begin{aligned} M \to M' &= p^{-1}(M^2-a^2)^{\frac{1}{2}}, \\ a \to a' &= qp^{-1}(M^2-a^2)^{\frac{1}{2}}, \\ \text{and} \qquad r \to r' &= r-M+p^{-1}(M^2-a^2)^{\frac{1}{2}}, \end{aligned}\right\} \tag{89}$$

we may, consistently with the choice of gauge made in §3, let

$$p = (M^2-a^2)^{\frac{1}{2}}/M \quad \text{and} \quad q = a/M. \tag{90}$$

With this choice of p and q, the expressions for Z_1 and Z_2 simplify and we are left with

$$Z_1 = (\Delta - a^2\delta)\,\rho^{-2} \quad \text{and} \quad Z_2 = 2aM\mu\rho^{-2}, \tag{91}$$

where $\rho^2 = r^2+a^2\mu^2$ (cf. equation (3)).

If we associate the foregoing solutions for Z_1 and Z_2 with the functions f and g considered in §6(b), then

$$f = (\chi^2-\omega^2)\,\mathrm{e}^{2\psi} = \mathrm{e}^{2\nu}-\omega^2\mathrm{e}^{2\psi} = (\Delta-a^2\delta)\,\rho^{-2} \tag{92}$$

and

$$g = 2aM\mu\rho^{-2}. \tag{93}$$

† I am again indebted to Mr B. Xanthopoulos for this observation.

The required solution for W follows from the equations (cf. equations (76))

$$g_{,2} = -\frac{4aMr\mu}{\rho^4} = \frac{(\Delta - a^2\delta)^2}{\rho^4\Delta} W_{,3} \quad \text{and} \quad g_{,3} = \frac{2aM(r^2 - a^2\mu^2)}{\rho^4} = -\frac{(\Delta - a^2\delta)^2}{\rho^4\delta} W_{,2}. \tag{94}$$

We find

$$W\left(= \frac{\omega}{\chi^2 - \omega^2}\right) = \frac{2aMr\delta}{\Delta - a^2\delta}. \tag{95}$$

Next, combining equations (92) and (95), we obtain

$$\omega\, e^{2\psi} = \frac{2aMr\delta}{\Delta - a^2\delta}(\chi^2 - \omega^2)\, e^{2\psi} = \frac{2aMr\delta}{\rho^2}. \tag{96}$$

Equations (92) and (96) can now be solved for the metric functions to give

$$e^{2\nu} = \rho^2\Delta/\Sigma^2, \quad e^{2\psi} = \delta\Sigma^2/\rho^2 \quad \text{and} \quad \omega = 2aMr/\Sigma^2, \tag{97}$$

where

$$\Sigma^2 = (\rho^4\Delta - 4a^2M^2r^2\delta)/(\Delta - a^2\delta) = (r^2 + a^2)^2 - a^2\Delta\delta; \tag{98}$$

and these are in agreement with the expressions, appropriate for the Kerr metric, given in equations (2) and (3).

It remains to determine $(\mu_2 + \mu_3)$ to complete the solution. First, we observe that the solutions for X and Y corresponding to the solutions (97) are

$$\left.\begin{aligned} X &= \frac{\Delta - a^2\delta}{\delta^{\frac{1}{2}}[\rho^2\Delta^{\frac{1}{2}} - 2aMr\delta^{\frac{1}{2}}]} = \frac{\Delta^{\frac{1}{2}} + a\delta^{\frac{1}{2}}}{\delta^{\frac{1}{2}}[(r^2 + a^2) + a(\Delta\delta)^{\frac{1}{2}}]} \\ \text{and} \quad Y &= \frac{\Delta - a^2\delta}{\delta^{\frac{1}{2}}[\rho^2\Delta^{\frac{1}{2}} + 2aMr\delta^{\frac{1}{2}}]} = \frac{\Delta^{\frac{1}{2}} - a\delta^{\frac{1}{2}}}{\delta^{\frac{1}{2}}[(r^2 + a^2) - a(\Delta\delta)^{\frac{1}{2}}]}. \end{aligned}\right\} \tag{99}$$

Inserting the foregoing solutions for X and Y in equation (43), we find after some reductions that we are left with the simple equation

$$(r - M)\frac{\partial}{\partial r}(\mu_2 + \mu_3) + \mu\frac{\partial}{\partial\mu}(\mu_2 + \mu_3) = 2 - \frac{(r - M)^2}{\Delta} - 2\frac{rM}{\rho^2}. \tag{100}$$

There is no difficulty in solving this elementary partial differential equation; and we find that

$$e^{\mu_2 + \mu_3} = \rho^2\Delta^{-\frac{1}{2}}, \tag{101†}$$

in agreement with the known solution for the Kerr metric.

With respect to the foregoing derivation of the Kerr metric two observations are apposite.

First, had we associated the solution (91) with the pair of functions (Ψ, Φ) (which is the more 'natural' pair to consider in the present treatment than (f, g)) we should

†There is another equation governing $\mu_2 + \mu_3$ which, together with equation (100), makes the solution (101) determinate. For that equation, see S. Chandrasekhar, *The Mathematical Theory of Black Holes* (Oxford: Clarendon Press, 1987), p. 288, eq.(130).

not have obtained the Kerr metric, but, rather, the metric derived from it after effecting the transformation (54)!

Second, it would have been more in the spirit of the present treatment to have sought directly the solutions of equations (41) and (42) for X and Y, instead of via Ernst's equation. But how could one have foreseen that the expressions for X and Y given in equations (99) represent solutions of equations (41) and (42)? Even to verify that they are requires some effort! However, that the deduced expressions for X and Y *are* solutions of equations (41) and (42) implies, in turn, that Ernst's equation allows solutions besides those which are expressible as ratios of polynomials of η and μ in the manner of Tomimatsu & Sato (1973). We consider this and related matters in §8 below.

8. A NEW CLASS OF SOLUTIONS ASSOCIATED WITH THE KERR–TOMIMATSU–SATO FAMILY

It was suggested in §7 that one might not have been able to foresee the solutions (99) for X and Y appropriate to the Kerr metric, in contrast to the solution (85) which did lead to the Kerr metric. On the other hand, from the close similarity of the real and the complex equations, for F and G on the one hand and $\mathscr{E}$ on the other, we can often write down, by simple inspection, solutions of equations (58) and (59) which, in some sense, correspond to the complex solutions of equation (71), and conversely. Thus, corresponding to the simple linear solution (85) for $\mathscr{E}$, we have the readily verifiable solution.

$$F = -p\eta - q\mu \quad \text{and} \quad G = -p\eta + q\mu \quad (p, q \text{ constants}, p^2 - q^2 = 1), \qquad (102)$$

of equations (58) and (59). We shall now show how these solutions for F and G enable us to write down a new solution of the field equations.

We shall find it convenient to consider for χ and ω solutions which are the negatives of those which follow from equations (52) and (102); this is permissible since equations (38) and (39) are invariant to simultaneous changes in the signs of χ and ω. Then, we obtain the solutions

$$\chi = \frac{p^2(\eta^2-1)+q^2(1-\mu^2)}{(p\eta+1)^2-q^2\mu^2} \quad \text{and} \quad \omega = \frac{2q\mu}{(p\eta+1)^2-q^2\mu^2}. \qquad (103)$$

The corresponding solutions for X and Y are

$$X = \frac{p\eta+q\mu-1}{p\eta+q\mu+1} \quad \text{and} \quad Y = \frac{p\eta-q\mu-1}{p\eta-q\mu+1}. \qquad (104)$$

Also, for F and G given in equation (102), equation (60) for $(\mu_2+\mu_3)$ becomes

$$2\eta\frac{\partial}{\partial\eta}(\mu_2+\mu_3)+2\mu\frac{\partial}{\partial\mu}(\mu_2+\mu_3) = \frac{4}{p^2\eta^2-q^2\mu^2-1}-\frac{3}{\eta^2-1}+\frac{1}{1-\mu^2}. \qquad (105)$$

The solution of this last equation, appropriate to the problem on hand, is given by

$$e^{\mu_2+\mu_3} = C\frac{p^2(\eta^2-1)+q^2(1-\mu^2)}{(\eta^2-1)^{\frac{3}{4}}(1-\mu^2)^{\frac{1}{4}}}, \tag{106}$$

where C is an adjustable scale factor. (Strictly, C can be an arbitrary function of the argument η/μ; see footnote on page 416.)

Reverting to the variable r, we find that the solutions given in equations (103) and (106) are:

$$\chi = \frac{\Delta + q^2p^{-2}(M^2-a^2)\,\delta}{[(r-M)+p^{-1}(M^2-a^2)^{\frac{1}{2}}]^2 - q^2p^{-2}(M^2-a^2)\,\mu^2}, \tag{107}$$

$$\omega = \frac{2qp^{-2}(M^2-a^2)\,\mu}{[(r-M)+p^{-1}(M^2-a^2)^{\frac{1}{2}}]^2 - q^2p^{-2}(M^2-a^2)\,\mu^2}, \tag{108}$$

and

$$e^{\mu_2+\mu_3} = \frac{p^2}{(M^2-a^2)^{\frac{1}{4}}}\,\frac{\Delta + q^2p^{-2}(M^2-a^2)\,\delta}{\Delta^{\frac{3}{4}}(1-\mu^2)^{\frac{1}{4}}}. \tag{109}$$

Combined with the known solutions (31) for $e^{\mu_3-\mu_2}$ and e^{β}, the foregoing equations provide a new solution of Einstein's equations.

As in §7, we can, consistently, with the choice of gauge made in §3, let

$$p = M/(M^2-a^2)^{\frac{1}{2}} \quad \text{and} \quad q = a/(M^2-a^2)^{\frac{1}{2}}, \tag{110}$$

in equations (107)–(109); but no special simplifications arise from this substitution.

We can similarly write down other solutions which stand in the same relation to the Tomimatsu–Sato family of solutions as the solution we have just written down stands in relation to Kerr's solution. Thus, Tomimatsu & Sato's '$\delta = 2$' solution is derived from the solution (cf. Tomimatsu & Sato 1973)

$$\mathscr{E} = \alpha/\beta,$$

where

$$\alpha = (p^2\eta^4+q^2\mu^4-1) - 2ipq\eta\mu(\eta^2-\mu^2) \quad \text{and} \quad \beta = 2p\eta(\eta^2-1) - 2iq\mu(1-\mu^2),$$
$$(p, q \text{ constants}, \; p^2+q^2 = 1) \tag{111}$$

of Ernst's equation. Corresponding to this complex solution of equation (71), we have the following real solutions of equations (58) and (59):

$$F = \alpha/\beta \quad \text{and} \quad G = \bar{\alpha}/\bar{\beta},$$

where

$$\alpha = (p^2\eta^4 - q^2\mu^4 - 1) + 2pq\eta\mu(\eta^2-\mu^2), \quad \beta = 2p\eta(\eta^2-1) + 2q\mu(1-\mu^2),$$
$$\bar{\alpha} = (p^2\eta^4 - q^2\mu^4 - 1) - 2pq\eta\mu(\eta^2-\mu^2), \quad \bar{\beta} = 2p\eta(\eta^2-1) - 2q\mu(1-\mu^2),$$
$$(p, q \text{ constants}, \; p^2-q^2 = 1). \tag{112}$$

It would not appear that at this stage any useful purpose will be served by writing out explicitly the solutions for the metric functions which follow from the foregoing solutions for F and G. It would rather appear that the central problem is to relate, quite generally, the real solutions of equations (58) and (59) with the complex solutions of equation (71). Some preliminary remarks bearing on this problem are made below.

(a) *The relations between the complex solutions for $\mathscr{E}$ and the real solutions for F and G*

A given stationary axisymmetric solution of the field equations specifies simultaneously solutions for $\mathscr{E}$ and for F and G. There must, accordingly, be a relation between the complex solutions of equation (71) and the real solutions of equations (58) and (59). A formal expression to this relation can be given.

The metric functions χ and ω are related to F and G by (cf. equation (52))

$$\chi = \frac{1-FG}{(1-F)(1-G)} \quad \text{and} \quad \omega = \frac{F-G}{(1-F)(1-G)}; \tag{113}$$

while the functions Ψ and Φ (defined in §6*a*) are related to $\mathscr{E}$ by

$$\left.\begin{aligned} \Psi &= (\Delta\delta)^{\frac{1}{2}}/\chi = \operatorname{Re} Z = (1-|\mathscr{E}|^2)/|1-\mathscr{E}|^2 \\ \text{and} \qquad \Phi &= \operatorname{Im} Z = -\mathrm{i}(\mathscr{E}-\mathscr{E}^*)/|1-\mathscr{E}|^2. \end{aligned}\right\} \tag{114}$$

Accordingly, one relation is

$$(1-|\mathscr{E}|^2)/|1-\mathscr{E}|^2 = (\Delta\delta)^{\frac{1}{2}}(1-F)(1-G)/(1-FG). \tag{115}$$

Further relations follow from equations (61); thus

$$\left.\begin{aligned} \mathrm{i}\left(\frac{\mathscr{E}-\mathscr{E}^*}{|1-\mathscr{E}|^2}\right)_{,2} &= +\delta\frac{(1-F)^2(1-G)^2}{(1-FG)^2}\left[\frac{F-G}{(1-F)(1-G)}\right]_{,3} \\ \text{and} \qquad \mathrm{i}\left(\frac{\mathscr{E}-\mathscr{E}^*}{|1-\mathscr{E}|^2}\right)_{,3} &= -\Delta\frac{(1-F)^2(1-G)^2}{(1-FG)^2}\left[\frac{F-G}{(1-F)(1-G)}\right]_{,2}. \end{aligned}\right\} \tag{116}$$

The foregoing relations are not particularly transparent; but in any given case they can be made explicit. Thus, we have seen how the linear solution of Ernst's equation leads to the Kerr metric (or, the one derived from it by the transformation (54)) and particular solutions for X and Y; and, conversely, how the linear solutions for F and G lead to the new metric (derived earlier in this section) and a particular solution for Z. The explicit relations are the following.

The solution, $\quad \mathscr{E} = p\eta + \mathrm{i}q\mu \quad (p, q \text{ constants}, p^2+q^2 = 1), \qquad (117)$

corresponds to

$$\left.\begin{aligned} X &= [(\eta^2-1)(1-\mu^2)]^{\frac{1}{2}}\frac{(p\eta+1)^2+q^2\mu^2}{p^2(\eta^2-1)-q^2(1-\mu^2)} + 2\frac{q}{p}\frac{(1-\mu^2)(p\eta+1)}{p^2(\eta^2-1)-q^2(1-\mu^2)} \\ \text{and} \quad Y &= [(\eta^2-1)(1-\mu^2)]^{\frac{1}{2}}\frac{(p\eta+1)^2+q^2\mu^2}{p^2(\eta^2-1)-q^2(1-\mu^2)} - 2\frac{q}{p}\frac{(1-\mu^2)(p\eta+1)}{p^2(\eta^2-1)-q^2(1-\mu^2)}; \end{aligned}\right\} \tag{118}$$

and, conversely, the solutions

$$F = p\eta + q\mu \quad \text{and} \quad G = p\eta - q\mu \quad (p, q \text{ constants}, p^2-q^2 = 1) \tag{119}$$

corresponds to the solution

$$Z = [(\eta^2-1)(1-\mu^2)]^{\frac{1}{2}} \frac{(p\eta+1)^2 - q^2\mu^2}{p^2(\eta^2-1)+q^2(1-\mu^2)} + \mathrm{i}2\frac{q}{p}\frac{(1-\mu^2)(p\eta+1)}{p^2(\eta^2-1)+q^2(1-\mu^2)}. \quad (120)$$

Analogous relations can be written for the remaining Tomimatsu–Sato solutions of Ernst's equations. It is possible that an examination of these relations in such special cases might reveal the basic connection that exists between the solutions of equations (58), (59) and (71).

9. Concluding remarks

While the principal motivation for this paper was to present an *ab initio* derivation of the Kerr metric based on an elementary treatment of the equations governing stationary axisymmetric fields, the treatment has revealed many of the inner relations that characterize the general solutions and opened an additional avenue towards the generation of new exact solutions. The simplicity, with which a new solution, associated in some sense with Kerr's solutions, was constructed, suggests that it will be useful to pursue this new avenue further and relate it to other known avenues that have been explored. Explorations along these lines are presently under active consideration by Mr B. Xanthopoulos.

I am greatly indebted to Dr A. Ashtekar and Mr B. Xanthopoulos for many useful discussions.

The research reported in this paper has in part been supported by a grant from the National Science Foundation under grant PHY 76-81102 with the University of Chicago.

References

Adler, A., Bazin, M. & Schiffer, M. 1975 *Introduction to relativity*. New York: McGraw-Hill.
Boyer, R. H. & Lindquist, R. W. 1967 *J. Math. Phys.* **8**, 265.
Carter, B. 1971 *Phys. Rev. Lett.* **26**, 331.
Carter, B. 1973 In *Black holes* (ed. C. DeWitt & B. S. DeWitt), p. 57. New York: Gordon and Breach.
Chandrasekhar, S. & Friedman, J. L. 1972 *Astrophys. J.* **175**, 379.
Condon, E. U. & Shortley, G. H. 1935 *The theory of atomic spectra*. Cambridge University Press.
Ernst, F. J. 1968 *Phys. Rev.* **167**, 1175.
Ernst, F. J. 1974 *J. Math. Phys.* **15**, 1409.
Kerr, R. P. 1963 *Phys. Rev. Lett.* **11**, 237.
Kerr, R. P. & Schild, A. 1965 *Comitato Nazionale per le Manifestazioni Celebrative del IV Centenario Della Nascita di Galileo Galilei, Atti del Convegno Sulla Relatività Generale: Problemi Dell'Energia E Onde Gravitazionali*, p. 1.
Kinnersley, W. 1975 In *General relativity and gravitation* (ed. G. Shaviv & J. Rosen), p. 109. New York: John Wiley.
Landau, L. D. & Lifschitz, E. M. 1975 *Classical fields*. Oxford: Pergamon.
Meinhardt, R. 1976 *Riv. nuovo Cimento* **6**, 405.
Papapetrou, A. 1953 *Annls Phys.* **12**, 309.
Robinson, D. C. 1975 *Phys. Rev. Lett.* **34**, 905.
Tomimatsu, A. & Sato, H. 1973 *Research Institute for Fundamental Physics*, no. 173. Kyoto: Kyoto University.

5

On the equations governing the axisymmetric perturbations of the Kerr black hole

By S. Chandrasekhar, F.R.S.
University of Chicago, Chicago, Illinois, 60637
and S. Detweiler
University of Maryland, College Park, Maryland, 20742

(*Received* 3 *February* 1975)

It is shown how Teukolsky's equation, governing the perturbations of the Kerr black hole, can be reduced, in the axisymmetric case, to a one-dimensional wave equation with four possible potentials. The potentials are implicitly, dependent on the frequency; and besides, depending on circumstances, they can be complex. In all cases (i.e. whether or not the potentials are real or complex), the problem of the reflexion and the transmission of gravitational waves by the potential barriers can be formulated, consistently, with the known conservation laws. It is, further, shown that all four potentials lead to the same reflexion and transmission coefficients.

1. Introduction

It is now generally believed that black holes occurring in nature must belong to the Kerr family. The Kerr metric (Kerr 1963) as written in the coordinate system, first introduced by Boyer & Lindquist (1967), is

$$\mathrm{d}s^2 = \frac{1}{r^2+a^2\cos^2\theta}\{-\Delta(\mathrm{d}t - a\sin^2\theta\,\mathrm{d}\varphi)^2 + [(r^2+a^2)\,\mathrm{d}\varphi - a\,\mathrm{d}t]^2\sin^2\theta\} + (r^2+a^2\cos^2\theta)\left[\frac{(\mathrm{d}r)^2}{\Delta} + (\mathrm{d}\theta)^2\right], \qquad (1)$$

where

$$\Delta = r^2 - 2Mr + a^2, \qquad (2)$$

M is the inertial mass, and a is the specific angular momentum of the black hole. The event horizon occurs at the larger of the two roots of the equation $\Delta = 0$ (so long as $a < M$).

The equations governing perturbations of the Kerr metric are, of course, basic for many problems in relativistic astrophysics concerned with what may happen in the neighbourhood of Kerr black holes (cf. Rees 1974). By considering a particular combination of the components of the Riemann tensor – the Newman–Penrose Ψ_0 – and restricting himself to perturbations with a dependence on t and φ of the form $\mathrm{e}^{\mathrm{i}(\sigma t + m\varphi)}$ (where m is an integer positive, negative, or zero), Teukolsky (1972, 1973) was able to separate the variables r and θ and obtain a pair of decoupled equations. The separation of the variables is a remarkable achievement in view of the initially axisymmetric situation – a situation in which one does not, normally, expect such a separation. And quite rightly, the derivation of Teukolsky's equation has been considered as providing the key to the solution of many problems in relativistic astrophysics.

Now Teukolsky's equation, when specialized to the Schwarzschild case (by setting $a = m = 0$) reduces to the Bardeen–Press equation (Bardeen & Press 1973). But it has been shown in a recent paper (Chandrasekhar 1975) that the Bardeen–Press equation can be transformed to either the Zerilli (1970) or the Regge–Wheeler (1957) equation governing the even and the odd parity perturbations of the Schwarzschild metric coefficients. Since the axisymmetric perturbations of rotating and non-rotating systems have many aspects in common in the Newtonian framework, the question arises whether Teukolsky's equation allows transformations to the forms of one dimensional wave equations which can be distinguished, for example, as governing the even and the odd parity perturbations. We shall show that such transformations are possible; indeed, it will appear that each class appears with a multiplicity two.

The basic ideas underlying the analysis of this paper are derived from an earlier paper (Chandrasekhar 1975; this paper will be referred to hereafter as I) in which explicit equations relating the functions satisfying the equations of Bardeen & Press, of Zerilli, and of Regge & Wheeler were derived.

2. Teukolsky's equation for axisymmetric perturbations

For axisymmetric perturbations ($m = 0$) of the Kerr metric, whose dependence on time is given by $e^{i\sigma t}$, Teukolsky's equation (Teukolsky 1973, equation (5.2)) reduces to the form

$$\frac{d^2\phi}{dr_*^2} + \sigma^2\phi - 4i\sigma\frac{2r\Delta-(r-M)(r^2+a^2)}{(r^2+a^2)^2}\phi - \left[\frac{\lambda\Delta}{(r^2+a^2)^2} + G^2 + \frac{dG}{dr_*}\right]\phi = 0, \tag{3}$$

where

$$\left.\begin{aligned} \frac{d}{dr_*} &= \frac{\Delta}{r^2+a^2}\frac{d}{dr}, \quad G = \frac{r\Delta}{(r^2+a^2)^2} + 2\frac{r-M}{r^2+a^2}, \\ \lambda &= \tau - 6, \quad \tau = E + a^2\sigma^2, \end{aligned}\right\} \tag{4}$$

and E is the separation constant which occurs in the equation

$$\frac{1}{\sin\theta}\frac{d}{d\theta}\left(\sin\theta\frac{dS}{d\theta}\right) - \frac{4S}{\sin^2\theta} + (a^2\sigma^2\cos^2\theta + 4a\sigma\cos\theta)S = -ES, \tag{5}$$

governing the angular dependence of the perturbation.

The boundary conditions with respect to which equation (3) must be solved are

$$\phi \to e^{i\sigma r_*}/\Delta \quad \text{as} \quad r_* \to -\infty \quad \text{and} \quad \phi \to e^{-i\sigma r_*}/r^2 \quad \text{as} \quad r_* \to +\infty, \tag{6}$$

appropriate for ingoing waves at the horizon ($r_* \to -\infty$) and outgoing waves at infinity ($r_* \to +\infty$). More generally, the asymptotic behaviours of the solutions of equation (3), for $r_* \to -\infty$ and $r_* \to +\infty$, are

$$\phi \to \phi_{\leftarrow} \frac{e^{i\sigma r_*}}{\Delta} + \phi_{\rightarrow} e^{-i\sigma r_*} \Delta \quad (r_* \to -\infty) \tag{7}$$

and

$$\phi \to \phi_{\leftarrow} e^{i\sigma r_*} r^2 + \phi_{\rightarrow} \frac{e^{-i\sigma r_*}}{r^2} \quad (r_* \to +\infty). \tag{8}$$

Analogous to the function Y introduced in the context of the Bardeen–Press equation in I (equation (62)), we shall now define the function

$$Y = \frac{\Delta}{(r^2+a^2)^2}\phi. \tag{9}$$

In terms of Y, (3) becomes

$$\Lambda^2 Y + P\Lambda_- Y - QY = 0. \tag{10}$$

where $\Lambda_\pm$ and Λ^2 denote the operators

$$\Lambda_\pm = \frac{d}{dr_*} \pm i\sigma \quad \text{and} \quad \Lambda^2 = \Lambda_+\Lambda_- = \Lambda_-\Lambda_+ = \frac{d^2}{dr_*^2} + \sigma^2, \tag{11}$$

and

$$P = 4\frac{r(r^2+a^2) - M(3r^2-a^2)}{(r^2+a^2)^2} = \frac{d}{dr_*}\left[\lg \frac{(r^2+a^2)^4}{\Delta^2}\right], \tag{12}$$

$$Q = \frac{\Delta}{(r^2+a^2)^2}\left[\nu + 3\frac{2Mr^3 - a^2(r^2+a^2)}{(r^2+a^2)^2}\right], \quad \text{and} \quad \tau = \nu + 2. \tag{13}$$

It can be readily verified that when we set

$$a = 0 \quad \text{and} \quad \nu = E - 2 = l(l+1) - 2 = 2n \qquad (14)\dagger$$

in equation (10), we recover the Bardeen–Press equation in the form given in I, equation (61).

3. The conditions for the equation for Y to be transformable to the form of a one dimensional wave equation

We now seek a transformation which will reduce equation (10) to a one dimensional wave equation of the form

$$\Lambda^2 Z = VZ, \tag{15}$$

where V is some potential function of r_* (or r) to be determined. It is clear that Z (on the assumption that one such exists) must be a linear combination of Y and its derivative since Y satisfies a second order differential equation. There is, therefore, no loss of generality in assuming that Z is related to Y in the manner (cf. I, equation (49))

$$Y = f\Lambda_+\Lambda_+ Z + W\Lambda_+ Z, \tag{16}$$

where f and W are certain functions of r_* (or r) which are, for the present, unspecified.

† The separation constant E in equation (5) has the value $l(l+1)$ when $a = 0$.

Applying the operator Λ_- to equation (16) and making use of the fact that Z has been assumed to satisfy equation (15), we find that

$$\Lambda_- Y = \left[\frac{d}{dr_*}(fV)+WV\right]Z+\left[fV+\frac{d}{dr_*}(W+2i\sigma f)\right]\Lambda_+ Z, \tag{17}$$

or, *defining* the functions

$$-\beta(r_*)\,M\frac{\Delta^2}{(r^2+a^2)^4}=\frac{d}{dr_*}(fV)+WV \tag{18}$$

and

$$R(r_*)=fV+\frac{d}{dr_*}(W+2i\sigma f), \tag{19}$$

we can write

$$\Lambda_- Y=-\beta M\frac{\Delta^2}{(r^2+a^2)^4}Z+R\Lambda_+ Z. \tag{20}$$

Also, we may note here that the relation (16), by virtue of equation (15), has the alternative form

$$Y=fVZ+(W+2i\sigma f)\,\Lambda_+ Z. \tag{21}$$

We must now require that equations (20) and (21) are compatible with the equation satisfied by Y. For this purpose, we must eliminate Y from equations (20) and (21). We proceed as follows.

First applying the operator Λ_- to equation (20), we obtain

$$\begin{aligned}\Lambda_-\Lambda_- Y &= \Lambda_-\left[-\beta M\frac{\Delta^2}{(r^2+a^2)^4}Z+R\Lambda_+ Z\right]\\ &= -\beta M\frac{\Delta^2}{(r^2+a^2)^4}(\Lambda_+-2i\sigma)Z-\beta MZ\frac{d}{dr_*}\frac{\Delta^2}{(r^2+a^2)^4}\\ &\quad -M\frac{\Delta^2}{(r^2+a^2)^4}Z\frac{d\beta}{dr_*}+R\Lambda^2 Z+\frac{dR}{dr_*}\Lambda_+ Z.\end{aligned} \tag{22}$$

On the other hand, by equation (10)

$$\Lambda_-\Lambda_- Y=-(2i\sigma+P)\,\Lambda_- Y+QY, \tag{23}$$

or, substituting for $\Lambda_- Y$ and Y from equations (20) and (21), we have

$$\begin{aligned}\Lambda_-\Lambda_- Y = -(2i\sigma+P)&\left[-\beta M\frac{\Delta^2}{(r^2+a^2)^4}Z+R\Lambda_+ Z\right]\\ &+Q[fVZ+(W+2i\sigma f)\,\Lambda_+ Z].\end{aligned} \tag{24}$$

By equating the right-hand sides of equations (22) and (24), we must obtain an *identity*. And the conditions, that the resulting equations may be an identity, are that the coefficients of Z and $\Lambda_+ Z$ in the equation, separately, vanish. In this manner, we obtain the pair of equations

$$-\beta M\frac{\Delta^2}{(r^2+a^2)^4}+\frac{dR}{dr_*}=QW-PR+2i\sigma(Qf-R) \tag{25}$$

and

$$RV-M\frac{\Delta^2}{(r^2+a^2)^4}\frac{d\beta}{dr_*}=QfV, \tag{26}$$

where it may be noted that the alternative form for P given in equation (12), namely $\{\lg[(r^2+a^2)^4/\Delta^2]\}_{,r_*}$, plays an essential role in the reductions. Making use of this same alternative form for P, we can rewrite equation (25) in the manner

$$\frac{d}{dr_*}\left[\frac{(r^2+a^2)^4}{\Delta^2}R\right] = \frac{(r^2+a^2)^4}{\Delta^2}[Q(W+2i\sigma f)-2i\sigma R]+\beta M. \tag{27}$$

It can now be directly verified that equations (18), (19), (26), and (27) admit the *integral*

$$Rf\frac{(r^2+a^2)^4}{\Delta^2}V+\beta M(W+2i\sigma f) = K = \text{a constant.} \tag{28}$$

(Note that K will be complex.) The integral (28) is the present counterpart of the relations (58) and (A 6) found in I in the context of the reduction of the Bardeen–Press equation.

(x) *The basic equations*

Rewriting equations (20) and (21) in the form

$$Y = fVZ+(W+2i\sigma f)\Lambda_+Z \tag{29}$$

and

$$\frac{(r^2+a^2)^4}{\Delta^2}\Lambda_-Y = -\beta MZ+\frac{(r^2+a^2)^4}{\Delta^2}R\Lambda_+Z, \tag{30}$$

we observe that the determinant of this system is precisely the integral (28). Accordingly, inverting the system, we obtain the pair of equations,

$$KZ = \frac{(r^2+a^2)^4}{\Delta^2}RY-\frac{(r^2+a^2)^4}{\Delta^2}(W+2i\sigma f)\Lambda_-Y \tag{31}$$

and

$$K\Lambda_+Z = \beta MY+\frac{(r^2+a^2)^4}{\Delta^2}fV\Lambda_-Y, \tag{32}$$

expressing Z and Λ_+Z in terms of Y and Λ_-Y. Again, equations (31) and (32) are the present counterparts of I, equations (56), (57), (A 10), and (A 11) derived in the context of the Bardeen–Press equation.

It can now be verified that *equations* (29)–(32), *by virtue of the relations* (18), (19), (26), *and* (27), *are necessary and sufficient for equation* (10) *to imply equation* (15), *and conversely*.

4. An explicit solution

Since the four equations (18), (19), (26), and (27) allow the integral (28), it will clearly suffice to supplement the integral with any three of the original equations; and we shall select equations (19), (26), and (27). In considering these equations, we shall find it convenient to define the following quantities:

$$\rho^2 = (r^2+a^2), \tag{33}$$

$$T = W+2i\sigma f, \tag{34}$$

and
$$q = \frac{\rho^8}{\Delta} Q = \nu\rho^4 - 3a^2\rho^2 + 6Mr^3. \tag{35}$$

In terms of these quantities, the equations we shall consider are

$$R = fV + \frac{\mathrm{d}T}{\mathrm{d}r_*}, \tag{36}$$

$$(R - Qf)\, V = M \frac{\Delta^2}{\rho^8} \frac{\mathrm{d}\beta}{\mathrm{d}r_*}, \tag{37}$$

$$\frac{\mathrm{d}}{\mathrm{d}r_*}\left(\frac{\rho^8}{\Delta^2} R\right) = \frac{\rho^8}{\Delta^2}(QT - 2\mathrm{i}\sigma R) + \beta M, \tag{38}$$

and
$$\frac{\rho^8}{\Delta^2} RfV + \beta MT = K. \tag{39}$$

Eliminating fV from equation (39) with the aid of equation (36), we obtain

$$R\left(R - \frac{\mathrm{d}T}{\mathrm{d}r_*}\right) + \frac{\Delta^2}{\rho^8}\beta MT = \frac{\Delta^2}{\rho^8} K. \tag{40}$$

Since we are provided with only two independent equations for the three unknown functions f, V, and W, it is clear that we have a considerable latitude in seeking solutions of equations (36)–(40). The guidelines we shall adopt in our search for solutions are the following: the *explicit* dependence of the various quantities on σ (i.e. apart from the *implicit* dependence through ν) is *linear* and, moreover, that V is *explicitly independent of* σ; and further, that an analytic solution is preferable to any that requires quadratures. It appears that a solution satisfying these requirements is possible only if *R and f, in addition to V, are also explicitly independent of σ*, while T and β, consistent with our requirements, are of the forms

$$T = T_1 + 2\mathrm{i}\sigma T_2 \quad \text{and} \quad \beta M = \beta_1 + 2\mathrm{i}\sigma\beta_2, \tag{41}$$

with T_2 and β_2, however, being *constants*. In making these various assumptions on the desired form of the solution, we are imposing more restrictions than we are permitted by the latitude that we have. Nevertheless, we shall verify that solutions of the stated forms exist by virtue of a remarkable identity (equation (55) below) that the function $F = q/\Delta$ satisfies.

Substituting the assumed forms for R, T, and βM in equation (38) and equating, separately, the terms which occur with the factor $\mathrm{i}\sigma$ and the terms which do not, we obtain

$$R = QT_2 + \frac{\Delta^2}{\rho^8}\beta_2, \tag{42}$$

and
$$\frac{\mathrm{d}}{\mathrm{d}r_*}\left(\frac{\rho^8}{\Delta^2} R\right) = \frac{\rho^8}{\Delta^2} QT_1 + \beta_1. \tag{43}$$

Next letting
$$K = \kappa_1 + 2\mathrm{i}\sigma\kappa_2, \tag{44}$$

we similarly obtain from equation (40) the pair of equations

$$\beta_1 T_2 + \beta_2 T_1 = \kappa_2 \tag{45}$$

and

$$R^2 - R\frac{\mathrm{d}T_1}{\mathrm{d}r_*} + \frac{\Delta^2}{\rho^8}\beta_1 T_1 = \frac{\Delta^2}{\rho^8}(\kappa_1 + 4\sigma^2\beta_2 T_2). \tag{46}$$

From equations (42), (43), (45), and (46), it is apparent that consistently with the assumptions already made, there is no loss of generality in setting

$$T_2 = 1 \tag{47}$$

since T_2 appears as a simple scaling factor which can be absorbed in the definitions of the various other quantities. Accordingly, setting $T_2 = 1$ and defining

$$F = \frac{q}{\Delta} \quad \left(\text{so that } Q = \frac{\Delta^2}{\rho^8}F\right) \tag{48}$$

and

$$\kappa = \kappa_1 + 4\sigma^2\beta_2, \tag{49}$$

we can rewrite equations (42), (43), (45), and (46) in the forms

$$R = \frac{\Delta^2}{\rho^8}(F + \beta_2), \tag{50}$$

$$\frac{\mathrm{d}F}{\mathrm{d}r_*} = T_1 F + \beta_1, \tag{51}$$

$$\beta_1 = \kappa_2 - \beta_2 T_1, \tag{52}$$

and

$$\frac{\Delta^2}{\rho^8}(F + \beta_2)^2 - (F + \beta_2)\frac{\mathrm{d}T_1}{\mathrm{d}r_*} + T_1(\kappa_2 - \beta_2 T_1) = \kappa. \tag{53}$$

Equations (51) and (52) can be combined to give

$$T_1 = \frac{1}{F - \beta_2}\left(\frac{\mathrm{d}F}{\mathrm{d}r_*} - \kappa_2\right). \tag{54}$$

Equations (50), (52), and (54) do, indeed, provide explicit analytical solutions for R, β_1, and T_1 (since F is a known function); but it remains to verify that with a suitable choice of the constants β_2, κ_2, and κ, which are at our disposal, equation (53) can be satisfied. We now proceed towards this verification.

Substituting for T_1 from equation (54) in equation (53), we obtain after some rearrangements

$$\left(\frac{\mathrm{d}F}{\mathrm{d}r_*}\right)^2 - F\frac{\mathrm{d}^2F}{\mathrm{d}r_*^2} + \frac{\Delta^2}{\rho^8}F^3 = (\kappa_2^2 - 2\beta_2\kappa) + \left(\kappa + 2\frac{\Delta^2}{\rho^8}\beta_2^2\right)F - \frac{\beta_2^2}{F}\left(\frac{\mathrm{d}^2F}{\mathrm{d}r_*^2} + \beta_2^2\frac{\Delta^2}{\rho^8} - \kappa\right). \tag{55}$$

In this last equation $F(= q/\Delta)$ is, of course, a known function. The various terms in the equation can, therefore, be directly evaluated. The evaluation of the terms, particularly those on the left-hand side, is cumbersome (through entirely straightforward). We find

$$\begin{aligned}&\rho^6[\nu^2(\nu+2)\rho^6-\nu(13\nu-6)a^2\rho^4+27\nu a^4\rho^2-27a^6]\\&\quad+M\rho^4[6\nu(\nu+2)r^3\rho^4+4\nu(5\nu-6)a^2r\rho^4-18(6\nu-5)a^4r\rho^2+126a^6r]\\&\quad+36M^2\rho^2[r^6\rho^2+3a^2r^4\rho^2+(3\nu-8)a^4r^2\rho^2-a^6r^2]-72M^3r^7(\rho^2+3a^2)\\&\quad=\Delta\rho^8(\kappa_2^2-2\beta_2\kappa)+q\rho^8\left(\kappa+2\frac{\Delta^2}{\rho^8}\beta_2^2\right)-\beta_2^2\Delta\rho^2\left[\frac{d}{dr}(\rho^2\Delta)\frac{d\Delta}{dr}-\rho^2\Delta\frac{d^2\Delta}{dr^2}\right]\\&\quad-\beta_2^2\frac{\Delta^2}{q}\left\{\rho^2\left[-\frac{d}{dr}(\rho^2\Delta)\frac{dq}{dr}+\rho^2\Delta\frac{d^2q}{dr^2}\right]+\beta_2^2\Delta^2-\kappa\rho^8\right\}.\end{aligned}\tag{56}$$

We shall now show that values for the constants κ, κ_2, and β_2 can be so chosen that equation (56) becomes an identity.

First, we observe that only $\kappa q\rho^8$ on the right-hand side of equation (56) can contribute a term in ρ^{12} to balance the term $\nu^2(\nu+2)\rho^{12}$ on the left-hand side. The equality of the coefficients of ρ^{12} on the two sides requires

$$\kappa=\nu(\nu+2).\tag{57}$$

Next, we observe that for equation (56) to be satisfied, the terms in braces in the last line on the right-hand side must be divisible by q; and the quantity in braces, after the substitution for κ its value (57), is found to be

$$\begin{aligned}\{\ \}=&-4\nu\rho^6(r^2-a^2)+6a^2\rho^4(3r^2-a^2)-12Mr\rho^2(3r^4+2a^2r^2-3a^4)\\&+36M^2r^2\rho^2(r^2-a^2)+\beta_2^2(\rho^4-4\rho^2Mr+4M^2r^2)-\nu(\nu+2)\rho^8.\end{aligned}\tag{58}$$

It is now readily verified that for this expression to be divisible by q it is necessary that

$$\beta_2^2=9a^4;\tag{59}$$

and when β_2^2 has this value, the expression is

$$\{\ \}=-q[\rho^2(\nu\rho^2+6r^2+a^2)-6Mr^3].\tag{60}$$

Accordingly, when κ and β_2^2 have the values assigned, the terms on the right-hand side of equation (56) reduce to

$$\begin{aligned}&\nu(\nu+2)q\rho^8+[\kappa_2^2-2\beta_2\nu(\nu+2)]\Delta\rho^8+9a^4\Delta^2[2q+\rho^2(\nu\rho^2+6r^2+a^2)-6Mr^3]\\&\qquad-9a^4\Delta\rho^2\left[\frac{d}{dr}(\rho^2\Delta)\frac{d\Delta}{dr}-\rho^2\Delta\frac{d^2\Delta}{dr^2}\right].\end{aligned}\tag{61}$$

We can now cancel from the left-hand side of equation (56) the term $\nu(\nu+2)q\rho^8$ on the right-hand side. After the cancellation, all the terms on the right-hand side occur with a factor Δ. The surviving terms on the left-hand side must, therefore, allow Δ as a factor; we find that they do and the equation that we are left to verify is

$$\begin{aligned}&[36M^2-2a^2\nu(5\nu-6)]\rho^8+27\nu a^4\rho^6-27a^6\rho^4-18Ma^4r\rho^2[(3\nu-5)\rho^2-4a^2]\\&\quad+36M^2(-6a^4\rho^4+8a^6\rho^2-3a^8)\\&\quad=[\kappa_2^2-2\beta_2\nu(\nu+2)]\rho^8+9a^4\Big\{\Delta[2q+\rho^2(\nu\rho^2+6r^2+a^2)-6Mr^3]\\&\qquad-\rho^2\left[\frac{d}{dr}(\rho^2\Delta)\frac{d\Delta}{dr}-\rho^2\Delta\frac{d^2\Delta}{dr^2}\right]\Big\}.\end{aligned}\tag{62}$$

The equality of the terms in ρ^8 on the two sides of this equation requires that

$$\kappa_2^2 - 2\beta_2 \nu(\nu+2) = 36M^2 - 2a^2\nu(5\nu-6); \tag{63}$$

and for equation (62) to be satisfied, the remaining terms must cancel which, in fact, they do. Thus, we have verified that equation (55) does indeed become an identity when

$$\kappa = \nu(\nu+2), \quad \beta_2 = \pm 3a^2,$$

and

$$\kappa_2 = \pm [36M^2 - 2a^2\nu(5\nu-6) + 2\beta_2 \nu(\nu+2)]^{\frac{1}{2}}. \tag{64}$$

The corresponding value of the constant K (cf. equations (44) and (49)) is

$$K = [\nu(\nu+2) - 4\sigma^2\beta_2] + 2i\sigma\kappa_2. \tag{65}$$

The coefficients β_2 and κ_2 can be assigned different signs independently of each other. There are thus four possible solutions; and it should be particularly noted that on account of the dependence of $\nu(= E - 2 + a^2\sigma^2)$ on the separation constant $E(= l(l+1)$ when $a = 0)$, κ_2 can become imaginary for a sufficiently large l (for a given σ) or for a sufficiently large σ (for a given a). We shall return to the meaning to be attributed to the four solutions in §8; and in §5 and the following sections we shall consider how to cope with the wave equation for Z when κ_2 becomes imaginary.

(*a*) *The solutions for V, f, and W*

With κ_2 and β_2 given by equations (64), the solutions for R, T_1, and β_1 are given by (cf. equations (48), (50), (52), and (54))

$$R = Q + \frac{\Delta^2}{\rho^8}\beta_2 = \frac{\Delta}{\rho^8}(q + \beta_2\Delta) = \frac{\Delta^2}{\rho^8}(F + \beta_2), \tag{66}$$

$$T_1 = \frac{1}{F - \beta_2}\left(\frac{dF}{dr_*} - \kappa_2\right), \tag{67}$$

and

$$\beta_1 = \kappa_2 - \beta_2 T_1. \tag{68}$$

The solution for V can now be obtained by combining equations (26), (36), and (68) in the manner

$$RV = Q\left(R - \frac{dT_1}{dr_*}\right) + \frac{\Delta^2}{\rho^8}\frac{d\beta_1}{dr_*}.$$

$$= QR - \left(Q + \frac{\Delta^2}{\rho^8}\beta_2\right)\frac{dT_1}{dr_*} = R\left(Q - \frac{dT_1}{dr_*}\right). \tag{69}$$

Therefore,

$$V = Q - \frac{dT_1}{dr_*}. \tag{70}$$

Substituting for T_1 in accordance with equation (67), we find after some reductions and rearrangements

$$\frac{1}{\Delta^2}(q - \beta_2\Delta)^2 V = \frac{\Delta^2}{\rho^8}F^3 + \left(\frac{dF}{dr_*}\right)^2 - F\frac{d^2F}{dr_*^2}$$
$$+ \beta_2\left\{\frac{d^2F}{dr_*^2} - \frac{\Delta^2}{\rho^8}F(2F - \beta_2)\right\} - \kappa_2\frac{dF}{dr_*}. \tag{71}$$

We observe that the terms in the first line on the right-hand side of this equation are the same as those we encountered earlier in equation (55). After some considerable simplification, equation (71) gives

$$V = \frac{\Delta}{\rho^8(q-\beta_2\Delta)^2}[\rho^6[\nu^2(\nu+2)\rho^6 - \nu(13\nu-6)a^2\rho^4 + 27\nu a^4\rho^2 - 27a^6]$$
$$+ M\rho^4[6\nu(\nu+2)r^3\rho^4 + 4\nu(5\nu-6)a^2 r\rho^4 - 18(6\nu-5)a^4 r\rho^2 + 126a^6 r]$$
$$+ 36M^2\rho^2[r^6\rho^2 + 3a^2r^4\rho^2 + (3\nu-8)a^4r^2\rho^2 - a^6r^2] - 72M^3r^7(\rho^2+3a^2)$$
$$- \kappa_2\,\rho^6\{2\nu r\rho^4 + 2M[-\nu\rho^2(3r^2-a^2) + 3r^4 + 3a^2(4r^2-a^2)] - 24M^2r^3\}$$
$$+ \beta_2 q\{-2(\nu-3)\rho^6 - 2a^2\rho^4 + 4Mr\rho^2[(\nu-7)\rho^2 + 2a^2] + 4M^2(9\rho^4 - 14a^2\rho^2 + 6a^4)\}$$
$$+ \beta_2\,\rho^2\Delta\{-4\nu\rho^4(r^2-a^2) + 6a^2\rho^2(3r^2-a^2) - 12Mr(3r^4+2a^2r^2-3a^4)$$
$$+ 36M^2r^2(r^2-a^2)\} + \beta_2^2\Delta^2 q]. \tag{72}$$

The potential given by equation (72) vanishes on the horizon (like Δ) and is $O(r^{-2})$ as $r \to \infty$; it is, therefore, a *short-range potential*. Also, in view of the ambiguity in the signs of β_2 and κ_2, *there are four possible potentials which characterize the problem.*

The solution for f follows from combining equations (19), (66), and (74) in the manner

$$fV = R - \frac{dT_1}{dr_*} = \frac{\Delta^2}{\rho^8}(F+\beta_2) - \frac{dT_1}{dr_*}$$
$$= Q - \frac{dT_1}{dr_*} + \beta_2\frac{\Delta^2}{\rho^8} = V + \beta_2\frac{\Delta^2}{\rho^8}. \tag{73}$$

Hence

$$f = 1 + \beta_2\frac{\Delta^2}{\rho^8 V}. \tag{74}$$

Accordingly, $f \to 1$ both at the horizon and at infinity.

Finally, the solution for W follows from the defining equations (34), (42), (43), and (67). Thus

$$W + 2i\sigma f = T_1 + 2i\sigma = \frac{1}{F-\beta_2}\left(\frac{dF}{dr_*} - \kappa_2\right) + 2i\sigma, \tag{75}$$

or

$$W = \frac{1}{F-\beta_2}\left(\frac{dF}{dr_*} - \kappa_2\right) + 2i\sigma(1-f). \tag{76}$$

Evaluating the term in F in equation (76), we obtain

$$W = \frac{1}{\rho^2(q-\beta_2\Delta)}\{2\nu r\rho^4 + 2M[-\nu\rho^2(3r^2-a^2) + 3r^4 + 3a^2(4r^2-a^2)]$$
$$- 24M^2r^3 - \kappa_2\,\rho^2\Delta\} + 2i\sigma(1-f). \tag{77}$$

(b) The potential in the Schwarzschild case, $a = 0$

By setting $a = 0$ in equation (72), we obtain

$$V_{a=0} = \frac{r-2M}{r^4(\nu r+6M)^2}\{\nu^2(\nu+2)r^3 + 6\nu(\nu+2)Mr^2 + 36M^2r - 72M^3$$
$$\mp 6M[2\nu r^2 - 6(\nu-1)Mr - 24M^2]\}. \tag{78}$$

Remembering that in this limit $\nu = 2n$, we verify that with the negative sign, the potential reduces to that of Zerilli while with the positive sign it reduces to that of Regge & Wheeler. We conclude that *the ambiguity in sign of κ_2 corresponds to the distinction between the odd and the even parity perturbations.*

(c) *An integral property of V*

A remarkable identity satisfied by the potential V follows from integrating equation (70) over the range of r_*. Thus,

$$\int_{-\infty}^{+\infty} V\,\mathrm{d}r_* = \int_{-\infty}^{+\infty} \frac{\Delta}{\rho^8} q\,\mathrm{d}r_* - T_1\Big|_{-\infty}^{+\infty}$$

$$= \int_{r_+}^{\infty} q\frac{\mathrm{d}r}{\rho^6} + (T_1)_{\Delta=0}, \tag{79}$$

where

$$r_+ = M + (M^2 - a^2)^{\frac{1}{2}} \tag{80}$$

is the radius of the horizon. Since, as may be verified (cf. equations (103) and (104) below),

$$(T_1)_{\Delta=0} = -\frac{r_+ - M}{r_+ M}, \tag{81}$$

$$\int_{-\infty}^{+\infty} V\,\mathrm{d}r_* = \int_{r_+}^{\infty} (\nu\rho^4 - 3a^2\rho^2 + 6Mr^3)\frac{\mathrm{d}r}{\rho^6} - \frac{r_+ - M}{r_+ M}; \tag{82}$$

or, after evaluating the elementary integral on the right-hand side, we have

$$\int_{-\infty}^{+\infty} V\,\mathrm{d}r_* = \frac{1}{a}(\nu - \tfrac{3}{2})\left(\tfrac{1}{2}\pi - \arctan\frac{r_+}{a}\right) + \frac{10M - r_+}{4Mr_+} - \frac{3a^2}{8Mr_+^2}. \tag{83}$$

Thus, *the value of the integral of V over the entire range of r_* is the same for all four cases*; *and, moreover, when the potential is complex, the integral over the imaginary part identically vanishes.*

Finally, we may note that when $a = 0$, equation (83) reduces to the known (equal) values of the integral over the Zerilli and the Regge–Wheeler potentials.

5. The transformation of the complex-conjugate equation for Y and the distinction between $Z^{(+\sigma)}$ and $Z^{(-\sigma)}$

So far we have been concerned with the transformation of Teukolsky's equation for Y, written in the particular form (10), to a one dimensional wave equation for Z with a suitably defined potential V. From the manner in which the transformation was effected in §4, it is clear that, had we started with the complex conjugate of equation (10) and followed the same procedure, we should not, necessarily, have obtained the complex conjugate of equation (15). The fact, that the expression for the potential V, which we finally arrived at, can, sometimes, be complex, makes it manifest that in seeking solutions of the forms (41) for T and β with the additional restrictions that V, R, and f are explicitly independent of σ, we are *not* (in spite of appearances) separating the real and the imaginary parts of T and β and restricting

V, R, and f to be real. On this account, we must carefully examine how the procedure of §4, when applied to the complex conjugate of equation (10), will affect the final outcome.

To distinguish equation (10) from its complex conjugate, we shall rewrite it in the form

$$\Lambda^2 Y^{(+\sigma)} + P\Lambda_- Y^{(+\sigma)} - Q Y^{(+\sigma)} = 0; \tag{84}$$

and write its complex conjugate as

$$\Lambda^2 Y^{(-\sigma)} + P\Lambda_+ Y^{(-\sigma)} - Q Y^{(-\sigma)} = 0. \tag{85}$$

And we shall distinguish the functions satisfying the one dimensional wave equations of the form (15), derived from equations (84) and (85), by $Z^{(+\sigma)}$ and $Z^{(-\sigma)}$, respectively. It should be explicitly noted that while $Y^{(+\sigma)}$ and $Y^{(-\sigma)}$ are complex conjugate functions $Z^{(+\sigma)}$ and $Z^{(-\sigma)}$ are not necessarily so.

A careful scrutiny of the result of applying the procedure of §4 to equation (85) reveals that $Z^{(+\sigma)}$ and $Z^{(-\sigma)}$ satisfy one dimensional wave equations with the *same* potential function:

$$\Lambda^2 Z^{(\pm\sigma)} = V Z^{(\pm\sigma)}. \tag{86}$$

In other words, *when V is complex, the equations governing $Z^{(+\sigma)}$ and $Z^{(-\sigma)}$ are not complex conjugate equations. Therefore, under these circumstances $Z^{(+\sigma)}$ and $Z^{(-\sigma)}$ are not complex conjugate functions.*

In addition to V, we also find that $R, f, T_1, \beta_1, \beta_2, \kappa$, and κ_2 have the same meanings in both transformations while the various other quantities are related in the manner

$$W^{(\pm\sigma)} \pm 2i\sigma f = T_1 \pm 2i\sigma, \tag{87}$$

$$K^{(\pm\sigma)} = [\nu(\nu+2) - 4\sigma^2\beta_2] \pm 2i\sigma\kappa_2, \tag{88}$$

$$Y^{(\pm\sigma)} = fVZ^{(\pm\sigma)} + (T_1 \pm 2i\sigma)\Lambda_\pm Z^{(\pm\sigma)}, \tag{89}$$

$$\Lambda_\mp Y^{(\pm\sigma)} = -(\beta_1 \pm 2i\sigma\beta_2)\frac{\Delta^2}{\rho^8} Z^{(\pm\sigma)} + R\Lambda_\pm Z^{(\pm\sigma)}, \tag{90}$$

$$K^{(\pm\sigma)} Z^{(\pm\sigma)} = \frac{\rho^8}{\Delta^2} R Y^{(\pm\sigma)} - \frac{\rho^8}{\Delta^2}(T_1 \pm 2i\sigma)\Lambda_\mp Y^{(\pm\sigma)}, \tag{91}$$

and

$$K^{(\pm\sigma)}\Lambda_\pm Z^{(\pm\sigma)} = (\beta_1 \pm 2i\sigma\beta_2) Y^{(\pm\sigma)} + \frac{\rho^8}{\Delta^2} f V \Lambda_\mp Y^{(\pm\sigma)}. \tag{92}$$

We may note here an equality which we shall need later. According to equation (88)

$$K^{(+\sigma)}K^{(-\sigma)} = \nu^2(\nu+2)^2 + 16\sigma^4\beta_2^2 - 4\sigma^2[2\nu(\nu+2)\beta_2 - \kappa_2^2], \tag{93}$$

or, substituting for β_2^2 and κ_2^2 their values in accordance with equations (64), we find

$$\begin{aligned} K^{(+\sigma)}K^{(-\sigma)} &= \nu^2(\nu+2)^2 - 8a^2\nu(5\nu-6)\sigma^2 + 144\sigma^2(M^2 + a^4\sigma^2) \\ &= |K|^2 \text{ (say)}. \end{aligned} \tag{94}$$

We observe that the value of $K^{(+\sigma)}K^{(-\sigma)} = |K|^2$ is independent of the ambiguity in the definitions of β_2 and κ_2; it is, therefore, the same in all four cases – a fact which will be of importance for later interpretations (§§6 and 7 below).

6. Conservation laws

Since the potential V in the wave equation (86) is of short range, we can formulate in its context, the standard problem of reflexion and transmission of waves by the potential barrier represented by V by seeking solutions of the equation having the asymptotic behaviours

$$\begin{aligned} Z^{(\pm\sigma)} &\to Z^{(\pm\sigma)}_{\text{inc.}}\,e^{\pm i\sigma r_*} + Z^{(\pm\sigma)}_{\text{ref.}}\,e^{\mp i\sigma r_*} \quad (r_* \to +\infty) \\ &\to \qquad Z^{(\pm\sigma)}_{\text{trans.}}\,e^{\pm i\sigma r_*} \quad (r_* \to -\infty). \end{aligned} \tag{95}$$

As we have already remarked, $Z^{(+\sigma)}$ and $Z^{(-\sigma)}$ will not be complex conjugate functions except when V is real. Nevertheless, with the *definitions*

$$\mathbb{R} = \frac{Z^{(+\sigma)}_{\text{ref.}}\,Z^{(-\sigma)}_{\text{ref.}}}{Z^{(+\sigma)}_{\text{inc.}}\,Z^{(-\sigma)}_{\text{inc.}}} \quad \text{and} \quad \mathbb{T} = \frac{Z^{(+\sigma)}_{\text{trans.}}\,Z^{(-\sigma)}_{\text{trans.}}}{Z^{(+\sigma)}_{\text{inc.}}\,Z^{(-\sigma)}_{\text{inc.}}}, \tag{96}$$

it follows from the usual arguments involving the Wronskians that (cf. Messiah 1964, pp. 106–108)

$$\mathbb{R} + \mathbb{T} = 1. \tag{97}$$

However, before we can attribute to $\mathbb{R}$ and $\mathbb{T}$ the physical meanings of reflexion and transmission coefficients, we must ensure that each of them, as defined, is real. And it must also be true that

$$\frac{(\text{flux of gravitational energy})_{\text{ref. or trans.}}}{(\text{flux of gravitational energy})_{\text{inc.}}} = \frac{Z^{(+\sigma)}_{\text{ref. or trans.}}\,Z^{(-\sigma)}_{\text{ref. or trans.}}}{Z^{(+\sigma)}_{\text{inc.}}\,Z^{(-\sigma)}_{\text{inc.}}}. \tag{98}$$

We shall now show that these requirements are met.

To prove that $\mathbb{R}$ and $\mathbb{T}$ as defined are real, we shall relate the various amplitudes $Z^{(+\sigma)}$ and $Z^{(-\sigma)}$, we have defined in equation (95), with the corresponding amplitudes of Y for $r_* \to +\infty$ and $r_* \to -\infty$.

Now, if we were solving the problem of reflexion and transmission in the context of equation (10), we should be seeking solutions of this equation with the asymptotic behaviours (cf. equations (7), (8), and (9))

$$\begin{aligned} Y^{(\pm\sigma)} &\to Y^{(\pm\sigma)}_{\text{inc.}}\,e^{\pm i\sigma r_*} + Y^{(\pm\sigma)}_{\text{ref.}}\,\frac{e^{\mp i\sigma r_*}}{r^4} \quad (r_* \to +\infty) \\ &\to \qquad \frac{Y^{(\pm\sigma)}_{\text{trans.}}}{\rho_+^3}\,e^{\pm i\sigma r_*} \quad (r_* \to -\infty), \end{aligned} \tag{99}$$

where we have introduced an additional factor $\rho_+^3 (= (r_+^2 + a^2)^{\frac{3}{2}})$ in the definition of the amplitude $Y^{(\pm\sigma)}_{\text{trans.}}$ to conform (for later comparisons) with the definitions of Teukolsky & Press (1974).

Considering equation (89) for incident waves from $+\infty$ and remembering that fV is $O(r^{-2})$ and T_1 is $O(r^{-1})$, we obtain

$$Y^{(\pm\sigma)}_{\text{inc.}} = -4\sigma^2 Z^{(\pm\sigma)}_{\text{inc.}}. \tag{100}$$

Similarly, considering equation (91) for the reflected wave at $+\infty$ and noting that $R\rho^8/\Delta^2$ is $O(r^2)$ and T_1 is $O(r^{-1})$, we obtain

$$K^{(\pm\sigma)} Z^{(\pm\sigma)}_{\text{ref.}} = -4\sigma^2 Y^{(\pm\sigma)}_{\text{ref.}}. \tag{101}$$

And finally considering equation (89) for the transmitted wave at $-\infty$ and remembering that fV vanishes on the horizon, we obtain

$$\frac{Y^{(\pm\sigma)}_{\text{trans.}}}{\rho_+^3} = (T_1 \pm 2i\sigma)_{\Delta=0}(\pm 2i\sigma) Z^{(\pm\sigma)}_{\text{trans.}}. \tag{102}$$

But it follows directly from equation (54) that

$$(T_1|_{\Delta=0} = -\left(\frac{1}{\rho^2}\frac{d\Delta}{dr}\right)_{\Delta=0} = -2\frac{r_+ - M}{\rho_+^2}, \tag{103}$$

where $$\rho_+^2 = r_+^2 + a^2 = 2Mr_+ \quad \text{and} \quad r_+ = M + (M^2 - a^2)^{\frac{1}{2}}. \tag{104}$$

Accordingly, $$Y^{(\pm\sigma)}_{\text{trans.}} = \pm 4i\sigma\left(\pm i\sigma - \frac{r_+ - M}{\rho_+^2}\right)\rho_+^3 Z^{(\pm\sigma)}_{\text{trans.}}. \tag{105}$$

From the relations (100) and (105) it is apparent that $Z^{(+\sigma)}_{\text{inc.}}$ and $Z^{(+\sigma)}_{\text{trans.}}$ are, respectively, the complex conjugates of $Z^{(-\sigma)}_{\text{inc.}}$ and $Z^{(-\sigma)}_{\text{trans.}}$. Also, since $K^{(+\sigma)}K^{(-\sigma)}$ is always real (cf. equation (93)), it follows from equation (101) that the product $Z^{(+\sigma)}_{\text{ref.}} Z^{(-\sigma)}_{\text{ref.}}$ is real. Hence $\mathbb{R}$ and $\mathbb{T}$ as defined in equations (96) are real, as required.

Considering next the requirement (98), we first observe that by virtue of the relations (100) and (101)

$$\frac{Z^{(+\sigma)}_{\text{ref.}} Z^{(-\sigma)}_{\text{ref.}}}{Z^{(+\sigma)}_{\text{inc.}} Z^{(-\sigma)}_{\text{inc.}}} = \frac{256\sigma^8}{K^{(+\sigma)}K^{(-\sigma)}}\frac{|Y_{\text{ref.}}|^2}{|Y_{\text{inc.}}|^2} = \frac{265\sigma^8}{|K|^2}\frac{|Y_{\text{ref.}}|^2}{|Y_{\text{inc.}}|^2}. \tag{106}$$

Now, Teukolsky & Press (1974, equation (4.12)) give for the ratio of the reflected and the incident fluxes of gravitational energy the same expressions in terms of the amplitudes of Y as in equation (106) except that, in place of $|K|^2$, they have a quantity, they have defined as, $|C|^2$. But a comparison of their definition of $|C|^2$ (Teukolsky & Press 1974, equation (3.23)), for the case $m = 0$, with equation (93) shows that

$$|C|^2_{\text{Teukolsky \& Press}; m=0} = |K|^2. \tag{107}$$

All the requirements for attributing to $\mathbb{R}$ and $\mathbb{T}$ the meanings of reflexion and transmission coefficients are, therefore, met.

We also observe that by virtue of the fact that $K^{(+\sigma)}K^{(-\sigma)}$ is the same, independently of the ambiguities in the signs of β_2 and κ_2, it would follow that the reflexion and the transmission coefficients derived from all four potentials must be the same. We shall give a direct demonstration of this fact in §7 below.

A further comparison with the results of Teukolsky & Press (1974; equation (5.8)) is possible by rewriting the expressions for $\mathbb{R}$ and $\mathbb{T}$ in equation (97) in terms of the amplitudes of Y with the aid of the relation (100), (101), and (105). We find

$$|Y_{\text{inc.}}|^2 = \frac{256\sigma^8}{|K|^2}|Y_{\text{ref.}}|^2 + \frac{\sigma^2}{(2Mr_+)^3[\sigma^2 + (M^2 - a^2)/4M^2r_+^2]}|Y_{\text{trans.}}|^2, \tag{108}$$

where we have made use of the relation (cf. equation (104))

$$(r_+ - M)^2 = M^2 - a^2. \tag{109}$$

Equation (108) is what Teukolsky & Press call their '*conservation law*'. It is now seen to be no more than the equality expressed by equation (97).

There is a reciprocal form of the conservation law (108). Instead of considering a wave incident on the barrier from $+\infty$ and getting partially reflected and transmitted (as we have done in writing equation (95)), we can, equally, consider a wave incident on the barrier from $-\infty$ and getting, similarly reflected and transmitted. In the latter case, the asymptotic behaviours of the solution at $-\infty$ and $+\infty$ will be given by

$$\begin{aligned} \tilde{Z}^{(\pm\sigma)} &\to \tilde{Z}^{(\pm\sigma)}_{\text{inc.}} e^{\mp i\sigma r_*} + \tilde{Z}^{(\pm\sigma)}_{\text{ref.}} e^{\pm i\sigma r_*} \quad (r_* \to -\infty) \\ &\to \tilde{Z}^{(\pm\sigma)}_{\text{trans.}} e^{\mp i\sigma r_*} \quad (r_* \to +\infty). \end{aligned} \tag{110}$$

The corresponding behaviours of the solution for Y are (cf. equations (7), (8), and (9))

$$\begin{aligned} \tilde{Y}^{(\pm\sigma)} &\to \frac{\tilde{Y}^{(\pm\sigma)}_{\text{inc.}}}{\rho_+^8} e^{\mp i\sigma r_*} \Delta^2 + \tilde{Y}^{(\pm\sigma)}_{\text{ref.}} e^{\pm i\sigma r_*} \quad (r_* \to -\infty) \\ &\to \tilde{Y}^{(\pm\sigma)}_{\text{trans.}} \frac{e^{\mp i\sigma r_*}}{r^4} \quad (r_* \to +\infty). \end{aligned} \tag{111}$$

The relations between the different reflected and the transmitted amplitudes, appropriate for this case, are the 'transposed' of the relations (102) and (105):

$$\tilde{Z}^{(\pm\sigma)}_{\text{trans.}} = -4\sigma^2 \frac{\tilde{Y}^{(\pm\sigma)}_{\text{trans.}}}{K^{(\pm\sigma)}} \quad \text{and} \quad \tilde{Z}^{(\pm\sigma)}_{\text{ref.}} = \frac{\tilde{Y}^{(\pm\sigma)}_{\text{ref.}}}{\pm 4i\sigma[\pm i\sigma - (r_+ - M)/\rho_+^2]}. \tag{112†}$$

To obtain the relation between the incident amplitudes, we consider equation (91) at the horizon. The term $R\rho^8/\Delta^2$ $(= \beta_2 + q/\Delta)$ does not contribute on account of the factor Δ^2 that occurs in the definition of $\tilde{Y}^{(\pm\sigma)}_{\text{inc.}}$. Accordingly,

$$K^{(\pm\sigma)} \tilde{Z}^{(\pm\sigma)}_{\text{inc.}} = -(T_1 \pm 2i\sigma)_{\Delta=0}\, \tilde{Y}^{(\pm\sigma)}_{\text{inc.}} e^{\pm i\sigma r_*} \left[\frac{1}{\Delta^2} \Lambda_\mp (\Delta^2 e^{\mp i\sigma r_*})\right]_{\Delta=0}; \tag{113}$$

or, after simplification (cf. equation (103)),

$$K^{(\pm\sigma)} \tilde{Z}^{(\pm\sigma)}_{\text{inc.}} = 4\left(\pm i\sigma - \frac{r_+ - M}{2r_+ M}\right)\left(\pm\sigma - \frac{r_+ - M}{r_+ M}\right) \tilde{Y}^{(\pm\sigma)}_{\text{inc.}}. \tag{114}$$

It is clear that the reflexion and the transmission coefficients defined in terms of $\tilde{Z}^{(\pm\sigma)}$ must be the same as those defined in terms of $Z^{(\pm\sigma)}$; and the conservation law (97) expressed in terms of $\tilde{Z}^{(\pm\sigma)}$, when rewritten in terms of $\tilde{Y}^{(\pm\sigma)}$ with the aid of equations (112) and (114), takes the form

$$\begin{aligned} \sigma^4 |\tilde{Y}_{\text{trans.}}|^2 + \frac{|K|^2}{256\sigma^2[\sigma^2 + (M^2 - a^2)/4M^2 r_+^2]} |\tilde{Y}_{\text{ref.}}|^2 \\ = \left(\sigma^2 + \frac{M^2 - a^2}{4M^2 r_+^2}\right)\left(\sigma^2 + \frac{M^2 - a^2}{M^2 r_+^2}\right) |\tilde{Y}_{\text{inc.}}|^2. \end{aligned} \tag{115}$$

† The factor ρ_+^3 is absent in the relation between the reflected amplitudes since we have not inserted this factor in our present definition of $\tilde{Y}_{\text{ref.}}$ (cf. equations (99) and (111)).

7. The relation between the functions satisfying the different wave equations

We shall distinguish the various quantities referring to the different potentials by superscripts i, j, etc. If we have a solution of equation (10) for Y which is appropriate to a particular situation, then equation (91) will enable us to deduce the different solutions Z^i which are appropriate to the same situation. We should, therefore, be able to relate the different solutions Z^i and Z^j with the aid of the relations (89)–(91). Thus, substituting for Y and $\Lambda_- Y$ from equations (89) and (90), appropriate for Z^j, in the relation,

$$K^i Z^i = \frac{\rho^8}{\Delta^2} R^i Y - \frac{\rho^8}{\Delta^2}(T_1^i + 2i\sigma)\Lambda_- Y, \tag{116}$$

appropriate for Z^i, we obtain

$$K^i Z^i = \left[\frac{\rho^8}{\Delta^2} R^i f^j V^j + (T_1^i + 2i\sigma)(\beta_1^j + 2i\sigma\beta_2^j)\right] Z^j + \frac{\rho^8}{\Delta^2}\left[R^i(T_1^j + 2i\sigma) - R^j(T_1^i + 2i\sigma)\right]\Lambda_+ Z^j. \tag{117}$$

Making use of the solutions for R and T_1 given in equations (66) and (67), we find that

$$\frac{\rho^8}{\Delta^2}[R^i(T_1^j + 2i\sigma) - R^j(T_1^i + 2i\sigma)] = 2i\sigma(\beta_2^i - \beta_2^j) + \frac{F^2 - \beta_2^2}{(F - \beta_2^i)(F - \beta_2^j)}(\kappa_2^i - \kappa_2^j). \tag{118}$$

Similarly, making use of the integral relation (39), we find that

$$\frac{\rho^8}{\Delta^2} R^i f^j V^j + (T_1^i + 2i\sigma)(\beta_1^j + 2i\sigma\beta_2^j) = \frac{R^i}{R^j} K^j - \frac{\beta_1^j + 2i\sigma\beta_2^j}{F + \beta_2^j}[R^i(T_1^j + 2i\sigma) - R^j(T_1^i + 2i\sigma)]. \tag{119}$$

Now combining equations (117)–(119), we obtain

$$K^i Z^i = \frac{F + \beta_2^i}{F + \beta_2^j} K^j Z^j + \left[2i\sigma(\beta_2^i - \beta_2^j) + (\kappa_2^i - \kappa_2^j)\frac{F^2 - \beta_2^2}{(F - \beta_2^i)(F - \beta_2^j)}\right] \times \left\{\Lambda_+ Z^j - \frac{\Delta}{q^2 - \beta_2^2 \Delta^2}[(\kappa_2^j + 2i\sigma\beta_2^j) q - \beta_2^j \Delta F_{,r_*} - 2i\sigma\beta_2^2 \Delta] Z\right\}; \tag{120†}$$

and this is the required relation between Z^i and Z^j. From this relation it is apparent that

$$K^i Z^i \to K^j Z^j + [2i\sigma(\beta_2^i - \beta_2^j) + \kappa_2^i - \kappa_2^j]\Lambda_+ Z^j \quad (r_* \to \pm\infty). \tag{121}$$

From the relation (121) it follows that if Z^j is a solution whose asymptotic behaviour is

$$Z^j \to e^{+i\sigma r_*} \quad (r_* \to \pm\infty), \tag{122}$$

† This equation represents a generalization of the relation (Chandrasekhar & Detweiler 1975, equation (61)) between the functions X and Z satisfying, respectively, the Regge–Wheeler and the Zerilli equation. In fact, equation (120) reduces to that relation, if we set $\beta_2^i = \beta_2^j = 0$, $\kappa_2^i = 6M$, and $\kappa_2^j = -6M$.

then the corresponding asymptotic behaviour of Z^i is given by

$$K^i Z^i \to [K^j - 4\sigma^2(\beta_2^i - \beta_2^j) + 2i\sigma(\kappa_2^i - \kappa_2^j)]\, Z^j \tag{123}$$

or, substituting for K^j its value,

$$K^j = \nu(\nu+2) - 4\sigma^2\beta_2^j + 2i\sigma\kappa_2^j, \tag{124}$$

we have

$$K^i Z^i \to K^i Z^j \quad \text{or} \quad Z^i \to Z^j \quad (Z^j \to e^{+i\sigma r_*};\, r_* \to \pm\infty). \tag{125}$$

On the other hand, if

$$Z^j \to e^{-i\sigma r_*} \quad (r_* \to \pm\infty), \tag{126}$$

then, it follows from the relation (121) that

$$K^i Z^i \to K^j Z^j \quad (Z^j \to e^{-i\sigma r_*};\, r_* \to \pm\infty). \tag{127}$$

The foregoing asymptotic relations between the solutions derived with the different potentials enable us to prove the following two theorems.

THEOREM 1. *The four potentials, appropriate to given ν and a, yield the same reflexion and transmission coefficients.*

Proof. Consider the solutions $Z^{(i,\pm\sigma)}$, derived with the potential V^i, whose asymptotic behaviours are given by

$$\begin{aligned} Z^{(i,+\sigma)} &\to e^{+i\sigma r_*} + A^{(+\sigma)} e^{-i\sigma r_*} \quad (r_* \to +\infty) \\ &\to B^{(+\sigma)} e^{+i\sigma r_*} \quad (r_* \to -\infty), \end{aligned} \tag{128}$$

and

$$\begin{aligned} Z^{(i,-\sigma)} &\to e^{-i\sigma r_*} + A^{(-\sigma)} e^{+i\sigma r_*} \quad (r_* \to +\infty) \\ &\to B^{(-\sigma)} e^{-i\sigma r_*} \quad (r_* \to -\infty). \end{aligned} \tag{129}$$

The reflexion and the transmission coefficients derived from this solution are

$$\mathbb{R} = A^{(+\sigma)} A^{(-\sigma)} \quad \text{and} \quad \mathbb{T} = B^{(+\sigma)} B^{(-\sigma)}. \tag{130}$$

In accordance with the relations (125) and (127), the solutions $Z^{(i,\pm\sigma)}$ with the asymptotic behaviours (128) and (129), can be transformed into solutions, appropriate for the potential V^j (via equation (120)), whose asymptotic behaviours will be given by

$$\begin{aligned} Z^{(j,+\sigma)} &\to e^{+i\sigma r_*} + A^{(+\sigma)} \frac{K^i}{K^j} e^{-i\sigma r_*} \quad (r_* \to +\infty) \\ &\to B^{(+\sigma)} e^{+i\sigma r_*} \quad (r_* \to -\infty), \end{aligned} \tag{131}$$

and

$$\begin{aligned} Z^{(j,-\sigma)} &\to \frac{K^i}{K^j} e^{-i\sigma r_*} + A^{(-\sigma)} e^{+i\sigma r_*} \quad (r_* \to +\infty) \\ &\to B^{(-\sigma)} \frac{K^i}{K^j} e^{-i\sigma r_*} \quad (r_* \to -\infty). \end{aligned} \tag{132}$$

It follows from these relations and the manner in which we have defined the reflexion and the transmission coefficients (cf. equation (96)) that these coefficients derived from the solutions $Z^{(i,\pm\sigma)}$ are the same as those derived from the solutions $Z^{(j,\pm\sigma)}$ And this is the theorem stated.

THEOREM 2. *If V^i is complex and $Z^{(i,+\sigma)}$ is a solution, derived with this potential, with the asymptotic behaviour*

$$\begin{aligned} Z^{(i,+\sigma)} &\to e^{+i\sigma r_*} + A^{(+\sigma)} e^{-i\sigma r_*} \qquad (r_* \to +\infty) \\ &\to \qquad B^{(+\sigma)} e^{+i\sigma r_*} \qquad (r_* \to -\infty), \end{aligned} \tag{133}$$

then, the reflexion and the transmission coefficients are given by

$$\mathbb{R} = |A^{(+\sigma)}|^2 \frac{(K^i)^2}{|K|^2} \quad \text{and} \quad \mathbb{T} = |B^{(+\sigma)}|^2. \tag{134}$$

Proof. First, we observe that the potential V can be complex only when κ_2 is purely imaginary (cf. equation (64)); and in this case, we can write

$$\kappa_2 = \pm ik_2 \quad \text{where} \quad k_2 \text{ is real.} \tag{135}$$

Let V^i and V^j be the potentials belonging to the same value of β_2 and appropriate to $+ik_2$ and $-ik_2$, respectively. It is manifest from equation (72) that, under these circumstances, V^i and V^j are complex conjugate functions. The constants K^i and K^j, associated with these potentials are

$$K^i = [\nu(\nu+2) - 4\sigma^2\beta_2] - 2\sigma k_2$$

and

$$K^j = [\nu(\nu+2) - 4\sigma^2\beta_2] + 2\sigma k_2. \tag{136}$$

Since V^j and V^i are complex conjugate functions, the equations satisfied by Z^j and the complex conjugate of Z^i are the same. Accordingly, the solution, derived with the potential V^j having the asymptotic behaviour (cf. equations (129) and (132)),

$$\begin{aligned} Z^{(j,-\sigma)} &\to e^{-i\sigma r_*} + A^{(-\sigma)} \frac{K^j}{K^i} e^{+i\sigma r_*} \qquad (r_* \to +\infty) \\ &\to \qquad B^{(-\sigma)} e^{-i\sigma r_*} \qquad (r_* \to -\infty), \end{aligned} \tag{137}$$

is the complex conjugate of the solution $Z^{(i,+\sigma)}$ having the asymptotic behaviour (133). Hence,

$$[A^{(+\sigma)}]^* = A^{(-\sigma)} \frac{K^j}{K^i} \quad \text{and} \quad [B^{(+\sigma)}]^* = B^{(-\sigma)}. \tag{138}$$

From these relations and equations (130) and (136), it follows that

$$\mathbb{R} = |A^{(+\sigma)}|^2 \frac{K^i}{K^j} = |A^{(+\sigma)}|^2 \frac{(K^i)^2}{|K|^2}, \tag{139}$$

and

$$\mathbb{T} = |B^{(+\sigma)}|^2; \tag{140}$$

and these are the relations that were to be established.

COROLLARY 1. *Quite generally, we may write*

$$\mathbb{R} = |A^{(+\sigma)}|^2 \frac{(K^i)(K^i)^*}{|K|^2} \quad \text{and} \quad \mathbb{T} = |B^{(+\sigma)}|^2. \tag{141}$$

The relations are manifest, since when V^i is real

$$(K^i)(K^i)^* = |K|^2; \tag{142}$$

and when V^i is complex K^i is real. In other words, the factor $(K^i)(K^i)^*/|K|^2$ which occurs in the definition of $\mathbb{R}$ is different from unity only when V^i is complex.

Corollary 2. *If instead of a solution $Z^{(i,+\sigma)}$ with the asymptotic behaviour* (133), *we consider a solution $\tilde{Z}^{(i,+\sigma)}$ with the behaviour*

$$\begin{aligned} \tilde{Z}^{(i,+\sigma)} &\to e^{-i\sigma r_*} + \tilde{A}^{(+\sigma)} e^{+\sigma r_*} \quad (r_* \to -\infty) \\ &\to \tilde{B}^{(+\sigma)} e^{-i\sigma r_*} \quad (r_* \to +\infty), \end{aligned} \tag{143}$$

then, the reflexion and the transmission coefficients are given by

$$\mathbb{R} = |\tilde{A}^{(+\sigma)}|^2 \frac{|K|^2}{(K^i)(K^i)^*} \quad \text{and} \quad \mathbb{T} = |\tilde{B}^{(+\sigma)}|^2. \tag{144}$$

The proof of this corollary follows along the same lines as the proofs of theorem 2 and corollary 1.

The importance of theorem 2 arises from the fact that, in order to evaluate the reflexion and the transmission coefficients, it is not necessary to integrate the wave equation for Z twice to obtain solutions with the asymptotic behaviours (128) and (129) appropriate for $Z^{(+\sigma)}$ and $Z^{(-\sigma)}$, separately: it will suffice to integrate the equation only once.

8. The nature of the potentials

We have seen how Teukolsky's equation, in the axisymmetric case, can be transformed into a one dimensional wave equation with four possible potentials. These potentials, unlike the potentials which appear in the equations governing the perturbations of the Schwarzschild metric, are frequency dependent. This frequency dependence occurs, implicitly, via

$$\nu = E - 2 + a^2\sigma^2, \tag{145}$$

where the separation constant E is further frequency-dependent. (Press & Teukolsky 1973) have provided a table from which E can be deduced for all $a\sigma \lesssim 3$.)

In addition to being frequency-dependent, the potentials can be real or complex: all four can be real, or a pair can be real and the other pair complex conjugate, or they can be two pairs of complex conjugates. According to equation (64), the condition for the occurrence of a complex-conjugate pair of potentials is

$$2a^2\nu(5\nu - 6) + 2\beta_2\nu(\nu+2) \geqslant 36M^2. \tag{146}$$

For the two possible values of β_2 (namely $+3a^2$ and $-3a^2$) the foregoing condition gives

$$\nu(\nu-6) > 9M^2/a^2 \quad \text{or} \quad \nu > \nu_1 = 3 + 3(1 + M^2/a^2)^{\frac{1}{2}}, \tag{147}$$

and

$$\nu > \nu_2 = 3M/2a. \tag{148}$$

Since the least value of ν, for a given l, is $(l-1)(1+2)$, it follows from (148) that all four potentials can be real if and only if

$$\frac{a}{M} \leqslant \frac{3}{2(l-1)(l+2)}. \tag{149}$$

Also, since

$$\nu_1 = 3 + 3\sqrt{2} = 7.243 \quad \text{for} \quad a/M = 1, \tag{150}$$

it follows that for the extreme case of the Kerr metric, all four potentials are complex for $l \geqslant 3$.

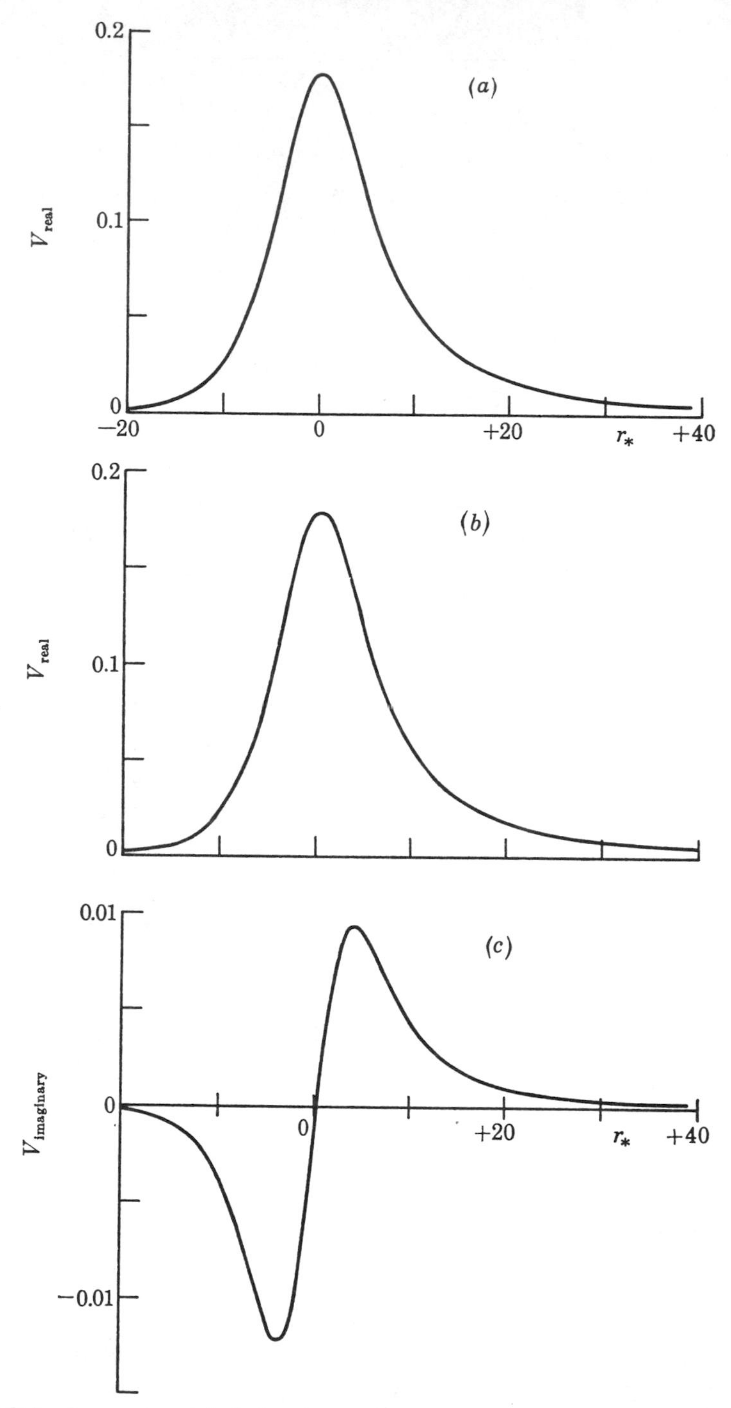

FIGURE 1. Of the four potentials belonging to $l = 2$, $a^2 = 0.81$ and $\sigma^2 = 0.13$, two are real (associated with the values $\kappa_2 = \pm 7.848$ and $\beta_2 = +2.43$); and two are complex conjugates (associated with the values $\kappa_2 = \pm 13.288i$ and $\beta_2 = -2.43$). The real potential belonging to $\kappa_2 = +7.848$ and $\beta_2 = +2.43$ is illustrated in (*a*); that belonging to $\kappa_2 = -7.848$ and $\beta_2 = +2.43$ cannot be distinguished from the one illustrated in the scale of the graph. In (*b*) and (*c*) the real and the imaginary parts of the potential belonging to $\kappa_2 = +13.288i$ and $\beta_2 = -2.43$ are illustrated.

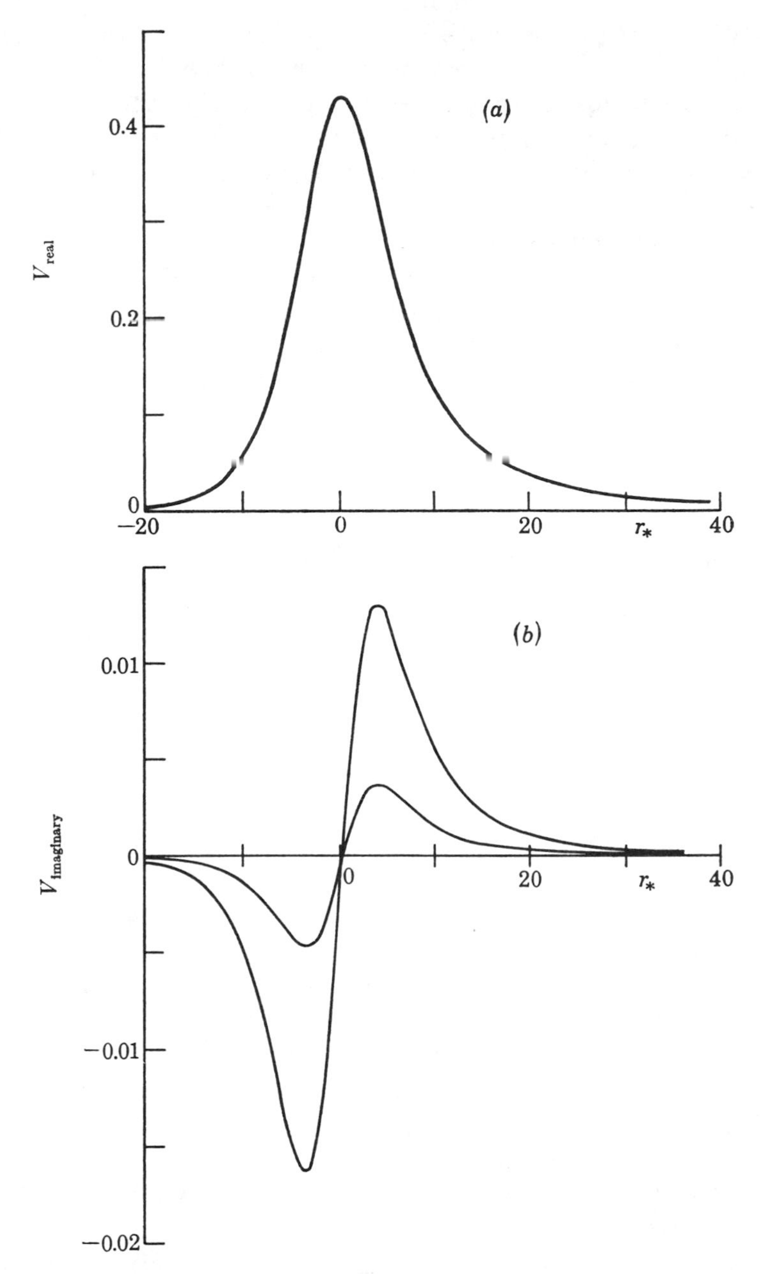

FIGURE 2. The four potentials belonging to $l = 3$, $a^2 = 0.81$ and $\sigma^2 = \frac{4}{9}$ occur as two complex-conjugate pairs belonging to $\kappa_2 = \pm 10.2333i$ and $\beta_2 = +2.43$, and $\kappa_2 = \pm 36.3761i$ and $\beta_2 = -2.43$. The real parts of the potentials (illustrated in (*a*)) in the two cases cannot be distinguished on the scale of the graph. The imaginary parts are illustrated in (*b*).

In figures 1 and 2, we illustrate the run of the potentials for some typical cases. We observe that for values of the parameters for which $\mathbb{R}$ and $\mathbb{T}$ are neither too close to unity nor too close to zero, the real parts of the potentials dominate and are nearly equal, their small differences being 'compensated' by their differing, (small) imaginary parts. It may be recalled in this connection that for a given ν, the integrals over r_* of the real parts of the potentials must be the same while those over the imaginary parts must be zero.

We have verified by direct numerical evaluation that all four potentials yield the same reflexion and transmission coefficients in accordance with either of the two equations (130) or (134). In particular, it was found that

$$\mathbb{R} = 0.13173 \quad \text{and} \quad \mathbb{T} = 0.86827 \quad \text{for} \quad l = 2, (a/M)^2 = 0.95, (\sigma M)^2 = 0.22028, \quad \text{and} \quad \nu = 4.09903;$$

$$\mathbb{R} = 0.34964 \quad \text{and} \quad \mathbb{T} = 0.65036 \quad \text{for} \quad l = 3, (a/M)^2 = 0.81, (\sigma M)^2 = 0.44444, \quad \text{and} \quad \nu = 10.241. \tag{151}$$

For comparison it may be noted that for the Schwarzschild black hole

$$\mathbb{R} = 0.0410 \quad \text{and} \quad \mathbb{T} = 0.9590 \quad \text{for} \quad l = 2 \quad \text{and} \quad (\sigma M)^2 = 0.22028;$$

$$\mathbb{R} = 0.1008 \quad \text{and} \quad \mathbb{T} = 0.8992 \quad \text{for} \quad l = 3 \quad \text{and} \quad (\sigma M)^2 = 0.44444. \tag{152}$$

We conclude that the Kerr black hole reflects radiation more efficiently than the Schwarzschild black hole of the same mass.

9. Concluding remarks

The transformation of Teukolsky's equation (for the case $m = 0$) to a one-dimensional wave equation has many aspects in common with the similar transformation (effected in (I)) of the Bardeen–Press equation to the wave equations governing the odd and the even parity perturbations of the Schwarzschild metric. In the Schwarzschild case the functions X and Z satisfying the one dimensional wave equations were explicitly related to the scalars defining the perturbations of the metric coefficients (cf. (I), equations (33) and (A 13)). It remains to relate, in a similar fashion, the different functions Z^i (satisfying the wave equations with the different potentials V^i) with the scalars defining the axisymmetric perturbations of the Kerr metric. These relations can be readily obtained with the aid of the equation (cf. equation (91))

$$\frac{\Delta^2}{\rho^8} K^i Z^i = R^i Y - (T_1^i + 2i\sigma)\,\Lambda_- Y \tag{153}$$

(where R^i and T_1^i are explicitly known functions), once Y has been expressed in terms of the perturbations of the metric coefficients. And to express Y in terms of the perturbed metric coefficients, we have only to evaluate the particular combination of the components of the perturbed Riemann tensor which makes up the Newman–Penrose $\delta\Psi_0$–a task which is, in principle, entirely straightforward. In any event, it is manifest that there exist four linear combinations of the perturbed metric coefficients which, after separating the angular dependence, satisfy four decoupled one dimensional wave equations. We shall return to these and related matters in a later communication.

The research reported in this paper has in part been supported by a grant from the Louis Block Fund and the National Science Foundation under grant MPS 74-17456 with the University of Chicago (S.C.) and by the National Aeronautics and Space Administration under grant NGR 21-002-010 with the University of Maryland (S.D.).

Note added in proof 6 *June* 1975. Meantime it has been possible to transform Teukolsky's equation for electromagnetic perturbations (in the general case) to a one dimensional wave equation with a short-range potential.

References

Bardeen, J. M. & Press, W. H. 1973 *J. Math. Phys.* **14**, 7.
Boyer, R. H. & Lindquist, R. W. 1967 *J. Math. Phys.* **8**, 265.
Chandrasekhar, S. 1975 *Proc. R. Soc. Lond.* A **343**, 289.
Chandrasekhar, S. & Detweiler, S. 1975 *Proc. R. Soc. Lond.* A **344**, 441.
Kerr, R. P. 1963 *Phys. Rev. Lett.* **11**, 237.
Messiah, A. 1964 *Quantum mechanics*, vol. I (translated by G. M. Temmer), Amsterdam: North Holland Publishing Co.
Press, W. H. & Teukolsky, S.A. 1973 *Astrophys. J.* **185**, 649.
Rees, M. 1974 *Observatory* **94**, 168.
Regge, T. & Wheeler, J. A. 1957 *Phys. Rev.* **108**, 1063.
Teukolsky, S. A. 1972 *Phys. Rev. Lett.* **29**, 114.
Teukolsky, S. A. 1973 *Astrophys. J.* **185**, 635.
Teukolsky, S. A. & Press, W. H. 1974 *Astrophys. J.* **193**, 443.
Zerilli, F. J. 1970 *Phys. Rev.* D **2**, 2141.

6

On a transformation of Teukolsky's equation and the electromagnetic perturbations of the Kerr black hole

BY S. CHANDRASEKHAR, F.R.S.

University of Chicago, Chicago, Illinois, 60637

(*Received* 11 *July* 1975)

Teukolsky's equation, governing the perturbations (scalar, electromagnetic, and gravitational) of the Kerr black hole, is transformed, by a simple change of variables, in a manner such that there is formally no difference in the treatments of the axisymmetric and the non-axisymmetric modes: the rôle of 'm' is effectively eliminated. By considering in detail the case of electromagnetic perturbations we show how, in all cases, the problems can be reduced to problems in the theory of penetration of one dimensional potential barriers with, however, certain novel features. The phenomenon of super-radiance, peculiar to the Kerr metric, emerges, for example, in an unexpected guise. The case of scalar perturbations is considered briefly in an appendix.

1. INTRODUCTION

In a recent paper (Chandrasekhar & Detweiler 1975; this paper will be referred to hereafter as (I)) Teukolsky's equation governing the axisymmetric (gravitational) perturbations of the Kerr black hole was transformed to a one dimensional equation as a natural generalization of the Zerilli and the Regge-Wheeler equations governing the perturbations of the Schwarzschild black hole. While this transformation was perhaps satisfying from a formal point of view, it must be conceded that only the non-axisymmetric perturbations – the m-modes with an $e^{im\varphi}$-dependence on the azimuthal angle φ (where m is a positive or a negative integer) – include the physically interesting aspects, such as the phenomenon of super-radiance, which are peculiar to the Kerr black hole. In this paper, we shall accordingly turn our attention to Teukolsky's equation governing the different m-modes and show how, by a single change of the independent variable, we can reduce it to a form in which m no longer appears explicitly and all the physically interesting features of the Kerr black hole emerge as novel aspects of the one dimensional barrier-penetration problem familiar in elementary quantum theory. We shall illustrate the particular appropriateness of the transformation by considering in detail the electromagnetic perturbations of the Kerr black hole. We shall return to the consideration of the gravitational perturbations in a later communication.

In an accompanying paper Detweiler (1976) has considered a transformation of Teukolsky's equation for the case $s = 1$ from a different point of view.

The screened portion of this paper (pp. 106–7) contains misstatements, and is erroneous. These errors are explained and corrected in the appendix to paper 9 in this volume. See below, pp. 134–37.

2. A transformation of Teukolsky's radial equation

Teukolsky's equation governing the radial function R is (cf. Teukolsky 1973, equation (4.9))

$$\Delta^{-s}\frac{\mathrm{d}}{\mathrm{d}r}\left(\Delta^{s+1}\frac{\mathrm{d}R}{\mathrm{d}r}\right)+\frac{1}{\Delta}[K^2+2\mathrm{i}s(r-M)K-\Delta(4\mathrm{i}s\sigma r+\lambda)]R=0, \tag{1}$$

where
$$K=(r^2+a^2)\sigma+am, \quad \Delta=r^2+a^2-2Mr, \tag{2}$$

$$\lambda=E+a^2\sigma^2+2a\sigma m-s(s+1)=\tau-s(s+1) \quad \text{(say)}, \tag{3}$$

$s=0$, 1, or 2 for scalar, electromagnetic, or gravitational perturbations, respectively, and E is the separation constant which occurs in the equation,

$$\frac{1}{\sin\theta}\frac{\mathrm{d}}{\mathrm{d}\theta}\left(\sin\theta\frac{\mathrm{d}S}{\mathrm{d}\theta}\right)+\left(a^2\sigma^2\cos^2\theta+2a\sigma s\cos\theta-\frac{m^2+s^2+2ms\cos\theta}{\sin^2\theta}+E\right)S=0, \tag{4}$$

governing the angular dependence of the perturbation. In deriving equations (1)–(4), it has been assumed that the t- and the φ-dependences of the perturbation is given by

$$\mathrm{e}^{\mathrm{i}(\sigma t+m\varphi)} \quad (\sigma>0). \tag{5}$$ †

It is apparent from equation (4) that the characteristic values of E are invariant to a simultaneous change in the signs of σ and m. This invariance has the consequence that if we wish to consider the complex conjugate of R by changing the sign of σ we must simultaneously change the sign of m: otherwise the value of λ in equation (1) will not be retained.

Now let
$$\alpha^2=a^2+am/\sigma, \tag{6}$$

so that
$$K=(r^2+\alpha^2)\sigma. \tag{7}$$

It is clear that for m negative and σ sufficiently small (and positive!) α^2 can become negative so that α can be imaginary. We shall consider the consequences of this latter possibility in §3; meantime, we shall note only that α^2 is invariant to a simultaneous change in the signs of σ and m.

We shall now introduce, in place of r, a new independent variable r_* defined by

$$\frac{\mathrm{d}}{\mathrm{d}r_*}=\frac{\Delta}{r^2+\alpha^2}\frac{\mathrm{d}}{\mathrm{d}r}. \tag{8}$$

In view of the admissibility of negative values for α^2, it is clear that the resulting $r_*(r)$-relation need not always be single-valued. Again, postponing to §§3 and 5 the consideration of this possible double-valuedness of the $r_*(r)$-relation, we shall, for the present, consider only the formal consequences of this change of variable. In addition to this change in the independent variable, we shall also let

$$\phi=\Delta^{\frac{1}{2}s}(r^2+\alpha^2)^{\frac{1}{2}}R. \tag{9}$$

† It should be noted that our convention with respect to the time-dependence is opposite to Teukolsky's: his ω is our minus σ; and σ will be assumed to be positive.

With these specified changes in the variables, we find that equation (1) reduces to

$$\frac{d^2\phi}{dr_*^2}+\sigma^2\phi-2is\sigma\frac{2r\Delta-(r-M)(r^2+\alpha^2)}{(r^2+\alpha^2)^2}\phi-\left[\frac{\lambda\Delta}{(r^2+\alpha^2)^2}+G^2+\frac{dG}{dr_*}\right]\phi=0, \quad (10)$$

where

$$G=\frac{s(r-M)}{r^2+\alpha^2}+\frac{r\Delta}{(r^2+\alpha^2)^2}. \quad (11)$$

We observe that equation (1) contains no explicit reference to m: *the non-axisymmetric case differs from the axisymmetric case only by the occurrence of* α^2 *in place of* a^2. (It should, however, be remembered that Δ involves a^2.)

3. The $r_*(r)$-relation and the manifestation of super-radiance

Integrating equation (8), we obtain the relation

$$r_*=r+\frac{2Mr_++am/\sigma}{r_+-r_-}\lg\left(\frac{r}{r_+}-1\right)-\frac{2Mr_-+am/\sigma}{r_+-r_-}\lg\left(\frac{r}{r_-}-1\right) \quad (r>r_+), \quad (12)$$

where

$$r_\pm=M\pm(M^2-a^2)^{\frac{1}{2}} \quad (13)$$

denote the radii of the outer and the inner horizons of the Kerr metric.

When considering the (external) perturbations of the Kerr black hole we are indifferent to 'what happens' inside $r<r_+$; and we need be concerned with the $r_*(r)$-relation only for $r>r_+$.

It is now apparent from equation (12), that the $r_*(r)$-relation is single-valued only so long as

$$r_+^2+\alpha^2=2Mr_+-a^2+\alpha^2=2Mr_++am/\sigma>0; \quad (14)$$

and when this inequality obtains

$$r_*\to+\infty \quad \text{when} \quad r\to\infty \quad \text{and} \quad r_*\to-\infty \quad \text{when} \quad r\to r_++0. \quad (15)$$

Letting

$$\sigma_s=-am/2Mr_+ \quad \text{(for } m \text{ negative)}, \quad (16)$$

and remembering that our convention with respect to σ is that it is positive, we conclude that *so long as* $\sigma>\sigma_s$, *the* $r_*(r)$*-relation is single-valued and the range of* r_* *is the entire interval* $(-\infty,+\infty)$. *But if* $0<\sigma<\sigma_s$ *and* $r_+^2+\alpha^2<0$, *the* $r_*(r)$*-relation is double-valued*: r_* *attains a minimum value when* $r=|\alpha|$ *and* $r_*\to+\infty$ *both when* $r\to\infty$ *and when* $r\to r_++0$. In the latter case, it can be readily shown that in the neighbourhood of $r=|\alpha|$, the $r_*(r)$-relation has the behaviour

$$r_*=r_*(|\alpha|)+\frac{|\alpha|}{\Delta_{|\alpha|}}(r-|\alpha|)^2+O[(r-|\alpha|)^3]. \quad (17)$$

Also, when $0<\sigma<\sigma_s$ and $r=|\alpha|$, the coefficients in equation (10) become singular before we arrive at the horizon; accordingly, the equation must be considered, separately, in the two branches of the $r_*(r)$-relation, namely for $\infty>r>|\alpha|$ and for $|\alpha|>r>r_+$.

Figure 1 illustrates the nature of the $r_*(r)$-relations for some typical situations that may arise.

As we shall show in detail in § 5, *the interval*, $0 < \sigma < \sigma_s$, *in* σ *defines precisely the range of frequencies for which the reflexion coefficient for incident waves exceeds unity and super-radiance occurs.*

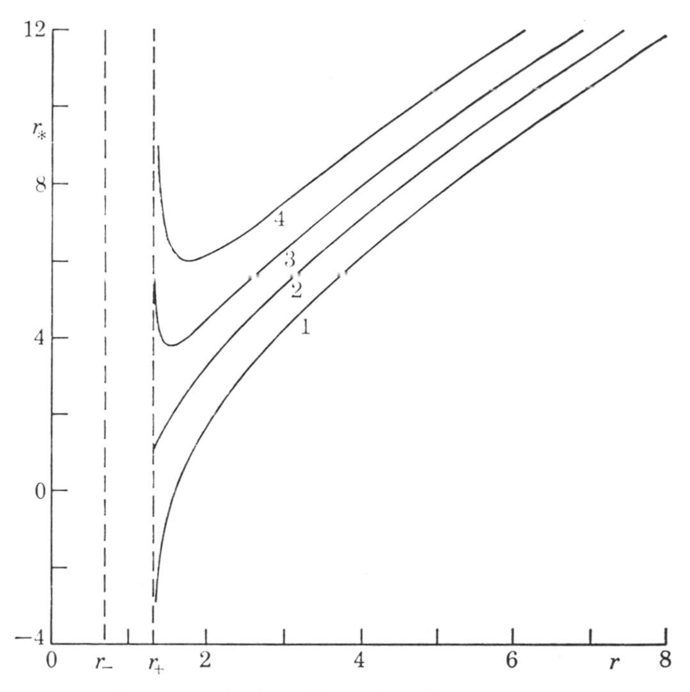

FIGURE 1. Illustrating the $r_*(r)$-relation. The unit of length is M; and the cases illustrated are for $a = 0.95$ when the outer horizon occurs at $r_+ = 1.312$ and the inner horizon at $r_- = 0.6878$. The curves labelled 1, 2, 3, and 4, respectively, are for values of α^2 equals -0.9975 (for a non-super-radiant frequency), -1.722 ($= -r_+^2$ for $\sigma = \sigma_s$), and -2.2975 and -3.0975 (for $\sigma < \sigma_s$).

It should be noted that α^2 becomes negative already before super-radiance occurs; thus

$$\alpha^2 \leqslant 0 \quad \text{for} \quad \sigma \leqslant \sigma_c = -m/a; \tag{18}$$

and $\alpha = 0$ when σ is '*co-rotational*'. In the interval, $\sigma_s < \sigma < \sigma_c$, while α^2 is negative, $r_+^2 + \alpha^2 > 0$; so that in this interval the $r_*(r)$-relation continues to be single-valued.

In contrast to the variable r_* which we have introduced, the conventional variable r_* (which we shall distinguish by $r_*^{(c)}$) which is introduced 'to push the horizon to minus infinity' is defined by

$$r_*^{(c)} = \int \frac{r^2 + a^2}{\Delta} \mathrm{d}r; \tag{19}$$

or, explicitly,

$$r_*^{(c)} = r + \frac{2Mr_+}{r_+ - r_-} \lg\left(\frac{r}{r_+} - 1\right) - \frac{2Mr_-}{r_+ - r_-} \lg\left(\frac{r}{r_-} - 1\right) \quad (r > r_+). \tag{20}$$

This $r_*^{(c)}(r)$-relation is always single-valued; and as required of it,

$$r_*^{(c)} \to +\infty \quad \text{when} \quad r \to \infty \quad \text{and} \quad r_*^{(c)} \to -\infty \quad \text{when} \quad r \to r_+ + 0. \tag{21}$$

The asymptotic relations between the variables r_* and $r_*^{(c)}$ may be noted here:

$$r_* \to r_*^{(c)} \quad \text{for} \quad r \to \infty, \quad \text{while} \quad r_* \to (k/\sigma)\, r_*^{(c)} \quad \text{as} \quad r \to r_+ + 0, \tag{22}$$

where

$$k = \sigma + am/2Mr_+. \tag{23}$$

We shall find that the relations (22) are crucial when we come to consider the relevant boundary conditions in §5.

4. The electromagnetic perturbations on the Kerr black hole; the axisymmetric case

To illustrate how, with the aid of equation (10), we may pass from solutions valid under conditions of axisymmetry to solutions valid quite generally, we shall consider in some detail the case of electromagnetic perturbations when $s = 1$. Considering then equation (10) under conditions of axisymmetry for the case $s = 1$, and letting

$$\phi = \frac{\rho^2}{\Delta^{\frac{1}{2}}}\, Y, \tag{24}$$

where

$$\rho^2 = r^2 + a^2, \tag{25}$$

we find that it becomes

$$\Lambda^2 Y + P\Lambda_- Y - QY = 0, \tag{26}$$

where

$$P = \frac{\mathrm{d}}{\mathrm{d}r_*}\left(\lg \frac{\rho^4}{\Delta}\right), \tag{27}$$

$$Q = \frac{\Delta}{\rho^4}\left(\tau - a^2 \frac{\Delta}{\rho^4}\right), \tag{28}$$

$$\Lambda_\pm = \frac{\mathrm{d}}{\mathrm{d}r_*} \pm \mathrm{i}\sigma \quad \text{and} \quad \Lambda^2 = \frac{\mathrm{d}^2}{\mathrm{d}r_*^2} + \sigma^2. \tag{29}$$

(In the case of axisymmetry our variable r_* coincides with $r_*^{(c)}$ and $\tau = E + a^2\sigma^2$.)

We observe that equation (26) is of the standard form that was first encountered when treating the perturbations of the Schwarzschild black hole (Chandrasekhar 1975, equation (61)) and, again, when treating the axisymmetric gravitational perturbations of the Kerr black hole ((I), equation (10)). And as in those cases, we shall now seek a transformation of equation (26) to the form of a one dimensional wave equation,

$$\Lambda^2 Z = VZ \tag{30}$$

with a short-range potential V. Accordingly, we set (cf. (I), equation (21))

$$Y = fVZ + (W + 2\mathrm{i}\sigma f)\Lambda_+ Z, \tag{31}$$

where f and W are certain functions of r_* (or r), unspecified for the present; and we express $\Lambda_- Y$ in the manner (cf. (I), equations (18)–(20))

$$\Lambda_- Y = -\beta M(\Delta/\rho^4)\, Z + R\Lambda_+ Z, \tag{32}$$

where β and R are now defined by

$$-\beta M\frac{\Delta}{\rho^4} = \frac{\mathrm{d}}{\mathrm{d}r_*}(fV)+WV \tag{33}$$

and

$$R = fV+\frac{\mathrm{d}}{\mathrm{d}r_*}(W+2\mathrm{i}\sigma f). \tag{34}$$

The requirement that Y satisfies equation (26), by virtue of equation (30) assumed to be satisfied by Z, leads to the pair of equations (cf. (I), equations (26) and (27))

$$RV-M\frac{\Delta}{\rho^4}\frac{\mathrm{d}\beta}{\mathrm{d}r_*} = QfV \tag{35}$$

and

$$\frac{\mathrm{d}}{\mathrm{d}r_*}\left(\frac{\rho^4}{\Delta}R\right) = \frac{\rho^4}{\Delta}[Q(W+2\mathrm{i}\sigma f)-2\mathrm{i}\sigma R]+\beta M. \tag{36}$$

As in Paper I (equation (28)), we now find that equations (33)–(36) admit the integral

$$\frac{\rho^4}{\Delta}RfV+\beta M(W+2\mathrm{i}\sigma f) = K = \text{constant}. \tag{37}$$

This integral enables us to invert equations (31) and (32) to give

$$(\Delta/\rho^4)\,KZ = RY-T\Lambda_-Y \tag{38}$$

and

$$K\Lambda_+Z = \beta MY+(\rho^4/\Delta)fV\Lambda_-Y, \tag{39}$$

where, for the sake of brevity, we have written

$$T = W+2\mathrm{i}\sigma f. \tag{40}$$

Since equations (33)–(36) allow the integral (37), it will suffice to consider the equations

$$fV = R-\frac{\mathrm{d}T}{\mathrm{d}r_*}, \tag{41}$$

$$\frac{\mathrm{d}}{\mathrm{d}r_*}\left(\frac{\rho^4}{\Delta}R\right) = \frac{\rho^4}{\Delta}(QT-2\mathrm{i}\sigma R)+\beta M, \tag{42}$$

$$R\left(R-\frac{\mathrm{d}T}{\mathrm{d}r_*}\right)+\frac{\Delta}{\rho^4}\beta MT = \frac{\Delta}{\rho^4}K \tag{43}$$

and

$$RV-QfV = M\frac{\Delta}{\rho^4}\frac{\mathrm{d}\beta}{\mathrm{d}r_*}, \tag{44}$$

where it may be noted that equation (43) is an alternative form of equation (37) in which fV has been replaced in accordance with equation (41).

(a) *An explicit solution*

It can be verified by direct substitution that

$$T = 2\mathrm{i}\sigma, \quad R = fV \tag{45}$$

$$R = \mp 2\sigma a\,\frac{\Delta}{\rho^4}, \quad \beta M = -2\mathrm{i}\sigma\left(\tau-a^2\frac{\Delta}{\rho^4}\pm 2\sigma a\right) \tag{46}$$

and

$$K = 4\sigma^2(\tau\pm 2\sigma a), \tag{47}$$

represent solutions of equations (41)–(43). Equation (44) then gives

$$V = \frac{\Delta}{\rho^4}\left[\tau - a^2\frac{\Delta}{\rho^4} \mp \mathrm{i}a\rho^2\frac{\mathrm{d}}{\mathrm{d}r}\left(\frac{\Delta}{\rho^4}\right)\right]. \tag{48}$$

The potential V given by equation (48) is clearly of short range: it vanishes on the horizon (in fact, exponentially as $r_* \to -\infty$) and falls off like r^{-2} as $r \to \infty$. Also, while V given by equation (48) is complex, its integral over r_* is always real (and finite).

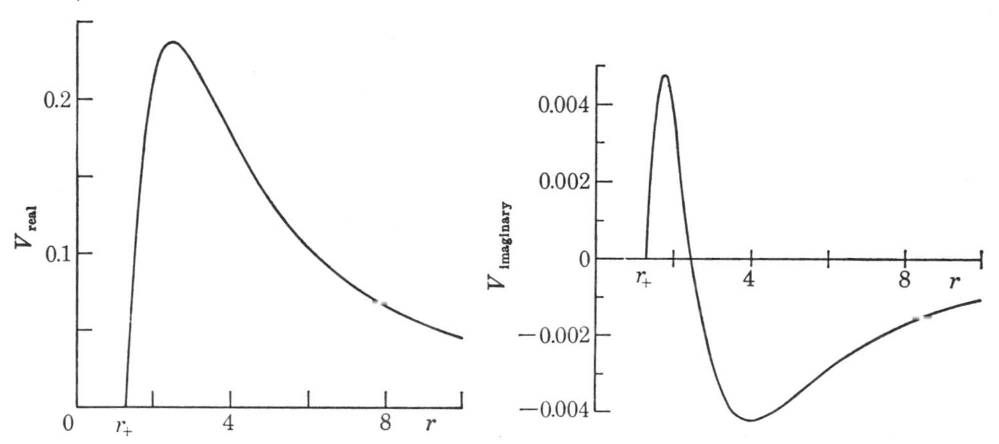

FIGURE 2. The potential appropriate for axisymmetric electromagnetic perturbations of a Kerr black hole ($a = 0.95$ and $\sigma = 1$).

Figure 2 illustrates the nature of the potential V given by equation (48) in a typical case.

With the solutions given in equations (45)–(47), equation (31), (32), (38), and (39) now take the forms

$$Y = \mp 2\sigma a\frac{\Delta}{\rho^4}Z + 2\mathrm{i}\sigma\Lambda_+ Z, \tag{49}$$

$$\Lambda_- Y = 2\mathrm{i}\sigma\frac{\Delta}{\rho^4}\left(\tau - a^2\frac{\Delta}{\rho^4} \pm 2\sigma a\right)Z \mp 2\sigma a\frac{\Delta}{\rho^4}\Lambda_+ Z, \tag{50}$$

$$KZ = \mp 2\sigma a Y - 2\mathrm{i}\sigma\frac{\rho^4}{\Delta}\Lambda_- Y \tag{51}$$

and

$$K\Lambda_+ Z = -2\mathrm{i}\sigma\left(\tau - a^2\frac{\Delta}{\rho^4} \pm 2\sigma a\right)Y \mp 2\sigma a\Lambda_- Y. \tag{52}$$

The two solutions, which we shall distinguish by Z_+ and Z_- and which follow from equations (49)–(52) by choosing the upper or the lower sign (wherever there is a choice) are simply related. Thus, by substituting in the relation,

$$K_+ Z_+ = -2\sigma a Y - 2\mathrm{i}\sigma(\rho^4/\Delta)\,\Lambda_- Y, \tag{53}$$

appropriate for Z_+, the expressions for Y and $\Lambda_- Y$ relating them to Z_-, we find that

$$K_+ Z_+ = 4\sigma^2(\tau - 2\sigma a - 2a^2(\Delta/\rho^4)) Z_- - 8\mathrm{i}a\sigma^2 \Lambda_+ Z_-. \tag{54}$$

Therefore,
$$K_+ Z_+ \to K_- Z_- - 8\mathrm{i}a\sigma^2 \Lambda_+ Z_- \quad (r_* \to \pm\infty). \tag{55}$$

From this last relation it follows that solutions for Z_- which have the asymptotic behaviours,

$$Z_- \to \mathrm{e}^{-\mathrm{i}\sigma r_*} \quad \text{and} \quad Z_- \to \mathrm{e}^{+\mathrm{i}\sigma r_*} \quad (r_* \to \pm\infty), \tag{56}$$

lead to solutions for Z_+ which have, respectively, the asymptotic behaviours,

$$Z_+ \to (K_-/K_+)\,\mathrm{e}^{-\mathrm{i}\sigma r_*} \quad \text{and} \quad Z_+ \to \mathrm{e}^{+\mathrm{i}\sigma r_*} \quad (r_* \to \pm\infty). \tag{57}$$

(*b*) *The distinction between* $Z^{(+\sigma)}$ *and* $Z^{(-\sigma)}$

Distinguishing Y and its complex conjugate by $Y^{(+\sigma)}$ and $Y^{(-\sigma)}$, we have the governing equations

$$\Lambda^2 Y^{(\pm\sigma)} + P\Lambda_\mp Y^{(\pm\sigma)} - Q Y^{(\pm\sigma)} = 0. \tag{58}$$

Similarly, distinguishing the functions satisfying the one dimensional wave equations of the form (30), derived from the equations for $Y^{(+\sigma)}$ and $Y^{(-\sigma)}$, by $Z^{(+\sigma)}$ and $Z^{(-\sigma)}$, respectively, we find by substitutions analogous to equations (45)–(47), that $Z^{(+\sigma)}$ and $Z^{(-\sigma)}$ satisfy wave equations with the *same* potential (48):

$$\Lambda^2 Z^{(\pm\sigma)} = V Z^{(\pm\sigma)}. \tag{59}$$

Since V is complex, it follows that $Z^{(+\sigma)}$ and $Z^{(-\sigma)}$, *unlike* $Y^{(+\sigma)}$ *and* $Y^{(-\sigma)}$, *are not complex-conjugate functions.*

If we select for $Z^{(+\sigma)}$ the solution which we had distinguished by Z_+ in § (*a*) above, we have the relations

$$T^{(\pm\sigma)} = \pm 2\mathrm{i}\sigma, \quad R^{(\pm\sigma)} = \mp 2\sigma a \frac{\Delta}{\rho^4}, \tag{60}$$

$$M\beta^{(\pm\sigma)} = \mp 2\mathrm{i}\sigma(\tau - a^2(\Delta/\rho^4) \pm 2\sigma a) \quad \text{and} \quad K^{(\pm\sigma)} = 4\sigma^2(\tau \pm 2\sigma a). \tag{61}$$

And the equations relating $Z^{(\pm\sigma)}$ and $Y^{(\pm\sigma)}$ are

$$Y^{(\pm\sigma)} = \mp 2\sigma a \frac{\Delta}{\rho^4} Z^{(\pm\sigma)} \pm 2\mathrm{i}\sigma \Lambda_\pm Z^{(\pm\sigma)}, \tag{62}$$

$$\Lambda_\mp Y^{(\pm\sigma)} = \pm 2\mathrm{i}\sigma\left(\tau - a^2\frac{\Delta}{\rho^4} \pm 2\sigma a\right)\frac{\Delta}{\rho^4} Z^{(\pm\sigma)} \mp 2\sigma a \frac{\Delta}{\rho^4}\Lambda_\pm Z^{(\pm\sigma)}, \tag{63}$$

$$K^{(\pm\sigma)} Z^{(\pm\sigma)} = \mp 2\sigma a Y^{(\pm\sigma)} \mp 2\mathrm{i}\sigma \frac{\rho^4}{\Delta}\Lambda_\mp Y^{(\pm\sigma)}, \tag{64}$$

and
$$K^{(\pm\sigma)}\Lambda_\pm Z^{(\pm\sigma)} = \mp 2\mathrm{i}\sigma(\tau - a^2\frac{\Delta}{\rho^4} \pm 2\sigma a)\, Y^{(\pm\sigma)} \mp 2\sigma a \Lambda_\mp Y^{(\pm\sigma)}. \tag{65}$$

From these relations it can be shown, by a procedure analogous to that which was used in § (*a*) above to relate the functions Z_+ and Z_-, that $Z^{(+\sigma)}$ is related to the complex conjugate of $Z^{(-\sigma)}$ by (cf. equation (54))

$$K^{(+\sigma)} Z^{(+\sigma)} = 4\sigma^2(\tau - 2\sigma a - 2a^2(\Delta/\rho^4))\,[Z^{(-\sigma)}]^* - 8\mathrm{i}a\sigma^2\Lambda_+ [Z^{(-\sigma)}]^*. \tag{66}$$

Therefore,

$$K^{(+\sigma)}Z^{(+\sigma)} \to K^{(-\sigma)}[Z^{(-\sigma)}]^* - 8ia\sigma^2\Lambda_+[Z^{(-\sigma)}]^* \quad (r_* \to \pm\infty). \tag{67}$$

From this last relation it follows that solutions for $Z^{(-\sigma)}$ which have the asymptotic behaviours,

$$Z^{(-\sigma)} \to e^{+i\sigma r_*} \quad \text{and} \quad Z^{(-\sigma)} \to e^{-i\sigma r_*} \quad (r_* \to \pm\infty), \tag{68}$$

lead to solutions for $Z^{(+\sigma)}$ which have, respectively, the asymptotic behaviours,

$$Z^{(+\sigma)} \to (K^{(-\sigma)}/K^{(+\sigma)})\, e^{-i\sigma r_*} \quad \text{and} \quad Z^{(+\sigma)} \to e^{+i\sigma r_*} \quad (r_* \to \pm\infty). \tag{69}$$

(c) *Reflexion and transmission coefficients*

In view of the short-range character of the potential (48), we can clearly seek solutions of equation (59) which have the asymptotic behaviours,

$$\begin{aligned} Z_+^{(+\sigma)} &\to e^{+i\sigma r_*} + A_+^{(+\sigma)} e^{-i\sigma r_*} \quad (r_* \to +\infty) \\ &\to \qquad B_+^{(+\sigma)} e^{+i\sigma r_*} \quad (r_* \to -\infty) \end{aligned} \tag{70}$$

and

$$\begin{aligned} Z_+^{(-\sigma)} &\to e^{-i\sigma r_*} + A_+^{(-\sigma)} e^{+i\sigma r_*} \quad (r_* \to +\infty) \\ &\to \qquad B_+^{(-\sigma)} e^{-i\sigma r_*} \quad (r_* \to -\infty). \end{aligned} \tag{71}$$

In writing the foregoing behaviours we have further distinguished the solutions by subscripts + to indicate that we are, in this instance, considering solutions appropriate to the potential (cf. equation (48))

$$V_+ = \frac{\Delta}{\rho^4}\left[\tau - a^2\frac{\Delta}{\rho^4} - ia\rho^2\frac{d}{dr}\left(\frac{\Delta}{\rho^4}\right)\right]. \tag{72}$$

Since $Z^{(+\sigma)}$ and $Z^{(-\sigma)}$ both satisfy wave equations with the same potential, the reflexion and the transmission coefficients defined in the manner,

$$\mathbb{R} = A_+^{(+\sigma)}A_+^{(-\sigma)} \quad \text{and} \quad \mathbb{T} = B_+^{(+\sigma)}B_+^{(-\sigma)} \tag{73}$$

satisfy *the conservation law*

$$\mathbb{R} + \mathbb{T} = 1. \tag{74}$$

On the other hand, by equation (69) relating the asymptotic behaviours of $Z_+^{(+\sigma)}$ and $[Z_+^{(-\sigma)}]^*$,

$$A_+^{(-\sigma)} = \frac{K_+^{(+\sigma)}}{K_+^{(-\sigma)}}[A_+^{(+\sigma)}]^* = \frac{\tau + 2a\sigma}{\tau - 2a\sigma}[A_+^{(+\sigma)}]^*$$

and

$$B_+^{(-\sigma)} = [B_+^{(+\sigma)}]^*. \tag{75}$$

The expressions (73) for $\mathbb{R}$ and $\mathbb{T}$ can, therefore, be rewritten in the forms

$$\mathbb{R} = \frac{\tau + 2a\sigma}{\tau - 2a\sigma}|A_+^{(+\sigma)}|^2 \quad \text{and} \quad \mathbb{T} = |B_+^{(+\sigma)}|^2. \tag{76}$$

These alternative forms for $\mathbb{R}$ and $\mathbb{T}$ show that, as defined in equations (73), they are indeed real. However, the importance of the expressions (76) for $\mathbb{R}$ and $\mathbb{T}$ consists in showing that for the evaluation of the reflexion and the transmission coefficients, it is not necessary to integrate the wave equation for Z twice in order to obtain the solutions with the different asymptotic behaviours (70) and (71); it will suffice to integrate the equation only once appropriate, for example, to the boundary conditions (70).

The expressions for $\mathbb{R}$ and $\mathbb{T}$ given in equation (76) apply for solutions of the wave equation with the potential V_+. If, instead, we had considered solutions of the wave equation with the potential

$$V_- = \frac{\Delta}{\rho^4}\left[\tau - a^2\frac{\Delta}{\rho^4} + \mathrm{i}a\rho^2\frac{\mathrm{d}}{\mathrm{d}r}\left(\frac{\Delta}{\rho^2}\right)\right] \tag{77}$$

we should have found

$$\mathbb{R} = \frac{\tau - 2a\sigma}{\tau + 2a\sigma}|A_-^{(+\sigma)}|^2 \quad \text{and} \quad \mathbb{T} = |B_-^{(+\sigma)}|^2. \tag{78}$$

On the other hand, according to equations (56) and (57),

$$A_+^{(+\sigma)} = \frac{\tau - 2a\sigma}{\tau + 2a\sigma}A_-^{(+\sigma)} \quad \text{and} \quad B_+^{(+\sigma)} = B_-^{(+\sigma)}. \tag{79}$$

Therefore equations (76) and (78) define the same reflexion and transmission coefficients.

Finally, by relating the solutions $Z^{(\pm\sigma)}$ with the corresponding solutions $Y^{(\pm\sigma)}$, in accordance with equations (62)–(65), we can show that the expressions for $\mathbb{R}$ and $\mathbb{T}$ we have given in terms of the solutions for $Z^{(\pm\sigma)}$ are in complete agreement with those given by Teukolsky & Press (1974, equations (4.7) and (5.7)) working directly with Teukolsky's equation.

5. The electromagnetic perturbations of the Kerr black hole; the non-axisymmetric case

Returning to equation (10) (in which r_* is defined as in equations (8) and (12)) and continuing the consideration of the case $s = 1$, we now set (cf. equations (24) and (25))

$$\phi = \frac{\rho^2}{\Delta^{\frac{1}{2}}}Y, \tag{80}$$

where

$$\rho^2 = r^2 + \alpha^2 = r^2 + (a^2 + am/\sigma). \tag{81}$$

With the substitution (80) in equation (10), we find that equations (26)–(28) continue to be applicable, as *written* provided ρ^2 is assigned its present meaning (81) and a^2, *wherever it occurs explicitly*,† is replaced by α^2. Moreover, since the reduction of equation (26) (with the aid of equations (49)–(52)) to the form of a one dimensional wave equation with a potential V given by equation (48), at no stage, made use of the explicit expression for Δ (except that its derivative with respect to r is $r - M$), it follows that equations (45)–(52) will again continue to be applicable provided ρ^2 is assigned its present meaning and a, wherever it occurs explicitly, is replaced by α.

† Thus, the meaning of Δ should not be changed.

In particular, we can now write (cf. equations (51) and (52))

$$KZ = \mp 2\sigma\alpha Y - 2i\sigma(\rho^4/\Delta)\Lambda_- Y \tag{82}$$

and

$$K\Lambda_+ Z = -2i\sigma(\tau - \alpha^2(\Delta/\rho^4) \pm 2\sigma\alpha)\, Y \mp 2\sigma\alpha\Lambda_- Y, \tag{83}$$

where

$$K = 4\sigma^2(\tau \pm 2\sigma\alpha), \tag{84}$$

and τ has the meaning (cf. equation (3))

$$\tau = E + a^2\sigma^2 + 2am\sigma. \tag{85}$$

And the one dimensional wave equation satisfied by Z is

$$\Lambda^2 Z = VZ, \tag{86}$$

where

$$V = \frac{\Delta}{\rho^4}\left[\tau - \alpha^2\frac{\Delta}{\rho^4} \mp i\alpha\rho^2\frac{\mathrm{d}}{\mathrm{d}r}\left(\frac{\Delta}{\rho^4}\right)\right]. \tag{87}$$

The reduction of equation (10) to equation (86) has so far been an exercise in formal manipulations; for, we have yet to consider how we are to allow for the two major facts: *first*, that the $r_*(r)$-relation becomes double-valued when σ is in the super-radiant interval $(0, \sigma_s)$ and *second*, that the potential V defined in equation (87) becomes singular at $r = |\alpha|$, outside of the horizon, if α^2 is negative and

$$r_+^2 - |\alpha|^2 < 0.$$

We consider these facts in §(*b*) below.

(*a*) *When σ is not super-radiant and $\sigma > \sigma_s$*

There is clearly no difficulty in using equation (86) with the potential (87) so long as $\sigma > \sigma_s$; for in this case, the $r_*(r)$-relation is single-valued and the potential is bounded and of short range. Consequently, we can seek solutions of the wave equation which have the behaviours,

$$\begin{aligned} Z_\pm &\to e^{+i\sigma r_*} + A_\pm e^{-i\sigma r_*} \quad (r_* \to +\infty) \\ &\to \qquad\quad B_\pm e^{+i\sigma r_*} \quad (r_* \to -\infty), \end{aligned} \tag{88}$$

where the subscript $\pm$ distinguishes the solutions determined with either the upper $(-)$ or the lower $(+)$ sign of the term $i\alpha\rho^2(\Delta/\rho^4)_{,r}$ in the expression for V. In terms of solutions satisfying the foregoing boundary conditions, the required reflexion and transmission coefficients are given by (cf. equations (76) and (79))

$$\mathbb{R} = \frac{\tau \pm 2\alpha\sigma}{\tau \mp 2\alpha\sigma}|A_\pm|^2 \quad \text{and} \quad \mathbb{T} = |B_\pm|^2. \tag{89}$$

The expressions for $\mathbb{R}$ and $\mathbb{T}$ given in equation (89) are applicable so long as $\alpha^2 > 0$, i.e. so long as $\sigma > \sigma_c(= -m/a)$. When, however, σ is in the interval $\sigma_s < \sigma < \sigma_c$, α^2 is negative and the potential (which is now real) can be written in the form

$$V_\pm = \frac{\Delta}{\rho^4}\left[\tau + |\alpha|^2\frac{\Delta}{\rho^4} \pm |\alpha|\,\rho^2\frac{\mathrm{d}}{\mathrm{d}r}\left(\frac{\Delta}{\rho^4}\right)\right], \tag{90}$$

where $$|\alpha|^2 = -\alpha^2 \quad \text{and} \quad \rho^2 = r^2 - |\alpha|^2; \tag{91}$$

and since $\sigma > \sigma_s$, $r_+^2 - |\alpha|^2 > 0$. The potential (90) continues to be bounded and of short range; and the theory of reflexion and transmission of one dimensional real potential barriers, familiar in elementary quantum theory, becomes directly applicable. It should, however, be noted that the two potentials included in the expression (90) both yield the same reflexion and transmission coefficients. The reason is that the corresponding amplitudes of the reflected and transmitted waves are related in the manner (cf. equation (79))

$$A_+ = \frac{\tau - 2i\sigma|\alpha|}{\tau + 2i\sigma|\alpha|} A_- \quad \text{and} \quad B_+ = B_- \tag{92}$$

and therefore, $$|A_+|^2 = |A_-|^2 \quad \text{and} \quad |B_+|^2 = |B_-|^2.$$

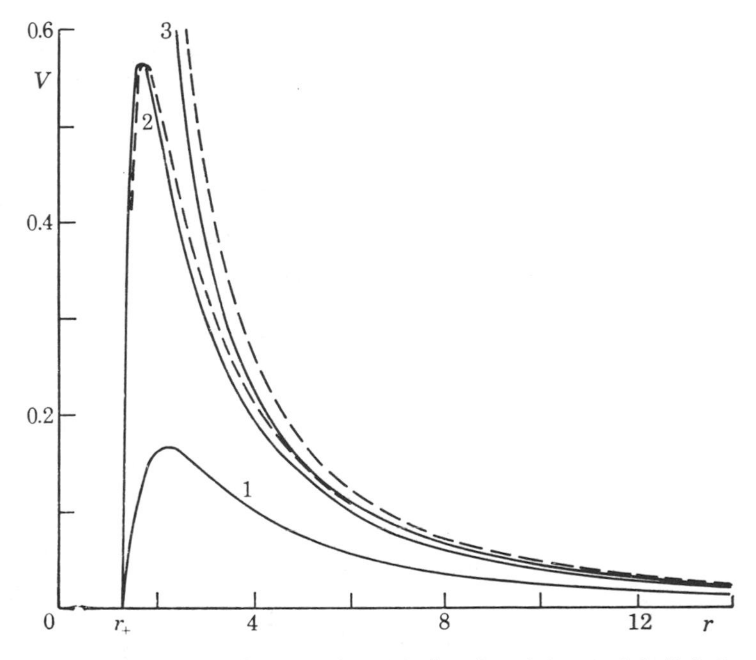

FIGURE 3. The potential given by equation (90) for $\alpha^2 = 0$ (curve labelled 1), $\alpha^2 = -0.9975$ (curves labelled 2, the solid curve for V_+ and the dashed curve for V_-), and $\alpha^2 = -1.7220$ (curves labelled 3, for the critical frequency σ_s); the solid curve for V_+ and the dashed curve for V_-). The potentials V_+ and V_- for the last case are singular at $r = r_+$.

Figure 3 illustrates the nature of the potentials which arise when treating these general non-axisymmetric modes for non-super-radiant frequencies in the interval $\sigma_s < \sigma \leqslant \sigma_c$. It may be noted in this connection that when σ has the co-rotational value $-m/a$ and $\alpha = 0$, the potential is particularly simple:

$$V_{\alpha=0} = \tau \Delta r^{-4}. \tag{93}$$

Two further aspects of the problem are worthy of notice. *First*, the asymptotic behaviour,

$$Z \to e^{\pm i\sigma r_*} \quad (r \to r_+ + 0), \tag{94}$$

required of the solutions of equation (86), as we approach the horizon, corresponds to the behaviour,

$$Z \to e^{\pm ikr_*^{(c)}} \quad (r_*^{(c)} \to -\infty), \tag{95}$$

(where $k = \sigma + am/2Mr_+$) when expressed in terms of the variable $r_*^{(c)}$, as usually defined (cf. equations (22) and (23)); and this is consistent with the boundary conditions as normally imposed on the solution (cf. Teukolsky 1973, equation (5.6)). This fact emphasizes the particular appropriateness of the choice of the variable r_*, as we have defined it. *Secondly*, we observe that as $\sigma \to \sigma_s + 0$, the potential becomes singular exactly at the horizon (see fig. 3). Thus, in this limit, the barrier presented to the incoming waves becomes infinite as the horizon is approached; consequently,

$$\mathbb{R} \to 1 \quad \text{and} \quad \mathbb{T} \to 0 \quad \text{as} \quad \sigma \to \sigma_s + 0. \tag{96}$$

This last fact is consistent with the expectation that super-radiance, with an accompanying reflexion coefficient $\mathbb{R} > 1$, begins *at* $\sigma = \sigma_s$.

(*b*) *When* $\sigma < \sigma_s$ *and super-radiance occurs*

As we have seen, when $\sigma < \sigma_s$ and $r_+^2 + \alpha^2 = r_+^2 - |\alpha|^2 < 0$, the $r_*(r)$-relation attains a minimum value at $r = |\alpha|$ and tends to $+\infty$ both when $r \to \infty$ and when $r \to r_+ + 0$. Therefore, we must consider, separately, the solutions along the two branches of the $r_*(r)$-relation, which starting from $r = |\alpha|$, progresses either towards $r \to \infty$ or towards the horizon at r_+. This requirement of separate considerations of the solution along the two branches is further compelled by the fact that the potential given by equation (90) has a singularity at $r = |\alpha|$. Thus, by rewriting the expression (90) for V in the manner,

$$V_\pm = \frac{\Delta}{\rho^4}\left[\tau + \frac{\Delta}{\rho^4}|\alpha|\,(|\alpha| \mp 4r) \pm 2|\alpha|\frac{r-M}{\rho^2}\right], \tag{97}$$

we find that in the neighbourhood of $r = |\alpha|$, $V_\pm$ has the behaviour,

$$V_\pm = (-3 \text{ or } 5)\frac{\Delta_{|\alpha|}^2}{16|\alpha|^2}\frac{1}{(r-|\alpha|)^4}, \tag{98}$$

where it may be noted that, in the same neighbourhood, the $r_*(r)$-relation has the behaviour (cf. equation (17))

$$r_* = r_*(|\alpha|) + \frac{|\alpha|}{\Delta_{|\alpha|}}(r-|\alpha|)^2 + \dots. \tag{99}$$

Postponing for the present the manner in which the singularity in $V_\pm$ at $r = |\alpha|$ is to be taken into account, we first observe that the solution to the problem of reflexion and transmission of waves of super-radiant frequencies requires that we seek solutions of the wave equation which satisfy the boundary conditions,

$$Z \to e^{+i\sigma r_*} + A\,e^{-i\sigma r_*} \quad \text{along the branch} \quad r \to \infty \quad \text{and} \quad r_* \to +\infty \tag{100}$$

and $$Z \to B e^{-i\sigma r_*} \quad \text{along the branch} \quad r \to r_+ \quad \text{and} \quad r_* \to +\infty. \tag{101}$$

Notice that we are now requiring that the transmitted wave has the behaviour $e^{-i\sigma r_*}$, in contrast to the behaviour $e^{+i\sigma r_*}$ which we required when $\sigma > \sigma_s$. The reason is that, in contrast to what obtained when $\sigma > \sigma_s$, when $\sigma < \sigma_s$, progressing *towards* the horizon is in the direction of *increasing* r_*.

From the constancy of the Wronskian, $W(Z, Z^*)$, along the solution, it now follows that with the definitions

$$\mathbb{R} = |A|^2 \quad \text{and} \quad \mathbb{T} = |B|^2, \tag{102}$$

the conservation law is

$$\mathbb{R} - \mathbb{T} = 1; \tag{103}$$

in other words, *we necessarily have super-radiance.*

Finally, it remains to clarify how a solution of the wave equation satisfying the boundary conditions (100) and (101) can be obtained, duly allowing for the singularity (98) in V at $r = |\alpha|$. For this purpose, we must examine the behaviour of the solution at $r = |\alpha|$.

Letting $$x = |(r - |\alpha|)|, \tag{104}$$

we find, with the aid of relations (98) and (99), that in the neighbourhood of $r = |\alpha|$, the wave equation for Z becomes

$$x^2 \frac{d^2 Z}{dx^2} - x \frac{dZ}{dx} = (-\tfrac{3}{4} \text{ or } \tfrac{5}{4}) Z \quad (x \sim 0). \tag{105}$$

From this equation it follows that

$$Z \sim (r - |\alpha|)^n, \tag{106}$$

where $$n = \tfrac{1}{2} \text{ or } \tfrac{3}{2} \text{ for } V_+ \quad \text{and} \quad n = -\tfrac{1}{2} \text{ or } \tfrac{5}{2} \text{ for } V_-. \tag{107}$$

In the neighbourhood of $r = |\alpha|$, the solution for Z must, therefore, have the form

$$Z = c_1 |(r - |\alpha|)|^{\frac{3}{2}} + c_2 |(r - |\alpha|)|^{\frac{1}{2}} \quad \text{for} \quad V_+,$$

and $$Z = c_3 |(r - |\alpha|)|^{\frac{5}{2}} + c_4 |(r - |\alpha|)|^{-\frac{1}{2}} \quad \text{for} \quad V_-, \tag{108}$$†

where $c_1, \ldots, c_4$ are constants. With this knowledge of the behaviour of Z at $r = |\alpha|$, a method of integrating the wave equation for Z, through the singularity at $r = |\alpha|$ and determining the reflexion and the transmission coefficients in accordance with the prescription (102), is the following.

We start with a solution for Z with the behaviour

$$Z \to e^{-i\sigma r_*} \quad \text{for} \quad r_* \to +\infty \quad \text{along the branch} \quad r \to r_+ + 0, \tag{109}$$

† Since the radial function R in Teukolsky's equation (1) is related to Y by

$$R = (r^2 - |\alpha|^2)^{\frac{1}{2}} Y/\Delta,$$

it follows from equation (82), for example, that these behaviours of Z at $r = |\alpha|$ are consistent with the continuity of R at $r = |\alpha|$.

and continue the integration forward from the horizon (but backward in r_*) to $r = |\alpha| - 0$. As we approach the singularity at $r = |\alpha|$, from the left, the solution will tend to a form specified in equations (108) with certain determinate values for the constants c_1 and c_2 (or c_3 and c_4). Since the behaviour of the solution specified in equations (108) applies to either side of $r = |\alpha|$, the form for Z determined by approaching $r = |\alpha| - 0$, applies equally to $r = |\alpha| + 0$. With the form of Z, for $r = |\alpha| + 0$, thus determined, we can continue the integration forward (in r and in r_*) beyond $r = |\alpha|$, along the branch $r \to \infty$. By such forward integration, we shall eventually find that, as $r \to \infty$, the solution tends to the limiting form

$$Z \to C_{\text{inc}} e^{+i\sigma r_*} + C_{\text{ref}} e^{-i\sigma r_*} \quad (r_* \to +\infty), \tag{110}$$

where C_{inc} and C_{ref} are certain constants that will be determined by the integration. Since the solution we started with corresponds to a transmitted wave of unit amplitude approaching the horizon, it is apparent that the required reflexion and transmission coefficients will be given by

$$\mathbb{R} = \frac{|C_{\text{ref}}|^2}{|C_{\text{inc}}|^2} \quad \text{and} \quad \mathbb{T} = \frac{1}{|C_{\text{inc}}|^2}. \tag{111}$$

The coefficients $\mathbb{R}$ and $\mathbb{T}$ derived in this fashion will satisfy the conservation law (103).

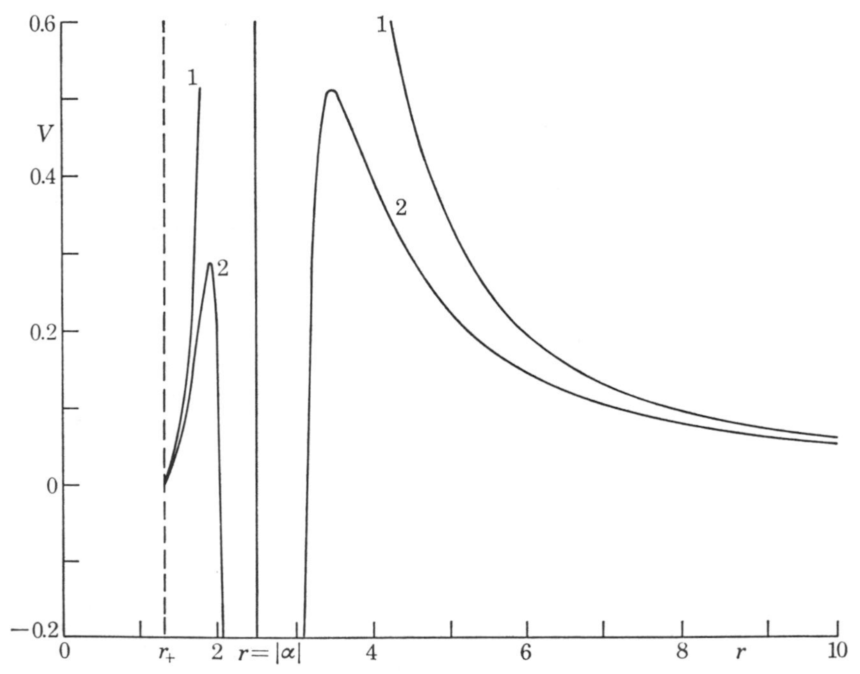

FIGURE 4. Illustrating the potential given by equation (97) for a super-radiant frequency ($a = 0.95$, $\alpha^2 = -6.25$, and $\sigma = 0.2656$). The curves labelled 1 and 2 refer to V_- and V_+, respectively.

6. Concluding remarks

The reduction of the problem of the electromagnetic perturbations of the Kerr black hole, in all cases, to problems – albeit novel problems – in the theory of penetration of one dimensional potential barriers, has revealed certain interesting aspects of the general perturbation problem as it pertains to the Kerr metric. Thus, the explicit appearance of m in the equations can be eliminated; the m-modes can be treated by methods which are formally the same as those which are applicable to axisymmetric modes; and phenomena such as super-radiance emerge cloaked in different guises. However, in order that we may transcribe the formal results obtained for the axisymmetric case, appropriate for the non-axisymmetric case, it is necessary that in obtaining the results for the axisymmetric case we nowhere make use of the explicit form of Δ, *except* that its derivative with respect to r is $r-M$. This was manifestly the case in the treatment of the axisymmetric electromagnetic perturbations in §4. But this was not the case in the treatment of the axisymmetric gravitational perturbations in paper I: in the course of the reductions, simplifications were effected which involved the explicit form of Δ. By a careful scrutiny of the analysis of paper I, it should be possible to rewrite the final expression for the potential (I, equation (72)) in a form in which in its transcription for the non-axisymmetric case is direct. We shall return to these matters pertaining to the gravitational perturbations of the Kerr black hole in a separate communication.

I am grateful to Miss Donna D. Elbert for her assistance in preparing the illustrations for this paper; and I have profited by discussions with Professor Peter G. O. Freund. The research reported in this paper has in part been supported by the National Science Foundation under grant MPS 74-17456 with the University of Chicago.

Appendix. Scalar perturbations of the Kerr black hole

For scalar perturbations $s = 0$ and equation (10), as it stands, already provides a one dimensional wave equation of the desired form. The equations can be written in the form

$$\Lambda^2\phi = V\phi, \tag{A 1}$$

where

$$V = \frac{\Delta}{\rho^4}\left\{\tau + \frac{1}{\rho^2}[\Delta + 2(r-M)] - 3\frac{r^2\Delta}{\rho^4}\right\}, \tag{A 2}$$

and ρ^2 and τ have the same meanings as in §5 (equations (81) and (85)).

For $\sigma > \sigma_s$, the potential given by equation (A 2) is of short range, real, and bounded; accordingly, the elementary theory of reflexion and transmission by one dimensional potential barriers is applicable. However, for $\sigma < \sigma_s$ the potential has a singularity at $r = |\alpha|$ where it has the behaviour

$$V = -3\frac{\Delta^2_{|\alpha|}}{16|\alpha|^2}\frac{1}{(r-|\alpha|)^4}. \tag{A 3}$$

It will be observed that this behaviour is the same as for the potential V_+ for electromagnetic perturbation (cf. equation (98)). Equation (A 1) can accordingly be integrated through the singularity at $r = |\alpha|$ to determine the reflexion and the transmission coefficients in the manner described in §5.

Note added in proof 26 *October* 1975. Meantime it has been possible to transcribe the results of paper I in a form that is applicable to a treatment of general non-axisymmetric gravitational perturbations of the Kerr black hole.

References

Chandrasekhar S., 1975 *Proc. R. Soc. Lond.* A **343**, 289.
Chandrasekhar, S. & Detweiler, S. 1975 *Proc. R. Soc. Lond.* A **345**, 145.
Detweiler, S. 1976 *Proc. R. Soc. Lond.* A (in the press).
Teukolsky, S. A. 1973 *Astrophys. J.* **185**, 635.
Teukolsky, S. A. & Press, W. H. 1974 *Astrophys. J.* **193**, 443.

7

The solution of Maxwell's equations in Kerr geometry

By S. Chandrasekhar, F.R.S.
University of Chicago, Chicago, Illinois, 60637

(*Received* 18 *September* 1975)

Explicit solutions for the vector-potential of a vacuum Maxwell-field in Kerr geometry are obtained in terms of Teukolsky's radial and angular functions.

1. Introduction

The consideration of electromagnetic perturbations of the Kerr black hole requires the solution of Maxwell's equations in the background Kerr geometry. In the Newman–Penrose formalism (Newman & Penrose 1962), four null vectors l^μ, n^μ, m^μ, and $m^{*\mu}$ are defined at each point in space-time; the vectors l^μ and n^μ are real while m^μ and $m^{*\mu}$ are complex conjugates of one another; and moreover

$$\boldsymbol{m}\cdot\boldsymbol{m}^* = -1, \quad \boldsymbol{l}\cdot\boldsymbol{n} = 1, \text{ and all other scalar products vanish.} \tag{1}$$

The null tetrad so chosen provides a quasi-orthonormal basis; and any tensor in which we may be interested is projected on to it. Equations are then derived for the scaars obtained by contracting the basic tensors – the antisymmetric tensor $F_{\mu\nu}$ when we are concerned with a Maxwell field and the Weyl tensor $C_{\alpha\beta\gamma\delta}$ when we are concerned with a gravitational field – with the vectors of the tetrad in all possible (non-trivial) ways. Thus, a Maxwell field, in the Newman–Penrose formalism, is characterized by the three scalars

$$\phi_{+1} = F_{\mu\nu} l^\mu m^\nu, \quad \phi_{-1} = F_{\mu\nu} m^{*\mu} n^\nu,$$

and

$$\phi_0 = \tfrac{1}{2} F_{\mu\nu}(l^\mu n^\nu + m^{*\mu} m^\nu); \tag{2}†$$

and a set of four coupled first-order differential equations are derived for them (equations (9)–(12), below). Starting from these equations, specialized appropriately for Kerr geometry, and seeking solutions in the Boyer–Lindquist (1967) coordinates (t, r, θ, φ), with a t- and a φ-dependence given by $e^{i(\sigma t + m\varphi)}$, Teukolsky (1972, 1973) obtained two decoupled equations for ϕ_{+1} and ϕ_{-1}; and he was further able to separate the variables r and θ in these equations and express the solutions for ϕ_{+1} and ϕ_{-1} in terms of certain basic radial and angular functions (equations (18) and (19) below). With the solutions for ϕ_{+1} and ϕ_{-1} specified in this fashion, the problem of solving Maxwell's equations in a background Kerr geometry may be considered

† Our notation is slightly different from the usual one: we are using ϕ_{+1} and ϕ_{-1}, in place of the customary ϕ_0 and ϕ_2, in order to conform to the 'spin weights' to which these functions belong (see equation (16) below).

as having been formally completed since all the other field quantities (except the Coulomb longitudinal-part) can in principle be determined in terms of ϕ_{+1} and ϕ_{-1}. Thus, Cohen & Kegeles (1974) and Chrzanowski (1975) have addressed themselves to precisely this last phase of the problem. In particular, Chrzanowski has derived formulae, in two different gauges, for the potentials, in the electromagnetic case, and for the perturbations of the metric coefficients, in the gravitational case. The present paper (and its sequel, in preparation) may be considered as providing an alternative treatment of the problem considered by Chrzanowski. It will appear that the method developed in this paper, in the context of electromagnetic perturbations, is direct and simple: it consists in simply writing out the expressions for ϕ_{+1} and ϕ_{-1}, explicitly, in terms of the potentials and observing that the resulting equations can be integrated by elementary methods and one is able to write down simple expressions for the different components of the potential in a general gauge.

2. The equations of the problem

In the Boyer–Lindquist coordinates, the Kerr metric can be written in the form

$$ds^2 = -e^{2\nu}(dt)^2 + e^{2\psi}(d\varphi - \omega\, dt)^2 + e^{2\mu_2}(dr)^2 + e^{2\mu_3}(d\theta)^2, \tag{3}$$

where

$$e^{\nu} = \frac{\rho\sqrt{\Delta}}{\Sigma}, \quad e^{\psi} = \frac{\sin\theta}{\rho}\Sigma, \quad e^{\mu_2} = \frac{\rho}{\sqrt{\Delta}}, \quad e^{\mu_3} = \rho, \quad \omega = \frac{2Mar}{\Sigma^2}, \tag{4}$$

$$\rho^2 = r^2 + a^2\cos^2\theta, \quad \Delta = r^2 + a^2 - 2Mr \quad \text{and} \quad \Sigma^2 = (r^2+a^2)^2 - a^2\Delta\sin^2\theta. \tag{5}$$

It may be parenthetically noted here that two elementary identities among the quantities we have defined and which contribute to much of the simplifications in the ensuing analysis are

$$a - (r^2+a^2)\,\omega = a\frac{\Delta\rho^2}{\Sigma^2} \quad \text{and} \quad 1 - a\omega\sin^2\theta = \frac{(r^2+a^2)\,\rho^2}{\Sigma^2}. \tag{6}$$

In the same Boyer–Lindquist coordinates, a null tetrad which satisfies the requirements (1) of the Newman–Penrose formalism and which is commonly used is that of Kinnersley (1969):

$$l^{\mu} = \frac{1}{\Delta}(r^2+a^2, \Delta, 0, a),$$

$$n^{\mu} = \frac{1}{2\rho^2}(r^2+a^2, -\Delta, 0, a),$$

$$m^{\mu} = \frac{1}{\bar{\rho}\sqrt{2}}(\mathrm{i}a\sin\theta, 0, 1, \mathrm{i}\,\mathrm{cosec}\,\theta), \tag{7}$$

where

$$\bar{\rho} = r + \mathrm{i}a\cos\theta \quad \text{and} \quad \bar{\rho}^* = r - \mathrm{i}a\cos\theta. \tag{8}$$

With this choice of the tetrad, the equations of Newman & Penrose for ϕ_{+1}, ϕ_{-1}, and ϕ_0 (with a t- and a φ-independence given by $e^{i(\sigma t+m\varphi)}$), are

$$\frac{1}{\bar{\rho}^*\sqrt{2}}\left(\mathscr{L}_1-\frac{ia\sin\theta}{\bar{\rho}^*}\right)\phi_{+1}=\left(\mathscr{D}+\frac{2}{\bar{\rho}^*}\right)\phi_0, \tag{9}$$

$$\frac{1}{\bar{\rho}^*\sqrt{2}}\left(\mathscr{L}_0+\frac{2ia\sin\theta}{\bar{\rho}^*}\right)\phi_0=\left(\mathscr{D}+\frac{1}{\bar{\rho}^*}\right)\phi_{-1}, \tag{10}$$

$$\frac{1}{\bar{\rho}\sqrt{2}}\left(\mathscr{L}_1^\dagger+\frac{ia\sin\theta}{\bar{\rho}^*}\right)\phi_{-1}=-\frac{\Delta}{2\rho^2}\left(\mathscr{D}^\dagger+\frac{2}{\bar{\rho}^*}\right)\phi_0, \tag{11}$$

$$\frac{1}{\bar{\rho}\sqrt{2}}\left(\mathscr{L}_0^\dagger+\frac{2ia\sin\theta}{\bar{\rho}^*}\right)\phi_0=-\frac{1}{2\rho^2}\left(\mathscr{D}^\dagger-\frac{1}{\bar{\rho}^*}\right)\Delta\phi_{+1}, \tag{12}$$

In the foregoing equations, $\mathscr{D}$, $\mathscr{D}^\dagger$, $\mathscr{L}_n$, and $\mathscr{L}_n^\dagger$ denote the operators (see equation (25) below)

$$\mathscr{D}=\partial_r+\frac{iK}{\Delta},\qquad \mathscr{D}^\dagger=\partial_r-\frac{iK}{\Delta}, \tag{13}$$

$$\mathscr{L}_n=\partial_\theta+Q+n\cot\theta \quad\text{and}\quad \mathscr{L}_n^\dagger=\partial_\theta-Q+n\cot\theta, \tag{14}$$

where
$$K=(r^2+a^2)\,\sigma+am \quad\text{and}\quad Q=a\sigma\sin\theta+m\operatorname{cosec}\theta. \tag{15}$$

Teukolsky's separated solutions for ϕ_{+1} and ϕ_{-1} are

$$\phi_{+1}=R_{+1}(r)\,S_{+1}(\theta) \quad\text{and}\quad \phi_{-1}=\frac{B}{(\bar{\rho}^*)^2}R_{-1}(r)\,S_{-1}(\theta), \tag{16}$$

where
$$B^2=(E+a^2\sigma^2+2a\sigma m)^2-4a^2\sigma^2-4a\sigma m, \tag{17}$$

and E is the 'separation constant' which occurs when the variables are separated (see equations (18) and (19) below). The radial, $R_{\pm1}(r)$, and the angular, $S_{\pm1}(\theta)$, functions are governed by the equations

$$\Delta^{-s}\frac{d}{dr}\left(\Delta^{s+1}\frac{dR_s}{dr}\right)+\frac{1}{\Delta}[K^2+2is(r-M)\,K-\Delta(4is\sigma r+\lambda_s)]\,R_s=0, \tag{18}$$

and

$$\frac{1}{\sin\theta}\frac{d}{d\theta}\left(\sin\theta\frac{dS_s}{d\theta}\right)+\left(a^2\sigma^2\cos^2\theta+2a\sigma s\cos\theta-\frac{m^2+s^2+2ms\cos\theta}{\sin^2\theta}+E\right)S_s=0, \tag{19}$$

where
$$\lambda_s=E+a^2\sigma^2+2a\sigma m-s(s+1),\quad s=\pm1, \tag{20}$$

and E is the separation constant – it is a characteristic value of equation (19) determined by the regularity conditions on S_s at $\theta=0$ and $\theta=\pi$.

It follows from equations (18) and (19) that ΔR_{+1} and $2BR_{-1}$ are complex conjugate functions while $S_{+1}(\theta)\equiv S_{-1}(\pi-\theta)$. Also, with respect to the operators $\mathscr{D}$, $\mathscr{D}^\dagger$, $\mathscr{L}_n$, and $\mathscr{L}_n^\dagger$, the functions $R_{\pm1}$ and $S_{\pm1}$ satisfy the following remarkable 'ladder' relations (cf. Teukolsky & Press 1974; Starobinsky 1973; and Starobinsky & Churoliv 1974):

$$\mathscr{D}\mathscr{D}R_{-1}=\tfrac{1}{2}R_{+1},\quad \Delta\mathscr{D}^\dagger\mathscr{D}^\dagger\Delta R_{+1}=2B^2R_{-1}, \tag{21}$$

$$\mathscr{L}_0\mathscr{L}_1S_{+1}=BS_{-1} \quad\text{and}\quad \mathscr{L}_0^\dagger\mathscr{L}_1^\dagger S_{-1}=BS_{+1}. \tag{22}$$

3. The solution for the vector potential

We shall express the Maxwell tensor $F_{\mu\nu}$ in terms of a vector potential A_μ in the usual manner,

$$F_{\mu\nu} = \partial_\nu A_\mu - \partial_\mu A_{\nu 1} \tag{23}$$

and evaluate ϕ_{+1}, ϕ_{-1}, and ϕ_0 directly in terms of it. Thus, considering ϕ_{+1}, we have

$$\phi_{+1} = l^\mu m^\nu (\partial_\nu A_\mu - \partial_\mu A_\nu). \tag{24}$$

Remembering that

$$l^\mu \partial_\mu = \mathscr{D} \quad \text{and} \quad m^\nu \partial_\nu = \frac{1}{\bar{\rho}\sqrt{2}} \mathscr{L}_0^\dagger, \tag{25}$$

and also that $\mathscr{D}$ operates on functions of r only while $\mathscr{L}_0^\dagger$ operates on functions of θ only, we readily verify that

$$\phi_{+1} = \frac{1}{\bar{\rho}\sqrt{2}} \mathscr{L}_0^\dagger \left(\frac{r^2+a^2}{\Delta} A_t + A_r + \frac{a}{\Delta} A_\varphi \right)$$

$$- \frac{1}{\bar{\rho}\sqrt{2}} \mathscr{D}(\text{i}a\, A_t \sin\theta + A_\theta + \text{i}A_\varphi \operatorname{cosec}\theta), \tag{26}$$

where we have made use of the explicit forms of l^μ and m^ν. Now letting

$$F_{+1} = A_r + \frac{a}{\Delta} A_\varphi + \frac{r^2+a^2}{\Delta} A_t \tag{27}$$

and

$$G_{+1} = A_\theta + \text{i}A_\varphi \operatorname{cosec}\theta + \text{i}a\, A_t \sin\theta, \tag{28}$$

and substituting for ϕ_{+1} Teukolsky's solution, we have

$$\sqrt{2}\,.\,(r + \text{i}a\cos\theta)\, R_{+1} S_{+1} = \mathscr{L}_0^\dagger F_{+1} - \mathscr{D} G_{+1}. \tag{29}$$

Similarly, the evaluation of ϕ_{-1} yields

$$2\sqrt{2}\,.\,\bar{\rho}(\bar{\rho}^*)^2 \phi_{-1} = 2\sqrt{2}\,.\,B(r + \text{i}a\cos\theta)\, R_{-1} S_{-1}$$

$$= \Delta(\mathscr{L}_0 F_{-1} - \mathscr{D}^\dagger G_{-1}), \tag{30}$$

where

$$F_{-1} = A_r - \frac{a}{\Delta} A_\varphi - \frac{r^2+a^2}{\Delta} A_t, \tag{31}$$

and

$$G_{-1} = A_\theta - \text{i}A_\varphi \operatorname{cosec}\theta - \text{i}a\, A_t \sin\theta. \tag{32}$$

The evaluation of ϕ_0 is equally straightforward; we find

$$\phi_0 = \frac{1}{2\rho^2} [\text{i}KA_r - (r^2+a^2) A_{t,r} - aA_{\varphi,r} - QA_\theta - \text{i}(a\, A_{t,\theta} \sin^2\theta + A_{\varphi,\theta}) \operatorname{cosec}\theta]. \tag{33}$$

(a) *The solutions of equations* (29) *and* (30)

Considering first equation (29), we observe that the general solution of the associated homogeneous equation is

$$F_{+1} = \mathscr{D} P_{+1} \quad \text{and} \quad G_{+1} = \mathscr{L}_0^\dagger P_{+1}, \tag{34}$$

where P_{+1} is an arbitrary function of r and θ. And a particular integral of the equation can be obtained by setting

$$F_{+1} = R_{+1}(r) f_{+1}(\theta) \quad \text{and} \quad G_{+1} = g_{+1}(r) S_{+1}(\theta), \tag{35}$$

where $f_{+1}(\theta)$ and $g_{+1}(r)$ are, as indicated, functions of θ and r, only. Inserting these forms for the solutions in equation (29), we obtain

$$\frac{1}{S_{+1}} \mathscr{L}_0^\dagger f_{+1} - \sqrt{2}.ia\cos\theta = \frac{1}{R_{+1}} \mathscr{D} g_{+1} + r\sqrt{2}. \tag{36}$$

The quantities on the left-hand side of this equation are functions of θ only, while the quantities on the right-hand side are functions of r only. Accordingly, the two sides must separately equal a constant. Since we are interested only in *some* particular integral, we can set each side of equation (36) to be equal to zero. We thus obtain the equations

$$\mathscr{L}_0^\dagger f_{+1} = \sqrt{2}.ia\, S_{+1}\cos\theta, \tag{37}$$

and

$$\mathscr{D} g_{+1} = -\sqrt{2}.rR_{+1}. \tag{38}$$

The solutions of these equations, as may be readily verified by making use of the relations (21) and (22), are

$$f_{+1}(\theta) = \sqrt{2}\frac{ia}{B}[(\cos\theta)\mathscr{L}_1^\dagger S_{-1} + (\sin\theta) S_{-1}] \tag{39}$$

and

$$g_{+1}(r) = -2\sqrt{2}.(r\mathscr{D}R_{-1} - R_{-1}). \tag{40}$$

The general solution of equation (29) is, therefore, given by

$$F_{+1} = f_{+1}R_{+1} + \mathscr{D}P_{+1} \quad \text{and} \quad G_{+1} = g_{+1}S_{+1} + \mathscr{L}_0^\dagger P_{+1}. \tag{41}$$

Similarly, we find that the general solution of equation (30) is given by

$$F_{-1} = f_{-1}\frac{R_{-1}}{\Delta} + \mathscr{D}^\dagger P_{-1} \quad \text{and} \quad G_{-1} = g_{-1}S_{-1} + \mathscr{L}_0 P_{-1}, \tag{42}$$

where P_{-1} is an arbitrary function of r and θ, and

$$f_{-1}(\theta) = 2\sqrt{2}.ia[(\cos\theta)\mathscr{L}_1 S_{+1} + (\sin\theta) S_{+1}] \tag{43}$$

and

$$g_{-1}(r) = -\frac{\sqrt{2}}{B}(r\mathscr{D}^\dagger\Delta R_{+1} - \Delta R_{+1}). \tag{44}$$

We may note here for future reference that $f_{-1}(\theta)$ and $g_{-1}(r)$ are solutions of the equations

$$\mathscr{L}_0 f_{-1} = 2\sqrt{2}.Bia\, S_{-1}\cos\theta, \tag{45}$$

and

$$\mathscr{D}^\dagger g_{-1} = -\frac{2\sqrt{2}}{\Delta}BrR_{-1}. \tag{46}$$

Equations (41) and (42) provide four equations which we can solve for the components of A_μ. We find

$$A_r = \tfrac{1}{2}\left(R_{+1} f_{+1} + \frac{R_{-1}}{\Delta} f_{-1}\right) + \tfrac{1}{2}(\mathscr{D} P_{+1} + \mathscr{D}^+ P_{-1}), \tag{47}$$

$$A_\theta = \tfrac{1}{2}(g_{+1} S_{+1} + g_{-1} S_{-1}) + \tfrac{1}{2}(\mathscr{L}_0^\dagger P_{+1} + \mathscr{L}_0 P_{-1}), \tag{48}$$

$$\rho^2 A_t = \tfrac{1}{2}(\Delta R_{+1} f_{+1} - R_{-1} f_{-1}) + \tfrac{1}{2} ia (g_{+1} S_{+1} - g_{-1} S_{-1}) \sin\theta$$
$$+ \tfrac{1}{2}\Delta(\mathscr{D} P_{+1} - \mathscr{D}^+ P_{-1}) + \tfrac{1}{2} ia (\mathscr{L}_0^\dagger P_{+1} - \mathscr{L}_0 P_{-1}) \sin\theta, \tag{49}$$

and

$$\rho^2 A_\phi = -\tfrac{1}{2} a (\Delta R_{+1} f_{+1} - R_{-1} f_{-1}) \sin^2\theta - \tfrac{1}{2} i (r^2 + a^2)(g_{+1} S_{+1} - g_{-1} S_{-1}) \sin\theta$$
$$- \tfrac{1}{2} a \Delta(\mathscr{D} P_{+1} - \mathscr{D}^\dagger P_{-1}) \sin^2\theta - \tfrac{1}{2} i (r^2 + a^2)(\mathscr{L}_0^\dagger P_{+1} - \mathscr{L}_0 P_{-1}) \sin\theta. \tag{50}$$

4. The verification of the solution and the freedom for a choice of gauge

It remains to verify that the solutions for the components of A_μ given in equations (47)–(50) are consistent with the solution for ϕ_0 implied by Teukolsky's solution for ϕ_{+1} and ϕ_{-1}. For this purpose, we shall first obtain an explicit solution for ϕ_0 by directly integrating the Newman–Penrose equations.

With the solutions for ϕ_{+1} and ϕ_{-1} given in equations (16), equations (9) and (11) can be written in the forms

$$\frac{R_{+1}}{\bar\rho^* \sqrt{2}}\left(\mathscr{L}_1 - \frac{ia\sin\theta}{\bar\rho^*}\right) S_{+1} = +\left(\mathscr{D} + \frac{2}{\bar\rho^*}\right)\phi_0 \tag{51}$$

and

$$\sqrt{2}\frac{BR_{-1}}{\Delta\bar\rho^*}\left(\mathscr{L}_1^\dagger - \frac{ia\sin\theta}{\bar\rho^*}\right) S_{-1} = -\left(\mathscr{D}^\dagger + \frac{2}{\bar\rho^*}\right)\phi_0. \tag{52}$$

Considering equation (51) and rewriting it, successively, in the manner

$$\mathscr{D}[(\bar\rho^*)^2 \phi_0] = \frac{1}{\sqrt{2}} R_{+1}[(r - ia\cos\theta)\mathscr{L}_1 S_{+1} - (ia\sin\theta) S_{+1}]$$
$$= \frac{1}{\sqrt{2}}\{(rR_{+1})\mathscr{L}_1 S_{+1} - ia[(\cos\theta)\mathscr{L}_1 S_{+1} + (\sin\theta) S_{+1}] R_{+1}\}, \tag{53}$$

and making use of equations (21), (38), and (43), we can write

$$\mathscr{D}[(\bar\rho^*)^2 \phi_0] = -\tfrac{1}{2}\mathscr{D}(g_{+1}\mathscr{L}_1 S_{+1} + f_{-1}\mathscr{D} R_{-1}). \tag{54}$$

This last equation can be integrated directly to give

$$(\bar\rho^*)^2 \phi_0 = -\tfrac{1}{2}(g_{+1}\mathscr{L}_1 S_{+1} + f_{-1}\mathscr{D} R_{-1}), \tag{55}$$

where, we have not included an arbitrary function of θ, we could have added to the solution on the right-hand side, since, as can be verified, any such addition to the solution would be incompatible with the other Newman–Penrose equation (10).

Similarly, we find from equations (12) and (52) that

$$(\bar{\rho}^*)^2\phi_0 = \tfrac{1}{2}(g_{-1}\mathscr{L}_1^\dagger S_{-1}+f_{+1}\mathscr{D}^\dagger\Delta R_{+1}). \tag{56}$$

A comparison of the solutions (55) and (56) leads to the interesting identity

$$f_{+1}\mathscr{D}^\dagger\Delta R_{+1}+f_{-1}\mathscr{D}R_{-1}+g_{+1}\mathscr{L}_1 S_{+1}+g_{-1}\mathscr{L}_1^\dagger S_{-1} = 0. \tag{57}$$

Now combining equations (55) and (56) we have the following alternative form for the solution for ϕ_0:

$$\phi_0 = \frac{1}{4(\bar{\rho}^*)^2}(f_{+1}\mathscr{D}^\dagger\Delta R_{+1}-f_{-1}\mathscr{D}R_{-1}-g_{+1}\mathscr{L}_1 S_{+1}+g_{-1}\mathscr{L}_1^\dagger S_{-1}). \tag{58}$$

Returning to the expression (33) for ϕ_0 and inserting in it the solutions for the components of A_μ given in equations (47)–(50), we find after some considerable simplifications that it can be reduced to the form

$$\begin{aligned}\phi_0 = \frac{1}{4\rho^2}\bigg[& -f_{+1}\left(\mathscr{D}^\dagger-\frac{2}{\bar{\rho}^*}\right)\Delta R_{+1}+f_{-1}\left(\mathscr{D}-\frac{2}{\bar{\rho}^*}\right)R_{-1} \\ & -g_{+1}\left(\mathscr{L}_1-\frac{2ia\sin\theta}{\bar{\rho}^*}\right)S_{+1}+g_{-1}\left(\mathscr{L}_1^\dagger-\frac{2ia\sin\theta}{\bar{\rho}^*}\right)S_{-1} \\ & -\left(\mathscr{D}^\dagger-\frac{2}{\bar{\rho}^*}\right)(\Delta\mathscr{D}P_{+1})+\left(\mathscr{D}-\frac{2}{\bar{\rho}^*}\right)(\Delta\mathscr{D}^\dagger P_{-1}) \\ & -\left(\mathscr{L}_1-\frac{2ia\sin\theta}{\bar{\rho}^*}\right)\mathscr{L}_0^\dagger P_{+1}+\left(\mathscr{L}_1^\dagger-\frac{2ia\sin\theta}{\bar{\rho}^*}\right)\mathscr{L}_0 P_{-1}\bigg].\end{aligned} \tag{59}$$

Equating now the two expressions (58) and (59) for ϕ_0 which we have found – one from a direct integration of the equations of Newman & Penrose and the other from the solution for A_μ which we have derived – we find that all the terms except those in P_{+1} and P_{-1} cancel; and the equality of the two expressions requires only

$$\begin{aligned}&\left(\mathscr{D}^\dagger-\frac{2}{\bar{\rho}^*}\right)(\Delta\mathscr{D}P_{+1})-\left(\mathscr{D}-\frac{2}{\bar{\rho}^*}\right)(\Delta\mathscr{D}^\dagger P_{-1}) \\ &\qquad +\left(\mathscr{L}_1-\frac{2ia\sin\theta}{\bar{\rho}^*}\right)\mathscr{L}_0^\dagger P_{+1}-\left(\mathscr{L}_1^\dagger-\frac{2ia\sin\theta}{\bar{\rho}^*}\right)\mathscr{L}_0 P_{-1} = 0.\end{aligned} \tag{60}$$

Equation (60) can be satisfied, trivially, by the choice

$$P_{+1} = P_{-1} = 0. \tag{61}$$

This choice is equivalent to a choice of gauge; in this gauge, the solutions, (47)–(50), for the components of the vector potential are particularly simple; they are given by

$$A_r = \tfrac{1}{2}\left(R_{+1}f_{+1}+\frac{R_{-1}}{\Delta}f_{-1}\right), \quad A_\theta = \tfrac{1}{2}(g_{+1}S_{+1}+g_{-1}S_{-1}), \tag{62}$$

$$\rho^2 A_t = \tfrac{1}{2}(\Delta R_{+1}f_{+1}-R_{-1}f_{-1})+\tfrac{1}{2}ia(g_{+1}S_{+1}-g_{-1}S_{-1})\sin\theta, \tag{63}$$

$$\rho^2 A_\phi = -\tfrac{1}{2}a(\Delta R_{+1}f_{+1}-R_{-1}f_{-1})\sin^2\theta-\tfrac{1}{2}i(r^2+a^2)(g_{+1}S_{+1}-g_{-1}S_{-1})\sin\theta. \tag{64}$$

However, in general, the terms in P_{+1} and P_{-1} must be included, as they are in equations (47)–(50). The arbitrariness in the solutions manifested by the appearance of the functions P_{+1} and P_{-1} in the solutions is clearly derived from the freedom we have in the choice of gauge. But this freedom is restricted by equation (60); in the alternative form,

$$\mathscr{D}^{\dagger}\frac{\Delta\mathscr{D}P_{+1}}{(\bar{\rho}^*)^2}+\mathscr{L}_1\frac{\mathscr{L}_0^{\dagger}P_{+1}}{(\bar{\rho}^*)^2}-\mathscr{D}\frac{\Delta\mathscr{D}^{\dagger}P_{-1}}{(\bar{\rho}^*)^2}-\mathscr{L}_1^{\dagger}\frac{\mathscr{L}_0P_{-1}}{(\bar{\rho}^*)^2}=0,$$

the equation has the appearance of a Coulomb gauge.

5. Concluding remarks

The direct and simple manner, in which it has been possible to specify the vector potential of a vacuum Maxwell-field in Kerr geometry in terms of Teukolsky's solution for ϕ_{+1} and ϕ_{-1}, suggests that a similar procedure should enable us to specify the gravitational perturbations of the Kerr metric coefficients in terms of Teukolsky's solutions for ϕ_{+2} and ϕ_{-2} (belonging to the spin weights $s = \pm 2$). This latter problem is now under consideration; it will be the subject of a forthcoming communication.

I am greatly indebted to Dr John L. Friedman and Dr Robert M. Wald for many clarifying discussions.

The research reported in this paper has in part been supported by the National Science Foundation under grant MPS 74–17456 with the University of Chicago.

References

Boyer, R. H. & Lindquist, R. W. 1967 *J. Math. Phys.* **8**, 265.
Chrzanowski, P. L. 1975 *Phys. Rev.* D **11**, 2042.
Cohen, J. M. & Kegeles, L. S. 1974 *Phys. Rev.* D **10**, 1070.
Kinnersley, W. 1969 *J. Math. Phys.* **10**, 1195.
Newman, E. T. & Penrose, R. 1963 *J. Math. Phys.* **3**, 566.
Starobinsky, A. A. 1973 *Zh. Eksp. Teor. Fiz.* **64**, 48 (also *Sov. Phys. – J.E.T.P.*, 1974, **37**, 28).
Starobinsky, A. A. & Churilov, S. M. 1973 *Zh. Eksp. Teor. Fiz.* **65**, 3 (also *Sov. Phys. - J.E.T.P.*, 1974, **38**, 1).
Teukolsky, S. A. 1972 *Phys. Rev. Lett.* **29**, 114.
Teukolsky, S. A. 1973 *Astrophys. J.* **185**, 635.
Teukolsky, S. A. & Press, W. H. 1974 *Astrophys. J.* **193**, 443.

8

The solution of Dirac's equation in Kerr geometry

BY S. CHANDRASEKHAR, F.R.S.
The School of Natural Sciences, The Institute for Advanced Study, Princeton, New Jersey, 08540‡

(*Received* 21 *April* 1976)

Dirac's equation for the electron in Kerr geometry is separated; and the general solution is expressed as a superposition of solutions derived from a purely radial and a purely angular equation.

1. INTRODUCTION

Teukolsky's (1972) separation of the variables of the equations governing the electromagnetic, the gravitational, and the two component neutrino-field perturbations of a Kerr black hole has been central to much of the later developments. But the lack of a similar separation of the variables of Dirac's equation for the electron has been an obstacle to progress along many desired directions (particularly, for the treatment of massive fields in the context of Hawking's (1975) quantal process of the evaporation of black holes). In this short paper, we shall show that Dirac's equation can also be separated and the solution expressed in terms of certain radial and angular functions satisfying decoupled equations; in consequence problems associated with an electron in the vicinity of Kerr black holes become amenable to treatment.

2. DIRAC'S EQUATION IN THE NEWMAN–PENROSE FORMALISM

In Penrose's (1972) spinor notation, the four components of the wave function which satisfy Dirac's equation are represented by two spinors P^A and Q^A (say); and Dirac's equation (in the units in which $c = \hbar = 1$) is written in the form

$$\nabla_{AA'}P^A + \mathrm{i}\mu_{\mathrm{e}}\bar{Q}_{A'} = 0 \tag{1}$$

and

$$\nabla_{AA'}Q^A + \mathrm{i}\mu_{\mathrm{e}}\bar{P}_{A'} = 0, \tag{2}$$

where $2^{\frac{1}{2}}\mu_{\mathrm{e}}$ is the mass of the electron (in the chosen units) and $\nabla_{AA'}$ is the symbol for covariant differentiation.

Following Newman & Penrose (1962), we introduce a basis ζ_a^A for the spinor space and a corresponding basis $\bar{\zeta}_{a'}^{A'}$ for the conjugate space. To the spinor basis is associated, at each point of the space-time, a null tetrad, $(\boldsymbol{l}, \boldsymbol{n}, \boldsymbol{m}, \overline{\boldsymbol{m}})$ satisfying the orthogonality relations, $\boldsymbol{l}\cdot\boldsymbol{n} = 1$, $\boldsymbol{m}\cdot\overline{\boldsymbol{m}} = -1$, and $\boldsymbol{l}\cdot\boldsymbol{m} = \boldsymbol{l}\cdot\overline{\boldsymbol{m}} = \boldsymbol{n}\cdot\boldsymbol{m} = \boldsymbol{n}\cdot\overline{\boldsymbol{m}} = 0$.

‡ Permanent address: the University of Chicago, Chicago, Illinois, 60637.
Reprinted with minor corrections.

The covariant derivative of a spinor ξ^A can then be expressed in terms of its components along the spinor basis, ζ_a^A, in the manner

$$\zeta_a^A \zeta_{a'}^{A'} \zeta_B^b \nabla_{AA'} \xi^B = \partial_{aa'} \xi^b + \Gamma^b_{caa'} \xi^c, \tag{3}$$

where $\Gamma^b_{caa'}$ are the spin coefficients and

$$\partial_{00'} = \mathrm{D}, \quad \partial_{11'} = \Delta, \quad \partial_{01'} = \delta, \quad \text{and} \quad \partial_{10'} = \bar{\delta} \tag{4}$$

are the directional derivatives along $\boldsymbol{l}$, $\boldsymbol{n}$, $\boldsymbol{m}$, and $\bar{\boldsymbol{m}}$.

Writing out, in accordance with equation (3), the components of equation (1), we obtain the pair of equations

$$(\mathrm{D} + \Gamma_{1000'} - \Gamma_{0010'}) P^0 + (\bar{\delta} + \Gamma_{1100'} - \Gamma_{0110'}) P^1 - \mathrm{i}\mu_e \bar{Q}^{1'} = 0 \tag{5}$$

and

$$(\delta + \Gamma_{1001'} - \Gamma_{0011'}) P^0 + (\Delta + \Gamma_{1101'} - \Gamma_{0111'}) P^1 + \mathrm{i}\mu_e \bar{Q}^{0'} = 0. \tag{6}$$

With the explicit nomenclature (Newman & Penrose 1962, equation (4.1*b*)) for the spin coefficients, equations (5) and (6) take the forms

$$(\mathrm{D} + \epsilon - \rho) P^0 + (\bar{\delta} + \pi - \alpha) P^1 = \mathrm{i}\mu_e \bar{Q}^{1'} \tag{7}$$

and

$$(\delta + \beta - \tau) P^0 + (\Delta + \mu - \gamma) P^1 = -\mathrm{i}\mu_e \bar{Q}^{0'}. \tag{8}$$

Equation (2) provides a similar pair of equations for Q^0 and Q^1. Letting

$$F_1 = P^0, \quad F_2 = P^1, \quad G_1 = \bar{Q}^{1'}, \quad \text{and} \quad G_2 = -\bar{Q}^{0'}, \tag{9}$$

we obtain the following set of four coupled equations:‡

$$(\mathrm{D} + \epsilon - \rho) F_1 + (\bar{\delta} + \pi - \alpha) F_2 = \mathrm{i}\mu_e G_1, \tag{10}$$

$$(\Delta + \mu - \gamma) F_2 + (\delta + \beta - \tau) F_1 = \mathrm{i}\mu_e G_2, \tag{11}$$

$$(\mathrm{D} + \bar{\epsilon} - \bar{\rho}) G_2 - (\delta + \bar{\pi} - \bar{\alpha}) G_1 = \mathrm{i}\mu_e F_2, \tag{12}$$

$$(\Delta + \bar{\mu} - \bar{\gamma}) G_1 - (\bar{\delta} + \bar{\beta} - \bar{\tau}) G_2 = \mathrm{i}\mu_e F_1. \tag{13}$$

3. Dirac's equation in Kerr geometry

In writing equations (10)–(13) appropriately in Kerr geometry and in the Boyer–Lindquist (1967) coordinates, we shall adopt for basis the null tetrad of Kinnersley (1969):

$$l^\mu = \frac{1}{\Delta}(r^2 + a^2, +\Delta, 0, a),$$

$$n^\mu = \frac{1}{2\rho^2}(r^2 + a^2, -\Delta, 0, a),$$

$$m^\mu = \frac{1}{\bar{\rho}\sqrt{2}}(\mathrm{i}a \sin\theta, 0, 1, \mathrm{i} \operatorname{cosec}\theta), \tag{14}$$

‡ I had originally derived these equations by starting with Dirac's equation in an orthonormal-tetrad frame (as given by Dirac 1958) and transforming it to a null-tetrad frame appropriate for the Newman–Penrose formalism. I am greatly indebted to Dr John L. Friedman for showing me how the equations can be derived directly and, indeed, very simply.

where
$$\Delta = r^2+a^2-2Mr, \quad \rho^2 = r^2+a^2\cos^2\theta,$$
$$\bar{\rho} = r+\mathrm{i}a\cos\theta, \quad \text{and} \quad \bar{\rho}^* = r-\mathrm{i}a\cos\theta. \tag{15}‡$$

We shall assume that the four components of the wave function have a dependence on the time t and the azimuthal angle φ given by
$$\mathrm{e}^{\mathrm{i}(\sigma t+m\varphi)}, \tag{16}$$
where $\sigma > 0$ (by convention). For this dependence on t and φ, the directional derivatives (4) have the values
$$\mathrm{D} = \mathscr{D}_0, \quad \Delta = -\frac{\Delta}{2\rho^2}\mathscr{D}_0^\dagger, \quad \delta = \frac{1}{\bar{\rho}\sqrt{2}}\mathscr{L}_0^\dagger, \quad \text{and} \quad \bar{\delta} = \frac{1}{\bar{\rho}^*\sqrt{2}}\mathscr{L}_0, \tag{17}$$
where
$$\mathscr{D}_n = \partial_r+\frac{\mathrm{i}K}{\Delta}+2n\frac{(r-M)}{\Delta}, \quad \mathscr{D}_n^\dagger = \partial_r-\frac{\mathrm{i}K}{\Delta}+2n\frac{(r-M)}{\Delta}, \tag{18}$$
$$\mathscr{L}_n = \partial_\theta+Q+n\cot\theta, \quad \mathscr{L}_n^\dagger = \partial_\theta-Q+n\cot\theta, \tag{19}$$
$$K = (r^2+a^2)\sigma+am, \quad \text{and} \quad Q = a\sigma\sin\theta+m\operatorname{cosec}\theta. \tag{20}$$

It should be noted that while $\mathscr{D}_n$ and $\mathscr{D}_n^\dagger$ are purely radial operators, $\mathscr{L}_n$ and $\mathscr{L}_n^\dagger$ are purely angular operators.

The spin coefficients appropriate for the chosen basis are known: they are listed, for example, by Teukolsky (1973, equation (4.5)). Substituting, then, for the spin coefficients and the directional derivatives, we find that equations (10)–(13) reduce to
$$\left(\mathscr{D}_0+\frac{1}{\bar{\rho}^*}\right)F_1+\frac{1}{\bar{\rho}^*\sqrt{2}}\mathscr{L}_{\frac{1}{2}}F_2 = +\mathrm{i}\mu_e G_1, \tag{21}$$
$$\frac{\Delta}{2\rho^2}\mathscr{D}_{\frac{1}{2}}^\dagger F_2-\frac{1}{\bar{\rho}\sqrt{2}}\left(\mathscr{L}_{\frac{1}{2}}^\dagger+\frac{\mathrm{i}a\sin\theta}{\bar{\rho}^*}\right)F_1 = -\mathrm{i}\mu_e G_2, \tag{22}$$
$$\left(\mathscr{D}_0+\frac{1}{\bar{\rho}}\right)G_2-\frac{1}{\bar{\rho}\sqrt{2}}\mathscr{L}_{\frac{1}{2}}^\dagger G_1 = +\mathrm{i}\mu_e F_2, \tag{23}$$
$$\frac{\Delta}{2\rho^2}\mathscr{D}_{\frac{1}{2}}^\dagger G_1+\frac{1}{\bar{\rho}^*\sqrt{2}}\left(\mathscr{L}_{\frac{1}{2}}-\frac{\mathrm{i}a\sin\theta}{\bar{\rho}}\right)G_2 = -\mathrm{i}\mu_e F_1. \tag{24}$$

4. The separation of the variables and the reduction of equations (21)–(24)

The form of equations (21–(24) suggest that in place of F_1 and G_2 we define
$$f_1 = \bar{\rho}^*F_1 = (r-\mathrm{i}a\cos\theta)F_1$$
and
$$g_2 = \bar{\rho}G_2 = (r+\mathrm{i}a\cos\theta)G_2. \tag{25}$$

‡ The spin coefficient ρ in equations (7), (10), and (11) should not be confused with ρ^2, $\bar{\rho}$, and $\bar{\rho}^*$ defined in the context of the Kerr metric; ρ with the meaning of the spin coefficient occurs only in equations (7), (10), and (11) and nowhere else in the paper.

Also, writing f_2 and g_1 in place of F_2 and G_1 (for symmetry in form of the resulting equations) we find that equations (21)–(24) become

$$\mathscr{D}_0 f_1 + 2^{-\frac{1}{2}}\mathscr{L}_{\frac{1}{2}} f_2 = +(\mathrm{i}\mu_e r + a\mu_e\cos\theta)\, g_1, \tag{26}$$

$$\varDelta\mathscr{D}^{\dagger}_{\frac{1}{2}} f_2 - 2^{+\frac{1}{2}}\mathscr{L}^{\dagger}_{\frac{1}{2}} f_1 = -2(\mathrm{i}\mu_e r + a\mu_e\cos\theta)\, g_2, \tag{27}$$

$$\mathscr{D}_0 g_2 - 2^{-\frac{1}{2}}\mathscr{L}^{\dagger}_{\frac{1}{2}} g_1 = +(\mathrm{i}\mu_e r - a\mu_e\cos\theta)\, f_2, \tag{28}$$

$$\varDelta\mathscr{D}^{\dagger}_{\frac{1}{2}} g_1 + 2^{+\frac{1}{2}}\mathscr{L}_{\frac{1}{2}} g_2 = -2(\mathrm{i}\mu_e r - a\mu_e\cos\theta)\, f_1. \tag{29}$$

It is now apparent that the variables can be separated by writing

$$f_1(r,\theta) = R_{-\frac{1}{2}}(r)\, S_{-\frac{1}{2}}(\theta), \quad f_2(r,\theta) = R_{+\frac{1}{2}}(r)\, S_{+\frac{1}{2}}(\theta),$$
$$g_1(r,\theta) = R_{+\frac{1}{2}}(r)\, S_{-\frac{1}{2}}(\theta), \quad \text{and} \quad g_2(r,\theta) = R_{-\frac{1}{2}}(r)\, S_{+\frac{1}{2}}(\theta), \tag{30}$$

where $R_{\pm\frac{1}{2}}(r)$ and $S_{\pm\frac{1}{2}}(\theta)$ are functions only of r and θ, respectively. For with these forms for the solutions, equations (26)–(29) become

$$(\mathscr{D}_0 R_{-\frac{1}{2}} - \mathrm{i}\mu_e r R_{+\frac{1}{2}})\, S_{-\frac{1}{2}} + [2^{-\frac{1}{2}}\mathscr{L}_{\frac{1}{2}} S_{+\frac{1}{2}} - (\alpha\mu_e\cos\theta)\, S_{-\frac{1}{2}}]\, R_{+\frac{1}{2}} = 0, \tag{31}$$

$$(\varDelta\mathscr{D}^{\dagger}_{\frac{1}{2}} R_{+\frac{1}{2}} + 2\mathrm{i}\mu_e R_{-\frac{1}{2}})\, S_{+\frac{1}{2}} - [2^{+\frac{1}{2}}\mathscr{L}^{\dagger}_{\frac{1}{2}} S_{-\frac{1}{2}} - 2(a\mu_e\cos\theta)\, S_{+\frac{1}{2}}]\, R_{-\frac{1}{2}} = 0, \tag{32}$$

$$(\mathscr{D}_0 R_{-\frac{1}{2}} - \mathrm{i}\mu_e r R_{+\frac{1}{2}})\, S_{+\frac{1}{2}} - [2^{-\frac{1}{2}}\mathscr{L}^{\dagger}_{\frac{1}{2}} S_{-\frac{1}{2}} - (a\mu_e\cos\theta)\, S_{+\frac{1}{2}}]\, R_{+\frac{1}{2}} = 0, \tag{33}$$

$$(\varDelta\mathscr{D}^{\dagger}_{\frac{1}{2}} R_{+\frac{1}{2}} + 2\mathrm{i}\mu_e R_{-\frac{1}{2}})\, S_{-\frac{1}{2}} + [2^{+\frac{1}{2}}\mathscr{L}_{\frac{1}{2}} S_{+\frac{1}{2}} - 2(a\mu_e\cos\theta)\, S_{-\frac{1}{2}}]\, R_{-\frac{1}{2}} = 0. \tag{34}$$

These equations imply that

$$\mathscr{D}_0 R_{-\frac{1}{2}} - \mathrm{i}\mu_e r R_{+\frac{1}{2}} = \lambda_1 R_{+\frac{1}{2}}; \qquad 2^{-\frac{1}{2}}\mathscr{L}_{\frac{1}{2}} S_{+\frac{1}{2}} - (a\mu_e\cos\theta)\, S_{-\frac{1}{2}} = -\lambda_1 S_{-\frac{1}{2}}, \tag{35}$$

$$\varDelta\mathscr{D}^{\dagger}_{\frac{1}{2}} R_{+\frac{1}{2}} + 2\mathrm{i}\mu_e R_{-\frac{1}{2}} = \lambda_2 R_{-\frac{1}{2}}; \quad 2^{+\frac{1}{2}}\mathscr{L}^{\dagger}_{\frac{1}{2}} S_{-\frac{1}{2}} - 2(a\mu_e\cos\theta)\, S_{+\frac{1}{2}} = +\lambda_2 S_{+\frac{1}{2}}, \tag{36}$$

$$\mathscr{D}_0 R_{-\frac{1}{2}} - \mathrm{i}\mu_e r R_{+\frac{1}{2}} = \lambda_3 R_{+\frac{1}{2}}; \qquad 2^{-\frac{1}{2}}\mathscr{L}^{\dagger}_{\frac{1}{2}} S_{-\frac{1}{2}} - (a\mu_e\cos\theta)\, S_{+\frac{1}{2}} = +\lambda_3 S_{+\frac{1}{2}}, \tag{37}$$

$$\varDelta\mathscr{D}^{\dagger}_{\frac{1}{2}} R_{+\frac{1}{2}} + 2\mathrm{i}\mu_e R_{-\frac{1}{2}} = \lambda_4 R_{-\frac{1}{2}}; \quad 2^{+\frac{1}{2}}\mathscr{L}_{\frac{1}{2}} S_{+\frac{1}{2}} - 2(a\mu_e\cos\theta)\, S_{-\frac{1}{2}} = -\lambda_4 S_{-\frac{1}{2}}, \tag{38}$$

where $\lambda_1, \ldots, \lambda_4$ are four constants of separation. However, it is manifest that the consistency equations (35)–(38) requires that

$$\lambda_1 = \lambda_3 = \tfrac{1}{2}\lambda_2 = \tfrac{1}{2}\lambda_4 = \lambda \quad \text{(say)}. \tag{39}$$

We are thus left with the two pairs of equations

$$\mathscr{D}_0 R_{-\frac{1}{2}} = (\lambda + \mathrm{i}\mu_e r)\, R_{+\frac{1}{2}},$$
$$\varDelta\mathscr{D}^{\dagger}_{\frac{1}{2}} R_{+\frac{1}{2}} = 2(\lambda - \mathrm{i}\mu_e r) R_{-\frac{1}{2}}; \tag{40}$$

and

$$\mathscr{L}_{\frac{1}{2}} S_{+\frac{1}{2}} = -2^{\frac{1}{2}}(\lambda - a\mu_e\cos\theta)\, S_{-\frac{1}{2}},$$
$$\mathscr{L}^{\dagger}_{\frac{1}{2}} S_{-\frac{1}{2}} = +2^{\frac{1}{2}}(\lambda + a\mu_e\cos\theta)\, S_{+\frac{1}{2}}. \tag{41}$$

From equations (40) and (41) it follows that

$$\varDelta^{\frac{1}{2}} R_{+\frac{1}{2}} \quad \text{and} \quad 2^{\frac{1}{2}} R_{-\frac{1}{2}} \tag{42}$$

are proportional to complex conjugate functions and

$$S_{+\frac{1}{2}}(\theta) \quad \text{is proportional to} \quad S_{-\frac{1}{2}}(\pi - \theta). \tag{43}$$

We can eliminate $S_{+\frac{1}{2}}$ from the pair of equations (41) to obtain an equation for $S_{-\frac{1}{2}}$; thus,

$$\left[\mathscr{L}_{\frac{1}{2}}\mathscr{L}^{\dagger}_{\frac{1}{2}}+\frac{a\mu_{\mathrm{e}}\sin\theta}{\lambda+a\mu_{\mathrm{e}}\cos\theta}\mathscr{L}^{\dagger}_{\frac{1}{2}}+2(\lambda^2-a^2\mu_{\mathrm{e}}^2\cos^2\theta)\right]S_{-\frac{1}{2}}=0; \tag{44}$$

and λ^2 is a characteristic value of this equation determined by the condition that $S_{+\frac{1}{2}}$ (and, therefore, also $S_{-\frac{1}{2}}$) is regular at $\theta = 0$ and $\theta = \pi$.

Similarly, we can eliminate $R_{+\frac{1}{2}}$ from the pair of equations (40) to obtain an equation for $R_{-\frac{1}{2}}$; thus,

$$\left[\Delta\mathscr{D}^{\dagger}_{\frac{1}{2}}\mathscr{D}_0-\frac{\mathrm{i}\mu_{\mathrm{e}}\Delta}{\lambda+\mathrm{i}\mu_{\mathrm{e}}r}\mathscr{D}_0-2(\lambda^2+\mu_{\mathrm{e}}^2r^2)\right]R_{-\frac{1}{2}}=0. \tag{45}$$

We have thus reduced the solution of Dirac's equation in Kerr geometry to the solution of a pair of decoupled equations; and the general solution can be expressed as a linear superposition of the various solutions belonging to the different characteristic values of λ^2.

It may be noted that by setting $\mu_{\mathrm{e}} = 0$ in equations (44) and (45) we recover Teukolsky's equations for the two-component neutrino-field; and, similarly, by setting $M = 0$, we obtain the solution of Dirac's equation in oblate spheroidal coordinates in Minkowskian space.

I am indebted to Dr S. A. Teukolsky, Dr N. J. M. Woodhouse, and most particularly to Dr J. L. Friedman for helpful remarks.

The present investigation was mostly carried out while the author was a visiting member in the School of Natural Sciences at the Institute for Advanced Study (Princeton) and supported by the National Science Foundation grant No. GP. 40768X. At the University of Chicago, the author's research is supported by the National Science Foundation grant MPS 74-17456.

References

Boyer, R. H. & Lindquist, R. W. 1967 *J. Math. Phys.* **14**, 7.

Dirac, P. A. M. 1958 *Max Planck Festschrift 1958, pp.* 339–344. Berlin: Veb. Deutscher Verlag der Wissenschaft.

Hawking, S. W. 1975 *Commun. Math. Phys.* **43**, 199; also in *Quantum gravity: an Oxford symposium* (eds C. J. Isham, R. Penrose & D. W. Sciama), pp. 78–135. Oxford: Clarendon Press.

Kinnersley, W. 1969 *J. Math. Phys.* **10**, 1195.

Newman, E. T. & Penrose, R. 1962 *J. Math. Phys.* **3**, 566.

Penrose, R. 1968 *Battelle Rencontres* (eds C. M. DeWitt & J. A. Wheeler), p. 121. New York: W. A. Benjamin, Inc.

Teukolsky, S. A. 1972 *Phys. Rev. Lett.* **29**, 114.

Teukolsky, S. A. 1973 *Astrophys. J.* **185**, 635.

9

On the reflexion and transmission of neutrino waves by a Kerr black hole

By S. Chandrasekhar, F.R.S.
University of Chicago, Chicago, Illinois, 60637

and S. Detweiler
University of Maryland, College Park, Maryland, 20742

(*Received 2 June* 1976)

The equations governing the two-component neutrino are reduced to the form of a one-dimensional wave equation. And it is shown how the absence of super-radiance (i.e. a reflexion coefficient in excess of one) for incident neutrino waves and its manifestation for incident electromagnetic and gravitational waves (of suitable frequencies) emerge very naturally from the character of the respective potential barriers that surround the Kerr black hole.

1. Introduction

The equations governing the two-component neutrino, in Kerr geometry, were separated by Teukolsky (1973) and by Unruh (1973, 1974). And Unruh further showed, on some very general grounds, that the neutrinos, unlike the photons and the gravitons, will not manifest the phenomenon of super-radiance. In this paper, we shall show how this difference between the neutrinos (with spin $\frac{1}{2}$) and the photons and the gravitons (with spins 1 and 2, respectively) arises very naturally from the differing characters of the potential barriers presented to them by the Kerr black hole.

The present paper is closely related to two recent ones (Chandrasekhar 1976*a* and Chandrasekhar & Detweiler 1976; these papers will be referred to hereafter as (I) and (II), respectively) devoted to the electromagnetic and the gravitational perturbations of the Kerr black hole; and it provides a supplement to them.

2. The equations of the problem

The equations governing the two-component neutrino can be obtained by putting $s = \pm\frac{1}{2}$ in Teukolsky's general equations

$$\Delta^{-s}\frac{\mathrm{d}}{\mathrm{d}r}\left(\Delta^{s+1}\frac{\mathrm{d}R}{\mathrm{d}r}\right)+\frac{1}{\Delta}[K^2+2\mathrm{i}s(r-M)\,K-\Delta(4\mathrm{i}s\sigma r+\lambda)]\,R = 0, \tag{1}$$

where
$$K = (r^2+a^2)\,\sigma+am, \quad \Delta = r^2+a^2-2Mr, \tag{2}$$

$$\lambda = \tau-s(s+1), \quad \tau = E+a^2\sigma^2+2a\sigma m, \tag{3}$$

and E is the separation constant which occurs in the equation,

$$\frac{1}{\sin\theta}\frac{\mathrm{d}}{\mathrm{d}\theta}\left(\sin\theta\frac{\mathrm{d}S}{\mathrm{d}\theta}\right)+\left(a^2\sigma^2\cos^2\theta+2a\sigma s\cos\theta-\frac{m^2+s^2+2ms\cos\theta}{\sin^2\theta}+E\right)S=0, \quad (4)$$

governing the angular dependence of the perturbation. In deriving equations (1)–(4), it has been assumed that the t- and φ-dependences of the perturbation are given by

$$e^{i(\sigma t+m\varphi)} \quad (\sigma>0). \quad (5)$$

For the particular cases $s=\pm\frac{1}{2}$, the foregoing equations follow from the elimination of $R_{-\frac{1}{2}}$ (or $R_{+\frac{1}{2}}$) and $S_{-\frac{1}{2}}$ (or $S_{+\frac{1}{2}}$) from the two pairs of equations

$$\mathscr{D}_0 R_{-\frac{1}{2}}=\nu^{\frac{1}{2}}R_{+\frac{1}{2}}, \quad \Delta^{\frac{1}{2}}\mathscr{D}_0^\dagger\Delta^{\frac{1}{2}}R_{+\frac{1}{2}}=\nu^{\frac{1}{2}}R_{-\frac{1}{2}}; \quad (6)$$

and

$$\mathscr{L}_{\frac{1}{2}}S_{+\frac{1}{2}}=-\nu^{\frac{1}{2}}S_{-\frac{1}{2}}, \quad \mathscr{L}_{\frac{1}{2}}^\dagger S_{-\frac{1}{2}}=+\nu^{\frac{1}{2}}S_{+\frac{1}{2}}, \quad (7)$$

where

$$\mathscr{D}_0=\frac{\mathrm{d}}{\mathrm{d}r}+\mathrm{i}\frac{K}{\Delta}, \quad \mathscr{D}_0^\dagger=\frac{\mathrm{d}}{\mathrm{d}r}-\mathrm{i}\frac{K}{\Delta},$$

$$\left.\begin{aligned}\mathscr{L}_n&=\partial_\theta+(a\sigma\sin\theta+m\operatorname{cosec}\theta)+n\cot\theta,\\ \mathscr{L}_n^\dagger&=\partial_\theta-(a\sigma\sin\theta+m\operatorname{cosec}\theta)+n\cot\theta,\end{aligned}\right\} \quad (8)$$

and

$$\nu=\tau+\tfrac{1}{4}. \quad (9)$$

Equations (6) and (7) are special cases of the equations derived in the context of separating Dirac's equation in Kerr geometry (by setting $\mu_e=0$ and replacing $2\lambda^2$ by ν and $2^{\frac{1}{2}}R_{-\frac{1}{2}}$ by $R_{-\frac{1}{2}}$ in Chandrasekhar (1976*b*, equations (40) and (41)). Equations (6) and (7) are also equivalent to the equations derived by Unruh (1973, equations (7*a*) and (7*b*) with the correspondence $R_1\to\Delta^{\frac{1}{2}}R_{+\frac{1}{2}}$ and $R_2\to R_{-\frac{1}{2}}$).

Returning to equation (1) for the particular case $s=+\frac{1}{2}$, we apply the transformations introduced in (I) to eliminate the explicit appearance of m. Thus, letting

$$\phi=\Delta^{\frac{1}{4}}(r^2+\alpha^2)^{\frac{1}{2}}R_{+\frac{1}{2}}, \quad (10)$$

where

$$\alpha^2=a^2+am/\sigma, \quad (11)$$

and introducing, in place of r, the variable r_* defined by

$$\frac{\mathrm{d}}{\mathrm{d}r_*}=\frac{\Delta}{r^2+\alpha^2}\frac{\mathrm{d}}{\mathrm{d}r}, \quad (12)$$

we reduce equation (1), for $s=+\frac{1}{2}$, to the form (cf. I, equations (10) and (11))

$$\frac{\mathrm{d}^2\phi}{\mathrm{d}r_*^2}+\sigma^2\phi-\mathrm{i}\sigma\frac{2r\Delta-(r-M)(r^2+\alpha^2)}{(r^2+\alpha^2)^2}\phi-\left[\frac{(\tau-\frac{3}{4})\Delta}{(r^2+\alpha^2)^2}+G^2+\frac{\mathrm{d}G}{\mathrm{d}r_*}\right]\phi=0, \quad (13)$$

where

$$G=\frac{r-M}{2(r^2+\alpha^2)}+\frac{r\Delta}{(r^2+\alpha^2)^2}. \quad (14)$$

And finally, we bring equation (13) to the standard form of this theory, namely,

$$\Lambda^2Y+P\Lambda_-Y-QY=0, \quad (15)$$

where $$\Lambda_{\pm} = \frac{d}{dr_*} \pm i\sigma \quad \text{and} \quad \Lambda^2 = \frac{d^2}{dr_*^2} + \sigma^2, \tag{16}$$

by the substitution $$Y = \frac{\Delta^{\frac{1}{4}}}{(r^2+\alpha^2)^{\frac{1}{2}}}\phi = \Delta^{\frac{1}{2}} R_{+\frac{1}{2}}, \tag{17}$$

and find $$P = \frac{d}{dr_*}\left(\lg\frac{\rho^2}{\Delta^{\frac{1}{2}}}\right) \quad \text{and} \quad Q = \nu\frac{\Delta}{\rho^4}, \tag{18}$$

where $$\rho^2 = r^2+\alpha^2. \tag{19}$$

For later reference, we may note here that in terms of the present variables the relations (6) take the forms,

$$\Lambda_+ R_{-\frac{1}{2}} = \frac{(\nu\Delta)^{\frac{1}{2}}}{\rho^2}\Delta^{\frac{1}{2}} R_{+\frac{1}{2}} = \frac{(\nu\Delta)^{\frac{1}{2}}}{\rho^2} Y \tag{20}$$

and $$\Lambda_- \Delta^{\frac{1}{2}} R_{+\frac{1}{2}} = \Lambda_- Y = \frac{(\nu\Delta)^{\frac{1}{2}}}{\rho^2} R_{-\frac{1}{2}}. \tag{21}$$

3. The reduction to a one-dimensional wave equation

As in the earlier papers on the electromagnetic and the gravitational perturbations, we shall now seek a transformation of equation (15) to the form of a one-dimensional wave equation

$$\Lambda^2 Z = VZ, \tag{22}$$

where the potential V is bounded and of short range for frequencies

$$\sigma > \sigma_s \quad (= -am/2Mr_+)$$

for which the $r_*(r)$-relation is single-valued (cf. (I), § 3).

By methods which are by now standard, the required transformation can be accomplished by the substitutions (cf. (I), equations (31) and (32))

$$Y = fVZ + T\Lambda_+ Z \tag{23}$$

and $$\Lambda_- Y = -\beta M\frac{\Delta^{\frac{1}{2}}}{\rho^2}Z + R\Lambda_+ Z, \tag{24}$$

where f, V, T, βM, and R are functions to be determined consistently with the equations (cf. (I), equations (41)–(44))

$$fV = R - \frac{dT}{dr_*}, \tag{25}$$

$$\frac{d}{dr_*}\left(\frac{\rho^2}{\Delta^{\frac{1}{2}}}R\right) = \frac{\rho^2}{\Delta^{\frac{1}{2}}}(QT - 2i\sigma R) + \beta M, \tag{26}$$

$$R\left(R - \frac{dT}{dr_*}\right) + \frac{\Delta^{\frac{1}{2}}}{\rho^2}\beta MT = \frac{\Delta^{\frac{1}{2}}}{\rho^2}K, \tag{27}$$

and
$$RV - QfV = M\frac{\Delta^{\frac{1}{2}}}{\rho^2}\frac{d\beta}{dr_*},\tag{28}$$

where K is a constant. And the relations, inverse to (23) and (24), are

$$\frac{\Delta^{\frac{1}{2}}}{\rho^2}KZ = RY - T\Lambda_- Y,\tag{29}$$

and
$$K\Lambda_+ Z = \beta MY + \frac{\rho^2}{\Delta^{\frac{1}{2}}}fV\Lambda_- Y.\tag{30}$$

(*a*) *An explicit solution*

It can be verified by direct substitution that

$$T = 2i\sigma, \quad R = fV,\tag{31}$$

$$R = \frac{\Delta^{\frac{1}{2}}}{\rho^2}q_\pm, \quad \beta M = -2i\sigma\left(\nu\frac{\Delta^{\frac{1}{2}}}{\rho^2} - q_\pm\right),\tag{32}$$

$$K = -4\sigma^2 q_\pm, \quad \text{and} \quad q_\pm = \pm 2i\sigma\nu^{\frac{1}{2}},\tag{33}$$

represent solutions of equations (25)–(27). Equation (28) then gives

$$V_\pm = \nu\frac{\Delta}{\rho^4} \mp \nu^{\frac{1}{2}}\frac{d}{dr_*}\left(\frac{\Delta^{\frac{1}{2}}}{\rho^2}\right).\tag{34}$$

With the solutions given in equations (31)–(33), the relations (23), (24), (29), and (30) now take the forms

$$Y = \frac{\Delta^{\frac{1}{2}}}{\rho^2}q_\pm Z + 2i\sigma\Lambda_+ Z,\tag{35}$$

$$\frac{\rho^2}{\Delta^{\frac{1}{2}}}\Lambda_- Y = 2i\sigma\left(\nu\frac{\Delta^{\frac{1}{2}}}{\rho^2} - q_\pm\right)Z + q_\pm\Lambda_+ Z,\tag{36}$$

and
$$KZ = q_\pm Y - 2i\sigma\frac{\rho^2}{\Delta^{\frac{1}{2}}}\Lambda_- Y,\tag{37}$$

$$K\Lambda_+ Z = -2i\sigma\left(\nu\frac{\Delta^{\frac{1}{2}}}{\rho^2} - q_\pm\right)Y + q_\pm\Lambda_- Y.\tag{38}$$

The two solutions, which we shall distinguish by Z_+ and Z_- and which follow from equations (35)–(38) by choosing the upper or the lower sign (wherever there is a choice), are simply related. Thus, by substituting in the relation

$$-8i\sigma^3\nu^{\frac{1}{2}}Z_+ = 2i\sigma\nu^{\frac{1}{2}}Y - 2i\sigma\frac{\rho^2}{\Delta^{\frac{1}{2}}}\Lambda_- Y,\tag{39}$$

appropriate for Z_+, the expressions for Y and $\Lambda_- Y$ relating them to Z_-, we find

$$Z_+ = \frac{i}{\sigma}\left[\frac{(\nu\Delta)^{\frac{1}{2}}}{\rho^2}Z_- - \frac{dZ_-}{dr_*}\right].\tag{40}$$

Therefore, given a solution Z_-, belonging to the potential V_-, we can obtain a solution Z_+, belonging to the potential V_+, with the aid of the relation (40). In particular, for solutions which are bounded at infinity and at the horizon,

$$Z_+ \to -\frac{i}{\sigma}\frac{dZ_-}{dr_*} \quad \text{for} \quad r \to \infty \quad \text{and} \quad r \to r_+ + 0. \tag{41}$$

4. Reflexion and transmission coefficients for $\sigma > \sigma_s$

There is clearly no ambiguity in using the one-dimensional wave equation, with either of the two potentials,

$$V_\pm = \nu\frac{\Delta}{\rho^4} \mp \frac{(\nu\Delta)^{\frac{1}{2}}}{\rho^4}\left[(r-M) - \frac{2r\Delta}{\rho^2}\right], \tag{42}‡$$

given by equation (34), so long as $\sigma > \sigma_s$; for, then, the underlying $r_*(r)$-relation is single-valued and the potentials are moreover bounded and of short range. We can, accordingly, seek solutions of the wave equation with the potential V_-, for example, which have the asymptotic behaviours,

$$\begin{aligned} Z_- &\to e^{+i\sigma r_*} + A\,e^{-i\sigma r_*} \quad (r_* \to +\infty) \\ &\to B\,e^{+i\sigma r_*} \quad (r_* \to -\infty). \end{aligned} \tag{43}$$

With the aid of the relation (40), we can derive from these solutions Z_-, solutions Z_+, of the wave equation with the potential V_+, which have the asymptotic behaviours (cf. equation (41))

$$\begin{aligned} Z_+ &\to e^{+i\sigma r_*} - A\,e^{-i\sigma r_*} \quad (r_* \to +\infty) \\ &\to B\,e^{+i\sigma r_*} \quad (r_* \to -\infty). \end{aligned} \tag{44}$$

Hence, both potentials will lead to the same reflexion $\mathbb{R}$ and transmission $\mathbb{T}$ coefficients given by

$$\mathbb{R} = |A|^2 \quad \text{and} \quad \mathbb{T} = |B|^2, \tag{45}$$

and satisfying the conservation law

$$\mathbb{R} + \mathbb{T} = 1. \tag{46}$$

It is, however, necessary to verify that the reflexion and the transmission coefficients, as we have defined them in terms of the solutions of the wave equations governing Z, are in accord with the physical definition of the number-current of the neutrinos flowing down the black hole.

According to Unruh (1973, equation (10)) the conserved number-current, $\partial N/\partial t$, of the neutrinos is given by

$$\partial N/\partial t = 4\pi\{|\Delta^{\frac{1}{2}}R_{+\frac{1}{2}}|^2 - |R_{-\frac{1}{2}}|^2\}. \tag{47}$$

‡ When $a = 0$, these potentials reduce to

$$V_\pm = \nu\left(1-\frac{2M}{r}\right)\frac{1}{r^2} \pm \nu^{\frac{1}{2}}\left(1-\frac{2M}{r}\right)^{\frac{1}{2}}\left(1-\frac{3M}{r}\right)\frac{1}{r^2}.$$

These potentials, appropriate to Schwarzschild geometry, occur in an early investigation of Brill & Wheeler (1957) on the equations governing the neutrino in curved space-time.

(This expression differs in sign from the one given by Unruh to allow for our present convention with regard to 'outgoing' and 'ingoing' waves being the opposite of his.) Substituting for $\Delta^{\frac{1}{2}}R_{+\frac{1}{2}}$ and $R_{-\frac{1}{2}}$, in equation (47), in terms of Y, in accordance with equations (17) and (21), we have

$$\partial N/\partial t = 4\pi\{|Y|^2 - |\rho^2 \Lambda_- Y/(\nu\Delta)^{\frac{1}{2}}|^2\}. \tag{48}$$

Now inserting for Y and $\Lambda_- Y$ the expressions given in equations (35) and (36), we find, after some simplifications, that we are left with

$$\partial N/\partial t = 32\pi \mathrm{i}\sigma^3(Z\,\mathrm{d}Z^*/\mathrm{d}r_* - Z^*\,\mathrm{d}Z/\mathrm{d}r_*) = 32\pi\mathrm{i}\sigma^3\,W(Z, Z^*). \tag{49}$$

Thus, the number-current of the neutrinos is proportional to the Wronskian, $W(Z, Z^*)$, of the solutions, Z and its complex-conjugate Z^*, of the one-dimensional wave equations which they satisfy; it is a constant consistent with physical requirements.

Evaluating $W(Z, Z^*)$ for $r_* \to +\infty$ and $r_* \to -\infty$, for solutions having the asymptotic behaviours specified in equations (43) and (44), we recover our earlier definitions of the reflexion and the transmission coefficients and the conservation law they satisfy:

$$(\partial N/\partial t)_{+\infty} = 32\pi\sigma^4(1-\mathbb{R}) = 32\pi\sigma^4\mathbb{T} = (\partial N/\partial t)_{-\infty}. \tag{50}$$

Figures 1*a, b* illustrate the nature of the potentials, $V_\pm$, for frequencies σ in the range $\sigma_s < \sigma \leqslant \sigma_c\ (= -m/a)$. In this range of frequences, $\alpha^2 < 0$ but $0 \leqslant |\alpha| < r_+$; and $|\alpha| \to r_+ + 0$ as $\sigma \to \sigma_s + 0$. As a consequence of this last fact, the potential barrier presented to the oncoming neutrino waves increases without bound as $\sigma \to \sigma_s + 0$. But the singularity of the potential, in this limit, at the horizon, is weaker for neutrinos then it is for photons or gravitons. Thus, considering the integral over V, as a measure of the barrier, we find that for both potentials given by equation (42),

$$\int_{-\infty}^{+\infty} V\,\mathrm{d}r_* = -\frac{\nu}{2|\alpha|}\lg\left[\frac{2Mr_+(1-\sigma_s/\sigma)}{(r_+ + |\alpha|)^2}\right] \quad (\text{for } \sigma_s < \sigma \leqslant \sigma_c \text{ and } 0 \leqslant |\alpha| < r_+). \tag{51}$$

This integral diverges logarithmically as $\sigma \to \sigma_s + 0$, in contrast to a divergence like $(1-\sigma_s/\sigma)^{-2}$ in the scalar, the electromagnetic, and the gravitational cases. It appears that this weaker divergence in the barrier for the neutrinos results in a *finite* transmission coefficient for them in the limit $\sigma \to \sigma_s + 0$. The numerical results given in table 1 strongly suggest that fact, though it remains to be established rigorously.

5. On the absence of super-radiance for $0 < \sigma < \sigma_s$

For $0 < \sigma < \sigma_s$, the $r_*(r)$-relation becomes double-valued and the potentials, further, become singular at $r = |\alpha|\ (> r_+$ when $\sigma < \sigma_s)$ (see figure 2). As explained in the appendix, we must now seek solutions of the wave equations for Z which satisfy the boundary conditions

$$Z \to \mathrm{e}^{+\mathrm{i}\sigma r_*} + A\,\mathrm{e}^{-\mathrm{i}\sigma r_*} \quad \text{along the branch} \quad r \to \infty \quad \text{and} \quad r_* \to +\infty. \tag{52}$$

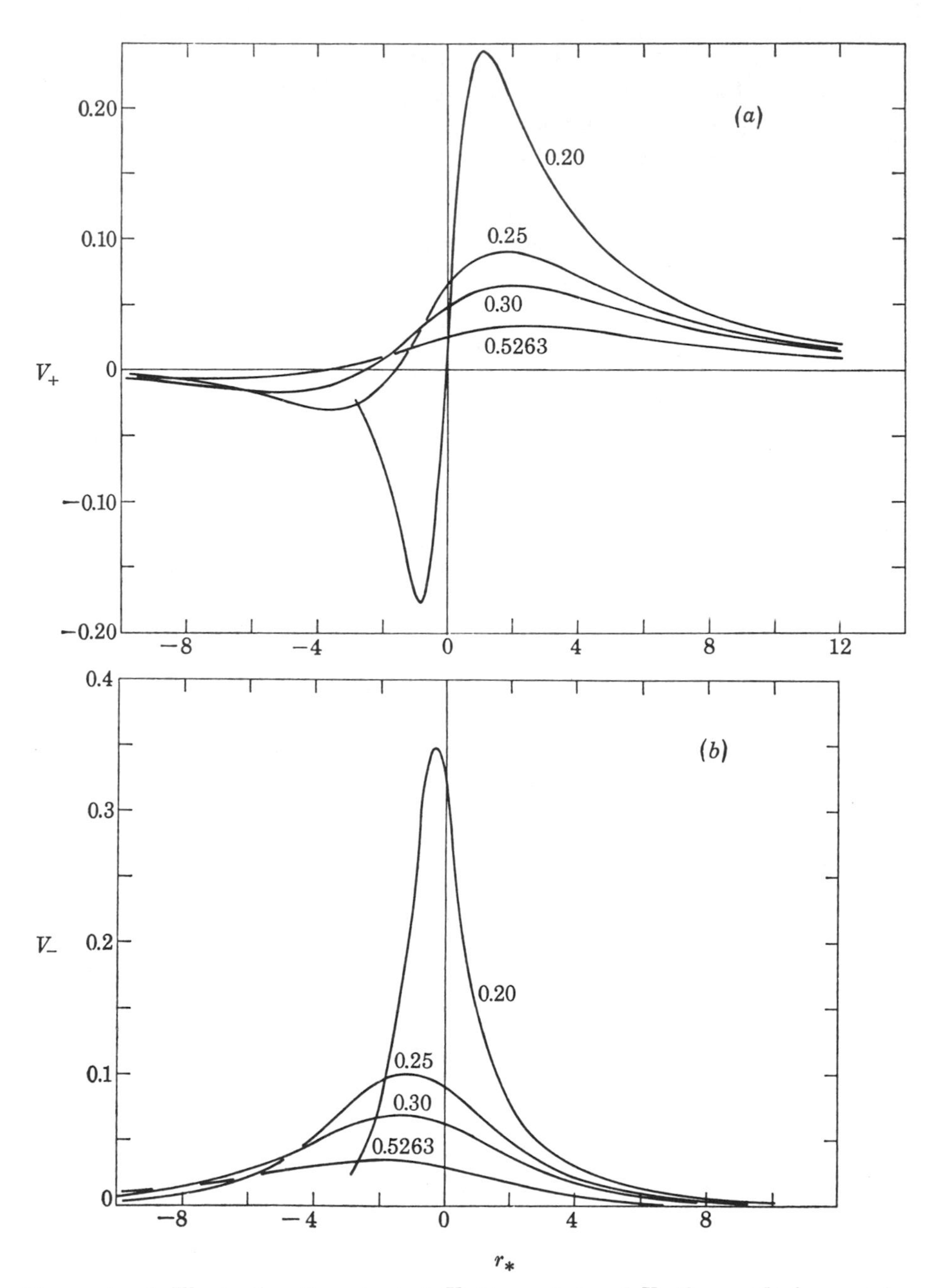

FIGURE 1*a*, *b*. Illustrating the potentials V_+ (figure 1*a*) and V_- (figure 1*b*) for neutrino waves, with $l = -m = 0.5$ and frequencies in the range $\sigma_s < \sigma \leqslant \sigma_c\ (= -m/a = 0.5263)$ for the values of the parameters considered, incident on a Kerr black hole ($a = 0.95$). The curves are labelled by the values of σ to which they belong.

and $\quad Z \rightarrow B\,e^{+i\sigma r_*}$ along the branch $r \rightarrow r_+ + 0$ and $r_* \rightarrow +\infty$. (53)

Also, by equation (A 13), for the case $s = \frac{1}{2}$, we are now considering,

$$W(Z, Z^*)_{r>|\alpha|} = W(Z, Z^*)_{r<|\alpha|}. \tag{54}$$

The Wronskian must therefore remain unchanged in the present case, in contrast to a change of sign that is demanded in the case of integral spins ($s = 0, 1$, and 2). On this account, the conservation law, in the case of the neutrinos, will continue to be given by

$$\mathbb{R} + \mathbb{T} = 1 \quad \text{also when} \quad 0 < \sigma < \sigma_s; \tag{55}$$

and super-radiance will not be manifested; and this is in agreement with the conclusion first reached by Unruh (1973).

TABLE 1. REFLEXION COEFFICIENTS FOR NEUTRINOS INCIDENT ON A KERR BLACK HOLE WITH $a = 0.95$

($l = 0.5$, $m = -0.5$)

σ	σ/σ_s	$\mathbb{R}$	σ	σ/σ_s	$\mathbb{R}$
0.181987	1.0055	0.97627	0.220987	1.2210	0.89187
0.182987	1.0111	0.97556	0.230987	1.2763	0.83457
0.183987	1.0166	0.97479	0.240987	1.3315	0.75356
0.184987	1.0221	0.97398	0.250000	1.3813	0.66005
0.185987	1.0276	0.97313	0.250987	1.3868	0.64885
0.186987	1.0332	0.97223	0.260987	1.4420	0.52777
0.187987	1.0387	0.97129	0.300000	1.6576	0.14125
0.188987	1.0442	0.97012	0.350000	1.9338	0.01622
0.189987	1.0497	0.96925	0.400000	2.2101	0.00197
0.200987	1.1105	0.95353	0.450000	2.4864	0.00027
0.210987	1.1658	0.92971			

Since there can be no super-radiance for any neutrino incidence and $\mathbb{R}$ is already very close to unity at $\sigma = \sigma_s$ (cf. table 1) there is not much interest to continue the calculations for $\sigma < \sigma_s$. But we may briefly indicate the character of the singularity in V at $r = |\alpha|$ and the behaviour of the solution for Z in its neighbourhood.

First, we observe that according to equation (42), $V_\pm$ has the behaviour,

$$V_\pm \simeq \nu^{\frac{1}{2}} \frac{\Delta^{\frac{3}{2}}_{|\alpha|}}{4\,|\alpha|^2} \frac{1}{(r-|\alpha|)^3}, \tag{56}$$

in the neighbourhood of $r = |\alpha|$. This behaviour of the potentials at $r = |\alpha|$ for $s = \frac{1}{2}$ differs from the corresponding behaviours of the potentials for $s = 0, 1$, and 2 in one important respect: the singularity is of lower order – a pole of order 3 instead of a pole of order 4.

Next, letting

$$x = r - |\alpha|, \tag{57}$$

we find that in the neighbourhood of $r = |\alpha|$, the equation governing Z becomes

$$x\frac{d^2 Z}{dx^2} - \frac{dZ}{dx} \mp \left(\frac{\nu}{\Delta_{|\alpha|}}\right)^{\frac{1}{2}} Z = 0. \tag{58}$$

And the solution of this equation in Watson's notation is (cf. Watson 1922, p. 97)

$$y\mathscr{C}_2(\mathrm{i}\sqrt{y}) \quad \text{for} \quad V_+ \quad \text{and} \quad y\mathscr{C}_2(\sqrt{y}) \quad \text{for} \quad V_-, \tag{59}$$

where

$$y = 4(\nu/\Delta_{|\alpha|})^{\frac{1}{2}}\,x, \tag{60}$$

and $\mathscr{C}_2$ denotes the general solution of Bessel's equation of order 2. With this known behaviour of Z at the singularity, there is no difficulty of principle in integrating the equation through $r = |\alpha|$ and completing the solution in the manner described in the appendix (in the electromagnetic context).

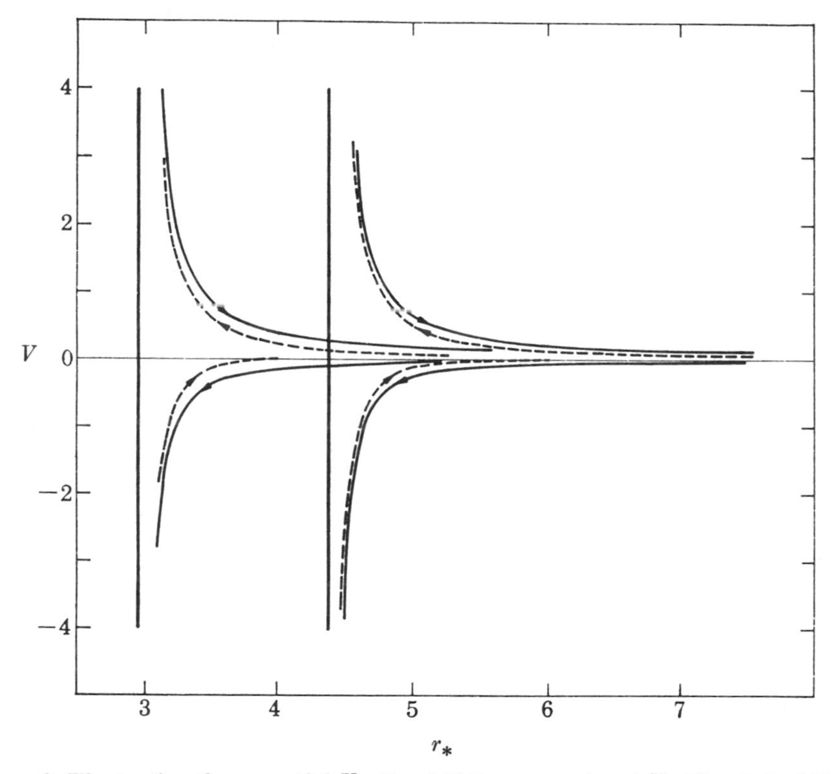

FIGURE 2. Illustrating the potential V_+ (the full-line curves) and V_- (the dashed-line curves) for neutrino waves, with $l = -m = 0.5$ and frequencies $\sigma < \sigma_s$ ($= 0.1810$ for the values of the parameters considered), incident on a Kerr black hole ($a = 0.95$). The curves to the left are for $\sigma = 0.16$ and the curves to the right are for $\sigma = 0.14$. The curves with arrows pointing to the left represent the potentials for $\infty > r > |\alpha|$ and the curves with arrows pointing to the right represent the potentials for $|\alpha| > r > r_+$. The vertical lines are positioned where the potentials become singular. (Note that for $\sigma < \sigma_s$, the $r_*(r)$-relation is double-valued and $r_* \to +\infty$ both when $r \to \infty$ and $r \to r_+$.)

6. Concluding remarks

In tables 2 and 3 we have listed the values of $\mathbb{R}$, for electromagnetic and gravitational waves incident on a Kerr black hole (with $a = 0.95$) and comparable to those listed in table 1 for incident neutrino waves. In figure 3, the dependence of the reflexion coefficient $\mathbb{R}$ on σ/σ_s is contrasted for the three cases $s = \frac{1}{2}$, 1, and 2. In many

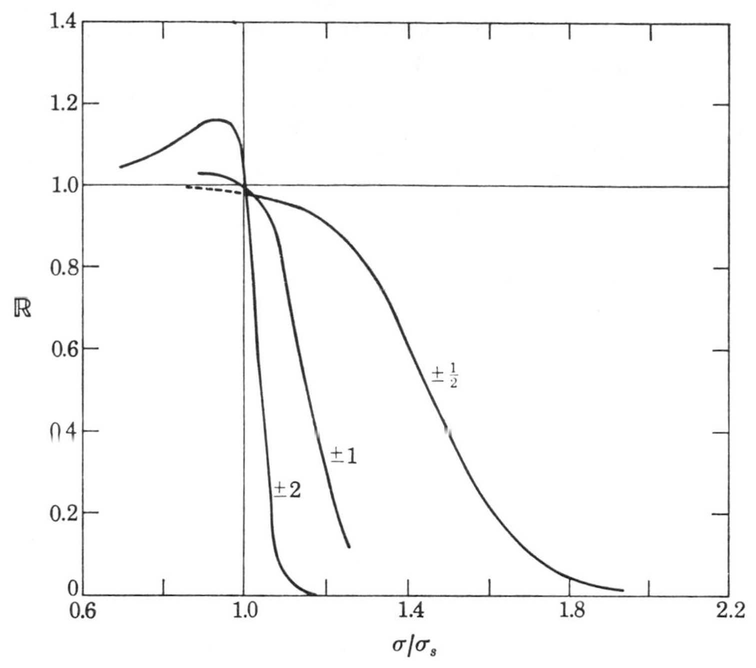

FIGURE 3. Illustrating the dependence of the reflexion coefficient ($\mathbb{R}$) on σ/σ_s for neutrinos ($l = -m = 0.5$), photons ($l = -m = 1$), and gravitons ($l = -m = 2$) incident on a Kerr black hole ($a = 0.95$). The curves are distinguished by the spins of the particles to which they refer. Notice that while the photons and the gravitons manifest super-radiance ($\mathbb{R} > 1$) for $\sigma < \sigma_s$, the neutrinos do not (for reasons explained in the test). The dashed-line part of the curve referring to the neutrinos represents a 'visual' extrapolation of the computed part drawn as a full-line curve.

TABLE 2. REFLEXION COEFFICIENTS FOR ELECTROMAGNETIC WAVES INCIDENT ON A KERR BLACK HOLE WITH $a = 0.95$

($l = 1$, $m = -1$)

σ	σ/σ_s	$\mathbb{R}$	σ	σ/σ_s	$\mathbb{R}$
0.325000	0.8979	1.02428	0.405593	1.1205	0.70998
0.345000	0.9531	1.01919	0.415593	1.1481	0.56810
0.350000	0.9669	1.01565	0.425593	1.1758	0.41943
0.365593	1.0100	0.99241	0.435593	1.2034	0.28686
0.375593	1.0376	0.96100	0.445593	1.2310	0.18435
0.385593	1.0653	0.90807	0.455593	1.2586	0.11332
0.395593	1.0929	0.82563			

ways it is remarkable how the manifestation of super-radiance for incident electromagnetic and gravitational waves and its absence for incident neutrino waves emerge so very naturally from the character of the respective potential barriers that surround the Kerr black hole.

We are grateful to Dr Don Page of the California Institute of Technology for providing us with his tabulation of the separation constant E of Teukolsky's equation for $s = \frac{1}{2}$. And one of us (S. C.) is greatly indebted to Dr Robert Wald and Dr John Friedman for many clarifying discussions.

The research reported in this paper has in part been supported by a grant from the National Science Foundation under grant MPS 74-17456 with the University of Chicago (S. C.) and by the National Science Foundation under grant MPS 74-18386-A01 and the University of Maryland Computer Science Center (S. D.).

Table 3. Reflexion coefficients for gravitational waves incident on a Kerr black hole with $a = 0.95$

$(l = 2,\ m = -2)$

σ	σ/σ_s	$\mathbb{R}$	σ	σ/σ_s	$\mathbb{R}$
0.50	0.6907	1.04422	0.77	1.0636	0.27057
0.55	0.7597	1.07031	0.78	1.0774	0.16754
0.60	0.8288	1.10693	0.79	1.0912	0.10007
0.65	0.8979	1.15358	0.80	1.1051	0.05844
0.70	0.9669	1.15101	0.81	1.1189	0.03372
0.73	1.0084	0.92530	0.82	1.1327	0.01932
0.74	1.0222	0.76882	0.83	1.1465	0.01103
0.75	1.0360	0.58828	0.84	1.1603	0.00628
0.76	1.0498	0.41376	0.85	1.1741	0.00357

Appendix. The proper boundary conditions for the equation $\Lambda^2 Z = VZ$ for $0 < \sigma < \sigma_s$

By S. Chandrasekhar, F.R.S.

When $0 < \sigma < \sigma_s$, the $r_*(r)$-relation becomes double-valued; and at the same time the potentials which occur in the wave equation for Z become singular at $r = |\alpha|$ $(> r_+$ when $\sigma < \sigma_s)$, i.e. precisely, when r_* attains its minimum value. In the context of the equations appropriate for electromagnetic perturbations, it was shown in (I), § 5, how one can consistently integrate the wave equation for Z through the singularity and deduce the emergence of super-radiance. But in formulating the boundary conditions for $r_* \to +\infty$, along the two branches of the $r_*(r)$-relation, it was assumed that as $r_* \to +\infty$ along the branch approaching the horizon at r_+, the wave progresses towards the horizon (even as in the case, when the $r_*(r)$-relation is single-valued); and, further, that the Wronskian $W(Z, Z^*)$ remains the same on either side of the singularity at $r = |\alpha|$. And it was shown, how solutions satisfying these conditions automatically lead to super-radiance in the form of the relation

$$\mathbb{R} - \mathbb{T} = 1. \tag{A 1}$$

However, the physical requirements of the problem are different: for $\sigma < \sigma_s$, the wave must appear as *emerging from* the horizon, not *progressing towards* it. The reason is that we have the right only to require that the group velocity of a wave packet as measured in any local frame is negative (cf. Teukolsky 1973, §5); and for

$\sigma < \sigma_s$ this requirement leads to the condition stated. Accordingly, we must, correctly, seek solutions of the wave equation for Z which satisfy the boundary conditions (in contrast to the conditions stated in (I), equations (100) and (101)),

$$Z \to e^{+i\sigma r_*} + A\,e^{-i\sigma r_*} \quad \text{along the branch} \quad r \to \infty \quad \text{and} \quad r_* \to +\infty \tag{A 2}$$

and
$$Z \to B e^{+i\sigma r_*} \quad \text{along the branch } r \to r_+ + 0 \quad \text{and} \quad r_* \to +\infty. \tag{A 3}$$

If we should require, at the same time, that $W(Z, Z^*)$ remains unchanged as we cross the singularity at $r = |\alpha|$, then we should not find any super-radiance. But, and this is the crucial point, *for integral spins* ($s = 0, 1$, and 2), *we must change the sign of the Wronskian* $W(Z, Z^*)$ *as we cross the singularity.* And the reason is the following.

Consider the Wronskian $W(\phi_1, \phi_2)$ of any two solutions of (I), equation (10). Since (cf. (I), equation (24); (II), equation (6); and equation (17) of this paper)

$$Y = \frac{\Delta^{\frac{1}{2}s}}{\rho^{2s}}\,\phi, \tag{A 4}$$

$$W(\phi_1, \phi_2) = \frac{\rho^{4s}}{\Delta^s}\,W(Y_1, Y_2). \tag{A 5}$$

Alternatively, we can also write

$$W(\phi_1, \phi_2) = \frac{\rho^{4s}}{\Delta^s}\,(Y_1 \Lambda_- Y_2 - Y_2 \Lambda_- Y_1). \tag{A 6}$$

Now substituting for Y and $\Lambda_- Y$ in terms of Z (in accordance with (I), equations (31) and (32), (II), equations (13) and (14), and equations (23) and (24) of this paper) we find

$$W(\phi_1, \phi_2) = \frac{\rho^{4s}}{\Delta^s}\left(f\,VR + \frac{\Delta^s}{\rho^{4s}}\beta MT\right)(Z_1 \Lambda_+ Z_2 - Z_2 \Lambda_+ Z_1), \tag{A 7}$$

or, by the integral which obtains in these cases (cf. (I), equation (37)), we have

$$W(\phi_1, \phi_2) = K\,W(Z_1, Z_2), \tag{A 8}$$

where K is a constant. It is this proportionality of the two Wronskians that underlies the equivalence of the conservation laws which follow from the constancy of $W(\phi_1, \phi_2)$, in the treatment of Teukolsky & Press (1974), and the constancy of $W(Z_1, Z_2)$ in our treatment.

Returning to the equation

$$W(Y_1, Y_2) = K\frac{\Delta^s}{\rho^{4s}}\,W(Z_1, Z_2), \tag{A 9}$$

we now observe that, by virtue of the relation,

$$Y = \frac{\Delta^s}{\rho^{2s-1}}\,R, \tag{A 10}$$

where R is the function which satisfies Teukolsky's radial equation for the spin-weight s,

$$W(Y_1, Y_2) = \frac{\Delta^{2s}}{\rho^{4s-2}}\,W(R_1, R_2). \tag{A 11}$$

In particular,

$$W(Y, Y^*) = \frac{\Delta^{2s}}{|r^2+\alpha^2|^{2s-1}} W(R, R^*) \tag{A 12}$$

Since $W(R, R^*)$ must be continuous across $r = |\alpha|$, it follows from equation (A 9) and (A 12) that

$$W(Z, Z^*)_{r>|\alpha|} = (-1)^{2s-1} W(Z, Z^*)_{r<|\alpha|}, \tag{A 13}$$

in case $\alpha^2 < 0$ and $r_+ < |\alpha|$, i.e. the case $\sigma < \sigma_s$. Therefore, *for integral spins*, $s = 0, 1,$ *and* 2, *we must change the sign of the Wronskian* $W(Z, Z^*)$ *as we cross the singularity at* $r = |\alpha|$; *and with this change of sign in the Wronskian, the solutions satisfying the boundary conditions* (A 2) *and* (A 3), *will again lead to super-radiance in the form of the relation* (A1).

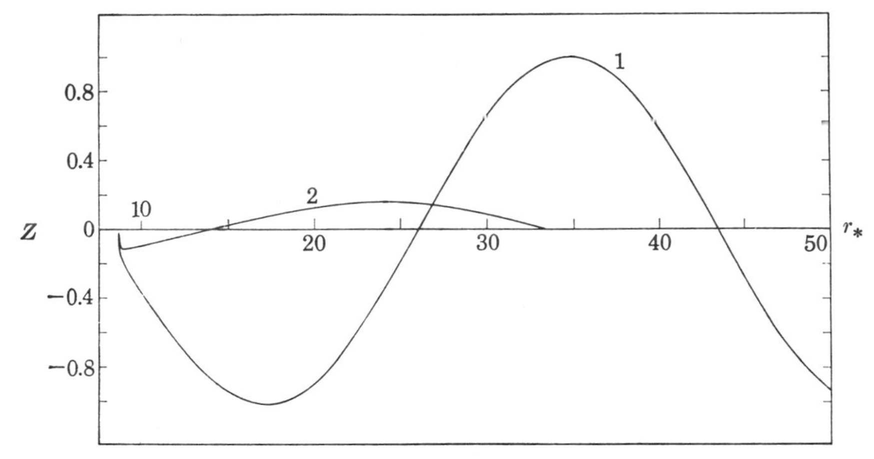

FIGURE A 1. Illustrating a solution of the wave equation for Z representing a standing electromagnetic wave ($l = -m = 1$) with a super-radiant frequency ($\sigma = \frac{1}{2}\sigma_s$) in the field of a Kerr black hole ($a = 0.95$). The part of the curve labelled '1' represents the *real* amplitude of the wave for $\infty > r > |\alpha|$ while the part of the curve labelled '2' represents the *imaginary* amplitude of the wave – imaginary on account of equation (A 16) – for $|\alpha| > r > r_+$. The turning point occurs at $r_* = 8.707$. (The solution was obtained by integrating the solution from the horizon as explained in the text; and it yielded a reflexion coefficient in agreement with what is obtained by more conventional methods.)

Solutions satisfying the altered boundary conditions can be found by the method described in (I), § 5 (*b*) with only minor modifications. Thus, we start with a solution for Z with the behaviour,

$$Z \to e^{+i\sigma r_*} \quad \text{for} \quad r_* \to +\infty \quad \text{along the branch} \quad r \to r_+ + 0, \tag{A 14}$$

and continue the integration forward from the horizon (but backward in r_*) to $r = |\alpha| - 0$. As we approach the singularity at $r = \alpha$, from the left, the solution will tend to a form specified in (I), equations (108) (for the case of electromagnetic perturbations considered) with certain determinate constants C_1 and C_2, for example.

More precisely, we may suppose that as $r \to |\alpha| - 0$, the solution for Z will tend to the form

$$Z \to C_1(|\alpha| - r)^{\frac{3}{2}} + C_2(|\alpha| - r)^{\frac{1}{2}} \quad (r \to |\alpha| - 0). \tag{A 15}$$

We conclude that as $r \to |\alpha| + 0$, the solution must tend to the form

$$Z \to \mathrm{i}C_1(r - |\alpha|)^{\frac{3}{2}} - \mathrm{i}C_2(r - |\alpha|)^{\frac{1}{2}} \quad (r \to |\alpha| + 0), \tag{A 16}$$

since only this choice will ensure the continuity of Teukolsky's radial function R at $r = |\alpha|$; and, consistent with our earlier requirement, the Wronskian $W(Z, Z^*)$ changes sign at $r = |\alpha|$. With the form of Z for $r = |\alpha| + 0$, thus determined, we can continue the integration forward (in r and in r_*) beyond $r = |\alpha|$, along the branch $r \to \infty$ and complete the solution of the problem in the manner described in (I), § 5 (*b*).

In figure A 1, we illustrate a solution for Z (a standing wave in this instance) obtained by a direct integration of the equation for Z, appropriate for electromagnetic perturbations, for a value of $\sigma = \frac{1}{2}\sigma_s$. I am grateful to Dr S. Detweiler for providing me with this numerical illustration of a solution directly integrated through the singularity.

I am greatly indebted to Dr R. Wald, Dr J. L. Friedman, and Dr S. Detweiler for discussions that clarified to me the various issues involved in a correct formulation of the boundary conditions for $\sigma < \sigma_s$.

References

Brill, D. R. & Wheeler, J. A. 1957 *Rev. mod. Phys.* **29**, 465.
Chandrasekhar, S. 1976*a* *Proc. R. Soc. Lond.* A **348**. 39.
Chandrasekhar, S. 1976*b* *Proc. R. Soc. Lond.* **A 349**, 571.
Chandrasekhar, S. & Detweiler, S. 1976 *Proc. R. Soc. Lond.* A **350**, 165.
Teukolsky, S. A. 1973 *Astrophys. J.* **185**, 635.
Teukolsky, S. A. & Press, W. H. 1974 *Astrophys. J.* **193**, 443.
Unruh, W. 1973 *Phys. Rev. Lett*, **31**, 1265.
Unruh, W. 1974 *Phys. Rev.* D2, **10**, 3194.
Watson, G. N. 1922 *Theory of Bessel functions.* Cambridge University Press.

The gravitational perturbations of the Kerr black hole

I. The perturbations in the quantities which vanish in the stationary state

BY S. CHANDRASEKHAR, F.R.S.
University of Chicago, Chicago, Illinois, 60637, U.S.A.

(*Received 2 May* 1977)

As a preliminary towards a complete integration of the Newman–Penrose equations governing the gravitational perturbations of the Kerr black hole, the perturbations in the spin coefficients and in the components of the Weyl tensor, which vanish in the stationary state, are considered. The manner of treatment of the basic equations yields Teukolsky's equations expressed directly in terms of the basic derivative operators of the theory and, further, suggests a preferred gauge in which two of the components of the Weyl tensor are governed by the same equations as a Maxwell field. Various identities and relations that are needed in subsequent work are assembled. In two appendixes, the solution of Maxwell's equations in Kerr geometry and the perturbations of the charged Kerr–Newman black hole are considered.

1. INTRODUCTION

An important advance towards our understanding of the gravitational perturbations of the Kerr black hole was made by Teukolsky in 1972. By considering two particular combinations of the components of the Weyl tensor – the Newman–Penrose quantities Ψ_0 and Ψ_4 (whose unperturbed values in the chosen frame are zero) – and analysing them into their normal modes, Teukolsky (1972, 1973) was able to express their solutions in terms of a pair of radial functions and an associated pair of angular functions, each satisfying a separate equation. In terms of the solutions of these separate equations one can determine, for example, the reflexion and the transmission coefficients of the Kerr black hole for incident gravitational waves (Press & Teukolsky 1973; Teukolsky & Press 1974; Chandrasekhar & Detweiler 1975, 1976; and Detweiler 1977). But the equations derived for Ψ_0 and Ψ_4 (even as the problems to which they are commonly addressed) do not depend upon, or require, any knowledge of the perturbations in the metric coefficients that are implied. In other words, in these common considerations, the basic question as to how the black hole is itself affected by the perturbation is side-stepped. And, while studies by Wald (1973), Cohen & Kegeles (1974), and Chrzanowski (1975) have dealt with the problem of expressing the perturbations in the metric coefficients in terms of Teukolsky's functions, a direct and a complete integration of the entire set of the Newman–Penrose equations governing the perturbations has not so far been attempted. The present paper is the first of two addressed to this larger problem. The precise aspects of the problems to which this paper is addressed are stated in §2*a*.

2. The Kerr geometry and the perturbation problem

In the Newman–Penrose formalism (1962; this paper will be referred to hereafter as N.P.) the description of space-time is in terms of a local tetrad-frame based on four null vectors ($\boldsymbol{l}, \boldsymbol{n}, \boldsymbol{m}, \overline{\boldsymbol{m}}$) of which $\boldsymbol{l}$ and $\boldsymbol{n}$ are real and $\boldsymbol{m}$ and $\overline{\boldsymbol{m}}$ are complex conjugates. It is further supposed that the chosen vectors satisfy the orthogonality relations $\boldsymbol{l}\cdot\boldsymbol{n} = 1$, $\boldsymbol{m}\cdot\overline{\boldsymbol{m}} = -1$, and $\boldsymbol{l}\cdot\boldsymbol{m} = \boldsymbol{l}\cdot\overline{\boldsymbol{m}} = \boldsymbol{n}\cdot\boldsymbol{m} = \boldsymbol{n}\cdot\overline{\boldsymbol{m}} = 0$. The associated spin-coefficients ($\kappa, \sigma, \lambda, \nu; \rho, \mu, \tau, \pi; \alpha, \beta, \gamma, \epsilon$; cf. N.P. equation (4.1*b*)) are no more than the standard Ricci rotation-coefficients defined in terms of the chosen null-tetrad (instead of in terms of the more conventional frame based on one time-like and three space-like vectors). Any tensor (in space-time) in which we may be interested is projected on to the chosen frame by contracting it with the basic null vectors in all possible non-trivial ways. Thus in place of the Weyl tensor, C_{abcd}, we consider the five complex scalars

$$
\begin{aligned}
\Psi_0 &= -C_{1313} = -C_{abcd}\, l^a m^b l^c m^d, \\
\Psi_1 &= -C_{1213} = -C_{abcd}\, l^a n^b l^c m^d, \\
\Psi_2 &= -C_{1342} = -C_{abcd}\, l^a m^b \overline{m}^c n^d, \\
\Psi_3 &= -C_{1242} = -C_{abcd}\, l^a n^b \overline{m}^c n^d,
\end{aligned}
$$

and

$$\Psi_4 = -C_{2424} = -C_{abcd}\, n^a \overline{m}^b n^c \overline{m}^d. \tag{1}$$

Since the Kerr metric is algebraically special and of Petrov type D, we can, in conformity with the Goldberg–Sachs theorem (cf. N.P., §3), choose the null basis in such a way that

$$\kappa = \sigma = \lambda = \nu = 0, \tag{2}$$

and the null vectors $\boldsymbol{l}$ and $\boldsymbol{n}$ form a shear-free null congruence of geodesics (not, necessarily, parametrized affinely). And when the basic tetrad is so chosen

$$\Psi_0 = \Psi_1 = \Psi_3 = \Psi_4 = 0; \tag{3}$$

and Ψ_2 is the only non-vanishing quantity.

Now the Kerr metric in the Boyer–Lindquist (1967) coordinates can be written in the form

$$ds^2 = -e^{2\nu}(dt)^2 + e^{2\psi}(d\varphi - \omega\, dt)^2 + e^{2\mu_2}(dr)^2 + e^{2\mu_3}(d\theta)^2, \tag{4}$$

where

$$e^{2\nu} = \frac{\rho^2 \Delta}{\Sigma^2}, \quad e^{2\psi} = \frac{\Sigma^2}{\rho^2}\sin^2\theta, \quad e^{2\mu_2} = \frac{\rho^2}{\Delta}, \quad e^{2\mu_3} = \rho^2, \quad \omega = \frac{2aMr}{\Sigma^2}, \tag{5}$$

and

$$\Delta = r^2 + a^2 - 2Mr, \quad \rho^2 = r^2 + a^2\cos^2\theta, \quad \Sigma^2 = (r^2+a^2)^2 - a^2\Delta\sin^2\theta. \tag{6}$$

A basic null-tetrad satisfying the requirements (2) and appropriate to the Kerr metric in the chosen coordinates was constructed by Kinnersley (1969) who further used the remaining freedom (after satisfying the requirements (2)) to make $\epsilon = 0$.

The Kinnersley tetrad is given by

$$\boldsymbol{l} = \frac{1}{\Delta}(r^2+a^2,\ +\Delta, 0, a),$$

$$\boldsymbol{n} = \frac{1}{2\rho^2}(r^2+a^2,\ -\Delta, 0, a),$$

and

$$\boldsymbol{m} = \frac{1}{\bar{\rho}\sqrt{2}}(\mathrm{i}a\sin\theta, 0, 1, \mathrm{i}\,\mathrm{cosec}\,\theta), \tag{7}$$

where

$$\bar{\rho} = r + a\mathrm{i}\cos\theta \quad \text{and} \quad \bar{\rho}^* = r - a\mathrm{i}\cos\theta. \tag{8}$$

The corresponding expressions for the spin-coefficients‡ are

$$\tilde{\rho} = -\frac{1}{\bar{\rho}^*}, \quad \beta = \frac{\cot\theta}{\bar{\rho}2\sqrt{2}}, \quad \pi = \frac{\mathrm{i}a\sin\theta}{(\bar{\rho}^*)^2\sqrt{2}}, \quad \tau = -\frac{\mathrm{i}a\sin\theta}{\rho^2\sqrt{2}},$$

$$\mu = -\frac{\Delta}{2\rho^2\bar{\rho}^*}, \quad \gamma = \mu + \frac{r-M}{2\rho^2}, \quad \alpha = \pi - \beta^*, \quad \text{and} \quad \epsilon = 0. \tag{9}$$

Also

$$\Psi_2 = -M(\bar{\rho}^*)^{-3}. \tag{10}$$

And, of course, κ, σ, λ, ν, Ψ_0, Ψ_1, Ψ_3 and Ψ_4 all vanish in the chosen frame.

This completes the description of Kerr geometry in the Newman–Penrose formalism.

(a) *The perturbation problem*

In considerations relative to the gravitational perturbations of a space-time, one is principally interested in the changes in the metric coefficients. In the Newman–Penrose formalism, the changes in the metric coefficients are directly related to (and expressible in terms of) the changes in the null vectors of the tetrad; the specification of these changes in the basic vectors is, therefore, the central problem of the theory. Besides, one is also interested in specifying the changes in the various spin-coefficients and in the projected components (the Ψs) of the Weyl tensor.

In the Newman–Penrose formalism, the perturbation problem is all of a piece. Yet, in the context of the Kerr metric, the problem divides itself, naturally, into two distinct parts in that the solution to the first part is a necessary prerequisite to obtaining the solution to the second part. The origin of this division is the fact that in the background Kerr geometry the unperturbed spin coefficients κ, σ, λ, and ν and the scalars Ψ_0, Ψ_1, Ψ_3, and Ψ_4 all vanish by virtue of the algebraic speciality of the Kerr metric and by the choice of the null basis that the speciality makes possible. As we shall presently see, under these circumstances, the Newman–Penrose formalism provides equations for the eight quantities (no longer zero, on account of the perturbation) which can be solved independently of the perturbations in all of the remaining quantities.

‡ To avoid ambiguity (with our usage of $\bar{\rho}$ and $\bar{\rho}^*$) we shall distinguish the spin-coefficient 'ρ' by denoting it $\tilde{\rho}$.

This paper is devoted to the solution of the quantities (κ, σ, λ, ν, Ψ_0, Ψ_1, Ψ_3, and Ψ_4) which vanish in the stationary state. The sequel to this paper (paper II) will be devoted to the solution of the quantities which are finite in the stationary state with the principal object of obtaining explicit expressions for the perturbations in the metric coefficients.

3. The derivative operators and some lemmas

With the intent of analysing the perturbations into their normal modes, we shall suppose that the perturbations to which the various quantities are subjected, have a t- and a φ-dependence given by

$$e^{i(\sigma t+m\varphi)}, \tag{11}$$

where σ is a constant (which we shall mostly consider as real and positive) and m is an integer (positive, negative, or zero). The basis vectors ($\boldsymbol{l}$, $\boldsymbol{n}$, $\boldsymbol{m}$, $\overline{\boldsymbol{m}}$) given in equation (7), when applied as tangent-vectors to functions with a t- and a φ-dependence given by (11), become the derivative operators

$$\boldsymbol{l} = \mathrm{D} = \mathscr{D}_0, \quad \boldsymbol{n} = \Delta = -\frac{\Delta}{2\rho^2}\mathscr{D}_0^\dagger,$$

$$\boldsymbol{m} = \delta = \frac{1}{\bar{\rho}\sqrt{2}}\mathscr{L}_0^\dagger, \quad \text{and} \quad \overline{\boldsymbol{m}} = \delta^* = \frac{1}{\bar{\rho}^*\sqrt{2}}\mathscr{L}_0; \tag{12}$$

where

$$\mathscr{D}_n = \partial_r + i\frac{K}{\Delta} + 2n\frac{r-M}{\Delta}, \quad \mathscr{D}_n^\dagger = \partial_r - i\frac{K}{\Delta} + 2n\frac{r-M}{\Delta};$$

$$\mathscr{L}_n = \partial_\theta + Q + n\cot\theta, \quad \mathscr{L}_n^\dagger = \partial_\theta - Q + n\cot\theta; \tag{13}$$

and

$$K = (r^2+a^2)\sigma + am, \quad Q = a\sigma\sin\theta + m\,\mathrm{cosec}\,\theta. \tag{14}$$

It will be noted that, while $\mathscr{D}_n$ and $\mathscr{D}_n^\dagger$ are purely radial operators, $\mathscr{L}_n$ and $\mathscr{L}_n^\dagger$ are purely angular operators.

The differential operators, we have defined, satisfy a number of elementary identities which we shall have occasions to use constantly in the subsequent analysis. For convenience, we collect them here as a series of lemmas.

Lemma 1. (i) $\qquad \mathscr{L}_n(\theta) = -\mathscr{L}_n^\dagger(\pi-\theta); \quad \mathscr{D}_n^\dagger = (\mathscr{D}_n)^*;$

(ii) $\qquad (\sin\theta)\mathscr{L}_{n+1} = \mathscr{L}_n\sin\theta; \quad (\sin\theta)\mathscr{L}_{n+1}^\dagger = \mathscr{L}_n^\dagger\sin\theta;$

(iii) $\qquad \Delta\mathscr{D}_{n+1} = \mathscr{D}_n\Delta; \quad \Delta\mathscr{D}_{n+1}^\dagger = \mathscr{D}_n^\dagger\Delta. \qquad (15)$

Lemma 2.

$$\left(\mathscr{D}+\frac{m}{\bar{\rho}^*}\right)\left(\mathscr{L}+\frac{mai\sin\theta}{\bar{\rho}^*}\right) = \left(\mathscr{L}+\frac{mai\sin\theta}{\bar{\rho}^*}\right)\left(\mathscr{D}+\frac{m}{\bar{\rho}^*}\right), \tag{16}$$

where $\mathscr{D}$ can be any $\mathscr{D}_n$ or $\mathscr{D}_n^\dagger$, $\mathscr{L}$ any $\mathscr{L}_{n'}$ or $\mathscr{L}_{n'}^\dagger$, and m is a constant (generally, an integer, positive or negative).

Lemma 3.

$$\mathscr{L}_{n+1}\mathscr{L}_{n+2}\ldots\mathscr{L}_{n+m}(f\cos\theta) = (\cos\theta)\mathscr{L}_{n+1}\ldots\mathscr{L}_{n+m}f - (m\sin\theta)\mathscr{L}_{n+2}\ldots\mathscr{L}_{n+m}f, \tag{17}$$

where f is any function of θ and the $\mathscr{L}$s can be replaced by $\mathscr{L}^\dagger$s.

Lemma 4. *If $f(\theta)$ and $g(\theta)$ are any two (bounded) functions regular at $\theta = 0$ and $\theta = \pi$, then*

$$\int_0^\pi g(\mathscr{L}_n f)\sin\theta\,d\theta = -\int_0^\pi f(\mathscr{L}^\dagger_{-(n-1)} g)\sin\theta\,d\theta. \tag{18}$$

Of the foregoing lemmas, the commutativity of the operators in lemma 2 provides the basis for the reductions leading to the Teukolsky separation of the variables (cf. §5).

And, finally, we may note the elementary relations

$$Q_{,\theta} + Q\cot\theta = 2a\sigma\cos\theta \quad \text{and} \quad K - aQ\sin\theta = \rho^2\sigma. \tag{19}$$

4. The Newman–Penrose equations that are already linearized

As we have stated, we shall be concerned in this paper only with the quantities κ, σ, λ, ν, Ψ_0, Ψ_1, Ψ_3, and Ψ_4, which vanish in the equilibrium stationary state. We find that, among the Newman–Penrose equations, four of the eight Bianchi identities (included in N.P. eq. (4.5)) and the two equations (N.P. eqns (4.2*b*) and (4.2*j*)) which give the components R_{1313} and R_{2424} (cf. the definitions of Ψ_0 snd Ψ_4 in equation (1)) of the Riemann tensor are *linear* and *homogeneous* in these quantities. They are

$$(\delta^* - 4\alpha + \pi)\,\Psi_0 - (\mathrm{D} - 2\epsilon - 4\tilde{\rho})\,\Psi_1 = 3\kappa\Psi_2,$$
$$(\Delta - 4\gamma + \mu)\,\Psi_0 - (\delta - 4\tau - 2\beta)\,\Psi_1 = 3\sigma\Psi_2,$$
$$(\mathrm{D} - \tilde{\rho} - \tilde{\rho}^* - 3\epsilon + \epsilon^*)\,\sigma - (\delta - \tau + \pi^* - \alpha^* - 3\beta)\,\kappa = \Psi_0; \tag{20}$$

and

$$(\mathrm{D} + 4\epsilon - \tilde{\rho})\,\Psi_4 - (\delta^* + 4\pi + 2\alpha)\,\Psi_3 = -3\lambda\Psi_2,$$
$$(\delta + 4\beta - \tau)\,\Psi_4 - (\Delta + 2\gamma + 4\mu)\,\Psi_3 = -3\nu\Psi_2,$$
$$(\Delta + \mu + \mu^* + 3\gamma - \gamma^*)\,\lambda - (\delta^* + 3\alpha + \beta^* + \pi - \tau^*)\,\nu = -\Psi_4. \tag{21}$$

The foregoing equations are already linearized in the sense that Ψ_0, Ψ_1, κ, and σ in equations (20) and Ψ_4, Ψ_3, λ, and ν in equations (21) are to be considered as quantities of the first order of smallness and that, accordingly, we may replace all the other quantities (including the derivative operators) which occur in these equations by their unperturbed values given in equations (9), (10), and (12). Equations (20) and (21) with these replacements are considered in §5 below; here we shall only make two observations concerning them. *First*, we are provided only six equations for the eight unknowns which occur in them; this implies that *their solution must involve two arbitrary functions. Second*, equations (20) governing Ψ_0, Ψ_1, κ, and σ decouple completely from equations (21) governing Ψ_4, Ψ_3, λ, and ν; this decoupling of the two sets of equations has, as we shall see, important consequences.

5. THE REDUCTION OF THE EQUATIONS AND THEIR SOLUTIONS

As we have stated in §4, we can replace the derivative operators, the various spin-coefficients (besides κ, σ, λ, and ν), and Ψ_2 in equations (20) and (21) by their values given in equations (9), (10), and (12). We find that the resulting equations take simple and symmetrical forms if we write them in terms of the variables

$$\Phi_0 = \Psi_0, \quad \Phi_1 = \Psi_1 \bar{\rho}^* \surd 2, \quad k = \kappa/(\bar{\rho}^*)^2 \surd 2, \quad \text{and} \quad s = \sigma \bar{\rho}/(\bar{\rho}^*)^2, \tag{22}$$

and
$$\Phi_4 = \Psi_4 (\bar{\rho}^*)^4, \quad \Phi_3 = \Psi_3 (\bar{\rho}^*)^3/\surd 2, \quad l = \lambda \bar{\rho}^*/2, \quad \text{and} \quad n = \nu \rho^2/\surd 2. \tag{23}$$

After the various replacements and substitutions, equations (20) and (21) reduce to the forms

$$\left(\mathscr{L}_2 - \frac{3ia\sin\theta}{\bar{\rho}^*}\right)\Phi_0 - \left(\mathscr{D}_0 + \frac{3}{\bar{\rho}^*}\right)\Phi_1 = -6Mk, \tag{24}$$

$$\Delta\left(\mathscr{D}_2^\dagger - \frac{3}{\bar{\rho}^*}\right)\Phi_0 + \left(\mathscr{L}_{-1}^\dagger + \frac{3ia\sin\theta}{\bar{\rho}^*}\right)\Phi_1 = +6Ms, \tag{25}$$

$$\left(\mathscr{D}_0 + \frac{3}{\bar{\rho}^*}\right)s - \left(\mathscr{L}_{-1}^\dagger + \frac{3ia\sin\theta}{\bar{\rho}^*}\right)k = +\frac{\bar{\rho}}{(\bar{\rho}^*)^2}\Phi_0, \tag{26}$$

and
$$\left(\mathscr{D}_0 - \frac{3}{\bar{\rho}^*}\right)\Phi_4 - \left(\mathscr{L}_{-1} + \frac{3ia\sin\theta}{\bar{\rho}^*}\right)\Phi_3 = +6Ml, \tag{27}$$

$$\left(\mathscr{L}_2^\dagger - \frac{3ia\sin\theta}{\bar{\rho}^*}\right)\Phi_4 + \Delta\left(\mathscr{D}_{-1}^\dagger + \frac{3}{\bar{\rho}^*}\right)\Phi_3 = +6Mn, \tag{28}$$

$$\Delta\left(\mathscr{D}_{-1}^\dagger + \frac{3}{\bar{\rho}^*}\right)l + \left(\mathscr{L}_{-1} + \frac{3ia\sin\theta}{\bar{\rho}^*}\right)n = +\frac{\bar{\rho}}{(\bar{\rho}^*)^2}\Phi_4. \tag{29}$$

It is evident that, by making use of the commutation relation (16), we can eliminate Φ_1 from equations (24) and (25) by operating equation (24) by

$$(\mathscr{L}_{-1}^\dagger + 3ia\sin\theta/\bar{\rho}^*)$$

and equation (25) by $(\mathscr{D}_0 + 3/\bar{\rho}^*)$ and adding. The right hand side of the resulting equation is, apart from a factor $6M$, precisely the quantity which occurs on the left hand side of equation (26). We, accordingly, obtain the decoupled equation

$$\left[\left(\mathscr{L}_{-1}^\dagger + \frac{3ia\sin\theta}{\bar{\rho}^*}\right)\left(\mathscr{L}_2 - \frac{3ia\sin\theta}{\bar{\rho}^*}\right) + \left(\mathscr{D}_0 + \frac{3}{\bar{\rho}^*}\right)\Delta\left(\mathscr{D}_2^\dagger - \frac{3}{\bar{\rho}^*}\right)\right]\Phi_0 = 6M\frac{\bar{\rho}}{(\bar{\rho}^*)^2}\Phi_0. \tag{30}$$

Similarly, we obtain from equations (27)–(30) the decoupled equation

$$\left[\left(\mathscr{L}_{-1} + \frac{3ia\sin\theta}{\bar{\rho}^*}\right)\left(\mathscr{L}_2^\dagger - \frac{3ia\sin\theta}{\bar{\rho}^*}\right) + \Delta\left(\mathscr{D}_{-1}^\dagger + \frac{3}{\bar{\rho}^*}\right)\left(\mathscr{D}_0 - \frac{3}{\bar{\rho}^*}\right)\right]\Phi_4 = 6M\frac{\bar{\rho}}{(\bar{\rho}^*)^2}\Phi_4. \tag{31}$$

On expanding equations (30) and (31), we find after some reductions (in which we make use of the elementary relations (19)) that we are left with

$$[\Delta\mathscr{D}_1\mathscr{D}_2^\dagger + \mathscr{L}_{-1}^\dagger\mathscr{L}_2 - 6i\sigma(r + ai\cos\theta)]\Phi_0 = 0 \tag{32}$$

and
$$[\Delta\mathscr{D}_{-1}^\dagger\mathscr{D}_0 + \mathscr{L}_{-1}\mathscr{L}_2^\dagger + 6i\sigma(r + ai\cos\theta)]\Phi_4 = 0. \tag{33}$$

Equations (32) and (33) clearly allow the separation of the variables. Thus, by the substitutions

$$\Phi_0 = R_{+2}(r)\,S_{+2}(\theta) \quad \text{and} \quad \Phi_4 = R_{-2}(r)\,S_{-2}(\theta), \tag{34}$$

where $R_{\pm 2}$ and $S_{\pm 2}$ are functions, respectively, of r and θ only, we obtain the two pairs of separated equations:

$$(\Delta \mathscr{D}_1 \mathscr{D}_2^\dagger - 6i\sigma r)\,R_{+2} = \bar{\lambda} R_{+2}, \tag{35}$$

$$(\mathscr{L}_{-1}^\dagger \mathscr{L}_2 + 6a\sigma\cos\theta)\,S_{+2} = -\bar{\lambda} S_{+2}; \tag{36}$$

and

$$(\Delta \mathscr{D}_{-1}^\dagger \mathscr{D}_0 + 6i\sigma r)\,R_{-2} = \bar{\lambda} R_{-2}, \tag{37}$$

$$(\mathscr{L}_{-1} \mathscr{L}_2^\dagger - 6a\sigma\cos\theta)\,S_{-2} = -\bar{\lambda} S_{-2}, \tag{38}$$

where $\bar{\lambda}$ is a separation constant.

It will be noticed that we have not distinguished between the separation constants that derive from equations (32) and (33). The reason is the following. Considering equation (36), we first observe that $\bar{\lambda}$ is a characteristic-value parameter that is to be determined by the requirement that $S_{+2}(\theta)$ be regular at $\theta = 0$ and $\theta = \pi$. On the other hand, since the operator acting on S_{-2} in equation (38) is the same as the operator acting on S_{+2} in equation (36) if we replace θ by $\pi - \theta$, it follows that a proper solution $S_{+2}(\theta; \bar{\lambda})$ of equation (36), belonging to a characteristic value $\bar{\lambda}$, provides a proper solution of equation (38) belonging to the same value $\bar{\lambda}$ if we replace θ by $\pi - \theta$. In other words, equations (36) and (38) determine the same set of characteristic values for $\bar{\lambda}$.

Further, by rewriting equation (35) in the alternative form

$$(\Delta \mathscr{D}_{-1} \mathscr{D}_0^\dagger - 6i\sigma r)\,\Delta^2 R_{+2} = \bar{\lambda} \Delta^2 R_{+2}, \tag{39}$$

and comparing it with equation (37), we conclude that if $\Delta^2 R_{+2}$ is a solution of equation (39) then its complex conjugate is a solution of equation (37).

It can be readily verified that equations (35)–(38) are equivalent to Teukolsky's equations; and our present characteristic-value parameter $\bar{\lambda}$ is related to his separation constant E by

$$\bar{\lambda} = E + a^2\sigma^2 + 2a\sigma m - 2. \tag{40}$$

It is, however, important to observe that, in the present treatment, Teukolsky's equations have emerged directly expressed in terms of the basic derivative operators $\mathscr{L}$ and $\mathscr{D}$. This emergence of the separated equations in the forms (35)–(37) is of significance from the point of view of the relations to be quoted in §7.

(a) *The solutions for the spin-coefficients κ, σ, λ, and ν and the choice of gauge*

Quite generally, the linearized equations governing the perturbations must be consistent with the freedoms we have in the choice of the tetrad and in the choice of the coordinates. Precisely, we have six degrees of freedom to make infinitesimal rotations of the local tetrad frame and four degrees of freedom to make infinitesimal coordinate transformations. We can exercise these available degrees of freedom to restrict the solutions of the linearized equations as convenience or occasion may dictate.

Returning to equations (24)–(29), we recall that the solutions of these equations must involve two arbitrary functions. Since $\Phi_0(=R_{+2}S_{+2})$ and $\Phi_4(=R_{-2}S_{-2})$ have already been determined (apart from normalization constants), we may let Φ_1 and Φ_3 be the two arbitrary functions. At this point, we may exercise four of the six available degrees of freedom in the choice of the local (perturbed) tetrad frame to make

$$\Phi_1 = \Phi_3 = 0 \quad \text{or, equivalently,} \quad \Psi_1 = \Psi_3 = 0. \tag{41}$$

In this gauge, the solutions for κ, σ, λ, and ν can be directly read off from equations (24)–(29); we have

$$\kappa = -\frac{\sqrt{2}}{6M}(\bar{\rho}^*)^2 R_{+2}\left(\mathscr{L}_2 - \frac{3ia\sin\theta}{\bar{\rho}^*}\right)S_{+2}, \tag{42}$$

$$\sigma = +\frac{1}{6M}\frac{(\bar{\rho}^*)^2}{\bar{\rho}}S_{+2}\Delta\left(\mathscr{D}_2^\dagger - \frac{3}{\bar{\rho}^*}\right)R_{+2}, \tag{43}$$

$$\lambda = +\frac{1}{6M}\frac{2}{\bar{\rho}^*}S_{-2}\left(\mathscr{D}_0 - \frac{3}{\bar{\rho}^*}\right)R_{-2}, \tag{44}$$

and

$$\nu = +\frac{\sqrt{2}}{6M}\frac{1}{\rho^2}R_{-2}\left(\mathscr{L}_2^\dagger - \frac{3ia\sin\theta}{\bar{\rho}^*}\right)S_{-2}, \tag{45}$$

where it should be noted that we have not so far specified the relative normalization of the functions $R_{+2}S_{+2}$ and $R_{-2}S_{-2}$.

Finally, it should be noted that the solutions for Φ_0 and Φ_4 are in no way dependent on the choice of the arbitrary functions that occur in the general solutions of equations (24)–(29). The reason is that Φ_0 and Φ_4 (i.e. Ψ_0 and Ψ_4) are invariant to infinitesimal rotations of the tetrad frame (cf. Teukolsky 1973; appendix A).

6. Is there a preferred gauge to express the solution?

In §5 we have seen how we may choose a gauge in which $\Psi_1 = \Psi_3 = 0$; this is in fact the gauge we shall adopt in paper II. But one may ask if there is a preferred gauge for the problem. That there may be one such is suggested by the following considerations.

The set of equations (24)–(26) governing Φ_0, Φ_1, k, and s, for example, appears strangely truncated: the symmetry of the equations in Φ_0, k and s is only partially present with respect to Φ_1, k, and s. Thus, as we have seen in §5, the equations permit the elimination of Φ_1 from equations (24) and (25), while equation (26) provides exactly the 'right' relation between k and s to obtain a decoupled equation for Φ_0. We may, similarly, eliminate Φ_0 from equations (24) and (25) by virtue of the commutativity of the operators $(\mathscr{L}_2 - 3ia\sin\theta/\bar{\rho}^*)$ and $\Delta(\mathscr{D}_2^\dagger - 3/\bar{\rho}^*)$; but we do not have the 'right' relation between k and s to obtain a decoupled equation for Φ_1. But, exercising the freedom we have to subject the local (perturbed) tetrad frame to an infinitesimal rotation, we can rectify the situation by supplying (*ad hoc*?) the needed equation. Thus, with the additional equation

$$\Delta\left(\mathscr{D}_2^\dagger-\frac{3}{\bar\rho^*}\right)k+\left(\mathscr{L}_2-\frac{3ia\sin\theta}{\bar\rho^*}\right)s = 2\frac{\bar\rho}{(\bar\rho^*)^2}\Phi_1, \tag{46}$$

we can then complete the elimination of Φ_0, k, and s to obtain the decoupled equation

$$\left[\Delta\left(\mathscr{D}_2^\dagger-\frac{3}{\bar\rho^*}\right)\left(\mathscr{D}_0+\frac{3}{\bar\rho^*}\right)+\left(\mathscr{L}_2-\frac{3ia\sin\theta}{\bar\rho^*}\right)\left(\mathscr{L}_{-1}^\dagger+\frac{3ia\sin\theta}{\bar\rho^*}\right)\right]\Phi_1 = 12M\frac{\bar\rho}{(\bar\rho^*)^2}\Phi_1. \tag{47}$$

On expanding this equation, we obtain

$$[\Delta\mathscr{D}_2^\dagger\mathscr{D}_0+\mathscr{L}_2\mathscr{L}_{-1}^\dagger-6i\sigma(r+ai\cos\theta)]\,\Phi_1 = 0. \tag{48}$$

This equation is clearly separable: by the substitution,

$$\Phi_1 = R_{+1}(r)\,S_{+1}(\theta), \tag{49}$$

– we justify the designation of these functions with the subscript $+1$ presently – we obtain the pair of equations:

$$(\Delta\mathscr{D}_2^\dagger\mathscr{D}_0-6i\sigma r)\,R_{+1} = +(\lambda^{(1)}-2)\,R_{+1} \tag{50}$$

and

$$(\mathscr{L}_2\mathscr{L}_{-1}^\dagger+6a\sigma\cos\theta)\,S_{+1} = -(\lambda^{(1)}-2)\,S_{+1}, \tag{51}$$

where $\lambda^{(1)}$ is a separation constant. By making use of the readily verifiable identities

$$\Delta\mathscr{D}_2^\dagger\mathscr{D}_0 = \Delta\mathscr{D}_1\mathscr{D}_1^\dagger+4ir\sigma-2 \tag{52}$$

and

$$\mathscr{L}_2\mathscr{L}_{-1}^\dagger = \mathscr{L}_0^\dagger\mathscr{L}_1-4a\sigma\cos\theta+2, \tag{53}$$

we can rewrite equations (50) and (51) in the forms

$$(\Delta\mathscr{D}_1\mathscr{D}_1^\dagger-2i\sigma r)\,R_{+1} = \lambda^{(1)}R_{+1} \tag{54}$$

and

$$(\mathscr{L}_0^\dagger\mathscr{L}_1+2a\sigma\cos\theta)\,S_{+1} = -\lambda^{(1)}S_{+1}. \tag{55}$$

With the equations written in these forms, we recognize that R_{+1} and S_{+1} are the same radial and angular functions which describe a Maxwell field (for spin $s = +1$) in Kerr geometry (see appendix A, equations (A 12) and (A 13)).

Similarly, supplementing equations (27)–(29) by the equation

$$\left(\mathscr{D}_0-\frac{3}{\bar\rho^*}\right)n-\left(\mathscr{L}_2^\dagger-\frac{3ia\sin\theta}{\bar\rho^*}\right)l = 2\frac{\bar\rho}{(\bar\rho^*)^2}\Phi_3, \tag{56}$$

we can obtain a decoupled equation for Φ_3 which on separation enables us to express Φ_3 as $R_{-1}(r)\,S_{-1}(\theta)$, where $R_{-1}(r)$ and $S_{-1}(\theta)$ satisfy the equations appropriate for a Maxwell field with spin $s = -1$ (cf. appendix A, equations (A 14) and (A 15)).

Thus, *in a gauge which restores to equations* (24)–(29) *complete symmetry in the relations among the various quantities which occur in them, the components* Ψ_1 *and* Ψ_3 *of the Weyl tensor are reflected and transmitted by the Kerr black hole exactly as though they were electromagnetic waves.*

Should one, on the foregoing grounds, consider that gauge as a 'preferred one'?

And, if it is a preferred one, what is its physical meaning? A later investigation pertaining to the perturbations of the charged Kerr–Newman black hole led to the same equations (46) and (56) which we supplied (*ad hoc*!) to restore the symmetry of the underlying equations. A brief outline of these circumstances is given in appendix B.

7. The Teukolsky–Starobinsky identities

Some very remarkable identities connecting the functions belonging to the spins $s = +2$ and $s = -2$ were discovered by Teukolsky (see Teukolsky & Press 1974) and by Starobinsky (1973; see also Starobinsky & Churilov 1973). We shall collect them here in a more systematic manner than they have been hitherto. (It should, however, be noted that the identities, as we shall state them, are not exactly as they were originally formulated.) We shall also include some additional relations that will be needed in part II.

The proofs of various identities and relations, though essentially straightforward, require a large amount of algebraic manipulations with the derivative operators $\mathscr{L}$ and $\mathscr{D}$ and involve frequent uses of lemmas 1 and 3 (in §3). The proofs will, accordingly, be omitted.

Theorem 1. $\Delta^2\mathscr{D}_0\mathscr{D}_0\mathscr{D}_0\mathscr{D}_0 R_{-2}$ *is a multiple of* $\Delta^2 R_{+2}$,

and $\Delta^2\mathscr{D}_0^\dagger\mathscr{D}_0^\dagger\mathscr{D}_0^\dagger\mathscr{D}_0^\dagger\Delta^2 R_{+2}$ *is a multiple of* R_{-2}.

In view of equations (37) and (39) governing R_{-2} and $\Delta^2 R_{+2}$, the theorem is equivalent to asserting the commutation relation

$$\Delta^2\mathscr{D}_0\mathscr{D}_0\mathscr{D}_0\mathscr{D}_0(\Delta\mathscr{D}_{-1}^\dagger\mathscr{D}_0 + 6\mathrm{i}\sigma r) = (\Delta\mathscr{D}_{-1}\mathscr{D}_0^\dagger - 6\mathrm{i}\sigma r)\,\Delta^2\mathscr{D}_0\mathscr{D}_0\mathscr{D}_0\mathscr{D}_0, \tag{57}$$

and its complex conjugate.

Corollary 1. *By a suitable choice of the relative normalization of the functions* $\Delta^2 R_{+2}$ *and* R_{-2} *we can arrange that*

$$\Delta^2\mathscr{D}_0\mathscr{D}_0\mathscr{D}_0\mathscr{D}_0 R_{-2} = \mathscr{C}\Delta^2 R_{+2} \tag{58}$$

and

$$\Delta^2\mathscr{D}_0^\dagger\mathscr{D}_0^\dagger\mathscr{D}_0^\dagger\mathscr{D}_0^\dagger\Delta^2 R_{+2} = \mathscr{C}^* R_{-2}, \tag{59}$$

where $\mathscr{C}$ *is some complex constant.*

We can arrange the relative normalizations in the way prescribed since $\Delta^2 R_{+2}$ and R_{-2} satisfy complex-conjugate equations.

Corollary 2. *The square of the absolute value of the constant* $\mathscr{C}$ *in equations* (58) *and* (59) *is given by*

$$|\mathscr{C}|^2 = \lambda^2(\lambda+2)^2 - 8\sigma^2\lambda[\alpha^2(5\lambda+6) - 12a^2] + 144\sigma^4\alpha^4 + 144\sigma^2 M^2, \tag{60}$$

where

$$\alpha^2 = a^2 + (am/\sigma). \tag{61}$$

The relation (60) is a consequence of the identity

$$\Delta^2\mathscr{D}_0^\dagger\mathscr{D}_0^\dagger\mathscr{D}_0^\dagger\mathscr{D}_0^\dagger\Delta^2\mathscr{D}_0\mathscr{D}_0\mathscr{D}_0\mathscr{D}_0 = |\mathscr{C}|^2 \bmod \Delta\mathscr{D}_{-1}^\dagger\mathscr{D}_0 + 6i\sigma r - \lambda = 0, \tag{62}$$

which directly follows from equations (58) and (59).

We shall designate $\Delta^2 R_{+2}$ and R_{-2} by P_{+2} and P_{-2} when their relative normalization is compatible with equations (58) and (59). With these designations, we may note here (for later use) that the explicit form of equation (58) is

$$\begin{aligned}
&\Delta^3\mathscr{D}_0\mathscr{D}_0\mathscr{D}_0\mathscr{D}_0 P_{-2} \\
&\quad = \{4i\,K(iK - r + M)(\lambda - 6i\sigma r) \\
&\qquad + [(\lambda - 6i\sigma r)(\lambda + 2 + 2i\sigma r) - 12i\sigma(iK - r + M)]\,\Delta\}P_{-2} \\
&\qquad + \{-8iK[K^2 + (r-M)^2] + [4iK(\lambda+2) + 8i\sigma r(r-M)]\,\Delta - 8i\sigma\Delta^2\}\mathscr{D}_0 P_{-2} \\
&\quad = \Delta\mathscr{C} P_{+2}.
\end{aligned} \tag{63}$$

Equation (59) provides the complex conjugate of equation (63).

Theorem 2 $\quad \mathscr{L}_{-1}\mathscr{L}_0\mathscr{L}_1\mathscr{L}_2 S_{+2}$ *is a multiple of* S_{-2},

and $\quad \mathscr{L}_{-1}^\dagger\mathscr{L}_0^\dagger\mathscr{L}_1^\dagger\mathscr{L}_2^\dagger S_{-2}$ *is a multiple of* S_{+2}.

In view of equations (36) and (38) governing S_{+2} and S_{-2}, the theorem is equivalent to asserting the commutation relation

$$\mathscr{L}_{-1}\mathscr{L}_0\mathscr{L}_1\mathscr{L}_2(\mathscr{L}_{-1}^\dagger\mathscr{L}_2 + 6a\sigma\cos\theta) = (\mathscr{L}_{-1}\mathscr{L}_2^\dagger - 6a\sigma\cos\theta)\mathscr{L}_{-1}\mathscr{L}_0\mathscr{L}_1\mathscr{L}_2, \tag{64}$$

and its 'adjoint' obtained by replacing θ by $\pi - \theta$.

Corollary 1. *If* S_{+2} *and* S_{-2} *are simultaneously normalized, then*

$$\mathscr{L}_{-1}\mathscr{L}_0\mathscr{L}_1\mathscr{L}_2 S_{+2} = DS_{-2} \tag{65}$$

and

$$\mathscr{L}_{-1}^\dagger\mathscr{L}_0^\dagger\mathscr{L}_1^\dagger\mathscr{L}_2^\dagger S_{-2} = DS_{+2}, \tag{66}$$

where D is some real constant.

The fact that the same constant D occurs in both equations (65) and (66) follows from writing the normalization integral for S_{-2} (say), expressing the integrand as a product of S_{-2} and $\mathscr{L}_{-1}\mathscr{L}_0\mathscr{L}_1\mathscr{L}_2 S_{+2}$, and after four successive applications of lemma 4 (§3), making use of the relation (66).

Corollary 2. *The constant D in the relations* (65) *and* (66) *is given by*

$$D^2 = \lambda^2(\lambda+2)^2 - 8\sigma^2\lambda[\alpha^2(5\lambda+6) - 12a^2] + 144\sigma^4\alpha^4, \tag{67}$$

where

$$\alpha^2 = a^2 + (am/\sigma).$$

The relation (67) is a consequence of the identity

$$\mathscr{L}_{-1}^\dagger\mathscr{L}_0^\dagger\mathscr{L}_1^\dagger\mathscr{L}_2^\dagger\mathscr{L}_{-1}\mathscr{L}_0\mathscr{L}_1\mathscr{L}_2 = D^2 \bmod \mathscr{L}_{-1}^\dagger\mathscr{L}_2 + (\lambda + 6a\sigma\cos\theta) = 0. \tag{68}$$

It will be noticed that (cf. equation (60))

$$|\mathscr{C}|^2 = D^2 + 144\sigma^2 M^2, \tag{69}$$

a relation which could not have been anticipated.

Again, we may note here (for later use) the explicit form of equation (65):

$$\mathscr{L}_{-1}\mathscr{L}_0\mathscr{L}_1\mathscr{L}_2 S_{+2} = [12a\sigma(Q\sin\theta - \cos\theta) + (\lambda + 6a\sigma\cos\theta)(\lambda + 2 - 4Q^2 + 4Q\cot\theta - 2a\sigma\cos\theta)]\, S_{+2}$$
$$+ [8a\sigma\operatorname{cosec}\theta + 2Q(-2\lambda - 4 + 4Q^2 - 4\cot^2\theta)]\mathscr{L}_{+2}S_{+2} = DS_{-2}. \tag{70}$$

Equation (66) provides the 'adjoint' of equation (70) obtained by replacing θ by $\pi - \theta$.

(a) *Alternative forms of the basic equations*

Equations (63) and (70) enable us to write the basic equations governing the radial and the angular functions in forms which we shall find essential for the analysis of the remaining Newman–Penrose equations in paper II.

Equation (63) and its complex conjugate when written out explicitly take the forms

$$+\mathrm{i}B_1\,\mathrm{d}P_{-2}/\mathrm{d}r = \Delta(\mathscr{C}_1 + \mathrm{i}\mathscr{C}_2)P_{+2} - [(A_1 - B_1 K/\Delta) + \mathrm{i}A_2]P_{-2}, \tag{71}$$

and

$$-\mathrm{i}B_1\,\mathrm{d}P_{+2}/\mathrm{d}r = \Delta(\mathscr{C}_1 - \mathrm{i}\mathscr{C}_2)P_{-2} - [(A_1 - B_1 K/\Delta) - \mathrm{i}A_2]P_{+2}, \tag{72}$$

where

$$\mathscr{C} = \mathscr{C}_1 + \mathrm{i}\mathscr{C}_2, \quad \Gamma_1 = \lambda(\lambda+2) - 12\sigma^2\alpha^2, \quad \Gamma_2 = 12\sigma M, \tag{73}$$

$$A_1 = +\Delta\Gamma_1 - 4\lambda K^2 + 24\sigma K(a^2 - Mr),$$
$$A_2 = -\Delta\Gamma_2 - 4\lambda[K(r-M) + r\sigma\Delta] + 24\sigma r K^2 = -\tfrac{1}{2}B_{1,r},$$

and

$$B_1 = -8K[K^2 + (r-M)^2] + 4\Delta[K(\lambda+2) + 2\sigma r(r-M)] - 8\sigma\Delta^2. \tag{74}$$

We shall find it convenient to consider

$$X = P_{+2} + P_{-2} \quad \text{and} \quad \mathrm{i}Y = P_{+2} - P_{-2}, \tag{75}$$

instead of P_{+2} and P_{-2}. From equations (71) and (72) we readily find that the equations governing X and Y are

$$B_1\,\mathrm{d}X/\mathrm{d}r = (\Delta\mathscr{C}_2 - A_2)X + (\Delta\mathscr{C}_1 + A_1 - B_1 K/\Delta)Y \tag{76}$$

and

$$B_1\,\mathrm{d}Y/\mathrm{d}r = (\Delta\mathscr{C}_1 - A_1 + B_1 K/\Delta)X - (\Delta\mathscr{C}_2 + A_2)Y. \tag{77}$$

It is important to observe that equations (76) and (77) involve the real and the imaginary parts of the constant $\mathscr{C}$. We do not, as yet, know them separately: we know only the sum of their squares. This lacuna in our information must be filled eventually. In a remarkable way, the real and the imaginary parts of $\mathscr{C}$ become determinate via an integrability condition which emerges when we come to analysing the Newman–Penrose equations in their entirety. But we may note, meantime, that, as a result of that analysis, we find that

$$\mathscr{C}_1 = D \quad \text{and} \quad \mathscr{C}_2 = -12\sigma M. \tag{78}$$

We can rewrite the equations governing S_{+2} and S_{-2} in a similar fashion. Thus equation (70) written out explicitly is

$$\beta_1 \mathscr{L}_2 S_{+2} = D S_{-2} - (\alpha_1 + \alpha_2) S_{+2}, \tag{79}$$

where
$$\beta_1 = 8Q^3 - 8Q\cot^2\theta - 4(\lambda+2)Q + 8a\sigma\operatorname{cosec}\theta,$$
$$\alpha_1 = \lambda(\lambda+2) - 12a^2\sigma^2 + 24a\sigma Q\operatorname{cosec}\theta - 4\lambda Q^2,$$
and
$$\alpha_2 = -24a\sigma Q^2\cos\theta + 4\lambda(Q\cot\theta + a\sigma\cos\theta). \tag{80}$$

Combining equation (79) with its 'adjoint' (obtained by replacing θ by $\pi-\theta$), we obtain the pair of equations

$$\beta_1 \,\mathrm{d}(S_{+2}+S_{-2})/\mathrm{d}\theta = -(\alpha_2 + 2\beta_1\cot\theta)(S_{+2}+S_{-2}) - (\alpha_1 + \beta_1 Q + D)(S_{+2}-S_{-2}) \tag{81}$$
and
$$\beta_1 \,\mathrm{d}(S_{+2}-S_{-2})/\mathrm{d}\theta = -(\alpha_1 + \beta_1 Q - D)(S_{+2}+S_{-2}) - (\alpha_2 + 2\beta_1\cot\theta)(S_{+2}-S_{-2}). \tag{82}$$

Equations (81) and (82) can be brought to forms very similar to equations (76) and (77) by rewriting them in terms of the variables

$$L = (S_{+2}+S_{-2})\sin^2\theta, \quad M = (S_{+2}-S_{-2})\sin^2\theta, \quad \text{and} \quad \mu = \cos\theta. \tag{83}$$

We find
$$\mathscr{B}_1 \,\mathrm{d}L/\mathrm{d}\mu = \mathscr{A}_2 L + [\mathscr{A}_1 + \mathscr{B}_1 \mathscr{Q}/(1-\mu^2) + D(1-\mu^2)]\,M, \tag{84}$$
and
$$\mathscr{B}_1 \,\mathrm{d}M/\mathrm{d}\mu = \mathscr{A}_2 M + [\mathscr{A}_1 + \mathscr{B}_1 \mathscr{Q}/(1-\mu^2) - D(1-\mu^2)]\,L, \tag{85}$$

where
$$\mathscr{A}_1 = \alpha_1\sin^2\theta = \Gamma_1(1-\mu^2) + 24a\sigma\mathscr{Q} - 4\lambda\mathscr{Q}^2,$$
$$\mathscr{A}_2 = \alpha_2\sin^2\theta = 4\lambda\mathscr{Q}\mu + 4\lambda a\sigma\mu(1-\mu^2) - 24a\sigma\mathscr{Q}^2\mu = \tfrac{1}{2}\mathscr{B}_{1,\mu},$$
$$\mathscr{B}_1 = \beta_1\sin^3\theta = 8\mathscr{Q}^3 - 8\mathscr{Q} - 4\lambda\mathscr{Q}(1-\mu^2) + 8a\sigma(1-\mu^2),$$
and
$$\mathscr{Q} = Q\sin\theta = a\sigma(1-\mu^2) + m. \tag{86}$$

8. Concluding remarks

Besides obtaining the solutions for Ψ_0, Ψ_1, Ψ_3, Ψ_4, κ, σ, λ, and ν (all of which vanish in the Kinnersley frame in the stationary state) in forms that will be needed for our subsequent analysis (in paper II) an attempt has been made to present a coherent treatment of the basic equations which is direct and simple. It is particularly important that Teukolsky's equations have emerged directly expressed in terms of the basic derivative operators $\mathscr{L}$ and $\mathscr{D}$. However, attention must be drawn to two significant lacunae in our information that remain to be filled. *First*, we do not as yet know the real and the imaginary parts of the Teukolsky–Starobinsky constant $\mathscr{C}$, separately, though we do know its absolute value. *Second*, even on the assumption that the radial functions $\Delta^2 R_{+2}$ and R_{-2} are normalized, relatively, so as to be compatible with the relations (58) and (59), the solutions for λ and ν given in equations (44) and (45) (in the gauge in which $\Psi_1 = \Psi_3 = 0$) yet need an additional numerical factor which is presently not known. These two lacunae in our information will be filled when we come to analysing (in paper II) the perturbations in the quantities which are finite in the stationary state.

The research reported in this paper has in part been supported by the National Science Foundation under grant PHY 76-81102 with the University of Chicago.

Appendix A. The solution of Maxwell's equations

We shall briefly outline a manner of solution of Maxwell's equations in which Teukolsky's equations emerge directly expressed in terms of the basic derivative operators (as they do in the treatment of the gravitational perturbations in §5).

In the Newman–Penrose formalism, the Maxwell tensor F. is replaced by the three scalars

$$\phi_0 = F_{ij} l^i m^j, \quad \phi_1 = \tfrac{1}{2} F_{ij}(l^i n^j + \overline{m}^i m^j), \quad \text{and} \quad \phi_2 = F_{ij} \overline{m}^i n^j. \tag{A 1}$$

The equations governing these scalars (N.P. equations (A 1)) in the Kinnersley frame take simple forms if they are written for

$$\Phi_0 = \phi_0, \quad \Phi_1 = \phi_1 \overline{\rho}^* \sqrt{2}, \quad \text{and} \quad \Phi_2 = 2\phi_2 (\overline{\rho}^*)^2. \tag{A 2}$$

We find

$$\left(\mathscr{L}_1 - \frac{\mathrm{i}a \sin\theta}{\overline{\rho}^*}\right)\Phi_0 = \left(\mathscr{D}_0 + \frac{1}{\overline{\rho}^*}\right)\Phi_1, \tag{A 3}$$

$$\left(\mathscr{L}_0 + \frac{\mathrm{i}a \sin\theta}{\overline{\rho}^*}\right)\Phi_1 = \left(\mathscr{D}_0 - \frac{1}{\overline{\rho}^*}\right)\Phi_2, \tag{A 4}$$

$$\left(\mathscr{L}_1^\dagger - \frac{\mathrm{i}a \sin\theta}{\overline{\rho}^*}\right)\Phi_2 = -\Delta\left(\mathscr{D}_0^\dagger + \frac{1}{\overline{\rho}^*}\right)\Phi_1, \tag{A 5}$$

and

$$\left(\mathscr{L}_0^\dagger + \frac{\mathrm{i}a \sin\theta}{\overline{\rho}^*}\right)\Phi_1 = -\Delta\left(\mathscr{D}_1^\dagger - \frac{1}{\overline{\rho}^*}\right)\Phi_0. \tag{A 6}$$

It is evident that the commutativity of the operators

$$(\mathscr{D}_0 + 1/\overline{\rho}^*) \quad \text{and} \quad (\mathscr{L}_0^\dagger + \mathrm{i}a\sin\theta/\overline{\rho}^*)$$

enables us to eliminate Φ_1 from equations (A 3) and (A 6) and to obtain a decoupled equation for Φ_0. We obtain

$$\left[\left(\mathscr{L}_0^\dagger + \frac{\mathrm{i}a \sin\theta}{\overline{\rho}^*}\right)\left(\mathscr{L}_1 - \frac{\mathrm{i}a \sin\theta}{\overline{\rho}^*}\right) + \Delta\left(\mathscr{D}_1 + \frac{1}{\overline{\rho}^*}\right)\left(\mathscr{D}_1^\dagger - \frac{1}{\overline{\rho}^*}\right)\right]\Phi_0 = 0. \tag{A 7}$$

A similar elimination of Φ_1 from equations (A 4) and (A 5) yields the equation

$$\left[\left(\mathscr{L}_0 + \frac{\mathrm{i}a \sin\theta}{\overline{\rho}^*}\right)\left(\mathscr{L}_1^\dagger - \frac{\mathrm{i}a \sin\theta}{\overline{\rho}^*}\right) + \Delta\left(\mathscr{D}_0^\dagger + \frac{1}{\overline{\rho}^*}\right)\left(\mathscr{D}_0 - \frac{1}{\overline{\rho}^*}\right)\right]\Phi_2 = 0. \tag{A 8}$$

On expanding equations (A 7) and (A 8) we are left with

$$[\Delta \mathscr{D}_1 \mathscr{D}_1^\dagger + \mathscr{L}_0^\dagger \mathscr{L}_1 - 2\mathrm{i}\sigma(r + a\mathrm{i}\cos\theta)]\,\Phi_0 = 0 \tag{A 9}$$

and

$$[\Delta \mathscr{D}_0^\dagger \mathscr{D}_0 + \mathscr{L}_0 \mathscr{L}_1^\dagger + 2\mathrm{i}\sigma(r + a\mathrm{i}\cos\theta)]\,\Phi_2 = 0. \tag{A 10}$$

These equations are clearly separable. Thus, with the substitutions

$$\Phi_0 = R_{+1}(r)\,S_{+1}(\theta) \quad \text{and} \quad \Phi_2 = R_{-1}(r)\,S_{-1}(\theta) \tag{A 11}$$

(where $R_{\pm 1}(r)$ and $S_{\pm 1}(\theta)$ are, respectively, functions of r and θ only), equations (A 9) and (A 10) separate to give the two pairs of equations

$$(\Delta\mathscr{D}_1\mathscr{D}_1^\dagger - 2i\sigma r)\,R_{+1} = \lambda R_{+1}, \tag{A 12}$$

$$(\mathscr{L}_0^\dagger\mathscr{L}_1 + 2a\sigma\cos\theta)\,S_{+1} = -\lambda S_{+1}, \tag{A 13}$$

$$(\Delta\mathscr{D}_0^\dagger\mathscr{D}_0 + 2i\sigma r)\,R_{-1} = \lambda R_{-1}, \tag{A 14}$$

and

$$(\mathscr{L}_0\mathscr{L}_1^\dagger - 2a\sigma\cos\theta)\,S_{-1} = -\lambda S_{-1}, \tag{A 15}$$

where λ is a separation constant. (We have not distinguished the separation constants derived from equations (A 9) and (A 10) since the characteristic values of λ determined by the two equations are the same.)

We readily verify that equations (A 12)–(A 15) are equivalent to Teukolsky's equations; and his separation constant E is related to our characteristic value parameter λ by (cf. equation (40))

$$\lambda = E + a^2\sigma^2 + 2am\sigma. \tag{A 16}$$

By rewriting equation (A 12) in the form

$$(\Delta\mathscr{D}_0\mathscr{D}_0^\dagger - 2i\sigma r)\,\Delta R_{+1} = \lambda\Delta R_{+1}, \tag{A 17}$$

and comparing it with equation (A 14), we conclude that, if ΔR_{+1} is a solution of equation (A 17), its complex conjugate is a solution of equation (A 14).

The functions belonging to spin one satisfy identities similar to those satisfied by the functions belonging to spin two. The identities appropriate for spin one (in contrast to those for spin two) are relatively easy to verify: the required manipulations with the derivative operators are, in many cases, minimal. We shall state the identities as theorems analogous to those enunciated in §7.

THEOREM 1. $\Delta\mathscr{D}_0\mathscr{D}_0 R_{-1}$ *is a multiple of* ΔR_{+1},

and $\Delta\mathscr{D}_0^\dagger\mathscr{D}_0^\dagger\Delta R_{+1}$ *is a multiple of* R_{-1}.

The theorem is equivalent to asserting the commutation relation

$$\Delta\mathscr{D}_0\,\mathscr{D}_0(\Delta\mathscr{D}_0^\dagger\mathscr{D}_0 + 2i\sigma r) = (\Delta\mathscr{D}_0\mathscr{D}_0^\dagger - 2i\sigma r)\,\Delta\mathscr{D}_0\mathscr{D}_0, \tag{A 18}$$

and its complex conjugate.

COROLLARY 1. *By a suitable choice of the relative normalization of the functions* ΔR_{+1} *and* R_{-1} *we can arrange that*

$$\Delta\mathscr{D}_0\mathscr{D}_0 R_{-1} = \mathscr{C}\Delta R_{+1} \tag{A 19}$$

and

$$\Delta\mathscr{D}_0^\dagger\mathscr{D}_0^\dagger\Delta R_{+1} = \mathscr{C}^* R_{-1}, \tag{A 20}$$

where C *is a constant (which can be complex‡).*

‡ We shall presently show that $\mathscr{C}$ is in fact real.

COROLLARY 2. *The square of the absolute value of the constant* $\mathscr{C}$ *is given by*

$$|\mathscr{C}|^2 = \lambda\!\!\!^{-2} - 4\alpha^2\sigma^2 \quad \text{where} \quad \alpha^2 = a^2 + (am/\sigma). \tag{A 21}$$

The relation (A 21) is a consequence of the identity

$$\Delta\mathscr{D}_0^\dagger \mathscr{D}_0^\dagger \Delta\mathscr{D}_0 \mathscr{D}_0 = |\mathscr{C}|^2 \bmod \Delta\mathscr{D}_0^\dagger\mathscr{D}_0 + 2\mathrm{i}\sigma r - \lambda\!\!\!^{-} = 0. \tag{A 22}$$

THEOREM 2. $\mathscr{L}_0\mathscr{L}_1 S_{+1}$ *is a multiple of* S_{-1},

and $\mathscr{L}_0^\dagger \mathscr{L}_1^\dagger S_{-1}$ *is a multiple of* S_{+1}.

This theorem is equivalent to asserting the commutation relation,

$$\mathscr{L}_0\mathscr{L}_1(\mathscr{L}_0^\dagger\mathscr{L}_1 + 2a\sigma\cos\theta) = (\mathscr{L}_0\mathscr{L}_1^\dagger - 2a\sigma\cos\theta)\,\mathscr{L}_0\mathscr{L}_1, \tag{A 23}$$

and its 'adjoint' obtained by replacing θ by $\pi - \theta$.

COROLLARY 1. *If* S_{+1} *and* S_{-1} *are simultaneously normalized, then*

$$\mathscr{L}_0\mathscr{L}_1 S_{+1} = DS_{-1} \quad \textit{and} \quad \mathscr{L}_0^\dagger\mathscr{L}_1^\dagger S_{-1} = DS_{+1}, \tag{A 24}$$

where D *is a real constant.*

COROLLARY 2. *The constant* D *is given by*

$$D^2 = \lambda\!\!\!^{-2} - 4\alpha^2\sigma^2, \quad \textit{where} \quad \alpha^2 = a^2 + (am/\sigma). \tag{A 25}$$

The relation (A 25) is a consequence of the identity

$$\mathscr{L}_0^\dagger\mathscr{L}_1^\dagger\mathscr{L}_0\mathscr{L}_1 = D^2 \bmod \mathscr{L}_0^\dagger\mathscr{L}_1 + (\lambda\!\!\!^{-} + 2a\sigma\cos\theta) = 0. \tag{A 26}$$

It will be observed that (in contrast to the case $s = \pm 2$, cf. equation (69))

$$D^2 = |\mathscr{C}|^2. \tag{A 27}$$

We shall now show that $\mathscr{C}$ is in fact equal to D, and, therefore, real.

In view of the commutativity of the operators $(\mathscr{D}_0 + 1/\bar{\rho}^*)$ and $(\mathscr{L}_0 + \mathrm{i}a\sin\theta/\bar{\rho}^*)$, we can eliminate Φ_1 from equations (A 3) and (A 4) to obtain a relation directly between Φ_0 and Φ_2. We find

$$\left(\mathscr{L}_0 + \frac{\mathrm{i}a\sin\theta}{\bar{\rho}^*}\right)\left(\mathscr{L}_1 - \frac{\mathrm{i}a\sin\theta}{\bar{\rho}^*}\right)\Phi_0 = \left(\mathscr{D}_0 + \frac{1}{\bar{\rho}^*}\right)\left(\mathscr{D}_0 - \frac{1}{\bar{\rho}^*}\right)\Phi_2. \tag{A 28}$$

On expanding this relation, we are left with

$$\mathscr{L}_0\mathscr{L}_1\Phi_0 = \mathscr{D}_0\mathscr{D}_0\Phi_2. \tag{A 29}$$

On inserting in equation (A 29) the solutions for Φ_0 and Φ_2 given in equations (A 11), we obtain

$$(\mathscr{L}_0\mathscr{L}_1 S_{+1})/S_{-1} = (\mathscr{D}_0\mathscr{D}_0 R_{-1})/R_{+1}. \tag{A 30}$$

If we now suppose that S_{+1} and S_{-1} are normalized, then it follows from equation (A 24) that

$$\mathscr{D}_0\mathscr{D}_0 R_{-1} = DR_{+1}. \tag{A 31}$$

Comparison of equations (A 19) and (A 31) shows that *the relative normalization of*

ΔR_{+1} and R_{-1} implied in writing equations (A 19) *and* (A 20) *must be associated with the simultaneous normalization of the angular functions S_{+1} and S_{-1}.* If follows at the same time that

$$\mathscr{C} = \mathscr{C}^* = D. \tag{A 32}$$

Finally, we may note that, with the stated normalizations of the radial and the angular functions, the solutions for ϕ_0 and ϕ_2 are given by (cf. equation (A 2))

$$\phi_0 = R_{+1}S_{+1} \quad \text{and} \quad \phi_2 = [1/2(\bar{\rho}^*)^2]\, R_{-1}S_{-1}. \tag{A33}$$

The corresponding solution for ϕ_1 is given in Chandrasekhar (1976).

Appendix B. The equations governing the perturbations of the Kerr–Newman black hole

It is not our intention in this appendix to go in any depth into the problem of the perturbations of the charged Kerr–Newman black hole. Our intention is, rather, to show how equations (46) and (56), by which we supplemented the original set of Newman–Penrose equations (24)–(29) to rectify their 'truncated' character, arises very naturally when we consider the coupled electromagnetic-gravitational perturbations of the Kerr–Newman black hole.

It is known (cf. Bose 1975; Lee 1976; Chitre 1976) that, in the Newman–Penrose formalism, the Kerr–Newman space-time allows itself to be described in *exactly* the same way as the Kerr space-time with the only formal difference that the 'horizon function' $\Delta(r)$ must be redefined with the meaning

$$\Delta(r) = r^2 + a^2 + Q_*^2 - 2Mr, \tag{B 1}$$

where Q_* denotes the charge of the black hole. In particular, the expressions (7) and (9) for the null basis and the spin coefficients continue to be valid with the new meaning for Δ. However, Ψ_2 has a different value now; it is given by (cf. equation (10))

$$\Psi_2 = -\frac{M}{(\bar{\rho}^*)^3} + \frac{Q_*^2}{\bar{\rho}(\bar{\rho}^*)^3}. \tag{B 2}$$

Also, ϕ_1 (one of the three scalars replacing Maxwell's tensor and defined in equation (A 1)) is non-vanishing and has the value

$$\phi_1 = \frac{Q_*}{2(\bar{\rho}^*)^2}. \tag{B 3}$$

The various Newman–Penrose equations, which govern the quantities which vanish in the stationary state, have been written down by Lee (1976). Using combinations of these equations (different from Lee's) and exercising the freedom to subject the (perturbed) null-tetrad frame to infinitesimal rotations, to make the changes in ϕ_0 and ϕ_2 equal to zero, we find that, with the same definitions (22) and

(23) as in the Kerr case, the relevant equations are

$$\left(\mathscr{L}_2 - \frac{3ia\sin\theta}{\bar{\rho}^*}\right)\Phi_0 - \left(\mathscr{D}_0 + \frac{3}{\bar{\rho}^*}\right)\Phi_1 = -2k\left[3\left(M - \frac{Q_*^2}{\bar{\rho}}\right) + Q_*^2\frac{\bar{\rho}^*}{(\bar{\rho})^2}\right], \tag{B 4}$$

$$\Delta\left(\mathscr{D}_2^\dagger - \frac{3}{\bar{\rho}^*}\right)\Phi_0 + \left(\mathscr{L}_{-1}^\dagger + \frac{3ia\sin\theta}{\bar{\rho}^*}\right)\Phi_1 = +2s\left[3\left(M - \frac{Q_*^2}{\bar{\rho}}\right) - Q_*^2\frac{\bar{\rho}^*}{(\bar{\rho})^2}\right], \tag{B 5}$$

$$\left(\mathscr{D}_0 + \frac{3}{\bar{\rho}^*}\right)s - \left(\mathscr{L}_{-1}^\dagger + \frac{3ia\sin\theta}{\bar{\rho}^*}\right)k = \frac{\bar{\rho}}{(\bar{\rho}^*)^2}\Phi_0, \tag{B 6}$$

and

$$\Delta\left(\mathscr{D}_2^\dagger - \frac{3}{\bar{\rho}^*}\right)k + \left(\mathscr{L}_2 - \frac{3ia\sin\theta}{\bar{\rho}^*}\right)s = 2\frac{\bar{\rho}}{(\bar{\rho}^*)^2}\Phi_1; \tag{B 7}$$

and a similar set of equations involving Φ_4, Φ_3, l, and n.

We observe that equation (B 7) is exactly the same as equation (46) which we supplied (*ad hoc*!) to restore the truncated symmetry of the equations (24)–(26). The remarkable feature of the foregoing equations is that equation (B 7) which arises so naturally when treating the perturbations of the charged Kerr–Newman black hole, persists, unaffected, even in the limit $Q_* \to 0$. In other words, the Kerr metric seems, in some sense, to be 'aware' that a charged version of itself exists!

By treating equations (B 4)–(B 7) in the same way that these equations were treated in §§5 and 6 when $Q_* = 0$, we obtain the pair of equations (cf. equations (32) and (48))

$$(\Delta\mathscr{D}_1\mathscr{D}_2^\dagger + \mathscr{L}_{-1}^\dagger\mathscr{L}_2 - 6i\sigma\bar{\rho})\Phi_0 = -2Q_*^2\left(\mathscr{L}_{-1}^\dagger\frac{k\bar{\rho}^*}{\bar{\rho}^2} + \mathscr{D}_0\frac{s\bar{\rho}^*}{\bar{\rho}^2}\right) \tag{B 8}$$

and

$$(\Delta\mathscr{D}_2^\dagger\mathscr{D}_0 + \mathscr{L}_2\mathscr{L}_{-1}^\dagger - 6i\sigma\bar{\rho})\Phi_1 = +2Q_*^2\left(\Delta\mathscr{D}_2^\dagger\frac{k\bar{\rho}^*}{\bar{\rho}^2} - \mathscr{L}_2\frac{s\bar{\rho}^*}{\bar{\rho}^2}\right). \tag{B 9}$$

But these equations do not seem to allow a separation of the variables in the same obvious way as when $Q_* = 0$.

References

Bose, S. K. 1975 *J. Math. Phys.* **16**, 772.
Boyer, R. H. & Lindquist, R. W. 1967 *J. Math. Phys.* **8**, 265.
Chandrasekhar, S. 1976 *Proc. R. Soc. Lond.* A **349**, 1.
Chandrasekhar, S. & Detweiler, S. 1975 *Proc. R. Soc. Lond.* A **345**, 145.
Chandrasekhar, S. & Detweiler, S. 1976 *Proc. R. Soc. Lond.* A **350**, 165.
Chitre, D. M. 1976 *Phys. Rev.* **13**, 2713.
Chrzanowski, P. L. 1975 *Phys. Rev.* D **11**, 2042.
Cohen, J. M. & Kegeles, L. S. 1974 *Phys. Rev.* D **10**, 1070.
Detweiler, S. 1977 *Proc. R. Soc. Lond.* A **352**, 381.
Kinnersley, W. 1969 *J. Math. Phys.* **10**, 1195.
Lee, C. H. 1976 *J. Math. Phys.* **17**, 1226.
Newman, E. & Penrose, R. 1962 *J. Math. Phys.* **3**, 566.
Press, W. H. & Teukolsky, S. A. 1973 *Astrophys. J.* **185**, 649.
Starobinsky, A. A. 1973 *Zh. eksp. teor. Fiz.* **64**, 48 (also 1974 *Soviet Phys. JETP* **37**, 28).

Starobinsky, A. A. & Churilov, S. M. 1973 *Zh. eksp. teor. Fiz.* **65**, 3 (also 1974 *Soviet Phys. JETP*, **38**, 1).
Teukolsky, S. A. 1972 *Phys. Rev. Lett.* **29**, 114.
Teukolsky, S. A. 1973 *Astrophys. J.* **185**, 638.
Teukolsky, S. A. & Press, W. H. 1974 *Astrophys. J.* **193**, 443.
Wald, R. 1973 *J. Math. Phys.* **14**, 1453.

11

The gravitational perturbations of the Kerr black hole. II. The perturbations in the quantities which are finite in the stationary state

BY S. CHANDRASEKHAR, F.R.S.
University of Chicago, Chicago, Illinois, 60637, U.S.A.

(*Received* 20 *June* 1977)

The present paper completes the integration of the linearized Newman–Penrose equations governing the gravitational perturbations of the Kerr black hole. The equations which determine the solutions are the four (complex) Bianchi identities (not used in part I) and the 24 equations which follow from the commutation relations. The principal results are (1) the demonstration that the perturbation in the Weyl scalar Ψ_2 must vanish in a gauge in which the scalars Ψ_1 and Ψ_3 are assumed to vanish identically; (2) the determination of the relative normalization of the radial functions (left unspecified in part I) through an integrability condition. Further, the solution to the integrability condition defines a function involving quadratures over Teukolsky's radial and angular functions; and it is in terms of this function that the perturbations in the metric coefficients are determined.

1. INTRODUCTION

The present paper is a continuation of an earlier one (Chandrasekhar 1977; this paper will be referred to hereafter as I). It completes the solution to the problem of the gravitational perturbations of the Kerr black hole.

We begin with a statement of the problem.

2. THE STATEMENT OF THE PROBLEM

We start with a description of Kerr geometry in a Newman–Penrose formalism with Kinnersley's choice of the basic null-tetrad frame (cf. I, §2). In the chosen frame, four of the five complex scalars, representing the Weyl tensor, vanish; and only one of them (namely, Ψ_2)‡ does not. Also, of the 12 complex spin coefficients, four of them (namely, κ, σ, λ, and ν) vanish, which reflects the fact that the two real null-vectors of the basis (namely, $\boldsymbol{l}$ and $\boldsymbol{n}$) form shear-free null congruences of geodesics. (With Kinnersley's choice of the basis, besides κ, σ, λ, and ν, the spin-coefficient ϵ also vanishes; but the vanishing of ϵ has no particular significance in our present context.)

‡ We shall continue to use the same notation as in I; and since the present paper is a continuation of I and must be read alongside it, we shall not repeat the definitions.

The screened portions of this paper (pp. 168, 169, 170, 179, and 180) contain misstatements, as explained in the footnote on page 183 of paper 12 in this volume. The erroneous statements are corrected in that paper.

When the Kerr black hole is subject to gravitational perturbations, the quantities which vanish in the stationary state will become quantities of the first order of smallness; and the quantities which are finite in the stationary state will, likewise, experience changes of the first order. As in I, with the intent of analysing the perturbations into their normal modes, we shall suppose that the changes which the various quantities experience have a t-and a φ-dependence given by

$$e^{i(\sigma t+m\varphi)}. \tag{1}$$

This common factor in all of the quantities expressing the perturbations will be suppressed; and the symbols representing them will be their corresponding amplitudes.

A general perturbation of the Kerr geometry, in the Newman–Penrose formalism, will be described by the two sets of amplitudes

$$\Psi_0,\quad \Psi_1,\quad \Psi_3,\quad \Psi_4,\quad \kappa,\quad \sigma,\quad \lambda,\quad \text{and}\quad \nu, \tag{2}$$

and

$$\Psi_2^{(1)},\quad \tilde{\rho}^{(1)},\quad \mu^{(1)},\quad \tau^{(1)},\quad \pi^{(1)},\quad \alpha^{(1)},\quad \beta^{(1)},\quad \gamma^{(1)},\quad \epsilon^{(1)},$$
$$\boldsymbol{l}^{(1)},\quad \boldsymbol{n}^{(1)},\quad \boldsymbol{m}^{(1)}\quad \text{and}\quad \boldsymbol{m}^{*(1)}, \tag{3}$$

where the superscripts '(1)', in the second set of amplitudes, distinguish them from their unperturbed values. (We have suppressed the corresponding distinguishing superscripts, in the first set of amplitudes, for the quantities which vanish in the stationary state.)

In I we have obtained explicit solutions for the quantities listed in (2). In particular, the solutions for the spin coefficients, κ, σ, λ and ν, in the gauge,

$$\Psi_1 \equiv \Psi_3 \equiv 0, \tag{4}$$

are given in I, eqns (42)–(45). And the solutions for Ψ_0 and Ψ_4 (which do not depend on the choice of gauge) are expressed in terms of Teukolsky's radial and angular functions, $R_{\pm 2}(r)$ and $S_{\pm 2}(\theta)$, by

$$\Psi_0 = R_{+2}S_{+2} \quad \text{and} \quad \Psi_4 = [1/(\bar{\rho}^*)^4]\,R_{-2}S_{-2}, \tag{5}$$

where it should be remembered that, if the radial and the angular functions are so chosen that $\Delta^2 R_{+2}$ and R_{-2} are compatible with I, eqns (58) and (59) and S_{+2} and S_{-2} are normalized to unity, then an additional numerical factor must be included in the expression for Ψ_4 – a factor which has not, as yet, been specified (cf. I, § 8). We shall return in § 9 to the question of specifying this undetermined factor and the related one of determining the real and the imaginary parts of the Teukolsky–Starobinsky constant $\mathscr{C}$ (I, eqn (60)).

(a) The representation of the perturbations in the basic tetrad by a matrix

It is convenient to introduce an index notation for the Kinnersley basis. Thus, letting

$$\boldsymbol{l}^1 = \boldsymbol{l},\quad \boldsymbol{l}^2 = \boldsymbol{n},\quad \boldsymbol{l}^3 = \boldsymbol{m}\quad \text{and}\quad \boldsymbol{l}^4 = \boldsymbol{m}^*, \tag{6}$$

we can express the perturbation $\boldsymbol{l}^{i(1)}$, in the vector $\boldsymbol{l}^i$, as a linear combination of the basis vectors $\boldsymbol{l}^j$, in the manner

$$\boldsymbol{l}^{i(1)} = A^i_j \boldsymbol{l}^j. \tag{7}$$

The perturbations in the basis vectors are then fully described by the matrix $\boldsymbol{A}$.

Since $\boldsymbol{l}^1$ and $\boldsymbol{l}^2$ are real and $\boldsymbol{l}^3$ and $\boldsymbol{l}^4$ are complex conjugates, it follows that the matrix elements A^1_1, A^1_2, A^2_2, and A^2_1 are real, while all the remaining elements are complex; also, that the elements in which the indices 3 and 4 are replaced one by the other are complex conjugates. It is, however, clear that the specification of the matrix $\boldsymbol{A}$ will require 16 real functions.

We shall find that the following multiples of A^i_j $(i \neq j)$ occur very naturally in the subsequent analysis:

$$\begin{aligned}
&F^1_2 = \frac{\Delta}{2\rho^2}A^1_2; \quad &&F^2_1 = \frac{2\rho^2}{\Delta}A^2_1; \quad &&F^3_1 = \frac{A^3_1}{\bar{\rho}^*}; \quad &&F^4_1 = \frac{A^4_1}{\bar{\rho}},\\
&F^1_3 = \frac{\Delta}{2\rho^2\bar{\rho}}A^1_3; \quad &&F^2_3 = \frac{A^2_3}{\bar{\rho}}; \quad &&F^3_2 = \frac{\Delta}{2\rho^2\bar{\rho}^*}A^3_2; \quad &&F^4_2 = \frac{\Delta}{2\rho^2\bar{\rho}}A^4_2,\\
&F^1_4 = \frac{\Delta}{2\rho^2\bar{\rho}^*}A^1_4; \quad &&F^2_4 = \frac{A^2_4}{\bar{\rho}^*}; \quad &&F^3_4 = \frac{A^3_4}{(\bar{\rho}^*)^2}; \quad &&F^4_3 = \frac{A^4_3}{(\bar{\rho})^2}.
\end{aligned} \tag{8}‡$$

We shall also find that the following combinations of F^i_j play important roles in the theory:

$$\begin{aligned}
F &= F^1_3 + F^1_4; \quad & B_1 &= (F^3_1 + F^3_2) + (F^4_1 + F^4_2),\\
G &= F^2_3 + F^2_4; \quad & B_2 &= (F^3_1 + F^3_2) - (F^4_1 + F^4_2),\\
H &= F^3_1 - F^3_2; \quad & C_1 &= (F^1_3 + F^2_3) - (F^1_4 + F^2_4),\\
J &= F^4_1 - F^4_2; \quad & C_2 &= (F^1_3 - F^2_3) - (F^1_4 - F^2_4).
\end{aligned} \tag{9}$$

(b) *The perturbations in the metric coefficients*

As we have stated in I, the central problem, in any theory of perturbations of a space-time geometry, is the specification of the normal modes of the perturbations in the metric coefficients. These metric perturbations are related to the perturbations in the basic null-tetrad by

$$g^{\mu\nu(1)} = l^\mu n^{\nu(1)} + l^{\mu(1)} n^\nu - m^\mu m^{*\nu(1)} - m^{*\mu(1)} m^\nu + l^\nu n^{\mu(1)} + l^{\nu(1)} n^\mu - m^{*\nu} m^{\mu(1)} - m^{\nu(1)} m^{*\mu}. \tag{10}$$

Evaluating the various components of $g^{\mu\nu(1)}$, in accordance with the foregoing formula and with the aid of equation (7) and I, eqn (7), we find:

$$\left.\begin{aligned}
g^{tt(1)} &= \frac{(r^2+a^2)^2}{\rho^2\Delta}(A^1_1 + A^2_2) - \frac{a^2\sin^2\theta}{\rho^2}(A^3_3 + A^4_4)\\
&\quad + \frac{(r^2+a^2)^2}{\rho^2\Delta}(F^1_2 + F^2_1) + a^2(F^3_4 + F^4_3)\sin^2\theta + \frac{\sqrt{2}}{\Delta}ia(r^2+a^2)(C_1 + B_2)\sin\theta,\\
g^{rr(1)} &= -\frac{\Delta}{\rho^2}(A^1_1 + A^2_2) + \frac{\Delta}{\rho^2}(F^1_2 + F^2_1),
\end{aligned}\right|$$

‡ As defined, F^1_2 and F^2_1 are real and $F^i_4 = (F^i_3)^*$, $F^4_i = (F^3_i)^*$ $(i = 1, 2)$ and $F^3_4 = (F^4_3)^*$.

$$
\left.
\begin{aligned}
g^{\theta\theta(1)} &= -\frac{1}{\rho^2}(A^3_3+A^4_4)-(F^3_4+F^4_3),\\
g^{\varphi\varphi(1)} &= \frac{a^2}{\rho^2\Delta}(A^1_1+A^2_2)-\frac{1}{\rho^2}(A^3_3+A^4_4)\operatorname{cosec}^2\theta\\
&\qquad +\frac{a^2}{\rho^2\Delta}(F^1_2+F^2_1)+(F^3_4+F^4_3)\operatorname{cosec}^2\theta+\frac{\sqrt{2}}{\Delta}\,\mathrm{i}a(C_1+B_2)\operatorname{cosec}\theta,\\
g^{tr(1)} &= -\frac{r^2+a^2}{\rho^2}(F^1_2-F^2_1)+\frac{\mathrm{i}a\sin\theta}{\sqrt{2}}(H-J-C_2),\\
g^{t\theta(1)} &= +\mathrm{i}a(F^3_4-F^4_3)\sin\theta+\frac{r^2+a^2}{\Delta\sqrt{2}}(F+G-B_1),\\
g^{t\varphi(1)} &= a\frac{r^2+a^2}{\Delta\rho^2}(A^1_1+A^2_2)-\frac{a}{\rho^2}(A^3_3+A^4_4)\\
&\quad +a\frac{r^2+a^2}{\Delta\rho^2}(F^1_2+F^2_1)+a(F^3_4+F^4_3)+\frac{\mathrm{i}}{\Delta\sqrt{2}}[(r^2+a^2)+a^2\sin^2\theta](C_1+B_2)\operatorname{cosec}\theta,\\
g^{r\theta(1)} &= -\frac{1}{\sqrt{2}}(F-G+J+H),\\
g^{r\varphi(1)} &= -\frac{a}{\rho^2}(F^1_2-F^2_1)+\frac{\mathrm{i}}{\sqrt{2}}(H-J-C_2)\operatorname{cosec}\theta,\\
g^{\theta\varphi(1)} &= \mathrm{i}(F^3_4-F^4_3)\operatorname{cosec}\theta+\frac{a}{\Delta\sqrt{2}}(F+G-B_1).
\end{aligned}
\right\}\tag{11}
$$

We observe that the perturbations in the metric coefficients depend only on the following ten combinations of the elements of $\boldsymbol{A}$:

$$
\begin{gathered}
A^1_1+A^2_2,\quad A^3_3+A^4_4,\quad F^1_2+F^2_1,\quad F^1_2-F^2_1,\quad F^3_4+F^4_3,\quad F^3_4-F^4_3,\\
C_1+B_2,\quad H-J-C_2,\quad F+G-B_1\quad\text{and}\quad F-G+J+H.
\end{gathered}\tag{12}
$$

(c) *The enumeration of the quantities which have to be determined, the equations which are available, and the degrees of gauge freedom which we have*

The quantities which have to be determined (listed in (2) and (3)) require ten real functions to specify the five (complex) Weyl scalars, twenty-four real functions to specify the twelve (complex) spin coefficients, and sixteen real functions to specify the matrix A (which defines the perturbed null tetrad): all together fifty real functions. These fifty functions are subject to ten degrees of gauge freedom. These ten degrees of freedom arise from the six degrees of freedom we have in setting up the local tetrad-frame and the four degrees of coordinate freedom we have from the general covariance of the theory.

In the Newman–Penrose formalism (as, indeed, in any equivalent tetrad formalism) one writes down *three* sets of equations: the *Bianchi identities*,

$$R_{ij[kl;\,m]} = 0, \tag{13}$$

expressed in terms of the spin coefficients and the Weyl and the Ricci scalars; the *commutation relations*,

$$[\boldsymbol{l}^i, \boldsymbol{l}^j] = \mathscr{C}^{ij}_{k}\boldsymbol{l}^k, \tag{14}$$

where the $\boldsymbol{l}$s (as tangent vectors) are to be interpreted as directional derivatives and the *structure constants*, $\mathscr{C}^{ij}_{k}$, are expressed in terms of the spin coefficients; and, finally, the components of the Riemann tensor derived from the *Ricci identity*,

$$l^{(a)}_{i;k;l} - l^{(a)}_{i;l;k} = l^{(a)m} R_{mikl}, \tag{15}$$

written in terms of the spin coefficients, their directional derivatives, and the Weyl and the Ricci scalars.

By counting each complex equation as equivalent to two (real) equations, we have 16 equations representing the Bianchi identities (when the Ricci scalars vanish as in the case of vacuum fields we are presently considering), 24 equations representing the commutation relations, and 36 equations representing the Ricci identities:‡ all together 76 equations. These 76 equations are ostensibly written down for determining the 50 real functions which we have enumerated and which are subject to 10 degrees of gauge freedom. It would, therefore, appear that among the 76 equations there can be no more than 40 independent equations.

Returning to the problem on hand, we take note that the quantities which remain to be determined are those listed in (3) (leaving aside the matters pertaining to the relative normalization of the functions R_{+2} and R_{-2} and the real and the imaginary parts of the Teukolsky–Starobinsky constant $\mathscr{C}$). The elements of $\boldsymbol{A}$, the perturbations in the eight (remaining) spin coefficients, and the perturbation in Ψ_2, all together require for their specification 34 real functions.

3. The linearization of the Bianchi identities

Four of the eight Bianchi identities listed by Newman & Penrose (1962, eqns (4.5)) have already been used in I (the first pair in each of eqns (20) and (21)) in obtaining the solutions for the quantities listed in (2). The remaining four identities, in the gauge $\Psi_1 \equiv \Psi_3 \equiv 0$ (which we are adopting permanently), are

$$\mathrm{D}\Psi_2 = 3\tilde{\rho}\Psi_2, \quad \Delta\Psi_2 = -3\mu\Psi_2, \quad \delta\Psi_2 = 3\tau\Psi_2, \quad \text{and} \quad \delta^*\Psi_2 = -3\pi\Psi_2, \tag{16}$$

when quantities of the second order of smallness (such as $\lambda\Psi_0$, $\kappa\Psi_4$, $\nu\Psi_0$, and $\sigma\Psi_4$) are neglected. The remarkable feature of equations (16) is that (in the gauge $\Psi_1 \equiv \Psi_3 \equiv 0$) *they are formally the same as in the stationary state: they are valid inclusive of quantities of the first order of smallness.*

On linearizing equations (16), we obtain

$$\left.\begin{aligned}
\tilde{\rho}^{(1)} &= -\left(\frac{A^1_1}{\bar{\rho}^*} + A^1_2\mu - A^1_3\tau + A^1_4\pi\right) + \mathscr{D}_0 X = -\frac{M^{(1)}}{\bar{\rho}^*} + \mathscr{D}_0 X,\\
\mu^{(1)} &= +\left(\frac{A^2_1}{\bar{\rho}^*} + A^2_2\mu - A^2_3\tau + A^2_4\pi\right) + \frac{\Delta}{2\rho^2}\mathscr{D}^\dagger_0 X = \frac{\Delta}{2\rho^2\bar{\rho}^*} M^{(2)} + \frac{\Delta}{2\rho^2}\mathscr{D}^\dagger_0 X,\\
\tau^{(1)} &= -\left(\frac{A^3_1}{\bar{\rho}^*} + A^3_2\mu - A^3_3\tau + A^3_4\pi\right) + \frac{1}{\bar{\rho}\sqrt{2}}\mathscr{L}^\dagger_0 X = -M^{(3)} + \frac{1}{\bar{\rho}\sqrt{2}}\mathscr{L}^\dagger_0 X,\\
\pi^{(1)} &= +\left(\frac{A^4_1}{\bar{\rho}^*} + A^4_2\mu - A^4_3\tau + A^4_4\pi\right) - \frac{1}{\bar{\rho}^*\sqrt{2}}\mathscr{L}_0 X = \frac{\bar{\rho}}{\bar{\rho}^*} M^{(4)} - \frac{1}{\bar{\rho}^*\sqrt{2}}\mathscr{L}_0 X,
\end{aligned}\right\} \tag{17}$$

‡ Of these 36 equations, 16 follow from the commutation relations. I am indebted to Dr Robert Wald for this observation.

where

$$\left.\begin{aligned} M^{(1)} &= +A_1^1 - F_2^1 + ia\sqrt{2}\frac{\rho^2}{\Delta}F\sin\theta,\\ M^{(2)} &= -A_2^2 + F_1^2 + ia\sqrt{2}\frac{\rho^2}{\Delta}G\sin\theta,\\ M^{(3)} &= H + \frac{ia\sin\theta}{\sqrt{2}}\left(F_4^3 + \frac{A_3^3}{\rho^2}\right),\\ M^{(4)} &= J + \frac{ia\sin\theta}{\sqrt{2}}\left(F_3^4 + \frac{A_4^4}{\rho^2}\right), \end{aligned}\right\} \tag{18}$$

and

$$X = -\frac{1}{3M}(\bar{\rho}^*)^3\Psi_2^{(1)}. \tag{19}$$

We may note here for future reference the following relations which are direct consequences of equations (17):

$$\left.\begin{aligned} -(\tau^* + \pi)^{(1)} &= -\frac{2ia\cos\theta}{\rho^2}A_1^4 + \frac{ia\Delta\cos\theta}{\rho^4}A_2^4\\ &\quad -\sqrt{2}\frac{iar\sin\theta}{\rho^2}\left(\frac{A_4^4}{\bar{\rho}^*} + \frac{A_3^4}{\bar{\rho}}\right) + \frac{1}{\bar{\rho}^*\sqrt{2}}\mathscr{L}_0 Y,\\ +(\mu^* - \mu)^{(1)} &= -\frac{2ia\cos\theta}{\rho^2}A_1^2 + \frac{ia\Delta\cos\theta}{\rho^4}A_2^2\\ &\quad -\sqrt{2}\frac{iar\sin\theta}{\rho^2}\left(\frac{A_4^2}{\bar{\rho}^*} + \frac{A_3^2}{\bar{\rho}}\right) - \frac{\Delta}{2\rho^2}\mathscr{D}_0^\dagger Y,\\ +(\rho^* - \rho)^{(1)} &= +\frac{2ia\cos\theta}{\rho^2}A_1^1 - \frac{ia\Delta\cos\theta}{\rho^4}A_2^1\\ &\quad +\sqrt{2}\frac{iar\sin\theta}{\rho^2}\left(\frac{A_4^1}{\bar{\rho}^*} + \frac{A_3^1}{\bar{\rho}}\right) - \mathscr{D}_0 Y,\\ (\tau - \pi^*)^{(1)} &= -\frac{2r}{\rho^2}A_1^3 + \frac{r\Delta}{\rho^4}A_2^3\\ &\quad +\sqrt{2}\frac{a^2\sin\theta\cos\theta}{\rho^2}\left(\frac{A_3^3}{\bar{\rho}} + \frac{A_4^3}{\bar{\rho}^*}\right) + \frac{1}{\bar{\rho}\sqrt{2}}\mathscr{L}_0^\dagger Z, \end{aligned}\right\} \tag{20}$$

where

$$Y = X - X^* \quad\text{and}\quad Z = X + X^*. \tag{21}$$

It is important to draw attention to the fact that the Bianchi identities have enabled us to express the perturbations in the spin-coefficients, $\tilde{\rho}$, μ, τ, and π, directly in terms of the perturbations in the basic tetrad and the perturbation in Ψ_2.

4. The linearization of the commutation relations; the three systems of equations

By expressing the commutation relations, as written down by Newman & Penrose (1962, eqns (4.4)), in conformity with equation (14), we find that the structure constants, $\mathscr{C}_k^{ij}$, are related to the spin coefficients as in the accompanying tabulation.

$$\left.\begin{array}{llll}
\mathscr{C}_1^{21} = \gamma+\gamma^*; & \mathscr{C}_1^{31} = \alpha^*+\beta-\pi^*; & \mathscr{C}_1^{32} = -\nu^*; & \mathscr{C}_1^{43} = \mu^*-\mu; \\
\mathscr{C}_2^{21} = \epsilon+\epsilon^*; & \mathscr{C}_2^{31} = \kappa; & \mathscr{C}_2^{32} = \tau-\alpha^*-\beta; & \mathscr{C}_2^{43} = \tilde{\rho}^*-\tilde{\rho}; \\
\mathscr{C}_3^{21} = -(\tau^*+\pi); & \mathscr{C}_3^{31} = -(\tilde{\rho}^*+\epsilon-\epsilon^*); & \mathscr{C}_3^{32} = \mu-\gamma+\gamma^*; & \mathscr{C}_3^{43} = \alpha-\beta^*; \\
\mathscr{C}_4^{21} = -(\tau+\pi^*); & \mathscr{C}_4^{31} = -\sigma; & \mathscr{C}_4^{32} = \lambda^*; & \mathscr{C}_4^{43} = \beta-\alpha^*.
\end{array}\right\} \quad (22)$$

For the stationary Kerr metric, described in Kinnersley's frame, the structure constants, C_k^{ij}, have the values given by

$$\left.\begin{array}{llll}
C_1^{21} = -\dfrac{r\Delta}{\rho^4}+\dfrac{r-M}{\rho^2}; & C_1^{31} = 0; \quad C_1^{32} = 0; & & C_1^{43} = \dfrac{ia\Delta\cos\theta}{\rho^4}, \\
C_2^{21} = 0; & C_2^{31} = 0; & C_2^{32} = \sqrt{2}\dfrac{a^2\sin\theta\cos\theta}{\rho^2\bar{\rho}}; & C_2^{43} = \dfrac{2ia\cos\theta}{\rho^2}, \\
C_3^{21} = -\sqrt{2}\dfrac{iar\sin\theta}{\rho^2\bar{\rho}^*}; & C_3^{31} = \dfrac{1}{\bar{\rho}}; & C_3^{32} = -\dfrac{\Delta}{2\rho^2\bar{\rho}}; & C_3^{43} = \dfrac{-1}{(\bar{\rho}^*)^2\sqrt{2}}(\bar{\rho}^*\cot\theta-ia\sin\theta), \\
C_4^{21} = +\sqrt{2}\dfrac{iar\sin\theta}{\rho^2\bar{\rho}}; & C_4^{31} = 0; \quad C_4^{32} = 0; & & C_4^{43} = \dfrac{+1}{(\bar{\rho})^2\sqrt{2}}(\bar{\rho}\cot\theta+ia\sin\theta).
\end{array}\right\} \quad (23)$$

Using equation (7), we can write the linearized version of equation (14) in the form

$$[A_k^i\, \boldsymbol{l}^k, \boldsymbol{l}^j]+[\boldsymbol{l}^i, A_k^j\, \boldsymbol{l}^k] = C_k^{ij} A_m^k\, \boldsymbol{l}^m + c_m^{ij}\, \boldsymbol{l}^m, \tag{24}$$

where c_m^{ij} is the perturbation in C_m^{ij}. The quantities c_m^{ij} are directly related to the perturbations in the particular combinations in the spin coefficients which occur in equation (22). By expanding equation (24), we obtain

$$A_k^i C_m^{kj}\, \boldsymbol{l}^m + A_k^j C_m^{ik}\, \boldsymbol{l}^m + (\boldsymbol{l}^i A_m^j)\, \boldsymbol{l}^m - (\boldsymbol{l}^j A_m^i)\, \boldsymbol{l}^m = C_k^{ij} A_m^k\, \boldsymbol{l}^m + c_m^{ij}\, \boldsymbol{l}^m, \tag{25}$$

or, since the $\boldsymbol{l}^m$s are linearly independent, we can write

$$\boldsymbol{l}^i A_m^j - \boldsymbol{l}^j A_m^i = A_k^i C_m^{jk} - A_k^j C_m^{ik} + C_k^{ij} A_m^k + c_m^{ij}. \tag{26}$$

Equation (26) is, in many ways, the basic equation of this theory: it provides a set of 24 simultaneous, linear, inhomogeneous, partial differential equations for the elements of $\boldsymbol{A}$; and the inhomogeneous terms are directly related to the perturbations in the spin coefficients.

The 24 equations, which equation (26) represents, can be grouped into three systems, of eight equations each, as follows:

It follows from the tabulation (22) that

$$\left.\begin{aligned} c_3^{21} &= -(\tau^*+\pi)^{(1)}; & c_1^{43} &= +(\mu^*-\mu)^{(1)},\\ c_4^{21} &= -(\tau+\pi^*)^{(1)}; & c_2^{43} &= +(\tilde{\rho}^*-\tilde{\rho})^{(1)},\\ c_1^{31}+c_2^{32} &= +(\tau-\pi^*)^{(1)}; & c_3^{31}+c_4^{41} &= -(\tilde{\rho}+\tilde{\rho}^*)^{(1)},\\ c_1^{41}+c_2^{42} &= +(\tau^*-\pi)^{(1)}; & c_3^{32}+c_4^{42} &= +(\mu+\mu^*)^{(1)}.\end{aligned}\right\} \tag{27}$$

Accordingly, we can write down a system of eight equations in which the inhomogeneous terms are directly related to the perturbations in the spin-coefficients $\tilde{\rho}$, τ, μ and π, but, as we have shown in § 3, the perturbations in these spin coefficients can be expressed (linearly) in terms of the elements of A and the perturbation in Ψ_2 (cf. equations (17) and (20)). Equation (26), therefore, provides a system of eight inhomogeneous equations (which we shall call system I) for the elements of A in which the inhomogeneous terms are expressible in terms of the perturbations in Ψ_2 only.

Next, we observe that

$$\left.\begin{aligned} c_2^{31} &= \kappa, & c_4^{31} &= -\sigma, & c_1^{32} &= -\nu^*, & c_4^{32} &= \lambda^*,\\ c_2^{41} &= \kappa^*, & c_3^{41} &= -\sigma^*, & c_1^{42} &= -\nu, & c_3^{42} &= \lambda.\end{aligned}\right\} \tag{28}$$

But we already have explicit solutions (I, eqns (42)–(45)) for κ, σ, λ and ν in terms of Teukolsky's functions (leaving aside the unspecified relative normalization of R_{+2} and R_{-2}). Consequently, equation (26) provides a further system of eight equations (which we shall call system II) for the elements of A in which we may consider the inhomogeneous terms as known.

The two systems of equations (systems I and II) provide a total of 16 equations for the 16 functions required to specify A and the two additional functions required to specify the perturbation in Ψ_2. However, we shall see presently that in these 16 equations the elements of A occur only in 14 different combinations, so that we have 16 equations to specify the 16 functions. But we shall find that not all of the 16 equations are independent.

Again, from the tabulation (22) it follows that

$$\left.\begin{aligned} c_1^{31}-c_2^{32} &= 2(\alpha^*+\beta)^{(1)}-(\tau+\pi^*)^{(1)},\\ c_1^{41}-c_2^{42} &= 2(\alpha+\beta^*)^{(1)}-(\tau^*+\pi)^{(1)},\\ c_4^{41}-c_3^{31} &= 2(\epsilon-\epsilon^*)^{(1)}-(\tilde{\rho}-\tilde{\rho}^*)^{(1)},\\ c_3^{42}-c_3^{32} &= 2(\gamma-\gamma^*)^{(1)}-(\mu-\mu^*)^{(1)},\\ c_1^{21} &= (\gamma+\gamma^*)^{(1)}, \quad c_3^{43} = (\alpha-\beta^*)^{(1)},\\ c_2^{21} &= (\epsilon+\epsilon^*)^{(1)}, \quad c_4^{43} = (\beta-\alpha^*)^{(1)}.\end{aligned}\right\} \tag{29}$$

Accordingly, we can derive from equation (26) a third system of eight equations in which the inhomogeneous terms are directly related to the perturbations in the four remaining spin-coefficients α, β, γ, and ϵ. Therefore, if A and $\Psi_2^{(1)}$ have already been determined, this last system of equations (which we shall call system III) will suffice to complete the solution.

We shall now write down the explicit forms of the three systems of equations.

SYSTEM I.

$$\left.\begin{aligned}
&\mathscr{D}_0^\dagger F_3^1+\mathscr{D}_0 F_3^2=-\mathrm{i}rT\sin\theta+\frac{2\mathrm{i}a\cos\theta}{\rho^2}J-\frac{1}{\rho^2\sqrt{2}}\mathscr{L}_0 Y \quad (21,3),\\
&\mathscr{D}_0^\dagger F_4^1+\mathscr{D}_0 F_4^2=+\mathrm{i}rT\sin\theta-\frac{2\mathrm{i}a\cos\theta}{\rho^2}H+\frac{1}{\rho^2\sqrt{2}}\mathscr{L}_0^\dagger Y \quad (21,4),\\
&\mathscr{L}_1 F_1^3-\mathscr{L}_1^\dagger F_1^4=\mathrm{i}\Delta T\cos\theta-\frac{2\mathrm{i}ar\sin\theta}{\rho^2}G-\frac{\Delta}{\rho^2\sqrt{2}}\mathscr{D}_0^\dagger Y \quad (43,1),\\
&\mathscr{L}_1 F_2^3-\mathscr{L}_1^\dagger F_2^4=\mathrm{i}\Delta T\cos\theta+\frac{2\mathrm{i}ar\sin\theta}{\rho^2}F-\frac{\Delta}{\rho^2\sqrt{2}}\mathscr{D}_0 Y \quad (43,2),\\
&\quad +\frac{1}{\sqrt{2}}\mathscr{L}_0^\dagger U-\rho^2\mathscr{D}_0 F_1^3+\rho^2\mathscr{D}_0^\dagger F_2^3=+2\mathrm{i}a(F_4^1+F_4^2)\cos\theta \quad (31,1)+(32,2),\\
&\quad +\frac{1}{\sqrt{2}}\mathscr{L}_0 U-\rho^2\mathscr{D}_0 F_1^4+\rho^2\mathscr{D}_0^\dagger F_2^4=-2\mathrm{i}a(F_3^1+F_3^2)\cos\theta \quad (41,1)+(42,2),\\
&\quad -\frac{1}{\sqrt{2}}\Delta\mathscr{D}_0 V+\rho^2\mathscr{L}_1 F_4^1+\rho^2\mathscr{L}_1^\dagger F_3^1=2\mathrm{i}ar(F_2^3-F_2^4)\sin\theta \quad (31,3)+(41,4),\\
&\quad +\frac{1}{\sqrt{2}}\Delta\mathscr{D}_0^\dagger V+\rho^2\mathscr{L}_1 F_4^2+\rho^2\mathscr{L}_1^\dagger F_3^2=2\mathrm{i}ar(F_1^4-F_1^3)\sin\theta \quad (32,3)+(42,4),
\end{aligned}\right\} \quad (30)$$

where $T=(\sqrt{2}/\rho^4)\,a(U-V)$, $U=A_1^1+A_2^2-Z$ and $V=A_3^3+A_4^4-Z$. (31)

SYSTEM II.

$$\left.\begin{aligned}
&\frac{1}{\sqrt{2}}\rho^2\mathscr{L}_0^\dagger F_2^1-\mathscr{D}_{-1}\rho^4F_2^3-2\mathrm{i}a\rho^2F_4^1\cos\theta=\tfrac{1}{2}\Delta\bar{\rho}\kappa \quad (31,2),\\
&\frac{1}{\sqrt{2}}\rho^2\mathscr{L}_0 F_2^1-\mathscr{D}_{-1}\rho^4F_2^4+2\mathrm{i}a\rho^2F_3^1\cos\theta=\tfrac{1}{2}\Delta\bar{\rho}^*\kappa^* \quad (41,2),\\
&\frac{1}{\sqrt{2}}\mathscr{L}_{-1}^\dagger\rho^4F_4^1-\tfrac{1}{2}\rho^2\Delta\,\mathscr{D}_0\rho^2F_4^3+\sqrt{2}\,\mathrm{i}ar\rho^2F_2^3\sin\theta=-\tfrac{1}{2}\Delta(\bar{\rho})^2\sigma \quad (31,4),\\
&\frac{1}{\sqrt{2}}\mathscr{L}_{-1}\rho^4F_3^1-\tfrac{1}{2}\rho^2\Delta\mathscr{D}_0\rho^2F_3^4-\sqrt{2}\,\mathrm{i}ar\rho^2F_2^4\sin\theta=-\tfrac{1}{2}\Delta(\bar{\rho}^*)^2\sigma^* \quad (41,3),\\
&\frac{1}{\sqrt{2}}\rho^2\mathscr{L}_0^\dagger F_1^2+\mathscr{D}_{-1}^\dagger\rho^4F_1^3-2\mathrm{i}a\rho^2F_4^2\cos\theta=-\frac{2\rho^4\bar{\rho}}{\Delta}\nu^* \quad (32,1),\\
&\frac{1}{\sqrt{2}}\rho^2\mathscr{L}_0 F_1^2+\mathscr{D}_{-1}^\dagger\rho^4F_1^4+2\mathrm{i}a\rho^2F_3^2\cos\theta=-\frac{2\rho^4\bar{\rho}^*}{\Delta}\nu \quad (42,1),\\
&\frac{1}{\sqrt{2}}\mathscr{L}_{-1}^\dagger\rho^4F_4^2+\tfrac{1}{2}\rho^2\Delta\mathscr{D}_0^\dagger\rho^2F_4^3-\sqrt{2}\,\mathrm{i}ar\rho^2F_1^3\sin\theta=\rho^2(\bar{\rho})^2\lambda^* \quad (32,4),\\
&\frac{1}{\sqrt{2}}\mathscr{L}_{-1}\rho^4F_3^2+\tfrac{1}{2}\rho^2\Delta\mathscr{D}_0^\dagger\rho^2F_3^4+\sqrt{2}\,\mathrm{i}ar\rho^2F_1^4\sin\theta=\rho^2(\bar{\rho}^*)^2\lambda \quad (42,3).
\end{aligned}\right\} \quad (32)$$

SYSTEM III.

$$\left.\begin{aligned}
&\frac{1}{\bar{\rho}^*\sqrt{2}}\mathscr{L}_0(A_1^1-A_2^2-Y)-\frac{1}{\bar{\rho}}\mathscr{D}_0(\bar{\rho}^2F_1^4)-\frac{1}{\rho^2(\bar{\rho}^*)^3}\mathscr{D}_{-2}^\dagger[\rho^4(\bar{\rho}^*)^2F_2^4]\\
&\qquad+\frac{2ia\cos\theta}{\bar{\rho}^*}(F_3^1-F_3^2)+\sqrt{2}\frac{ia\sin\theta}{\bar{\rho}^*}\bar{\rho}F_3^4+\sqrt{2}\frac{ia\sin\theta}{(\bar{\rho}^*)^2}A_4^4\\
&\qquad\qquad= 2(\alpha+\beta^*)^{(1)}\quad (41,1)\text{–}(42,2)\\
&\frac{\sqrt{2}}{\Delta\rho^2}\mathscr{L}_{-1}\rho^4F_4^1-\frac{\sqrt{2}}{\Delta\rho^2}\mathscr{L}_{-1}^\dagger\rho^4F_3^1-\mathscr{D}_0(A_4^4-A_3^3-Y)\\
&\qquad+2\sqrt{2}\frac{iar\sin\theta}{\Delta}(F_2^3+F_2^4)=2(\epsilon-\epsilon^*)^{(1)}\quad (41,4)\text{–}(31,3),\\
&\frac{1}{(\bar{\rho}^*)^4\sqrt{2}}\mathscr{L}_{-1}(\bar{\rho}^*)^4F_4^2-\frac{1}{(\bar{\rho})^4\sqrt{2}}\mathscr{L}_{-1}^\dagger(\bar{\rho})^4F_3^2-\frac{\Delta}{2\rho^2}\mathscr{D}_0^\dagger(A_3^3-A_4^4-Y)\\
&\qquad+\frac{2ia\Delta\cos\theta}{\rho^4}(F_1^2-A_2^1)-\sqrt{2}\frac{iar\sin\theta}{\rho^2}(F_1^3+F_1^4)=2(\gamma-\gamma^*)^{(1)}\quad (42,3)\text{–}(32,3),\\
&\frac{\Delta}{2\rho^2}\mathscr{D}_0^\dagger A_1^1+\mathscr{D}_0A_1^0=A_2^1\frac{d}{dr}\left(\frac{\Delta}{2\rho^2}\right)+\sqrt{2}\frac{iar\sin\theta}{\rho^2}(F_1^3-F_1^4)-(\gamma+\gamma^*)^{(1)}\quad (21,1),\\
&-\tfrac{1}{2}\Delta\mathscr{D}_1^\dagger\frac{A_2^1}{\rho^2}-\mathscr{D}_0A_2^2=2\sqrt{2}\frac{a^2\sin\theta\cos\theta}{\Delta}F-2\sqrt{2}\frac{iar\sin\theta}{\Delta}(F_2^3-F_2^4)\\
&\qquad\qquad+(\epsilon+\epsilon^*)^{(1)}\quad (21,2),\\
&+\frac{1}{\bar{\rho}^*\sqrt{2}}\mathscr{L}_0A_3^3-\frac{1}{\sqrt{2}}\mathscr{L}_1^\dagger\frac{A_3^4}{\bar{\rho}}=J+\frac{2ia\cos\theta}{\bar{\rho}^*}(F_3^1+F_3^2)\\
&\qquad+\frac{A_4^4}{(\bar{\rho}^*)^2\sqrt{2}}(\bar{\rho}^*\cot\theta-ia\sin\theta)+(\alpha-\beta^*)^{(1)}\quad (43,3).
\end{aligned}\right\}\quad(33)$$

(In the foregoing, we have not included the equations for $(\alpha^*+\beta)^{(1)}$ and $(\alpha^*-\beta)^{(1)}$ since they can be written down directly from the equations for $(\alpha+\beta^*)^{(1)}$ and $(\alpha-\beta^*)^{(1)}$.)

5. Further identities

By combining the Bianchi identities (16) (in the gauge $\Psi_1\equiv\Psi_3\equiv 0$ and correct inclusive of quantities of the first order of smallness, only) with the *exact* commutation relations, we can derive some further identities which we shall find useful. Thus, applying the commutation relation (cf. equations (14) and (22)),

$$\begin{aligned}(\Delta D-D\Delta)&=[l^2,l^1]=\mathscr{C}_k^{21}l^k\\
&=(\gamma+\gamma^*)D+(\epsilon+\epsilon^*)\Delta-(\tau^*+\pi)\delta-(\tau+\pi^*)\delta^*,\end{aligned}\tag{34}$$

to Ψ_2 (inclusive of first-order perturbations), the left hand side of the equation gives

$$\begin{aligned}(\Delta D-D\Delta)\Psi_2&=3\Delta(\tilde{\rho}\Psi_2)+3D(\mu\Psi_2)\\
&=3\Psi_2(\Delta\tilde{\rho}+D\mu)+3\tilde{\rho}\Delta\Psi_2+3\mu D\Psi_2\\
&=3\Psi_2(\Delta\tilde{\rho}+D\mu);\end{aligned}\tag{35}$$

while the right hand side gives

$$3[(\gamma+\gamma^*)\tilde{\rho}-(\epsilon+\epsilon^*)\mu+\pi\pi^*-\tau\tau^*]\Psi_2. \tag{36}$$

We thus obtain the identity

$$\Delta\tilde{\rho}+\mathrm{D}\mu = (\gamma+\gamma^*)\tilde{\rho}-(\epsilon+\epsilon^*)\mu+\pi\pi^*-\tau\tau^*. \tag{37}$$

We similarly obtain from the other commutation relations the further identities

$$\delta^*\tau+\delta\pi = \mu^*\tilde{\rho}-\tilde{\rho}^*\mu+(\alpha^*-\beta)\pi+(\alpha-\beta^*)\tau, \tag{38}$$

$$\delta\tilde{\rho}-\mathrm{D}\tau = (\alpha^*+\beta-\pi^*)\tilde{\rho}-\kappa\mu+\sigma\pi-(\tilde{\rho}^*+\epsilon-\epsilon^*)\tau, \tag{39}$$

$$\Delta\tau+\delta\mu = \nu^*\tilde{\rho}+(\tau-\alpha^*-\beta)\mu+\lambda^*\pi-(\mu-\gamma+\gamma^*)\tau, \tag{40}$$

$$\Delta\pi+\delta^*\mu = -\nu\tilde{\rho}-(\tau^*-\alpha-\beta^*)\mu+\lambda\tau-(\mu^*+\gamma-\gamma^*)\pi \tag{41}$$

We shall return to a consideration of these identities in § 7.

6. The reduction of system I

By eliminating T from the first four equations of system I (equations (30)) we obtain the pair of equations,

$$\mathscr{D}_0^\dagger F+\mathscr{D}_0 G = \frac{2\mathrm{i}a\cos\theta}{\rho^2}(J-H)-\frac{\sqrt{2}}{\rho^2}QY \tag{42}$$

and

$$\mathscr{L}_1 H-\mathscr{L}_1^\dagger J = -\frac{2\mathrm{i}ar\sin\theta}{\rho^2}(F+G)+\frac{\sqrt{2}}{\rho^2}\mathrm{i}KY, \tag{43}$$

where we may recall that F, G, J, and H are defined in equations (9). We find that this pair of equations plays a central role in the subsequent analysis.

Next, by subtracting equation (21, 4) from equation (21, 3) and by adding equations (43, 1) and (43, 2), we obtain the pair of equations

$$\mathscr{D}_0^\dagger(F_3^1-F_4^1)+\mathscr{D}_0(F_3^2-F_4^2) = -2\mathrm{i}rT\sin\theta+\frac{2\mathrm{i}a\cos\theta}{\rho^2}(J+H)-\frac{\sqrt{2}}{\rho^2}\frac{\partial Y}{\partial\theta}, \tag{44}$$

$$\mathscr{L}_1(F_1^3+F_2^3)-\mathscr{L}_1^\dagger(F_1^4+F_2^4) = 2\mathrm{i}\Delta T\cos\theta+\frac{2\mathrm{i}ar\sin\theta}{\rho^2}(F-G)-\frac{\sqrt{2}}{\rho^2}\Delta\frac{\partial Y}{\partial r}. \tag{45}$$

By similar additions and subtractions we can replace the second four equations of system I by the equivalent set:

$$\left.\begin{aligned}
\mathscr{D}_0(F_1^3+F_1^4)-\mathscr{D}_0^\dagger(F_2^3+F_2^4) &= \frac{2\mathrm{i}a\cos\theta}{\rho^2}C_1+\frac{\sqrt{2}}{\rho^2}\frac{\partial U}{\partial\theta},\\
\mathscr{D}_0(F_1^3-F_1^4)-\mathscr{D}_0^\dagger(F_2^3-F_2^4) &= -\frac{2\mathrm{i}a\cos\theta}{\rho^2}(F+G)-\frac{\sqrt{2}}{\rho^2}QU,\\
\mathscr{L}_1(F_4^1+F_4^2)+\mathscr{L}_1^\dagger(F_3^1+F_3^2) &= \frac{2\mathrm{i}ar\sin\theta}{\rho^2}(J-H)+\frac{\sqrt{2}}{\rho^2}\mathrm{i}KV,\\
\mathscr{L}_1(F_4^1-F_4^2)+\mathscr{L}_1^\dagger(F_3^1-F_3^2) &= \frac{2\mathrm{i}ar\sin\theta}{\rho^2}B_2+\frac{\sqrt{2}}{\rho^2}\Delta\frac{\partial V}{\partial r}.
\end{aligned}\right\} \tag{46}$$

It is useful to rewrite equations (42), (43), and (46) in the alternative forms:

$$\frac{iK}{\Delta}C_2 = \frac{\partial C_1}{\partial r} + 2irT\sin\theta - \frac{2ia\cos\theta}{\rho^2}(J+H) + \frac{\sqrt{2}}{\rho^2}\frac{\partial Y}{\partial \theta}, \tag{47}$$

$$QB_1 = -\left(\frac{\partial}{\partial\theta} + \cot\theta\right)B_2 + 2i\Delta T\cos\theta + \frac{2iar\sin\theta}{\rho^2}(F-G) - \frac{\sqrt{2}}{\rho^2}\Delta\frac{\partial Y}{\partial r}, \tag{48}$$

$$\frac{iK}{\Delta}B_1 = \frac{2ia\cos\theta}{\rho^2}C_1 - \frac{\partial}{\partial r}(J+H) + \frac{\sqrt{2}}{\rho^2}\frac{\partial U}{\partial\theta}, \tag{49}$$

$$\frac{iK}{\Delta}B_2 = -\frac{2ia\cos\theta}{\rho^2}(F+G) + \frac{\partial}{\partial r}(J-H) - \frac{\sqrt{2}}{\rho^2}QU, \tag{50}$$

$$QC_1 = -\frac{2iar\sin\theta}{\rho^2}(J-H) + \left(\frac{\partial}{\partial\theta} + \cot\theta\right)(F+G) - \frac{\sqrt{2}}{\rho^2}iKV, \tag{51}$$

$$QC_2 = -\frac{2iar\sin\theta}{\rho^2}B_2 + \left(\frac{\partial}{\partial\theta} + \cot\theta\right)(F-G) - \frac{\sqrt{2}}{\rho^2}\Delta\frac{\partial V}{\partial r}, \tag{52}$$

where B_1, B_2, C_1, and C_2 are defined in equations (9).

It can be shown that not all of the eight equations (42), (43), and (47)–(52) are independent. Thus, it can be verified that equations (51) and (52) are dependent on equations (42), (43), (47), and (50); and, similarly, equations (49) and (50) are dependent on equations (42), (43), (48), and (51). We conclude that *only six of the eight equations* (42), (43), *and* (47)–(52) *are independent*. We shall take equations (49)–(52) together with equations (42) and (43) as our basic set of equations. The latter two equations can be written alternatively in the forms

$$\frac{iK}{\Delta}(F-G) = \frac{\partial}{\partial r}(F+G) - \frac{2ia\cos\theta}{\rho^2}(J-H) + \frac{\sqrt{2}}{\rho^2}QY, \tag{53}$$

$$Q(J+H) = \left(\frac{\partial}{\partial\theta} + \cot\theta\right)(J-H) - \frac{2iar\sin\theta}{\rho^2}(F+G) + \frac{\sqrt{2}}{\rho^2}iKY. \tag{54}$$

From equations (49)–(54), it is clear that the quantities B_1, B_2, C_1, C_2, $F-G$, and $J+H$ can all be expressed in terms of $F+G$, $J-H$, U, and V. (It will eventually turn out that U and V remain as arbitrary functions at our disposal.)

7. The vanishing of the perturbation in Ψ_2

We consider in this section the identities derived in §5. First, we observe that some of the identities are equivalent to linear combinations of some of the Ricci identities (15). Thus, it can be readily verified that the sum of the Newman & Penrose (1962) eqns (4.2h) and (4.2q) is identical with the difference of equations (37) and (38).

Again the difference of the Newman & Penrose (1962) eqns (4.2k) and (4.2c) is given by

$$\delta\tilde{\rho} - \delta^*\sigma - D\tau + \Delta\kappa = \tilde{\rho}(\alpha^* + \beta) - \sigma(3\alpha - \beta^*) + \tau(\tilde{\rho} - \tilde{\rho}^*) + \kappa(\mu - \mu^*) \\ - \tilde{\rho}(\tau + \pi^*) - \sigma(\tau^* + \pi) - \tau(\epsilon - \epsilon^*) + \kappa(3\gamma + \gamma^*). \tag{55}$$

But, by equation (39), the right hand side of equation (55) must equal

$$\tilde{\rho}(\alpha^* + \beta - \pi^*) - \kappa\mu + \sigma\pi - \tau(\tilde{\rho}^* + \epsilon - \epsilon^*) + \Delta\kappa - \delta^*\sigma; \tag{56}$$

and the equality requires

$$\Delta\kappa - \delta^*\sigma = (2\mu - \mu^* + 3\gamma + \gamma^*)\kappa - (\tau^* + 2\pi + 3\alpha - \beta^*)\sigma \tag{57}$$

– an equation which can be verified to be consistent with the solutions for κ and σ given in I, eqns (42) and (43). Similarly, it can be verified that the difference of the Newman & Penrose (1962) eqns (4.2m) and (4.2i) together with equation (41) is consistent with the solutions for λ and ν given in I, eqns (44) and (45). [It should be noted that the differences of the Newman–Penrose equations, (4.2k) – (4.2c) and (4.2m) – (4.2i), are *not* equations which can be derived from the commutation relations (see footnote on p. 445).

It can also be shown, after some considerable reductions, that the linearized version of equation (40) is equivalent to a combination of equations (32, 1) and (32, 4) of system II; and, similarly, equation (39) is equivalent to a combination of equations (31, 2) and (31, 4) of system II.

From the foregoing remarks, it should not be concluded that the identities derived in §5 do not provide any useful information. Indeed, we shall presently show that the linearized versions of equations (37) and (38) provide a most important information: in the gauge in which $\Psi_1 \equiv \Psi_3 \equiv 0$, the perturbation in Ψ_2 must vanish identically. In view of the surprising (and unexpected) nature of this result, we shall give some details of the steps leading to it.

The linearized version of equation (37) is

$$D\mu^{(1)} + D^{(1)}\mu + \Delta\tilde{\rho}^{(1)} - \Delta^{(1)}\frac{1}{\bar{\rho}^*} = (\pi^{(1)}\pi^* - \tau^{*(1)}\tau) \\ + (\pi^{*(1)}\pi - \tau^{(1)}\tau^*) + \tilde{\rho}^{(1)}(\gamma + \gamma^*) - \frac{1}{\bar{\rho}^*}(\gamma + \gamma^*)^{(1)} - (\epsilon + \epsilon^*)^{(1)}\mu. \tag{58}$$

Making use of the equations in §3 as well as the expressions for $(\gamma + \gamma^*)^{(1)}$ and $(\epsilon + \epsilon^*)^{(1)}$ included in system III, we find that the various groups of terms in equation (58) have the values

$$\pi^{(1)}\pi^* - \tau^{*(1)}\tau + \pi^{*(1)}\pi - \tau^{(1)}\tau^* \\ = \frac{2ia\sin\theta}{\rho^4}Q[rZ - (ia\cos\theta)\,Y] + \sqrt{2}\,\frac{ia\sin\theta}{\rho^2}(H - J), \tag{59}$$

$$\rho^{(1)}(\gamma + \gamma^*) = -\frac{1}{\bar{\rho}^*}\frac{d}{dr}\left(\frac{\Delta}{2\rho^2}\right)\left[A_1^1 - \frac{\Delta}{2\rho^2}A_2^1 + \sqrt{2}\,\frac{ia\sin\theta}{\Delta}\rho^2 F\right] + \frac{d}{dr}\left(\frac{\Delta}{2\rho^2}\right)\mathscr{D}_0 X, \tag{60}$$

$$
\begin{aligned}
&-\mu(\epsilon+\epsilon^*)^{(1)}-\frac{1}{\bar{\rho}^*}(\gamma+\gamma^*)^{(1)} \\
&\quad=\frac{1}{\bar{\rho}^*}\left[-\frac{\Delta^2}{4\rho^2}\mathscr{D}_1^\dagger\frac{A_2^1}{\rho^2}-\frac{\Delta}{2\rho^2}\mathscr{D}_0A_2^2-\sqrt{2}\,\frac{a^2\sin\theta\cos\theta}{\rho^2}F+\sqrt{2}\,\frac{\mathrm{i}ar\sin\theta}{\rho^2}(F_2^3-F_2^4)\right. \\
&\qquad\left.+\frac{\Delta}{2\rho^2}\mathscr{D}_0^\dagger A_1^1+\mathscr{D}_0A_1^2-\frac{\mathrm{d}}{\mathrm{d}r}\left(\frac{\Delta}{2\rho^2}\right)A_2^2-\sqrt{2}\,\frac{\mathrm{i}ar\sin\theta}{\rho^2}(F_1^3-F_1^4)\right]
\end{aligned}
\tag{61}
$$

$$
\begin{aligned}
&\mathrm{D}\mu^{(1)}+\mathrm{D}^{(1)}\mu+\Delta\rho^{(1)}-\Delta^{(1)}\frac{1}{\bar{\rho}^*}=\frac{1}{\bar{\rho}^*}\mathscr{D}_0\left(A_1^2-\frac{\Delta}{2\rho^2}A_2^2+\frac{\mathrm{i}a\sin\theta}{\sqrt{2}}G\right) \\
&\quad+\frac{\Delta}{2\rho^2\bar{\rho}^*}\mathscr{D}_0^\dagger\left(A_1^1-\frac{\Delta}{2\rho^2}A_2^1+\sqrt{2}\,\frac{\mathrm{i}a\sin\theta}{\Delta}\rho^2F\right)-\sqrt{2}\,\frac{a^2\sin\theta\cos\theta}{\rho^2\bar{\rho}^*}F \\
&\quad+\frac{\Delta}{2\rho^2}(\mathscr{D}_0\mathscr{D}_0^\dagger-\mathscr{D}_0^\dagger\mathscr{D}_0)X+\frac{\mathrm{d}}{\mathrm{d}r}\left(\frac{\Delta}{2\rho^2}\right)\mathscr{D}_0^\dagger X-\frac{1}{\bar{\rho}^*}\left(A_1^1-\frac{\Delta}{2\rho^2}A_2^1\right)\frac{\mathrm{d}}{\mathrm{d}r}\left(\frac{\Delta}{2\rho^2}\right).
\end{aligned}
\tag{62}
$$

Combining the foregoing expressions, we find that we are left with

$$Q(rZ-\mathrm{i}aY\cos\theta)\sin\theta=0, \tag{63}$$

after some lengthy reductions and remarkable cancellations.

A similar reduction of the linearized version of equation (38) yields

$$K(rY-\mathrm{i}aZ\cos\theta)=0. \tag{64}$$

(It might be noted that in the reductions leading to equations (63) and (64), equations (42) and (43) play essential roles.)

Since neither K nor Q can vanish except when both σ and m are zero, it follows from equations (63) and (64) that

$$rZ=\mathrm{i}aY\cos\theta \quad\text{and}\quad rY=\mathrm{i}aZ\cos\theta \tag{65}$$

if *either* σ *or* m *is different from zero*. But the two equations in (65) are incompatible unless

$$Z\equiv Y\equiv 0 \quad (\sigma \text{ or } m\neq 0) \tag{66}$$

We conclude that *the perturbation* $\Psi_2^{(1)}$ *in* Ψ_2 *must vanish identically except for quasi-stationary axisymmetric perturbations*. The exception noted is consistent with the expectation that we should be able to slide along the Kerr sequence quasi-stationarily by axisymmetric deformations which change the mass and, therefore, change Ψ_2 (cf. Wald 1973).

In concluding this section, it is worth emphasizing, once again, that the vanishing of $\Psi_2^{(1)}$ has been derived with no assumptions except that we are in the gauge $\Psi_1\equiv\Psi_3\equiv 0$. It is not clear why this should be so.

8. The reduction of system II; a fundamental integrability condition

By addition and subtraction, we can replace the equations of system II (equations (32)) by the following equivalent set of eight equations:

$$\left.\begin{aligned}
&\left.\begin{aligned}
&\sqrt{2}\,Q\rho^2F^1_2+\mathscr{D}_{-1}\rho^4(F^3_2-F^4_2)+2\mathrm{i}a\rho^2F\cos\theta=\tfrac{1}{2}\Delta(\bar{\rho}^*\kappa^*-\bar{\rho}\kappa),\\
&\sqrt{2}\,Q\rho^2F^2_1-\mathscr{D}^\dagger_{-1}\rho^4(F^3_1-F^4_1)+2\mathrm{i}a\rho^2G\cos\theta=-\frac{2\rho^4}{\Delta}(\bar{\rho}^*\nu-\bar{\rho}\nu^*),
\end{aligned}\right\}\\
&\left.\begin{aligned}
&\sqrt{2}\,\rho^2\frac{\partial F^1_2}{\partial\theta}-\mathscr{D}_{-1}\rho^4(F^3_2+F^4_2)+2\mathrm{i}a\rho^2(F^1_3-F^1_4)\cos\theta=\tfrac{1}{2}\Delta(\bar{\rho}^*\kappa^*+\bar{\rho}\kappa),\\
&\sqrt{2}\,\rho^2\frac{\partial F^2_1}{\partial\theta}+\mathscr{D}^\dagger_{-1}\rho^4(F^3_1+F^4_1)+2\mathrm{i}a\rho^2(F^2_3-F^2_4)\cos\theta=-\frac{2\rho^4}{\Delta}(\bar{\rho}^*\nu+\bar{\rho}\nu^*),
\end{aligned}\right\}\\
&\left.\begin{aligned}
&-\mathrm{i}K\rho^4F^3_4+\frac{1}{\sqrt{2}}\mathscr{L}^\dagger_{-1}\rho^4(F^1_4+F^2_4)-\sqrt{2}\,\mathrm{i}ar\rho^2H\sin\theta=(\bar{\rho})^2(\rho^2\lambda^*-\tfrac{1}{2}\Delta\sigma),\\
&-\mathrm{i}K\rho^4F^4_3+\frac{1}{\sqrt{2}}\mathscr{L}_{-1}\rho^4(F^1_3+F^2_3)+\sqrt{2}\,\mathrm{i}ar\rho^2J\sin\theta=(\bar{\rho}^*)^2(\rho^2\lambda-\tfrac{1}{2}\Delta\sigma^*),
\end{aligned}\right\}\\
&\left.\begin{aligned}
&-\rho^2\Delta\frac{\partial}{\partial r}\rho^2F^3_4+\frac{1}{\sqrt{2}}\mathscr{L}^\dagger_{-1}\rho^4(F^1_4-F^2_4)+\sqrt{2}\,\mathrm{i}ar\rho^2(F^3_1+F^3_2)\sin\theta=-(\bar{\rho})^2(\rho^2\lambda^*+\tfrac{1}{2}\Delta\sigma),\\
&-\rho^2\Delta\frac{\partial}{\partial r}\rho^2F^4_3+\frac{1}{\sqrt{2}}\mathscr{L}_{-1}\rho^4(F^1_3-F^2_3)-\sqrt{2}\,\mathrm{i}ar\rho^2(F^4_1+F^4_2)\sin\theta=-(\bar{\rho}^*)^2(\rho^2\lambda+\tfrac{1}{2}\Delta\sigma^*).
\end{aligned}\right\}
\end{aligned}\right\}\tag{67}$$

Next, by adding the equations in each of the four pairs of equations included in the foregoing set and making use of equations (49)–(52) for further reductions, we obtain the four equations,

$$\frac{Q}{\sqrt{2}}(F^1_2+F^2_1-U)+\sqrt{\Delta}\frac{\partial}{\partial r}\rho^2\frac{J-H}{\sqrt{\Delta}}=\frac{1}{2\rho^2}\left[\tfrac{1}{2}\Delta(\bar{\rho}^*\kappa^*-\bar{\rho}\kappa)-\frac{2\rho^4}{\Delta}(\bar{\rho}^*\nu-\bar{\rho}\nu^*)\right],\tag{68}$$

$$\frac{1}{\sqrt{2}}\frac{\partial}{\partial\theta}(F^1_2+F^2_1-U)+\sqrt{\Delta}\frac{\partial}{\partial r}\rho^2\frac{J+H}{\sqrt{\Delta}}=\frac{1}{2\rho^2}\left[\tfrac{1}{2}\Delta(\bar{\rho}^*\kappa^*+\bar{\rho}\kappa)-\frac{2\rho^4}{\Delta}(\bar{\rho}^*\nu+\bar{\rho}\nu^*)\right],\tag{69}$$

$$-\frac{\mathrm{i}K}{\sqrt{2}}[\rho^2(F^3_4+F^4_3)+V]+\frac{\partial}{\partial\theta}\rho^2(F+G)=\frac{+1}{\rho^2\sqrt{2}}[(\bar{\rho})^2(\rho^2\lambda^*-\tfrac{1}{2}\Delta\sigma)+(\bar{\rho}^*)^2(\rho^2\lambda-\tfrac{1}{2}\Delta\sigma^*)],\tag{70}$$

$$\frac{\Delta}{\sqrt{2}}\frac{\partial}{\partial r}[\rho^2(F^3_4+F^4_3)+V]-\frac{\partial}{\partial\theta}\rho^2(F-G)=\frac{+1}{\rho^2\sqrt{2}}[(\bar{\rho})^2(\rho^2\lambda^*+\tfrac{1}{2}\Delta\sigma)+(\bar{\rho}^*)^2(\rho^2\lambda+\tfrac{1}{2}\Delta\sigma^*)].\tag{71}$$

Eliminating $(F^1_2+F^2_1-U)$ from equations (68) and (69) and $[\rho^2(F^3_4+F^4_3)+V]$ from equations (70) and (71) we find (after some reductions in which we make use of equations (53) and (54)),

$$\sqrt{\Delta}\frac{\partial}{\partial r}\rho^2\frac{J+H}{\sqrt{\Delta}}-\frac{\partial}{\partial\theta}\frac{\sqrt{\Delta}}{Q}\frac{\partial}{\partial r}\rho^2\frac{J-H}{\sqrt{\Delta}} = \frac{1}{2\rho^2}\left[\tfrac{1}{2}\Delta(\bar{\rho}^*\kappa^*+\bar{\rho}\kappa)-\frac{2\rho^4}{\Delta}(\bar{\rho}^*\nu+\bar{\rho}\nu^*)\right]$$
$$-\frac{\partial}{\partial\theta}\left\{\frac{1}{2\rho^2 Q}\left[\tfrac{1}{2}\Delta(\bar{\rho}^*\kappa^*-\bar{\rho}\kappa)-\frac{2\rho^4}{\Delta}(\bar{\rho}^*\nu-\bar{\rho}\nu^*)\right]\right\} \quad (72)$$

and

$$\frac{\partial}{\partial\theta}\rho^2(F-G)+i\Delta\frac{\partial}{\partial r}\frac{1}{K}\frac{\partial}{\partial\theta}\rho^2(F+G) = \frac{-1}{\rho^2\sqrt{2}}[(\bar{\rho})^2(\rho^2\lambda^*+\tfrac{1}{2}\Delta\sigma)+(\bar{\rho}^*)^2(\rho^2\lambda+\tfrac{1}{2}\Delta\sigma^*)]$$
$$+\Delta\frac{\partial}{\partial r}\left\{\frac{i}{\rho^2 K\sqrt{2}}[(\bar{\rho})^2(\rho^2\lambda^*-\tfrac{1}{2}\Delta\sigma)+(\bar{\rho}^*)^2(\rho^2\lambda-\tfrac{1}{2}\Delta\sigma^*)]\right\}. \quad (73)$$

The foregoing equations can be further reduced to the forms (again, making use of equations (53) and (54)):

$$\frac{2a\sqrt{\Delta}}{Q}\frac{\partial}{\partial r}\frac{1}{\sqrt{\Delta}}[K(J-H)\cos\theta-irQ(F+G)\sin\theta]$$
$$=\frac{Q}{2\rho^2}\left[\tfrac{1}{2}\Delta(\bar{\rho}^*\kappa^*+\bar{\rho}\kappa)-\frac{2\rho^4}{\Delta}(\bar{\rho}^*\nu+\bar{\rho}\nu^*)\right]$$
$$-Q\frac{\partial}{\partial\theta}\left\{\frac{1}{2\rho^2 Q}\left[\tfrac{1}{2}\Delta(\bar{\rho}^*\kappa^*-\bar{\rho}\kappa)-\frac{2\rho^4}{\Delta}(\bar{\rho}^*\nu-\bar{\rho}\nu^*)\right]\right\}. \quad (74)$$

and

$$-\frac{2ia}{K}\frac{\partial}{\partial\theta}[K(J-H)\cos\theta-irQ(F+G)\sin\theta]$$
$$=-\frac{iK}{\rho^2\Delta\sqrt{2}}[(\bar{\rho})^2(\rho^2\lambda^*+\tfrac{1}{2}\Delta\sigma)+(\bar{\rho}^*)^2(\rho^2\lambda+\tfrac{1}{2}\Delta\sigma^*)]$$
$$+iK\frac{\partial}{\partial r}\left\{\frac{i}{\rho^2 K\sqrt{2}}[(\bar{\rho})^2(\rho^2\lambda^*-\tfrac{1}{2}\Delta\sigma)+(\bar{\rho}^*)^2(\rho^2\lambda-\tfrac{1}{2}\Delta\sigma^*)]\right\}. \quad (75)$$

Alternative forms of these equations are:

$$2a\frac{\partial}{\partial\theta}[K(J-H)\cos\theta-irQ(F+G)\sin\theta]$$
$$=\frac{iK^2}{\sqrt{2}}\left\{\mathscr{D}_0^\dagger\frac{1}{K}\left[\tfrac{1}{2}\Delta\left(\frac{\bar{\rho}}{\bar{\rho}^*}\sigma+\frac{\bar{\rho}^*}{\bar{\rho}}\sigma^*\right)\right]-\mathscr{D}_0\frac{1}{K}[(\bar{\rho})^2\lambda^*+(\bar{\rho}^*)^2\lambda]\right\} \quad (76)$$

and

$$2a\sqrt{\Delta}\frac{\partial}{\partial r}\frac{1}{\sqrt{\Delta}}[K(J-H)\cos\theta-irQ(F+G)\sin\theta]$$
$$=-\tfrac{1}{2}Q^2\left\{\mathscr{L}_0^\dagger\frac{1}{Q}\left(\tfrac{1}{2}\Delta\frac{\kappa^*}{\bar{\rho}}-\frac{2\rho^2\bar{\rho}^*\nu}{\Delta}\right)-\mathscr{L}_0\frac{1}{Q}\left(\tfrac{1}{2}\Delta\frac{\kappa}{\bar{\rho}^*}-\frac{2\rho^2\bar{\rho}\nu^*}{\Delta}\right)\right\}. \quad (77)$$

Equations (76) and (77) manifestly require that an integrability condition be satisfied: *the application of the operators,*

$$\sqrt{\Delta}\frac{\partial}{\partial r}\frac{1}{\sqrt{\Delta}} \quad \textit{and} \quad \frac{\partial}{\partial\theta}, \quad (78)$$

respectively, to the right hand sides of equations (76) *and* (77) *must yield an equality.* This condition is a crucial one in this theory: it determines (as we shall see in § 9 below) the relative normalization of the radial functions R_{+2} and R_{-2} and the real and the imaginary parts of the Teukolsky–Starobinsky constant.

9. The solution of the integrability condition

It will be observed that on the right hand side of equation (76) both σ and λ occur. But according to the solutions for these quantities obtained in I (equations (43) and (44)) σ is expressed in terms of R_{+2} and S_{+2} while λ is expressed in terms of R_{-2} and S_{-2}; and, as we have emphasized, the relative normalization of the functions R_{+2} and R_{-2} (when the angular functions S_{+2} and S_{-2} are both normalized to unity) has yet to be specified. In consequence of this lacuna in our information, the expression on the right hand side of equation (76) is not well defined. The same remark applies equally to the expression on the right hand side of equation (77).

From the discussion of the Teukolsky–Starobinsky identities in I, § 7, it is apparent that the problem of specifying the relative normalization of the functions R_{+2} and R_{-2} is closely related to the determination of the real and the imaginary parts of the Teukolsky–Starobinsky constant, $\mathscr{C}$ $(= \mathscr{C}_1 + i\mathscr{C}_2)$. Thus, if we should choose to normalize $\Delta^2 R_{+2} (= P_{+2})$ and $R_{-2} (= P_{-2})$ so as to be compatible with I, eqns (58) and (59), then only a numerical factor in the solution for Ψ_4 needs to be specified. But the problem of specifying the constants $\mathscr{C}_1$ and $\mathscr{C}_2$ in I, eqns (76) and (77), will still remain.

The nature of the problem we are presented here, becomes clearer when we examine the treatment of Maxwell's equations in I, appendix A. There, it was found that with the choice of ΔR_{+1} and R_{-1} that is compatible with I, eqns (A 19) and (A 20), an additional factor $\frac{1}{2}$ had still to be allowed for in the solution for ϕ_2 (I, eqn (A 33)); and the determination of this factor $\frac{1}{2}$ was simultaneous with the establishment of the reality of the corresponding constant $\mathscr{C}$ (I, eqn (A 32)).

Turning to the integrability condition with which we are presented, we find very early in the analysis that, with the choice of P_{+2} and P_{-2} as our basic solutions for the radial functions, the solutions for Ψ_0 and Ψ_4 must, in fact, be

$$\Delta^2 \Psi_0 = P_{+2} S_{+2} \quad \text{and} \quad \Psi_4 = \frac{1}{4(\bar{\rho}^*)^4} P_{-2} S_{-2} \tag{79}$$

– i.e. a factor $\frac{1}{4}$ as against a factor $\frac{1}{2}$ in the spin-1 case. It is convenient to assume the validity of the solutions (79) from the outset though it is strictly not necessary: we can include an additional factor – q, say – in the expression for Ψ_4 (in which case, the subsequent analysis will show that $q = 1$). Nothing essential is lost in suppressing the factor q: its restoration is manifest at all stages; but it will destroy the symmetry of the formulae we wish to display.

With Ψ_0 and Ψ_4 given by equations (79) (with the understanding that S_{+2} and S_{-2} are to be normalized to unity and $P_{+2} = \Delta^2 R_{+2}$ and $P_{-2} = R_{-2}$ are to be compatible with I, eqns (58) and (59) and satisfy I, eqns (76) and (77)), the solutions for

κ, σ, λ, and ν given in I, eqns (42)–(45), become

$$\left.\begin{aligned}
\kappa &= -\frac{1}{6M}\frac{\sqrt{2}}{\Delta^2}(\bar{\rho}^*)^2 P_{+2}\left(\mathscr{L}_2 - \frac{3ia\sin\theta}{\bar{\rho}^*}\right)S_{+2},\\
\sigma &= +\frac{1}{6M}\frac{1}{\Delta}\frac{(\bar{\rho}^*)^2}{\bar{\rho}}S_{+2}\left(\mathscr{D}_0^\dagger - \frac{3}{\bar{\rho}^*}\right)P_{+2},\\
\lambda &= +\frac{1}{6M}\frac{1}{2\bar{\rho}^*}S_{-2}\left(\mathscr{D}_0 - \frac{3}{\bar{\rho}^*}\right)P_{-2},\\
\nu &= +\frac{1}{6M}\frac{\sqrt{2}}{4\rho^2}P_{-2}\left(\mathscr{L}_2^\dagger - \frac{3ia\sin\theta}{\bar{\rho}^*}\right)S_{-2}.
\end{aligned}\right\} \tag{80}$$

Inserting the foregoing solutions for κ, σ, λ, and ν in the expressions on the right hand sides of equations (76) and (77), we find (after some considerable reductions in which we make use of the equations satisfied by $P_{\pm 2}$ and $S_{\pm 2}$ in the forms obtained in I):

$$\begin{aligned}
\frac{i}{12M\sqrt{2}}\frac{K}{\Delta}\Big\{\bar{\rho}^*\Big[&\left(\lambda + 6i\sigma r + \frac{6r\sigma\Delta}{K\bar{\rho}^*}\right)P_{+2} + 2\left(r - M - iK - \frac{r\sigma\Delta}{K} - \frac{\Delta}{\bar{\rho}^*}\right)\mathscr{D}_0^\dagger P_{+2}\Big]S_{+2}\\
+\bar{\rho}\Big[&\left(\lambda + 6i\sigma r + \frac{6r\sigma\Delta}{K\bar{\rho}}\right)P_{+2} + 2\left(r - M - iK - \frac{r\sigma\Delta}{K} - \frac{\Delta}{\bar{\rho}}\right)\mathscr{D}_0^\dagger P_{+2}\Big]S_{-2}\\
-\bar{\rho}^*\Big[&\left(\lambda - 6i\sigma r + \frac{6r\sigma\Delta}{K\bar{\rho}^*}\right)P_{-2} + 2\left(r - M + iK - \frac{r\sigma\Delta}{K} - \frac{\Delta}{\bar{\rho}^*}\right)\mathscr{D}_0 P_{-2}\Big]S_{-2}\\
-\bar{\rho}\Big[&\left(\lambda - 6i\sigma r + \frac{6r\sigma\Delta}{K\bar{\rho}}\right)P_{-2} + 2\left(r - M + iK - \frac{r\sigma\Delta}{K} - \frac{\Delta}{\bar{\rho}}\right)\mathscr{D}_0 P_{-2}\Big]S_{+2}\Big\}.
\end{aligned} \tag{81}$$

and

$$\begin{aligned}
\frac{1}{12M\sqrt{2}}\frac{Q}{\Delta}\Big\{-\bar{\rho}^*\Big[&-\left(\lambda + 6a\sigma\cos\theta - \frac{6ia^2\sigma\sin\theta\cos\theta}{Q\bar{\rho}^*}\right)S_{+2}\\
&+2\left(\cot\theta + Q - \frac{ia\sin\theta}{\bar{\rho}^*} - \frac{a\sigma\cos\theta}{Q}\right)\mathscr{L}_2 S_{+2}\Big]P_{+2}\\
+\bar{\rho}\Big[&-\left(\lambda - 6a\sigma\cos\theta + \frac{6ia^2\sigma\sin\theta\cos\theta}{Q\bar{\rho}}\right)S_{-2}\\
&+2\left(\cot\theta - Q + \frac{ia\sin\theta}{\bar{\rho}} - \frac{a\sigma\cos\theta}{Q}\right)\mathscr{L}_2^\dagger S_{-2}\Big]P_{+2}\\
+\bar{\rho}^*\Big[&-\left(\lambda - 6a\sigma\cos\theta - \frac{6ia^2\sigma\sin\theta\cos\theta}{Q\bar{\rho}^*}\right)S_{-2}\\
&+2\left(\cot\theta - Q - \frac{ia\sin\theta}{\bar{\rho}^*} - \frac{a\sigma\cos\theta}{Q}\right)\mathscr{L}_2^\dagger S_{-2}\Big]P_{-2}\\
-\bar{\rho}\Big[&-\left(\lambda + 6a\sigma\cos\theta + \frac{6ia^2\sigma\sin\theta\cos\theta}{Q\bar{\rho}}\right)S_{+2}\\
&+2\left(\cot\theta + Q + \frac{ia\sin\theta}{\bar{\rho}} - \frac{a\sigma\cos\theta}{Q}\right)\mathscr{L}_2 S_{+2}\Big]P_{-2}\Big\}.
\end{aligned} \tag{82}$$

After some further reductions, the foregoing expressions can be transformed to

$$\frac{\mathrm{i}}{12M\sqrt{2}}\{[r(K\mathscr{D}_0^\dagger\mathscr{D}_0^\dagger P_{+2}-2r\sigma\mathscr{D}_0^\dagger P_{+2})+2(3r\sigma P_{+2}-K\mathscr{D}_0^\dagger P_{+2})$$
$$-r(K\mathscr{D}_0\mathscr{D}_0 P_{-2}-2r\sigma\mathscr{D}_0 P_{-2})-2(3r\sigma P_{-2}-K\mathscr{D}_0 P_{-2})]\,(S_{+2}+S_{-2})$$
$$-\mathrm{i}a[(K\mathscr{D}_0^\dagger\mathscr{D}_0^\dagger P_{+2}-2r\sigma\mathscr{D}_0^\dagger P_{+2})+(K\mathscr{D}_0\mathscr{D}_0 P_{-2}-2r\sigma\mathscr{D}_0 P_{-2})]\,(S_{+2}-S_{-2})\cos\theta\} \quad (83)$$

and

$$\frac{1}{12M\Delta\sqrt{2}}[\![\mathrm{i}a\{[Q\mathscr{L}_1\mathscr{L}_2 S_{+2}-(2a\sigma\cos\theta)\mathscr{L}_2 S_{+2}]\cos\theta-2[(3a\sigma\cos\theta)S_{+2}-Q\mathscr{L}_2 S_{+2}]\sin\theta$$
$$+[Q\mathscr{L}_1^\dagger\mathscr{L}_2^\dagger S_{-2}-(2a\sigma\cos\theta)\mathscr{L}_2^\dagger S_{-2}]\cos\theta-2[(3a\sigma\cos\theta)S_{-2}-Q\mathscr{L}_2^\dagger S_{-2}]\sin\theta\}(P_{+2}-P_{-2})$$
$$-\{[Q\mathscr{L}_1\mathscr{L}_2 S_{+2}-(2a\sigma\cos\theta)\mathscr{L}_2 S_{+2}]-[Q\mathscr{L}_1^\dagger\mathscr{L}_2^\dagger S_{-2}-(2a\sigma\cos\theta)\mathscr{L}_2^\dagger S_{-2}]\}r(P_{+2}+P_{-2})]\!] \quad (84)$$

We have to require now that the result of applying the operator $\Delta^{\frac{1}{2}}\partial_r\Delta^{-\frac{1}{2}}$ to the expression (83) yields the same result as applying the operator ∂_θ to the expression (84). In the first instance, one is at a loss to know as to how one is to proceed: the expressions which have to be equated involve up to the third derivatives of the radial and the angular functions defined only by the differential equations they satisfy. However, one soon realizes that the only way in which the required conditions can be found is to replace, at each stage, the derivatives of the functions $P_{\pm 2}(r)$ and $S_{\pm 2}(\theta)$ by appropriate combinations of the functions themselves with the aid of I, eqns (76), (77), (81), and (82). By such replacements, the expressions (83) and (84) become

$$\frac{1}{12MB_1\sqrt{2}}[\![\{2K^2(\mathscr{C}_2-\Gamma_2-4\lambda\sigma r)-2[K(r-M)-\sigma r\Delta](\mathscr{C}_1-\Gamma_1)\}rX$$
$$+\{2K^2(\mathscr{C}_1-\Gamma_1)+2[K(r-M)-\sigma r\Delta](\mathscr{C}_2-\Gamma_2-4\lambda\sigma r)+8\lambda\sigma K\Delta\}rY$$
$$+2K(\Delta\mathscr{C}_1-A_1)X-2[3r\sigma B_1+K(\Delta\mathscr{C}_2+A_2)]\,Y]\!]\,(S_{+2}+S_{-2})$$
$$+\frac{a}{12MB_1\sqrt{2}}[\![\{2K^2(\mathscr{C}_1+\Gamma_1)+2[K(r-M)-\sigma r\Delta](\mathscr{C}_2+\Gamma_2+4\lambda\sigma r)-8\lambda\sigma K\Delta\}X$$
$$+\{-2K^2(\mathscr{C}_2+\Gamma_2+4\lambda\sigma r)+2[K(r-M)-\sigma r\Delta](\mathscr{C}_1+\Gamma_1)\}\,Y]\!]\,(S_{+2}-S_{-2})\cos\theta \quad (85)$$

and

$$-\frac{1}{12M\Delta\beta_1\sqrt{2}}[\![\{2(Q\cot\theta-a\sigma\cos\theta)(D-\Gamma_1)+8a\sigma\lambda Q^2\cos\theta\}(S_{+2}+S_{-2})$$
$$+\{-2Q^2(D-\Gamma_1)-8a\sigma\lambda(Q\operatorname{cosec}\theta-a\sigma\cos^2\theta)\}(S_{+2}-S_{-2})]\!]\,rX$$
$$-\frac{a}{12M\Delta\beta_1\sqrt{2}}[\![\{2Q^2(D+\Gamma_1)-8a\sigma\lambda(Q\operatorname{cosec}\theta-a\sigma\cos^2\theta)\}(S_{+2}+S_{-2})\cos\theta$$
$$+\{-2(Q\cot\theta-a\sigma\cos\theta)(D+\Gamma_1)+8a\sigma\lambda Q^2\cos\theta\}(S_{+2}-S_{-2})\cos\theta$$
$$-2\{[(3a\sigma\cos\theta)\beta_1+\alpha_2 Q](S_{+2}+S_{-2})+Q(\alpha_1+D)(S_{+2}-S_{-2})\}\sin\theta]\!]\,Y, \quad (86)$$

where it may be recalled that (cf. I, eqns (60), (67), (69), (73), and (75))

$$X = P_{+2}+P_{-2}, \quad iY = P_{+2}-P_{-2}, \tag{87}$$

$$\mathscr{C} = \mathscr{C}_1+i\mathscr{C}_2, \quad \Gamma_1 = \lambda(\lambda+2)-12\sigma^2\alpha^2, \quad \Gamma_2 = 12\sigma M,$$
$$D^2 = \lambda^2(\lambda+2)^2-8\sigma^2\lambda[\alpha^2(5\lambda+6)-12a^2]+144\sigma^4\alpha^4,$$
$$|\mathscr{C}|^2 = D^2+144\sigma^2M^2, \quad \alpha^2 = a^2+(am/\sigma), \tag{88}$$

and the remaining symbols (A_1, A_2, B_1, α_1, α_2, and β_1) have the same meanings as in I, eqns (74) and (80). It might also be noted that, in obtaining the expressions (85) and (86), besides I, eqns (76), (77), (81) and (82), the following equations (and their complex conjugates and adjoints) were found useful:

$$\left.\begin{aligned} \Delta\mathscr{D}_0\mathscr{D}_0P_{-2} &= 2(iK+r-M)\mathscr{D}_0P_{-2}+(\lambda-6i\sigma r)P_{-2} \\ \Delta^2\mathscr{D}_0\mathscr{D}_0\mathscr{D}_0P_{-2} &= [4iK(iK+r-M)+\Delta(\lambda+2-2i\sigma r)]\mathscr{D}_0P_{-2} \\ &\quad +[2iK(\lambda-6i\sigma r)-6i\sigma\Delta]P_{-2}, \end{aligned}\right\} \tag{89}$$

and

$$\left.\begin{aligned} \mathscr{L}_1\mathscr{L}_2S_{+2} &= -(\lambda+6a\sigma\cos\theta)S_{+2}+2(Q+\cot\theta)\mathscr{L}_2S_{+2} \\ \mathscr{L}_0\mathscr{L}_1\mathscr{L}_2S_{+2} &= [6a\sigma\sin\theta-2Q(\lambda+6a\sigma\cos\theta)]S_{+2} \\ &\quad -[(\lambda+2)+2a\sigma\cos\theta-4Q(Q+\cot\theta)]\mathscr{L}_2S_{+2}. \end{aligned}\right\} \tag{90}$$

Now applying, respectively, the operators $\Delta^{\frac{1}{2}}\partial_r\Delta^{-\frac{1}{2}}$ and ∂_θ to the expressions (85) and (86) and making the same replacements of the derivatives in terms of the functions, we find, after some very considerable reductions and equally remarkable cancellations, that we are finally left with

$$\begin{aligned} &\sqrt{\Delta}\frac{\partial}{\partial r}\frac{2a}{\sqrt{\Delta}}\frac{\partial}{\partial\theta}[K(J-H)\cos\theta-irQ(F+G)\sin\theta] \\ &= \frac{1}{48M\Delta\sqrt{2}}[\![a\{[(\mathscr{C}_2+\Gamma_2)+4\lambda\sigma r]X+(\mathscr{C}_1+\Gamma_1)Y\}(S_{+2}-S_{-2})\cos\theta \\ &+\{-(\mathscr{C}_1-\Gamma_1)rX+[r(\mathscr{C}_2+\Gamma_2)+4\sigma(\lambda\alpha^2-6a^2)Y\}(S_{+2}+S_{-2})]\!] \end{aligned} \tag{91}$$

and

$$\begin{aligned} &2a\frac{\partial}{\partial\theta}\sqrt{\Delta}\frac{\partial}{\partial r}\frac{1}{\sqrt{\Delta}}[K(J-H)\cos\theta-irQ(F+G)\sin\theta] \\ &= \frac{1}{48M\Delta\sqrt{2}}[\![a\{(D+\Gamma_1)(S_{+2}-S_{-2})\cos\theta+(4\sigma/a)(\lambda\alpha^2-6a^2)(S_{+2}+S_{-2})\}Y \\ &+\{-(D-\Gamma_1)(S_{+2}+S_{-2})+4a\sigma\lambda(S_{+2}-S_{-2})\cos\theta\}rX]\!]. \end{aligned} \tag{92}$$

A comparison of the expressions on the right hand sides of equations (91) and (92) shows that their equality only (!) requires that

$$\mathscr{C}_1 = D \quad \text{and} \quad \mathscr{C}_2 = -\Gamma_2 = -12\sigma M \tag{93}$$

– remarkably simple results to arrive at, after such a long road. [The reductions described in this section alone took the author over 100 pages!]

Now defining the functions

$$\mathscr{S}_+ = \int (S_{+2}+S_{-2})\,d\theta, \quad \mathscr{S}_- = \int (S_{+2}-S_{-2})\cos\theta\,d\theta,$$
$$\mathscr{R}_+ = \Delta^{\frac{1}{2}}\int \frac{rX}{\Delta^{\frac{3}{2}}}\,dr, \quad \text{and} \quad \mathscr{R}_- = \Delta^{\frac{1}{2}}\int \frac{Y}{\Delta^{\frac{3}{2}}}\,dr, \tag{94}$$

we can write the integral of equations (76) and (77) in the form

$$K(J-H)\cos\theta - irQ(F+G)\sin\theta = \Psi, \tag{95}$$

where
$$\Psi = \frac{1}{96M\sqrt{2}}\Big[(\mathscr{C}_1+\Gamma_1)\mathscr{R}_-\mathscr{S}_- + \frac{4\sigma}{a}(\lambda\alpha^2-6a^2)\mathscr{R}_-\mathscr{S}_+$$
$$+4\lambda\sigma\mathscr{R}_+\mathscr{S}_- - \frac{1}{a}(\mathscr{C}_1-\Gamma_1)\mathscr{R}_+\mathscr{S}_+\Big]. \tag{96}$$

We conclude this section by noting that with the aid of the integral (95) we can rewrite equations (53) and (54) in the alternative forms

$$\rho^2(J+H) = \frac{\partial}{\partial\theta}\rho^2\frac{J-H}{Q} + \frac{2a\Psi}{Q^2}, \tag{97}$$

and

$$i\rho^2\frac{F-G}{\Delta} = \frac{\partial}{\partial r}\rho^2\frac{F+G}{K} - \frac{2ia\Psi}{K^2}. \tag{98}$$

10. The completion of the reduction of system II

In §§ 8 and 9, our considerations were restricted to the four equations (68)–(71) derived from the eight equations (67) (which are equivalent to those of system II). After satisfying the integrability condition, which follows from these equations, we are left with three independent equations which we may take to be equations (68), (70), and the integral (95). It remains to consider a further set of four equations, from system II, which are independent of the ones already considered.

By subtracting the second equation from the first (instead of adding them as in § 8) in each of the four pairs of equations in (69), we obtain

$$\sqrt{2}\,Q\rho^2(F^1_2-F^2_1) + \Delta\frac{\partial}{\partial r}\frac{\rho^4}{\Delta}B_2 + \frac{iK}{\Delta}\rho^4(J-H)$$
$$+2ia\rho^2(F-G)\cos\theta = \tfrac{1}{2}\Delta(\bar\rho^*\kappa^*-\bar\rho\kappa) + \frac{2\rho^4}{\Delta}(\bar\rho^*\nu-\bar\rho\nu^*), \tag{99}$$

$$\sqrt{2}\,\rho^2\frac{\partial}{\partial\theta}(F^1_2-F^2_1) - \Delta\frac{\partial}{\partial r}\frac{\rho^4}{\Delta}B_1 + \frac{iK}{\Delta}\rho^4(J+H)$$
$$+2ia\rho^2C_2\cos\theta = \tfrac{1}{2}\Delta(\bar\rho^*\kappa^*+\bar\rho\kappa) + \frac{2\rho^4}{\Delta}(\bar\rho^*\nu+\bar\rho\nu^*), \tag{100}$$

$$-iK\rho^4(F^3_4-F^4_3) - \frac{1}{\sqrt{2}}\left(\frac{\partial}{\partial\theta}-\cot\theta\right)\rho^4C_1 - \frac{Q}{\sqrt{2}}\rho^4(F+G)$$
$$-\sqrt{2}\,iar\rho^2(J+H)\sin\theta = (\bar\rho)^2(\rho^2\lambda^*-\tfrac{1}{2}\Delta\sigma) - (\bar\rho^*)^2(\rho^2\lambda-\tfrac{1}{2}\Delta\sigma^*), \tag{101}$$

$$-\rho^2\Delta\frac{\partial}{\partial r}\rho^2(F^3_4-F^4_3) - \frac{1}{\sqrt{2}}\left(\frac{\partial}{\partial\theta}-\cot\theta\right)\rho^4C_2 - \frac{Q}{\sqrt{2}}\rho^4(F-G)$$
$$+\sqrt{2}\,iar\rho^2B_1\sin\theta = -(\bar\rho)^2(\rho^2\lambda^*+\tfrac{1}{2}\Delta\sigma) + (\bar\rho^*)^2(\rho^2\lambda+\tfrac{1}{2}\Delta\sigma^*). \tag{102}$$

On eliminating $(F_2^1 - F_1^2)$ from equations (99) and (100) and $\rho^2(F_4^3 - F_3^4)$ from equations (101) and (102) and making use of equations (95), (97), and (100) in reducing the eliminants, we find that we are finally left with

$$\frac{2ia\Delta}{Q^2K}\left[\frac{\partial}{\partial r}K\frac{\partial}{\partial r}\frac{\Psi}{K}+\left(\frac{4\alpha^2\sigma^2}{K^2}-\frac{K^2}{\Delta^2}\right)\Psi\right]$$
$$=\frac{\partial}{\partial\theta}\frac{1}{Q\rho^2}\left[\tfrac{1}{2}\Delta(\bar{\rho}^*\kappa^*-\bar{\rho}\kappa)+\frac{2\rho^4}{\Delta}(\bar{\rho}^*\nu-\bar{\rho}\nu^*)\right]$$
$$-\frac{\Delta}{2\rho^2}(\bar{\rho}^*\kappa^*+\bar{\rho}\kappa)-\frac{2\rho^2}{\Delta}(\bar{\rho}^*\nu+\bar{\rho}\nu^*) \qquad (103)$$

and
$$\frac{2a\Delta}{K^2Q}\left[\frac{\partial}{\partial\theta}Q\frac{\partial}{\partial\theta}\frac{\Psi}{Q}-\left(\frac{4\alpha^2\sigma^2}{Q^2}-Q^2\right)\Psi\right]$$
$$=\Delta\frac{\partial}{\partial r}\frac{i\sqrt{2}}{K\rho^2}[(\bar{\rho})^2(\rho^2\lambda^*-\tfrac{1}{2}\Delta\sigma)-(\bar{\rho}^*)^2(\rho^2\lambda-\tfrac{1}{2}\Delta\sigma^*)]$$
$$+\frac{\sqrt{2}}{\rho^2}[-(\bar{\rho})^2(\rho^2\lambda^*+\tfrac{1}{2}\Delta\sigma)+(\bar{\rho}^*)^2(\rho^2\lambda+\tfrac{1}{2}\Delta\sigma^*)]. \qquad (104)$$

Since both sides of equations (103) and (104) are expressible solely in terms of Teukolsky's functions, it is clear that what we have here are identities among them. We conclude that equations (99) and (100) and, similarly, equations (101) and (102) are not independent equations; and that we may take equations (99) and (101) as the two independent equations of the set.

We have now completed the reduction of all eight equations of system II to an irreducible set comprised of the five equations (68), (70), (95), (99), and (101).

11. Recapitulation

We have completed the solution of all the equations derived from the Bianchi identities and the commutation relations (as well as some of the equations derived from the Ricci identities). Before we tabulate the solutions, it will be useful to enumerate the different independent equations (out of a total of 76) which have been solved explicitly, the quantities for which solutions have been found, and the degree of arbitrariness which is finally left in the solutions.

The four (complex) Bianchi identities as well as the two (complex) Ricci identities which are already linearized (in the sense of I, § 4) enabled the solution of Ψ_0, Ψ_1, Ψ_3, Ψ_4, κ, σ, λ, and ν in terms of Teukolsky's radial and angular functions. The solutions involved two arbitrary (complex) functions which we eliminated by permanently adopting the gauge in which $\Psi_1 \equiv \Psi_3 \equiv 0$. In this gauge the solutions for the remaining six (complex) quantities became definite apart from the relative normalization of the radial functions which was left unspecified.

The remaining four (complex) Bianchi identities served to express (uniquely) the perturbations in the spin-coefficients $\tilde{\rho}$, μ, τ, and π in terms of the elements of the matrix A (specifying the perturbed null-tetrad) and the perturbation in Ψ_2, or equivalently X (cf. equation (19)).

Turning to the 24 equations derived from the linearized commutation-relation (equation (26)), we found that eight of them (equations (33)) served to express (again, uniquely) the perturbations in the spin-coefficients α, β, γ, and ϵ in terms of the elements of A and X. The remaining 16 equations were grouped into two systems (systems I and II; equations (30) and (32)) of eight equations each. With the known solutions for κ, σ, λ, and ν (in terms of Teukolsky's functions) and for $\tilde{\rho}^{(1)}$, $\mu^{(1)}$, $\tau^{(1)}$, and $\pi^{(1)}$ (in terms of A and X), the equations belonging to systems I and II become equations for the elements of A and X.

It was found that only six equations of system I were independent. These six equations (equations (49)–(54)), together with two of the identities (equations (37) and (38)) derived from the Bianchi identities (equations (16)) and the commutator relations, required that $X \equiv 0$; in other words, the perturbation in Ψ_2 must vanish identically – a result both surprising and unexpected.

The eight equations of system II (in their alternative forms (67)) provided two sets of equations of four each (equations (68)–(71) and (99)–(102)). The first set led to a crucial integrability condition (cf. equations (76) and (77)) whose satisfaction determined the relative normalization of the radial functions and the real and the imaginary parts of the Teukolsky–Starobinsky constant. And the end result was the replacement of the four equations (68)–(71) by the two equations (68) and (70) and the solution (95) of the integrability condition. With the aid of the solution (95), it was then shown that the second set of four equations (99)–(102) includes only two independent equations which we took to be equations (99) and (101).

In summary, then, the solution of the 16 equations included in systems I and II leaves us with the 11 equations (49)–(52), (68), (70), (95), (97), (98), (99), and (101). These 11 equations provide *explicit* solutions for 11 particular combinations of the elements of A. But the solutions leave unspecified the 4 diagonal elements of A and *one* of the two functions $(F+G)$ and $(J-H)$. Including the 12 spin coefficients, the Weyl scalars Ψ_0 and Ψ_4, and the 11 combinations of the elements of A, we have all together the solution for 39 real quantities. And these 39 quantities have been obtained as the solutions of 39 independent equations taken out of the entire set of the Newman–Penrose equations.‡

The solution for the 39 quantities (which we have enumerated) have been obtained in the gauge $\Psi_1 \equiv \Psi_3 \equiv 0$; and, as we have seen, they involve five arbitrary functions. Besides, it has been shown that the imposition of the gauge, $\Psi_1 \equiv \Psi_3 \equiv 0$, has required, as one of its consequences, the vanishing of the perturbation in Ψ_2.

From the foregoing account, it would appear that we have in fact obtained the general solution to the problem of the gravitational perturbations of the Kerr black hole. And, if that is the case, we may without loss of generality set

$$A_1^1 = A_2^2 = A_3^3 = A_4^4 = 0 \quad \text{and} \quad F+G=0 \quad \text{or} \quad J-H=0, \tag{105}$$

‡ Besides the 24 commutation relations and the 16 Bianchi identities, 8 equations derived from the Ricci identities (not included in the set derivable from the commutation relations) – i.e. a total of 48 equations – have been explicitly satisfied. At least four other additional equations have also been directly verified.

and obtain two alternative forms for the solutions. The solutions for the principal quantities are given below:

$$\left.\begin{aligned} &\textit{if}\quad F+G=0:\quad J-H=\frac{\Psi}{K\cos\theta},\quad F-G=-\frac{2a\Delta}{\rho^2K^2}\Psi,\\ &J+H=\frac{1}{\rho^2}\frac{\partial}{\partial\theta}\frac{\rho^2\Psi}{KQ\cos\theta}+\frac{2a}{\rho^2Q^2}\Psi,\\ &\frac{\mathrm{i}K}{\Delta}B_2=\frac{\partial}{\partial r}\frac{\Psi}{K\cos\theta},\quad QC_1=-\frac{2\mathrm{i}ar\sin\theta}{\rho^2K\cos\theta}\Psi; \end{aligned}\right\}\tag{106}$$

and

$$\left.\begin{aligned} &\textit{if}\quad J-H=0:\quad F+G=\mathrm{i}\frac{\Psi}{rQ\sin\theta},\quad J+H=\frac{2a}{\rho^2Q^2}\Psi,\\ &F-G=\frac{\Delta}{\rho^2}\frac{\partial}{\partial r}\frac{\rho^2\Psi}{rKQ\sin\theta}-\frac{2a\Delta}{\rho^2K^2}\Psi,\\ &\frac{\mathrm{i}K}{\Delta}B_2=\frac{2a\cos\theta}{r\rho^2Q\sin\theta}\Psi,\quad QC_1=\mathrm{i}\left(\frac{\partial}{\partial\theta}+\cot\theta\right)\frac{\Psi}{rQ\sin\theta}. \end{aligned}\right\}\tag{107}$$

The expressions for B_1 and C_2 are formally the same in both cases and are given by

$$\frac{\mathrm{i}K}{\Delta}B_1=\frac{2\mathrm{i}a\cos\theta}{\rho^2}C_1-\frac{\partial}{\partial r}(J+H)\tag{108}$$

and

$$QC_2=-\frac{2\mathrm{i}ar\sin\theta}{\rho^2}B_2+\left(\frac{\partial}{\partial\theta}+\cot\theta\right)(F-G).\tag{109}$$

The corresponding solutions for $(F^1_2+F^2_1)$ and $(F^3_4+F^4_3)$ follow directly from equations (68) and (70); while the solutions for $(F^1_2-F^2_1)$ and $(F^3_4-F^4_3)$ follow, similarly, from equations (99) and (101).

And finally, by setting the diagonal elements of A equal to zero in the expressions given in equations (11), we can read off the solutions for the metric perturbations.

12. Concluding remarks

The problem to which this paper is addressed and the manner in which its solution was accomplished are fully described in §§ 2 and 11. But one aspect of the matter deserves some comment. It is the fact that the entire linearized set of the Newman–Penrose equations allows an explicit integration in the case of Kerr geometry. Not only that; except for the simple quadratures needed to specify the functions $\mathscr{R}_\pm$ and $\mathscr{S}_\pm$ (in terms of which the solutions are expressed) one was not called upon to solve any differential equation as such: the solution was accomplished by methods almost wholly algebraic. Clearly, there must be some 'symmetry' of the Kerr space-time which makes all this possible. Indeed, there is so much else that has the aura of the miraculous about the Kerr metric that one wonders.

Turning to more mundane questions which can be answered with the aid of the solutions for the metric perturbations obtained in this paper, we shall list three.

First, the solutions obtained in this paper must apply to the Schwarzschild metric by simply setting $a = 0$. But how are the solutions so obtained related to those found in earlier treatments (such as in Chandrasekhar 1975)?

Second, how are the functions $Z^{(\pm\sigma)}$, which have been found by Chandrasekhar & Detweiler (1975, 1976) to satisfy one-dimensional wave equations, related to the metric perturbations?

And third, how is the notion of parity to be grafted on to the present treatment?

We shall return to these and related questions on other occasions.

During the past 12 months when the subject of this paper has been my principal concern, I have profited greatly by frequent discussions with Dr S. Teukolsky, Dr R. Wald, and Mr B. Xanthopoulos. I am particularly grateful to Mr Xanthopoulos for his patience in checking a large part of the massive reductions involved in this work.

The present work was started during the summer (June–August) of 1976 when I was a visitor at the Department of Applied Mathematics, Cambridge University (England) as a guest of Dr S. Hawking, F.R.S.; my work during that period was supported in part by the United Kingdom Science Research Council under grant BRG 61662 'Mathematical problems in the general theory of relativity'. I am extremely grateful to Dr Hawking for his hospitality. At Chicago, my research is in part supported by a grant from the National Science Foundation under grant PHY76-81102 with the University of Chicago.

Note added in proof, *22 November* 1977. Note that the finiteness of Ψ (as defined in equation (96)) for $a \to 0$ requires that D $(= \mathscr{C}_1)$ be defined as the positive square root of the expression given in equation (88). Further, it can be shown that not only does Ψ become separable in the variables, it can also be expressed directly in terms of the Teukolsky's radial and angular functions.

References

Chandrasekhar, S. 1975 *Proc. R. Soc. Lond.* A **343**, 289.
Chandrasekhar, S. 1977 *Proc. R. Soc. Lond.* A **358**, 421.
Chandrasekhar, S. & Detweiler, S. 1975 *Proc. R. Soc. Lond.* A **345**, 145.
Chandrasekhar, S. & Detweiler, S. 1976 *Proc. R. Soc. Lond.* A **350**, 165.
Newman, E. & Penrose, R. 1962 *J. Math. Phys.* **3**, 566.
Wald, R. 1973 *J. Math. Phys.* **14**, 1453.

12

The gravitational perturbations of the Kerr black hole III. Further amplifications†

By S. Chandrasekhar, F.R.S.
University of Chicago, Chicago, Illinois 60637, U.S.A.

(*Received* 22 *August* 1978)

The present paper is devoted to an amplification of the solution of the Newman–Penrose equations considered in the two earlier papers of this series. The principal amplification consists in showing that the function Ψ, in terms of which the metric perturbations are most simply expressed (and which was thought to require quadratures), besides being separable in its variables, is expressible directly in terms of the Teukolsky functions (and eliminates the need for quadratures). It is further shown that the completion of the solution for the metric perturbations requires the consideration of four additional equations which follow from four Ricci identities (not hitherto considered); and the solution of these equations is found. It is also pointed out that while the perturbation in the Weyl scalar, Ψ_2, *can* be set equal to zero, it cannot be *deduced* to be zero. (The contrary statement in the earlier paper arose from an error of a factor 2 in one of the equations.)

Numerical verification of some of the principal equations and identities of the theory is provided.

An important aspect of the analysis contained in this paper is the emergence of several crucial identities among the Teukolsky functions which one might despair of verifying directly.

1. Introduction

The present paper amplifies and completes the investigation of the perturbation of the Kerr black hole begun in two earlier papers (Chandrasekhar 1978*a* and *b*; these papers will be referred to hereafter as I and II, respectively). But, first, it is important to correct an error in paper II: it is *not* true as claimed in II, § 7 that the vanishing of the perturbation, $\Psi_2^{(1)}$, in the Weyl scalar, Ψ_2, is *deducible* in the gauge in which $\Psi_1 \equiv \Psi_3 \equiv 0$ (which is allowed by the freedom we have in the choice of the perturbed *tetrad frame*). As was emphasized to the author by Dr John Friedman, while the perturbation in Ψ_2 *can* be set equal to zero, it *must* be left unspecified in a solution of the Newman–Penrose equation‡ (by the freedom we have in the choice of the perturbed *coordinate frame*). And at Friedman's suggestion, Mrs Eliane

† Part II appeared in *Proc. R. Soc. Lond.* A **358**, 441–465 (1978).

‡ Dr Robert Wald had also made similar remarks to the author.

Reprinted with minor corrections.

Lessner undertook a careful scrutiny of the author's work sheets and discovered an error of a factor 2 in II, equation (59): it should correctly read

$$\pi^{(1)}\pi^* - \tau^{*(1)}\tau + \pi^{*(1)}\pi - \tau^{(1)}\tau^* = \frac{\mathrm{i}a\sin\theta}{\rho^4}Q\,[rZ - (\mathrm{i}a\cos\theta)\,Y] + \surd 2\frac{\mathrm{i}a\sin\theta}{\rho^2}(H-J); \quad (1)$$

that is without the factor 2 in front of the first term on the right hand side of the equation. When this corrected equation is substituted in II, equation (58), along with II, equations (60), (61), and (62), we find that the equation is, in fact, *identically* satisfied. A similar reduction of II, equation (38), after correcting for a 'corresponding' error, shows that it too is identically satisfied. Thus, in agreement with what one ought to have expected, the solution of the Newman–Penrose equations does leave $\Psi_2^{(1)}$ unspecified. And leaving $\Psi_2^{(1)}$ unspecified in no way affects the subsequent analysis in II, §§ 8, 9, and 10: the only change that is required is to redefine the function Ψ (II, equation (95)) by

$$\Psi = K(J-H)\cos\theta - \mathrm{i}rQ(F+G)\sin\theta + \mathrm{i}KQY/a\surd 2, \quad (2)$$

where Y is related to $\Psi_2^{(1)}$ as in II, equation (21). The integrability condition (derived in § 8) and its solution (obtained in § 9) as well as the subsequent analysis (in § 10) are unaffected by this redefinition. And, moreover, as we have stated, we *can* set $\Psi_2^{(1)} = 0$, by a choice of the perturbed coordinate frame.

Again, in II § 11 it was stated (correctly) that the solution of the Newman–Penrose equations that were considered (namely, the entire set of the 24 commutation relations, the 16 Bianchi identities, and some 8 assorted Ricci identities) left one of the two functions, $F+G$ and $J-H$, unspecified. But the (tentative) inference that was drawn, namely, that one, or the other of these two functions, can be set equal to zero (cf. II, equation (105)) is wrong.† We shall show, instead, that four of the Ricci identities, not considered in II, do provide an additional relation between $(F+G)$ and $(J-H)$ and make them determinate. (But obtaining this second relation turned out to be a rather formidable task.)

The principal amplifications that are made in the present paper are the following. It is shown that the function Ψ (in terms of which the solutions for the metric perturbations are most simply expressed) is separable in its variables and is expressible as a product of a radial and an angular function, $\mathscr{R}(r)$ and $\mathscr{S}(\theta)$; and, further, that these functions are explicitly expressible in terms of the Teukolsky functions. Also, as we have already stated, a second relation between the metric functions, $F+G$ and $J-H$, is obtained.

Perhaps the most important fact that emerges from the analysis of the present paper is that a systematic solution of the Newman–Penrose equations leads to many remarkable identities among the Teukolsky functions, identities which

† To be specific, besides the incorrect statements made in the abstract, II, §§ 7 and 11, that the vanishing of $\Psi_2^{(1)}$ is deducible, the only equations which are affected are II, equations (59), (63)–(66), and (105)–(109). All the other equations in II are presumed to be correct (cf. §§ 3 and 4 below where numerical verification is provided for some of the key equations).

would appear well-nigh impossible of direct verification. And this fact raises the question whether, in a similar way, the Newman–Penrose equations may not be used to generate a new class of identities among some of the special functions of mathematical physics.

2. A bracket notation

We shall find that the following notation is helpful in abbreviating the equations of the theory. We let

$$[P]^{\pm} = P_{+2} \pm P_{-2}; \quad [S]^{\pm} = S_{+2} \pm S_{-2}; \tag{3}$$

$$[\mathscr{D}P]^{\pm} = \mathscr{D}_0^{\dagger} P_{+2} \pm \mathscr{D}_0 P_{-2}; \quad [\mathscr{D}\mathscr{D}P]^{\pm} = \mathscr{D}_0^{\dagger}\mathscr{D}_0^{\dagger} P_{+2} \pm \mathscr{D}_0 \mathscr{D}_0 P_{-2}; \tag{4}$$

$$[\mathscr{L}S]^{\pm} = \mathscr{L}_2 S_{+2} \pm \mathscr{L}_2^{\dagger} S_{-2}; \quad [\mathscr{L}\mathscr{L}S]^{\pm} = \mathscr{L}_1 \mathscr{L}_2 S_{+2} \pm \mathscr{L}_1^{\dagger}\mathscr{L}_2^{\dagger} S_{-2}. \tag{5}$$

(Note that we have defined earlier, $X = P_{+2}+P_{-2}$ and $\mathrm{i}Y = P_{+2}-P_{-2}$. We shall continue to use these alternative definitions.)

Elementary consequences of the foregoing definitions are:

$$\frac{\mathrm{d}}{\mathrm{d}r}[P]^{\pm} = [\mathscr{D}P]^{\pm} + \frac{\mathrm{i}K}{\Delta}[P]^{\mp}; \quad \frac{\mathrm{d}}{\mathrm{d}r}[\mathscr{D}P]^{\pm} = [\mathscr{D}\mathscr{D}P]^{\pm} + \frac{\mathrm{i}K}{\Delta}[\mathscr{D}P]^{\mp}, \tag{6}$$

$$\frac{\mathrm{d}}{\mathrm{d}\theta}[S]^{\pm} = [\mathscr{L}S]^{\pm} - Q[S]^{\mp} - 2[S]^{\pm}\cot\theta, \tag{7}$$

and

$$\frac{\mathrm{d}}{\mathrm{d}\theta}[\mathscr{L}S]^{\pm} = [\mathscr{L}\mathscr{L}S]^{\pm} - Q[\mathscr{L}S]^{\mp} - [\mathscr{L}S]^{\pm}\cot\theta. \tag{8}$$

Also, making use of II, equations (89) and (90), we may write in our present notation

$$\Delta[\mathscr{D}\mathscr{D}P]^{\pm} = -2\mathrm{i}K[\mathscr{D}P]^{\mp} + 2(r-M)[\mathscr{D}P]^{\pm} + \lambda[P]^{\pm} + 6\mathrm{i}\sigma r[P]^{\mp}, \tag{9}$$

and

$$[\mathscr{L}\mathscr{L}S]^{\pm} = 2Q[\mathscr{L}S]^{\mp} + 2[\mathscr{L}S]^{\pm}\cot\theta - \lambda[S]^{\pm} - 6a\sigma[S]^{\mp}\cos\theta. \tag{10}$$

3. The separability of $\Psi(r, \theta)$

The solution for

$$\Psi = K(J-H)\cos\theta - \mathrm{i}rQ(F+G)\sin\theta + \mathrm{i}KQY/a\surd 2, \tag{11}$$

given by II, equations (88) and (94)–(96), is

$$\Psi = \frac{1}{96M\surd 2}\Big[(\mathscr{C}_1+\Gamma_1)\,\mathscr{R}_-\mathscr{S}_- + \frac{4\sigma}{a}(\lambda\alpha^2-6a^2)\,\mathscr{R}_-\mathscr{S}_+ + 4\lambda\sigma\mathscr{R}_+\mathscr{S}_- - \frac{1}{a}(\mathscr{C}_1-\Gamma_1)\,\mathscr{R}_+\mathscr{S}_+\Big], \tag{12}$$

where

$$\mathscr{S}_+ = \int [S]^+ \mathrm{d}\theta, \quad \mathscr{S}_- = \int [S]^- \cos\theta\,\mathrm{d}\theta, \tag{13}$$

$$\mathscr{R}_+ = \Delta^{\frac{1}{2}}\int \frac{rX}{\Delta^{\frac{3}{2}}}\mathrm{d}r, \quad \mathscr{R}_- = \Delta^{\frac{1}{2}}\int \frac{Y}{\Delta^{\frac{3}{2}}}\mathrm{d}r, \tag{14}$$

$$\mathscr{C}_1^2 = \lambda^2(\lambda+2)^2 - 8\sigma^2\lambda[\alpha^2(5\lambda+6)-12a^2] + 144\sigma^4\alpha^4, \tag{15}$$

$$\Gamma_1 = \lambda(\lambda+2) - 12\sigma^2\alpha^2, \quad \text{and} \quad \alpha^2 = a^2 + (am/\sigma). \tag{16}$$

We first observe that the finiteness of Ψ, for $a \to 0$, requires that $\mathscr{C}_1$ be defined as the positive square root of the quantity on the right hand side of (15); for only then will $\mathscr{C}_1 - \Gamma_1 \to 0$, as $a \to 0$. We have, in fact, the identity

$$\mathscr{C}_1^2 - \Gamma_1^2 = -16\sigma^2\lambda(\lambda\alpha^2 - 6a^2). \tag{17}$$

Accordingly, we may write

$$\frac{1}{a}(\mathscr{C}_1 - \Gamma_1) = \frac{\mathscr{C}_1^2 - \Gamma_1^2}{a(\mathscr{C}_1 + \Gamma_1)} = -\frac{16\sigma^2\lambda(\lambda\alpha^2 - 6a^2)}{a(\mathscr{C}_1 + \Gamma_1)}. \tag{18}$$

Making use of this last relation, we can rewrite the solution (12) for Ψ in the form:

$$\begin{aligned}\Psi &= \frac{1}{96M\sqrt{2}}\left\{[(\mathscr{C}_1 + \Gamma_1)\mathscr{R}_- + 4\lambda\sigma\mathscr{R}_+]\mathscr{S}_- + \frac{4\sigma}{a}(\lambda\alpha^2 - 6a^2)\left[\mathscr{R}_- + \frac{4\lambda\sigma}{\mathscr{C}_1 + \Gamma_1}\mathscr{R}_+\right]\mathscr{S}_+\right\}\\ &= \frac{\mathscr{C}_1 + \Gamma_1}{96M\sqrt{2}}\left[\mathscr{R}_- + \frac{4\lambda\sigma}{\mathscr{C}_1 + \Gamma_1}\mathscr{R}_+\right]\left[\mathscr{S}_- + \frac{4\sigma(\lambda\alpha^2 - 6a^2)}{a(\mathscr{C}_1 + \Gamma_1)}\mathscr{S}_+\right].\end{aligned} \tag{19}$$

The separability of Ψ is now manifest:

$$\Psi = \frac{\mathscr{C}_1 + \Gamma_1}{96M\sqrt{2}}\mathscr{R}(r)\mathscr{S}(\theta), \tag{20}$$

where $$\mathscr{R}(r) = \mathscr{R}_- + \frac{4\lambda\sigma}{\mathscr{C}_1 + \Gamma_1}\mathscr{R}_+ \quad \text{and} \quad \mathscr{S}(\theta) = \mathscr{S}_- + \frac{4\sigma(\lambda\alpha^2 - 6a^2)}{a(\mathscr{C}_1 + \Gamma_1)}\mathscr{S}_+. \tag{21}$$

From the definitions of $\mathscr{R}$ and $\mathscr{S}$ it follows that

$$\sqrt{\Delta}\frac{d}{dr}\left(\frac{\mathscr{R}}{\sqrt{\Delta}}\right) = \frac{1}{\Delta}\left(Y + \frac{4\lambda\sigma}{\mathscr{C}_1 + \Gamma_1}rX\right) \tag{22}$$

and $$\frac{d\mathscr{S}}{d\theta} = [S]^- \cos\theta + \frac{4\sigma(\lambda\alpha^2 - 6a^2)}{a(\mathscr{C}_1 + \Gamma_1)}[S]^+. \tag{23}$$

(a) *The expression of $\mathscr{R}$ and $\mathscr{S}$ in terms of the Teukolsky functions*

Since the expression given in II, equation (83) is $2a\Psi_{,\theta}$ (cf. II, equation (76)), we can now write, in accordance with equations (20) and (22),

$$\begin{aligned}2a\frac{\partial\Psi}{\partial\theta} &= \frac{a(\mathscr{C}_1 + \Gamma_1)}{48M\sqrt{2}}\mathscr{R}\left\{[S]^- \cos\theta + \frac{4\sigma(\lambda\alpha^2 - 6a^2)}{a(\mathscr{C}_1 + \Gamma_1)}[S]^+\right\},\\ &= \frac{i}{12M\sqrt{2}}[\![\{rK[\mathscr{D}\mathscr{D}P]^- - 2(r^2\sigma + K)[\mathscr{D}P]^- + 6r\sigma[P]^-\}[S]^+\\ &\qquad - ia\{K[\mathscr{D}\mathscr{D}P]^+ - 2r\sigma[\mathscr{D}P]^+\}[S]^- \cos\theta]\!].\end{aligned} \tag{24}$$

Equating now the coefficients of $[S]^- \cos\theta$ and $[S]^+$ on the two sides of this equation we obtain two alternative expressions for $\mathscr{R}$:

$$\tfrac{1}{4}(\mathscr{C}_1 + \Gamma_1)\mathscr{R} = K[\mathscr{D}\mathscr{D}P]^+ - 2r\sigma[\mathscr{D}P]^+, \tag{25}$$

and $$\sigma(\lambda\alpha^2 - 6a^2)\mathscr{R} = i\{rK[\mathscr{D}\mathscr{D}P]^- - 2(r^2\sigma + K)[\mathscr{D}P]^- + 6r\sigma[P]^-\}. \tag{26}$$

As explained in II, § 9, the expressions on the right hand sides of (25) and (26) can be written out explicitly in terms of the functions X and Y by making use of I, equations (76) and (77). The required expressions can be read off from II, equation (85). We have

$$\tfrac{1}{4}(\mathscr{C}_1+\Gamma_1)\,\mathscr{R} = \frac{1}{B_1}[\![\{2K^2(\mathscr{C}_1+\Gamma_1)+8\lambda\sigma r[K(r-M)-\sigma r\Delta]-8\lambda\sigma K\Delta\}X + \{-8K^2\lambda\sigma r+2(\mathscr{C}_1+\Gamma_1)\ [K(r-M)-\sigma r\Delta]\}\,Y]\!] \tag{27}$$

and

$$\sigma(\lambda\alpha^2-6a^2)\,\mathscr{R} = \frac{1}{B_1}[\![\{-8K^2\sigma(6M+\lambda r)-2[K(r-M)-\sigma r\Delta]\,(\mathscr{C}_1-\Gamma_1)\}rX + \{2K^2(\mathscr{C}_1-\Gamma_1)-8\sigma(6M+\lambda r)\,[K(r-M)-\sigma r\Delta]+8\lambda\sigma K\Delta\}\,r\,Y + 2K(\Delta\mathscr{C}_1-A_1)\,X-2[3r\sigma B_1+K(\Delta\mathscr{C}_2+A_2)]\,Y]\!]. \tag{28}$$

Again, since the expression given in II, equation (84) is $2a(\Psi/\sqrt{\Delta})_{,r}\sqrt{\Delta}$, we can now write in accordance with (20) and (23)

$$\begin{aligned}2a\sqrt{\Delta}\frac{\partial}{\partial r}\frac{\Psi}{\sqrt{\Delta}} &= \frac{a(\mathscr{C}_1+\Gamma_1)}{48M\Delta\sqrt{2}}\mathscr{S}\left(Y+\frac{4\lambda\sigma}{\mathscr{C}_1+\Gamma_1}rX\right)\\ &= -\frac{1}{12M\Delta\sqrt{2}}[\![\{Q[\mathscr{L}\mathscr{L}S]^- -(2a\sigma\cos\theta)\,[\mathscr{L}S]^-\}rX\\ &\quad +a\{Q[\mathscr{L}\mathscr{L}S]^+\cos\theta+2(Q\sin\theta-a\sigma\cos^2\theta)\,[\mathscr{L}S]^+\\ &\quad -6a\sigma\,(\sin\theta\cos\theta)\,[S]^+\}\,Y]\!].\end{aligned} \tag{29}$$

Equating now the coefficients of Y and rX on the two sides of the equation, we obtain the following two alternative expressions for $\mathscr{S}$:

$$-\lambda a\sigma\mathscr{S} = Q[\mathscr{L}\mathscr{L}S]^- - 2a\sigma[\mathscr{L}S]^-\cos\theta \tag{30}$$

and

$$-\tfrac{1}{4}(\mathscr{C}_1+\Gamma_1)\mathscr{S} = Q[\mathscr{L}\mathscr{L}S]^+\cos\theta+2(Q\sin\theta-a\sigma\cos^2\theta)\,[\mathscr{L}S]^+-6a\sigma[S]^+\sin\theta\cos\theta. \tag{31}$$

And according to II, equation (86), the corresponding expressions for $\mathscr{S}$ in terms of the Teukolsky functions are

$$\begin{aligned}-\tfrac{1}{4}(\mathscr{C}_1+\Gamma_1)\mathscr{S} &= \frac{1}{\beta_1}[\![\{2Q^2(\mathscr{C}_1+\Gamma_1)-8a\sigma\lambda(Q\operatorname{cosec}\theta-a\sigma\cos^2\theta)\}\,[S]^+\cos\theta\\ &\quad +\{-2(Q\cot\theta-a\sigma\cos\theta)\,(\mathscr{C}_1+\Gamma_1)+8a\sigma\lambda Q^2\cos\theta\}\,[S]^-\cos\theta\\ &\quad -2\{[(3a\sigma\cos\theta)\,\beta_1+\alpha_2 Q]\,[S]^+ + Q(\alpha_1+\mathscr{C}_1)\,[S]^-\}\sin\theta]\!],\end{aligned} \tag{32}$$

and

$$-\lambda a\sigma\mathscr{S} = \frac{1}{\beta_1}[\![\{2(Q\cot\theta-a\sigma\cos\theta)\,(\mathscr{C}_1-\Gamma_1)+8a\sigma\lambda Q^2\cos\theta\}\,[S]^+ + \{-2Q^2(\mathscr{C}_1-\Gamma_1)-8a\sigma\lambda(Q\operatorname{cosec}\theta-a\sigma\cos^2\theta)\}\,[S]^-]\!]. \tag{33}$$

We shall find that the relations (25), (26), (30), and (31) play crucial roles in the subsequent developments.

TABLE 1. VERIFICATION OF EQUATIONS (27) AND (28)

r	$\sigma = 0.25$				$\sigma = 0.50$			
	X	Y	$\mathscr{R}$ by eqn (27)	$\mathscr{R}$ by eqn (28)	X	Y	$\mathscr{R}$ by eqn (27)	$\mathscr{R}$ by eqn (28)
2.5	16.972	0.41131	−8.3126	−8.3127	+0.32142	+6.3237	+2.3381	+2.3381
3.0	52.753	15.546	−8.9470	−8.9471	−8.0250	+18.950	+5.3234	+5.3233
4.0	208.48	171.03	+4.8439	+4.8421	−68.375	+43.073	+12.998	+12.998
5.0	442.33	672.60	+50.52	+50.53	−188.77	−4.7272	+16.683	+16.683
7.0	278.24	3529.2	+264.75	+264.76	−188.92	−545.15	−4.6649	−4.6650

TABLE 2. VERIFICATION OF EQUATIONS (32) AND (33)

$\cos\theta$	$\sigma = 0.25$				$\sigma = 0.50$			
	$[S]^+$	$[S]^-$	$\mathscr{S}$ by eqn (32)	$\mathscr{S}$ by eqn (33)	$[S]^+$	$[S]^-$	$\mathscr{S}$ by eqn (32)	$\mathscr{S}$ by eqn (33)
0.80	1.31885	1.29393	0.243779	0.24377	1.33806	1.31866	0.23362	0.23362
0.60	1.05031	0.94742	0.23945	0.23944	1.02204	0.93939	0.22480	0.22480
0.40	0.85966	0.62068	0.18052	0.18051	0.80114	0.60288	0.16691	0.16691
0.20	0.74574	0.30707	0.09574	0.09573	0.67051	0.29447	0.08769	0.08769
0.04	0.70936	0.06121	0.01948	0.01947	0.62901	0.05845	0.01779	0.01779

(b) Numerical verification of the identities

The various identities among the radial and angular Teukolsky functions, that follow from (25)–(28) and (30)–(33) are, to say the least, unexpected. For example, could one have foretold that the quantities on the right hand sides of (25) and (26), or of equations (30) and (31), are proportional to each other and that the constants of proportionalities are precisely those that are specified? Or, to consider a different aspect of the matter: what we have, in effect, shown is that the indefinite integrals which initially defined the functions $\mathscr{R}$ and $\mathscr{S}$, namely,

$$\mathscr{R} = \sqrt{\Delta}\int^{r}\left(Y+\frac{4\lambda\sigma}{\mathscr{C}_1+\Gamma_1}rX\right)\frac{\mathrm{d}r}{\Delta^{\frac{3}{2}}} \tag{34}$$

and
$$\mathscr{S} = \int^{\theta}\left\{[S]^{-}\cos\theta+\frac{4\sigma(\lambda\alpha^2-6a^2)}{a(\mathscr{C}_1+\Gamma_1)}[S]^{+}\right\}\mathrm{d}\theta, \tag{35}$$

are expressible algebraically in terms of the Teukolsky functions. Could one have expected this? It is difficult, even, to imagine how one might begin to verify these various differential and integral equalities, *ab initio*, from the differential equations which define the Teukolsky functions.

In view of the massive reductions which were needed to arrive at the various relations we have quoted, it was felt that it would be useful to confirm the correctness of the reductions by a direct numerical evaluation of the functions $\mathscr{R}$ and $\mathscr{S}$ by the two alternative sets of formulae: equations (27) and (28) for $\mathscr{R}$, and (32) and (33) for $\mathscr{S}$. Dr S. Detweiler kindly provided the author, for this purpose, with tables of the functions $X(r)$ and $Y(r)$ (in the normalization specified in I, equations (58) and (59)) and $S_{+2}(\theta)$ and $S_{-2}(\theta)$ (in their standard normalizations) and their derivatives. Dr Detweiler, further, checked his numerically evaluated derivatives of the four functions against the values which follow from I, equations (74)–(77) and (80)–(82); and he found that the agreement was entirely within the accuracy of the numerical integrations.

In tables 1 and 2, we present selected entries from an extensive tabulation of the functions $\mathscr{R}$ and $\mathscr{S}$, with the aid of the two alternative sets of formulae. The agreement in the two sets is an assurance for the correctness of the formulae and the reductions leading to them.

4. The differential equations satisfied by $\mathscr{R}$ and $\mathscr{S}$

In completing the reduction of the eight equations included in II, equations (67) – the equations of system II – it was stated in II, § 10 that the eliminants (II, equations (103) and (104)) of II, equations (99) and (100) and of II, equations (101) and (102), must be considered as identities 'since both sides of the equations are expressible solely in terms of the Teukolsky functions'. But what the meaning or the content of these identities may be, was not considered. We shall now show that II, equations (103) and (104) provide equations which express the Teukolsky functions in terms of $\mathscr{R}$ and $\mathscr{S}$ and their derivatives.

First, we observe that II, equations (103) and (104) can be written alternatively in the forms

$$\frac{ia\Delta}{K}\left\{\frac{\partial}{\partial r}\left(K\frac{\partial}{\partial r}\frac{\Psi}{K}\right)+\left(\frac{4\alpha^2\sigma^2}{K^2}-\frac{K^2}{\Delta^2}\right)\Psi\right\}$$
$$=\tfrac{1}{2}Q^2\left\{\mathscr{L}_0^\dagger\frac{1}{Q}\left(\tfrac{1}{2}\Delta\frac{\kappa^*}{\bar{\rho}}+\frac{2\rho^2\bar{\rho}^*}{\Delta}\nu\right)-\mathscr{L}_0\frac{1}{Q}\left(\tfrac{1}{2}\Delta\frac{\kappa}{\bar{\rho}^*}+\frac{2\rho^2\bar{\rho}}{\Delta}\nu^*\right)\right\}, \tag{36}$$

and

$$\frac{a}{Q}\left\{\frac{\partial}{\partial\theta}\left(Q\frac{\partial}{\partial\theta}\frac{\Psi}{Q}\right)-\left(\frac{4\alpha^2\sigma^2}{Q^2}-Q^2\right)\Psi\right\}$$
$$=\frac{i}{\sqrt{2}}K^2\left\{\mathscr{D}_0^\dagger\frac{1}{K}\left[-\tfrac{1}{2}\Delta\frac{\bar{\rho}}{\bar{\rho}^*}\sigma+\tfrac{1}{2}\Delta\frac{\bar{\rho}^*}{\bar{\rho}}\sigma^*\right]-\mathscr{D}_0\frac{1}{K}[-\bar{\rho}^2\lambda^*+(\bar{\rho}^*)^2\lambda]\right\}. \tag{37}$$

Considering first equation (36), we find that on substituting for the spin coefficients, κ and ν, their solutions given in II, equations (80), the terms on the right hand side of the equation, by reductions analogous to those used in II in the context of the right hand sides of II, equations (76) and (77) (and briefly explained in II, § 9), can be expressed in terms of $\mathscr{S}$, X, and Y. Thus, we find

$$\frac{i}{12M\Delta\sqrt{2}}[\![\{Q[\mathscr{L}\mathscr{L}S]^- - 2a\sigma[\mathscr{L}S]^-\cos\theta\}rY$$
$$-a\{Q[\mathscr{L}\mathscr{L}S]^+\cos\theta+2(Q\sin\theta-a\sigma\cos^2\theta)[\mathscr{L}S]^+-6a\sigma[S]^+\cos\theta\sin\theta\}X]\!]$$
$$=ia\frac{\mathscr{C}_1+\Gamma_1}{48M\Delta\sqrt{2}}\left(X-\frac{4\lambda\sigma}{\mathscr{C}_1+\Gamma_1}rY\right)\mathscr{S}, \tag{38}$$

where we have made use of both the relations (30) and (31). On the other hand, the fact that Ψ is separable (as expressed in (20)) enables us to write the left hand side of equation (36) in the form

$$\frac{ia\Delta}{K}\frac{\mathscr{C}_1+\Gamma_1}{96M\sqrt{2}}\mathscr{S}\left[\frac{d}{dr}\left(K\frac{d}{dr}\frac{\mathscr{R}}{K}\right)+\left(\frac{4\alpha^2\sigma^2}{K^2}-\frac{K^2}{\Delta^2}\right)\mathscr{R}\right]. \tag{39}$$

Accordingly, (36) reduces to the following equation for the radial function, $\mathscr{R}$:

$$\frac{\Delta^2}{K}\left[\frac{d}{dr}\left(K\frac{d}{dr}\frac{\mathscr{R}}{K}\right)+\left(\frac{4\alpha^2\sigma^2}{K^2}-\frac{K^2}{\Delta^2}\right)\mathscr{R}\right]=2\left(X-\frac{4\lambda\sigma}{\mathscr{C}_1+\Gamma_1}rY\right). \tag{40}$$

Equation (40), in combination with (22), will enable us to express X and Y, each, in terms of $\mathscr{R}$ and its derivatives. Could one have foretold that the radial functions, P_{+2} and P_{-2}, can both be expressed in terms of a same radial function $\mathscr{R}$?

Considering next (37), we similarly find that the terms on the right hand side of the equation can be reduced to give

$$\frac{\mathrm{i}}{12M\sqrt{2}}[\![-\{rK[\mathscr{D}\mathscr{D}P]^- - 2(r^2\sigma+K)[\mathscr{D}P]^- + 6r\sigma[P]^-\}[S]^- + \mathrm{i}a\{K[\mathscr{D}\mathscr{D}P]^+ - 2r\sigma[\mathscr{D}P]^+\}[S]^+\cos\theta]\!]$$
$$= \frac{\mathrm{i}}{12M\sqrt{2}}\{\mathrm{i}\sigma(\lambda\alpha^2-6a^2)\,\mathscr{R}[S]^- + \tfrac{1}{4}\mathrm{i}a(\mathscr{C}_1+\Gamma_1)\,\mathscr{R}[S]^+\cos\theta\}$$
$$= -\frac{a(\mathscr{C}_1+\Gamma_1)}{48M\sqrt{2}}\mathscr{R}\left\{[S]^+\cos\theta + \frac{4\sigma(\lambda\alpha^2-6a^2)}{a(\mathscr{C}_1+\Gamma_1)}[S]^-\right\}, \tag{41}$$

where we have made use of both the relations (25) and (26); while the left hand side gives

$$\frac{a}{Q}\frac{\mathscr{C}_1+\Gamma_1}{96M\sqrt{2}}\mathscr{R}\left[\frac{\mathrm{d}}{\mathrm{d}\theta}\left(Q\frac{\mathrm{d}}{\mathrm{d}\theta}\frac{\mathscr{S}}{Q}\right) - \left(\frac{4\alpha^2\sigma^2}{Q^2} - Q^2\right)\mathscr{S}\right]. \tag{42}$$

Accordingly, (37) reduces to the following equation for the angular function, $\mathscr{S}$:

$$\frac{1}{Q}\left[\frac{\mathrm{d}}{\mathrm{d}\theta}\left(Q\frac{\mathrm{d}}{\mathrm{d}\theta}\frac{\mathscr{S}}{Q}\right) - \left(\frac{4\alpha^2\sigma^2}{Q^2} - Q^2\right)\mathscr{S}\right] = -2\left\{[S]^+\cos\theta + \frac{4\sigma(\lambda\alpha^2-6a^2)}{a(\mathscr{C}_1+\Gamma_1)}[S]^-\right\}. \tag{43}$$

Equation (43), in combination with (23), will enable us to express $[S]^+$ and $[S]^-$, each, in terms of $\mathscr{S}$ and its derivatives. Again, could one have foretold that both the angular functions, S_{+2} and S_{-2}, can both be expressed in terms of a same angular function $\mathscr{S}$?

The real 'depth' of the identities implied by (40) and (43) becomes clear only when we ask how we can verify them. For this purpose it is convenient to rewrite the equations in the equivalent forms

$$\frac{\Delta^2}{K}\left(\frac{\mathrm{d}^2\mathscr{R}}{\mathrm{d}r^2} - \frac{2r\sigma}{K}\frac{\mathrm{d}\mathscr{R}}{\mathrm{d}r} + \frac{2\sigma\Delta^2-K^3}{K\Delta^2}\mathscr{R}\right) = 2\left(X - \frac{4\lambda\sigma}{\mathscr{C}_1+\Gamma_1}rY\right), \tag{44}$$

and

$$\frac{1}{Q}\left[\frac{\mathrm{d}^2\mathscr{S}}{\mathrm{d}\theta^2} - \left(\frac{2a\sigma\cos\theta}{Q} - \cot\theta\right)\frac{\mathrm{d}\mathscr{S}}{\mathrm{d}\theta} + \left(Q^2 - \operatorname{cosec}^2\theta - \frac{2a\sigma}{Q\sin\theta}\right)\mathscr{S}\right]$$
$$= -2\left\{[S]^+\cos\theta + \frac{4\sigma(\lambda\alpha^2-6a^2)}{a(\mathscr{C}_1+\Gamma_1)}[S]^-\right\}. \tag{45}$$

The terms on the right hand sides of the foregoing equations can be expressed in terms of X and Y or $[S]^+$ and $[S]^-$ with the aid of the equations

$$\frac{\mathrm{d}\mathscr{R}}{\mathrm{d}r} = \frac{1}{\Delta}\left(Y + \frac{4\lambda\sigma}{\mathscr{C}_1+\Gamma_1}rX\right) + \frac{r-M}{\Delta}\mathscr{R}, \tag{46}$$

$$\frac{\mathrm{d}^2\mathscr{R}}{\mathrm{d}r^2} = \frac{1}{\Delta}\left(\frac{\mathrm{d}Y}{\mathrm{d}r} + \frac{4\lambda\sigma}{\mathscr{C}_1+\Gamma_1}r\frac{\mathrm{d}X}{\mathrm{d}r} + \frac{4\lambda\sigma}{\mathscr{C}_1+\Gamma_1}X\right) - \frac{r-M}{\Delta^2}\left(Y + \frac{4\lambda\sigma}{\mathscr{C}_1+\Gamma_1}rX\right) + \frac{a^2-M^2}{\Delta^2}\mathscr{R}, \tag{47}$$

$$\frac{\mathrm{d}\mathscr{S}}{\mathrm{d}\theta} = [S]^-\cos\theta + \frac{4\sigma(\lambda\alpha^2-6a^2)}{a(\mathscr{C}_1+\Gamma_1)}[S]^+, \tag{48}$$

and

$$\frac{\mathrm{d}^2\mathscr{S}}{\mathrm{d}\theta^2} = -[S]^-\sin\theta + \cos\theta\frac{\mathrm{d}[S]^-}{\mathrm{d}\theta} + \frac{4\sigma(\lambda\alpha^2-6a^2)}{a(\mathscr{C}_1+\Gamma_1)}\frac{\mathrm{d}[S]^+}{\mathrm{d}\theta}, \tag{49}$$

since $\mathscr{R}$ and $\mathscr{S}$ are known in terms of X and Y and $[S]^+$ and $[S]^-$ (by equations (27) (or (28)) and (32) (or (33)) and the derivatives of X and Y and of $[S]^+$ and $[S]^-$ are expressible in terms of the functions themselves (by I, equations (74)–(77) and (80)–(82)). The complexity of these various relations which must be inserted in (44) and (45) to reduce them to identities is a measure of their 'depth'.

Again, since (44) and (45) have emerged after so much involved reductions, it was felt that it would be useful to confirm their correctness by a direct numerical evaluation of the quantities on the left-hand sides of both these equations by the substitutions we have enumerated and comparing the values so obtained with the values of the simple quantities on the right hand sides. The comparisons exhibited in tables 3 and 4 ensure the correctness of the equations and the reductions leading to them.

Finally, it is worth noticing that we can write differential equations for $\mathscr{R}$ and $\mathscr{S}$ which contain no reference to the Teukolsky functions. Thus, by expressing the derivatives of X and Y, which occur in (47), in terms of the functions themselves (by making use of I, equations (74)–(77)) we obtain a second equation relating X and Y to the derivatives of $\mathscr{R}$; this second equation, in combination with (46) will enable us to express X and Y, separately, in terms of $\mathscr{R}$ and its derivatives. The substitution of these latter relations in (44) will provide us with an equation governing $\mathscr{R}$ with no references to X and Y. The equation which we obtain in this manner is

$$\frac{\Delta^2}{K}\left[\frac{\mathrm{d}}{\mathrm{d}r}\left(K\frac{\mathrm{d}}{\mathrm{d}r}\frac{\mathscr{R}}{K}\right)+\left(\frac{4\alpha^2\sigma^2}{K^2}-\frac{K^2}{\Delta^2}\right)\mathscr{R}\right]$$

$$=-\frac{2}{[\det]_r}\left\{(\not{p}^2r^2+1)\left[\Delta^2\frac{\mathrm{d}^2\mathscr{R}}{\mathrm{d}r^2}-(a^2-M^2)\mathscr{R}\right]\right.$$

$$-\left[2\not{p}r\frac{\Delta^2\mathscr{C}_1}{B_1}+(\not{p}^2r^2-1)\frac{\Delta^2\mathscr{C}_2}{B_1}-(\not{p}^2r^2+1)\left(\frac{\Delta A_2}{B_1}+M\right)+r(\not{p}^2a^2-1)\right]$$

$$\left.\times\left[\Delta\frac{\mathrm{d}\mathscr{R}}{\mathrm{d}r}-(r-M)\mathscr{R}\right]\right\}, \tag{50}$$

where $$[\det]_r=\Delta\left[(\not{p}^2r^2-1)\frac{\Delta\mathscr{C}_1}{B_1}+(\not{p}^2r^2+1)\left(\frac{A_1}{B_1}-\frac{K}{\Delta}\right)-2\not{p}r\frac{\Delta\mathscr{C}_2}{B_1}-\not{p}\right]. \tag{51}$$

and $\not{p}=4\lambda\sigma/(\mathscr{C}_1+\Gamma_1)$.

Similarly, the equation for $\mathscr{S}$ which we obtain is

$$\frac{1}{Q}\left[\frac{\mathrm{d}}{\mathrm{d}\theta}\left(Q\frac{\mathrm{d}}{\mathrm{d}\theta}\frac{\mathscr{S}}{Q}\right)-\left(\frac{4\alpha^2\sigma^2}{Q^2}-Q^2\right)\mathscr{S}\right]$$

$$=\frac{2}{[\det]_\theta}\left\{(\not{q}^2-\cos^2\theta)\frac{\mathrm{d}^2\mathscr{S}}{\mathrm{d}\theta^2}+\left[\frac{1}{\beta_1}(\alpha_2+2\beta_1\cot\theta)(\not{q}^2-\cos^2\theta)\right.\right.$$

$$\left.\left.-\frac{2\mathscr{C}_1}{\beta_1}\not{q}\cos\theta-\cos\theta\sin\theta\right]\frac{\mathrm{d}\mathscr{S}}{\mathrm{d}\theta}\right\}, \tag{52}$$

where

$$[\det]_\theta = [\varphi \sin\theta + (\alpha_1 + \beta_1 Q)(\varphi^2 - \cos^2\theta)/\beta_1 + \mathscr{C}_1(\varphi^2 + \cos^2\theta)/\beta_1], \tag{53}$$

and $\varphi = 4\sigma(\lambda\alpha^2 - 6a^2)/a(\mathscr{C}_1 + \Gamma_1)$.

While (50) and (52) are, perhaps, too complicated to be of any practical use, they do serve to emphasize that, in principle, we can eliminate any direct reference to the Teukolsky functions when writing the solutions for the metric perturbations. While this may be too 'extreme' a view, it is certainly true that the functions $\mathscr{R}$ and $\mathscr{S}$ play crucial roles in the solution of the Newman–Penrose equations (as will become abundantly clear in §§ 6 and 7 below). And it does appear that the properties of the Teukolsky functions as expressed in the Teukolsky–Starobinsky identities and in the identities involving $\mathscr{R}$ and $\mathscr{S}$ are the significant ones of the theory.

TABLE 3. VERIFICATION OF EQUATION (44)

	$\sigma = 0.25$		$\sigma = 0.50$	
r	l.h.s. of eqn (44)	r.h.s. of eqn (44)	l.h.s. of eqn (44)	r.h.s. of eqn (44)
2.5	33.75	33.79	−1.7234	−1.7236
3.0	98.67	98.67	−24.558	−24.560
4.0	316.7	316.7	−162.53	−162.54
5.0	391.9	391.9	−374.00	−374.00
7.0	−3063.4	−3063.6	+193.38	+193.36

TABLE 4. VERIFICATION OF EQUATION (45)

	$\sigma = 0.25$		$\sigma = 0.50$	
$\cos\theta$	l.h.s. of eqn (45)	r.h.s. of eqn (45)	l.h.s. of eqn (45)	r.h.s. of eqn (45)
0.80	−0.32882	−0.32880	−0.27035	−0.27036
0.60	+0.04393	+0.04395	+0.10609	+0.10609
0.40	+0.16675	+0.16676	+0.21427	+0.21428
0.20	+0.12444	+0.12444	+0.14949	+0.14951
0.04	+0.02753	+0.02752	+0.03257	+0.03259

5. FOUR LINEARIZED RICCI IDENTITIES

As we have explained in II, § 2 (c), we have ten degrees of gauge freedom which arise from the six degrees of freedom, we have in the choice of the perturbed tetrad frame, and the four degrees of freedom, we have in the choice of the coordinate frame. The freedoms which have origins in these two sources are, in the first instance, independent of one another. Thus, in letting $\Psi_1 \equiv \Psi_3 \equiv 0$, we have used four of the six degrees of freedom available in the choice of the perturbed tetrad frame; and if we let $\Psi_2^{(1)} = 0$, we use two of the four degrees of freedom available in the choice of the perturbed coordinate frame. These two choices, $\Psi_1 \equiv \Psi_3 \equiv 0$ and $\Psi_2^{(1)} = 0$, do not interfere with one another and exhaust six of the ten degrees of available freedom. The further fact, that $U = A_1^1 + A_2^2$ and $V = A_3^3 + A_4^4$ are left unspecified when the solutions for the metric perturbations are completed, accounts for the remaining four degrees of gauge freedom. Consequently, there is no room left for leaving one of the two functions, $F+G$ and $J-H$, unspecified, as presumed in II, § 11. Clearly, then, there must be some equations among the Ricci identities (since *all* the equations derived from the commutation relations and the Bianchi identities have been satisfied), which have not been considered and which will provide the missing relation and complete the solution. An examination of the various Ricci identities suggests that the equations we must consider are the following ones from the Newman–Penrose set of equations (Newman & Penrose 1962, equations (4.2a), (4.2n), (4.2g), and (4.2p)):

$$\left.\begin{aligned} D\tilde{\rho}-\delta^*\kappa &= (\tilde{\rho}^2+\sigma\sigma^*)+(\epsilon+\epsilon^*)\,\tilde{\rho}-\kappa^*\tau-\kappa(3\alpha+\beta^*-\pi), \\ \delta\nu-\Delta\mu &= (\mu^2+\lambda\lambda^*)+(\gamma+\gamma^*)\,\mu-\nu^*\pi+\nu(\tau-3\beta-\alpha^*), \\ D\lambda-\delta^*\pi &= (\tilde{\rho}\lambda+\sigma^*\mu)+\pi^2+(\alpha-\beta^*)\,\pi-\kappa^*\nu-\lambda(3\epsilon-\epsilon^*), \\ \delta\tau-\Delta\sigma &= (\mu\sigma+\lambda^*\tilde{\rho})+\tau^2+(\beta-\alpha^*)\,\tau-\nu^*\kappa-\sigma(3\gamma-\gamma^*). \end{aligned}\right\} \tag{54}$$

The linearized versions of these equations are

$$\left.\begin{aligned} D\tilde{\rho}^{(1)}+2\frac{\tilde{\rho}^{(1)}}{\bar{\rho}^*}-D^{(1)}\frac{1}{\bar{\rho}^*}+\frac{(\epsilon+\epsilon^*)^{(1)}}{\bar{\rho}^*} &= \delta^*\kappa-\kappa(3\alpha+\beta^*-\pi)-\kappa^*\tau, \\ \Delta\mu^{(1)}+(2\mu+\gamma+\gamma^*)\,\mu^{(1)}+\Delta^{(1)}\mu+(\gamma+\gamma^*)^{(1)}\mu &= \delta\nu-\nu(\tau-3\beta-\alpha^*)+\nu^*\pi, \\ \delta^*\pi^{(1)}+(2\pi+\alpha-\beta^*)\,\pi^{(1)}+\delta^{*(1)}\pi+(\alpha-\beta^*)^{(1)}\pi &= D\lambda+\lambda/\bar{\rho}^*-\sigma^*\mu, \\ \delta\tau^{(1)}-(2\tau+\beta-\alpha^*)\,\tau^{(1)}+\delta^{(1)}\tau+(\alpha^*-\beta)^{(1)}\tau &= \Delta\sigma+\sigma\mu-\lambda^*/\bar{\rho}^*-\sigma(3\gamma-\gamma^*). \end{aligned}\right\} \tag{55}$$

Inserting in the foregoing equations, the solutions for κ, σ, λ, and ν given in II, equations (80), for $\tilde{\rho}^{(1)}$, $\tau^{(1)}$, $\mu^{(1)}$, and $\pi^{(1)}$ given in II, equations (20), and for $\alpha^{(1)}$, $\beta^{(1)}$, $\gamma^{(1)}$, and $\epsilon^{(1)}$ given in II, equations (33) (the equations of 'system III'), we find, after some considerable reductions, that they yield

$$\begin{aligned} -\mathrm{i}KF_2^1+\tfrac{1}{2}\Delta\mathscr{D}_0\,U+\frac{1}{\sqrt{2}}(\mathrm{i}a\sin\theta)\,\mathscr{D}_{-1}\rho^2F+\sqrt{2}\,.\,(a^2\sin\theta\cos\theta)\,F& \\ -\sqrt{2}\,.\,(\mathrm{i}ar\sin\theta)\,(F_2^3-F_2^4) = -\frac{\Delta}{2\sqrt{2}}\left[(\bar{\rho}^*)^2\mathscr{L}_1\frac{\kappa}{(\bar{\rho}^*)^2}+\frac{\mathrm{i}a\sin\theta}{\bar{\rho}}\kappa^*\right]&, \end{aligned} \tag{56}$$

$$\begin{aligned} \mathrm{i}KF_1^2+\tfrac{1}{2}\Delta\mathscr{D}_0^\dagger\,U-\frac{1}{\sqrt{2}}(\mathrm{i}a\sin\theta)\,\mathscr{D}_{-1}^\dagger\rho^2G-\sqrt{2}\,.\,(a^2\sin\theta\cos\theta)\,G& \\ -\sqrt{2}\,.\,(\mathrm{i}ar\sin\theta)\,(F_1^3-F_1^4) = \sqrt{2}\frac{\rho^4\bar{\rho}^*}{\Delta}\left[\frac{1}{\rho^2\bar{\rho}}\mathscr{L}_1^\dagger(\rho^2\nu)+\frac{\mathrm{i}a\sin\theta}{(\bar{\rho}^*)^2}\nu^*\right]&, \end{aligned} \tag{57}$$

$$\begin{aligned} (\mathrm{i}aQ\sin\theta)\,F_3^4+\frac{\mathrm{i}a\sin\theta}{2\rho^2}\mathscr{L}_0\,V+\frac{1}{\sqrt{2}}\left(\mathscr{L}_{-1}J-\frac{2\mathrm{i}a\sin\theta}{\bar{\rho}}J\right)& \\ +\sqrt{2}\frac{a^2\sin\theta\cos\theta}{\rho^2}(F_3^1+F_3^2) = \frac{\bar{\rho}^*}{\bar{\rho}}\left[\mathscr{D}_0(\bar{\rho}^*\lambda)+\frac{\Delta}{2\rho^2}\sigma^*\right]&, \end{aligned} \tag{58}$$

$$\begin{aligned} (\mathrm{i}aQ\sin\theta)\,F_4^3-\frac{\mathrm{i}a\sin\theta}{2\rho^2}\mathscr{L}_0^\dagger\,V-\frac{1}{\sqrt{2}}\left(\mathscr{L}_{-1}^\dagger H-\frac{2\mathrm{i}a\sin\theta}{\bar{\rho}}H\right)& \\ +\sqrt{2}\frac{a^2\sin\theta\cos\theta}{\rho^2}(F_4^1+F_4^2) = -\left[\frac{\Delta\bar{\rho}^*}{2\bar{\rho}}\mathscr{D}_1^\dagger\frac{\bar{\rho}\sigma}{(\bar{\rho}^*)^2}+\frac{\bar{\rho}}{\bar{\rho}^*}\lambda^*\right]&. \end{aligned} \tag{59}$$

By subtracting (56) from (57) and by adding (58) and (59), we obtain the following pair of complex equations:

$$\mathrm{i}K(F^1_2+F^2_1-U)-\sqrt{2}\,.\,(a^2\sin\theta\cos\theta)\,(F+G)+\sqrt{2}\,.\,(\mathrm{i}ar\sin\theta)\,(J-H)$$

$$-\frac{1}{\sqrt{2}}\,\mathrm{i}a\sin\theta(\mathscr{D}_{-1}\rho^2F+\mathscr{D}^\dagger_{-1}\rho^2G)=\sqrt{2}\frac{\rho^4\bar{\rho}^*}{\Delta}\left[\frac{1}{\rho^2\bar{\rho}}\mathscr{L}^\dagger_1(\rho^2\nu)+\frac{\mathrm{i}a\sin\theta}{(\bar{\rho}^*)^2}\nu^*\right]$$

$$+\frac{\Delta}{2\sqrt{2}}\left[(\bar{\rho}^*)^2\mathscr{L}_1\frac{\kappa}{(\bar{\rho}^*)^2}+\frac{\mathrm{i}a\sin\theta}{\bar{\rho}}\kappa^*\right],\qquad(60)$$

and

$$\frac{\mathrm{i}aQ\sin\theta}{\rho^2}\left[\rho^2(F^3_4+F^4_3)+V\right]+\sqrt{2}\,\frac{a^2\sin\theta\cos\theta}{\rho^2}(F+G)$$

$$+\frac{1}{\sqrt{2}}\left[\mathscr{L}_{-1}J-\mathscr{L}^\dagger_{-1}H-\frac{2\mathrm{i}a\sin\theta}{\bar{\rho}}(J-H)\right]=\frac{\bar{\rho}^*}{\bar{\rho}}\left[\mathscr{D}_0(\bar{\rho}^*\lambda)+\frac{\Delta}{2\rho^2}\sigma^*\right]$$

$$-\left[\frac{\Delta\bar{\rho}^*}{2\bar{\rho}}\mathscr{D}^\dagger_1\frac{\bar{\rho}\sigma}{(\bar{\rho}^*)^2}+\frac{\bar{\rho}}{\bar{\rho}^*}\lambda^*\right].\qquad(61)$$

(*a*) *The elimination of* $(F^1_2+F^2_1-U)$ *and* $[\rho^2(F^3_4+F^4_3)+V]$

We shall now show how, by eliminating $(F^1_2+F^2_1-U)$ and $[\rho^2(F^3_4+F^4_3)+V]$ from (60) and (61), we obtain a remarkably simple set of equations ((92)–(95) below) governing $F+G$ and $J-H$.

Considering first the imaginary parts of (60) and (61), we have

$$(\mathrm{i}a\sin\theta)\left\{\sqrt{\Delta}\frac{\partial}{\partial r}\left(\rho^2\frac{F+G}{\sqrt{\Delta}}\right)-[r(F+G)+(\mathrm{i}a\cos\theta)(J-H)]\right\}=[\kappa,\nu]-[\kappa,\nu]^*\qquad(62)$$

and $$\frac{\partial}{\partial\theta}\rho^2(J-H)-(\mathrm{i}a\sin\theta)\,[r(F+G)+(\mathrm{i}a\cos\theta)(J-H)]=[\lambda,\sigma]-[\lambda,\sigma]^*,\qquad(63)$$

where $$[\kappa,\nu]=\frac{\rho^4\bar{\rho}}{2\Delta}\left[\frac{1}{\rho^2\bar{\rho}^*}\mathscr{L}_1(\rho^2\nu^*)-\frac{\mathrm{i}a\sin\theta}{\bar{\rho}^2}\nu\right]+\tfrac{1}{8}\Delta\left[\bar{\rho}^2\mathscr{L}^\dagger_1\frac{\kappa^*}{\bar{\rho}^2}-\frac{\mathrm{i}a\sin\theta}{\bar{\rho}^*}\kappa\right]\qquad(64)$$

and $$[\lambda,\sigma]=\frac{1}{2\sqrt{2}}(\bar{\rho}^*)^2\left[\mathscr{D}_0(\bar{\rho}^*\lambda)+\frac{\Delta}{2\rho^2}\sigma^*\right]-\frac{1}{2\sqrt{2}}\rho^2\left[\frac{\Delta\bar{\rho}^*}{2\bar{\rho}}\mathscr{D}^\dagger_1\frac{\bar{\rho}\sigma}{(\bar{\rho}^*)^2}+\frac{\bar{\rho}}{\bar{\rho}^*}\lambda^*\right].\qquad(65)$$

We observe that in both (62) and (63) the combination, $[r(F+G)+\mathrm{i}a(J-H)\cos\theta]$, occurs. In (62) we can replace it by a combination of $F+G$ and Ψ, while in (63) we can replace it by a combination of $J-H$ and Ψ by making use of one or the other of the two relations included in

$$r(F+G)+(\mathrm{i}a\cos\theta)(J-H)=\frac{r\sigma}{K}\rho^2(F+G)+\frac{\mathrm{i}a\Psi}{K}$$

$$=\frac{-\mathrm{i}\sigma}{Q\sin\theta}\rho^2(J-H)\cos\theta+\frac{\mathrm{i}\Psi}{Q\sin\theta}.\qquad(66)$$

After these replacements, (62) and (63) can be brought to the forms

$$(\mathrm{i}a\sin\theta)\left[\sqrt{(\Delta K)}\frac{\partial}{\partial r}\left(\rho^2\frac{F+G}{\sqrt{(\Delta K)}}\right)-\frac{\mathrm{i}a\Psi}{K}\right]=[\kappa,\nu]-[\kappa,\nu]^* \tag{67}$$

and

$$\sqrt{(Q\sin\theta)}\frac{\partial}{\partial\theta}\left(\rho^2\frac{J-H}{\sqrt{(Q\sin\theta)}}\right)+\frac{a\Psi}{Q}=[\lambda,\sigma]-[\lambda,\sigma]^*. \tag{68}$$

Returning to (60) and (61), we next consider their real parts:

$$\frac{1}{\sqrt{2}}\mathrm{i}K(F_2^1+F_1^2-U)+(a\sin\theta)[\mathrm{i}r(J-H)-a(F+G)\cos\theta]=[\kappa,\nu]+[\kappa,\nu]^*, \tag{69}$$

and

$$\frac{1}{\sqrt{2}}\mathrm{i}Q(a\sin\theta)[\rho^2(F_4^3+F_3^4)+V]-(a\sin\theta)[\mathrm{i}r(J-H)-a(F+G)\cos\theta]$$
$$=[\lambda,\sigma]+[\lambda,\sigma]^*. \tag{70}$$

By making use of one or the other of the two relations included in

$$\begin{aligned}\mathrm{i}r(J-H)-a(F+G)\cos\theta&=\frac{\mathrm{i}\alpha^2\sigma}{arQ\sin\theta}\rho^2(J-H)-\frac{\mathrm{i}a\cos\theta}{rQ\sin\theta}\Psi,\\&=-\frac{\alpha^2\sigma}{aK\cos\theta}\rho^2(F+G)+\frac{\mathrm{i}r}{K\cos\theta}\Psi,\end{aligned} \tag{71}$$

we can bring (69) and (70) to the forms

$$\frac{1}{\sqrt{2}}\mathrm{i}K(F_2^1+F_1^2-U)+\frac{\mathrm{i}\alpha^2\sigma}{rQ}\rho^2(J-H)-\frac{\mathrm{i}a^2\cos\theta}{rQ}\Psi=[\kappa,\nu]+[\kappa,\nu]^* \tag{72}$$

and

$$\frac{1}{\sqrt{2}}\mathrm{i}Q(a\sin\theta)[\rho^2(F_4^3+F_3^4)+V]+\frac{\alpha^2\sigma\sin\theta}{K\cos\theta}\rho^2(F+G)-\frac{\mathrm{i}ar\sin\theta}{K\cos\theta}\Psi=[\lambda,\sigma]+[\lambda,\sigma]^*. \tag{73}$$

We can now eliminate $(F_2^1+F_1^2-U)$ and $[\rho^2(F_4^3+F_3^4)+V]$ from (71) and (72) with the aid of II, equations (68) and (69). We write these latter equations in the forms

$$\frac{1}{\sqrt{2}}Q(F_2^1+F_1^2-U)+\sqrt{\Delta}\frac{\partial}{\partial r}\left(\rho^2\frac{J-H}{\sqrt{\Delta}}\right)=(\kappa,\nu), \tag{74}$$

and

$$-\frac{1}{\sqrt{2}}\mathrm{i}K[\rho^2(F_4^3+F_3^4)+V]+\frac{\partial}{\partial\theta}(\rho^2(F+G))=(\lambda,\sigma), \tag{75}$$

where

$$(\kappa,\nu)=\tfrac{1}{2}[\tfrac{1}{2}\Delta(\bar\rho^*\kappa^*-\bar\rho\kappa)-2\rho^4(\bar\rho^*\nu-\bar\rho\nu^*)/\Delta]/\rho^2 \tag{76}$$

and

$$(\lambda,\sigma)=\frac{1}{\sqrt{2}}[(\bar\rho)^2(\rho^2\lambda^*-\tfrac{1}{2}\Delta\sigma)+(\bar\rho^*)^2(\rho^2\lambda-\tfrac{1}{2}\Delta\sigma^*)]/\rho^2. \tag{77}$$

Carrying out the stated eliminations, we obtain the pair of equations

$$K\sqrt{\Delta}\frac{\partial}{\partial r}\left(\rho^2\frac{J-H}{\sqrt{\Delta}}\right)-\frac{\alpha^2\sigma}{r}\rho^2(J-H)+\frac{a^2\cos\theta}{r}\Psi = \mathrm{i}Q\{[\kappa,\nu]+[\kappa,\nu]^*\}+K(\kappa,\nu), \quad (78)$$

and

$$(Qa\sin\theta)\frac{\partial}{\partial\theta}(\rho^2(F+G))+\frac{\alpha^2\sigma\sin\theta}{\cos\theta}\rho^2(F+G)-\frac{\mathrm{i}ar\sin\theta}{\cos\theta}\Psi$$
$$= K\{[\lambda,\sigma]+[\lambda,\sigma]^*\}+(Qa\sin\theta)(\lambda,\sigma), \quad (79)$$

where it may be noted that the left hand sides of these equations have the alternative forms

$$r\sqrt{(\Delta K)}\left[\frac{\partial}{\partial r}\left(\frac{\sqrt{K}}{r\sqrt{\Delta}}\rho^2(J-H)\right)+\frac{a^2\cos\theta}{r^2\sqrt{(\Delta K)}}\Psi\right], \quad (80)$$

and $$(a\cos\theta)\sqrt{(Q\sin\theta)}\left[\frac{\partial}{\partial\theta}\left(\frac{\sqrt{(Q\sin\theta)}}{\cos\theta}\rho^2(F+G)\right)-\frac{\mathrm{i}r\sin\theta}{(\cos^2\theta)\sqrt{(Q\sin\theta)}}\Psi\right]. \quad (81)$$

The four equations governing $F+G$ and $J-H$, which we have derived are

$$\frac{\partial}{\partial r}\left(\frac{1}{\sqrt{(\Delta K)}}\rho^2(F+G)\right)-\frac{\mathrm{i}a}{\sqrt{(K^3\Delta)}}\Psi = \frac{1}{(\mathrm{i}a\sin\theta)\sqrt{(\Delta K)}}\{[\kappa,\nu]-[\kappa,\nu]^*\}. \quad (82)$$

$$\frac{\partial}{\partial\theta}\left(\frac{\sqrt{(Q\sin\theta)}}{\cos\theta}\rho^2(F+G)\right)-\frac{\mathrm{i}r\sin\theta}{(\cos^2\theta)\sqrt{(Q\sin\theta)}}\Psi$$
$$= \frac{1}{(a\cos\theta)\sqrt{(Q\sin\theta)}}[\![K\{[\lambda,\sigma]+[\lambda,\sigma]^*\}+(Qa\sin\theta)(\lambda,\sigma)]\!], \quad (83)$$

$$\frac{\partial}{\partial r}\left(\frac{\sqrt{K}}{r\sqrt{\Delta}}\rho^2(J-H)\right)+\frac{a^2\cos\theta}{r^2\sqrt{(\Delta K)}}\Psi = \frac{1}{r\sqrt{(\Delta K)}}[\![\mathrm{i}Q\{[\kappa,\nu]+[\kappa,\nu]^*\}+K(\kappa,\nu)]\!], \quad (84)$$

$$\frac{\partial}{\partial\theta}\left(\frac{1}{\sqrt{(Q\sin\theta)}}\rho^2(J-H)\right)+\frac{a}{Q^{\frac{3}{2}}\sqrt{\sin\theta}}\Psi = \frac{1}{\sqrt{(Q\sin\theta)}}\{[\lambda,\sigma]-[\lambda,\sigma]^*\}. \quad (85)$$

We consider the further reduction of these equations in § 6 below.

6. The reduction of equations (82)–(85)

In our further consideration of (82)–(85) we shall adopt a coordinate gauge in which $\Psi_2^{(1)} = 0$. In this gauge

$$\Psi = K(J-H)\cos\theta-\mathrm{i}rQ(F+G)\sin\theta. \quad (86)$$

Accordingly, it would appear that we should seek a transformation of (82)–(85) such that $J-H$ and $F+G$ occur in the combinations

$$Z_1 = K(J-H)\cos\theta \quad \text{and} \quad Z_2 = -\mathrm{i}rQ(F+G)\sin\theta, \quad (87)$$

so that $$\Psi = Z_1+Z_2. \quad (88)$$

The desired transformation can be effected by multiplying (82), (83), (84), and (85) by $-\mathrm{i}\sqrt{(Q\sin\theta)}/\cos\theta$, $-\mathrm{i}/\sqrt{(\Delta K)}$, $\mathrm{i}/\sqrt{(Q\sin\theta)}$, and $\sqrt{K}/r\sqrt{\Delta}$, respectively. Thus, with the definitions

$$E = \frac{\rho^2}{(r\cos\theta)\sqrt{(\Delta KQ\sin\theta)}}, \tag{89}$$

$$\left.\begin{aligned} A_1 &= \frac{r^2\sin\theta}{\rho^2\cos\theta}; \quad B_1 = \frac{aK\cos\theta}{\rho^2 Q}, \\ A_2 &= \frac{arQ\sin\theta}{K\rho^2}; \quad B_2 = \frac{a^2\cos^2\theta}{r\rho^2}, \end{aligned}\right\} \tag{90}$$

and

$$\left.\begin{aligned} X_1 &= \frac{rQ}{a\rho^2}\{[\kappa,\nu]-[\kappa,\nu]^*\}, \\ Y_1 &= \frac{K\cos\theta}{\rho^2}\{[\lambda,\sigma]-[\lambda,\sigma]^*\}, \\ X_2 &= \frac{\cos\theta}{\rho^2}[\![\mathrm{i}Q\{[\kappa,\nu]+[\kappa,\nu]^*\}+K(\kappa,\nu)]\!], \\ Y_2 &= \mathrm{i}\frac{r}{a\rho^2}[\![K\{[\lambda,\sigma]+[\lambda,\sigma]^*\}+(Qa\sin\theta)(\lambda,\sigma)]\!], \end{aligned}\right\} \tag{91}$$

we find, after the stated multiplications, that (82)–(85) take the remarkably simple forms

$$\frac{\partial}{\partial r}EZ_2 - A_2 E\Psi = -EX_1, \tag{92}$$

$$\frac{\partial}{\partial\theta}EZ_2 - A_1 E\Psi = -EY_2, \tag{93}$$

$$\frac{\partial}{\partial r}EZ_1 + B_2 E\Psi = +EX_2, \tag{94}$$

and

$$\frac{\partial}{\partial\theta}EZ_1 + B_1 E\Psi = +EY_1. \tag{95}$$

We may note here, for future reference, that the various bracket-expressions involving the spin-coefficients, which occur in the definitions of X_1, X_2, Y_1, and Y_2 are given by:

$$\left.\begin{aligned} (\kappa,\nu) &= \frac{1}{12M\Delta\sqrt{2}}[\![rX[\mathscr{L}S]^- + aY\{[\mathscr{L}S]^+\cos\theta + 3[S]^+\sin\theta\}]\!], \\ [\kappa,\nu]-[\kappa,\nu]^* &= \frac{1}{24M\Delta\sqrt{2}}[\![X\{(r^2-a^2\cos^2\theta)[\mathscr{L}\mathscr{L}S]^- - 2a^2[\mathscr{L}S]^-\sin\theta\cos\theta\} \\ &\quad + 2arY\{[\mathscr{L}\mathscr{L}S]^+\cos\theta + [\mathscr{L}S]^+\sin\theta\}]\!]. \\ [\kappa,\nu]+[\kappa,\nu]^* &= \frac{\mathrm{i}}{24M\Delta\sqrt{2}}[\![2arX\{[\mathscr{L}\mathscr{L}S]^-\cos\theta + 2[\mathscr{L}S]^-\sin\theta\} \\ &\quad + Y\{-(r^2-a^2\cos^2\theta)[\mathscr{L}\mathscr{L}S]^+ + 4a^2[\mathscr{L}S]^+\sin\theta\cos\theta \\ &\quad + 6a^2[S]^+\sin^2\theta\}]\!], \end{aligned}\right\} \tag{96}$$

$$
\left.
\begin{aligned}
(\lambda, \sigma) &= \frac{1}{12M\sqrt{2}} [\![\{-r[\mathscr{D}P]^- + 3[P]^-\}[S]^+ + ia[\mathscr{D}P]^+[S]^- \cos\theta]\!], \\
[\lambda, \sigma] - [\lambda, \sigma]^* &= \frac{1}{24M\sqrt{2}} [\![\{-(r^2 - a^2\cos^2\theta)[\mathscr{D}\mathscr{D}P]^+ + 2r[\mathscr{D}P]^+\}[S]^- \\
&\qquad + 2ia\{r[\mathscr{D}\mathscr{D}P]^- - [\mathscr{D}P]^-\}[S]^+ \cos\theta]\!]. \\
[\lambda, \sigma] + [\lambda, \sigma]^* &= \frac{1}{24M\sqrt{2}} [\![2ia\{r[\mathscr{D}\mathscr{D}P]^+ - 2[\mathscr{D}P]^+\}[S]^- \cos\theta \\
&\qquad + \{-(r^2 - a^2\cos^2\theta)[\mathscr{D}\mathscr{D}P]^- + 4r[\mathscr{D}P]^- - 6[P]^-\}[S]^+]\!].
\end{aligned}
\right\} \tag{97}
$$

There are some unexpected relations among the coefficients E, A_1, A_2, B_1, and B_2, which play crucial roles for the solvability of (92)–(95). We enumerate these relations in the following lemma:

LEMMA.

$$
\left.
\begin{aligned}
&\text{(i)} && A_1 B_2 = A_2 B_1; \quad (A_1 + B_1)(A_2 - B_2) = (A_1 - B_1)(A_2 + B_2); \\
&\text{(ii)} && A_1 + B_1 = \alpha^2\sigma/aQ \cos\theta \quad (\textit{a function of } \theta \textit{ only}); \\
&\text{(iii)} && A_2 + B_2 = \alpha^2\sigma/rK \quad (\textit{a function of } r \textit{ only}); \\
&\text{(iv)} && \partial A_1/\partial r = \partial A_2/\partial\theta; \quad \partial B_1/\partial r = \partial B_2/\partial\theta; \\
&\text{(v)} && \partial \lg E/\partial r = A_2 - B_2 - (r - M)/\Delta; \quad \partial \lg E/\partial\theta = A_1 - B_1; \\
&\text{(vi)} && A_1 \partial \lg E/\partial r - A_2 \partial \lg E/\partial\theta = -A_1(r - M)/\Delta; \\
\text{and}\ &\text{(vii)} && B_1 \partial \lg E/\partial r - B_2 \partial \lg E/\partial\theta = -B_1(r - M)/\Delta.
\end{aligned}
\right\} \tag{98}
$$

7. The solution of equations (92)–(95)

By adding (92) and (94), and similarly (93) and (95), we obtain the pair of equations (since $Z_1 + Z_2 = \Psi$)

$$\partial(E\Psi)/\partial r - (A_2 - B_2)E\Psi = -E(X_1 - X_2) \tag{99}$$

and

$$\partial(E\Psi)/\partial\theta - (A_1 - B_1)E\Psi = +E(Y_1 - Y_2). \tag{100}$$

By making use of relations included in (98), the foregoing equations can be reduced to the forms

$$\sqrt{\Delta}\frac{\partial}{\partial r}\left(\frac{\Psi}{\sqrt{\Delta}}\right) = -(X_1 - X_2) \quad \text{and} \quad \frac{\partial\Psi}{\partial\theta} = (Y_1 - Y_2). \tag{101}$$

By virtue of the known properties of Ψ these equations imply that

$$X_1 - X_2 = -\frac{\mathscr{C}_1 + \Gamma_1}{96M\Delta\sqrt{2}}\mathscr{S}\left(Y + \frac{4\lambda\sigma}{\mathscr{C}_1 + \Gamma_1} rX\right) \tag{102}$$

and

$$Y_1 - Y_2 = +\frac{\mathscr{C}_1 + \Gamma_1}{96M\sqrt{2}}\mathscr{R}\left\{[S]^- \cos\theta + \frac{4\sigma(\lambda\alpha^2 - 6a^2)}{a(\mathscr{C}_1 + \Gamma_1)}[S]^+\right\}. \tag{103}$$

In view of the complexity of the equations defining $X_1, \ldots, Y_2$, the foregoing relations are remarkable identities: they could not have been foreseen. However, in this instance, they can be verified directly (with some effort) if proper use is made of the relations (25), (26), (30), and (31).

Again, the consistency of (92) and (93), and of (94) and (95), requires that the following integrability conditions be satisfied:

$$\frac{\partial}{\partial r}A_1 E\Psi - \frac{\partial}{\partial \theta}A_2 E\Psi = \frac{\partial}{\partial r}EY_2 - \frac{\partial}{\partial \theta}EX_1, \tag{104}$$

and

$$\frac{\partial}{\partial r}B_1 E\Psi - \frac{\partial}{\partial \theta}B_2 E\Psi = \frac{\partial}{\partial r}EY_1 \quad \frac{\partial}{\partial \theta}EX_2. \tag{105}$$

By making use of relations included in (98), we find that (104) and (105) can be brought to the forms

$$A_1\sqrt{\Delta}\frac{\partial}{\partial r}\left(\frac{\Psi}{\sqrt{\Delta}}\right) - A_2\frac{\partial \Psi}{\partial \theta} = \frac{1}{E}\left(\frac{\partial}{\partial r}EY_2 - \frac{\partial}{\partial \theta}EX_1\right) \tag{106}$$

and

$$B_1\sqrt{\Delta}\frac{\partial}{\partial r}\left(\frac{\Psi}{\sqrt{\Delta}}\right) - B_2\frac{\partial \Psi}{\partial \theta} = \frac{1}{E}\left(\frac{\partial}{\partial r}EY_1 - \frac{\partial}{\partial \theta}EX_2\right). \tag{107}$$

Now, substituting from (101) for the quantities on the left hand sides of (106) and (107), we obtain the pair of equations

$$\sqrt{\Delta}\frac{\partial}{\partial r}\left(\frac{Y_1}{\sqrt{\Delta}}\right) - \frac{\partial X_2}{\partial \theta} = -B_1X_1 + A_1X_2 - A_2Y_1 + B_2Y_2 \tag{108}$$

and

$$\sqrt{\Delta}\frac{\partial}{\partial r}\left(\frac{Y_2}{\sqrt{\Delta}}\right) - \frac{\partial X_1}{\partial \theta} = -B_1X_1 + A_1X_2 - A_2Y_1 + B_2Y_2. \tag{109}$$

The difference of (108) and (109) gives

$$\sqrt{\Delta}\frac{\partial}{\partial r}\left(\frac{Y_1 - Y_2}{\sqrt{\Delta}}\right) + \frac{\partial}{\partial \theta}(X_1 - X_2) = 0; \tag{110}$$

and this relation is no more than what is required by (101). But the sum of (108) and (109) (after some reductions making use, once again, of relations included in (98)) provides a new relation:

$$\frac{\partial}{\partial r}E[(Y_1 + Y_2) - (A_1 + B_1)\Psi] = \frac{\partial}{\partial \theta}E[(X_1 + X_2) - (A_2 + B_2)\Psi]. \tag{111}$$

This relation must be an identity since all the quantities which occur in this equation are known functions. We shall presently consider the implications of this identity.

Returning to (92)–(95), we now find that the differences of (92) and (94), and of (93) and (95), yield the pair of equations

$$\frac{\partial}{\partial r}E(Z_1 - Z_2) = E[(X_1 + X_2) - (A_2 + B_2)\Psi] \tag{112}$$

and

$$\frac{\partial}{\partial \theta}E(Z_1 - Z_2) = E[(Y_1 + Y_2) - (A_1 + B_1)\Psi]. \tag{113}$$

Equation (111) guarantees the integrability of these equations, so that the solution for $E(Z_1 - Z_2)$ is correctly given by

$$E(Z_1 - Z_2) = \int^r E[(X_1 + X_2) - (A_2 + B_2)\Psi]\,\mathrm{d}r$$
$$= \int^\theta E[(Y_1 + Y_2) - (A_1 + B_1)\Psi]\,\mathrm{d}\theta. \tag{114}$$

This solution involves integrals over many functions and the question arises whether the various (indefinite) integrals, which occur in the solution, can be expressed explicitly in terms of known functions, even as the integrals defining the functions $\mathscr{R}$ and $\mathscr{S}$ were so expressed in § 3. We now turn to this question.

(a) *The implications of the integrability condition* (111) *and the solution for* $Z_1 - Z_2$

We recall that the explicit expression of the integrals defining the functions $\mathscr{R}$ and $\mathscr{S}$, in terms of known functions, was accomplished only via the identities implied by the integrability condition considered in II, §§ 8 and 9. It will accordingly appear that we must now seek the identities implied by the present integrability condition (111). For this purpose, we need explicit expressions for $X_1 + X_2$ and $Y_1 + Y_2$. Evaluating them, in accordance with the definitions (91) and the expressions for the various 'brackets' given in (96) and (97), we find, after some considerable reductions, that they can be written in the forms:

$$X_1 + X_2 = \frac{rX}{a\Delta}\left[G_1(\theta) - \frac{4a^2 Q\cos\theta}{\rho^2}G_2(\theta)\right] + \frac{Y}{\Delta}\left[G_3(\theta) - \frac{4a^2 Q\cos\theta}{\rho^2}G_4(\theta)\right] \tag{115}$$

and

$$Y_1 + Y_2 = [S]^- \cos\theta\left[F_1(r) - \frac{4rKF_2(r)}{\rho^2}\right] + \frac{\mathrm{i}[S]^+}{a}\left[F_3(r) - \frac{4rK}{\rho^2}F_4(r)\right]. \tag{116}$$

where a common factor $1/(24M\sqrt{2})$ has been suppressed, and

$$\left.\begin{aligned}
&F_1(r) = K[\mathscr{D}\mathscr{D}P]^+ + 2r\sigma[\mathscr{D}P]^+; \quad F_2(r) = r[\mathscr{D}\mathscr{D}P]^+ - [\mathscr{D}P]^+;\\
&F_3(r) = 3Kr[\mathscr{D}\mathscr{D}P]^- - 2\alpha^2\sigma[\mathscr{D}P]^- - 6r\sigma[P]^-; \quad F_4(r) = r^2[\mathscr{D}\mathscr{D}P]^- - r[\mathscr{D}P]^-;\\
&G_1(\theta) = Q[\mathscr{L}\mathscr{L}S]^- + 2a\sigma[\mathscr{L}S]^-\cos\theta; \quad G_2(\theta) = [\mathscr{L}\mathscr{L}S]^-\cos\theta + [\mathscr{L}S]^-\sin\theta;\\
&G_3(\theta) = 3Q[\mathscr{L}\mathscr{L}S]^+\cos\theta + (2\alpha^2\sigma/a)[\mathscr{L}S]^+ + 6a\sigma[S]^+\cos\theta\sin\theta;\\
&G_4(\theta) = [\mathscr{L}\mathscr{L}S]^+\cos^2\theta + [\mathscr{L}S]^+\cos\theta\sin\theta.
\end{aligned}\right\} \tag{117}$$

It can be verified that among the functions $F_1, \ldots, G_4$ the following relations exist:

$$\left.\begin{aligned}
&F_1 - 2K[\mathscr{D}\mathscr{D}P]^+ = -\tfrac{1}{4}(\mathscr{C}_1 + \Gamma_1)\,\mathscr{R}; \quad G_1 - 2Q[\mathscr{L}\mathscr{L}S]^- = \lambda a\sigma\mathscr{S};\\
&F_3 - 4KF_4/r = \mathrm{i}\sigma(\lambda\alpha^2 - 6a^2)\,\mathscr{R}; \quad G_3 - 4QG_4/\cos\theta = \tfrac{1}{4}(\mathscr{C}_1 + \Gamma_1)\mathscr{S}.
\end{aligned}\right\} \tag{118}$$

The appearance of the functions $\mathscr{R}$ and $\mathscr{S}$ in these relations is noteworthy.

Turning to the integrability condition (111), we observe that it requires the equality of the two expressions

$$\frac{1}{E}\frac{\partial}{\partial r}\left[\frac{\partial}{\partial\theta}E(Z_1-Z_2)\right]$$
$$=\sqrt{\Delta}\frac{\partial}{\partial r}\left(\frac{Y_1+Y_2}{\sqrt{\Delta}}\right)+(A_2-B_2)(Y_1+Y_2)+(A_2+B_2)\frac{\partial\Psi}{\partial\theta}$$
$$-\left\{(A_1+B_1)\sqrt{\Delta}\frac{\partial}{\partial r}\left(\frac{\Psi}{\sqrt{\Delta}}\right)+(A_2+B_2)\frac{\partial\Psi}{\partial\theta}+(A_1+B_1)(A_2-B_2)\Psi\right\} \quad (119)$$

and

$$\frac{1}{E}\frac{\partial}{\partial\theta}\left[\frac{\partial}{\partial r}E(Z_1-Z_2)\right]$$
$$=\frac{\partial}{\partial\theta}(X_1+X_2)+(A_1-B_1)(X_1+X_2)+(A_1+B_1)\sqrt{\Delta}\frac{\partial}{\partial r}\left(\frac{\Psi}{\sqrt{\Delta}}\right)$$
$$-\left\{(A_1+B_1)\sqrt{\Delta}\frac{\partial}{\partial r}\left(\frac{\Psi}{\sqrt{\Delta}}\right)+(A_2+B_2)\frac{\partial\Psi}{\partial\theta}+(A_1-B_1)(A_2+B_2)\Psi\right\}. \quad (120)$$

We notice that the quantities in the second lines of both equations are the same. Consequently, it will suffice to consider the required equality of the quantities in the first lines only. (But this fact, as we shall see in § 7 (*b*) below, entails an essential lacuna in the information derived from the integrability condition.)

On substituting for X_1+X_2 and Y_1+Y_2 from (115) and (116), we find that the terms in the first lines on the right hand sides of (119) and (120) can be brought to the following forms if appropriate use is made of the various relations included in (98) and (118), as well as the identities involving the functions $\mathscr{R}$ and $\mathscr{S}$:

$$\left[\left\{\sqrt{\Delta}\frac{\mathrm{d}}{\mathrm{d}r}\left(\frac{F_1}{\sqrt{\Delta}}\right)-\frac{2}{r}F_1+\frac{2\alpha^2\sigma}{r}[\mathscr{D}\mathscr{D}P]^+\right\}-\frac{4r}{\rho^2}\left\{\sqrt{\Delta}\frac{\mathrm{d}}{\mathrm{d}r}\left(\frac{KF_2}{\sqrt{\Delta}}\right)-r\sigma F_2-\tfrac{1}{2}F_1\right\}\right][S]^-\cos\theta$$
$$+\left[\left\{\sqrt{\Delta}\frac{\mathrm{d}}{\mathrm{d}r}\left(\frac{F_3}{\sqrt{\Delta}}\right)-\frac{2}{r}F_3+\frac{4\alpha^2\sigma}{r^2}F_4\right\}-\frac{4r}{\rho^2}\left\{\sqrt{\Delta}\frac{\mathrm{d}}{\mathrm{d}r}\left(\frac{KF_4}{\sqrt{\Delta}}\right)-r\sigma F_4-\tfrac{1}{2}F_3\right\}\right]\frac{\mathrm{i}[S]^+}{a}; \quad (121)$$

and

$$\left[\left\{\frac{\mathrm{d}G_1}{\mathrm{d}\theta}+2\frac{\sin\theta}{\cos\theta}G_1-\frac{2\alpha^2\sigma}{a\cos\theta}[\mathscr{L}\mathscr{L}S]^-\right\}-\frac{4a^2\cos\theta}{\rho^2}\left\{\frac{\mathrm{d}}{\mathrm{d}\theta}(QG_2)-a\sigma G_2\cos\theta+\tfrac{1}{2}G_1\sin\theta\right\}\right]\frac{rX}{a\Delta}$$
$$+\left[\left\{\frac{\mathrm{d}G_3}{\mathrm{d}\theta}+2\frac{\sin\theta}{\cos\theta}G_3-\frac{4\alpha^2\sigma}{a\cos^2\theta}G_4\right\}-\frac{4a^2\cos\theta}{\rho^2}\left\{\frac{\mathrm{d}}{\mathrm{d}\theta}(QG_4)-a\sigma G_4\cos\theta+\tfrac{1}{2}G_3\sin\theta\right\}\right]\frac{Y}{\Delta}. \quad (122)$$

Our aim is now to reduce the expressions (121) and (122) to forms in which (apart from simple factors such as r^2 or $\cos^2\theta$) only the functions X, Y, $[S]^+$, and $[S]^-\cos\theta$ occur and none of their derivatives. The analogous reductions, in the context of the integrability condition considered in II, § 9, were arduous since they required three successive applications of the (complicated) equations (I, equations (76), (77), (81), and (82)) which express the derivatives of the Teukolsky functions in terms of the functions themselves. Fortunately, in the present context the application of these equations can be avoided if judicious use is made of the relation included in (118) and the identities involving $\mathscr{R}$ and $\mathscr{S}$; but the reductions to the forms given below are not straightforward. We find

$$\frac{[S]^- \cos\theta}{\rho^2\Delta}\left[\!\!\left[\tfrac{1}{4}(\mathscr{C}_1+\Gamma_1)(-3r^2+a^2\cos^2\theta)\left(Y+\frac{4\lambda\sigma}{\mathscr{C}_1+\Gamma_1}rX\right)\right.\right.$$
$$\left.\left.+2\sigma r(\alpha^2+a^2\cos^2\theta)(\lambda X-6\sigma r Y)\right]\!\!\right]$$
$$+\frac{[S]^+}{a\rho^2\Delta}\left[\!\!\left[\sigma(\lambda\alpha^2-6a^2)(-r^2+3a^2\cos^2\theta)\left(Y+\frac{4\lambda\sigma}{\mathscr{C}_1+\Gamma_1}rX\right)\right.\right.$$
$$\left.\left.-2a^2\sigma\{6\sigma r(r^2-\alpha^2)X+[(\lambda\alpha^2-6a^2)+3(\lambda+2)r^2]\,Y\}\cos^2\theta\right]\!\!\right] \tag{123}$$

and

$$\frac{rX}{a\rho^2\Delta}\left[\!\!\left[\lambda a\sigma(-r^2+3a^2\cos^2\theta)\left\{[S]^-\cos\theta+\frac{4\sigma(\lambda\alpha^2-6a^2)}{a(\mathscr{C}_1+\Gamma_1)}[S]^+\right\}\right.\right.$$
$$\left.\left.-2a\sigma(r^2-\alpha^2)\{\lambda[S]^-+6a\sigma[S]^+\cos\theta\}\cos\theta\right]\!\!\right]$$
$$+\frac{Y}{\rho^2\Delta}\left[\!\!\left[\tfrac{1}{4}(\mathscr{C}_1+\Gamma_1)(-3r^2+a^2\cos^2\theta)\left\{[S]^-\cos\theta+\frac{4\sigma(\lambda\alpha^2-6a^2)}{a(\mathscr{C}_1+\Gamma_1)}[S]^+\right\}\right.\right.$$
$$\left.\left.-(2\sigma/a)r^2\{6a\sigma(\alpha^2+a^2\cos^2\theta)[S]^-\cos\theta-[(\lambda\alpha^2-6a^2)-3a^2(\lambda+2)\cos^2\theta][S]^+\}\right]\!\!\right]. \tag{124}$$

A comparison of these expressions shows that, as required, they are indeed the same.

After some rearrangements of the terms in (123) (or (124)), we can now write, in place of (119) and (120), the single equation

$$\frac{1}{E}\frac{\partial^2}{\partial r\partial\theta}E(Z_1-Z_2)$$
$$=(rX[S]^-\cos\theta)\lambda\sigma[-3(r^2-a^2\cos^2\theta)+2\alpha^2]/\rho^2\Delta$$
$$+(Y[S]^-\cos\theta)\{\tfrac{1}{4}(\mathscr{C}_1+\Gamma_1)(-3r^2+a^2\cos^2\theta)-12\sigma^2r^2(\alpha^2+a^2\cos^2\theta)\}/\rho^2\Delta$$
$$+rX[S]^+\left\{\frac{4\lambda\sigma^2(\lambda\alpha^2-6a^2)}{\mathscr{C}_1+\Gamma_1}(-r^2+3a^2\cos^2\theta)-12a^2\sigma^2(r^2-\alpha^2)\cos^2\theta\right\}\Big/a\rho^2\Delta$$
$$+Y[S]^+\{\sigma(\lambda\alpha^2-6a^2)(-r^2+a^2\cos^2\theta)-6a^2\sigma(\lambda+2)r^2\cos^2\theta\}/a\rho^2\Delta$$
$$-\left\{(A_1+B_1)\sqrt{\Delta}\frac{\partial}{\partial r}\left(\frac{\Psi}{\sqrt{\Delta}}\right)+(A_2+B_2)\frac{\partial\Psi}{\partial\theta}+(A_1\pm B_1)(A_2\mp B_2)\Psi\right\}. \tag{125}$$

An equivalent form of this equation is

$$\begin{aligned}
&\frac{1}{E}\frac{\partial^2}{\partial r\partial\theta}E(Z_1-Z_2)\\
&\quad=-\tfrac{3}{4}(\mathscr{C}_1+\Gamma_1)\,r^2(Y+\not{p}rX)\{[S]^-\cos\theta+\tfrac{1}{3}\not{q}[S]^+\}/\rho^2\Delta\\
&\qquad+\tfrac{1}{4}a^2(\mathscr{C}_1+\Gamma_1)(Y+3\not{p}rX)\{[S]^-\cos\theta+\not{q}[S]^+\}\cos^2\theta/\rho^2\Delta\\
&\qquad-12\sigma^2r^2Y(\alpha^2+a^2\cos^2\theta)[S]^-\cos\theta/\rho^2\Delta-12a\sigma^2(r^2-\alpha^2)\,rX[S]^+\cos^2\theta/\rho^2\Delta\\
&\qquad+2\alpha^2\sigma\lambda rX[S]^-\cos\theta/\rho^2\Delta-6a\sigma(\lambda+2)\,r^2Y[S]^+\cos^2\theta/\rho^2\Delta\\
&\qquad-\tfrac{1}{4}(\mathscr{C}_1+\Gamma_1)\left[\!\!\left[\frac{\alpha^2\sigma\mathscr{S}}{aQ\cos\theta}\frac{1}{E}\frac{\partial}{\partial r}E\mathscr{R}+\frac{\alpha^2\sigma}{rK}\mathscr{R}\{[S]^-\cos\theta+\not{q}[S]^+\}\right]\!\!\right]\\
&\qquad\text{or}\\
&\qquad-\tfrac{1}{4}(\mathscr{C}_1+\Gamma_1)\left[\!\!\left[\frac{\alpha^2\sigma\mathscr{R}}{rK}\frac{1}{E}\frac{\partial}{\partial\theta}E\mathscr{S}+\frac{\alpha^2\sigma}{aQ\cos\theta}\frac{\mathscr{S}}{\Delta}(Y+\not{p}rX)\right]\!\!\right],
\end{aligned}\tag{126}$$

where, for brevity, we have written

$$\not{p}=\frac{4\lambda\sigma}{\mathscr{C}_1+\Gamma_1}\quad\text{and}\quad\not{q}=\frac{4\sigma(\lambda\alpha^2-6a^2)}{a(\mathscr{C}_1+\Gamma_1)}.\tag{127}$$

Also, the alternative forms, in (126), for the terms in the last line of (125) follow from relations included in (98).

The solution for Z_1-Z_2 can now be written down by directly integrating, over r and θ, the expression on the right hand side of (126), after multiplication by E. Since, apart from the factor $\rho^2(=r^2+a^2\cos^2\theta)$, E is a product of a function of r and a function of θ, it is clear that the solution for Z_1-Z_2, obtained by integration, is a sum of products of an integral over r and an integral over θ. The radial integrals which occur in the solution are of six kinds: over rX, r^3X, Y, r^2Y, $\mathscr{R}$, and $r^2\mathscr{R}$. The integrals over the first four are with the weight function $[r\sqrt{(K\Delta^3)}]^{-1}$ while the last two are with the weight function $[r^2\sqrt{(K^3\Delta)}]^{-1}$. Similarly, the angular integrals which occur are also six kinds: over $[S]^+$, $[S]^+\cos^2\theta$, $[S]^-\cos\theta$, $[S]^-\cos^3\theta$, $\mathscr{S}$, and $\mathscr{S}\cos^2\theta$. The integrals over the first four are with the weight function $[\cos\theta\sqrt{(Q\sin\theta)}]^{-1}$ while the last two are with the weight function $[a\cos^2\theta\sqrt{(Q^3\sin\theta)}]^{-1}$. We shall denote the various integrals (with the specified weight functions) with angular brackets enclosing the quantity which is integrated. Thus, in this notation,

$$\left.\begin{aligned}
\langle r^3X\rangle&=\int\frac{r^3X}{r\sqrt{(K\Delta^3)}}\,dr;\quad \langle[S]^+\cos^2\theta\rangle=\int\frac{[S]^+\cos^2\theta}{\cos\theta\sqrt{(Q\sin\theta)}}\,d\theta,\\
\langle\mathscr{R}\rangle&=\int\frac{\mathscr{R}}{r^2\sqrt{(K^3\Delta)}}\,dr;\quad \langle\mathscr{S}\cos^2\theta\rangle=\int\frac{\mathscr{S}\cos^2\theta}{a\cos^2\theta\sqrt{(Q^3\sin\theta)}}\,d\theta.
\end{aligned}\right\}\tag{128}$$

The solution for Z_1-Z_2, in the notation adopted, can now be written in the form

$$\begin{aligned}
&E(Z_1-Z_2)\\
&\quad=-\tfrac{3}{4}(\mathscr{C}_1+\Gamma_1)\{\langle r^2Y\rangle+\not{p}\langle r^3X\rangle\}\{\langle[S]^-\cos\theta\rangle+\tfrac{1}{3}\not{q}\langle[S]^+\rangle\}\\
&\qquad+\frac{a^2}{4}(\mathscr{C}_1+\Gamma_1)\{\langle Y\rangle+3\not{p}\langle rX\rangle\}\{\langle[S]^-\cos^3\theta\rangle+\not{q}\langle[S]^+\cos^2\theta\rangle\}\\
&\qquad-12\sigma^2\langle r^2Y\rangle\{\alpha^2\langle[S]^-\cos\theta\rangle+a^2\langle[S]^-\cos^3\theta\rangle\}
\end{aligned}$$

$$- 12a\sigma^2\langle[S]^+\cos^2\theta\rangle\{\langle r^3X\rangle - \alpha^2\langle rX\rangle\}$$
$$+ 2\lambda\alpha^2\sigma\langle rX\rangle\langle[S]^-\cos\theta\rangle - 6a\sigma(\lambda+2)\langle r^2Y\rangle\langle[S]^+\cos^2\theta\rangle$$
$$- \tfrac{1}{4}\alpha^2\sigma(\mathscr{C}_1+\Gamma_1)\left\{\frac{\mathscr{R}}{r\sqrt{(\Delta K)}}[r^2\langle\mathscr{S}\rangle + a^2\langle\mathscr{S}\cos^2\theta\rangle\right.$$
$$\left.+\langle r^2\mathscr{R}\rangle[\langle[S]^-\cos\theta\rangle + \not{q}\langle[S]^+\rangle] + a^2\langle\mathscr{R}\rangle[\langle[S]^-\cos^3\theta\rangle + \not{q}\langle[S]^+\cos^2\theta\rangle]\right\}$$

or

$$- \tfrac{1}{4}\alpha^2\sigma(\mathscr{C}_1+\Gamma_1)\left\{\frac{\mathscr{S}}{\cos\theta\sqrt{(Q\sin\theta)}}[\langle r^2\mathscr{R}\rangle + a^2\langle\mathscr{R}\rangle\cos^2\theta]\right.$$
$$\left.+\langle\mathscr{S}\rangle[\langle r^2Y\rangle + \not{p}\langle r^3X\rangle] + a^2\langle\mathscr{S}\cos^2\theta\rangle[\langle Y\rangle + \not{p}\langle rX\rangle]\right\}. \tag{129}$$

(b) The identities which follow from the solution (129)

The question now arises, whether the various integrals which occur in the solution (129) can all be expressed in terms of known functions, even as the integrals defining $\mathscr{R}$ and $\mathscr{S}$ were expressed in § 3 in terms of the Teukolsky functions. To answer this question, we proceed in a similar way.

We differentiate the solution (129) with respect to r, and with respect to θ, and equate the resulting expressions with what we know them to be, namely, those given by (112) and (113). In this manner, we obtain the equations

$$\frac{1}{E}\frac{\partial}{\partial r}E(Z_1-Z_2) = X_1+X_2-\frac{\alpha^2\sigma}{rK}\Psi$$
$$= \frac{rX}{a\Delta\rho^2}(\rho^2G_1-4a^2QG_2\cos\theta) + \frac{Y}{\Delta\rho^2}(\rho^2G_3-4a^2QG_4\cos\theta) - \frac{\alpha^2\sigma(\mathscr{C}_1+\Gamma_1)}{4rK}\mathscr{R}\mathscr{S}$$
$$= \frac{(Q\sin\theta)^{\frac{1}{2}}\cos\theta}{\Delta\rho^2}[\![-\tfrac{3}{4}(\mathscr{C}_1+\Gamma_1)r^2(Y+\not{p}rX)\{\langle[S]^-\cos\theta\rangle + \tfrac{1}{3}\not{q}\langle[S]^+\rangle\}$$
$$+ \tfrac{1}{4}a^2(\mathscr{C}_1+\Gamma_1)(Y+3\not{p}rX)\{\langle[S]^-\cos^3\theta\rangle + \not{q}\langle[S]^+\cos^2\theta\rangle\}$$
$$- 12\sigma^2r^2Y\{\alpha^2\langle[S]^-\cos\theta\rangle + a^2\langle[S]^-\cos^3\theta\rangle\} + 2\lambda\alpha^2\sigma rX\langle[S]^-\cos\theta\rangle$$
$$- 12a\sigma^2(r^2-\alpha^2)rX\langle[S]^+\cos^2\theta\rangle - 6a\sigma(\lambda+2)r^2Y\langle[S]^+\cos^2\theta\rangle$$
$$- \tfrac{1}{4}\alpha^2\sigma(\mathscr{C}_1+\Gamma_1)(Y+\not{p}rX)\{r^2\langle\mathscr{S}\rangle + a^2\langle\mathscr{S}\cos^2\theta\rangle\}]\!] - \frac{\alpha^2\sigma(\mathscr{C}_1+\Gamma_1)}{4rK}\mathscr{R}\mathscr{S}. \tag{130}$$

and

$$\frac{1}{E}\frac{\partial}{\partial\theta}E(Z_1-Z_2) = Y_1+Y_2-\frac{\alpha^2\sigma}{aQ\cos\theta}\Psi$$
$$= \frac{[S]^-\cos\theta}{\rho^2}(\rho^2F_1-4rKF_2) + \frac{\mathrm{i}[S]^+}{a\rho^2}(\rho^2F_3-4rKF_4) - \frac{\alpha^2\sigma(\mathscr{C}_1+\Gamma_1)}{4aQ\cos\theta}\mathscr{R}\mathscr{S}$$
$$= \frac{r\sqrt{(\Delta K)}}{\rho^2}[\![-\tfrac{3}{4}(\mathscr{C}_1+\Gamma_1)\{\langle r^2Y\rangle + \not{p}\langle r^3X\rangle\}\{[S]^-\cos\theta + \tfrac{1}{3}\not{q}[S]^+\}$$
$$+ \tfrac{1}{4}a^2(\mathscr{C}_1+\Gamma_1)\{\langle Y\rangle + 3\not{p}\langle rX\rangle\}\{[S]^-\cos\theta + \not{q}[S]^+\}\cos^2\theta$$
$$- 12\sigma^2\langle r^2Y\rangle(\alpha^2+a^2\cos^2\theta)[S]^-\cos\theta + 2\lambda\alpha^2\sigma\langle rX\rangle[S]^-\cos\theta$$
$$- 12a\sigma^2\{\langle r^3X\rangle - \alpha^2\langle rX\rangle\}[S]^+\cos^2\theta - 6a\sigma(\lambda+2)\langle r^2Y\rangle[S]^+\cos^2\theta$$
$$- \tfrac{1}{4}\alpha^2\sigma(\mathscr{C}_1+\Gamma_1)\{\langle r^2\mathscr{R}\rangle + a^2\langle\mathscr{R}\rangle\cos^2\theta\}\{[S]^-\cos\theta + \not{q}[S]^+\}]\!] - \frac{\alpha^2\sigma(\mathscr{C}_1+\Gamma_1)}{4aQ\cos\theta}\mathscr{R}\mathscr{S}. \tag{131}$$

We observe that the terms in $\mathscr{R}\mathscr{S}$ cancel in both (130) and (131). The remaining terms in (130) consist of functions of θ which occur, respectively, with the factors r^3X, rX, Y, and r^2Y. We may, accordingly, equate the functions of θ which occur with these four radial factors. Similarly, we may equate in (131) the functions of r which occur with the four angular factors, $[S]^+$, $[S]^+\cos^2\theta$, $[S]^-\cos\theta$, and $[S]^-\cos^3\theta$. In this manner, we obtain the following eight identities:

$$\frac{G_1}{(Q\sin\theta)^{\frac{1}{2}}\cos\theta} = -3\lambda a\sigma\{\langle[S]^-\cos\theta\rangle+\tfrac{1}{3}q\langle[S]^+\rangle+\tfrac{1}{3}\alpha^2\sigma\langle\mathscr{S}\rangle\}-12a^2\sigma^2\langle[S]^+\cos^2\theta\rangle, \tag{132}$$

$$\begin{aligned}\frac{G_1\cos^2\theta-4QG_2\cos\theta}{(Q\sin\theta)^{\frac{1}{2}}\cos\theta} &= 3\lambda a\sigma\{\langle[S]^-\cos^3\theta\rangle+q\langle[S]^+\cos^2\theta\rangle-\tfrac{1}{3}\alpha^2\sigma\langle\mathscr{S}\cos^2\theta\rangle\}\\ &\quad+12\alpha^2\sigma^2\langle[S]^+\cos^2\theta\rangle+2\lambda(\alpha^2\sigma/a)\langle[S]^-\cos\theta\rangle.\end{aligned} \tag{133}$$

$$\begin{aligned}\frac{G_3}{(Q\sin\theta)^{\frac{1}{2}}\cos\theta} &= -\tfrac{3}{4}(\mathscr{C}_1+\Gamma_1)\{\langle[S]^-\cos\theta\rangle+\tfrac{1}{3}q\langle[S]^+\rangle+\tfrac{1}{3}\alpha^2\sigma\langle\mathscr{S}\rangle\}\\ &\quad-6a\sigma(\lambda+2)\langle[S]^+\cos^2\theta\rangle-12\sigma^2\{\alpha^2\langle[S]^-\cos\theta\rangle+a^2\langle[S]^-\cos^3\theta\rangle\}\end{aligned} \tag{134}$$

$$\begin{aligned}\frac{G_3\cos^2\theta-4QG_4\cos\theta}{(Q\sin\theta)^{\frac{1}{2}}\cos\theta} &= \frac{\mathscr{C}_1+\Gamma_1}{4(Q\sin\theta)^{\frac{1}{2}}\cos\theta}\mathscr{S}\cos^2\theta\\ &= \tfrac{1}{4}(\mathscr{C}_1+\Gamma_1)\{\langle[S]^-\cos^3\theta\rangle+q\langle[S]^+\cos^2\theta\rangle-\alpha^2\sigma\langle\mathscr{S}\cos^2\theta\rangle\},\end{aligned} \tag{135}$$

$$F_1/r\surd(\Delta K) = \tfrac{1}{4}(\mathscr{C}_1+\Gamma_1)\{\langle Y\rangle+3p\langle rX\rangle-\alpha^2\sigma\langle\mathscr{R}\rangle\}-12\sigma^2\langle r^2Y\rangle, \tag{136}$$

$$\begin{aligned}\frac{r^2F_1-4rKF_2}{r\surd(\Delta K)} &= -\tfrac{3}{4}(\mathscr{C}_1+\Gamma_1)\{\langle r^2Y\rangle+p\langle r^3X\rangle+\tfrac{1}{3}\alpha^2\sigma\langle r^2\mathscr{R}\rangle\}\\ &\quad-12\alpha^2\sigma^2\langle r^2Y\rangle+2\lambda\alpha^2\sigma\langle rX\rangle,\end{aligned} \tag{137}$$

$$\begin{aligned}iF_3/r\surd(\Delta K) &= \tfrac{1}{4}a(\mathscr{C}_1+\Gamma_1)q\{\langle Y\rangle+3p\langle rX\rangle-\alpha^2\sigma\langle\mathscr{R}\rangle\}\\ &\quad-12\sigma^2\{\langle r^3X\rangle-\alpha^2\langle rX\rangle\}-6\sigma(\lambda+2)\langle r^2Y\rangle,\end{aligned} \tag{138}$$

and

$$\begin{aligned}i(r^2F_3-4rKF_4)/r\surd(\Delta K) &= -\tfrac{1}{4}a(\mathscr{C}_1+\Gamma_1)qr^2\mathscr{R}/r\surd(\Delta K)\\ &= -\tfrac{1}{4}a(\mathscr{C}_1+\Gamma_1)q\{\langle r^2Y\rangle+p\langle r^3X\rangle+\alpha^2\sigma\langle r^2\mathscr{R}\rangle\}.\end{aligned} \tag{139}$$

It may be noted that in (135) and (139), we have incorporated two of the relations included in (118).

The eight identities, (132)–(139), will not suffice to express the 12 integrals, which occur in the solution for Z_1-Z_2, in terms of known functions. But additional identities follow from integrating the equations

$$\begin{aligned}\frac{\partial}{\partial\theta}E\mathscr{S} &= E\left[\frac{d\mathscr{S}}{d\theta}+(A_1-B_1)\mathscr{S}\right]\\ &= E\left[\{[S]^-\cos\theta+q[S]^+\}+\frac{\sigma}{\rho^2aQ\cos\theta}(r^2\alpha^2-a^2\alpha^2\cos^2\theta-2a^2r^2\cos^2\theta)\mathscr{S}\right]\end{aligned} \tag{140}$$

and

$$\begin{aligned}\frac{\partial}{\partial r}E\mathscr{R} &= E\left[\surd\Delta\frac{d}{dr}\left(\frac{\mathscr{R}}{\surd\Delta}\right)+(A_2-B_2)\mathscr{R}\right]\\ &= E\left[\frac{1}{\Delta}(Y+prX)+\frac{\sigma}{\rho^2rK}(r^2\alpha^2-a^2\alpha^2\cos^2\theta-2a^2r^2\cos^2\theta)\mathscr{R}\right].\end{aligned} \tag{141}$$

We find

$$\frac{\mathscr{S}}{(Q\sin\theta)^{\frac{1}{2}}\cos\theta} = \langle [S]^- \cos\theta\rangle + q\langle [S]^+\rangle + \sigma[\alpha^2\langle \mathscr{S}\rangle - 2a^2\langle \mathscr{S}\cos^2\theta\rangle]. \tag{142}$$

$$\frac{\mathscr{S}\cos^2\theta}{(Q\sin\theta)^{\frac{1}{2}}\cos\theta} = \langle [S]^- \cos^3\theta\rangle + q\langle [S]^+ \cos^2\theta\rangle - \alpha^2\sigma\langle \mathscr{S}\cos^2\theta\rangle. \tag{143}$$

$$\mathscr{R}/r\surd(\Delta K) = \langle Y\rangle + p\langle rX\rangle - \sigma[\alpha^2\langle \mathscr{R}\rangle + 2\langle r^2\mathscr{R}\rangle], \tag{144}$$

and

$$r^2\mathscr{R}/r\surd(\Delta K) = \langle r^2 Y\rangle + p\langle r^3 X\rangle + \alpha^2\sigma\langle r^2\mathscr{R}\rangle. \tag{145}$$

But two of these equations, namely (143) and (145), are the same as (135) and (139), included in the earlier set of eight identities. This duplication appears coincidental. But it is, perhaps, to be traced to the circumstance that the terms in $\mathscr{R}\mathscr{S}$ in (130) and (131) cancel and to the still earlier fact that parts of (119) and (120) are manifestly the same; and that, in consequence, the integrability condition fails to provide all the information that is needed to make the solution fully explicit. As it turns out, it *just* fails: it leaves one radial and one angular integral unevaluated.

Finally, we may note the following two alternative ways in which the solution (129) for $Z_1 - Z_2$ can be written down by making use of relations (132) (130) (and restoring also the factor $1/(24M\surd 2)$ which we had suppressed in (115) and (116)):

$$\begin{aligned} Z_1 - Z_2 = {} & \frac{r\surd(\Delta K)}{(24M\surd 2)\rho^2}\Big\{\frac{G_1}{a}\langle r^3X\rangle + a(G_1\cos^2\theta - 4QG_2\cos\theta)\langle rX\rangle \\ & + G_3\langle r^2Y\rangle + \tfrac{1}{4}a^2(\mathscr{C}_1 + \Gamma_1)\langle Y\rangle\mathscr{S}\cos^2\theta \\ & - \tfrac{1}{4}\alpha^2\sigma(\mathscr{C}_1+\Gamma_1)\mathscr{S}[\langle r^2\mathscr{R}\rangle + a^2\langle\mathscr{R}\rangle\cos^2\theta] \\ = {} & \frac{(Q\sin\theta)^{\frac{1}{2}}\cos\theta}{(24M\surd 2)\rho^2}\{a^2F_1\langle [S]^-\cos^3\theta\rangle + (r^2F_1 - 4rKF_2)\langle [S]^-\cos\theta\rangle \\ & + iaF_3\langle [S]^+\cos^2\theta\rangle - \tfrac{1}{4}(\mathscr{C}_1+\Gamma_1)q\langle [S]^+\rangle r^2\mathscr{R} \\ & - \tfrac{1}{4}\alpha^2\sigma(\mathscr{C}_1+\Gamma_1)\,\mathscr{R}[r^2\langle\mathscr{S}\rangle + a^2\langle\mathscr{S}\cos^2\theta\rangle]\}. \end{aligned} \tag{146}$$

For the six integrals which appear in each of the two foregoing alternative forms of the solution, we have only five equations relating them to known functions. One further relation (either radial or angular) is needed to make the solution fully explicit. The solution of the Newman–Penrose equations, to the extent that we have carried out, does not provide it. In § 8 we consider how this sole lacuna in our information may be filled.

8. Concluding remarks

It was made clear in II that the solutions for the perturbations in the metric coefficients (listed in II, equations (11)) can be written, simply, in terms of the solutions for the two functions $F+G$ and $J-H$. The solutions for these two functions have now been obtained (apart from a single lacuna in our information needed to make the solutions fully explicit); and the solutions for the metric perturbations which follow are fully consistent with the available ten degrees of gauge freedom: a gauge in which $\Psi_1 \equiv \Psi_3 \equiv 0$ was explicitly chosen at the outset and the solutions obtained leave $\Psi_2^{(1)}$ and the diagonal elements of the matrix $\boldsymbol{A}$ unspecified; and these fully account for the ten degrees of freedom.

Apart from the fact that the solution of the Newman–Penrose equations has required an inordinate amount of reductions, the analysis has disclosed aspects of solving a perturbation problem in this formalism which are, perhaps, worthy of some note.

In the context, specifically, of the perturbations of the Kerr metric, the unexpected features relate to the identities among the Teukolsky functions which have emerged. The most striking of these identities are those related to the function Ψ: its definition and its expression as a product of a radial and an angular function, $\mathscr{R}(r)$ and $\mathscr{S}(\theta)$; the existence of two alternative sets of formulae for $\mathscr{R}$ and $\mathscr{S}$ relating them directly to the Teukolsky functions and the fact that these formulae evaluate certain indefinite integrals over the Teukolsky functions; and, finally, the feasibility of these functions replacing the Teukolsky functions, altogether, for the purposes of specifying the metric perturbations.

We may now ask what the general implications may be of the features which have emerged in the context of solving the problem of the perturbations of the Kerr metric. It would appear that the features we have described originate in the very superfluity of the equations which the Newman–Penrose formalism provides for solving Einstein's equations. As we have stated in II, § 2 (*c*), the Newman–Penrose formalism provides a total of 76 real equations; and 'these equations have ostensibly been written down for determining 50 real functions subject to 10 degrees of gauge freedom.' One might infer that there can be 'no more than 40 independent equations' among the 76. But can one unambiguously select a particular 40?

In the context of the problem we have solved, the 16 Bianchi identities do provide independent information. But of the 24 equations (grouped under three systems of 8 equations each) which follow from the commutation relations, 6 equations belonging to system I (II, equations (30)), 5 equations belonging to system II (II, equations (32)), and all 8 equations belonging to system III (II, equations (33)) provide independent information, i.e. a total of 19 equations out of 24. But is this number 19 peculiar to the problem considered, or, is it of more general significance? In this connection, it is worth noticing that it is an integrability condition which arose from the consideration of four of the equations of system II (II, equations (68)–(91)) that (in addition to determining the relative normalizations of Ψ_0 and Ψ_4 and the real and the imaginary parts of the Teukolsky–Starobinsky constant $\mathscr{C}$) required the definition of Ψ and, eventually, led to the identities associated with it; and, further, that it was the consideration of the remaining four equations of the same system (II, equations (99)–(102)) that led to the differential equations (40) and (43) governing $\mathscr{R}$ and $\mathscr{S}$.

Turning to the 18 (complex) Ricci identities, we know that 8 of these follow from the commutation relations (as was pointed out by R. Wald). In the problem we have considered, six of the Ricci identities have played essential roles in the solution of the spin coefficients, κ, σ, λ, and ν and in the solution of $Z_1 - Z_2$. (Of the four real equations considered in the latter context, two of them lead to identities and, of the remaining two, one provides the integrability condition for the solvability of the other.) We have not so far considered six of the Ricci identities which involve the derivatives of the spin coefficients α, β, γ, and ϵ. It may be expected that a consideration of these Ricci identities will lead to further relations among the Teukolsky functions and that two of them will enable us to fill the last remaining lacuna in our information (mentioned in § 7). It will appear, then, that a consideration of these Ricci identities should be undertaken, though their linearization, which will require the differentiation of the equations of system III (II, equations (33)), is not likely to be 'easy.'

The major question which emerges from the present investigation is this: Can the superfluity of the Newman–Penrose equations enable us to discover new classes of identities among the special functions† of mathematical physics when they occur as solutions of Einstein's equations?

In the course of this investigation, I have profited by the interest shown by several colleagues. I am particularly grateful to Dr John L. Friedman for his critical judgement; his perceptive comments have proved invaluable to me. I am also indebted to Mrs Eliane Lessner for her discovery of an important error of a factor 2 in one of my earlier formulae (to which reference has been made in § 1); to Dr S. Detweiler for providing me with tables of the functions X and Y (in my chosen normalizations) and S_{+2} and S_{-2} (in their standard normalizations) as well as their derivatives; to Mr B. Xanthopoulos for many helpful discussions and for his checking some of the reductions; and to Miss Donna Elbert for her assistance in the numerical verification of some of the key equations of the theory (included in tables 1–4).

The research reported in this paper has in part been supported by a grant from the National Science Foundation under grant PHY 76-81102 A01 with the University of Chicago.

References

Chandrasekhar, S. 1978*a* *Proc. R. Soc. Lond.* A **358**, 421.
Chandrasekhar, S. 1978*b* *Proc. R. Soc. Lond.* A **358**, 441.
Newman, E. T. & Penrose, R. 1962 *J. Math. Phys.* **3**, 566.

† We may include Teukolsky's functions among the 'special' functions of mathematical physics since they reduce to conventional ones in the limit $M \to 0$.

The gravitational perturbations of the Kerr black hole IV. The completion of the solution†

By S. Chandrasekhar, F.R.S.
University of Chicago, Chicago, Illinois 60637, *U.S.A.*

(*Received* 25 *February* 1980)

This paper eliminates the last remaining lacuna in the information that was needed to make the solution for the perturbations in the metric coefficients of the Kerr space–time fully explicit. The requisite information is obtained from a pair of equations which is complementary to the one considered in paper III; and the solution of the Newman–Penrose equations governing the perturbations is, thus, completed.

1. Introduction

The solutions, for the perturbations of the metric coefficients of the Kerr space–time, via the Newman–Penrose formalism, provided in the earlier papers of this series (Chandrasekhar 1978*a*, *b*, and 1979)‡ were complete except for one lacuna in the information needed to make them fully explicit: for the six indefinite integrals (radial or angular), in terms of which the solutions were expressed (cf. III, eqn (146)), we had obtained only five identities relating them to known functions. In this paper, we shall show how this last remaining defect in the solution can be remedied: by a procedure which departs from that followed in paper III (starting from the verification of the integrability condition, III, eqn (111)), we shall obtain, in a fairly straightforward fashion, explicit solutions for Z_1 and Z_2 (leading to one more remarkable identity). While we are thus able to circumvent the need to establish the various integral relations which follow, namely III, eqns (131)–(139), it was fortunate, in some ways, that they *were* established: they might have been overlooked otherwise!

Finally, in the Appendix, we consider the form the solutions take in the Schwarz-

† Part III appeared in *Proc. R. Soc. Lond.* A **365**, 425–451 (1979).

‡ These papers will be referred to hereafter as papers I, II, and III, respectively. When equations from one or other of the papers are referred to, they will be prefixed by the corresponding Roman numerals.

schild limit, $a \to 0$, for two reasons: *first*, the passage to the limit, $a \to 0$, is by no means a straightforward one; and, *second*, the various identities which have appeared in the course of the analysis are exhibited in less opaque guises in this limit.

2. A brief recapitulation and some preliminary relations

It was shown, already in paper II (§ 6), that the solution to the metric perturbations can be completed, once the functions $F+G$ and $J-H$ are determined. And it was shown, further, in papers II and III that, in the combinations (II, eqn (95), and III, eqns (86) and (87)),

$$Z_1 = K(J-H)\cos\theta \quad \text{and} \quad Z_2 = -\mathrm{i}rQ(F+G)\sin\theta, \tag{1}$$

their sum, Ψ, can be expressed in the form,

$$\Psi = Z_1 + Z_2 = \frac{\mathscr{C}_1 + \Gamma_1}{96M\sqrt{2}}\mathscr{R}(r)\mathscr{S}(\theta), \tag{2}$$

where the functions $\mathscr{R}(r)$ and $\mathscr{S}(\theta)$, initially defined as indefinite integrals over the radial and the angular Teukolsky functions (III, eqns (34) and (35)), are in fact directly expressible in terms of the Teukolsky functions themselves by virtue of some remarkable identities which emerged at this stage (III, eqns (25)–(28) and (30)–(33)).

Next, by considering a certain pair of combinations of the linearized versions of four particular Ricci identities (listed in III, eqns (54)) we showed how the derivatives of Z_1 and Z_2, with respect to r and with respect to θ, can be expressed in terms of Z_1 and Z_2 themselves (III, eqns (92)–(95)). These equations further led to an integrability condition (namely III, eqn (111)) for the solvability of these equations for $(Z_1 - Z_2)$. The rest of paper III was largely devoted to deriving the implications of the integrability condition. But, as we have stated, the integrability condition did not provide all the information that was needed to make the solution for $(Z_1 - Z_2)$ fully explicit: a single lacuna remained. In this paper, we shall show how, by making use of the same equations in a different way, we can obtain the solutions for Z_1 and Z_2 in the explicit forms that we seek.

The fact, that the derivatives of Z_1 and Z_2 are expressible as linear combinations of the functions themselves, means that, in any equation in which the derivatives of Z_1 and Z_2 (no matter of how high orders) occur, we can replace them by the functions. As examples, consider II, eqns (53) and (54):

$$\frac{\mathrm{i}K}{\Delta}(F-G) = \frac{\partial}{\partial r}(F+G) - \frac{2\mathrm{i}a\cos\theta}{\rho^2}(J-H) \tag{3}$$

and

$$Q(J+H) = \left(\frac{\partial}{\partial\theta} + \cot\theta\right)(J-H) - \frac{2\mathrm{i}ar\sin\theta}{\rho^2}(F+G). \tag{4}$$

By rewriting the terms on the right-hand sides of these equations in terms of Z_1 and Z_2, we obtain

$$\frac{rKQ\sin\theta}{\Delta}(F-G)=\frac{\partial Z_2}{\partial r}-\frac{Z_2}{r}-\frac{2arQ\sin\theta}{K\rho^2}Z_1 \tag{5}$$

and
$$KQ(J+H)\cos\theta=\frac{\partial Z_1}{\partial\theta}+\frac{Z_1}{\cos\theta\sin\theta}+\frac{2aK\cos\theta}{\rho^2 Q}Z_2. \tag{6}$$

Now, substituting for the derivatives of Z that occur on the right-hand sides of equations (5) and (6) from III, eqns (92)–(95), we find, after some further reductions, that

$$\frac{rKQ\sin\theta}{\Delta}(F-G)=-\frac{arQ\sin\theta}{K\rho^2}Z_1+\left(\frac{r-M}{\Delta}-\frac{r}{\rho^2}\right)Z_2-X_1 \tag{7}$$

and
$$KQ(J+H)\cos\theta=\frac{(r^2+a^2)\cot\theta}{\rho^2}Z_1+\frac{aK\cos\theta}{\rho^2 Q}Z_2+Y_1, \tag{8}$$

where (cf. III, eqns (91), (96), and (97))

$$X_1=\frac{rQ}{a\rho^2}\{[\kappa,\nu]-[\kappa,\nu]^*\} \tag{9}$$

and
$$Y_1=\frac{K\cos\theta}{\rho^2}\{[\lambda,\sigma]-[\lambda,\sigma]^*\}. \tag{10}$$

In reducing equations (5) and (6) to the forms (7) and (8), we have replaced Ψ by (Z_1+Z_2) and made use of some of the relations included in the lemma (III, eqn (98)).

3. The equations for determining Z_1 and Z_2

We return to III, eqns (56)–(59), which represent the linearized versions of the Ricci identities, III, eqns (54). Instead of considering the *difference* of eqns (56) and (57) and the *sum* of eqns (58) and (59), as we did in paper III, we shall now consider the *sum* of eqns (56) and (57) and the *difference* of eqns (58) and (59). The complementary pair of equations are

$$-\mathrm{i}K(F^1_2-F^2_1)+\tfrac{1}{2}\Delta(\mathscr{D}_0+\mathscr{D}^\dagger_0)U+\tfrac{1}{\sqrt{2}}(\mathrm{i}a\sin\theta)(\mathscr{D}_{-1}\rho^2F-\mathscr{D}^\dagger_{-1}\rho^2G)$$
$$+\sqrt{2}\,(a^2\sin\theta\cos\theta)(F-G)-\sqrt{2}\,(\mathrm{i}ar\sin\theta)B_2$$
$$=-\frac{\Delta}{2\sqrt{2}}\left[(\bar\rho^*)^2\mathscr{L}_1\frac{\kappa}{(\bar\rho^*)^2}+\frac{\mathrm{i}a\sin\theta}{\bar\rho}\kappa^*\right]+\sqrt{2}\frac{\rho^4\bar\rho^*}{\Delta}\left[\frac{1}{\rho^2\bar\rho}\mathscr{L}^\dagger_1(\rho^2\nu)+\frac{\mathrm{i}a\sin\theta}{(\bar\rho^*)^2}\nu^*\right], \tag{11}$$

and

$$-(\mathrm{i}aQ\sin\theta)(F^3_4-F^4_3)+\frac{\mathrm{i}a\sin\theta}{2\rho^2}(\mathscr{L}_0+\mathscr{L}^\dagger_0)V+\sqrt{2}\frac{a^2\sin\theta\cos\theta}{\rho^2}C_1$$
$$+\tfrac{1}{\sqrt{2}}\left(\mathscr{L}_{-1}J+\mathscr{L}^\dagger_{-1}H-\frac{2\mathrm{i}a\sin\theta}{\bar\rho}J-\frac{2\mathrm{i}a\sin\theta}{\bar\rho}H\right)$$
$$=\frac{\bar\rho^*}{\bar\rho}\left[\mathscr{D}_0(\bar\rho^*\lambda)+\frac{\Delta}{2\rho^2}\sigma^*\right]+\left[\frac{\Delta\bar\rho^*}{2\bar\rho}\mathscr{D}^\dagger_1\frac{\bar\rho\sigma}{(\bar\rho^*)^2}+\frac{\bar\rho}{\bar\rho^*}\lambda^*\right]. \tag{12}$$

Our basic equations, for our further considerations, are the imaginary part of equation (11) and the real part of equation (12). These equations can be written in the forms

$$\mathscr{D}_{-1}\rho^2 F - \mathscr{D}^\dagger_{-1}\rho^2 G = \frac{2}{ia\sin\theta}[\{\kappa,\nu\}-\{\kappa,\nu\}^*], \tag{13}$$

and

$$\rho^2\left[\mathscr{L}_{-1}J + \mathscr{L}^\dagger_{-1}H - \frac{2a^2\sin\theta\cos\theta}{\rho^2}(J+H)\right] = 2\,[\{\lambda,\sigma\}+\{\lambda,\sigma\}^*], \tag{14}$$

where

$$\{\kappa,\nu\} = \tfrac{1}{8}\Delta\left[\bar\rho^2\mathscr{L}^\dagger_1\frac{\kappa^*}{\bar\rho^2} - \frac{ia\sin\theta}{\bar\rho^*}\kappa\right] - \frac{\rho^4\bar\rho}{2\Delta}\left[\frac{1}{\rho^2\bar\rho^*}\mathscr{L}_1(\rho^2\nu^*) - \frac{ia\sin\theta}{\bar\rho^2}\nu\right] \tag{15}$$

and

$$\{\lambda,\sigma\} = \frac{1}{2\sqrt{2}}(\bar\rho^*)^2\left[\mathscr{D}_0(\bar\rho^*\lambda) + \frac{\Delta}{2\rho^2}\sigma^*\right] + \frac{1}{2\sqrt{2}}\rho^2\left[\frac{\Delta\bar\rho^*}{2\bar\rho}\mathscr{D}^\dagger_1\frac{\bar\rho\sigma}{(\bar\rho^*)^2} + \frac{\bar\rho}{\bar\rho^*}\lambda^*\right]. \tag{16}$$

Rewriting equation (13) in the form

$$\Delta\frac{\partial}{\partial r}\frac{\rho^2(F-G)}{\Delta} + \frac{iK\rho^2}{\Delta}(F+G) = \Delta\frac{\partial}{\partial r}\frac{\rho^2}{rKQ\sin\theta}\left[\frac{rKQ\sin\theta}{\Delta}(F-G)\right] - \frac{K\rho^2}{r\Delta Q\sin\theta}Z_2$$

$$= -\frac{2i}{a\sin\theta}[\{\kappa,\nu\}-\{\kappa,\nu\}^*], \tag{17}$$

and making use of equation (7), we obtain

$$rK\Delta\frac{\partial}{\partial r}\frac{\rho^2}{rK}\left[-\frac{arQ\sin\theta}{K\rho^2}Z_1 + \left(\frac{r-M}{\Delta} - \frac{r}{\rho^2}\right)Z_2 - X_1\right] - \frac{K^2\rho^2}{\Delta}Z_2$$

$$= -2i\frac{rKQ}{a}[\{\kappa,\nu\}-\{\kappa,\nu\}^*]. \tag{18}$$

Similarly, rewriting equation (14) in the form

$$\sin\theta\frac{\partial}{\partial\theta}\frac{\rho^2(J+H)}{\sin\theta} + Q\rho^2(J-H) = 2\,[\{\lambda,\sigma\}+\{\lambda,\sigma\}^*], \tag{19}$$

and making use of equation (8), we obtain

$$\sin\theta\frac{\partial}{\partial\theta}\frac{\rho^2}{KQ\cos\theta\sin\theta}\left[\frac{(r^2+a^2)\cot\theta}{\rho^2}Z_1 + \frac{aK\cos\theta}{\rho^2 Q}Z_2 + Y_1\right] + \frac{\rho^2 Q}{K\cos\theta}Z_1$$

$$= 2[\{\lambda,\sigma\}+\{\lambda,\sigma\}^*]. \tag{20}$$

Now expanding the left-hand sides of equations (18) and (20) and substituting, once again, for the derivatives of Z which occur, from III, eqns (92)–(95), we obtain, after some considerable reductions, the pair of equations:

$$\left[\frac{\rho^2}{\Delta}(a^2-M^2-K^2)-\frac{\sigma}{K}\{\Delta a^2\cos^2\theta+2r\rho^2(r-M)-2r^2\Delta\}\right]Z_2$$
$$+\frac{3a\sigma}{K^2}(r^2\Delta Q\sin\theta)Z_1=\frac{arQ\sin\theta}{K}\Delta X_2+[\rho^2(r-M)-r\Delta]X_1$$
$$+\frac{rKQ}{a}\left\{\Delta\frac{\partial}{\partial r}\frac{[\kappa,\nu]-[\kappa,\nu]^*}{K}-2i[\{\kappa,\nu\}-\{\kappa,\nu\}^*]\right\} \qquad (21)$$

and

$$\left[\rho^2(Q^2-\operatorname{cosec}^2\theta)+\frac{a\sigma}{Q\sin\theta}(3r^2\sin^2\theta-2\rho^2)\right]Z_1-\frac{3a^2\sigma}{Q^2}(K\cos^2\theta)Z_2$$
$$=-[(r^2+a^2)\cot\theta]Y_1+\frac{aK\cos\theta}{Q}Y_2$$
$$+QK\cos\theta\left\{-\sin\theta\frac{\partial}{\partial\theta}\frac{[\lambda,\sigma]-[\lambda,\sigma]^*}{Q\sin\theta}+2[\{\lambda,\sigma\}+\{\lambda,\sigma\}^*]\right\}. \qquad (22)$$

It is now manifest that either of these two equations will suffice to determine Z_1 and Z_2 (since their sum is known to be Ψ). We shall, however, find it more convenient to treat the two equations symmetrically as in §4 below.

4. The solutions for Z_1 and Z_2

In equation (21), we replace Z_1 on the left-hand side by $\Psi-Z_2$ while, in equation (22), we replace Z_2 by $\Psi-Z_1$. Also, we replace X_2 and Y_2 (on the right-hand sides of these equations) by X_1 and Y_1, respectively, by making use of the relations (III, eqns (101)–(103)):

$$X_2=X_1+\sqrt{\Delta}\frac{\partial}{\partial r}\frac{\Psi}{\sqrt{\Delta}}=X_1+\frac{\mathscr{C}_1+\Gamma_1}{96M\Delta\sqrt{2}}\mathscr{S}\left(Y+\frac{4\lambda\sigma}{\mathscr{C}_1+\Gamma_1}rX\right) \qquad (23)$$

and
$$Y_2=Y_1-\frac{\partial\Psi}{\partial\theta}=Y_1-\frac{\mathscr{C}_1+\Gamma_1}{96M\sqrt{2}}\mathscr{R}\left\{[S]^-\cos\theta+\frac{4\sigma(\lambda\alpha^2-6a^2)}{a(\mathscr{C}_1+\Gamma_1)}[S]^+\right\}. \qquad (24)$$

In this manner, we reduce equations (21) and (22) to the forms

$$\rho^2\left[\frac{1}{\Delta}(a^2-M^2-K^2)-\frac{\sigma\Delta}{K}\left\{1+\frac{2r(r-M)}{\Delta}-\frac{3r^2\sigma}{K}\right\}\right]Z_2$$
$$=-\frac{arQ\sin\theta}{K}\Delta\left\{\sqrt{\Delta}\frac{\partial}{\partial r}\frac{\Psi}{\sqrt{\Delta}}-\frac{3r\sigma}{K}\Psi\right\}$$
$$+\frac{rKQ}{a}\left\{\frac{\Delta}{K}\left(\frac{\partial}{\partial r}+\frac{r-M}{\Delta}-\frac{3r\sigma}{K}\right)\{[\kappa,\nu]-[\kappa,\nu]^*\}-2i[\{\kappa,\nu\}-\{\kappa,\nu\}^*]\right\}, \qquad (25)$$

and

$$\rho^2\left[Q^2-\operatorname{cosec}^2\theta+\frac{a\sigma}{Q\sin\theta}\left(3\sin^2\theta-2+\frac{3a\sigma\cos^2\theta\sin\theta}{Q}\right)\right]Z_1$$

$$=-\frac{aK\cos\theta}{Q}\left(\frac{\partial\Psi}{\partial\theta}-\frac{3a\sigma\cos\theta}{Q}\right)\Psi+KQ\cos\theta$$

$$\times\left\{\left[-\frac{1}{Q}\left(\frac{\partial}{\partial\theta}+\cot\theta\right)+\frac{3a\sigma\cos\theta}{Q^2}\right]\{[\lambda,\sigma]-[\lambda,\sigma]^*\}+2\left[\{\lambda,\sigma\}+\{\lambda,\sigma\}^*\right]\right\}. \quad (26)$$

The right-hand sides of equations (25) and (26) can be further simplified by making use of the following expressions for $\{[\kappa,\nu]-[\kappa,\nu]^*\}$, etc. (cf. III, eqns (96) and (97)):

$$[\kappa,\nu]-[\kappa,\nu]^*=\frac{1}{24M\Delta\sqrt{2}}[\![X\{(r^2-a^2\cos^2\theta)[\mathscr{L}\mathscr{L}S]^--2a^2[LS]^-\sin\theta\cos\theta\}$$
$$+2arY\{[\mathscr{L}\mathscr{L}S]^+\cos\theta+[\mathscr{L}S]^+\sin\theta\}]\!], \quad (27)$$

$$\{\kappa,\nu\}-\{\kappa,\nu\}^*=\frac{\mathrm{i}}{24M\Delta\sqrt{2}}[\![Y\{(r^2-a^2\cos^2\theta)[\mathscr{L}\mathscr{L}S]^--2a^2[\mathscr{L}S]^-\sin\theta\cos\theta\}$$
$$-2arX\{[\mathscr{L}\mathscr{L}S]^+\cos\theta+[\mathscr{L}S]^+\sin\theta\}]\!], \quad (28)$$

$$[\lambda,\sigma]-[\lambda,\sigma]^*=\frac{-1}{24M\sqrt{2}}[\![\{(r^2-a^2\cos^2\theta)[\mathscr{D}\mathscr{D}P]^+-2r[\mathscr{D}P]^+\}[S]^-$$
$$-2\mathrm{i}a\{r[\mathscr{D}\mathscr{D}P]^--[\mathscr{D}P]^-\}[S]^+\cos\theta]\!], \quad (29)$$

$$\{\lambda,\sigma\}+\{\lambda,\sigma\}^*=\frac{+1}{24M\sqrt{2}}[\![\{(r^2-a^2\cos^2\theta)[\mathscr{D}\mathscr{D}P]^+-2r[\mathscr{D}P]^+\}[S]^+$$
$$-2\mathrm{i}a\{r[\mathscr{D}\mathscr{D}P]^--[\mathscr{D}P]^-\}[S]^-\cos\theta]\!]. \quad (30)$$

We thus obtain the following explicit solutions for Z_1 and Z_2:

$$\rho^2\left[(Q^2-\operatorname{cosec}^2\theta)+\frac{a\sigma}{Q\sin\theta}\left(3\sin^2\theta-2+\frac{3a\sigma}{Q}\cos^2\theta\sin\theta\right)\right]Z_1$$

$$=-\tfrac{1}{4}(\mathscr{C}_1+\Gamma_1)\frac{aK\cos\theta}{Q}\left\{\mathscr{R}\left[[S]^-\cos\theta+\frac{4\sigma(\lambda\alpha^2-6a^2)}{a(\mathscr{C}_1+\Gamma_1)}[S]^+\right]-\frac{3a\sigma\cos\theta}{Q}\mathscr{R}\mathscr{S}\right\}$$

$$+K\cos\theta\Big[\!\Big[(2a^2\sin\theta\cos\theta)[\mathscr{D}\mathscr{D}P]^+[S]^-$$

$$+\{(r^2-a^2\cos^2\theta)[\mathscr{D}\mathscr{D}P]^+-2r[\mathscr{D}P]^+\}$$

$$\times\left\{[\mathscr{L}S]^--\left(\cot\theta+\frac{3a\sigma\cos\theta}{Q}\right)[S]^-+Q[S]^+\right\}-2\mathrm{i}a\cos\theta$$

$$\times\{r[\mathscr{D}\mathscr{D}P]^--[\mathscr{D}P]^-\}\left\{[\mathscr{L}S]^+-\left(\frac{1}{\sin\theta\cos\theta}+\frac{3a\sigma}{Q}\cos\theta\right)[S]^++Q[S]^-\right\}\Big]\!\Big], \quad (31)$$

and

$$\rho^2\left[\frac{1}{\Delta}(a^2-M^2-K^2)-\frac{\sigma\Delta}{K}\left\{1+\frac{2r(r-M)}{\Delta}-\frac{3r^2\sigma}{K}\right\}\right]Z_2$$

$$= \tfrac{1}{4}(\mathscr{C}_1+\Gamma_1)\frac{arQ\sin\theta}{K}\Delta\left\{\frac{\mathscr{S}}{\Delta}\left(Y+\frac{4\lambda\sigma}{\mathscr{C}_1+\Gamma_1}rX\right)-\frac{3r\sigma}{K}\mathscr{R}\mathscr{S}\right\}+\frac{rQ}{a}\Big[\!\Big[2r[\mathscr{L}\mathscr{L}S]^- X$$

$$+\{(r^2-a^2\cos^2\theta)[\mathscr{L}\mathscr{L}S]^- -(2a^2\sin\theta\cos\theta)[\mathscr{L}S]^-\}$$

$$\times\left\{[\mathscr{D}P]^+ -\left(\frac{r-M}{\Delta}+\frac{3r\sigma}{K}\right)X+K\frac{Y}{\Delta}\right\}$$

$$-2ar\{[\mathscr{L}\mathscr{L}S]^+\cos\theta+[\mathscr{L}S]^+\sin\theta\}$$

$$\times\left\{\mathrm{i}\,[\mathscr{D}P]^- +\left(\frac{Mr-a^2}{r\Delta}+\frac{3r\sigma}{K}\right)Y+K\frac{X}{\Delta}\right\}\Big]\!\Big], \qquad (32)$$

where a common factor, $1/(24M\sqrt{2})$, in the solutions for Z_1 and Z_2 has been suppressed.

It is, of course, a necessary identity that Z_1 and Z_2, determined in accordance with equations (31) and (32), are consistent with the requirement

$$Z_1+Z_2 = \tfrac{1}{4}(\mathscr{C}_1+\Gamma_1)\mathscr{R}(r)\mathscr{S}(\theta). \qquad (33)$$

In view of the massive reductions involved in obtaining the solutions (31) and (32), an assurance that nó algebraic (or other) error has, inadvertently, crept into the analysis, was sought in a direct numerical verification of the equality (33) for Z_1 and Z_2 determined (also numerically) with the aid of equations (31) and (32). For a wide range of the variables r and θ (for $\sigma = 0{\cdot}25$ and $\sigma = 0{\cdot}5$ and $a = 0{\cdot}95$) the required equality was verified. Moreover, in the appendix, we shall be able to verify directly the equality (33) in the Schwarzschild limit, $a \to 0$.

5. The completion of the solution

With Z_1 and Z_2 determined by equations (31) and (32), the solutions for the remaining functions (in terms of which the metric perturbations are listed in paper II, eqns (11)) can be readily found. Thus, equations (7) and (8) already provide the solutions for $F-G$ and $J+H$. Similarly, making use of II, eqns (49) and (50), we now obtain

$$\frac{\mathrm{i}K}{\Delta}B_2 = \frac{a(K+\rho^2\sigma)\cos\theta}{r\rho^2KQ\sin\theta}Z_2+\frac{1}{K\cos\theta}\left[X_2-\left(\frac{r\sigma}{K}+\frac{r}{\rho^2}-\frac{r-M}{\Delta}\right)Z_1\right] \qquad (34)$$

and

$$QC_1 = \frac{\mathrm{i}r(\rho^2\sigma-aQ\sin\theta)}{\rho^2KQ\sin\theta}Z_1-\frac{\mathrm{i}}{rQ\sin\theta}\left[Y_2+\left(\frac{a\sigma\cos\theta}{Q}-\frac{r^2+a^2}{\rho^2}\cot\theta\right)Z_2\right]. \qquad (35)$$

With B_2 and C_1 given by these equations, the solutions for B_1 and C_2 follow from (II, eqns (49) and (52)):

$$\frac{\mathrm{i}K}{\Delta}B_1 = +\frac{2\mathrm{i}a\cos\theta}{\rho^2}C_1 - \frac{\partial}{\partial r}(J+H) \tag{36}$$

and

$$QC_2 = -\frac{2\mathrm{i}ar\sin\theta}{\rho^2}B_2 + \frac{1}{\sin\theta}\frac{\partial}{\partial\theta}(F-G)\sin\theta. \tag{37}$$

And finally III, eqns (74) and (75), and II, eqns (99) and (100), complete the solution:

$$\frac{\mathrm{i}K}{\sqrt{2}}(F^1_2+F^2_1) = -\frac{\mathrm{i}a}{rKQ\cos\theta}[(r^2Q\sin\theta)Z_1-(aK\cos^2\theta)Z_2]+[\kappa,\nu]+[\kappa,\nu]^*, \tag{38}$$

$$\frac{\mathrm{i}aQ\sin\theta}{\sqrt{2}}\rho^2(F^4_3+F^3_4) = +\frac{\mathrm{i}a}{rKQ\cos\theta}[(r^2Q\sin\theta)Z_1-(aK\cos^2\theta)Z_2]+[\lambda,\sigma]+[\lambda,\sigma]^*, \tag{39}$$

$$\frac{Q}{\sqrt{2}}(F^1_2-F^2_1)+\mathrm{i}\frac{K^2\rho^4-2a^2\Delta^2\cos^2\theta}{2\Delta K^2\rho^2\cos^2\theta}Z_1+\frac{\mathrm{i}a\Delta\cos\theta}{rKQ\sin\theta}\left[\left(\frac{r-M}{\Delta}-\frac{r}{\rho^2}\right)Z_2-X_1\right] + \frac{\Delta}{2\rho^2}\frac{\partial}{\partial r}\frac{\rho^4}{\Delta}B_2 = \langle\kappa,\nu\rangle, \tag{40}$$

$$\frac{\mathrm{i}K}{\sqrt{2}}\rho^2(F^4_3-F^3_4)-\mathrm{i}\frac{Q^2\rho^4+2a^2r^2\sin^2\theta}{2r\rho^2Q^2\sin^2\theta}Z_2-\frac{\mathrm{i}ar\sin\theta}{KQ\cos\theta}\left[\frac{(r^2+a^2)\cot\theta}{\rho^2}Z_1+Y_1\right] - \frac{1}{2\rho^2}\left(\frac{\partial}{\partial\theta}-\cot\theta\right)\rho^4C_1 = \langle\lambda,\sigma\rangle. \tag{41}$$

The functions of the spin coefficients which occur on the right-hand sides of equations (38) and (39) have already been listed in paper III (eqns (37) and (38)); and we note below the additional functions that appear in equations (40) and (41):

$$\langle\kappa,\nu\rangle = \frac{1}{2\rho^2}\left[\tfrac{1}{2}\Delta(\bar\rho^*\kappa^*-\bar\rho\kappa)+\frac{2\rho^4}{\Delta}(\bar\rho^*\nu-\bar\rho\nu^*)\right] = \frac{\mathrm{i}}{12M\Delta\sqrt{2}}[\![rY[\mathscr{L}S]^- - aX\{[\mathscr{L}S]^+\cos\theta+3[S]^+\sin\theta\}]\!] \tag{42}$$

and

$$\langle\lambda,\sigma\rangle = \frac{1}{\rho^2\sqrt{2}}[\bar\rho^2(\rho^2\lambda^*-\tfrac{1}{2}\Delta\sigma)-(\bar\rho^*)^2(\rho^2\lambda-\tfrac{1}{2}\Delta\sigma^*)] = \frac{1}{12M\sqrt{2}}[\![\{-r[\mathscr{D}P]^-+3[P]^-\}[S]^-+\mathrm{i}a[\mathscr{D}P]^+[S]^+\cos\theta]\!]. \tag{43}$$

6. Concluding remarks

Deleting the references to the 'single lacuna in the information', we can let the 'Concluding remarks' of paper III (§8) stand unchanged. Perhaps, it may be stated, from the vantage point of the present, that the Bianchi identities, the commutation relations, and six particular Ricci identities suffice to determine all the quantities and provide, in addition, ten integral identities (included in III, eqns (34), (35), and (132)–(139)). And while the manner of the final outcome was unknown, even at the stage of paper III, there has been, through the course of these papers, no circumlocution in the approach towards the complete solution: the analysis has been, almost, self-propelled.

The research reported in this paper has in part been supported by a grant from the National Science Foundation under grant PHY 78-24275 with the University of Chicago.

Appendix. The form of the solutions in the Schwarzschild limit, $a \to 0$

The form of the solutions in the Schwarzschild limit, $a \to 0$, is of some general interest; among other things, it will enable us to present the basic identities in their simplest and, in some instances, in their directly verifiable forms. Moreover, as it is manifest from some of the equations of the theory (e.g. III, eqns (26) and (30)) the passage to the limit, $a \to 0$, will require some care.

First, we observe that, in the limit $a \to 0$,

$$\left.\begin{aligned} &Q = m \operatorname{cosec}\theta, \quad K = r^2\sigma, \quad \bar{\rho} = \bar{\rho}^* = r, \quad \rho^2 = r^2, \\ &\Delta = r^2 - 2Mr, \quad \text{and} \quad \mathscr{C}_1 = \Gamma_1 = \lambda(\lambda+2). \end{aligned}\right\} \tag{A 1}$$

And from III, eqn (18), we conclude that

$$\mathscr{C}_1 - \Gamma_1 = -\frac{8\sigma\lambda m}{\lambda+2}a + O(a^2). \tag{A 2}$$

Also, the limiting forms of the coefficients, listed in I, eqns (74) and (80), are

$$\left.\begin{aligned} A_1 &= \Delta\lambda(\lambda+2) - 4\lambda r^4\sigma^2 - 24M\sigma^2 r^3, \\ A_2 &= -12\Delta M\sigma - 4\lambda\sigma r^2(2r-3M) + 24\sigma^3 r^5, \\ B_1 &= 4r^2\sigma[\lambda\Delta + 2M(r-3M) - 2r^4\sigma^2]; \end{aligned}\right\} \tag{A 3}$$

and

$$\left.\begin{aligned} \alpha_1 &= \lambda(\lambda+2) - 4\lambda m^2 \operatorname{cosec}^2\theta, \\ \alpha_2 &= 4\lambda m \cos\theta \operatorname{cosec}^2\theta, \\ \beta_1 &= 4m[2(m^2-1) - \lambda\sin^2\theta]\operatorname{cosec}^3\theta. \end{aligned}\right\} \tag{A 4}$$

Greatest interest is clearly attached to the limiting forms of the expressions for $\mathscr{R}$, $\mathscr{S}$, Z_1 and Z_2. By making use of equations (A 3) and (A 4), we find that III, eqns (27) and (28), and similarly III, eqns (32) and (33), provide, consistently with each other, the following expressions for $\mathscr{R}$ and $\mathscr{S}$:

$$\tfrac{1}{2}[\lambda\Delta + 2M(r-3M) - 2r^4\sigma^2]\mathscr{R} = \left[r - \frac{2}{\lambda+2}(r-3M)\right] r\sigma X + \left[M - \frac{2\sigma^2}{\lambda+2}r^3\right] Y, \tag{A 5}$$

and

$$\tfrac{1}{2}[\lambda\sin^2\theta - 2(m^2-1)]\mathscr{S} = \frac{m\lambda}{\lambda+2}[S]^+\cos\theta\sin\theta + \left(\frac{2m^2}{\lambda+2} - 1\right)[S]^-\sin\theta. \tag{A 6}$$

These expressions for $\mathscr{R}$ and $\mathscr{S}$ evaluate the integrals

$$\mathscr{R} = \sqrt{\Delta}\int\left(Y+\frac{2r\sigma}{\lambda+2}X\right)\frac{\mathrm{d}r}{\Delta^{\frac{3}{2}}}, \tag{A 7}$$

and

$$\mathscr{S} = \int\left\{[S]^{-}\cos\theta+\frac{2m}{\lambda+2}[S]^{+}\right\}\mathrm{d}\theta. \tag{A 8}$$

It is, as we have stated before, a striking aspect of the Newman–Penrose formalism, when applied to the theory of perturbations of the type-D space–times, that it yields novel integral relations, such as (A 7) and (A 8), among the known classical functions.

Turning next to the solutions for Z_1 and Z_2 given by equations (31) and (32), we find after some careful analysis that, in the limit, they give (omitting again the factor $1/(24M\sqrt{2})$)

$$Z_1 = -\tfrac{1}{2}\lambda(\lambda+2)\mathscr{R}\frac{2\sin\theta\cos\theta}{\lambda\sin^2\theta-2(m^2-1)}\{m[S]^{+}-[S]^{-}\cos\theta\} \tag{A 9}$$

and

$$Z_2 = \mathscr{R}\frac{2\lambda\sin\theta}{\lambda\sin^2\theta-2(m^2-1)}\{m(\lambda+1)[S]^{+}\cos\theta+[m^2-\tfrac{1}{2}(\lambda+2)(1+\cos^2\theta)][S]^{-}\}. \tag{A 10}$$

In obtaining the foregoing formulae, the following relations, valid in the limit $a\to 0$, were used:

$$m[\mathscr{L}\mathscr{L}S]^{-}\operatorname{cosec}\theta = -\sigma(\lambda\mathscr{S}-2[\mathscr{L}S]^{-}\cos\theta)a+O(a^2), \tag{A 11}$$

$$m\{[\mathscr{L}\mathscr{L}S]^{+}\cos\theta+2[\mathscr{L}S]^{+}\sin\theta\}\operatorname{cosec}\theta = -\tfrac{1}{2}\lambda(\lambda+2)\mathscr{S}, \tag{A 12}$$

$$r\sigma\{r[\mathscr{D}\mathscr{D}P]^{+}-2[\mathscr{D}P]^{+}\} = \tfrac{1}{2}\lambda(\lambda+2)\mathscr{R}, \tag{A 13}$$

and

$$\begin{aligned}\tfrac{1}{2}\lambda(\lambda+2)\mathscr{S}+m[\mathscr{L}S]^{+} &= -\tfrac{1}{4}(\lambda+2)\{-\lambda\mathscr{S}+2[\mathscr{L}S]^{-}\cos\theta\}\\ &= \frac{\lambda\sin\theta}{\lambda\sin^2\theta-2(m^2-1)}\{m(\lambda+1)[S]^{+}\cos\theta+[m^2-\tfrac{1}{2}(\lambda+2)(1+\cos^2\theta)][S]^{-}\}.\end{aligned} \tag{A 14}$$

For Z_1 and Z_2, given by equations (A 9) and (A 10), it can be directly verified that, indeed,

$$Z_1+Z_2 = \tfrac{1}{2}\lambda(\lambda+2)\mathscr{R}\mathscr{S}. \tag{A 15}$$

References

Chandrasekhar, S. 1978a *Proc. R. Soc. Lond.* A **358**, 421–439. (Paper I.)
Chandrasekhar, S. 1978b *Proc. R. Soc. Lond.* A **358**, 441–465. (Paper II.)
Chandrasekhar, S. 1979 *Proc. R. Soc. Lond.* A **365**, 425–451. (Paper III.)

3. *The Reissner-Nordström Black Hole*

14

On the equations governing the perturbations of the Reissner–Nordström black hole

By S. Chandrasekhar, F.R.S.
University of Chicago, Chicago, Illinois, 60637, U.S.A.

(*Received* 28 *August* 1978)

By considering suitable combinations of the Weyl scalars and the spin coefficients, the basic equations governing the perturbations of the Reissner–Nordström black hole, in the Newman–Penrose formalism, are decoupled; a fundamental pair of decoupled equations are obtained. It is then shown how this pair of decoupled equations can be transformed into one dimensional wave equations which are appropriate for describing the perturbations of odd and of even parity. A simple relation is obtained which will allow derivation of a solution belonging to one parity from a solution belonging to the opposite parity. Finally, equations are derived in terms of which one can readily ascertain how an arbitrary superposition of gravitational and electromagnetic waves, incident on the black hole, will be reflected and absorbed.

1. Introduction

The spherically symmetric charged black hole, described by the Riessner–Nordström metric, provides the simplest context in which one can study the general relativistic phenomenon of the conversion of gravitational energy into electromagnetic energy (and vice versa) in spacetimes described by the Einstein–Maxwell equations. For this reason the perturbations of the Reissner–Nordström black holes have been studied, among others, by Zerilli (1974), Moncrief (1974*a*, *b*, 1975), Lun (1974), Chitre (1976), Lee (1976), and Matzner (1976). But the basic theory, as currently developed, is incomplete in several respects, as we shall presently detail.

The perturbations of the black-hole solutions of general relativity† can be studied from two different stand-points and from two different bases: *either*, directly for the metric perturbations via the Einstein or the Einstein–Maxwell equations linearized about the unperturbed solution, *or* for the Weyl and the Maxwell scalars via the equations of the Newman–Penrose formalism. The theories developed from the two bases complement each other very effectively, as the study of the perturbations of the Schwarzschild metric has shown (cf. Chandrasekhar 1975; Chandrasekhar & Detweiler 1975*a*) and the present study devoted to the Reissner–Nordström metric will show.

In his study of the metric perturbations of the Reissner–Nordström black hole, Moncrief (1974*a*, *b*, 1975) was able to reduce the various governing equations to a

† There are only four of these, associated with the names of Schwarzschild, Reissner–Nordström, Kerr, and Kerr–Newman.

pair of decoupled one dimensional wave equations both for the odd and for the even parity perturbations. But it would appear from Moncrief's *separate* investigations, devoted to the stability of the Reissner–Nordström black hole for odd and for even parity perturbations (and from later investigations by others, as well), that it is not generally realized that the equations governing the odd and the even parity perturbations are not really independent, for we shall show that, as in the case of the perturbations of the Schwarzschild black hole, so also in the case of the perturbations of the Reissner–Nordström black hole, the solutions appropriate to one parity can be deduced, very simply, from the solutions appropriate to the opposite parity.

Again, while it should be expected on general grounds that the wave equations, derived from an investigation of the metric perturbations, ought to be deducible from the equations of the Newman–Penrose formalism (as was, indeed, possible in the context of the Schwarzschild metric), attempts, even, to decouple the basic equations have so far failed (cf. Chitre 1976; Lee 1976).

In this paper we shall show how the Newman–Penrose equations, governing the basic scalars and spin coefficients, can be decoupled to yield a pair of (decoupled) equations, and how these equations can be transformed to one dimensional wave equations appropriate for the odd and for the even parity perturbations by transformations which are dual (in a sense which we shall define).

2. Moncrief's equations for odd parity perturbations

It will be convenient, for future reference, to write Moncrief's equations for odd parity perturbations in the forms we shall need them. (We shall not, however, need his equations for even parity perturbations.)

Let H_1 and H_2 denote the incident electromagnetic and gravitational perturbations, with a time dependence $e^{i\sigma t}$, and belonging to spherical harmonics of order l. (These are the quantities, denoted by H and Q by Matzner (1976, equations (2.10) and (2.12).) Moncrief's equations for H_1 and H_2 are comprised in

$$(\Lambda^2 - \overline{V}) \begin{bmatrix} H_1 \\ H_2 \end{bmatrix} = \frac{\Delta}{r^5} \begin{bmatrix} 3M & 2Q_* \mu \\ 2Q_* \mu & -3M \end{bmatrix} \begin{bmatrix} H_1 \\ H_2 \end{bmatrix}, \qquad (1)^*$$

where

$$\left.\begin{aligned} \Lambda^2 &= \frac{d^2}{dr_*^2} + \sigma^2, \quad \frac{d}{dr_*} = \frac{\Delta}{r^2}\frac{d}{dr}, \\ \Delta &= r^2 - 2Mr + Q_*^2, \quad \mu^2 = (l-1)(l+2), \end{aligned}\right\} \qquad (2)$$

and

$$\overline{V} = \frac{\Delta}{r^5}\left[(\mu^2+2)r - 3M + \frac{4Q_*^2}{r}\right]. \qquad (3)$$

Also, M denotes the mass and Q_* the charge of the black hole.

* This equation was also derived, independently, by Lun (1974) from the Newman–Penrose formalism using a null basis different from Kinnersley's. But Lun did not obtain either the even parity equations or the relations between the solutions belonging to the two parities. Dr E. D. Fackerell drew my attention to Lun's paper (after the present investigation had been completed).

By letting

$$\left.\begin{aligned} Z_1 &= +[\surd(9M^2+4Q_*^2\mu^2)+3M]H_1+2Q_*\mu H_2, \\ \text{and} \qquad Z_2 &= -[\surd(9M^2+4Q_*^2\mu^2)+3M]H_2+2Q_*\mu H_1, \end{aligned}\right\} \tag{4}$$

we can decouple the equations for H_1 and H_2, to obtain the one dimensional wave equations

$$\Lambda^2 Z_i = V_i Z_i \quad (i = 1, 2), \tag{5}$$

where

$$V_i = \Delta r^{-5}[(\mu^2+2)r - q_j(1+q_i/\mu^2 r)] \quad (i,j = 1,2; i \neq j), \tag{6}$$

$$q_1 = 3M + \surd(9M^2+4Q_*^2\mu^2), \quad \text{and} \quad q_2 = 3M - \surd(9M^2+4Q_*^2\mu^2). \tag{7}$$

The convention, that i and j take the values 1 *and* 2, *but* $i \neq j$, *will be strictly adhered to in this paper; it will not be restated on every occasion.*

It is worth noting that

$$q_1 + q_2 = 6M \quad \text{and} \quad q_1 q_2 = -4Q_*^2\mu^2; \tag{8}$$

and that in the Schwarzschild limit ($Q_* = 0$), $q_1 = 6M$ and $q_2 = 0$. Also note that $q_2 < 0$.

3. The decoupling of the Newman–Penrose equations

In an appendix to an earlier paper (Chandrasekhar 1978, appendix B) the basic equations governing the perturbations of the Kerr–Newman black hole, in the Newman–Penrose formalism, were written down in a gauge in which the Maxwell scalars, ϕ_0 and ϕ_2, vanish identically. Reasons were adduced in that paper why the chosen gauge may be specially appropriate for considerations pertaining to the perturbations of charged black holes. We shall accordingly adopt this gauge for considering the perturbations of the Reissner–Nordström black hole. The relevant equations can now be written down by simply setting the Kerr parameter, a, equal to zero in the equations derived for the Kerr–Newman black hole (Chandrasekhar 1978, equations (B 4)–(B 9)). The equations are

$$\left.\begin{aligned} \mathscr{L}_2 \Phi_0 - (\mathscr{D}_0 + 3/r)\Phi_1 &= -2k(3M - 2Q_*^2/r), \\ \Delta(\mathscr{D}_2^\dagger - 3/r)\Phi_0 + \mathscr{L}_{-1}^\dagger \Phi_1 &= +2s(3M - 4Q_*^2/r), \\ (\mathscr{D}_0 + 3/r)s - \mathscr{L}_{-1}^\dagger k &= \Phi_0/r, \\ \text{and} \qquad \Delta(\mathscr{D}_2^\dagger - 3/r)k + \mathscr{L}_2 s &= 2\Phi_1/r, \end{aligned}\right\} \tag{9}$$

where

$$\left.\begin{aligned} \mathscr{D}_n &= \partial_r + \mathrm{i}(r^2\sigma/\Delta) + 2n(r-M)/\Delta, \quad \mathscr{D}_n^\dagger = \partial_r - \mathrm{i}(r^2\sigma/\Delta) + 2n(r-M)/\Delta, \\ \mathscr{L}_n &= \partial_\theta + m\operatorname{cosec}\theta + n\cot\theta, \quad \text{and} \quad \mathscr{L}_n^\dagger = \partial_\theta - m\operatorname{cosec}\theta + n\cot\theta. \end{aligned}\right\} \tag{10}$$

The variables in equations (9) can be separated by the substitutions

$$\left.\begin{aligned} \Phi_0(r,\theta) &= R_2(r)S_2(\theta), \quad \Phi_1(r,\theta) = R_1(r)S_1(\theta), \\ k(r,\theta) &= k(r)S_1(\theta), \quad \text{and} \quad s(r,\theta) = s(r)S_2(\theta), \end{aligned}\right\} \tag{11}$$

where $R_1(r)$ and $R_2(r)$ are two radial functions, and the angular functions, $S_2(\theta)$ and $S_1(\theta)$ are defined by the equations (cf. Chandrasekhar 1978, equations (36) and (51))

$$\mathscr{L}_{-1}^{\dagger}\mathscr{L}_2 S_2 = -\mu^2 S_2 \quad \text{and} \quad \mathscr{L}_2\mathscr{L}_{-1}^{\dagger} S_1 = -\mu^2 S_1. \tag{12}$$

The functions S_1 and S_2, when normalized, are further related by

$$\mathscr{L}_2 S_2 = \mu S_1 \quad \text{and} \quad \mathscr{L}_{-1}^{\dagger} S_1 = -\mu S_2. \tag{13}$$

With the substitutions (11), equations (9) become

$$\left.\begin{aligned} \mu R_2 - (\mathscr{D}_0 + 3/r) R_1 &= -2k(3M - 2Q_*^2/r), \\ \Delta(\mathscr{D}_2^{\dagger} - 3/r) R_2 - \mu R_1 &= +2s(3M - 4Q_*^2/r), \\ (\mathscr{D}_0 + 3/r) s + \mu k &= R_2/r, \\ \text{and} \qquad \Delta(\mathscr{D}_2^{\dagger} - 3/r) k + \mu s &= 2R_1/r. \end{aligned}\right\} \tag{14}$$

Now consider the following combinations of the Weyl scalars and spin coefficients:

$$\left.\begin{aligned} F_1 = R_2 + q_1 k/\mu, \quad F_2 = R_2 + q_2 k/\mu, \\ G_1 = R_1 + q_1 s/\mu, \quad \text{and} \quad G_2 = R_1 + q_2 s/\mu. \end{aligned}\right\} \tag{15}$$

The equations governing these particular combinations of the Weyl scalars and spin coefficients can be readily found from suitable combinations of equations (14). They suffice to decouple the equations; and we find

$$\Delta(\mathscr{D}_2^{\dagger} - 3/r) F_i = \mu(1 + 2q_i/\mu^2 r) G_j \tag{16}$$

and
$$(\mathscr{D}_0 + 3/r) G_i = \mu(1 + q_i/\mu^2 r) F_j \quad (i,j = 1,2; i \neq j). \tag{17}$$

Now, letting
$$F_i = r^3 Y_i/\Delta^2 \quad \text{and} \quad G_i = X_i/r^3, \tag{18}$$

we find that equations (16) and (17) become

$$\mathscr{D}_0^{\dagger} Y_i = \mu \frac{\Delta}{r^6}\left(1 + \frac{2q_i}{\mu^2 r}\right) X_j, \tag{19}$$

and
$$\mathscr{D}_0 X_j = \mu \frac{r^6}{\Delta}\left(1 + \frac{q_j}{\mu^2 r}\right) Y_i \quad (i,j = 1,2; i \neq j). \tag{20}$$

Since
$$\frac{\Delta}{r^2}\mathscr{D}_0 = \frac{d}{dr_*} + i\sigma = \Lambda_+ \quad \text{and} \quad \frac{\Delta}{r^2}\mathscr{D}_0^{\dagger} = \frac{d}{dr_*} - i\sigma = \Lambda_-, \tag{21}$$

we can rewrite equations (19) and (20) in the forms

$$\Lambda_- Y_i = \mu \frac{\Delta^2}{r^8}\left(1 + \frac{2q_i}{\mu^2 r}\right) X_j, \tag{22}$$

and
$$\Lambda_+ X_j = \mu \frac{r^4}{\Delta}\left(1 + \frac{q_j}{\mu^2 r}\right) Y_i. \tag{23}$$

Finally, eliminating the X's in favour of the Y's, we obtain the pair of basic equations

$$\Lambda^2 Y_i + P_i \Lambda_- Y_i - Q_i Y_i = 0 \quad (i = 1, 2), \tag{24}$$

where

$$P_i = \frac{\mathrm{d}}{\mathrm{d}r_*} \lg \left(\frac{r^8}{D_i}\right); \quad D_i = \Delta^2 \left(1 + \frac{2q_i}{\mu^2 r}\right), \tag{25}$$

and

$$Q_i = \mu^2 \frac{\Delta}{r^4} \left(1 + \frac{2q_i}{\mu^2 r}\right) \left(1 + \frac{q_j}{\mu^2 r}\right) \quad (i, j = 1, 2; i \neq j). \tag{26}$$

4. The transformation of equation (24) to the form of a one dimensional wave equation

Our aim is to transform equation (24) to the form

$$\Lambda^2 Z_i = V_i Z_i, \tag{27†}$$

where V_i is some potential function.

Equation (24) is of the standard form that has recently been considered in several papers (e.g. Chandrasekhar & Detweiler 1975*b*); and the methods developed in those papers for transforming such equations to wave equations of the form (27), are readily available.

We assume, then, that Y_i is related to Z_i in the manner

$$Y_i = f_i V_i Z_i + (W_i + 2i\sigma f_i) \Lambda_+ Z_i, \quad (i = 1, 2), \tag{28}$$

where f_i, V_i, and W_i are certain functions unspecified for the present. Applying the operator Λ_- to equation (28) and making use of the fact that Z_i has been assumed to satisfy equation (27), we find that

$$\Lambda_- Y_i = -r^{-8} D_i \beta_i Z_i + R_i \Lambda_+ Z_i, \tag{29}$$

where

$$-r^{-8} D_i \beta_i = \mathrm{d}(f_i V_i)/\mathrm{d}r_* + W_i V_i, \tag{30}$$

and

$$R_i = f_i V_i + \mathrm{d}(W_i + 2i\sigma f_i)/\mathrm{d}r_*. \tag{31}$$

The requirement that equations (28) and (29) are compatible with equations (24) satisfied by Y_i, leads to the pair of equations

$$\frac{\mathrm{d}}{\mathrm{d}r_*}\left(\frac{r^8}{D_i} R_i\right) = \frac{r^8}{D_i} [Q_i (W_i + 2i\sigma f_i) - 2i\sigma R_i] + \beta_i, \tag{32}$$

and

$$-\frac{D_i}{r^8} \frac{\mathrm{d}\beta_i}{\mathrm{d}r_*} = (Q_i f_i - R_i) V_i. \tag{33}$$

It can be verified that equations (30)–(33) allow the integral

$$r^8 R_i f_i V_i / D_i + \beta_i (W_i + 2i\sigma f_i) = K_i = \text{constant}. \tag{34}$$

† The subscript i is superfluous in this section. But we are retaining it to avoid any subsequent ambiguity and also to emphasize that we are, in fact, dealing simultaneously, with two equations.

Accordingly, it will suffice to consider any four of the five equations (30)–(34). (For more details concerning the derivation of the foregoing equations, see Chandrasekhar & Detweiler 1975*b*, §3.)

5. The transformation of equation (24) to the wave equation governing the odd parity perturbations

In the context of the theory of the perturbations of the Schwarzschild metric it is known that the transformation equations of §4 are satisfied consistently with the assumptions β = constant and $f = 1$ (cf. Chandrasekhar 1975). We shall now show that the same assumptions also lead to consistent solutions in the case we are considering.

We ask then whether equations (30)–(34) will allow solutions consistently with the assumptions

$$\beta_i = \text{constant}, \quad \text{and} \quad f_i = 1, \tag{35}$$

and for the potentials

$$V_i = r^{-5}\Delta[(\mu^2+2)r - q_j(1+q_i/\mu^2 r)] \quad (i,j = 1,2; i \neq j), \tag{36}$$

appropriate for odd parity perturbations (cf. equation (6)).

We first observe that the assumptions (35), in conjunction with (33), require that (cf. equation (26))

$$R_i = Q_i = \mu^2 \frac{\Delta}{r^4}\left(1+\frac{2q_i}{\mu^2 r}\right)\left(1+\frac{q_j}{\mu^2 r}\right) \quad (i,j = 1,2; i \neq j). \tag{37}$$

From (31), (36), and (37) it now follows that

$$\frac{\mathrm{d}W_i}{\mathrm{d}r_*} = Q_i - V_i = \frac{\Delta}{r^5}\left[-2r + 2(q_1+q_2) + \frac{3q_1 q_2}{\mu^2 r}\right]; \tag{38}$$

or, making use of (2) and (8), we obtain

$$\frac{\mathrm{d}W_i}{\mathrm{d}r} = -\frac{2}{r^2} + \frac{12M}{r^3} - \frac{12Q_*^2}{r^4}. \tag{39}$$

The required solution of this equation is

$$W_i = W = 2r^{-2}(r - 3M + 2Q_*^2/r). \tag{40}$$

(Notice that this solution is the same for $i = 1$ and 2.)

We next verify that when the expressions (36) and (40) for V_i and W_i are substituted in equation (30), we find, consistently with our assumptions, that

$$\beta_i = -q_j = \text{constant} \quad (i \neq j = 1 \text{ or } 2). \tag{41}$$

We complete the verification of the entire set of the transformation equations by observing that, with the expressions (36) and (40) for V_i and W_i, (34) yields for the quantity on the left hand side the constant value

$$K_i = \mu^2(\mu^2+2) - 2\mathrm{i}\sigma q_j. \tag{42}$$

We conclude that (24) can be transformed to Moncrief's one dimensional wave equations, governing the odd parity perturbations, by the substitutions

$$Y_i = V_i^{(-)} Z_i^{(-)} + (W^{(-)} + 2i\sigma) \Lambda_+ Z_i^{(-)}, \tag{43}$$

$$\Lambda_- Y_i = r^{-8} D_i q_j Z_i^{(-)} + Q_i \Lambda_+ Z_i^{(-)}, \tag{44}$$

$$V_i^{(-)} = r^{-5}\Delta[(\mu^2+2) r - q_j(1+q_i/\mu^2 r)], \tag{45}$$

and

$$W^{(-)} = 2r^{-2}(r - 3M + 2Q_*^2/r), \tag{46}$$

where, we have now distinguished by the superscript $(-)$ the various quantities pertaining to the odd parity perturbations.

The inverse of the relations (43) and (44) are

$$r^{-8} D_i K_i^{(-)} Z_i^{(-)} = Q_i Y_i - (W^{(-)} + 2i\sigma) \Lambda_- Y_i, \tag{47}$$

and

$$r^{-8} D_i K_i^{(-)} \Lambda_+ Z_i^{(-)} = -q_j r^{-8} D_i Y_i + V_i^{(-)} \Lambda_- Y_i, \tag{48}$$

where

$$K_i^{(-)} = \mu^2(\mu^2+2) - 2i\sigma q_j. \tag{49}$$

6. The dual transformation and the wave equations governing the even parity perturbations

It is shown in the appendix that, *if the transformation equations of* § 4 *are satisfied, consistently with the assumptions* β = *constant and* $f = 1$, *then they are equally satisfied with the assumptions* $\tilde{\beta} = -\beta$ *and* $f = 1$. The transformations which follow from the latter assumptions are *dual* to the original in the sense recently defined by Heading (1977).

We shall now show that the transformations dual to the ones considered in § 5 lead to the wave equations appropriate for the even parity perturbations.

We assume then that (cf. equation (41))

$$\beta_i^{(+)} = +q_j \quad \text{and} \quad f_i^{(+)} = 1, \tag{50}$$

where we are now distinguishing the quantities, pertaining to the dual transformations, by the superscript $(+)$.

Now, letting (cf. equation (A 5))

$$F_i = \frac{r^8}{D_i} Q_i = \mu^2 \frac{r^4}{\Delta}\left(1 + \frac{q_j}{\mu^2 r}\right) \quad (i, j = 1, 2; i \neq j), \tag{51}$$

we deduce from (A 6) that

$$W_i^{(+)} = \frac{d}{dr_*} \lg F_i - \frac{\beta_i^{(+)}}{F_i} = \frac{d}{dr_*} \lg F_i - \frac{q_j}{F_i}. \tag{52}$$

Since (as may be readily verified),

$$\frac{d}{dr_*}\lg F_i = \frac{2}{r^2}\left(r-3M+\frac{2Q_*^2}{r}\right)-\frac{q_j\Delta}{r^4(\mu^2+q_j/r)},$$

$$= W^{(-)}-\frac{q_j\Delta}{r^4(\mu^2+q_j/r)}, \tag{53}$$

it follows that

$$W_i^{(+)} = W^{(-)}-2q_j\Delta/r^4(\mu^2+q_j/r). \tag{54}$$

The corresponding expression for the potential $V_i^{(+)}$ is given by (cf. equation (38))

$$V_i^{(+)} = Q_i-\frac{dW_i^{(+)}}{dr_*} = Q_i-\frac{dW^{(-)}}{dr_*}+2q_j\frac{d}{dr_*}\left(\frac{\Delta}{r^4(\mu^2+q_j/r)}\right), \tag{55}$$

or

$$V_i^{(+)} = V_i^{(-)}+2q_j\frac{d}{dr_*}\left(\frac{\Delta}{r^4(\mu^2+q_j/r)}\right). \tag{56}$$

The equality of the integrals,

$$\int_{-\infty}^{+\infty} V_i^{(+)}\,dr_* = \int_{-\infty}^{+\infty} V_i^{(-)}\,dr_*, \tag{57}$$

which is manifest from (56), is a necessary condition for the fact (which we shall prove in §8 below) that the two potentials, $V_i^{(+)}$ and $V_i^{(-)}$, yield the same reflexion and transmission coefficients.

An equivalent form of the potential $V_i^{(+)}$ is

$$V_i^{(+)} = V_i^{(-)}+\frac{2q_j\Delta}{r^4(\mu^2+q_j/r)}\left[\frac{q_j\Delta}{r^4(\mu^2+q_j/r)}-W^{(-)}\right]. \tag{58}$$

It can be verified that this expression for $V_i^{(+)}$ agrees with what Moncrief (1974*b*, eqns (13) and (19)) has obtained from his study of the even parity metric perturbations.

Finally, we may note that in agreement with (A 4) we find (cf. equation (49))

$$r^8Q_i V_i^{(+)}/D_i+q_j(W_i^{(+)}+2i\sigma) = K_i^{(+)} = \mu^2(\mu^2+2)+2i\sigma q_j. \tag{59}$$

The transformation equations, appropriate for the even parity perturbations (that are analogous to equations (43), (44), (47), and (48), for the odd parity perturbations), are

$$Y_i = V_i^{(+)}Z_i^{(+)}+(W_i^{(+)}+2i\sigma)\Lambda_+Z_i^{(+)}, \tag{60}$$

$$\Lambda_-Y_i = -r^{-8}D_iq_jZ_i^{(+)}+Q_i\Lambda_+Z_i^{(+)}, \tag{61}$$

$$r^{-8}D_iK_i^{(+)}Z_i^{(+)} = Q_iY_i-(W_i^{(+)}+2i\sigma)\Lambda_-Y_i, \tag{62}$$

and

$$r^{-8}D_iK_i^{(+)}\Lambda_+Z_i^{(+)} = +r^{-8}q_jD_iY_i+V_i^{(+)}\Lambda_-Y_i. \tag{63}$$

7. An explicit relation between the solutions for $Z_i^{(-)}$ and $Z_i^{(+)}$

We have seen how equation (24) for Y_i $(i = 1, 2)$ can be transformed to the one dimensional wave equations

$$\Lambda^2 Z_i^{(\pm)} = V_i^{(\pm)} Z_i^{(\pm)} \quad (i = 1, 2), \tag{64}$$

where $V_i^{(\pm)}$ are given by equations (45) and (58). All four of these potentials are real and are of *short range*: their integrals over the range of r_*, $(-\infty < r_* < +\infty)$, are finite. Moreover, the potentials tend to zero exponentially as $r_x \to -\infty$, and tend to zero like r_*^{-2} as $r_* \to +\infty$.

With the aid of the transformation equations (43), (44), (47), (48), and (60)–(63) we shall now show that there is a simple relation which enables us to obtain a solution $Z_i^{(-)}$ of odd parity from a solution $Z_i^{(+)}$ of even parity (and conversely).

Let $Z_i^{(-)}$ $(i = 1, 2)$ denote solutions of the wave equations appropriate for odd parity. Then making use of (47), relating $Z_i^{(-)}$ to Y_i and $\Lambda_- Y_i$, and then using (60) and (61), relating Y_i and $\Lambda_- Y_i$ to $Z_i^{(+)}$, we obtain

$$\begin{aligned} r^{-8} D_i K_i^{(-)} Z_i^{(-)} &= Q_i Y_i - (W^{(-)} + 2i\sigma) \Lambda_- Y_i \\ &= Q_i [V_i^{(+)} Z_i^{(+)} + (W_i^{(+)} + 2i\sigma) \Lambda_+ Z_i^{(+)}] \\ &\qquad - (W^{(-)} + 2i\sigma)[-r^{-8} D_i q_j Z_i^{(+)} + Q_i \Lambda_+ Z_i^{(+)}], \end{aligned} \tag{65}$$

or, after some regrouping of the terms, we have

$$\begin{aligned} r^{-8} D_i K_i^{(-)} Z_i^{(-)} &= r^{-8} D_i [r^8 Q_i V_i^{(+)}/D_i + q_j (W^{(-)} + 2i\sigma)] Z_i^{(+)} \\ &\qquad + Q_i (W_i^{(+)} - W^{(-)}) \Lambda_+ Z_i^{(+)}. \end{aligned} \tag{66}$$

The expression on the right hand side of this equation can be simplified considerably by making use of (42), (54), and (59). We find

$$[\mu^2(\mu^2+2) - 2i\sigma q_j] Z_i^{(-)} = \left[\mu^2(\mu^2+2) + \frac{2q_j^2 \Delta}{r^4(\mu^2 + q_j/r)}\right] Z_i^{(+)} - 2q_j \frac{dZ_i^{(+)}}{dr_*}. \tag{67}$$

Using equation (67) we can derive a solution $Z_i^{(-)}$ (appropriate for describing odd parity perturbations) from a solution $Z_i^{(+)}$ (appropriate for even parity perturbations). It is difficult to see how this relation between $Z_i^{(-)}$ and $Z_i^{(+)}$ could have been found except via the Newman–Penrose equations.

Since the second term in square brackets on the right hand side vanishes both on the horizon ($r_* \to -\infty$ and $\Delta = 0$) and at infinity ($r_* \to +\infty$), it follows that

$$[\mu^2(\mu^2+2) - 2i\sigma q_j] Z_i^{(-)} \to \mu^2(\mu^2+2) Z_i^{(+)} - 2q_j \, dZ_i^{(+)}/dr_* \; (r_* \to \pm\infty). \tag{68}$$

In particular the solutions $Z_i^{(-)}$, derived from solutions $Z_i^{(+)}$ (in accordance with equations (67)), having the asymptotic behaviours

$$Z_i^{(+)} \to e^{+i\sigma r_*} \quad \text{and} \quad Z_i^{(+)} \to e^{-i\sigma r_*} \quad \text{as} \quad r_* \to \pm\infty, \tag{69}$$

have, respectively, the asymptotic behaviours

$$Z_i^{(-)} \to e^{+i\sigma r_*} \quad \text{and} \quad Z_i^{(-)} \to \frac{\mu^2(\mu^2+2) + 2i\sigma q_j}{\mu^2(\mu^2+2) - 2i\sigma q_j} e^{-i\sigma r_*}. \tag{70}$$

8. The reflexion and transmission of incident gravitational and electromagnetic waves

Since the potentials $V_i^{(\pm)}$ that occur in the one dimensional wave equation for $Z_i^{(\pm)}$ are of short range, we can seek solutions of the wave equations which have the asymptotic behaviours

$$\begin{aligned} Z_i^{(\pm)} &\rightarrow e^{+i\sigma r_*} + A_i^{(\pm)} e^{-i\sigma r_*} \quad \text{as} \quad r_* \rightarrow +\infty, \\ &\rightarrow B_i^{(\pm)} e^{+i\sigma r_*} \quad \text{as} \quad r_* \rightarrow -\infty, \end{aligned} \tag{71}$$

and since the potentials are real, we have

$$|A_i^{(\pm)}|^2 + |B_i^{(\pm)}|^2 = 1. \tag{72}$$

From equations (69) and (70), relating the asymptotic behaviours of the solutions of odd and even parity, it follows that

$$A_i^{(+)} = A_i^{(-)} e^{i\Delta_i} \quad \text{and} \quad B_i^{(+)} = B_i^{(-)}, \tag{73}$$

where

$$e^{i\Delta_i} = \frac{\mu^2(\mu^2+2) - 2i\sigma q_j}{\mu^2(\mu^2+2) + 2i\sigma q_j}; \tag{74}$$

and in particular,

$$|A_i^{(+)}|^2 = |A_i^{(-)}|^2 \quad \text{and} \quad |B_i^{(+)}|^2 = |B_i^{(-)}|^2. \tag{75}$$

Thus, the reflexion ($\mathbb{R}_i$) and the transmission ($\mathbb{T}_i$) coefficients are the same for the odd and the even parity solutions, $Z_i^{(\pm)}$. Also, while there is no difference in the relative phases of the transmitted amplitudes, there is a difference in the relative phases of the reflected amplitudes which is given by (74).

In view of the relations (73) and (75), we can write the asymptotic behaviours of the solutions for $Z_1^{(\pm)}$ and $Z_2^{(\pm)}$ in the following forms:

$$\left.\begin{aligned} Z_1^{(\pm)} &\rightarrow \exp(+i\sigma r_*) + \mathbb{R}_1^{\frac{1}{2}} \exp(i\delta_1^{r,\pm} - i\sigma r_*) \quad \text{as} \quad r_* \rightarrow +\infty, \\ &\rightarrow \mathbb{T}_1^{\frac{1}{2}} \exp(i\delta_1^{t} + i\sigma r_*) \quad \text{as} \quad r_* \rightarrow -\infty, \\ \text{and} \quad Z_2^{(\pm)} &\rightarrow \exp(+i\sigma r_*) + \mathbb{R}_2^{\frac{1}{2}} \exp(i\delta_2^{r,\pm} - i\sigma r_*) \quad \text{as} \quad r_* \rightarrow +\infty, \\ &\rightarrow \mathbb{T}_2^{\frac{1}{2}} \exp(i\delta_2^{t} + i\sigma r_*) \quad \text{as} \quad r_* \rightarrow -\infty, \end{aligned}\right\} \tag{76}$$

where $\mathbb{R}_i$ and $\mathbb{T}_i$ $(i = 1, 2)$ denote the (common) reflexion and transmission coefficients for either $Z_i^{(+)}$ or $Z_i^{(-)}$. It will be noticed that in accordance with equations (73) we have not distinguished the phases δ_i^t of the transmitted amplitudes by '$(\pm)$', since they are the same for both parities. But we have distinguished the phases of the reflected amplitudes by '$(\pm)$'; their difference is, in fact, given by

$$\exp i(\delta_i^{r,+} - \delta_i^{r,-}) = \frac{\mu^2(\mu^2+2) - 2i\sigma q_j}{\mu^2(\mu^2+2) + 2i\sigma q_j}, \quad (i, j = 1, 2; i \neq j). \tag{77}$$

We now turn to the physical problem of how a superposition of gravitational and electromagnetic waves incident on the Reissner–Nordström black hole will be reflected and transmitted. For the solution of this problem, we need to know how the amplitudes of the incident waves of the two kinds are to be related to $Z_i^{(\pm)}$. The required relations have been given by Olson & Unruh (1974) and by Matzner (1976). According to these authors, the amplitudes, $H_1^{(\pm)}$ and $H_2^{(\pm)}$, of the electromagnetic and the gravitational perturbations, are related to $Z_i^{(\pm)}$ by the equations

$$\left.\begin{aligned} H_2^{(\pm)} &= Z_2^{(\pm)}\cos\psi + Z_1^{(\pm)}\sin\psi, \quad Z_2^{(\pm)} = H_2^{(\pm)}\cos\psi - H_1^{(\pm)}\sin\psi,\\ H_1^{(\pm)} &= Z_1^{(\pm)}\cos\psi - Z_2^{(\pm)}\sin\psi, \quad Z_1^{(\pm)} = H_2^{(\pm)}\sin\psi + H_1^{(\pm)}\cos\psi,\end{aligned}\right\} \tag{78}$$

where
$$\sin(2\psi) = \mp 2\sqrt{\frac{-q_1 q_2}{(q_1-q_2)^2}} = \mp\frac{2Q_*\mu}{\sqrt{(9M^2+4Q_*^2\mu^2)}}. \tag{79}$$

The manner in which a given superposition of gravitational and electromagnetic waves incident on the black hole will be reflected and transmitted can be readily ascertained with the aid of equations (76) and (78). Thus, let $H_1^{(\text{inc})}$ and $H_2^{(\text{inc})}$ be the amplitudes of the incident electromagnetic and gravitational waves of an assigned parity. (We are, for the present, suppressing the distinguishing superscript '(±)'.) In accordance with equations (8) we may now write

$$\left.\begin{aligned} H_1^{(\text{inc})} &= |Z_1|\,e^{i\sigma r_*}\cos\psi - |Z_2|\,e^{i\sigma r_*}\sin\psi,\\ \text{and}\qquad H_2^{(\text{inc})} &= |Z_2|\,e^{i\sigma r_*}\cos\psi + |Z_1|\,e^{i\sigma r_*}\sin\psi.\end{aligned}\right\} \tag{80}$$

By equations (76) the corresponding expressions for the reflected amplitudes are

$$H_1^{(\text{ref})} = \mathbb{R}_1^{\frac{1}{2}}(|Z_1|\cos\psi)\exp(i\delta_1^r - i\sigma r_*) - \mathbb{R}_2^{\frac{1}{2}}(|Z_2|\sin\psi)\exp(i\delta_2^r - i\sigma r_*) \tag{81}$$

and
$$H_2^{(\text{ref})} = \mathbb{R}_2^{\frac{1}{2}}(|Z_2|\cos\psi)\exp(i\delta_2^r - i\sigma r_*) + \mathbb{R}_1^{\frac{1}{2}}(|Z_1|\sin\psi)\exp(i\delta_1^r - i\sigma r_*). \tag{82}$$

From these expressions it follows that the fluxes of energy in the reflected waves are given by

$$|H_1^{(\text{ref})}|^2 = \mathbb{R}_1|Z_1|^2\cos^2\psi + \mathbb{R}_2|Z_2|^2\sin^2\psi - (\mathbb{R}_1\mathbb{R}_2)^{\frac{1}{2}}|Z_1|\,|Z_2|\sin(2\psi)\cos(\Delta_{12}^r) \tag{83}$$

and

$$|H_2^{(\text{ref})}|^2 = \mathbb{R}_2|Z_2|^2\cos^2\psi + \mathbb{R}_1|Z_1|^2\sin^2\psi + (\mathbb{R}_1\mathbb{R}_2)^{\frac{1}{2}}|Z_1|\,|Z_2|\sin(2\psi)\cos(\Delta_{12}^r), \tag{84}$$

where
$$\Delta_{12}^r = \delta_1^r - \delta_2^r. \tag{85}$$

The corresponding expressions for the transmitted fluxes are

$$|H_1^{(\text{trans})}|^2 = \mathbb{T}_1|Z_1|^2\cos^2\psi + \mathbb{T}_2|Z_2|^2\sin^2\psi - (\mathbb{T}_1\mathbb{T}_2)^{\frac{1}{2}}|Z_1|\,|Z_2|\sin(2\psi)\cos(\Delta_{12}^t) \tag{86}$$

and

$$|H_2^{(\text{trans})}|^2 = \mathbb{T}_2|Z_2|^2\cos^2\psi + \mathbb{T}_1|Z_1|^2\sin^2\psi + (\mathbb{T}_1\mathbb{T}_2)^{\frac{1}{2}}|Z_1|\,|Z_2|\sin(2\psi)\cos(\Delta_{12}^t), \tag{87}$$

where
$$\Delta_{12}^t = \delta_1^t - \delta_2^t. \tag{88}$$

The above expressions are, of course, consistent with the conservation of total energy. Thus

$$\left.\begin{aligned} &|H_2^{(\text{ref})}|^2 + |H_1^{(\text{ref})}|^2 = \mathbb{R}_2|Z_2|^2 + \mathbb{R}_1|Z_1|^2,\\ &|H_2^{(\text{trans})}|^2 + |H_1^{(\text{trans})}|^2 = \mathbb{T}_2|Z_2|^2 + \mathbb{T}_1|Z_1|^2\\ \text{and}\quad &|H_2^{(\text{ref})}|^2 + |H_2^{(\text{trans})}|^2 + |H_1^{(\text{ref})}|^2 + |H_1^{(\text{trans})}|^2 = |Z_1|^2 + |Z_2|^2.\end{aligned}\right\} \tag{89}$$

It is evident from our earlier remarks that the expressions (86) and (87), for the transmitted fluxes, are the same for both parities. But the reflected fluxes do depend on parity by virtue of (77).

A question of particular interest is when the incident flux of energy is entirely of one kind (gravitational or electromagnetic) and we ask: What fraction of the incident energy is reflected in the other kind?

Consider the case when the incident energy is entirely gravitational so that $H_1^{(\mathrm{inc})} = 0$. We may then write (cf. equations (78))

$$Z_2^{(\mathrm{inc})} = |Z| \cos \psi \quad \text{and} \quad Z_1^{(\mathrm{inc})} = |Z| \sin \psi. \tag{90}$$

Then, by equation (83) the fraction of the incident energy that will be reflected as electromagnetic waves is given by

$$|H_1^{(\mathrm{ref})}|^2/|Z|^2 = [\mathbb{R}_2 + \mathbb{R}_1 - 2(\mathbb{R}_1 \mathbb{R}_2)^{\frac{1}{2}} \cos (\Delta_{12}^{\mathrm{r};\pm})] \sin^2 \psi \cos^2 \psi, \tag{91}$$

where we have restored the distinguishing superscript '(±)' to emphasize that this *conversion factor* does depend on parity.

From the symmetry of equation (91) in the indices 1 and 2, it follows that *the conversion factor is the same whether the incident energy is entirely gravitational or entirely electromagnetic*. (This consequence of the relations derived in this section was first noted by Gunter (1979) who, in a separate paper, has examined in detail the various aspects of the reflexion and the transmission of waves incident on the Reissner–Nordström black hole.)

9. Concluding remarks

It is noteworthy that the treatments of the perturbations of the Schwarzschild and the Reissner–Nordström black holes, on the Newman–Penrose formalism, are so closely parallel. In a later paper we shall present a treatment of the metric perturbations of the Reissner–Nordström black hole which will be similarly parallel to the treatment of the metric perturbations of the Schwarzschild black hole given earlier (Chandrasekhar (1975).

I am indebted to Mr B. Xanthopoulos for many useful discussions. The research reported in this paper has in part been supported by the National Science Foundation under grant PHY 76-81102 A01 with the University of Chicago.

Appendix. On dual transformations

We shall consider, quite generally, when the transformation equations (30)–(34) will allow solutions consistently with the assumptions

$$\beta = \text{constant} \quad \text{and} \quad f = 1. \tag{A 1}$$

(We are suppressing the subscript i as superfluous in the present context.)

The assumptions (A 1), in conjunction with the equations (33), (31), (32), and (34) successively, require

$$R = Q; \quad V = Q - \mathrm{d}W/\mathrm{d}r_*, \tag{A 2}$$

$$\mathrm{d}(r^8 Q/D)/\mathrm{d}r_* = (r^8 QW/D) + \beta, \tag{A 3}$$

and

$$r^8 QV/D + \beta W = K - 2\mathrm{i}\sigma\beta = \kappa \quad \text{(say)}. \tag{A 4}$$

(Note that κ is to be a constant.)

With the definition

$$F = r^8 Q/D, \tag{A 5}$$

equations (A 3) and (A 4) give

$$W = \frac{1}{F}\frac{\mathrm{d}F}{\mathrm{d}r_*} - \frac{\beta}{F}, \tag{A 6}$$

and

$$FV + \beta W = \kappa. \tag{A 7}$$

From equations (A 2) and (A 7) it now follows that

$$F(Q - \mathrm{d}W/\mathrm{d}r_*) + \beta W = \kappa. \tag{A 8}$$

Finally, eliminating W from equation (A 8) with the aid of equation (A 6), we obtain the basic condition

$$\frac{1}{F}\left(\frac{\mathrm{d}F}{\mathrm{d}r_*}\right)^2 - \frac{\mathrm{d}^2 F}{\mathrm{d}r_*^2} + \frac{D}{r^8}F^2 = \frac{\beta^2}{F} + \kappa. \tag{A 9}$$

Thus a necessary and sufficient condition for the transformation equations (30)–(34) to allow solutions consistently with the assumptions (A 1) is that equation (A 9) be satisfied by the *given* F for some suitably chosen constants β and κ. On the other hand, since β occurs as β^2 in equation (A 9), it follows that if equation (A 9) is satisfied for some values of the constants β^2 and κ then two transformations associated with $+\beta$ and $-\beta$ (but the same κ) are possible; and they lead to the dual transformations of Heading (1977).

It is apparent from equations (A 2) and (A 8) that the potentials derived from dual transformations are simply related (as we have seen in detail in the particular example considered in the text).

References

Chandrasekhar, S. 1975 *Proc. R. Soc. Lond.* A **343**, 289.
Chandrasekhar, S. 1978 *Proc. R. Soc. Lond.* A **358**, 421.
Chandrasekhar, S. & Detweiler, S. 1975*a* *Proc. R. Soc. Lond.* A **344**, 441.
Chandrasekhar, S. & Detweiler, S. 1975*b* *Proc. R. Soc. Lond.* A **345**, 145.
Chitre, D. M. 1976 *Phys. Rev.* D13, 2713.
Gunter, D. 1979 (in preparation).
Heading, J. 1977 *J. Phys.* A **10**, 885.
Lee, C. H. 1976 *J. Math. Phys.* **17**, 1226.
Lun, A. W. C. 1974 *Nuovo Cim. Lett.* **10**, 681.
Matzner, R. 1976 *Phys. Rev.* D**14**, 3724.
Moncrief, V. 1974*a* *Phys. Rev.* D**9**, 2707.
Moncrief, V. 1974*b* *Phys. Rev.* D**10**, 1057.
Moncrief, V. 1975 *Phys. Rev.* D**12**, 1526.
Olson, D. W. & Unruh, W. G. 1974 *Phys. Rev. Lett.* **33**, 1116.
Zerilli, F. J. 1973 *Phys. Rev.* D**9**, 860.

15

On the metric perturbations of the Reissner–Nordström black hole

By S. Chandrasekhar, F.R.S. and B. C. Xanthopoulos
University of Chicago, Chicago, Illinois 60637, U.S.A.

(*Received* 31 *October* 1978)

The two pairs of one-dimensional wave equations which govern the odd and the even-parity perturbations of the Reissner–Nordström black hole are derived directly from a treatment of its metric perturbations. The treatment closely parallels the corresponding treatment in the context of the Schwarzschild black hole.

1. Introduction

In a recent paper (Chandrasekhar 1979) the equations governing the perturbations of the Reissner–Nordström black hole were considered in the framework of the Newman–Penrose formalism; it was shown how the relevant equations can be reduced to either of two pairs of one-dimensional wave equations (appropriate for the two parities) by transformations that exactly parallel those which accomplish a similar reduction in the context of the perturbations of the Schwarzschild black hole. In this paper, we shall present a treatment of the metric perturbations of the Reissner–Nordström black hole which similarly parallels the treatment of the metric perturbations of the Schwarzschild black hole (given in Chandrasekhar 1975; this paper will be referred to hereafter as Sch.). The directness and the conceptual simplicity of the present treatment should be contrasted with the complexity of the earlier treatments by Zerilli (1973) and by Moncrief (1974*a*, *b*, 1975).

2. The statement of the problem

The Reissner–Nordström metric written in its standard form is

$$ds^2 = -e^{2\nu}(dt)^2 + r^2 \sin^2\theta (d\varphi)^2 + e^{2\lambda}(dr)^2 + r^2(d\theta)^2, \tag{1}$$

where

$$e^{2\nu} = e^{-2\lambda} = 1 - (2m/r) + (Q_*^2/r^2), \tag{2}$$

and the only non-vanishing component of the antisymmetric Maxwell-tensor is

$$-F_{tr} = F_{rt} = Q_*/r^2. \tag{3}$$

In the foregoing equations m denotes the mass and Q_* the charge of the black hole.

In considering the perturbations of this spherically symmetric space time, we can, without loss of generality,† restrict outselves to time-dependent axisymmetric

† Since solutions with a dependence $e^{im\varphi}$ (where m is an integer, positive or negative) can be obtained by a rotation from those with $m = 0$.

perturbations. And as has been shown (Chandrasekhar & Friedman 1972) a form of the metric adequate for treating time-dependent axisymmetric systems in general relativity is

$$ds^2 = -e^{2\nu}(dt)^2 + e^{2\psi}(d\varphi - \omega\, dt - q_2\, dx^2 - q_3\, dx^3)^2 + e^{2\mu_2}(dx^2)^2 + e^{2\mu_3}(dx^3)^2, \tag{4}$$

where ν, ψ, μ_2, μ_3, ω, q_2, and q_3 are functions of $t(=x^0)$ and the two spatial coordinates x^2 and x^3 but independent of the azimuthal angle $\varphi(=x^1)$.

With the identifications,

$$x^1 = \varphi, \quad x^2 = r \quad \text{and} \quad x^3 = \theta, \tag{5}$$

the Reissner–Nordström metric corresponds to the various metric coefficients having the values

$$e^{2\nu} = e^{-2\mu_2} = 1 - (2m/r) + (Q_*^2/r^2), \quad e^{\mu_2} = e^{\lambda}, \quad e^{\mu_3} = r, \quad e^{\psi} = r \sin\theta, \tag{6}$$

and

$$\omega = q_2 = q_3 = 0. \tag{7}$$

Also,

$$F_{02} = -F_{20} = -Q_*/r^2 \quad \text{and} \quad F_{ij} = 0 \quad (ij \neq 02 \text{ and } 20). \tag{8}$$

When the Reissner–Nordström solution is perturbed, we shall write

$$\nu + \delta\nu, \quad \lambda + \delta\lambda, \quad \psi + \delta\psi, \quad \mu_3 + \delta\mu_3 \quad \text{and} \quad F_{02} + \delta F_{02} \tag{9}$$

for the corresponding metric functions and the $(0, 2)$-component of the Maxwell tensor. Besides, ω, q_2, q_3 and $F_{ij}(ij \neq 02$ or $20)$ need no longer be zero; and since these quantities vanish in the background geometry, we shall let the same symbols stand for the perturbations (i.e. without distinguishing them by a prefix 'δ'). Equations governing these perturbations can be readily written down by linearizing the field equations, valid for the metric (4), about the static solution given by equations (6)–(8).

Expressions for the various components of the Ricci and the Einstein tensor for a metric of the form (4) are given in Chandrasekhar & Friedman (1972). In the present context, we shall need, in addition, the explicit forms of Maxwell's equations for the same metric. We shall provide these equations in § 3 and their linearized versions about the Reissner–Nordström solution in § 4.

3. Maxwell's equations in a non-stationary axisymmetric space time

Perhaps the simplest way of obtaining the explicit forms of Maxwell's equations in a space time with the metric (4) is to use a standard tetrad formalism with the basis,

$$e_{(a)i} = \begin{vmatrix} -e^{+\nu} & 0 & 0 & 0 \\ -\omega e^{\psi} & e^{\psi} & -q_2 e^{\psi} & -q_3 e^{\psi} \\ 0 & 0 & e^{\mu_2} & 0 \\ 0 & 0 & 0 & e^{\mu_3} \end{vmatrix}, \tag{10}$$

and
$$e^i_{(a)} = \begin{vmatrix} e^{-\nu} & \omega e^{-\nu} & 0 & 0 \\ 0 & e^{-\psi} & 0 & 0 \\ 0 & q_2 e^{-\mu_2} & e^{-\mu_2} & 0 \\ 0 & q_3 e^{-\mu_3} & 0 & e^{-\mu_3} \end{vmatrix} \tag{11}$$

which satisfy the ortho-normal requirement,

$$\eta^{(a)(b)} = \eta_{(a)(b)} = e_{(a)i} e^i_{(b)} = (-1, +1, +1, +1). \tag{12}$$

(When there is likelihood of an ambiguity, we shall enclose the tetrad indices in parentheses to distinguish them from the tensor indices, but we shall dispense with the enclosing parentheses when no ambiguity is likely.)

Maxwell's equations governing the tetrad components, $F_{(a)(b)}$, of the Maxwell tensor are

$$\eta^{(a)(b)} F_{(c)(a)|(b)} = 0 \quad \text{and} \quad F_{[(a)(b)|(c)]} = 0, \tag{13}$$

where
$$F_{(a)(b)|(c)} = F_{(a)(b),(c)} - \eta^{(m)(n)}[\gamma_{(m)(a)(c)} F_{(n)(b)} - \gamma_{(m)(b)(c)} F_{(a)(n)}]. \tag{14}$$

and a vertical rule and a comma denote, respectively, the intrinsic and directional derivative with respect to the index following. Also,

$$\gamma_{(a)(b)(c)} = e^i_{(a)} e_{(b)i;j} e^j_{(c)} \tag{15}$$

are the Ricci rotation coefficients.

With the rotation coefficients listed in the appendix, we find that Maxwell's equations which follow can be grouped into two sets of four equations each which we shall describe as *odd* and as *even*. The odd equations are

$$\left.\begin{aligned} &(e^{\psi+\mu_2}F_{12})_{,3} + (e^{\psi+\mu_3}F_{31})_{,2} = 0, \\ &(e^{\psi+\nu}F_{01})_{,2} + (e^{\psi+\mu_2}F_{12})_{,0} = 0, \\ &(e^{\psi+\nu}F_{01})_{,3} + (e^{\psi+\mu_3}F_{13})_{,0} = 0, \\ \text{and}\quad &(e^{\mu_2+\mu_3}F_{01})_{,0} + (e^{\nu+\mu_3}F_{12})_{,2} + (e^{\nu+\mu_2}F_{13})_{,3} \\ &\quad = e^{\psi+\mu_3}(\omega_{,2} - q_{2,0}) F_{02} + e^{\psi+\mu_2}(\omega_{,3} - q_{3,0}) F_{03} + e^{\psi+\nu}(q_{3,2} - q_{2,3}) F_{23}. \end{aligned}\right\} \tag{16}$$

And the even equations are

$$\left.\begin{aligned} &(e^{\psi+\mu_3}F_{02})_{,2} + (e^{\psi+\mu_2}F_{03})_{,3} = 0, \\ &(e^{\psi+\nu}F_{32})_{,2} + (e^{\psi+\mu_2}F_{03})_{,0} = 0, \\ &(e^{\psi+\nu}F_{23})_{,3} + (e^{\psi+\mu_3}F_{02})_{,0} = 0, \\ \text{and}\quad &(e^{\nu+\mu_2}F_{02})_{,3} + (e^{\nu+\mu_3}F_{30})_{,2} + (e^{\mu_2+\mu_3}F_{23})_{,0} \\ &\quad = -e^{\psi+\nu}(q_{3,2} - q_{2,3}) F_{01} + e^{\psi+\mu_2}(\omega_{,3} - q_{3,0}) F_{12} + e^{\psi+\mu_3}(\omega_{,2} - q_{2,0}) F_{31} \end{aligned}\right\} \tag{17}$$

It will be observed that in each of the two groups of equations, we can dispense with the first equation since it provides only the integrability condition for the two following equations.

4. Maxwell's equations linearized about the Reissner–Nordström solution

Remembering that $F_{02} = -F_{20} = F_{(0)(2)}$ is the only non-vanishing component of the Maxwell tensor in the static spherically symmetric background, we readily find that the linearized versions of equations (16) and (17) (dispensing with the first equation in each case) which govern the perturbations are

$$(r\,e^{\nu}F_{01}\sin\theta)_{,r} + r\,e^{-\nu}F_{12,0}\sin\theta = 0, \tag{18}$$

$$r\,e^{\nu}(F_{01}\sin\theta)_{,\theta} + r^2F_{13,0}\sin\theta = 0, \tag{19}$$

$$e^{-\nu}rF_{01,0} + (r\,e^{\nu}F_{12})_{,r} + F_{13,\theta} = -Q_*(\omega_{,2} - q_{2,0})\sin\theta, \tag{20}$$

and

$$r\,e^{-\nu}F_{03,0} = (r\,e^{\nu}F_{23})_{,r}, \tag{21}$$

$$\delta F_{02,0} - \frac{Q_*}{r^2}(\delta\psi + \delta\mu_3)_{,0} + \frac{e^{\nu}}{r\sin\theta}(F_{23}\sin\theta)_{,\theta} = 0, \tag{22}$$

$$\left[\delta F_{02} - \frac{Q_*}{r^2}(\delta\nu + \delta\lambda)\right]_{,\theta} + (r\,e^{\nu}F_{30})_{,r} + r\,e^{-\nu}F_{23,0} = 0. \tag{23}$$

It will be observed that the odd equations (18)–(20) do not involve the perturbations of any of the quantities which are non-vanishing in the background. In this respect, F_{01}, F_{12} and F_{13} (together with ω, q_2 and q_3, as we shall see in § 5 below) behave like a *test field*. Moreover, we can eliminate F_{12} and F_{13} from equation (20) to obtain a single second-order equation for F_{01} relating it only to $\omega_{,20} - q_{2,00}$. Thus, letting

$$B = F_{01}\sin\theta, \tag{24}$$

differentiating equation (20) with respect to $x^0(=t)$, and making use of equations (18) and (19), we obtain the equation

$$[e^{2\nu}(r\,e^{\nu}B)_{,r}]_{,r} + \frac{e^{\nu}}{r}\left(\frac{B_{,\theta}}{\sin\theta}\right)_{,\theta}\sin\theta - r\,e^{-\nu}B_{,00} = Q_*(\omega_{,20} - q_{2,00})\sin^2\theta. \tag{25}$$

To proceed further, we must supplement the foregoing equations with the linearized versions of the Einstein–Maxwell equations. For this purpose, it is first necessary to relate the perturbations in the Ricci tensor with the perturbations in the Maxwell tensor as required by the field equations, namely,

$$R_{(a)(b)} = 2[F_{(a)}{}^{(m)}F_{(b)(m)} - \tfrac{1}{4}\eta_{(a)(b)}F^{(m)(n)}F_{(m)(n)}]. \tag{26}$$

Again, remembering that $F_{02}(=F_{(0)(2)})$ is the only non-vanishing component of the Maxwell tensor in the static background, we obtain from equation (26) the following expression for $\delta R_{(a)(b)}$:

$$\delta R_{(a)(b)} = 2[\delta F_{(a)}{}^{(m)}F_{(b)(m)} + F_{(a)}{}^{(m)}\delta F_{(b)(m)} - \eta_{(a)(b)}Q_*\,\delta F_{02}/r^2]. \tag{27}$$

From this last equation, we find

$$\delta R_{00} = \delta R_{11} = -\delta R_{22} = \delta R_{33} = -2Q_*\,\delta F_{02}/r^2, \tag{28}$$

$$\left.\begin{aligned}\delta R_{01} &= -2Q_* F_{12}/r^2, & \delta R_{03} &= 2Q_* F_{23}/r^2,\\ \delta R_{12} &= +2Q_* F_{01}/r^2, & \delta R_{23} &= +2Q_* F_{03}/r^2,\end{aligned}\right\} \tag{29}$$

and

$$\delta R_{02} = \delta R_{13} = 0. \tag{30}$$

5. The reduction of the equations governing the odd-parity perturbations

The equations governing the various quantities can be readily written down by transcribing in the present context the various equations given by Chandrasekhar & Friedman (1972; see particularly equations in part I, § III and part III). Thus, considering the (1, 3)-, (1, 2)- and the (0, 1)-components of the field equations, we have (cf. Chandrasekhar & Friedman 1972, eqn (18) and eqns (149), et seq.),

$$[r^2 e^{2\nu}(q_{2,3}-q_{3,2})\sin^3\theta]_{,2} - r^2 e^{-2\nu}(\omega_{,3}-q_{3,0})_{,0}\sin^3\theta = 0, \tag{31}$$

$$\begin{aligned}&[r^2 e^{2\nu}(q_{2,3}-q_{3,2})\sin^3\theta]_{,3} + r^4(\omega_{,2}-q_{2,0})_{,0}\sin^3\theta\\ &\qquad = 2(r^3 e^{\nu}\sin^2\theta)\,\delta R_{12} = 4Q_*\, r e^{\nu} B\sin\theta,\end{aligned} \tag{32}$$

and

$$\begin{aligned}&[r^4(\omega_{,2}-q_{2,0})\sin^3\theta]_{,2} + [r^2 e^{-2\nu}(\omega_{,3}-q_{3,0})\sin^3\theta]_{,3}\\ &\qquad = 2(r^3 e^{-\nu}\sin^2\theta)\,\delta R_{01} = -4Q_*\, r e^{-\nu} F_{12}\sin^2\theta.\end{aligned} \tag{33}$$

Now letting (as in Sch. eqn (11a))

$$Q = r^2 e^{2\nu}(q_{2,3}-q_{3,2})\sin^3\theta, \tag{34}$$

we can rewrite equations (31) and (32) in the forms

$$e^{2\nu}\frac{Q_{,2}}{r^2\sin^3\theta} = +(\omega_{,3}-q_{3,0})_{,0}, \tag{35}$$

and

$$\frac{Q_{,3}}{r^4\sin^3\theta} = -(\omega_{,2}-q_{2,0})_{,0} + \frac{4Q_*}{r^3\sin^2\theta}e^{\nu}B. \tag{36}$$

Eliminating ω from the foregoing equations, we obtain

$$\frac{1}{\sin^3\theta}\left(\frac{e^{2\nu}}{r^2}Q_{,r}\right)_{,r} + \frac{1}{r^4}\left(\frac{Q_{,\theta}}{\sin^3\theta}\right)_{,\theta} = (q_{2,3}-q_{3,2})_{,00} + 4Q_*\frac{e^{\nu}}{r^3}\left(\frac{B}{\sin^2\theta}\right)_{,\theta}, \tag{37}$$

or, assuming that all the quantities have a time dependence $e^{i\sigma t}$, we have

$$\frac{1}{\sin^3\theta}\left(\frac{e^{2\nu}}{r^2}Q_{,r}\right)_{,r} + \frac{1}{r^4}\left(\frac{Q_{\theta}}{\sin^3\theta}\right)_{,\theta} + \frac{\sigma^2 Q}{r^2 e^{2\nu}\sin^3\theta} = 4Q_*\frac{e^{\nu}}{r^3}\left(\frac{B}{\sin^2\theta}\right)_{,\theta}. \tag{38}$$

Equation (38) provides the generalization of the equation (Sch. eqn (11b)) which governs the odd parity perturbations in the context of the Schwarzschild metric.

Equation (38) must now be supplemented by equation (25). Eliminating $(\omega_{,20}-q_{2,00})$, which occurs on the right hand side of this equation, with the aid of equation (36), we obtain

$$[e^{2\nu}(r e^{\nu}B)_{,r}]_{,r} + (\sin\theta)\frac{e^{\nu}}{r}\left(\frac{B_{,\theta}}{\sin\theta}\right)_{,\theta} + \left(\sigma^2 r e^{-\nu} - \frac{4Q_*^2}{r^3}e^{\nu}\right)B = -Q_*\frac{Q_{,\theta}}{r^4\sin\theta}. \tag{39}$$

(a) The separation of the variables

It is known that the left hand side of equation (38) is separable in the variables r and θ by the substitution (cf. Sch. appendix)

$$Q(r,\theta) = Q(r)\,P_{l+2}(\cos\theta|-3) = Q(r)\,Y(\theta) \quad \text{(say)}, \tag{40}$$

where $P_{l+2}(\cos\theta|-3) = Y(\theta)$ is the Gegenbauer polynomial of order $(l+2)$ and index -3 and satisfies the equation

$$\frac{d}{d\theta}\left(\frac{1}{\sin^3\theta}\frac{dY}{d\theta}\right) = -\mu^2\frac{Y}{\sin^3\theta}\ [\mu^2 = 2n = (l-1)(l+2)]. \tag{41}$$

It is clear that the substitution (40) will separate equation (38) provided that $B(r,\theta)$ is of the form

$$B(r,\theta) = B(r)\,Y_{,\theta}/\sin\theta. \tag{42}$$

This form for B is consistent with the simultaneous separability of equation (39) by virtue of the identity (readily verified)

$$\sin\theta\frac{d}{d\theta}\left(\frac{1}{\sin\theta}\frac{d}{d\theta}\frac{Y_{,\theta}}{\sin\theta}\right) = -(\mu^2+2)\frac{Y_{,\theta}}{\sin\theta}. \tag{43}$$

With the substitutions (40) and (42), equations (38) and (39) provide the coupled radial equations

$$\left(\frac{e^{2\nu}}{r^2}Q_{,r}\right)_{,r} + \left(\frac{\sigma^2}{r^2 e^{2\nu}} - \frac{\mu^2}{r^4}\right)Q = -\frac{4Q_*\mu^2}{r^3}e^{\nu}B, \tag{44}$$

and

$$[e^{2\nu}(r\,e^{\nu}B)_{,r}]_{,r} + \left[\sigma^2 r\,e^{-\nu} - (\mu^2+2)\frac{e^{\nu}}{r} - \frac{4Q_*^2}{r^3}e^{\nu}\right]B = -Q_*\frac{Q}{r^4}. \tag{45}$$

With the further substitutions

$$Q = rH_2^{(-)} \quad \text{and} \quad r\,e^{\nu}B = -H_1^{(-)}/2\mu, \tag{46}$$

and the transformation to the variable r_* defined by

$$\frac{d}{dr_*} = e^{2\nu}\frac{d}{dr}, \tag{47}$$

equations (44) and (45) can be brought to the forms

$$\frac{d^2H_2^{(-)}}{dr_*^2} + \sigma^2 H_2^{(-)} = \frac{e^{2\nu}}{r^3}\left[(\mu^2+2)\,r - 6m + \frac{4Q_*^2}{r}\right]H_2^{(-)} + \frac{2Q_*\mu}{r^3}e^{2\nu}H_1^{(-)}, \tag{48}$$

and

$$\frac{d^2H_1^{(-)}}{dr_*^2} + \sigma^2 H_1^{(-)} = \frac{e^{2\nu}}{r^3}\left[(\mu^2+2)\,r + \frac{4Q_*^2}{r}\right]H_1^{(-)} + \frac{2Q_*\mu}{r^3}e^{2\nu}H_2^{(-)}. \tag{49}$$

These equations are the same as the equations which Moncrief (1974*a*) has derived by very different methods, and they can be decoupled to provide the same pair of one-dimensional wave equations to which the Newman–Penrose formalism directly leads (Chandrasekhar 1979).

6. The equations governing the even-parity perturbations

We now turn to the even-parity perturbations. For this purpose, it will suffice to consider the (0, 2)-, (0, 3)-, (2, 3)- and the (2, 2)-components of the Einstein–Maxwell equations supplemented by the Maxwell equations (21)–(23). The Einstein–Maxwell equations are (cf. Sch. eqns (12)–(15))

$$(\delta\psi+\delta\mu_3)_{,r}+\left(\frac{1}{r}-\nu_{,r}\right)(\delta\psi+\delta\mu_3)-\frac{2}{r}\delta\lambda=0, \tag{50}$$

$$[(\delta\psi+\delta\lambda)_{,\theta}+(\delta\psi-\delta\mu_3)\cot\theta]_{,0}=-\mathrm{e}^{\nu+\mu_3}\delta R_{03}=-2Q_*\,\mathrm{e}^{\nu}F_{23}/r, \tag{51}$$

$$(\delta\psi+\delta\nu)_{,r\theta}+(\delta\psi-\delta\mu_3)_{,r}\cot\theta+\left(\nu_{,r}-\frac{1}{r}\right)\delta\nu_{,\theta}-\left(\nu_{,r}+\frac{1}{r}\right)\delta\lambda_{,\theta}$$
$$=-\mathrm{e}^{\mu_2+\mu_3}\delta R_{23}=-2Q_*\,\mathrm{e}^{-\nu}F_{03}/r, \tag{52}$$

and
$$\mathrm{e}^{2\nu}\left[\frac{2}{r}\mathrm{d}\nu_{,r}+\left(\frac{1}{r}+\nu_{,r}\right)(\delta\psi+\delta\mu_3)_{,r}-2\delta\lambda\left(\frac{1}{r^2}+2\frac{\nu_{,r}}{r}\right)\right]$$
$$+\frac{1}{r^2}[(\delta\psi+\delta\nu)_{,\theta\theta}+(2\delta\psi+\delta\nu-\delta\mu_3)_{,\theta}\cot\theta+2\delta\mu_3]$$
$$-\mathrm{e}^{-2\nu}(\delta\psi+\delta\mu_3)_{,00}=\delta G_{22}=\delta R_{22}=2Q_*\,\delta F_{02}/r^2. \tag{53}$$

In addition, we shall find the following (1, 1)-component of the Einstein–Maxwell's equations useful (cf. Sch. eqn (16)):

$$\mathrm{e}^{2\nu}\left[\delta\psi_{,rr}+2\left(\frac{1}{r}+\nu_{,r}\right)\delta\psi_{,r}+\frac{1}{r}(\delta\psi+\delta\nu+\delta\mu_3-\delta\lambda)_{,r}-2\frac{\delta\lambda}{r}\left(\frac{1}{r}+2\nu_{,r}\right)\right]$$
$$+\frac{1}{r^2}[\delta\psi_{,\theta\theta}+\delta\psi_{,\theta}\cot\theta+(\delta\psi+\delta\nu-\delta\mu_3+\delta\lambda)_{,\theta}\cot\theta+2\delta\mu_3]$$
$$-\mathrm{e}^{-2\nu}\delta\psi_{,00}=-\delta R_{11}=2Q_*\,\delta F_{02}/r^2. \tag{54}$$

The right hand sides of the foregoing equations are separable by the substitutions (due to Friedman 1973)

$$\left.\begin{aligned}\delta\nu&=N(r)\,P_l(\cos\theta)\,\mathrm{e}^{i\sigma t},\\ \delta\lambda&=L(r)\,P_l(\cos\theta)\,\mathrm{e}^{i\sigma t},\\ \delta\mu_3&=[T(r)\,P_l+V(r)\,P_{l,\theta\theta}]\,\mathrm{e}^{i\sigma t},\\ \delta\psi&=[T(r)\,P_l+V(r)\,P_{l,\theta}\cot\theta]\,\mathrm{e}^{i\sigma t},\\ \text{and}\qquad \delta\psi+\delta\mu_3&=[2T-l(l+1)\,V]\,P_l\,\mathrm{e}^{i\sigma t},\end{aligned}\right\} \tag{55}$$

where N, L, T and V are four radial functions, and $P_l(\cos\theta)$ $(\equiv P_l)$ is the Legendre function. With these substitutions equations (50)–(53) become

$$\left[\frac{\mathrm{d}}{\mathrm{d}r}+\left(\frac{1}{r}-\nu_{,r}\right)\right][2T-l(l+1)\,V]-\frac{2}{r}L=0, \tag{56}$$

$$(T-V+L)\,P_{l,\theta}=\frac{2iQ_*}{\sigma}\frac{\mathrm{e}^{\nu}}{r}F_{23}, \tag{57}$$

$$\left[(T-V+N)_{,r}-\left(\frac{1}{r}-\nu_{,r}\right)N-\left(\frac{1}{r}+\nu_{,r}\right)L\right]P_{l,\theta}=-2Q_*\frac{\mathrm{e}^{-\nu}}{r}F_{03}, \tag{58}$$

and

$$\left\{ e^{2\nu} \left[\!\left[\frac{2}{r} N_{,r} + \left(\frac{1}{r} + \nu_{,r}\right) [2T - l(l+1)\, V]_{,r} - \frac{2}{r}\left(\frac{1}{r} + 2\nu_{,r}\right) L \right]\!\right] - \frac{l(l+1)}{r^2} N - \frac{2n}{r^2} T \right.$$
$$\left. + \sigma^2 e^{-2\nu} [2T - l(l+1)\, V] \right\} P_l = 2Q_* \, \delta F_{02}/r^2, \quad (59)$$

where we have assumed that δF_{02}, F_{03}, and F_{23} have the same time dependence $e^{i\sigma t}$. The consistency of equations (56)–(59) requires that δF_{02}, F_{03}, and F_{23} are of the forms

$$\delta F_{02} = \frac{r^2 e^{2\nu}}{2Q_*} B_{02}(r)\, P_l, \quad F_{03} = -\frac{r\, e^{+\nu}}{2Q_*} B_{03}(r)\, P_{l,\theta},$$

and

$$F_{23} = -\frac{i\sigma}{2Q_*} r\, e^{-\nu} B_{23}(r)\, P_{l,\theta}. \quad (60)$$

We shall presently verify that these forms for δF_{02}, F_{03} and F_{23} are consistent with the simultaneous separability Maxwell's equations (21)–(23). But meantime we note that equations (57)–(59) reduce to the forms

$$T - V + L = B_{23}, \quad (61)$$

$$(T - V + N)_{,r} - \left(\frac{1}{r} - \nu_{,r}\right) N - \left(\frac{1}{r} + \nu_{,r}\right) L = B_{03}, \quad (62)$$

and

$$\frac{2}{r} N_{,r} + \left(\frac{1}{r} + \nu_{,r}\right) [2T - l(l+1)\, V] - \frac{2}{r}\left(\frac{1}{r} + 2\nu_{,r}\right) L$$
$$- \frac{l(l+1)}{r^2} e^{-2\nu} N - \frac{2n}{r^2} e^{-2\nu} T + \sigma^2 e^{-4\nu} [2T - l(l+1)\, V] = B_{02}. \quad (63)$$

Turning next to Maxwell's equations (21)–(23), we now find that the substitutions (60) yield the further radial equations

$$B_{03} = \frac{1}{r^2} (r^2 B_{23})_{,r} = B_{23,r} + \frac{2}{r} B_{23}, \quad (64)$$

$$r^4 e^{2\nu} B_{02} = 2Q_*^2 [2T - l(l+1)\, V] - l(l+1)\, r^2 B_{23}, \quad (65)$$

and

$$(r^2 e^{2\nu} B_{03})_{,r} + r^2 e^{2\nu} B_{02} + \sigma^2 r^2 e^{-2\nu} B_{23} = 2Q_*^2 (N + L)/r^2. \quad (66)$$

We shall find it convenient to write

$$X = nV = \tfrac{1}{2}(l-1)(l+2)\, V, \quad (67)$$

in which case (by virtue of equation (61))

$$2T - l(l+1)\, V = -2(L + X - B_{23}). \quad (68)$$

Also, we may note here that we may use the linear relations (61) and (65) to eliminate T and B_{02} from the remaining equations and consider N, L, $V(=X/n)$, B_{23} and B_{03} as the principal functions in terms of which to express the perturbations.

Making use of equations (64) and (68), we can rewrite equations (56) and (62) in the forms

$$(L+X-B_{23})_{,r} = -\left(\frac{1}{r}-\nu_{,r}\right)(L+X-B_{23})-\frac{1}{r}L, \tag{69}$$

and

$$N_{,r}-L_{,r} = \left(\frac{1}{r}-\nu_{,r}\right)N+\left(\frac{1}{r}+\nu_{,r}\right)L+\frac{2}{r}B_{23}. \tag{70}$$

Returning to equations (53) and (54), we observe that the difference of these equations gives

$$e^{2\nu}\left[\delta\psi_{,rr}+2\left(\frac{1}{r}+\nu_{,r}\right)\delta\psi_{,r}-\frac{1}{r}(\delta\lambda+\delta\nu)_{,r}-\nu_{,r}(\delta\psi+\delta\mu_3)_{,r}\right]$$
$$+\frac{1}{r^2}(\delta\lambda_{,\theta}\cot\theta-\delta\nu_{,\theta\theta})+e^{-2\nu}\delta\mu_{3,00} = 0. \tag{71}$$

By virtue of equations (55), the left hand side of equation (71) is a linear combination of terms with the angular factors P_l and $P_{l,\theta}\cot\theta$. We can accordingly equate, separately, the terms with these two factors; and the terms with the factor $P_{l,\theta}\cot\theta$ yield the equation

$$V_{,rr}+2\left(\frac{1}{r}+\nu_{,r}\right)V_{,r}+\frac{e^{-2\nu}}{r^2}(N+L)+\sigma^2 e^{-4\nu}V = 0. \tag{72}$$

We observe that equation (72) is formally the same as that which obtains in the context of the Schwarzschild metric (cf. Sch. eqn (24)).

Equations (63), (69) and (70) provide three linear first-order equations for the three radial functions L, N and $V(=X/n)$. By suitably combining these equations, we can express the derivative of each of them as linear combinations of L, N, V, B_{23}, and B_{03}. Thus (cf. Sch. eqns (27)–(31))

$$N_{,r} = aN+bL+c(X-B_{23}), \tag{73}$$

$$L_{,r} = \left(a-\frac{1}{r}+\nu_{,r}\right)N+\left(b-\frac{1}{r}-\nu_{,r}\right)L+c(X-B_{23})-\frac{2}{r}B_{23}, \tag{74}$$

and
$$X_{,r} = -\left(a-\frac{1}{r}+\nu_{,r}\right)N-\left(b+\frac{1}{r}-2\nu_{,r}\right)L-\left(c+\frac{1}{r}-\nu_{,r}\right)(X-B_{23})+B_{03}, \tag{75}$$

where
$$a = \frac{n+1}{r}e^{-2\nu}, \tag{76}$$

$$b = -\frac{1}{r}-\frac{n}{r}e^{-2\nu}+\nu_{,r}+r\nu_{,r}^2+\sigma^2 r\,e^{-4\nu}-2\frac{Q_*^2}{r^3}e^{-2\nu},$$
$$= -\frac{1}{r}-\frac{n}{r}e^{-2\nu}+\frac{m}{r^2}e^{-2\nu}+\frac{m^2}{r^3}e^{-4\nu}+\sigma^2 r\,e^{-4\nu}-\frac{Q_*^2}{r^3}(1+2e^{2\nu})e^{-4\nu}, \tag{77}$$

$$c = -\frac{1}{r}+\frac{e^{-2\nu}}{r}+r\nu_{,r}^2+\sigma^2 r\,e^{-4\nu}-2\frac{Q_*^2}{r^3}e^{-2\nu},$$
$$= -\frac{1}{r}+\frac{1}{r}e^{-2\nu}+\frac{m^2}{r^3}e^{-4\nu}+\sigma^2 r\,e^{-4\nu}-\frac{Q_*^2}{r^3}(1+e^{2\nu})e^{-4\nu}. \tag{78}$$

We shall find that of the three equations (73)–(75), the equation for $X_{,r}$ is the one that is most used in the subsequent reductions. We accordingly note that the following relations,

$$b-c-\nu_{,r} = -a, \tag{79}$$

$$-b-\frac{1}{r}+2\nu_{,r} = \frac{n}{r}\,\mathrm{e}^{-2\nu}+\frac{m}{r^2}\,\mathrm{e}^{-2\nu}-\frac{1}{r^3}(m^2-Q_*^2)\,\mathrm{e}^{-4\nu}-\sigma^2 r\,\mathrm{e}^{-4\nu}, \tag{80}$$

and

$$a-\frac{1}{r}+\nu_{,r} = \frac{1}{r^2}\mathrm{e}^{-2\nu}\left(nr+3m-2\frac{Q_*^2}{r}\right) = -\frac{\varpi}{r^2}\mathrm{e}^{-2\nu}\quad(\text{say}), \tag{81}$$

where

$$\varpi = -\left(nr+3m-2\frac{Q_*^2}{r}\right) = m-(n+2)\,r+2r\,\mathrm{e}^{2\nu}, \tag{82}$$

enable us to write equation (75) in the form

$$X_{,r} = \frac{\varpi}{r^2}\mathrm{e}^{-2\nu}N+\left[-\frac{1}{r}\,\mathrm{e}^{-2\nu}+\frac{m}{r^2}\,\mathrm{e}^{-2\nu}-\frac{1}{r^3}(m^2-Q_*^2)\,\mathrm{e}^{-4\nu}-\sigma^2 r\,\mathrm{e}^{-4\nu}\right](L+X-B_{23}) + aL+B_{03}. \tag{83}$$

Equations (64), (66), (69), (70), (72) and (83) are the basic equations of the theory.

7. The reduction of the equations governing the even-parity perturbations

Our aim is now to find two linear combinations $H_1^{(+)}$ and $H_2^{(+)}$ of the scalars L, N, X, B_{23} and B_{03} which will satisfy coupled equations of the form (cf. equations (48) and (49))

$$\mathrm{e}^{-2\nu}\Lambda^2\left|\begin{matrix} H_1^{(+)} \\ H_2^{(+)}\end{matrix}\right| = U\left|\begin{matrix} H_1^{(+)} \\ H_2^{(+)}\end{matrix}\right| + W\left|\begin{matrix} +3m & -2Q_*\mu \\ -2Q_*\mu & -3m\end{matrix}\right|\left|\begin{matrix} H_1^{(+)} \\ H_2^{(+)}\end{matrix}\right|, \tag{84}$$

where

$$\Lambda^2 = \frac{\mathrm{d}^2}{\mathrm{d}r_*^2}+\sigma^2, \tag{85}$$

and U and W are certain functions of r (as yet, unspecified); for, it is only when the equations can be reduced to this form can we obtain decoupled one-dimensional wave equations of the forms which Moncrief (1974*b*) has derived and which the Newman–Penrose formalism yields very directly (Chandrasekhar 1979).

The search for suitable functions $H_1^{(+)}$ and $H_2^{(+)}$ is fraught with ambiguity; and the only quide we have is that in the context of the Schwarzschild perturbations, the Zerilli function Z (which replaces $H_2^{(+)}$) is given by (Sch. eqn (33))

$$Z = \frac{r}{n}X-\frac{r^2}{nr+3m}(L+X). \tag{86}$$

After some trials, it appeared that we should replace

$$\left.\begin{aligned} L+X \quad &\text{by}\quad L+X-B_{23}, \\ \text{and}\qquad nr+3m \quad &\text{by}\quad nr+3m-2Q_*^2/r = -\varpi. \end{aligned}\right\} \tag{87}$$

We were led to consider $(L+X-B_{23})$ as the proper replacement for $L+X$ since the two functions satisfy, formally, the same equations (cf. eqn (69) and Sch. eqn (32)). In any event, with the replacements (87), we were led to try

$$H_2^{(+)} = \frac{r}{n}X + \frac{r^2}{\varpi}(L+X-B_{23}) \tag{88}$$

as the right combination; and as we shall now verify, with this definition of $H_2^{(+)}$, we are able to determine, uniquely, U, W and $H_1^{(+)}$ consistently with equation (84).

First, differentiating the expression for H_2 with respect to r_* and making use of equations (69) and (70), we find

$$\begin{aligned}\frac{\mathrm{d}H_2^{(+)}}{\mathrm{d}r_*} = \frac{r}{n}\mathrm{e}^{2\nu}X_{,r} + \mathrm{e}^{2\nu}\left(\frac{1}{n}+\frac{r}{\varpi}\right)X - \frac{r}{\varpi}\mathrm{e}^{2\nu}B_{23}\\ + \left[\frac{r}{\varpi^2}(2nr+3m)\,\mathrm{e}^{2\nu} + \frac{r-m}{\varpi}\right](L+X-B_{23}).\end{aligned} \tag{89}$$

We may note here that the following elementary relations were found to be useful in the foregoing and in the subsequent reductions:

$$\left.\begin{aligned}2r\,\mathrm{e}^{2\nu} = \varpi + (n+2)\,r - m, \quad 2Q_*^2 = r(\varpi+nr+3m),\\ \varpi_{,r} = -(\varpi+2nr+3m)/r \quad \text{and} \quad (\mathrm{e}^{2\nu})_{,r} = -(\varpi+nr+m)/r^2.\end{aligned}\right\} \tag{90}$$

Differentiating equation (89) once again with respect to r_* and making use of equations (64), (69), (70) and (72), we find, successively,

$$\begin{aligned}\mathrm{e}^{-2\nu}\Lambda^2 H_2^{(+)} = {}&\left[\mathrm{e}^{2\nu}\left(\frac{1}{n}+\frac{r}{\varpi}\right)\right]_{,r}X - \left[\left(\mathrm{e}^{2\nu}\frac{r}{\varpi}\right)_{,r} - 2\frac{\mathrm{e}^{2\nu}}{\varpi}\right]B_{23}\\ &+\mathrm{e}^{2\nu}\frac{r}{\varpi}\left[\left(-\frac{1}{r}\mathrm{e}^{2\nu}+\frac{m}{r^2}\mathrm{e}^{-2\nu}-\frac{m^2}{r^3}\mathrm{e}^{-4\nu}+\frac{Q_*^2}{r^3}\mathrm{e}^{-4\nu}\right)(L+X-B_{23})+aL\right]\\ &+\left\{\left[\mathrm{e}^{2\nu}\frac{r}{\varpi^2}(2nr+3m)+\frac{r-m}{\varpi}\right]_{,r}\right.\\ &\left.-\left(\frac{1}{r}-\nu_{,r}\right)\left[\mathrm{e}^{2\nu}\frac{r}{\varpi^2}(2nr+3m)+\frac{r-m}{\varpi}\right]\right\}(L+X-B_{23})\\ &-\frac{1}{r}L-\frac{1}{r}\left[\mathrm{e}^{2\nu}\frac{r}{\varpi^2}(2nr+3m)+\frac{r-m}{\varpi}\right]L,\\ = {}&\left\{\left[\mathrm{e}^{2\nu}\frac{r}{\varpi^2}(2nr+3m)+\frac{r-m}{\varpi}\right]_{,r}-\left(\frac{1}{r}-\nu_{,r}\right)\mathrm{e}^{2\nu}\frac{r}{\varpi^2}(2nr+3m)\right.\\ &\left.+\frac{1}{r\varpi}(3m-2r)-\frac{\mathrm{e}^{2\nu}}{\varpi^2}(2nr+3m)+\frac{nr+m-\varpi}{\varpi r}\right\}(L+X-B_{23})\\ &+\left\{\frac{\mathrm{e}^{2\nu}}{\varpi^2}(2nr+3m)-\frac{nr+m-\varpi}{\varpi r}+\left[\left(\mathrm{e}^{2\nu}\frac{r}{\varpi}\right)_{,r}-\frac{2}{\varpi}\mathrm{e}^{2\nu}\right]\right\}(X-B_{23})\\ &+\left[\frac{1}{n}(\mathrm{e}^{2\nu})_{,r}+\frac{2}{\varpi}\mathrm{e}^{2\nu}\right]X.\end{aligned} \tag{91}$$

After some further simplifications, we find that the foregoing equation can be reduced to the form

$$e^{-2\nu}\Lambda^2 H_2^{(+)} = \left[2(\varpi+2nr+3m)\,W - \frac{1}{r^3}(\varpi+m+nr) + \frac{2n}{r\varpi}e^{2\nu}\right]\frac{r^2}{\varpi}(L+X-B_{23})$$

$$+\left[2nrW - \frac{1}{r^3}(\varpi+m+nr) + \frac{2n}{r\varpi}e^{2\nu}\right]\frac{r}{n}X - 2r^2WB_{23}, \quad (92)$$

where
$$W = \frac{e^{2\nu}}{r^2\varpi^2}(2nr+3m) - \frac{m+nr}{r^3\varpi}. \quad (93)$$

(In defining W as we have, we are anticipating that this function will play the role of 'W' in equation (84).)

We now compare the right hand side of equation (92) with what we should expect on equation (84), namely,

$$e^{-2\nu}\Lambda^2 H_2^{(+)} = (U-3mW)\left[\frac{r}{n}X + \frac{r^2}{\varpi}(L+X-B_{23})\right] - 2Q_*\mu W H_1^{(+)}. \quad (94)$$

The compatibility of equations (93) and (94) requires that we identify

$$U - 3mW = 2nrW - \frac{1}{r^3}(\varpi+m+nr) + \frac{2n}{r\varpi}e^{2\nu}, \quad (95)$$

or
$$U = (2nr+3m)\,W - \frac{1}{r^3}(\varpi+m+nr) + \frac{2n}{r\varpi}e^{2\nu}, \quad (96)$$

and, further, that

$$H_1^{(+)} = \frac{1}{Q_*\mu}\left[r^2B_{23} - (\varpi+nr+3m)\frac{r^2}{\varpi}(L+X-B_{23})\right]$$

$$= \frac{1}{Q_*\mu}\left[r^2B_{23} - 2Q_*^2\frac{r}{\varpi}(L+X-B_{23})\right]. \quad (97)$$

An equivalent form of $H_1^{(+)}$ is

$$H_1^{(+)} = \frac{1}{Q_*\mu}\left(r^2B_{23} + 2\frac{Q_*^2}{n}X - 2\frac{Q_*^2}{r}H_2^{(+)}\right). \quad (98)$$

It remains to verify that, with the definitions (88), (93), (96) and (98) of $H_2^{(+)}$, W, U, and $H_1^{(+)}$, the equation,

$$e^{-2\nu}\Lambda^2 H_1^{(+)} = (U+3mW)\,H_1^{(+)} - 2Q_*\mu W H_2^{(+)}, \quad (99)$$

is satisfied. Inserting in this equation, the expression (98) for $H_1^{(+)}$, we require to verify

$$e^{-2\nu}\Lambda^2\left(r^2B_{23} + 2\frac{Q_*^2}{n}X - 2\frac{Q_*^2}{r}H_2^{(+)}\right)$$

$$= (U+3mW)\left(r^2B_{23} + 2\frac{Q_*^2}{n}X\right) - 2\frac{Q_*^2}{r}[U+(3m+2nr)\,W]\,H_2^{(+)}. \quad (100)$$

The effect of the operator Λ^2 on $(H_2^{(+)}/r)$, on the left hand side of equation (100) can be readily written down with our present knowledge of $\Lambda^2 H_2^{(+)}$; and we find after some further reductions (in which we make use of equations (95) and (96)) that equation (100) reduces to the form,

$$\begin{aligned} e^{-2\nu}\Lambda^2\left(r^2 B_{23}+2\frac{Q_*^2}{n}X\right) \\ = -\left[2\varpi W+\frac{1}{r^3}(\varpi+m+nr)-\frac{2n}{r\varpi}e^{2\nu}\right]\left(r^2B_{23}+2\frac{Q_*^2}{n}X\right) \\ +2\frac{Q_*^2}{r}\left\{\left[2\varpi W+\frac{1}{r^3}(\varpi+m+nr)+\frac{2}{r^2}e^{2\nu}\right]H_2^{(+)}-\frac{2}{r}\frac{dH_2^{(+)}}{dr_*}\right\}, \end{aligned} \quad (101)$$

where we may substitute for $dH_2^{(+)}/dr_*$ from equation (89). The left hand side of equation (101) can be readily evaluated with the aid of equations (64), (65), (66) and (72). We find

$$e^{-2\nu}\Lambda^2\left(r^2B_{23}+2\frac{Q_*^2}{n}X\right)=\frac{4Q_*^2}{r^2}(L+X-B_{23})+2(n+1)B_{23}-4\frac{Q_*^2}{nr}X_{,r}. \quad (102)$$

It is now a simple matter to verify that the right hand sides of equations (101) and (102) agree. Thus, with $H_1^{(+)}$, $H_2^{(+)}$, U and W as defined, equation (99) is in fact satisfied.

Letting

$$\left.\begin{aligned} Z_1 &= +[\sqrt{(9m^2+4Q_*^2\mu^2)}+3m]H_1^{(+)}-2Q_*\mu H_2^{(+)}, \\ Z_2 &= -[\sqrt{(9m^2+4Q_*^2\mu^2)}+3m]H_2^{(+)}-2Q_*\mu H_1^{(+)}, \end{aligned}\right\} \quad (103)$$

we can decouple the equations for $H_1^{(+)}$ and $H_2^{(+)}$ to obtain a pair of one-dimensional wave equations for Z_1 and Z_2; and these equations agree with the equations derived by Moncrief (1974*b*; see also Chandrasekhar 1979).

While we have derived the basic equations governing the odd- and the even-parity perturbations directly from a consideration of the metric perturbations, it is important to observe that the derivation of these same equations from the Newman–Penrose formalism has the great advantage that both pairs of equations follow from the same basic pair of equations, by transformations which are dual, and that in consequence the solutions appropriate to one parity can be deduced, simply, from solutions appropriate to the opposite parity (cf. Chandrasekhar 1979, §§6 and 7).

And, finally, the manner, in which an arbitrary superposition of gravitational and electromagnetic waves, incident on the black hole, will be reflected and absorbed, has been described in Chandrasekhar (1972, §8).

8. Concluding remarks

The present treatment of the metric perturbations of the Reissner–Nordström black hole, as well as the earlier treatment of the perturbations on the Newman–Penrose formalism, exhibit remarkable similarities (almost to the point of identity) with the corresponding treatments in the context of the Schwarzschild black hole.

This far-reaching similarity suggests that the perturbations of the Kerr–Newman black hole should allow a treatment which parallels the treatment (now complete) of the perturbations of the Kerr black hole; but whether this is in fact possible, must await future investigations.

The research reported in this paper has in part been supported by National Science Foundation under grant PHY 76-81102 A01 with the University of Chicago.

Appendix

For convenience we list below the various Ricci rotation-coefficients for the non-stationary axisymmetric metric (4) described in the ortho-normal basis defined in equations (10) and (11).

$$\gamma_{100} = 0;$$
$$\gamma_{101} = \psi_{,0}\,\mathrm{e}^{-\nu};$$
$$\gamma_{102} = +\tfrac{1}{2}\mathrm{e}^{\psi-\nu-\mu_2}(\omega_{,2} - q_{2,0});$$
$$\gamma_{103} = +\tfrac{1}{2}\mathrm{e}^{\psi-\nu-\mu_3}(\omega_{,3} - q_{3,0});$$
$$\gamma_{120} = -\tfrac{1}{2}\,\mathrm{e}^{\psi-\nu-\mu_2}(\omega_{,2} - q_{2,0});$$
$$\gamma_{121} = \psi_{,2}\,\mathrm{e}^{-\mu_2};$$
$$\gamma_{122} = 0;$$
$$\gamma_{123} = -\tfrac{1}{2}\mathrm{e}^{-\psi-\mu_2-\mu_3}(q_{3,2} - q_{2,3});$$
$$\gamma_{130} = -\tfrac{1}{2}\mathrm{e}^{\psi-\nu-\mu_3}(\omega_{,3} - q_{3,0});$$
$$\gamma_{131} = \psi_{,3}\,\mathrm{e}^{-\mu_3};$$
$$\gamma_{132} = +\tfrac{1}{2}\mathrm{e}^{\psi-\mu_2-\mu_3}(q_{3,2} - q_{2,3});$$
$$\gamma_{133} = 0;$$
$$\gamma_{200} = \nu_{,2}\,\mathrm{e}^{-\mu_2};$$
$$\gamma_{201} = +\tfrac{1}{2}\mathrm{e}^{\psi-\nu-\mu_2}(\omega_{,2} - q_{2,0});$$
$$\gamma_{202} = \mu_{2,0}\,\mathrm{e}^{-\nu};$$
$$\gamma_{203} = 0;$$
$$\gamma_{230} = 0;$$
$$\gamma_{231} = +\tfrac{1}{2}\mathrm{e}^{\psi-\mu_2-\mu_3}(q_{3,2} - q_{2,3});$$
$$\gamma_{232} = +\mu_{2,3}\,\mathrm{e}^{-\mu_3};$$
$$\gamma_{233} = -\mu_{3,2}\,\mathrm{e}^{-\mu_2};$$
$$\gamma_{300} = \nu_{,3}\,\mathrm{e}^{-\mu_3};$$
$$\gamma_{301} = +\tfrac{1}{2}\mathrm{e}^{\psi-\nu-\mu_3}(\omega_{,3} - q_{3,0});$$
$$\gamma_{302} = 0;$$
$$\gamma_{303} = \mu_{3,0}\,\mathrm{e}^{-\nu}.$$

References

Chandrasekhar, S. 1975 *Proc. R. Soc. Lond.* A **343**, 289.
Chandrasekhar, S. 1979 *Proc. R. Soc. Lond.* A **365**, 453.
Chandrasekhar, S. & Friedman, J. L. 1972 *Astrophys. J.* **175**, 379.
Friedman, J. L. 1973 *Proc. R. Soc. Lond.* A **335**, 163.
Moncrief, V. 1974*a* *Phys. Rev.* D **9**, 2707.
Moncrief, V. 1974*b* *Phys. Rev.* D **10**, 1057.
Moncrief, V. 1975 *Phys. Rev.* D **12**, 1526.
Zerilli, F. J. 1973 *Phys. Rev.* D **9**, 860.

16

On one-dimensional potential barriers having equal reflexion and transmission coefficients

By S. Chandrasekhar, F.R.S.

University of Chicago, Chicago, Illinois 60637, U.S.A.

(*Received* 30 *July* 1979)

Based on the results of earlier studies on the perturbations of the Schwarzschild and the Reissner–Nordström black holes, it is shown that there exists a very general class of potential pairs ($V^{(+)}$ and $V^{(-)}$) which yield the same reflexion and transmission coefficients. It is further shown that these potentials, $V^{(+)}$ and $V^{(-)}$, satisfy an infinite hierarchy of integral equalities which are, formally, the same as the conserved quantities allowed by the Korteweg–deVries equation.

1. Introduction

The study of the perturbations of the black-hole solutions of general relativity has disclosed several novel aspects in the theory of penetration of one dimensional potential barriers familiar in elementary quantum theory (for an account of these aspects, see Chandrasekhar 1979*a*). In particular, in the context of the perturbations of the Schwarzschild and the Reissner–Nordström black holes, it was found (cf. Chandrasekhar & Detweiler 1975 and Chandrasekhar 1979*b*) that the one dimensional wave equations, governing the perturbations of opposite parities, yield the same reflexion and transmission coefficients. A general question in the theory of 'inverse-scattering' which this result raises is how one may distinguish such potential barriers. This question is a more general one than the celebrated one of establishing the uniqueness (or, otherwise) of a potential from a knowledge of the resulting S-matrix. In this paper, we shall show how the known results in the theory of the perturbations of the Schwarzschild and the Reissner–Nordström black holes suggest a very general class of pairs of potentials which yield the same reflexion and transmission coefficients and disclose a connection between the present considerations and the theory of solitons and the Korteweg–deVries equation.

2. A summary of the known results

Since the equations governing the perturbations of the Schwarzschild black hole are special cases of the equations governing the perturbations of the Reissner–Nordström black hole, it will suffice to consider the latter.

The two pairs of wave equations, appropriate to the perturbations belonging to the opposite parities, are given by (Chandrasekhar 1979*b*); see particularly the equations in §§ 5 and 6 and in the appendix)

$$\Lambda^2 Z_i^{(\pm)} = V_i^{(\pm)} Z_i^{(\pm)} \quad \left(\Lambda^2 = \frac{d^2}{dr_*^2} + \sigma^2; \ \frac{d}{dr_*} = \frac{\Delta}{r^2}\frac{d}{dr}\right), \tag{1}$$

where

$$V_i^{(\pm)} = Q_i - \frac{dW_i^{(\pm)}}{dr_*}, \tag{2}$$

$$Q_i = \mu^2 \frac{\Delta}{r^4}\left(1 + \frac{2q_i}{\mu^2 r}\right)\left(1 + \frac{q_j}{\mu^2 r}\right), \tag{3}$$

$$W_i^{(\pm)} = \frac{1}{F_i}\frac{dF_i}{dr_*} - \frac{\beta_i^{(\pm)}}{F_i}, \quad \beta_i^{(\pm)} = \pm q_j, \tag{4}$$

$$F_i = \frac{r^8}{D_i} Q_i, \quad D_i = \Delta^2\left(1 + \frac{2q_i}{\mu^2 r}\right) \quad (i,j = 1, 2; \ i \neq j), \tag{5}$$

$$q_1 = 3M + \sqrt{(9M^2 + 4Q_*^2 \mu^2)}, \quad q_2 = 3M - \sqrt{(9M^2 + 4Q_*^2 \mu^2)}, \tag{6}$$

$$\Delta = r^2 - 2Mr + Q_*^2, \quad \mu^2 = 2n = (l-1)(l+2), \tag{7}$$

and M and Q_* are the mass and the charge of the black hole.

Substituting for $W_i^{(\pm)}$ from equation (4) into equation (2), we find

$$V_i^{(\pm)} = \frac{D_i}{r^8} F_i - \frac{1}{F_i}\frac{d^2 F_i}{dr_*^2} + \frac{1}{F_i^2}\left(\frac{dF_i}{dr_*}\right)^2 + \beta_i^{(\pm)} \frac{d}{dr_*}\left(\frac{1}{F_i}\right), \tag{8}$$

or, making use of the differential equation satisfied by F_i (namely, equation (A 9) in Chandrasekhar 1979*b*) – and this is the key step – we obtain the basic formula,

$$V_i^{(\pm)} = \frac{\beta_i^2}{F_i^2} + \frac{\kappa}{F_i} + \beta_i^{(\pm)} \frac{d}{dr_*}\left(\frac{1}{F_i}\right), \tag{9}$$

where

$$\kappa = \mu^2(\mu^2 + 2). \tag{10}$$

Now letting

$$f_i = \frac{1}{F_i} = \frac{D_i}{r^8 Q_i} = \frac{\Delta}{r^3(\mu^2 r + q_j)} \quad (i,j = 1, 2; \ i \neq j), \tag{11}$$

we arrive at the potentials (cf. equation (4))

$$V_i^{(\pm)} = \pm \beta_i \frac{df_i}{dr_*} + \beta_i^2 f_i^2 + \kappa f_i. \tag{12}†$$

These potentials, appropriate for describing the perturbations of the opposite parities, differ only in the sign of the term in β_i. And, as we have seen, these potentials yield the same reflexion and transmission coefficients.

† The reduction of the potentials to this form, in the simpler context of the Schwarzschild black hole, was accomplished by Heading (1977) by a different route but using the equivalent relations derived in Chandrasekhar (1975).

3. Wave equations yielding the same reflexion and transmission coefficients

From the foregoing summary of the known results, it would appear that the wave equations

$$\frac{d^2Z^{(+)}}{dx^2}+\sigma^2 Z^{(+)} = V^{(+)}Z^{(+)} = \left(+\beta\frac{df}{dx}+\beta^2 f^2+\kappa f\right) Z^{(+)} \tag{13}$$

and

$$\frac{d^2Z^{(-)}}{dx^2}+\sigma^2 Z^{(-)} = V^{(-)}Z^{(-)} = \left(-\beta\frac{df}{dx}+\beta^2 f^2+\kappa f\right) Z^{(-)}, \tag{14}$$

where κ and β are some real constants and f is an arbitrary continuous function, which together with all its derivatives vanishes for both $x \to +\infty$ and $x \to -\infty$ and whose integral over the entire range of x is finite, determine equal values for the reflexion and the transmission coefficients. Indeed, we shall show that while the relative amplitudes of the transmitted waves are the same for both potentials, the corresponding amplitudes of the reflected waves differ only in their phases. (It will be noted that, for convenience, we have replaced r_* by x as the independent variable.)

First, we observe that a relation between $Z^{(-)}$ and $Z^{(+)}$ of the form,

$$Z^{(-)} = p(x)\, Z^{(+)} + q(x)\, Z^{(+)}_{,x}, \tag{15}$$

requires, as necessary and sufficient conditions, that the functions p and q satisfy the equations (cf. Heading 1977)

$$2p_{,x}+q_{,xx} = [V^{(-)}-V^{(+)}]\,q \tag{16}$$

and

$$[\sigma^2-V^{(+)}]\,q^2+p^2+pq_{,x}-qp_{,x} = C^2, \tag{17}$$

where C^2 is some real constant. For $V^{(+)}$ and $V^{(-)}$ given in equations (13) and (14), it can be readily verified that the foregoing equations are satisfied by

$$p = \kappa+2\beta^2 f, \quad q = -2\beta(=\text{constant}) \tag{18}$$

and

$$C^2 = \kappa^2+4\sigma^2\beta^2. \tag{19}$$

(It may be noted, parenthetically here, that C^2 is the 'Starobinsky constant' of the problem!)

We may accordingly write

$$(\kappa-2i\sigma\beta)\, Z^{(-)} = (\kappa+2\beta^2 f)\, Z^{(+)} - 2\beta Z^{(+)}_{,x}, \tag{20}$$

where we have chosen a relative normalization of $Z^{(-)}$ and $Z^{(+)}$ such that the inverse relation between $Z^{(+)}$ and $Z^{(-)}$ is given (in the same normalization) by

$$(\kappa+2i\sigma\beta)\, Z^{(+)} = (\kappa+2\beta^2 f)\, Z^{(-)} + 2\beta Z^{(-)}_{,x}. \tag{21}$$

From the relations (20) and (21), it now follows, from our assumption, that f vanishes for $x \to \pm\infty$, that solutions for $Z^{(-)}$, derived from solutions for $Z^{(+)}$ having the asymptotic behaviours,

$$Z^{(+)} \to \mathrm{e}^{+\mathrm{i}\sigma x} \quad \text{and} \quad Z^{(+)} \to \mathrm{e}^{-\mathrm{i}\sigma x} \quad (x \to \pm\infty) \tag{22†}$$

have, respectively, the asymptotic behaviours,

$$Z^{(-)} \to \mathrm{e}^{+\mathrm{i}\sigma x} \quad \text{and} \quad Z^{(-)} \to \frac{\kappa + 2\mathrm{i}\sigma\beta}{\kappa - 2\mathrm{i}\sigma\beta}\mathrm{e}^{-\mathrm{i}\sigma x} \quad (x \to \pm\infty). \tag{23}$$

Turning next to the problem of barrier-penetration, we must seek solutions of the wave equations (13) and (14) which have the asymptotic behaviours,

$$\left.\begin{aligned} Z^{(\pm)} &\to \mathrm{e}^{+\mathrm{i}\sigma x} + A^{(\pm)}(\sigma)\,\mathrm{e}^{-\mathrm{i}\sigma x} \quad (x \to +\infty), \\ &\to \qquad B^{(\pm)}(\sigma)\,\mathrm{e}^{+\mathrm{i}\sigma x} \quad (x \to -\infty). \end{aligned}\right\} \tag{24}$$

From the relations (22) and (23), it now follows that

$$B^{(-)}(\sigma) = B^{(+)}(\sigma) \quad \text{and} \quad A^{(-)}(\sigma) = \frac{\kappa + 2\mathrm{i}\sigma\beta}{\kappa - 2\mathrm{i}\sigma\beta}A^{(+)}(\sigma). \tag{25}$$

Thus, as stated, while the transmitted amplitudes, $B^{(\pm\sigma)}$, are identically the same, the reflected amplitudes, $A^{(\pm)}(\sigma)$, differ only in their phases. The equality of the reflexion and the transmission coefficients is now an immediate consequence.

4. The infinite hierarchy of integral relations between $V^{(+)}$ and $V^{(-)}$

It was noted early in these studies devoted to the perturbations of the black-hole solutions of general relativity, that the equality of the reflexion and the transmission coefficients is always associated with an equality of the integrals of the potentials over the range of r_* (or x). It was soon pointed out to the writer by Professor K. M. Case that the equality of the integrals of $V^{(+)}$ and $V^{(-)}$ must be the first of an infinite hierarchy of integral equalities one should expect if $V^{(+)}$ and $V^{(-)}$ are to yield the same reflexion and transmission coefficients. Indeed, as is shown in the Appendix, the various combinations $V^{(\pm)}$ and their derivatives, $V^{(\pm)\prime}$, $V^{(\pm)\prime\prime}$, etc., whose integrals must be equal are, formally, the same as the conserved quantities allowed by the Korteweg–deVries equation, namely (cf. Miura *et al.* 1968),

$$\left.\begin{aligned} &\text{(i) } V; \quad \text{(ii) } V^2; \quad \text{(iii) } 2V^3 + V'^2; \quad \text{(iv) } 5V^4 + 10VV'^2 + V''^2; \\ &\text{(v) } 14V^5 + 70V^2V'^2 + 14VV''^2 + V'''^2; \quad \text{etc.} \quad (V = V^{(\pm)}). \end{aligned}\right\} \tag{26}$$

(Five additional members of this hierarchy can be read off from the list given by Miura *et al.* by replacing their u by $-6V$.)

† Solutions with these asymptotic behaviours exist, since the potentials $V^{(+)}$ and $V^{(-)}$, by virtue of the restrictions imposed on f, are of short range.

The fact that the potentials $V^{(+)}$ and $V^{(-)}$, given in equations (13) and (14) do satisfy the various integral equalities can be verified, individually, by evaluating the expressions (26) (and those not listed) for

$$V = \beta f' + \beta^2 f^2 + \kappa f, \tag{27}$$

and showing that the terms, odd in β, are expressible as derivatives of combinations of f and its derivatives, f', f'', etc., and must, therefore, vanish on integration (since f and its derivatives have been assumed to vanish for $x \to \pm\infty$).

The equality of the integrals of $V^{(+)}$ and $V^{(-)}$ follows, for example, from the fact that the integral of $\beta f'$ vanishes.

Considering next V^2, for the terms odd in β, we have

$$2\beta(\beta^2 f^2 + \kappa f) f' = \tfrac{2}{3}\beta^3 (f^3)' + \kappa\beta (f^2)'. \tag{28}$$

The integral of this expression clearly vanishes; and the remaining terms, even in β, are the same for $V^{(+)}$ and $V^{(-)}$.

Similarly, considering the remaining three quantities listed in (26), we find that the terms, odd in β, can be reduced, respectively, to the forms:

(iii) $$2\beta^3(ff'^2)' + \kappa\beta(f'^2)' + 6\beta(\beta^2 f^2 + \kappa f)^2 f'; \tag{29}$$

(iv) $$20\beta^5(f^3 f'^2)' + 30\kappa\beta^3(f^2 f'^2)' + 10\kappa^2\beta(ff'^2)' + 2\beta^3[(ff''^2)' + 2(f'^2 f'')'] + \kappa\beta(f''^2)' + 40\beta(\beta^2 f^2 + \kappa f)^3 f', \tag{30}$$

and

(v) $$\begin{aligned} &140\beta^7(f^5 f'^2)' + \beta^5[70(ff'^4)' + 56(f^2 f'^2 f'')' + 28(f^3 f''^2)'] \\ &+ \beta^3[2(ff'''^2)' + 12(f' f'' f''')' - 4(f''^3)'] + 350\kappa\beta^5(f^4 f'^2)' \\ &+ \kappa\beta^3[56(ff'^2 f'')' + 35(f'^4)' + 42(f^2 f''^2)'] + \kappa\beta(f'''^2)' + 280\kappa^2\beta^2(f^3 f'^2)' \\ &+ 14\kappa^2\beta(ff''^2)' + 70\kappa^3\beta(f^2 f'^2)' + 70\beta(\beta^2 f^2 + \kappa f)^4 f'. \end{aligned} \tag{31}$$

Clearly, the integrals of all these expressions vanish; and the remaining terms in the expansions of the respective quantities, being even in β, are the same when they are evaluated for $V = V^{(+)}$ and $V = V^{(-)}$; and the equality of the integrals follows.

The integral equalities of order higher than the fifth, we have considered, can, one may be confident, be verified, individually, in similar fashion. It does not appear that an inductive proof that will establish *all* the equalities can be devised since no such proof has been devised even for establishing the existence of the infinite set of conserved quantities allowed by the Korteweg–deVries equation. (In the latter case, an ingenious procedure, involving the inversion of an infinite series, was devised by Miura *et al.* 1968.) But the explicit verification of the first five equalities strengthens our belief in the correctness of the results of the analysis given in the appendix, even if the analysis itself does not meet the requisite standards of mathematical rigour.

Finally, we may note here the value of the integrals over the five quantities listed in equation (26) for the potentials appropriate for the Schwarzschild metric. It is known that (Chandrasekhar 1975, eq. (38))

(i) $$\int_{-\infty}^{+\infty} V^{(\pm)}\,dr_* = \frac{1}{4M}(4n+1). \tag{32}$$

I am indebted to Miss Donna Elbert for the following evaluation of the integrals over the four remaining quantities:

$$\left.\begin{aligned}
&\text{(ii)} \quad \frac{1}{480M^3}(5p^2-18p+18);\\
&\text{(iii)} \quad \frac{1}{26\,880M^5}(16p^3-83p^2+150p-87);\\
&\text{(iv)} \quad \frac{1}{128M^7}\left(\frac{1}{168}p^4-\frac{53}{1386}p^3+\frac{263}{2772}p^2-\frac{147}{1430}p+\frac{444}{10\,010}\right);\\
&\text{(v)} \quad \frac{1}{512M^9}\left(\frac{14}{6435}p^5-\frac{41}{2574}p^4+\frac{56}{1155}p^3-\frac{2557}{34\,320}p^2+\frac{1203}{19\,448}p-\frac{723}{38\,896}\right),
\end{aligned}\right\} \qquad (33)$$

where $p = 2(n+1)$.

5. Concluding remarks

The ideas underlying this paper require to be generalized to allow for complex potentials of the kind encountered in the treatment of the perturbations of the Kerr metric. In the context of the Kerr metric, we are led, besides, to *four* potentials all of which yield the same reflexion and transmission coefficients. It is not clear whether the four potentials can be reduced to certain standard forms as has been possible for the pair of real potentials considered in this paper. These are questions that remain to be investigated; but the relation, already disclosed to the theory of the solitons and the Korteweg–deVries equation, is perhaps of some interest in itself.

The research reported in this paper has in part been supported by a grant from the National Science Foundation under grant PHY 78-24275 with the University of Chicago.

Appendix. Necessary conditions for different potentials to yield the same transmission amplitude

The analysis which follows is largely derived from a paper by Faddeev (1964). While Faddeev's paper originated in his studies on the theory of 'inverse scattering', it is equally relevant to the problem considered in this paper even though, for its strict applicability, some of the restrictive assumptions of Faddeev's analysis must be relaxed.

The one-dimensional wave equation

$$\psi_{,xx}+(\sigma^2-V)\psi = 0 \quad (-\infty < x < +\infty), \qquad \text{(A 1)}$$

where $V(x)$ is a potential whose integral over the range of x is bounded, allows solutions with the asymptotic behaviours,

$$\left.\begin{aligned}
\psi &\to e^{+i\sigma x}+A(\sigma)\,e^{-i\sigma x} && (x\to+\infty)\\
&\to B(\sigma)\,e^{+i\sigma x} && (x\to-\infty).
\end{aligned}\right\} \qquad \text{(A 2)}$$

We now ask for the conditions on V which will yield the same complex amplitude $B(\sigma)$.

In all discussions bearing on the question of the extent to which the potential V is uniquely determined by a knowledge of *both* $A(\sigma)$ and $B(\sigma)$, it is customary to restrict V by the requirement that

$$\int_{-\infty}^{+\infty} V(x)\,(1+|x|)\,dx \quad \text{is bounded.} \tag{A 3}$$

When $V(x)$ satisfies this requirement, it has been shown that the amplitude $B(\sigma)$, for complex σ, is an analytic function of σ in the lower half of the complex σ-plane. The potentials in which we are interested have an inverse square behaviour for $x \to +\infty$ and, therefore, do not satisfy the requirement (A 3). We shall, nevertheless, assume that $\lg B(\sigma)$ allows a Laurent expansion in inverse powers of σ for $\text{Im}(\sigma) < 0$ – an assumption whose validity remains to be established.

Writing

$$\psi = e^{i\sigma x + w}, \tag{A 4}$$

we find that w satisfies the differential equation

$$w_{,xx} + 2i\sigma w_{,x} + w_{,x}^2 - V = 0. \tag{A 5}$$

By the further substitutions,

$$w = -\int_x^{\infty} v(x', \sigma)\,dx' \quad \text{and} \quad w_{,x} = -v, \tag{A 6}$$

we obtain the Riccati equation

$$v_{,x} + 2i\sigma v + v^2 - V = 0. \tag{A 7}$$

The boundary conditions (A 2) imposed on ψ, now require that

$$\left.\begin{aligned} w &\to \lg B \quad \text{for} \quad x \to -\infty \\ &\to 0 \quad\;\; \text{for} \quad x \to +\infty \end{aligned}\right\} \quad (\text{Im}\,(\sigma) < 0); \tag{A 8}$$

and also, that

$$\lg B = -\int_{-\infty}^{+\infty} v(x, \sigma)\,dx. \tag{A 9}$$

And finally, it follows from equation (A 7) that

$$v \to V/2i\sigma \quad (|\sigma| \to \infty). \tag{A 10}$$

We now suppose that v can be expanded in a Laurent series of the form

$$v = \sum_{n=1}^{\infty} \frac{v_n(x)}{(2i\sigma)^n}, \tag{A 11}$$

where, according to equation (A 10)

$$v_1 = V. \tag{A 12}$$

Inserting the series expansion (A 11) into equation (A 9), we obtain

$$\lg B = -\sum_{n=1}^{\infty} \frac{1}{(2i\sigma)^n} \int_{-\infty}^{+\infty} v_n(x)\,dx, \tag{A 13}$$

where we have assumed that integration, term by term, is permissible. Thus, we have an expansion for $\lg B$ of the form

$$\lg B = -\sum_{n-1}^{\infty} c_n \sigma^{-n}, \tag{A 14}$$

where

$$(2i)^n c_n = \int_{-\infty}^{+\infty} v_n dx. \tag{A 15}$$

Now inserting the expansion (A 11) in equation (A 7) and assuming that the coefficients of the different inverse powers of σ can be equated, we obtain the recurrence relation,

$$v_n = -v_{n-1,x} - \sum_{l=1}^{n-1} v_l v_{n-1-l}. \tag{A 16}$$

Using this recurrence relation, we can solve for the v_n's, successively, starting with (cf. equation A 10))

$$v_1 = V. \tag{A 17}$$

We find

$$\left.\begin{aligned} v_2 &= -v_{1,x} = -V_{,x}, \\ v_3 &= -v_{2,x} - v_1^2 = V_{,xx} - V^2, \\ v_4 &= -v_{3,x} - 2v_1 v_2 = -V_{,xxx} + 2(V^2)_{,x}, \\ v_5 &= -v_{4,x} - 2v_1 v_3 - v_2^2 = V_{,xxxx} - 3(V^2)_{,xx} + (V_{,x})^2 + 2V^3, \end{aligned}\right\} \tag{A 18}$$

etc.

The coefficients of odd order, c_{2n+1}, in the expansion for $\lg B$ are, therefore, given by

$$2ic_1 = \int_{-\infty}^{+\infty} V\,dx; \; -(2i)^3 c_3 = \int_{-\infty}^{+\infty} V^2\,dx; \; (2i)^5 c_5 = \int_{-\infty}^{+\infty} (2V^3 + V_{,x})^2\,dx; \text{ etc.,} \tag{A 19}$$

while all the coefficients of even order, c_{2n}, vanish.

That the further terms in the hierarchy (A 19) are precisely the conserved quantities of the Korteweg–deVries equation follows from the fact that its solution is expressible in terms of the reflexion and the transmission amplitudes of the associated one dimensional wave equation (cf. Whitham 1974).

Now if two different potentials should yield the same transmission amplitude $B(\sigma)$, then we should expect that the derived Laurent expansions of $\lg B$, for both potentials, must coincide term by term; and we may conclude that the integrals listed in (A 19) (and the others of the infinite sequence to which they belong) must be the same when evaluated for the two potentials. We have explicitly verified in

§ 4 that this is the case for the integrals of the first five orders for the potentials, $V^{(+)}$ and $V^{(-)}$, considered, even though, they were required to satisfy less stringent conditions than those imposed on V in the present analysis.

References

Chandrasekhar, S. 1975 *Proc. R. Soc. Lond.* A **343**, 289.

Chandrasekhar, S. 1979*a* *General relativity – an Einstein centenary survey* (ed. W. Israel & S. Hawking), ch. 7. Cambridge University Press.

Chandrasekhar, S. 1979*b* *Proc. R. Soc. Lond.* A **365**, 453.

Chandrasekhar, S. & Detweiler, S. 1975 *Proc. R. Soc. Lond.* A **344**, 441.

Faddeev, L. D. 1964 *Tr. Mat. Inst. V. A. Steklova*, **73**, 314.

Heading, J. 1977 *J. Phys.* A **10**, 885.

Miura, R. M., Gardner, C. S. & Kruskal, M. D. 1968 *J. Math. Phys.* **9**, 1262.

Whitham, G. B. 1974 *Linear and nonlinear waves*, §§ 16.14–16.16 and 17.2–17.4. New York: J. Wiley.

17

On crossing the Cauchy horizon of a Reissner–Nordström black-hole

By S. Chandrasekhar, F.R.S., and J. B. Hartle
University of Chicago, Chicago, Illinois 60637, U.S.A.

(*Received* 18 *June* 1982)

The behaviour on the Cauchy horizon of a flux of gravitational and/or electromagnetic radiation crossing the event horizon of a Reissner–Nordström black-hole is investigated as a problem in the theory of one-dimensional potential-scattering. It is shown that the flux of radiation received by an observer crossing the Cauchy horizon, along a radial time-like geodesic, diverges for all physically reasonable perturbations crossing the event horizon, even including those with compact support.

1. Introduction

The geometry of the analytically completed Reissner–Nordström space–time suggests that one may emancipate oneself from the past and escape into 'new worlds' by crossing the inner Cauchy horizon of the black hole. But as Simpson & Penrose (1972) have pointed out the 'projected journey is likely to be a dangerous undertaking'. The anticipated 'danger' consists of the following.

Consider an observer describing a time-like geodesic crossing the Cauchy horizon across C′E in the Penrose diagram (figure 1) of the space–time. Physical considerations (cf. §7) would suggest that the flux of radiation, derived from a smallest source in the 'outside world', that is incident on the observer will diverge at the instant of his crossing the horizon. Since Simpson & Penrose's early preliminary considerations, several authors, including McNamara (1978*a*, *b*), Gürsel *et al.* (1979*a*, *b*), and Matzner *et al.* (1979) have considered this problem. While the considerations by these various authors have contributed to a basic understanding of the problem, no simple and complete demonstration of the divergence of the flux of radiation incident on an observer crossing the Cauchy horizon has been provided on the basis of the standard equations of the underlying perturbation theory. In this paper we provide such a demonstration.

We begin with a brief summary of the basic equations governing the perturbations of the Reissner–Nordström black-hole as presented by Chandrasekhar (1979, 1983) and Chandrasekhar & Xanthopoulos (1979).

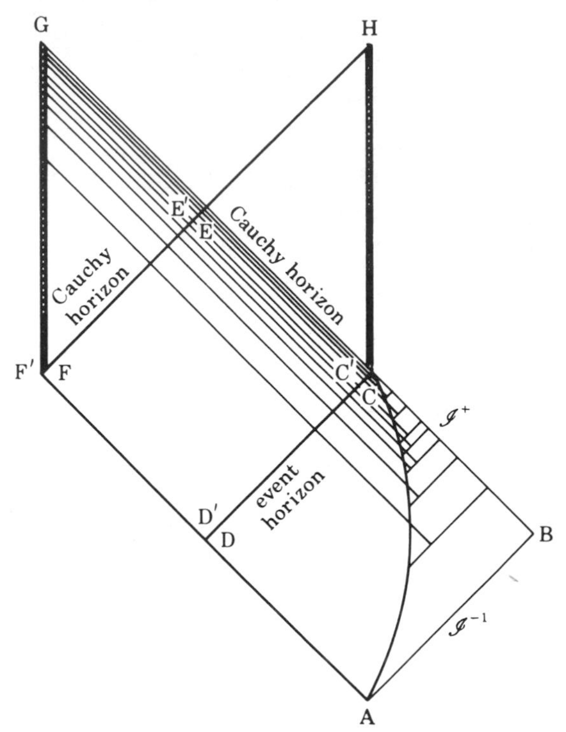

FIGURE 1. The Penrose diagram for the Reissner–Nordström space–time illustrating how the secondary scattering of perturbations along the arc AC will result in an accumulation of wavefronts at $\mathscr{I}^+$.

2. THE BASIC EQUATIONS OF THE PROBLEM

The perturbations of the Reissner–Nordström black-hole (of mass M and charge Q_* in the units in which $c = G = 1$) analysed into their normal modes with a time dependence $e^{i\sigma t}$ and angular dependences given by appropriate Legendre functions (belonging to given l and m) fall into two classes, which we shall distinguish as *axial* and *polar*. The equations governing these two classes of perturbations (distinguished by superscripts + and −) can be reduced to a pair of one-dimensional Schrödinger wave equations of the standard form. We have

$$\frac{d^2Z_i^{(\pm)}}{dr_*^2} + \sigma^2 Z_i^{(\pm)} = V_i^{(\pm)} Z_i^{(\pm)} \quad (i = 1, 2), \tag{1}$$

where

$$r_* = r + \frac{1}{2\kappa_+} \lg|r_+ - r| - \frac{1}{2\kappa_-} \lg|r - r_-|, \tag{2}$$

$$\kappa_+ = \frac{r_+ - r_-}{2r_+^2}, \quad \kappa_- = \frac{r_+ - r_-}{2r_-^2}, \tag{3}$$

$$V_i^{(-)} = \frac{\Delta}{r^5}\left[(\mu^2 + 2)\,r - q_j + \frac{4Q_*^2}{r}\right] \quad (i, j = 1, 2;\ i \neq j), \tag{4}$$

$$V_i^{(+)} = V_i^{(-)} + 2q_j \frac{\mathrm{d}}{\mathrm{d}r_*} \frac{\Delta}{r^3(\mu^2 r + q_j)} \quad (i,j = 1,2;\ i \neq j), \tag{5}$$

$$\Delta = (r - r_+)(r - r_-) = r^2 - 2Mr + Q_*^2, \quad \mu^2 = (l-1)(l+2), \tag{6}$$

$$q_1 = 3M + \sqrt{(9M^2 + 4Q_*^2 \mu^2)} \quad \text{and} \quad q_2 = 3M - \sqrt{(9M^2 + 4Q_*^2 \mu^2)}. \tag{7}$$

An important relation between the solutions belonging to the axial and the polar perturbations, which simplifies the further discussion of these perturbations considerably, is that given a solution $Z_i^{(+)}$ (or $Z_i^{(-)}$) belonging to one parity we can derive a solution belonging to the opposite parity. With a suitable normalization, the relation in question is

$$[\mu^2(\mu^2+2) \pm 2\mathrm{i}\sigma q_j]\, Z_i^{(\pm)} = \left[\mu^2(\mu^2+2) + \frac{2q_j^2 \Delta}{r^3(\mu^2 r + q_j)}\right] Z_i^{(\mp)} \pm 2q_j \frac{\mathrm{d}Z_i^{(\mp)}}{\mathrm{d}r_*} \quad (i,j = 1,2;\ i \neq j). \tag{8}$$

In the context of equation (1), the standard problem is, of course, that of the reflexion and transmission of incident waves. Precisely, the problem is to obtain solutions of the equations satisfying the boundary conditions

$$\left.\begin{aligned} Z_i^{(\pm)} &\to \mathrm{e}^{+\mathrm{i}\sigma r_*} + R_i^{(\pm)}(\sigma)\, \mathrm{e}^{-\mathrm{i}\sigma r_*} \quad && (r_* \to +\infty), \\ Z_i^{(\pm)} &\to T_i^{(\pm)}(\sigma)\, \mathrm{e}^{+\mathrm{i}\sigma r_*} && (r_* \to -\infty), \end{aligned}\right\} \tag{9}$$

where $R_i^{(\pm)}(\sigma)$ and $T_i^{(\pm)}(\sigma)$ denote the amplitudes of the reflected and the transmitted waves resulting from the incident wave, $\mathrm{e}^{+\mathrm{i}\sigma r_*}$, of unit amplitude.

By virtue of the relation (8), the reflexion and the transmission amplitudes, belonging to the axial and the polar perturbations, are very simply related:

$$T_i^{(+)}(\sigma) = T_i^{(-)}(\sigma) \quad \text{and} \quad R_i^{(+)}(\sigma) = \frac{\mu^2(\mu^2+2) - 2\mathrm{i}\sigma q_j}{\mu^2(\mu^2+2) + 2\mathrm{i}\sigma q_j} R_i^{(-)}(\sigma) \quad (i,j = 1,2;\ i \neq j). \tag{10}$$

Finally, we may note that the functions $Z_i^{(\pm)}$ in terms of which the perturbations have been expressed are gauge-invariant: they can, for example, be expressed explicitly in terms of the Weyl and the Maxwell scalars of the Newman–Penrose formalism.

(*a*) *Some known relations in the theory of one-dimensional scattering*

For later reference, we shall collect here some elementary relations in the theory of one-dimensional potential-scattering. The relations in various guises can be found in a variety of places. A recent useful reference is Deift & Trubowitz (1979).

We shall suppress the distinguishing subscripts and superscripts and replace r_* by x. The range of x (as of r_*) is $(-\infty, +\infty)$. We shall consider then equation (1) in the form

$$\left(\frac{\mathrm{d}^2}{\mathrm{d}x^2} + \sigma^2\right) Z = VZ; \tag{11}$$

and we shall further suppose that V (even as $V^{(\pm)}$ is) is bounded and that its integral over the range of r_* is finite.

We consider two solutions, $f_1(x, \sigma)$ and $f_2(x, \sigma)$, of equation (11) with the asymptotic behaviours

$$\left.\begin{aligned} &f_1(x, \sigma) \to e^{-i\sigma x} \quad (x \to +\infty) \\ \text{and} \qquad &f_2(x, \sigma) \to e^{+i\sigma x} \quad (x \to -\infty). \end{aligned}\right\} \tag{12}$$

(Solutions with these behaviours exist when $V(x)$ is integrable.) It can be shown (by evaluating their Wronskians) that the pairs, $\{f_1(x, \sigma), f_1(x, -\sigma)\}$ and $\{f_2(x, \sigma), f_2(x, -\sigma)\}$, represent independent solutions. It follows that there exist unique functions $R_1(\sigma)$ and $T_1(\sigma)$ such that

$$f_2(x, \sigma) = \frac{R_1(\sigma)}{T_1(\sigma)} f_1(x, \sigma) + \frac{1}{T_1(\sigma)} f_1(x, -\sigma). \tag{13}$$

From a comparison of the asymptotic behaviour,

$$f_2(x, \sigma) \to \frac{R_1(\sigma)}{T_1(\sigma)} e^{-i\sigma x} + \frac{1}{T_1(\sigma)} e^{+i\sigma x} \quad (x \to +\infty), \tag{14}$$

with equation (9) we conclude that $R_1(\sigma)$ and $T_1(\sigma)$ have the same meanings as reflexion and transmission amplitudes as we have defined them.

From equation (14) and the known asymptotic behaviours of $f_1(x, \pm\sigma)$ and $f_2(x, \pm\sigma)$ we can express $1/T_1(\sigma)$ and $R_1(\sigma)/T_1(\sigma)$ in terms of the Wronskians of f_1 and f_2; thus

$$\frac{1}{T_1(\sigma)} = -\frac{1}{2i\sigma}[f_1(x, \sigma), f_2(x, \sigma)] \tag{15}$$

and

$$\frac{R_1(\sigma)}{T_1(\sigma)} = -\frac{1}{2i\sigma}[f_2(x, \sigma), f_1(x, -\sigma)], \tag{16}$$

where

$$[f, g] = f'g - fg' \tag{17}$$

(the prime denoting differentiation).

Finally, writing

$$f_2(x, \sigma) = e^{i\sigma x} + \psi(x) \quad \text{and} \quad \psi(x) \to 0 \quad \text{as} \quad x \to -\infty, \tag{18}$$

and solving for $\psi(x)$ in terms of a suitably defined Green function, we obtain for $f_2(x, \sigma)$ the integral equation

$$f_2(x, \sigma) = e^{+i\sigma x} + \int_{-\infty}^{x} \frac{\sin\sigma(x - x_1)}{\sigma} V(x_1) f_2(x_1, \sigma)\, dx_1. \tag{19}$$

In similar fashion, we find for $f_1(x, \sigma)$ the integral equation

$$f_1(x, \sigma) = e^{-i\sigma x} - \int_{x}^{\infty} \frac{\sin\sigma(x - x_1)}{\sigma} V(x_1) f_1(x_1, \sigma)\, dx_1. \tag{20}$$

3. The equations governing the dispersion of waves in the domain between the two horizons

The equations of § 2 are entirely general. We shall now specialize them to the particular problem in the perturbations of the Reissner–Nordström black-hole that we wish to consider. The range of r in which we are now interested is

$$r_- < r < r_+; \tag{21}$$

the corresponding range of r_* is

$$+\infty > r_* > -\infty \quad (r_* \to +\infty \text{ for } r \to r_- + 0 \text{ and } r_* \to -\infty \text{ for } r \to r_+ - 0). \tag{22}$$

Also, Δ as defined in equation (6) is now negative. By virtue of this last fact, all four potentials, $V_i^{(\pm)}$ $(i = 1, 2)$, are negative in the entire interval (21). We are thus dealing with *potential wells* (in contrast to *potential barriers* outside the event horizon). Since the considerations in this section are applicable to all four potentials, we shall suppress the distinguishing subscripts and superscripts.

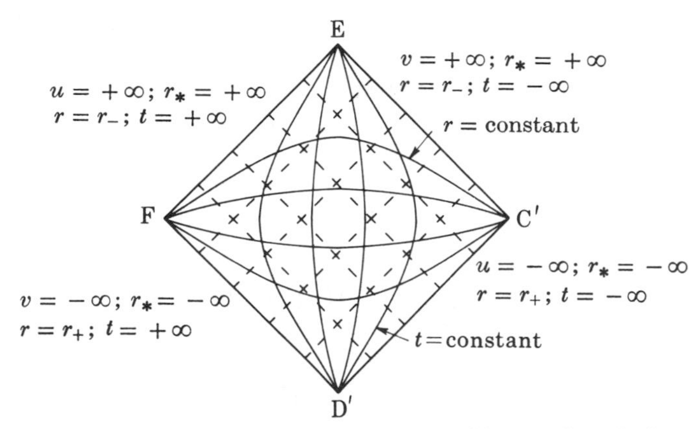

Figure 2. The different coordinate systems used in the domain between the two horizons of the Reissner–Nordström black-hole.

The physical problem that is of interest in the context of perturbations between the two horizons is the dispersion of waves entering the domain across the event horizon at D′C′ (see figure 1). We must therefore seek solutions of the basic equations that satisfy the boundary conditions

$$\left.\begin{aligned} Z(r^*, \sigma) &\to A(\sigma)\,\mathrm{e}^{-\mathrm{i}\sigma r_*} + B(\sigma)\,\mathrm{e}^{+\mathrm{i}\sigma r_*} \quad && (r_* \to +\infty,\ r \to r_- + 0) \\ Z(r^*, \sigma) &\to e^{-\mathrm{i}\sigma r_*} && (r_* \to -\infty,\ r \to r_+ - 0). \end{aligned}\right\} \tag{23}$$

We can readily convince ourselves that the amplitudes $A(\sigma)$ and $B(\sigma)$ are related to the reflexion and the transmission amplitudes, as defined in § 2, in the manner

$$A(\sigma) = \frac{1}{T_1(-\sigma)} \quad \text{and} \quad B(\sigma) = \frac{R_1(-\sigma)}{T_1(-\sigma)}. \tag{24}$$

In analysing the radiation arriving at the Cauchy horizon at r_- we must distinguish the edges EC′ and EF in the Penrose diagram (figure 1). For this reason, we restore the time dependence $e^{i\sigma t}$ of the solutions and introduce the null coordinates(see figure 2)

$$u = r_* + t \quad \text{and} \quad v = r_* - t, \tag{25}$$

and write in place of equation (23)

$$Z(r_*, t) \to e^{-i\sigma v} + [A(\sigma) - 1] e^{-i\sigma v} + B(\sigma) e^{+i\sigma u} \quad (r_* \to +\infty; r \to r_- + 0) \tag{26}$$

and

$$Z(r_*, t) \to e^{-i\sigma v} \quad (r_* \to -\infty; r \to r_+ - 0). \tag{27}$$

We now suppose that the form of the perturbation crossing the event horizon at D′C′ is $\hat{Z}(v)$ so that

$$\mathscr{Z}(\sigma) = \frac{1}{2\pi} \int_{-\infty}^{+\infty} \hat{Z}(v) e^{i\sigma v} dv. \tag{28}$$

In writing equation (28), we are supposing that the physical restrictions on the radiation crossing the event horizon will allow us to take the Fourier transform of $\hat{Z}(v)$. (We return to this matter in § 6.)

The radiation crossing the event horizon will disperse in the domain between the two horizons; and at the Cauchy horizon the amplitude of the scattered wave will be determined by (cf. equation (26))

$$\mathscr{Z}_{\text{scattered}}(r^*, t) \to X(v) + Y(u) \quad (r_* \to +\infty), \tag{29}$$

where

$$X(v) = \int_{-\infty}^{+\infty} \mathscr{Z}(\sigma) [A(\sigma) - 1] e^{-i\sigma v} d\sigma \tag{30}$$

and

$$Y(u) = \int_{-\infty}^{+\infty} \mathscr{Z}(\sigma) B(\sigma) e^{+i\sigma u} d\sigma. \tag{31}$$

As we have explained in § 1, we are interested in the flux of radiation received by an observer crossing the Cauchy horizon along a radial time-like geodesic. If U^j denotes the four-velocity of the observer along the geodesic, the flux of radiation received by the observer is proportional to the square of the *amplitude*,

$$\mathscr{F} = U^j Z_{,j}. \tag{32}$$

It can be shown that, at the edges EF and EC′ of the Cauchy horizon, this amplitude is given by (cf. Matzner *et al.* 1979; Chandrasekhar 1983)

$$\mathscr{F}_{\text{EF}} = -\frac{2r_-^2}{r_+ - r_-} E \lim_{u \to +\infty} e^{\kappa_- u} Y_{,u}, \tag{33}$$

and

$$\mathscr{F}_{\text{EC}'} = +\frac{2r_-^2}{r_+ - r_-} |E| \lim_{v \to +\infty} e^{\kappa_- v} X_{,-v}, \tag{34}$$

where E denotes the constant time-component of the covariant four-velocity U_j of the observer. We conclude from equations (24), (30), and (31) that the divergence, or otherwise, of the received fluxes on the Cauchy horizon depend on

$$Y_{,u} = \int_{-\infty}^{+\infty} \mathrm{i}\sigma \frac{R_1(-\sigma)}{T_1(-\sigma)} \mathscr{Z}(\sigma)\,\mathrm{e}^{+\mathrm{i}\sigma u}\,\mathrm{d}\sigma \tag{35}$$

and

$$X_{,-v} = \int_{-\infty}^{+\infty} \mathrm{i}\sigma \left[\frac{1}{T_1(-\sigma)} - 1\right] \mathscr{Z}(\sigma)\,\mathrm{e}^{-\mathrm{i}\sigma v}\,\mathrm{d}\sigma, \tag{36}$$

where we may substitute for $R_1(-\sigma)/T_1(-\sigma)$ and $1/T_1(-\sigma)$ in accordance with equations (15) and (16). Thus, apart from constants, the quantities to be considered are

$$\mathscr{F}_u = \mathrm{e}^{\kappa_- u} \int_{-\infty}^{+\infty} [f_2(x, -\sigma), f_1(x, +\sigma)]\, \mathscr{Z}(\sigma)\,\mathrm{e}^{+\mathrm{i}\sigma u}\,\mathrm{d}\sigma \tag{37}$$

and

$$\mathscr{F}_v = \mathrm{e}^{\kappa_- v} \int_{-\infty}^{+\infty} \{[f_1(x, -\sigma), f_2(x-\sigma)] - 2\mathrm{i}\sigma\}\, \mathscr{Z}(\sigma)\,\mathrm{e}^{-\mathrm{i}\sigma v}\,\mathrm{d}\sigma; \tag{38}$$

and, as we have stated, we are primarily interested in the asymptotic behaviours of $\mathscr{F}_u$ for $u \to +\infty$ and $\mathscr{F}_v$ for $v \to +\infty$. The required asymptotic behaviours are most conveniently ascertained by integration along suitably defined contours in the complex σ-plane. But the application of this method requires a knowledge of the analytic properties of the functions $f_1(x, \pm\sigma)$ and $f_2(x, -\infty)$ for complex arguments; and these properties in turn depend on the analytic properties of the potentials. We now turn to a consideration of these matters.

4. The analytic properties of the potentials

In view of the relations (10) between the reflexion and the transmission amplitudes belonging to the axial and the polar perturbations, it will clearly suffice to restrict our consideration to the analytic properties of the functions $f_1(x, \pm\sigma)$ and $f_2(x, -\sigma)$ belonging to the axial perturbations: the integrand of $X_{,-v}$ is the same for the two classes of perturbations while the integrand of $Y_{,u}$ differs only by a simple factor. This is a fortunate circumstance since the potentials $V_i^{(-)}$ $(i = 1, 2)$, as complex functions of r, are analytic in the entire complex plane except at $r = 0$; in contrast, $V_i^{(+)}$ $(i = 1, 2)$ has singularities besides the one at $r = 0$. The remaining considerations in this section will therefore be restricted to $V_i^{(-)}$. For convenience we shall suppress the distinguishing superscript and subscript. We shall also write (cf. equation (2))

$$r_* = x = r + \frac{1}{2\kappa_+} \lg\left(1 - \frac{r}{r_+}\right) - \frac{1}{2\kappa_-} \lg\left(\frac{r}{r_-} - 1\right) + x_0, \tag{39}$$

where

$$x_0 = \frac{1}{2\kappa_+} \lg r_+ - \frac{1}{2\kappa_-} \lg r_- \quad \text{and} \quad r_- < r < r_+. \tag{40}$$

First, we observe that $V(x)$, as defined in equation (4), has the asymptotic behaviours

$$\left.\begin{aligned} V(x) &\to \text{constant} \times e^{-2\kappa_- x} \quad (x \to +\infty), \\ V(x) &\to \text{constant} \times e^{+2\kappa_+ x} \quad (x \to -\infty). \end{aligned}\right\} \tag{41}$$

and

We shall now show, by considering V as a complex function either of r or of

$$z = x - x_0 + iy, \tag{42}$$

that $V(x)$ has the representations

$$V(x) = \int_{2\kappa_+}^{\infty} \mathscr{V}_+(\xi)\, e^{+x\xi}\, d\xi \tag{43}$$

and

$$V(x) = \int_{2\kappa_-}^{\infty} \mathscr{V}_-(\xi)\, e^{-x\xi}\, d\xi. \tag{44}$$

To establish the foregoing representations and to obtain explicit expressions for $\mathscr{V}_\pm(\xi)$, it is necessary to evaluate the inverse Laplace-transform,

$$\mathscr{V}_\pm(\xi) = \frac{1}{2\pi i} \int_{\mathscr{C}_\pm} V(z)\, e^{\mp\xi z}\, dz, \tag{45}$$

where $\mathscr{C}_\pm$ are contours that run parallel to the imaginary z-axis with $\mathscr{C}_+$ lying entirely to the left of any singularity of $V(z)$ and $\mathscr{C}_-$ lying entirely to the right of the singularities.

As we have seen, $V(r)$ is an analytic function of r except at $r = 0$. It follows that $V(z)$ is an analytic function of z except at the corresponding poles located along the imaginary z-axis with periods π/κ_+ and π/κ_-. There is, however, no pole at $z = 0$. Therefore, in the complex r-plane, $\mathscr{C}_+$ is a closed contour about the point r_+ not including r_-. This means that $V(z)$ evaluated on $\mathscr{C}_+$ will be periodic in y with the period π/κ_+. For the integral defining $\mathscr{V}_+(\xi)$, we shall thus have

$$\mathscr{V}_+(\xi) = \frac{e^{\xi x_+}}{2\pi} \int_{-\infty}^{+\infty} V(x_+ + iy)\, e^{-i\xi y}\, dy, \tag{46}$$

where x_+ is the constant real part of z along $\mathscr{C}_+$. The integral on the right-hand side of equation (46) clearly reduces to an infinite sum of δ-functions. We may thus write

$$\mathscr{V}_+(\xi) = \sum_{-\infty}^{+\infty}{}' C_m^{(+)}\, \delta(\xi - 2m\kappa_+). \tag{47}$$

The prime in the summation sign in equation (47) indicates that there is no term with $m = 0$ since the coefficient $C_0^{(+)}$ is identically zero: the integral (46) can then be evaluated along a contour in the left-hand plane where no poles are enclosed.

From equations (43) and (47) we now obtain the representation

$$V(x) = \sum_{m=1}^{\infty} C_m^{(+)}\, e^{2m\kappa_+ x}. \tag{48}$$

We observe that the leading term in (48) agrees with known asymptotic behaviour of $V(x)$ for $x \to -\infty$ (cf. equation (41)).

In similar fashion, starting with equation (44), we shall find

$$\mathscr{V}_-(\xi) = \sum_{m=-\infty}^{+\infty}{}' C_m^{(-)}\, \delta(\xi - 2m\kappa_-), \tag{49}$$

and

$$V(x) = \sum_{m=1}^{\infty} C_m^{(-)}\, e^{-2m\kappa_- x}. \tag{50}$$

The constants, $C_m^{(\pm)}$, that occur in the expansions (48) and (50) for $V(x)$ can be bounded by considering them (as we may) as the coefficients in the Fourier expansions of $V(x_\pm + iy)$ where x_+ and x_- are to the right and to the left, respectively, of any singularity of V. Thus, by writing

$$C_m^{(+)} = \frac{\kappa_+}{\pi} e^{-2m\kappa_+ x_+} \int_0^{\pi/\kappa_+} V(x_+ + iy)\, e^{-2im\kappa_+ y}\, dy, \tag{51}$$

we conclude that

$$C_m^{(+)} \leqslant M^{(+)}(x_+)\, e^{-2m\kappa_+ x_+}, \tag{52}$$

where

$$M^{(+)}(x_+) = \frac{\kappa_+}{\pi} \int_0^{\pi/\kappa_+} |V(x_+ + iy)|\, dy. \tag{53}$$

Similarly, we have

$$C_m^{(-)} \leqslant M^{(-)}(x_-)\, e^{2m\kappa_- x_-}, \tag{54}$$

where

$$M^{(-)}(x_-) = \frac{\kappa_-}{\pi} \int_0^{\pi/\kappa_-} |V(x_- + iy)|\, dy. \tag{55}$$

5. Analytic properties of $f_1(x, \sigma)$ and $f_2(x, \sigma)$

We shall now show that with the representations (48) and (50) for the potentials belonging to the axial perturbations and the bounds (52) and (54) on the coefficients, $C_m^{(\pm)}$, we can specify completely the analytic properties of the scattering functions $f_1(x, \sigma)$ and $f_2(x, \sigma)$. In the analysis, we shall, in the main, follow the treatment in Hartle & Wilkins (1972) of a related problem in the theory of the perturbations of the Kerr black-hole. The method of treatment, originally due to Martin (1959), is a standard one in the theory of potential scattering (see for example de Alfairo & Regge (1965) for a review).

Considering first $f_2(x, -\sigma)$, we can obtain a formal solution of the integral equation (cf. equation (19))

$$f_2(x, -\sigma) = e^{-i\sigma x} + \int_{-\infty}^{x} \frac{\sin \sigma(x - x_1)}{\sigma} V(x_1) f_2(x_1, -\sigma)\, dx \tag{56}$$

by successive iteration. Thus, we write

$$f_2(x, -\sigma) = e^{-i\sigma x} + \sum_{n=1}^{\infty} f_2^{(n)}(x, -\sigma), \tag{57}$$

where

$$f_2^{(n)}(x, -\sigma) = \int_{-\infty}^{x} dx_1 \frac{\sin \sigma(x - x_1)}{\sigma} V(x_1) f_2^{(n-1)}(x_1, -\sigma). \tag{58}$$

The explicit expression for the nth iterate is

$$f_2^{(n)}(x,-\sigma)=\int_{-\infty}^{x_0}\mathrm{d}x_1\int_{-\infty}^{x_1}\mathrm{d}x_2\ldots\int_{-\infty}^{x_{n-1}}\mathrm{d}x_n\prod_{i=1}^{n}\frac{\sin\sigma(x_{i-1}-x_i)}{\sigma}V(x_i)\,\mathrm{e}^{-\mathrm{i}\sigma x_n}$$

$$=\frac{\mathrm{e}^{-\mathrm{i}\sigma x}}{(2\mathrm{i}\sigma)^n}\int_{-\infty}^{x_0}\mathrm{d}x_1\ldots\int_{-\infty}^{x_{n-1}}\mathrm{d}x_n\prod_{i=1}^{n}\{[\mathrm{e}^{2\mathrm{i}\sigma(x_{i-1}-x_i)}-1]\,V(x_i)\},\tag{59}$$

where $x_0=x$. Letting

$$y_i=x_{i-1}-x_i,\tag{60}$$

we can rewrite equation (59), alternatively, in the form

$$f_2^{(n)}(x,-\sigma)=\frac{\mathrm{e}^{-\mathrm{i}\sigma x}}{(2\mathrm{i}\sigma)^n}\int_0^{\infty}\mathrm{d}y_1\ldots\int_0^{\infty}\mathrm{d}y_n\prod_{i=1}^{n}\{(1-\mathrm{e}^{2\mathrm{i}\sigma y_i})\,V(x_i)\}.\tag{61}$$

Now substituting for $V(x_i)$ its representation (48) and performing the integrations over $y_n,\ldots,y_1$, successively, we find

$$f_2^{(n)}(x,-\sigma)=\sum_{m_n>m_{n-1}>\ldots>m_1}C^{(+)}_{m_n-m_{n-1}}C^{(+)}_{m_{n-1}-m_{n-2}}\ldots C^{(+)}_{m_2-m_1}C^{(+)}_{m_1}\times d^{(+)}_{m_n}d^{(+)}_{m_{n-1}}\ldots d^{(+)}_{m_1}\,\mathrm{e}^{2m_n\kappa_+x-\mathrm{i}\sigma x},\tag{62}$$

where
$$d^{(+)}_m=-[4m\kappa_+(m\kappa_+-\mathrm{i}\sigma)]^{-1}.\tag{63}$$

In equation (62) the summation, as indicated, is over positive integers such that $m_n>\ldots>m_1$.

We observe that each $f_2^{(n)}(x,-\sigma)$, as defined in equation (62), is an analytic function of σ except for isolated poles at

$$\sigma=-\mathrm{i}m\kappa_+\quad(m=1,2,\ldots).\tag{64}$$

Further, by making use of the bounds (52) on the coefficients $C^{(+)}_m$ we can now bound $f_2^{(n)}(x,-\sigma)$. Thus, from equation (62) we conclude that

$$|f_2^{(n)}(x,-\sigma)|\leqslant[M^{(+)}(x_+)]^n\sum_{m_n>\ldots>m_1}|d^{(+)}_{m_n}|\ldots|d^{(+)}_{m_1}|\,\mathrm{e}^{2m_n\kappa_+(x-x_+)}.\tag{65}$$

Noting that $m_n>n$ and symmetrizing the summand, we obtain

$$|f_2^{(n)}(x,-\sigma)|\leqslant[M^{(+)}(x_+)\,\mathrm{e}^{2\kappa_+(x-x_+)}\phi(\sigma)]^n/n!\tag{66}$$

where
$$\phi(\sigma)=\sum_{m=1}^{\infty}|d^{(+)}_m|.\tag{67}$$

The series for $\phi(\sigma)$ converges for all σ in any bounded subdomain of the σ-plane that excludes the points $\sigma=-\mathrm{i}m\kappa_+$ $(m=1,2,\ldots)$. In any such bounded subdomain, $\phi(\sigma)$ may be replaced by its maximum value in equation (66). The bound on $f_2^{(n)}(x,-\sigma)$, thus obtained, ensures the uniform convergence (in the same subdomains) of the infinite series in terms of which the solution for $f_2(x,-\sigma)$ has been expressed in equation (57). Also, since each of the terms, $f_2^{(n)}(x,-\sigma)$, is analytic, except at $\sigma=-\mathrm{i}m\kappa_+$ $(m=1,2,\ldots)$, the uniform convergence of the infinite series ensures, in turn, that $f_2(x,-\sigma)$ is analytic in the same subdomains. We conclude that $f_2(x, -\sigma)$ *is an analytic function of* σ *except for poles at* $\sigma = -\mathrm{i}m\kappa_+$ $(m = 1, 2, \ldots)$. (See figure 3.)

In an entirely analogous fashion, we can show that $f_1(x, +\sigma)$ *is an analytic function of* σ *except for poles at* $\sigma = +\mathrm{i}m\kappa_-$ $(m = 1, 2, \ldots)$. (See figure 3.)

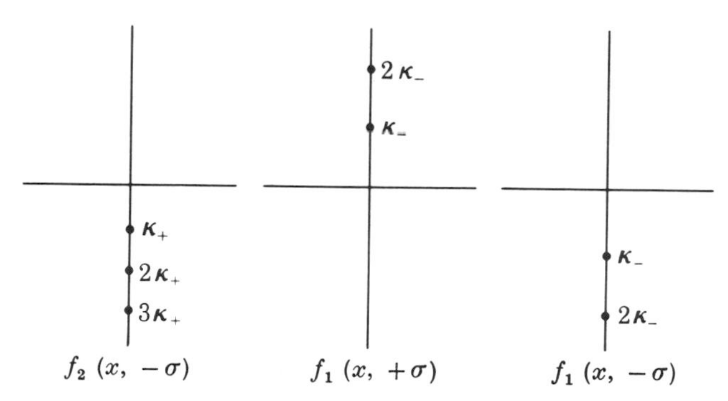

FIGURE 3. The disposition of the poles of the scattering amplitudes $f_1(x, \pm\sigma)$ and $f_2(x, -\sigma)$.

6. The asymptotic behaviours of $\mathscr{F}_u$ and $\mathscr{F}_v$ on the Cauchy horizon

We now turn to the consideration of our principal inquiry, namely, the asymptotic behaviours, on the Cauchy horizon, of $\mathscr{F}_u$ and $\mathscr{F}_v$ defined as Fourier integrals in equations (37) and (38). The asymptotic behaviours of such Fourier integrals can be ascertained by integration along contours distorted into that half σ-plane in which the integrand is bounded in the asymptotic limit. The required asymptotic behaviours then depend on the singularity of the integrand closest to the real axis.

The singularities of the integrands in equations (37) and (38) are related to those of $f_1(x, \sigma)$ and $f_2(x, \sigma)$, which we have specified in § 5 for axial perturbations, and also to those of $\mathscr{Z}(\sigma)$, which depend on the spectral distribution of the radiation crossing the event horizon specified by $\hat{Z}(v)$.

Considering first the expected behaviour of $\hat{Z}(v)$, we shall suppose that it vanishes at least as rapidly as v^{-1} for $v \to +\infty$. This behaviour for $v \to +\infty$ encompasses the power-law decays that are typical of perturbations resulting from slightly aspherical gravitational collapse (cf. Price 1972). As for the behaviour of $\hat{Z}(v)$ for $v \to -\infty$, we do not expect the perturbations on the Cauchy horizon to depend sensitively on this behaviour, since in any realistic collapse it will be obscured by the collapsing body. On this last account, we may take $\hat{Z}(v)$ to vanish for v less than some specified v_0. An inverse power-law behaviour for $\hat{Z}(v)$ for $v \to +\infty$ and a cut-off at some v_0 for $v \to -\infty$ will ensure that its Fourier transform, $\mathscr{Z}(\sigma)$, exists. However, an inverse power-law behaviour of $\hat{Z}(v)$ for $v \to +\infty$ will give rise to singularities in $\mathscr{Z}(\sigma)$ along the real axis and in the lower half-plane; but the cut-off at some v_0 for $v \to -\infty$ will ensure that $\mathscr{Z}(\sigma)$ is free of singularities and analytic in the upper half-plane.

Returning to the consideration of the asymptotic behaviours of $\mathscr{F}_u$ and $\mathscr{F}_v$, we shall, in the first instance, restrict ourselves to axial perturbations for which we have determined the analytic properties of $f_1(x, \sigma)$ and $f_2(x, \sigma)$ in § 5. Considering first the integral representing $\mathscr{F}_u$ in equation (37), we distort the contour into the upper half σ-plane where the singularity nearest the real axis, of $f_1(x, +\sigma)$ and $f_2(x, -\sigma)$, is the pole of $f_1(x, +\sigma)$ at $\sigma = +\mathrm{i}\kappa_-$ (see figure 3). Since $\mathscr{Z}(\sigma)$, as we have seen, is analytic in the upper half-plane, the pole at $\sigma = +\mathrm{i}\kappa_-$ will control the asymptotic behaviour of the integral in equation (37); the integral, in fact, will have the behaviour $\mathrm{e}^{-\kappa_- u}$. Therefore, by equation (37)

$$\mathscr{F}_u \to \text{constant}, \quad \text{as} \quad u \to +\infty. \tag{68}$$

The flux of radiation received by the geodesic observer, being proportional to the square of $\mathscr{F}_u$, will, therefore, be finite as he (or she) crosses the Cauchy horizon at EF.

Considering next the integral representing $\mathscr{F}_v$ in equation (38), we now distort the contour into the lower half σ-plane. The singularity nearest the real axis, of $f_1(x, -\sigma)$ and $f_2(x, -\sigma)$, is that of $f_2(x, -\sigma)$ at $\sigma = -\mathrm{i}\kappa_+$ (see figure 3) since $\kappa_+ < \kappa_-$ (cf. equation (37)). There may, however, be singularities nearer to the real axis derived from $\mathscr{Z}(\sigma)$: a circumstance that will arise if $\hat{Z}(v)$ vanishes less rapidly than $\mathrm{e}^{-\kappa_+ v}$ for $v \to +\infty$. In the last event, the integral in equation (38) will behave like $\mathrm{d}\hat{Z}/\mathrm{d}v$ and

$$\mathscr{F}_v \to \text{constant} \times \mathrm{e}^{\kappa_- v}\, \mathrm{d}\hat{Z}/\mathrm{d}v, \quad \text{as} \quad v \to +\infty. \tag{69}$$

If on the other hand $\hat{Z}(v)$ vanishes more rapidly than $\mathrm{e}^{-\kappa_+ v}$ for $v \to +\infty$, then the singularity of $f_2(x, -\sigma)$ at $\sigma = -\mathrm{i}\kappa_+$ will control the asymptotic behaviour of the integral in equation (38): it will behave like $\mathrm{e}^{-\kappa_+ v}$ and we shall have

$$\mathscr{F}_v \to \text{constant}\ \mathrm{e}^{(\kappa_- - \kappa_+) v}, \quad \text{as} \quad v \to +\infty. \tag{70}$$

It is important to emphasize that the behaviour predicted by equation (70) represents the *slowest* rate of divergence of $\mathscr{F}_v$. The flux of radiation received by the geodesic observer is therefore *always* divergent as he (or she) crosses the Cauchy horizon at EC′.

The behaviours of $\mathscr{F}_u$ and $\mathscr{F}_v$ on the Cauchy horizon that we have established for axial perturbations apply equally for polar perturbations. This follows from equations (10) relating the reflexion and the transmission amplitudes for the two classes of perturbations. Thus, from the equality of the transmission amplitudes for the two classes of perturbations, we conclude from equations (34) and (36) that our earlier discussion of the integral defining $\mathscr{F}_u$ is unaltered. But the integrand defining $Y_{,u}$ in equation (35) has the additional factor

$$\frac{\mu^2(\mu^2+2) + 2\mathrm{i}\sigma q_j}{\mu^2(\mu^2+2) - 2\mathrm{i}\sigma q_j}. \tag{71}$$

However, this additional factor does not give rise to any singularity in the upper half-plane nearer to the real axis than that of $f_1(x, +\sigma)$ at $\sigma = +\mathrm{i}\kappa_-$. The finiteness of $\mathscr{F}_u$ for $u \to +\infty$, therefore, follows equally for polar perturbations.

7. Concluding remarks

We have seen that the theory of one-dimensional potential-scattering provides an adequate base for a simple and a mathematically complete description of the behaviours one may expect on the Cauchy horizon of the Reissner–Nordström space–time (for $Q_*^2 < M$). The description we have arrived at is in accord with a simple picture due to Penrose (private communication). The picture is the following.

Gravitational (and/or electromagnetic) perturbations of the exterior region will generate waves; these will cross the horizon either directly or after secondary scattering by the curvature of the background space–time. In general, we may expect an infinite number of such scattered waves between the source and $\mathscr{I}^+$ (see figure 1). In the conformal Penrose diagram, this means that the scattered waves will accumulate on the compactified $\mathscr{I}^+$. By continuity, there must be a similar accumulation on the Cauchy horizon GEC resulting in an infinite density of wave fronts. In consequence, an observer crossing the null surface GEC will experience the impact of an infinite flux of radiation. But there is no such cause for an infinite flux on the null surface EF. These expected behaviours are precisely what we have established on the basis of the scattering theory.

The basic facts, that the flux of radiation becomes infinite for $v \to +\infty$ and remains finite for $u \to +\infty$, do not depend on any particular feature of the perturbation crossing the event horizon: they are true for perturbations with inverse power-law behaviours and they are true for perturbations with compact support. The manner of the divergence of $\mathscr{F}_v$ for $v \to +\infty$ does depend on the character of the perturbation; but there is always a slowest rate of approach to infinity given by equation (70).

The divergence of the radiation flux on the Cauchy horizon has been established on the basis of the first-order linear perturbation-theory. It is important to emphasize that there is no divergence in any of the functions described by the linear theory. What are divergent are the derived fluxes. This fact suggests that a perturbation theory that includes terms of the second and higher orders will diverge since the sources for the second-order theory, for example, will be provided by the fluxes of the first-order theory; and these, as we have seen, diverge on the Cauchy horizon. If this expectation is justified, we should conclude that a perturbation theory is, strictly, not adequate for describing the Reissner–Nordström geometry in the presence of external sources even though we have a finite first-order theory.

The research reported in this paper has in part been supported by grants from the National Science Foundation under grants PHY 80-26043 with the University of Chicago and PHY 81-07384 with the University of California at Santa Barbara.

APPENDIX. ON THE FINITENESS OF $A(\sigma)$ AND $B(\sigma)$ IN THE LIMIT $\sigma = 0$

The analysis in § 6 establishes the analytic properties of the scattering amplitudes $A(\sigma)$ and $B(\sigma)$ (in the combinations $\sigma A(\sigma)$ and $\sigma B(\sigma)$) by making use of equations (15), (16), and (24), expressing them in terms of the Wronskians of the functions $f_1(x, \pm\sigma)$ and $f_2(x, -\sigma)$, and of our knowledge of the analytic properties of these functions. But the analysis leaves open the question whether or not $A(\sigma)$ and $B(\sigma)$ are singular at $\sigma = 0$. While this question is not relevant for the purposes of ascertaining the behaviours of $\mathscr{F}_u$ and $\mathscr{F}_v$ on the Cauchy horizon, it is relevant for ascertaining the behaviours of the perturbations themselves.

In reality neither $A(\sigma)$ nor $B(\sigma)$ is singular at $\sigma = 0$. This fact derives from a result of Gürsel *et al.* (1979*b*) that the Reissner–Nordström geometry allows static perturbations that are regular at both horizons.

To deduce the behaviours of $A(\sigma)$ and $B(\sigma)$ at $\sigma = 0$ from the known existence of regular static perturbations, consider equation (11) in the limit $\sigma \to 0$. In this limit, the solutions, for both $x \to +\infty$ and $x \to -\infty$, either tend to constants or tend towards a linear variation with x. When seeking solutions that are regular at infinity, we need not concern ourselves with solutions that tend towards a linear variation with x. The solutions that tend to a constant for $x \to +\infty$ must be proportional to

$$\lim_{\sigma\to 0}[f_1(x, +\sigma)+f_1(x, -\sigma)]. \tag{A 1}$$

Similarly, solutions that tend to a constant for $x \to -\infty$ must be proportional to

$$\lim_{\sigma\to 0}[f_2(x, +\sigma)+f_2(x, -\sigma)]. \tag{A 2}$$

The existence of solutions that are regular both for $x \to +\infty$ and for $x \to -\infty$ in the zero-frequency limit, implies that

$$[f_1(x, +\sigma)+f_1(x, -\sigma)] = K[f_2(x, +\sigma)+f_2(x, -\sigma)] \quad (\sigma \to 0), \tag{A 3}$$

where K is a constant. Taking the Wronskian of both sides of equation (A 3) with $f_2(x, +\sigma)$, making use of equations (15) and (16), and the further relation

$$[f_2(x, +\sigma), f_2(x, -\sigma)] = 2i\sigma, \tag{A 4}$$

and passing to the limit $\sigma = 0$, we find

$$A(0)-B(0) = K. \tag{A 5}$$

On the other hand, from equation (23) it follows that

$$A(0)+B(0) \quad \text{is finite} \tag{A 6}$$

since the solution of equation (11) satisfying the boundary condition specified must extend to $\sigma = 0$. From equations (A 5) and (A 6) it is manifest that $A(\sigma)$ *and* $B(\sigma)$ *are both finite.*

Now making use of the finiteness of $A(\sigma)$ and $B(\sigma)$ at $\sigma = 0$, we can, by a straightforward extension of the arguments of § 6, determine, quite generally, the asymptotic behaviours of $X(v)$ and $Y(u)$ defined in equations (30) and (31). We find that for $v \to +\infty$, $X(v)$ vanishes like $e^{-\kappa_+ v}$ or $d\hat{Z}/dv$ whichever is slower, and for $u \to +\infty$, $Y(u)$ vanishes like $e^{-\kappa_- u}$. Thus, while the flux associated with $\mathscr{F}_v$ (by virtue of the additional factor $e^{\kappa_- v}$ in its definition) diverges on the Cauchy horizon, the linear perturbations themselves are smooth.

References

Chandrasekhar, S. 1979 *Proc. R. Soc. Lond.* A **365**, 453.

Chandrasekhar, S. 1983 *The mathematical theory of black holes*, ch. 5. Oxford: Clarendon Press.

Chandrasekhar, S. & Xanthopoulos, B. C. 1979 *Proc. R. Soc. Lond.* A **367**, 1.

de Alfaro, V. & Regge, T. 1965 *Potential scattering*. Amsterdam: North Holland.

Deift, P. & Trubowitz, E. 1979 *Communs. pure appl. Math.* **32**, 121.

Gürsel, Y., Novikov, I. D., Sandberg, V. D. & Starobinsky, A. A. 1979*b* *Phys. Rev.* D **20** 1260.

Gürsel, Y., Sandberg, V. D., Novikov, I. D. & Starobinsky, A. A. 1979*a* *Phys. Rev.* D **19**, 413.

Hartle, J. B. & Wilkins, D. C. 1974 *Communs. math. Phys.* **38**, 47.

McNamara, J. M. 1978*a* *Proc. R. Soc. Lond.* A **358**, 499.

McNamara, J. M. 1978*b* *Proc. R. Soc. Lond.* A **364**, 121.

Martin, A. 1959 *Nuovo Cim.* **14**, 403.

Matzner, R. A., Zamorano, N. & Sandberg, V. D. 1979 *Phys. Rev.* D **19**, 2821.

Price, R. 1972 *Phys. Rev.* D **5**, 2419.

Simpson, M. & Penrose, R. 1973 *Int. J. theor. Phys.* **7**, 183.

On algebraically special perturbations of black holes

By Subrahmanyan Chandrasekhar, F.R.S.
University of Chicago, Chicago, Illinois 60637, U.S.A.

(*Received* 23 *May* 1983)

Algebraically special perturbations of black holes excite gravitational waves that are either purely ingoing or purely outgoing. Solutions, appropriate to such perturbations of the Kerr, the Schwarzschild, and the Reissner–Nordström black-holes, are obtained in explicit forms by different methods. The different methods illustrate the remarkable inner relations among different facets of the mathematical theory. In the context of the Kerr black-hole they derive from the different ways in which the explicit value of the Starobinsky constant emerges, and in the context of the Schwarzschild and the Reissner–Nordström black-holes they derive from the potential barriers surrounding them belonging to a special class.

1. Introduction

The Kerr space–time, being of Petrov-type D, allows two shear-free congruences of null geodesics. And in a Newman–Penrose formalism, with a null basis provided by these geodesics and a suitably selected pair of complex-conjugate null vectors, Ψ_2 is the only non-vanishing one among the five Weyl-scalars, $\Psi_0, \Psi_1, \ldots, \Psi_4$.

Perturbations of the Kerr space–time, conceived as caused by the incidence of gravitational waves which are reflected and absorbed by the black hole, are expressed in terms of Ψ_0 and Ψ_4, whose relative amplitudes are uniquely determined by the theory. Also, by availing ourselves of the allowed gauge and tetrad freedoms, we can arrange that Ψ_1 and Ψ_3 continue to vanish in the perturbed space–time, while Ψ_2 retains its original value unchanged. The question occurs whether there may not be special perturbations in which only one of Ψ_0 and Ψ_4 is non-zero while the other vanishes, i.e. perturbations in which we have only incoming ($\Psi_0 \neq 0$) or only outgoing ($\Psi_4 \neq 0$) waves. Such *algebraically special perturbations* of the Kerr space–time were considered (in passing) by Wald (1973). Wald's result generalized an earlier one of Couch & Newman (1973) in the context of the Schwarzschild black-hole. In this paper, we shall consider these algebraically special perturbations in somewhat greater detail and show that they are related to other aspects of the general mathematical theory of black holes.

The plan of the paper is the following. In §2, we obtain the explicit form of the algebraically special perturbations of the Kerr black-hole. In §3, we show how these same solutions can be derived from the 'transformation theory' as described in *The Mathematical Theory of Black Holes* (Chandrasekhar 1983, §97; this book will be referred to hereafter as *M.T.*). In §4, we specialize the results of §§2 and 3 to the case of the Schwarzschild black-hole; and in §5, we show how the algebraically special perturbations of the Schwarzschild and the Reissner–Nordström black-holes follow in a very natural way from the special nature of the potential barriers surrounding them. And, finally, in §6, we consider briefly the broader implications of the various relations that are derived.

2. The algebraically special perturbations of the Kerr black-hole

In the theory of the gravitational perturbations of the Kerr black-hole, the equations governing the Weyl scalars Ψ_0 and Ψ_4 (with a time, t, and an azimuth angle, φ, dependence given by $e^{i(\sigma t+m\varphi)}$, where m is an integer, positive, negative, or zero) decouple from the rest and separate; and their solutions are expressed in the forms,

$$\Psi_0 = R_{+2}(r)\,S_{+2}(\theta) \quad \text{and} \quad \Psi_4 = (r-\mathrm{i}a\cos\theta)^{-4}R_{-2}(r)\,S_{-2}(\theta), \tag{1}$$

where $S_{+2}(\theta)\,(\equiv S_{-2}(\pi-\theta))$ are the normalized angular functions of Teukolsky (belonging to a common characteristic value, λ) and $\Delta^2 R_{+2}$ and R_{-2} satisfy the complex-conjugate pair of equations,

$$(\Delta\mathscr{D}_{-1}\mathscr{D}_0^\dagger - 6\mathrm{i}\sigma^+ r - \lambda)\,(\Delta^2 R_{+2}) = 0 \tag{2}$$

and

$$(\Delta\mathscr{D}_{-1}^\dagger\mathscr{D}_0 + 6\mathrm{i}\sigma^+ r - \lambda)\,(R_{-2}) = 0, \tag{3}$$

where

$$\mathscr{D}_n^\dagger = (\mathscr{D}_n)^* = \frac{\mathrm{d}}{\mathrm{d}r} - \mathrm{i}\frac{\sigma^+\varpi^2}{\Delta} + 2n\frac{(r-M)}{\Delta}, \tag{4}$$

$$\varpi^2 = r^2+\alpha^2, \quad \alpha^2 = a^2 + am/\sigma^+ \quad \text{and} \quad \Delta = r^2 - 2Mr + a^2. \tag{5}$$

The functions $\Delta^2 R_{+2}$ and R_{-2} are related by the Teukolsky–Starobinsky identities,

$$\left.\begin{aligned} &\Delta^2\mathscr{D}_0^\dagger\mathscr{D}_0^\dagger\mathscr{D}_0^\dagger\mathscr{D}_0^\dagger(\Delta^2 R_{+2}) = \mathscr{C}_1 R_{-2}, \\ \text{and}\qquad &\Delta^2\mathscr{D}_0\mathscr{D}_0\mathscr{D}_0\mathscr{D}_0 R_{-2} = \mathscr{C}_2\Delta^2 R_{+2}, \end{aligned}\right\} \tag{6}$$

where

$$\mathscr{C}_1\mathscr{C}_2 = \lambda^2(\lambda+2)^2 - 8\sigma^{+2}\lambda[\alpha^2(5\lambda+6)-12a^2] + 144\sigma^{+2}(M^2+\sigma^{+2}\alpha^4) = |\mathscr{C}|^2 \tag{7}$$

(say), is the *Starobinsky constant*. Since $\Delta^2 R_{+2}$ and R_{-2} satisfy complex-conjugate equations, we can normalize them, relatively, in such a way that $\Delta^2 R_{+2} = P_{+2}$ and $R_{-2} = P_{-2}$ satisfy the complex conjugate relations,

$$\left.\begin{aligned} &\Delta^2\mathscr{D}_0^\dagger\mathscr{D}_0^\dagger\mathscr{D}_0^\dagger\mathscr{D}_0^\dagger P_{+2} = \mathscr{C}^* P_{-2}, \\ \text{and}\qquad &\Delta^2\mathscr{D}_0\mathscr{D}_0\mathscr{D}_0\mathscr{D}_0 P_{-2} = \mathscr{C}P_{+2}, \end{aligned}\right\} \tag{8}$$

where $|\mathscr{C}|^2$ has the value specified in equation (7). With P_{+2} and P_{-2} defined in this manner, the solutions for Ψ_0 and Ψ_4, required for the problem of the reflection and transmission of incident gravitational waves, are given by

$$\Psi_0 = \frac{1}{\Delta^2}P_{+2}S_{+2} \quad \text{and} \quad \Psi_4 = \frac{1}{4(r-\mathrm{i}a\cos\theta)^4}P_{-2}S_{-2}, \tag{9}$$

without any ambiguity (apart from a constant of proportionality). (All of the foregoing relations can be found in *M.T.*, Chapter 9.)

From the foregoing equations, in particular from equations (6), it follows that we cannot in general ensure that $R_{-2} = 0$ without at the same time requiring $R_{+2} = 0$; and conversely. And the only way we can ensure a non-vanishing R_{+2} with a vanishing R_{-2} is to let $\mathscr{C}_2 = 0$ so that $\Delta^2 R_{+2} = P_{+2}$ satisfies the equation,

$$\mathscr{D}_0^\dagger \mathscr{D}_0^\dagger \mathscr{D}_0^\dagger \mathscr{D}_0^\dagger P_{+2} = 0. \tag{10}$$

A non-trivial solution of this equation, satisfying equation (2) at the same time, can be found as follows.

Since

$$\mathscr{D}_0^\dagger = \frac{\varpi^2}{\Delta}\left(\frac{\mathrm{d}}{\mathrm{d}r_*} - \mathrm{i}\sigma^+\right), \tag{11}$$

where r_* is a new radial coordinate defined by the equation,

$$\frac{\mathrm{d}r_*}{\mathrm{d}r} = \frac{\varpi^2}{\Delta}, \tag{12}$$

it is manifest that

$$\mathscr{D}_0^\dagger(r^n \,\mathrm{e}^{+\mathrm{i}\sigma^+ r_*}) = nr^{n-1}\,\mathrm{e}^{+\mathrm{i}\sigma^+ r_*}. \tag{13}$$

Accordingly, the general solution of equation (10) is given by

$$P_{+2} = (Ar^3 + Br^2 + Cr + D)\,\mathrm{e}^{+\mathrm{i}\sigma^+ r_*}, \tag{14}$$

where A, B, C, and D are constants of integration. The requirement, now, that P_{+2} given by equation (14) satisfies equation (2) also, leads to the condition

$$2[\mathrm{i}\sigma^+ r^2 - r + (M + \mathrm{i}\sigma^+ \alpha^2)](3Ar^2 + 2Br + C) + 2(r^2 - 2Mr + a^2)(3Ar + B) \\ - (6\mathrm{i}\sigma^+ r + \lambda)(Ar^3 + Br^2 + Cr + D) = 0. \tag{15}$$

The coefficients of the different powers of r in this equation must vanish separately; and we are led to the following equations:

$$\left.\begin{aligned} \lambda A + 2\mathrm{i}\sigma^+ B &= 0, \\ 6(M - \mathrm{i}\sigma^+\alpha^2)A + (\lambda + 2)B + 4\mathrm{i}\sigma^+ C &= 0, \\ 6a^2 A + 4\mathrm{i}\sigma^+\alpha^2 B - (\lambda + 2)C - 6\mathrm{i}\sigma^+ D &= 0, \\ 2a^2 B + 2(M + \mathrm{i}\sigma^+\alpha^2)C - \lambda D &= 0. \end{aligned}\right\} \tag{16}$$

The determinant of this system of equations, namely,

$$\begin{vmatrix} \lambda & 2\mathrm{i}\sigma^+ & 0 & 0 \\ 6(M - \mathrm{i}\sigma^+\alpha^2) & \lambda + 2 & 4\mathrm{i}\sigma^+ & 0 \\ 6a^2 & 4\mathrm{i}\sigma^+\alpha^2 & -(\lambda + 2) & -6\mathrm{i}\sigma^+ \\ 0 & 2a^2 & 2(M + \mathrm{i}\sigma^+\alpha^2) & -\lambda \end{vmatrix}, \tag{17}$$

must vanish; and the vanishing of this determinant leads precisely to the condition

$$|\mathscr{C}|^2 = 0. \tag{18}$$

It is manifest that the same condition will follow from requiring that $\Delta^2 R_{+2} = 0$ and that P_{-2} is a solution of equation (3) compatible with

$$\mathscr{D}_0 \mathscr{D}_0 \mathscr{D}_0 \mathscr{D}_0 P_{-2} = 0. \tag{19}$$

The vanishing of the Starobinsky constant, $|\mathscr{C}|^2$, therefore, is a necessary and sufficient condition for the existence of solutions that correspond only to ingoing or only to outgoing waves.

When $|\mathscr{C}|^2 = 0$, the constants A, B, C, and D in the solution (14) become determinate, apart from a constant of proportionality. We shall find it convenient to set

$$A = 2\mathrm{i}\sigma^+, \tag{20}$$

in which case,

$$B = -\lambda, \quad C = -\frac{\mathrm{i}}{4\sigma^+}\mathscr{Q} \quad \text{and} \quad D = 2a^2 - \tfrac{2}{3}\lambda\alpha^2 + \frac{\lambda+2}{24\sigma^{+2}}\mathscr{Q}, \tag{21}$$

where

$$\mathscr{Q} = \lambda(\lambda+2) - 12\sigma^{+2}\alpha^2 - 12\mathrm{i}\sigma^+ M, \tag{22}$$

and σ^+ is a root of equation (18). As a result a solution leading only to ingoing waves (from infinity) and a non-vanishing Ψ_0 is given by

$$P_{+2} = f(r)\,\mathrm{e}^{+\mathrm{i}\sigma r_*}, \tag{23}$$

where

$$f(r) = 2\mathrm{i}\sigma^+ r^3 - \lambda r^2 - \frac{\mathrm{i}}{4\sigma^+}\mathscr{Q}r + 2a^2 - \tfrac{2}{3}\lambda\alpha^2 + \frac{\lambda+2}{24\sigma^{+2}}\mathscr{Q}. \tag{24}$$

Solutions leading only to outgoing waves and non-vanishing P_{-2} can be obtained simply by reversing the sign of σ^+ in the foregoing equations (without, however, altering the sign of σ^+ in the definition of α^2). These are the algebraically special solutions. The condition, $|\mathscr{C}|^2 = 0$, for the existence of such solutions is stated in Wald's paper (Wald 1973); but the explicit solution (24) is not given.

We may note here for future reference the following elementary relation. Defining the complex-conjugate of $\mathscr{Q}$, namely

$$\mathscr{Q}^* = \lambda(\lambda+2) - 12\sigma^{+2}\alpha^2 + 12\mathrm{i}\sigma^+ M, \tag{25}$$

we find,

$$|\mathscr{Q}|^2 = \lambda^2(\lambda+2)^2 + 144\sigma^{+2}(M^2 + \sigma^{+2}\alpha^4) - 24\lambda(\lambda+2)\,\sigma^{+2}\alpha^2, \tag{26}$$

or, by virtue of the vanishing of $|\mathscr{C}|^2$,

$$|\mathscr{Q}|^2 = 16\sigma^{+2}\lambda(\lambda\alpha^2 - 6a^2). \tag{27}$$

3. Relation to the transformation theory

The algebraically special perturbations derived in §2 can also be obtained from the transformation theory described in *M.T.*, §97; but first, we shall summarize the basic equations of the theory.

With the definition

$$Y^{(\pm\sigma^+)} = |\varpi^2|^{-\frac{3}{2}} P_{\pm 2}, \tag{28}$$

we have, in place of equations (2) and (3), the equation,

$$\Lambda^2 Y^{(\pm\sigma^+)} + P\Lambda_\mp Y^{(\pm\sigma^+)} - QY^{(\pm\sigma^+)} = 0, \tag{29}$$

where

$$\Lambda_\pm = \frac{\mathrm{d}}{\mathrm{d}r_*} \pm \mathrm{i}\sigma^+, \quad \Lambda^2 = \frac{\mathrm{d}^2}{\mathrm{d}r_*^2} + \sigma^{+2}, \tag{30}$$

$$P = \frac{\mathrm{d}}{\mathrm{d}r_*} \lg \frac{\varpi^8}{\Delta^2}, \quad \text{and} \quad Q = \frac{\Delta}{\varpi^8}[\lambda\varpi^4 + 3\varpi^2(r^2 - a^2) - 3r^2\Delta]. \tag{31}$$

Equation (29) is then transformed to the one-dimensional wave-equation,

$$\Lambda^2 Z^{(\pm\sigma^+)} = VZ^{(\pm\sigma^+)}, \tag{32}$$

by the substitutions,

$$Y^{(\pm\sigma^+)} = \left(R - \frac{\mathrm{d}T_1}{\mathrm{d}r_*}\right) Z^{(\pm\sigma^+)} + (T_1 \pm 2\mathrm{i}\sigma^+)\Lambda_\pm Z^{(\pm\sigma^+)}, \tag{33}$$

$$\frac{\Delta^2}{\varpi^8} K^{(\pm\sigma^+)} Z^{(\pm\sigma^+)} = RY^{(\pm\sigma^+)} - (T \pm 2\mathrm{i}\sigma^+)\Lambda_\mp Y^{(\pm\sigma^+)}, \tag{34}$$

where

$$R = Q + \frac{\Delta^2}{\varpi^8}\beta_2 = \frac{\Delta^2}{\varpi^8}(F + \beta_2), \quad F = \frac{\varpi^8}{\Delta^2}Q, \tag{35}$$

$$T_1 = \frac{1}{F - \beta_2}\left(\frac{\mathrm{d}F}{\mathrm{d}r_*} - \kappa_2\right), \quad \beta_2 = \pm 3\alpha^2, \tag{36}$$

$$\kappa_2 = \pm\{36M^2 - 2\lambda[\alpha^2(5\lambda + 6) - 12a^2] + 2\beta_2\lambda(\lambda + 2)\}^{\frac{1}{2}}, \tag{37}$$

and

$$K^{(\pm\sigma^+)} = \lambda(\lambda + 2) - 4\sigma^{+2}\beta_2 \pm 2\mathrm{i}\sigma^+\kappa_2. \tag{38}$$

The potential, V, in equation (32), which is the same for $Z^{(\pm\sigma^+)}$, is given by

$$V = Q - \frac{\mathrm{d}T_1}{\mathrm{d}r_*}. \tag{39}$$

Since

$$K^{(+\sigma^+)}K^{(-\sigma^+)} = |\mathscr{C}|^2, \tag{40}$$

it follows that we can recover the algebraically special solutions by setting $K^{(+\sigma^+)}$ (or $K^{(-\sigma^+)}$) $= 0$ in equation (34). Thus, setting

$$K^{(+\sigma^+)} = \lambda(\lambda + 2) - 4\sigma^{+2}\beta_2 + 2\mathrm{i}\sigma^+\kappa_2 = 0, \tag{41}$$

we obtain from equation (34)

$$\left(\frac{R}{T_1 + 2\mathrm{i}\sigma^+}\right)_{K^{(+\sigma^+)}=0} = \frac{1}{Y^{(+\sigma^+)}}\Lambda_- Y^{(+\sigma^+)}. \tag{42}$$

To be in accord with the solution obtained in §2, it is necessary that equation (42) be compatible with

$$Y^{(+\sigma^+)} = \frac{f(r)}{|\varpi^2|^{\frac{3}{2}}} \mathrm{e}^{+\mathrm{i}\sigma^+ r_*}, \tag{43}$$

where $f(r)$ is given by equation (24). In other words, we must have

$$\left(\frac{R}{T_1+2i\sigma^+}\right)_{K^{(+\sigma^+)}=0} = \frac{\Delta}{\varpi^2}\frac{d}{dr}\lg\frac{f}{|\varpi^2|^{\frac{3}{2}}} = \frac{\Delta}{\varpi^4}\frac{\varpi^2 f_{,r}-3rf}{f}, \tag{44}$$

where, according to equations (35) and (36),

$$\frac{R}{T_1+2i\sigma^+} = \frac{\Delta^2}{\varpi^8}\frac{F^2-\beta_2^2}{F_{,r_*}+2i\sigma^+F-(\kappa_2+2i\sigma^+\beta_2)}. \tag{45}$$

Equation (44), therefore, requires that

$$\frac{\varpi^4}{\Delta}\left[\frac{F_{,r_*}+2i\sigma^+F-(\kappa_2+2i\sigma^+\beta_2)}{F^2-\beta_2^2}\right]_{K^{(+\sigma^+)}=0} = \frac{f}{\varpi^2 f_{,r}-3rf}. \tag{46}$$

Now, letting

$$F = q/\Delta, \tag{47}$$

where

$$\begin{aligned} q &= \lambda\varpi^4+3\varpi^2(r^2-a^2)-3r^2\Delta \\ &= \lambda\varpi^4+3r^2(\alpha^2-2a^2)+6Mr^3-3\alpha^2a^2, \end{aligned} \tag{48}$$

we find,

$$\begin{aligned} F^2-\beta_2^2 &= \frac{1}{\Delta^2}(q^2-\beta_2^2\Delta^2) \\ &= \frac{\varpi^2}{\Delta^2}(\lambda\varpi^2+6Mr-6a^2)\,[\lambda\varpi^4+6r^2(\alpha^2-a^2)+6Mr(r^2-\alpha^2)], \end{aligned} \tag{49}$$

and

$$\begin{aligned} F_{,r_*} &= \frac{\Delta}{\varpi^2}F_{,r} = \frac{1}{\Delta\varpi^2}(q_{,r}\Delta-\Delta_{,r}q) \\ &= \frac{1}{\Delta\varpi^2}[-2(r-M)\lambda\varpi^4+2\varpi^2(2\lambda r\Delta-3Mr^2-3Ma^2+6a^2r) \\ &\qquad +12r\Delta(Mr-a^2)]. \end{aligned} \tag{50}$$

The equality (46), that has to be verified, now takes the form

$$\left\{\frac{\varpi^2}{(q^2-\beta_2^2\Delta^2)}[(q_{,r}\Delta-\Delta_{,r}q)+2i\sigma^+\varpi^2q-(\kappa_2+2i\sigma^+\beta_2)\varpi^2\Delta]\right\}_{K^{(+\sigma^+)}=0} = \frac{f}{\varpi^2 f_{,r}-3rf}, \tag{51}$$

where, on the right side, for f given by equation (24),

$$\varpi^2 f_{,r}-3rf = \lambda r^3+\frac{i}{2\sigma^+}(\mathscr{Q}+12\sigma^{+2}\alpha^2)r^2-\left(6a^2+\frac{\lambda+2}{8\sigma^{+2}}\mathscr{Q}\right)r-i\frac{\alpha^2}{4\sigma^+}\mathscr{Q}. \tag{52}$$

In verifying the equality (51), we must, of course, bear in mind that σ^+ is not arbitrary, but must be assigned a value compatible with $K^{(+\sigma^+)}=0$, i.e.

$$\lambda(\lambda+2)-4\sigma^{+2}\beta_2+2i\sigma^+\kappa_2 = 0, \tag{53}$$

or, alternatively,

$$\kappa_2+2i\sigma^+\beta_2 = \frac{i}{2\sigma^+}\lambda(\lambda+2). \tag{54}$$

Also, rewriting equation (53) in the form,

$$\beta_2(2i\sigma^+)^2+\kappa_2(2i\sigma^+)+\lambda(\lambda+2) = 0, \tag{55}$$

we may solve for $2i\sigma^+$ to obtain,

$$2i\sigma^+ = -\frac{1}{2\beta_2}(\kappa-\bar{\kappa}_2), \tag{56}$$

where (cf. equation (37))

$$\bar{\kappa}_2 = \pm\{36M^2-2\lambda[\alpha^2(5\lambda+6)-12a^2]-2\beta_2\lambda(\lambda+2)\}^{\frac{1}{2}}. \tag{57}$$

So far, we have not restricted β_2 to either of the two permissible values $\pm 3\alpha^2$. However, it now appears that for the validity of equation (51), with f given by equation (24), it is necessary that we assign to β_2 the value

$$\beta_2 = +3\alpha^2, \tag{58}†$$

in which case,

$$K^{(+\sigma^+)} = \lambda(\lambda+2)-12\sigma^{+2}\alpha^2+2i\sigma^+\kappa_2 = 0, \tag{59}$$

$$2i\sigma^+ = -\frac{\kappa_2-\bar{\kappa}_2}{6\alpha^2}, \quad \mathscr{Q} = -2i\sigma^+(\kappa_2+6M), \quad \mathscr{Q}^* = -2i\sigma^+(\kappa_2-6M), \tag{60}$$

$$\kappa_2 = \pm\{36M^2-2\lambda[\alpha^2(5\lambda+6)-12a^2]+6\alpha^2\lambda(\lambda+2)\}^{\frac{1}{2}}, \tag{61}$$

and

$$\bar{\kappa}_2 = \pm\{36M^2-2\lambda[\alpha^2(5\lambda+6)-12a^2]-6\alpha^2\lambda(\lambda+2)\}^{\frac{1}{2}}. \tag{62}$$

From these equations it follows that (cf. equation (27))

$$|\mathscr{Q}|^2 = 4\sigma^{+2}(36M^2-\kappa_2^2) = 16\sigma^{+2}\lambda(\lambda\alpha^2-6a^2), \tag{63}$$

or

$$\kappa_2^2 = 36M^2-4\lambda(\lambda\alpha^2-6a^2). \tag{64}$$

And since

$$\kappa_2^2-\bar{\kappa}_2^2 = 12\alpha^2\lambda(\lambda+2), \tag{65}$$

$$\bar{\kappa}_2^2 = 36M^2-16\alpha^2\lambda^2-24\lambda(\alpha^2-a^2). \tag{66}$$

Additional relations that follow from equations (56), (58), (60), and (65) are

$$\left.\begin{aligned} \frac{1}{\kappa_2-\bar{\kappa}_2} &= \frac{\kappa_2+\bar{\kappa}_2}{12\alpha^2\lambda(\lambda+2)}, \\ \frac{1}{2i\sigma^+} &= -\frac{6\alpha^2}{\kappa_2-\bar{\kappa}_2} = -\frac{\kappa_2+\bar{\kappa}_2}{2\lambda(\lambda+2)}, \\ \mathscr{Q} &= \frac{1}{6\alpha^2}(\kappa_2-\bar{\kappa}_2)(\kappa_2+6M). \end{aligned}\right\} \tag{67}$$

† This choice of β_2^2, in this context, has no bearing on the four distinct potentials that the transformation theory provides for each assigned value of σ^+. This follows from the equation $2i\sigma^+ = -(\kappa_2-\bar{\kappa}_2)/6\alpha^2$ together with equations (61) and (62) (which replace $|\mathscr{C}|^2 = 0$) since we may *independently* associate with κ_2 and $\bar{\kappa}_2$ *either* sign; and the ensemble of roots that we obtain in this fashion must coincide with the roots provided by $|\mathscr{C}|^2 = 0$.

By making use of the foregoing relations, we can eliminate the explicit appearance of σ^+ in equations (24) and (53); we find

$$f = -\frac{\kappa_2 - \bar{\kappa}_2}{6\alpha^2} r^3 - \lambda r^2 - \tfrac{1}{2}(\kappa_2 + 6M) r + 2a^2 - \tfrac{2}{3}\lambda a^2 - \frac{1}{12\lambda}(\kappa_2 + \bar{\kappa}_2)(\kappa_2 + 6M), \tag{68†}$$

and

$$\varpi^2 f_{,r} - 3rf = \lambda r^3 + [\tfrac{1}{2}(\kappa_2 + \bar{\kappa}_2) + 6M] r^2 + [-6a^2 + \frac{1}{4\lambda}(\kappa_2 + \bar{\kappa}_2)(\kappa_2 + 6M)] r - \tfrac{1}{2}\alpha^2(\kappa_2 + 6M). \tag{69}$$

Therefore, equation (51) rewritten in the form,

$$\begin{aligned}
&[\lambda\varpi^2 + 6Mr - 6a^2][\lambda\varpi^4 + 6r^2(\alpha^2 - a^2) + 6Mr(r^2 - \alpha^2)] \times \Big\{ -\frac{\kappa_2 - \bar{\kappa}_2}{6\alpha^2} r^3 \\
&\qquad - \lambda r^2 - \tfrac{1}{2}(\kappa_2 + 6M) r + 2a^2 - \tfrac{2}{3}\lambda\alpha^2 - \frac{1}{12\lambda}(\kappa_2 + \bar{\kappa}_2)(\kappa_2 + 6M) \Big\} \\
&= \Big\{ -2(r - M)\lambda\varpi^4 + 2\varpi^2[2\lambda r(r^2 - 2Mr + a^2) - 3Mr^2 + 6a^2 r - 3Ma^2] \\
&\qquad + 12(r^2 - 2Mr + a^2)(Mr^2 - a^2 r) - \tfrac{1}{2}(\kappa_2 + \bar{\kappa}_2)(r^2 + \alpha^2)(r^2 - 2Mr + a^2) \\
&\qquad - \frac{\kappa_2 - \bar{\kappa}_2}{6\alpha^2}\varpi^2 [\lambda\varpi^4 + 3r^2(\alpha^2 - 2a^2) + 6Mr^3 - 3\alpha^2 a^2] \Big\} \\
&\qquad \times \Big\{ \lambda r^3 + [\tfrac{1}{2}(\kappa_2 + \bar{\kappa}_2) + 6M] r^2 + \Big[-6a^2 + \frac{1}{4\lambda}(\kappa_2 + \bar{\kappa}_2)(\kappa_2 + 6M) \Big] r \\
&\qquad - \tfrac{1}{2}\alpha^2(\kappa_2 + 6M) \Big\},
\end{aligned} \tag{70}$$

must be an identity. Since the two sides of equation (70) are both polynomials of degree nine in r, it is necessary that the coefficients of each of the ten powers of $r, r^9, \ldots, r^0$, agree. It can be directly verified that this is indeed the case by making use *only* of the *quadratic* identities (64)–(66); but the verification is long and tedious. The agreement of the solution, obtained in § 2, with that which follows from the transformation theory, though necessary, is not so immediately manifest as one might have hoped: it requires the ratio of two polynomials of degree six to be factorizable to a ratio of two polynomials of degree three!

(a) *The solutions for* $Z^{(\pm\sigma)}$

The solutions for $Z^{(\pm\sigma)}$, which directly give the amplitudes of the incoming and the outgoing waves, can now be obtained from equation (33). Thus, considering $Z^{(+\ )}$, equation (33) with $Y^{(+\sigma)}$ given by equation (43), we have

$$Y^{(+\sigma)} = \frac{f(r)}{|\varpi^2|^{\frac{3}{2}}} e^{+i\sigma r_*} = \left(R - \frac{dT_1}{dr_*} \right) Z^{(+\sigma)} + (T_1 + 2i\sigma^+) \Lambda_+ Z^{(+\sigma)}. \tag{71}$$

The associated homogeneous equation allows a solution of the form

$$Z^{(+\sigma)} = \mathscr{Z}(r) e^{-i\sigma r_*}, \tag{72}$$

† The coefficient of r^3 in this expression for f has the alternative form $-2\lambda(\lambda + 2)/(\kappa_2 + \bar{\kappa}_2)$.

where

$$\frac{d}{dr_*}\lg \mathscr{Z} = \frac{1}{T_1+2i\sigma^+}\frac{dT_1}{dr_*} - \frac{R}{T_1+2i\sigma^+}. \tag{73}$$

Remembering that we are at present considering the case $K^{(+\sigma^+)}$, we can make use of equation (44) and rewrite equation (73) in the form

$$\frac{d}{dr_*}\lg \mathscr{Z} = \frac{d}{dr_*}\lg(T_1+2i\sigma^+) - \frac{d}{dr_*}\lg\frac{f}{|\varpi^2|^{\frac{3}{2}}}. \tag{74}$$

Accordingly,

$$\mathscr{Z}(r) = \frac{|\varpi^2|^{\frac{3}{2}}}{f}(T_1+2i\sigma^+); \tag{75}$$

or again, by making use of equation (44), we can write

$$\mathscr{Z}(r) = R\frac{|\varpi^2|^{\frac{3}{2}}}{f}\frac{1}{[\lg(f/|\varpi^2|^{\frac{3}{2}})]_{,r_*}}. \tag{76}$$

The solution of the inhomogeneous equation can now be completed. We find

$$Z^{(+\sigma^+)} = \text{constant}\ \chi\int\frac{dr_*}{\chi^2}, \tag{77}$$

where

$$\chi = \frac{q+3\alpha^2\Delta}{|\varpi^2|^{\frac{1}{2}}}\frac{e^{-i\sigma r_*}}{\varpi^2 f_{,r}-3rf}, \tag{78}$$

$$q+3\alpha^2\Delta = \lambda\varpi^4+6r^2(\alpha^2-a^2)+6Mr(r^2-\alpha^2), \tag{79}$$

and $(\varpi^2 f_{,r}-3rf)$ is given by equation (69).

4. Specialization to the Schwarzschild black-hole

The solutions for the algebraically special perturbations of the Schwarzschild black-hole can be obtained by setting

$$a^2 = \alpha^2 = 0, \quad \lambda = \mu^2 = (l-1)(l+2), \tag{80}$$

$$\mathscr{Q} = \mu^2(\mu^2+2) - 12i\sigma^+ M, \tag{81}$$

and

$$\begin{aligned}|\mathscr{C}|^2 &= \mu^4(\mu^2+2)^2+144\sigma^{+2}M^2\\ &= [\mu^2(\mu^2+2)+12i\sigma^+ M][\mu^2(\mu^2+2)-12i\sigma^+ M] = 0.\end{aligned} \tag{82}$$

Thus, we find from equations (24) and (28) that

$$Y = \frac{f(r)}{r^3}e^{+i\sigma r_*}, \tag{83}$$

where

$$f(r) = r^2(2i\sigma^+ r-\mu^2)+\tfrac{1}{4}\mathscr{Q}\left(\frac{r}{i\sigma^+}+\frac{\mu^2+2}{6\sigma^{+2}}\right). \tag{84}$$

Two cases arise depending on which of the two factors of $|\mathscr{C}|^2$ we set equal to zero; case (i):

$$\mu^2(\mu^2+2)+12\mathrm{i}\sigma^+ M = 0, \quad \mathscr{Q} = -24\mathrm{i}\sigma^+ M, \quad \text{and} \quad \mathrm{i}\sigma^+ = -\frac{\mu^2(\mu^2+2)}{12M}, \tag{85}$$

when $Y = Y^{(+)} = -\dfrac{e^{+\mathrm{i}\sigma^+ r_*}}{6M\mu^2 r^3}[\mu^4(\mu^2+2)r^3+6M\mu^4 r^2+36M^2\mu^2 r+72M^3]$; (86)

and case (ii):

$$\mu^2(\mu^2+2)-12\mathrm{i}\sigma^+ M = 0, \quad \mathscr{Q} = 0, \quad \text{and} \quad \mathrm{i}\sigma^+ = +\frac{\mu^2(\mu^2+2)}{12M}, \tag{87}$$

when

$$Y = Y^{(-)} = \frac{\mu^2}{6Mr}[(\mu^2+2)r-6M]e^{+\mathrm{i}\sigma^+ r_*}. \tag{88}$$

It can be readily verified that the solutions for $Y^{(\pm)}$ given by equations (86) and (88) are consistent with the equation

$$\pm 6M + \frac{r^8}{\Delta^2}V^{(\pm)}\frac{1}{Y^{(\pm)}}\Lambda_- Y^{(\pm)} = 0, \tag{89}$$

which follows from the transformation theory for the axial and the polar perturbations (cf. *M.T.*, equation (319), p. 187).

Turning to the corresponding solutions for $Z^{(\pm)}$ given by equation (77), we first observe that

$$\chi = \frac{r^2(\mu^2 r+6M)}{(\varpi^2 f_{,r}-3rf)_{a=0}}e^{-\mathrm{i}\sigma^+ r_*}. \tag{90}$$

We verify that the polar and the axial perturbations arise from setting, respectively,

$$\text{(i)}\ \kappa_2 = \bar{\kappa}_2 = +6M, \quad \text{and} \quad \text{(ii)}\ \kappa_2 = \bar{\kappa}_2 = -6M. \tag{91}$$

From equation (69) we find that for these two cases,

$$\begin{aligned}(\varpi^2 f_{,r}-3rf)_{a=0} &= \mu^2 r^3+12Mr^2+\frac{36}{\mu^2}M^2 r \\ &= \frac{r}{\mu^2}(\mu^2 r+6M)^2 \quad \text{(case (i))},\end{aligned} \tag{92}$$

and

$$\chi^{(+)} = \frac{\mu^2 r}{\mu^2 r+6M}e^{-\mathrm{i}\sigma^+ r_*}; \tag{93}$$

and

$$(\varpi^2 f_{,r}-3rf)_{a=0} = \mu^2 r^3 \quad \text{(case (ii))}, \tag{94}$$

and

$$\chi^{(-)} = \frac{\mu^2 r+6M}{\mu^2 r}e^{-\mathrm{i}\sigma^+ r_*}. \tag{95}$$

The solution for $Z^{(\pm)}$ now takes the form,

$$Z^{(\pm)} = \text{constant}\left(\frac{r}{\mu^2 r+6M}\right)^{\pm 1}e^{\pm\sigma^+{}_0 r_*}\int\left(\frac{\mu^2 r+6M}{r}\right)^{\pm 2}e^{\mp 2\sigma^+{}_0 r_*}\,dr_*, \tag{96}$$

where

$$\sigma^+{}_0 = \frac{\mu^2(\mu^2+2)}{12M}. \tag{97}$$

5. An alternative derivation of the algebraically special perturbations of the Schwarzschild and the Reissner–Nordström black-holes

It is known that the equations governing the perturbations of the Schwarzschild and the Reissner–Nordström black-holes can be reduced to one-dimensional wave-equations with potentials which belong to a particular special class. Thus, in the case of the Schwarzschild black-hole the polar and the axial perturbations are governed by the equation,

$$\Lambda^2 Z^{(\pm)} = \left(\pm\beta\frac{\mathrm{d}f}{\mathrm{d}r_*}+\beta^2 f^2+\kappa f\right) Z^{(\pm)}, \tag{98}$$

where

$$\beta = 6M, \quad \kappa = \mu^2(\mu^2+2), \quad \text{and} \quad f = \frac{\Delta}{r^3(\mu^2 r+\beta)}. \tag{99}$$

Similarly, in the case of the Reissner–Nordström black-hole, the equations are reducible to the pair of equations,

$$\Lambda^2 Z_i^{(\pm)} = \left(\pm\beta_i\frac{\mathrm{d}f_i}{\mathrm{d}r_*}+\beta_i^2 f_i^2+\kappa f_i\right) Z_i^{(\pm)} \quad (i = 1, 2), \tag{100}$$

where

$$\left.\begin{aligned}\beta_i = q_j(i,j = 1,2;\, i \neq j), \quad \kappa = \mu^2(\mu^2+2), \quad f_i = \frac{\Delta}{r^3(\mu^2 r+\beta_i)},\\ q_1 = 3M+\sqrt{(9M^2+4Q_*^2\mu^2)}, \quad q_2 = 3M-\sqrt{(9M^2+4Q_*^2\mu^2)},\end{aligned}\right\} \tag{101}$$

and Q_* is the charge of the black hole.

We shall consider equation (98) as a prototype of these equations. The 'Riccati form' of the terms in f on the right side of equation (98) suggests the following sequence of transformations:

$$\psi = \beta f+\frac{\kappa}{2\beta} \quad \text{and} \quad \pm\psi = \frac{\mathrm{d}}{\mathrm{d}r_*}\lg\chi; \tag{102}$$

for, by these substitutions the terms in f become

$$\begin{aligned}\pm\beta\frac{\mathrm{d}f}{\mathrm{d}r_*}+\beta^2 f^2+Kf &= \pm\frac{\mathrm{d}\psi}{\mathrm{d}r_*}+\psi^2-\frac{\kappa^2}{4\beta^2}\\ &= \frac{1}{\chi}\frac{\mathrm{d}^2\chi}{\mathrm{d}r_*^2}-\frac{\kappa^2}{4\beta^2};\end{aligned} \tag{103}$$

and the equation for $Z^{(\pm)}$ takes the form,

$$\frac{1}{Z^{(\pm)}}\frac{\mathrm{d}^2 Z^{(\pm)}}{\mathrm{d}r_*^2}+\left(\sigma^{+2}+\frac{\kappa^2}{4\beta^2}\right) = \frac{1}{\chi}\frac{\mathrm{d}^2\chi}{\mathrm{d}r_*^2}. \tag{104}$$

In this form, it is plainly manifest that the case,

$$\sigma^{+2}+\frac{\kappa^2}{4\beta^2} = 0, \tag{105}$$

is distinguished, since the equation is then directly integrable. This is precisely the case of the vanishing Starobinsky constant and the algebraically special perturbations.

The general solution of equation (104), when equality (105) holds and

$$\sigma^{+} = \pm\, \mathrm{i}\frac{\kappa}{2\beta} = \pm\, \mathrm{i}\frac{\mu^2(\mu^2+2)}{2\beta}, \tag{106}$$

is given by (cf. equation (77))

$$Z^{(\pm)} = \chi \int \frac{\mathrm{d}r_*}{\chi^2}. \tag{107}$$

For the particular form of f that occurs in equations (98) and (100), we find that

$$\chi = \left(\frac{r}{\mu^2 r+\beta}\right)^{\pm 1} \mathrm{e}^{\pm\sigma_0 r_*} \quad \left(\sigma^{+}{}_0 = \frac{\mu^2(\mu^2+2)}{2\beta}\right); \tag{108}$$

and the solution for $Z^{(\pm)}$ becomes

$$Z^{(\pm)} = \text{constant}\left(\frac{r}{\mu^2 r+\beta}\right)^{\pm 1} \mathrm{e}^{\pm\sigma_0 r_*} \int \left(\frac{\mu^2 r+\beta}{r}\right)^{\pm 2} \mathrm{e}^{\mp 2\sigma_0 r_*}\,\mathrm{d}r_*, \tag{109}$$

in agreement with the solution (96) obtained in §4. But a fact which emerges (which had not been known before) is that solutions for 'algebraically special perturbations',† of *exactly* the same forms as for the Schwarzschild black-hole, exist also for the Reissner–Nordström black-hole.

6. Concluding remarks

The consideration of algebraically special perturbations of the Kerr, the Schwarzschild, and the Reissner–Nordström black-holes has revealed, once again, the remarkable relations that exist among the different facets of the mathematical theory. The relations are exemplified, for example, by the three different contexts in which the explicit value (7) of the Starobinsky constant, $|\mathscr{C}|^2$, emerges: *first*, in the direct evaluation of the Teukolsky–Starobinsky identities as set out in *M.T.*, §81 pp. 436–438) – an evaluation shrouded in complexity till the very last step of miraculous cancellations when the veil is lifted; *second*, as the condition that the radial Teukolsky-equations allow solutions of the form of an outgoing or of an ingoing wave with an amplitude which is a cubic polynomial in r – an evaluation that is direct and simple; and *third*, as the factored $K^{(+\sigma)}K^{(-\sigma)}$ in the transformation theory, independently of the Teukolsky–Starobinsky identities but requiring the 'formidable task' of verifying that the coefficient Q in equation (29) satisfies a strange nonlinear second-order differential equation (*M.T.*, p. 505, equation (403)). These different manners of evaluations of $|\mathscr{C}|^2$ must have a common origin. But what

† These perturbations, in the case of the Reissner–Nordström black-hole, will consist of a superposition (with determinate weights) of purely ingoing (or purely outgoing) gravitational and electromagnetic waves (cf. *M.T.* p. 259, equation (347)).

is it? And why, indeed, is the simple matter of taking the complex-conjugate of P_{+2} equivalent to applying the fourth-order differential operator $\Delta^2 \mathscr{D}_0^\dagger \mathscr{D}_0^\dagger \mathscr{D}_0^\dagger \mathscr{D}_0^\dagger$? These questions occur in the context of the Kerr black-hole. In the context of the Schwarzschild and the Reissner–Nordstörm black-holes, they are included in the single question as to the origin of the potential barriers surrounding them belonging to a very special class. These questions, like so many others in the mathematical theory of black holes, are sheathed in enigmas.

I am grateful to Professor Robert Wald for discussions which clarified certain essential issues.

The research reported in this paper has in part been supported by a grant from the National Science Foundation under grant PHY 80-26043 with the University of Chicago.

References

Chandrasekhar, S. 1983 *The mathematical theory of black holes*. Oxford: Clarendon Press.
Couch, W. E. & Newman, E. T. 1973 *J. math. Phys.* **14**, 285.
Wald, R. M. 1973 *J. math. Phys.* **14**, 1453.

Appendix. The characteristic values of σ^+ belonging to the algebraically special perturbations of the Kerr black-hole

I am grateful to Mr S. K. Chakrabarti for obtaining for me the roots of the equation,

$$|\mathscr{C}|^2 = 0, \tag{A 1}$$

for $l = 2$ and $m = 0, \pm 1, \pm 2$ and different values of a. The results of his calculations are summarized in table A 1.

The tabulated values of σ^+ and λ are consistent with equations (61) and (62) for κ_2 and $\bar{\kappa}_2$ with the positive sign chosen for the right sides of both equations *except* for $a = 0.5$ and $m = 0$ for which case the negative sign for $\bar{\kappa}_2$ should be chosen (while retaining the positive sign for κ_2).

Table A 1. The roots of $|\mathscr{C}|^2 = 0$: the characteristic frequencies σ^+ and the corresponding value of λ

	$l = 2; m = 0$		$l = 2; m = \pm 1$		$l = 2; m = \pm 2$	
a	σ^+	λ	σ^+	λ	σ^+	λ
0	2i	4.0	0 + 2i	4.0	0 + 2i	4.0
0.1	2.0155i	3.9806	∓0.2609 + 1.9609i	3.8973 ± 0.6484i	∓0.4592 + 1.8310i	3.6862 ± 1.2154i
0.2	2.0652i	3.9174	∓0.4886 + 1.8528i	3.6241 ± 1.1986i	∓0.6831 + 1.5375i	3.0747 ± 2.0250i
0.3	2.1620i	3.7917	∓0.6608 + 1.7013i	3.2620 ± 1.6014i	∓0.7602 ± 1.2887i	2.4688 ± 2.5282i
0.4	2.3453i	3.5476	∓0.7749 + 1.5369i	2.8875 ± 1.8666i	∓0.7754 + 1.0993i	1.9336 ± 2.8606i
0.5	2.9019i	2.8285	∓0.8434 + 1.3800i	2.5416 ± 2.0315i	∓0.7652 + 0.9552i	1.4681 ± 3.0954i
0.6	—	—	∓0.8813 + 1.2383i	2.2367 ± 2.1298i	—	—
0.7	—	—	∓0.8994 + 1.1118i	1.1927 ± 2.1836i	—	—

PART TWO

The Mathematical Theory of Colliding Waves

19

On the Nutku–Halil solution for colliding impulsive gravitational waves

BY SUBRAHMANYAN CHANDRASEKHAR,[1] F.R.S., AND VALERIA FERRARI[2]

[1] *University of Chicago, Chicago, Illinois* 60637, *U.S.A.*
[2] *Dipartimento di Fisica 'G. Marconi', Universita di Roma, Rome, Italy*

(*Received* 27 *March* 1984)

The equations appropriate for space–times with two space-like Killing-vectors are set up, *ab initio*, and explicit expressions for the components of the Riemann, the Ricci, and the Einstein tensors in a suitable tetrad frame are written. The equations for the vacuum are reduced to a single equation of the Ernst type. It is then shown that the simplest linear solution of the Ernst equation leads directly to the Nutku–Halil solution for two colliding impulsive gravitational waves with uncorrelated polarizations. Thus, in some sense, the Nutku–Halil solution occupies the same place in space–times with two space-like Killing-vectors as the Kerr solution does in space–times with one time-like and one space-like Killing-vector.

The Nutku–Halil solution is further described in a Newman–Penrose formalism; and the expressions for the Weyl scalars, in particular, make the development of curvature singularities manifest.

Finally, a theorem analogous to Robinson's theorem (but much less strong) is proved for space–times with two space-like Killing-vectors.

1. INTRODUCTION

The study of colliding impulsive gravitational waves was initiated by Khan & Penrose (1971). They obtained the metric describing the most general space–time representing the collision of two parallel polarized impulsive gravitational waves; and they drew attention to certain generic singularities that result from such collisions. In spite of the brevity of Khan & Penrose's paper, it is remarkably complete in the description of the space–time they had considered, both in its physical and in its mathematical aspects.

The nature of the singularities of the Khan–Penrose space–time has been further examined by Szekeres (1972) and by Matzner & Tipler (1984). Also, the choice of a pure δ-function form (for the waves before collision) requires consideration of the consistency of the junction conditions that one imposes at the interface between the curved space–time, in which the interaction occurs, and the flat space–time external to the region of the interaction. These junction conditions have been examined in considerable detail by Bel & Szekeres (1977) following O'Brien & Synge (1952) and Pirani (1964). In a different direction, Nutku & Halil (1977), in a paper of considerable virtuosity, removed the restriction to parallel polarizations in the Kahn–Penrose solution.

Reprinted with minor corrections.

This paper is addressed, principally, to a more standard derivation of the Nutku–Halil solution than the one sketched by the authors; and to draw attention to certain reciprocal relations that exist between stationary axisymmetric space–times and space–times with two space-like Killing-vectors. It will appear that the Nutku–Halil solution occupies the same central place in space–times with two space-like Killing-vectors as the Kerr solution does in stationary axisymmetric space–times.

The present paper derives from an earlier paper on the Kerr metric (Chandrasekhar (1978); see also Chandrasekhar (1983), *The Mathematical Theory of Black Holes* (§§52 and 53), which will be referred to hereafter as M.T.); and is similarly motivated.

The reciprocal relation, between stationary axisymmetric space–times and space–times with two space-like Killing-vectors, to which reference was made, arises in the following way.

Rewriting the metric,

$$ds^2 = e^{2\nu}(dt)^2 - e^{2\psi}(d\varphi - \omega\, dt)^2 - e^{2\mu_2}(dx^2)^2 - e^{2\mu_3}(dx^3)^2, \tag{1}$$

appropriate to stationary axisymmetric space–times (where ν, ψ, ω, μ_2, and μ_3 are functions only of the two space-like coordinates x^2 and x^3) in the form

$$ds^2 = e^{\psi+\nu}\left[\chi(dt)^2 - \frac{1}{\chi}(d\varphi - \omega\, dt)^2\right] - e^{2\mu_2}(dx^2)^2 - e^{2\mu_3}(dx^3)^2, \tag{2}$$

where

$$\chi = e^{-\psi+\nu}, \tag{3}$$

one finds that the entire solution to Einstein's equations reduces to solving for the pair of functions χ and ω. And for χ and ω one can derive equations of two alternative forms: a real pair of equations (M.T., p. 280, equations (57) and (58)) for the functions

$$X = \chi + \omega \quad \text{and} \quad Y = \chi - \omega, \tag{4}$$

or a single complex equation – the *Ernst equation* – for a complex function $Z = \Psi + i\Phi$, where Φ is a potential for ω and Ψ is a function that we associate with Φ. One can obtain, by simple inspection, simple solutions for either X and Y or for Z. The simplest of these solutions for Z (or, rather, its *adjoint*, $\tilde{Z}$; cf. M.T., p. 285) leads to the Kerr solution. The analogous simplest solution of the X- and Y-equations (derived in Chandrasekhar (1978), §8) did not, at the time, seem to have any special significance. However, soon after its publication, Nutku, in a personal communication, drew the attention of one of us (S.C.) to the fact that the solution, obtained via the simplest solution of the X- and Y-equations, is exactly the one he had discovered for the collision of two impulsive gravitational waves with non-parallel polarizations, if one allows for the 'complexification' required for the replacement of the time-like Killing-vector, dt, by a space-like Killing-vector, dx^3, say. In this paper, we shall show in some detail how this reciprocity arises. We shall further describe the Nutku–Halil solution in a Newman–Penrose formalism. And finally, we shall prove a theorem analogous to Robinson's theorem for stationary axisymmetric space–times.

2. The chosen form of the metric; and the resulting field equations

From the discussion in M.T., chapter 2, §§11–13, it follows that the metric appropriate for a space–time with two space-like Killing-vectors, dx^1 and dx^2, can be written in the form (cf. M.T., p. 72, equation (38))

$$ds^2 = e^{2\nu}(dt)^2 - e^{2\mu_3}(dx^3)^2 - e^{2\psi}(dx^1 - q_2 dx^2)^2 - e^{2\mu_2}(dx^2)^2, \tag{5}$$

where ν, ψ, q_2, μ_2 and μ_3 are functions of t (equal to x^0) and x^3, only; and we have the freedom to impose any desired coordinate condition on ν and μ_3.

Writing the metric (5) in the form

$$ds^2 = \frac{e^{\mu_3+\nu}}{\sqrt{\Delta}}[(dt)^2 - \Delta(dx^3)^2] - e^{\beta}\left[\chi(dx^2)^2 + \frac{1}{\chi}(dx^1 - q_2 dx^2)^2\right], \tag{6}$$

where
$$\Delta = e^{2\mu_3-2\nu}, \quad \chi = e^{-\psi+\mu_2}, \quad \text{and} \quad e^{\beta} = e^{\psi+\mu_2}, \tag{7}$$

and interchanging the variables x^1 and x^2, we conclude that if the pair (χ, q_2) represents a solution then so does the pair $(\tilde{\chi}, \tilde{q}_2)$ where

$$\tilde{\chi} = \frac{\chi}{\chi^2 + q_2^2} \quad \text{and} \quad \tilde{q}_2 = \frac{q_2}{\chi^2 + q_2^2}. \tag{8}$$

This is the analogue, in the present context, of the process of *conjugation* described in M.T., §52 (*a*).

We can now write down the components of the Riemann tensor, appropriate for the chosen form (5) of the metric, in a tetrad frame with the basis one-forms

$$\omega^0 = e^{\nu}\,dt, \quad \omega^1 = e^{\psi}(dx^1 - q_2 dx^2), \quad \omega^2 = e^{\mu_2}\,dx^2$$

and
$$\omega^3 = e^{\mu_3}\,dx^3, \tag{9}$$

by suitably specializing the expressions listed in M.T., pp. 78–79. We find that the non-vanishing components of the Riemann tensor, in our present context are

$$R_{1212} = e^{-2\mu_3}\psi_{,3}\mu_{2,3} - e^{-2\nu}\psi_{,0}\mu_{2,0} - \tfrac{1}{4}e^{2\psi-2\mu_2}[e^{-2\mu_3}(q_{2,3})^2 - e^{-2\nu}(q_{2,0})^2],$$

$$R_{1313} = e^{-\psi-\mu_3}(e^{\psi-\mu_3}\psi_{,3})_{,3} - e^{-2\nu}\psi_{,0}\mu_{3,0} - \tfrac{1}{4}e^{2\psi-2\mu_3-2\mu_2}(q_{2,3})^2,$$

$$R_{1010} = e^{-\psi-\nu}(e^{\psi-\nu}\psi_{,0})_{,0} - e^{-2\mu_3}\psi_{,3}\nu_{,3} - \tfrac{1}{4}e^{2\psi-2\nu-2\mu_2}(q_{2,0})^2,$$

$$R_{2323} = e^{-\mu_2-\mu_3}(e^{\mu_2-\mu_3}\mu_{2,3})_{,3} - e^{-2\nu}\mu_{3,0}\mu_{2,0} + \tfrac{3}{4}e^{2\psi-2\mu_2-2\mu_3}(q_{2,3})^2,$$

$$R_{2020} = e^{-\mu_2-\nu}(e^{\mu_2-\nu}\mu_{2,0})_{,0} - e^{-2\mu_3}\nu_{,3}\mu_{2,3} + \tfrac{3}{4}e^{2\psi-2\mu_2-2\nu}(q_{2,0})^2,$$

$$R_{3030} = e^{-\mu_3-\nu}[(e^{\mu_3-\nu}\mu_{3,0})_{,0} - (e^{\nu-\mu_3}\nu_{,3})_{,3}],$$

$$R_{1310} = e^{-\nu-\mu_3}[-\psi_{,3,0} - \psi_{,3}(\psi-\mu_3)_{,0} + \psi_{,0}\nu_{,3} + \tfrac{1}{4}e^{2\psi-2\mu_2}q_{2,0}q_{2,3}],$$

$$R_{1332} = e^{\psi-2\mu_3-\mu_2}q_{2,3}(\psi_{,3} - \tfrac{1}{2}\mu_{2,3}) + \tfrac{1}{2}e^{-\mu_3-\mu_2}(e^{\psi-\mu_3}q_{2,3})_{,3} - \tfrac{1}{2}e^{\psi-2\nu-\mu_2}q_{2,0}\mu_{3,0},$$

$$R_{1002} = e^{\psi-2\nu-\mu_2}q_{2,0}(\psi_{,0} - \tfrac{1}{2}\mu_{2,0}) + \tfrac{1}{2}e^{-\nu-\mu_2}(e^{\psi-\nu}q_{2,0})_{,0} - \tfrac{1}{2}e^{\psi-2\mu_3-\mu_2}q_{2,3}\nu_{,3},$$

$$R_{3220} = +e^{-\mu_3-\nu}[\mu_{2,3,0} + \mu_{2,3}(\mu_2-\mu_3)_{,0} - \mu_{2,0}\nu_{,3} + \tfrac{3}{4}e^{2\psi-2\mu_2}q_{2,3}q_{2,0}],$$

$$R_{1230} = +\tfrac{1}{2}\mathrm{e}^{-\mu_3-\nu}[(\mathrm{e}^{\psi-\mu_2}q_{2,0})_{,3}-(\mathrm{e}^{\psi-\mu_2}q_{2,3})_{,0}],$$

$$R_{1023} = +\tfrac{1}{2}\mathrm{e}^{-\mu_2-\mu_3}(\mathrm{e}^{\psi-\nu}q_{2,0})_{,3}+\tfrac{1}{2}\mathrm{e}^{\psi-\mu_2-\mu_3-\nu}q_{2,3}(2\psi-\mu_2-\mu_3)_{,0},$$

$$R_{1302} = -\tfrac{1}{2}\mathrm{e}^{-\nu-\mu_2}(\mathrm{e}^{\psi-\mu_3}q_{2,3})_{,0}-\tfrac{1}{2}\mathrm{e}^{\psi-\mu_2-\mu_3-\nu}q_{2,0}(2\psi-\nu-\mu_2)_{,3}. \tag{10}$$

The components which vanish identically are

$$R_{1213} = R_{1210} = R_{1223} = R_{1220} = R_{1330} = R_{1003} = R_{2330} = R_{3002} = 0. \tag{11}$$

The components of the Ricci and the Einstein tensors are given by certain linear combinations of the components of the Riemann tensor (specified in M.T., p. 80, equations (78)). For our present purposes it will suffice to consider the following equations (cf. M.T., pp. 274–75, for the corresponding equations appropriate for stationary axisymmetric space–times):

$$\begin{aligned}\mathrm{e}^{-2\nu}[\mu_{2,0,0}+\mu_{2,0}(\psi+\mu_2+\mu_3-\nu)_{,0}]-\mathrm{e}^{-2\mu_3}[\mu_{2,3,3}+\mu_{2,3}(\psi+\mu_2+\nu-\mu_3)_{,3}]\\ = -\tfrac{1}{2}\mathrm{e}^{2\psi-2\mu_2}[\mathrm{e}^{-2\nu}(q_{2,0})^2-\mathrm{e}^{-2\mu_3}(q_{2,3})^2] \quad (R_{22}=0),\end{aligned} \tag{12}$$

$$\begin{aligned}\mathrm{e}^{-2\nu}[\psi_{,0,0}+\psi_{,0}(\psi+\mu_2+\mu_3-\nu)_{,0}]-\mathrm{e}^{-2\mu_3}[\psi_{,3,3}+\psi_{,3}(\psi+\mu_2+\nu-\mu_3)_{,3}]\\ = +\tfrac{1}{2}\mathrm{e}^{2\psi-2\mu_2}[\mathrm{e}^{-2\nu}(q_{2,0})^2-\mathrm{e}^{-2\mu_3}(q_{2,3})^2] \quad (R_{11}=0),\end{aligned} \tag{13}$$

$$(\mathrm{e}^{3\psi+\nu-\mu_2-\mu_3}q_{2,3})_{,3}-(\mathrm{e}^{3\psi+\mu_3-\mu_2-\nu}q_{2,0})_{,0} = 0 \quad (R_{12}=0), \tag{14}$$

$$\begin{aligned}-(\psi+\mu_2)_{,0,3}+(\psi+\mu_2)_{,0}\nu_{,3}+(\psi+\mu_2)_{,3}\mu_{3,0}-\psi_{,0}\psi_{,3}-\mu_{2,3}\mu_{2,0}\\ = \tfrac{1}{2}\mathrm{e}^{2\psi-2\mu_2}q_{2,0}q_{3,0} \quad (R_{03}=0),\end{aligned} \tag{15}$$

$$\begin{aligned}&\mathrm{e}^{-2\mu_3}[(\psi+\mu_2)_{,3,3}+(\mu_2-\mu_3)_{,3}(\psi+\mu_2)_{,3}+\psi_{,3}\psi_{,3}]\\ &-\mathrm{e}^{-2\nu}[(\mu_2+\mu_3)_{,0}\psi_{,0}+\mu_{3,0}\mu_{2,0}] = -\tfrac{1}{4}\mathrm{e}^{2\psi-2\mu_2}[\mathrm{e}^{-2\nu}(q_{2,0})^2+\mathrm{e}^{-2\mu_3}(q_{2,3})^2]\\ &\qquad (G_{00}=0),\end{aligned} \tag{16}$$

$$\begin{aligned}&\mathrm{e}^{-2\nu}[(\psi+\mu_2)_{,0,0}+(\mu_2-\nu)_{,0}(\psi+\mu_2)_{,0}+\psi_{,0}\psi_{,0}]\\ &-\mathrm{e}^{-2\mu_3}[(\mu_2+\nu)_{,3}\psi_{,3}+\nu_{,3}\mu_{2,3}] = -\tfrac{1}{4}\mathrm{e}^{2\psi-2\mu_2}[\mathrm{e}^{-2\nu}(q_{2,0})^2+\mathrm{e}^{-2\mu_3}(q_{2,3})^2]\\ &\qquad (G_{33}=0).\end{aligned} \tag{17}$$

Letting (cf. equation (7)),

$$\beta = \psi+\mu_2, \tag{18}$$

we can rewrite equations (12) and (13) in the forms

$$(\mathrm{e}^{\beta+\mu_3-\nu}\mu_{2,0})_{,0}-(\mathrm{e}^{\beta+\nu-\mu_3}\mu_{2,3})_{,3} = -\tfrac{1}{2}\mathrm{e}^{3\psi-\mu_2}[\mathrm{e}^{\mu_3-\nu}(q_{2,0})^2-\mathrm{e}^{\nu-\mu_3}(q_{2,3})^2], \tag{19}$$

$$(\mathrm{e}^{\beta+\mu_3-\nu}\psi_{,0})_{,0}-(\mathrm{e}^{\beta+\nu-\mu_3}\psi_{,3})_{,3} = +\tfrac{1}{2}\mathrm{e}^{3\psi-\mu_2}[\mathrm{e}^{\mu_3-\nu}(q_{2,0})^2-\mathrm{e}^{\nu-\mu_3}(q_{2,3})^2]. \tag{20}$$

The sum and the difference of these equations are

$$[\mathrm{e}^{\mu_3-\nu}(\mathrm{e}^{\beta})_{,0}]_{,0}-[\mathrm{e}^{\nu-\mu_3}(\mathrm{e}^{\beta})_{,3}]_{,3} = 0, \tag{21}$$

$$[\mathrm{e}^{\beta+\mu_3-\nu}(\psi-\mu_2)_{,0}]_{,0}-[\mathrm{e}^{\beta+\nu-\mu_3}(\psi-\mu_2)_{,3}]_{,3} = \mathrm{e}^{3\psi-\mu_2}[\mathrm{e}^{\mu_3-\nu}(q_{2,0})^2-\mathrm{e}^{\nu-\mu_3}(q_{2,3})^2]. \tag{22}$$

Subtraction of equations (16) and (17) yields the same equation (21), while addition gives

$$\begin{aligned}4\,\mathrm{e}^{\nu-\mu_3}(\beta_{,3}\nu_{,3}+\psi_{,3}\mu_{2,3})+4\,\mathrm{e}^{\mu_3-\nu}(\beta_{,0}\mu_{3,0}+\psi_{,0}\mu_{2,0})\\ = 2\,\mathrm{e}^{-\beta}\{[\mathrm{e}^{\nu-\mu_3}(\mathrm{e}^{\beta})_{,3}]_{,3}+[\mathrm{e}^{\mu_3-\nu}(\mathrm{e}^{\beta})_{,0}]_{,0}\}\\ +\mathrm{e}^{2\psi-2\mu_2}[\mathrm{e}^{\mu_3-\nu}(q_{2,0})^2+\mathrm{e}^{\nu-\mu_3}(q_{2,3})^2].\end{aligned} \tag{23}$$

3. The choice of gauge and the reduction of the equations to standard forms

Considering first equation (21), we can, without loss of generality, suppose that the solution for e^{β} is separable in the variables x^0 and x^3 and is expressible in the form

$$e^{\beta} = f(x^3)\sqrt{\Delta(x^0)}, \tag{24}$$

where $f(x^3)$ and $\Delta(x^0)$ are functions only of the variable specified; and, further, that by making use of the freedom we have to impose a coordinate condition on μ_3 and ν, we can arrange that

$$e^{2(\mu_3-\nu)} = \Delta(x^0). \tag{25}$$

With these assumptions, equation (21) reduces to

$$\tfrac{1}{2}\Delta_{,0,0} - f_{,3,3}/f = 0. \tag{26}$$

From this equation, we may conclude, again without loss of generality that

$$f = \sin x^3 \quad \text{and} \quad \Delta = -(x^0)^2 + 2Tx^0 + a^2, \tag{27}$$

where T and a^2 are constants. By letting

$$x^0 - T = (T^2+a^2)^{\frac{1}{2}}\eta, \tag{28}$$

we find that displacing the origin of x^0 to T and measuring time (x^0) in the unit $(T^2+a^2)^{\frac{1}{2}}$, we can write

$$\Delta = 1-\eta^2. \tag{29}$$

We shall also find it convenient to let

$$\mu = \cos x^3 \tag{30}$$

in place of x^3 and write

$$\delta = \sin^2 x^3 = 1-\mu^2. \tag{31}$$

With these definitions,

$$e^{2(\mu_3-\nu)} = \Delta = 1-\eta^2 \quad \text{and} \quad e^{2\beta} = e^{2(\psi+\mu_2)} = (1-\eta^2)(1-\mu^2) = \Delta\delta. \tag{32}$$

Hereafter, we shall consistently use η and μ as the variables in place of x^0 and x^3; they play, respectively, the roles of time and spatial distance.

With the solutions for $e^{\mu_3-\nu}$ and e^{β} given in equations (32), we can bring equations (22) and (14), respectively, to the forms

$$[\Delta(\psi-\mu_2)_{,0}]_{,0} - [\delta(\psi-\mu_2)_{,3}]_{,3} = e^{2(\psi-\mu_2)}[\Delta(q_{2,0})^2 - \delta(q_{2,3})^2], \tag{33}$$

and

$$[\Delta\, e^{2(\psi-\mu_2)} q_{2,0}]_{,0} - [\delta\, e^{2(\psi-\mu_2)} q_{2,3}]_{,3} = 0. \tag{34}$$

Letting χ have the same meaning as in equation (7), we can rewrite equations (33) and (34) in the forms

$$\left[\frac{\Delta}{\chi}\chi_{,0}\right]_{,0} - \left[\frac{\delta}{\chi}\chi_{,3}\right]_{,3} = -\frac{1}{\chi^2}[\Delta(q_{2,0})^2 - \delta(q_{2,3})^2], \tag{35}$$

and

$$\left(\frac{\Delta}{\chi^2}q_{2,0}\right)_{,0} - \left(\frac{\delta}{\chi^2}q_{2,3}\right)_{,3} = 0; \tag{36}$$

or, alternatively,

$$\chi[(\Delta\chi_{,0})_{,0}-(\delta\chi_{,3})_{,3}] = \Delta[(\chi_{,0})^2-(q_{2,0})^2]-\delta[(\chi_{,3})^2-(q_{2,3})^2], \tag{37}$$

and

$$\chi[(\Delta q_{2,0})_{,0}-(\delta q_{2,3})_{,3}] = 2\Delta\chi_{,0}q_{2,0}-2\delta\chi_{,3}q_{2,3}. \tag{38}$$

Now, expressing χ and q_2 as the real and the imaginary part of a complex function

$$Z = \chi+iq_2, \tag{39}$$

we can combine equations (37) and (38) into the single complex equation,

$$\mathscr{R}e\,(Z)\{[(1-\eta^2)\,Z_{,\eta}]_{,\eta}-[(1-\mu^2)\,Z_{,\mu}]_{,\mu}\} = (1-\eta^2)\,(Z_{,\eta})^2-(1-\mu^2)\,(Z_{,\mu})^2. \tag{40}$$

It is noteworthy that at this stage we are *not* led to a pair of real equations as we were for $\chi+\omega$ and $\chi-\omega$ in the stationary axisymmetric case; we are led, instead, directly to *Ernst's equation*.

By the substitution

$$Z = \frac{1+E}{1-E}, \tag{41}$$

equation (40) becomes

$$\begin{aligned}(1-EE^*)\{[(1-\eta^2)\,E_{,\eta}]_{,\eta}-[(1-\mu^2)\,E_{,\mu}]_{,\mu}\}\\ = -2E^*[(1-\eta^2)\,(E_{,\eta})^2-(1-\mu^2)\,(E_{,\mu})^2].\end{aligned} \tag{42}$$

The metric functions χ and q_2 are given in terms of the solution of equation (42) by

$$\chi = \frac{1-|E|^2}{|1-E|^2} \quad \text{and} \quad iq_2 = \frac{E-E^*}{|1-E|^2}. \tag{43}$$

Finally, equations (15) and (23) (which determine the last remaining metric function $\nu+\mu_3$), after some elementary reductions, become

$$\frac{\mu}{1-\mu^2}(\nu+\mu_3)_{,\eta}+\frac{\eta}{1-\eta^2}(\nu+\mu_3)_{,\mu} = -\frac{1}{2(\mathscr{R}e\,Z)^2}(Z_{,\eta}Z^*_{,\mu}+Z^*_{,\eta}Z_{,\mu}) \tag{44}$$

and

$$\begin{aligned}2\eta(\nu+\mu_3)_{,\eta}+2\mu(\nu+\mu_3)_{,\mu} &= \frac{3}{1-\eta^2}+\frac{1}{1-\mu^2}\\ &\quad -\frac{1}{(\mathscr{R}e\,Z)^2}[(1-\eta^2)|Z_{,\eta}|^2+(1-\mu^2)|Z_{,\mu}|^2],\end{aligned} \tag{45}$$

or, equivalently, in terms of E,

$$\frac{\mu}{1-\mu^2}(\nu+\mu_3)_{,\eta}+\frac{\eta}{1-\eta^2}(\nu+\mu_3)_{,\mu} = -\frac{2}{(1-|E|^2)^2}(E_{,\eta}E^*_{,\mu}+E^*_{,\eta}E_{,\mu}) \tag{46}$$

and

$$\begin{aligned}2\eta(\nu+\mu_3)_{,\eta}+2\mu(\nu+\mu_3)_{,\mu} &= \frac{3}{1-\eta^2}+\frac{1}{1-\mu^2}\\ &\quad -\frac{4}{(1-|E|^2)^2}[(1-\eta^2)|E_{,\eta}|^2+(1-\mu^2)|E_{,\mu}|^2].\end{aligned} \tag{47}$$

The equations derived from a potential for q_2

Even as in the case of stationary axisymmetric space–times, we can now define a potential for q_2, consistently with equation (34), in the manner,

$$\Phi_{,0} = \frac{\delta}{\chi^2} q_{2,3} \quad \text{and} \quad \Phi_{,3} = \frac{\Delta}{\chi^2} q_{2,0}. \tag{48}$$

The potential Φ will be governed by the equation

$$\left(\frac{\chi^2}{\delta}\Phi_{,0}\right)_{,0} - \left(\frac{\chi^2}{\Delta}\Phi_{,3}\right)_{,3} = 0; \tag{49}$$

and expressed in terms of Φ, equation (35) takes the form

$$[\Delta(\lg\chi)_{,0}]_{,0} - [\delta(\lg\chi)_{,3}]_{,3} = \frac{\chi^2}{\delta}(\Phi_{,0})^2 - \frac{\chi^2}{\Delta}(\Phi_{,3})^2. \tag{50}$$

Letting

$$\Psi = \sqrt{(\Delta\delta)}/\chi, \tag{51}$$

we find that equations (49) and (50) can be reduced to the forms

$$\Psi[(\Delta\Psi_{,0})_{,0} - (\delta\Psi_{,3})_{,3}] = \Delta[(\Psi_{,0})^2 - (\Phi_{,0})^2] - \delta[(\Psi_{,3})^2 - (\Phi_{,3})^2], \tag{52}$$

and

$$\Psi[(\Delta\Phi_{,0})_{,0} - (\delta\Phi_{,3})_{,3}] = 2\Delta\Psi_{,0}\Phi_{,0} - 2\delta\Psi_{,3}\Phi_{,3}. \tag{53}$$

From a comparison of these equations with equations (37) and (38) it is apparent that by combining Ψ and Φ in the manner

$$Z^\dagger = \Psi + i\Phi, \tag{54}$$

we shall find that $Z^\dagger$ satisfies the same equation (40). Thus, in contrast to the stationary axisymmetric case, we obtain an Ernst equation both from χ and q_2 and from Ψ and Φ.

4. The derivation of the Nutku–Halil solution

It is well known that the simplest solution of Ernst's equation (42) is given by

$$E = p\eta + iq\mu, \tag{55}$$

where p and q are real constants satisfying the condition

$$p^2 + q^2 = 1. \tag{56}$$

As we shall presently verify, the Nutku–Halil solution follows *directly* from this simplest solution. (It will be recalled that the Kerr solution also follows from this same solution of Ernst's equation; but the Ernst equation considered in that context is that appropriate for the functions $\tilde{\Phi}$ and $\tilde{\Psi}$ belonging to the conjugate metric (cf. M.T., §54).)

For E given by equation (55) the metric functions χ and q_2 are given by (cf. equations (43))

$$\chi = e^{-\psi+\mu_2} = \frac{1-p^2\eta^2-q^2\mu^2}{(1-p\eta)^2+q^2\mu^2} = \frac{p^2(1-\eta^2)+q^2(1-\mu^2)}{(1-p\eta)^2+q^2\mu^2} \tag{57}$$

and
$$q_2 = \frac{2q\mu}{(1-p\eta)^2+q^2\mu^2}. \tag{58}$$

Equations (46) and (47) governing $(\nu+\mu_3)$ are
$$\frac{\mu}{1-\mu^2}(\nu+\mu_3)_{,\eta}+\frac{\eta}{1-\eta^2}(\nu+\mu_3)_{,\mu} = 0 \tag{59}$$
and
$$2\eta(\nu+\mu_3)_{,\eta}+2\mu(\nu+\mu_3)_{,\mu} = \frac{3}{1-\eta^2}+\frac{1}{1-\mu^2}-\frac{4}{1-p^2\eta^2-q^2\mu^2}. \tag{60}$$

We readily verify that the solution of these equations is given by (cf. Chandrasekhar (1978), equation (106))
$$e^{\mu_3+\nu} = \text{constant}\,\frac{1-p^2\eta^2-q^2\mu^2}{(1-\eta^2)^{\frac{3}{4}}(1-\mu^2)^{\frac{1}{4}}}. \tag{61}$$

Now combining the solutions (32), (57), (58) and (61), we obtain for the various metric coefficients the solution
$$e^{2\nu} = \frac{1-p^2\eta^2-q^2\mu^2}{(1-\eta^2)^{\frac{5}{4}}(1-\mu^2)^{\frac{1}{4}}}, \quad e^{2\mu_3} = \frac{1-p^2\eta^2-q^2\mu^2}{(1-\eta^2)^{\frac{1}{4}}(1-\mu^2)^{\frac{1}{4}}}, \tag{62}$$
$$e^{2\psi} = (1-\eta^2)^{\frac{1}{2}}(1-\mu^2)^{\frac{1}{2}}\frac{(1-p\eta)^2+q^2\mu^2}{1-p^2\eta^2-q^2\mu^2}, \tag{63}$$
and
$$e^{2\mu_2} = (1-\eta^2)^{\frac{1}{2}}(1-\mu^2)^{\frac{1}{2}}\frac{1-p^2\eta^2-q^2\mu^2}{(1-p\eta)^2+q^2\mu^2}. \tag{64}$$

Before writing down the resulting expression for the metric, we may first note that (cf. equation (6))
$$\begin{aligned}-e^{\beta}\left[\chi(dx^2)^2+\frac{1}{\chi}(dx^1-q_2\,dx^2)^2\right] &= -e^{\beta}\left[\frac{\chi^2+q_2^2}{\chi}(dx^2)^2+\frac{1}{\chi}(dx^1)^2-\frac{2q_2}{\chi}dx^1\,dx^2\right]\\ &= -e^{\beta}\left[\frac{|Z|^2}{\chi}(dx^2)^2+\frac{1}{\chi}(dx^1)^2-\frac{2q_2}{\chi}dx^1\,dx^2\right].\end{aligned} \tag{65}$$

and that for E given by equations (55)
$$|Z|^2 = \left|\frac{1+E}{1-E}\right|^2 = \frac{(1+p\eta)^2+q^2\mu^2}{(1-p\eta)^2+q^2\mu^2}. \tag{66}$$

Inserting for χ, q_2, and $|Z|^2$ from equations (57), (58), and (66) in the last line of (65) and making use of the solutions for e^{β}, e^{ν}, and $e^{2\mu_3}$ given in equations (32) and (62), we can now write down the resulting metric in the form
$$\begin{aligned}ds^2 = {}&\frac{1-p^2\eta^2-q^2\mu^2}{(1-\eta^2)^{\frac{1}{4}}(1-\mu^2)^{\frac{1}{4}}}\left[\frac{(d\eta)^2}{1-\eta^2}-\frac{(d\mu)^2}{1-\mu^2}\right]\\ &-\frac{(1-\eta^2)^{\frac{1}{2}}(1-\mu^2)^{\frac{1}{2}}}{1-p^2\eta^2-q^2\mu^2}\{[(1+p\eta)^2+q^2\mu^2](dx^2)^2\\ &+[(1-p\eta)^2+q^2\mu^2](dx^1)^2-4q\mu\,dx^1\,dx^2\},\end{aligned} \tag{67}$$

where we have set the constant in equation (61) equal to unity.

The metric (67) can be written in a more symmetrical form by using in place of η and μ the variables ψ and θ defined by

$$\eta = \cos\psi \quad \text{and} \quad \mu = \cos\theta, \tag{68}$$

so that the range of both η and μ (i.e. of 't' and 'z') is $(0, 1)$. In these variables,

$$E = p\cos\psi + iq\cos\theta, \tag{69}$$

and

$$1-|E|^2 = 1-p^2\cos^2\psi - q^2\cos^2\theta = p^2\sin^2\psi + q^2\sin^2\theta = \Delta \text{ (say)}, \tag{70}$$

and the metric becomes

$$\begin{aligned} ds^2 = {} & \frac{\Delta}{(\sin\psi\sin\theta)^{\frac{1}{2}}}[(d\psi)^2-(d\theta)^2] \\ & -\frac{\sin\psi\sin\theta}{\Delta}[|1+E|^2(dx^2)^2+|1-E|^2(dx^1)^2-4q\cos\theta\, dx^1\,dx^2]. \end{aligned} \tag{71}$$

The contravariant form of this metric is

$$\begin{aligned} \partial s^2 = {} & \frac{(\sin\psi\sin\theta)^{\frac{1}{2}}}{\Delta}(\partial_\psi^2-\partial_\theta^2) \\ & -\frac{1}{\Delta\sin\psi\sin\theta}[|1-E|^2\partial_2^2+|1+E|^2\partial_1^2-4q\cos\theta\,\partial_1\partial_2]. \end{aligned} \tag{72}$$

We observe that the metric (71) is of the original form (5) with the definitions

$$\left.\begin{aligned} & e^{2\nu} = e^{2\mu_3} = \frac{\Delta}{(\sin\psi\sin\theta)^{\frac{1}{2}}}, \quad e^{2\psi} = \frac{\Sigma}{\Delta}\sin\psi\sin\theta, \quad e^{2\mu_2} = \frac{\Delta}{\Sigma}\sin\psi\sin\theta, \\ & \text{and} \qquad q_2 = \frac{2q}{\Sigma}\cos\theta, \end{aligned}\right\} \tag{73}$$

where we have written

$$\Sigma = |1-E|^2 = (1-p\cos\psi)^2+q^2\cos^2\theta. \tag{74}$$

Finally, we may note here for future reference the following identity:

$$|1+E|^2\,|1-E|^2-4q^2\cos^2\theta = \Delta^2. \tag{75}$$

The tetrad components of the Riemann tensor

There is in principle no difficulty in inserting the metric coefficients (73) in the expressions listed in equation (10) to obtain the components of the Riemann tensor. The required reductions can be simplified by noting that apart from the components (11), which vanish identically, and the cyclic identity

$$R_{1230}+R_{1023}+R_{1302} = 0, \tag{76}$$

we have the following relations, which are consequences of the vanishing of the Ricci tensor:

$$\left.\begin{aligned} & R_{1332} = R_{1002}, \quad R_{0223} = -R_{0113}, \quad R_{0202} = -R_{1313}, \\ & R_{0303} = -R_{1212}, \quad R_{2323} = -R_{1010} = R_{0202}+R_{0303}. \end{aligned}\right\} \tag{77}$$

We find

$$R_{0303} = \frac{1}{\Delta^3}(\sin\psi\,\sin\theta)^{\frac{1}{2}}[2p^2q^2(\cos^2\psi\,\sin^2\theta - \sin^2\psi\,\cos^2\theta) - \Delta(p^2-q^2) + \tfrac{1}{4}\Delta^2(\operatorname{cosec}^2\psi - \operatorname{cosec}^2\theta)],$$

$$R_{1230} = \frac{2pq}{\Delta^3}(\sin\psi\,\sin\theta)^{\frac{3}{2}},$$

$$R_{1023} - R_{1302} = \frac{3q}{2\Sigma\Delta^3}(\sin\psi\,\sin\theta)^{\frac{3}{2}}[\![8pq^2\cos^2\theta + 4p^2[(1-p\cos\psi)^2 - 3q^2\cos^2\theta]\cos\psi, - \Delta\{[(1-p\cos\psi)^2 - q^2\cos^2\theta]\cos\psi\,\operatorname{cosec}^2\psi, + 2p(1-p\cos\psi)\cos^2\theta\,\operatorname{cosec}^2\theta\}]\!],$$

$$R_{1332} = R_{1002} = -\frac{(\sin\psi\,\sin\theta)^{\frac{1}{2}}}{4\Sigma\Delta^2}3q\cos\theta\left\{3\Delta - 4q^2\sin^2\theta - \frac{4}{\Delta}[2p(1-p\cos\psi)(p^2\sin^2\psi - q^2\sin^2\theta)]\right\},$$

$$-2R_{3220} = -2R_{1310}$$

$$= \frac{1}{\Delta}(\sin\psi\,\sin\theta)^{\frac{1}{2}}\left\{\frac{24p^2q^2}{\Sigma\Delta^2}(1-p\cos\psi)\sin\psi\,\cos\psi\,\sin\theta\,\cos\theta - \frac{6}{\Sigma\Delta}pq^2\sin\psi\,\sin\theta\,\cos\theta - \frac{1}{\Sigma\Delta}[7(1-p\cos\psi) - 2\Delta](p^2\sin\psi\,\cos\psi \times \cot\theta + q^2\sin\theta\,\cos\theta\,\cot\psi) + \frac{3p}{2\Sigma}\sin\psi\,\cot\theta + 2\cot\theta\,\cot\psi\right\},$$

$$-8R_{2323}e^{2\nu} = \frac{1}{\Delta^2}[-4p^2q^2(3\cos^2\psi\,\sin^2\theta + \sin^2\psi\,\cos^2\theta) + 20p^4\sin^2\psi\,\cos^2\psi + 12q^4\sin^2\theta\,\cos^2\theta]$$
$$+ \frac{1}{\Sigma\Delta}[-32p^3(1-p\cos\psi)\sin^2\psi\,\cos\psi + 16q^4\sin^2\theta\,\cos^2\theta]$$
$$+ \frac{1}{\Sigma}(6p^2 + 2q^2 - 20p^2\cos^2\psi - 8q^2\cos^2\theta + 14p\cos\psi)$$
$$+ \frac{1}{\Delta}(6p^2 + 2q^2 - 16p^2\cos^2\psi - 12q^2\cos^2\theta) - 8 - (\cot^2\psi - \cot^2\theta)$$

$$-8R_{0202}e^{2\nu} = \frac{1}{\Delta^2}[-4p^2q^2(3\sin^2\psi\,\cos^2\theta + \cos^2\psi\,\sin^2\theta) + 12p^4\sin^2\psi\,\cos^2\psi + 20q^4\sin^2\theta\,\cos^2\theta]$$
$$+ \frac{1}{\Sigma\Delta}[-16p^3(1-p\cos\psi)\sin^2\psi\,\cos\psi + 32q^4\sin^2\theta\,\cos^2\theta]$$
$$+ \frac{1}{\Sigma}(6q^2 + 2p^2 - 8p^2\cos^2\psi - 20q^2\cos^2\theta + 6p\cos\psi)$$
$$+ \frac{1}{\Delta}(6q^2 + 2p^2 - 12p^2\cos^2\psi - 16q^2\cos^2\theta) - 8 + (\cot^2\psi - \cot^2\theta). \quad (78)$$

The remaining components can be obtained by making use of the relations (76) and (77).

[64]

5. The description of the Nutku–Halil solution in a Newman–Penrose formalism

First, we observe that by virtue of the identity

$$\begin{aligned}&|1-E|^2\,(\mathrm{d}x^1)^2+|1+E|^2\,(\mathrm{d}x^2)^2-(4q\cos\theta)\,\mathrm{d}x^1\,\mathrm{d}x^2\\&=|1-E|^2\,(\mathrm{d}x^1)^2+|1+E|^2\,(\mathrm{d}x^2)^2+2i(E-E^*)\,\mathrm{d}x^1\,\mathrm{d}x^2\\&=|(1-E)\,\mathrm{d}x^1+i(1+E)\,\mathrm{d}x^2|^2,\end{aligned}\tag{79}$$

we can rewrite the metric (71) in the alternative form

$$\mathrm{d}s^2=\frac{\Delta}{(\sin\psi\sin\theta)^{\frac{1}{2}}}[(\mathrm{d}\psi)^2-(\mathrm{d}\theta)^2]-\frac{\sin\psi\sin\theta}{\Delta}|(1-E)\,\mathrm{d}x^1+i(1+E)\,\mathrm{d}x^2|^2.\tag{80}$$

It is now manifest that a null tetrad-basis is provided by the vectors,

$$\begin{array}{llllll} & & \psi & \theta & x^1 & x^2\\ +l_i=e^2= & \dfrac{\sqrt{\Delta}}{(\sin\psi\sin\theta)^{\frac{1}{4}}\sqrt{2}} & \{1, & -1, & 0, & 0\ \},\\ +n_i=e^1= & \dfrac{\sqrt{\Delta}}{(\sin\psi\sin\theta)^{\frac{1}{4}}\sqrt{2}} & \{1, & +1, & 0, & 0\ \},\\ -m_i=e^4= & \dfrac{(\sin\psi\sin\theta)^{\frac{1}{2}}}{\sqrt{(2\Delta)}} & \{0, & 0, & (1-E^*), & -i(1+E^*)\},\\ -\bar{m}_i=e^3= & \dfrac{(\sin\psi\sin\theta)^{\frac{1}{2}}}{\sqrt{(2\Delta)}} & \{0, & 0, & (1-E), & +i(1+E)\ \}.\end{array}\tag{81}$$

The contravariant form of these basis vectors is given by

$$\begin{array}{llllll} l^i= & \dfrac{(\sin\psi\sin\theta)^{\frac{1}{4}}}{\sqrt{(2\Delta)}} & \{1, & +1, & 0, & 0\ \},\\ n^i= & \dfrac{(\sin\psi\sin\theta)^{\frac{1}{4}}}{\sqrt{(2\Delta)}} & \{1, & -1, & 0, & 0\ \},\\ \bar{m}^i= & \dfrac{1}{\sqrt{(2\Delta\sin\psi\sin\theta)}} & \{0, & 0, & 1+E^*, & -i(1-E^*)\},\\ m^i= & \dfrac{1}{\sqrt{(2\Delta\sin\psi\sin\theta)}} & \{0, & 0, & 1+E, & +i(1-E)\ \},\end{array}\tag{82}$$

where, in deriving these expressions, we have made use of the identity (75).

(a) *The spin coefficients*

The spin coefficients (as defined in M.T., chapter 1, equations (286)) with respect to the chosen basis are most conveniently evaluated via the λ-symbols (defined in M.T., chapter 1, equations (266)). By virtue of the facts that the matrix formed by the column vectors $(l^i, n^i, m^i, \bar{m}^i)$ is reducible and that its elements are functions of ψ and θ only, many of the λ-symbols vanish; in particular,

$$\lambda_{311}=\lambda_{322}=\lambda_{231}=\lambda_{312}=\lambda_{321}=\lambda_{314}=\lambda_{324}=\lambda_{344}=0.\tag{83}$$

As a result the following spin coefficients vanish:

$$\kappa=\nu=\tau=\pi=\alpha=\beta=0.\tag{84}$$

(We may note that the vanishing of κ and ν is consistent with the fact that (l^i) and (n^i) are null geodesic-congruences.) The remaining spin coefficients are given by

$$\left.\begin{aligned}\sigma &= \lambda_{331}, \quad \lambda = \lambda_{244}, \qquad \rho = \mathscr{R}e\,\lambda_{431}, \quad \mu = \mathscr{R}e\,\lambda_{243},\\ \epsilon &= \tfrac{1}{2}(\lambda_{211} - i\mathscr{I}m\,\lambda_{431}), \quad \nu = \tfrac{1}{2}(\lambda_{221} - i\mathscr{I}m\,\lambda_{243});\end{aligned}\right\} \tag{85}$$

and we find

$$\begin{aligned}
\sigma &= -\frac{(\sin\psi\,\sin\theta)^{\frac{1}{4}}}{(2\Delta^3)^{\frac{1}{2}}}(p\sin\psi - iq\sin\theta),\\
\lambda &= +\frac{(\sin\psi\,\sin\theta)^{\frac{1}{4}}}{(2\Delta^3)^{\frac{1}{2}}}(p\sin\psi - iq\sin\theta),\\
\rho &= -\frac{(\sin\psi\,\sin\theta)^{\frac{1}{4}}}{(8\Delta)^{\frac{1}{2}}}(\cot\psi + \cot\theta),\\
\mu &= +\frac{(\sin\psi\,\sin\theta)^{\frac{1}{4}}}{(8\Delta)^{\frac{1}{2}}}(\cot\psi - \cot\theta),\\
\epsilon &= \frac{(\sin\psi\,\sin\theta)^{\frac{1}{4}}}{(2\Delta)^{\frac{3}{2}}}\{+p^2\sin\psi\,\cos\psi + q^2\sin\theta\,\cos\theta - \tfrac{1}{4}\Delta(\cot\theta + \cot\psi)\\
&\quad + ipq(\sin\psi\,\cos\theta - \cos\psi\,\sin\theta)\},\\
\gamma &= \frac{(\sin\psi\,\sin\theta)^{\frac{1}{4}}}{(2\Delta)^{\frac{3}{2}}}\{-p^2\sin\psi\,\cos\psi + q^2\sin\theta\,\cos\theta - \tfrac{1}{4}\Delta(\cot\theta - \cot\psi)\\
&\quad + ipq(\sin\psi\,\cos\theta + \cos\psi\,\sin\theta)\}.
\end{aligned} \tag{86}$$

(b) *The Weyl scalars*

With the spin coefficients listed in equations (83) and (86), the Weyl scalars can be obtained by making use of the Ricci identities (M.T., chapter 1, equations (310)); thus from equations (310, c) and (310, i), namely

$$\boldsymbol{D}\tau - \boldsymbol{\Delta}\kappa = \rho(\tau + \pi^*) + \sigma(\tau^* + \pi) + \tau(\epsilon - \epsilon^*) - \kappa(3\gamma + \gamma^*) + \Psi_1, \tag{87}$$

$$\boldsymbol{D}\nu - \boldsymbol{\Delta}\pi = \mu(\pi + \tau^*) + \lambda(\pi^* + \tau) + \pi(\gamma - \gamma^*) - \nu(3\epsilon + \epsilon^*) + \Psi_3, \tag{88}$$

and the vanishing of the spin coefficients (83), we conclude that

$$\Psi_1 = \Psi_3 = 0. \tag{89}$$

Similarly, it follows from M.T., equation (310, l) p. 46, that

$$\Psi_2 = \mu\rho - \lambda\sigma; \tag{90}$$

and inserting the values of μ, ρ, λ and σ from equations (86), we find

$$\Psi_2 = \frac{(\sin\psi\,\sin\theta)^{\frac{1}{2}}}{2\Delta^3}[(p\sin\psi - iq\sin\theta)^2 - \tfrac{1}{4}\Delta^2(\operatorname{cosec}^2\psi - \operatorname{cosec}^2\theta)]. \tag{91}$$

Next, by M.T., equation (310, b), p. 46, we have, in the present context,

$$\begin{aligned}
\frac{\Psi_0}{\sigma} &= \boldsymbol{D}\,\lg\sigma - (\rho + \rho^* + 3\epsilon - \epsilon^*)\\
&= \frac{(\sin\psi\,\sin\theta)^{\frac{1}{4}}}{(2\Delta)^{\frac{1}{2}}}\left(\frac{\partial}{\partial\psi} + \frac{\partial}{\partial\theta}\right)\lg\sigma - (\rho + \rho^* + 3\epsilon - \epsilon^*);
\end{aligned} \tag{92}$$

and we find

$$\Psi_0 = -\frac{3}{2\Delta^2}(\sin\psi \sin\theta)^{\frac{1}{2}}(p\sin\psi - iq\sin\theta) \times\left[\tfrac{1}{2}(\cot\psi+\cot\theta)-\frac{p\cos\psi+iq\cos\theta}{p\sin\psi+iq\sin\theta}\right], \quad (93)$$

or, alternatively,

$$\Psi_0 = \frac{3}{4}\frac{\sin(\theta-\psi)}{(\sin\psi\sin\theta)^{\frac{1}{2}}(p\sin\psi+iq\sin\theta)^3}. \quad (94)$$

Similarly, from M.T., equation (310, j), p. 46, we find

$$\Psi_4 = \frac{3}{4}\frac{\sin(\theta+\psi)}{(\sin\psi\sin\theta)^{\frac{1}{2}}(p\sin\psi+iq\sin\theta)^3}. \quad (95)$$

With the specification of the null basis, the spin coefficients, and the Weyl scalars, we have completed the description of the Nutku–Halil space–time in a Newman–Penrose formalism. We shall return in §6 to the implications of the expressions (91), (94) and (95) for the non-vanishing Weyl-scalars.

6. The introduction of null coordinates and the completion of the space–time to describe the collision of two impulsive gravitational waves

So far we have concerned ourselves only with the region of the space–time in which the interaction between the colliding waves occurs. To complete the solution, we must extend the domain of validity of the solution to include the flat space–time that prevailed before the instant of the collision at $\eta = 0$. For this purpose, it is convenient to express the metric (80) in terms of two null coordinates, u and v, in place of ψ and θ; and this is readily accomplished by setting

$$\psi = \xi - \zeta \quad \text{and} \quad \theta = \xi + \zeta,$$

and

$$u = \cos\xi \quad \text{and} \quad v = \sin\zeta, \quad (96)$$

Then

$$(d\psi)^2 - (d\theta)^2 = -4d\xi\, d\zeta = 4\frac{du\, dv}{(1-u^2)^{\frac{1}{2}}(1-v^2)^{\frac{1}{2}}}; \quad (97)$$

and the equations relating ψ and θ and u and v are

$$\begin{aligned}\eta = \cos\psi &= \cos(\xi-\zeta) = \cos\xi\cos\zeta + \sin\xi\sin\zeta \\ &= u\sqrt{(1-v^2)} + v\sqrt{(1-u^2)},\end{aligned} \quad (98)$$

and

$$\begin{aligned}\mu = \cos\theta &= \cos(\xi+\zeta) = \cos\xi\cos\zeta - \sin\xi\sin\zeta \\ &= u\sqrt{(1-v^2)} - v\sqrt{(1-u^2)}.\end{aligned} \quad (99)$$

Also,

$$\begin{aligned}\sin\psi\sin\theta &= \sin^2\xi\cos^2\zeta - \cos^2\xi\sin^2\zeta \\ &= (1-u^2)(1-v^2) - u^2v^2 = 1-u^2-v^2,\end{aligned} \quad (100)$$

and

$$\begin{aligned}E &= p\cos\psi + iq\cos\theta \\ &= (p+iq)u\sqrt{(1-v^2)} + (p-iq)v\sqrt{(1-u^2)}.\end{aligned} \quad (101)$$

Since $p^2+q^2 = 1$, it is convenient to define

$$p = \cos\alpha \quad \text{and} \quad q = \sin\alpha, \tag{102}$$

in which case the expression for E takes the form

$$E = \mathrm{e}^{i\alpha} u\sqrt{(1-v^2)}+\mathrm{e}^{-i\alpha} v\sqrt{(1-u^2)}. \tag{103}$$

The metric (80) in terms of the null coordinates u and v thus takes the form

$$\begin{aligned} \mathrm{d}s^2 = {} & 4\frac{1-|E|^2}{[(1-u^2-v^2)(1-u^2)(1-v^2)]^{\frac{1}{2}}}\mathrm{d}u\,\mathrm{d}v \\ & -\frac{1-u^2-v^2}{1-|E|^2}[(1-E)\,\mathrm{d}x^1+i(1+E)\,\mathrm{d}x^2][(1-E^*)\,\mathrm{d}x^1-i(1+E^*)\,\mathrm{d}x^2]. \end{aligned} \tag{104}$$

(*a*) *The effect of a rotation of the* (x^1, x^2)*-axes*

It is of interest to notice that if x^1 and x^2 are subject to the transformation,

$$\left.\begin{aligned} & x^1 = x\cos\varphi-y\sin\varphi \\ \text{and} \qquad & x^2 = x\sin\varphi+y\cos\varphi. \end{aligned}\right\} \tag{105}$$

corresponding to a rotation of the (x^1, x^2)-axes in the (x^1, x^2)-plane by an angle φ, the metric (104) retains its form if E is replaced by

$$E_\varphi = \mathrm{e}^{i(\alpha-2\varphi)} u\sqrt{(1-v^2)}+\mathrm{e}^{-i(\alpha+2\varphi)} v\sqrt{(1-u^2)}. \tag{106}$$

Accordingly, we may suppose that in general E is given by

$$E = \mathrm{e}^{i\beta} u\sqrt{(1-v^2)}+\mathrm{e}^{i\gamma} v\sqrt{(1-u^2)}; \tag{107}$$

and the angle α (equal to $\cos^{-1} p$) determining the planes of polarizations of the colliding waves is given by

$$\alpha = \tfrac{1}{2}(\beta-\gamma). \tag{108}$$

(*b*) *The completion of the metric* (104) *to correspond to the space–time of two colliding impulsive gravitational waves*

The ranges of the coordinates u and v as originally defined are restricted to the domain (cf. equation (96) and the earlier definitions of the variables η and μ)

$$0 \leqslant u \leqslant 1 \quad \text{and} \quad 0 \leqslant v \leqslant 1. \tag{109}$$

But it is manifest from the form of the metric (104) that it is singular on the surface

$$u^2+v^2 = 1. \tag{110}$$

Indeed, on this surface the curvature scalars become infinite. This is evident from the expressions (91), (94), and (95) for the Weyl scalars Ψ_2, Ψ_0, and Ψ_4. Thus, Ψ_2, expressed in terms of u and v, is

$$\begin{aligned} \Psi_2 = \frac{(1-u^2-v^2)^{\frac{1}{2}}}{2(1-|E|^2)^3}\Big\{ & (p^2-q^2)(1-u^2-v^2+2u^2v^2)-2uv(1-u^2)^{\frac{1}{2}}(1-v^2)^{\frac{1}{2}} \\ & -2ipq(1-u^2-v^2)-(1-|E|^2)^2\frac{uv(1-u^2)^{\frac{1}{2}}(1-v^2)^{\frac{1}{2}}}{(1-u^2-v^2)^2}\Big\}; \end{aligned} \tag{111}$$

and Ψ_2 has the behaviour $(1-u^2-v^2)^{-\frac{3}{2}}$ when $u^2+v^2 \to 1$. The scalars Ψ_0 and Ψ_4 also diverge for $u^2+v^2 \to 1$; but they have the weaker behaviour $(1-u^2-v^2)^{-\frac{1}{2}}$ on the singular surface. We conclude that *the space–time described by the metric* (104) *develops an essential curvature singularity on* $u^2+v^2=1$.

In contrast to what happens on $u^2+v^2=1$, on the surfaces,

$$u=0, \quad 0 \leqslant v < 1 \quad \text{and} \quad v=0, \quad 0 \leqslant u < 1, \tag{112}$$

we have only coordinate singularities, as is evident when the metric is written in terms of the coordinates ψ and θ: $\sin\psi \sin\theta$ does not vanish except on $u^2+v^2=1$ (cf. equation (80)). In view of the coordinate nature of the singularities that characterize the surfaces (112), we can clearly extend the space–time described by the metric (104), in the region designated by I in figure 1, across these surfaces into the regions designated by II and III. The extension can be accomplished in more than one way if we are not restricted to analytic continuations. In particular, to extend the space–time to represent the collision of two plane-fronted impulsive gravitational waves, we simply write (following Khan & Penrose (1971) and Nutku & Halil (1977))

$$uH(u) \quad \text{and} \quad vH(v), \tag{113}$$

in place of u and v, where $H(u)$ and $H(v)$ denote the Heaviside step-functions that are unity for positive values of the argument and zero otherwise. This manner of extension raises several questions. Is the space–time in the regions II ($u<0$, $0 \leqslant v < 1$) and III ($v<0$, $0 \leqslant u < 1$) flat as required? Are the discontinuities in the metric functions so introduced justified? And, finally, how does the occurrence of the curvature singularity, on $u^2+v^2=1$, affect the space–time on the surfaces, $v=1$ for $u<0$ and $u=1$ for $v<0$? These questions have been considered in detail by Penrose (unpublished work), Bell & Szekeres (1974), and by Matzner & Tipler (1984). The following comments are made essentially for the sake of completeness in the context of the present analysis.

First, we observe that by the substitutions (113), the metric (104) in the region II (after a suitable rotation of the (x^1, x^2)-axes) becomes

$$ds^2 = 4\,du\,dv - (1-v)^2\,(dx^1)^2 - (1+v)^2\,(dx^2)^2. \tag{114}$$

By the transformation,

$$x^1 = \frac{x}{1-v}, \quad x^2 = \frac{y}{1+v}, \quad v = V,$$

and

$$u = U + \frac{1}{4}\left(\frac{x^2}{1-v} - \frac{y^2}{1+v}\right), \tag{115}$$

the metric (114) takes the manifestly Minkowskian form

$$ds^2 = 4\,dU\,dV - (dx)^2 - (dy)^2, \tag{116}$$

verifying its flatness.

Considering next the nature of the singularity on the surface $u=0$ ($0 \leqslant v < 1$) introduced by the substitutions (113), we find that approaching the surface from the positive side of the u-axis, the non-vanishing Weyl-scalars Ψ_2, Ψ_0, and Ψ_4, given by equations (91), (94) and (95) tend to the values

$$\Psi_2 \to \frac{1}{2}\frac{(p-iq)^2}{(1-v^2)^{\frac{3}{2}}}, \quad \Psi_0 \to \tfrac{3}{2}(p-iq)^3 \frac{v}{(1-v^2)^{\frac{3}{2}}} \quad \text{and} \quad \Psi_4 \to 0$$
$$(u \to +0, \theta+\psi \to \pi - 0). \quad (117)$$

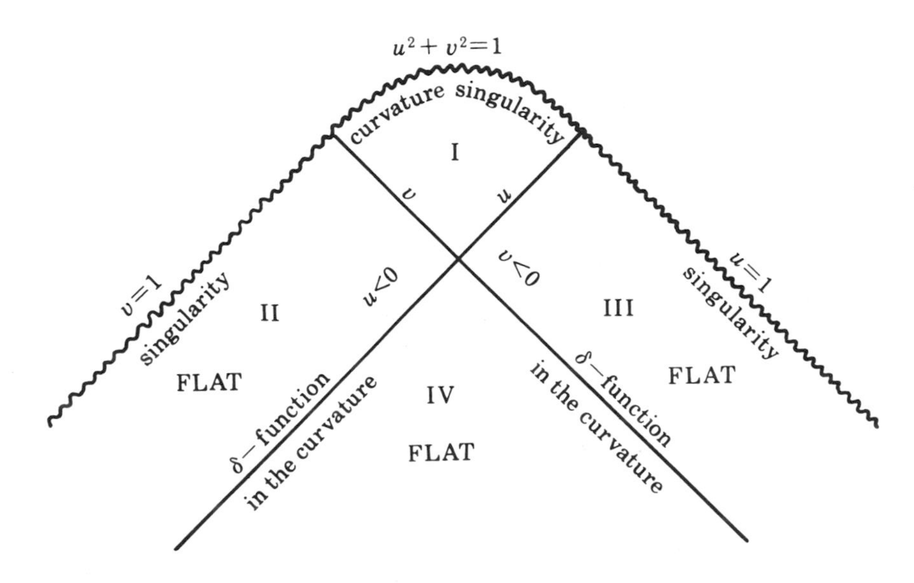

FIGURE 1. To illustrate the space–time resulting from the collision of impulsive gravitational waves.

Similarly, by approaching the surface $v = 0 (0 \leqslant u < 1)$ from the positive side of v-axis, they tend to the values

$$\Psi_2 \to \frac{1}{2}\frac{(p-iq)^2}{(1-u^2)^{\frac{3}{2}}}, \quad \Psi_0 \to 0 \quad \text{and} \quad \Psi_4 \to \tfrac{3}{2}(p-iq)^3 \frac{u}{(1-u^2)^{\frac{3}{2}}}$$
$$(v \to +0, \theta-\psi \to +0). \quad (118)$$

The discontinuities introduced by the substitutions (113), make the expressions for the Weyl scalars to be the following:

$$\left.\begin{aligned} \Psi_0 &= \frac{(p-iq)}{2(1-v^2)}\delta(u) + H(u)\,\Psi_0^{(\mathrm{I})}, \quad \Psi_4 = \frac{(p-iq)}{2(1-u^2)}\delta(v) + H(v)\,\Psi_4^{(\mathrm{I})} \\ \text{and} \qquad \Psi_2 &= H(u)\,H(v)\,\Psi_2^{(\mathrm{I})}, \end{aligned}\right\} \quad (119)$$

where $\Psi_2^{(\mathrm{I})}$, $\Psi_0^{(\mathrm{I})}$, and $\Psi_4^{(\mathrm{I})}$ denote the expressions for these scalars in region I, given by equations (91), (94), and (95). The simple discontinuities in the Weyl scalars, resulting from the presence of the Heaviside functions, do not conflict with any of the requirements of the field equations since they occur on null surfaces. The considerations, leading to the absence of a δ-function in the expression for Ψ_2 and its presence, with the factors specified, in the expressions for Ψ_0 and Ψ_4, are somewhat subtle. They are explained in a paper by Dr B. Xanthopoulos and one of us (S.C.) (on the analogue of the Kerr–Newman solution for a space–time with two space-like Killing vectors) now in the process of being submitted to the Royal Society. The presence of these δ-function singularities in the expressions for Ψ_0 and Ψ_4 reflect the assumed (idealized) impulsive character of the waves, which collide at $\eta = 0$ and $\mu = 0$.

Finally, we note that the restrictions of the space–times in regions II and III by the requirements $(u < 0, 0 \leqslant v < 1)$ and $(v < 0, 0 \leqslant u < 1)$ make $v = 1 (u < 0)$ and $u = 1$ $(v < 0)$ singular. This fact becomes manifest, when we consider, for example, the null rays (cf. equation (115)),

$$x^2 = 0, \quad x^1 = \frac{x_0}{1-u} \quad \text{and} \quad u = u_0 + \frac{1}{4}\frac{x_0^2}{1-v} \quad (u_0 < 0), \tag{120}$$

in region II. All these rays, except the one for $x_0 = 0$, intersect the v-axis at

$$v = 1 - x_0^2/4(-u_0) < 1, \tag{121}$$

cross into the region I smoothly, and continue to intersect the curve $u^2 + v^2 = 1$ in finite time. The disposition of these null rays (for $x_0 \neq 0$) and, in particular, the manner of their avoidance of $v = 1$ (while in region II) and reaching the singularity at $u^2 + v^2 = 1$ in finite time are clear manifestations of the singularity of $u = 1$ $(v < 0)$; and, similarly, of $v = 1 (u < 0)$.

The extension of the space–times of regions II and III into the flat-region IV requires, of course, no special comment!

7. The analogue or Robinson's theorem for space–times with two space-like Killing-vectors

In stationary axisymmetric space–times we have Robinson's (1975) theorem (for an account see M.T., §55), which guarantees the uniqueness of a solution, external to a smooth event-horizon and asymptotically flat, for any assigned pair of values for the mass, M, and the angular momentum J (so long as $M^2/J^2 < 1$). In this section we consider whether an analogous theorem can be formulated in the framework of space–times with two space-like Killing-vectors ($\mathrm{d}x^1$ and $\mathrm{d}x^2$). But some important similarities and differences should first be noted.

As we have seen, in space–times with two space-like Killing-vectors, we can always choose a gauge in which the duration in time t and the spatial extent in x^3 are both finite: this fact corresponds to the allowed range, (0, 1), for both the variables η and μ. (This choice of gauge is entirely analogous to the choice that endows every stationary axisymmetric space–time with a 'spherical' null-surface, $\eta = 1$ (cf. M.T., §53).) However, the space–time is extensible for negative values for both η and μ; in the context of colliding gravitational-waves, this extension is accomplished by the introduction of the Heaviside step-function and by making the extended part of the space–time flat. But, in general, it may not be possible to extend beyond $\eta = 1$ and $\mu = 1$: curvature singularities may develop.

Turning to the formal similarities in the equations governing stationary axisymmetric space–times and space–times with two space-like Killing-vectors, we find that the basic functions that describe the (x^1, x^2)-part of the metric, namely (cf. equation (6))

$$-e^{\psi+\mu_2}\left[\chi(dx^2)^2+\frac{1}{\chi}(dx^1-q_2\,dx^2)^2\right], \tag{122}$$

are the functions χ and q_2. (The factor $e^{\psi+\mu_2}$ in front depends on the choice of gauge; in the chosen gauge it is $(1-\eta^2)^{\frac{1}{2}}\,(1-\mu^2)^{\frac{1}{2}}$.) Letting

$$\chi = X \quad \text{and} \quad q_2 = Y, \tag{123}$$

the equations governing them ((35) and (36)) can be rewritten in the forms,

$$E = 0 \quad \text{and} \quad F = 0, \tag{124}$$

where

$$E(X, Y) = \left[(\eta^2-1)\frac{X_{,\eta}}{X}\right]_{,\eta}+(\eta^2-1)\frac{(Y_{,\eta})^2}{X^2}$$
$$+\left[(1-\mu^2)\frac{X_{,\mu}}{X}\right]_{,\mu}+(1-\mu^2)\frac{(Y_{,\mu})^2}{X^2}, \tag{125}$$

and

$$F(X, Y) = \left[(\eta^2-1)\frac{Y_{,\eta}}{X^2}\right]_{,\eta}+\left[(1-\mu^2)\frac{Y_{,\mu}}{X^2}\right]_{,\mu}. \tag{126}$$

These equations are formally identical with those underlying Robinson's theorem (cf. M.T., §55, equations (143)–(145), p. 292). But there are important differences: the range of η is now $(0, 1)$ and not $(1, \infty)$, and there are differences in the boundary conditions as well. In the case considered by Robinson (cf. M.T., §55, equations (146) and (147)),

$$X \to (1-\mu^2)\,\eta^2+O(\eta) \quad \text{and} \quad Y \to 2J\mu(3-\mu^2)+O(\eta^{-1}) \quad \text{for} \quad \eta \to \infty; \tag{127}$$

while in the present context, we shall require that both X and Y are bounded for all values of η and μ in the allowed range $(0, 1)$, as they are for the solutions (57) and (58) we have found for them. More particularly, we shall require that

$$X \to 1+O(\eta, \mu^2) \quad \text{and} \quad Y = C\mu+O(\eta, \mu^3) \quad \text{for} \quad \eta, \mu \to 0, \tag{128}$$

where C is a constant; and, further, that X and Y are not singular at $\eta = 1$ and $\mu = 1$.

We may note here that for the solutions obtained in §4,

$$X = 1+2p\eta+2p^2\eta^2-2q^2\mu^2+\dots$$

and

$$Y = 2q\mu(1+2p\eta+3p^2\eta^2-q^2\mu^2+\dots), \tag{129}$$

where $p^2+q^2 = 1$. It is of interest to inquire in this connection, whether unique solutions for X and Y can be obtained by assuming that expansions in powers of η and μ, valid near $\eta = \mu = 0$, are possible. It appears that they may not be possible. Thus, we find, for example, that on insertion of the expansions

$$X = \chi = 1+2p\eta+a\eta^2+b\mu^2+O(\mu^4, \eta^3),$$

and

$$Y = q_2 = 2q\mu(1+\alpha\eta+\beta\eta^2+\nu\mu^2)+O(\mu^5, \eta^3) \tag{130}$$

(where $p, a, b, q, \alpha, \beta$, and ν are unspecified constants), in equations (37) and (38), the equality of the *leading terms*, already, requires that

$$1-3\nu+\beta = 2(p\alpha-b) \quad \text{and} \quad (a-b) = 2(p^2+q^2); \tag{131}$$

and the equality of the next higher terms requires many further terms in the expansions (130); and the process appears unending. This apparent impossibility of obtaining determinate solutions by series expansions (known to be applicable to the particular solutions (57) and (58) that we have found) must be related to the hyperbolic character of the underlying equations.

Keeping the foregoing facts in mind, we consider two solution pairs (X_1, Y_1) and (X_2, Y_2), which are bounded and continuous in the ranges $0 \leqslant \eta \leqslant 1$ and $0 \leqslant \mu \leqslant 1$. Since, along with (X, Y), (cX, cY), where c is an arbitrary constant, also represents a solution, there is no loss of generality in supposing that $X_1 = X_2$ at $\eta = 0$ and $\mu = 0$. And we shall also assume that both Y_1 and Y_2 vanish at the origin. Our assumptions, then, are:

$$X_1 = X_2 = 1 \quad \text{and} \quad Y_1 = Y_2 = 0 \quad \text{at} \quad \eta = 0 \quad \text{and} \quad \mu = 0. \tag{132}$$

Since X and Y formally satisfy the same equations as in Robinson's theorem, we can define the same functional $\mathbb{R}$ (M.T., equation (148), p. 293) and proceed similarly by making use of the same lemmas. We shall then obtain the identity (M.T., pp. 295–297)

$$\begin{aligned} &\frac{1}{2}\int_0^1 d\mu \left\{(1-\eta^2)\left[\frac{(Y_2-Y_1)^2+(X_2-X_1)^2}{X_1X_2}\right]\right\}_{\eta=0}^{\eta=1} \\ &-\frac{1}{2}\int_0^1 d\eta \left\{(1-\mu^2)\left[\frac{(Y_2-Y_1)^2+(X_2-X_1)^2}{X_1X_2}\right]\right\}_{\mu=0}^{\mu=1} \\ &+\int_0^1\int_0^1 \frac{d\eta\, d\mu}{4X_1^3X_2^3}[\![-(1-\eta^2)\,\mathbb{R}_1+(1-\mu^2)\,\mathbb{R}_2]\!] = 0, \end{aligned} \tag{133}$$

where $\mathbb{R}_1$ and $\mathbb{R}_2$ are the same positive-definite expressions that occur in M.T., equation (156), p. 297.

We first observe that the surface integrals do not survive by virtue of the assumed boundedness of the solutions at $\eta = 1$ and $\mu = 1$, and the boundary conditions (132) that we have stipulated at the origin. We are thus left with the volume integral:

$$\int_0^1\int_0^1 \frac{d\eta\, d\mu}{4X_1^3X_2^3}[\![(1-\eta^2)\,\mathbb{R}_1-(1-\mu^2)\,\mathbb{R}_2]\!] = 0. \tag{134}$$

The principal and the decisive difference between the present case and the case considered by Robinson is that we now have a *difference* of two positive-definite terms and not a *sum*. Therefore, Robinson's conclusion that

$$X_1 = X_2 \quad \text{and} \quad Y_1 = Y_2 \tag{135}$$

does *not* follow. But if we *assume* that $\mathbb{R}_1 = 0$, *then* $\mathbb{R}_2 = 0$ *and conversely*. Proceeding further, as in the proof of Robinson's theorem, we may now concude that

$$\left(\frac{Y_2 - Y_1}{X_1 X_2}\right)_{,\eta} = 0 \quad \textit{and} \quad \left(\frac{X_1}{X_2}\right)_{,\eta} = 0 \tag{136}$$

implies that

$$\left(\frac{Y_2 - Y_1}{X_1 X_2}\right)_{,\mu} = 0 \quad \textit{and} \quad \left(\frac{X_1}{X_2}\right)_{,\mu} = 0, \tag{137}$$

and conversely: a non-trivial conclusion, perhaps; but its meaning is not clear.

8. Concluding remarks

Perhaps the most distinctive feature of space–times with two space-like Killing-vectors ($\mathrm{d}x^1$, $\mathrm{d}x^2$) is their finite duration in time and finite extent in the third spatial direction, x^3. This last feature is the principal characteristic of space–times representing the collision of two planar gravitational waves. The origins of time and of the coordinate x^3 refer to the instant and the location of the collision; and the finite duration in t and the finite spatial extent in x^3 are consequences of the simultaneous focusing of the scattered waves to form a singularity of the space–time.

On the more formal side, particular interest is attached to the exact correspondence that is revealed between the Nutku–Halil solution in space–times with two space-like Killing-vectors and the Kerr solution in space–times that are stationary and axisymmetric. A special case of this correspondence is that between the Khan–Penrose solution and the Schwarzschild solution.

Finally, the analogue of Robinson's theorem proved in §7 indicates clearly what may result from replacing a time-like killing-vector by a space-like one.

The research reported in this paper has, in part, been supported by grants from the National Science Foundation under grant PHY 80-26043 with the University of Chicago.

References

Bell, P. & Szekeres, P. 1974 *Gen. Rel. Grav.* **5**, 275.
Chandrasekhar, S. 1978 *Proc. R. Soc. Lond.* A **358**, 406.
Chandrasekhar, S. 1983 *The mathematical theory of black holes*. Oxford: Clarendon Press.
Khan, K. & Penrose, R. 1971 *Nature, Lond.* **229**, 185.
Matzner, R. A. & Tipler, F. J. 1984 *Phys. Rev.* D **29**, 1575.
Nutku, Y. & Halil, M. 1977 *Phys. Rev. Lett.* **39**, 1379.
O'Brien, S. & Synge, J. L. 1952 *Communs Dubl. Inst. advd Stud.* A **9**.
Pirani, F. A. E. 1964 In *Brandeis Lectures on General Relativity*, pp. 269–275. New Jersey: Prentice-Hall.
Robinson, D. C. 1972 *Phys. Rev. Lett.* **34**, 905.
Szekeres, P. 1972 *J. math. Phys.* **13**, 286.

20

On colliding waves in the Einstein–Maxwell theory

By Subrahmanyan Chandrasekhar[1], F.R.S.,
and Basilis C. Xanthopoulos[2]

[1] *University of Chicago, Chicago, Illinois* 60637, *U.S.A.*
[2] *Department of Physics, University of Crete and Research Centre of Crete, Iraklion, Greece*

(*Received* 2 *October* 1984)

An exact solution of the Einstein–Maxwell equations is obtained that represents a space–time which describes consistently the collision between two plane impulsive gravitational waves, each supporting an electromagnetic shock-wave. In obtaining the solution, the relationship, which had been established earlier, between the solutions describing stationary black-holes and solutions describing colliding plane-waves, is extended to the Einstein–Maxwell equations (and exploited).

The case when the colliding waves are parallelly polarized is analysed in detail to exhibit the singularities and the discontinuities that occur on the null boundaries characteristic of this problem. It is found that the passage of the waves, prior to collision, produces a spray of gravitational and electromagnetic radiation and the collision results in the scattering and the focusing of the waves and the development of a space–time singularity.

The solution that is obtained avoids in a natural way certain conceptual difficulties (such as the occurrence of the 'square root' of a δ-function and current sheets) that had been anticipated.

1. Introduction

Khan & Penrose (1971; see also Szekeres 1970, 1972) showed that space–time singularities develop when two parallelly polarized plane impulsive gravitational waves collide and scatter off each other. On the strength of this example, they argued that the scattering of waves by waves in general relativity may be as generic a cause for the development of space–time singularities as gravitational collapse. In view of the continuing obscurity which appears to shroud the nature of space–time singularities in general relativity, the arguments of Khan & Penrose would amply justify a broader study of related problems involving the scattering of waves than has hitherto been accorded. Thus, apart from an investigation by Nutku & Halil (1977) in which the restriction to parallel polarizations in the example considered by Khan & Penrose was removed, some of the basic questions which their example raises have remained unresolved. One such question concerns the collision of gravitational waves in the exact framework of the Einstein–Maxwell equations. In this latter framework one cannot have gravitational waves without accompanying electromagnetic waves, and conversely. The question then occurs: would an impulsive gravitational wave with its associated δ-function singularity in the Weyl tensor imply a similar δ-function singularity in the Maxwell stress-tensor? If that should occur, then the expression for the electromagnetic field variables would involve the square root of the δ-function; and one would be at a loss to know how to interpret such a function. Since the coupling of whatever field may be present with gravitation (which it must entail) is an essential feature of general relativity, a resolution of the conceptual questions raised by as simple a notion as an impulsive gravitational wave would appear useful if not necessary. And from past experience in such matters, one may conclude that nothing less than *exact* solutions describing physically realizable situations will suffice.

A recent paper (Chandrasekhar & Ferrari 1984; this paper will be referred to hereafter as Paper I), in which a new derivation of the Nutku–Halil solution was given, suggests a direct approach to the problem of colliding waves in the exact framework of the Einstein–Maxwell equations. In that paper it was pointed out how, in a very definite mathematical sense, the solutions of Khan & Penrose and of Nutku & Halil (in the region of the space–time in which the interaction between the colliding waves occurs) are, for space–times with two space-like Killing vectors, the exact analogues of the Schwarzschild and the Kerr solutions for space–times with one space-like and one time-like Killing vector. The question, as to what the corresponding analogue of the charged Kerr–Newman solution is, is an immediate and a definite one; and as we shall show, the methods of Paper I provide a prescription for finding that analogue. In this paper we shall obtain that solution. Further, by considering in detail the particular solution which is the analogue of the Reissner–Nordström solution (or, equivalently, of the Khan–Penrose solution in the Einstein–Maxwell theory for parallelly polarized colliding waves), we shall show that none of the anticipated difficulties arise: they resolve themselves naturally – a further example of the inner consistency of the general theory of relativity!

2. The basic equations

As in Paper I, we start with the metric

$$\mathrm{d}s^2 = \mathrm{e}^{2\nu}(\mathrm{d}x^0)^2 - \mathrm{e}^{2\mu_3}(\mathrm{d}x^3)^2 - \mathrm{e}^{2\psi}(\mathrm{d}x^1 - q_2\,\mathrm{d}x^2)^2 - \mathrm{e}^{2\mu_2}(\mathrm{d}x^2)^2, \tag{1}$$

where ν, μ_3, ψ, q_2, and μ_2 are functions of x^0 and x^3 only, with the freedom to impose a coordinate condition on ν and μ_3. (As explained in Paper I, §2, the metric of the form (1) is appropriate for space–times with two commuting space-like Killing vectors.)

Since we are now concerned with the Einstein–Maxwell equations, we shall consider, first, Maxwell's equations in a space–time with a metric of the chosen form. These equations can be directly written down by specializing the equations given in *The mathematical theory of black holes* (Chandrasekhar 1983, chapter 2, §15, equations (95); this book will be referred to hereafter as M.T.). Thus, we find that

$$F_{12} = F_{03} = 0, \tag{2}$$

and the remaining equations are

$$(e^{\psi+\nu} F_{01})_{,3} + (e^{\psi+\mu_3} F_{13})_{,0} = 0, \tag{3}$$

$$(e^{\psi+\nu} F_{23})_{,3} + (e^{\psi+\mu_3} F_{02})_{,0} = 0, \tag{4}$$

$$(e^{\mu_2+\mu_3} F_{01})_{,0} + (e^{\nu+\mu_2} F_{13})_{,3} = -e^{\psi+\mu_3} F_{02}\, q_{2,0} - e^{\psi+\nu} F_{23}\, q_{2,3}, \tag{5}$$

$$(e^{\mu_2+\mu_3} F_{23})_{,0} + (e^{\nu+\mu_2} F_{02})_{,3} = +e^{\psi+\nu} F_{01}\, q_{2,3} + e^{\psi+\mu_3} F_{13}\, q_{2,0}. \tag{6}$$

(Here and in the rest of this paper the indices refer to orthonormal tetrad-components and commas denote coordinate derivatives.) From equations (3) and (4), it is apparent that the two pairs of components, (F_{01}, F_{13}) and (F_{23}, F_{02}), of the Maxwell tensor can be derived from two potentials A and B in the manner

$$\left.\begin{aligned} e^{\psi+\nu} F_{01} &= +A_{,0}; \quad e^{\psi+\nu} F_{23} = +B_{,0}, \\ e^{\psi+\mu_3} F_{13} &= -A_{,3}; \quad e^{\psi+\mu_3} F_{02} = -B_{,3}. \end{aligned}\right\} \tag{7}$$

In terms of the potentials A and B, equations (5) and (6) take the forms

$$(e^{-\psi-\nu+\mu_2+\mu_3} A_{,0})_{,0} - (e^{-\psi-\mu_3+\nu+\mu_2} A_{,3})_{,3} = B_{,3}\, q_{2,0} - B_{,0}\, q_{2,3}, \tag{8}$$

$$(e^{-\psi-\nu+\mu_2+\mu_3} B_{,0})_{,0} - (e^{-\psi-\mu_3+\nu+\mu_2} B_{,3})_{,3} = A_{,0}\, q_{2,3} - A_{,3}\, q_{2,0}. \tag{9}$$

Equations (8) and (9) can be combined into a single equation for the *complex potential*

$$H = A + iB, \tag{10}$$

in the form

$$(e^{-\psi-\nu+\mu_2+\mu_3} H_{,0})_{,0} - (e^{-\psi-\mu_3+\nu+\mu_2} H_{,3})_{,3} = i(H_{,0}\, q_{2,3} - H_{,3}\, q_{2,0}). \tag{11}$$

With the energy–momentum stress-tensor of the Maxwell field, in an orthonormal tetrad frame, given by

$$T_{ab} = \eta^{cd} F_{ac} F_{bd} - \tfrac{1}{4}\eta_{ab} F_{ef} F^{ef}, \tag{12}$$

where $\eta^{cd} = (1, -1, -1, -1)$ is the local Minkowskian metric, the Einstein field-equations require

$$\begin{aligned} R_{ab} &= -2T_{ab} \\ &= -2\eta^{cd} F_{ac} F_{bd} - \eta_{ab}[(F_{01})^2 + (F_{02})^2 - (F_{13})^2 - (F_{23})^2]. \end{aligned} \tag{13}$$

Evaluating the various components of R_{ab} with the aid of equations (7), we find

$$+R_{00} = R_{33} = e^{-2\psi-2\nu}[(A_{,0})^2 + (B_{,0})^2] + e^{-2\psi-2\mu_3}[(A_{,3})^2 + (B_{,3})^2], \tag{14}$$

$$-R_{11} = R_{22} = e^{-2\psi-2\nu}[(A_{,0})^2 + (B_{,0})^2] - e^{-2\psi-2\mu_3}[(A_{,3})^2 + (B_{,3})^2], \tag{15}$$

$$R_{12} = 2\, e^{-2\psi-\nu-\mu_3}(A_{,0} B_{,3} - A_{,3} B_{,0}), \tag{16}$$

$$R_{03} = 2\, e^{-2\psi-\nu-\mu_3}(A_{,0} A_{,3} + B_{,0} B_{,3}). \tag{17}$$

The required field equations now follow from equating the expressions on the right-hand sides of the foregoing equations with the corresponding expressions for the Ricci (or the Einstein) tensor in terms of the metric. The expressions for the latter can be read off from Paper I, equations (12)–(18); and we find that these equations are now replaced by

$$e^{-2\nu}[\mu_{2,0,0}+\mu_{2,0}(\psi+\mu_2+\mu_3-\nu)_{,0}]-e^{-2\mu_3}[\mu_{2,3,3}+\mu_{2,3}(\psi+\mu_2+\nu-\mu_3)_{,3}]$$
$$=-\tfrac{1}{2}e^{2\psi-2\mu_2}[e^{-2\nu}(q_{2,0})^2-e^{-2\mu_3}(q_{2,3})^2]$$
$$+e^{-2\psi-2\nu}[(A_{,0})^2+(B_{,0})^2]-e^{-2\psi-2\mu_3}[(A_{,3})^2+(B_{,3})^2], \tag{18}$$

$$e^{-2\nu}[\psi_{,0,0}+\psi_{,0}(\psi+\mu_2+\mu_3-\nu)_{,0}]-e^{-2\mu_3}[\psi_{,3,3}+\psi_{,3}(\psi+\mu_2+\nu-\mu_3)_{,3}]$$
$$=+\tfrac{1}{2}e^{2\psi-2\mu_2}[e^{-2\nu}(q_{2,0})^2-e^{-2\mu_3}(q_{2,3})^2]$$
$$-e^{-2\psi-2\nu}[(A_{,0})^2+(B_{,0})^2]+e^{-2\psi-2\mu_3}[(A_{,3})^2+(B_{,3})^2], \tag{19}$$

$$(e^{3\psi+\nu-\mu_2-\mu_3}q_{2,3})_{,3}-(e^{3\psi+\mu_3-\mu_2-\nu}q_{2,0})_{,0}=4(A_{,0}B_{,3}-A_{,3}B_{,0}), \tag{20}$$

$$-(\psi+\mu_2)_{,0,3}+(\psi+\mu_2)_{,0}\nu_{,3}+(\psi+\mu_2)_{,3}\mu_{3,0}-\psi_{,0}\psi_{,3}-\mu_{2,3}\mu_{2,0}$$
$$=\tfrac{1}{2}e^{2\psi-2\mu_2}q_{2,0}q_{2,3}+2e^{-2\psi}(A_{,0}A_{,3}+B_{,0}B_{,3}), \tag{21}$$

$$e^{-2\mu_3}[(\psi+\mu_2)_{,3,3}+(\mu_2-\mu_3)_{,3}(\psi+\mu_2)_{,3}+\psi_{,3}\psi_{,3}]$$
$$-e^{-2\nu}[(\mu_2+\mu_3)_{,0}\psi_{,0}+\mu_{3,0}\mu_{2,0}]=-\tfrac{1}{4}e^{2\psi-2\mu_2}[e^{-2\nu}(q_{2,0})^2+e^{-2\mu_3}(q_{2,3})^2]$$
$$-e^{-2\psi-2\nu}[(A_{,0})^2+(B_{,0})^2]-e^{-2\psi-2\mu_3}[(A_{,3})^2+(B_{,3})^2], \tag{22}$$

$$e^{-2\nu}[(\psi+\mu_2)_{,0,0}+(\mu_2-\nu)_{,0}(\psi+\mu_2)_{,0}+\psi_{,0}\psi_{,0}]$$
$$-e^{-2\mu_3}[(\mu_2+\nu)_{,3}\psi_{,3}+\nu_{,3}\mu_{2,3}]=-\tfrac{1}{4}e^{2\psi-2\mu_2}[e^{-2\nu}(q_{2,0})^2+e^{-2\mu_3}(q_{2,3})^2]$$
$$-e^{-2\psi-2\nu}[(A_{,0})^2+(B_{,0})^2]-e^{-2\psi-2\mu_3}[(A_{,3})^2+(B_{,3})^2]. \tag{23}$$

Now, letting

$$\beta=\psi+\mu_2, \tag{24}$$

we can rewrite equations (18) and (19) in the forms

$$(e^{\beta+\mu_3-\nu}\mu_{2,0})_{,0}-(e^{\beta+\nu-\mu_3}\mu_{2,3})_{,3}=-\tfrac{1}{2}e^{3\psi-\mu_2}[e^{\mu_3-\nu}(q_{2,0})^2-e^{\nu-\mu_3}(q_{2,3})^2]$$
$$+e^{-\psi+\mu_2}\{e^{\mu_3-\nu}[(A_{,0})^2+(B_{,0})^2]-e^{\nu-\mu_3}[(A_{,3})^2+(B_{,3})^2]\}, \tag{25}$$

$$(e^{\beta+\mu_3-\nu}\psi_{,0})_{,0}-(e^{\beta+\nu-\mu_3}\psi_{,3})_{,3}=+\tfrac{1}{2}e^{3\psi-\mu_2}[e^{\mu_3-\nu}(q_{2,0})^2-e^{\nu-\mu_3}(q_{2,3})^2]$$
$$-e^{-\psi+\mu_2}\{e^{\mu_3-\nu}[(A_{,0})^2+(B_{,0})^2]-e^{\nu-\mu_3}[(A_{,3})^2+(B_{,3})^2]\}. \tag{26}$$

The sum and difference of these equations are

$$[e^{\mu_3-\nu}(e^{\beta})_{,0}]_{,0}-[e^{\nu-\mu_3}(e^{\beta})_{,3}]_{,3}=0, \tag{27}$$

$$[e^{\beta+\mu_3-\nu}(\psi-\mu_2)_{,0}]_{,0}-[e^{\beta+\nu-\mu_3}(\psi-\mu_2)_{,3}]_{,3}$$
$$=e^{3\psi-\mu_2}[e^{\mu_3-\nu}(q_{2,0})^2-e^{\nu-\mu_3}(q_{2,3})^2]$$
$$-2e^{-\psi+\mu_2}\{e^{\mu_3-\nu}[(A_{,0})^2+(B_{,0})^2]-e^{\nu-\mu_3}[(A_{,3})^2+(B_{,3})^2]\}. \tag{28}$$

The subtraction of equations (22) and (23) yields the same equation (27), while addition gives

$$4e^{\nu-\mu_3}(\beta_{,3}\nu_{,3}+\psi_{,3}\mu_{2,3})+4e^{\mu_3-\nu}(\beta_{,0}\mu_{3,0}+\psi_{,0}\mu_{2,0})$$
$$=2e^{-\beta}\{[e^{\nu-\mu_3}(e^{\beta})_{,3}]_{,3}+[e^{\mu_3-\nu}(e^{\beta})_{,0}]_{,0}\}$$
$$+e^{2\psi-2\mu_2}[e^{\nu-\mu_3}(q_{2,3})^2+e^{\mu_3-\nu}(q_{2,0})^2]$$
$$+4e^{\nu+\mu_3}\{e^{-2\psi-2\mu_3}[(A_{,3})^2+(B_{,3})^2]+e^{-2\psi-2\nu}[(A_{,0})^2+(B_{,0})^2]\}. \tag{29}$$

Finally, we may note that equation (20) has the alternative form

$$[e^{3\psi+\mu_3-\mu_2-\nu}q_{2,0}+2(AB_{,3}-A_{,3}B)]_{,0} \qquad\qquad -[e^{3\psi+\nu-\mu_2-\mu_3}q_{2,3}-2(A_{,0}B-AB_{,0})]_{,3}=0. \quad (30)$$

3. The choice of gauge and the reduction of the equations to standard forms

As in Paper I (equations (29)–(32)), by exercising the gauge freedom we have to impose a coordinate-condition on x^0 and x^3, we can, without loss of generality, choose

$$e^{2(\mu_3-\nu)}=\Delta=1-\eta^2 \quad \text{and} \quad e^{2\beta}=\Delta\delta=(1-\eta^2)(1-\mu^2), \quad (31)$$

where η measures time from a suitable origin and in a suitable unit and

$$\mu=\cos x^3, \quad (32)$$

replaces the coordinate x^3. *In the rest of this paper, we shall use η and μ as the variables in place of x^0 and x^3; and the subscripts '0' and '3', from now on, will refer to η and μ. The variables η and μ play the roles of time and horizontal distance in our present problem.*

Again, as in Paper I, we shall further define

$$\chi=e^{-\psi+\mu_2}. \quad (33)$$

With the foregoing definitions and choice of variables, equations (11), (28), and (30) take the forms

$$\left(\frac{\chi}{\sqrt{(\Delta\delta)}}\Delta H_{,0}\right)_{,0}-\left(\frac{\chi}{\sqrt{(\Delta\delta)}}\delta H_{,3}\right)_{,3}=i(H_{,3}q_{2,0}-H_{,0}q_{2,3}), \quad (34)$$

$$\left(\frac{\Delta}{\chi^2}q_{2,0}+2\,\mathscr{I}m\,HH^*_{,3}\right)_{,0}-\left(\frac{\delta}{\chi^2}q_{2,3}+2\,\mathscr{I}m\,HH^*_{,0}\right)_{,3}=0, \quad (35)$$

and

$$\left(\frac{\Delta}{\chi}\chi_{,0}\right)_{,0}-\left(\frac{\delta}{\chi}\chi_{,3}\right)_{,3}=-\frac{1}{\chi^2}[\Delta(q_{2,0})^2-\delta(q_{2,3})^2] \qquad +\frac{2\chi}{\sqrt{(\Delta\delta)}}[\Delta|H_{,0}|^2-\delta|H_{,3}|^2], \quad (36)$$

where, in equations (28) and (30), we have replaced the real potentials A and B by the complex potential H.

Following the treatment of the corresponding equations in the stationary axisymmetric case in M.T. (§109(b), pp. 567–70), we shall rewrite equation (35) in terms of a potential Φ which it allows; thus, we can let

$$\Phi_{,0}=\frac{\delta}{\chi^2}q_{2,3}+2\,\mathscr{I}m\,HH^*_{,0} \quad \text{and} \quad \Phi_{,3}=\frac{\Delta}{\chi^2}q_{2,0}+2\,\mathscr{I}m\,HH^*_{,3}. \quad (37)$$

Solving these equations for $q_{2,0}$ and $q_{2,3}$, we obtain

$$q_{2,0} = \frac{\delta}{\Psi^2}(\Phi_{,3} - 2\,\mathscr{I}m\,HH^*_{,3}) \quad \text{and} \quad q_{2,3} = \frac{\Delta}{\Psi^2}(\Phi_{,0} - 2\,\mathscr{I}m\,HH^*_{,0}), \tag{38}$$

where we have written

$$\Psi = \frac{\sqrt{(\Delta\delta)}}{\chi}. \tag{39}$$

The integrability condition for equations (38) is

$$\left[\frac{\Delta}{\Psi^2}(\Phi_{,0} - 2\,\mathscr{I}m\,HH^*_{,0})\right]_{,0} - \left[\frac{\delta}{\Psi^2}(\Phi_{,3} - 2\,\mathscr{I}m\,HH^*_{,3})\right]_{,3} = 0; \tag{40}$$

and equations (34) and (36) written in terms of Ψ and Φ are

$$\Psi[(\Delta H_{,0})_{,0} - (\delta H_{,3})_{,3}] = \Delta H_{,0}[\Psi_{,0} - \mathrm{i}(\Phi_{,0} - 2\,\mathscr{I}m\,HH^*_{,0})] - \delta H_{,3}[\Psi_{,3} - \mathrm{i}(\Phi_{,3} - 2\,\mathscr{I}m\,HH^*_{,3})], \tag{41}$$

and

$$\Psi[(\Delta\Psi_{,0})_{,0} - (\delta\Psi_{,3})_{,3}] = \Delta(\Psi_{,0})^2 - \delta(\Psi_{,3})^2 - 2\Psi[\Delta|H_{,0}|^2 - \delta|H_{,3}|^2] - \Delta(\Phi_{,0} - 2\,\mathscr{I}m\,HH^*_{,0})^2 + \delta(\Phi_{,3} - 2\,\mathscr{I}m\,HH^*_{,3})^2. \tag{42}$$

By a sequence of transformations similar to those described in M.T. (pp. 569–70), we can reduce equations (40)–(42) to the following pair of coupled equations:

$$(\mathscr{R}e\,Z - |H|^2)[(\Delta Z_{,0})_{,0} - (\delta Z_{,3})_{,3}] = \Delta(Z_{,0})^2 - \delta(Z_{,3})^2 - 2H^*(\Delta Z_{,0}H_{,0} - \delta Z_{,3}H_{,3}) \tag{43}$$

and

$$(\mathscr{R}e\,Z - |H|^2)[(\Delta H_{,0})_{,0} - (\delta H_{,3})_{,3}] = \Delta H_{,0}Z_{,0} - \delta H_{,3}Z_{,3} - 2H^*[\Delta(H_{,0})^2 - \delta(H_{,3})^2], \tag{44}$$

where

$$Z = \Psi + |H|^2 - \mathrm{i}\Phi. \tag{45}$$

Once equations (43) and (44) have been solved for Ψ, Φ and H, χ follows at once from the defining equation (39), while to obtain q_2 we must integrate equations (38). Finally, the last remaining metric function, $(\nu + \mu_3)$, will be determined by equations (21) and (29) which, after some reductions, take the forms

$$-\frac{\mu}{1-\mu^2}(\nu+\mu_3)_{,0} - \frac{\eta}{1-\eta^2}(\nu+\mu_3)_{,3} = \frac{1}{\chi^2}(\chi_{,0}\chi_{,3} + q_{2,0}q_{2,3}) + \frac{2\chi}{\sqrt{(\Delta\delta)}}(H_{,0}H^*_{,3} + H^*_{,0}H_{,3}), \tag{46}$$

$$2\eta(\nu+\mu_3)_{,0} + 2\mu(\nu+\mu_3)_{,3} = \frac{3}{1-\eta^2} + \frac{1}{1-\mu^2} - \frac{1}{\chi^2}\{\Delta[(\chi_{,0})^2 + (q_{2,0})^2] + \delta[(\chi_{,3})^2 + (q_{2,3})^2]\} - \frac{4\chi}{\sqrt{(\Delta\delta)}}(\Delta H_{,0}H^*_{,0} + \delta H_{,3}H^*_{,3}). \tag{47}$$

Since our present interest in the foregoing equations is to obtain a solution which would be the analogue of the Kerr–Newman solution in the stationary axisymmetric case (even as the Khan–Penrose and the Nutku–Halil solutions are the analogues of the Schwarzschild and the Kerr solutions, in the sense clarified in Paper I), we shall seek the desired solution of equations (43) and (44) in the special class for which (cf. M.T., §110, p. 573, equation (76))

$$H = Q(Z+1), \tag{48}$$

where Q is a constant which can, in general, be complex. However, since equations (43) and (44) are invariant to an arbitrary (constant) change in the phase of H, there is no loss of generality in assuming that Q is real. When H is related to Z as in equation (48), *both* equations (43) and (44) reduce to the single equation

$$\tfrac{1}{2}[(1-2Q^2)(Z+Z^*)-2Q^2(|Z|^2+1)]\,[(\Delta Z_{,0})_{,0}-(\delta Z_{,3})_{,3}]$$
$$= [1-2Q^2(Z^*+1)]\,[\Delta(Z_{,0})^2-\delta(Z_{,3})^2]. \tag{49}$$

By the substitution

$$Z = \frac{1+E}{1-E}, \tag{50}$$

equation (49) becomes

$$(1-4Q^2-|E|^2)[(\Delta E_{,0})_{,0}-(\delta E_{,3})_{,3}] = -2E^*[\Delta(E_{,0})^2-\delta(E_{,3})^2]; \tag{51}$$

and it is the solution of this equation that we shall be concerned with.

In terms of a solution of equation (51), the required solutions for Ψ, Φ, and H are given by

$$\Psi = \frac{1}{|1-E|^2}(1-4Q^2-|E|^2), \quad \Phi = \mathrm{i}\frac{E-E^*}{|1-E|^2}, \tag{52}$$

and

$$H = \frac{2Q}{1-E}.$$

4. The specification of the solution that is sought

As we have stated in the Introduction, we wish to obtain a solution of the equations derived in §3 that will be a natural generalization of the Nutku–Halil solution of the vacuum equations considered in Paper I even as the Kerr–Newman solution is a natural generalization of the Kerr solution. Or, equivalently, we wish to obtain a solution that will reduce to the Nutku–Halil solution when we set $H = 0$. This general requirement will have to be made more precise.

We have seen in Paper I that the Nutku–Halil solution follows from the simplest solution

$$E = p\eta + \mathrm{i}q\mu \quad (p^2+q^2 = 1), \tag{53}$$

of the Ernst equation (I, equation (42)) for

$$Z = \frac{1+E}{1-E}. \tag{54}$$

But I, equation (42) satisfied by E is *not* comparable to the present equation (51) since Z as now defined in equation (45) does not reduce to I, equation (39), namely

$$Z = \chi + \mathrm{i} q_2; \tag{55}$$

it reduces instead to $Z^{\dagger}$ as defined in I, equation (54) (with an inconsequential change in the sign of Φ), namely

$$Z^{\dagger} = \Psi - \mathrm{i}\Phi, \tag{56}$$

where

$$\Psi = \sqrt{(\Delta\delta)}/\chi \tag{57}$$

and Φ is the associated potential for q_2 given by

$$\Phi_{,0} = \frac{\delta}{\chi^2} q_{2,3} \quad \text{and} \quad \Phi_{,3} = \frac{\Delta}{\chi^2} q_{2,0}, \tag{58}$$

or, equivalently,

$$q_{2,3} = \frac{\Delta}{\Psi^2} \Phi_{,0} \quad \text{and} \quad q_{2,0} = \frac{\delta}{\Psi^2} \Phi_{,3}. \tag{59}$$

The corresponding Ernst equation,

$$(1 - |E^{\dagger}|^2)\,[(\Delta E^{\dagger}_{,0})_{,0} - (\delta E^{\dagger}_{,3})_{,3}] = -2(E^{\dagger})^*[\Delta(E^{\dagger}_{,0})^2 - \delta(E^{\dagger}_{,3})^2], \tag{60}$$

for

$$E^{\dagger} = \frac{Z^{\dagger} - 1}{Z^{\dagger} + 1}, \tag{61}$$

is comparable to equation (51).

Now the expressions for χ and q_2, appropriate for the Nutku–Halil solution, are (cf. Paper I, equations (57) and (58)),

$$\chi = \frac{1 - p^2\eta^2 - q^2\mu^2}{(1 - p\eta)^2 + q^2\mu^2} \quad \text{and} \quad q_2 = \frac{2q\mu}{(1 - p\eta)^2 + q^2\mu^2} \quad (p^2 + q^2 = 1). \tag{62}$$

The corresponding solutions for Ψ and Φ are

$$\Psi = \sqrt{(\Delta\delta)}\,\frac{(1 - p\eta)^2 + q^2\mu^2}{1 - p^2\eta^2 - q^2\mu^2} = \frac{1 - |E^{\dagger}|^2}{|1 - E^{\dagger}|^2}, \tag{63}$$

and

$$\Phi = -\frac{2q}{p}\,\frac{(1 - \mu^2)(1 - p\eta)}{1 - p^2\eta^2 - q^2\mu^2} = \mathrm{i}\,\frac{E^{\dagger} - (E^{\dagger})^*}{|1 - E^{\dagger}|^2}. \tag{64}$$

Our aim then is to find a solution of equation (49) which for $Q = 0$ reduces to

$$Z^{\dagger} = \Psi - \mathrm{i}\Phi \tag{65}$$

with Ψ and Φ given by equations (63) and (64). In the following section we shall show how such a solution can be obtained.

5. The required solution for the metric coefficients

We begin by stating a lemma that is directly verifiable.

Lemma. *If $E^\dagger$ is a solution of equation* (60), *then $\alpha E^\dagger$ is a solution of equation* (51) *for*

$$\alpha = \sqrt{(1-4Q^2)}. \tag{66}$$

Since both equations (51) and (60) are invariant to arbitrary changes in the phases of E and $E^\dagger$, there is clearly no loss of generality in supposing that α is real and $4Q^2 < 1$; and the range of α we shall consider is $(0, 1)$. It may be recalled here that a similar restriction on Q^2 applies to the Reissner–Nordström solution if we are to avoid a naked singularity.

The lemma provides the key to obtaining the solution of equation (49) that we seek. It is, in fact, given by

$$E = \alpha E^\dagger = \alpha \frac{Z^\dagger - 1}{Z^\dagger + 1}, \tag{67}$$

where

$$Z^\dagger = \Psi - i\Phi, \tag{68}$$

and Ψ and Φ are the solutions (63) and (64), appropriate to the vacuum and which lead to the Nutku–Halil solution.

To avoid ambiguity, we shall distinguish the quantities χ, q_2 Ψ, and Φ (introduced in §3 in the context of the Einstein–Maxwell equations) from the same quantities defined in §4 (in the context of the vacuum equations) by inserting a subscript 'e' to the former set. Thus, *we shall write*

$$\chi_e, q_{2e}, \Psi_e \quad \textit{and} \quad \Phi_e$$

for the quantities defined in the context of the Einstein–Maxwell equations; *and χ, q_2, Ψ, and Φ will denote the corresponding quantities for the vacuum* (and whose values are given in equations (62)–(64).

With the solution for E given by equation (67), the solution for Ψ_e follows from equation (50); thus

$$\Psi_e = \alpha^2 \frac{(1-|E^\dagger|^2)}{|1-\alpha E^\dagger|^2} = \alpha^2 \frac{(1-|E^\dagger|^2)}{|1-E^\dagger|^2} \frac{|1-E^\dagger|^2}{|1-\alpha E^\dagger|^2}$$

$$= \alpha^2 \Psi \frac{|1-E^\dagger|^2}{|1-\alpha E^\dagger|^2}. \tag{69}$$

But

$$\frac{1-\alpha E^\dagger}{1-E^\dagger} = \tfrac{1}{2}[(1-\alpha) Z^\dagger + (1+\alpha)]. \tag{70}$$

Accordingly, we may write

$$\Psi_e = \frac{4\alpha^2}{\varpi^2} \Psi, \tag{71}$$

where

$$\varpi^2 = |(1-\alpha)Z^\dagger + (1+\alpha)|^2 = |(1-\alpha)(\Psi - \mathrm{i}\Phi) + (1+\alpha)|^2$$
$$= (1-\alpha)^2(\Psi^2 + \Phi^2) + 2(1-\alpha^2)\Psi + (1+\alpha)^2. \tag{72}$$

Similarly, from the remaining expressions for Φ_e and H given in equations (50), we find

$$\Phi_e = \mathrm{i}\alpha \frac{E^\dagger - (E^\dagger)^*}{|1-\alpha E^\dagger|^2} = \frac{4\alpha}{\varpi^2}\Phi \tag{73}$$

and

$$H = \frac{2Q}{1-\alpha E^\dagger} = \sqrt{(1-\alpha^2)}\frac{Z^\dagger + 1}{(1-\alpha)Z^\dagger + (1+\alpha)}. \tag{74}$$

Finally,

$$\chi_e = \frac{\sqrt{(\Delta\delta)}}{\Psi_e} = \frac{\varpi^2}{4\alpha^2}\chi. \tag{75}$$

It remains to solve for q_{2e} and $(\nu + \mu_3)$ to complete the solution.

Equations (71) and (73), relating Ψ_e and Φ_e with the vacuum solutions Ψ and Φ, can be deduced from an 'Ehlers–Harrison' (Ehlers 1957; Harrison 1968) transformation with suitably chosen parameters. The same transformation can also be used to infer the solution for $(\nu+\mu_3)$ given in equation (113) below. But it does not appear that there is any short-cut to obtaining the solution (117) for q_{2e}.

(a) The solution for q_{2e}

To obtain the solution for q_{2e}, we must return to the defining equations (38) and integrate them with our present knowledge of H and Φ_e. Thus, from equation (74) we find that

$$H_{,0} = \frac{4\alpha Q Z^\dagger_{,0}}{(Z^\dagger+1)^2(1-\alpha E^\dagger)^2} \tag{76}$$

and

$$HH^*_{,0} = \frac{8\alpha Q^2}{\varpi^4}(Z^\dagger)^*_{,0}(Z^\dagger+1)[(1-\alpha)Z^\dagger + (1+\alpha)]. \tag{77}$$

From this last equation we deduce that

$$2\,\mathscr{I}m\, HH^*_{,0} = \frac{4\alpha(1-\alpha^2)}{\varpi^4}\{(1-\alpha)[(\Psi^2 - \Phi^2 + \Psi)\Phi_{,0} - \Phi(1+2\Psi)\Psi_{,0}]$$
$$+ (1+\alpha)[(\Psi+1)\Phi_{,0} - \Phi\Psi_{,0}]\}. \tag{78}$$

Similarly, by combining the equations,

$$(\varpi^2)_{,0} = 2(1-\alpha)^2(\Psi\Psi_{,0} + \Phi\Phi_{,0}) + 2(1-\alpha^2)\Psi_{,0}, \tag{79}$$

and

$$\Phi_{e,0} = \frac{4\alpha}{\varpi^4}[\varpi^2\Phi_{,0} - \Phi(\varpi^2)_{,0}], \tag{80}$$

which follow from equations (72) and (73), we find

$$\Phi_{e,0} = \frac{4\alpha}{\varpi^4}\{[(1-\alpha)^2(\Psi^2-\Phi^2) + 2(1-\alpha^2)\Psi + (1+\alpha)^2]\Phi_{,0}$$
$$- 2\Phi\Psi_{,0}[(1-\alpha)^2\Psi + (1-\alpha^2)]\}. \tag{81}$$

Inserting equations (78) and (81) in the defining equation

$$q_{2e,3} = \frac{\Delta}{\Psi_e^2}(\Phi_{e,0} - 2\,\mathscr{I}m\,HH^*_{,0}), \tag{82}$$

we find after some considerable reductions that

$$q_{2e,3} = \frac{\Delta}{4\alpha^2\Psi^2}\{[(1-\alpha)^2(\Phi^2 - \Psi^2) + (1+\alpha)^2]\Phi_{,0} + 2(1-\alpha)^2\Phi\Psi\Psi_{,0}\}. \tag{83}$$

Similarly, we find

$$q_{2e,0} = \frac{\delta}{4\alpha^2\Psi^2}\{[(1-\alpha)^2(\Phi^2 - \Psi^2) + (1+\alpha)^2]\Phi_{,3} + 2(1-\alpha)^2\Phi\Psi\Psi_{,3}\}. \tag{84}$$

Alternative forms of these equations are:

$$q_{2e,3} = \frac{(1+\alpha)^2}{4\alpha^2}\frac{\Delta}{\Psi^2}\Phi_{,0} + \frac{(1-\alpha)^2}{4\alpha^2}\frac{\Delta}{\Psi^2}[-(\Psi^2 - \Phi^2)\Phi_{,0} + 2\Phi\Psi\Psi_{,0}], \tag{85}$$

$$q_{2e,0} = \frac{(1+\alpha)^2}{4\alpha^2}\frac{\delta}{\Psi^2}\Phi_{,3} + \frac{(1-\alpha)^2}{4\alpha^2}\frac{\delta}{\Psi^2}[-(\Psi^2 - \Phi^2)\Phi_{,3} + 2\Phi\Psi\Psi_{,3}]. \tag{86}$$

It is now apparent from equations (58) that apart from the factor $(1+\alpha)^2/4\alpha^2$, the first terms, on the right-hand sides of equations (85) and (86), integrate to give the vacuum solution q_2. Accordingly, we may write

$$q_{2e} = \frac{(1+\alpha)^2}{4\alpha^2}q_2 + \frac{(1-\alpha)^2}{4\alpha^2}q_2^{(e)}, \tag{87}$$

where $q_2^{(e)}$ is to be determined from the equations

$$q^{(e)}_{2,0} = -\frac{\delta}{\Psi^2}(\Psi^2 - \Phi^2)\Phi_{,3} + 2\delta\Phi\frac{\Psi_{,3}}{\Psi}, \tag{88}$$

and

$$q^{(e)}_{2,3} = -\frac{\Delta}{\Psi^2}(\Psi^2 - \Phi^2)\Phi_{,0} + 2\Delta\Phi\frac{\Psi_{,0}}{\Psi}. \tag{89}$$

Again, making use of equations (58), (59), and (63), we can reduce equations (88) and (89) to the forms:

$$q^{(e)}_{2,0} = -(\Psi^2 - \Phi^2)q_{2,0} - 2\Phi\left\{\mu + (1-\mu^2)\left[\ln\left(\frac{1-p^2\eta^2-q^2\mu^2}{(1-p\eta)^2+q^2\mu^2}\right)\right]_{,3}\right\}, \tag{90}$$

$$q^{(e)}_{2,3} = -(\Psi^2 - \Phi^2)q_{2,3} - 2\Phi\left\{\eta + (1-\eta^2)\left[\ln\left(\frac{1-p^2\eta^2-q^2\mu^2}{(1-p\eta)^2+q^2\mu^2}\right)\right]_{,0}\right\}, \tag{91}$$

where, according to the solution for q_2 given in equation (62),

$$q_{2,0} = \frac{4pq\mu(1-p\eta)}{[(1-p\eta)^2+q^2\mu^2]^2}; \quad q_{2,3} = \frac{2q[(1-p\eta)^2-q^2\mu^2]}{[(1-p\eta)^2+q^2\mu^2]^2}. \tag{92}$$

The reduction of equations (90) and (91) with the known solutions (63) and (64) for Ψ and Φ can be considerably lightened by defining the quantities

$$\boldsymbol{\alpha} = 1 - p\eta, \quad \boldsymbol{\beta} = q\mu, \quad \text{and} \quad \Delta = 1 - p^2\eta^2 - q^2\mu^2, \tag{93}$$

rewriting the basic equations in the forms,

$$\chi = \frac{\Delta}{\alpha^2+\beta^2}, \quad \Phi = -\frac{2q}{p}(1-\mu^2)\frac{\alpha}{\Delta}, \quad \Psi = [(1-\eta^2)(1-\mu^2)]^{\frac{1}{2}}\frac{\alpha^2+\beta^2}{\Delta},$$

$$q_{2,0} = \frac{4p\alpha\beta}{(\alpha^2+\beta^2)^2}, \quad q_{2,3} = \frac{2q(\alpha^2-\beta^2)}{(\alpha^2+\beta^2)^2}, \tag{94}$$

and making use of the identities

$$\Delta^2 = 4\alpha^2 - 2(\alpha^2+\beta^2)\Delta - (\alpha^2+\beta^2)^2,$$

and (95)

$$\alpha^2+\beta^2 = 2\alpha - \Delta.$$

In this manner we find that equation (90) eventually reduces to the simple equation

$$q^{(e)}_{2,0} = \frac{4q^3}{p}\mu(1-\mu^2)^2\frac{(1-p\eta)}{[(1-p\eta)^2+q^2\mu^2]^2}, \tag{96}$$

which immediately integrates to give

$$q^{(e)}_2 = \frac{2q^3}{p^2}\frac{\mu(1-\mu^2)^2}{(1-p\eta)^2+q^2\mu^2}+f(\mu), \tag{97}$$

where $f(\mu)$ is a function of μ which is left undetermined at this stage.

Simplifying equation (91) in similar fashion, we find

$$q^{(e)}_{2,3} = \frac{2q(1-\mu^2)}{p^2}\left[-1+\frac{4\alpha^2}{\alpha^2+\beta^2}+\frac{q^2(1-\mu^2)}{(\alpha^2+\beta^2)}-2\frac{q^4\mu^2(1-\mu^2)}{(\alpha^2+\beta^2)^2}\right]. \tag{98}$$

On the other hand, differentiating the solution (97) with respect to μ, we find

$$q^{(e)}_{2,3} = \frac{2q(1-\mu^2)}{p^2}\left[-\frac{4q^2\mu^2}{\alpha^2+\beta^2}+\frac{q^2(1-\mu^2)}{(\alpha^2+\beta^2)}-2\frac{q^4\mu^2(1-\mu^2)}{(\alpha^2+\beta^2)^2}\right]+f_{,3}. \tag{99}$$

From a comparison of equations (98) and (99) we find,

$$f_{,3} = \frac{6q}{p^2}(1-\mu^2). \tag{100}$$

Therefore,

$$f = \frac{2q}{p^2}(3-\mu^2)\mu; \tag{101}$$

and the required solution for $q^{(e)}_2$ is given by

$$q^{(e)}_2 = \frac{2q\mu}{p^2}\left[\frac{q^2(1-\mu^2)^2}{(1-p\eta)^2+q^2\mu^2}+(3-\mu^2)\right]. \tag{102}$$

The complete solution for q_{2e} can now be written down by combining q_2 (given by equation (62)) and $q^{(e)}_2$ (given by equation (102)) in accordance with equation (87) (see equation (117) below).

(b) The solution for $(\nu+\mu_3)$

We now turn to equations (46) and (47) for the solution of $(\nu+\mu_3)$. These equations in our present notation are

$$-\frac{\mu}{1-\mu^2}(\nu+\mu_3)_{,0}-\frac{\eta}{1-\eta^2}(\nu+\mu_3)_{,3}=\frac{1}{\chi_e^2}(\chi_{e,0}\chi_{e,3}+q_{2e,0}q_{2e,3})$$
$$+2\frac{\chi_e}{\sqrt{(\Delta\delta)}}(H_{,0}H^*_{,3}+H^*_{,0}H_{,3}) \quad (103)$$

and

$$2\eta(\nu+\mu_3)_{,0}+2\mu(\nu+\mu_3)_{,3}=\frac{3}{1-\eta^2}+\frac{1}{1-\mu^2}$$
$$-\frac{1}{\chi_e^2}\{\Delta[(\chi_{e,0})^2+(q_{2e,0})^2]+\delta[(\chi_{e,3})^2+(q_{2e,3})^2]\}$$
$$-\frac{4\chi_e}{\sqrt{(\Delta\delta)}}[\Delta H_{,0}H^*_{,0}+\delta H_{,3}H^*_{,3}]. \quad (104)$$

Considering first equation (103), we can reduce the terms on the right-hand side of this equation by making use of the various definitions and equations in §4 and in the earlier parts of this section and in particular equations (56)–(59) of §4 and equations (72), (74), (75), (76), (83), and (84) of this section. Thus, we find

$$2\frac{\chi_e}{\sqrt{(\Delta\delta)}}[H_{,0}H^*_{,3}+H^*_{,0}H_{,3}]$$
$$=2\left[\frac{\varpi^2\chi}{4\alpha^2\sqrt{(\Delta\delta)}}\right]\frac{16Q^2\alpha^2}{|Z^\dagger+1|^4|1-\alpha E^\dagger|^4}[Z^\dagger_{,0}(Z^\dagger)^*_{,3}+(Z^\dagger)^*_{,0}Z^\dagger_{,3}]$$
$$=\frac{4(1-\alpha^2)}{\varpi^2\Psi}(\Psi_{,0}\Psi_{,3}+\Phi_{,0}\Phi_{,3}), \quad (105)$$

and

$$\frac{1}{\chi_e^2}(\chi_{e,0}\chi_{e,3}+q_{2e,0}q_{2e,3})$$
$$=(\ln\chi_e)_{,0}(\ln\chi_e)_{,3}+\frac{16\alpha^4}{\varpi^4\chi^2}q_{2e,0}q_{2e,3}$$
$$=\left(\frac{\chi_{,0}}{\chi}+\frac{\varpi^2{}_{,0}}{\varpi^2}\right)\left(\frac{\chi_{,3}}{\chi}+\frac{\varpi^2{}_{,3}}{\varpi^2}\right)+\frac{q_{2,0}q_{2,3}}{\chi^2}-\frac{\Phi_{,0}\Phi_{,3}}{\Psi^2}$$
$$+\frac{1}{\varpi^4\Psi^2}\{[(1-\alpha)^2(\Phi^2-\Psi^2)+(1+\alpha)^2]\Phi_{,3}+2(1-\alpha)^2\Phi\Psi\Psi_{,3}\}$$
$$\times\{[(1-\alpha)^2(\Phi^2-\Psi^2)+(1+\alpha)^2]\Phi_{,0}+2(1-\alpha)^2\Phi\Psi\Psi_{,0}\}$$
$$=\frac{\chi_{,0}\chi_{,3}+q_{2,0}q_{2,3}}{\chi^2}-\left(\frac{\Psi_{,0}}{\Psi}+\frac{\eta}{1-\eta^2}\right)\frac{\varpi^2{}_{,3}}{\varpi^2}-\left(\frac{\Psi_{,3}}{\Psi}+\frac{\mu}{1-\mu^2}\right)\frac{\varpi^2{}_{,0}}{\varpi^2}$$

$$+\frac{1}{\varpi^4\Psi^2}[\![\varpi^2{}_{,0}\varpi^2{}_{,3}\Psi^2-\varpi^4\Phi_{,0}\Phi_{,3}$$

$$+\{[(1-\alpha)^2(\Phi^2-\Psi^2)+(1+\alpha)^2]\Phi_{,3}+2(1-\alpha)^2\Phi\Psi\Psi_{,3}\}$$

$$+\{[(1-\alpha)^2(\Phi^2-\Psi^2)+(1+\alpha)^2]\Phi_{,0}+2(1-\alpha)^2\Phi\Psi\Psi_{,0}\}]\!], \tag{106}$$

where in the third line we have added and subtracted $q_{2,0}q_{2,3}/\chi^2$ $(=\Phi_{,0}\Phi_{,3}/\Psi^2)$ as needed in the subsequent reductions. The terms in the double square brackets with the factor $1/\varpi^4\Psi^2$ can be further reduced by making use of the definition of ϖ^2 in equation (72) and the expression

$$\varpi^2{}_{,i}=2(1-\alpha)[(1-\alpha)(\Psi\Psi_{,i}+\Phi\Phi_{,i})+(1+\alpha)\Psi_{,i}] \quad (i=0,3), \tag{107}$$

for its derivatives. We find

$$\frac{1}{\varpi^4\Psi^2}[\![\;]\!]=\frac{2(1-\alpha)}{\varpi^2\Psi}[2(1-\alpha)\Psi\Psi_{,0}\Psi_{,3}+(1-\alpha)(\Psi_{,0}\Phi_{,3}+\Psi_{,3}\Phi_{,0})\Phi$$

$$-2(1+\alpha)\Phi_{,0}\Phi_{,3}]. \tag{108}$$

This expression combined with the one on the right-hand side of equation (105) (in the last line) gives

$$\frac{2(1-\alpha)}{\varpi^2\Psi}[2(1-\alpha)\Psi\Psi_{,0}\Psi_{,3}+2(1+\alpha)\Psi_{,0}\Psi_{,3}+(1-\alpha)\Phi(\Psi_{,0}\Phi_{,3}+\Psi_{,3}\Phi_{,0})]$$

$$=\frac{2(1-\alpha)}{\varpi^2\Psi}\{\Psi_{,3}[(1+\alpha)\Psi_{,0}+(1-\alpha)(\Psi\Psi_{,0}+\Phi\Phi_{,0})]$$

$$+\Psi_{,0}[(1+\alpha)\Psi_{,3}+(1-\alpha)(\Psi\Psi_{,3}+\Phi\Phi_{,3})]\}$$

$$=\frac{1}{\varpi^2\Psi}(\Psi_{,3}\varpi^2{}_{,0}+\Psi_{,0}\varpi^2{}_{,3}), \tag{109}$$

where, in arriving at the last step in the reductions we have made use of equation (107). The final result of the reduction of equation (103) is, therefore,

$$-\frac{\mu}{1-\mu^2}(\nu+\mu_3)_{,0}-\frac{\eta}{1-\eta^2}(\nu+\mu_3)_{,3}=\frac{1}{\chi^2}(\chi_{,0}\chi_{,3}+q_{2,0}q_{2,3})$$

$$-\frac{\mu}{1-\mu^2}\frac{\varpi^2{}_{,0}}{\varpi^2}-\frac{\eta}{1-\eta^2}\frac{\varpi^2{}_{,3}}{\varpi^2}, \tag{110}$$

where it may be recalled that χ and q_2 are the solutions appropriate to the vacuum.

A similar reduction of equation (104) yields

$$2\eta(\nu+\mu_3)_{,0}+2\mu(\nu+\mu_3)_{,3}=\frac{3}{1-\eta^2}+\frac{1}{1-\mu^2}$$

$$-\frac{1}{\chi^2}\{\Delta[(\chi_{,0})^2+(q_{2,0})^2]+\delta[(\chi_{,3})^2+(q_{2,3})^2]\}+2\eta\frac{\varpi^2{}_{,0}}{\varpi^2}+2\mu\frac{\varpi^2{}_{,3}}{\varpi^2}. \tag{111}$$

We now observe that equations (110) and (111), apart from the terms in ϖ^2 on the right-hand sides of these equations, are the same as the equations for the vacuum in Paper I, equations (39), (44), and (45). We conclude that the solution of equations (110) and (111) is given by

$$\nu+\mu_3 = \text{the solution for the vacuum} + \ln \varpi^2 + \text{constant}. \tag{112}$$

With the solution for the vacuum given in Paper I, equation (61), we can now write the solution for $(\nu+\mu_3)$ in the form

$$e^{\nu+\mu_3} = \frac{\varpi^2}{4\alpha^2}\frac{\Delta}{(1-\eta^2)^{\frac{3}{4}}(1-\mu^2)^{\frac{1}{4}}}, \tag{113}$$

where Δ is defined in equation (93) and the constant of integration in equation (112) has been so chosen that when $\alpha = 1$, the solution reduces to that for the vacuum.

(c) The expression for the metric

With the solution for the metric coefficients completed, we can write the metric in the form

$$ds^2 = \frac{\varpi^2}{4\alpha^2}\frac{\Delta}{(1-\eta^2)^{\frac{1}{4}}(1-\mu^2)^{\frac{1}{4}}}\left[\frac{(d\eta)^2}{1-\eta^2}-\frac{(d\mu)^2}{1-\mu^2}\right]$$

$$-(1-\eta^2)^{\frac{1}{2}}(1-\mu^2)^{\frac{1}{2}}\left[\chi_e(dx^2)^2+\frac{1}{\chi_e}(dx^1-q_{2e}\,dx^2)^2\right], \tag{114}$$

where

$$\Delta = 1-p^2\eta^2-q^2\mu^2, \tag{115}$$

$$\chi_e = \frac{\varpi^2}{4\alpha^2}\frac{\Delta}{(1-p\eta)^2+q^2\mu^2}, \tag{116}$$

$$q_{2e} = \frac{(1+\alpha)^2}{4\alpha^2}\frac{2q\mu}{(1-p\eta)^2+q^2\mu^2}+\frac{(1-\alpha)^2}{4\alpha^2}\frac{2q\mu}{p^2}\left[\frac{q^2(1-\mu^2)^2}{(1-p\eta)^2+q^2\mu^2}+(3-\mu^2)\right], \tag{117}$$

$$\varpi^2 = (1-\alpha)^2(\Psi^2+\Phi^2)+2(1-\alpha^2)\Psi+(1+\alpha)^2, \tag{118}$$

$$\Psi = \frac{\sqrt{(\Delta\delta)}}{\Delta}[(1-p\eta)^2+q^2\mu^2], \quad \text{and} \quad \Phi = -\frac{2q}{p\Delta}(1-\mu^2)(1-p\eta). \tag{119}$$

We shall find it convenient to rewrite the metric (114) by replacing η and μ by ψ and θ where

$$\eta = \cos\psi \quad \text{and} \quad \mu = \cos\theta, \tag{120}$$

and defining

$$\chi_e + iq_{2e} = \mathscr{Z}, \quad \mathscr{Z} = \frac{1+\mathscr{E}}{1-\mathscr{E}},$$

$$\chi_e = \frac{1-|\mathscr{E}|^2}{|1-\mathscr{E}|^2} \quad \text{and} \quad iq_{2e} = \frac{\mathscr{E}-\mathscr{E}^*}{|1-\mathscr{E}|^2}, \tag{121}$$

where $\mathscr{Z}$ and $\mathscr{E}$ are formally the same as 'Z' and 'E' defined in Paper I, equations (39) and (41), and reduce to them, in the limit $H = 0$. With the foregoing definitions, the metric (114) takes a form similar to what we had for the vacuum (cf. Paper I, equation (80)); thus

$$ds^2 = \frac{\varpi^2}{4\alpha^2}\frac{\Delta}{(\sin\psi\sin\theta)^{\frac{1}{2}}}[(d\psi)^2-(d\theta)^2] - \frac{\sin\psi\sin\theta}{1-|\mathscr{E}|^2}|(1-\mathscr{E})\,dx^1+i(1+\mathscr{E})\,dx^2|^2, \quad (122)$$

where

$$\Delta = p^2\sin^2\psi+q^2\sin^2\theta, \quad (p^2+q^2=1). \quad (123)$$

6. The description of the space–time in a Newman–Penrose formalism. The Weyl and the Maxwell scalars

It is manifest from the metric written in the form (122) that a null-tetrad basis is provided by the vectors:

$$\begin{array}{lllll} & \psi & \theta & x^1 & x^2 \\ \boldsymbol{e}_{(1)} = +\boldsymbol{e}^{(2)} = (l_i) = +\frac{1}{\sqrt{2}} & (U, & -U, & 0, & 0 \quad), \\ \boldsymbol{e}_{(2)} = +\boldsymbol{e}^{(1)} = (n_i) = +\frac{1}{\sqrt{2}} & (U, & +U, & 0, & 0 \quad), \\ \boldsymbol{e}_{(3)} = -\boldsymbol{e}^{(4)} = (m_i) = -\frac{1}{\sqrt{2}} & (0, & 0, & V\mathscr{Q}^*_-, & -iV\mathscr{Q}^*_+ \quad), \\ \boldsymbol{e}_{(4)} = -\boldsymbol{e}^{(3)} = (\bar{m}_i) = -\frac{1}{\sqrt{2}} & (0, & 0, & V\mathscr{Q}_-, & +iV\mathscr{Q}_+ \quad), \end{array} \quad (124)$$

where, for the sake of brevity, we have introduced

$$U = \frac{\sqrt{\Delta}}{(\sin\psi\sin\theta)^{\frac{1}{4}}}\frac{\varpi}{2\alpha}, \quad V = (\sin\psi\sin\theta)^{\frac{1}{2}},$$

and

$$\mathscr{Q}_\pm = \frac{1\pm\mathscr{E}}{(1-|\mathscr{E}|^2)^{\frac{1}{2}}}. \quad (125)$$

The quantities $\mathscr{Q}_\pm$, as defined, satisfy the identity

$$\mathscr{Q}_-\mathscr{Q}^*_+ + \mathscr{Q}^*_-\mathscr{Q}_+ = 2. \quad (126)$$

With the aid of the null basis (124), we find that the non-vanishing spin-coefficients are given by

$$\sigma = \frac{\mathscr{E}^*_{,\psi}+\mathscr{E}^*_{,\theta}}{U(1-|\mathscr{E}|^2)\sqrt{2}}, \quad \lambda = -\frac{\mathscr{E}_{,\psi}-\mathscr{E}_{,\theta}}{U(1-|\mathscr{E}|^2)\sqrt{2}},$$

$$\rho = -\frac{\cot\psi+\cot\theta}{(2\sqrt{2})\,U}, \quad \mu = +\frac{\cot\psi-\cot\theta}{(2\sqrt{2})\,U},$$

$$\epsilon = \frac{1}{(2\sqrt{2})\,U}\left[(\ln U)_{,\theta} + (\ln U)_{,\psi} + \frac{1}{2(1-|\mathscr{E}|^2)}(\mathscr{E}^*\mathscr{E}_{,\psi} - \mathscr{E}\mathscr{E}^*_{,\psi} + \mathscr{E}^*\mathscr{E}_{,\theta} - \mathscr{E}\mathscr{E}^*_{,\theta})\right],$$

$$\gamma = \frac{1}{(2\sqrt{2})\,U}\left[(\ln U)_{,\theta} - (\ln U)_{,\psi} + \frac{1}{2(1-|\mathscr{E}|^2)}(\mathscr{E}^*\mathscr{E}_{,\psi} - \mathscr{E}\mathscr{E}^*_{,\psi} - \mathscr{E}^*\mathscr{E}_{,\theta} + \mathscr{E}\mathscr{E}^*_{,\theta})\right]. \quad (127)$$

With the foregoing spin-coefficients as the only non-vanishing ones, the Weyl scalars Ψ_1 and Ψ_3 and the Maxwell scalar ϕ_1 vanish identically, a consequence, really, of the existence of the two commuting Killing fields. The remaining scalars follow from the Ricci identities (M.T. Chapter 1, equations 310 (f), (b), (j), (a), (n), and (g), respectively) by noting that in the present context the directional derivatives, $\boldsymbol{D}$ and $\boldsymbol{\Delta}$, are given by

$$\boldsymbol{e}^{(2)} = \boldsymbol{D} = \frac{1}{U\sqrt{2}}\left(\frac{\partial}{\partial\psi} + \frac{\partial}{\partial\theta}\right) \quad \text{and} \quad \boldsymbol{e}^{(1)} = \boldsymbol{\Delta} = \frac{1}{U\sqrt{2}}\left(\frac{\partial}{\partial\psi} - \frac{\partial}{\partial\theta}\right). \quad (128)$$

We find

$$\Psi_2 = \mu\rho - \lambda\sigma$$

$$+\frac{\Psi_0}{\sigma} = \frac{1}{U\sqrt{2}}\left(\frac{\partial}{\partial\psi} + \frac{\partial}{\partial\theta}\right)\ln\sigma - (\rho + \rho^* + 3\epsilon - \epsilon^*),$$

$$-\frac{\Psi_4}{\lambda} = \frac{1}{U\sqrt{2}}\left(\frac{\partial}{\partial\psi} - \frac{\partial}{\partial\theta}\right)\ln\lambda + (\mu + \mu^* + 3\gamma - \gamma^*),$$

$$+\frac{\Phi_{00}}{\rho} = \frac{1}{U\sqrt{2}}\left(\frac{\partial}{\partial\psi} + \frac{\partial}{\partial\theta}\right)\ln\rho - (\rho + \epsilon + \epsilon^*) - \frac{\sigma\sigma^*}{\rho},$$

$$-\frac{\Phi_{22}}{\mu} = \frac{1}{U\sqrt{2}}\left(\frac{\partial}{\partial\psi} - \frac{\partial}{\partial\theta}\right)\ln\mu + (\mu + \gamma + \gamma^*) + \frac{\lambda\lambda^*}{\mu},$$

$$+\frac{\Phi_{20}}{\lambda} = \frac{1}{U\sqrt{2}}\left(\frac{\partial}{\partial\psi} + \frac{\partial}{\partial\theta}\right)\ln\lambda - (\rho - 3\epsilon + \epsilon^*) - \frac{\sigma^*\mu}{\lambda}. \quad (129)$$

On evaluating the scalars with the aid of the spin coefficients (127), we find

$$\Psi_2 = -\frac{1}{8U^2}(\cot^2\psi - \cot^2\theta) + \frac{(\mathscr{E}_{,\psi} - \mathscr{E}_{,\theta})(\mathscr{E}^*_{,\psi} + \mathscr{E}^*_{,\theta})}{2U^2(1-|\mathscr{E}|^2)^2}, \quad (130)$$

$$\Psi_0 = \frac{\mathscr{E}^*_{,\psi} + \mathscr{E}^*_{,\theta}}{2U^2(1-|\mathscr{E}|^2)}\left\{(\cot\psi + \cot\theta) - 2\left(\frac{U_{,\theta}}{U} + \frac{U_{,\psi}}{U}\right) + \frac{\mathscr{E}^*_{,\psi,\psi} + \mathscr{E}^*_{,\theta,\theta} + 2\mathscr{E}^*_{,\psi,\theta}}{\mathscr{E}^*_{,\psi} + \mathscr{E}^*_{,\theta}} + 2\,\frac{\mathscr{E}(\mathscr{E}^*_{,\psi} + \mathscr{E}^*_{,\theta})}{1-|\mathscr{E}|^2}\right\}, \quad (131)$$

$$\Psi_4 = \frac{\mathscr{E}_{,\psi} - \mathscr{E}_{,\theta}}{2U^2(1-|\mathscr{E}|^2)}\left\{(\cot\psi - \cot\theta) + 2\left(\frac{U_{,\theta}}{U} - \frac{U_{,\psi}}{U}\right) + \frac{\mathscr{E}_{,\psi,\psi} + \mathscr{E}_{,\theta,\theta} - 2\mathscr{E}_{,\psi,\theta}}{\mathscr{E}_{,\psi} - \mathscr{E}_{,\theta}} + 2\,\frac{\mathscr{E}^*(\mathscr{E}_{,\psi} - \mathscr{E}_{,\theta})}{1-|\mathscr{E}|^2}\right\}, \quad (132)$$

$$\Phi_{00} = -\frac{\cot\psi+\cot\theta}{4U^2}\left\{-\frac{\operatorname{cosec}^2\psi+\operatorname{cosec}^2\theta}{\cot\psi+\cot\theta}+\tfrac{1}{2}(\cot\psi+\cot\theta)-2\left(\frac{U_{,\theta}}{U}+\frac{U_{,\psi}}{U}\right)\right.$$
$$\left.+2\frac{|\mathscr{E}_{,\psi}+\mathscr{E}_{,\theta}|^2}{(\cot\psi+\cot\theta)(1-|\mathscr{E}|^2)^2}\right\}, \tag{133}$$

$$\Phi_{22} = -\frac{\cot\psi-\cot\theta}{4U^2}\left\{-\frac{\operatorname{cosec}^2\psi+\operatorname{cosec}^2\theta}{\cot\psi-\cot\theta}+\tfrac{1}{2}(\cot\psi-\cot\theta)+2\left(\frac{U_{,\theta}}{U}-\frac{U_{,\psi}}{U}\right)\right.$$
$$\left.+2\frac{|\mathscr{E}_{,\psi}-\mathscr{E}_{,\theta}|^2}{(\cot\psi-\cot\theta)(1-|\mathscr{E}|^2)^2}\right\}, \tag{134}$$

$$\Phi_{20} = -\frac{\mathscr{E}_{,\psi}-\mathscr{E}_{,\theta}}{2U^2(1-|\mathscr{E}|^2)}\left\{\tfrac{1}{2}(\cot\psi+\cot\theta)+2\frac{\mathscr{E}^*(\mathscr{E}_{,\psi}+\mathscr{E}_{,\theta})}{1-|\mathscr{E}|^2}\right.$$
$$\left.+\tfrac{1}{2}(\cot\psi-\cot\theta)\frac{\mathscr{E}_{,\psi}+\mathscr{E}_{,\theta}}{\mathscr{E}_{,\psi}-\mathscr{E}_{,\theta}}+\frac{\mathscr{E}_{,\psi,\psi}-\mathscr{E}_{,\theta,\theta}}{\mathscr{E}_{,\psi}-\mathscr{E}_{,\theta}}\right\}. \tag{135}$$

With the specification of the null basis, the spin coefficients, and the Weyl and the Maxwell scalars, we have completed the description of the space–time in a Newman–Penrose formalism.

7. The introduction of the null-coordinates and the extension of the space–time into regions II, III, and IV

So far, we have concerned ourselves only with the region of the space–time in which the interaction between the colliding waves – gravitational and electromagnetic – occurs; this is region I in the space–time diagram illustrated in figure 1 and to which the metric (122) we have derived applies. To specify the limits of this region and to extend the space–time beyond these limits, it is convenient to rewrite the metric (122) in terms of two null coordinates, u and v, in place of ψ and θ, defined by the equations (cf. I, equation (96))

$$\eta = \cos\psi = u\surd(1-v^2)+v\surd(1-u^2)$$

and

$$\mu = \cos\theta = u\surd(1-v^2)-v\surd(1-u^2), \tag{136}$$

or, equivalently,

$$u = \cos\tfrac{1}{2}(\theta+\psi) \quad \text{and} \quad v = \sin\tfrac{1}{2}(\theta-\psi). \tag{136'}$$

By this transformation,

$$(\mathrm{d}\psi)^2-(\mathrm{d}\theta)^2 = \frac{4\,\mathrm{d}u\,\mathrm{d}v}{\surd(1-u^2)\surd(1-v^2)}, \tag{137}$$

and the metric takes the form

$$\mathrm{d}s^2 = \frac{\varpi^2\Delta}{\alpha^2[(1-u^2-v^2)(1-u^2)(1-v^2)]^{\frac{1}{2}}}\,\mathrm{d}u\,\mathrm{d}v$$
$$-\frac{1-u^2-v^2}{1-|\mathscr{E}|^2}|(1-\mathscr{E})\,\mathrm{d}x^1+\mathrm{i}(1+\mathscr{E})\,\mathrm{d}x^2|^2, \tag{138}$$

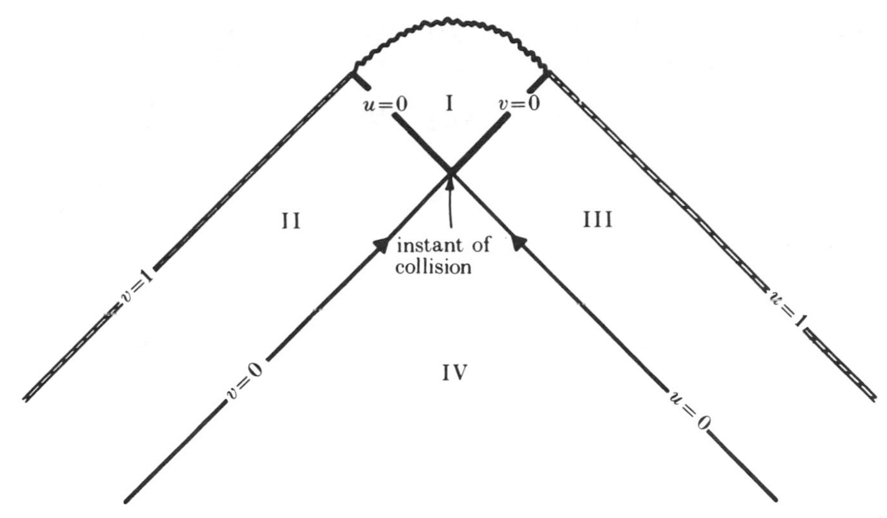

FIGURE 1. The different regions of the space–time that have to be distinguished. In §5, the metric appropriate to region I, in which the interaction between the colliding waves (propagated along the null directions u and v) occurs, is derived. In §§7 and 8, the metric appropriate to regions II, III, and IV is considered. (For a more detailed description of the space–time see figure 2 in §12.)

where ϖ^2, Δ, and $\mathscr{E}$, given in terms of η and μ in equations (115)–(118) and (121), must be rewritten in terms of u and v with the aid of equations (136).

The ranges of the coordinates u and v, as we have presently defined them by equations (136), are restricted to the domain

$$0 \leqslant u \leqslant 1 \quad \text{and} \quad 0 \leqslant v \leqslant 1. \tag{139}$$

However, it is manifest from the form of the metric (138) that it is singular on the surface

$$u^2 + v^2 = 1. \tag{140}$$

Indeed, a curvature singularity develops on this surface as may be verified by considering $\mathscr{R}e\,\Psi_2$, since (M.T., p. 44, equation (299))

$$-2\,\mathscr{R}e\,\Psi_2 = C_{ijkl}\,l^i n^j l^k n^l = C_{ijkl}\,m^i\,\overline{m}^j\,m^k\,\overline{m}^l, \tag{141}$$

is invariant to scale transformations, i.e. to tetrad transformations belonging to class III (cf. M.T., §8(g), equation (347)). Thus, by equation (130),

$$\mathscr{R}e\,\Psi_2 = \frac{2\alpha^2}{\varpi^2\Delta}\left[\sqrt{(1-u^2-v^2)}\,\frac{|\mathscr{E}_{,\psi}|^2 - |\mathscr{E}_{,\theta}|^2}{(1-|\mathscr{E}|^2)^2} - \frac{uv\sqrt{(1-u^2)}\sqrt{(1-v^2)}}{(1-u^2-v^2)^{\frac{3}{2}}}\right]; \tag{142}$$

and this manifestly diverges on the surface (140).

In contrast to what happens on $u^2+v^2 = 1$, we have only coordinate singularities on the surfaces,

$$u = 0, \quad 0 \leqslant v < 1 \quad \text{and} \quad v = 0, \quad 0 \leqslant u < 1, \tag{143}$$

separating region I from regions II and III. Following the original prescription of Penrose, we shall extend the metric (138) across these surfaces by the substitutions

$$uH(u) \quad \text{and} \quad vH(v), \tag{144}$$

in place of u and v in the metric coefficients, where $H(u)$ and $H(v)$ are the Heaviside functions that are unity for positive *and* zero values of the argument and zero for negative values of the argument. By this substitution, we shall obtain a C^0-extension of the metric; but δ-function singularities or H-function discontinuities, or both, will occur in the Weyl and the Maxwell scalars. Postponing to § 10 the consideration of the singularities and the discontinuities that occur on the surfaces (143), we shall pass on now to the metrics one obtains for regions II, III, and IV by the substitutions (144).

We consider first the metrics one obtains for regions II and III. Since the metric (138) is entirely symmetric in u and v, it will suffice to consider the metric in region II. In this region,

$$\eta \to v, \quad \mu \to -v, \quad \Delta \to 1-v^2, \quad \text{and} \quad \delta \to 1-v^2, \tag{145}$$

by the substitution (144) and the various functions defined in equations (115)–(119) become

$$\Delta = 1-v^2, \quad \Psi = 1-2pv+v^2, \quad \Phi = -2q(1-pv)/p,$$

$$\varpi^2 = \frac{(1-\alpha)^2}{p^2}[4+p^2(v^4+6v^2-3)-4pv(1+q^2+p^2v^2)]$$
$$+2(1-\alpha^2)(1-2pv+v^2)+(1+\alpha)^2,$$

$$\chi_e = \frac{\varpi^2}{4\alpha^2}\frac{1-v^2}{1-2pv+v^2},$$

and

$$q_{2e} = -\frac{qv}{2\alpha^2p^2(1-2pv+v^2)}\{(1+\alpha)^2p^2$$
$$+(1-\alpha)^2[4+2pv(v^2-3)-p^2(1-v^2)^2]\}. \tag{146}$$

With these definitions, the metric takes the form

$$\mathrm{d}s^2 = \frac{\varpi^2}{\alpha^2}\,\mathrm{d}u\,\mathrm{d}v-(1-v^2)\left[\chi_e(\mathrm{d}x^2)^2+\frac{1}{\chi_e}(\mathrm{d}x^1-q_{2e}\,\mathrm{d}x^2)^2\right], \tag{147}$$

where ϖ^2, χ_e, and q_{2e} are functions of the one null-coordinate v only.

The metric appropriate to region III can be written down by similarly setting $\eta = \mu = u$ in equations (115)–(119).

It is clear that, unlike in the case of the vacuum, the space–time in regions II and III is *not* flat. As to what their precise nature is, we shall consider in the following section.

The extension of the space–time into region IV

Since both u and v are negative in region IV, the effect of the substitutions (144) on the metric is to set both u and v equal to zero in the expressions for the different metric coefficients. The resulting metric can therefore be obtained by simply putting $v = 0$ in the expressions listed in (146). We thus find that

$$\left.\begin{aligned} \varpi^2 &= \frac{4}{p^2}[(1-\alpha)^2 q^2 + p^2], \quad q_{2e} = 0, \\ \chi_e &= \varpi^2/4\alpha^2 = C, \end{aligned}\right\} \qquad (148)$$

and

where C is a positive constant. Therefore, by the extension, the metric in region IV becomes

$$ds^2 = 4C\,du\,dv - \left[C(dx^2)^2 + \frac{1}{C}(dx^1)^2\right], \qquad (149)$$

which is manifestly Minkowskian. We conclude that the space–time that is causally to the past of the instant of collision is flat; clearly a requirement of the circumstances we are envisaging.

Finally, it should be noted that by the substitutions (144), we have achieved a unique C^0-extension of the space–time in region I to the entire space–time. It is, of course, conceivable that other manners of extension are possible.

8. The nature of the space–time in regions II and III

As we have seen in §7, the space–time in regions II and III is described by a metric of the form

$$ds^2 = e^{2\nu}[(dx^0)^2 - (dx^3)^2] - e^{2\psi}(dx^1 - q_2\,dx^2)^2 - e^{2\mu_2}(dx^2)^2, \qquad (150)$$

where ν, ψ, q_2, and μ_2 are functions only of one or the other of the two null-coordinates $x^0 \pm x^3$; and it will clearly suffice to consider only one of the two cases. Choosing $x^0 - x^3 (= v)$ as the independent variable, we shall indicate by primes the derivatives with respect to the argument; thus

$$\nu' = \nu_{,0} = -\nu_{,3}. \qquad (151)$$

Since the metric (150) is of the standard form considered in M.T., Chapter 2, we can at once write down the tetrad components of the Riemann tensor by specializing appropriately the expressions listed on pp. 78–79. We thus find that the non-vanishing components of the Riemann tensor, in the present context, are given by

$$\begin{aligned} R_{1313} &= R_{1010} = R_{1310} \\ &= e^{-2\nu}[\psi'' + \psi'(\psi' - 2\nu')] - \tfrac{1}{4}e^{2\psi - 2\nu - 2\mu_2}(q_2')^2 = L \quad \text{(say)}, \end{aligned} \qquad (152)$$

$$\begin{aligned} R_{1332} &= R_{1002} = -R_{1023} = R_{1302} \\ &= \tfrac{1}{2}e^{\psi - 2\nu - \mu_2}[q_2'' + q_2'(3\psi' - 2\nu' - \mu_2')] = N \quad \text{(say)}, \end{aligned} \qquad (153)$$

and

$$R_{2323} = R_{2020} = -R_{3220}$$
$$= e^{-2\nu}[\mu_2'' + \mu_2'(\mu_2' - 2\nu')] + \tfrac{3}{4}\, e^{2\psi - 2\mu_2 - 2\nu}(q_2')^2 = M \quad \text{(say)}. \tag{154}$$

Using these expressions, we find that the only non-vanishing components of the Ricci tensor are

$$R_{00} = R_{33} = R_{03} = -(L+M). \tag{155}$$

Besides, $R = 0$ and accordingly,

$$R_{00} = G_{00} \quad \text{and} \quad R_{33} = G_{33}. \tag{156}$$

The solutions (2) and (7) of Maxwell's equations in the earlier context of the metric (1) can be directly specialized to our present context when the metric coefficients are functions of $(x^0 - x^3)$ only and $\nu = \mu_3$. We find

$$F_{12} = F_{03} = 0, \tag{157}$$

$$F_{01} = F_{13} = e^{-\psi-\nu} A' \quad \text{and} \quad F_{23} = F_{02} = e^{-\psi-\nu} B'. \tag{158}$$

The corresponding expressions for the Maxwell stress-tensor are:

$$T_{00} = T_{33} = T_{03} = -e^{2\psi-2\nu}[(A')^2 + (B')^2]; \tag{159}$$

and the remaining components vanish.

From equations (155) and (159) it follows that the Einstein–Maxwell equations provide only the single equation

$$L + M = e^{-2\nu}[(\psi + \mu_2)'' + (\psi')^2 + (\mu_2')^2 - 2\nu'(\psi' + \mu_2')]$$
$$+ \tfrac{1}{2}\, e^{2\psi - 2\mu_2 - 2\nu}(q_2')^2$$
$$= -2\, e^{2\psi - 2\nu}[(A')^2 + (B')^2], \tag{160}$$

while N is left unrestricted. We conclude that *any metric of the form* (150), *in which the metric coefficients are functions only of a single null-coordinate, is a solution of the Einstein–Maxwell equations provided* $L+M \leqslant 0$; and among the six functions ν, ψ, q_2, μ_2, A, and B there is only the one constraint (160). (This result seems to be known; see Misner *et al.* (1970) and Bell & Szekeres (1974).) The requirement $L+M \leqslant 0$ follows from equation (179) (below) since $\Phi_{22} = |\phi_2|^2 \geqslant 0$. We shall presently identify the two invariants of this space–time.

For a direct comparison with the equations obtained for region I, it is convenient to define

$$\beta = \psi + \mu_2, \quad \chi = e^{-\psi + \mu_2}, \tag{161}$$

and

$$\mathscr{Z} = \chi + i q_2 = \frac{1+\mathscr{E}}{1-\mathscr{E}}. \tag{162}$$

With these definitions, the metric (150) takes the form

$$ds^2 = e^{2\nu}[(dx^0)^2 - (dx^3)^2] - \frac{e^{\beta}}{1-|\mathscr{E}|^2}\,|(1-\mathscr{E})\,dx^1 + i(1+\mathscr{E})\,dx^2|^2; \tag{163}$$

and is directly comparable to the metric (122) for region I.

In terms of the variables (161) and (162) the expression for $M+L$ given in equation (161) becomes

$$M+L = e^{-2\nu}\left[\beta'' + \tfrac{1}{2}(\beta')^2 - 2\nu'\beta' + \frac{1}{2\chi^2}|\mathscr{Z}'|^2\right]; \tag{164}$$

or, since

$$\mathscr{Z}' = \frac{2\mathscr{E}'}{(1-\mathscr{E})^2} \quad \text{and} \quad \chi = \frac{1-|\mathscr{E}|^2}{|1-\mathscr{E}|^2}, \tag{165}$$

we can write

$$M+L = e^{-2\nu}\left[\beta'' + \tfrac{1}{2}(\beta')^2 - 2\nu'\beta' + \frac{2|\mathscr{E}'|^2}{(1-|\mathscr{E}|^2)^2}\right]. \tag{166}$$

Also, we may note here for future reference that

$$\begin{aligned} M-L+2iN = e^{-2\nu}\bigg\{&(\mu_2-\psi)'' + (\mu_2'+\psi')(\mu_2'-\psi') - 2\nu'(\mu_2'-\psi') \\ &+ \frac{(q_2')^2}{\chi^2} + \frac{i}{\chi}[q_2'' + q_2'(3\psi' - 2\nu' - \mu_2')]\bigg\} \\ = e^{-2\nu}&\left[\frac{1}{\chi}\mathscr{Z}'' + \frac{1}{\chi}(\beta'-2\nu')\mathscr{Z}' - \frac{1}{\chi^2}(\mathscr{Z}')^2\right]; \end{aligned} \tag{167}$$

or, substituting for $\mathscr{Z}$ and χ in terms of $\mathscr{E}$, we have

$$M-L+2iN = \frac{2\,e^{-2\nu}(1-\mathscr{E}^*)}{(1-\mathscr{E})(1-|\mathscr{E}|^2)}\left[\mathscr{E}'' + (\beta'-2\nu')\mathscr{E}' + 2\frac{\mathscr{E}^*(\mathscr{E}')^2}{1-|\mathscr{E}|^2}\right]. \tag{168}$$

The spin coefficients and the Weyl and the Maxwell scalars

From a comparison of the metrics (122) and (163), it is evident that with the definitions

$$U = e^{\nu}, \quad V = e^{\beta/2} \quad \text{and} \quad \mathscr{Q}_\pm = \frac{1 \pm \mathscr{E}}{\sqrt{(1-|\mathscr{E}|^2)}}, \tag{169}$$

we can set up for the space–time, we are presently considering, a null-tetrad basis of the same form as (124) in §6. With the basis so chosen, we find that the non-vanishing spin-coefficients are

$$\lambda = -(\sqrt{2})\,e^{-\nu}\frac{\mathscr{E}'}{1-|\mathscr{E}|^2}, \quad \mu = \frac{1}{\sqrt{2}}e^{-\nu}\beta'$$

and

$$\gamma = \frac{e^{-\nu}}{\sqrt{2}}\left[-\nu' + \frac{\mathscr{E}^*\mathscr{E}' - \mathscr{E}\mathscr{E}^{*\prime}}{2(1-|\mathscr{E}|^2)}\right]; \tag{170}$$

and, besides, these are functions of $(x^0 - x^3)$ only. From these facts and from equations (129) (in which ∂_ψ and ∂_θ are now to be replaced by ∂_{x^0} and ∂_{x^3}) we find that only Ψ_4 and Φ_{22} do not vanish; and they are given by

$$-[\Psi_4] = e^{-\nu}\lambda'\sqrt{2} + \lambda(\mu + \mu^* + 3\gamma - \gamma^*) \tag{171}$$

and

$$-[\Phi_{22}] = e^{-\nu}\mu'\sqrt{2} + \mu(\mu + \gamma + \gamma^*) + \lambda\lambda^*. \tag{172}$$

Evaluating these expressions, we find, on further comparison with equations (166) and (168), that

$$[\Psi_4] = \frac{2\,e^{-2\nu}}{1-|\mathscr{E}|^2}\left[\mathscr{E}'' + (\beta' - 2\nu')\,\mathscr{E}' + 2\frac{\mathscr{E}^*(\mathscr{E}')^2}{1-|\mathscr{E}|^2}\right]$$

$$= \frac{1-\mathscr{E}}{1-\mathscr{E}^*}\,(M - L + 2iN), \tag{173}$$

and

$$[\Phi_{22}] = -e^{-2\nu}\left[\beta'' + \tfrac{1}{2}\beta'(\beta' - 4\nu') + \frac{2|\mathscr{E}'|^2}{(1-|\mathscr{E}|^2)^2}\right]$$

$$= -(M+L). \tag{174}$$

We have enclosed the expressions for Ψ_4 and Φ_{22} in square brackets to emphasize that they are not comparable to the expressions listed in equations (130)–(135) for region I: the basis null-vectors adopted in the two cases are differently scaled. In fact, if we are to make the tetrads chosen in regions I and II agree, then the basis vectors must be rescaled so that

$$[n_i] = -(1-v^2)^{-\frac{1}{4}} n_i, \tag{175}$$

and

$$[l_i] = -(1-v^2)^{+\frac{1}{4}} l_i. \tag{176}$$

Since both Ψ_4 and Φ_{22} are quadratic in n_i, the factor, by which the expressions (173) and (174) must be multiplied to make then comparable to the expressions (130)–(135), is

$$\sqrt{(1-v^2)}. \tag{177}$$

We shall accordingly write

$$\Psi_4 = \frac{1-\mathscr{E}}{1-\mathscr{E}^*}(M - L + 2iN)\sqrt{(1-v^2)} \tag{178}$$

and

$$\Phi_{22} = -(M+L)\sqrt{(1-v^2)}. \tag{179}$$

From the non-vanishing of only Ψ_4 and Φ_{22}, we conclude that in region II we have a pure field of gravitational and electromagnetic radiation propagated in the direction of increasing v. Similarly in region III only Ψ_0 and Φ_{00} will be non-vanishing and we shall, again, have a pure field of gravitational and electromagnetic radiation, now, propagated in the direction of increasing u.

Finally, we may recall that, as we have remarked earlier, a metric of the form (150) in which ν, ψ, q_2, and μ_2 are functions only of one of the two null-coordinates, $(x^0 \pm x^3)$, is a solution of the Einstein–Maxwell equations; and, as we have now found, it represents a space–time in which a pure radiation field, propagating in the chosen null direction, prevails. The two gauge invariant combinations of the metric functions are $M+L$ and $M-L+2iN$; and these are, in fact, proportional to the Maxwell and the Weyl scalars describing the radiation field.

9. The explicit forms of the Weyl and the Maxwell scalars for the case $q = 0$

It will be convenient to have the explicit forms of the Weyl and the Maxwell scalars, in the different regions for the case $q = 0$, before we proceed to the consideration of the singularities or the discontinuities, or both, that occur along the null boundaries separating them. (We shall not consider in detail the case $q \neq 0$ in this paper: the relevant expressions for the Weyl and the Maxwell scalars are far too complicated; and it does not appear that they will change any of the principal conclusions.)

(a) *The Weyl and the Maxwell scalars in region* I

When $q = 0$ and $p = 1$, the expressions for the various metric functions simplify considerably. Thus, in place of equations (115)–(119), we now have

$$\Delta = \varDelta = 1-\eta^2 = \sin^2\psi,$$

$$\Psi = \sin\theta \sin\psi \tan^2 \tfrac{1}{2}\psi, \quad \Phi = 0$$

$$\varpi/2\alpha = a + b\sin\theta\sin\psi\tan^2\tfrac{1}{2}\psi = X \quad \text{(say)},$$

$$\chi_e = X^2 \cot^2\tfrac{1}{2}\psi \quad \text{and} \quad q_{2e} = 0, \tag{180}$$

where we have written

$$a = (1+\alpha)/2\alpha \quad \text{and} \quad b = (1-\alpha)/2\alpha. \tag{181}$$

Also (cf. equations (121) and (125))

$$\mathscr{E} = \frac{X^2\cot^2\tfrac{1}{2}\psi - 1}{X^2\cot^2\tfrac{1}{2}\psi + 1} \quad \text{and} \quad U = \frac{\sin^{\frac{3}{4}}\psi}{\sin^{\frac{1}{4}}\theta} X. \tag{182}$$

And the metric takes the simple form

$$\mathrm{d}s^2 = \left(\frac{\sin^3\psi}{\sin\theta}\right)^{\frac{1}{2}} X^2[(\mathrm{d}\psi)^2 - (\mathrm{d}\theta)^2]$$

$$-\sin\theta\sin\psi\left[\frac{(\mathrm{d}x^1)^2}{X^2\cot^2\tfrac{1}{2}\psi} + (X^2\cot^2\tfrac{1}{2}\psi)(\mathrm{d}x^2)^2\right]. \tag{183}$$

The task of the reduction and the simplification of the expressions for the Weyl and the Maxwell scalars (which is considerable even in this simpler case) can be lightened if appropriate use is made of the following relations:

$$X_{,\theta} = b\cos\theta\sin\psi\tan^2\tfrac{1}{2}\psi; \quad X_{,\psi} = b(2+\cos\psi)\sin\theta\tan^2\tfrac{1}{2}\psi,$$

$$U_{,\theta}/U = -\tfrac{1}{4}\cot\theta + X_{,\theta}/X; \quad U_{,\psi}/U = \tfrac{3}{4}\cot\psi + X_{,\psi}/X, \tag{184}$$

and

$$\mathscr{E}_{,\psi} \pm \mathscr{E}_{,\theta} = \frac{4X}{(X^2\cot^2\tfrac{1}{2}\psi+1)^2}\{-a\,\mathrm{cosec}\,\psi\cot^2\tfrac{1}{2}\psi + b[\sin\theta + \sin(\theta\pm\psi)]\}. \tag{185}$$

We find:

$$\Psi_0 = \frac{3}{4}\frac{\sin^{\frac{1}{2}}\theta}{X^4\cot^4\frac{1}{2}\psi\,\sin^{\frac{3}{2}}\psi}\left[\!\!\left[a^2\frac{\sin(\theta-\psi)}{\sin^2\psi\sin\theta}\cot^4\tfrac{1}{2}\psi\right.\right.$$

$$+ab\frac{\cot^2\frac{1}{2}\psi}{\sin\theta\sin\psi}[\sin(\theta+\psi)+2\sin\theta]^2$$

$$\left.\left.-b^2\{[\sin(\theta+\psi)+2\sin\theta]^2+\sin\theta\sin(\theta-\psi)\}\right]\!\!\right],$$

$$\Psi_4 = \frac{3}{4}\frac{\sin^{\frac{1}{2}}\theta}{X^4\cot^4\frac{1}{2}\psi\,\sin^{\frac{3}{2}}\psi}\left[\!\!\left[a^2\frac{\sin(\theta+\psi)}{\sin^2\psi\sin\theta}\cot^4\tfrac{1}{2}\psi\right.\right.$$

$$+ab\frac{\cot^2\frac{1}{2}\psi}{\sin\theta\sin\psi}[\sin(\theta-\psi)+2\sin\theta]^2$$

$$\left.\left.-b^2\{[\sin(\theta-\psi)+2\sin\theta]^2+\sin\theta\sin(\theta+\psi)\}\right]\!\!\right],$$

$$\Psi_2 = \frac{1}{8}\frac{\sin^{\frac{1}{2}}\theta}{X^4\cot^4\frac{1}{2}\psi\,\sin^{\frac{3}{2}}\psi}\left[\!\!\left[a^2(3\,\mathrm{cosec}^2\psi+\mathrm{cosec}^2\theta)\cot^4\tfrac{1}{2}\psi\right.\right.$$

$$-2ab\frac{\cot^2\frac{1}{2}\psi}{\sin\theta\sin\psi}[(\sin^2\theta-\sin^2\psi)+4(1+\cos\psi)\sin^2\theta]$$

$$\left.\left.+b^2[3(\sin^2\theta-\sin^2\psi)+4(1+2\cos\psi)\sin^2\theta]\right]\!\!\right],$$

$$\Phi_{00} = +\frac{ab}{2(\sin\theta\sin^5\psi)^{\frac{1}{2}}X^4\cot^2\frac{1}{2}\psi}[\sin(\theta+\psi)+2\sin\theta]^2,$$

$$\Phi_{22} = +\frac{ab}{2(\sin\theta\sin^5\psi)^{\frac{1}{2}}X^4\cot^2\frac{1}{2}\psi}[\sin(\theta-\psi)+2\sin\theta]^2,$$

$$\Phi_{02} = -\frac{ab}{2(\sin\theta\sin^5\psi)^{\frac{1}{2}}X^4\cot^2\frac{1}{2}\psi}[\sin(\theta+\psi)+2\sin\theta][\sin(\theta-\psi)+2\sin\theta]. \tag{186}$$

The limiting values to which the various scalars tend as we approach the null boundary,

$$u=0,\quad 0\leqslant v<1, \tag{187}$$

can be readily ascertained by noting that on this boundary

$$\begin{aligned}\theta+\psi\to\pi-0:\quad &\sin\theta=\sin\psi=\sqrt{(1-v^2)};\quad \cos\psi=-\cos\theta=v,\\ &\sin(\theta+\psi)=0;\quad \sin(\theta-\psi)=2v\sqrt{(1-v^2)},\\ &\cot^2\tfrac{1}{2}\psi=(1+v)/(1-v),\quad\text{and}\quad X=a+b(1-v)^2.\end{aligned} \tag{188}$$

We find

$$\Phi_{00} \to \frac{2ab}{X^4} \frac{(1-v)^{\frac{1}{2}}}{(1+v)^{\frac{3}{2}}}, \quad \Phi_{22} \to \frac{2ab}{X^4} (1-v^2)^{\frac{1}{2}},$$

$$\Phi_{02} \to -\frac{2ab}{X^4} \left(\frac{1-v}{1+v}\right)^{\frac{1}{2}};$$

$$\Psi_0 \to \frac{3}{2X^4(1-v^2)^{\frac{3}{2}}} [a-b(1-v)^2]\,[av+b(2+v)\,(1-v)^2],$$

$$\Psi_4 \to \frac{3b}{X^4} (1-v^2)^{\frac{1}{2}} [a-b(1-v)^2],$$

$$\Psi_2 \to \frac{1}{2X^4(1-v^2)^{\frac{3}{2}}} [a-b(1-v)^2]\,[a-b(1-v)^2(1+2v)]. \tag{189}$$

An important consequence of the solution for Φ_{22} included in the foregoing equations should be noted: since $ab = (1-\alpha^2)/4\alpha^2$, it follows from the positive-definiteness of Φ_{22} that $\alpha^2 < 1$, an inequality which we have hitherto assumed implicitly.

The corresponding limits on the boundary

$$v = 0, \quad 0 \leqslant u < 1, \tag{190}$$

can be obtained by simply replacing v by u and interchanging Ψ_0 and Ψ_4 and Φ_{00} and Φ_{22}.

(*b*) *The Weyl and the Maxwell scalars in regions* II *and* III

From equations (146) it follows that when $q = 0$ and $p = 1$,

$$\chi_e = \frac{1+v}{1-v} X^2 \quad \text{and} \quad q_{2e} = 0, \tag{191}$$

where

$$X = a+b(1-v)^2, \tag{192}$$

is the same expression that we defined earlier in equations (181) and (188); and the metric consistent with the notation of §8 is

$$\mathrm{d}s^2 = 4X^2\,[(\mathrm{d}x^0)^2-(\mathrm{d}x^3)^2]-\left[X^2(1+v)^2\,(\mathrm{d}x^2)^2+\frac{1}{X^2}(1-v)^2\,(\mathrm{d}x^1)^2\right], \tag{193}$$

where $v = (x^0-x^3)$. Comparison of the metrics (150) and (193) shows that

$$\mathrm{e}^{\nu} = 2X, \quad \mathrm{e}^{\beta} = (1-v^2), \quad \mathrm{e}^{\psi} = (1-v)/X, \quad \text{and} \quad \mathrm{e}^{\mu_2} = X(1+v). \tag{194}$$

Remembering that now, $N = 0$, the expressions for $M+L$ and $M-L$ given in equations (164) and (167) give

$$M-L = \mathrm{e}^{-2\nu}\left[\left(\frac{\chi'}{\chi}\right)' + (\beta'-2\nu')\frac{\chi'}{\chi}\right] = \frac{3b}{X^4}\,[a-b(1-v)^2], \tag{195}$$

and

$$M+L = e^{-2\nu}\left[\beta''+\tfrac{1}{2}(\beta')^2+\frac{1}{2}\left(\frac{\chi'}{\chi}\right)^2-2\nu'\beta'\right] = -\frac{2ab}{X^4}. \tag{196}$$

Inserting these expressions in equations (178) and (179) and remembering that $\mathscr{E}$ is now real, we obtain

$$\Psi_4 = \frac{3b}{X^4}[a-b(1-v)^2]\surd(1-v^2) \quad \text{and} \quad \Phi_{22} = \frac{2ab}{X^4}\surd(1-v^2). \tag{197}$$

We observe that these expressions for Ψ_4 and Φ_{22} agree with the values they attain, as $u \to +0$ (from the side of region I), on the null boundary separating regions I and II. They are, therefore, *continuous* on the boundary. The remaining scalars, Ψ_0, Ψ_2, Φ_{00}, and Φ_{02}, listed in (189) experience discontinuities or singularities, or both, as we cross the boundary. We shall determine the nature of these in §10.

It follows in similar fashion that in region III only, Ψ_0 and Φ_{00} do not vanish; and they will be given by

$$\Psi_0 = \frac{3b}{Y^4}[a-b(1-u)^2]\surd(1-u^2) \quad \text{and} \quad \Phi_{00} = \frac{2ab}{Y^4}\surd(1-u^2) \tag{198}$$

where

$$Y = a+b(1-u)^2; \tag{199}$$

and these will be continuous on the null boundary ($v = 0$, $0 \leqslant u < 1$), separating regions I and III.

We also note that as we approach the boundary ($v \to +0$ (from the side of region II), $u < 0$), separating regions II and IV, Ψ_4 and Φ_{22} attain the limiting values

$$\Psi_4 = \tfrac{3}{2}(1-\alpha)\,\alpha^3 \quad \text{and} \quad \Phi_{22} = \tfrac{1}{2}(1-\alpha^2)\,\alpha^2. \tag{200}$$

And Ψ_0 and Φ_{00} will attain the same limiting values (as Ψ_4 and Φ_{22}) as the boundary, $u \to +0$, $v < 0$, separating regions III and IV, is approached.

Since region IV is flat, all the Weyl and the Maxwell scalars will vanish in this region. Therefore, we shall have singularities or discontinuities, or both, as we cross into regions II and III; and their nature will be considered in §10.

10. THE CHARACTERIZATION OF THE SINGULARITIES AND THE DISCONTINUITIES ALONG THE NULL BOUNDARIES

Singularities or discontinuities, or both, along the null boundaries separating the different regions will arise from the substitutions,

$$u \to u\,H(u) \quad \text{and} \quad v \to vH(v), \tag{201}$$

that we have made for extending the metric, derived for region I, into the other regions. An exact description of the resulting singularities and discontinuities can be obtained from the following simple considerations.

Let $f(u)$ be a function of u, which at $u = 0$, has the Taylor expansion

$$f(u) = a+bu+cu^2+\ldots. \tag{202}$$

Further, let

$$\bar{f}(u) = f(u\,H(u)) \tag{203}$$

be the function obtained from $f(u)$ by the substitution $u \to u\,H(u)$. Then $\bar{f}(u)$, at $u = 0$, has the expansion

$$\bar{f}(u) = a + bu\,H(u) + cu^2\,H^2(u) + \dots. \tag{204}$$

By differentiating $\bar{f}(u)$ successively and making use of the identities

$$H'(u) = \delta(u), \quad u\delta(u) = 0 \quad \text{and} \quad u\delta'(u) = -\delta(u) \tag{205}$$

(in the sense of distributions), we find

$$\bar{f}'_+(0) = b \quad \text{and} \quad \bar{f}''_+(0) = b\delta(u) + 2c, \tag{206}$$

where it should be noted that the validity of these relations depends crucially on the definition of the Heaviside function requiring $H(0) = 1$ and $H(u) = 0$ only for u *strictly* negative. From the relations (206) it follows that $f(u\,H(u))$ will have a δ-function singularity in its second derivative at $u = 0$ and that

$$\bar{f}''_+(0) = f''(0) + f'(0)\,\delta(u); \tag{207}$$

and, in contrast, the first derivative of $\bar{f}(u)$ will exhibit only discontinuity at $u = 0$. We shall use these results in deciding which among the Weyl and the Maxwell scalars will exhibit a δ-function singularity and which a simple H-function discontinuity along the null boundaries separating the different regions.

We shall consider first the boundary

$$(u = 0; 0 \leqslant v < 1) \quad \text{and} \quad \theta + \psi \to \pi - 0, \tag{208}$$

separating regions I and II. By the transformation (136)

$$u = \cos\tfrac{1}{2}(\psi + \theta), \tag{209}$$

$$\left(\frac{\partial}{\partial\psi} + \frac{\partial}{\partial\theta}\right) f(u) = -f'(u)\sin\tfrac{1}{2}(\psi + \theta) \tag{210}$$

and

$$\left(\frac{\partial}{\partial\psi} - \frac{\partial}{\partial\theta}\right) f(u) = 0. \tag{211}$$

On $u = 0$, equations (210) and (211) give

$$\left[\left(\frac{\partial}{\partial\psi} + \frac{\partial}{\partial\theta}\right) f(u)\right]_{u=0} = -f'(0) \quad \text{and} \quad \left[\left(\frac{\partial}{\partial\psi} - \frac{\partial}{\partial\theta}\right) f(u)\right]_{u=0} = 0. \tag{212}$$

Turning now to the Weyl and the Maxwell scalars listed in equations (130)–(135), we observe that Ψ_2 and Φ_{22} do not involve any second derivatives of the metric coefficients; they cannot therefore involve any δ-function singularity: they can at most suffer an H-function discontinuity. The Weyl scalar Ψ_4 and the Maxwell scalar Φ_{20} involve second derivatives only in the combinations, $(\partial_\psi - \partial_\theta)^2 \mathscr{E}$ and $(\partial_\psi - \partial_\theta)(\partial_\psi + \partial_\theta)\mathscr{E}$. By equations (212) these scalars cannot also involve any δ-function singularity. However, Ψ_0 involves the term

$$\frac{1}{2U^2(1 - |\mathscr{E}|^2)}\left(\frac{\partial}{\partial\psi} + \frac{\partial}{\partial\theta}\right)^2 \mathscr{E}^*. \tag{213}$$

Therefore, by equations (207) and (212) it must have a δ-function singularity given by

$$-\left[\frac{\mathscr{E}^*_{,\psi}+\mathscr{E}^*_{,\theta}}{2U^2(1-|\mathscr{E}|^2)}\right]_{\theta+\psi\to\pi-0}\delta(u)\quad(0\leqslant v<1).\tag{214}$$

Similarly, along

$$(v=0,\, 0\leqslant u<1)\quad\text{and}\quad\theta-\psi\to+0,\tag{215}$$

we shall have a δ-function singularity in Ψ given by

$$-\left[\frac{\mathscr{E}_{,\psi}-\mathscr{E}_{,\theta}}{2U^2(1-|\mathscr{E}|^2)}\right]_{\theta-\psi\to+0}\delta(v)\quad(0\leqslant u<1).\tag{216}$$

For the case $q=0$, the factors of the δ-functions in (214) and (216) can be explicitly evaluated by making use of equations (182) and (185); we find that the δ-function singularity in Ψ_0, on the null boundary separating regions I and II, is given by

$$\frac{a-b(1-v)^2}{2(1-v^2)\,[a+b(1-v^2)]^3}\,\delta(u);\tag{217}$$

and the complete solution for Ψ_0 in regions I and II can be written in the form

$$\Psi_0^{(\mathrm{I})}H(u)+\frac{a-b(1-v)^2}{2(1-v^2)\,[a+b(1-v^2)]^3}\,\delta(u),\tag{218}$$

where the superscript (I) indicates that it represents the solution for region I given in equations (186). Similarly, we may write for Ψ_2, Φ_{00}, and Φ_{20} the solutions

$$\Psi_2^{(\mathrm{I})}H(u),\quad \Phi_{00}^{(\mathrm{I})}H(u),\quad\text{and}\quad \Phi_{20}^{(\mathrm{I})}H(u);\tag{219}$$

and, as we have directly verified, Ψ_4 and Φ_{22} are continuous across the boundary (208).

The corresponding solutions that obtain in regions I and III can be written down by simply replacing u by v and interchanging Ψ_0 and Ψ_4 and Φ_{22} and Φ_{00} in equations (218) and (219).

The singularities and the discontinuities that occur along the null boundaries separating region IV and regions II and III can be ascertained similarly. Thus, for the case $q=0$, considered in §9, Ψ_4 has a δ-function singularity given by

$$\left[\frac{1}{2X^3}X'\right]_{v=0}\delta(v)=-\tfrac{1}{2}(1-\alpha)\,\alpha^2\delta(v),\tag{220}$$

along the null boundary, $v=0$ and $u<0$, separating regions II and IV. In addition Ψ_4 also experiences an H-function discontinuity of amount specified in equation (200). The Maxwell scalar, Φ_{22}, however, experiences only an H-function discontinuity of amount also specified in equation (200). Finally, the singularity in Ψ_0 and the discontinuity in Φ_{00} along the null boundary, $u=0$ and $v<0$, separating regions III and IV, is the same as for Ψ_4 and Φ_{22} along $(v=0,\ u<0)$ with u replaced by v.

11. The consistency of the derived solution with the requirements of Maxwell's equations and the absence of current sheets

The characterization of the discontinuities in the Maxwell scalars in §10 leaves open the question whether the resulting discontinuities in the Maxwell field across the null boundaries are consistent with the requirements of Maxwell's equations. The requirements are of two kinds: the *mathematical requirement* that the 'jump-conditions' across these surfaces of discontinuity, that follow from Maxwell's equation,

$$\nabla_{[i} F_{jk]} = 0, \tag{221}$$

are satisfied; and the *physical requirement* that no current sheets occur on the surfaces of discontinuity in accord with the equation

$$J_i = \nabla^j F_{ij} = 0. \tag{222}$$

(The subscripts, i, j, k, etc. (from the latter part of the alphabet) are tensor indices (in conformity with the notation in M.T.).) The reason for this latter requirement is that, since Maxwell's equations, without charges and currents, have been satisfied in the rest of space–time, the occurrence of current sheets, on the null boundaries separating the different regions, would imply the creation of charges from nowhere; and this we cannot tolerate!

As we have seen in §10, the Maxwell tensor F_{ij} suffers H-function discontinuities across the null boundaries separating the different regions. Following Pirani (1964; see also O'Brien & Synge 1952 and Bell & Szekeres 1974) we shall express this fact that, across the boundary separating regions I and II, and III and IV, for example, F_{ij} has the form

$$F_{ij} = f_{ij} + \psi_{ij} H(u), \tag{223}$$

where ψ_{ij} is that part of F_{ij} that discontinuously becomes zero for $u < 0$ and f_{ij} that part which remains continuous at $u = 0$. We have shown that across this boundary $u = 0$, ϕ_0 suffers an H-function discontinuity. Therefore, only ϕ_0 contributes to ψ_{ij} and we may write (cf. M.T., p. 51, equation (324))

$$\psi_{ij} = (\phi_0^* m_i + \phi_0 \overline{m}_i) n_j - (\phi_0 \overline{m}_j + \phi_0^* m_j) n_i. \tag{224}$$

Therefore, Maxwell's equation (221) now requires

$$f_{[ij,k]} + \psi_{[ij,k]} H(u) + \psi_{[ij} u_{,k]} \delta(u) = 0. \tag{225}$$

Since Maxwell's equations have been satisfied in all four regions I, II, III, and IV,

$$\left.\begin{aligned} &f_{[ij,k]} + \psi_{[ij,k]} = 0 \quad (u > 0;\ \text{in regions I and III}) \\ \text{and}\qquad &\psi_{[ij,k]} = 0 \quad (u < 0;\ \text{in regions II and IV}); \end{aligned}\right\} \tag{226}$$

therefore, the required '*jump-condition*', on $u = 0$, is

$$\psi_{[ij} u_{,k]} = 0. \tag{227}$$

Now, for the null tetrad chosen (cf. equation (124))

$$l_i\,\mathrm{d}x^i = \frac{1}{\sqrt{2}}U(\mathrm{d}\psi - \mathrm{d}\theta) = \mathrm{d}v$$

and

$$n_i\,\mathrm{d}x^i = \frac{1}{\sqrt{2}}U(\mathrm{d}\psi + \mathrm{d}\theta) = \mathrm{d}u. \tag{228}$$

Therefore,

$$u_{,i} \text{ is a vector parallel to } n_i, \tag{229}$$

and

$$v_{,i} \text{ is a vector parallel to } l_i. \tag{230}$$

Returning to the jump-condition (227) we have the requirement

$$\psi_{[ij}\,n_{k]} = 0; \tag{231}$$

and this requirement is manifestly met by ψ_{ij} given by (224).

Turning next to the condition that must be met across the boundary separating regions I and III, and II and IV, we must now require

$$\psi_{[ij}\,v_{,k]} = 0, \tag{232}$$

where

$$\psi_{ij} = -(\phi_2 m_i + \phi_2^* \bar{m}_i)\,l_j + (\phi_2 m_j + \phi_2^* \bar{m}_j)\,l_i, \tag{233}$$

since it is ϕ_2 that experiences an H-function discontinuity (in v) while ϕ_0 is continuous. By (230), the jump condition is equivalent to the requirement

$$\psi_{[ij}\,l_{k]} = 0; \tag{234}$$

and this requirement is, again, met by ψ_{ij} given by (233).

Turning next to the examination of whether or not current sheets are present at the interfaces separating the different regions, we shall first write down Maxwell's equations, including currents, in a Newman–Penrose formalism.

In a general tetrad formalism, the current vector $J_{(a)}$ is given by

$$J_{(a)} = \eta^{(b)(c)}\,F_{(a)(b)|(c)}, \tag{235}$$

where the vertical rule signifies 'intrinsic differentiation' (as defined, for example, in M.T., p. 37). In terms of the Maxwell scalars ϕ_0, ϕ_1, and ϕ_2, equation (235) gives

$$J_{(1)} = (\phi_1 + \phi_1^*)_{|1} - \phi_{0|4} - \phi^*_{0\,|3},$$

$$J_{(2)} = \phi_{2|3} + \phi^*_{2|4} - (\phi_1 + \phi_1^*)_{|2},$$

$$J_{(3)} = \phi_{1|3} + \phi^*_{2|1} - \phi_{0|2} - \phi^*_{1|3} = J^*_{(4)}. \tag{236}$$

Substituting for the intrinsic derivatives their expressions in terms of directional derivatives and spin coefficients, we obtain

$$J_{(1)} = [\boldsymbol{D}\phi_1 - \boldsymbol{\delta}^*\phi_0 + (2\alpha - \pi)\,\phi_0 - 2\rho\phi_1 + \kappa\phi_2] + \text{complex conjugate}, \tag{237}$$

$$J_{(2)} = [\boldsymbol{\delta}\phi_2 - \boldsymbol{\Delta}\phi_1 + \nu\phi_0 - 2\mu\phi_1 + (2\beta - \tau)\,\phi_2] + \text{complex conjugate}, \tag{238}$$

$$J_{(3)} = [\boldsymbol{\delta}\phi_1 - \boldsymbol{\Delta}\phi_0 + (2\gamma - \mu)\,\phi_0 - 2\tau\phi_1 + \sigma\phi_2]$$

$$+ [\boldsymbol{D}\phi_2 - \boldsymbol{\delta}^*\phi_1 + \lambda\phi_0 - 2\pi\phi_1 + (2\epsilon - \rho)\,\phi_2]^* = J^*_{(4)}. \tag{239}$$

Since we have satisfied all of Maxwell's equations, without currents, in regions I, II, III, and IV except on the boundaries separating them, only current sheets can be present (if at all) at the interfaces as δ-function distributions. By our discussion in §10, δ-function distributions can arise only if the expressions (237)–(239) for the currents involve the second derivatives $\partial^2/\partial u^2$ or $\partial^2/\partial v^2$ with respect to u and v; but no such singularity will arise by the presence of the mixed derivative $\partial^2/\partial u\,\partial v$. Since none of the expressions (127) for the spin coefficients involve any second derivative of the metric coefficients, their appearance in equations (237)–(239) cannot lead to any δ-function distribution. Also, since ϕ_1 vanishes identically (for the problem on hand) and the directional derivatives $\boldsymbol{\delta}$ and $\boldsymbol{\delta}^*$ (which occur in the expressions for $J_{(1)}$ and $J_{(2)}$) operate only in the (x^1, x^2)-plane and none of the functions (including ϕ_0 and ϕ_2) are dependent on x^1 and x^2, it follows directly from equations (237) and (238) that

$$J_{(1)} = J_{(2)} = 0. \tag{240}$$

To show that $J_{(3)}$ and $J_{(4)}$ also vanish, we must examine the limits

$$\lim_{u\to 0 \text{ or } v\to 0} [\boldsymbol{D}\phi_2^* - \boldsymbol{\Delta}\phi_0]. \tag{241}$$

Before we examine the limits (241), we first remark that the expressions for the Maxwell scalars given in equations (133)–(135) determine only

$$\Phi_{00} = |\phi_0|^2, \quad \Phi_{22} = |\phi_2|^2 \quad \text{and} \quad \Phi_{20} = \phi_2\phi_0^*. \tag{242}$$

Therefore, if we write

$$\phi_0 = |\phi_0|\,\mathrm{e}^{\mathrm{i}\zeta_0} \quad \text{and} \quad \phi_2 = |\phi_2|\,\mathrm{e}^{\mathrm{i}\zeta_2}, \tag{243}$$

the known solutions for Φ_{00}, Φ_{22}, and Φ_{20} (or Φ_{02}) determine only the absolute values $|\phi_0|$ and $|\phi_2|$ of ϕ_0 and ϕ_2 and the difference $(\zeta_0-\zeta_2)$ in their phases; the sum of their phases $(\zeta_0+\zeta_2)$ is left undetermined. The phases ζ_0 and ζ_2, separately, (required for the evaluation of the limits (241)) can be determined by considering the equations

$$\boldsymbol{D}\phi_2 = -\lambda\phi_0 + (\rho - 2\epsilon)\,\phi_2$$

and

$$\boldsymbol{\Delta}\phi_0 = +\sigma\phi_2 - (\mu - 2\gamma)\,\phi_0, \tag{244}$$

that follow from M.T., p. 52, equations (331) and (332) for the problem on hand. Rewriting the foregoing equations in the forms

$$\mathrm{i}\boldsymbol{D}\zeta_2 = -\boldsymbol{D}\ln|\phi_2| + (\rho-2\epsilon) - \lambda\frac{|\phi_0|}{|\phi_2|}\,\mathrm{e}^{\mathrm{i}(\zeta_0-\zeta_2)},$$

$$\mathrm{i}\boldsymbol{\Delta}\zeta_0 = -\boldsymbol{\Delta}\ln|\phi_0| - (\mu-2\gamma) + \sigma\frac{|\phi_2|}{|\phi_0|}\,\mathrm{e}^{\mathrm{i}(\zeta_2-\zeta_0)}, \tag{245}$$

we observe that all the quantities on the right-hand sides of these equations are known; they can, therefore, be used to determine ζ_0 and ζ_2, separately.

Returning to the expression (241), we have to examine

$$\lim_{u\to 0 \text{ or } v\to 0} [e^{-i\zeta_2}(\boldsymbol{D}|\phi_2|-i|\phi_2|\boldsymbol{D}\zeta_2)-e^{i\zeta_0}(\boldsymbol{\Delta}|\phi_0|+i|\phi_0|\boldsymbol{\Delta}\zeta_0)], \tag{246}$$

A careful examination of the quantity in square brackets in (246), with the known expressions for $\boldsymbol{\Phi}_{00}$ and $\boldsymbol{\Phi}_{22}$, given in equations (133) and (135), and equations (245) governing ζ_2 and ζ_0, shows that it involves second derivatives only in the combination

$$\left(\frac{\partial}{\partial\psi}+\frac{\partial}{\partial\theta}\right)\left(\frac{\partial}{\partial\psi}-\frac{\partial}{\partial\theta}\right) \quad \text{or} \quad \frac{\partial^2}{\partial u\,\partial v}. \tag{247}$$

But as we have seen, mixed derivatives cannot give rise to δ-function distributions. We therefore conclude that

$$J_{(3)} = J_{(4)} = 0 \quad \text{on} \quad (u = 0 \quad \text{or} \quad v = 0). \tag{248}$$

The demonstration is now complete that the discontinuities in the Maxwell field, that arise from the metric we have derived, satisfy the necessary jump conditions and do not also require any current sheets at the null interfaces between the different regions. The physical interpretation of these facts must be that the discontinuities in the Maxwell field are in some sense 'supported' by the impulsive gravitational waves.

12. Concluding remarks

In view of the complexity of the analysis and of the many issues that had to be resolved, we shall, in this concluding section, restate the essential physical and mathematical content of the problem that has been studied.

The problem originated in an observation of Penrose that colliding waves in general relativity may provide further examples of situations which, generically, may lead to space–time singularities; and a further remark of his in a personal communication that he had "thought a little about Einstein–Maxwell impulsive waves [and] the curious feature that in order that the Ricci curvatures be a δ-function, the Maxwell field must be a kind of 'square-root of the δ-function'; and it was never clear to me how to make sense of such things".

This paper is concerned with the extension of earlier treatments of the problem of colliding impulsive pure gravitational waves to the case when gravitational and electromagnetic waves are coupled via the Einstein–Maxwell equations; but the restriction to plane-fronted impulsive waves and space–times with two commuting space-like Killing vectors is retained.

An exact solution of the Einstein–Maxwell equations is obtained that is a natural generalization (in a sense we shall presently clarify) of the solutions of Khan & Penrose and of Nutku & Halil. The physical content of the solution is best described by the space–time diagram illustrated in figure 2 and in the detailed legend given below it. It is especially noteworthy that the discontinuities in the Maxwell field, which occur along the null boundaries separating the different regions, are consistent with the 'jump conditions' that must be satisfied and do not also require the presence of current sheets as δ-function distributions at the

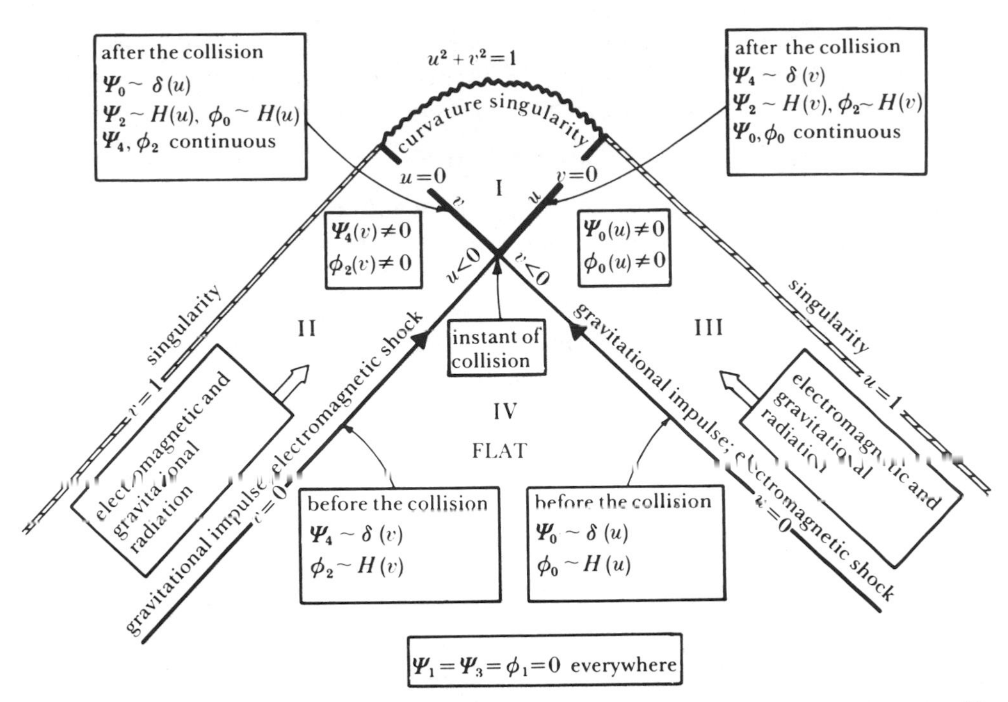

FIGURE 2. The space–time diagram for two colliding plane impulsive gravitational waves. The colliding waves are propagated along the null directions, u and v. The time coordinate is along the vertical, and the spatial direction of propagation (in space) is along the horizontal. The plane of the wavefronts (on which the geometry is invariant) is orthogonal to the plane of the diagram. The instant of the collision is at the origin of the (u, v)-coordinates.

The flat portion of the space–time, prior to the arrival of either wave, is region IV. These waves produce a spray of gravitational and electromagnetic radiation which fills regions II and III; and the region of the space–time in which the waves scatter off each other and focus is region I. The result of the collision is the development of a curvature space–time singularity on $u^2+v^2 = 1$; and $v = 1$ or $u = 1$ are space–time singularities for observers who do *not* observe the collision. Singular behaviours in the Weyl scalars, Ψ_4, Ψ_2, and Ψ_0 (as Dirac δ-functions or Heaviside (H) step-functions, or both, expressing the shock-wave character of the colliding electromagnetic waves) occur, as indicated, along the null boundaries separating the different regions.

interfaces. The analysis did not require the introduction of any inadmissible notion such as the 'square root of a δ-function'.

The two null 3-surfaces, $(v = 1,\ -\infty < u < 0)$ and $(u = 1,\ -\infty < v < 0)$, represented by lines in figure 2, are, as indicated, space–time singularities. In the case of the vacuum (when regions II and III are flat) the origin and the nature of these singularities have been amply discussed by Matzner & Tipler (1984). That they continue to be space–time singularities in our present context (when regions II and III are no longer flat) can be visualized most simply by tracing the time-like and the null geodesics that enter regions II and III by crossing the coordinate-axes $v = 0$ or $u = 0$ at some negative values of u or v.

The construction of the geodesics in regions II and III, in which the metric coefficients are functions of only one of the two null-coordinates, v or u, is simple since the space–time allows three conserved momenta, p_{x^1}, p_{x^2}, and p_u or p_v; and the energy integral can be immediately integrated. Thus, in the case $q = 0$, the three conserved momenta in region II are (cf. equation (150))

$$p_{x^1} = e^{2\psi}(x^1)^{\cdot} = \alpha, \quad p_{x^2} = e^{2\mu_2}(x^2)^{\cdot} = \beta, \quad \text{and} \quad p_u = \tfrac{1}{2}e^{2\nu}\dot{v} = \gamma, \tag{249}$$

where α, β, and γ are constants and the dots denote differentiation with respect to some affine parameter; and the energy integral gives

$$e^{2\nu}\dot{u}\dot{v} - e^{2\psi}[(x^1)^{\cdot}]^2 - e^{2\mu_2}[(x^2)^{\cdot}]^2 = E, \tag{250}$$

where $E > 0$ for time-like geodesics and $E = 0$ for null geodesics. From the foregoing four conservation laws, we readily obtain the equation

$$4\gamma^2\frac{du}{dv} = \alpha^2 e^{-2\psi+2\nu} + \beta^2 e^{-2\mu_2+2\nu} + E e^{2\nu}. \tag{251}$$

With $e^{2\nu}$, $e^{2\psi}$, and $e^{2\mu_2}$ given in equations (194), equation (251) can be integrated to give

$$\gamma^2(u+c) = \alpha^2\int_0^v [a+b(1-v)^2]^4\frac{dv}{(1-v)^2} + \beta^2\frac{v}{1+v} + E\int_0^v [a+b(1-v)^2]^2\,dv, \tag{252}$$

where $u = -c$ is the point on the negative u-axis at which the geodesic crosses into region II. Accordingly $c > 0$. We now observe that the right-hand side of equation (252) is positive-definite (for $v > 0$); and further that the first term diverges to $+\infty$ for $v \to 1-0$ with the behaviour

$$\alpha^2 a^4\frac{v}{1-v}. \tag{253}$$

Therefore, *no time-like or null geodesic with* $\alpha \neq 0$ *can avoid crossing the v-axis into region* I *for some* $0 < v < 1$ *and hitting the singularity at* $u^2+v^2 = 1$. But for $\alpha = 0$, there exist geodesics, both time-like and null, which can avoid crossing into region I: these geodesics hit the singularity at $v = 1$. The situation is exactly the same as in the vacuum; and we may pictorially represent the space–time in the manner of Penrose (see figure 7 in Matzner & Tipler 1984).

On the mathematical side, the procedure that was followed in obtaining the solution departs from earlier attempts in which solutions were sought for the metric, for regions of the space–time in which the colliding waves scatter off each other, that will be compatible with carefully formulated initial conditions on characteristic surfaces (cf. Bell & Szekeres 1974). We have started, instead, at the 'opposite end' by first selecting (*ad hoc*!) the metric, for the region of interaction, by an 'aesthetic' criterion and then systematically extending it across null boundaries, where coordinate singularities occur, with the aid of the Heaviside step-function. The fact that by this procedure, we have been able to obtain a physically consistent solution, is to be credited to the criterion for the choice of the solution in the interacting region. In the choice, we were guided by the observation that both the Khan–Penrose and the Nutku–Halil solutions follow from the *same* simplest solution (cf. Paper I, equation (55)),

$$E = p\eta + iq\mu \quad (p^2 + q^2 = 1), \tag{254}$$

of the Ernst equation (for space–times with two commuting space-like Killing vectors) as the Schwarzschild and the Kerr solution do from, formally, the *same* Ernst equation (now for space–times with one space-like and one time-like Killing vector). We, therefore, sought (and found) for the region of interaction, a solution of the Einstein–Maxwell equations that can be obtained from the Nutku–Halil solution by following the same procedure by which one obtains the Kerr–Newman solution, for the charged black-hole, from the Kerr solution; and, in both cases, the *same* solution of (formally) the *same* coupled Ernst equations is involved. The fact that this 'inverted' procedure leads to a physically consistent solution is a further manifestation of the firm aesthetic base of the general theory of relativity.

While we have obtained in this paper the solution for the general case when the polarizations of the colliding waves are not parallel, we have analysed in full detail (in §9) only the case when the polarizations are parallel. In a later paper, we intend to analyse in equal detail the more general case of non-parallel polarizations. But there are other directions in which the methods of this paper and of paper I can be extended; for example, to the problem of colliding gravitational waves coupled with hydrodynamic shocks (a problem which we are presently investigating).

We are grateful to Professor R. Geroch for some useful discussions. The research reported in this paper has, in part, been supported by grants from the National Science Foundation under grant PHY 80-26043 with the University of Chicago. B. C. Xanthopoulos's tenure at the University of Chicago during June–September 1984 was supported by a grant from the Senior Scholar Fulbright Program; he also wishes to express his thanks to the Relativity Group at the Fermi Institute of the University of Chicago for their hospitality.

References

Bell, P. & Szekeres, P. 1974 *Gen. Rel. Grav.* **5**, 275.
Chandrasekhar, S. 1983 *The mathematical theory of black holes* (referred to as M.T.). Oxford: Clarendon Press.
Chandrasekhar, S. & Ferrari, V. 1984 *Proc. R. Soc. Lond.* A **396**, 55–74.
Ehlers, J. 1957 Dissertation. University of Hamburg.
Harrison, B. K. 1968 *J. math. Phys.* **9**, 1744.
Khan, K. & Penrose, R. 1971 *Nature, Lond.* **229**, 185.
Matzner, R. A. & Tipler, F. J. 1984 *Phys. Rev.* D **29**, 1575.
Misner, C. W., Thorne, K. S. & Wheeler, J. W. 1970 *Gravitation.* San Francisco: W. H. Freeman & Co.
Nutku, Y. & Halil, M. 1977 *Phys. Rev. Lett.* **39**, 1379.
O'Brien, S. & Synge, J. L. 1952 *Communs Dubl. Inst. advd. Stud.* A **9**.
Pirani, F. A. E. 1964 In *Brandeis Lectures on General Relativity*, pp. 269–275. New Jersey: Prentice-Hall.
Szekeres, P. 1970 *Nature, Lond.* **228**, 1183.
Szekeres, P. 1972 *J. math. Phys.* **13**, 286.

21

On the collision of impulsive gravitational waves when coupled with fluid motions

By Subrahmanyan Chandrasekhar[1], F.R.S., and Basilis C. Xanthopoulos[2]

[1] *University of Chicago, Chicago, Illinois* 60637, *U.S.A.*

[2] *Department of Physics, University of Crete and Research Centre of Crete, Iraklion, Greece*

(*Received* 15 *May* 1985)

An exact solution of Einstein's equations, with a source derived from a perfect fluid in which the energy density, ϵ, is equal to the pressure, p, is obtained. The solution describes the space–time following the collision of plane impulsive gravitational waves and is the natural generalization of the Nutku–Halil solution of the vacuum equations, in the region of interaction under similar basic conditions. A consistent extension of the solution, prior to the instant of collision, requires that the fluid in the region of interaction is the direct result of a transformation of incident null-dust (i.e. of massless particles describing null trajectories). The ultimate result of the collision is the development of a space–time singularity, the nature of which is strongly dependent on the amplitude and the character of the sound waves that are present. The distribution of ϵ that follows the collision has many intriguing features.

The solution obtained in this paper provides the first example of an induced transformation of a massless into a massive particle.

1. Introduction

In a recent paper (Chandrasekhar & Ferrari 1984; this paper will be referred to hereafter as Paper I) a new method of derivation was given of the then known solutions of Khan & Penrose (1971) and of Nutku & Halil (1977) for colliding plane impulsive gravitational waves. In a later paper (Chandrasekhar & Xanthopoulos 1985; this paper will be referred to hereafter as Paper II) the method was extended to obtain the complete solution of the same underlying problem when the gravitational waves are coupled with electromagnetic waves through the Einstein–Maxwell equations. In this paper, we shall extend the study, further, to the case when the gravitational waves are coupled with fluid motions; and we shall obtain the general solution of the basic equations, in the region of interaction, in the special case when the energy density (ϵ) is equal to the pressure (p) and the velocity of sound equals the velocity of light. We shall find that the extension of the solution to include the entire space–time, prior to the instant of collision, requires that, at collision, an initial 'spray' of massless particles, describing null trajectories, is transformed into a perfect fluid with $\epsilon = p$ (in which the stream lines describe time-like trajectories).

2. The hydrodynamic equations

In the conventional tensor-notation, the equations of hydrodynamics, governing a perfect fluid, follow from the identity

$$T^{ij}{}_{;j} = 0, \tag{1}$$

where the stress–energy tensor, T^{ij}, is given by

$$T^{ij} = (\epsilon + p)\, u^i\, u^j - p g^{ij}, \tag{2}$$

where ϵ denotes the energy density and p the pressure. Equation (1), written out in full for T^{ij} given by equation (2), is

$$(\epsilon + p)\, u^i{}_{;j}\, u^j + u^i[(\epsilon + p)_{,j}\, u^j + (\epsilon + p)\, u^j{}_{;j}] = p_{,j}\, g^{ij}. \tag{3}$$

Contracting this equation with u_i, we obtain

$$(\epsilon + p)_{,j}\, u^j + (\epsilon + p)\, u^j{}_{;j} = p_{,j}\, u^j. \tag{4}$$

Using this last equation to eliminate the terms in square brackets in equation (3), we obtain

$$(\epsilon + p)\, u_{i;j}\, u^j + u_i\, u^j\, p_{,j} = p_{,i}. \tag{5}$$

For the special case $\epsilon = p$, with which we shall be principally concerned in the present paper, equation (4) takes the form

$$(u^j \surd \epsilon)_{;j} = \frac{1}{\surd -g} [u^j \surd(-\epsilon g)]_{,j} = 0 \quad (\epsilon = p). \tag{6}$$

In our present study, as in our earlier studies on the collision of impulsive gravitational waves in Papers I and II, we shall restrict ourselves to space-times with two space-like Killing-vectors ($\partial/\partial x^1$ and $\partial/\partial x^2$); and for the metric appropriate to such space–times, we shall assume the same form, namely (cf. Paper I, equation (5))

$$ds^2 = e^{2\nu}(dx^0)^2 - e^{2\mu_3}(dx^3)^2 - e^{2\psi}(dx^1 - q_2\, dx^2)^2 - e^{2\mu_2}(dx^2)^2, \tag{7}$$

where the various functions, ν, μ_3, ψ, q_2, and μ_2, introduced depend only on the variables x^0 and x^3. Also, in the present context, we shall suppose that the only non-vanishing components of the four-velocity are u^0 and u^3. Under these circumstances, equation (6), applicable to the case $\epsilon = p$, takes the form

$$(e^{\psi + \mu_2 + \mu_3 + \nu}\, u^0 \surd \epsilon)_{,0} + (e^{\psi + \mu_2 + \mu_3 + \nu}\, u^3 \surd \epsilon)_{,3} = 0 \quad (\epsilon = p). \tag{8}$$

3. The components of the stress-energy tensor and the equations of motion in a tetrad frame

In the rest of the paper we shall write the various equations in a tetrad frame with the basis one-forms

$$\omega^{(0)} = e^{\nu}\, dx^0, \quad \omega^{(1)} = e^{\psi}(dx^1 - q_2\, dx^2), \quad \omega^{(2)} = e^{\mu_2}\, dx^2,$$

and

$$\omega^{(3)} = e^{\mu_3}\, dx^3; \tag{9}$$

and to distinguish the tetrad components from the tensor components used in §2, we shall (in this section) enclose the tetrad indices in parentheses. Thus, writing

$$T_{(a)(b)} = (\epsilon+p)\,u_{(a)}\,u_{(b)} - p\eta_{(a)(b)}, \tag{10}$$

where $\eta_{(a)(b)}$ is a diagonal matrix with the elements $(1, -1, -1, -1)$, we readily find that the components of the Ricci and the Einstein tensors, as required by the Einstein field-equations, are

$$\begin{aligned}
T_{(0)(0)} - \tfrac{1}{2}T &= (\epsilon+p)\,u_{(0)}^2 - \tfrac{1}{2}(\epsilon-p) = -\tfrac{1}{2}R_{(0)(0)},\\
T_{(3)(3)} + \tfrac{1}{2}T &= (\epsilon+p)\,u_{(3)}^2 + \tfrac{1}{2}(\epsilon-p) = -\tfrac{1}{2}R_{(3)(3)},\\
T_{(2)(2)} + \tfrac{1}{2}T &= T_{(1)(1)} + \tfrac{1}{2}T = \tfrac{1}{2}(\epsilon-p) = -\tfrac{1}{2}R_{(1)(1)} = -\tfrac{1}{2}R_{(2)(2)},\\
T_{(0)(3)} &= (\epsilon+p)\,u_{(0)}\,u_{(3)} = -\tfrac{1}{2}R_{(0)(3)},\\
G_{(0)(0)} &= -2(\epsilon+p)\,u_{(0)}^2 + 2p, \quad G_{(3)(3)} = -2(\epsilon+p)\,u_{(3)}^2 - 2p,
\end{aligned} \tag{11}$$

where

$$T = \epsilon - 3p = \tfrac{1}{2}R, \tag{12}$$

denotes the trace of $T_{(a)(b)}$. We also find that

$$R_{(1)(2)} = R_{(1)(3)} = R_{(0)(1)} = R_{(0)(2)} = 0. \tag{13}$$

The hydrodynamic equation, in tetrad notation, is given by (cf. equation (5)),

$$u^{(b)}u_{(a)|(b)} = \frac{1}{\epsilon+p}\,[p_{,(a)} - u_{(a)}\,u^{(b)}p_{,(b)}], \tag{14}$$

where the vertical rule in $u_{(a)|(b)}$ denotes intrinsic differentiation as defined in *The mathematical theory of black holes* (Chandrasekhar 1983, p. 37; this book will be referred to hereafter as M.T.). Making use of the Ricci rotation-coefficients listed in M.T. (equation (91), p. 82), we find that the equations governing $u_{(0)}$ and $u_{(3)}$ are

$$+e^{-\nu}u_{(0)}\,u_{(0),0} - e^{-\mu_3}u_{(3)}\,u_{(0),3} - e^{-\mu_3}\nu_{,3}\,u_{(0)}\,u_{(3)} + e^{-\nu}\mu_{3,0}\,u_{(3)}^2$$
$$= \frac{1}{\epsilon+p}\,(-e^{-\nu}u_{(3)}^2\,p_{,0} + u_{(0)}\,u_{(3)}\,e^{-\mu_3}p_{,3}), \tag{15}$$

$$-e^{-\mu_3}u_{(3)}\,u_{(3),3} + e^{-\nu}u_{(0)}\,u_{(3),0} + e^{-\nu}\mu_{3,0}\,u_{(3)}\,u_{(0)} - e^{-\mu_3}\nu_{,3}\,u_{(0)}^2$$
$$= \frac{1}{\epsilon+p}\,(+e^{-\mu_3}u_{(0)}^2\,p_{,3} - u_{(3)}\,u_{(0)}\,e^{-\nu}p_{,0}). \tag{16}$$

Also, since the tetrad components, $u_{(0)}$ and $u_{(3)}$, of the four-velocity are related to the tensor components, u^0 and u^3 by

$$u^{(0)} = u_{(0)} = e^{\nu}u^0 = e^{-\nu}u_0,$$

and

$$u^{(3)} = -u_{(3)} = e^{\mu_3}u^3 = -e^{-\mu_3}u_3, \tag{17}$$

equation (8), applicable to the case $\epsilon = p$, now takes the form

$$(e^{\psi+\mu_2+\mu_3}u_{(0)}\sqrt{\epsilon})_{,0} - (e^{\psi+\mu_2+\nu}u_{(3)}\sqrt{\epsilon})_{,3} = 0. \tag{18}$$

4. THE FIELD EQUATIONS

The relevant field equations for the problem on hand can be written down by inserting on the right-hand sides of equations (12)–(17) of Paper I the expressions for the components of the Ricci and the Einstein tensors, listed in equations (11)–(13), indicated in the margins of these same equations (I, (12)–(17)). The resulting equations can be combined in the same fashion as in Paper I and we find that in place of I, equations (21), (14), (22), (15), and (23) we now have, respectively,

$$[e^{\mu_3-\nu}(e^{\beta})_{,0}]_{,0}-[e^{\nu-\mu_3}(e^{\beta})_{,3}]_{,3} = -2\,e^{\psi+\mu_2+\mu_3+\nu}(\epsilon-p), \tag{19}$$

$$(e^{3\psi+\nu-\mu_2-\mu_3}\,q_{2,3})_{,3}-(e^{3\psi+\mu_3-\mu_2-\nu}\,q_{2,0})_{,0} = 0, \tag{20}$$

$$[e^{\beta+\mu_3-\nu}(\psi-\mu_2)_{,0}]_{,0}-[e^{\beta+\nu-\mu_3}(\psi-\mu_2)_{,3}]_{,3} = e^{3\psi-\mu_2}[e^{\mu_3-\nu}(q_{2,0})^2-e^{\nu-\mu_3}(q_{2,3})^2], \tag{21}$$

$$\begin{aligned}-(\psi+\mu_2)_{,0,3}+(\psi+\mu_2)_{,0}\,\nu_{,3}+(\psi+\mu_2)_{,3}\,\mu_{3,0}-\psi_{,0}\,\psi_{,3}-\mu_{2,3}\,\mu_{2,0}&\\ = \tfrac{1}{2}\,e^{2\psi-2\mu_2}\,q_{2,0}\,q_{2,3}-2(\epsilon+p)\,u_0\,u_3\,e^{\nu+\mu_3},&\end{aligned} \tag{22}$$

and

$$\begin{aligned}&4\,e^{\nu-\mu_3}(\beta_{,3}\,\nu_{,3}+\psi_{,3}\,\mu_{2,3})+4\,e^{\mu_3-\nu}(\beta_{,0}\,\mu_{3,0}+\psi_{,0}\,\mu_{2,0})\\ &\quad = 2\,e^{-\beta}\{[e^{\nu-\mu_3}(e^{\beta})_{,3}]_{,3}+[e^{\mu_3-\nu}(e^{\beta})_{,0}]_{,0}\}\\ &\qquad +e^{2\psi-2\mu_2}[e^{\nu-\mu_3}(q_{2,3})^2+e^{\mu_3-\nu}(q_{2,0})^2]-4(\epsilon+p)\,(u_0^2+u_3^2)\,e^{\nu+\mu_3},\end{aligned} \tag{23}$$

where we have dispensed with the parentheses distinguishing the tetrad indices (since we shall have no further occasions to use tensor indices). Also, as in Papers I and II, we have written

$$\beta = \psi+\mu_2. \tag{24}$$

5. THE SPECIAL CASE $\epsilon = p$

We observe that when $\epsilon = p$, equations (19), (20) and (21) become identical with the equations governing the vacuum (cf. I, equations (14), (21) and (22)). Accordingly, in this special case we can make the same choice of gauge and coordinates as in the vacuum, namely (cf. I, equations (30)–(32))

$$\left.\begin{aligned} e^{2(\mu_3-\nu)} &= \Delta = 1-\eta^2\\ \text{and}\qquad e^{2\beta} &= e^{2(\psi+\mu_2)} = (1-\eta^2)\,(1-\mu^2) = \Delta\delta,\end{aligned}\right\} \tag{25}$$

where μ ($= \cos x^3$) is a measure of the distance between the plane wavefronts and η is a measure of the time from the instant of the collision. Also, since χ ($= e^{-\psi+\mu_2}$) and q_2 satisfy the same equations as for the vacuum, it would be natural to assume for them the same solutions that lead to the Nutku–Halil solution, namely

$$\chi = e^{-\psi+\mu_2} = \frac{p^2(1-\eta^2)+q^2(1-\mu^2)}{(1-p\eta)^2+q^2\mu^2},\quad q_2 = \frac{2q\mu}{(1-p\eta)^2+q^2\mu^2}\quad (p^2+q^2 = 1), \tag{26}$$

since we should expect to recover these solutions when $\epsilon = p = 0$. For the present, we shall *not*, however, make this assumption. But it is essential that the absence of terms on the right-hand side of equation (19), when $\epsilon = p$, enables us to make the same choice of coordinates η and μ as in Papers I and II. In what follows *we shall use η and μ in place of x^0 and x^3; and the subscripts '0' and '3' will refer, from now on, to η and μ.*

Returning to equations (22) and (23), which determine the last remaining metric function $(\nu+\mu_3)$, we find on further simplification (cf. I, equations (44) and (45) and II, equations (46) and (47)):

$$-\frac{\mu}{1-\mu^2}(\nu+\mu_3)_{,0}-\frac{\eta}{1-\eta^2}(\nu+\mu_3)_{,3} = \frac{1}{\chi^2}(\chi_{,0}\chi_{,3}+q_{2,0}q_{2,3})+\frac{8\epsilon}{\sqrt{(1-\mu^2)}}u_0 u_3 e^{\nu+\mu_3}, \tag{27}$$

$$2\eta(\nu+\mu_3)_{,0}+2\mu(\nu+\mu_3)_{,3} = -\frac{1}{\chi^2}\{\Delta[(\chi_{,0})^2+(q_{2,0})^2]+\delta[(\chi_{,3})^2+(q_{2,3})^2]\} +\frac{3}{1-\eta^2}+\frac{1}{1-\mu^2}+8\epsilon(u_0^2+u_3^2)e^{2\mu_3}. \tag{28}$$

(a) *The reduction of the hydrodynamic equations*

Equations (27) and (28) must be supplemented by the hydrodynamic equations (18) and (15) and (16) for $\epsilon = p$. Considering first equation (18) and substituting for $e^{\psi+\mu_2}$ and $e^{\nu-\mu_3}$ from equations (25), we obtain

$$[e^{\mu_3}u_0\sqrt{(\Delta\epsilon)}]_{,0}+[e^{\mu_3}u_3\sqrt{(\delta\epsilon)}]_{,3} = 0. \tag{29}$$

With the definitions,

$$P = e^{\mu_3}u_0\sqrt{\epsilon} \quad \text{and} \quad Q = e^{\mu_3}u_3\sqrt{\epsilon}, \tag{30}$$

we can, in accordance with equation (29), derive P and Q from a potential ϕ in the manner,

$$\left.\begin{aligned} & e^{\mu_3}u_0\sqrt{(\Delta\epsilon)} = +P\sqrt{\Delta} = +\phi_{,3}, \\ \text{and} \quad & e^{\mu_3}u_3\sqrt{(\delta\epsilon)} = +Q\sqrt{\delta} = -\phi_{,0}. \end{aligned}\right\} \tag{31}$$

From equations (30) it follows that

$$u_0/u_3 = P/Q; \tag{32}$$

combining this equation with the identity

$$u_0^2-u_3^2 = 1, \tag{33}$$

we obtain

$$u_0 = P/\sqrt{(P^2-Q^2)} \quad \text{and} \quad u_3 = Q/\sqrt{(P^2-Q^2)}. \tag{34}$$

Considering next equations (15) and (16) (for the case $\epsilon = p$), we find that in the chosen system of coordinates, η and μ, they become

$$\Delta^{\frac{1}{2}}(u_0 u_{0,0}+u_3^2\mu_{3,0})+\delta^{\frac{1}{2}}u_3(u_{0,3}+u_0\nu_{,3})+\frac{u_3}{2\epsilon}(u_3\epsilon_{,0}\sqrt{\Delta}+u_0\epsilon_{,3}\sqrt{\delta}) = 0, \tag{35}$$

$$\delta^{\frac{1}{2}}(u_3 u_{3,3}+u_0^2\nu_{,3})+\Delta^{\frac{1}{2}}u_0(u_{3,0}+u_3\mu_{3,0})+\frac{u_0}{2\epsilon}(u_0\epsilon_{,3}\sqrt{\delta}+u_3\epsilon_{,0}\sqrt{\Delta}) = 0. \tag{36}$$

It is clear that equations (35) and (36) allow $u_0^2 - u_3^2 = 1$ as an integral. Accordingly, it will suffice to consider only one of them, say (35). So rewriting equation (35) in the form

$$u_3^2\left(\mu_{3,0}+\frac{\epsilon_{,0}}{2\epsilon}\right)\sqrt{\Delta}+u_0\,u_3\left(\nu_{,3}+\frac{\epsilon_{,3}}{2\epsilon}\right)\sqrt{\delta}+u_0\,u_{0,0}\sqrt{\Delta}+u_3\,u_{0,3}\sqrt{\delta}=0, \quad (37)$$

and making use of the relations

$$\left.\begin{aligned}\mu_{3,0}+\frac{\epsilon_{,0}}{2\epsilon}&=\frac{P_{,0}}{P}-\frac{u_{0,0}}{u_0},\\ \nu_{,3}+\frac{\epsilon_{,3}}{2\epsilon}&=\frac{Q_{,3}}{Q}-\frac{u_{3,3}}{u_3},\end{aligned}\right\} \quad (38)$$

which follow from the definitions (30), we obtain after some simplifications,

$$\sqrt{\Delta}\,\frac{P_{,0}}{P}+\sqrt{\delta}\,\frac{u_0\,Q_{,3}}{u_3\,Q}+\sqrt{\Delta}\,\frac{u_{0,0}}{u_0\,u_3^2}+\sqrt{\delta}\left(\frac{u_0}{u_3}\right)_{,3}=0. \quad (39)$$

Now, substituting for u_0 and u_3 from equations (34), we find that we are left with

$$Q_{,0}\sqrt{\Delta}+P_{,3}\sqrt{\delta}=0. \quad (40)$$

This last equation combined with equations (31) provides the basic equation

$$\Delta\phi_{,0,0}-\delta\phi_{,3,3}=0. \quad (41)$$

In terms of a solution of this equation, $u_0\sqrt{\epsilon}$ and $u_3\sqrt{\epsilon}$ are given by

$$u_0\sqrt{\epsilon}=\frac{e^{-\mu_3}}{\sqrt{\Delta}}\,\phi_{,3} \quad \text{and} \quad u_3\sqrt{\epsilon}=-\frac{e^{-\mu_3}}{\sqrt{\delta}}\,\phi_{,0}. \quad (42)$$

(b) *The reduction of the equations governing* $\nu+\mu_3$

We shall write

$$\nu+\mu_3=(\nu+\mu_3)_{\text{vac.}}+f. \quad (43)$$

Since, by the choice of gauge,

$$\nu-\mu_3=(\nu-\mu_3)_{\text{vac.}}, \quad (44)$$

it is clear that

$$\mu_3=(\mu_3)_{\text{vac.}}+\tfrac{1}{2}f \quad \text{and} \quad \nu=(\nu)_{\text{vac.}}+\tfrac{1}{2}f. \quad (45)$$

Since the solution for $\nu+\mu_3$, appropriate for the vacuum, satisfies equations (27) and (28) when $\epsilon=0$, it is clear that the equations governing f are

$$\frac{\mu}{1-\mu^2}f_{,0}+\frac{\eta}{1-\eta^2}f_{,3}=-\frac{8\epsilon}{\sqrt{(1-\mu^2)}}\,u_0\,u_3\,e^{\nu+\mu_3}, \quad (46)$$

$$2\eta f_{,0}+2\mu f_{,3}=8\epsilon(u_0^2+u_3^2)\,e^{2\mu_3}. \quad (47)$$

Now, substituting for $u_0\sqrt{\epsilon}$ and $u_3\sqrt{\epsilon}$ from equations (42), we obtain for f the equations

$$\frac{\mu}{1-\mu^2}f_{,0}+\frac{\eta}{1-\eta^2}f_{,3} = \frac{8}{\Delta\delta}\phi_{,0}\phi_{,3}, \tag{48}$$

$$\eta f_{,0}+\mu f_{,3} = \frac{4}{\Delta\delta}[\Delta(\phi_{,0})^2+\delta(\phi_{,3})^2]. \tag{49}$$

Solving these equations for $f_{,0}$ and $f_{,3}$, we obtain

$$f_{,0} = \frac{4}{\eta^2-\mu^2}\left\{-2\mu\phi_{,0}\phi_{,3}+\frac{\eta}{\Delta}[\Delta(\phi_{,0})^2+\delta(\phi_{,3})^2]\right\}, \tag{50}$$

$$f_{,3} = \frac{4}{\eta^2-\mu^2}\left\{+2\eta\phi_{,0}\phi_{,3}-\frac{\mu}{\delta}[\Delta(\phi_{,0})^2+\delta(\phi_{,3})^2]\right\}. \tag{51}$$

The integrability condition for these equations is guaranteed by the field equations; in fact it leads us exactly to the equation (41) governing ϕ.

In the future, we shall dispense with the distinguishing subscript 'vac.' for μ_3 and ν and retain for them (as we already have for β, χ, and q_2) the meaning that they refer to solutions appropriate for the vacuum so that for the problem on hand

$$\mu_3+\tfrac{1}{2}f \quad \text{and} \quad \nu+\tfrac{1}{2}f \tag{52}$$

replace what we have hitherto denoted by μ_3 and ν.

Finally, we may note that from equations (33) and (42), we find that the energy density, ϵ, is given by

$$\epsilon = -\frac{e^{-2\mu_3-f}}{\Delta\delta}[\Delta(\phi_{,0})^2-\delta(\phi_{,3})^2]. \tag{53}$$

6. The transformation to null coordinates

We shall find it convenient to consider the basic equations (41), (50), and (51) in a system of null coordinates u and v defined in the same manner as in Paper I, equations (96)–(100) and Paper II, equations (136) and (136′). The defining equations are

$$\left.\begin{aligned}\eta = \cos\psi = \cos(\xi-\zeta); \quad \mu = \cos\theta = \cos(\xi+\zeta),\\ u = \cos\xi \quad \text{and} \quad v = \sin\zeta.\end{aligned}\right\} \tag{54}$$

By virtue of these definitions

$$\left.\begin{aligned}\eta = u\sqrt{(1-v^2)}+v\sqrt{(1-u^2)},\\ \mu = u\sqrt{(1-v^2)}-v\sqrt{(1-u^2)},\end{aligned}\right\} \tag{55}$$

and
$$\sin\psi\sin\theta = \sin(\xi+\zeta)\sin(\xi-\zeta) = 1-u^2-v^2. \tag{56}$$

By the transformation of variables,

$$\left.\begin{aligned}
\phi_{,\eta} &= \phi_{,0} = -\frac{\phi_{,\xi}-\phi_{,\zeta}}{2\sqrt{(1-\eta^2)}}; \quad \phi_{,\mu} = \phi_{,3} = -\frac{\phi_{,\xi}+\phi_{,\zeta}}{2\sqrt{(1-\mu^2)}};\\
\phi_{,\eta,\eta} &= \frac{1}{4(1-\eta^2)}(\phi_{,\xi,\xi}-2\phi_{,\xi,\zeta}+\phi_{,\zeta,\zeta})-\frac{\eta}{2(1-\eta^2)^{\frac{3}{2}}}(\phi_{,\xi}-\phi_{,\zeta});\\
\phi_{,\mu,\mu} &= \frac{1}{4(1-\mu^2)}(\phi_{,\xi,\xi}+2\phi_{,\xi,\zeta}+\phi_{,\zeta,\zeta})-\frac{\mu}{2(1-\mu^2)^{\frac{3}{2}}}(\phi_{,\xi}+\phi_{,\zeta});
\end{aligned}\right\} \tag{57}$$

and equation (41), in the (ξ, ζ)-variables, becomes

$$2\phi_{,\xi,\zeta}+\frac{\cos(\xi-\zeta)}{\sin(\xi-\zeta)}(\phi_{,\xi}-\phi_{,\zeta})-\frac{\cos(\xi+\zeta)}{\sin(\xi+\zeta)}(\phi_{,\xi}+\phi_{,\zeta}) = 0. \tag{58}$$

On the other hand, by equations (54),

$$\phi_{,\xi} = -\phi_{,u}\sin\xi, \quad \phi_{,\zeta} = \phi_{,v}\cos\zeta, \quad \phi_{,\xi,\zeta} = -\phi_{,u,v}\sin\xi\cos\zeta. \tag{59}$$

With these substitutions, equation (58), in the (u, v)-variables, takes the form

$$\phi_{,u,v}+\frac{v\phi_{,u}+u\phi_{,v}}{1-u^2-v^2} = 0. \tag{60}$$

For future reference, we may note here the following relations, which are further consequences of equations (57)–(59):

$$2\phi_{,0}\phi_{,3} = \frac{(1-u^2)(\phi_{,u})^2-(1-v^2)(\phi_{,v})^2}{2(1-u^2-v^2)}, \tag{61}$$

$$\Delta(\phi_{,0})^2+\delta(\phi_{,3})^2 = \tfrac{1}{2}[(1-u^2)(\phi_{,u})^2+(1-v^2)(\phi_{,v})^2]. \tag{62}$$

Returning to equations (50) and (51) governing f, we first observe that by the change of variables adopted

$$\left.\begin{aligned}
f_{,u} &= (f_{,0}+f_{,3})\sqrt{(1-v^2)}-(f_{,0}-f_{,3})\frac{uv}{\sqrt{(1-u^2)}},\\
f_{,v} &= (f_{,0}-f_{,3})\sqrt{(1-u^2)}-(f_{,0}+f_{,3})\frac{uv}{\sqrt{(1-v^2)}}.
\end{aligned}\right\} \tag{63}$$

Now, substituting for $f_{,0}$ and $f_{,3}$ in the foregoing equations from equations (50) and (51) and simplifying with the aid of equations (61) and (62), we find that we are left with

$$f_{,u} = \frac{2(\phi_{,u})^2}{u(1-u^2-v^2)}, \quad f_{,v} = \frac{2(\phi_{,v})^2}{v(1-u^2-v^2)}. \tag{64}$$

Again, the integrability condition for these equations is guaranteed by equation (60) governing ϕ.

We may note that the expression (53) for ϵ in the (u, v)-variables becomes

$$\epsilon = -\frac{(1-u^2)^{\frac{1}{2}}(1-v^2)^{\frac{1}{2}}}{(1-u^2-v^2)^2}e^{-2\mu_3-f}\phi_{,u}\phi_{,v}. \tag{65}$$

Finally, it is important to observe that *equations* (60), (64), *and* (65) *are entirely general: they are in no way dependent on the choice we may make for the solutions* χ, q_2, *and* $\nu+\mu_3$ *for the vacuum.*

7. The solution for ϕ

We shall first show that equation (60) is separable in the variables

$$r = u^2 - v^2 \quad \text{and} \quad s = 1 - u^2 - v^2. \tag{66}$$

By this change of variables,

$$\phi_{,u} = 2u(\phi_{,r} - \phi_{,s}) \quad \text{and} \quad \phi_{,v} = -2v(\phi_{,r} + \phi_{,s}) \tag{67}$$

and

$$\phi_{,u,v} = -4uv(\phi_{,r,r} - \phi_{,s,s}); \tag{68}$$

and equation (60) reduces to

$$\phi_{,r,r} - \left(\phi_{,s,s} - \frac{1}{s}\phi_{,s}\right) = 0. \tag{69}$$

Equation (69) is clearly separable; thus, writing

$$\phi(r, s) = g(r)\,h(s), \tag{70}$$

we find

$$\frac{\mathrm{d}^2 g}{\mathrm{d}r^2} = \alpha^2 g \quad \text{and} \quad \frac{\mathrm{d}^2 h}{\mathrm{d}s^2} - \frac{1}{s}\frac{\mathrm{d}h}{\mathrm{d}s} - \alpha^2 h = 0, \tag{71}$$

where α^2 is the separation constant. (One could also have chosen a negative constant of separation; but that choice, as one can easily verify, does not lead to acceptable solutions for ϵ.) The fundamental solutions of equations (71) are:

$$g = \mathrm{e}^{\pm\alpha r} \quad \text{and} \quad h = s\mathscr{C}_1(\mathrm{i}\alpha s), \tag{72}$$

where $\mathscr{C}_1$ is Bessel's function of order 1 for a purely imaginary argument. (The case when $\alpha = 0$ is considered separately; see equation (81) below.)

We shall presently verify that the separable solution for ϕ, which ensures the positive definiteness of the energy density ϵ in its domain of definition, is given by

$$\phi = C\,\mathrm{e}^{\pm\alpha r}\,\alpha s K_1(\alpha s), \tag{73}$$

where K_1 is the Bessel function of order 1 for a purely imaginary argument in Watson's notation (cf. G. N. Watson 1944, p. 78) and C is a constant. (The general solution for ϕ can be expressed as an arbitrary linear combination of the solution (73) for differing Cs and αs; see equation (82) below.)

For ϕ given by equation (73),

$$\phi_{,r} = \pm C\alpha^2\,\mathrm{e}^{\pm\alpha r}\,sK_1(\alpha s) \tag{74}$$

and

$$\phi_{,s} = -C\alpha^2\,\mathrm{e}^{\pm\alpha r}\,sK_0(\alpha s), \tag{75}$$

where we have made use of the identity

$$[zK_1(z)]' = -zK_0(z). \tag{76}$$

We shall now verify that the solution (73) for ϕ does provide a positive definite expression for ϵ. Noting that (cf. equations (67))

$$\phi_{,u}\phi_{,v} = -4uv[(\phi_{,r})^2-(\phi_{,s})^2], \tag{77}$$

and substituting for $\phi_{,r}$ and $\phi_{,s}$ from equations (74) and (75), we find from equation (65) that

$$\epsilon = 4C^2\alpha^4\, e^{-2\mu_3-f}\, uv(1-u^2)^{\frac{1}{2}}(1-v^2)^{\frac{1}{2}}\, e^{\pm 2\alpha r}[K_1^2(\alpha s)-K_0^2(\alpha s)], \tag{78}$$

which is indeed positive definite since $K_1(z) > K_0(z)$.

An alternative separable solution for ϕ, which provides a positive-definite expression for ϵ and is further symmetrical in u and v, is

$$\phi = C\alpha s K_1(\alpha s)\sinh \alpha r, \tag{79}$$

when

$$\epsilon = 4C^2\alpha^4\, e^{-2\mu_3-f}\, uv(1-u^2)^{\frac{1}{2}}(1-v^2)^{\frac{1}{2}}[K_1^2(\alpha s)\cosh^2\alpha r - K_0^2(\alpha s)\sinh^2\alpha r], \tag{80}$$

which is again manifestly positive-definite.

The general solution for ϕ when $\alpha = 0$ is

$$\phi = C[\beta_1+(u^2-v^2)][\beta_2+(1-u^2-v^2)^2], \tag{81}$$

where C, β_1 and β_2 are constants. The expression for ϵ that follows can be made positive-definite for suitable choices of C, β_1, and β_2.

As we have already remarked the general solution for ϕ can be expressed as a linear superposition (as a discrete sum or as an integral) of the fundamental solution (73). Restricting ourselves to a discrete sum and ignoring various special cases that may arise, we can write

$$\phi = \sum_i C_i\alpha_i\, e^{\alpha_i r} sK_1(\alpha_i s). \tag{82}$$

The corresponding expression for ϵ is

$$\epsilon = 4uv(1-u^2)^{\frac{1}{2}}(1-v^2)^{\frac{1}{2}}\, e^{-2\mu_3-f}\{\sum_i C_i^2\alpha_i^4\, e^{2\alpha_i r}[K_1^2(\alpha_i s)-K_0^2(\alpha_i s)]$$
$$+2\sum_{i>j} C_iC_j\alpha_i^2\alpha_j^2\, e^{+(\alpha_i+\alpha_j)r}[K_1(\alpha_i s)K_1(\alpha_j s)-K_0(\alpha_i s)K_0(\alpha_j s)]\}. \tag{83}$$

8. The solution for f

We now turn to the solution of equations (64) for f. First, we observe that by changing the independent variables to r and s, the first of equations (64) gives

$$f_{,u} = 2u(f_{,r}-f_{,s}) = \frac{2[2u(\phi_{,r}-\phi_{,s})]^2}{u(1-u^2-v^2)} = \frac{8u}{s}(\phi_{,r}-\phi_{,s})^2, \tag{84}$$

or

$$f_{,r}-f_{,s} = +\frac{4}{s}(\phi_{,r}-\phi_{,s})^2. \tag{85}$$

Similarly, the second of equations (64) gives

$$f_{,r}+f_{,s} = -\frac{4}{s}(\phi_{,r}+\phi_{,s})^2. \tag{86}$$

From equations (85) and (86) it follows:

$$f_{,r} = -\frac{8}{s}\phi_{,r}\phi_{,s} \quad \text{and} \quad f_{,s} = -\frac{4}{s}[(\phi_{,r})^2+(\phi_{,s})^2]. \tag{87}$$

Now, substituting for $\phi_{,r}$ and $\phi_{,s}$ from equations (74) and (75) we find

$$f_{,r} = \pm 8C^2\alpha^4 \, e^{\pm 2\alpha r} sK_1(\alpha s)\, K_0(\alpha s) \tag{88}$$

and

$$f_{,s} = -4C^2\alpha^4 \, e^{\pm 2\alpha r} s[K_1^2(\alpha s)+K_0^2(\alpha s)]. \tag{89}$$

Making use of equation (76) and the further relation

$$K_0'(z) = -K_1(z), \tag{90}$$

we readily verify that the required solution for f is

$$f = 4C^2\alpha^3 \, e^{\pm 2\alpha r} sK_1(\alpha s)\, K_0(\alpha s). \tag{91}$$

Similarly, for ϕ given by equation (79) the corresponding solution for f is

$$f = 2C^2\alpha^2\{2\alpha s K_1(\alpha s)\, K_0(\alpha s)\cosh^2(\alpha r) - \alpha^2 s^2[K_1^2(\alpha s)-K_0^2(\alpha s)]\}. \tag{92}$$

Finally, for ϕ given by equation (82) the solution for f is

$$f = 4\sum_i C_i^2\alpha_i^3 \, e^{2\alpha_i r} sK_1(\alpha_i s)\, K_0(\alpha_i s)$$

$$+8\sum_{i>j} C_i C_j \frac{\alpha_i^2\alpha_j^2}{\alpha_i+\alpha_j} e^{(\alpha_i+\alpha_j)r} s[K_1(\alpha_i s)\, K_0(\alpha_j s)+K_1(\alpha_j s)\, K_0(\alpha_i s)]. \tag{93}$$

(In writing the solution (93) we are ignoring special cases such as may arise when for particular choices of α_i and α_j, $\alpha_i+\alpha_j = 0$.)

9. The choice of the solutions for χ and q_2

As we have remarked earlier, the determination of the metric of the space–time for the problem on hand consists in selecting any solution of the vacuum equations (for the functions χ and q_2 and the solution for $\nu+\mu_3$ that follows from them) and supplementing it with any solution for ϕ and f derived in §§7 and 8.

Since we have complete freedom of choice in the selection of solutions for χ and q_2, it would, in the first instance, appear most natural to select for them the solutions (26) that lead to the Nutku–Halil solution: for, it represents the solution we wish to generalize when $\epsilon \neq 0$ and its selection affects in no way what we may wish to assume about ϕ and f. So by making this selection, the corresponding solution for $\nu+\mu_3$ is (cf. I, equation (61))

$$e^{\mu_3+\nu} = \frac{1-p^2\eta^2-q^2\mu^2}{(1-\eta^2)^{\frac{3}{4}}(1-\mu^2)^{\frac{1}{4}}} \quad (p^2+q^2 = 1). \tag{94}$$

Combined with equations (25) and (26), we find (cf. I, equations (62)–(64))

$$e^{2\nu} = \frac{1-p^2\eta^2-q^2\mu^2}{(1-\eta^2)^{\frac{5}{4}}(1-\mu^2)^{\frac{1}{4}}}, \quad e^{2\mu_3} = \frac{1-p^2\eta^2-q^2\mu^2}{(1-\eta^2)^{\frac{1}{4}}(1-\mu^2)^{\frac{1}{4}}}, \tag{95}$$

$$e^{2\psi} = (1-\eta^2)^{\frac{1}{2}}(1-\mu^2)^{\frac{1}{2}}\frac{(1-p\eta)^2+q^2\mu^2}{1-p^2\eta^2-q^2\mu^2}, \tag{96}$$

$$e^{2\mu_2} = (1-\eta^2)^{\frac{1}{2}}(1-\mu^2)^{\frac{1}{2}}\frac{1-p^2\eta^2-q^2\mu^2}{(1-p\eta)^2+q^2\mu^2}. \tag{97}$$

In the standard form in which we write the metric (as in I, equation (6)), '$e^{\mu_3+\nu}$' is the sole term that differs, in our present context, from its vacuum value (by the factor e^f). We may accordingly write the metric in the form (cf. I, equation (80))

$$ds^2 = \frac{\Delta\, e^f}{(\sin\psi\,\sin\theta)^{\frac{1}{2}}}[(d\psi)^2-(d\theta)^2] - \frac{\sin\psi\,\sin\theta}{\Delta}\,|(1-E)\,dx^1+i(1+E)\,dx^2|^2, \tag{98}$$

where

$$\left.\begin{aligned} E &= p\eta+iq\mu = p\cos\psi+iq\cos\theta, \\ \Delta &= 1-|E|^2 = p^2\sin^2\psi+q^2\sin^2\theta, \end{aligned}\right\} \tag{99}$$

and f is given by any of the solutions (91), (92), or (93) (depending on our choice).

10. The description of the space–time in a Newman–Penrose formalism

From a comparison of the metric (98) with those derived in Papers I and II (I, equation (80) and II, equation (122)), it is evident that we can set up a null tetrad-basis in exactly the same way and derive the relevant spin coefficients. Thus, the definitions of Paper II, equations (124) and (125) can be taken over by identifying $\mathscr{E}$ with our present (simpler) E and replacing

$$\left.\begin{aligned} &U \quad \text{by} \quad \tilde{U} = e^{f/2}\,U = e^{f/2}\frac{\sqrt{\Delta}}{(\sin\psi\,\sin\theta)^{\frac{1}{4}}}, \\ &\mathbf{D} \quad \text{by} \quad \tilde{\mathbf{D}} = e^{-f/2}\,\mathbf{D} = \frac{e^{-f/2}}{U\sqrt{2}}(\partial_\psi+\partial_\theta), \\ &\mathbf{\Delta} \quad \text{by} \quad \tilde{\mathbf{\Delta}} = e^{-f/2}\,\mathbf{\Delta} = \frac{e^{-f/2}}{U\sqrt{2}}(\partial_\psi-\partial_\theta). \end{aligned}\right\} \tag{100}$$

With the foregoing replacements, we deduce from II (equations (127)) that the non-vanishing spin-coefficients for the present problem (distinguished by a tilde) are given by

$$\left.\begin{aligned} &\tilde{\sigma} = e^{-f/2}\,\sigma; \quad \tilde{\lambda} = e^{-f/2}\,\lambda, \\ &\tilde{\rho} = e^{-f/2}\,\rho; \quad \tilde{\mu} = e^{-f/2}\,\mu, \\ &\tilde{\epsilon} = e^{-f/2}(\epsilon+\tfrac{1}{4}\mathbf{D}f), \quad \tilde{\gamma} = e^{-f/2}(\gamma-\tfrac{1}{4}\mathbf{\Delta}f), \end{aligned}\right\} \tag{101}$$

where σ, λ, ρ, μ, ϵ, and γ are the spin coefficients for the vacuum given in Paper I, equations (86) (see also Chandrasekhar & Ferrari 1985, where some misprints in these equations are noted). With the spin coefficients given in equations (101), the various Weyl and Ricci scalars can be found by systematically going through the 'Ricci identities' listed in M.T., equations (310), pp. 46–47. We find that the Weyl scalars are given by

$$\left.\begin{aligned}&\tilde{\Psi}_1 = \tilde{\Psi}_3 = 0,\\ &\tilde{\Psi}_0 = e^{-f}(\Psi_0 - \sigma\,\mathbf{D}f); \quad \tilde{\Psi}_4 = e^{-f}(\Psi_4 + \lambda\,\mathbf{\Delta}f).\end{aligned}\right\} \tag{102}$$

$$\begin{aligned}\tilde{\Psi}_2 &= e^{-f}\left[\Psi_2 + \frac{1}{12U^2}(f_{,\theta,\theta} - f_{,\psi,\psi})\right]\\ &= e^{-f}\,\Psi_2 - \tfrac{1}{12}R = e^{-f}\,\Psi_2 + \tfrac{1}{3}\epsilon,\end{aligned} \tag{103}$$

where the quantities distinguished by tildes refer to the problem on hand, while those not so distinguished refer to the vacuum.

Similarly, we find that the non-vanishing Ricci scalars are given by

$$\begin{aligned}\tilde{\Phi}_{00} &= -\rho\,e^{-f}\,\mathbf{D}f = -\tfrac{1}{2}R_{11},\\ \tilde{\Phi}_{22} &= +\mu\,e^{-f}\,\mathbf{\Delta}f = -\tfrac{1}{2}R_{22},\\ \tilde{\Phi}_{11} &= \frac{e^{-f}}{8U^2}(f_{,\theta,\theta} - f_{,\psi,\psi}) = -\tfrac{1}{4}R_{12},\\ \tilde{\Lambda} &= -\frac{e^{-f}}{24U^2}(f_{,\theta,\theta} - f_{,\psi,\psi}) = +\tfrac{1}{12}R_{12} = \tfrac{1}{24}R = -\tfrac{1}{6}\epsilon.\end{aligned} \tag{104}$$

11. The character of the fluid motions in region I and the nature of the space–time singularity at $u^2 + v^2 = 1$

We have seen in Papers I and II that a space–time singularity develops on $u^2 + v^2 = 1$ by a focusing of the colliding waves when they scatter off one another. We shall find that while a space–time singularity continues to manifest itself in the present context as well, its character depends in a complicated way on the amplitude of the sound waves present, on the particular superposition of the separable solutions that occur, and on whether the colliding gravitational waves are parallelly polarized or not. To illustrate the nature of these effects, we shall, for the sake of simplicity, restrict our considerations to the solution (79), which leads to a symmetrical distribution of ϵ in u and v in region I.

Substituting for $e^{-2\mu_3}$ from equation (95) in the expression (80) for ϵ, we find

$$\epsilon = 4C^2\alpha^4 uv(1-u^2)^{\frac{1}{2}}(1-v^2)^{\frac{1}{2}}\frac{(1-u^2-v^2)^{\frac{1}{2}}}{\Delta}e^{-f} \times [K_1^2(\alpha s)\cosh^2(\alpha r) - K_0^2(\alpha s)\sinh^2(\alpha r)], \tag{105}$$

where it may be recalled that the corresponding solution for f is given by equation (92) and that

$$r = u^2 - v^2 \quad\text{and}\quad s = 1 - u^2 - v^2. \tag{106}$$

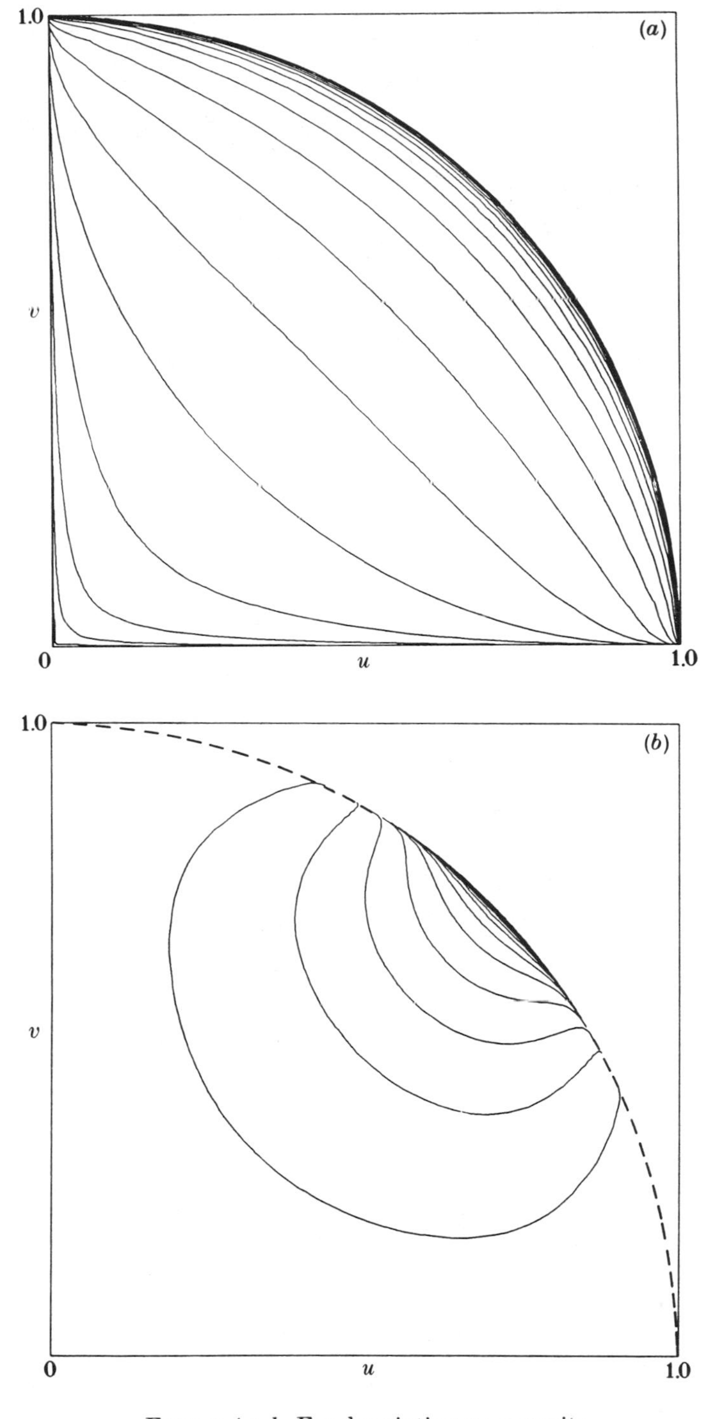

FIGURE 1 a, b. For description see opposite.

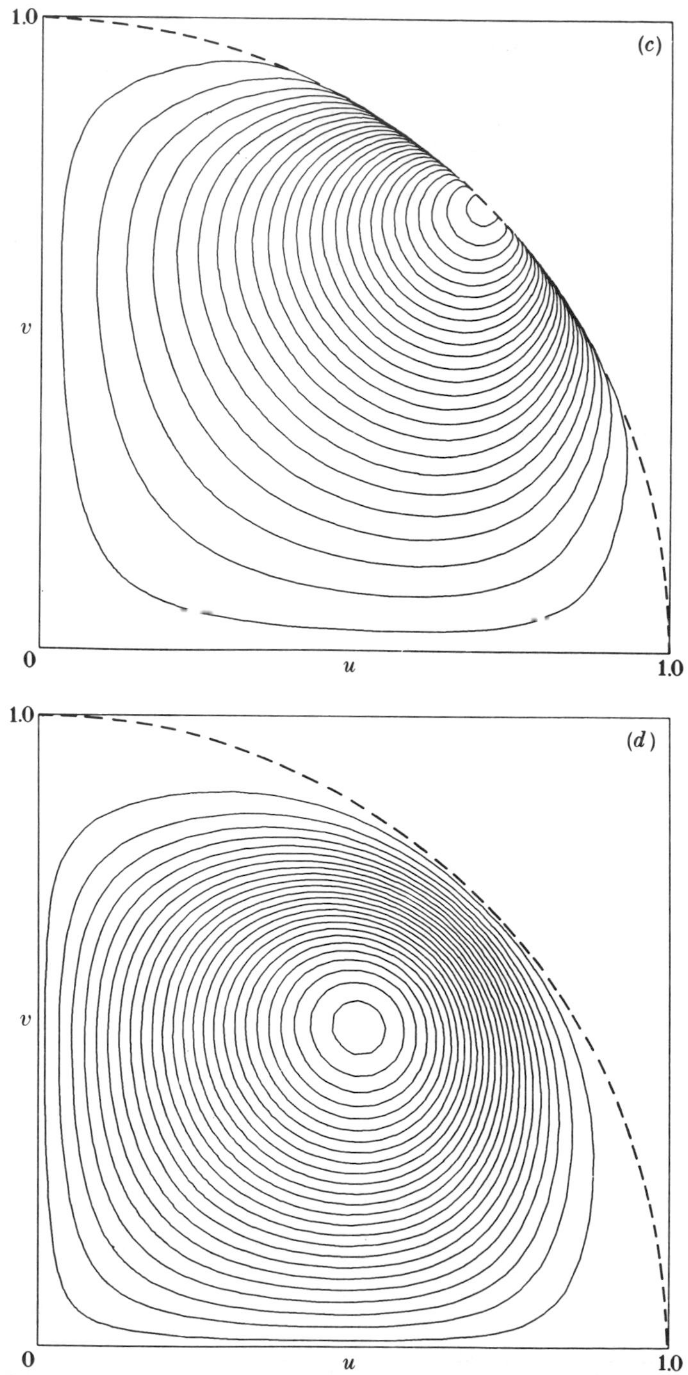

FIGURE 1. The curves of constant ϵ for $q = 0$, $\alpha = 0.5$ and for various values of C. Along the null boundaries, $u = 0$ $(0 \leqslant v \leqslant 1)$ and $v = 0$ $(0 \leqslant u \leqslant 1)$, ϵ vanishes in all cases. The various cases illustrated are: (a) $C = 0.2$, which corresponds to case (i) (see text) when ϵ diverges along the entire arc, $u^2+v^2 = 1$; (b) $C = 1.80$, which corresponds to case (iii) when ϵ diverges only along the part of the arc, $u^2+v^2 = 1$; (c) $C = \sqrt{3.5}$, which corresponds to case (iv) when ϵ vanishes along the entire arc, $u^2+v^2 = 1$, except at the midpoint; and (d) $C = 2.20$, which corresponds to case (v) when ϵ vanishes along the entire arc, $u^2+v^2 = 1$.

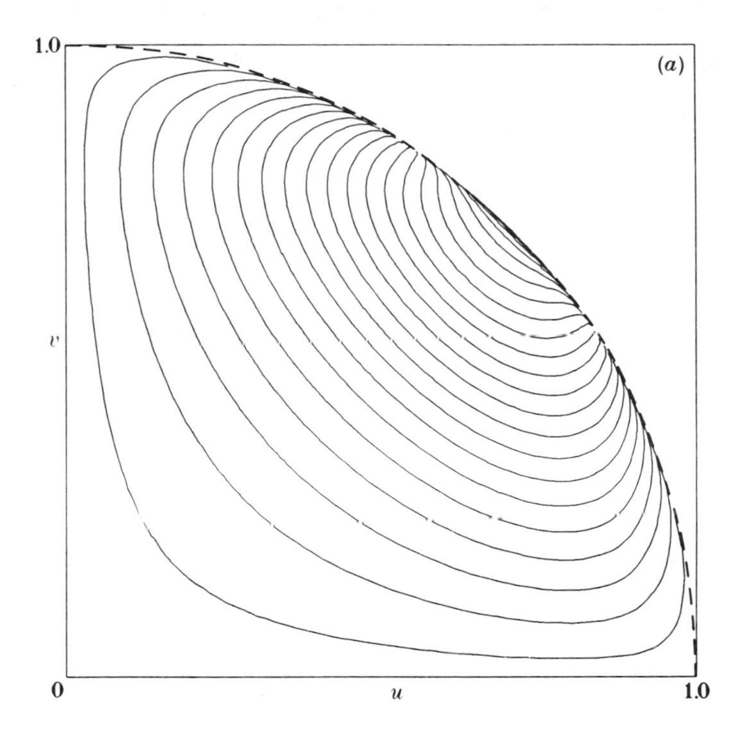

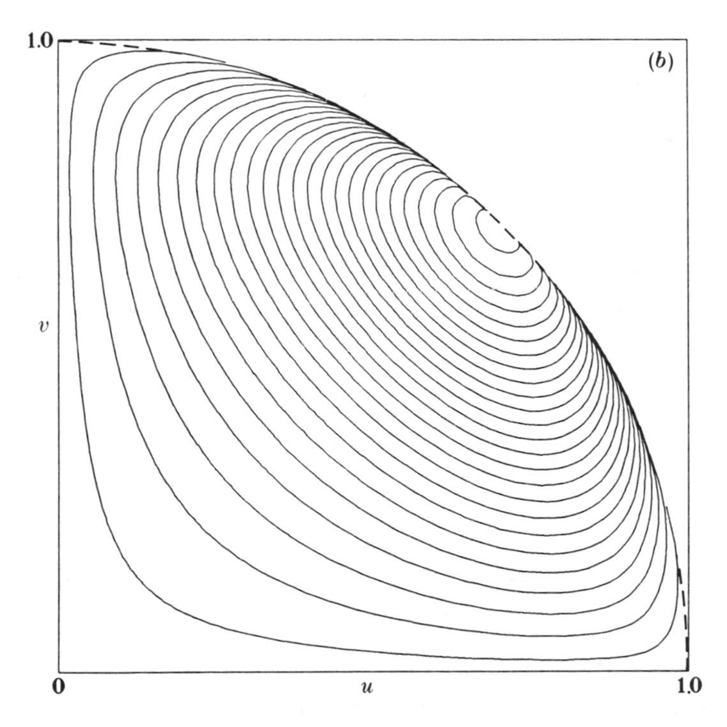

FIGURE $2a, b$. For description see opposite.

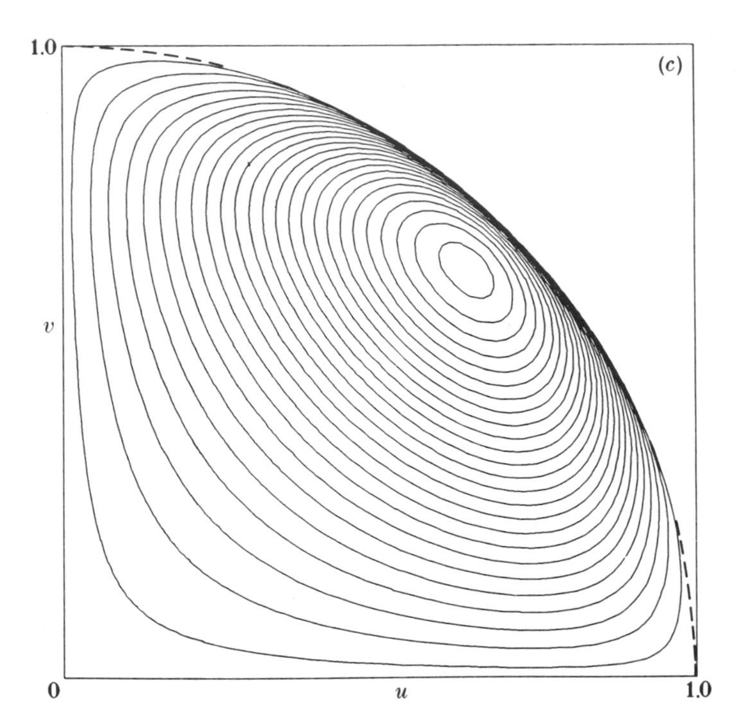

FIGURE 2. The curves of constant ϵ for $q^2 = p^2 = 0.5$, $\alpha = 0.5$ and for various values of C. Along the null boundaries, $u = 0$ $(0 \leqslant v \leqslant 1)$ and $v = 0$ $(0 \leqslant u \leqslant 1)$, ϵ vanishes in all cases. The various cases illustrated are: (a) $C = 1.2$, which corresponds to case (iii) when ϵ diverges only along a part of the arc, $u^2+v^2 = 1$; (b) $C = C_1 = \sqrt{1.5}$, which corresponds to case (iv) when ϵ vanishes along the entire arc, $u^2+v^2 = 1$, except at the midpoint; and (c) $C = 1.3$, which corresponds to case (v) when ϵ vanishes along the entire arc, $u^2+v^2 = 1$.

Figures 1 and 2 (drawn for us by computer tracing by Mr S. K. Chakrabarti) exhibit contours of constant ϵ for $\alpha = 0.5$ and some selected values of C for $q = 0$ (figure 1) and $q \neq 0$ (figure 2). In figure 3 the curves of constant ϕ – the 'streamlines' – are illustrated for a typical case. Notice, in particular, that the streamlines are normal to the u and v axes.

The features exhibited in figures 1 and 2 can be understood in terms of the behaviour of ϵ as we approach the bounding arc $u^2+v^2 = 1-0$, i.e. as $s \to 0$. Making use of the behaviours

$$K_0(z) \to -(\gamma_{\mathrm{E}} + \ln \tfrac{1}{2}z) \quad \text{and} \quad K_1(z) \to z^{-1} \quad \text{as} \quad z \to 0, \tag{107}$$

of the Bessel functions, where γ_{E} $(= 0.57721\ldots)$ is Euler's constant, we readily deduce from the expression (92) for f that

$$\begin{aligned} f &\to 2C^2\alpha^2\{-2[\gamma_{\mathrm{E}} - \ln 2 + \ln \alpha + \ln(1-u^2-v^2)]\cosh^2 \alpha r - 1\} \\ &= -4C^2\alpha^2\{(\gamma_{\mathrm{E}} - \ln 2 + \ln \alpha)\cosh^2 \alpha r + \tfrac{1}{2} + \ln(1-u^2-v^2)\cosh^2 \alpha r\} \end{aligned}$$

$$(u^2+v^2 \to 1-0). \tag{108}$$

Hence,

$$e^{-f} \to A(1-u^2-v^2)^k, \tag{109}$$

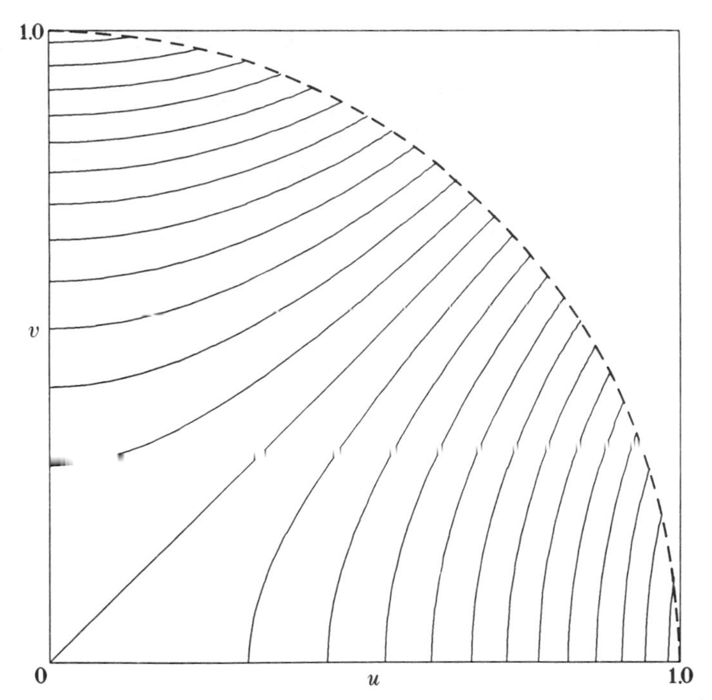

FIGURE 3. The curves of constant ϕ (stream lines) for $q = 0$, $\alpha = 0.5$ and $C = \sqrt{3.5}$.

where
$$A = \exp\left[4C^2\alpha^2(\gamma_E - \ln 2 + \ln\alpha)\cosh^2(\alpha r) + \tfrac{1}{2}\right], \tag{110}$$
and
$$k = 4C^2\alpha^2\cosh^2[\alpha(u^2 - v^2)]. \tag{111}$$

Accordingly,
$$\epsilon \to 4C^2\alpha^2 uv(1-u^2)^{\frac{1}{2}}(1-v^2)^{\frac{1}{2}}\frac{\cosh^2\alpha(u^2-v^2)}{\Delta}\frac{A}{(1-u^2-v^2)^{\frac{3}{2}-k}}. \tag{112}$$

The behaviour of Δ for $u^2+v^2 \to 1-0$ is different when $q = 0$ and $q \neq 0$. Thus, from equations (55) and the definition of Δ, we find
$$\Delta = 1 - u^2 - v^2 + 2uv[uv - (p^2-q^2)(1-u^2)^{\frac{1}{2}}(1-v^2)^{\frac{1}{2}}]. \tag{113}$$
Therefore,
$$\Delta = 4q^2u^2v^2 \quad \text{when} \quad u^2+v^2 = 1; \tag{114}$$
and it is finite when $q \neq 0$. When $q = 0$, a more detailed examination is necessary; and we find that
$$\Delta \sim \frac{1}{4u^2v^2}(1-u^2-v^2)^2 \quad (u^2+v^2 \to 1-0;\, q = 0). \tag{115}$$

We conclude that the behaviour of ϵ for $u^2+v^2 \to 1-0$ is different for $q = 0$ and $q \neq 0$; thus
$$\left.\begin{aligned} &\epsilon \sim (1-u^2-v^2)^{k-3.5} \quad (q = 0)\\ \text{and}\quad &\epsilon \sim (1-u^2-v^2)^{k-1.5} \quad (q \neq 0).\end{aligned}\right\} \tag{116}$$

Therefore, at a point v $(= \sqrt{(1-u^2)})$ on the arc $u^2+v^2 = 1$ (bounding region I), ϵ will diverge only if
$$k = 4C^2\alpha^2\cosh^2[\alpha(1-2v^2)] < k_c, \tag{117}$$

where $k_c = 3.5$ for $q = 0$ and $k_c = 1.5$ for $q \neq 0$. Defining for a given α (> 0) the two amplitudes

$$\left.\begin{aligned} C_0 &= \sqrt{k_c}/(2\alpha \cosh \alpha) \quad (v = 1 \text{ or } 0) \\ \text{and} \qquad C_1 &= \sqrt{k_c}/(2\alpha) \qquad (v = 1/\sqrt{2}), \end{aligned}\right\} \tag{118}$$

we can distinguish the following cases:

(i) $C < C_0$, when ϵ diverges along the entire arc $u^2+v^2 = 1$ including ($u = 1, v = 0$) and ($u = 0$, $v = 1$);

(ii) $C = C_0$, when ϵ diverges along the arc $u^2+v^2 = 1$ except at its end points, where it is finite;

(iii) $C_0 < C < C_1$, when ϵ diverges only along the part of the arc included between

$$v_+ = \sqrt{[(1+x)/2]} \quad \text{and} \quad v_- = \sqrt{[(1-x)/2]}, \tag{119}$$

where x is the positive root of the equation

$$2C\alpha \cosh(\alpha x) = \sqrt{k_c} \tag{120}$$

and vanishes outside these limits: at $v = v_+$ and $v = v_-$, ϵ is finite;

(iv) $C = C_1$, when ϵ vanishes along the entire arc $u^2+v^2 = 1$ except at the midpoint, $v = 1/\sqrt{2}$, where it is finite; and finally

(v) $C > C_1$, when ϵ vanishes along the entire arc $u^2+v^2 = 1$.

Since the Weyl scalar, Ψ_2, for the vacuum has the behaviour (cf. I, equation (111))

$$\Psi_2 \sim -\frac{1}{8q^2 s^{\frac{3}{2}}} \; (q \neq 0) \quad \text{and} \quad \Psi_2 \sim \frac{6u^4v^4}{s^{\frac{7}{2}}} \; (q = 0), \tag{121}$$

for $u^2+v^2 \to 1-0$, it follows that the term $e^{-f}\,\Psi_2$ included in $\tilde{\Psi}_2$ (besides the term $\frac{1}{3}\epsilon$ in equation (103)) has the same behaviour as ϵ (cf. equation (112)) as we approach $u^2+v^2 = 1-0$. (It should be noted that the order of the divergence of Ψ_2 is also different for the Khan–Penrose and the Nutku–Halil solutions.)

From the foregoing discussion of the behaviour of $\tilde{\Psi}_2$ and ϵ as $u^2+v^2 \to 1-0$, it follows that they do not exhibit any divergence when the colliding waves are coupled with sound waves of sufficient amplitude (in the case $\epsilon = p$ we have considered). Nevertheless, the singularity of the space–time on $u^2+v^2 = 1$ is not eliminated: it continues to be manifested in the derivatives of the curvature invariants of sufficiently high order since the exponent k determining their behaviour for $u^2+v^2 \to 1-0$ is in general an irrational number.

12. The extension of the space–time into regions II, III, and IV

So far, we have concerned ourselves only with the region of the space–time in which the interaction between the colliding waves – gravitational and hydrodynamic – occurs; this is region I in the space–time diagram illustrated in figure 4. The boundaries of region I are

$$u = 0, \quad 0 \leqslant v \leqslant 1; \quad v = 0, \quad 0 \leqslant u \leqslant 1, \quad \text{and} \quad u^2+v^2 \leqslant 1. \tag{122}$$

To complete the solution, we must extend the domain of validity of the solution to include regions II, III, and IV, prior to the instant of collision. Following the

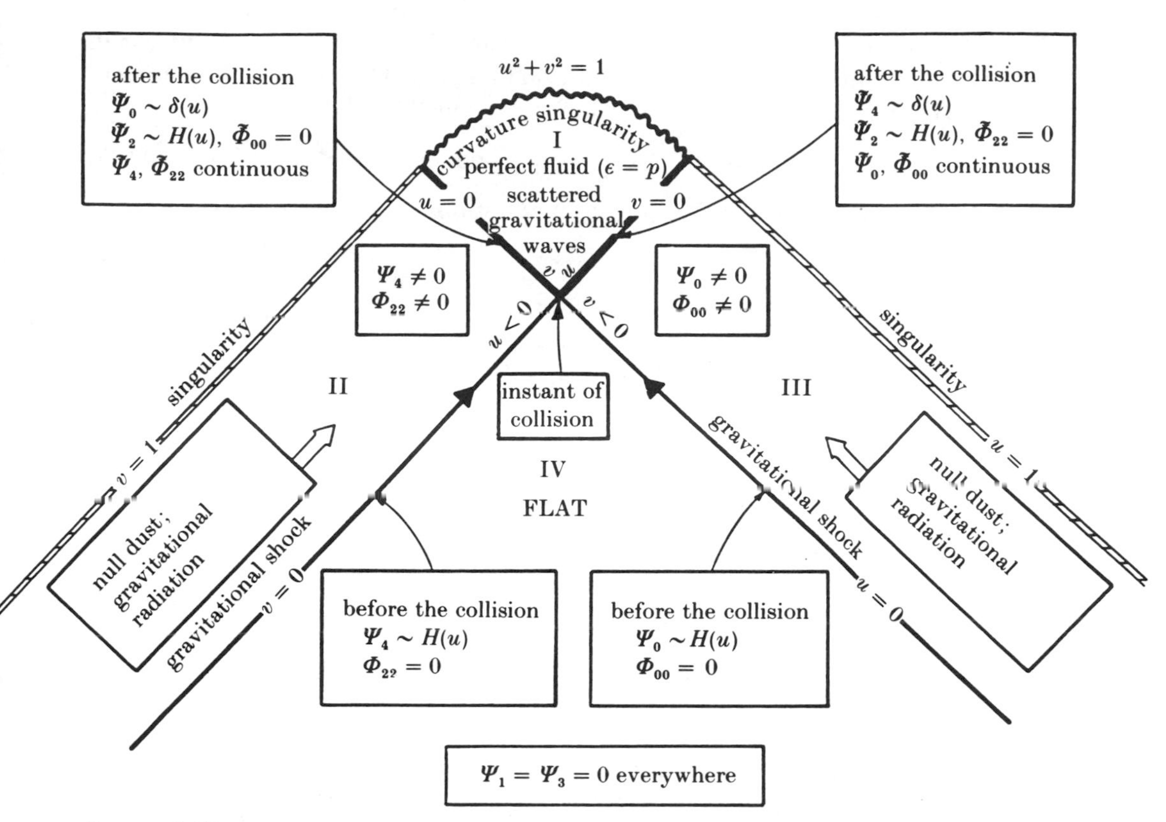

FIGURE 4. The space–time diagram for two colliding plane impulsive gravitational waves when coupled with a perfect fluid. The colliding waves are propagated along the null directions, u and v. The time coordinate is along the vertical, and the spatial direction of propagation (in space) is along the horizontal. The plane of the wavefronts (on which the geometry is invariant) is orthogonal to the plane of the diagram. The instant of collision is at the origin of the (u, v)-coordinates.

The flat portion of the space–time, prior to the arrival of either wave, is region IV. In regions II (III) we have gravitational radiation characterized by a non-vanishing Ψ_4 (Ψ_0) and a field of pure radiation characterized by a non-vanishing Φ_{22} (Φ_{00}). As a result of the collision the pure radiation in regions II and III is transformed into a perfect fluid with the equation of state $\epsilon = p$. Subsequently, a curvature singularity develops on the arc, $u^2+v^2 = 1$. The nature of the fluid motions in region I is illustrated in figures 1 and 2. Along the null boundaries $u = 0$ $(0 \leqslant v \leqslant 1)$ and $v = 0$ $(0 \leqslant u \leqslant 1)$ the Ricci scalars are all continuous; but the Weyl scalars Ψ_4, Ψ_0, and Ψ_2 are characterized by Dirac δ-function singularities or Heaviside (H) step-function discontinuities, or both, as indicated in the diagram.

original prescription of Penrose, we shall extend the metric (98) across the null boundaries,

$$u = 0, \quad 0 \leqslant v \leqslant 1 \quad \text{and} \quad v = 0, \quad 0 \leqslant u \leqslant 1, \tag{123}$$

by the substitutions,

$$u\,H(u) \quad \text{and} \quad v\,H(v), \tag{124}$$

in place of u and v in the various metric functions, where $H(u)$ and $H(v)$ are the Heaviside functions that are unity for positive and zero values of the arguments and zero otherwise.

From the form of the metric (98), it is clear that it will suffice to consider it in region II: the results of a similar consideration in region III can be readily written. In region II, by the substitutions (124) (cf. equations (55)),

$$\eta \to v, \quad \mu \to -v, \quad \Delta \to 1-v^2, \quad \text{and} \quad \delta \to 1-v^2; \tag{125}$$

and the metric functions χ and q_2 (given in equations (26)) become

$$\chi = \frac{1-v^2}{1-2pv+v^2} \tag{126}$$

and

$$q_2 = -\frac{2qv}{1-2pv+v^2}, \tag{127}$$

and $f(u, v)$ is replaced by $f(0, v)$. With these definitions, the metric (98) reduces to

$$ds^2 = 4\,e^{f(0,\,v)}\,du\,dv - (1-v^2)\left[\chi(dx^2)^2 + \frac{1}{\chi}(dx^1 - q_2\,dx^2)^2\right]; \tag{128}$$

or, by letting

$$u = x^0 + x^3 \quad \text{and} \quad v = x^0 - x^3, \tag{129}$$

the metric takes the standard form,

$$ds^2 = e^{2\nu}[(dx^0)^2 - (dx^3)^2] - e^{2\psi}(dx^1 - q_2\,dx^2)^2 - e^{2\mu_2}(dx^2)^2, \tag{130}$$

where

$$\left.\begin{aligned} e^{2\nu} = 4\,e^{f(0,\,v)}, \quad e^{2\psi} = 1-2pv+v^2, \quad e^{2\mu_2} = \frac{(1-v^2)^2}{1-2pv+v^2}, \\ e^{\beta} = e^{\psi+\mu_2} = 1-v^2, \quad \text{and} \quad \chi = e^{-\psi+\mu_2}. \end{aligned}\right\} \tag{131}$$

In Paper II, it was shown that a metric of the form (130) in which ν, ψ, q_2, and μ_2 are functions of v $(= x^0 - x^3)$ only, the sole non-vanishing components of the Ricci tensor are

$$R_{00} = R_{33} = R_{03} = -(L+M), \tag{132}$$

where $L+M$ is defined in Paper II, equation (160) (see equation (139) below); and, besides

$$R = 0. \tag{133}$$

On the other hand, if the equation of state, $\epsilon = p$, assumed in region I, continues to hold in region II, then we should have (cf. equation (11))

$$R_{00} = -4\epsilon u_0^2, \quad R_{33} = -4\epsilon u_3^2, \quad R_{03} = -4\epsilon u_0 u_3, \quad \text{and} \quad R = -4\epsilon; \tag{134}$$

and these requirements are clearly incompatible with equations (132) and (133). Therefore, the extension of the metric (98) by the substitutions (124) leads to inconsistencies if it is assumed that the character of the fluid remains unchanged in regions II and III. Since we know that the space–time in regions II and III is flat when the Nutku–Halil solution (with the same χ and q_2) is extended into these regions by the substitutions (124), we can avoid the mutual inconsistency of equations (132), (133) and (134) by stipulating that

$$f(0, v) = f(u, 0) = \text{a constant}. \tag{135}$$

From equations (64) it now follows that the constancy of f along the null boundaries (123) implies a similar constancy of ϕ. But it can be shown that a solution of equation (60), governing ϕ, which is constant along the characteristics (123) is identically a constant in the entire region I (see Appendix A). The solution thus becomes trivial: $\epsilon = 0$!

An alternative method of extension, in which all the metric functions, *except* f, are extended by the substitutions (124) and $f(u, v)$ is replaced by $f(u, v)\, H(u)\, H(v)$, while mathematically self-consistent, also fails since the resulting solution violates the positive-energy condition on the null boundaries (see Appendix B).

Since the extension of the metric (98) into regions II and III thus appears impossible if we insist on the inviolability of the underlying equation of state $\epsilon = p$, we shall inquire, without prejudice, the character of a fluid source which will be compatible with equations (133) and (134).

The components of the Ricci tensor (as required by the Einstein field-equations) derived from a perfect fluid-source, without making the assumption that the four-velocity is time-like and $u_0^2 - u_3^2 = 1$, are (cf. equations (11))

$$\left.\begin{aligned} (\epsilon+p)\,(u_0^2+u_3^2)+2p &= -R_{00} = L+M, \\ (\epsilon+p)\,(u_0^2+u_3^2)-2p &= -R_{33} = L+M, \\ 2(\epsilon+p)\,u_0\,u_3 &= -R_{03} = L+M. \end{aligned}\right\} \tag{136}$$

These equations clearly require that

$$p = 0, \quad u_0^2 = u_3^2, \quad \text{and} \quad 2\epsilon u_0^2 = L+M. \tag{137}$$

In other words, *the metric* (130) *is compatible with a null dust or 'pure radiation'* (in the nomenclature of Kramer *et al.* (1980, p. 22)) *provided*

$$L+M > 0. \tag{138}^*$$

(Notice that this last requirement is the opposite of what was required in the context of Paper II, equation (160).) We had noticed this compatibility at an early stage in our investigation of this problem, but we dismissed the possibility as 'manifestly untenable': we simply did not contemplate the possibility of a change in the character of the source. However, in discussing with Roger Penrose the impasse we thought we had reached in the matter of extending our solution in region I into regions II and III, he suggested (without any indication from us) that the extension could possibly be accomplished with pure radiation in regions II and III. Moreover, he considered the possibility as a 'very reasonable one', pointing out that the essential difference between a perfect fluid with $\epsilon = p$ and a null dust is that while in the former case, the Ricci tensor R^{ij} is proportional to $u^i\, u^j$, where u^i is a time-like vector, in the latter case, it is proportional to $k^i\, k^j$, where k^i is a null vector. But before we accept the interpretation of the metric (130), with the metric functions defined as in equations (131), as derived from pure radiation, it is necessary that we verify a number of requirements including the positivity of $L+M$.

Now associated with a metric of the form (130), there are two important

*The difference in the senses of the inequality is due to the difference in our conventions regarding the energy momentum tensor in the two cases. It is not a significant one.

combinations of the metric functions, namely (cf. Paper II, equations (160) and (167))

$$L+M = e^{-2\nu}[\beta'' + (\psi')^2 + (\mu_2')^2 - 2\nu'\beta' + \tfrac{1}{2}(q_2'/\chi)^2], \tag{139}$$

and

$$M-L+2iN = e^{-2\nu}\left[\frac{1}{\chi}\mathscr{Z}'' + \frac{1}{\chi}(\beta' - 2\nu')\mathscr{Z}' - \frac{1}{\chi^2}(\mathscr{Z}')^2\right], \tag{140}$$

where primes denote differentiations with respect to v and

$$\mathscr{Z} = \chi + iq_2 = \frac{1+\mathscr{E}}{1-\mathscr{E}}. \tag{141}$$

The importance of these combinations arises from the fact that they are simply related to the only Ricci and Weyl scalars that do not vanish, namely Φ_{22} and Ψ_4; thus (cf. Paper II, equations (178) and (179))

$$\Phi_{22} = -(L+M)\sqrt{(1-v^2)}, \tag{142}$$

and

$$\Psi_4 = \frac{1-\mathscr{E}}{1-\mathscr{E}^*}(M-L+2iN)\sqrt{(1-v^2)}. \tag{143}$$

Since the metric (130), with the metric functions as defined in equations (131), is flat when ν = a constant, it follows that for the problem on hand,

$$L+M = -2\,e^{-2\nu}\nu'\beta' = (e^{-2\nu})_{,v}\beta_{,v}, \tag{144}$$

or

$$L+M = \frac{v}{2(1-v^2)}\,e^{-f(0,v)}f_{,v}(0,v), \tag{145}$$

which is positive as required since equations (64) guarantee that

$$f_{,v} > 0. \tag{146}$$

Similarly, from equation (140) we conclude that

$$M-L+2iN = \frac{\chi' + iq_2'}{\chi}(e^{-2\nu})_{,v}. \tag{147}$$

On evaluating χ'/χ and q_2'/χ with the aid of equations (126) and (127), we find

$$\frac{\chi' + iq_2'}{\chi} = \frac{2}{1-2pv+v^2}\left[-2\frac{v(1-pv)}{1-v^2} + p - iq\right], \tag{148}$$

or

$$\frac{\chi' + iq_2'}{\chi} = \frac{2(p-iq)}{(1-v^2)(1-2pv+v^2)}[1-v(p+iq)]^2. \tag{149}$$

Thus,

$$M-L+2iN = -\frac{(p-iq)}{2(1-v^2)(1-2pv+v^2)}[1-v(p+iq)]^2\,e^{-f(0,v)}f_{,v}(0,v). \tag{150}$$

Returning to the expressions (142) and (143) for Φ_{22} and Ψ_4, we first note that

$$\frac{1-\mathscr{E}}{1-\mathscr{E}^*} = \frac{\chi+1-iq_2}{\chi+1+iq_2} = \frac{(\chi+1)^2 - q_2^2 - 2iq_2(\chi+1)}{(\chi+1)^2+q_2^2}, \tag{151}$$

which, on further simplification, becomes

$$\frac{1-\mathscr{E}}{1-\mathscr{E}^*} = \frac{[1-v(p-iq)]^2}{1-2pv+v^2}. \tag{152}$$

From equations (142), (143), (145), (150), and (152) we now find

$$\Phi_{22} = -\frac{v}{2\sqrt{(1-v^2)}}\,e^{-f(0,v)}f_{,v}(0,v), \tag{153}$$

and

$$\Psi_4 = -\frac{p-iq}{2\sqrt{(1-v^2)}}\,e^{-f(0,v)}f_{,v}(0,v). \tag{154}$$

In similar fashion, we shall find that in region III,

$$\Phi_{00} = -\frac{u}{2\sqrt{(1-u^2)}}\,e^{-f(u,0)}f_{,u}(u,0), \tag{155}$$

and

$$\Psi_0 = -\frac{p-iq}{2\sqrt{(1-u^2)}}\,e^{-f(u,0)}f_{,u}(u,0). \tag{156}$$

The Weyl and the Ricci scalars in region I and the characterization of the singularities and the discontinuities along the null boundaries

As we have seen in §10, equations (104), the Ricci and the Weyl scalars in region I are simply related to those that obtain in the vacuum. Noting that

$$\mathbf{D}f = -\frac{(\sin\psi\sin\theta)^{\frac{1}{4}}}{(2\Delta)^{\frac{1}{2}}}(1-u^2)^{\frac{1}{2}}f_{,u} \tag{157}$$

and

$$\Delta f = -\frac{(\sin\psi\sin\theta)^{\frac{1}{4}}}{(2\Delta)^{\frac{1}{2}}}(1-v^2)^{\frac{1}{2}}f_{,v}, \tag{158}$$

and making use of the spin coefficients listed in Paper I, equations (86) (see also the errata in Chandrasekhar & Ferrari (1985)), we find that

$$\tilde{\Phi}_{00} = -\frac{(\sin\psi\sin\theta)^{\frac{1}{2}}}{4\Delta}(\cot\psi+\cot\theta)\,e^{-f}f_{,u}\sqrt{(1-u^2)}, \tag{159}$$

$$\tilde{\Phi}_{22} = -\frac{(\sin\psi\sin\theta)^{\frac{1}{2}}}{4\Delta}(\cot\psi-\cot\theta)\,e^{-f}f_{,v}\sqrt{(1-v^2)}, \tag{160}$$

$$\tilde{\Psi}_0 = e^{-f}\Psi_0 - \frac{(\sin\psi\sin\theta)^{\frac{1}{2}}}{2\Delta^2}(p\sin\psi - iq\sin\theta)(1-u^2)^{\frac{1}{2}}e^{-f}f_{,u}, \tag{161}$$

$$\tilde{\Psi}_4 = e^{-f}\Psi_4 - \frac{(\sin\psi\sin\theta)^{\frac{1}{2}}}{2\Delta^2}(p\sin\psi - iq\sin\theta)(1-v^2)^{\frac{1}{2}}e^{-f}f_{,v} \tag{162}$$

and

$$\tilde{\Psi}_2 = e^{-f}\Psi_2 + \tfrac{1}{3}\epsilon, \tag{163}$$

where, it may be recalled that in accordance with equations (65), (67) and (95),

$$\epsilon = 4\frac{uv(1-u^2)^{\frac{1}{2}}(1-v^2)^{\frac{1}{2}}}{\Delta(1-u^2-v^2)^{\frac{3}{2}}}\,e^{-f}[(\phi_{,r})^2-(\phi_{,s})^2]. \tag{164}$$

From the foregoing equations it follows that on the null boundary $u = 0$ where

$$\sin\psi = \sin\theta = \sqrt{(1-v^2)}, \quad \cos\psi = -\cos\theta = v \quad \text{and} \quad \Delta = \sin^2\psi, \tag{165}$$

$$\left.\begin{aligned} \tilde{\Phi}_{22} = -\frac{v}{2\sqrt{(1-v^2)}}\,\mathrm{e}^{-f(0,v)} f_{,v}(0,v), \quad \tilde{\Psi}_4 = -\frac{p-\mathrm{i}q}{2\sqrt{(1-v^2)}}\,\mathrm{e}^{-f(0,v)} f_{,v}(0,v), \\ \tilde{\Phi}_{00} = 0 \quad \text{and} \quad R = -4\epsilon = 0. \end{aligned}\right\} \tag{166}$$

A comparison with equations (153) and (154) now shows that on $u = 0$ $(0 \leqslant v \leqslant 1)$, $\tilde{\Psi}_4$ *and all the Ricci scalars are continuous*. But the Weyl scalars $\tilde{\Psi}_2$ and $\tilde{\Psi}_0$ suffer discontinuities and, or, singularities given by

$$\tilde{\Psi}_0 = \mathrm{e}^{-f(0,v)}\left[\frac{p-\mathrm{i}q}{2(1-v^2)}\,\delta(u) + \tfrac{3}{2}(p-\mathrm{i}q)^3\,\frac{v}{(1-v^2)^{\frac{3}{2}}}\,H(u)\right] \\ -\frac{p-\mathrm{i}q}{2(1-v^2)}\,\mathrm{e}^{-f(0,v)} f_{,u}(0,v)\,H(u), \tag{167}$$

and

$$\tilde{\Psi}_2 = \mathrm{e}^{-f(0,v)}\,\frac{(p-\mathrm{i}q)^2}{2(1-v^2)^{\frac{3}{2}}}\,H(u), \tag{168}$$

where we have incorporated the results for the vacuum given in Paper I, equations (117)–(119).

Similarly, on the null boundary $v = 0$ $(0 \leqslant u \leqslant 1)$, $\tilde{\Psi}_0$ and all the Ricci scalars will be continuous while $\tilde{\Psi}_2$ and $\tilde{\Psi}_4$ will suffer discontinuities and, or, singularities analogous to $\tilde{\Psi}_2$ and $\tilde{\Psi}_0$ on $u = 0$ $(0 \leqslant v \leqslant 1)$.

Considering next the null boundary separating region II and region IV, which is, of course, flat, we conclude from equations (153) and (154) that

$$\Phi_{22} = 0 \quad \text{and} \quad \Psi_4 = -\tfrac{1}{2}(p-\mathrm{i}q)\,\mathrm{e}^{-f(0,0)} f_{,v}(0,0). \tag{169}$$

Therefore, along $v = 0$ $(u < 0)$, Φ_{22} vanishes and is continuous while Ψ_4 experiences an H-function discontinuity of an amount specified in equation (169). Similarly, along $u = 0$ $(v < 0)$ (separating regions III and IV) Φ_{00} vanishes and Ψ_0 experiences an H-function discontinuity of amount

$$\Psi_0 = -\tfrac{1}{2}(p-\mathrm{i}q)\,\mathrm{e}^{-f(0,0)} f_{,u}(0,0). \tag{170}$$

13. Concluding remarks

In this paper, we have considered the interaction of gravitational waves and sound waves in a perfect fluid with the equation of state $\epsilon = p$, in a space–time that admits two space-like Killing vectors. We show that Einstein's field-equations, under these circumstances and for the equation of state considered, permit the same choice of gauge and coordinates that has been used earlier in considering the collision of impulsive gravitational waves. The coordinates η and μ chosen limit the region of the space–time, in which the gravitational waves and the sound waves scatter off each other, to a finite duration $(0 \leqslant \eta \leqslant 1)$ and to a finite spatial extent $(0 \leqslant \mu \leqslant 1)$ (normal to the plane of stratification). With the coordinates, so chosen, the coupled Einstein hydrodynamic-equations enable us to reduce the metric (7) to the particular form

$$ds^2 = e^{\mu_3+\nu+f}(1-\eta^2)^{\frac{1}{2}}\left[\frac{(d\eta)^2}{1-\eta^2}-\frac{(d\mu)^2}{1-\mu^2}\right]$$
$$-(1-\eta^2)^{\frac{1}{2}}(1-\mu^2)^{\frac{1}{2}}\left[\chi(dx^2)^2+\frac{1}{\chi}(dx^1-q_2\,dx^2)^2\right], \quad (171)$$

where χ, q_2, and $\mu_3+\nu$ represent solutions of the vacuum equations and f is the only function that derives from the presence of the fluid.

The fluid motions, themselves, can be derived from a potential, ϕ, which in a suitable system of null coordinates, u and v (derived from η and μ), is governed by the equation

$$\phi_{,u,v}+\frac{v\phi_{,u}+u\phi_{,v}}{1-u^2-v^2}=0. \quad (172)$$

The function f is determined, in terms of a solution of equation (172), by

$$f_{,u}=\frac{2(\phi_{,u})^2}{u(1-u^2-v^2)} \quad \text{and} \quad f_{,v}=\frac{2(\phi_{,v})^2}{v(1-u^2-v^2)}. \quad (173)$$

The general solution of equations (172) and (173), which ensures only the positive-definiteness of ϵ, is obtained in explicit forms in §§7 and 8.

The problem of solving the coupled Einstein hydrodynamic-equations is thus reduced to selecting *any* desired solution of the vacuum equations, for χ, q_2, and $\mu_3+\nu$, and supplementing it with the general solution for f we have found.

The explicit form of the metric (98) derived in §9, is limited to the domain

$$0\leqslant u\leqslant 1, \quad 0\leqslant v\leqslant 1 \quad \text{and} \quad u^2+v^2\leqslant 1. \quad (174)$$

On $u^2+v^2=1$, the space–time develops a curvature singularity. The nature of this singularity (exhibited by the contours of constant ϵ in figures 1 and 2) is strongly affected by the amplitudes and the superpositions of the sound waves that are present.

The extension of the space–time described by the metric (98) to regions II and III provides perhaps the most unexpected element of the analysis: the transformation of massless particles describing null trajectories into a perfect fluid in which the stream lines describe time-like trajectories. It is remarkable that such a transformation is possible in the framework of the general theory of relativity; it adds to the already abundant richness of the theory.

There is a further point to be noted. In discussions relating to the equation of state that may prevail at ultra-high densities there are two views: an 'ancient view' that it is given by $p=\frac{1}{3}\epsilon$ (attributed to J. von Neumann; see, for example, Chandrasekhar (1935, p. 693)) and a more modern view that it is given by $p=\epsilon$ (attributed to Zeldovich 1962). The prevailing view appears to be non-committal. But the fact that a perfect fluid with the equation of state $p=\epsilon$ emerges as the result of an induced transformation of pure radiation gives credence to this equation of state.

Besides the theoretical insights provided by the example considered, it may have wider implications. Thus, the distribution of ϵ that develops, particularly, the occurrence of density maxima in the 'interior', away from the boundaries, when acoustic waves of sufficient amplitudes are present, is highly suggestive. It is not unreasonable to expect that the phenomena that are exhibited in figures 1 and 2 may be retained when the problem is reconsidered under more 'realistic' conditions. If this expectation is confirmed, then the density fluctuations in ϵ that arise under the idealized circumstances considered in this paper may have relevance to the physics of the early universe when those circumstances may have been approximated. Also, the fact that prior to the instant of collision one has pure radiation adds a further intriguing element.

We are grateful to Professor R. Geroch, Dr L. Lindblom, and Mr C. Cutler for many helpful discussions. We are also most indebted to Professor Roger Penrose, F.R.S., for convincing us that the extension of the solution in region I into regions II and III, on the assumption that these regions are filled with pure radiation, is physically reasonable. The research reported in this paper has, in part, been supported by grants from the National Science Foundation under grant PHY 84-16691 with the University of Chicago.

References

Chandrasekhar, S. 1935 *Mon. Not. R. astr. Soc.* **95**, 678.
Chandrasekhar, S. 1983 *The mathematical theory of black holes*. Oxford: Clarendon Press.
Chandrasekhar, S. & Ferrari, V. 1984 *Proc. R. Soc. Lond.* A **396**, 55.
Chandrasekhar, S. & Ferrari, V. 1985 *Proc. R. Soc. Lond.* A **398**, 429. (Errata to Chandrasekhar & Ferrari 1984.)
Chandrasekhar, S. & Xanthopoulos, B. C. 1985 *Proc. R. Soc. Lond.* A **398**, 223.
Khan, K. & Penrose, R. 1971 *Nature, Lond.* **229**, 185.
Kramer, D., Stephani, H., Herlt, E. & MacCallum, M. 1980 *Exact solutions of Einstein's field equations*. Cambridge University Press.
Nutku, Y. & Halil, M. 1977 *Phys. Rev. Lett.* **39**, 1379.
Watson, G. N. 1944 *A treatise on the theory of Bessel functions*. Cambridge University Press.
Zel'dovich, Ya. B. 1962 *Soviet Phys. JETP* **14**, 1143.

Appendix A

In this appendix, we shall show that if a solution ϕ of equation (60) is required to take a constant value along the null boundaries $u = 0$ $(0 \leqslant v \leqslant 1)$ and $v = 0$ $(0 \leqslant u \leqslant 1)$, then it is identically a constant in the entire region I.

It is convenient to introduce the variables

$$x = 1-2u^2 \quad \text{and} \quad y = 1-2v^2, \tag{A 1}$$

whence equation (60) takes the form

$$\phi_{,x,y} - \frac{1}{2}\frac{\phi_{,x}+\phi_{,y}}{x+y} = 0; \tag{A 2}$$

and region I in the (u, v)-plane is mapped onto the triangle CAB in the (x, y)-plane

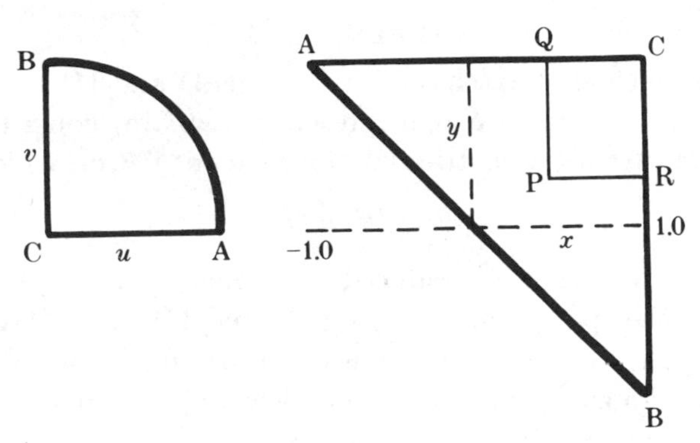

FIGURE A 1. The mapping of region I in the (u, v)-plane onto the (x, y)-plane.

(see figure A 1). We seek a solution of equation (A 2) with the boundary conditions that ϕ is a constant along AC and CB.

Let $P = (x_0, y_0)$ be any point in the interior of the triangle CAB. Through P draw the characteristics PQ (x = constant = x_0) and PR (y = constant = y_0). Multiply equation (A 2) by $(\phi_{,x}+\phi_{,y})$ and integrate over the rectangular domain PQCR. We obtain

$$\int_{x_0}^{1}\int_{y_0}^{1}(\phi_{,x}+\phi_{,y})\,\phi_{,x,y}\,\mathrm{d}x\,\mathrm{d}y-\frac{1}{2}\int_{x_0}^{1}\int_{y_0}^{1}\frac{(\phi_{,x}+\phi_{,y})^2}{x+y}\,\mathrm{d}x\,\mathrm{d}y=0. \tag{A 3}$$

Rewriting the first of the two integrals on the left-hand side in the form

$$\frac{1}{2}\int_{x_0}^{1}\int_{y_0}^{1}\{[(\phi_{,x})^2]_{,y}+[(\phi_{,y})^2]_{,x}\}\,\mathrm{d}x\,\mathrm{d}y, \tag{A 4}$$

we can effect one of the two integrations; and we obtain

$$\frac{1}{2}\int_{x_0}^{1}(\phi_{,x})^2\Big|_{y_0}^{1}\,\mathrm{d}x+\frac{1}{2}\int_{y_0}^{1}(\phi_{,y})^2\Big|_{x_0}^{1}\,\mathrm{d}y. \tag{A 5}$$

Since, by assumption,

$$\phi_{,x}(x,1)=0 \quad\text{and}\quad \phi_{,y}(1,y)=0, \tag{A 6}$$

equation (A 3) reduces to

$$\frac{1}{2}\int_{x_0}^{1}[\phi_{,x}(x,y_0)]^2\,\mathrm{d}x+\frac{1}{2}\int_{y_0}^{1}[\phi_{,y}(x_0,y)]^2\,\mathrm{d}y+\frac{1}{2}\int_{x_0}^{1}\int_{y_0}^{1}\frac{(\phi_{,x}+\phi_{,y})^2}{x+y}\,\mathrm{d}x\,\mathrm{d}y=0. \tag{A 7}$$

Since all three integrands in equation (A 7) are positive definite, it follows that

$$\left.\begin{aligned}\phi_{,x}(x,y_0)&=0, \quad x_0\leqslant x\leqslant 1,\\ \phi_{,y}(x_0,y)&=0, \quad y_0\leqslant y<1.\end{aligned}\right\} \tag{A 8}$$

Since (x_0, y_0) is any point in the interior of the triangle CAB, the constancy of ϕ over the entire domain CAB is proved.

APPENDIX B

An alternative method of extension into regions II and III that was considered (and abandoned for reasons we shall presently explain) consists in making the substitutions (124) in all the metric functions except $f(u, v)$, which is replaced by

$$f(u, v)\, H(u)\, H(v). \tag{B 1}$$

By this manner of extension, f is reduced to zero, as we cross the null boundaries (123), by fiat and the space–time in regions II and III made flat. But the metric will be discontinuous at the null boundaries (123) and cease to be C^0. However, as C. Cutler has shown (personal communication) the discontinuities can be eliminated by the coordinate transformation

$$\left.\begin{aligned} \tilde{u}(u, v) &= \int_0^u \exp\left[H(u')f(u', vH(v))\right] \mathrm{d}u', \\ \tilde{v}(u, v) &= \int_0^v \exp\left[H(v')f(uH(u), v')\right] \mathrm{d}v', \end{aligned}\right\} \tag{B 2}$$

and the metric made C^0 in the $(\tilde{u}, \tilde{v})$-variables. The discontinuity of the metric along the null boundaries is therefore, only, a coordinate singularity. This inference is in accord with the behaviour of the Ricci and the Weyl scalars that results from the substitutions (124) and (B 1). At $u = 0$, we find, for example:

$$\tilde{\Psi}_0 = \mathrm{e}^{-f(0, v)} \left[\frac{p - \mathrm{i}q}{2(1-v^2)}\,(1 - f(0, v))\,\delta(u) + \tfrac{3}{2}(p - \mathrm{i}q)^3 \frac{v}{(1-v^2)^{\frac{3}{2}}} H(u)\right],$$

$$\tilde{\Psi}_4 = -\frac{p - \mathrm{i}q}{2\sqrt{(1-v^2)}}\, \mathrm{e}^{-f(0, v)} f_{,v}(0, v)\, H(u),$$

$$\tilde{\Psi}_2 = +\mathrm{e}^{-f(0, v)} \left[\frac{(p - \mathrm{i}q)^2}{2(1-v^2)^{\frac{3}{2}}} H(u) - \tfrac{1}{12} f_{,v}(0, v)\, \delta(u)\right],$$

$$\tilde{\Phi}_{22} = -\frac{v}{2\sqrt{(1-v^2)}}\, \mathrm{e}^{-f(0, v)} f_{,v}(0, v)\, H(u), \quad \tilde{\Phi}_{00} = 0,$$

$$\tilde{\Lambda} = \tfrac{1}{12} R_{12} = \tfrac{1}{24} R = \tfrac{1}{24}\, \mathrm{e}^{-f(0, v)} f_{,v}(0, v)\, \delta(u). \tag{B 3}$$

We observe that none of the scalars exhibit a singularity worse than a δ-function. We conclude that the proposed manner of extension encounters no mathematical inconsistency.

In spite of the freedom from mathematical inconsistency, the resulting solution has a serious flaw: it derives from the fact that the δ-function singularity (expressing the impulsive character of the sound waves) appears only in R and in R_{12} $(= \frac{1}{2}R)$, a fact that implies that $T_{11} = T_{22} = 0$ and the δ-function singularity occurs only in the spatial components T_{33} and T_{44}. In other words, only the spatial components of the energy–momentum tensor have an impulsive character; the temporal components do not. The positive-energy condition is, therefore, violated on the null boundaries; and the proposed manner of extension is untenable.

On the collision of impulsive gravitational waves when coupled with null dust

BY SUBRAHMANYAN CHANDRASEKHAR[1], F.R.S., AND BASILIS C. XANTHOPOULOS[2]

[1] *The University of Chicago, Chicago, Illinois* 60637, *U.S.A.*
[2] *Department of Physics, University of Crete, and Research Centre of Crete, Iraklion, Greece*

(*Received* 12 *August* 1985)

The problem of colliding impulsive gravitational waves is considered when the region of space–time, after the instant of collision, is filled with a mixture of null dusts moving in opposite directions. The extension of the resulting space–time, to regions before the instant of collision, shows that null dust follows the leading edges of the impulsive waves, and, further, that one can arrange that the space–time in these regions is identical with what prevails when a perfect fluid with $\epsilon = p$ fills the region after the instant of collision. This ambiguity in the space–time, after the instant of collision, must be traced to an inherent ambiguity in the nature of null dust and its relation with a perfect fluid with $\epsilon = p$.

1. INTRODUCTION

In an earlier paper (Chandrasekhar & Xanthopoulos 1985; this paper will be referred to hereafter as Paper I), it was shown how the presence of a perfect fluid with an equation of state ϵ (the energy density) $= p$ (the pressure), in regions of space–time after the instant of collision of two impulsive gravitational waves, implies that it is the result of a transformation of null dust following the leading edges of the impulsive waves. Two aspects of this transformation have been clarified since Paper I was written. First, it has been drawn to our attention independently by Professor Roger Penrose and by Dr L. Lindblom (private communications) that the transformation in question can take place in flat space–time without the intervention of gravity; and second, that it is related to an inherent ambiguity in the notion of a null dust that is concealed in the expression for its energy–momentum tensor, namely,

$$T^{ij} = Ek^i k^j, \tag{1}$$

where E is some positive scalar function and $\boldsymbol{k}$ is a null vector. Because $\boldsymbol{k}$, as a null vector, is undetermined to the extent of a multiplicative factor, the magnitude of E is undetermined without some normalization convention (as in the Newman–Penrose formalism). Even after such a normalization of $\boldsymbol{k}$, the expression (1) for T^{ij} does not specify the substance uniquely. In this sense, null dust is, as Robert

Geroch has aptly described, 'secretive'; but, in reality, it is no more 'secretive' than a perfect fluid is by the specification of *its* energy–momentum tensor,

$$T^{ij} = (\epsilon+p)\, u^i u^j - p g^{ij}, \tag{2}$$

without an equation of state relating ϵ and p.

It will be recalled that in region I, in the context of the problem considered in Paper I (see figure 4), the energy density, ϵ, vanishes on the null boundaries ($u = 0$ and $v = 0$), the four-velocity, u^i, becomes null, and the energy–momentum tensor reduces to the form (1). The fluid, accordingly, acquires the character of a 'null dust' at the null boundaries. Because the Ricci tensor (now equal to the energy–momentum tensor) is continuous across the null boundaries, it is reasonable to suppose that it is this same limiting form of the perfect fluid with $\epsilon = p$ that prevails in regions II and III. However, it would appear that once one is in region II (or III), there is no way in which this fact can be discerned or ascertained. For an observer in region II (or III), the transition to a perfect fluid with $\epsilon = p$, at the instant of collision, when and if it occurs, must appear as a *phase transition*. It would clearly be useful in this connection to re-examine the problem considered in Paper I, as Dr Lindblom suggested to us, when null dust by assumption prevails in region I. It is to this problem that the present paper is addressed. The solution turns out to be surprisingly simple; and it sheds some light on the nature of null dust and its relation to a perfect fluid with $\epsilon = p$.

2. The characterization of null dust and the equations of motion in a tetrad frame

In this study, as in our earlier studies on the collision of impulsive gravitational waves, we shall restrict ourselves to space–times with two space-like Killing vectors ($\partial/\partial x^1$ and $\partial/\partial x^2$); and we shall assume for the metric the same form, namely,

$$ds^2 = e^{2\nu}(dx^0)^2 - e^{2\mu_3}(dx^3)^2 - e^{2\psi}(dx^1 - q_2\, dx^2)^2 - e^{2\mu_2}(dx^2)^2, \tag{3}$$

where ν, μ_3, ψ, q_2 and μ_2 are functions of x^0 and x^3 only. And we shall write the various equations in a tetrad frame with the basis vectors,

$$\omega^{(0)} = e^{\nu}\, dx^0, \quad \omega^{(1)} = e^{\psi}(dx^1 - q_2\, dx^2), \quad \omega^{(2)} = e^{\mu_2}\, dx^2,$$
$$\text{and} \quad \omega^{(3)} = e^{\mu_3}\, dx^3. \tag{4}$$

To distinguish the tetrad components from the tensor components (used in §1), we shall (in this section only) enclose the tetrad indices in parentheses.

The energy–momentum tensor for null dust, in a tetrad frame, has the form

$$T^{(a)(b)} = E k^{(a)} k^{(b)}. \tag{5}$$

Under the present conditions, the non-vanishing components of $k^{(a)}$ are

$$k^{(0)} = k_{(0)} \quad \text{and} \quad k^{(3)} = -k_{(3)}, \tag{6}$$

with the requirement

$$(k_{(0)})^2 - (k_{(3)})^2 = 0. \tag{7}$$

Accordingly, we must distinguish the two cases,

$$k_{(0)} = +k_{(3)} \quad \text{and} \quad k_{(0)} = -k_{(3)}, \tag{8}$$

and consider the null dust as composed of two constituents distinguished by $k_{(0)} = +k_{(3)}$ and $k_{(0)} = -k_{(3)}$. Consistently with this distinction, we shall write the energy–momentum tensor as a sum of two terms in the form

$$T^{(a)(b)} = 2(\epsilon k^{(a)} k^{(b)} + \bar{\epsilon} \bar{k}^{(a)} \bar{k}^{(b)}), \tag{9}$$

where

$$k_{(0)} = +k_{(3)} = k \quad \text{and} \quad \bar{k}_{(0)} = -\bar{k}_{(3)} = \bar{k}. \tag{10}$$

The corresponding non-vanishing components of the Ricci tensor (as required by Einstein's field equations) are

$$R_{(0)(0)} = R_{(3)(3)} = -4(\epsilon k^2 + \bar{\epsilon}\bar{k}^2) \tag{11}$$

and

$$R_{(0)(3)} = -4(\epsilon k^2 - \bar{\epsilon}\bar{k}^2); \tag{12}$$

and the scalar curvature is, of course, zero.

The required equations of motion now follow from

$$T^{(a)(b)}{}_{|(b)} = 2(\epsilon k^{(a)} k^{(b)} + \bar{\epsilon}\bar{k}^{(a)}\bar{k}^{(b)})_{|(b)} = 0. \tag{13}$$

In evaluating equation (13) appropriately for this problem, we shall, for the sake of simplicity, consider only the first of the two terms included in the expression for $T^{(a)(b)}$, but distinguish the two cases (8). We have,

$$(\epsilon k^{(a)} k^{(b)})_{|(b)} = \epsilon k^{(a)} k^{(b)}{}_{|(b)} + k^{(b)} (\epsilon k^{(a)})_{|(b)}. \tag{14}$$

On evaluating the terms on the right-hand side of equation (14) with the aid of the Ricci rotation coefficients listed in Chandrasekhar (1983, p. 82), we find

$$k^{(b)}{}_{|(b)} = e^{-\nu}[k_{,0} + (\psi + \mu_2 + \mu_3)_{,0} k] \mp e^{-\mu_3}[k_{,3} + (\psi + \mu_2 + \nu)_{,3} k], \tag{15}$$

where the upper sign applies to the case, $k_{(0)} = +k_{(3)} = k$, and the lower sign to the case, $k_{(0)} = -k_{(3)} = k$. Similarly, we find,

$$\begin{aligned} k^{(b)}(\epsilon k^{(0)})_{|(b)} = {} & e^{-\nu}[k(\epsilon k)_{,0} + \mu_{3,0}(\epsilon k^2)] \\ & \mp e^{-\mu_3}[k(\epsilon k)_{,3} + \nu_{,3}(\epsilon k^2)], \end{aligned} \tag{16}$$

and

$$\begin{aligned} k^{(b)}(\epsilon k^{(3)})_{|(b)} = {} & \mp e^{-\nu}[k(\epsilon k)_{,0} + \mu_{3,0}(\epsilon k^2)] \\ & + e^{-\mu_3}[k(\epsilon k)_{,3} + \nu_{,3}(\epsilon k^2)]. \end{aligned} \tag{17}$$

Combining equations (14)–(17) appropriately, we find that equation (13) leads to the pair of equations,

$$\begin{aligned} & e^{-\nu}[(\epsilon k^2 + \bar{\epsilon}\bar{k}^2)_{,0} + (\psi + \mu_2 + 2\mu_3)_{,0}(\epsilon k^2 + \bar{\epsilon}\bar{k}^2)] \\ & \qquad - e^{-\mu_3}[(\epsilon k^2 - \bar{\epsilon}\bar{k}^2)_{,3} + (\psi + \mu_2 + 2\nu)_{,3}(\epsilon k^2 - \bar{\epsilon}\bar{k}^2)] = 0 \end{aligned} \tag{18}$$

and

$$e^{-\nu}[(\epsilon k^2-\bar{\epsilon}\bar{k}^2)_{,0}+(\psi+\mu_2+2\mu_3)_{,0}(\epsilon k^2-\bar{\epsilon}\bar{k}^2)]$$
$$-e^{-\mu_3}[(\epsilon k^2+\bar{\epsilon}\bar{k}^2)_{,3}+(\psi+\mu_2+2\nu)_{,3}(\epsilon k^2+\bar{\epsilon}\bar{k}^2)]=0; \tag{19}$$

or, equivalently,

$$e^{\nu-\mu_3}[e^{\psi+\mu_2+2\mu_3}(\epsilon k^2+\bar{\epsilon}\bar{k}^2)]_{,0}-[e^{\psi+\mu_2+2\nu}(\epsilon k^2-\bar{\epsilon}\bar{k}^2)]_{,3}=0 \tag{20}$$

and

$$e^{\nu-\mu_3}[e^{\psi+\mu_2+2\mu_3}(\epsilon k^2-\bar{\epsilon}\bar{k}^2)]_{,0}-[e^{\psi+\mu_2+2\nu}(\epsilon k^2+\bar{\epsilon}\bar{k}^2)]_{,3}=0. \tag{21}$$

3. The choice of gauge and of coordinates

Because R_{11} and R_{22} both vanish for null dust (even as they do for a perfect fluid with $\epsilon=p$), it is clear that we can reduce the field equations in the same manner as in Paper I (see particularly §§5 and 13). Thus, we can make the same choice of gauge and of coordinates, as for the vacuum, namely,

$$e^{2(\mu_3-\nu)}=\Delta=1-\eta^2 \tag{22}$$

and

$$e^{2\beta}=e^{2(\psi+\mu_2)}=(1-\eta^2)(1-\mu^2)=\Delta\delta, \tag{23}$$

where μ ($=\cos x^3$) is a measure of the distance between the plane wave-fronts and η is a measure of the time from the instant of the collision. Further, with the metric written in the form

$$ds^2=e^{\mu_3+\nu}\sqrt{\Delta}\left[\frac{(d\eta)^2}{\Delta}-\frac{(d\mu)^2}{\delta}\right]-\sqrt{(\Delta\delta)}\left[\chi(dx^2)^2+\frac{1}{\chi}(dx^1-q_2\,dx^2)^2\right], \tag{24}$$

we may choose for the functions χ and q_2 any solution appropriate to the vacuum. The equations governing $(\nu+\mu_3)$ involve the sources explicitly through R_{03} and $(R_{00}+R_{33})$. We shall consider these equations in §6 below. Meanwhile, it is convenient to define other pairs of coordinates (besides η and μ) which we shall find useful:

$$\left.\begin{aligned}&\psi=\arccos\eta,\quad \theta=\arccos\mu;\\ &\xi=\tfrac{1}{2}(\theta+\psi),\quad \zeta=(\theta-\psi);\\ \text{and}\quad &u=\cos\xi,\quad v=\sin\zeta.\end{aligned}\right\} \tag{25}$$

The last pair of variables, (u, v), define the null coordinates. The relevant transformation equations for passing from one pair of variables to another will be found in Paper I (§6, equations (57) and (59)).

4. The reduction and the solution of the equations of motion

Returning to equations (20) and (21), substituting from equations (22) and (23) and changing to the variables η and μ, we find that they reduce to the forms

$$[(\epsilon k^2+\bar{\epsilon}\bar{k}^2)e^{2\mu_3}\sqrt{\Delta}]_{,\eta}+[(\epsilon k^2-\bar{\epsilon}\bar{k}^2)e^{2\mu_3}\sqrt{\delta}]_{,\mu}=0 \tag{26}$$

and

$$[(\epsilon k^2-\bar{\epsilon}\bar{k}^2)e^{2\mu_3}\sqrt{\Delta}]_{,\eta}+[(\epsilon k^2+\bar{\epsilon}\bar{k}^2)e^{2\mu_3}\sqrt{\delta}]_{,\mu}=0. \tag{27}$$

With the definitions,

$$(\epsilon k^2+\bar{\epsilon}\bar{k}^2)\,\mathrm{e}^{2\mu_3}=+\frac{\phi_{,\mu}}{\sqrt{\Delta}} \tag{28}$$

and

$$(\epsilon k^2-\bar{\epsilon}\bar{k}^2)\,\mathrm{e}^{2\mu_3}=-\frac{\phi_{,\eta}}{\sqrt{\delta}}, \tag{29}$$

equation (26) is identically satisfied and equation (27) leads to the equation,

$$(\phi_{,\eta}\sqrt{\Delta})_{,\eta}\sqrt{\Delta}-(\phi_{,\mu}\sqrt{\delta})_{,\mu}\sqrt{\delta}=0. \tag{30}$$

Making use of the transformation equations (57) and (59) of Paper I, we find that in terms of the variables, (θ,ψ), (ξ,ζ), and (u,v), equation (30) becomes

$$\phi_{,\psi,\psi}-\phi_{,\theta,\theta}=0,\quad \phi_{,\xi,\zeta}=0\quad\text{and}\quad \phi_{,u,v}=0. \tag{31}$$

We can therefore write the general solution for ϕ in the form

$$\phi(u,v)=\Phi(v)+\bar{\Phi}(u), \tag{32}$$

where Φ and $\bar{\Phi}$ are functions of the arguments specified, arbitrary and unrestricted for the present.

5. The completion of the solution

It remains to determine $(\nu+\mu_3)$ to complete the solution of the field equations. As in Paper I, §4(*b*), we shall write

$$\nu+\mu_3=(\nu+\mu_3)_{\text{vac}}+f, \tag{33}$$

where $(\nu+\mu_3)_{\text{vac}}$ is a solution of the vacuum equations that belongs with χ and q_2. The equations governing f (similar to I, equations (46) and (47)) are

$$\frac{\mu}{\delta}f_{,\eta}+\frac{\eta}{\Delta}f_{,\mu}=\frac{2R_{03}}{\sqrt{\delta}}\,\mathrm{e}^{\nu+\mu_3} \tag{34}$$

and

$$\eta f_{,\eta}+\mu f_{,\mu}=-(R_{00}+R_{33})\,\mathrm{e}^{2\mu_3}. \tag{35}$$

Now, substituting for R_{03} and $(R_{00}+R_{33})$ in accordance with equations (11), (12), (28), and (29), we obtain

$$\frac{\mu}{\delta}f_{,\eta}+\frac{\eta}{\Delta}f_{,\mu}=8\frac{\phi_{,\eta}}{\delta\sqrt{\Delta}}, \tag{36}$$

and

$$\eta f_{,\eta}+\mu f_{,\mu}=8\frac{\phi_{,\mu}}{\sqrt{\Delta}}. \tag{37}$$

In the future, we shall dispense with the distinguishing subscript 'vac' for μ_3 and ν and retain for them (as we already have for β, χ, and q_2) the meaning that they refer to solutions appropriate to the vacuum, so that for the problem on hand,

$$\mu_3+\tfrac{1}{2}f\quad\text{and}\quad\nu+\tfrac{1}{2}f \tag{38}$$

replace what we have hitherto denoted by μ_3 and ν.

Returning to equations (36) and (37), we find that in the (ξ, ζ) variables they reduce to

$$(f_{,\xi}-f_{,\zeta})\mu\sqrt{\Delta}+(f_{,\xi}+f_{,\zeta})\eta\sqrt{\delta} = 8(\phi_{,\xi}-\phi_{,\zeta}) \tag{39}$$

and

$$(f_{,\xi}-f_{,\zeta})\eta\sqrt{\delta}+(f_{,\xi}+f_{,\zeta})\mu\sqrt{\Delta} = 8(\phi_{,\xi}+\phi_{,\zeta}), \tag{40}$$

where it may be noted that

$$\mu\sqrt{\Delta} = \sin(\xi-\zeta)\cos(\xi+\zeta) \quad \text{and} \quad \eta\sqrt{\delta} = \sin(\xi+\zeta)\cos(\xi-\zeta). \tag{41}$$

Now solving equations (39) and (40) for $f_{,\xi}$ and $f_{,\zeta}$, we find

$$f_{,\xi} = 8\frac{\phi_{,\xi}}{\sin 2\xi} \quad \text{and} \quad f_{,\zeta} = -8\frac{\phi_{,\zeta}}{\sin 2\zeta}; \tag{42}$$

or, in terms of the null variables u and v,

$$f_{,u} = 4\frac{\overline{\Phi}_{,u}}{u\sqrt{(1-u^2)}} \quad \text{and} \quad f_{,v} = -4\frac{\Phi_{,v}}{v\sqrt{(1-v^2)}}, \tag{43}$$

where we have substituted for ϕ its solution (32). Therefore f, like ϕ, is a sum of a function of u only and a function of v only; and we may write

$$f(u,v) = \overline{F}(u)+F(v). \tag{44}$$

Because we can impose on f conditions of continuity normally required of metric functions, we can invert equations (43) to determine $\overline{\Phi}(u)$ and $\Phi(v)$ in terms of $\overline{F}(u)$ and $F(v)$; thus,

$$\left.\begin{aligned} \overline{\Phi}(u) &= +\frac{1}{4}\int^u \mathrm{d}u\, \overline{F}_{,u}\, u\sqrt{(1-u^2)}, \\ \text{and} \qquad \Phi(v) &= -\frac{1}{4}\int^v \mathrm{d}v\, F_{,v}\, v\sqrt{(1-v^2)}. \end{aligned}\right\} \tag{45}$$

(a) The form of the metric

The solutions for ϕ and f which follow from equations (32), (44), and (45), can be associated with any solution of the vacuum equations for the metric functions χ, q_2, and $\nu+\mu_3$. For our purposes, it will suffice to choose the solutions for χ, q_2, and $\nu+\mu_3$ that lead to the Nutku–Halil solution for colliding impulsive gravitational waves. With this choice, the metric takes the form (cf. I, equations (98) and (99))

$$\mathrm{d}s^2 = \frac{\Delta \exp[\overline{F}(u)+F(v)]}{(\sin\psi\sin\theta)^{\frac{1}{2}}}[(\mathrm{d}\psi)^2-(\mathrm{d}\theta)^2] - \frac{\sin\psi\sin\theta}{\Delta}|(1-E)\,\mathrm{d}x^1+\mathrm{i}(1+E)\,\mathrm{d}x^2|^2, \tag{46}$$

where

$$\left.\begin{aligned} E &= p\eta+\mathrm{i}q\mu = p\cos\psi+\mathrm{i}q\cos\theta \quad (p^2+q^2=1), \\ \text{and} \qquad \Delta &= 1-|E|^2 = p^2\sin^2\psi+q^2\sin^2\theta = 1-p^2\eta^2-q^2\mu^2. \end{aligned}\right\} \tag{47}$$

(*b*) *The description of the space–time in a Newman–Penrose formalism*

In view of the formal identity of the metric (46) with the one considered in Paper I (equation (98)), it is clear that we can relate (as in Paper I, equations (101)–(104)) the spin coefficients and the Weyl and the Ricci scalars (distinguished by tildes) with those (without tildes) that obtain for the vacuum (listed in Chandrasekhar & Ferrari (1984), equations (86) and (93)–(95); for some misprints in these equations see Chandrasekhar & Ferrari 1985). However, because in the present context (cf. equations (31)),

$$f_{,\theta,\theta}-f_{,\psi,\psi}=0, \tag{48}$$

some of the expressions simplify while others vanish. Thus the non-vanishing Weyl and Ricci scalars are now given by

$$\left.\begin{aligned}&\tilde{\Psi}_0=e^{-f}(\Psi_0-\sigma D\bar{F}),\quad \tilde{\Psi}_4=e^{-f}(\Psi_4+\lambda \varDelta F),\\ &\tilde{\Psi}_2=e^{-f}\,\Psi_2, \tilde{\Phi}_{00}=-\rho\,e^{-f}D\bar{F},\quad\text{and}\quad \tilde{\Phi}_{22}=+\mu\,e^{-f}\varDelta F,\end{aligned}\right\} \tag{49}$$

where

$$\left.\begin{aligned}D\bar{F}(u)&=-\frac{(\sin\psi\sin\theta)^{\frac{1}{4}}}{(2\Delta)^{\frac{1}{2}}}\bar{F}_{,u}\surd(1-u^2),\\ \text{and}\qquad \varDelta F(v)&=-\frac{(\sin\psi\sin\theta)^{\frac{1}{4}}}{(2\Delta)^{\frac{1}{2}}}F_{,v}\surd(1-v^2).\end{aligned}\right\} \tag{50}$$

And, explicitly, we find (cf. Paper I, equations (159), (162))

$$\left.\begin{aligned}\tilde{\Phi}_{00}&=-\frac{(\sin\psi\sin\theta)^{\frac{1}{2}}}{4\Delta}(\cot\psi+\cot\theta)\,e^{-f}\bar{F}_{,u}\surd(1-u^2),\\ \tilde{\Phi}_{22}&=-\frac{(\sin\psi\sin\theta)^{\frac{1}{2}}}{4\Delta}(\cot\psi-\cot\theta)\,e^{-f}F_{,v}\surd(1-v^2),\\ \tilde{\Psi}_0&=e^{-f}\Psi_0-\frac{(\sin\psi\sin\theta)^{\frac{1}{2}}}{2\Delta^2}(p\sin\psi-iq\sin\theta)\,e^{-f}\bar{F}_{,u}\surd(1-u^2),\\ \tilde{\Psi}_4&=e^{-f}\Psi_4-\frac{(\sin\psi\sin\theta)^{\frac{1}{2}}}{2\Delta^2}(p\sin\psi-iq\sin\theta)\,e^{-f}F_{,v}\surd(1-v^2),\end{aligned}\right\} \tag{51}$$

where it may be recalled that $f=\bar{F}(u)+F(v)$.

It should be noted that if we should choose $\bar{F}(u)=0$, then $\tilde{\Phi}_{00}=0$; and similarly, $\tilde{\Phi}_{22}=0$ if we should choose $F(v)=0$.

6. The nature of the space–time singularity at $u^2+v^2=1$ and the distribution of null dust in region I

Earlier studies on the collision of impulsive gravitational waves have shown that a space–time singularity develops at $u^2+v^2=1$ by a focusing of the waves when they scatter off each other. The coupling of the gravitational waves with null dust does not prevent the development of such a singularity, though its character can be altered. Thus, the sets of behaviour (cf. Paper I, equations (121))

$$\left.\begin{aligned} \Psi_2 \sim -\frac{1}{8q^2(1-u^2-v^2)^{\frac{3}{2}}} \quad (q \neq 0) \quad \text{and} \quad \Psi_2 \sim \frac{6u^4v^4}{(1-u^2-v^2)^{\frac{7}{2}}} \quad (q=0) \\ (u^2+v^2 \to 1-0), \end{aligned}\right\} \tag{52}$$

of the Weyl scalar Ψ_2 in the vacuum are modified by a factor e^{-f} by the presence of the null dust; and because $f = \bar{F}(u)+F(v)$, we can affect the nature of the singularity to any extent we may desire.

Turning next to the distribution of the null dust, we first solve for ϵk^2 and $\bar{\epsilon}\bar{k}^2$ from equations (28) and (29) and find

$$\epsilon k^2 = \frac{\phi_{,\mu}\sqrt{\delta}-\phi_{,\eta}\sqrt{\Delta}}{2\sqrt{(\Delta\delta)}} e^{-2\mu_3-f} \tag{53}$$

and

$$\bar{\epsilon}\bar{k}^2 = \frac{\phi_{,\mu}\sqrt{\delta}+\phi_{,\eta}\sqrt{\Delta}}{2\sqrt{(\Delta\delta)}} e^{-2\mu_3-f}, \tag{54}$$

where we have replaced $2\mu_3$ by $2\mu_3+f$ in accordance with our present understanding (38). Now making use of the transformation equations (54), (57), and (59) of Paper I, we can rewrite the expressions for ϵk^2 and $\bar{\epsilon}\bar{k}^2$ as

$$\epsilon k^2 = -\frac{\phi_{,\zeta}}{2\sqrt{(\Delta\delta)}} e^{-2\mu_3-f} = -\frac{\sqrt{(1-v^2)}}{2(1-u^2-v^2)} e^{-2\mu_3-f}\Phi_{,v}, \tag{55}$$

and

$$\bar{\epsilon}\bar{k}^2 = -\frac{\phi_{,\xi}}{2\sqrt{(\Delta\delta)}} e^{-2\mu_3-f} = +\frac{\sqrt{(1-u^2)}}{2(1-u^2-v^2)} e^{-2\mu_3-f}\bar{\Phi}_{,u}. \tag{56}$$

Now, substituting for $e^{-2\mu_3}$ its expression (given in I, equation (95)),

$$e^{-2\mu_3} = \frac{1}{\mathbf{\Delta}}(1-u^2-v^2)^{\frac{1}{2}}, \tag{57}$$

we obtain

$$\epsilon k^2 = -\frac{1}{2}\left(\frac{1-v^2}{1-u^2-v^2}\right)^{\frac{1}{2}} \frac{e^{-f}}{\mathbf{\Delta}} \Phi_{,v} \tag{58}$$

and

$$\bar{\epsilon}\bar{k}^2 = +\frac{1}{2}\left(\frac{1-u^2}{1-u^2-v^2}\right)^{\frac{1}{2}} \frac{e^{-f}}{\mathbf{\Delta}} \bar{\Phi}_{,u}; \tag{59}$$

or, expressed entirely in terms of f, by making use of equations (43), we have

$$\epsilon k^2 = \frac{v(1-v^2)}{8(1-u^2-v^2)^{\frac{1}{2}}} \frac{e^{-f}}{\mathbf{\Delta}} F_{,v}, \tag{60}$$

and

$$\bar{\epsilon}\bar{k}^2 = \frac{u(1-u^2)}{8(1-u^2-v^2)^{\frac{1}{2}}} \frac{e^{-f}}{\mathbf{\Delta}} \bar{F}_{,u}. \tag{61}$$

In view of the positivity of ϵk^2 and $\bar{\epsilon}\bar{k}^2$, we must, in our choice of the functions $\bar{F}(u)$ and $F(v)$, require that

$$\bar{F}_{,u} > 0 \quad (0 \leqslant u \leqslant 1) \quad \text{and} \quad F_{,v} > 0 \quad (0 \leqslant v \leqslant 1). \tag{62}$$

Also, without loss of generality, we may suppose that

$$F(0) = \bar{F}(0) = 0. \tag{63}$$

According to equations (60) and (61), ϵk^2 and $\bar{\epsilon}\bar{k}^2$ vanish on the null boundaries $u = 0$ and $v = 0$, respectively. Their behaviour on $u^2+v^2 = 1$ will depend on the choice of $F(v)$ and $\bar{F}(u)$; but will be modified by the factors

$$(1-u^2-v^2)^{-\frac{1}{2}} \quad \text{when} \quad q \neq 0 \quad \text{and} \quad (1-u^2-v^2)^{-\frac{5}{2}} \quad \text{when} \quad q = 0. \tag{64}$$

7. The extension of the space–time into regions II, III and IV

To complete the solution obtained in region I, we must extend it to include regions II, III, and IV. The required extension can be accomplished by Penrose's standard prescription of substituting $uH(u)$ and $vH(v)$ in place of u and v in the various metric functions, where $H(u)$ and $H(v)$ are the Heaviside functions which are unity for positive and zero values of the arguments and zero otherwise. We find that in region II, for example (cf. Paper I, equation (128)), the metric reduces to

$$\mathrm{d}s^2 = 4\mathrm{e}^{F(v)}\,\mathrm{d}u\,\mathrm{d}v - (1-v^2)\left[\chi(\mathrm{d}x^2)^2 + \frac{1}{\chi}(\mathrm{d}x^1 - q_2\,\mathrm{d}x^2)^2\right], \tag{65}$$

where χ and q_2 have the same meanings as in Paper I, equations (126) and (127). (Also, we have set $\bar{F}(0) = 0$ in accordance with equation (63).) In view of the restrictions (62), the metric (65) is compatible with null dust prevailing in region II. And we find, as in Paper I, equations (153) and (154), that the non-vanishing Ricci and Weyl scalars are given by

$$\Phi_{22} = -\frac{v}{2\sqrt{(1-v^2)}}\mathrm{e}^{-F(v)}\,F_{,v} \tag{66}$$

and

$$\Psi_4 = -\frac{p-\mathrm{i}q}{2\sqrt{(1-v^2)}}\mathrm{e}^{-F(v)}\,F_{,v}. \tag{67}$$

Similarly, we find that in region III the non-vanishing Ricci and Weyl scalars are

$$\Phi_{00} = -\frac{u}{2\sqrt{(1-u^2)}}\mathrm{e}^{-\bar{F}(u)}\,\bar{F}_{,u} \tag{68}$$

and

$$\Psi_0 = -\frac{p-\mathrm{i}q}{2\sqrt{(1-v^2)}}\mathrm{e}^{-\bar{F}(u)}\,\bar{F}_{,u}. \tag{69}$$

We can now readily verify that the Ricci and the Weyl scalars given by the foregoing equations join continuously with the values that follow from equations (51) on the respective common null boundaries.

The extension of the space–time to region IV and the characterization of the singularities on the various null boundaries do not call for any special comments because the discussion of these matters in Paper I is unchanged.

One important observation with respect to the present problem requires to be made. Because we are permitted to set $\bar{F}(u)$ or $F(v)$ equal to zero, it follows that *we can have null dust and the accompanying gravitational radiation in region* II *or in region* III *and have region* III *or region* II *flat*. Indeed, the fact that the solutions for f and ϕ are expressible as sums of a function of u only and a function of v only, implies that the streams of null dust crossing the null boundaries $u = 0$ and $v = 0$ do not influence one another in region I: they pass by unconcerned!

8. Concluding remarks

The solutions, for the two problems of colliding impulsive gravitational waves obtained in Paper I and in the present paper, provide a striking illustration of the ambiguous nature of null dust. Thus, because in the solution (44) for f obtained in this paper the functions $\bar{F}(u)$ and $F(v)$ are arbitrary (except for requirements of continuity and the restrictions (62) and (63)), we can specify that they are the same as the functions $f(u, 0)$ and $f(0, v)$ where $f(u, v)$ is the solution for the problem considered in Paper I (§8). We should then have *solutions in regions* II, III, *and* IV *which are identical for the two problems, but which are entirely different in region* I: *appropriate to a perfect fluid with* $\epsilon = p$ *in one case and a mixture of null dusts moving in opposite direction in the other*. The reason for this (initially) surprising ambiguity in the evolution of space–time must reside in the nature of null dust and in its special relation to a perfect fluid with $\epsilon = p$, a relation that must be traced, in the last analysis, to the fact that the Ricci tensor in one case is the tensor product of two time-like vectors, while in the other case it is the tensor product of two null vectors. The consequences that emerge from this fact clearly require further study.

We are grateful to Professor R. Geroch and Dr L. Lindblom for many fruitful discussions; in particular, we owe to Dr Lindblom the initial suggestion for investigating the problem considered in this paper. We are also indebted to Professor Roger Penrose for some useful correspondence. The research reported in this paper has, in part, been supported by grants from the National Science Foundation under grant PHY 84-16691 with the University of Chicago.

References

Chandrasekhar, S. 1983 *The mathematical theory of black holes*. Oxford: Clarendon Press.
Chandrasekhar, S. & Ferrari, V. 1984 *Proc. R. Soc. Lond.* A **396**, 55–74.
Chandrasekhar, S. & Ferrari, V. 1985 *Proc. R. Soc. Lond.* A **398**, 429.
Chandrasekhar, S. & Xanthopoulos, B. C. 1985 *Proc. R. Soc. Lond.* A **402**, 205–224.

23

A new type of singularity created by colliding gravitational waves

By Subrahmanyan Chandrasekhar, F.R.S., and Basilis C. Xanthopoulos†

The University of Chicago, Chicago, Illinois 60637, *U.S.A.*

(*Received* 14 *April* 1986)

An exact solution is obtained for colliding plane impulsive gravitational waves accompanied by shock waves, which, in contrast to other known solutions, results in the development of a null surface which acts like an event horizon. The analytic extension of the solution across the null surface reveals the existence of time-like curvature singularities along two hyperbolic arcs in the extended domain, reminiscent of the ring singularity of the Kerr metric. Besides, the space–time, in the region of the interaction of the colliding waves, is of Petrov-type D and locally isometric to the Kerr space–time in a region interior to the ergosphere. Various other aspects of the solution are also discussed.

1. Introduction

In some recent papers (Chandrasekhar & Ferrari 1984; Chandrasekhar & Xanthopoulos 1985*a*, *b*, *c*, 1986; these papers will be referred to hereafter as Papers I, II, III, IV and V, respectively) the collision of plane-fronted gravitational waves has been considered under a variety of circumstances. The solutions that have been found are all generalizations of the basic ones of Khan & Penrose (1971) and of Nutku & Halil (1977) to include sources derived from a Maxwell field (Paper II), a perfect fluid with the equation of state, energy-density = pressure (Paper III), or a null dust (Paper V).

A common feature of the solutions that have been considered so far (including those of Szekeres (1970, 1972), Halil (1979) and Ferrari & Ibañez (1986)) is the development of a curvature singularity, a finite time after the instant of collision and at a finite distance from the plane on which the collision occurred. Following Penrose, the development of the singularity is attributed to the self-focusing of the gravitational waves on scattering (though, as we have seen in Paper III, the singularity can be 'weakened' considerably by coupling with acoustic waves of sufficient amplitude).

In this paper, we return to the vacuum equations; and we shall obtain a solution which represents the collision of impulsive waves accompanied by shock waves. The solution has many properties that are not shared by the other solutions for colliding waves that have been considered so far. The most important among these

† Permanent address: Department of Physics, University of Crete, and Research Centre of Crete, Iraklion, Greece.

is the occurrence of a null surface (which acts like an event horizon) and an associated coordinate singularity in place of a space–time curvature singularity. The analytic extension of the space–time, beyond the null surface, reveals, in turn, in the extended manifold, a time-like curvature-singularity along two-dimensional hyperbolic arcs reminiscent of the ring singularity of the Kerr space–time. Indeed, the region of the space–time in which the colliding waves interact is of Petrov-type D and locally isometric to a part of the Kerr space–time interior to the ergo-sphere where it has also two space-like Killing fields.

The plan of the paper is the following. In §2, we assemble the basic equations and clarify the origin of the solution we shall be considering. In §3, we consider in some detail the requirements that must be met for a consistent description of colliding waves. In §4 we derive the new solution which is the principal subject of this paper (§§4–9).

2. The basic equations

In this paper we shall be concerned only with the vacuum equations. The relevant equations as derived by Chandrasekhar & Ferrari (Paper I; see also the errata in Chandrasekhar & Ferrari 1985) are the following.

In a gauge and in a coordinate system that is appropriate for describing the collision of plane-fronted gravitational waves, the metric of the space–time, in the region of interaction of the colliding waves, can be written in the form

$$ds^2 = e^{\nu+\mu_3}\sqrt{\Delta}\left[\frac{(d\eta)^2}{\Delta}-\frac{(d\mu)^2}{\delta}\right]-\sqrt{(\Delta\delta)}\left[\chi(dx^2)^2+\frac{1}{\chi}(dx^1-q_2\,dx^2)^2\right], \tag{1}$$

where η measures time from the instant of collision, μ measures the distance normal to the planes (x^1, x^2) spanned by the two Killing fields,

$$\Delta = 1-\eta^2 \quad \text{and} \quad \delta = 1-\mu^2. \tag{2}$$

Letting

$$\chi+iq_2 = Z = \frac{1+E}{1-E}, \tag{3}$$

we have the equations

$$(Z+Z^*)\,[(\Delta Z_{,\eta})_{,\eta}-(\delta Z_{,\mu})_{,\mu}] = 2[\Delta(Z_{,\eta})^2-\delta(Z_{,\mu})^2] \tag{4}$$

and

$$(1-|E|^2)\,[(\Delta E_{,\eta})_{,\eta}-(\delta E_{,\mu})_{,\mu}] = -2E^*[\Delta(E_{,\eta})^2-\delta(E_{,\mu})^2], \tag{5}$$

of which the latter is an Ernst equation. In terms of a solution, E, of the Ernst equation (5), the equations governing $(\nu+\mu_3)$ are

$$\frac{\mu}{\delta}(\nu+\mu_3)_{,\eta}+\frac{\eta}{\Delta}(\nu+\mu_3)_{,\mu} = -\frac{1}{\chi^2}(\chi_{,\eta}\chi_{,\mu}+q_{2,\eta}q_{2,\mu}) = -2\frac{E_{,\eta}E^*_{,\mu}+E^*_{,\eta}E_{,\mu}}{(1-|E|^2)^2}, \tag{6}$$

and

$$\begin{aligned}2\eta(\nu+\mu_3)_{,\eta}+2\mu(\nu+\mu_3)_{,\mu} &= \frac{3}{\Delta}+\frac{1}{\delta}-\frac{1}{\chi^2}\{\Delta[(\chi_{,\eta})^2+(q_{2,\eta})^2]+\delta[(\chi_{,\mu})^2+(q_{2,\mu})^2]\}\\ &= \frac{3}{\Delta}+\frac{1}{\delta}-\frac{4}{(1-|E|^2)^2}[\Delta|E_{,\eta}|^2+\delta|E_{,\mu}|^2].\end{aligned} \tag{7}$$

Alternatively, by deriving q_2 from a potential Φ in the manner (Paper I, equations (48) and (54)),

$$\Phi_{,\eta} = \frac{\delta}{\chi^2} q_{2,\mu} \quad \text{and} \quad \Phi_{,\mu} = \frac{\Delta}{\chi^2} q_{2,\eta}, \tag{8}$$

and letting

$$Z^\dagger = \Psi + \mathrm{i}\Phi = \frac{1+E^\dagger}{1-E^\dagger}, \tag{9}$$

where

$$\Psi = \sqrt{(\Delta\delta)}/\chi \tag{10}$$

we obtain equations for $Z^\dagger$ and $E^\dagger$ which are identical in forms with equations (4) and (5).

Equations (6) and (7) governing $(\nu+\mu_3)$ can now be expressed in terms of $E^\dagger$ as follows. By making use of the definitions (8) and (10), we find

$$\begin{aligned}\frac{1}{\chi^2}(\chi_{,\eta}\chi_{,\mu} + q_{2,\eta} q_{2,\mu}) &= (\ln\chi)_{,\eta}(\ln\chi)_{,\mu} + \frac{\Phi_{,\eta}\Phi_{,\mu}}{\Psi^2} \\ &= \left(\frac{\eta}{\Delta} + \frac{\Psi_{,\eta}}{\Psi}\right)\left(\frac{\mu}{\delta} + \frac{\Psi_{,\mu}}{\Psi}\right) + \frac{\Phi_{,\eta}\Phi_{,\mu}}{\Psi^2} \\ &= \frac{\mu}{\delta}\left[\ln\frac{\Psi}{\sqrt[4]{(\Delta\delta)}}\right]_{,\eta} + \frac{\eta}{\Delta}\left[\ln\frac{\Psi}{\sqrt[4]{(\Delta\delta)}}\right]_{,\mu} + \frac{\Phi_{,\eta}\Phi_{,\mu} + \Psi_{,\eta}\Psi_{,\mu}}{\Psi^2}.\end{aligned} \tag{11}$$

Thus equation (6) reduces to the form

$$\begin{aligned}&\frac{\mu}{\delta}\left[(\nu+\mu_3) + \ln\frac{\Psi}{\sqrt[4]{(\Delta\delta)}}\right]_{,\eta} + \frac{\eta}{\Delta}\left[(\nu+\mu_3) + \ln\frac{\Psi}{\sqrt[4]{(\Delta\delta)}}\right]_{,\mu} \\ &\qquad = -\frac{1}{\Psi^2}(\Phi_{,\eta}\Phi_{,\mu} + \Psi_{,\eta}\Psi_{,\mu}) = -2\frac{E^\dagger_{,\eta}(E^\dagger)^*_{,\mu} + (E^\dagger)^*_{,\eta}E^\dagger_{,\mu}}{(1-|E^\dagger|^2)^2}.\end{aligned} \tag{12}$$

Similarly, equation (7) reduces to the form

$$\begin{aligned}&2\eta\left[(\nu+\mu_3) + \ln\frac{\Psi}{\sqrt[4]{(\Delta\delta)}}\right]_{,\eta} + 2\mu\left[(\nu+\mu_3) + \ln\frac{\Psi}{\sqrt[4]{(\Delta\delta)}}\right]_{,\mu} \\ &\qquad = \frac{3}{\Delta} + \frac{1}{\delta} - \frac{4}{(1-|E^\dagger|^2)^2}[\Delta|E^\dagger_{,\eta}|^2 + \delta|E^\dagger_{,\mu}|^2].\end{aligned} \tag{13}$$

As shown in Paper I, the Nutku–Halil solution for colliding impulsive gravitational waves follows from the simplest solution,

$$E = p\eta + \mathrm{i}q\mu \quad (p^2+q^2 = 1), \tag{14}$$

of the Ernst equation (5). In this paper, we shall consider the solution which follows from the same solution of the Ernst equation for $E^\dagger$.

It will be recalled that in the analogous considerations pertaining to stationary axisymmetric space–times in *The mathematical theory of black holes* (Chandrasekhar 1983, §§53 and 54), the functions χ and ω (corresponding to the present pair χ and q_2) do *not* combine to yield an Ernst equation; they combine instead to yield a pair of real equations for $X = \chi + \omega$ and $Y = \chi - \omega$. The simplest solution of the X- and Y-equations does not lead to a solution of physical interest (cf. Chandrasekhar

1978, §8). But we do have an Ernst equation at the level of Ψ and Φ (the potential for ω); and we also have an Ernst equation for the functions $\tilde{\Psi}$ and $\tilde{\Phi}$ of the 'conjugate metric' obtained via the transformations $t \rightarrow i\varphi$ and $\varphi \rightarrow -it$. It is the simplest solution of this latter Ernst equation that leads to the Kerr solution. In space–times with two space-like Killing vectors, the distinction between a metric and its conjugate is trivial since it corresponds to a simple interchange of the Killing-vectors ∂_{x^1} and ∂_{x^2}. Therefore, the solution we shall obtain in this paper, by considering the simplest solution for $E^{\dagger}$, is the strict analogue of the Kerr solution. Indeed, as we have stated earlier, the solution is of Petrov-type D and is locally isometric to the Kerr solution.

3. The requirements on a solution for a consistent description of colliding gravitational waves

An observer witnessing the collision of two plane-fronted gravitational waves will divide his experience into three distinct intervals: an interval during which neither of the two oncoming wavefronts has crossed his world-line and the space–time is flat; an interval during which only one of two wavefronts has crossed his world-line (either by meeting him head-on or overtaking him from behind); and, finally, an interval during which both wavefronts have crossed his world-line and he witnesses the scattering of the colliding waves, one by the other. The corresponding division of the space–time into four non-overlapping regions (allowing for the two alternative sequences in which the colliding wavefronts can cross the observer's world-line) is most conveniently visualized in a system of null coordinates, u and v, which are related to η and μ (in the domains described by them) by

$$\left.\begin{aligned} \eta &= u\sqrt{(1-v^2)}+v\sqrt{(1-u^2)} \\ \text{and} \qquad \mu &= u\sqrt{(1-v^2)}-v\sqrt{(1-u^2)}. \end{aligned}\right\} \tag{15}$$

The coordinate patch, $0 \leqslant \eta \leqslant 1$ and $-1 \leqslant \mu \leqslant +1$, is mapped on to the patch, $0 \leqslant u \leqslant 1$ and $0 \leqslant v \leqslant 1$, of the (u, v)-plane.

In terms of the null coordinates, the metric (1) takes the form,

$$\mathrm{d}s^2 = 4\sqrt{\Delta}\frac{\mathrm{d}u\,\mathrm{d}v}{\sqrt{(1-u^2)}\sqrt{(1-v^2)}}\,\mathrm{e}^{\mu_3+\nu} - (1-u^2-v^2)\left[\chi(\mathrm{d}x^2)^2+\frac{1}{\chi}(\mathrm{d}x^1-q_2\,\mathrm{d}x^2)^2\right], \tag{16}$$

where it may be noted that

$$\sqrt{(\Delta\delta)} = 1-u^2-v^2, \tag{17}$$

and $\mu_3+\nu$, χ, and q_2 are now expressed as functions of u and v in accordance with equation (15).

As was first shown by Khan & Penrose (1971), and amply confirmed since, the space–time resulting from the collision of plane-fronted gravitational waves develops a singularity on the arc,

$$u^2+v^2 = 1, \tag{18}$$

in the (u, v)-plane. In all cases considered hitherto, the singularity that develops is a curvature singularity. For the solution obtained in this paper, we shall find that no curvature singularity develops on $u^2+v^2 = 1$; but the determinant of the metric does vanish. A coordinate singularity thus obtains and the extension of the metric beyond $1-u^2-v^2 = 0$ must be considered (see §7 below). In any event, the equations of §2 suffice only to determine the space–time in the region

$$0 \leqslant u < 1, \quad 0 \leqslant v < 1 \quad \text{and} \quad u^2+v^2 < 1 \tag{19}$$

after the instant of collision at $u = 0$ and $v = 0$. This is region I of the space–time (see figure 1 in Paper I and in Paper II). It remains to ascertain the nature of the space–time in regions II, III, and IV.

If we consider the collision of two impulsive gravitational waves, the space–times in regions II and III (where only one of the two colliding wavefronts has crossed the world-line of the observer) must continue to remain flat (even as it was in region IV before the instant of collision). It will not, however, suffice to simply state these requirements: the postulated flatness of the space–times of regions II, III, and IV must be compatible with the metric (16) determined in region I by the equations of §2. More generally, we need an algorism for extending the metric in region I to regions II, III, and IV *and* the requirements that must be satisfied for the metric, so obtained, to be physically and mathematically acceptable.

The algorithm we shall consider for extending the metric (16) to negative values of u and v is by the Penrose–Khan substitutions,

$$u \to uH(u) \quad \text{and} \quad v \to vH(v), \tag{20}$$

where $H(u)$ and $H(v)$ are the Heaviside functions that are unity for positive and zero values of the argument and zero for negative values of the argument. Whereas alternative methods of extension are conceivable (cf. Paper III, Appendix B and Szekeres 1972), the substitutions (20) have several advantages: the resulting metric is C^0; at the null boundaries, $u = 0$ and $v = 0$, the curvature scalars behave entirely consistently with the assumed character of the colliding fronts by being continuous or by having an H-function discontinuity and/or a δ-function singularity. In particular, we find that the radiative parts of the gravitational field are continuous across the null boundaries in accord with the requirements of the conservation of energy.

In view of the symmetry of the problem in u and v, it will suffice to consider the extension of the metric into regions II and IV. The extension of the metric (16) into region II is effected by the substitutions (cf. equation (15)),

$$\eta = v \quad \text{and} \quad \mu = -v, \tag{21}$$

when the metric becomes

$$ds^2 = 4\, e^{g(v)}\, du\, dv - (1-v^2)\left[\chi(v)\,(dx^2)^2 + \frac{1}{\chi(v)}\,(dx^1 - q_2(v)\, dx^2)^2\right], \tag{22}$$

where

$$\left.\begin{aligned} g(v) &= (\nu+\mu_3)_{(\eta = v\,;\,\mu = -v)}, \quad \chi(v) = \chi(\eta = v\,;\mu = -v), \\ \text{and} \qquad q_2(v) &= q_2(\eta = v\,;\mu = -v). \end{aligned}\right\} \tag{23}$$

The metric (22) must satisfy several requirements to be acceptable: it must not be degenerate; it must be free of singularities; and it must represent a solution of Einstein's vacuum equations. That the extended metric (22) will satisfy these requirements is by no means evident. Thus the solution,

$$Z = \chi = e^{\alpha\eta\mu} \quad \text{and} \quad q_2 = 0 \quad (\alpha = \text{constant}), \tag{24}$$

of equation (4) leads to the solution,

$$e^{\mu_3+\nu} = \frac{\eta^2-\mu^2}{(1-\eta^2)^{\frac{3}{4}}(1-\mu^2)^{\frac{1}{4}}} \exp\left[\alpha^2(\eta^2\mu^2-\eta^2-\mu^2)\right]/4, \tag{25}$$

of equations (6) and (7). This solution, when extended by the substitutions (21), makes

$$e^{\mu_3+\nu} = 0 \quad (\text{in region II}). \tag{26}$$

The extended metric is therefore degenerate; and the solutions (24) and (25) for the metric coefficients in region I are therefore unacceptable.

More generally, we find that the solution,

$$\ln\chi = \sum_n A_n P_n(\mu) P_n(\eta) \quad (A_n = \text{const.}), \tag{27}$$

of the equation (cf. Paper I, equation (35) for $q_2 = 0$)

$$[\Delta(\ln\chi)_{,\eta}]_{,\eta} - [\delta(\ln\chi)_{,\mu}]_{,\mu} = 0, \tag{28}$$

where P_n denotes the Legendre polynomial, does not also lead to an acceptable metric when extended to regions II and III. On the other hand, the solution,

$$\ln\chi = \ln\frac{1+\eta}{1-\eta} + \sum_n A_n P_n(\mu) P_n(\eta) + \sum_n B_n Q_n(\mu) Q_n(\eta), \tag{29}$$

of equation (28), where the first term represents the solution appropriate to the Khan–Penrose metric (cf. Paper I, equation (57), $q_2 = 0$) and Q_n denotes the Legendre function of the second kind, does lead to acceptable solutions for special values of the constants A_n and B_n (cf. Ferrari & Ibañez 1986). (The solution (29) for $\ln\chi$ is the analogue, in the present context of the solution that leads to static distorted black holes, cf. *The mathematical theory of black holes* (Chandrasekhar 1983, §112).)

Examples of solutions in region I, which when extended into regions II and III exhibit singularities, are provided by the Tomimatsu & Sato family of solutions. Thus, considering the '$(\delta = 2)$'-member, we have the solution

$$e^{\nu+\mu_3} = [(\eta^2-\mu^2)^3(1-\mu^2)^{\frac{1}{4}}(1-\eta^2)^{\frac{3}{4}}]^{-1}\{[p^2(\eta^2-1)^2+q^2(1-\mu^2)^2]^2 - 4p^2q^2(\eta^2-1)(1-\mu^2)(\eta^2-\mu^2)^2\}, \tag{30}$$

which belongs to the solution (cf. Tomimatsu & Sato 1973),

$$E = \frac{p^2\eta^4+q^2\mu^4-1-2ipq\eta\mu(\eta^2-\mu^2)}{2p\eta(\eta^2-1)-2iq\mu(1-\mu^2)} \quad (p^2+q^2=1), \tag{31}$$

of the Ernst equation. We observe that the expression (30) diverges by the substitutions (21) and, therefore, is unacceptable.

It remains to ascertain the conditions when the metric (22) will represent a solution of Einstein's vacuum equations. The metric (22) is of the form,

$$ds^2 = e^{2\nu(v)}[(dx^0)^2-(dx^3)^2]-e^{2\psi(v)}[dx^1-q_2(v)\,dx^2]^2-e^{2\mu_2(v)}(dx^2)^2 \quad (v = x^0-x^3), \tag{32}$$

considered in Paper II (§8) and Paper III (§12). Also, it has been shown that the metric (32) is a solution of Einstein's equation for a null Maxwell field or a null dust depending on whether,

$$L+M<0 \quad \text{or} \quad L+M>0, \tag{33}^*$$

respectively, where (cf. Paper II, equation (160))

$$L+M = e^{-2\nu}[(\psi+\mu_2)''+(\psi')^2+(\mu_2')^2-2\nu'(\psi+\mu_2)']+\tfrac{1}{2}e^{2\psi-2\mu_2-2\nu}(q_2')^2, \tag{34}$$

and primes denote differentiations with respect to v; and, further, that the metric (32) will be a solution of Einstein's vacuum equations if and only if

$$L+M = 0. \tag{35}$$

A comparison of the metrics (22) and (32) shows that in the present context,

$$e^{2\nu} = 4\,e^{g}, \quad e^{2\psi} = (1-v^2)/\chi, \quad e^{2\mu_2} = (1-v^2)\,\chi,$$

and

$$q_2 = q_2, \tag{36}$$

and the condition (35) takes the form

$$L+M = \tfrac{1}{4}e^{-g}\left\{-\frac{2}{(1-v^2)^2}+\frac{2v}{1-v^2}g'+\frac{1}{2\chi^2}[(\chi')^2+(q_2')^2]\right\} = 0. \tag{37}$$

This requirement is an essential one since the extension of the metric across the null boundary is not a smooth one.

Because the determinant of the metric (22) is given by

$$-4\,e^{2g(v)}\,(1-v^2)^2, \tag{38}$$

we can combine the requirements of non-degeneracy, freedom from singularities, and compatability with Einstein's vacuum equations in the following single statement.

A solution of equations (5), (6), *and* (7) *will describe the collision of two plane-fronted gravitational waves, after the instant of collision, provided that the limits* $g(v)$, $\chi(v)$, *and* $q_2(v)$, *as defined in equations* (23), *and the limits* $g(u)$, $\chi(u)$, *and* $q_2(u)$ *as defined by the equations,*

$$g(u) = (\nu+\mu_3)_{(\eta=u\,;\,\mu=u)}, \quad \chi(u) = \chi(\eta=u;\mu=u),$$

and

$$q_2(u) = q_2(\eta=u;\mu=u), \tag{39}$$

exist on the null boundaries, ($u = 0, 0 \leqslant v \leqslant 1$) *and* ($v = 0, 0 \leqslant u \leqslant 1$), $\chi(u)$ *and* $\chi(v)$ *are different from zero, and the condition* (37) *and a similar condition involving* $g(u)$, $\chi(u)$, $q_2(u)$ *are satisfied.*

Assuming then that the various requirements on the solution derived for

* The difference in the senses of the inequality is due to the difference in our conventions regarding the energy momentum tensor in the two cases. It is not a significant one.

region I are satisfied, we can completely characterize the gravitational field in region II (for example) by the Weyl scalar Ψ_4 given by (Paper II, equation (178))

$$\Psi_4 = \frac{1-\mathscr{E}}{1-\mathscr{E}^*}(M-L+2\mathrm{i}N)\sqrt{(1-v^2)}, \tag{40}$$

where in our present context,

$$M-L+2\mathrm{i}N = \tfrac{1}{4}\,\mathrm{e}^{-g(v)}\left[\frac{\mathscr{X}''}{\chi}-\frac{\mathscr{X}'}{\chi}\left(\frac{2v}{1-v^2}+g'\right)-\frac{(\mathscr{X}')^2}{\chi^2}\right], \tag{41}$$

and

$$\mathscr{X} = \chi(v)+\mathrm{i}q_2(v) = \frac{1+\mathscr{E}(v)}{1-\mathscr{E}(v)}. \tag{42}$$

The further extension of the metric into region IV is accomplished by simply setting $v = 0$ in the coefficients of the metric (22). The resulting metric is clearly one of constant coefficients; and the space–time will be flat as required provided the substitution $v = 0$ in equations (23) leads to a regular non-degenerate metric. (We are not aware of any instance in which this last extension fails.)

4. The metric for region I derived from the simplest solution of the Ernst equation for $E^\dagger$

It is clear that the solution for Ψ and Φ for

$$E^\dagger = p\eta+\mathrm{i}q\mu \quad (p^2+q^2 = 1), \tag{43}$$

must be the same as for χ and q_2 for the same solution of the Ernst equation for E. Therefore by Paper I, equations (57) and (58),

$$\Psi = \frac{1-p^2\eta^2-q^2\mu^2}{(1-p\eta)^2+q^2\mu^2} \quad \text{and} \quad \Phi = \frac{2q\mu}{(1-p\eta)^2+q^2\mu^2}. \tag{44}$$

From the defining equations (8) and (10), we now find

$$\chi = \sqrt{(\Delta\delta)}\,\frac{(1-p\eta)^2+q^2\mu^2}{1-p^2\eta^2-q^2\mu^2} \tag{45}$$

and

$$q_2 = \frac{2q}{(1+p)}\,\frac{(1-\eta)(p\mu^2+p\eta+\mu^2-1)}{1-p^2\eta^2-q^2\mu^2}, \tag{46}$$

where, for convenience, a constant of integration in the solution for q_2 has been so chosen that q_2 vanishes for $\eta = 1$. (The reasons for this choice are explained in §7 below.)

An alternative form of the expression (46) for q_2 which we shall find useful is

$$q_2 = -\frac{2q\delta(1-p\eta)}{p(1-p^2\eta^2-q^2\mu^2)}+\frac{2q}{p(1+p)}. \tag{47}$$

Turning next to the solution of $\nu+\mu_3$, we conclude from a comparison of equations (6) and (7) with equations (12) and (13) that it is given by

$$\nu+\mu_3+\ln\frac{\Psi}{\sqrt{(\Delta\delta)}} = (\nu+\mu_3)_{\text{Nutku-Halil}} \tag{48}$$

because E and $E^\dagger$ are the same for the two solutions. Therefore,

$$e^{\nu+\mu_3} = \frac{\sqrt[4]{(\Delta\delta)}}{\Psi} \frac{1-p^2\eta^2-q^2\mu^2}{(1-\eta^2)^{\frac{3}{4}}(1-\mu^2)^{\frac{1}{4}}}, \tag{49}$$

where we have substituted for $(\nu+\mu_3)_{\text{Nutku-Halil}}$ from Paper I, equation (61). With the solution for Ψ given by equation (44), we find

$$e^{\nu+\mu_3} = \frac{(1-p\eta)^2+q^2\mu^2}{\sqrt{(1-\eta^2)}}. \tag{50}$$

With the solutions for χ, q_2, and $e^{\nu+\mu_3}$ given by equations (45), (47), and (50), we have the metric (cf. equation (1))

$$ds^2 = X\left[\frac{(d\eta)^2}{1-\eta^2}-\frac{(d\mu)^2}{1-\mu^2}\right]-\Delta\delta\frac{X}{Y}(dx^2)^2-\frac{Y}{X}(dx^1-q_2\,dx^2)^2, \tag{51}$$

where

$$X = (1-p\eta)^2+q^2\mu^2, \quad Y = 1-|E^\dagger|^2 = 1-p^2\eta^2-q^2\mu^2 = p^2\Delta+q^2\delta, \tag{52}$$

and

$$q_2 = -\frac{2q\delta(1-p\eta)}{pY}+\frac{2q}{p(1+p)}; \tag{53}$$

or, alternatively, by setting

$$\eta = \cos\psi \quad \text{and} \quad \mu = \cos\theta, \tag{54}$$

we have

$$ds^2 = X[(d\psi)^2-(d\theta)^2]-\sin^2\psi\,\sin^2\theta\frac{X}{Y}(dx^2)^2-\frac{Y}{X}(dx^1-q_2\,dx^2)^2, \tag{55}$$

where X, Y, and q_2 must now be rewritten in terms of ψ and θ.

For the purposes of setting up a null tetrad-frame (cf. Appendix A), it is convenient to write the metric in the form

$$ds^2 = X[(d\psi)^2-(d\theta)^2]-\frac{\sin\psi\,\sin\theta}{1-|\mathscr{E}|^2}|(1-\mathscr{E})\,dx^1+i(1+\mathscr{E})\,dx^2|^2, \tag{56}$$

where

$$\mathscr{E} = \frac{\chi+iq_2-1}{\chi+iq_2+1} = \frac{\mathscr{Z}-1}{\mathscr{Z}+1}. \tag{57}$$

(It should be noted that we have used $\mathscr{E}$ and $\mathscr{Z}$ in equation (57), and not E and Z, to emphasize that the solution for $\chi+iq_2$, considered, is *not directly* derived as a solution of the Ernst equation for $(\chi+iq_2-1)/(\chi+iq_2+1)$ but *indirectly* (as in the case presently considered) *via* equations (8) and (10) from a solution of the Ernst equation for $(\Psi+i\Phi-1)/(\Psi+i\Phi+1)$. The reason for making this distinction is that a simple solution of the Ernst equation derived for $\Psi+i\Phi$ often leads to a quite *complicated* expression for $\mathscr{E}$. This convention for the usage of $\mathscr{E}$ and E will be adhered to in the future.)

Still another form of the metric (51) from which it is manifest that no singularity arises from a zero of the function Y is the following. Considering the last two terms in the (x^1, x^2)-part of the metric and substituting for q_2 from equation (53), we obtain

$$\Delta\delta\frac{X}{Y}(\mathrm{d}x^2)^2+\frac{Y}{X}\left\{\left[\mathrm{d}x^1-\frac{2q}{p(1+p)}\mathrm{d}x^2\right]+2\frac{q\delta(1-p\eta)}{pY}\mathrm{d}x^2\right\}^2$$

$$=\frac{\delta}{p^2XY}[p^2\Delta X^2+4q^2\delta(1-p\eta)^2](\mathrm{d}x^2)^2$$

$$+\frac{Y}{X}\left[\mathrm{d}x^1-\frac{2q}{p(1+p)}\mathrm{d}x^2\right]^2+4q\delta\frac{1-p\eta}{pX}\left[\mathrm{d}x^1-\frac{2q}{p(1+p)}\mathrm{d}x^2\right]\mathrm{d}x^2. \quad (58)$$

Now making use of the identity (cf. equation (B 12))

$$p^2\Delta X^2+4q^2\delta(1-p\eta)^2 = Y\{[(1-p\eta)^2+q^2]^2+p^2q^2\Delta\delta\}, \quad (59)$$

we find that the metric takes the form

$$\mathrm{d}s^2 = X\left[\frac{(\mathrm{d}\eta)^2}{1-\eta^2}-\frac{(\mathrm{d}\mu)^2}{1-\mu^2}\right]-\frac{\delta}{p^2X}\{[(1-p\eta)^2+q^2]^2+p^2q^2\Delta\delta\}(\mathrm{d}x^2)^2$$

$$-\frac{Y}{X}\left[\mathrm{d}x^1-\frac{2q}{p(1+p)}\mathrm{d}x^2\right]^2-4q\delta\frac{1-p\eta}{pX}\left[\mathrm{d}x^1-\frac{2q}{p(1+p)}\mathrm{d}x^2\right]\mathrm{d}x^2, \quad (60)$$

in which Y no longer appears in a denominator.

5. The extension of the space–time in region I_0 into regions II, III and IV

By the algorism explained in §3, the extension of the metric (51) in region I (or, rather region I_0 bounded by the null boundaries, $u = 0$, $v = 0$, and $u^2+v^2 = 1$) is effected by the substitutions (21) in the metric coefficients (52) and (53). We then obtain, in the notation of §3, equation (22),

$$\mathrm{d}s^2 = \frac{4\chi}{\sqrt{(1-v^2)}}\mathrm{d}u\,\mathrm{d}v-(1-v^2)\left[\chi(\mathrm{d}x^2)^2+\frac{1}{\chi}(\mathrm{d}x^1-q_2\,\mathrm{d}x^2)^2\right], \quad (61)$$

where

$$\chi = \chi(v) = 1-2pv+v^2, \quad q_2 = q_2(v) = \frac{2q}{(1+p)}[(1+p)v-1], \quad (62)$$

and

$$g(v) = \ln[(1-2pv+v^2)/\sqrt{(1-v^2)}]. \quad (63)$$

With the present definitions (62) and (63), we readily verify that the condition (37), for the metric (61) to represent a solution of Einstein's equations for the vacuum, is satisfied. That this condition is satisfied is, of course, necessary for the solution we have obtained to have a physical meaning.

As we have stated in §3, the only non-vanishing Weyl scalar in region II is Ψ_4. Its value can now be found by inserting the present solution for $g(v)$, $\mathscr{Z}(=\chi+\mathrm{i}q_2)$ and $\mathscr{E} = (\mathscr{Z}-1)/(\mathscr{Z}+1)$ in equations (40) and (41). We find

$$\Psi_4 = -\frac{3}{2}\frac{\chi-\mathrm{i}q_2+1}{\chi+\mathrm{i}q_2+1}\frac{(p-\mathrm{i}q)^2}{(1-pv+\mathrm{i}qv)^3}. \quad (64)$$

We shall verify in §6 that this value of Ψ_4 agrees with what we find on approaching the common boundary from the side of region I. Thus, Ψ_4 is continuous across the null-boundary $u = 0$ and $0 \leqslant v \leqslant 1$ separating regions I and II.

The further extension of the metric into region IV is accomplished by simply setting $v = 0$ in the coefficients (62) and (63). We thus obtain

$$ds^2 = 4\,du\,dv - \left[(dx^2)^2 + \left(dx^1 + \frac{2q}{1+p}\,dx^2\right)^2\right], \tag{65}$$

which manifestly represents a flat space–time (as required). On the other hand, as we approach the null-boundary $v = 0$ and $-\infty < u < 0$, from the side of region II, Ψ_4 attains the value

$$\Psi_4 = -\tfrac{3}{2}(p - iq). \tag{66}$$

Therefore, Ψ_4 experiences an H-function discontinuity of amount (66) on the null-boundary separating the regions II and IV. In addition, Ψ_4 also experiences a δ-function singularity given by

$$\left[\frac{(1-\mathscr{E})\,\mathscr{E}'}{(1+\mathscr{E})^3}\right]_{(v=0)} \delta(v) = -\frac{(1+p)^3\,(p-iq)}{(1+p-2iq)^3}\,\delta(v). \tag{67}$$

And finally, by interchanging the roles of u and v and of Ψ_4 and Ψ_0, we can ascertain the circumstances prevailing in region III and of the discontinuities and the singularities that are present on $u = 0$ and $-\infty < v < 0$.

It is on account of the simultaneous presence of an H-function discontinuity and a δ-function singularity on the null boundaries $(v = 0, -\infty < u < 0)$ and $(u = 0, -\infty < v < 0)$ that *we interpret the space–time as resulting from the collision of impulsive gravitational waves accompanied by gravitational shock-waves.*

(a) *The description of the geodesics in region II and the nature of the singularity along $v = 1$ and $-\infty < u < 0$*

The nature of the singularity that develops on $v = 1$ $(-\infty < u < 0)$ manifests itself in the behaviour of the geodesics in region II (cf. Matzner & Tipler 1984 for an analogous discussion in the context of the Khan–Penrose solution). The delineation of the geodesics in this region is simple because the space–time, by virtue of the metric coefficients depending only on the one variable v, allows three conserved momenta, p_u, p_{x^1}, and p_{x^2} besides the energy integral. These are:

$$p_u = \frac{2\chi}{\sqrt{(1-v^2)}}\dot{v} = \gamma; \quad p_{x^1} = \frac{1-v^2}{\chi}(\dot{x}^1 - q_2\,\dot{x}^2) = p_1; \tag{68}$$

$$p_{x^2} = -\frac{1-v^2}{\chi}(\dot{x}^1 - q_2\,\dot{x}^2)\,q_2 + (1-v^2)\,\chi\dot{x}^2 = -p_1 q_2 + (1-v^2)\,\chi\dot{x}^2 = p_2; \tag{69}$$

and

$$\begin{aligned}\epsilon &= \frac{4\chi}{\sqrt{(1-v^2)}}\,\dot{u}\dot{v} - \frac{1-v^2}{\chi}(\dot{x}^1 - q_2\,\dot{x}^2)^2 - (1-v^2)\,\chi(\dot{x}^2)^2 \\ &= 2\gamma\dot{u} - p_1^2\frac{\chi}{1-v^2} - \frac{1}{\chi(1-v^2)}(p_2 + p_1 q_2)^2,\end{aligned} \tag{70}$$

where γ, p_1, p_2, and ϵ (greater than 0 for time-like geodesics and equal to 0 for null geodesics) are constants of the motion. From the foregoing conservation laws, we readily obtain the equation

$$\gamma^2 \frac{du}{dv} = p_1^2 \frac{\chi^2}{(1-v^2)^{\frac{3}{2}}} + (p_2 + p_1 q_2)^2 \frac{1}{(1-v^2)^{\frac{3}{2}}} + \epsilon \frac{\chi}{(1-v^2)^{\frac{1}{2}}}; \tag{71}$$

or, after integration

$$\gamma^2(u+c) = \int_0^v \left\{ \frac{1}{(1-v^2)^{\frac{3}{2}}} [p_1^2 \chi^2 + (p_2 + p_1 q_2)^2] + \epsilon \frac{\chi}{\surd(1-v^2)} \right\} dv, \tag{72}$$

where $u = -c$ $(c > 0)$ is the point on the negative u-axis at which the geodesic crosses into region II. Because according to equation (72)

$$\gamma^2(u+c) \to \frac{1}{\surd(1-v^2)} \left[4(1-p)^2 p_1^2 + \left(p_2 + \frac{2p_1 pq}{1+p} \right)^2 \right] \quad (\text{as } v \to 1-0), \tag{73}$$

it follows that no time-like or null geodesics with p_1 and/or $p_2 \neq 0$, can avoid crossing the v-axis into region I_0 for some $0 < v < 1$. The only exceptions are time-like geodesics with $p_1 = p_2 = 0$ and the null geodesics $u =$ constant. The situation is essentially the same as in the other contexts that have been considered (Matzner & Tipler 1984; paper II, §12).

6. The Weyl scalars, the type-D character of the space–time, and the local isometry with the Kerr space–time

The quantities in which we are principally interested are the Weyl scalars, Ψ_0, Ψ_2, and Ψ_4, which do not vanish identically in the null-tetrad frame chosen in Appendix A. The required scalars can be evaluated directly, without any difficulty, from equations (A 20), (A 23), and (A 26) because the present metric has been derived from the simplest solution,

$$E = p \cos\psi + iq \cos\theta \quad (p^2 + q^2 = 1), \tag{74}$$

of the Ernst equation governing $Z = \Psi + i\Phi$, where it will be noted that we have suppressed the superscript '†' to distinguish E.

For the metric (56), the quantities U and V, as introduced in equation (A 1) have the meanings (cf. equations (52) and (54)),

$$V^2 = \sin\psi \sin\theta \quad \text{and} \quad U^2 = (1 - p\cos\psi)^2 + q^2 \cos^2\theta = |1-E|^2. \tag{75}$$

Further relations that are useful in the simplifications of the expressions for the Weyl scalars are

$$D = \cot\psi - \cot\theta, \quad \overline{D} = -\left(\frac{E_{,\psi} - E_{,\theta}}{1-E} + \frac{E^*_{,\psi} - E^*_{,\theta}}{1-E^*} \right); \tag{76}$$

$$S = \cot\psi + \cot\theta, \quad \overline{S} = -\left(\frac{E_{,\psi} + E_{,\theta}}{1-E} + \frac{E^*_{,\psi} + E^*_{,\theta}}{1-E^*} \right); \tag{77}$$

$$|E_{,\psi} \pm E_{,\theta}|^2 = 1 - |E|^2, \quad \text{and} \quad E_{,\psi,\psi} + E_{,\theta,\theta} \pm 2E_{,\theta,\psi} = -E. \tag{78}$$

With the aid of these relations, we readily find that equation (A 20) gives

$$\Psi_2 = \frac{1}{2(1-E)^3} = \frac{1}{2(1-p\cos\psi - \mathrm{i}q\cos\theta)^3}. \tag{79}$$

Turning next to the expression (A 23) for Ψ_4, we first find that, by making use of equations (76) and (78) only, it can be simplified to the form

$$\begin{aligned} 4U^2\frac{\chi+1+\mathrm{i}q_2}{\chi+1-\mathrm{i}q_2}\Psi_4 &= D^2 + D_{,\psi} - D_{,\theta} \\ &\quad + \frac{D}{1-|E|^2}[E^*(E_{,\psi}-E_{,\theta}) + E(E^*_{,\psi}-E^*_{,\theta})] \\ &\quad - \Theta\left[3\frac{(E_{,\psi}-E_{,\theta})^2}{1-E} + \frac{|E_{,\psi}-E_{,\theta}|^2}{1-E^*} - E\right] + \frac{4}{1-|E|^2}. \end{aligned} \tag{80}$$

Now substituting for E its expression (74), we find

$$\frac{\chi+1+\mathrm{i}q_2}{\chi+1-\mathrm{i}q_2}\Psi_4 = -\frac{3}{2}\frac{(p\sin\psi - \mathrm{i}q\sin\theta)^2}{(1-E)^3(1-|E|^2)}. \tag{81}$$

An analogous reduction of the expression (A 26) yields,

$$\frac{\chi+1-\mathrm{i}q_2}{\chi+1+\mathrm{i}q_2}\Psi_0 = -\frac{3}{2}\frac{(p\sin\psi + \mathrm{i}q\sin\theta)^2}{(1-E)^3(1-|E|^2)}. \tag{82}$$

First, we observe that on the null boundary $u = 0\ (0 \leqslant v \leqslant 1)$, where $\eta = -\mu = v$ and $E = (p-\mathrm{i}q)v$, Ψ_4 given by equation (81) agrees with the value (64) it has in region II; *Ψ_4 is, therefore, continuous across this boundary.*

Because

$$E = p + \mathrm{i}q\cos\theta \quad\text{and}\quad q_2 = 0, \quad\text{when}\quad u^2+v^2=1 \quad\text{and}\quad \eta = \cos\psi = +1, \tag{83}$$

it follows from equations (79), (81), and (82) that

$$\Psi_4 = \Psi_0 = \frac{3}{2(1-p-\mathrm{i}q\cos\theta)^3} = 3\Psi_2, \quad\text{when}\quad u^2+v^2=1 \quad\text{and}\quad \eta = 1. \tag{84}$$

We conclude that, *except when $q = 0$ and $p = 1$, the Weyl scalars Ψ_0, Ψ_4, and (most importantly) Ψ_2 do not diverge when $u^2+v^2 \to 1$.* (It may be noted that for $q = 0$ the Weyl scalars diverge for $s = 1-u^2-v^2 \to 0$ like s^{-6}; and this is stronger than the rates of divergence encountered with the Khan–Penrose ($s^{-\frac{7}{2}}$) and the Nutku–Halil ($s^{-\frac{3}{2}}$) solutions.) Therefore, in contrast to all other examples of colliding gravitational waves that have been considered hitherto, *no curvature singularity develops in the present instance when $u^2+v^2 \to 1$.* Because, however, the determinant of the metric (56) vanishes on $u^2+v^2 = 1$ when $\sin\psi\sin\theta = 0$, a coordinate singularity certainly obtains here. The question of the analytic continuation of the metric beyond $u^2+v^2 = 1$ and the characterization of this surface are taken up in §§7 and 8. We turn now to important aspects of the space–time that follow from equations (79), (81), and (82).

(a) The type-D character of the space–time

It follows directly from equations (79), (81), and (82) that

$$\Psi_0 \Psi_4 = 9\Psi_2^2. \tag{85}$$

In the context of this relation, we shall now prove the following theorem:

THEOREM. *If in some chosen null-tetrad frame,* $(l, n, m, \overline{m})$

$$\Psi_1 = \Psi_3 = 0 \quad \text{and} \quad \Psi_0 \Psi_4 = 9\Psi_2^2, \tag{86}$$

then, the space–time is of Petrov-type D†.

Proof. We prove the theorem by showing that by a suitable rotation of the null tetrad-frame we can make all the Weyl scalars, except Ψ_2, to vanish.

First, we recall that rotations of a chosen null tetrad-frame, $(\boldsymbol{l}, \boldsymbol{n}, \boldsymbol{m}, \overline{\boldsymbol{m}})$ can be decomposed into three classes. (For a description of these classes of rotations and their effects on the Weyl scalars see *The mathematical theory of black holes* (Chandrasekhar 1983, chapter 1, pp. 53–55).) In particular, a rotation of class I (which leaves the vector $\boldsymbol{l}$ unchanged) alters an initial set of Weyl scalars, $\Psi_0, \ldots, \Psi_4$, into

$$\begin{aligned}
&\tilde{\Psi}_0 = \Psi_0, \quad \tilde{\Psi}_1 = \Psi_1 + a^*\Psi_0, \quad \tilde{\Psi}_2 = \Psi_2 + 2a^*\Psi_1 + (a^*)^2\,\Psi_0,\\
&\tilde{\Psi}_3 = \Psi_3 + 3a^*\Psi_2 + 3(a^*)^2\,\Psi_1 + (a^*)^3\,\Psi_0,\\
&\tilde{\Psi}_4 = \tilde{\Psi}_4 + 4a^*\Psi_3 + 6(a^*)^2\,\Psi_2 + 4(a^*)^3\,\Psi_1 + (a^*)^4\,\Psi_0,
\end{aligned} \tag{87}$$

where the parameter a denotes a complex scalar. A subsequent rotation of class II (which leaves the vector $\boldsymbol{n}$ unchanged) alters the set, $\tilde{\Psi}_1, \ldots, \tilde{\Psi}_4$, into

$$\begin{aligned}
&\tilde{\tilde{\Psi}}_0 = \tilde{\Psi}_0 + 4b\tilde{\Psi}_1 + 6b^2\tilde{\Psi}_2 + 4b^3\tilde{\Psi}_3 + b^4\tilde{\Psi}_4,\\
&\tilde{\tilde{\Psi}}_1 = \tilde{\Psi}_1 + 3b\tilde{\Psi}_2 + 3b^2\tilde{\Psi}_3 + b^3\tilde{\Psi}_4,\\
&\tilde{\tilde{\Psi}}_2 = \tilde{\Psi}_2 + 2b\tilde{\Psi}_3 + b^2\tilde{\Psi}_4, \quad \tilde{\tilde{\Psi}}_3 = \tilde{\Psi}_3 + b\tilde{\Psi}_4, \quad \tilde{\tilde{\Psi}}_4 = \tilde{\Psi}_4,
\end{aligned} \tag{88}$$

where the parameter b is another complex scalar.

Let initially $\Psi_1 = \Psi_3 = 0$ (as in equation (86)). Then a rotation of class I with a parameter a, will result in a set $\tilde{\Psi}_1, \ldots, \tilde{\Psi}_4$ given by

$$\begin{aligned}
&\tilde{\Psi}_0 = \Psi_0, \quad \tilde{\Psi}_1 = a^*\Psi_0, \quad \tilde{\Psi}_2 = \Psi_2 + (a^*)^2\,\Psi_0,\\
&\tilde{\Psi}_3 = 3a^*\Psi_2 + (a^*)^3\,\Psi_0, \quad \tilde{\Psi}_4 = \Psi_4 + 6(a^*)^2\,\Psi_2 + (a^*)^4\,\Psi_0.
\end{aligned} \tag{89}$$

We now choose

$$(a^*)^2 = -3\Psi_2/\Psi_0, \tag{90}$$

so that

$$\tilde{\Psi}_3 = 0. \tag{91}$$

At the same time, by virtue of the relation $\Psi_0 \Psi_4 = 9\Psi_2^2$ among the initial set,

$$\tilde{\Psi}_4 = \Psi_4 - 18\frac{\Psi_2^2}{\Psi_0} + 9\frac{\Psi_2^2}{\Psi_0^2}\,\Psi_0 = \frac{1}{\Psi_0}\,(\Psi_0\,\Psi_4 - 9\Psi_2^2) = 0. \tag{92}$$

† We are grateful to the referee for pointing out that this theorem follows directly from a discussion of the 'invariants' as described in Kramer *et al.* (1980, p. 64).

We are thus left with

$$\tilde{\Psi}_0 = \Psi_0, \quad \tilde{\Psi}_1 = a^*\Psi_0, \quad \tilde{\Psi}_2 = -2\Psi_2, \quad \tilde{\Psi}_3 = \tilde{\Psi}_4 = 0. \tag{93}$$

We now subject the tetrad to a rotation of class II with parameter b. The set of Weyl scalars (93) will become

$$\left.\begin{aligned} &\tilde{\tilde{\Psi}}_4 = 0, \quad \tilde{\tilde{\Psi}}_3 = 0, \quad \tilde{\tilde{\Psi}}_2 = \tilde{\Psi}_2 = -2\Psi_2, \\ &\tilde{\tilde{\Psi}}_1 = \tilde{\Psi}_1 + 3b\tilde{\Psi}_2, \quad \tilde{\tilde{\Psi}}_0 = \Psi_0 + 4b\tilde{\Psi}_1 + 6b^2\tilde{\Psi}_2. \end{aligned}\right\} \tag{94}$$

We now choose,

$$b = -\frac{\tilde{\Psi}_1}{3\tilde{\Psi}_2} = \tfrac{1}{6}a^*\frac{\Psi_0}{\Psi_2}, \tag{95}$$

so that

$$\tilde{\tilde{\Psi}}_1 = 0, \quad \text{and} \quad \tilde{\tilde{\Psi}}_0 = \Psi_0 + \tfrac{1}{3}(a^*)^2\frac{\Psi_0^2}{\Psi_2} = 0, \tag{96}$$

by our choice of a^* (see equation (90)). Thus, as a result of the two rotations, we are left with

$$\tilde{\tilde{\Psi}}_0 = \tilde{\tilde{\Psi}}_1 = \tilde{\tilde{\Psi}}_3 = \tilde{\tilde{\Psi}}_4 = 0 \quad \text{and} \quad \tilde{\tilde{\Psi}}_2 = -2\Psi_2. \tag{97}$$

The type-D character of the metric now follows.

(*b*) *Local isometry with the Kerr space–time*

The facts, that the space–time we are presently considering is of type-D and that the Kerr space–time, also of type-D, has, interior to its ergo-sphere, a region in which it has two space-like Killing fields, and further that both metrics derive from the same solution of the Ernst equation, suggest that the two space–times may be locally isometric. We shall find that this is indeed the case.

With the substitutions

$$\left.\begin{aligned} &p = \mp\sqrt{(M^2-a^2)}/M, \quad q = \pm a/M, \\ &\eta = \mp(M-r)/\sqrt{(M^2-a^2)}, \quad \mu = \cos\theta, \end{aligned}\right\} \tag{98}$$

consistent with the requirement $p^2+q^2 = 1$, where M and a ($< |M|$) are constants, we have the correspondence,

$$1-p\eta = r/M \quad \text{and} \quad 1-\eta^2 = -\Delta/(M^2-a^2), \tag{99}$$

where Δ now stands for the 'horizon function'

$$\Delta = r^2-2Mr+a^2. \tag{100}$$

In the (r,θ)-coordinates, the metric functions X, Y, and q_2, defined in equations (52) and (53) become

$$X = \frac{1}{M^2}(r^2+a^2\cos^2\theta) = \frac{\rho^2}{M^2}, \quad Y = -\frac{1}{M^2}(\Delta-a^2\delta) = -\frac{1}{M^2}(\rho^2-2Mr), \tag{101}$$

and

$$q_2 = \frac{2q}{p(1+p)} - \frac{2q\delta(1-p\eta)}{pY} = \frac{2q}{p(1+p)} - \frac{2aMr\sin^2\theta}{(\Delta-a^2\delta)\sqrt{(M^2-a^2)}}. \tag{102}$$

Inserting these expressions in equation (51), we find

$$ds^2 = \frac{\Delta - a^2\delta}{\rho^2}\left[\left(dx^1 - \frac{2q}{p(1+p)}dx^2\right) + \frac{2aMr\sin^2\theta}{(\Delta - a^2\delta)\sqrt{(M^2-a^2)}}dx^2\right]^2$$
$$-\frac{\Delta\rho^2\sin^2\theta}{M^2(\Delta - a^2\delta)}(d\varphi)^2 - \frac{\rho^2}{M^2\Delta}[(dr)^2 + \Delta(d\theta)^2]. \quad (103)$$

With the further substitutions,

$$t = M\left[x^1 - \frac{2q}{p(1+p)}x^2\right] \quad \text{and} \quad \varphi = \frac{M}{\sqrt{(M^2-a^2)}}x^2, \quad (104)$$

the metric becomes

$$M^2\,ds^2 = \frac{\Delta - a^2\delta}{\rho^2}\left[dt + \frac{2aMr\sin^2\theta}{\Delta - a^2\delta}d\varphi\right]^2 - \frac{\rho^2}{\Delta}[(dr)^2 + \Delta(d\theta)^2] - \frac{\Delta\rho^2\sin^2\theta}{\Delta - a^2\delta}(d\varphi)^2. \quad (105)$$

By making use of the relations,

$$\rho^4\Delta - 4a^2M^2r^2\sin^2\theta = \Sigma^2(\Delta - a^2\delta), \quad (106)$$

$$\Sigma^2 = (r^2+a^2)^2 - a^2\Delta\delta = \rho^2(r^2+a^2) + 2a^2Mr\sin^2\theta, \quad (107)$$

and

$$\Delta - a^2\delta = \rho^2 - 2Mr, \quad (108)$$

we readily verify that the metric (105) is, indeed, the Kerr metric in its standard form. It should, however, be emphasized that the isometry that is established is only *local*: in particular, it maps a linear combination of the Killing vectors ∂_{x^1} and ∂_{x^2}, with open orbits, to ∂_φ with closed orbits (cf. equation (112) below).

It is of interest to inquire what the null boundaries of region I_0, namely,

$$s = 1 - u^2 - v^2 = 0, \quad u = 0, \quad \text{and} \quad v = 0. \quad (109)$$

correspond to in the isometry. Because

$$\eta = 1 \quad \text{when} \quad s = 0, \quad \eta = -\mu \quad \text{when} \quad u = 0,$$
$$\text{and} \quad \eta = +\mu \quad \text{when} \quad v = 0, \quad (110)$$

it follows that these boundaries correspond, respectively, to

$$r = M \pm \sqrt{(M^2-a^2)}, \ (s=0); \quad r = M \mp (M^2-a^2)^{\frac{1}{2}}\cos\theta, \ (u=0),$$

and

$$r = M \pm (M^2-a^2)^{\frac{1}{2}}\cos\theta, \ (v=0). \quad (111)$$

Thus, the region of the Kerr space–time that is mapped by the isometry is to the interior of the two ergo-spheres (the internal and the external) i.e. in the region where the Kerr metric does have two space-like Killing vectors. It is noteworthy that the boundary $s = 0$, where the coordinate singularity of the metric (51) occurs, coincides, in the isometry, with the event horizon (for the + sign) or the Cauchy horizon (for the − sign).

We may notice here that by virtue of the transformation equations (104), the Killing vectors, ∂_{x^1} and ∂_{x^2}, which have the contravariant components (0, 0, 1, 0) and (0, 0, 0, 1) in the (η, μ, x^1, x^2)-coordinates, have in the (t, r, θ, φ)-coordinates the components

$$\partial_{x^1} = (M, 0, 0, 0) \quad \text{and} \quad \partial_{x^2} = \left(-\frac{2qM}{p(1+p)}, 0, 0, \frac{M}{\sqrt{(M^2-a^2)}}\right) \tag{112}$$

where

$$\frac{2q}{p(1+p)} = -\frac{2M[M \pm \sqrt{(M^2-a^2)}]}{a\sqrt{(M^2-a^2)}}. \tag{113}$$

Equations (112) express the Killing vectors ∂_{x^1} and ∂_{x^2} in terms of the time-translational and the rotational Killing vectors of the Kerr metric in the isometry established.

7. The extension of the space–time across $1-u^2-v^2 = 0$ in region I

In the null coordinates, u and v, defined in equation (15), the metric (51) takes the form

$$ds^2 = X\frac{4\,du\,dv}{\sqrt{[(1-u^2)(1-v^2)]}} - (1-u^2-v^2)^2\frac{X}{Y}(dx^2)^2 - \frac{Y}{X}(dx^1 - q_2\,dx^2)^2, \tag{114}$$

where, it may be recalled that

$$X = (1-p\eta)^2 + q^2\mu^2, \quad Y = 1-p^2\eta^2-q^2\mu^2 = p^2\Delta + q^2\delta, \tag{115}$$

and

$$q_2 = \frac{2q}{(1+p)}\frac{(1+p)\mu^2-(1-p\eta)}{(1+\eta)\delta Y}(1-u^2-v^2)^2. \tag{116}$$

Written in this form, it is apparent that the determinant of the metric vanishes on the surface $u^2+v^2 = 1$. Further, the norm of the Killing vector ∂_{x^2} and its scalar product with the other Killing vector ∂_{x^1}, given by

$$|\partial_{x^2}|^2 = -\frac{1}{XY}\left\{X^2 + \left(\frac{2q}{1+p}\right)^2\left[\frac{(1+p)\mu^2-(1-p\eta)}{(1+\eta)\delta}\right]^2(1-u^2-v^2)^2\right\}(1-u^2-v^2)^2, \tag{117}$$

and

$$(\partial_{x^1}\cdot\partial_{x^2}) = \frac{2q}{1+p}\frac{(1+p)\mu^2-(1-p\eta)}{(1+\eta)\delta X}(1-u^2-v^2)^2, \tag{118}$$

also vanish on $u^2+v^2 = 1$. Therefore, ∂_{x^2} becomes null when $u^2+v^2 = 1$; but ∂_{x^1} remains space-like. This 'asymmetry' in the vectors ∂_{x^1} and ∂_{x^2} is by choice: it is a direct consequence of the choice of the constant of integration that was made in selecting the solution for q_2 in §4 (see equation (46) and the remark following it). A different choice would only have meant that some linear combination of the two Killing vectors will have played the present role of ∂_{x^2}; but it is convenient to have assigned to ∂_{x^2} the distinguishing role.

We have seen in §6 that no curvature singularity develops on $u^2+v^2 = 1$. But the vanishing of the determinant of the metric (114) on this surface implies that at the least a coordinate singularity obtains here. We shall accordingly seek a transformation of coordinates which will free the metric of this singularity.

First, it is convenient to rewrite the metric (114) in terms of the variables,

$$s = 1-u^2-v^2 \quad \text{and} \quad r = u^2-v^2, \tag{119}$$

which we have found useful in other contexts (in Papers III and V, for example). The metric becomes

$$ds^2 = \frac{X}{4uv\sqrt{[(1-u^2)(1-v^2)]}}[(\mathrm{d}s)^2-(\mathrm{d}r)^2]-(1-u^2-v^2)^2\frac{X}{Y}(\mathrm{d}x^2)^2 - \frac{Y}{X}(\mathrm{d}x^1-q_2\,\mathrm{d}x^2)^2. \tag{120}$$

To motivate the transformation we shall presently adopt, it is useful to consider, first, the limiting form of the metric (120) when $s \to 0$. Because

$$\eta \to 1, \quad \mu \to r, \quad \text{and} \quad \delta \to 1-r^2 \quad \text{when} \quad s \to 0, \tag{121}$$

we find

$$X \to (1-p)[(1-p)+(1+p)r^2], \quad Y \to q^2(1-r^2), \tag{122}$$

and

$$q_2 \to \frac{1}{q(1+p)}\frac{(1+p)r^2-(1-p)}{(1-r^2)^2}s^2, \tag{123}$$

and the metric assumes the form

$$ds^2 \simeq (1-p)\frac{(1-p)+(1+p)r^2}{(1-r^2)}\left[(\mathrm{d}s)^2-\frac{s^2}{q^2}(\mathrm{d}x^2)^2-(\mathrm{d}r)^2\right] + \frac{2}{q}\frac{(1+p)r^2-(1-p)}{(1-r^2)[(1-p)+(1+p)r^2]}s^2\,\mathrm{d}x^1\,\mathrm{d}x^2 - \frac{(1+p)(1-r^2)}{(1-p)+(1+p)r^2}(\mathrm{d}x^1)^2. \tag{124}$$

Before we had reduced the metric to its limiting form (124), one of us (S.C.) had the opportunity to discuss this matter of the extension of the metric (120) across $u^2+v^2 = 1$ with Professor Roger Penrose. With astonishing insight, he suggested a manner of transformation which proved successful. In some measure, the suggested transformation is made 'obvious' by the metric in its reduced limiting form (124). We set

$$\xi = s\,\mathrm{e}^{+cx^2} \quad \text{and} \quad \zeta = s\,\mathrm{e}^{-cx^2}, \tag{125}$$

where c is, for the present, an unspecified constant. By this transformation

$$2\,\mathrm{d}s = s\left(\frac{\mathrm{d}\xi}{\xi}+\frac{\mathrm{d}\zeta}{\zeta}\right) \quad \text{and} \quad 2\,\mathrm{d}x^2 = \frac{1}{c}\left(\frac{\mathrm{d}\xi}{\xi}-\frac{\mathrm{d}\zeta}{\zeta}\right). \tag{126}$$

Therefore, with the choice

$$c = 1/q, \tag{127}$$

$$(\mathrm{d}s)^2-\frac{s^2}{q^2}(\mathrm{d}x^2)^2 = \mathrm{d}\xi\,\mathrm{d}\zeta, \tag{128}$$

and the metric (124) becomes

$$ds^2 \simeq (1-p)\frac{(1-p)+(1+p)r^2}{1-r^2}[\mathrm{d}\xi\,\mathrm{d}\zeta-(\mathrm{d}r)^2] + \frac{(1+p)r^2-(1-p)}{(1-r^2)[(1-p)+(1+p)r^2]}(\zeta\,\mathrm{d}\xi-\xi\,\mathrm{d}\zeta)\,\mathrm{d}x^1 - \frac{(1+p)(1-r^2)}{(1-p)+(1+p)r^2}(\mathrm{d}x^1)^2. \tag{129}$$

It is manifest that by the transformation the coordinate singularity has been eliminated.

More generally, with the substitutions

$$\xi = s\,\mathrm{e}^{+x^2/q} \quad \text{and} \quad \zeta = s\,\mathrm{e}^{-x^2/q} \quad (\xi\zeta = s^2;\, \xi/\zeta = \mathrm{e}^{2x^2/q}), \tag{130}$$

the exact metric (120) becomes

$$\mathrm{d}s^2 = \frac{X}{4\xi\zeta}\left[\frac{1}{H}(\zeta\,\mathrm{d}\xi+\xi\,\mathrm{d}\zeta)^2 - \frac{q^2}{Y}(\zeta\,\mathrm{d}\xi-\xi\,\mathrm{d}\zeta)^2\right] - \frac{Y}{X}\left[\mathrm{d}x^1-(1-p)\frac{Q}{Y}(\zeta\,\mathrm{d}\xi-\xi\,\mathrm{d}\zeta)\right]^2 - \frac{X}{H}(\mathrm{d}r)^2, \tag{131}$$

where

$$H = 4uv\surd[(1-u^2)(1-v^2)] = \eta^2-\mu^2 = \delta-\Delta, \tag{132}$$

$$Q = \frac{(1+p)\mu^2-(1-p\eta)}{(1+\eta)(1-\mu^2)}, \tag{133}$$

and η and μ, expressed in terms of ξ, ζ, and r, are given by

$$\left.\begin{aligned}\eta &= \tfrac{1}{2}\{\surd[(1+r)^2-\xi\zeta]+\surd[(1-r)^2-\xi\zeta]\},\\ \mu &= \tfrac{1}{2}\{\surd[(1+r)^2-\xi\zeta]-\surd[(1-r)^2-\xi\zeta]\}.\end{aligned}\right\} \tag{134}$$

We may note here for future reference that H, as defined in equation (132), has the alternative forms

$$H = \surd[1-(s+r)^2]\surd[1-(s-r)^2] = \surd[(1+r)^2-\xi\zeta]\surd[(1-r)^2-\xi\zeta] = \surd[(1+\xi\zeta-r^2)^2-4\xi\zeta]. \tag{135}$$

With the aid of the relations,

$$Y-q^2H = \Delta = \frac{s^2}{\delta}, \quad Y+q^2H = 2q^2\delta+(p^2-q^2)\Delta, \tag{136}$$

we can rewrite the metric (131) in the following form in which the absence of any singularity when $\xi = 0$ and/or $\zeta = 0$ is made manifest:

$$\mathrm{d}s^2 = \frac{X}{4HY\delta}\{\zeta^2(\mathrm{d}\xi)^2+\xi^2(\mathrm{d}\zeta)^2+2[2q^2\delta^2+(p^2-q^2)\Delta\delta]\,\mathrm{d}\xi\,\mathrm{d}\zeta\} - \frac{Y}{X}\left[\mathrm{d}x^1-(1-p)\frac{Q}{Y}(\zeta\,\mathrm{d}\xi-\xi\,\mathrm{d}\zeta)\right]^2 - \frac{X}{H}(\mathrm{d}r)^2. \tag{137}$$

The determinant of this metric is given by

$$-\frac{X^2q^2}{4H^2}. \tag{138}$$

A singularity when $X = 0$ is indicated. We shall show in §8 below that, in fact, a curvature singularity develops in the extended domain when $X = 0$.

8. The extended space–time

The correspondence of the space–time manifold (with coordinates ξ, ζ, r and x^1) described by the metric (137) and the manifold (with coordinates u, v, x^1, and x^2) described by the metric (120), is illustrated in figures 1 and 2.

In the (ξ, ζ)-plane, the curves of constant x^2 are the radial lines through the origin, the ξ-axis ($\zeta = 0$) and the ζ-axis ($\xi = 0$) corresponding, respectively, to $x_2 \to +\infty$ and $x_2 \to -\infty$; and $s = 0$ along the two axes. The curves of constant $s > 0$ are hyperbolae in the positive (ξ, ζ)-quadrant. In particular, the hyperbola $\xi\zeta = 1$ defines the instant of collision, $u = 0, v = 0$, and $-\infty < x^2 < +\infty$. It should, however, be noted that the description of the space–time in the (ξ, ζ)-coordinates breaks down on the null boundaries, separating regions I from regions II and III, when $u = 0$ and $0 \leqslant v \leqslant 1$ *or* $v = 0$ and $0 \leqslant u \leqslant 1$: for, then *either* $s - r\,(= 1-2u^2) = 1$ *or* $s + r\,(= 1-2v^2) = 1$; and in either case $H = 0$ (by the first of equations (135)). In other words, the space–time described by the metric (137) includes the space–time described by the metric (120) *exclusive* of the null boundaries at $u = 0$ and $v = 0$.

The manifold described by the (ξ, ζ, r, x^1)-coordinates includes, besides the region I_0 in the positive (ξ, ζ)-quadrant, three additional quadrants. Considering first the negative (ξ, ζ)-quadrant ($\xi < 0$ and $\zeta < 0$), we observe that the metric (137) is invariant to a simultaneous change in the signs of ξ and ζ; therefore, *the space–time represented in the two quadrants*, $(\xi \geqslant 0, \zeta \geqslant 0)$ *and* $(\xi \leqslant 0, \zeta \leqslant 0)$ *is isometric*. With the same transformation, the coordinates in terms of which the manifold is now described, namely, $s < 0$ and $-1 < r < +1$, again break down when either $r + s = -1$ or $r - s = +1$ and H vanishes (by one of equations (135)). It is, however, clear that by the coordinate transformation,

$$\xi = (\tilde{u}^2 + \tilde{v}^2 - 1)\, e^{+x^2/q}, \quad \zeta = (\tilde{u}^2 + \tilde{v}^2 - 1)\, e^{-x^2/q},$$

and
$$r = \tilde{v}^2 - \tilde{u}^2, \quad (0 \leqslant \tilde{u} \leqslant 1, 0 \leqslant \tilde{v} \leqslant 1), \tag{139}$$

we can eliminate the singularity; and the transformed metric in the $(\tilde{u}, \tilde{v})$-variables will be isometric to (120) with the correspondence

$$\tilde{u} \rightleftharpoons \surd(1 - v^2) \quad \text{and} \quad \tilde{v} \rightleftharpoons \surd(1 - u^2), \tag{140}$$

or, equivalently,

$$u \rightleftharpoons \surd(1 - \tilde{v}^2) \quad \text{and} \quad v \rightleftharpoons \surd(1 - \tilde{u}^2). \tag{141}$$

At the same time the variables η and μ, defined in terms of u and v (as in equation (15)) will correspond to $\tilde{\eta}$ and $\tilde{\mu}$, defined in identical manner in terms of $\tilde{u}$ and $\tilde{v}$.

The dissimilarity (in appearance) of the regions I_0 and I_e in the (u, v)-plane, in spite of their isometry, can be rectified by the choice of the coordinates (see Paper III, Appendix A),

$$\left.\begin{aligned} &x = 1 - 2u^2, \quad y = 1 - 2v^2, \\ \text{and} \quad &\tilde{x} = 1 - 2\tilde{u}^2 = 2v^2 - 1 = -y \quad \text{and} \quad \tilde{y} = 1 - 2\tilde{v}^2 = 2u^2 - 1 = -x, \end{aligned}\right\} \tag{142}$$

in place of u and v and $\tilde{u}$ and $\tilde{v}$ (see figure 3).

We now turn to the nature of the space–time in the domains, $(\xi > 0, \zeta < 0)$ and

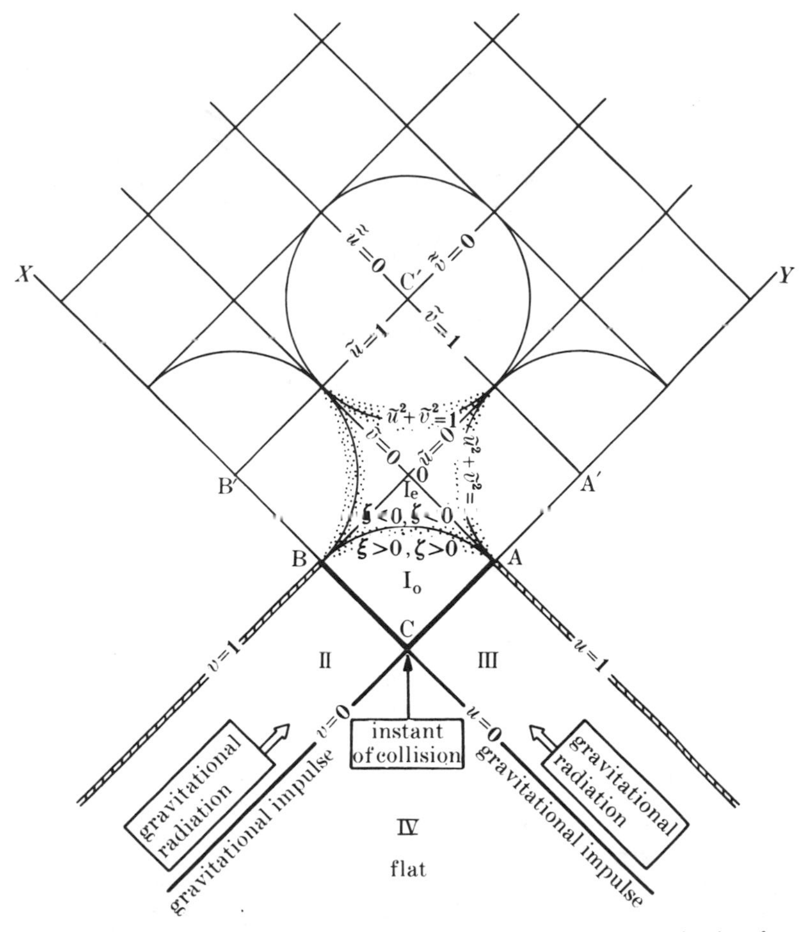

FIGURE 1. The space–time diagram for colliding plane impulsive gravitational waves each accompanied by a gravitational shock-wave. The colliding waves are propagated along the null directions u and v. The time coordinate is along the vertical and the directions of wave-propagation are along the horizontal. The plane of the wavefronts on which the geometry is invariant is orthogonal to the plane of the diagram.

The collision occurs at C. Regions IV, II, and III represent the space–time before the arrival of either of the two wavefronts or after the passage of one or the other of them. The scattering of the waves (one off the other) occurs to the future of C.

In the problem considered, the surface, $s = 1-u^2-v^2 = 0$, represented by the arc BA, is null; further, a coordinate singularity occurs in the (u, v)-coordinate system when $s = 0$. The extension of the metric across $s = 0$ is achieved by the (ξ, ζ, r, x^1)-chart described in §7 (see figure 2). The extended metric includes two additional regions, $(\xi > 0, \zeta < 0)$ and $(\xi < 0, \zeta > 0)$ indicated in the diagram by the distribution of dots along BA. By the extension, the regions I_o and I_e are isometric; and they are naturally represented by the null coordinates (u, v) and $(\tilde{u} = \sqrt{(1-v^2)}, \tilde{v} = \sqrt{(1-u^2)})$, respectively.

An analytic extension of the space–time beyond the domain of dependence of the initial data is obtained by first adjoining region I_e by three additional regions by letting $\tilde{u}$ and $\tilde{v}$ take negative values. These regions, isometric to each other and to I_e, can be extended similarly to complete the square CB′C′A′. The maximal extension is now obtained by tiling the entire region (to the future of C) by panels which are replicas of the panel CB′C′A′.

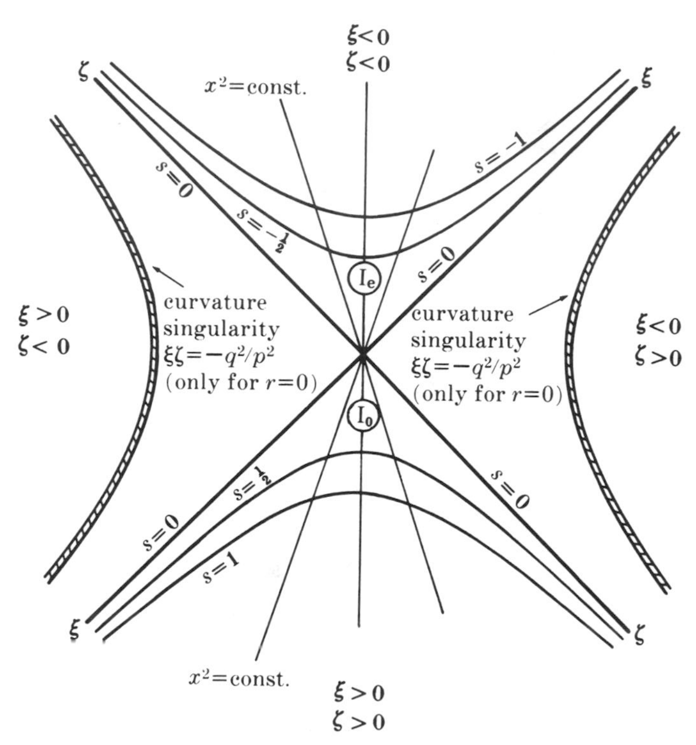

FIGURE 2. The projection of the space–time in the (ξ, ζ)-plane. By the transformation to the (ξ, ζ, r, x^1)-chart, described in §7, the curves of constant x^2 are radial lines through the origin while the curves of constant s are hyperbolae. Regions $\mathrm{I_0}$ and $\mathrm{I_e}$ of figure 1 are represented by areas in the positive and in the negative quadrants that are included between the axes and the hyperbolic arcs $\xi\zeta = +1$. In the regions, $(\xi > 0, \zeta < 0)$ and $(\xi < 0, \zeta > 0)$, included in the extended manifold, time-like curvature singularities occur when $\xi\zeta = -q^2/p^2$ and for $r = 0$ only.

$(\xi < 0, \zeta > 0)$. In these domains, the function H cannot vanish because $\xi\zeta < 0$ (cf. equation (135)). However, the function X and, together with it, the determinant (138) of the metric (137) can vanish. By the definition of X given in equations (115), the conditions for the vanishing of X are

$$\mu = 0 \quad \text{and} \quad \eta = 1/p; \tag{143}$$

and by equations (134) these conditions are equivalent to

$$r = 0 \quad \text{and} \quad \eta = \surd(1-\xi\zeta) = 1/p, \tag{144}$$

or

$$r = 0 \quad \text{and} \quad \xi\zeta = 1-1/p^2 = -q^2/p^2. \tag{145}$$

Because (cf. equations (79), (81), and (82))

$$|\Psi_0|^2 = |\Psi_4|^2 = 9|\Psi_2|^2 = \tfrac{9}{4}X^{-3}, \tag{146}$$

it follows that along the hyperbolic arcs $\xi\zeta = -q^2/p^2$ in the (ξ, ζ)-plane for $r = 0$, we have a time-like curvature-singularity reminiscent of the ring singularity in the Kerr space–time.

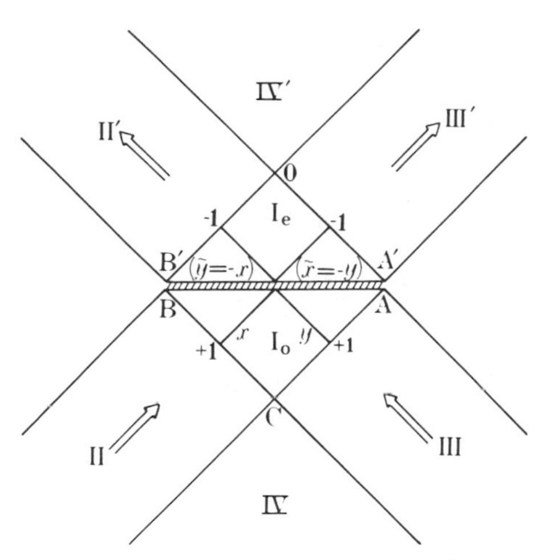

FIGURE 3. An alternative representation of the space–time diagram, which achieves manifest symmetry between the region of the space–time before and after the formation of the null-surface, is provided by the transformation to the coordinates $x = 1-2u^2$ and $y = 1-2v^2$. The regions marked $\mathrm{I_0}$, II, III, IV in this figure and in figure 1 correspond. A permissible extension beyond the domain of dependence, described in §8, is by simply attaching to region $\mathrm{I_e}$, regions II′, III′, and IV′ isometric to regions II, III, and IV. By this C^0 extension the two worlds joined together by the (ξ, ζ, r, x^1)-chart are mirror images of one another.

(a) *The disposition of the light-cones on* $\xi = 0$ *and* $\zeta = 0$

An important aspect of the extended space–time is the nature of the partitioning surfaces at $\xi = 0$ and $\zeta = 0$. First, we conclude, from the vanishing of the contravariant components,

$$g^{\xi\xi} = -\xi^2/(\delta X q^2) \quad \text{and} \quad g^{\zeta\zeta} = -\zeta^2/(\delta X q^2), \tag{147}$$

of the metric tensor on $\xi = 0$ and $\zeta = 0$, that these are *null surfaces*. More generally, the disposition of the light-cones on these surfaces can be ascertained from the two directions, $(\mathrm{d}\xi/\mathrm{d}\zeta)$, of the null-cone projected on the surfaces of constant r and constant x^1. From equation (137), we deduce that the equation determining the required null-directions are given by the roots of the equation,

$$\frac{X}{4HY\delta}\left\{\zeta^2\left(\frac{\mathrm{d}\xi}{\mathrm{d}\zeta}\right)^2 + \xi^2 + 2[2q^2\delta^2 + (p^2-q^2)\,\xi\zeta]\frac{\mathrm{d}\xi}{\mathrm{d}\zeta}\right\} - (1-p)^2\frac{Q^2}{XY}\left(\zeta\frac{\mathrm{d}\xi}{\mathrm{d}\zeta} - \xi\right)^2 = 0. \tag{148}$$

On $\xi = 0$, the equation reduces to

$$[X^2 - 4(1-p)^2 Q^2 H\delta]\,\zeta^2\left(\frac{\mathrm{d}\xi}{\mathrm{d}\zeta}\right)^2 = -4q^2\delta^2 X^2\left(\frac{\mathrm{d}\xi}{\mathrm{d}\zeta}\right), \tag{149}$$

where, by equations (122), (123) and (133), X, Y, Q, H, and δ are given by

$$\left.\begin{aligned} &X = (1-p)\,[(1+p)\,r^2 + (1-p)], \quad Q = [(1+p)\,r^2 - (1-p)]/2(1-r^2), \\ &Y = q^2(1-r^2) \quad \text{and} \quad H = \delta = 1-r^2. \end{aligned}\right\} \tag{150}$$

From equation (149), it now follows that the two required null-directions are

$$\frac{d\xi}{d\zeta}=0 \quad \text{and} \quad \frac{d\xi}{d\zeta}=-\frac{(1-r^2)^2}{r^2\zeta^2}[(1+p)\,r^2+(1-p)]^2. \tag{151}$$

Similarly, on $\zeta = 0$, the two null directions are

$$\frac{d\zeta}{d\xi}=0 \quad \text{and} \quad \frac{d\zeta}{d\xi}=-\frac{(1-r^2)^2}{r^2\xi^2}[(1+p)\,r^2+(1-p)]^2. \tag{152}$$

The disposition of the null cones on $\xi = 0$ and $\zeta = 0$, that follows from equations (151) and (152), is illustrated in figure 4. An important feature, which emerges from

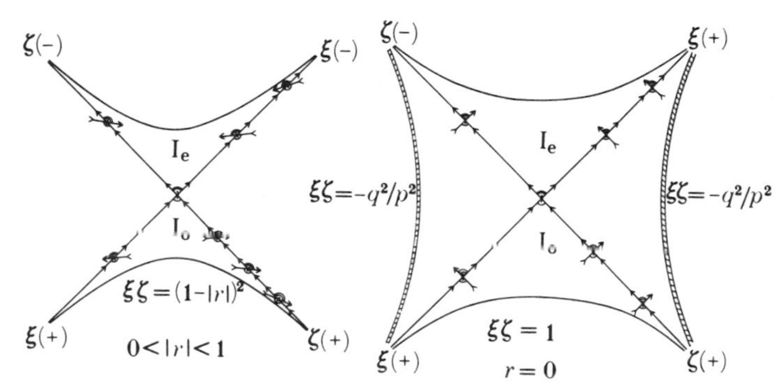

FIGURE 4. The disposition of the light-cones for two sections, r = constant and x^1 = constant, of the space–time: for $0 < |r| < 1$ (on the left) and for $r = 0$ (on the right).

For $0 < |r| < 1$, the light cones have the normal opening of 90° only at the origin ($\xi = \zeta = 0$); and they open out increasingly as one recedes from the origin along the axes and asymptotically tend to 180°. (Note that the regions I_0 and I_e are bounded by the coordinate axes and the hyperbolic arcs $\xi\zeta = (1-|r|)^2$).

For $r = 0$, the light cones have a constant opening of 90° along the entire lengths of the two axes. The section $r = 0$ is distinguished by the occurrence of the curvature singularity along the hyperbolic arcs $\xi\zeta = -q^2/p^2$.

The null surfaces $\xi\zeta = s^2 = 0$ are horizons since no future-directed time-like or null curve starting from the region I_e can cross into the right or the left quadrants of the extended manifold.

these orientations of the light cones on these surfaces, is that while future-directed time-like trajectories, starting from region I_0 (in the positive (ξ, ζ)-quadrant) can enter the region, $(\xi > 0, \zeta < 0)$, or $(\xi < 0, \zeta > 0)$, it is not possible to enter these regions, starting from region I_e (in the negative (ξ, ζ)-quadrant) by future-directed time-like trajectories. In these respects the surfaces, $(\xi = 0, \zeta < 0)$ and $(\xi < 0, \zeta = 0)$ act very much like the event horizon of a black hole. (See note added in proof at end of paper.)

The space–time in the (ξ, ζ, r, x^1)-chart can be visualized, as illustrated and described in figure 5, in the three dimensions ξ, ζ, and r.

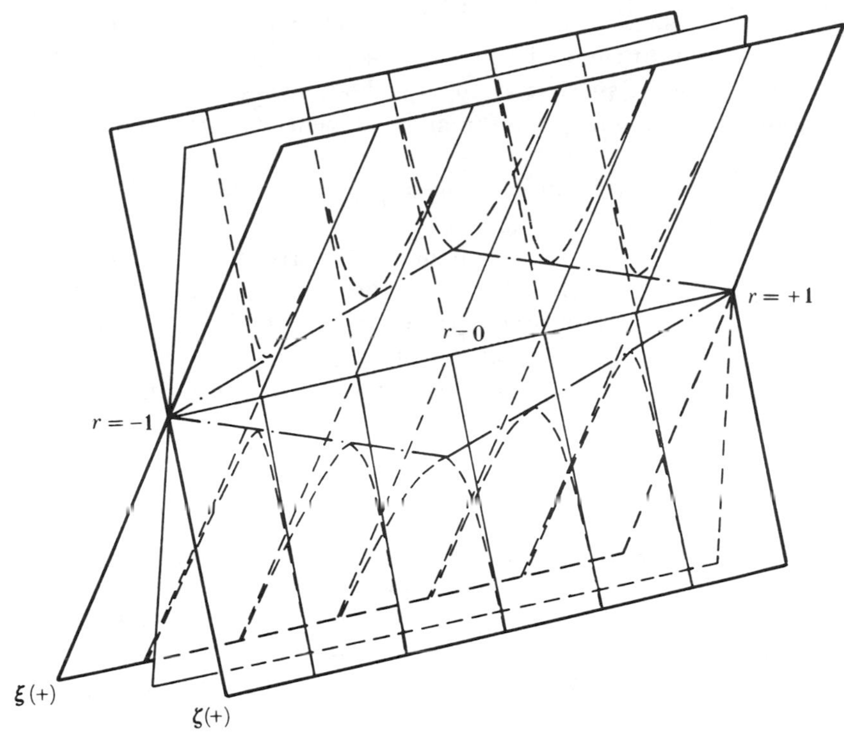

FIGURE 5. A three-dimensional visualization of the extended space–time in the (ξ, ζ, r, x^1)-chart. The coordinate x^1 is suppressed and the range of r is $(-1, +1)$.

The regions included, in each section, $r =$ const., between the two hyperbolic arcs and the ξ- and the ζ-axes, are the regions that we have designated earlier as I_o (in the lower half below the r-axis) and I_e (in the upper half above the r-axis). The space–time in the region CB′C′A′ of figure 1 is thus represented by the two infinite quadrants $(\xi > 0, \zeta < 0)$ and $(\xi < 0, \zeta > 0)$ joined together by the regions I_o and I_e.

(*b*) *The extension of the space–time beyond the domain of dependence of the initial data*

The space–time and its extensions that we have considered in §§4–8, complete the description of the time-development within the entire domain of dependence of the initial data providing for the collision of a pair of impulsive gravitational waves, each accompanied by a shock wave, approaching each other from plus infinity and minus infinity. But the space–time that has unfolded is geodesically incomplete. Its further extension remains, therefore, to be considered.

Since the region I_e is isometric to the region I_o, the simplest permissible extension is to attach to the null boundaries, $\tilde{u} = 0$ and $\tilde{v} = 0$, regions II′ and III′ isometric to regions II and III and include between II′ and III′ a flat region IV′ isometric to IV. By this extension, the world across $s = 0$ is literally a mirror image of the world that was left behind. Besides, the extended space–time manifold includes, as we have seen, the regions, $(\xi > 0, \zeta < 0)$ and $(\xi < 0, \zeta > 0)$, in which curvature singularities along hyperbolic arcs occur.

The extension (beyond the domain of dependence) considered in the preceding paragraph is not a smooth one: it includes H-function discontinuities and δ-function singularities at the null boundaries separating the regions $\mathrm{I_e}$, II′, III′, and IV′, even as they are present at the null boundaries separating the regions $\mathrm{I_o}$, II, III, and IV (and no more!). But a more conventional smooth extension can be obtained as follows.

We first extend the region $\mathrm{I_e}$ smoothly by adjoining to it three additional regions which we obtain by simply allowing $\tilde{u}$ and $\tilde{v}$ to take negative values. It is clear that the three regions adjoined will be bounded by arcs that are *exact* replicas of the arc $s = 0$ separating the regions $\mathrm{I_o}$ and $\mathrm{I_e}$: the arcs will represent null surfaces; and the associated coordinate singularities that will occur on them can be eliminated by coordinate transformations of identical (ξ, ζ)-type. The extended space–times in the new charts will include regions (with curvature singularities) isometric to the regions, $(\xi > 0, \zeta < 0)$ and $(\xi < 0, \zeta > 0)$, that we have described earlier in this section. In this manner, we complete the specification of the space–time in the square CB′C′A′ in figure 1. The maximal analytical extension of the space–time, beyond the domain of dependence, can now be obtained by tiling the entire region XCY (to the future of the instant of collision) by laying side by side (both in the u- and in the v-directions) replicas of the panel CB′C′A′.

It is remarkable that the space–time, resulting from the collision of the gravitational waves that we have considered, should bear such close resemblance to Alice's anticipations with respect to the world *through the looking glass*: 'it (the passage in the looking-glass house) is very like our passage as far as you can see, only it may be quite different on beyond.'

9. Concluding remarks

As we have remarked more than once, the space–time for colliding gravitational waves considered in this paper differs from all earlier ones considered in similar or related contexts, in that a null surface and an associated coordinate singularity occur where (in the other cases) a space-like curvature-singularity sealed off the space–time. However, as we have seen, the analytic extension of the space–time across the null surface reveals the occurrence of time-like curvature singularities along two-dimensional hyperbolic arcs. From the vantage point of the present, the important question appears to be *not* one as to when and where the singularity occurs as a result of the collision, but rather one as to its generic nature: whether it is space-like or time-like. In gravitational collapse, the generic singularity is that of the Kerr black-hole; and it is time-like. It is possible that the same thing is true of colliding waves as well, in which case the example considered in this paper is likely to be the generic one.

We are most indebted to Professor Roger Penrose for the key suggestion with regard to the extension of the space–time considered in §7. We are also grateful to Professor R. Geroch and Professor R. Wald for numerous discussions. The research reported in this paper has, in part, been supported by grants from the National Science Foundation under grant PHY-84-16691 with the University of Chicago. We are also grateful for a grant from the Division of the Physical Sciences of the University of Chicago which made the present collaboration possible.

References

Chandrasekhar, S. 1978 *Proc. R. Soc. Lond.* A **358**, 405–420.
Chandrasekhar, S. 1983 *The mathematical theory of black holes.* Oxford: Clarendon Press.
Chandrasekhar, S. & Ferrari, V. 1984 *Proc. R. Soc. Lond.* A **396**, 55–74.
Chandrasekhar, S. & Ferrari, V. 1985 *Proc. R. Soc. Lond.* A **398**, 429.
Chandrasekhar, S. & Xanthopoulos, B. C. 1985*a* *Proc. R. Soc. Lond.* A **398**, 223–259.
Chandrasekhar, S. & Xanthopoulos, B. C. 1985*b* *Proc. R. Soc. Lond.* A **402**, 37–65.
Chandrasekhar, S. & Xanthopoulos, B. C. 1985*c* *Proc. R. Soc. Lond.* A **402**, 205–224.
Chandrasekhar, S. & Xanthopoulos, B. C. 1986 *Proc. R. Soc. Lond.* A **403**, 189–198.
Ferrari, V. & Ibañez, J. 1986 *A new exact solution for colliding gravitational plane waves.* Preprint.
Halil, M. 1979 *J. math. Phys.* **20**, 120–125.
Khan, K. & Penrose, R. 1971 *Nature, Lond.* **229**, 185–186.
Kramer, D., Stephani, H., Herlt, E. & MacCallum, M. 1980 *Exact solutions of Einstein's field equations.* Cambridge University Press.
Matzner, R. A. & Tipler, F. J. 1984 *Phys. Rev.* D **29**, 1575–1583.
Nutku, Y. & Halil, M. 1977 *Phys. Rev. Lett.* **39**, 1379–1382.
Szekeres, P. 1970 *Nature, Lond.* **228**, 1183–84.
Szekeres, P. 1972 *J. math. Phys.* **13**, 286–294.
Tomimatsu, A. & Sato, H. 1973 *Prog. theor. Phys.* **50**, 95–110.

Appendix A. Expressions for the Weyl scalars explicitly in terms of a solution of the Ernst equation for $\Psi + i\Phi$

A metric suitable for a space–time with two commuting space-like Killing vectors (∂_{x^1} and ∂_{x^2}) can, in general, be cast in the form

$$ds^2 = U^2[(d\psi)^2 - (d\theta)^2] - \frac{V^2}{1-|\mathscr{E}|^2}|(1-\mathscr{E})\,dx^1 + i(1+\mathscr{E})\,dx^2|^2, \tag{A 1}$$

where

$$\left.\begin{aligned} \mathscr{E} &= \frac{\mathscr{X}-1}{\mathscr{X}+1}, \quad \mathscr{X} = \chi + iq_2 = \frac{1+\mathscr{E}}{1-\mathscr{E}}; \\ 1-|\mathscr{E}|^2 &= \frac{4\chi}{(\chi+1)^2+q_2^2}, \quad \chi = \frac{1-|\mathscr{E}|^2}{|1-\mathscr{E}|^2}, \quad iq_2 = \frac{\mathscr{E}-\mathscr{E}^*}{|1-\mathscr{E}|^2}, \end{aligned}\right\} \tag{A 2}$$

and U, V, and $\mathscr{E}$ are all functions of ψ and θ only. A convenient null tetrad-frame for the description of the space–time in a Newman–Penrose formalism is provided by (cf. Paper II, equation (124))

$$\left.\begin{array}{rlcccc} & & \psi & \theta & x^1 & x^2 \\ (l_i) = & +\dfrac{1}{\sqrt{2}}(& U, & -U, & 0, & 0), \\ (n_i) = & +\dfrac{1}{\sqrt{2}}(& U, & +U, & 0, & 0), \\ (m_i) = & -\dfrac{1}{\sqrt{2}}(& 0, & 0, & V\mathscr{Q}_-^*, & -iV\mathscr{Q}_+^*), \\ (\overline{m}_i) = & -\dfrac{1}{\sqrt{2}}(& 0, & 0, & V\mathscr{Q}_-, & +iV\mathscr{Q}_+), \end{array}\right\} \tag{A 3}$$

where

$$\mathscr{Q}_\pm = \frac{1 \pm \mathscr{E}}{\sqrt{(1-|\mathscr{E}|^2)}}. \tag{A 4}$$

With the null frame chosen in this manner, the non-vanishing Weyl scalars are given by (cf. Paper II, equations (130)–(132) and Paper IV, equations (A 4)

$$\Psi_2 = \frac{1}{2U^2}\left\{-[(\ln V)_{,\psi}]^2+[(\ln V)_{,\theta}]^2+\frac{(\mathscr{E}_{,\psi}-\mathscr{E}_{,\theta})(\mathscr{E}^*_{,\psi}+\mathscr{E}^*_{,\theta})}{(1-|\mathscr{E}|^2)^2}\right\}, \qquad (\text{A } 5)$$

$$\Psi_4 = \frac{1}{2U^2(1-|\mathscr{E}|^2)}\left\{2(\mathscr{E}_{,\psi}-\mathscr{E}_{,\theta})[(\ln V)_{,\psi}-(\ln V)_{,\theta}-(\ln U)_{,\psi}+(\ln U)_{,\theta}]\right.$$
$$\left.+\left(\frac{\partial}{\partial\psi}-\frac{\partial}{\partial\theta}\right)^2\mathscr{E}+\frac{2\mathscr{E}^*(\mathscr{E}_{,\psi}-\mathscr{E}_{,\theta})^2}{1-|\mathscr{E}|^2}\right\}, \qquad (\text{A } 6)$$

$$\Psi_0^* = \frac{1}{2U^2(1-|\mathscr{E}|^2)}\left\{2(\mathscr{E}_{,\psi}+\mathscr{E}_{,\theta})[(\ln V)_{,\psi}+(\ln V)_{,\theta}-(\ln U)_{,\psi}-(\ln U)_{,\theta}]\right.$$
$$\left.+\left(\frac{\partial}{\partial\psi}+\frac{\partial}{\partial\theta}\right)^2\mathscr{E}+\frac{2\mathscr{E}^*(\mathscr{E}_{,\psi}+\mathscr{E}_{,\theta})^2}{1-|\mathscr{E}|^2}\right\}. \qquad (\text{A } 7)$$

These expressions are far too complicated for a direct evaluation of the Weyl scalars if $\mathscr{E}$ is not *directly* derived from the Ernst equation for $\chi+\mathrm{i}q_2$ but *indirectly* via the equations

$$\left.\begin{aligned} \frac{\chi_{,\psi}}{\chi} &= 2(\ln V)_{,\psi}-\frac{\Psi_{,\psi}}{\Psi}; \quad \frac{q_{2,\psi}}{\chi}=\frac{\Phi_{,\theta}}{\Psi}, \\ \frac{\chi_{,\theta}}{\chi} &= 2(\ln V)_{,\theta}-\frac{\Psi_{,\theta}}{\Psi}; \quad \frac{q_{2,\theta}}{\chi}=\frac{\Phi_{,\psi}}{\Psi}, \end{aligned}\right\} \qquad (\text{A } 8)$$

where Ψ and Φ are expressed directly in terms of a solution E of an Ernst equation obtained by the substitutions

$$Z = \Psi+\mathrm{i}\Phi = \frac{1+E}{1-E}, \quad \Psi = \frac{1-|E|^2}{|1-E|^2}, \quad \text{and} \quad \mathrm{i}\Phi = \frac{E-E^*}{|1-E|^2}. \qquad (\text{A } 9)$$

In these cases the evaluation of the Weyl scalars will be much simpler if equations (A 5)–(A 7) are expressed in terms of E rather than in terms of $\mathscr{E}$. We shall now show how this can be accomplished.

First, with the aid of the following directly verifiable relations

$$\mathscr{Z}_{,\psi} = \frac{2\mathscr{E}_{,\psi}}{(1-\mathscr{E})^2}, \quad \mathscr{Z}_{,\theta} = \frac{2\mathscr{E}_{,\theta}}{(1-\mathscr{E})^2}, \quad \mathscr{Z}_{,\psi,\theta} = \frac{2\mathscr{E}_{,\psi,\theta}}{(1-\mathscr{E})^2}+\frac{4\mathscr{E}_{,\psi}\mathscr{E}_{,\theta}}{(1-\mathscr{E})^3}; \qquad (\text{A } 10)$$

$$\mathscr{Z}_{,\psi,\psi} = \frac{2\mathscr{E}_{,\psi,\psi}}{(1-\mathscr{E})^2}+\frac{4(\mathscr{E}_{,\psi})^2}{(1-\mathscr{E})^3}, \quad \mathscr{Z}_{,\theta,\theta} = \frac{2\mathscr{E}_{,\theta,\theta}}{(1-\mathscr{E})^2}+\frac{4(\mathscr{E}_{,\theta})^2}{(1-\mathscr{E})^3}; \qquad (\text{A } 11)$$

$$\mathscr{E}_{,\psi}\pm\mathscr{E}_{,\theta} = \tfrac{1}{2}(1-\mathscr{E})^2(\mathscr{Z}_{,\psi}\pm\mathscr{Z}_{,\theta}). \qquad (\text{A } 12)$$

and

$$\left(\frac{\partial}{\partial\psi}\pm\frac{\partial}{\partial\theta}\right)^2\mathscr{E}+2\frac{\mathscr{E}^*(\mathscr{E}_{,\psi}\pm\mathscr{E}_{,\theta})^2}{1-|\mathscr{E}|^2} = \tfrac{1}{2}(1-\mathscr{E})^2\left[\left(\frac{\partial}{\partial\psi}\pm\frac{\partial}{\partial\theta}\right)^2\mathscr{Z}-\frac{1}{\chi}(\mathscr{Z}_{,\psi}\pm\mathscr{Z}_{,\theta})^2\right], \qquad (\text{A } 13)$$

we can rewrite equations (A 5)–(A 7) in the forms:

$$\Psi_2 = \frac{1}{2U^2}\left\{-[(\ln V)_{,\psi}]^2 + [(\ln V)_{,\theta}]^2 + \frac{1}{4\chi^2}(\mathscr{Z}_{,\psi} - \mathscr{Z}_{,\theta})(\mathscr{Z}^*_{,\psi} + \mathscr{Z}^*_{,\theta})\right\}, \tag{A 14}$$

$$\Psi_4 = \frac{(1-\mathscr{E})^2}{4U^2(1-|\mathscr{E}|^2)}\left\{2[(\ln V)_{,\psi} - (\ln V)_{,\theta} - (\ln U)_{,\psi} + (\ln U)_{,\theta}](\mathscr{Z}_{,\psi} - \mathscr{Z}_{,\theta}) + \left(\frac{\partial}{\partial\psi} - \frac{\partial}{\partial\theta}\right)^2 \mathscr{Z} - \frac{1}{\chi}(\mathscr{Z}_{,\psi} - \mathscr{Z}_{,\theta})^2\right\}, \tag{A 15}$$

$$\Psi_0^* = \frac{(1-\mathscr{E})^2}{4U^2(1-|\mathscr{E}|^2)}\left\{2[(\ln V)_{,\psi} + (\ln V)_{,\theta} - (\ln U)_{,\psi} - (\ln U)_{,\theta}](\mathscr{Z}_{,\psi} + \mathscr{Z}_{,\theta}) + \left(\frac{\partial}{\partial\psi} + \frac{\partial}{\partial\theta}\right)^2 \mathscr{Z} - \frac{1}{\chi}(\mathscr{Z}_{,\psi} + \mathscr{Z}_{,\theta})^2\right\}. \tag{A 16}$$

Now, with the aid of the following relations (which follow directly from equations (A 7)–(A 9))

$$\left.\begin{aligned} \frac{1}{\chi}(\mathscr{Z}_{,\psi} - \mathscr{Z}_{,\theta}) &= 2[(\ln V)_{,\psi} - (\ln V)_{,\theta}] - \frac{1}{\Psi}(Z_{,\psi} - Z_{,\theta}), \\ \frac{1}{\chi}(\mathscr{Z}_{,\psi} + \mathscr{Z}_{,\theta}) &= 2[(\ln V)_{,\psi} + (\ln V)_{,\theta}] - \frac{1}{\Psi}(Z^*_{,\psi} + Z^*_{,\theta}), \end{aligned}\right\} \tag{A 17}$$

and

$$Z_{,\psi} = \frac{2E_{,\psi}}{(1-E)^2}, \quad Z_{,\theta} = \frac{2E_{,\theta}}{(1-E)^2}, \tag{A 18}$$

as well as equation (A 13) in which $\mathscr{Z}$ and $\mathscr{E}$ are replaced by Z and E (which is permissible by virtue of equations (A 18)), we can express, for example, all the quantities on the right-hand side of equation (A 14) in terms of E. Thus, by making use of the equation,

$$\frac{1}{4\chi^2}(\mathscr{Z}_{,\psi} - \mathscr{Z}_{,\theta})(\mathscr{Z}^*_{,\psi} + \mathscr{Z}^*_{,\theta}) = [(\ln V)_{,\psi}]^2 - [(\ln V)_{,\theta}]^2 + \frac{1}{4\Psi^2}[(Z_{,\psi})^2 - (Z_{,\theta})^2] - \frac{1}{\Psi}[(\ln V)_{,\psi} Z_{,\psi} - (\ln V)_{,\theta} Z_{,\theta}], \tag{A 19}$$

(which is an immediate consequence of equations (A 17)) we find that equation (A 14) for Ψ_2 becomes

$$\Psi_2 = \frac{1}{2U^2}\left\{\frac{(1-E^*)^2}{(1-E)^2(1-|E|^2)^2}[(E_{,\psi})^2 - (E_{,\theta})^2] - 2\frac{(1-E^*)}{(1-E)(1-|E|^2)}[(\ln V)_{,\psi} E_{,\psi} - (\ln V)_{,\theta} E_{,\theta}]\right\}. \tag{A 20}$$

Similarly, rewriting equation (A 15) in the form

$$4U^2 \frac{\chi+1+\mathrm{i}q_2}{\chi+1-\mathrm{i}q_2}\Psi_4 = [(\ln V)_{,\psi} - (\ln V)_{,\theta} - (\ln U)_{,\psi} + (\ln U)_{,\theta}]\frac{2}{\chi}(\mathscr{Z}_{,\psi} - \mathscr{Z}_{,\theta}) + \frac{1}{\chi}\left(\frac{\partial}{\partial\psi} - \frac{\partial}{\partial\theta}\right)(\mathscr{Z}_{,\psi} - \mathscr{Z}_{,\theta}) - \frac{1}{\chi^2}(\mathscr{Z}_{,\psi} - \mathscr{Z}_{,\theta})^2, \tag{A 21}$$

and making use of the identity

$$\frac{(1-\mathscr{E})^2}{1-|\mathscr{E}|^2} = \frac{1-\mathscr{E}}{\chi(1-\mathscr{E}^*)} = \frac{\mathscr{Z}^*+1}{\chi(\mathscr{Z}+1)} = \frac{\chi+1-iq_2}{\chi(\chi+1+iq_2)}, \tag{A 22}$$

we find after some reductions that

$$\begin{aligned} 4U^2\frac{\chi+1+iq_2}{\chi+1-iq_2}\Psi_4 = {} & D^2+(D_{,\psi}-D_{,\theta})-D\overline{D} \\ & -\tfrac{1}{2}D[\Theta(E_{,\psi}-E_{,\theta})+\Theta^*(E^*_{,\psi}-E^*_{,\theta})]+\overline{D}\Theta(E_{,\psi}-E_{,\theta}) \\ & -\Theta\left[E_{,\psi,\psi}+E_{,\theta,\theta}-2E_{,\theta,\psi}+2\frac{(E_{,\psi}-E_{,\theta})^2}{1-E}\right] \\ & +\frac{4}{(1-|E|^2)^2}|E_{,\psi}-E_{,\theta}|^2, \end{aligned} \tag{A 23}$$

where

$$\Theta = \frac{2(1-E^*)}{(1-E)(1-|E|^2)}, \tag{A 24}$$

$$D = 2(\ln V)_{,\psi}-2(\ln V)_{,\theta} \quad \text{and} \quad \overline{D} = 2(\ln U)_{,\psi}-2(\ln U)_{,\theta}. \tag{A 25}$$

By analogous reductions we find

$$\begin{aligned} 4U^2\frac{\chi+1+iq_2}{\chi+1-iq_2}\Psi^*_0 = {} & S^2+(S_{,\psi}+S_{,\theta})-S\overline{S} \\ & -\tfrac{1}{2}S[\Theta(E_{,\psi}+E_{,\theta})+\Theta^*(E^*_{,\psi}+E^*_{,\theta})]+\overline{S}\Theta^*(E^*_{,\psi}+E^*_{,\theta}) \\ & -\Theta^*\left[E^*_{,\psi,\psi}+E^*_{,\theta,\theta}+2E^*_{,\theta,\psi}+2\frac{(E^*_{,\psi}+E^*_{,\theta})^2}{1-E^*}\right] \\ & +\frac{4}{(1-|E|^2)^2}|E_{,\psi}+E_{,\theta}|^2, \end{aligned} \tag{A 26}$$

where

$$S = 2(\ln V)_{,\psi}+2(\ln V)_{,\theta} \quad \text{and} \quad \overline{S} = 2(\ln U)_{,\psi}+2(\ln U)_{,\theta}. \tag{A 27}$$

Because the solutions of the Ernst equations that we use are generally very simple, the evaluation of the expressions (A 20), (A 23), and (A 26) presents no formidable problem.

Appendix B. The separation of the Hamilton–Jacobi equation

The separation of the geodesic equations of motion in the space–time considered in this paper is to be expected in view of its local isometry with the Kerr space–time. But in view of the complexity of the coordinate transformation needed to exhibit the isometry, it is simpler to separate the Hamilton–Jacobi equation *ab initio*. In this Appendix we shall show how this can be accomplished.

The existence of the two space-like Killing vectors ∂_{x^1} and ∂_{x^2} implies that the geodesic equations allow the two conserved momenta (cf. equations (68) and (69))

$$p_{x^1} = \frac{Y}{X}(\dot{x}^1-q_2\dot{x}^2) = p_1 = \text{const.}, \tag{B 1}$$

$$p_{x^2} = \Delta\delta\frac{X}{Y}\dot{x}^2-q_2\frac{Y}{X}(\dot{x}^1-q_2\dot{x}^2) = p_2 = \text{const.}, \tag{B 2}$$

besides the energy integral,

$$\delta_1 = X\left(\frac{\dot{\eta}^2}{\Delta}-\frac{\dot{\mu}^2}{\delta}\right)-\frac{Y}{X}(\dot{x}^1-q_2\dot{x}^2)^2-\Delta\delta\frac{X}{Y}(\dot{x}^2)^2, \tag{B 3}$$

where
$$\delta_1 = \begin{cases} 1 & \text{for time-like geodesics,} \\ 0 & \text{for null geodesics.} \end{cases} \tag{B 4}$$

From the contravariant form of the metric (55), namely

$$g^{ij} = \begin{vmatrix} \frac{\Delta}{X} & 0 & 0 & 0 \\ 0 & -\frac{\delta}{X} & 0 & 0 \\ 0 & 0 & -\left(\frac{X}{Y}+q_2^2\frac{Y}{\Delta\delta X}\right) & -q_2\frac{Y}{\Delta\delta X} \\ 0 & 0 & -q_2\frac{Y}{\Delta\delta X} & -\frac{Y}{\Delta\delta X} \end{vmatrix} \quad (\text{columns } \eta,\ \mu,\ x^1,\ x^2) \tag{B 5}$$

we obtain the Hamilton–Jacobi equation:

$$\begin{aligned} 2\frac{\partial S}{\partial \tau} &= g^{ij}\frac{\partial S}{\partial x^i}\frac{\partial S}{\partial x^j} \\ &= \frac{\Delta}{X}\left(\frac{\partial S}{\partial \eta}\right)^2-\frac{\delta}{X}\left(\frac{\partial S}{\partial \mu}\right)^2-\left(\frac{X}{Y}+q_2^2\frac{Y}{\Delta\delta X}\right)\left(\frac{\partial S}{\partial x^1}\right)^2 \\ &\quad -2q_2\frac{Y}{\Delta\delta X}\frac{\partial S}{\partial x^1}\frac{\partial S}{\partial x^2}-\frac{Y}{\Delta\delta X}\left(\frac{\partial S}{\partial x^2}\right)^2. \end{aligned} \tag{B 6}$$

Assuming that the variables can be separated, we seek a solution of equation (B 6) of the form,

$$S = \tfrac{1}{2}\delta_1\tau+p_1x^1+p_2x^2+S_\eta(\eta)+S_\mu(\mu), \tag{B 7}$$

where, as the notation implies, S_η and S_μ are functions only of the variables specified. For this chosen form of the solution, equation (B 6) becomes

$$\delta_1 X = \Delta\left(\frac{dS_\eta}{d\eta}\right)^2-\delta\left(\frac{dS_\mu}{d\mu}\right)^2-\left\{\left(\frac{X^2}{Y}+q_2^2\frac{Y}{\Delta\delta}\right)p_1^2+2q_2\frac{Y}{\Delta\delta}p_1p_2+\frac{Y}{\Delta\delta}p_2^2\right\}. \tag{B 8}$$

Now, considering the terms

$$\frac{X^2}{Y}p_1^2+\frac{Y}{\Delta\delta}(q_2p_1+p_2)^2, \tag{B 9}$$

and substituting for q_2 from equation (53), we obtain

$$\begin{aligned} &\frac{X^2}{Y}p_1^2+\frac{Y}{\Delta\delta}\left[-\frac{2q}{pY}\delta(1-p\eta)p_1+c\right]^2 \\ &= \frac{p_1^2}{Y}\left[X^2-4(1-p\eta)^2+4\frac{Y}{p^2\Delta}(1-p\eta)^2\right]+\frac{c^2p^2}{\delta}+\frac{c^2q^2}{\Delta}-4\frac{p_1cq}{p\Delta}(1-p\eta), \end{aligned} \tag{B 10}$$

where
$$c = p_2 + \frac{2q}{p(1+p)} p_1 = \text{const.} \tag{B 11}$$

Now making use of the identity,
$$X^2 - 4(1-p\eta)^2 = -Y[X + 2(1-p\eta)], \tag{B 12}$$

the first terms on the right-hand side of equation (B 10) can be rewritten in the form
$$\begin{aligned} p_1^2 & \left\{ \frac{4}{p^2 \Delta} (1-p\eta)^2 - [X + 2(1-p\eta)] \right\} \\ & = p_1^2 \left[p^2 \Delta - 4(1-p\eta) + \frac{4}{p^2 \Delta} (1-p\eta)^2 + q^2 \delta \right] \\ & = + \frac{p_1^2}{p^2 \Delta} [(1-p\eta)^2 + q^2]^2 + p_1^2 q^2 \delta. \end{aligned} \tag{B 13}$$

By these reductions, equation (B 8) takes the form
$$\begin{aligned} \delta_1[(1-p\eta)^2 + q^2\mu^2] = & - \left[\delta \left(\frac{dS_\mu}{d\mu} \right)^2 + p_1^2 q^2 \delta + \frac{c^2 p^2}{\delta} \right] \\ & + \Delta \left(\frac{dS_\eta}{d\eta} \right)^2 - \frac{p_1^2}{p^2 \Delta} [(1-p\eta)^2 + q^2]^2 + \frac{4 p_1 c q}{p \Delta} (1-p\eta) - \frac{c^2 q^2}{\Delta}. \end{aligned} \tag{B 14}$$

The separability of the equation is manifest and we infer that
$$\Delta \left(\frac{dS_\eta}{d\eta} \right)^2 - \frac{1}{\Delta} \left\{ \frac{p_1^2}{p^2} [(1-p\eta)^2 + q^2]^2 - \frac{4 p_1 c q}{p} (1-p\eta) + c^2 q^2 \right\} - \delta_1 (1-p\eta)^2 = \mathscr{Q}, \tag{B 15}$$

and
$$\delta \left(\frac{dS_\mu}{d\mu} \right)^2 + p_1^2 q^2 \delta + \frac{c^2 p^2}{\delta} + \delta_1 q^2 \mu^2 = \mathscr{Q}, \tag{B 16}$$

where $\mathscr{Q}$ is a separation constant. With the abbreviations
$$R = \Delta[\mathscr{Q} + \delta_1 (1-p\eta)^2] + \frac{p_1^2}{p^2} [(1-p\eta)^2 + q^2]^2 - \frac{4 p_1 c q}{p} (1-p\eta) + c^2 q^2. \tag{B 17}$$

and
$$\Theta = \delta(\mathscr{Q} - \delta_1 q^2 \mu^2) - c^2 p^2 - p_1^2 q^2 \delta^2 \tag{B 18}$$

the solution for S is given by
$$S = \tfrac{1}{2} \delta_1 \tau + p_1 x^1 + p_2 x^2 + \int^\eta \frac{\sqrt{R}}{\Delta} d\eta + \int^\mu \frac{\sqrt{\Theta}}{\delta} d\mu. \tag{B 19}$$

We may note that by making use of the relations
$$\frac{dS_\eta}{d\eta} = \frac{\sqrt{R}}{\Delta} = p_\eta = \frac{X}{\Delta} \dot{\eta}; \quad \frac{dS_\mu}{d\mu} = \frac{\sqrt{\Theta}}{\delta} = p_\mu = \frac{X}{\delta} \dot{\mu}, \tag{B 20}$$

we obtain from equations (B 15) and (B 16) two alternative expressions for the separation constant $\mathscr{Q}$, namely,
$$\mathscr{Q} = \frac{X^2}{\Delta} \dot{\eta}^2 - \delta_1 (1-p\eta)^2 - \frac{1}{\Delta} \left\{ \frac{p_1^2}{p^2} [(1-p\eta)^2 + q^2]^2 - \frac{4 p_1 c q}{p} (1-p\eta) + c^2 q^2 \right\}, \tag{B 21}$$

and
$$\mathscr{Q} = \frac{X^2}{\delta}\mu^2 + \delta_1 q^2\mu^2 + \frac{c^2p^2}{\delta} + p_1^2 q^2\delta. \tag{B 22}$$

The equivalence of these two expressions follows from the energy integral (B 3).

The equations governing the geodesic motion can be deduced from the solution (B 19) by the standard procedure of setting to zero the partial derivatives of S with respect to the different constants of the motion $-\mathscr{Q}$, δ_1, p_1, and p_2 in this instance. Of the resulting equations the most important is

$$\frac{\partial S}{\partial \mathscr{Q}} = \frac{1}{2}\int^{\eta} \frac{1}{\Delta\sqrt{R}}\frac{\partial R}{\partial \mathscr{Q}}\mathrm{d}\eta + \frac{1}{2}\int^{\mu}\frac{1}{\delta\sqrt{\Theta}}\frac{\partial \Theta}{\partial \mathscr{Q}}\mathrm{d}\mu = 0, \tag{B 23}$$

because it leads to the equation,

$$\int^{\eta}\frac{\mathrm{d}\eta}{\sqrt{R}} = -\int^{\mu}\frac{\mathrm{d}\mu}{\sqrt{\Theta}}, \tag{B 24}$$

which determines the projection of the orbit in the (η, μ)-plane. Indeed, simple considerations relating to the required positivity of Θ already give considerable information concerning the disposition of the orbits.

Note added in proof (11 *August* 1986). A quantity of some significance that characterizes a horizon is the prevailing surface-gravity, κ. Apart from a constant of proportionality, it is determined by evaluating the equation,

$$A^i A_{j;i} = -\tfrac{1}{2}(A^i A_i)_{,j} = \kappa A_j, \tag{i}$$

on the horizon, where A^i denotes the Killing-vector that becomes null on the horizon. For the case on hand, we shall carry out the evaluation in the coordinate system (ξ, r, x^1, ζ) that is smooth on the horizon, $s^2 = \xi\zeta = 0$. The Killing-vector, $\boldsymbol{A}$, which becomes null on the horizon, has the components,

$$A^i = \partial_{x^2} = \frac{1}{q}(\xi, 0, 0, -\zeta). \tag{ii}$$

With the metric given by (137) the evaluation of κ is straightforward; we find

$$\kappa = -\frac{1}{q} \quad \text{on } \xi = 0 \quad \text{and } \kappa = +\frac{1}{q} \text{ on } \zeta = 0. \tag{iii}$$

Therefore *the absolute value of the surface gravity is constant on the horizon*; *but it takes values of opposite signs on* $\xi = 0$ *and* $\zeta = 0$.

The vector $\boldsymbol{A}$ is unspecified to the extent of a constant of proportionality. It can be made determinate by requiring that it is normalized to -1 in the flat region-IV where it becomes space-like. The vector $\boldsymbol{A}$, which in the (ξ, r, x^1, ζ)-coordinate system is given by equation (ii), in the (η, μ, x^1, x^2)-coordinate system is given by

$$\boldsymbol{A} = (0, 0, 0, 1). \tag{iv}$$

The corresponding covariant and contravariant components of $\boldsymbol{A}$ in region IV can be ascertained by the requirement that (cf. (51)),

$$S = A_j\,\mathrm{d}x^j = g_{ij}\,A^i\,\mathrm{d}x^j = q_2\frac{Y}{X}\mathrm{d}x^1 - \left(\Delta\delta\frac{X}{Y} + q_2^2\frac{Y}{X}\right)\mathrm{d}x^2, \tag{v}$$

retains its value when the metric (51) is extended to region IV and the metric takes the form (65) and

$$X = Y = \Delta = \delta = 1 \quad \text{and} \quad q_2 = -2q/(1+p). \tag{vi}$$

Therefore,

$$S = q_2\,dx^1 - (1+q_2^2)\,dx^2 = -\frac{2q}{1+p}\,dx^1 - \frac{5-3p}{1+p}\,dx^2. \tag{vii}$$

We conclude that

$$A_1 = -\frac{2q}{1+p}, \quad A_2 = -\frac{5-3p}{1+p}, \quad A^1 = 0 \quad \text{and} \quad A^2 = 1. \tag{viii}$$

Therefore,

$$A^i A_i = -(5-3p)/(1+p), \tag{ix}$$

which is negative as required by the space-like character of $\boldsymbol{A}$. We can normalize this vector to -1 and consider

$$\tilde{\boldsymbol{A}} = \epsilon\left(\frac{1+p}{5-3p}\right)^{\frac{1}{2}}\boldsymbol{A} \quad (\epsilon = \pm 1), \tag{x}$$

in place of $\boldsymbol{A}$. With this normalized vector, we shall obtain for the surface-gravity on the horizon,

$$\kappa = -\frac{\epsilon}{q}\left(\frac{1+p}{5-3p}\right)^{\frac{1}{2}} \quad \text{on } \xi = 0 \quad \text{and} \quad \kappa = +\frac{\epsilon}{q}\left(\frac{1+p}{5-3p}\right)^{\frac{1}{2}} \quad \text{on } \zeta = 0. \tag{xi}$$

24

On colliding waves that develop time-like singularities: a new class of solutions of the Einstein–Maxwell equations

By Subrahmanyan Chandrasekhar, F.R.S., and Basilis C. Xanthopoulos†

The University of Chicago, Chicago, Illinois 60637, *U.S.A.*

(*Received* 21 *July* 1986)

Solutions of the Einstein–Maxwell equations are found that provide generalizations of a solution discovered by Bell and Szekeres, which represents the collision of impulsive gravitational waves coupled with electromagnetic shock-waves in a conformally flat space-time. Starting with the Bell–Szekeres solution in a form more general than their original one (though equivalent to it) and applying to it a so-called Ehlers transformation, we obtain a new family of Petrov type-D space-times in which horizons form and subsequently two-dimensional time-like singularities develop. A second solution provides a generalization of the Bell–Szekeres solution in the same way as the axisymmetric distorted static black-hole solutions provide a generalization of the Schwarzschild solution. This second solution also forms a horizon but the time-like singularity that develops is three-dimensional.

The mathematical theory that is developed seems specially adapted to the solution of these and related problems.

1. Introduction

In an earlier paper (Chandrasekhar & Xanthopoulos 1986 referred to hereafter as Paper II) we presented an example of a collision of plane-fronted gravitational waves that led to the development of a horizon and subsequent time-like singularity. In this paper, we shall present further examples, in the context of the Einstein–Maxwell equations, which exhibit similar behaviours: a one-parameter family of type-D space-times and a space-time with a metric that is diagonal and as many parameters as one may wish. The solutions that are obtained are, in some sense, generalizations of the one discovered by Bell & Szekeres (1974) that represents the collision of impulsive gravitational waves coupled with electromagnetic waves in a conformally flat space-time.

The plan of the paper is the following. In §2, we outline the theory which provides the base for the solutions obtained in this paper. It derives from the observation (briefly explored in the context of cylindrical waves in a recent paper, Chandrasekhar (1986, Appendix C) that on a certain assumption, the equation governing a complex potential H, describing the electromagnetic field, reduces to

† Permanent address: Department of Physics, University of Crete, and Research Centre of Crete, Iraklion, Greece.

an Ernst equation. The equations determining a one-parameter family of space-times that one may obtain by subjecting an Ernst function to a so-called 'Ehlers transformation' are also derived in this section. In §3, we show that the simplest linear solution of the Ernst equation for H leads to the Bell–Szekeres solution. In §4, we consider the space-times that are obtained by subjecting the linear solution of the Ernst equation to the Ehlers transformation. We show that the one-parameter family of space-times derived in this fashion are of Petrov-type D; that they develop horizons; and that in the manifold extended beyond the horizon, two-dimensional time-like singularities occur (exactly as in the solution considered in Paper II). In §5, we consider the general solution for the case when the two space-like commuting Killing-vectors are hypersurface orthogonal. This solution represents a generalization of the Bell–Szekeres solution even as the solution describing distorted static axisymmetric black-holes is a generalization of the Schwarzschild solution. It is shown that this general solution also develops a horizon; and that in the manifold extended beyond the horizon, *three-dimensional* time-like singularities occur. Finally, in two Appendixes we consider some related matters.

2. The general theory

The Einstein–Maxwell equations, for a space-time with two commuting space-like Killing-vectors, are fully written out in Chandrasekhar & Xanthopoulos (1985 Paper I). We shall quote the principal equations of the theory from that paper without much explanation; indeed, familiarity with the methods and the notations of Paper I (as well as of Paper II) will be assumed.

First, we observe that in a gauge and in a coordinate system, appropriate for describing the collision of plane-fronted gravitational waves, the metric of the space-time, in the region of the interaction of the colliding waves, can be written in the form,

$$\mathrm{d}s^2 = \mathrm{e}^{\nu+\mu_3}\sqrt{\Delta}\left[\frac{(\mathrm{d}\eta)^2}{\Delta}-\frac{(\mathrm{d}\mu)^2}{\delta}\right]-\sqrt{(\Delta\delta)}\left[\chi(\mathrm{d}x^2)^2+\frac{1}{\chi}(\mathrm{d}x^1-q_2\,\mathrm{d}x^2)^2\right], \tag{1}$$

where η measures the time from the instant of collision, μ measures the distance normal to the planes, (x^1, x^2), spanned by the two Killing-vectors,

$$\Delta = 1-\eta^2 = \sin^2\psi \quad \text{and} \quad \delta = 1-\mu^2 = \sin^2\theta. \tag{2}$$

The electromagnetic field is expressed in terms of a complex-potential H; and the field equations relate H to the metric coefficients (I, equation (11)).

One then defines the functions Ψ and Φ, in place of χ and q_2, by

$$\Psi = \frac{\sqrt{(\Delta\delta)}}{\chi}, \tag{3}$$

$$q_{2,0} = \frac{\delta}{\Psi^2}(\Phi_{,3}-2\,\mathscr{Im}\,HH^*_{,3}) \quad \text{and} \quad q_{2,3} = \frac{\Delta}{\Psi^2}(\Phi_{,0}-2\,\mathscr{Im}\,HH^*_{,0}), \tag{4}$$

where the subscripts '0' and '3' refer to η and μ, respectively. (On occasions, we shall revert to η and μ in place of '0' and '3' to avoid ambiguity.) One then finds that the complex functions H and

$$Z = \Psi + |H|^2 - \mathrm{i}\Phi, \tag{5}$$

satisfy the following pair of coupled equations (I, equations (43) and (44))

$$(\mathscr{R}e\, Z - |H|^2)[(\Delta Z_{,0})_{,0} - (\delta Z_{,3})_{,3}] = \Delta (Z_{,0})^2 - \delta (Z_{,3})^2 - 2H^*(\Delta Z_{,0} H_{,0} - \delta Z_{,3} H_{,3}). \tag{6}$$

$$(\mathscr{R}e\, Z - |H|^2)[(\Delta H_{,0})_{,0} - (\delta H_{,3})_{,3}] = \Delta H_{,0} Z_{,0} - \delta H_{,3} Z_{,3} - 2H^*[\Delta (H_{,0})^2 - \delta (H_{,3})^2]. \tag{7}$$

The observation that leads to the new class of solutions obtained in this paper is, simply, that if Z were assumed to be a constant (real or complex), equation (6) governing Z is identically satisfied and we are left with the single equation (7) for H. A constant imaginary part of Z does not affect equation (7); it only makes Φ a constant as is evident from equation (5). Also, a constant Φ does not affect the metric functions as is apparent from equations (4). We can, therefore, assume that Z is a real constant. And since replacing Z by $\alpha^2 Z$ and H by αH, where α is a real constant, leaves equation (7) unchanged, no loss of generality is entailed by the assumption (in present context) that

$$Z = 1. \tag{8}$$

Equation (7) then reduces to

$$(1 - |H|^2)[(\Delta H_{,0})_{,0} - (\delta H_{,3})_{,3}] = -2H^*[\Delta (H_{,0})^2 - \delta (H_{,3})^2]. \tag{9}$$

In other words, H satisfies the Ernst equation for the vacuum in its standard form, namely

$$(1 - |E|^2)[(\Delta E_{,0})_{,0} - (\delta E_{,3})_{,3}] = -2E^*[\Delta (E_{,0})^2 - \delta (E_{,3})^2]. \tag{10}$$

When this observation was made and briefly explored in the context of cylindrical waves (Chandrasekhar 1986, appendix C), we were not aware that it had already been noted by Kramer *et al.* (1980, p. 330); but the observation does not seem to have been pursued.

We assume, then, that

$$H = E, \tag{11}$$

where E is *any* solution of the Ernst equation (10). Then,

$$\Psi = 1 - |H|^2 = 1 - |E|^2 \quad \text{and} \quad \chi = \frac{\sqrt{(\Delta\delta)}}{\Psi} = \frac{\sqrt{(\Delta\delta)}}{1 - |E|^2}, \tag{12}$$

and equation (4) governing q_2 takes the form:

$$q_{2,0} = -\frac{\delta}{\Psi^2}(2\,\mathscr{I}m\, HH^*_{,3}) = \mathrm{i}\frac{\delta}{(1-|E|^2)^2}(EE^*_{,3} - E^*E_{,3}), \tag{13}$$

$$q_{2,3} = -\frac{\Delta}{\Psi^2}(2\,\mathscr{I}m\, HH^*_{,0}) = \mathrm{i}\frac{\Delta}{(1-|E|^2)^2}(EE^*_{,0} - E^*E_{,0}). \tag{14}$$

The remaining metric function, $\nu+\mu_3$, now follows from I, equations (46) and (47); by making use of equations (11)–(14), they can be reduced to the forms (cf. II, equations (12) and (13)):

$$\frac{\mu}{\delta}\left[(\nu+\mu_3)+\ln\frac{\Psi}{\sqrt[4]{(\Delta\delta)}}\right]_{,0}+\frac{\eta}{\Delta}\left[(\nu+\mu_3)+\ln\frac{\Psi}{\sqrt[4]{(\Delta\delta)}}\right]_{,3}$$
$$=-2\,\frac{E_{,0}E^*_{,3}+E^*_{,0}E_{,3}}{(1-|E|^2)^2},\qquad(15)$$

$$2\eta\left[(\nu+\mu_3)+\ln\frac{\Psi}{\sqrt[4]{(\Delta\delta)}}\right]_{,0}+2\mu\left[(\nu+\mu_3)+\ln\frac{\Psi}{\sqrt[4]{(\Delta\delta)}}\right]_{,3}$$
$$=\frac{3}{\Delta}+\frac{1}{\delta}-4\,\frac{\Delta|E_{,0}|^2+\delta|E_{,3}|^2}{(1-|E|^2)^2}.\qquad(16)$$

(a) The Ehlers-transformed solution

It is known that associated with any Ernst equation, we have the equation

$$\mathscr{Re}\,(Z)[(\Delta Z_{,0})_{,0}-(\delta Z_{,3})_{,3}]=\Delta(Z_{,0})^2-\delta(Z_{,3})^2,\qquad(17)$$

where

$$Z=\frac{1+E}{1-E}\quad\text{or}\quad E=\frac{Z-1}{Z+1}.\qquad(18)$$

From equation (17), it follows that *if Z is a solution, then so are Z^{-1} and $Z+2i\beta$ where β is any real constant*. Therefore, $\tilde{Z}$ given by

$$\frac{1}{\tilde{Z}}=\frac{1}{Z}+2i\beta\quad\text{or}\quad\tilde{Z}=\frac{Z}{1+2i\beta Z}\qquad(19)$$

is also a solution of equation (17). The corresponding solution of the Ernst equation is

$$\tilde{E}=\frac{\tilde{Z}-1}{\tilde{Z}+1}=\frac{E-i\beta(1+E)}{1+i\beta(1+E)}=\frac{J}{K}\quad\text{(say)},\qquad(20)$$

where

$$J=E-i\beta(1+E)\quad\text{and}\quad K=1+i\beta(1+E).\qquad(21)$$

We shall call $\tilde{E}$ the *Ehlers transform* of E. By this transformation, we shall obtain, from any given E, a one-parameter family of solutions distinguished by β.

It is a remarkable fact that the right-hand sides of equations (15) and (16) are invariant to Ehler's transformation (20): a fact that is readily verified by making use of the relations,

$$\tilde{E}_{,i}=\frac{E_{,i}}{K^2},\quad|\tilde{E}_{,i}|^2=\frac{|E_{,i}|^2}{|K|^4}\quad(i=0,3);\qquad\tilde{E}_{,0}\tilde{E}^*_{,3}=\frac{E_{,0}E^*_{,3}}{|K|^4}\qquad(22)$$

and

$$\tilde{\Psi}=1-|\tilde{E}|^2=\frac{1}{|K|^2}(1-|E|^2)=\frac{\Psi}{|K|^2}.\qquad(23)$$

From the stated invariance, it follows from equations (15) and (16) that

$$(\tilde{\nu}+\tilde{\mu}_3)+\ln\frac{\tilde{\Psi}}{\sqrt[4]{(\Delta\delta)}}=(\nu+\mu_3)+\ln\frac{\Psi}{\sqrt[4]{(\Delta\delta)}},\qquad(24)$$

or

$$e^{\tilde{\nu}+\tilde{\mu}_3}=\frac{\Psi}{\tilde{\Psi}}e^{\nu+\mu_3}=|K|^2e^{\nu+\mu_3}.\qquad(25)$$

Finally, from equations (13), (14), (22) and (23), we find that the equations governing the Ehlers transform, $\tilde{q}_2$, of q_2 are:

$$\tilde{q}_{2,0} = q_{2,0} - 2\delta\, \mathscr{I}m \left\{ \frac{E^*_{,3}}{(1-|E|^2)^2} [\beta^2(1+E)^2 + i\beta(E^2-1)] \right\}, \tag{26}$$

$$\tilde{q}_{2,3} = q_{2,3} - 2\Delta\, \mathscr{I}m \left\{ \frac{E^*_{,0}}{(1-|E|^2)^2} [\beta^2(1+E)^2 + i\beta(E^2-1)] \right\}. \tag{27}$$

(b) A digression on the relation between vacuum solutions belonging to an Ernst function and its Ehlers transform

From the invariance of the right-hand sides of equations (12) and (13), Paper II, to Ehlers transformation, we conclude that $\tilde{\nu}+\tilde{\mu}_3$ and $\nu+\mu_3$, belonging to an Ernst function and to its Ehlers transform, are related by (cf. equation (25))

$$e^{\tilde{\nu}+\tilde{\mu}_3} = \frac{\Psi}{\tilde{\Psi}} e^{\nu+\mu_3}. \tag{28}$$

However, the functions Ψ and $\tilde{\Psi}$ are now related differently than in equation (23). For, when we are concerned with solutions of the vacuum equations, then, by definition (cf. II, equation (A 9))

$$\Psi = \frac{1-|E|^2}{|1-E|^2} \quad \text{and} \quad \tilde{\Psi} = \frac{1-|\tilde{E}|^2}{|1-\tilde{E}|^2}, \tag{29}$$

where by equations (20) and (21),

$$1-\tilde{E} = \frac{1}{K}[1-E+2i\beta(1+E)] \quad \text{and} \quad 1-|\tilde{E}|^2 = \frac{1}{|K|^2}(1-|E|^2). \tag{30}$$

(Note that we have suppressed the 'dagger' distinguishing E in the equations of II, §2.) Accordingly,

$$\frac{\Psi}{\tilde{\Psi}} = \left| 1 + 2i\beta \frac{1+E}{1-E} \right|^2 = |1+2i\beta Z|^2 = |1+2i\beta(\Psi + i\Phi)|^2. \tag{31}$$

Equation (28) now gives

$$e^{\tilde{\nu}+\tilde{\mu}_3} = [(1-2\beta\Phi)^2 + 4\beta^2\Psi^2]\, e^{\nu+\mu_3}. \tag{32}$$

Relations equivalent to (28) and (32), in the more general context of space-times with one Killing field, were first derived by Geroch (1971); and they were included in Kramer *et al.* (1980, p. 324, statement following equation (30.5)).

Finally, we may observe that given any vacuum solution derived from a particular solution E of the Ernst equation, we can derive a solution of the Einstein–Maxwell equations belonging to the same E $(= H)$. We provide an illustration in Appendix B.

3. The solution derived from $E = p\eta + iq\mu$, $(p^2+q^2 = 1)$; The Bell–Szekeres solution

We now consider the simplest solution of the Ernst equation for E, namely

$$E = p\eta + iq\mu, \tag{33}$$

where p and q are two real constants restricted by the requirement,

$$p^2+q^2 = 1. \tag{34}$$

Then, by (12),

$$\Psi = 1-|E|^2 = 1-p^2\eta^2-q^2\mu^2. \tag{35}$$

It is convenient to introduce the abbreviation (as in II, equation (52))

$$Y = 1-p^2\eta^2-q^2\mu^2 = p^2\Delta+q^2\delta. \tag{36}$$

Then,

$$\Psi = Y \quad \text{and} \quad \chi = \sqrt{(\Delta\delta)}/Y. \tag{37}$$

Next, we observe that for the chosen E, the right-hand sides of (15) and (16) are, respectively,

$$0 \quad \text{and} \quad \frac{3}{\Delta}+\frac{1}{\delta}-\frac{4}{Y}; \tag{38}$$

and these are the same that occur on the right-hand sides of equations (59) and (60) in Chandrasekhar & Ferrari (1984) (in their derivation of the Nutku–Halil solution). We conclude that

$$\nu+\mu_3+\ln\frac{\Psi}{\sqrt[4]{(\Delta\delta)}} = (\nu+\mu_3)_{\text{Nutku–Halil}}. \tag{39}$$

Therefore

$$e^{\nu+\mu_3} = \frac{1-p^2\eta^2-q^2\mu^2}{(1-\eta^2)^{\frac{3}{4}}(1-\mu^2)^{\frac{1}{4}}}\frac{\sqrt[4]{(\Delta\delta)}}{\Psi} = \frac{1}{\sqrt{\Delta}}. \tag{40}$$

Turning next to equations (13) and (14), we obtain

$$q_{2,0} = \frac{2pq}{Y^2}\eta(1-\mu^2) \quad \text{and} \quad q_{2,3} = -\frac{2pq}{Y^2}\mu(1-\eta^2); \tag{41}$$

and the solution of these equations is

$$q_2 = -\frac{p\Delta}{qY}, \tag{42}$$

where the constant of integration has been so chosen that $q_2 = 0$ for $\eta = 1$.

With the solution for the metric coefficients completed, we can write the metric in the form

$$ds^2 = \left[\frac{(d\eta)^2}{\Delta}-\frac{(d\mu)^2}{\delta}\right]-\frac{\Delta\delta}{Y}(dx^2)^2-Y(dx^1-q_2\,dx^2)^2. \tag{43}$$

Even though the metric, as written, gives the impression that the two commuting Killing-vectors ∂_{x^1} and ∂_{x^2} are not hypersurface orthogonal, the impression is illusory! As we shall presently show, the metric can be brought to a diagonal form by choosing as basis vectors, suitable constant linear combinations of ∂_{x^1} and ∂_{x^2}, as we may, because they are also Killing vectors. Thus, considering the (x^1, x^2), part of the metric (43), we may rewrite it, successively, in the manner:

$$\frac{\Delta}{q^2 Y}(p^2\Delta+q^2\delta)(\mathrm{d}x^2)^2+Y(\mathrm{d}x^1)^2+2\frac{p}{q}\Delta\,\mathrm{d}x^1\,\mathrm{d}x^2$$

$$=\frac{\Delta}{q^2}(\mathrm{d}x^2)^2+p^2\Delta(\mathrm{d}x^1)^2+2\frac{p\Delta}{q}\mathrm{d}x^1\,\mathrm{d}x^2+q^2\delta(\mathrm{d}x^1)^2$$

$$=\frac{\Delta}{q^2}(\mathrm{d}x^2+pq\,\mathrm{d}x^1)^2+q^2\delta(\mathrm{d}x^1)^2. \tag{44}$$

Therefore, by the replacements,

$$q\,\mathrm{d}x^1 \quad \text{by} \quad \mathrm{d}x^1 \quad \text{and} \quad \frac{1}{q}\mathrm{d}x^2+p\,\mathrm{d}x^1 \quad \text{by} \quad \mathrm{d}x^2, \tag{45}$$

the metric takes the diagonal form

$$\mathrm{d}s^2=\left[\frac{(\mathrm{d}\eta)^2}{\Delta}-\frac{(\mathrm{d}\mu)^2}{\delta}\right]-\delta(\mathrm{d}x^1)^2-\Delta(\mathrm{d}x^2)^2, \tag{46}$$

or, alternatively, in terms of the coordinates (ψ, θ) defined by (2),

$$\mathrm{d}s^2=(\mathrm{d}\psi)^2-(\mathrm{d}\theta)^2-(\mathrm{d}x^1)^2\sin^2\theta-(\mathrm{d}x^2)^2\sin^2\psi. \tag{46$'$}$$

This is the Bell–Szekeres solution. We could have derived it directly by considering *ab initio* the solution,

$$H=\eta, \quad p=1 \quad \text{and} \quad q=0. \tag{47}$$

In our experience, this is the first instance in which what one might have thought of as the analogue of the Kerr solution reduces to the Schwarzschild solution!

The Weyl and the Maxwell scalars for the space-time described by the metric (46) can be inferred as special cases of equations (87)–(90), derived in §4 below, by setting $\beta=0$ and $p=1$. We thus find that

$$\left.\begin{aligned}&\Psi_0=\Psi_1=\Psi_2=\Psi_3=\Psi_4=0\\ \text{and}\qquad &\Phi_{00}=\Phi_{22}=-\Phi_{20}=\tfrac{1}{2} \quad\text{and}\quad \Phi_{11}=0.\end{aligned}\right\} \tag{48}$$

Thus, the space-time is conformally flat with a prevailing electromagnetic field characterized by constant Maxwell scalars. This is in agreement with what Bell and Szekeres found.

It is of interest to write the metric (46′) in a form that is manifestly conformally flat. By making the coordinate transformation

$$\left.\begin{aligned}&\cos\psi=\frac{1}{2r}(r^2-t^2+1), \quad \tanh x^2=\frac{1}{2t}(r^2-t^2-1),\\ &t=\frac{\sin\psi\cosh x^2}{\cos\psi-\sin\psi\sinh x^2} \quad\text{and}\quad r=\frac{1}{\cos\psi-\sin\psi\sinh x^2},\end{aligned}\right\} \tag{48$'$}$$

we find that the metric (46′) reduces to the required form:

$$\mathrm{d}s^2=\frac{1}{r^2}[(\mathrm{d}t)^2-(\mathrm{d}r)^2-r^2(\mathrm{d}\theta)^2-r^2\sin^2\theta\,(\mathrm{d}x^1)^2]. \tag{48$''$}$$

Hence the Bell–Szekeres solution is equivalent to the Bertotti–Robinson homogeneous solution and it admits six independent Killing fields (see Kramer *et al.* 1980, p. 120). The transformation (48′) is equivalent to one given by Griffiths (1985).

(a) The extension of the space-time into regions II and III

In the terms of the null coordinates, u and v, related to η and μ by the equations

$$\eta = u\surd(1-v^2)+v\surd(1-u^2) \quad \text{and} \quad \mu = u\surd(1-v^2)-v\surd(1-u^2), \tag{49}$$

the metric takes the form

$$\mathrm{d}s^2 = \frac{4\,\mathrm{d}u\,\mathrm{d}v}{\surd[(1-u^2)(1-v^2)]}-(1-\mu^2)(\mathrm{d}x^1)^2-(1-\eta^2)(\mathrm{d}x^2)^2, \tag{50}$$

where it may be recalled that

$$\varDelta\delta = (1-\eta^2)(1-\mu^2) = (1-u^2-v^2)^2. \tag{51}$$

The extension of the space-time across the null boundaries at $u=0$ and $v=0$ is made by the standard Penrose-algorism of replacing u and v in the metric coefficients by $uH(u)$ and $vH(v)$, where $H(u)$ and $H(v)$ are the Heaviside step-functions that are unity for non-negative values of the argument and zero for negative values. Thus, in region II, the metric becomes

$$\mathrm{d}s^2 = \frac{4\,\mathrm{d}u\,\mathrm{d}v}{\surd(1-v^2)}-(1-v^2)[(\mathrm{d}x^1)^2+(\mathrm{d}x^2)^2]. \tag{52}$$

The designations we are adopting for the different regions of the space-time are the same as in the earlier papers (see figure 1 in Paper I and figures 1 and 3 in Paper II). For this metric, we find from the formalism described in Paper I, §8, that

$$L+M = -\frac{1}{2\surd(1-v^2)} < 0, \tag{53}$$

where

$$N = M-L \equiv 0. \tag{54}$$

Accordingly, the space-time continues to be conformally flat in region II whereas the Maxwell scalar $\varPhi_{22}$ (cf. I, equation (179)) has the same constant value $\frac{1}{2}$ as in region I. On the other hand, because $\varPhi_{00}$ and $\varPhi_{22}$ are non-vanishing in region I, we have electromagnetic shocks at the null boundaries, $u=0$ and $v=0$.

The singularity of the space-time along $v=1$ and $u=1$, in regions II and III, can be deduced (as in earlier contexts) by considering the geodesics in these regions. Thus, in place of II, equations (72), we now have

$$\gamma^2(u+c) = \int_0^v \left(\frac{p_1^2+p_2^2}{1-v^2}+\epsilon\right)\frac{\mathrm{d}v}{\surd(1-v^2)}, \tag{55}$$

where γ, p_1, p_2, and ϵ (greater than zero for time-like geodesics and zero for null geodesics) are constants of the motion and $u=-c$ $(c>0)$ is the point on the u-axis at which the geodesic crosses into region II. From equation (55), we conclude that no time-like or null geodesic, with p_1 and/or $p_2 \neq 0$, can avoid crossing the v-axis into region I for some $0 < v < 1$. The situation is the same as for the space-time considered in Paper II.

(*b*) *The extension of the space-time across* $s = 0$

From the metric written in the form (50), it is manifest that its determinant vanishes on the surface $1-u^2-v^2 = s = 0$. A coordinate singularity thus occurs when $s = 0$; and because $s = 0$ is a null surface on which ∂_{x^2} further becomes null, it is in fact a horizon. The extension of the space-time across $s = 0$ can be accomplished as in Paper II, §7.

First, we rewrite the metric in terms of the variables,

$$r = u^2-v^2 \quad \text{and} \quad s = 1-u^2-v^2, \tag{56}$$

in the form

$$\mathrm{d}s^2 = \frac{(\mathrm{d}s)^2-(\mathrm{d}r)^2}{4uv\sqrt{(1-u^2)}\sqrt{(1-v^2)}} - \frac{s^2}{\delta}(\mathrm{d}x^2)^2 - \delta(\mathrm{d}x^1)^2. \tag{57}$$

(The occurrence, here and elsewhere, of s in the line-element $\mathrm{d}s^2$ and as the variable s, is unfortunate, but unlikely to cause any confusion.) The extension across $s = 0$ is now achieved by the substitutions (cf. II, equation (130))

$$\xi = s\,\mathrm{e}^{+x^2} \quad \text{and} \quad \zeta = s\,\mathrm{e}^{-x^2} \quad (\xi\zeta = s^2), \tag{58}$$

when the metric becomes

$$\mathrm{d}s^2 = \frac{1}{4H\delta^2}[4\delta^2\,\mathrm{d}\xi\,\mathrm{d}\zeta + (\zeta\,\mathrm{d}\xi - \xi\,\mathrm{d}\zeta)^2] - \frac{(\mathrm{d}r)^2}{H} - \delta(\mathrm{d}x^1)^2, \tag{59}$$

where

$$\begin{aligned} H &= 4uv\sqrt{(1-u^2)}\sqrt{(1-v^2)} = \delta - \Delta \\ &= \sqrt{[(1+r)^2-\xi\zeta]}\sqrt{[(1-r)^2-\xi\zeta]}. \end{aligned} \tag{60}$$

It is clear that, unlike as in the space-time considered in Paper II, no singularity occurs in the extended domain – a fact not in itself surprising, because no scattering of gravitational waves takes place in the space-time and gravitation (as manifested by a non-vanishing Weyl tensor) is confined, exclusively, to the null boundaries, $u = 0$ and $v = 0$, as impulsive waves.

The complete analytic extension of the space-time can be carried out as in Paper II, §8(*b*). The simplest C^0-extension is provided by that illustrated in figure 3 of Paper II.

4. The space-time derived from the Ehlers transform of $E = p\eta + \mathrm{i}q\mu$

Because the solution derived from $E = p\eta + \mathrm{i}q\mu$ degenerated to the Bell–Szekeres solution, we shall consider its Ehlers transform. For this choice of E, the functions J and K defined in equations (21) become

$$J = p\eta + \beta q\mu - \mathrm{i}[\beta(1+p\eta) - q\mu], \tag{61}$$

and

$$K = 1 - \beta q\mu + \mathrm{i}\beta(1+p\eta). \tag{62}$$

Accordingly, by equations (23)

$$\tilde{\Psi} = \frac{\Psi}{|K|^2} = \frac{Y}{\Pi} \quad \text{and} \quad \tilde{\chi} = \sqrt{(\Delta\delta)}\frac{\Pi}{Y}, \tag{63}$$

where, in addition to Y (defined in equation (36)), we have introduced the abbreviation

$$\Pi = |K|^2 = (1-\beta q\mu)^2 + \beta^2(1+p\eta)^2. \tag{64}$$

By equations (25) and (40), the corresponding solution for $(\tilde{\nu}+\tilde{\mu}_3)$ is given by

$$e^{\tilde{\nu}+\tilde{\mu}_3} = |K|^2\, e^{\nu+\mu_3} = \frac{\Pi}{\sqrt{\Delta}}. \tag{65}$$

Turning next to equations (26) and (27) governing $\tilde{q}_2$ and writing

$$\tilde{q}_2 = q_2 + Q, \tag{66}$$

we obtain

$$Q_{,0} = \beta\frac{2q\delta}{Y^2}\{\beta[(1+p\eta)^2 - q^2\mu^2] - 2pq\eta\mu\}$$

and

$$Q_{,3} = \beta\frac{2p\Delta}{Y^2}\{(1-p^2\eta^2+q^2\mu^2) - 2\beta q\mu(1+p\eta)\}. \tag{67}$$

The solution of these equations, apart from a constant of integration, is given by

$$Q = \frac{2}{Y}\left[\beta\mu p\Delta + \beta^2\frac{q}{p}(1+p\eta)\,\delta\right]. \tag{68}$$

Combined with the solution (42) for q_2, we obtain

$$\tilde{q}_2 = -\frac{(1-\eta)}{qY}\{p[1-2\beta q\mu + 2\beta^2(1+p)](1+\eta) + 2\beta^2q^2\delta\}, \tag{69}$$

where the constant of integration has now been chosen to make $\tilde{q}_2$ vanish for $\eta = 1$.

With the solution for the metric coefficients completed, we can now write the metric in the form

$$ds^2 = \Pi\left[\frac{(d\eta)^2}{\Delta} - \frac{(d\mu)^2}{\delta}\right] - \Delta\delta\frac{\Pi}{Y}(dx^2)^2 - \frac{Y}{\Pi}(dx^1 - \tilde{q}_2\,dx^2)^2, \tag{70}$$

where it may be recalled that

$$Y = 1 - p^2\eta^2 - q^2\mu^2 \quad \text{and} \quad \Pi = (1-\beta q\mu)^2 + \beta^2(1+p\eta)^2. \tag{71}$$

The metric (70) is of exactly the same form as that derived in Paper II, equation (51) with the only difference being that Π and $\tilde{q}_2$ replace X and q_2. (In both cases, $\tilde{q}_2$ and q_2 have been arranged to vanish for $\eta = 1$.)

It is important to observe that while the solution $E = p\eta + iq\mu$ with $q \neq 0$ led to the Bell–Szekeres solution in its original form, *when* a suitable choice of the basis vectors, ∂_{x^1} and ∂_{x^2}, was made, the Ehlers transformation removes the degeneracy. It is essential for our purposes to have started with the solution, $E = p\eta + iq\mu$ with $q \neq 0$; otherwise, we should not have obtained the non-degenerate metric (70) in which *both* β and q are essential parameters.

(a) The extension of the space-time into regions II and III

The metric (70) in region I, when extended into region II by the standard Penrose-procedure, becomes

$$ds^2 = \frac{4\Pi_0}{\sqrt{(1-v^2)}}\,du\,dv - (1-v^2)\left[\Pi_0(dx^2)^2 + \frac{1}{\Pi_0}(dx^1 - \bar{q}_2\,dx^2)^2\right], \tag{72}$$

where

$$\Pi_0 = 1 + 2\beta qv + \beta^2(1 + 2pv + v^2), \tag{73}$$

and

$$\bar{q}_2 = -\frac{1}{q}[p(1+2\beta qv) + 2\beta^2(1+p) - 2\beta^2 q^2 v]. \tag{74}$$

The metric (72) is of the standard form considered in Paper I, §8 equation (150), with the correspondence

$$e^{2\nu} = \frac{4\Pi_0}{\sqrt{(1-v^2)}}, \quad e^{\beta} = (1-v^2), \quad \chi = \Pi_0,$$

$$e^{2\mu_2} = (1-v^2)\Pi_0 \quad \text{and} \quad e^{2\psi} = \frac{1-v^2}{\Pi_0}. \tag{75}$$

With this correspondence, we find from I, equations (152)–(154), (160) and (167) that

$$M + L = -\frac{1}{2\Pi_0^2\sqrt{(1-v^2)}}, \tag{76}$$

$$M - L = -\frac{3}{4\Pi_0^3\sqrt{(1-v^2)}}[\Pi_0(\Pi_0 - 1) - \beta^2(1-v^2)\{3\Pi_0 - 4[q + \beta(p+v)]^2\}], \tag{77}$$

$$N = -\frac{3\beta(\beta q - p)}{4\Pi_0^3\sqrt{(1-v^2)}}[v + 2\beta q + \beta^2(2p + 3v - v^3)]. \tag{78}$$

In deriving these formulae we have made use of the following identities, which are readily verified.

$$(\Pi_{0,v})^2 + (\bar{q}_{2,v})^2 = 4\beta^2\Pi_0 \quad \text{and} \quad v\Pi_{0,v} + \beta^2(1-v^2) = \Pi_0 - 1. \tag{79}$$

It follows from equations (76)–(78) that in region II we have both gravitational and electromagnetic radiation characterized, respectively, by a Weyl scalar Ψ_4 and a Maxwell scalar Φ_{22} (cf. I, equations (178) and (179)). We also observe that in the limit, $\beta = 0$ and $\Pi_0 = 1$, $M + L$ reduces to the expression (53) for the Bell–Szekeres solution whereas $M - L$ and N both vanish consistently with the requirement of conformal flatness in this limit.

The singularity of the space-times in regions II and III, along $v = 1$ and $u = 1$, follows as in §3(*b*) and in Paper II, §5: the geodesics in these regions have exactly the same dispositions.

(b) The Weyl and the Maxwell scalars in region I; the type-D character of the space-time

The expressions for the Weyl and the Maxwell scalars given in Appendix A (equations (A 6)) can be used, also, when, as in the present instance, we are considering the Ehlers transform of a particular solution for E: we have only to replace E, Ψ, and U^2, where they occur, by

$$\tilde{E} = \frac{J}{K}, \quad \tilde{\Psi} = \frac{\Psi}{|K|^2} \quad \text{and} \quad \tilde{U}^2 = |K|^2, \tag{80}$$

where J and K are given in equations (21); and, in the reductions make use of the relations (cf. equations (22))

$$\tilde{E}_{,i} = \frac{E_{,i}}{K^2}, \quad \tilde{E}_{,i,j} = \frac{1}{K^2}\left[E_{,i,j} - 2\frac{\mathrm{i}\beta}{K}E_{,i}E_{,j}\right], \tag{81}$$

$$\tilde{U}^2_{,i} = |K|^2_{,i} = -\mathrm{i}\beta(KE^*_{,i} - K^*E_{,i}) \quad (i,j = \psi, \theta), \tag{82}$$

and

$$\left(\frac{\partial}{\partial\psi} \pm \frac{\partial}{\partial\theta}\right)^2 \tilde{E} = \frac{1}{K^2}\left[\left(\frac{\partial}{\partial\psi} \pm \frac{\partial}{\partial\theta}\right)^2 E - 2\frac{\mathrm{i}\beta}{K}(E_{,\psi} \pm E_{,\theta})^2\right]. \tag{83}$$

The reductions are greatly simplified if use is made of the following identities:

$$J - EK^* = -\mathrm{i}\beta\Psi \quad \text{and} \quad |J|^2 - |E|^2|K|^2 = \Psi(|K|^2 - 1). \tag{84}$$

For the particular case we are presently considering,

$$E = p\eta + \mathrm{i}q\mu \quad (p^2 + q^2 = 1), \tag{85}$$

$$V^2 = \sin\psi\sin\theta \quad \text{and} \quad D_{\pm} = -2(1 \mp \cot\psi\cot\theta). \tag{86}$$

We find that the expressions for the Weyl and the Maxwell scalars simplify considerably and we are left with

$$\Psi_2 = -\tfrac{1}{2}\mathrm{i}\beta\frac{J}{|K|^2(K^*)^2}, \tag{87}$$

$$\frac{1-\mathscr{E}}{1-\mathscr{E}^*}\Psi_0 = +\tfrac{3}{2}\mathrm{i}\beta\frac{J}{|K|^2(K^*)^2}\frac{p\sin\psi - \mathrm{i}q\sin\theta}{p\sin\psi + \mathrm{i}q\sin\theta}, \tag{88}$$

$$\frac{1-\mathscr{E}^*}{1-\mathscr{E}}\Psi_4 = +\tfrac{3}{2}\mathrm{i}\beta\frac{J}{|K|^2(K^*)^2}\frac{p\sin\psi + \mathrm{i}q\sin\theta}{p\sin\psi - \mathrm{i}q\sin\theta}, \tag{89}$$

$$-\frac{1-\mathscr{E}^*}{1-\mathscr{E}}\Phi_{20} = \frac{1}{2|K|^4}\frac{p\sin\psi + \mathrm{i}q\sin\theta}{p\sin\psi - \mathrm{i}q\sin\theta}; \quad \Phi_{22} = \Phi_{00} = \frac{1}{2|K|^4}, \tag{90}$$

where J and K have their present values given in (61) and (62) and $|K|^2 = \Pi$.

First, we observe that when $\beta = 0$, the Weyl scalars Ψ_0, Ψ_2, and Ψ_4 vanish (in addition to Ψ_1 and Ψ_3 which vanish identically by symmetry requirements). Also, in the same limit, when $K = 1$, the Maxwell scalars Φ_{00} and Φ_{22} take the value $\frac{1}{2}$; and the phase by which Φ_{20} differs from $-\frac{1}{2}$ can be made zero by a linear transformation, with constant coefficients, of the x^1 and the x^2 coordinates, that will bring the metric (43) to its diagonal form (46).

From equations (87)–(89) it is evident that

$$9\Psi_2^2 = \Psi_0\Psi_4. \tag{91}$$

It follows (as in II, §6(a)) that the space-time is of Petrov-type D. It does not appear that the present one-parameter family of type-D metrics (distinguished by β) is included in any known family: it is not included, for example, in the Plebanski–Demianski family (cf. Kramer *et al.* 1980, §19.1.2, p. 210).

The question now arises whether the principal null directions of the Weyl and the Maxwell tensors coincide. The answer will be in the affirmative if the rotation that brings the Weyl tensor to its normal form (when Ψ_2 is the only non-vanishing scalar) also brings the Maxwell tensor to its normal form (when Φ_{11} will be the only non-vanishing scalar).

In Paper II, §6 (a) we showed that if

$$\Psi_1 = \Psi_3 = 0 \quad \text{and} \quad 9\Psi_2^2 = \Psi_0 \Psi_4, \tag{92}$$

the Weyl tensor can be brought to its normal form by applying, successively, two rotations belonging, respectively, to classes I and II with parameters a^* and b given by (II, equations (90) and (95))

$$(a^*)^2 = -3\frac{\Psi_2}{\Psi_0}, \quad b = \tfrac{1}{6}a^*\frac{\Psi_0}{\Psi_2} \quad \text{and} \quad a^*b = -\tfrac{1}{2}. \tag{93}$$

We shall now determine the effect of these rotations on the Maxwell scalars. Quite generally, the effect of rotations belonging to classes I and II applied, successively, to the Maxwell scalars, is given by (Chandrasekhar 1983, p. 54, equations (344))

$$\left.\begin{aligned} \tilde{\phi}_0 = \phi_0, \quad \tilde{\phi}_1 = \phi_1 + a^*\phi_0 \quad &\text{and} \quad \tilde{\phi}_2 = \phi_2 + 2a^*\phi_1 + (a^*)^2\phi_0; \\ \tilde{\tilde{\phi}}_0 = \tilde{\phi}_0 + 2b\tilde{\phi}_1 + b^2\tilde{\phi}_2, \quad \tilde{\tilde{\phi}}_1 = \tilde{\phi}_1 + b\tilde{\phi}_2 \quad &\text{and} \quad \tilde{\tilde{\phi}}_2 = \tilde{\phi}_2. \end{aligned}\right\} \tag{94}$$

Therefore, if initially $\phi_1 = 0$ (as it is in the case we are presently considering),

$$\tilde{\tilde{\phi}}_0 = (1+a^*b)^2\phi_0 + b^2\phi_2, \quad \tilde{\tilde{\phi}}_1 = a^*(1+a^*b)\phi_0 + b\phi_2,$$

and

$$\tilde{\tilde{\phi}}_2 = \phi_2 + (a^*)^2\phi_0. \tag{95}$$

For parameters a^* and b given by equations (93), we find that

$$\tilde{\tilde{\phi}} = \frac{1}{12\Psi_2}(3\phi_0\Psi_2 - \phi_2\Psi_0) \quad \text{and} \quad \tilde{\tilde{\phi}}_2 = -\frac{1}{\Psi_0}(3\phi_0\Psi_2 - \phi_2\Psi_0). \tag{96}$$

Therefore, the rotations considered will bring the Maxwell tensor to its normal form if, and only if,

$$3\frac{\phi_0\Psi_2}{\phi_2\Psi_0} = 3\frac{\Phi_{00}\Psi_2}{\Phi_{20}\Psi_0} = 1. \tag{97}$$

We conclude that *the necessary and sufficient conditions, that a space-time in which*

$$\Psi_1 = \Psi_3 = \Phi_{11} = 0, \tag{98}$$

is of type D and the principal null-directions of the Weyl and the Maxwell tensors coincide, are

$$9\Psi_2^2 = \Psi_0\Psi_4 \quad \text{and} \quad 3\Phi_{00}\Psi_2 = \Phi_{20}\Psi_0. \tag{99}$$

For the space-time we are presently considering, we have already noted that the first of these conditions is satisfied; we readily verify that the second of these conditions is also satisfied by the expressions for the Weyl and the Maxwell scalars we have listed.

It is a remarkable fact that we should obtain a family of space-times (distinguished by the parameters β and q) with this abundant structure by applying the Ehlers transformation to the Bell–Szekeres conformally flat space-time in its *non*-degenerate form.

(c) The extension of the space-time across the horizon at $s = 0$

In view of the formal identity of the metric (70) with the one considered in Paper II (equation (51)), we can accomplish its extension across the horizon at $s = 0$ in an identical manner. Thus, by rewriting the metric (70) in the (r, s)-variables (defined in equation (56)) in the form

$$ds^2 = \frac{\Pi}{H}[(ds)^2-(dr)^2]-s^2\frac{\Pi}{Y}(dx^2)^2-\frac{Y}{\Pi}(dx^1-\tilde{q}_2\,dx^2)^2, \tag{100}$$

we achieve its extension by the substitutions

$$\xi = s\,e^{+x^2/q} \quad\text{and}\quad \zeta = s\,e^{-x^2/q} \quad (\xi\zeta = s^2), \tag{101}$$

when the metric becomes (cf. II, equation (137))

$$ds^2 = \frac{\Pi}{4HY\delta}\{\zeta^2(d\xi)^2+\xi^2(d\zeta)^2+2[2q^2\delta^2+(p^2-q^2)\xi\zeta]\,d\xi\,d\zeta\}$$
$$-\frac{Y}{\Pi}\left[dx^1-\frac{Q}{Y}(\zeta\,d\xi-\xi\,d\zeta)\right]^2-\frac{\Pi}{H}(dr)^2, \tag{102}$$

where
$$Q = -\frac{1}{2\delta}\left\{p[1-2\beta q\mu+2\beta^2(1+p)]+\frac{2\beta^2q^2\delta}{1+\eta}\right\}, \tag{103}$$

and η and μ, expressed in terms of ξ, ζ, and r, are given by

$$\left.\begin{aligned}\eta &= \tfrac{1}{2}\{\sqrt{[(1+r)^2-\xi\zeta]}+\sqrt{[(1-r)^2-\xi\zeta]}\}\\ \text{and}\qquad \mu &= \tfrac{1}{2}\{\sqrt{[(1+r)^2-\xi\zeta]}-\sqrt{[(1-r)^2-\xi\zeta]}\}.\end{aligned}\right\} \tag{104}$$

The determinant of the metric (102) is

$$-\frac{\Pi^2q^2}{4H^2}; \tag{105}$$

and it is smooth across $s = 0$, as required.

It is now evident from the expressions (87)–(90) for the Weyl and the Maxwell scalars that the metric develops a curvature singularity when

$$|K|^2 = \Pi = (1-\beta q\mu)^2+\beta^2(1+p\eta)^2 = 0, \tag{106}$$

i.e. when
$$\beta q\mu = 1 \quad\text{and}\quad p\eta = -1. \tag{107}$$

The corresponding locus in the (ξ, ζ)-plane follows from the equations

$$\xi\zeta = s^2 = \Delta\delta = (1-\eta^2)(1-\mu^2) \quad\text{and}\quad r = u^2-v^2 = \eta\mu. \tag{108}$$

We find

$$\xi\zeta = -\frac{1}{p^2\beta^2}(\beta^2q^2-1) \quad\text{and}\quad r = -\frac{1}{\beta qp}. \tag{109}$$

Because $|r| \leqslant 1$,

$$|\beta q| > |\beta qp| \geqslant 1 \quad \text{and} \quad \xi\zeta < 0. \tag{110}$$

We conclude that a time-like curvature-singularity develops along hyperbolic arcs, in the quadrants $\xi\zeta < 0$, for $r = -1/(\beta qp)$ and $-\infty < x_1 < +\infty$. Depending on the sign of β, the singularity occurs for $-1 < r < 0$ or $0 < r < 1$. It will be recalled that for the space-time considered in Paper II, the singularity similarly occurred along hyperbolic arcs but for $r = 0$. (Note that no singularity in the extended domain occurs when $|\beta q| < 1$.)

This completes our consideration of the space-times derived by the Ehlers transformation.

5. The general solution for the case, $Z = 1$ and $q_2 = 0$

In this section, we shall investigate the case when $Z = 1$ and the basis vectors, ∂_{x^1} and ∂_{x^2}, are assumed, *ab initio*, to be hypersurface orthogonal.

First, we observe that, when $q_2 = 0$, the equation (I, equation (35)),

$$(\mathscr{I}m\, HH^*_{,3})_{,0} - (\mathscr{I}m\, HH^*_{,0})_{,3} = 0, \tag{111}$$

governing the complex-potential H, enables us to conclude, by effecting a suitable 'duality transformation' that we may assume, without loss of generality, that

$$H \quad \text{is real.} \tag{112}$$

Then, with the definition,

$$Z = \Psi + H^2, \tag{113}$$

and our present assumption that $q_2 = \Phi = 0$, equations (6) and (7) can be cast in the forms

$$\left(\frac{\Delta}{\Psi} H_{,0}\right)_{,0} - \left(\frac{\delta}{\Psi} H_{,3}\right)_{,3} = 0; \quad \left(\frac{\Delta}{\Psi} Z_{,0}\right)_{,0} - \left(\frac{\delta}{\Psi} Z_{,3}\right)_{,3} = 0. \tag{114}$$

If we now assume that

$$Z = 1, \tag{115}$$

(there is no loss of generality in assuming that $Z = 1$ rather than some other real constant) equations (114) reduce to the single equation

$$\left(\frac{\Delta}{1-H^2} H_{,0}\right)_{,0} - \left(\frac{\delta}{1-H^2} H_{,3}\right)_{,3} = 0. \tag{116}$$

With the substitution

$$\tfrac{1}{2}\ln\frac{1+H}{1-H} = F, \tag{117}$$

equation (116) becomes

$$[(1-\eta^2)F_{,\eta}]_{,\eta} - [(1-\mu^2)F_{,\mu}]_{,\mu} = 0. \tag{118}$$

In terms of a solution F of this equation,

$$H = \tanh F \quad \text{and} \quad \Psi = 1 - H^2 = \operatorname{sech}^2 F. \tag{119}$$

Turning next to equations (15) and (16) governing $(\nu+\mu_3)$, we find

$$\frac{\mu}{\delta}\left[(\nu+\mu_3)+\ln\frac{\Psi}{\sqrt[4]{(\Delta\delta)}}\right]_{,0}+\frac{\eta}{\Delta}\left[(\nu+\mu_3)+\ln\frac{\Psi}{\sqrt[4]{(\Delta\delta)}}\right]_{,3}=-4\frac{H_{,0}H_{,3}}{(1-H^2)^2}$$

$$=-\left(\ln\frac{1+H}{1-H}\right)_{,0}\left(\ln\frac{1+H}{1-H}\right)_{,3}=-4F_{,0}F_{,3}, \quad (120)$$

$$2\eta\left[(\nu+\mu_3)+\ln\frac{\Psi}{\sqrt[4]{(\Delta\delta)}}\right]_{,0}+2\mu\left[(\nu+\mu_3)+\ln\frac{\Psi}{\sqrt[4]{(\Delta\delta)}}\right]_{,3}$$

$$=-\frac{1}{\Delta}+\frac{1}{\delta}+\frac{4}{\Delta}-\Delta\left[\left(\ln\frac{1+H}{1-H}\right)_{,0}\right]^2-\delta\left[\left(\ln\frac{1+H}{1-H}\right)_{,3}\right]^2$$

$$=-\frac{1}{\Delta}+\frac{1}{\delta}+\frac{4}{\Delta}-4[\Delta(F_{,0})^2+\delta(F_{,3})^2]. \quad (121)$$

From a comparison with equations (59) and (60) in Chandrasekhar & Ferrari (1984) we conclude that

$$\exp[(\nu+\mu_3)_{F=0}]=\frac{\sqrt[4]{(\Delta\delta)}}{\Psi}\exp[(\nu+\mu_3)_{\text{Nutku-Halil};\,q=0}]$$

$$=\frac{\sqrt[4]{(\Delta\delta)}}{\Psi}\frac{\Delta^{\frac{1}{4}}}{\delta^{\frac{1}{4}}}=\frac{\sqrt{\Delta}}{\Psi}. \quad (122)$$

Therefore, by writing

$$\nu+\mu_3=(\nu+\mu_3)_{F=0}+f \quad (123)$$

the equations governing f are:

$$\frac{\mu}{\delta}f_{,0}+\frac{\eta}{\Delta}f_{,3}=-4F_{,0}F_{,3}, \quad (124)$$

$$\eta f_{,0}+\mu f_{,3}=\frac{2}{\Delta}-2[\Delta(F_{,0})^2+\delta(F_{,3})^2]. \quad (125)$$

With f determined by these equations, the solution for $(\nu+\mu_3)$ takes the form (cf. equations (119) and (122))

$$e^{\nu+\mu_3}=e^f(\cosh^2 F)\sqrt{\Delta}. \quad (126)$$

It is convenient to separate from F the special solution,

$$\tfrac{1}{2}\ln\frac{1+\eta}{1-\eta}, \quad (127)$$

of equation (118) and write F as a sum in the form

$$F=\tfrac{1}{2}\ln\frac{1+\eta}{1-\eta}+\Pi, \quad (128)$$

where Π is governed by the same equation,

$$[(1-\eta^2)\Pi_{,\eta}]_{,\eta}-[(1-\mu^2)\Pi_{,\mu}]_{,\mu}=0. \quad (129)$$

The special solution (127), that we have thus separated, corresponds to the solution, $H = \eta$ (cf. equation (117)), which, as we know, leads to the *original* Bell–Szekeres solution.

The general separable solution for Π is given by

$$[P_n(\eta) \quad \text{and/or} \quad Q_n(\eta)] \times [P_n(\mu) \quad \text{and/or} \quad Q_n(\mu)], \tag{130}$$

where P_n and Q_n denote the Legendre functions of the first and the second kinds, respectively, of order n; and the general solution for Π is expressible as arbitrary linear combinations of these basic solutions. However, in our present context, the extendibility of the metric into regions II and III is *assured* if the solution for Π includes only terms in the Legendre functions of the first kind (cf. remarks made in a similar context in Paper II following equation (29)). On this account, the solution for Π we shall consider in this paper is

$$\Pi = \sum_n A_n P_n(\eta) P_n(\mu), \tag{131}$$

where the A_n's are arbitrary constants. The solution for F which we then adopt is

$$F = \tfrac{1}{2} \ln \frac{1+\eta}{1-\eta} + \sum_n A_n P_n(\eta) P_n(\mu). \tag{132}$$

This expression for F as the sum of two parts – a part leading to the Bell–Szekeres solution and a part (with as many free parameters, A_n, as one may wish) leading to its generalization – corresponds exactly to a similar separation of the solution for '$\ln \Psi$' for static axisymmetric black-holes as the sum of two parts: a part leading to the Schwarzschild solution and a part leading to its distortion by a general distribution of 'distant' matter (cf. Chandrasekhar 1983, §112).

In terms of Π the equations governing f are:

$$\frac{\mu \Delta}{\delta} f_{,\eta} + \eta f_{,\mu} = -4\Pi_{,\mu} - 4\Delta \Pi_{,\eta} \Pi_{,\mu}, \tag{133}$$

$$\eta f_{,\eta} + \mu f_{,\mu} = -4\Pi_{,\eta} - 2[\Delta(\Pi_{,\eta})^2 + \delta(\Pi_{,\mu})^2]; \tag{134}$$

and the solutions for Ψ and $(\nu + \mu_3)$ take the forms

$$\frac{1}{\Psi} = \cosh^2 F = \frac{1}{\Delta}(\cosh \Pi + \eta \sinh \Pi)^2 \tag{135}$$

and

$$e^{\nu + \mu_3} = \frac{\Sigma^2}{\sqrt{\Delta}} e^f, \tag{136}$$

where we have written

$$\Sigma = \cosh \Pi + \eta \sinh \Pi. \tag{137}$$

There is, in principle, no difficulty in writing down the solution of equations (133) and (134) for any finite sum of terms in the solution (131) for Π. Thus, for

$$\Pi = A_1 P_1(\eta) P_1(\mu) + A_2 P_2(\eta) P_2(\mu), \tag{138}$$

we find

$$f = -4A_1\mu + A_1^2\Delta\delta + 6A_1A_2\Delta\delta\eta\mu + 6A_2\eta\delta + \tfrac{9}{8}A_2^2\Delta\delta(9\eta^2\mu^2 - \eta^2 - \mu^2 + 1) + \text{const.} \quad (139)$$

The problem of solving equations (133) and (134) in general will be considered in a separate paper (Chandrasekhar 1987, in preparation).

With the solution for the metric coefficients completed, we can now write the metric in the form

$$ds^2 = e^f\Sigma^2\left[\frac{(d\eta)^2}{\Delta} - \frac{(d\mu)^2}{\delta}\right] - \frac{\Delta}{\Sigma^2}(dx^1)^2 - \delta\Sigma^2(dx^2)^2. \quad (140)$$

Equation (140) represents a complete generalization of the original Bell–Szekeres solution (to which it reduces when $\Pi = 0$, $f = 0$, and $\Sigma = 1$).

The extension of the metric (140) into regions II and III presents no difficulty. We find that in these regions we have coupled gravitational and electromagnetic radiations; and the boundaries $v = 1$ and $u = 1$ are singular in the same way as in the examples considered in Paper II and §§3 and 4.

(a) The absence of a curvature singularity on $s = 0$

From the form of the metric (140), it is apparent that, as in all the other cases we have considered hitherto in the context of the collision of plane-fronted gravitational waves, its determinant vanishes when $\Delta\delta = s^2 = 0$. Whether a curvature singularity develops on this surface can be ascertained by the behaviour of the Weyl scalar Ψ_2 for $s \to 0$. In the present instance, when $q_2 = 0$, Ψ_2 is given by (cf. I, equation (130))

$$8U^2\Psi_2 = -(\cot^2\psi - \cot^2\theta) + \frac{1}{\chi^2}[(\chi_{,\psi})^2 - (\chi_{,\theta})^2], \quad (141)$$

where

$$U^2 = e^f\Sigma^2. \quad (142)$$

Rewriting (141) in terms of $\Psi\ (= \sqrt{(\Delta\delta)}/\chi)$, we find

$$8U^2\Psi_2 = \frac{1}{\Psi^2}[\Delta(\Psi_{,\eta})^2 - \delta(\Psi_{,\mu})^2] + \frac{2}{\Psi}(\eta\Psi_{,\eta} - \mu\Psi_{,\mu}); \quad (143)$$

or, substituting for Ψ from equation (119), we have

$$2U^2\Psi_2 = \{[\Delta(F_{,\eta})^2 - \delta(F_{,\mu})^2]\tanh F - (\eta F_{,\eta} - \mu F_{,\mu})\}\tanh F. \quad (144)$$

Finally, making use of equation (128), we obtain

$$\Psi_2 = \frac{e^{-f}}{2\Sigma^2}\left\{[\Delta(\Pi_{,\eta})^2 + \Pi_{,\eta} - \delta(\Pi_{,\mu})^2]\tanh F + \frac{1}{\Sigma}(\Delta\Pi_{,\eta} + 1)\sinh\Pi + \mu\Pi_{,\mu}\right\}\tanh F, \quad (145)$$

where it may be noted that

$$\tanh F = \frac{1}{\Sigma}(\sinh\Pi + \eta\cosh\Pi). \quad (146)$$

It is clear that for Π given by equation (131), Ψ_2 remains finite for $s \to 0$. Accordingly, $s = 0$ represents a coordinate singularity, and for the same reasons as in §3 it is a null-surface that is also a horizon. A smooth extension of the space-time across $s = 0$ is considered in §(*b*) below.

(b) The extension of the space-time across $s = 0$

First, we rewrite the metric (140) in terms of the variables r and s (defined in equation (56)) in the form

$$ds^2 = \frac{e^f \Sigma^2}{H}[(ds)^2 - (dr)^2] - \frac{s^2}{\delta \Sigma^2}(dx^1)^2 - \delta \Sigma^2 (dx^2)^2. \tag{147}$$

In contrast to the metric (57) considered in §3(*b*), it is the vector ∂_{x^1} (not ∂_{x^2}) that becomes null when $s = 0$. Therefore, to obtain a smooth extension across $s = 0$, we now make, instead of (58), the substitutions

$$\xi = s\, e^{+x^1} \quad \text{and} \quad \zeta = s\, e^{-x^1} \quad (\xi\zeta = s^2). \tag{148}$$

The metric (147) then becomes

$$ds^2 = \frac{1}{4H\Sigma^2 s^2}(e^f \Sigma^4 - H\delta^{-1})[\zeta^2(d\xi)^2 + \xi^2(d\zeta)^2] + \frac{1}{2H\Sigma^2}(e^f \Sigma^4 + H\delta^{-1})\, d\xi\, d\zeta - \frac{e^f \Sigma^2}{H}(dr)^2 - \delta\Sigma^2(dx^2)^2. \tag{149}$$

The *appearance* of singularity in the first term of the metric, for $s \to 0$, can be eliminated and its smoothness across $s = 0$ made manifest by first noting that

$$\Sigma = e^{\Pi}\left[1 - \frac{s^2}{\delta(1+\eta)} e^{-\Pi} \sinh \Pi\right], \quad H = \delta\left(1 - \frac{s^2}{\delta^2}\right), \tag{150}$$

rewriting the metric in the form

$$ds^2 = \frac{1}{4H\Sigma^2 s^2}\left\{e^{f+4\Pi}\left[1 - \frac{s^2}{\delta(1+\eta)} e^{-\Pi} \sinh \Pi\right]^4 - 1 + \frac{s^2}{\delta^2}\right\}[\zeta^2(d\xi)^2 + \xi^2(d\zeta)^2] + \frac{1}{2H\Sigma^2}(e^f \Sigma^4 + H/\delta)\, d\xi\, d\zeta - \frac{e^f \Sigma^2}{H}(dr)^2 - \delta\Sigma^2(dx^2)^2, \tag{151}$$

and observing that by virtue of equation (133)

$$f + 4\Pi = O(s^2) \quad \text{for} \quad s \to 0. \tag{152}$$

(c) The development of three-dimensional time-like singularity in the extended manifold

The determinant of the metric, namely

$$-\frac{e^{2f} \Sigma^4}{4H^2}, \tag{153}$$

vanishes when $\Sigma = 0$; at the same time, the Weyl scalar Ψ_2, given by equation (145), diverges. A curvature singularity, therefore, develops when

$$\Sigma = \cosh \Pi + \eta \sinh \Pi = 0, \tag{154}$$

or

$$\coth \Pi = -\eta. \tag{155}$$

From equation (104), expressing η and μ in terms of r, s, ξ and ζ it follows that the singularity develops when

$$\tfrac{1}{2}\{\sqrt{[(1+r)^2-\xi\zeta]}+\sqrt{[(1-r)^2-\xi\zeta]}\} = -\coth \Pi. \tag{156}$$

From the fact that $|\coth \Pi| \geqslant 1$, we conclude that solutions of equation (156) exist only when

$$\xi\zeta < 0 \quad \text{and} \quad \Pi = \sum_n A_n P_n(\eta) P_n(\mu) < 0. \tag{157}$$

Clearly, equation (156) determines a *three-dimensional time-like surface*. The difference from the cases considered in Paper II and in §4 arises from the fact that, while in these latter cases, *both* η and μ were determined by the condition for the singularity (as by II, equation (143), and equation (107)), in the present case only a relation between them (namely, equation (155)) is required.

6. Concluding remarks

The solution for the vacuum equations found in Paper I, together with the large class of solutions of the Einstein–Maxwell equations found in this paper, show that the collision of plane-fronted impulsive gravitational waves, either by themselves or coupled with electromagnetic shock waves, leading to the development of horizons and subsequent time-like singularities, is not an isolated phenomenon. In any event, the development of space-like curvature singularities (as in the Khan–Penrose and in the Nutku–Halil solutions) can no longer be maintained as universal. It is, however, more difficult to decide when one or the other of the two alternatives will result under particular circumstances. On the basis of the experience gained, one may perhaps draw, tentatively, the following conclusions.

Collision of pure gravitational waves (i.e. in the framework of the Einstein vacuum equations) with parallel polarizations will, perhaps, always result in strong three-dimensional space-like curvature singularities. When the polarizations of the colliding waves are not parallel, one may expect weaker space-like curvature singularities unless the space-time has some special structure such as a type-D character. In the latter case, horizons may form and subsequently time-like singularities (as is the case with the space-time considered in Paper II). In the framework of the Einstein–Maxwell equations it appears that we must distinguish two cases: the case when the source of the gravitation is not entirely attributable to the prevailing electromagnetic field and the Weyl tensor does not vanish when the Maxwell tensor does (as is the case with the space-time considered in Paper I); and the case when the gravitation is entirely attributable to the prevailing electromagnetic field, when the Weyl tensor vanishes when the Maxwell tensor does (as is the case with the space-times considered in this paper). In the former case, the development of space-like curvature singularities may be the rule; and in the latter case, the development of horizons and time-like singularities may likewise prevail. It is important to observe in this connection that it is the latter alternative that occurs for the space-times considered in §5 which are not of type-D and in which the commuting space-like Killing vectors *are* hypersurface orthogonal.

Our experience is more limited when sources other than a Maxwell field are coupled with the gravitational field. In the special case, when gravitation is coupled with a perfect fluid with the equation of state, the energy density equals the pressure, the space-like curvature singularity that occurs is weakened considerably by the acoustic waves that are present. At the same time novel phenomena such as the transformation of the perfect-fluid into null-dust can also take place.

Turning next to other aspects of the class of solutions of the Einstein–Maxwell equations considered in this paper, namely those that are obtained on a special assumption which reduces the entire set of field equations to a single Ernst-equation, we recognize that the Bell–Szekeres solution plays for this class the same role as the Khan–Penrose solution, for colliding plane-fronted impulsive gravitational waves, and the Schwarzschild solution, for stationary axisymmetric space-times. The role of the type-D character for all of these space-times is noteworthy.

The fact that horizons can form under circumstances different from those one normally considers (namely as the result of gravitational collapse) raises the question whether Hawking radiation will be emitted under these new circumstances.

And finally, at a more technical level, we may ask whether solutions of the Einstein–Maxwell equations, of the class considered in this paper, exist also for space-times which are stationary and axisymmetric. It is clear that, analogous to the separable solution with cylindrical symmetry (describing monochromatic waves) considered in Chandrasekhar (1986, appendix C), we may consider separable solutions of the form

$$H = \mathrm{e}^{\mathrm{i}kz}\, F(\varpi) \tag{158}$$

(in a system of coordinates, t, z, ϖ and ϕ), when the radial function $F(\varpi)$ will satisfy the equation,

$$-F(1+F^2)\,k^2+(1-F^2)\,\frac{1}{\varpi}\,(\varpi F_{,\,\varpi})_{,\,\varpi}+2F(F_{,\,\varpi})^2 = 0 \tag{159}$$

(in place of equation (63) in Chandrasekhar (1986)); in other words, a similar equation belonging to Painlevé transcendent III but for a 'purely imaginary argument'. A preliminary examination indicates that a solution of (159) leading to an 'acceptable' space-time does not exist; but the matter requires further investigation.

We are grateful to Dr Yavuz Nutku for stimulating our interest in the Bell–Szekeres solution. The research reported in this paper has, in part, been supported by a grant from the National Science Foundation under grant

PHY-84-16691 with the University of Chicago. We are also grateful for a grant from the Division of the Physical Sciences of the University of Chicago that made the present collaboration possible.

References

Bell, P. & Szekeres, P. 1974 *Gen. Rel. Grav.* **5**, 275.
Chandrasekhar, S. 1983 *The mathematical theory of black holes*. Oxford: Clarendon Press.
Chandrasekhar, S. 1986 *Proc. R. Soc. Lond.* A **408**, 209–232.
Chandrasekhar, S. 1987 (In preparation.)
Chandrasekhar, S. & Ferrari, V. 1984 *Proc. R. Soc. Lond.* A **396**, 55.
Chandrasekhar, S. & Xanthopoulos, B. C. 1985 *Proc. R. Soc. Lond.* A **398**, 223.
Chandrasekhar, S. & Xanthopoulos, B. C. 1986 *Proc. R. Soc. Lond.* A **408**, 175–208.
Geroch, R. 1971 *J. math. Phys.* **12**, 918.
Griffiths, J. B. 1985 In *Galaxies, axisymmetric systems, and relativity* (ed. M. A. H. MacCallum), p. 199. Cambridge University Press.
Kramer, D., Stephani, H., Herlt, E. & MacCallum, M. 1980 *Exact solutions of Einstein's field equations*. Cambridge University Press.

Appendix A. Expressions for the Weyl and the Maxwell scalars explicitly in terms of a solution E of the Ernst equation for H

In Appendix A of Paper II the expressions for the Weyl scalars, given in Paper I (equations (130)–(132)) in terms of

$$\mathscr{E} = \frac{\mathscr{X}-1}{\mathscr{X}+1}, \quad \text{where} \quad \mathscr{X} = \chi + \mathrm{i}q_2 \tag{A 1}$$

for a space-time, with a metric written in its standard form (II, equation (A 1)), were re-expressed in terms of the Ernst function,

$$E = \frac{\Psi + \mathrm{i}\Phi - 1}{\Psi + \mathrm{i}\Phi + 1}. \tag{A 2}$$

This transformation applies only to the vacuum solutions; it does not apply to the solutions which are our present concern, namely, those derived from the Ernst equation for H when $Z = 1$; for, in place of II, equation (A 8), we now have

$$\chi = \frac{V^2}{\Psi}, \quad \Psi = 1 - |E|^2, \tag{A 3}$$

$$q_{2,\psi} = \mathrm{i}\,\frac{V^2}{\Psi^2}(EE^*_{,\theta} - E^*E_{,\theta}) \quad \text{and} \quad q_{2,\theta} = \mathrm{i}\,\frac{V^2}{\Psi^2}(EE^*_{,\psi} - E^*E_{,\psi}). \tag{A 4}$$

By virtue of these equations, the reductions of Appendix A of Paper II have to be modified considerably. Thus, in place of II, equations (A 17), we now have

$$\frac{1}{\chi}(\mathscr{X}_{,\psi} + \mathscr{X}_{,\theta}) = 2\left[(\ln V)_{,\psi} + (\ln V)_{,\theta} + \frac{E^*}{\Psi}(E_{,\psi} + E_{,\theta})\right],$$

$$\frac{1}{\chi}(\mathscr{X}_{,\psi} - \mathscr{X}_{,\theta}) = 2\left[(\ln V)_{,\psi} - (\ln V)_{,\theta} + \frac{E}{\Psi}(E^*_{,\psi} - E^*_{,\theta})\right]. \tag{A 5}$$

Proceeding along similar lines, we find that the expressions for the Weyl and the Maxwell scalars listed in Paper I (equations (130)–(135)) are now replaced by

$$2U^2\Psi_2 = \frac{E^2}{\Psi^2}[(E^*_{,\psi})^2-(E^*_{,\theta})^2]+2\frac{E}{\Psi}[(\ln V)_{,\psi}E^*_{,\psi}-(\ln V)_{,\theta}E^*_{,\theta}],$$

$$4U^2\frac{1-\mathscr{E}^*}{1-\mathscr{E}}\Psi_0^* = D_+-2\left[\frac{(U^2)_{,\psi}}{U^2}+\frac{(U^2)_{,\theta}}{U^2}\right]\left[(\ln V)_{,\psi}+(\ln V)_{,\theta}+\frac{E^*}{\Psi}(E_{,\psi}+E_{,\theta})\right]$$

$$+\frac{2}{\Psi}|E_{,\psi}+E_{,\theta}|^2+4\frac{|E|^2}{\Psi^2}|E_{,\psi}+E_{,\theta}|^2+2\frac{E^*}{\Psi}\left(\frac{\partial}{\partial\psi}+\frac{\partial}{\partial\theta}\right)^2E$$

$$+\frac{2}{\Psi}[(\ln V)_{,\psi}+(\ln V)_{,\theta}][E(E^*_{,\psi}+E^*_{,\theta})+E^*(E_{,\psi}+E_{,\theta})],$$

$$4U^2\frac{1-\mathscr{E}^*}{1-\mathscr{E}}\Psi_4 = D_--2\left[\frac{(U^2)_{,\psi}}{U^2}-\frac{(U^2)_{,\theta}}{U^2}\right]\left[(\ln V)_{,\psi}-(\ln V)_{,\theta}+\frac{E}{\Psi}(E^*_{,\psi}-E^*_{,\theta})\right]$$

$$+\frac{2}{\Psi}|E_{,\psi}-E_{,\theta}|^2+4\frac{|E|^2}{\Psi^2}|E_{,\psi}-E_{,\theta}|^2+2\frac{E}{\Psi}\left(\frac{\partial}{\partial\psi}-\frac{\partial}{\partial\theta}\right)^2E^*$$

$$+\frac{2}{\Psi}[(\ln V)_{,\psi}-(\ln V)_{,\theta}][E(E^*_{,\psi}-E^*_{,\theta})+E^*(E_{,\psi}-E_{,\theta})],$$

$$4U^2\Phi_{00} = -D_++4\left(\frac{U_{,\psi}}{U}+\frac{U_{,\theta}}{U}\right)[(\ln V)_{,\psi}+(\ln V)_{,\theta}]-2\frac{|E|^2}{\Psi^2}|E_{,\psi}+E_{,\theta}|^2$$

$$-\frac{2}{\Psi}[(\ln V)_{,\psi}+(\ln V)_{,\theta}][E(E^*_{,\psi}+E^*_{,\theta})+E^*(E_{,\psi}+E_{,\theta})],$$

$$4U^2\Phi_{22} = -D_-+4\left(\frac{U_{,\psi}}{U}-\frac{U_{,\theta}}{U}\right)[(\ln V)_{,\psi}-(\ln V)_{,\theta}]-2\frac{|E|^2}{\Psi^2}|E_{,\psi}-E_{,\theta}|^2$$

$$-\frac{2}{\Psi}[(\ln V)_{,\psi}-(\ln V)_{,\theta}][E(E^*_{,\psi}-E^*_{,\theta})+E^*(E_{,\psi}-E_{,\theta})],$$

$$-2U^2\frac{1-\mathscr{E}^*}{1-\mathscr{E}}\Phi_{20} = \frac{E}{\Psi}(E^*_{,\psi,\psi}-E^*_{,\theta,\theta})+\frac{1}{\Psi}(E_{,\psi}+E_{,\theta})(E^*_{,\psi}-E^*_{,\theta})$$

$$+2\frac{E^2}{\Psi^2}[(E^*_{,\psi})^2-(E^*_{,\theta})^2]+2\frac{E}{\Psi}[(\ln V)_{,\psi}E^*_{,\psi}-(\ln V)_{,\theta}E^*_{,\theta}], \tag{A 6}$$

where

$$D_+ = 2\left(\frac{\partial}{\partial\psi}+\frac{\partial}{\partial\theta}\right)^2\ln V+4[(\ln V)_{,\psi}+(\ln V)_{,\theta}]^2$$

and

$$D_- = 2\left(\frac{\partial}{\partial\psi}-\frac{\partial}{\partial\theta}\right)^2\ln V+4[(\ln V)_{,\psi}-(\ln V)_{,\theta}]^2. \tag{A 7}$$

Appendix B. Solutions of the Einstein–Maxwell equation that are the analogues of vacuum solutions

Because a solution of the Einstein–Maxwell equations, belonging to the class considered in this paper, follows from any solution of the Ernst equation that we may choose, we can clearly assign to H the value $E^\dagger$, where $E^\dagger$ is a solution of the Ernst equation from which solutions for the metric functions, Ψ and Φ, appropriate to a vacuum space-time, are obtained via the equation (cf. I, equations (56) and (61)),

$$E^\dagger = \frac{\Psi - \mathrm{i}\Phi - 1}{\Psi - \mathrm{i}\Phi + 1} = \frac{Z^\dagger - 1}{Z^\dagger + 1}. \tag{B 1}$$

With the assignment

$$H = E^\dagger, \tag{B 2}$$

the metric function Ψ_e, belonging to a solution of the Einstein–Maxwell equations, is given by

$$\Psi_e = 1 - |E^\dagger|^2 = \frac{4\Psi}{(\Psi+1)^2 + \Phi^2}; \tag{B 3}$$

or, letting

$$\Sigma = \tfrac{1}{4}[(\Psi+1)^2 + \Phi^2], \tag{B 4}$$

$$\Psi_e = \frac{\Psi}{\Sigma} \quad \text{and} \quad \chi_e = \frac{\surd(\Delta\delta)}{\Psi}\Sigma = \chi\Sigma. \tag{B 5}$$

From (15) and (16) it further follows that

$$\mathrm{e}^{(\nu+\mu_3)_e} = \frac{\Psi}{\Psi_e}\mathrm{e}^{\nu+\mu_3} = \Sigma\, \mathrm{e}^{\nu+\mu_3}. \tag{B 6}$$

Turning next to (13) and (14) we must substitute for E, on the right-hand sides of these equations, its present value $E^\dagger$. With this substitution, we find on simplification,

$$q_{2e,3} = \frac{\Delta}{4\Psi^2}[(1+\Phi^2-\Psi^2)\Phi_{,0} + 2\Phi\Psi\Psi_{,0}], \tag{B 7}$$

and

$$q_{2e,0} = \frac{\delta}{4\Psi^2}[(1+\Phi^2-\Psi^2)\Phi_{,3} + 2\Phi\Psi\Psi_{,3}]. \tag{B 8}$$

Writing q_{2e} as the sum,

$$q_{2e} = \tfrac{1}{4}q_2 + Q, \tag{B 9}$$

where q_2 is the solution of the vacuum equations,

$$q_{2,0} = \frac{\delta}{\Psi^2}\Phi_{,3} \quad \text{and} \quad q_{2,3} = \frac{\Delta}{\Psi^2}\Phi_{,0}, \tag{B 10}$$

we find that the equations determining Q are

$$Q_{,3} = \tfrac{1}{4}\Delta\frac{\Phi^2}{\Psi^2}\left(\frac{\Psi^2+\Phi^2}{\Phi}\right)_{,0}, \quad Q_{,0} = \tfrac{1}{4}\delta\frac{\Phi^2}{\Psi^2}\left(\frac{\Psi^2+\Phi^2}{\Phi}\right)_{,3}. \tag{B 11}$$

These equations can, in principle, be solved by elementary quadratures. We shall, thus, have obtained a solution of the Einstein–Maxwell equations from a known solution of the vacuum equations.

The analogue of the Nutku–Halil solution

As an illustration of the procedure we have outlined, we shall obtain the analogue of the Nutku–Halil solution. The solutions for Ψ, Φ, q_2 and $\nu+\mu_3$, belonging to the Nutku–Halil solution, are (cf. Paper I, equations (62)–(64), and Chandrasekhar & Ferrari 1984, equation (61))

$$\Psi = \sqrt{(\Delta\delta)}\frac{X}{Y}, \quad \Phi = -\frac{2q}{p}\frac{\delta(1-p\eta)}{Y}, \quad q_2 = \frac{2q\mu}{X} \quad \text{and} \quad e^{\nu+\mu_3} = \frac{Y}{\Delta^{\frac{3}{4}}\delta^{\frac{1}{4}}}, \tag{B 12}$$

where X and Y have their current meanings:

$$X = (1-p\eta)^2+q^2\mu^2 \quad \text{and} \quad Y = 1-p^2\eta^2-q^2\mu^2. \tag{B 13}$$

The corresponding solutions for Ψ_e, χ_e and $(\nu+\mu_3)_e$ now follow directly from equations (B 5) and (B 6).

To evaluate q_{2e} in accordance with equations (B 9) and (B 11), we first note that, for Ψ and Φ given in equations (B 12),

$$\frac{\Phi^2+\Psi^2}{\Phi} = -\frac{1}{2pq}\{(1-p\eta)^3+2q^2(1-p\eta)+q^2(1+p\eta)\}+\tfrac{1}{2}pq\frac{\Delta\mu^2}{1-p\eta}. \tag{B 14}$$

Equation (B 11) now gives

$$Q_{,\eta} = \frac{q^3}{p}\frac{1-p\eta}{X^2}\delta^2\mu, \tag{B 15}$$

$$Q_{,\mu} = \frac{3q\delta}{2p^2}+\frac{q^3\delta}{2p^2X^2}[(1-5\mu^2)X-2q^2\mu^2\delta]. \tag{B 16}$$

The solution of these equations, apart from a constant of integration, is

$$Q = \frac{q\mu}{2p^2X}[(3-\mu^2)X+q^2\delta^2]. \tag{B 17}$$

With the solution for q_2 given in equation (B 12), we now find, in accordance with (B 9), that

$$q_{2e} = \frac{q\mu}{2p^2X}[1+(1-p\eta)^2(3-\mu^2)+q^2\mu^2]. \tag{B 18}$$

With the solution for the metric functions completed, we can now write the resulting metric in the form,

$$ds^2 = \Sigma\frac{Y}{\sqrt[4]{(\Delta\delta)}}\left[\frac{(d\eta)^2}{\Delta}-\frac{(d\mu)^2}{\delta}\right]-\Delta\delta\frac{\Sigma}{\Psi}(dx^2)^2-\frac{\Psi}{\Sigma}(dx^1-q_{2e}dx^2)^2. \tag{B 19}$$

We find that the Weyl scalar Ψ_2 for the metric (B 19) is given by

$$\Psi_2^* = \frac{\Psi^2 - \Phi^2 - 1 + 2i\Phi\Psi}{16X\Sigma^2(\sin\psi\sin\theta)^{\frac{1}{2}}}\Big\{Z^\dagger_{,\psi}\cot\psi - Z^\dagger_{,\theta}\cot\theta$$

$$+\frac{Y}{8\Sigma X(\sin\psi\sin\theta)}(\Psi^2 - \Phi^2 - 1 + 2i\Phi\Psi)[(Z^\dagger_{,\psi})^2 - (Z^\dagger_{,\theta})^2]\Big\}. \quad \text{(B 20)}$$

It diverges for s ($= \sin\psi\sin\theta$) $\to 0$ like $s^{-\frac{3}{2}}$. The space-time described by the metric (B 19) develops a curvature singularity on $s = 0$ exactly like the Nutku–Halil solution. Besides, as can be directly verified, the metric is extendible into regions II and III by the standard procedures.

25

The effect of sources on horizons that may develop when plane gravitational waves collide

By Subrahmanyan Chandrasekhar[1], F.R.S., and Basilis C. Xanthopoulos[2]

[1] *The University of Chicago, Chicago, Illinois 60637, U.S.A.*
[2] *Department of Physics, University of Crete, and Research Centre of Crete, Iraklion, Greece*

(*Received* 16 *March* 1987)

Colliding plane gravitational waves that lead to the development of a horizon and a subsequent time-like singularity are coupled with an electromagnetic field, a perfect fluid (whose energy density, ϵ, equals the pressure, p), and null dust (consisting of massless particles). The coupling of the gravitational waves with an electromagnetic field does not affect, in any essential way, the development of the horizon or the time-like singularity if the polarizations of the colliding gravitational waves are not parallel. If the polarizations are parallel, the space-like singularity which occurs in the vacuum is transformed into a horizon followed by a three-dimensional time-like singularity by the merest presence of the electromagnetic field. The coupling of the gravitational waves with an $(\epsilon = p)$-fluid and null dust affect the development of horizons and singularities in radically different ways: the $(\epsilon = p)$-fluid affects the development *decisively* in *all* cases but qualitatively in the same way, while null dust prevents the development of horizons and allows only the development of space-like singularities. The contrasting behaviours of an $(\epsilon = p)$-fluid and of null dust in the framework of general relativity is compared with the behaviours one may expect, under similar circumstances, in the framework of special relativity.

1. Introduction

In a series of papers published in these *Proceedings* during the past three years (Chandrasekhar & Ferrari 1984; Chandrasekhar & Xanthopoulos 1985*a*, *b*, 1986*a*, *b*, 1987) the collision of plane-fronted waves has been addressed under a variety of circumstances. In this paper, we consider some further amplifications, primarily with the object of completing the main outlines of the theory that has been developed.

The paper is divided into two parts. In part I, we obtain the analogue of the vacuum solution obtained in Chandrasekhar & Xanthopoulos (1986*b*, referred to hereafter as Paper III) in the framework of the Einstein–Maxwell equations, as set out in Chandrasekhar & Xanthopoulos (1985*a*, referred to hereafter as Paper I). In part II, we consider, in the manner of Chandrasekhar & Xanthopoulos (1985*b* and 1986*a*, referred to hereafter as Papers II*a* and II*b*) the effect of coupling to the same vacuum solution (of Paper III) sources derived from a perfect fluid in which the energy density is equal to the pressure, and from null dust. We shall find that these provide some additional insights as to how the development of horizons can be affected when the colliding gravitational waves are coupled with souces of electrodynamic or hydrodynamic origins.

Some related matters are considered in three Appendixes.

Familiarity with the definitions, notations, and methods of the earlier papers will be assumed; only a minimum of explanation will be provided. An adequate summary of the earlier papers is given in Chandrasekhar & Xanthopoulos (1987, §6); and for a more extensive review of the entire field see Chandrasekhar (1987, §VI and particularly tables 1 and 2 on pp. 42–44).

Part I. The Einstein–Maxwell analogue of the vacuum solution of Paper III

2. The basic equations

For the metric of the space–time, in the region of interaction of the colliding waves, we choose our standard form,

$$ds^2 = e^{\nu+\mu_3}\sqrt{\Delta}\left[\frac{(d\eta)^2}{\Delta}-\frac{(d\mu)^2}{\delta}\right]-\sqrt{(\Delta\delta)}\left[\chi(dx^2)^2+\frac{1}{\chi}(dx^1-q_2\,dx^2)^2\right], \tag{1}$$

where η measures time from the instant of collision, μ measures the distance normal to the planes, (x^1, x^2), spanned by the two Killing-fields, ∂_{x^1} and ∂_{x^2},

$$\Delta = 1-\eta^2 \quad \text{and} \quad \delta = 1-\mu^2. \tag{2}$$

It is known that a solution of Einstein's vacuum equations can be derived from a solution, E, of the Ernst equation (III, equation (5)) for

$$E = \frac{Z-1}{Z+1} \quad \text{where} \quad Z = \Psi - i\Phi; \tag{3}$$

Φ is a potential (as defined in III, equation (8)) for the metric function q_2, and Ψ $(= \sqrt{(\Delta\delta)}/\chi)$ is a function associated with Φ. In terms of E, the functions Ψ and Φ are given by

$$\Psi = \frac{1-|E|^2}{|1-E|^2} \quad \text{and} \quad \Phi = i\,\frac{E-E^*}{|1-E|^2}. \tag{4}$$

In writing the foregoing equations, we have suppressed the superscript '†' (used in earlier papers) to distinguish the solutions derived via Ψ and Φ from those derived via χ and q_2. Also, we have defined $Z = \Psi - i\Phi$ (instead of $Z = \Psi + i\Phi$ as in Paper III) to be in accord with the definitions of Paper I.

As we have explained in Paper I, §5, a solution of the Einstein–Maxwell equations for

$$Z_e = \Psi_e - i\Phi_e + |H|^2 = \frac{1+E_e}{1-E_e}, \tag{5}$$

for the special case,

$$H = Q(Z_e+1) \tag{6}$$

where Q is some non-zero real constant, follows from the solution,

$$E_e = \alpha E, \tag{7}$$

of the associated Einstein–Maxwell–Ernst equation (Paper I, equation (49)), where E is a solution of the vacuum Ernst equation and

$$\alpha = \sqrt{(1-4Q^2)} \quad (0 < \alpha < 1). \tag{8}$$

In terms of E, the required solutions for Ψ_e, χ_e, Φ_e, and H are given by (Paper I, equations (69) and (73)–(75))

$$\Psi_e = \frac{\alpha^2(1-|E|^2)}{|1-\alpha E|^2} = \alpha^2 \frac{|1-E|^2}{|1-\alpha E|^2}\Psi, \tag{9}$$

$$\chi_e = \frac{\sqrt{(\Delta\delta)}}{\alpha^2}\frac{|1-\alpha E|^2}{1-|E|^2} = \frac{|1-\alpha E|^2}{\alpha^2|1-E|^2}\chi, \tag{10}$$

$$\Phi_e = \mathrm{i}\alpha\frac{E-E^*}{|1-\alpha E|^2} \quad \text{and} \quad H = \frac{2Q}{1-\alpha E}. \tag{11}$$

The solution for q_{2e} follows from the equations (I, (38))

$$q_{2e,\eta} = \frac{\delta}{\Psi_e^2}(\Phi_{e,\mu} - 2\,\mathrm{Im}\,HH^*_{,\mu}) \tag{12}$$

and

$$q_{2e,\mu} = \frac{\Delta}{\Psi_e^2}(\Phi_{e,\eta} - 2\,\mathrm{Im}\,HH^*_{,\eta}); \tag{13}$$

or, substituting for Φ_e and H from (11), we obtain:

$$q_{2e,\eta} = \frac{\mathrm{i}\delta}{\alpha^2(1-|E|^2)^2}[(1-\alpha E)(E-\alpha)E^*_{,\mu} - (1-\alpha E^*)(E^*-\alpha)E_{,\mu}], \tag{14}$$

and

$$q_{2e,\mu} = \frac{\mathrm{i}\Delta}{\alpha^2(1-|E|^2)^2}[(1-\alpha E)(E-\alpha)E^*_{,\eta} - (1-\alpha E^*)(E^*-\alpha)E_{,\eta}], \tag{15}$$

Finally, by an analysis similar to the one set out in Paper I, §5*b*, we conclude that the solution for the last remaining metric function $(\nu+\mu_3)_e$ is determined in terms of the corresponding vacuum-solution, $(\nu+\mu_3)$, by

$$e^{(\nu+\mu_3)_e} = e^{(\nu+\mu_3)_{\mathrm{vac}}}\frac{|1-\alpha E|^2}{\alpha^2|1-E|^2}. \tag{16}$$

In Appendix A, we derive expression for the Weyl and the Maxwell scalars explicitly in terms of the solution E of the vacuum equation.

3. The solution of the Einstein–Maxwell equations derived from the solution, $E = p\eta - \mathrm{i}q\mu$ $(p^2+q^2=1)$ of the vacuum equation

The solution of the vacuum equations obtained in Paper III follows from the simplest solution, $E = p\eta \pm \mathrm{i}q\mu$ $(p^2+q^2=1)$, of the Ernst equation. In conformity with our present convention, we shall adopt

$$E = p\eta - \mathrm{i}q\mu \quad (p^2+q^2=1), \tag{17}$$

as the solution from which we wish to derive a solution of the Einstein–Maxwell equations.

The solutions for Ψ, χ, Φ and $(\nu+\mu_3)$, appropriate to the vacuum, given in equations III, (44), (45) and (50), are

$$\Psi = \frac{1-p^2\eta^2-q^2\mu^2}{(1-p\eta)^2+q^2\mu^2}, \quad \chi = \sqrt{(\Delta\delta)}\,\frac{(1-p\eta)^2+q^2\mu^2}{1-p^2\eta^2-q^2\mu^2}, \tag{18}$$

$$\Phi = \frac{2q\mu}{(1-p\eta)^2+q^2\mu^2} \quad \text{and} \quad e^{(\mu_3+\nu)_{\text{vac}}} = \frac{(1-p\eta)^2+q^2\mu^2}{\sqrt{\Delta}}. \tag{19}$$

The corresponding solutions, Ψ_e, χ_e and $(\nu+\mu_3)_e$, appropriate to the Einstein–Maxwell equations, are, by equations (9), (10) and (11),

$$\Psi_e = \alpha^2\,\frac{1-p^2\eta^2-q^2\mu^2}{(1-\alpha p\eta)^2+\alpha^2q^2\mu^2}, \tag{20}$$

$$\chi_e = \frac{\sqrt{(\Delta\delta)}}{\alpha^2}\,\frac{(1-\alpha p\eta)^2+\alpha^2q^2\mu^2}{1-p^2\eta^2-q^2\mu^2}, \tag{21}$$

and

$$e^{(\mu_3+\nu)_e} = \frac{1}{\alpha^2\sqrt{\Delta}}\,[(1-\alpha p\eta)^2+\alpha^2q^2\mu^2]. \tag{22}$$

Equations (14) and (15), governing q_{2e}, now become

$$\left.\begin{aligned} q_{2e,\mu} &= \frac{2pq\mu\Delta}{\alpha^2(1-p^2\eta^2-q^2\mu^2)^2}\,(1+\alpha^2-2\alpha p\eta), \\ \text{and}\quad q_{2e,\eta} &= -\frac{2q\delta(1+\alpha^2-2\alpha p\eta)\,p\eta}{\alpha^2(1-p^2\eta^2-q^2\mu^2)^2}+\frac{2q\delta}{\alpha(1-p^2\eta^2-q^2\mu^2)}.\end{aligned}\right\} \tag{23}$$

We readily verify that the solution of these equations is given by

$$q_{2e} = \frac{p\Delta}{\alpha^2 q}\,\frac{1+\alpha^2-2\alpha p\eta}{1-p^2\eta^2-q^2\mu^2}+\frac{2\eta}{\alpha q}+\text{const.} \tag{24}$$

or, equivalently,

$$q_{2e} = -\frac{q\delta}{\alpha^2 p}\,\frac{1+\alpha^2-2\alpha p\eta}{1-p^2\eta^2-q^2\mu^2}+\text{const.} \tag{25}$$

It is convenient to determine the constant in the solution (24) for q_{2e} by requiring that it vanishes for $\eta = 1$, for reasons explained in Paper III following equation (118). We thus find,

$$q_{2e} = \frac{(1-\eta)}{\alpha^2 q(1-p^2\eta^2-q^2\mu^2)}\,[p(1+\alpha^2)\,(1+\eta)-2\alpha(1+p^2\eta-q^2\mu^2)]. \tag{26}$$

With the solution for the metric functions completed, we can now write the metric in the form,

$$ds^2 = X\left[\frac{(d\eta)^2}{1-\eta^2}-\frac{(d\mu)^2}{1-\mu^2}\right]-\Delta\delta\,\frac{X}{Y}\,(dx^2)^2-\frac{Y}{X}\,(dx^1-q_{2e}\,dx^2)^2, \tag{27}$$

where

$$X = \frac{1}{\alpha^2}\,[(1-\alpha p\eta)^2+\alpha^2q^2\mu^2] \quad \text{and} \quad Y = 1-p^2\eta^2-q^2\mu^2 = p^2\Delta+q^2\delta. \tag{28}$$

We observe that, except for the generalizations of the definitions of the functions X and q_2, the metric (27) is of exactly the same form as the vacuum metric, equation III (51).

In terms of the null-coordinates, u and v, defined in the manner,

$$\eta = u\sqrt{(1-v^2)}+v\sqrt{(1-u^2)} \quad \text{and} \quad \mu = u\sqrt{(1-v^2)}-v\sqrt{(1-u^2)}, \tag{29}$$

the metric takes the form (cf. equation III (114))

$$ds^2 = \frac{4X\,du\,dv}{\sqrt{[(1-u^2)(1-v^2)]}}-(1-u^2-v^2)^2\frac{X}{Y}(dx^2)^2-\frac{Y}{X}(dx^1-q_{2e}\,dx^2)^2. \tag{30}$$

(a) The extension of the space-time into regions II and III

The extension of the space-time across the null boundaries, at $u = 0$ and $v = 0$, is accomplished by the standard Penrose-algorism described in Paper III, §3. Thus in region II (in the designation adopted in Papers I and III) the metric becomes

$$ds^2 = 4e^{g(v)}\,du\,dv-(1-v^2)\left\{\chi(v)(dx^2)^2+\frac{1}{\chi(v)}[dx^1-q_{2e}(v)\,dx^2]^2\right\}, \tag{31}$$

where

$$\left.\begin{aligned} e^{g(v)} &= \frac{1-2\alpha pv+\alpha^2v^2}{\alpha^2\sqrt{(1-v^2)}}, \quad \chi(v) = \frac{1}{\alpha^2}(1-2\alpha pv+\alpha^2v^2), \\ \text{and} \qquad q_{2e}(v) &= \frac{1}{\alpha^2 q}[(1+\alpha^2)p-2\alpha(1-q^2v)]. \end{aligned}\right\} \tag{32}$$

For this metric we find, from the formalism described in Paper III, §3 and in particular from equations III (37) and (41), that

$$L+M = -\frac{\alpha^2(1-\alpha^2)}{2(1-2\alpha pv+\alpha^2v^2)^2\sqrt{(1-v^2)}}, \tag{33}$$

and

$$M-L+2iN = \frac{3\alpha^3[(p+iq)v-\alpha][1-\alpha(p+iq)v]^2}{2(1-2\alpha pv+\alpha^2v^2)^3\sqrt{(1-v^2)}}(p-iq)^2. \tag{34}$$

Moreover, it follows from equations I (178) and (179) that the non-vanishing Maxwell and Weyl scalars in region II are given by

$$\Phi_{22} = \frac{\alpha^2(1-\alpha^2)}{2(1-2\alpha pv+\alpha^2v^2)^2}, \tag{35}$$

and

$$\frac{1-\mathscr{E}^*}{1-\mathscr{E}}\Psi_4 = \frac{3\alpha^3[(p+iq)v-\alpha][1-\alpha(p+iq)v]^2}{2(1-2\alpha pv+\alpha^2v^2)^3}(p-iq)^2. \tag{36}$$

We shall find in part (*b*) below that Φ_{22} and Ψ_4, given by these equations, agree with the values they attain, in region I, as $u \to +0$. These scalars are, therefore, continuous across the null boundary.

By interchanging the roles of u and v, of Φ_{22} and Φ_{00}, and of Ψ_4 and Ψ_0, we can ascertain the circumstances prevailing in region III.

And, finally, the extension of the space-time into region IV, by the same algorism, will lead to a flat space-time.

(*b*) *The Weyl and the Maxwell scalars and the type-*D *character of the space-time*

In Appendix A, we have listed the Weyl and the Maxwell scalars for a space-time that is derived from a solution, E, of the vacuum Ernst-equation for $\Psi - \mathrm{i}\Phi$. In the present instance, the required scalars can, therefore, be evaluated directly from equations (A 15)–(A 20) by substituting for E the simplest solution

$$E = p\cos\psi - \mathrm{i}q\cos\theta \quad (p^2+q^2=1). \tag{37}$$

For the metric (27),

$$V^2 = \sin\psi\sin\theta, \quad U^2 = X = \frac{1}{\alpha^2}[(1-\alpha p\cos\psi)^2+\alpha^2q^2\cos^2\theta] = \frac{1}{\alpha^2}|1-\alpha E|^2, \tag{38}$$

$$D = \cot\psi-\cot\theta; \quad \bar{D} = -\alpha\left[\frac{E_{,\psi}-E_{,\theta}}{1-\alpha E}+\frac{E^*_{,\psi}-E^*_{,\theta}}{1-\alpha E^*}\right],$$

and

$$S = \cot\psi+\cot\theta; \quad \bar{S} = -\alpha\left[\frac{E_{,\psi}+E_{,\theta}}{1-\alpha E}+\frac{E^*_{,\psi}+E^*_{,\theta}}{1-\alpha E^*}\right]. \tag{39}$$

The evaluation of the scalars is considerably simplified if appropriate use is made of the following elementary identities:

$$E^*_{,\psi}-E^*_{,\theta} = E_{,\psi}+E_{,\theta} = -p\sin\psi+\mathrm{i}q\sin\theta,$$

$$E^*_{,\psi}+E^*_{,\theta} = E_{,\psi}-E_{,\theta} = -p\sin\psi-\mathrm{i}q\sin\theta,$$

$$(E_{,\psi}+E_{,\theta})(E_{,\psi}-E_{,\theta}) = (E^*_{,\psi}+E^*_{,\theta})(E^*_{,\psi}-E^*_{,\theta}) = |E_{,\psi}\pm E_{,\theta}|^2 = 1-|E|^2,$$

$$\left(\frac{\partial}{\partial\psi}\pm\frac{\partial}{\partial\theta}\right)^2 E = -E;$$

$$D^2+D_{,\psi}-D_{,\theta} = -2(1+\cot\psi\cot\theta),$$

$$S^2+S_{,\psi}+S_{,\theta} = -2(1-\cot\psi\cot\theta),$$

$$SD+D_{,\psi}+D_{,\theta} = 0;$$

$$S(E^*_{,\psi}-E^*_{,\theta})+D(E^*_{,\psi}+E^*_{,\theta}) = -2E,$$

$$D[E^*(E_{,\psi}-E_{,\theta})+E(E_{,\psi}+E_{,\theta})] = 2[(1+\cot\psi\cot\theta)(1-|E|^2)-1],$$

$$S[E^*(E_{,\psi}+E_{,\theta})+E(E_{,\psi}-E_{,\theta})] = 2[(1-\cot\psi\cot\theta)(1-|E|^2)-1]. \tag{40}$$

We find:

$$\Psi_2 = -\tfrac{1}{2}\alpha^3\frac{(E-\alpha)}{(1-\alpha E)(1-\alpha E^*)^3}, \tag{41}$$

$$\frac{1-\mathscr{E}^*}{1-\mathscr{E}}\Psi_4 = +\tfrac{3}{2}\alpha^3\frac{(E-\alpha)(E_{,\psi}+E_{,\theta})^2}{(1-\alpha E)(1-\alpha E^*)^3(1-|E|^2)}, \tag{42}$$

$$\frac{1-\mathscr{E}}{1-\mathscr{E}^*}\Psi_0 = +\tfrac{3}{2}\alpha^3\frac{(E-\alpha)(E^*_{,\psi}+E^*_{,\theta})^2}{(1-\alpha E)(1-\alpha E^*)^3(1-|E|^2)}, \tag{43}$$

$$\Phi_{00} = \Phi_{22} = \frac{1}{2}\frac{\alpha^2(1-\alpha^2)}{|1-\alpha E|^4}, \tag{44}$$

and
$$\frac{1-\mathscr{E}^*}{1-\mathscr{E}}\,\Phi_{20} = -\tfrac{1}{2}\alpha^2(1-\alpha^2)\,\frac{(E_{,\psi}+E_{,\theta})^2}{|1-\alpha E|^4\,(1-|E|^2)}. \tag{45}$$

We verify that the foregoing expressions for the Weyl and the Maxwell scalars satisfy the relations,
$$9\Psi_2^2 = \Psi_0\,\Psi_4 \quad\text{and}\quad 3\Phi_{00}\,\Psi_2 = \Phi_{20}\,\Psi_0. \tag{46}$$
Besides, by symmetry requirements,
$$\Psi_1 = \Psi_3 = \Phi_{11} = 0. \tag{47}$$
It follows (as in Chandrasekhar & Xanthopoulos 1987, §4(*b*), equations (98) and (99)) that *the space-time is of Petrov type-*D *and that the principal null directions of the Weyl and the Maxwell tensors coincide.*

We observe that none of the Weyl and the Maxwell scalars diverge on the boundary $u^2+v^2 = 1$ where $\eta = 1$, $E = p-\mathrm{i}q\mu$ (p and $\alpha \neq 1$); and that on the null boundary, $u = 0$, where $\eta = -\mu = v$, the Maxwell scalar Φ_{22} and the Weyl scalar Ψ_4, in region I, agree with the values (35) and (36) they have in region II.

(*c*) *Local isometry with the Kerr–Newman space-time*

The facts, that the space-time we are presently considering is of type-D and that the Kerr–Newman space-time (also of type-D) has, interior to its ergosphere, a region in which it has two space-like Killing vectors and, further, that both the space-times derive from the same solution of the Ernst equation, strongly suggest that the two space-times may be locally isometric. As in III §(6*b*), we shall verify that this is indeed the case.

For our present purposes, it is important to choose for q_{2e} the alternative form of the solution,
$$q_{2e} = -\frac{q\delta}{\alpha^2 p}\,\frac{1+\alpha^2-2\alpha p\eta}{1-p^2\eta^2-q^2\mu^2}, \tag{48}$$
given by (25) so that q_{2e} may have a smooth behaviour as $q\to 0$. Now, defining
$$p = \frac{\lambda}{\alpha M}\,\surd(M^2-a^2-Q_*^2),\quad q = -\frac{\lambda}{\alpha M}\,a,\quad Q_*^2 = (1-\alpha^2)\,M^2 \quad\text{and}\quad \lambda = \pm 1, \tag{49}$$
consistently with the requirement, $p^2+q^2 = 1$, and letting
$$\eta = \frac{\lambda(M-r)}{\surd(M^2-a^2-Q_*^2)},\quad \mu = \cos\theta \quad\text{and}\quad \Delta = r^2-2Mr+a^2+Q_*^2. \tag{50}$$
we find
$$1-\alpha p\eta = r/M,\quad 1-\eta^2 = -\Delta/(M^2-a^2-Q_*^2)$$
$$X = (r^2+a^2\cos^2\theta)/(\alpha^2M^2),\quad Y = -(\Delta-a^2\delta)/(\alpha^2M^2)$$
and
$$q_{2e} = -\frac{a(r^2+a^2-\Delta)\sin^2\theta}{(\Delta-a^2\delta)\surd(M^2-a^2-Q_*^2)}. \tag{51}$$
The substitutions,
$$t = \alpha Mx^1 \quad\text{and}\quad \phi = \frac{\alpha M}{\surd(M^2-a^2-Q_*^2)}\,x^2, \tag{52}$$
now transform the metric (27) into the standard Kerr–Newman form, apart from a factor $(\alpha M)^2$.

(d) The extension of the space-time across $s = 1-u^2-v^2 = 0$

First, we rewrite the metric (30) in terms of the variables,

$$r = u^2-v^2 \quad \text{and} \quad s = 1-u^2-v^2, \tag{53}$$

when it becomes

$$\mathrm{d}s^2 = \frac{X}{H}[(\mathrm{d}s)^2-(\mathrm{d}r)^2]-s^2\frac{X}{Y}(\mathrm{d}x^2)^2-\frac{Y}{X}(\mathrm{d}x^1-q_{2e}\,\mathrm{d}x^2)^2, \tag{54}$$

where
$$H = 4uv\sqrt{[(1-u^2)(1-v^2)]} = \eta^2-\mu^2 = \delta-\Delta. \tag{55}$$

The determinant of the metric (54) vanishes when $s = 0$. However, as we have seen in part (*b*), none of the Weyl and the Maxwell scalars diverge when $s = 0$: they remain finite. We can conclude, as in Paper III, that only a coordinate singularity obtains here. Indeed, in view of the formal identity of the metric (54) with the one considered in Paper III, equation (120), we can eliminate the coordinate singularity by the substitutions

$$\xi = s\,\mathrm{e}^{+x^2/q} \quad \text{and} \quad \zeta = s\,\mathrm{e}^{-x^2/q}, \tag{56}$$

when the metric becomes (cf. equation III (137))

$$\mathrm{d}s^2 = \frac{X}{4HY\delta}\{\zeta^2(\mathrm{d}\xi)^2+\xi^2(\mathrm{d}\zeta)^2+2[2q^2\delta^2+(p^2-q^2)\Delta\delta]\,\mathrm{d}\xi\,\mathrm{d}\zeta\}$$
$$-\frac{Y}{X}\left[\mathrm{d}x^1-\frac{Q}{Y}(\zeta\,\mathrm{d}\xi-\xi\,\mathrm{d}\zeta)\right]^2-\frac{X}{H}(\mathrm{d}r)^2, \tag{57}$$

where

$$Q = \tfrac{1}{2}q\frac{q_{2e}Y}{\Delta\delta} = \frac{1}{2\alpha^2(1+\eta)\delta}[p(1+\alpha^2)(1+\eta)-2\alpha(1+p^2\eta-q^2\mu^2)]. \tag{58}$$

It is manifest from (41)–(45) that all the Weyl and the Maxwell scalars diverge in the extended domain, when

$$X = |1-\alpha E|^2 = (1-\alpha p\eta)^2+\alpha^2q^2\mu^2 = 0, \tag{59}$$

i.e. when
$$\eta = 1/(\alpha p) \quad \text{and} \quad \mu = 0, \tag{60}$$

if $q \neq 0$. (The case $q = 0$ is treated separately in part (*e*) below.) Since

$$\left.\begin{aligned}\eta &= \tfrac{1}{2}\{\sqrt{[(1+r)^2-\xi\zeta]}+\sqrt{[(1-r)^2-\xi\zeta]}\}\\ \text{and} \quad \mu &= \tfrac{1}{2}\{\sqrt{[(1+r)^2-\xi\zeta]}-\sqrt{[(1-r)^2-\xi\zeta]}\},\end{aligned}\right\} \tag{61}$$

it follows that the conditions (60) are equivalent to

$$r = 0 \quad \text{and} \quad \xi\zeta = 1-(\alpha p)^{-2} < 0, \tag{62}$$

since both α^2 and p^2 are required, by definitions, to be less than 1. We conclude that time-like singularities occur along the hyperbolic arcs, $\xi\zeta = -(1-\alpha^2p^2)/(\alpha^2p^2)$ for $r = 0$. The situation is the same as for the space-time considered in Paper III (see in particular figures 2 and 5).

(e) The hypersurface orthogonal case, $q = 0$

From (23) it follows that when $q = 0$, we may set

$$q_{2e} = \text{a constant}. \tag{63}$$

Without loss of generality, we may set

$$q_{2e} = 0; \tag{64}$$

for, if q_{2e} were not zero, we can restore the *status quo ante* by replacing $\mathrm{d}x^1 -$ constant $\mathrm{d}x^2$ by $\mathrm{d}x^1$, which we may, since ∂_{x^1} and ∂_{x^2} represent commuting Killing-fields. Therefore, when $q = 0$,

$$X = \frac{1}{\alpha^2}(1-\alpha\eta)^2 \quad \text{and} \quad Y = 1-\eta^2 = \Delta; \tag{65}$$

and the metric reduces to

$$\mathrm{d}s^2 = X\left[\frac{(\mathrm{d}\eta)^2}{\Delta} - \frac{(\mathrm{d}\mu)^2}{\delta}\right] - \delta X(\mathrm{d}x^2)^2 - \frac{\Delta\delta}{\delta X}(\mathrm{d}x^1)^2, \tag{66}$$

or, in the (r, s)-variables,

$$\mathrm{d}s^2 = \frac{X}{H}[(\mathrm{d}s)^2 - (\mathrm{d}r)^2] - \delta X(\mathrm{d}x^2)^2 - \frac{s^2}{\delta X}(\mathrm{d}x^1)^2. \tag{67}$$

At the same time, the Weyl and the Maxwell scalars, listed in (41)–(45), become

$$\left.\begin{aligned} \Psi_0 = \Psi_4 = -3\Psi_2 = \tfrac{3}{2}\alpha^3\frac{\eta-\alpha}{(1-\alpha\eta)^4}, \\ \text{and} \qquad \Phi_{00} = \Phi_{22} = -\Phi_{20} = \frac{1}{2}\frac{\alpha^2(1-\alpha^2)}{(1-\alpha\eta)^4}, \end{aligned}\right\} \tag{68}$$

and none of these diverge when $s = 0$ and $\eta = 1$ so long as $\alpha < 1$, as required by definition. Clearly, a coordinate singularity obtains when $s = 0$. To determine the substitutions that will eliminate this coordinate singularity, we first observe that when $s \to 0$,

$$\eta \to 1, \quad \mu \to r, \quad \delta \to 1 - r^2 \quad \text{and} \quad H \to 1-r^2, \tag{69}$$

and the metric (67) tends to the form

$$\mathrm{d}s^2 \sim \frac{X_0}{1-r^2}\left[(\mathrm{d}s)^2 - \frac{s^2}{X_0^2}(\mathrm{d}x^1)^2\right] - \frac{X_0}{1-r^2}(\mathrm{d}r)^2 - X_0(1-r^2)(\mathrm{d}x^2)^2, \tag{70}$$

where

$$X_0 = (1-\alpha)^2/\alpha^2. \tag{71}$$

Comparison with equation III (124) shows that the substitutions, that will eliminate the coordinate singularity, are

$$\xi = s\,\mathrm{e}^{+x^1/X_0} \quad \text{and} \quad \zeta = s\,\mathrm{e}^{-x^1/X_0} \quad (\xi\zeta = s^2). \tag{72}$$

With these substitutions, the metric (67) becomes

$$\mathrm{d}s^2 = \frac{Xs^2}{4H}\left(\frac{\mathrm{d}\xi}{\xi} + \frac{\mathrm{d}\zeta}{\zeta}\right)^2 - \frac{s^2X_0^2}{4X\delta}\left(\frac{\mathrm{d}\xi}{\xi} - \frac{\mathrm{d}\zeta}{\zeta}\right)^2 - \frac{X}{H}(\mathrm{d}r)^2 - \delta X(\mathrm{d}x^2)^2 \tag{73}$$

or equivalently,

$$\mathrm{d}s^2 = \frac{X}{4H\delta s^2}\left\{\left(\delta - \frac{X_0^2}{X^2}H\right)[\zeta^2(\mathrm{d}\xi)^2 + \xi^2(\mathrm{d}\zeta)^2] + 2s^2\left(\delta + \frac{X_0^2}{X^2}H\right)\mathrm{d}\xi\,\mathrm{d}\zeta\right\} - \frac{X}{H}(\mathrm{d}r)^2 - \delta X(\mathrm{d}x^2)^2. \quad (74)$$

On the other hand, by making use of (55), we have

$$\delta - \frac{X_0^2}{X^2}H = \delta\left(1 - \frac{X_0^2}{X^2}\right) + \frac{X_0^2}{X^2}\Delta = \frac{s^2}{\delta X^2}\left[\frac{\delta}{1+\eta}(X+X_0)(\sqrt{X}+\sqrt{X_0}) + X_0^2\right]. \quad (75)$$

We can, accordingly, rewrite the metric (73) in the form,

$$\mathrm{d}s^2 = \frac{1}{4HX\delta^2}\left\{\left[X_0^2 + \frac{\delta}{1+\eta}(X+X_0)(\sqrt{X}+\sqrt{X_0})\right][\zeta^2(\mathrm{d}\xi)^2 + \xi^2(\mathrm{d}\zeta)^2] + 2X^2\delta\left[\delta + \frac{X_0^2}{X^2}(\delta - \Delta)\right]\mathrm{d}\xi\,\mathrm{d}\zeta\right\} - \frac{X}{H}(\mathrm{d}r)^2 - \delta X(\mathrm{d}x^2)^2, \quad (76)$$

in which the absence of any singularity, when $\xi = 0$ and/or $\zeta = 0$, is made manifest.

Again, in the extended space-time, the Weyl and the Maxwell scalars, given by (68), diverge when

$$X = 1 - \alpha\eta = 0 \quad \text{or} \quad \eta = 1/\alpha > 1; \quad (77)$$

or by (61), when

$$\sqrt{[(1+r)^2 - \xi\zeta]} + \sqrt{[(1-r)^2 - \xi\zeta]} = 2/\alpha > 2. \quad (78)$$

Since r is now unrestricted, *the time-like singularity which occurs in the extended domain is three-dimensional.*

Part II. The effect of a perfect fluid (with the equation of state, $\epsilon = p$) and of null dust on the development of singularities

4. The solution of the Einstein–hydrodynamic equations derived from the solution, $E = p\eta + \mathrm{i}q\mu$ ($p^2 + q^2 = 1$), of the vacuum equations

In Chandrasekhar & Xanthopoulos (1985*b*, referred to hereafter as Paper II*a*) it was shown that given *any* vacuum space-time, described by a metric of the standard form (1), the only effect of coupling the gravitational field with a perfect fluid, with the equation of state $\epsilon = p$, is to alter the metric to the form,

$$\mathrm{d}s^2 = \mathrm{e}^{\mu_3+\nu+f}\sqrt{\Delta}\left[\frac{(\mathrm{d}\eta)^2}{\Delta} - \frac{(\mathrm{d}\mu)^2}{\delta}\right] - \sqrt{(\Delta\delta)}\left[\chi(\mathrm{d}x^2)^2 + \frac{1}{\chi}(\mathrm{d}x^1 - q_2\,\mathrm{d}x^2)^2\right], \quad (79)$$

where, to repeat, χ, q_2, and $\mu_3 + \nu$ represent *any* solution of the vacuum equations and f is the *only* function that derives from the presence of the fluid. The fluid motions themselves are derived from a potential, ϕ, which, in the null coordinates defined in equation (29), is governed by the equation (IIa, (60)),

$$\phi_{,u,v}+\frac{v\phi_{,u}+u\phi_{,v}}{1-u^2-v^2}=0; \tag{80}$$

and the function f is determined in terms of a solution of (80) by (cf. equation IIa (64))

$$f_{,u}=\frac{2(\phi_{,u})^2}{u(1-u^2-v^2)},\quad f_{,v}=\frac{2(\phi_{,v})^2}{v(1-u^2-v^2)}, \tag{81}$$

while the energy density ϵ $(=p)$ is given by (equation IIa (65))

$$\epsilon=-\frac{(1-u^2)^{\frac{1}{2}}(1-v^2)^{\frac{1}{2}}}{(1-u^2-v^2)^2\sqrt{\Delta}}\,\mathrm{e}^{-\mu_3-\nu-f}\,\phi_{,u}\phi_{,v}. \tag{82}$$

In Paper IIa we chose for the vacuum space-time the Nutku–Halil solution. We shall now consider for the vacuum solution that obtained in Paper III; this solution, in contrast to the Nutku–Halil solution, develops a horizon in place of a space-like curvature-singularity. In the (r,s)-variables of §3(d), the metric will then be given by (cf. equations III (115), (120), and (132))

$$\mathrm{d}s^2=\mathrm{e}^f\frac{X}{H}[(\mathrm{d}s)^2-(\mathrm{d}r)^2]-s^2\frac{X}{Y}(\mathrm{d}x^2)^2-\frac{Y}{X}(\mathrm{d}x^1-q_2\,\mathrm{d}x^2)^2, \tag{83}$$

where $$X=(1-p\eta)^2+q^2\mu^2;\quad q_2=-\frac{2q\delta(1-p\eta)}{pY}+\frac{2q}{p(1+p)}, \tag{84}$$

and $$H=4uv\sqrt{[(1-u^2)(1-v^2)]}=\delta-\Delta. \tag{85}$$

The equations that determine ϕ, f, and ϵ, in the same (r,s)-variables, take the forms:

$$\phi_{,r,r}-\left(\phi_{,s,s}-\frac{1}{s}\phi_{,s}\right)=0, \tag{86}$$

$$f_{,r}=-\frac{8}{s}\phi_{,r}\phi_{,s},\quad f_{,s}=-\frac{4}{s}[(\phi_{,r})^2+(\phi_{,s})^2], \tag{87}$$

and $$\epsilon=\frac{H}{s^2}\frac{\mathrm{e}^{-f}}{X}[(\phi_{,r})^2-(\phi_{,s})^2]. \tag{88}$$

It may be noted here (for future reference) that for the solution considered in Paper IIa, the expression for ϵ is

$$\epsilon=\frac{H}{s^{\frac{3}{2}}}\frac{\mathrm{e}^{-f}}{Y}[(\phi_{,r})^2-(\phi_{,s})^2]. \tag{89}$$

It was shown in Paper IIa that the general solution of (86), which ensures the positivity of ϵ in its domain of definition, can be written as a linear superposition (as sums or integrals) of the fundamental separable solution,

$$\phi=C\,\mathrm{e}^{\pm\alpha r}\,\alpha sK_1(\alpha s). \tag{90}$$

where α^2 is the separation constant and K_1 is Bessel's function of order 1 for a purely imaginary argument. The essential features of the general solution are sufficiently illustrated by the particular solution,

$$\phi = C\alpha s K_1(\alpha s) \sinh(\alpha r), \tag{91}$$

which was analysed in detail in Paper IIa. The corresponding solutions for f and ϵ are:

$$f = 2C^2\alpha^2\{2\alpha s K_1(\alpha s) K_0(\alpha s) \cosh^2(\alpha r) - \alpha^2 s^2[K_1^2(\alpha s) - K_0^2(\alpha s)]\}, \tag{92}$$

$$\epsilon = C^2\alpha^4 \frac{e^{-f}H}{X} [K_1^2(\alpha s) \cosh^2(\alpha r) - K_0^2(\alpha s) \sinh^2(\alpha r)]. \tag{93}$$

For the Nutku–Halil solution, considered in Paper IIa (equation (105)),

$$\epsilon = C^2\alpha^4 e^{-f} \frac{Hs^{\frac{1}{2}}}{Y} [K_1^2(\alpha s) \cosh^2(\alpha r) - K_0^2(\alpha s) \sinh^2(\alpha r)]. \tag{94}$$

In view of the occurrence of e^{-f} as a factor in the metric, in the expression for ϵ, and in the expressions for the Weyl and the Ricci scalars (see (103) and (104) below), the amplitude of the sound waves present, as measured by $C^2\alpha^2$, is decisive in determining the behaviour of the space-time as $s \to 0$. Since sufficient attention was not paid in Paper IIa to the delicate issues that arise in this context, we shall consider both solutions – that derived in the present paper and that derived in Paper IIa – in parallel.

It was shown in Paper IIa (equations (109)–(111)) that for the solution (92), the behaviour of e^{-f} as $s \to 0$ is given by

$$e^{-f} \to A(r)\, s^{k(r)}, \tag{95}$$

where

$$A(r) = \exp[4C^2\alpha^2(\gamma_E - \ln 2 + \ln\alpha) \cosh^2(\alpha r) + \tfrac{1}{2}] \quad (\gamma_E = 0.57721\ldots = \text{Euler's constant}), \tag{96}$$

and

$$k(r) = 4C^2\alpha^2 \cosh^2(\alpha r). \tag{97}$$

From (93) it now follows that for $s \to 0$,

$$\epsilon \sim \begin{cases} C^2\alpha^2 \dfrac{A(r)(1-r^2)\cosh^2(\alpha r)}{(1-p)^2 + q^2 r^2} s^{k-2} & (q \neq 0), \\ 4C^2\alpha^2[A(r)(1-r^2)^3 \cosh^2(\alpha r)] s^{k-6} & (q = 0). \end{cases} \tag{98}$$

The corresponding behaviours for the solution (94) are:

$$\epsilon \sim \begin{cases} \dfrac{C^2\alpha^2}{q^2} [A(r) \cosh^2(\alpha r)] s^{k-1.5} & (q \neq 0), \\ C^2\alpha^2[A(r)(1-r^2)^2 \cosh^2(\alpha r)] s^{k-3.5} & (q = 0). \end{cases} \tag{99}$$

From the similarity of the behaviours (98) and (99), we conclude that the disposition of the contours of constant ϵ for the solution (93) must qualitatively be the same as for the solution (94) illustrated in figures 1 and 2 in Paper IIa. In particular, the various cases distinguished in the context of the earlier solution apply equally to the present solution if the critical exponent k_c is assigned the values 2 and 6 in place of 1.5 and 3.5 (as follows from the behaviours (98) and (99)). More precisely, defining for each α the two amplitudes,

$$\left.\begin{aligned} C_0 &= \sqrt{k_c}/(2\alpha\cosh\alpha) \quad (s=0,\, r=\pm 1), \\ \text{and}\qquad C_1 &= \sqrt{k_c}/(2\alpha) \qquad\qquad (s=r=0), \end{aligned}\right\} \tag{100}$$

we distinguish the following cases:

(i) $C < C_0$, when ϵ diverges along the entire arc $u^2+v^2 = 1$ including the end points, $(u = 1, v = 0)$ and $(u = 0, v = 1)$;

(ii) $C = C_0$, when ϵ diverges along the entire arc $u^2+v^2 = 1$ except at its end points where it is finite;

(iii) $C_0 < C < C_1$, when ϵ diverges only along the part of the arc $u^2+v^2 = 1$, included between

$$v_+ = \sqrt{[(1+x)/2]} \quad \text{and} \quad v_- = \sqrt{[(1-x)/2]},$$

where x is the positive root of the equation,

$$2C\alpha\cosh(\alpha x) = \sqrt{k_c}, \tag{101}$$

and vanishes outside these limits (at $v = v_+$ and $v = v_-$, ϵ is finite);

(iv) $C = C_1$, when ϵ vanishes along the entire arc $u^2+v^2 = 1$ except at the mid point, $u = v = 1/\sqrt{2}$, where it is finite; and finally

(v) $C > C_1$, when ϵ vanishes along the entire arc $u^2+v^2 = 1$.

The different cases distinguished lead to different behaviours of the space-time as $s \to 0$. To elucidate the nature of these behaviours, we must first ascertain how the Weyl and the Ricci scalars depend on f and ϵ.

We have shown in Paper IIa, §10, how the Weyl and the Ricci scalars when $f \neq 0$ can be expressed very simply in terms of the Weyl scalars and spin-coefficients that obtain for the vacuum (when $f = 0$). Thus, by making use of the identities,

$$\partial_\psi + \partial_\theta = -2u(1-u^2)^{\frac{1}{2}}(\partial_r - \partial_s) \quad \text{and} \quad \partial_\psi - \partial_\theta = +2v(1-v^2)^{\frac{1}{2}}(\partial_r + \partial_s), \tag{102}$$

we can rewrite equations IIa (102)–(104) in the forms,

$$\begin{aligned} \tilde{\Psi}_1 &= \tilde{\Psi}_3 = 0, \\ \tilde{\Psi}_0 &= e^{-f}\,\Psi_0 - \sqrt{2}\,\frac{u\sqrt{(1-u^2)}}{U}\,\sigma(\partial_r - \partial_s)\,e^{-f}, \\ \tilde{\Psi}_4 &= e^{-f}\,\Psi_4 - \sqrt{2}\,\frac{v\sqrt{(1-v^2)}}{U}\,\lambda(\partial_r + \partial_s)\,e^{-f}, \\ \tilde{\Psi}_2 &= e^{-f}\,\Psi_2 + \tfrac{1}{3}\epsilon; \end{aligned} \tag{103}$$

and

$$\begin{aligned} \tilde{\Phi}_{00} &= -\sqrt{2}\,\frac{u\sqrt{(1-u^2)}}{U}\,\rho(\partial_r - \partial_s)\,e^{-f}, \\ \tilde{\Phi}_{22} &= -\sqrt{2}\,\frac{v\sqrt{(1-v^2)}}{U}\,\mu(\partial_r + \partial_s)\,e^{-f}, \\ \tilde{\Phi}_{11} &= +\tfrac{1}{2}\epsilon; \quad \tilde{\Lambda} = -\tfrac{1}{6}\epsilon, \end{aligned} \tag{104}$$

where the quantities distinguished by tildes refer to the problem on hand, while those not so distinguished refer to the vacuum. Also,

$$U^2 = e^{\mu_3+\nu}\sqrt{\Delta} \tag{105}$$

quite generally; and for the two solutions considered,

$$U^2 = \begin{cases} X & \text{(for the solution considered in Paper III)}, \\ Y/\sqrt[4]{(\Delta\delta)} & \text{(for the Nutku–Halil solution)}. \end{cases} \tag{106}$$

To determine the behaviours of the Weyl and the Ricci scalars given in (103) and (104), it is necessary that we know the spin-coefficients, ρ, μ, σ and λ, besides the Weyl scalars Ψ_0, Ψ_2 and Ψ_4, for the vacuum space-time. The requisite information is available for the Nutku–Halil solution in Chandrasekhar & Ferrari (1984, equations (86), (91), (94) and (95)). But for the vacuum solution considered in Paper III, only the Weyl scalars were explicitly evaluated; the spin-coefficients were not.

In Appendix C, we have listed the spin-coefficients for a vacuum space-time derived from a solution E of the Ernst equation for $\Psi + i\Phi$. In the present instance, the required spin-coefficients can be obtained by making the following substitutions in (C 1)–(C 5):

$$\begin{aligned} E &= p\cos\psi + iq\cos\theta, \quad U^2 = (1-p\cos\psi)^2 + q^2\cos^2\theta = X; \\ S &= \cot\psi + \cot\theta, \quad D = \cot\psi - \cot\theta; \\ \bar{S} &= \frac{2}{X}[p(1-p\cos\psi)\sin\psi - q^2\cos\theta\sin\theta], \\ \bar{D} &= \frac{2}{X}[p(1-p\cos\psi)\sin\psi + q^2\cos\theta\sin\theta]. \end{aligned} \tag{107}$$

We find:

$$\begin{aligned} (2\sqrt{2})\rho &= -\frac{1}{\sqrt{X}}(\cot\psi + \cot\theta); \quad (2\sqrt{2})\mu = +\frac{1}{\sqrt{X}}(\cot\psi - \cot\theta) \\ 2\sqrt{2}\,\frac{1-\mathscr{E}}{1-\mathscr{E}^*}\,\sigma &= \frac{1}{\sqrt{X}}\left[\cot\psi + \cot\theta + 2\frac{1-p\cos\psi + iq\cos\theta}{(1-p\cos\psi - iq\sin\theta)(p\sin\psi - iq\sin\theta)}\right], \\ -2\sqrt{2}\,\frac{1-\mathscr{E}^*}{1-\mathscr{E}}\,\lambda &= \frac{1}{\sqrt{X}}\left[\cot\psi - \cot\theta + 2\frac{1-p\cos\psi + iq\cos\theta}{(1-p\cos\psi - iq\sin\theta)(p\sin\psi + iq\sin\theta)}\right]. \end{aligned} \tag{108}$$

The Weyl scalars for the vacuum solution we are presently considering have been evaluated in Paper III (equatons (79), (81) and (82)); and according to these equations, their behaviours, for $s \to 0$, are given by:

$$\left.\begin{aligned} \Psi_0 = \Psi_4 = 3\Psi_2 &= \tfrac{3}{2}(1-p-iqr)^{-3} \quad (q \neq 0), \\ -\Psi_0 \sim -\Psi_4 \sim 3\Psi_2 &\sim 12(1-r^2)^3 s^{-6} \quad (q = 0). \end{aligned}\right\} \tag{109}$$

Similarly, from (108), we conclude that the spin-coefficients, σ, λ, ρ, and μ have the behaviours:

$$\left.\begin{aligned} \sigma \sim -\lambda \sim -\rho \sim \mu \sim \frac{1}{2\sqrt{2}} \frac{\sqrt{(1-r^2)}}{\sqrt{[(1-p)^2+q^2r^2]}} s^{-1} \quad (q \neq 0), \\ \sigma \sim -\lambda \sim -3\rho \sim 3\mu \sim \frac{3}{\sqrt{2}} (1-r^2)^{\frac{3}{2}} s^{-3} \quad (q = 0). \end{aligned}\right\} \tag{110}$$

Combining the foregoing behaviours in accordance with (98), (103) and (104), we find:

$$\tilde{\Psi}_0 \sim \tilde{\Psi}_4 \sim -\tilde{\Phi}_{00} \sim -\tilde{\Phi}_{22} \sim \frac{(1-r^2)\,k(r)\,A(r)}{4[(1-p)^2+q^2r^2]} s^{k-2},$$

$$\tilde{\Psi}_2 \sim \tfrac{1}{3}C^2\alpha^2 \frac{(1-r^2)\,A(r)\cosh^2(\alpha r)}{(1-p)^2+q^2r^2} s^{k-2}, \tag{111}$$

for $q \neq 0$; and

$$\tilde{\Psi}_0 = \tilde{\Psi}_4 \sim -3(1-r^2)^3\,[4-k(r)]\,A(r)\,s^{k-6},$$

$$\tilde{\Psi}_2 \sim 4(1-r^2)^3\,[1+\tfrac{1}{3}C^2\alpha^2\cosh^2(\alpha r)]\,A(r)\,s^{k-6},$$

$$\tilde{\Phi}_{00} \sim \tilde{\Phi}_{22} \sim -(1-r^2)^3\,k(r)\,A(r)\,s^{k-6}. \tag{112}$$

for $q = 0$.

With the Weyl scalars and spin-coefficients for the Nutku–Halil solution given in Chandrasekhar & Ferrari (1984, equations (86), (91), (94) and (95)) we similarly find:

$$\Psi_0 \sim \Psi_4 \begin{cases} \sim \dfrac{3i}{4q^3(1-r^2)} s^{-\frac{1}{2}}; \\ \sim \frac{3}{4}(1-r^2)^2 s^{-\frac{7}{2}}; \end{cases} \quad \Psi_2 \left\{\begin{aligned} &\sim -\frac{1}{8q^2} s^{-\frac{3}{2}} && (q \neq 0), \\ &\sim \tfrac{3}{8}(1-r^2)^2 s^{-\frac{7}{2}} && (q = 0); \end{aligned}\right\} \tag{113}$$

and

$$\sigma \sim -\lambda \begin{cases} \sim \dfrac{i}{q^2(1-r^2)\sqrt{2}} s^{\frac{1}{4}}; \\ \sim -\dfrac{1}{\sqrt{2}} (1-r^2)\,s^{-\frac{7}{4}}; \end{cases} \quad -\rho \sim \mu \left\{\begin{aligned} &\sim \frac{1}{2q\sqrt{2}} s^{-\frac{3}{4}} && (q \neq 0), \\ &\sim \frac{1}{2\sqrt{2}} (1-r^2)\,s^{-\frac{7}{4}} && (q = 0). \end{aligned}\right\} \tag{114}$$

Now combining these behaviours in accordance with equation (99), (103) and (104) we find:

$$\tilde{\Psi}_0 \sim \tilde{\Psi}_4 \sim \frac{i}{4q^3(1-r^2)} [3+2k(r)]\,A(r)\,s^{k-0.5},$$

$$\tilde{\Psi}_2 \sim -\frac{1}{24q^2} [3-8C^2\alpha^2\cosh^2(\alpha r)]\,A(r)\,s^{k-1.5},$$

$$\tilde{\Phi}_{00} \sim \tilde{\Phi}_{22} \sim -\frac{1}{4q^2} k(r)\,A(r)\,s^{k-1.5}, \tag{115}$$

for $q \neq 0$; and

$$\tilde{\Psi}_0 \sim \tilde{\Psi}_4 \sim \tfrac{1}{4}(1-r^2)^2\,[3-2k(r)]\,A(r)\,s^{k-3.5},$$

$$\tilde{\Psi}_2 \sim (1-r^2)^2\,[\tfrac{3}{8}+\tfrac{1}{3}C^2\alpha^2\cosh^2(\alpha r)]\,A(r)\,s^{k-3.5},$$

$$\tilde{\Phi}_{00} \sim \tilde{\Phi}_{22} \sim -\tfrac{1}{4}(1-r^2)^2\,k(r)\,A(r)\,s^{k-3.5} \quad (\text{for } q = 0), \tag{116}$$

where it may be noted that use has been made of the asymptotic relations,

$$U \sim [q\sqrt{(1-r^2)}]\,s^{-\frac{1}{4}} \quad (q \neq 0), \qquad U \sim \frac{s^{\frac{3}{4}}}{\sqrt{(1-r^2)}} \quad (q = 0), \tag{117}$$

for $s \to 0$.

It is manifest from the foregoing behaviours of the Weyl and the Ricci scalars for $s \to 0$, that the presence of the fluid, ever so evanescent, is the decisive factor. This fact is particularly striking in the context of the solution considered in Paper III: in the absence of the fluid, a horizon develops: but the horizon is destroyed and replaced by a space-like curvature-singularity so long as the amplitude C (see (91)) is less than C_0 $(= \sqrt{2}/(2\alpha \cosh \alpha)$ and ϵ diverges along the entire arc $u^2+v^2 = 1$. (This is the first of the five cases distinguished in our description of the contours of constant ϵ in region I. A reference to figures 1 and 2 in Paper II*a* may be helpful in visualizing the circumstances.) As the amplitude C increases, the space-like singularity is gradually lifted: for $C_0 < C < C_1$ – the third of the five cases distinguished – the Weyl and the Ricci scalars diverge along a part of the arc $u^2+v^2 = 1$ (included between the points v_+ and v_-), while they vanish along the rest of the arc. For amplitudes $C > C_1$, ϵ and, along with it, all the Weyl and the Ricci scalars, vanish along the entire arc. In this last case, the space-time tends to become flat, smoothly, as $s \to 0$. Since, further, the contour of constant ϵ approaches being null as $\epsilon \to 0$, one may, perhaps, conjecture that even though the space-time becomes flat in the limit, the surface $s = 0$ may, in fact, be a 'singularity' not unlike the one that occurs along the null boundaries, $v = 1$ and $u = 1$, of regions II and III. But the matter clearly requires further investigation.

In the context of the foregoing discussion, it is important to observe that the non-vanishing components for the four-velocity, u_0 and u_3, have smooth behaviours for $s \to 0$. For, from equation II*a* (42), (65), (67) we find

$$(\delta-\Delta)[(\phi_{,r})^2-(\phi_{,s})^2]\,(u_3)^2 = \{[u\sqrt{(1-u^2)}-v\sqrt{(1-v^2)}]\,\phi_{,r}$$
$$-[u\sqrt{(1-u^2)}+v\sqrt{(1-v^2)}]\,\phi_{,s}\}^2; \tag{118}$$

and since in general $\phi_{,r}$ remains finite while $\phi_{,s} \to 0$ for $s \to 0$ (cf. equation II*a* (74), (75) and (107)), it follows that $u_3 \to 0$ while $u_0 \to 1$ as $s \to 0$.

It remains to consider the extension of the space-time into regions II, III and IV by the standard Penrose algorism. It is clear that in the present instance, the metric of the space-time extended into region II, for example, will be of the same form as that obtained for the vacuum in Paper III, equations (22) and (61)–(63) with the only difference that g must be replaced by $g+f$.

Since the Ricci scalars vanish for the vacuum space-time, it follows from equation III (37) that the only contribution to $L+M$ will derive from the term in g' so that we shall now have

$$L+M = \frac{v}{2(1-v^2)}\,e^{-(g+f)}\,f_{,v}(0,v)$$
$$= \frac{v}{2(1-2pv+v^2)\sqrt{(1-v^2)}}\,e^{-f(0,v)}\,f_{,v}(0,v). \tag{119}$$

The resulting expression for the Ricci scalar $\tilde{\Phi}_{22}$ is

$$\tilde{\Phi}_{22} = -\frac{v}{2(1-2pv+v^2)}\,e^{-f(0,v)}\,f_{,v}(0,v). \tag{120}$$

The requirement that $\tilde{\Phi}_{22} < 0$ for the prevalence of null dust in region II (as was the case for the solution considered in Paper II*a*) is assured by the positivity of $f_{,v}$, guaranteed by (81).

Turning next to the Weyl scalar Ψ_4 given by equation III (40), we first observe that according to equation III (41),

$$M - L + 2iN = e^{-f}\left[(M-L+2iN)_{vac} - \frac{1}{4}\frac{\chi' + iq_2'}{\chi}\,e^{-g}f'\right], \tag{121}$$

where primes denote differentiation with respect to v. Using the definitions of g, χ and q_2 given in equations III (62) and (63), we find:

$$\frac{1}{4}\frac{\chi' + iq_2'}{\chi}\,e^{-g}f' = \frac{\sqrt{(1-v^2)}}{2(1-2pv+v^2)^2}\,(v-p+iq)f'. \tag{122}$$

Adding this contribution to $(M-L+2iN)_{vac}$, we shall find for $\tilde{\Psi}_4$ the expression (cf. equations I (178) and III (64))

$$\tilde{\Psi}_4 = -\tfrac{1}{2}\,e^{-f(0,v)}\,\frac{\chi - iq_2 + 1}{\chi + iq_2 - 1}\left[\frac{3(p-iq)^2}{(1-pv+iqv)^3} + \frac{1-v^2}{(1-2pv+v^2)^2}\,(v-p+iq)f_{,v}(0,v)\right]. \tag{123}$$

It can be verified that $\tilde{\Phi}_{22}$ and $\tilde{\Psi}_4$ given by (120) and (123) agree with those that follow from equations (103) and (104) on the common null-boundary separating regions I and II.

The extension of the metric into regions III and IV calls for no special comment.

5. The effect of null dust on the horizon that develops in the vacuum space-time considered in Paper III

Null dust as defined in Chandrasekhar & Xanthopoulos (1986*a*; referred to hereafter as Paper II*b*) consists of massless particles (with no internal pressure) describing null trajectories. Its energy momentum tensor is given by

$$T^{ij} = Ek^ik^j \quad (T^i_i = 0), \tag{124}$$

where E is some positive scalar function and $\boldsymbol{k}$ is a null vector. More precisely, null dust can be considered as a mixture of two sorts of massless particles which, at any point of space-time, move in opposite directions. Accordingly, the energy momentum tensor, in a tetrad frame suitable for a space-time with two space-like Killing vectors ∂_{x^1} and ∂_{x^2}, is expressible as a sum of two terms in the form (cf. equations II*b* (9) and (10)),

$$T^{(a)(b)} = 2(\epsilon k^{(a)}k^{(b)} + \bar{\epsilon}\bar{k}^{(a)}\bar{k}^{(b)}), \tag{125}$$

where $$k_{(0)} = +k_{(3)} = k \quad\text{and}\quad \bar{k}_{(0)} = -\bar{k}_{(3)} = \bar{k}. \tag{126}$$

It was shown in Paper IIb, that the coupling of the gravitational field with null dust modifies a vacuum metric into exactly the same form (79) as a perfect fluid with the equation of state $\epsilon = p$, with the difference that the function f now derives from the presence of null dust; and, further, that the only requirement on f is that it satisfy the equation

$$f_{,u,v} = 0. \tag{127}$$

where u and v are the null coordinates defined in (29). In other words,

$$f(u, v) = \bar{F}(u) + F(v), \tag{128}$$

where $\bar{F}$ and F are, for the present, arbitrary functions of u and v, respectively. In terms of F and $\bar{F}$, ϵk^2 and $\bar{\epsilon}\bar{k}^2$, which specify the null dust, are given by

$$\epsilon k^2 = \frac{v(1-v^2)}{8s} e^{-2\mu_3 - f} F_{,v} \quad \text{and} \quad \bar{\epsilon}\bar{k}^2 = \frac{u(1-u^2)}{8s} e^{-2\mu_3 - f} \bar{F}_{,u}. \tag{129}$$

For the vacuum space-time considered in Paper III,

$$e^{2\mu_3} = X = (1-p\eta)^2 + q^2\mu^2, \tag{130}$$

and we have

$$\epsilon k^2 = \frac{v(1-v^2)}{8X} e^{-f} F_{,v} s^{-1} \quad \text{and} \quad \bar{\epsilon}\bar{k}^2 = \frac{u(1-u^2)}{8X} e^{-f} \bar{F}_{,u} s^{-1}. \tag{131}$$

For comparison, we may note here that for the vacuum space-time considered in Paper IIb,

$$\epsilon k^2 = \frac{v(1-v^2)}{8Y} e^{-f} F_{,v} s^{-\frac{1}{2}} \quad \text{and} \quad \bar{\epsilon}\bar{k}^2 = \frac{u(1-u^2)}{8Y} e^{-f} \bar{F}_{,u} s^{-\frac{1}{2}}. \tag{132}$$

For consistent extendability of the space-time into regions II and III, it is necessary to require that (cf. equation IIb (62))

$$\bar{F}_{,u} > 0 \quad (0 \leqslant u \leqslant 1) \quad \text{and} \quad F_{,v} > 0 \quad (0 \leqslant v \leqslant 1). \tag{133}$$

Also, since the addition of constants to $\bar{F}$ and F can be absorbed into the definitions of $\bar{\epsilon}$ and ϵ, there is no loss of generality in supposing in addition that

$$\bar{F}(0) = F(0) = 0, \quad \bar{F}(u) \geqslant 0 \quad (0 \leqslant u \leqslant 1), \quad \text{and} \quad F(v) \geqslant 0 \quad (0 \leqslant v \leqslant 1). \tag{134}$$

The Weyl and the Ricci scalars, for the problem we are presently considering, can be directly written down from (103) and (104) by simply replacing f by $\bar{F}(u)+F(v)$ and ϵ by 0 (cf. equations IIa (102), (103), and IIb (49)). In rewriting the equations, it is convenient to introduce the variables (cf. equation IIa (A 1))

$$x = 1-2u^2 = s-r \quad \text{and} \quad y = 1-2v^2 = s+r$$
$$(-1 \leqslant x \leqslant 1,\ -1 \leqslant y \leqslant +1,\ 0 \leqslant x+y \leqslant 2), \tag{135}$$

and write

$$f(u, v) \equiv \bar{F}(x) + F(y). \tag{136}$$

Equations (103) and (104) now give:

$$\tilde{\Psi}_0 = e^{-f}\left[\Psi_0 - 2\sqrt{2}\,\frac{u\sqrt{(1-u^2)}}{U}\,\sigma\bar{F}_{,x}\right], \quad \tilde{\Psi}_4 = e^{-f}\left[\Psi_4 + 2\sqrt{2}\,\frac{v\sqrt{(1-v^2)}}{U}\,\lambda F_{,y}\right]$$
$$\tilde{\Psi}_2 = e^{-f}\Psi_2;\ \tilde{\Phi}_{00} = -2\sqrt{2}\,\frac{u\sqrt{(1-u^2)}}{U}\,e^{-f}\rho\bar{F}_{,x}, \quad \tilde{\Phi}_{22} = 2\sqrt{2}\,\frac{v\sqrt{(1-v^2)}}{U}\,e^{-f}\mu F_{,y}, \tag{137}$$

where, as in §4, the quantities without tildes refer to the vacuum space-time while the quantities with tildes refer to the problem on hand.

The behaviours of the Weyl and the Ricci scalars can now be written down by combining the behaviours of the various scalars and spin-coefficients given in (109) and (110) in accordance with (137). We find:

$$\tilde{\Psi}_0 \sim -\tilde{\Phi}_{00} \sim -\frac{1-r^2}{2[(1-p)^2+q^2r^2]}\,\mathrm{e}^{-f}\,\bar{F}_{,x}\,s^{-1},$$

$$\tilde{\Psi}_4 \sim -\tilde{\Phi}_{22} \sim -\frac{1-r^2}{2[(1-p)^2+q^2r^2]}\,\mathrm{e}^{-f}\,F_{,y}\,s^{-1},$$

and

$$\tilde{\Psi}_2 = \tfrac{1}{2}(1-p-\mathrm{i}qr)^{-3}\,\mathrm{e}^{-f}, \tag{138}$$

for $q \neq 0$; and

$$\tilde{\Psi}_0 \sim \tilde{\Psi}_4 \sim -3\tilde{\Psi}_2 \sim -12(1-r^2)^3\,\mathrm{e}^{-f}\,s^{-6},$$

$$\tilde{\Phi}_{00} \sim 2(1-r^2)^3\,\mathrm{e}^{-f}\,\bar{F}_{,x}\,s^{-5}, \quad \text{and} \quad \tilde{\Phi}_{22} \sim 2(1-r^2)^3\,\mathrm{e}^{-f}\,F_{,y}\,s^{-5}, \tag{139}$$

for $q = 0$.

For comparison, the corresponding behaviours for the space-time considered in Paper II*b* are:

$$\tilde{\Psi}_0 \sim \tilde{\Psi}_4 \sim \frac{3\mathrm{i}\mathrm{e}^{-f}}{4q^3(1-r^2)}\,s^{-\frac{1}{2}}, \quad \tilde{\Psi}_2 \sim -\frac{1}{8q^2}\,\mathrm{e}^{-f}\,s^{-\frac{3}{2}}$$

and

$$\tilde{\Phi}_{00} \sim \frac{\mathrm{e}^{-f}}{2q^2}\,\bar{F}_{,x}\,s^{-\frac{1}{2}}, \quad \tilde{\Phi}_{22} \sim \frac{\mathrm{e}^{-f}}{2q^2}\,F_{,y}\,s^{-\frac{1}{2}}, \tag{140}$$

for $q \neq 0$; and

$$\tilde{\Psi}_0 \sim \tilde{\Psi}_4 \sim 2\tilde{\Psi}_2 \sim \tfrac{3}{4}(1-r^2)^2\,\mathrm{e}^{-f}\,s^{-\frac{7}{2}},$$

and

$$\tilde{\Phi}_{00} \sim \tfrac{1}{2}(1-r^2)^2\,\mathrm{e}^{-f}\,\bar{F}_{,x}\,s^{-\frac{5}{2}}, \quad \tilde{\Phi}_{22} \sim \tfrac{1}{2}(1-r^2)^2\,\mathrm{e}^{-f}\,F_{,y}\,s^{-\frac{5}{2}}, \tag{141}$$

for $q = 0$.

Before we consider the implications of the foregoing behaviours of the Weyl and the Ricci scalars, it is useful to state explicitly the following, otherwise obvious, lemma.

If $\bar{F}(x)$ and $F(y)$ are positive bounded functions defined in a triangular domain,

$$-1 \leqslant x \leqslant +1, \quad -1 \leqslant y \leqslant +1, \quad \text{and} \quad 0 \leqslant x+y \leqslant 2, \tag{142}$$

then, for $m > 0$,

$$I = \frac{\mathrm{e}^{-\bar{F}(x)-F(y)}}{(x+y)^m} \to \infty \quad \text{as} \quad (x+y) \to 0. \tag{143}$$

Conversely, if I should be bounded in a finite segment, S, along the hypotenuse, $x+y=0$, then either $\bar{F}(x)$ or $F(y)$ (positive by definition) must be unbounded for the projected range of the segment S along the x- or the y-coordinate axes. *Therefore, under no physically reasonable conditions can I remain bounded as $x+y = 2s \to +0$.*

In view of the foregoing lemma, it is manifest from the behaviours (138)–(140) of the Weyl and the Ricci scalars that, in *all* the cases considered, a space-like singularity develops when $s \to 0$. In particular, this is the case even for the vacuum solution considered in Paper III which develops a horizon, with the Weyl scalars remaining finite, when $s \to 0$. The behaviours specified in (138) apply to this case; and as is manifest, the merest presence of null dust destroys the horizon and replaces it with a space-like singularity. In this particular respect, the null dust has the same effect as the $(\epsilon = p)$-fluid. But unlike the $(\epsilon = p)$-fluid, increasing intensity of the prevailing null dust does not eliminate the development of the curvature singularity: in *all* cases considered, the curvature singularity is eliminated (and replaced by a space-time that becomes flat when $s \to 0$) if the $(\epsilon = p)$-fluid is present in sufficient amount.

When a space-like singularity develops in the absence of the null dust (as in the cases to which the behaviours (139)–(141) apply), the presence of the dust does not affect the negative power of s with which the singularity develops, again in contrast to the $(\epsilon = p)$-fluid whose presence is *always* decisive.

6. A comparison of the behaviours of null dust and an $(\epsilon = p)$-fluid in the frameworks of special and general relativity

When the transformation of null dust into an $(\epsilon = p)$-fluid first emerged from the analysis of Paper II*a*, the phenomenon was initially described as 'gravitationally induced'. Roger Penrose (in a personal communication) objected, with the remark that 'the same behaviour could also occur for the corresponding equations for fluids in special relativity'; and Penrose explained as follows.

Consider the collision of two plane pulses of null dust. We can obtain their behaviour in the region of interaction by adding the null vectors, $k^i_{(1)}\sqrt{\epsilon_1}$ and $k^i_{(2)}\sqrt{\epsilon_2}$ of the two pulses. The resulting vector, being the sum of two non-proportional null-vectors, will be time-like and will describe the streamlines of a *non-rotating* $(\epsilon = p)$-fluid. Figure 1*a* is Penrose's illustration of the phenomenon.

In the notation and in the frame-work of the present and the earlier Papers II*a* and II*b*, the equations appropriate to a special-relativistic treatment of the problem are the following.

In a flat space-time with the minkowskian metric,

$$ds^2 = (dx^0)^2 - (dx^3)^2 - (dx^1)^2 - (dx^2)^2. \tag{144}$$

equations II*a* (41) and (42), governing an $(\epsilon = p)$-fluid, reduce to

$$U^0 = u^0\sqrt{\epsilon} = \phi_{,3} \quad \text{and} \quad U^3 = u^3\sqrt{\epsilon} = -\phi_{,0} \tag{145}$$

and

$$\phi_{,0,0} - \phi_{,3,3} = 0. \tag{146}$$

In terms of the null coordinates,

$$u = x^0 + x^3 \quad \text{and} \quad v = x^0 - x^3, \tag{147}$$

the equations become

$$U^u = (U^0 + U^3) = -2\phi_{,v}, \quad U^v = (U^0 - U^3) = +2\phi_{,u} \tag{148}$$

and

$$\phi_{,u,v} = 0. \tag{149}$$

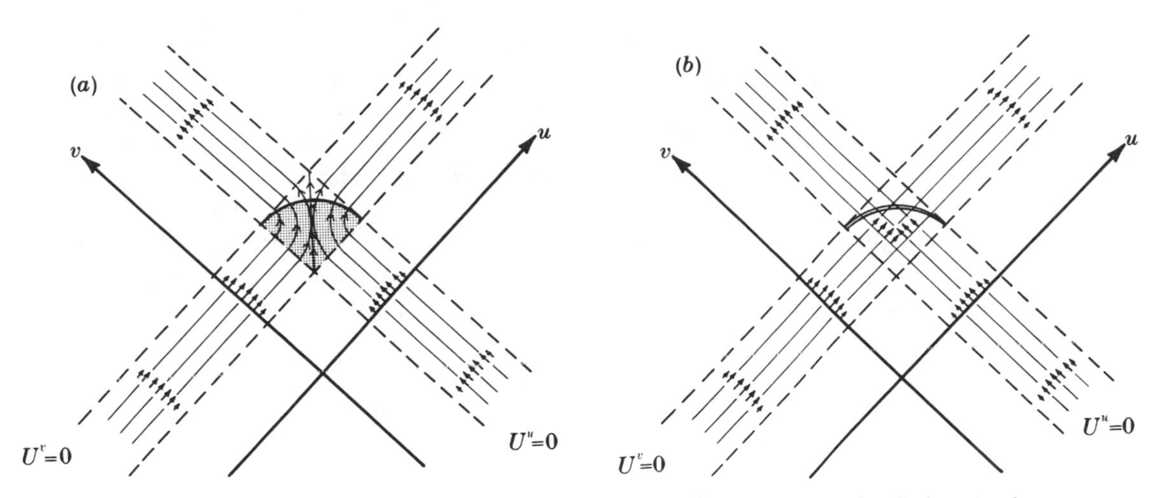

FIGURE 1. Comparison of the results of the collision of plane waves of null dust in the frameworks of special and general relativity. (For explanation see text.)

With the general solution of equation (149) given by

$$\phi(u,v) = \bar{\Phi}(u)+\Phi(v), \tag{150}$$

where $\bar{\Phi}(u)$ and $\Phi(v)$ are functions only of the arguments specified, we have

$$U^u = -2\Phi_{,v} \quad \text{and} \quad U^v = 2\bar{\Phi}_{,u}. \tag{151}$$

In other words, *the velocity of fluid elements, following streamlines in the u-direction, depend only on v, and conversely.*

Turning next to null dust, we envisaged it as consisting of two sorts of massless particles (distinguished by ϵ and $\bar{\epsilon}$) characterized by $k^0+k^3 = 0$ and $\bar{k}^0-\bar{k}^3 = 0$, i.e. by their following null trajectories in the v- or the u-directions, respectively. Denoting, in analogy with the definitions (145) and (148) for the $(\epsilon = p)$-fluid,

$$\epsilon k^2 = (U^v)^2 \quad \text{and} \quad \bar{\epsilon}\bar{k}^2 = (U^u)^2, \tag{152}$$

we find that equations II*b* (28) and (29) in Minkowskian space-time, become

$$(U^v)^2 = \phi_{,u} \quad (U^u)^2 = -\phi_{,v} \tag{153}$$

and

$$\phi_{,u,v} = 0. \tag{154}$$

Again, writing the solution of ϕ as a sum of two functions $\bar{N}(u)$ and $N(v)$, we obtain,

$$(U^u)^2 = -N_{,v} \quad \text{and} \quad (U^v)^2 = +\bar{N}_{,u}. \tag{155}$$

In other words, *the massless particles describing null trajectories in the u-direction depend only on v, and conversely.*

In view of the similar behaviours of the streamlines of fluid elements in the $(\epsilon = p)$-fluid and of the null trajectories of the massless particles constituting null dust, Penrose's explanation of the transformation of null dust into an $(\epsilon = p)$-fluid becomes transparent. An alternative possibility, not mentioned by Penrose, is that the two streams of null dust simply pass through each other without the one affecting the other. This second possibility is illustrated in figure 1*b*.

The two possible results of collision of two plane-waves of null dust, in the framework of special relativity, correspond exactly to the two behaviours of colliding waves of null dust, in the framework of general relativity, that we have described in §§4 and 5. But apart from this correspondence, what happens as a result of the collision in the two frameworks is radically different. The presence only of null dust in the region of interaction results, in all cases considered, in an impenetrable barrier, in the form of a space-like singularity, developing on the surface $u^2+v^2 = 1$. If, on the other hand, we have only an $(\epsilon = p)$-fluid in the region of interaction, the fluid is trapped behind $u^2+v^2 = 1$, in *all* cases: either by a space-like singularity developing (for small amplitudes) or by the space-time becoming flat, smoothly, as $u^2+v^2 \to +1$ (for large amplitudes). The results of the collision in the two cases, which emerge from general relativity, cannot really have been anticipated from any approximate version of the theory.

7. Concluding remarks

In the present series of papers the collision of gravitational waves, either by themselves or coupled with sources of electromagnetic or hydrodynamic origin, has been studied in a variety of cases. The principal results of these studies (with the exception of those of part II of this paper) have been summarized recently by one of us (Chandrasekhar 1987: see particularly tables 1 and 2 on pp. 42–44). Besides, the results of part II have been adequately summarized in §4 (following (117)), §5 (following (143)), and in §6. We shall, therefore, confine our present remarks, principally, to the results of this paper.

The solution of the Einstein–Maxwell equations, analogous to the vacuum solution obtained in Paper III, shows that the coupling of the gravitational waves to electromagnetic waves does not affect the essential features of the problem, namely, the development of a horizon and, subsequently, of a time-like curvature-singularity along hyperbolic arcs. However, in the hypersurface-orthogonal case, the strong space-like singularity that occurs in the vacuum is transformed into a horizon followed by a three-dimensional time-like singularity by the merest presence of the electromagnetic field. This essential difference in the two cases (of parallel or non-parallel polarizations of the colliding gravitational waves) is reminiscent of the replacement of the space-like singularity in the Schwarzschild space-time by a Cauchy horizon followed by a time-like singularity in the Reissner–Nordström space-time. (The completion of the analysis of Paper I, in Appendix B, of the Einstein-Maxwell analogue of the Nutku–Halil solution, does not disclose any new feature.)

The results of part II reveal the very different ways in which an $(\epsilon = p)$-fluid and null dust affect the development of horizons and singularities: the entirely decisive way in which the $(\epsilon = p)$-fluid affects the development in *all cases* but *in qualitatively the same way*; and the *invariable* development of a singularity when null dust is present even most evanescently. And the comparison of these behaviours with what one may expect in the framework of special relativity shows the danger of extrapolating to general relativity what one knows otherwise.

In the context of the problems considered in part II, we may allow ourselves one speculative conjecture. As was pointed out in Paper II*a* (§13), Zel'dovich (1962) appears to have established that the ultimate equation of state of neutral matter, consisting of charged constituents of opposite signs, is one in which equality between the energy-density and the pressure is attained and the velocity of sound equals the velocity of light. If the reality of this ultimate state of matter is accepted, then the fact that such matter can be converted into null dust, both in the frameworks of special and general relativity, gives credence to the possibility that null dust, as (classically) defined in this paper, may have an equal reality. If one goes one step further and regards null dust as the earliest constituent of the Universe, then its conversion into an $(\epsilon = p)$-fluid, in the environment then prevailing, may hold the clue to the origin of matter in the early Universe.

Independently of the foregoing speculation, the present studies suggest that 'exploring general relativity, sensitive to its aesthetic base, may lead to a deepening of our understanding of the physical content of the theory'.

The research reported in this paper has, in part, been supported by grants from the National Science Foundation under grant PHY-84-16691 with the University of Chicago. We are also grateful for a grant from the Division of Physical Sciences of the University of Chicago which has enabled our continued collaboration by making possible periodic visits by B.C.X. to the University of Chicago.

References

Chandrasekhar, S. 1987 *Karl Schwarzschild Lecture: The aesthetic base of the general theory of relativity*. Hamburg: Mitteilungen der Astronomischen Gesellschaft Nr. 67.
Chandrasekhar, S. & Ferrari, V. 1984 *Proc. R. Soc. Lond.* A **396**, 55–74.
Chandrasekhar, S. & Xanthopoulos, B. C. 1985*a* *Proc. R. Soc. Lond.* A **398**, 223–259 (Paper I).
Chandrasekhar, S. & Xanthopoulos, B. C. 1985*b* *Proc. R. Soc. Lond.* A **402**, 37–65 (Paper II*a*).
Chandrasekhar, S. & Xanthopoulos, B. C. 1986*a* *Proc. R. Soc. Lond.* A **403**, 189–198 (Paper II*b*).
Chandrasekhar, S. & Xanthopoulos, B. C. 1986*b* *Proc. R. Soc. Lond.* A **408**, 175–208 (Paper III).
Chandrasekhar, S. & Xanthopoulos, B. C. 1987 *Proc. R. Soc. Lond.* A **410**, 311–336.
Zel'dovich, Ya. B. 1962 *Soviet Phys. JETP* **14**, 1143.

Appendix A. Expressions for the Weyl and the Maxwell scalars explicitly in terms of the solution, E, of the vacuum Ernst-equation

We start with the same form of the metric and described in the same null tetrad-frame as set out in Paper III, Appendix A. The expressions for the Weyl scalars, given in equations III (A 14)–(A 16), ostensibly for the vacuum, are equally valid for an Einstein–Maxwell space-time as is evident from a comparison of equations I (130)–(132) and equations III (A 5)–(A 7). The required expressions are:

$$8U^2\Psi_2 = -SD + \frac{1}{\chi^2}(\mathscr{X}_{,\psi} - \mathscr{X}_{,\theta})(\mathscr{X}^*_{,\psi} + \mathscr{X}^*_{,\theta}), \tag{A 1}$$

$$4U^2\frac{1-\mathscr{E}^*}{1-\mathscr{E}}\Psi_4 = (D-\bar{D})\frac{1}{\chi}(\mathscr{Z}_{,\psi}-\mathscr{Z}_{,\theta})+\frac{1}{\chi}\left(\frac{\partial}{\partial\psi}-\frac{\partial}{\partial\theta}\right)^2\mathscr{Z}-\frac{1}{\chi^2}(\mathscr{Z}_{,\psi}-\mathscr{Z}_{,\theta})^2$$
$$= (D-\bar{D})\frac{1}{\chi}(\mathscr{Z}_{,\psi}-\mathscr{Z}_{,\theta})+\left(\frac{\partial}{\partial\psi}-\frac{\partial}{\partial\theta}\right)\left[\frac{1}{\chi}(\mathscr{Z}_{,\psi}-\mathscr{Z}_{,\theta})\right]$$
$$+\frac{1}{\chi}(\mathscr{Z}_{,\psi}-\mathscr{Z}_{,\theta})\left(\frac{\chi_{,\psi}}{\chi}-\frac{\chi_{,\theta}}{\chi}\right)-\frac{1}{\chi^2}(\mathscr{Z}_{,\psi}-\mathscr{Z}_{,\theta})^2, \qquad \text{(A 2)}$$

and

$$4U^2\frac{1-\mathscr{E}^*}{1-\mathscr{E}}\Psi_0^* = (S-\bar{S})\frac{1}{\chi}(\mathscr{Z}_{,\psi}+\mathscr{Z}_{,\theta})+\left(\frac{\partial}{\partial\psi}+\frac{\partial}{\partial\theta}\right)\left[\frac{1}{\chi}(\mathscr{Z}_{,\psi}+\mathscr{Z}_{,\theta})\right]$$
$$+\frac{1}{\chi}(\mathscr{Z}_{,\psi}+\mathscr{Z}_{,\theta})\left(\frac{\chi_{,\psi}}{\chi}+\frac{\chi_{,\theta}}{\chi}\right)-\frac{1}{\chi^2}(\mathscr{Z}_{,\psi}+\mathscr{Z}_{,\theta})^2, \qquad \text{(A 3)}$$

where, to repeat for the sake of convenience, the definitions

$$\mathscr{Z} = \chi+iq_2 = \frac{1+\mathscr{E}}{1-\mathscr{E}}, \quad \chi = \frac{1-|\mathscr{E}|^2}{|1-\mathscr{E}|^2}, \quad iq_2 = \frac{\mathscr{E}-\mathscr{E}^*}{|1-\mathscr{E}|^2}; \qquad \text{(A 4)}$$

$$S = 2(\ln V)_{,\psi}+2(\ln V)_{,\theta}; \quad \bar{S} = 2(\ln U)_{,\psi}+2(\ln U)_{,\theta}; \qquad \text{(A 5)}$$

$$D = 2(\ln V)_{,\psi}-2(\ln V)_{,\theta}; \quad \bar{D} = 2(\ln U)_{,\psi}-2(\ln U)_{,\theta}; \qquad \text{(A 6)}$$

where χ and q_2 now refer to the solutions of the Einstein–Maxwell equations. (We are here suppressing the distinguishing subscript e.)

The expressions for the Maxwell scalars, given in equations I (133)–(135), similarly written in terms of $\mathscr{Z}$, with the aid of equations III (A 12) and (A 13), are:

$$-4U^2\Phi_{00} = S_{,\psi}+S_{,\theta}+\tfrac{1}{2}S^2-S\bar{S}+\frac{1}{2\chi^2}|\mathscr{Z}_{,\psi}+\mathscr{Z}_{,\theta}|^2, \qquad \text{(A 7)}$$

$$-4U^2\Phi_{22} = D_{,\psi}-D_{,\theta}+\tfrac{1}{2}D^2-D\bar{D}+\frac{1}{2\chi^2}|\mathscr{Z}_{,\psi}-\mathscr{Z}_{,\theta}|^2, \qquad \text{(A 8)}$$

and

$$-4U^2\frac{1-\mathscr{E}^*}{1-\mathscr{E}}\Phi_{20} = \frac{1}{2\chi}[S(\mathscr{Z}_{,\psi}-\mathscr{Z}_{,\theta})+D(\mathscr{Z}_{,\psi}+\mathscr{Z}_{,\theta})]$$
$$+\left(\frac{\partial}{\partial\psi}+\frac{\partial}{\partial\theta}\right)\left[\frac{1}{\chi}(\mathscr{Z}_{,\psi}-\mathscr{Z}_{,\theta})\right]+\frac{1}{\chi}(\mathscr{Z}_{,\psi}-\mathscr{Z}_{,\theta})\left(\frac{\chi_{,\psi}}{\chi}+\frac{\chi_{,\theta}}{\chi}\right)$$
$$-\frac{1}{\chi^2}(\mathscr{Z}_{,\psi}-\mathscr{Z}_{,\theta})(\mathscr{Z}_{,\psi}+\mathscr{Z}_{,\theta}). \qquad \text{(A 9)}$$

In our present context (cf. (10)),

$$\chi = \frac{V^2}{\alpha^2}\frac{|1-\alpha E|^2}{1-|E|^2} \quad [V^2 = \surd(\Delta\delta) = \sin\psi\sin\theta]. \qquad \text{(A 10)}$$

Accordingly,

$$\frac{\chi_{,i}}{\chi} = 2(\ln V)_{,i}+\frac{1}{1-|E|^2}\left[\frac{(E^*-\alpha)E_{,i}}{1-\alpha E}+\frac{(E-\alpha)E^*_{,i}}{1-\alpha E^*}\right] \quad (i=\psi,\theta). \qquad \text{(A 11)}$$

Also, (14) and (15) can be rewritten, as a single equation, in the form,

$$\frac{q_{2,i}}{\chi} = \frac{i}{|1-\alpha E|^2(1-|E|^2)}[(1-\alpha E)(E-\alpha)E^*_{,j}-(1-\alpha E^*)(E^*-\alpha)E_{,j}]$$
$$(i,j=\psi,\theta;\, i\neq j). \qquad \text{(A 12)}$$

With the aid of equations (A 11) and (A 12), we obtain

$$\frac{1}{\chi}(\mathscr{Z}_{,\psi}-\mathscr{Z}_{,\theta}) = D+2\frac{(E-\alpha)(E^*_{,\psi}-E^*_{,\theta})}{(1-\alpha E^*)(1-|E|^2)} \qquad \text{(A 13)}$$

and

$$\frac{1}{\chi}(\mathscr{Z}_{,\psi}+\mathscr{Z}_{,\theta}) = S+2\frac{(E^*-\alpha)(E_{,\psi}+E_{,\theta})}{(1-\alpha E)(1-|E|^2)}. \qquad \text{(A 14)}$$

The required expressions for the Weyl and the Maxwell scalars now follow from substituting the expressions (A 11), (A 13) and (A 14) in equations (A 1)–(A 3) and (A 7)–(A 9). After some lengthy, but straightforward, reductions, we find:

$$4U^2\Psi_2' - \frac{E-\alpha}{(1-|E|^2)(1-\alpha E^*)}\bigg[S(E^*_{,\psi}-E^*_{,\theta})+D(E^*_{,\psi}+E^*_{,\theta})$$
$$+2(E-\alpha)\frac{(E^*_{,\psi}+E^*_{,\theta})(E^*_{,\psi}-E^*_{,\theta})}{(1-|E|^2)(1-\alpha E^*)}\bigg], \qquad \text{(A 15)}$$

$$4U^2\frac{1-\mathscr{E}^*}{1-\mathscr{E}}\Psi_4 = D^2-D\bar{D}+D_{,\psi}-D_{,\theta}+\frac{2(E-\alpha)}{(1-|E|^2)(1-\alpha E^*)}\left(\frac{\partial}{\partial\psi}-\frac{\partial}{\partial\theta}\right)^2 E^*$$
$$+\frac{1}{1-|E|^2}\left[D\frac{(E^*-\alpha)(E_{,\psi}-E_{,\theta})}{1-\alpha E}+(D-2\bar{D})\frac{(E-\alpha)(E^*_{,\psi}-E^*_{,\theta})}{1-\alpha E^*}\right]$$
$$+2\frac{(1+\alpha^2)(1+|E|^2)-2\alpha(E+E^*)}{(1-|E|^2)^2|1-\alpha E|^2}|E_{,\psi}-E_{,\theta}|^2$$
$$+4\alpha\frac{(E-\alpha)(E^*_{,\psi}-E^*_{,\theta})^2}{(1-|E|^2)(1-\alpha E^*)^2}, \qquad \text{(A 16)}$$

$$4U^2\frac{1-\mathscr{E}^*}{1-\mathscr{E}}\Psi_0^* = S^2-S\bar{S}+S_{,\psi}+S_{,\theta}+\frac{2(E^*-\alpha)}{(1-|E|^2)(1-\alpha E)}\left(\frac{\partial}{\partial\psi}+\frac{\partial}{\partial\theta}\right)^2 E$$
$$+\frac{1}{1-|E|^2}\left[S\frac{(E-\alpha)(E^*_{,\psi}+E^*_{,\theta})}{1-\alpha E^*}+(S-2\bar{S})\frac{(E^*-\alpha)(E_{,\psi}+E_{,\theta})}{1-\alpha E}\right]$$
$$+2\frac{(1+\alpha^2)(1+|E|^2)-2\alpha(E+E^*)}{(1-|E|^2)^2|1-\alpha E|^2}|E_{,\psi}+E_{,\theta}|^2$$
$$+4\alpha\frac{(E^*-\alpha)(E_{,\psi}+E_{,\theta})^2}{(1-|E|^2)(1-\alpha E)^2}; \qquad \text{(A 17)}$$

$$-4U^2\Phi_{00} = S_{,\psi}+S_{,\theta}+S^2-S\bar{S}+2\frac{|E-\alpha|^2|E_{,\psi}+E_{,\theta}|^2}{(1-|E|^2)^2|1-\alpha E|^2}$$
$$+\frac{S}{1-|E|^2}\left[\frac{E^*-\alpha}{1-\alpha E}(E_{,\psi}+E_{,\theta})+\frac{E-\alpha}{1-\alpha E^*}(E^*_{,\psi}+E^*_{,\theta})\right], \qquad \text{(A 18)}$$

$$-4U^2\Phi_{22} = D_{,\psi}-D_{,\theta}+D^2-D\bar{D}+2\,\frac{|E-\alpha|^2\,|E_{,\psi}-E_{,\theta}|^2}{(1-|E|^2)^2\,|1-\alpha E|^2}$$
$$+\frac{D}{1-|E|^2}\left[\frac{E-\alpha}{1-\alpha E^*}\,(E^*_{,\psi}-E^*_{,\theta})+\frac{E^*-\alpha}{1-\alpha E}\,(E_{,\psi}-E_{,\theta})\right] \qquad \text{(A 19)}$$

$$-4U^2\,\frac{1-\mathscr{E}^*}{1-\mathscr{E}}\,\Phi_{20} = SD+D_{,\psi}+D_{,\theta}+2\,\frac{(E-\alpha)\,(E^*_{,\psi,\psi}-E^*_{,\theta,\theta})}{(1-\alpha E^*)\,(1-|E|^2)}$$
$$+4E(E-\alpha)\,\frac{(E^*_{,\psi}-E^*_{,\theta})\,(E^*_{,\psi}+E^*_{,\theta})}{(1-\alpha E^*)\,(1-|E|^2)^2}+2(1-\alpha^2)\,\frac{(E_{,\psi}+E_{,\theta})\,(E^*_{,\psi}-E^*_{,\theta})}{|1-\alpha E|^2\,(1-|E|^2)}$$
$$+\frac{E-\alpha}{(1-\alpha E^*)\,(1-|E|^2)}\,[S(E^*_{,\psi}-E^*_{,\theta})+D(E^*_{,\psi}+E^*_{,\theta})]. \qquad \text{(A 20)}$$

Appendix B. The Weyl and the Maxwell scalars for the solution obtained in Paper I

In Paper I, the Einstein–Maxwell analogue of the Nutku–Halil vacuum solution was obtained. But only the hypersurface orthogonal case was analysed in detail: the Weyl and the Maxwell scalars for the general case were not evaluated; and the character of the singularity that develops on the surface $u^2+v^2 = 1$ was not fully specified. In this Appendix, we shall show how the various scalars can be expressed directly in terms of certain basic functions that determine them for the vacuum together with some additional terms; and how the character of the singularities that develop on $u^2+v^2 = 1$ can be inferred without difficulty.

As is manifest from (A 1)–(A 3) and (A 7)–(A 9) of Appendix A, all the required scalars can be evaluated in terms of the functions

$$\mathscr{X}_e = \chi_e+\mathrm{i}q_{2e} \quad \text{and} \quad U_e^2, \qquad \text{(B 1)}$$

where, to avoid ambiguity, we have used the subscript e to distinguish the quantities that refer to the solution of the Einstein–Maxwell equations from the corresponding ones that refer to the vacuum solution. For the particular solution considered in Paper I (see equations I (75), (125), (87) and (72), respectively),

$$\chi_e = \frac{\varpi^2}{4\alpha^2}\,\chi, \quad U_e^2 = \frac{\varpi^2}{4\alpha^2}\,U^2. \qquad \text{(B 2)}$$

and

$$q_{2e} = \frac{(1+\alpha)^2}{4\alpha^2}\,(q_2+\epsilon^2 q_2^{(e)}), \qquad \text{(B 3)}$$

where

$$\alpha = \surd(1-4Q^2), \quad \epsilon = (1-\alpha)/(1+\alpha) \qquad \text{(B 4)}$$

$$\frac{\varpi^2}{4\alpha^2} = \frac{(1+\alpha)^2}{4\alpha^2}\,[1+2\epsilon\Psi+\epsilon^2(\Psi^2+\Phi^2)] = \frac{(1+\alpha)^2}{4\alpha^2}\,(1+F) \qquad \text{(B 5)}$$

and

$$F = 2\epsilon\Psi+\epsilon^2(\Psi^2+\Phi^2). \qquad \text{(B 6)}$$

In the foregoing equations, the unsuffixed quantities, χ, q_2, Ψ, Φ and U refer to the vacuum solution and are given by equations I (62), (63) and (64); thus,

$$\chi = \frac{Y}{(1-p\eta)^2+q^2\mu^2}, \quad q_2 = \frac{2q\mu}{(1-p\eta)^2+q^2\mu^2}, \quad U^2 = \frac{Y}{V}, \tag{B 7}$$

$$\Psi = \frac{\sqrt{(\Delta\delta)}}{Y}[(1-p\eta)^2+q^2\mu^2] \quad \text{and} \quad \Phi = -\frac{2q\delta}{pY}(1-p\eta), \tag{B 8}$$

where, as hitherto,

$$\Delta = 1-\eta^2, \quad \delta = 1-\mu^2, \quad \text{and} \quad Y = 1-p^2\eta^2-q^2\mu^2 \quad (p^2+q^2=1). \tag{B 9}$$

And, finally, the equations determining $q_2^{(e)}$ are

$$q^{(e)}_{2,0} = \frac{4q^3\mu\delta^2}{p}\frac{1-p\eta}{[(1-p\eta)^2+q^2\mu^2]^2}, \tag{B 10}$$

and

$$q^{(e)}_{2,3} = \frac{2q\delta}{p^2[(1-p\eta)^2+q^2\mu^2]}\left\{3(1-p\eta)^2-q^2\mu^2+q^2\delta\frac{(1-p\eta)^2-q^2\mu^2}{(1-p\eta)^2+q^2\mu^2}\right\}. \tag{B 11}$$

The expressions, (A 1)–(A 3) and (A 7)–(A 9), for the Weyl and the Maxwell scalars, are all expressed in terms of

$$\frac{1}{\chi_e}(\mathscr{X}_{e,\psi} \pm \mathscr{X}_{e,\theta}). \tag{B 12}$$

Making use of (B 2) and (B 3), we can write

$$\frac{\mathscr{X}_{e,i}}{\chi_e} = \frac{1}{1+F}\left[\left(\frac{\mathscr{X}_{,i}}{\chi}\right)_{\text{vac}} + \mathscr{X}_i^{(e)}\right], \tag{B 13}$$

where

$$\mathscr{X}_i^{(e)} = \frac{1}{\chi}[(\chi F)_{,i} + \mathrm{i}\epsilon^2 q^{(e)}_{2,i}] \quad (i = 0, 3). \tag{B 14}$$

Inserting for F its expression (B 6), we find,

$$\mathscr{X}_i^{(e)} = \epsilon\Psi[\ln(\Delta\delta)]_{,i} + \epsilon^2\left[\frac{\chi_{,i}}{\chi}(\Psi^2+\Phi^2)+(\Psi^2+\Phi^2)_{,i}+\frac{q^{(e)}_{2,i}}{\chi}\right], \tag{B 15}$$

where we have used the relation $\chi\Psi = \sqrt{(\Delta\delta)}$. It should be noted that consistent with the notation adopted in (B 14),

$$\mathscr{X}_\psi^{(e)} = -\mathscr{X}_0^{(e)}\sin\psi \quad \text{and} \quad \mathscr{X}_\theta^{(e)} = -\mathscr{X}_3^{(e)}\sin\theta. \tag{B 16}$$

Now making use of our knowledge of the functions χ, Ψ and Φ, we find:

$$\Psi = \sqrt{(\Delta\delta)}\frac{\alpha^2+\beta^2}{Y}, \quad \Psi^2+\Phi^2 = \frac{4\alpha\delta}{p^2 Y}(\alpha-p^2\Delta)+\Delta\delta, \tag{B 17}$$

$$\Psi_{,0} = -\left[\frac{\eta}{\Delta}+\frac{2p(\alpha^2-\beta^2)}{(\alpha^2+\beta^2)Y}\right]\Psi, \quad \Psi_{,3} = -\left[\frac{\mu}{\delta}-\frac{4q\alpha\beta}{(\alpha^2+\beta^2)Y}\right]\Psi, \tag{B 18}$$

$$\tfrac{1}{2}(\Psi^2+\Phi^2)_{,0} = -\frac{\delta}{Y^2}\left[\eta(\alpha^2+\beta^2)^2+\frac{2}{p}(\alpha^2-\beta^2)(q^2+\alpha^2)\right], \tag{B 19}$$

$$\tfrac{1}{2}(\Psi^2+\Phi^2)_{,3} = -\frac{\Delta}{Y^2}[\mu(\boldsymbol{\alpha}^2+\boldsymbol{\beta}^2)^2+4q\delta\boldsymbol{\alpha\beta}], \tag{B 20}$$

$$\frac{1}{\chi}(\chi F)_{,0} = -2\epsilon\frac{\eta\Psi}{\Delta}+\epsilon^2\left[\frac{2\delta(\boldsymbol{\alpha}^2-\boldsymbol{\beta}^2)}{p(\boldsymbol{\alpha}^2+\boldsymbol{\beta}^2)Y}(4\boldsymbol{\alpha}-p^2\Delta)-\frac{8\boldsymbol{\alpha}^2(\boldsymbol{\alpha}^2-\boldsymbol{\beta}^2)\delta}{p(\boldsymbol{\alpha}^2+\boldsymbol{\beta}^2)Y^2}-\frac{2\eta\delta(\boldsymbol{\alpha}^2+\boldsymbol{\beta}^2)^2}{Y^2}\right], \tag{B 21}$$

$$\frac{1}{\chi}(\chi F)_{,3} = -2\epsilon\frac{\mu\Psi}{\delta}+\epsilon^2\left[\frac{4q\Delta\delta\boldsymbol{\alpha\beta}}{(\boldsymbol{\alpha}^2+\boldsymbol{\beta}^2)Y}-\frac{16q\delta\boldsymbol{\alpha}^3\boldsymbol{\beta}}{p^2(\boldsymbol{\alpha}^2+\boldsymbol{\beta}^2)Y^2}-\frac{2\mu\Delta(\boldsymbol{\alpha}^2+\boldsymbol{\beta}^2)^2}{Y^2}\right], \tag{B 22}$$

$$\frac{q^{(e)}_{2,0}}{\chi} = \frac{4q^2\delta^2}{pY}\frac{\boldsymbol{\alpha\beta}}{\boldsymbol{\alpha}^2+\boldsymbol{\beta}^2}, \quad \frac{q^{(e)}_{2,3}}{\chi} = \frac{2q\delta}{p^2Y}\left(3\boldsymbol{\alpha}^2-\boldsymbol{\beta}^2+q^2\delta\frac{\boldsymbol{\alpha}^2-\boldsymbol{\beta}^2}{\boldsymbol{\alpha}^2+\boldsymbol{\beta}^2}\right), \tag{B 23}$$

where, for brevity, we have written

$$\boldsymbol{\alpha} = 1-p\eta \quad \text{and} \quad \boldsymbol{\beta} = q\mu. \tag{B 24}$$

Also, we find that

$$\left(\frac{\mathscr{Z}_{,0}}{\chi}\right)_{\text{vac}} = Ap \quad \text{and} \quad \left(\frac{\mathscr{Z}_{,3}}{\chi}\right)_{\text{vac}} = \text{i}Aq, \tag{B 25}$$

where

$$A = \frac{2}{(\boldsymbol{\alpha}^2+\boldsymbol{\beta}^2)Y}(\boldsymbol{\alpha}^2-\boldsymbol{\beta}^2+2\text{i}\boldsymbol{\alpha\beta}) \quad \text{and} \quad |A|^2 = \frac{4}{Y^2}. \tag{B 26}$$

From the foregoing equations, it is evident that none of the quantities which enter into the definitions of $\mathscr{Z}_{e,\psi}/\chi_e$ and $\mathscr{Z}_{e,\theta}/\chi_e$ diverge when

$$s \to 0, \quad \eta \to 1, \quad \boldsymbol{\alpha} \to 1-p \quad \text{and} \quad Y \to q^2\delta. \tag{B 27}$$

Indeed, when $s \to 0$, we find from (B 16)–(B 23) that the various quantities tend to the following limiting values:

$$F = \frac{4\epsilon^2\boldsymbol{\alpha}^2}{p^2q^2}, \quad F_{,\psi} = 2\epsilon\frac{\boldsymbol{\alpha}^2+\boldsymbol{\beta}^2}{q^2\sqrt{\delta}}, \quad F_{,\theta} = 0;$$

$$\Psi_{,\psi} = \frac{\boldsymbol{\alpha}^2+\boldsymbol{\beta}^2}{q^2\sqrt{\delta}}, \quad \Psi_{,\theta} = 0; \quad \mathscr{Z}_{,\psi} = 0, \quad \frac{\mathscr{Z}_{,\theta}}{\chi} = -\text{i}Aq\sin\theta;$$

$$Z^{(e)}_{\psi} = 2\epsilon\frac{\boldsymbol{\alpha}^2+\boldsymbol{\beta}^2}{q^2\sqrt{\delta}} = \epsilon\mathscr{Z}^{(e)}_{\psi}(0) \quad \text{(say)},$$

and

$$\mathscr{Z}^{(e)}_{\theta} = \epsilon^2\left[\frac{16\boldsymbol{\alpha}^3\boldsymbol{\beta}}{p^2q^3(\boldsymbol{\alpha}^2+\boldsymbol{\beta}^2)\delta}-\frac{2\text{i}}{p^2q}\left(3\boldsymbol{\alpha}^2-\boldsymbol{\beta}^2+q^2\delta\frac{\boldsymbol{\alpha}^2-\boldsymbol{\beta}^2}{\boldsymbol{\alpha}^2+\boldsymbol{\beta}^2}\right)\right]\sin\theta$$

$$= \epsilon^2\mathscr{Z}^{(e)}_{\theta}(0)\sin\theta \quad \text{(say)}, \tag{B 28}$$

while

$$U_e^2 \sim \frac{(1+\alpha)^2}{4\alpha^2}\left(1+\frac{4\epsilon^2\boldsymbol{\alpha}^2}{p^2q^2}\right)q^2\delta s^{-\frac{1}{2}} \quad (s \to 0). \tag{B 29}$$

We can now determine the behaviours of the Weyl scalars as $s \to 0$. Thus, considering Ψ_2 given by (A 1) we have,

$$8U_e^2\Psi_2 = -(\operatorname{cosec}^2\psi-\operatorname{cosec}^2\theta)+\frac{1}{\chi_e^2}(\mathscr{Z}_{e,\psi}-\mathscr{Z}_{e,\theta})(\mathscr{Z}^*_{e,\psi}+\mathscr{Z}^*_{e,\theta}). \tag{B 30}$$

It is evident that the only term on the right-hand side that can lead to a divergence of Ψ_2 as $s \to 0$ is the first; and from the behaviour (B 29) of U_e^2 for $s \to 0$, it follows that

$$\Psi_2 \sim -\frac{\alpha^2 p^2}{2(1+\alpha)^2} \frac{s^{-\frac{3}{2}}}{(p^2q^2+4\mathbf{a}^2\epsilon^2)}. \tag{B 31}$$

Similarly, from (A 3) it is clear that the only term on the right-hand side that can lead to a divergence of Ψ_0 is

$$(S-\bar{S})\frac{1}{\chi_e}(\mathscr{X}^*_{e,\psi}+\mathscr{X}^*_{e,\theta}). \tag{B 32}$$

Since

$$S-\bar{S} = \tfrac{3}{2}(\cot\psi+\cot\theta)-\frac{Y_{,\psi}+Y_{,\theta}}{Y}-\frac{F_{,\psi}+F_{,\theta}}{1+F} \sim \frac{3}{2}\frac{\sin\theta}{s} \quad (s\to 0), \tag{B 33}$$

the behaviour of Ψ_0, apart from a phase-factor, $e^{i\phi}$, is given by

$$e^{i\phi}\,\Psi_0 \sim \frac{3\alpha^2}{2(1+\alpha)^2}\frac{1}{(1+4\mathbf{a}^2\epsilon^2/p^2q^2)^2}\,|(iAq+\epsilon^2\mathscr{X}_\theta^{(e)*}(0))\sin\theta+\epsilon\mathscr{X}_\psi^{(e)}(0)|\,\frac{s^{-\frac{1}{2}}}{q^2\sqrt{\delta}}. \tag{B 34}$$

Turning next to the Maxwell scalars, it will clearly suffice to consider Φ_{00} given by (A 7). We have

$$-4U_e^2\Phi_{00} = S_{,\psi}+S_{,\theta}+\tfrac{1}{2}S^2-S\bar{S}+\frac{1}{2\chi_e^2}|\mathscr{X}_{e,\psi}+\mathscr{X}_{e,\theta}|^2. \tag{B 35}$$

Simplifying the four terms in S on the right-hand side of this equation, we find

$$S_{,\psi}+S_{,\theta}+S^2-(\cot\psi+\cot\theta)\left(\frac{Y_{,\psi}+Y_{,\theta}}{Y}+\frac{F_{,\psi}+F_{,\theta}}{1+F}\right)$$
$$= -\frac{2}{Y}-(\cot\psi+\cot\theta)\frac{F_{,\psi}+F_{,\theta}}{1+F}; \tag{B 36}$$

and the expression for Φ_{00} reduces to

$$-4U_e^2\Phi_{00} = -\frac{2}{Y}-(\cot\psi+\cot\theta)\frac{F_{,\psi}+F_{,\theta}}{1+F}$$
$$+\frac{1}{2(1+F)^2}|-A(p\sin\psi+iq\sin\theta)+\mathscr{X}_\psi^{(e)}+\mathscr{X}_\theta^{(e)}|^2. \tag{B 37}$$

We observe that the expression on the right-hand side vanishes identically when $F=0$ by virtue of the identity (cf. (B 26)):

$$\tfrac{1}{2}|A(p\sin\psi+iq\sin\theta)|^2 = \tfrac{1}{2}|A|^2\,Y = \frac{2}{Y}. \tag{B 38}$$

More generally, when $s\to 0$, the divergence in Φ_{00} can arise only from the term,

$$-(\cot\psi+\cot\theta)\frac{F_{,\psi}}{1+F} \sim -\frac{\sin\theta}{s}\left(2\epsilon\frac{\mathbf{a}^2+\boldsymbol{\beta}^2}{q^2\sqrt{\delta}}\right)\frac{1}{1+4\epsilon^2\mathbf{a}^2/p^2q^2}. \tag{B 39}$$

Accordingly

$$\Phi_{00} \sim 2\epsilon\frac{\alpha^2}{(1+\alpha)^2}\frac{p^4(\mathbf{a}^2+\boldsymbol{\beta}^2)}{(p^2q^2+4\epsilon^2\mathbf{a}^2)^2}\frac{s^{-\frac{1}{2}}}{\delta}. \tag{B 40}$$

It is of interest to contrast the behaviours for $s\to 0$, that we have found, with what obtains for the Nutku–Halil solution and for the solution for the hypersurface orthogonal case analysed in Paper I. A comparison of the present behaviours (B 31) and (B 34) with the behaviours that follow from Chandrasekhar & Ferrari (1984, equations (93)–(95)), shows that the two are the same. On the other hand, for the hypersurface orthogonal-case analysed in Paper I, the behaviours of the Weyl and the Maxwell scalars that follow from (186) are

$$\left.\begin{aligned} \Psi_0 \sim \Psi_4 \sim 2\Psi_2 &\sim \frac{3\alpha^2\delta^2}{(1+\alpha)^2} s^{-\frac{7}{2}}, \\ \text{and} \qquad \Phi_{00} \sim \Phi_{22} \sim -\Phi_{02} &\sim \frac{9\alpha^2(1-\alpha)\,\delta}{2(1+\alpha)^3} s^{-\frac{1}{2}}; \end{aligned}\right\} \tag{B 41}$$

in other words, the Weyl scalars have a much stronger divergence – Ψ_0 and Ψ_4 even stronger than Ψ_2 – while the behaviours of the Maxwell scalars remain unchanged.

Appendix C. Expressions for the spin-coefficients explicitly in terms of a solution of the Ernst equation for $\Psi + i\Phi$

In Appendix A to Paper III, the Weyl scalars for a vacuum space-time, derived via a solution E of the Ernst equation for $\Psi + i\Phi$, were expressed in terms of E. The spin-coefficients, ρ, μ, σ, λ, ϵ and γ (which are the non-vanishing ones) can, likewise, be expressed in terms of E. Thus, by making use, successively, of equations III A(12), A(17) and A(18), the expressions for the spin-coefficients given in equations I (127) can be reduced to the forms:

$$(2\surd 2)\,U\rho = -S; \quad (2\surd 2)\,U\mu = +D, \tag{C 1}$$

$$+2\surd 2\,\frac{1-\mathscr{E}}{1-\mathscr{E}^*}\,U\sigma = S - 2\,\frac{1-E^*}{1-E}\,\frac{E_{,\psi}+E_{,\theta}}{1-|E|^2}, \tag{C 2}$$

$$-2\surd 2\,\frac{1-\mathscr{E}^*}{1-\mathscr{E}}\,U\lambda = D - 2\,\frac{1-E^*}{1-E}\,\frac{E_{,\psi}-E_{,\theta}}{1-|E|^2}, \tag{C 3}$$

$$(4\surd 2)\,U\epsilon = \bar{S} - 2\mathrm{i}S\,\frac{q_2\chi}{(\chi+1)^2+q_2^2} - \left[\frac{\mathscr{X}^*-1}{\mathscr{X}+1}\,\frac{1-E}{1-E^*}\,\frac{E^*_{,\psi}+E^*_{,\theta}}{1-|E|^2} - \text{c.c.}\right], \tag{C 4}$$

$$(4\surd 2)\,U\gamma = \bar{D} - 2\mathrm{i}D\,\frac{q_2\chi}{(\chi+1)^2+q_2^2} - \left[\frac{\mathscr{X}^*-1}{\mathscr{X}+1}\,\frac{1-E^*}{1-E}\,\frac{E_{,\psi}-E_{,\theta}}{1-|E|^2} - \text{c.c.}\right]. \tag{C 5}$$

26

A perturbation analysis of the Bell–Szekeres space-time

By Subrahmanyan Chandrasekhar[1], F.R.S.,
and Basilis C. Xanthopoulos[2]

[1] *The University of Chicago, Chicago, Illinois* 60637, *U.S.A.*
[2] *Department of Physics, University of Crete, and Research Centre of Crete, Iraklion, Greece*

(*Received* 22 *April* 1988)

A perturbation analysis of the Bell–Szekeres space-time is provided both in the region where the scattering of the colliding plane-fronted waves occurs (Region I) and in the regions prior to the instant of collision (Regions II and III). A complete set of normal modes, expressed in terms of spin-weighted spherical harmonics and bounded in the entire Region I, is obtained. These modes exhibit a behaviour of ever-increasing frequency as the horizon (that develops as a result of the collision) is approached. Region II, bounded by the null lines, $u = 0 (0 \leqslant v \leqslant 1)$ and $v = 1 (u \leqslant 0)$ (where u and v define the directions of propagation of the colliding waves) does not allow u-independent perturbations to *all* orders; while all the u-dependent perturbations exhibit strong divergences along the *entire* line, $v = 1, u < 0$. This same description applies to Region III with the roles of u and v interchanged. These behaviours of the perturbations, in the regions of the space-time prior to the instant of collision, raise some serious conceptual questions.

1. Introduction

Several solutions representing the collision of plane-fronted gravitational waves that lead to the development of horizons have recently been found (Chandrasekhar & Xanthopoulos 1986, 1987*a*, *b*; Ferrari & Ibañez 1987, 1988). The stability of these solutions is, of course, a matter of particular interest. In the context of the solution obtained in Chandrasekhar & Xanthopoulos (1986) it has been shown (Chandrasekhar & Xanthopoulos 1987*b*) in the framework of the exact theory, that the merest presence of an $(\epsilon = p)$-fluid or of null dust, in the region of the interaction of the colliding waves, will destroy the horizon. In a perturbative treatment of the same problem, Yurtsever (1987) has argued that space-times representing colliding waves which develop Killing Cauchy-horizons are unstable. (We comment on Yurtsever's formulation of the problem in §11.)

In this paper, we shall consider the simplest of these space-times, namely, that of Bell & Szekeres (1974; see also Chandrasekhar & Xanthopoulos 1987*a*, §3), which represents the collision of impulsive gravitational waves with accompanying electromagnetic waves in a conformally flat space-time. The theory of the perturbations of this space-time, apart from its intrinsic interest, is noteworthy in that it provides a simple illustrative example of the method described in *The* *mathematical theory of black holes* (Chandrasekhar 1983, Chapter IX; this book will be referred to hereafter as *M.T.*) for treating the gravitational perturbations of the Kerr space-time.

2. The description of the Bell–Szekeres space-time in a shear-free null-tetrad basis

The metric of the Bell–Szekeres space-time is given by

$$ds^2 = (d\psi)^2 - (d\theta)^2 - (dx^1)^2 \sin^2\theta - (dx^2)^2 \sin^2\psi, \tag{1}$$

where $\eta(=\cos\psi)$ measures the time from the instant of collision of the waves and $\mu(=\cos\theta)$ measures the spatial distance normal to the planes, (x^1, x^2) spanned by two Killing fields ∂_{x^1} and ∂_{x^2}. The space-time is conformally flat, the Ricci part of the Riemann tensor deriving solely from the presence of the electromagnetic field. The space-time has been described in a Newman–Penrose formalism in Chandrasekhar & Xanthopoulos (1987*a*); but the basis of the null vectors chosen, adapted to the existence of the two Killing fields, is not a convenient one for a perturbation analysis. We shall find the following shear-free basis as a more suitable one:

$$\begin{array}{lllll} & \psi & \theta & x^1 & x^2 \\ (l^i) = \dfrac{1}{\sqrt{2}}(1, & 0, & 0, & -\operatorname{cosec}\psi), \\ (n^i) = \dfrac{1}{\sqrt{2}}(1, & 0, & 0, & +\operatorname{cosec}\psi), \\ (m^i) = \dfrac{1}{\sqrt{2}}(0, & -1, & -\mathrm{i}\operatorname{cosec}\theta, & 0), \\ (\bar{m}^i) = \dfrac{1}{\sqrt{2}}(0, & -1, & +\mathrm{i}\operatorname{cosec}\theta, & 0). \end{array} \tag{2}$$

With this choice of null-basis, $\boldsymbol{l}$ and $\boldsymbol{n}$ are the principal null-directions of the Maxwell tensor.

Relative to the basis (2), the spin-coefficients

$$\kappa, \sigma, \lambda, \nu, \rho, \mu, \tau \text{ and } \pi \tag{3}$$

vanish; and the only non-vanishing spin-coefficients are

$$\epsilon = -\gamma = \frac{\cot\psi}{2\sqrt{2}} \quad \text{and} \quad \alpha = -\beta = \frac{\cot\theta}{2\sqrt{2}}. \tag{4}$$

And consistent with the conformal flatness of the space-time,

$$\Psi_0 = \Psi_1 = \Psi_2 = \Psi_3 = \Psi_4 = 0. \tag{5}$$

The only non-vanishing Maxwell scalar is ϕ_1:

$$\Phi_{11} = \tfrac{1}{2} \quad \text{and} \quad \Phi_{00} = \Phi_{01} = \Phi_{22} = 0. \tag{6}$$

From the vanishing of Φ_{00}, Φ_{01} and Φ_{22}, we conclude that

$$\phi_0 = \phi_2 = 0; \tag{7}$$

while going back to the full Maxwell equations (*M.T.* p. 52, equations (330)–(333)) we find that when $\phi_0 = \phi_2 = 0$,

$$\mathrm{D}\phi_1 = \delta^*\phi_1 = \delta\phi_1 = \Delta\phi_1 = 0; \tag{8}$$

and that in consequence,

$$\phi_1 = \text{constant} = \frac{1}{\sqrt{2}}, \tag{9}$$

where we have ignored a physically irrelevant phase factor.

3. The perturbation problem: formulation and definitions

When a space-time is perturbed, the various quantities describing it will experience first-order changes. Since the eight spin-coefficients (3) and the five Weyl-scalars (5) vanish in the unperturbed space-time, they will in general cease to vanish in the perturbed space-time and become quantities of the first order. We shall retain for them the same symbols with the understanding that they are all quantities of the first order of smallness.

As regards the Maxwell scalars, since ϕ_0 and ϕ_2 vanish in the unperturbed space-time, they will be expected to become quantities of the first order in the perturbed space-time. However, by availing ourselves of the freedom to subject the null tetrad to an infinitesimal rotation, we can make ϕ_0 and ϕ_2 vanish without affecting ϕ_1 (cf. *M.T.* p. 54, equation (344)). Therefore, without loss of generality, we may assume that ϕ_0 and ϕ_2 continue to vanish in the perturbed space-time; and that the only Maxwell scalar subject to a first-order change is ϕ_1:

$$\phi_1 \rightarrow \phi_1 + \phi_1^{(1)}. \tag{10}$$

We conclude, then, that the perturbed space-time will be characterized by

$$\Psi_0, \Psi_1, \Psi_2, \Psi_3, \Psi_4, \phi_1^{(1)},$$

$$\kappa, \sigma, \lambda, \nu, \rho, \mu, \tau \text{ and } \pi, \tag{11}$$

as well as the first-order changes,

$$\epsilon^{(1)}, \gamma^{(1)}, \alpha^{(1)} \text{ and } \beta^{(1)}, \tag{12}$$

in the non-vanishing spin-coefficients, ϵ, γ, α and β; and $\boldsymbol{l}^{(1)}$, $\boldsymbol{n}^{(1)}$, $\boldsymbol{m}^{(1)}$ and $\bar{\boldsymbol{m}}^{(1)}$ in the basis vectors $\boldsymbol{l}$, $\boldsymbol{n}$, $\boldsymbol{m}$ and $\bar{\boldsymbol{m}}$.

As is customary, we shall analyse the perturbations into normal modes by supposing that all the perturbed quantities have an x^1- and an x^2-dependence given by

$$e^{i(k_1 x^1 + k_2 x^2)}, \tag{13}$$

where k_1 and k_2 are real constants. Since x^1 and x^2 both range over the entire real axis, this dependence on x^1 and x^2 corresponds to a Fourier analysis of the perturbations. The common factor (13) in all the quantities describing the perturbations will be suppressed; and the symbols representing them will be their amplitudes.

The basis vectors $\boldsymbol{l}$, $\boldsymbol{n}$, $\boldsymbol{m}$ and $\bar{\boldsymbol{m}}$, defined in equation (2), when applied as tangent vectors to functions having an x^1- and x^2-dependence given by (13), become the derivative operators

$$\boldsymbol{l} = \mathrm{D} = +\frac{1}{\sqrt{2}}\mathscr{D}_0^\dagger, \quad \boldsymbol{n} = \Delta = +\frac{1}{\sqrt{2}}\mathscr{D}_0,$$

$$\boldsymbol{m} = \delta = -\frac{1}{\sqrt{2}}\mathscr{L}_0^\dagger, \quad \text{and} \quad \bar{\boldsymbol{m}} = \delta^* = -\frac{1}{\sqrt{2}}\mathscr{L}_0, \tag{14}$$

where,

$$\begin{aligned} \mathscr{D}_n &= \partial_\psi + \mathrm{i}k_2 \operatorname{cosec}\psi + n\cot\psi, \\ \mathscr{D}_n^\dagger &= \partial_\psi - \mathrm{i}k_2 \operatorname{cosec}\psi + n\cot\psi, \\ \mathscr{L}_n &= \partial_\theta + k_1 \operatorname{cosec}\theta + n\cot\theta, \\ \mathscr{L}_n^\dagger &= \partial_\theta - k_1 \operatorname{cosec}\theta + n\cot\theta. \end{aligned} \tag{15}$$

It will be noticed that while $\mathscr{D}_n$ and $\mathscr{D}_n^\dagger$ are purely temporal operators, $\mathscr{L}_n$ and $\mathscr{L}_n^\dagger$ are purely spatial operators; and as such the operators $\mathscr{D}$ and $\mathscr{L}$ commute.

The derivative operators we have defined satisfy some elementary identities. For later reference, we shall state them here as lemmas.

Lemma 1.

$$\begin{aligned} \mathscr{L}_n(\theta) &= -\mathscr{L}_n^\dagger(\pi-\theta), \quad \sin\theta\,\mathscr{L}_n = \mathscr{L}_{n-1}\sin\theta, \\ \mathscr{D}_n(\psi) &= -\mathscr{D}_n^\dagger(\pi-\psi), \quad \text{and} \quad \mathscr{D}_n^\dagger = (\mathscr{D}_{n)})^*. \end{aligned} \tag{16}$$

Lemma 2.

$$\begin{aligned} \mathscr{L}_0^\dagger\mathscr{L}_1 &= \partial_{\theta\theta} + \cot\theta\,\partial_\theta - 2k_1\operatorname{cosec}\theta\cot\theta - (k_1^2+1)\operatorname{cosec}^2\theta \\ &= \mathscr{L}_2\mathscr{L}_{-1}^\dagger - 2 \end{aligned} \tag{17}$$

and

$$\begin{aligned} \mathscr{D}_0^\dagger\mathscr{D}_1 &= \partial_{\psi\psi} + \cot\psi\,\partial_\psi - 2\mathrm{i}k_2\operatorname{cosec}\psi\cot\psi + (k_2^2-1)\operatorname{cosec}^2\psi \\ &= \mathscr{D}_2\mathscr{D}_{-1}^\dagger - 2. \end{aligned} \tag{18}$$

We also have the 'adjoint' relations,

$$\mathscr{L}_0\mathscr{L}_1^\dagger - \mathscr{L}_2^\dagger\mathscr{L}_{-1} = \mathscr{D}_0\mathscr{D}_1^\dagger - \mathscr{D}_2^\dagger\mathscr{D}_{-1} = -2. \tag{19}$$

Lemma 3. *If $f(\theta)$ and $g(\theta)$ are any two bounded differentiable functions in the closed interval* $[0, \pi]$, *then*

$$\int_0^\pi g(\mathscr{L}_n f)\sin\theta\,\mathrm{d}\theta = -\int_0^\pi f(\mathscr{L}_{-n+1}^\dagger g)\sin\theta\,\mathrm{d}\theta. \tag{20}$$

4. THE EQUATIONS GOVERNING κ, σ, ρ, τ, Ψ_0, Ψ_1, AND Ψ_2 AND THEIR SOLUTION

In the treatment of the gravitational perturbations of the Kerr space-time in *The mathematical theory of black holes* (Chapter IX), the first problem that is addressed is to the solution of the quantities which vanish in the background (κ, σ, λ, ν, Ψ_0, Ψ_1, Ψ_3 and Ψ_4 in that context) by considering those equations which are linear and homogeneous in them. These equations are 'already linearized' in the sense that in them we can replace *all* the other quantities (including the basis vectors) by their unperturbed values. For the same reasons, in treating the perturbation of the Bell–Szekeres space-time, we shall first consider those among the Newman–Penrose equations which are linear and homogeneous in the spin coefficients κ, σ, ρ, τ, ν, λ, μ and π, the Weyl scalars, Ψ_0, Ψ_1, Ψ_2, Ψ_3 and Ψ_4, and the Maxwell scalars ϕ_0 and ϕ_2 (which in the chosen gauge continue to vanish in the first order). There are two sets of ten equations each. Thus, in the present context, when the non-vanishing spin-coefficients, ϵ, γ, α and β have the values (4) in the background geometry, the Ricci identities,

M.T. pages 46 and 47, equations (310), a, b, c, k, p and q, and the Bianchi identities,

M.T. pages 49 and 50, equations (321) and (321′), a, b, e and f,

provide, respectively, equations (i), (vi), (ix), (x), (ii), (iii), (vii), (v), (viii) and (iv) of the following set:

$$
\begin{aligned}
&\mathscr{D}^\dagger_{-1}\rho+\mathscr{L}_1\kappa = 0, &&\text{(i)} &&\mathscr{D}^\dagger_{-1}\sigma+\mathscr{L}^\dagger_{-1}\kappa = \Psi_0\sqrt{2}, &&\text{(vi)}\\
&\mathscr{L}^\dagger_{-1}\tau+\mathscr{D}_1\sigma = 0, &&\text{(ii)} &&\mathscr{D}^\dagger_{-1}\Psi_1+\mathscr{L}_2\Psi_0 = \kappa\sqrt{2}, &&\text{(vii)}\\
&\mathscr{D}_1\rho+\mathscr{L}_1\tau = -\Psi_2\sqrt{2}, &&\text{(iii)} &&\mathscr{L}^\dagger_{-1}\Psi_1+\mathscr{D}_2\Psi_0 = \sigma\sqrt{2}, &&\text{(viii)}\\
&\mathscr{D}_1\Psi_1+\mathscr{L}^\dagger_0\Psi_2 = +\tau\sqrt{2}, &&\text{(iv)} &&\mathscr{D}^\dagger_0\tau-\mathscr{D}_2\kappa = \Psi_1\sqrt{2}, &&\text{(ix)}\\
&\mathscr{L}_1\Psi_1+\mathscr{D}^\dagger_0\Psi_2 = +\rho\sqrt{2}, &&\text{(v)} &&\mathscr{L}^\dagger_0\rho-\mathscr{L}_2\sigma = \Psi_1\sqrt{2}. &&\text{(x)}
\end{aligned}
\qquad (21)
$$

The adjoint set of equations governing ν, λ, μ, π, Ψ_2, Ψ_3 and Ψ_4 are considered in §5 below (cf. equations (63)).

We observe that equations (21)(i) and (ii) enable us to express ρ, κ, τ and σ in terms of two potentials P and Q in the manner:

$$\rho = +\mathscr{L}_1 P\sqrt{2};\quad \kappa = -\mathscr{D}^\dagger_{-1}P\sqrt{2}, \qquad (22)$$

and

$$\tau = -\mathscr{D}_1 Q\sqrt{2};\quad \sigma = +\mathscr{L}^\dagger_{-1}Q\sqrt{2}. \qquad (23)$$

In terms of these expressions for ρ, κ, τ and σ the remaining eight equations of (31) take the forms:

$$
\begin{aligned}
&\mathscr{L}_1\mathscr{D}_1(P-Q) = -\Psi_2, &&\text{(iii)} &&\mathscr{D}^\dagger_{-1}\Psi_1+\mathscr{L}_2\Psi_0 = -2\mathscr{D}^\dagger_{-1}P, &&\text{(vii)}\\
&\mathscr{D}_1\Psi_1+\mathscr{L}^\dagger_0\Psi_2 = -2\mathscr{D}_1 Q, &&\text{(iv)} &&\mathscr{L}^\dagger_{-1}\Psi_1+\mathscr{D}_2\Psi_0 = +2\mathscr{L}^\dagger_{-1}Q, &&\text{(viii)}\\
&\mathscr{L}_1\Psi_1+\mathscr{D}^\dagger_0\Psi_2 = +2\mathscr{L}_1 P, &&\text{(v)} &&\mathscr{D}_2\mathscr{D}^\dagger_{-1}P-\mathscr{D}^\dagger_0\mathscr{D}_1 Q = \Psi_1, &&\text{(ix)}\\
&\mathscr{D}^\dagger_{-1}\mathscr{L}^\dagger_{-1}(P-Q) = -\Psi_0, &&\text{(vi)} &&\mathscr{L}^\dagger_0\mathscr{L}_1 P-\mathscr{L}_2\mathscr{L}^\dagger_{-1}Q = \Psi_1. &&\text{(x)}
\end{aligned}
\qquad (24)
$$

Rewriting equation (24)(iv) in the form

$$\mathscr{D}_1(\Psi_1+2Q) = -\mathscr{L}_0^\dagger \Psi_2 \tag{25}$$

and making use of equation (24)(iii), we obtain

$$\mathscr{D}_1(\Psi_1+2Q) = \mathscr{D}_1 \mathscr{L}_0^\dagger \mathscr{L}_1(P-Q). \tag{26}$$

In the present context we may infer from equation (26), without loss of generality, that

$$\Psi_1+2Q = \mathscr{L}_0^\dagger \mathscr{L}_1(P-Q). \tag{27}$$

(For an explanation of the reasoning behind this inference see footnote on p. 394 in *M.T.*) From equations (24)(iii) and (v), (vi) and (vii) and (vi) and (viii), we similarly infer that

$$\Psi_1-2P = \mathscr{D}_0^\dagger \mathscr{D}_1(P-Q), \tag{28}$$

$$\Psi_1+2P = \mathscr{L}_2 \mathscr{L}_{-1}^\dagger(P-Q) \quad \text{and} \quad \Psi_1-2Q = \mathscr{D}_2 \mathscr{D}_{-1}^\dagger(P-Q). \tag{29}$$

Subtracting equation (28) from equation (27), we obtain

$$(\mathscr{L}_0^\dagger \mathscr{L}_1 - \mathscr{D}_0^\dagger \mathscr{D}_1)(P-Q) = 2(P+Q). \tag{30}$$

From equations (29), we similarly obtain,

$$(\mathscr{L}_2 \mathscr{L}_{-1}^\dagger - \mathscr{D}_2 \mathscr{D}_{-1}^\dagger)(P-Q) = 2(P+Q). \tag{31}$$

It can be readily verified that by virtue of the identities (19) of Lemma 2 (§3), equations (30) and (31) are consistent with each other and with equations (24)(ix) and (x). Finally, we rewrite equation (30) in the form

$$(\mathscr{L}_0^\dagger \mathscr{L}_1 - \mathscr{D}_0^\dagger \mathscr{D}_1 - 2) P = (\mathscr{L}_0^\dagger \mathscr{L}_1 - \mathscr{D}_0^\dagger \mathscr{D}_1 + 2) Q. \tag{32}$$

So far, we have not included Maxwell's equations in our considerations. Equations that are already linearized (in our present context, as well) are provided by equations (205) and (206) in *M.T.* p. 239. Remembering that in the gauge we have chosen, ϕ_0 and ϕ_2 continue to vanish in the first order, *M.T.* (p. 239) equation (205) gives

$$(\Delta - 4\gamma)\kappa - (\delta^* - 4\alpha)\sigma + 2\Psi_1 = 0; \tag{33}$$

or, substituting for γ, α, Δ and δ^* from equations (4) and (14), we obtain,

$$\mathscr{D}_2 \kappa + \mathscr{L}_2 \sigma + 2\Psi_1 \sqrt{2} = 0. \tag{34}$$

Inserting for κ, σ and Ψ_1 from equations (22), (23), (27) amd (28), we obtain,

$$-\mathscr{D}_2 \mathscr{D}_{-1}^\dagger P + \mathscr{L}_2 \mathscr{L}_{-1}^\dagger Q + \mathscr{L}_0^\dagger \mathscr{L}_1(P-Q) - 2Q + \mathscr{D}_0^\dagger \mathscr{D}_1(P-Q) + 2P = 0, \tag{35}$$

or, on simplifying, we have

$$\mathscr{L}_0^\dagger \mathscr{L}_1 P = \mathscr{D}_0^\dagger \mathscr{D}_1 Q. \tag{36}$$

Now combining equations (32) and (36) in the manner,

$$\begin{aligned} \mathscr{D}_0^\dagger \mathscr{D}_1(\mathscr{L}_0^\dagger \mathscr{L}_1 - \mathscr{D}_0^\dagger \mathscr{D}_1 - 2) P &= \mathscr{D}_0^\dagger \mathscr{D}_1(\mathscr{L}_0^\dagger \mathscr{L}_1 - \mathscr{D}_0^\dagger \mathscr{D}_1 + 2) Q \\ &= (\mathscr{L}_0^\dagger \mathscr{L}_1 - \mathscr{D}_0^\dagger \mathscr{D}_1 + 2)\mathscr{D}_0^\dagger \mathscr{D}_1 Q \\ &= (\mathscr{L}_0^\dagger \mathscr{L}_1 - \mathscr{D}_0^\dagger \mathscr{D}_1 + 2)\mathscr{L}_0^\dagger \mathscr{L}_1 P, \end{aligned} \tag{37}$$

we find
$$[(\mathscr{L}_0^\dagger \mathscr{L}_1 - \mathscr{D}_0^\dagger \mathscr{D}_1)^2 + 2(\mathscr{L}_0^\dagger \mathscr{L}_1 + \mathscr{D}_0^\dagger \mathscr{D}_1)]P = 0. \tag{38}$$
In similar fashion, we find that Q satisfies the same equation.

(a) The separability of equation (38) *and the expression of the solutions in terms of the spin-weighted spherical harmonics*

The substitutions
$$P = f(\theta)\, g(\psi), \tag{39}$$
$$\mathscr{L}_0^\dagger \mathscr{L}_1 f = (a+b) f \quad \text{and} \quad \mathscr{D}_0^\dagger \mathscr{D}_1 g = (a-b) g, \tag{40}$$
where a and b are constants, are consistent with equation (38) provided
$$a + b^2 = 0 \quad \text{or} \quad b = \surd(-a). \tag{41}$$
In other words, equation (38) allows separable solutions of the form (39) with $f(\theta)$ and $g(\psi)$ satisfying the equations,
$$\mathscr{L}_0^\dagger \mathscr{L}_1 f = (a + \surd -a) f, \tag{42}$$
and
$$\mathscr{D}_0^\dagger \mathscr{D}_1 g = (a - \surd -a) g, \tag{43}$$
where $a \pm \surd(-a)$ are separation constants.

Since $\mu = \cos\theta$ measures distance normal to the planes, (x^1, x^2), we must clearly require f to be smooth in the entire interval $0 \leqslant \theta \leqslant \pi$ (or $-1 \leqslant \mu \leqslant +1$). By lemma 3 (equation (20)) of §3, it follows that in the present context,
$$\int_0^\pi f(\mathscr{L}_0^\dagger \mathscr{L}_1 f) \sin\theta \, d\theta = -\int_0^\pi (\mathscr{L}_1 f)^2 \sin\theta \, d\theta, \tag{44}$$
we conclude that the separation constant, $a + \surd(-a)$, on the right-hand side of equation (42) must be a negative real constant. Consistent with this last requirement, we can rewrite equations (42) and (43) in the forms,
$$\mathscr{L}_0^\dagger \mathscr{L}_1 f = -\lambda(\lambda+1) f, \tag{45}$$
and
$$\mathscr{D}_0^\dagger \mathscr{D}_1 g = -(\lambda+1)(\lambda+2) g \tag{46}$$
where $a = -(\lambda+1)^2$ and $\lambda+1$ is a real positive constant.

Equation (45) written out explicitly is (cf. equation (17))
$$\left[\frac{1}{\sin\theta}\frac{d}{d\theta}\sin\theta\frac{d}{d\theta} - \frac{(k_1^2+1) + 2k_1\cos\theta}{\sin^2\theta} + \lambda(\lambda+1)\right] f = 0. \tag{47}$$

In this form, we recognize that the operator acting on f is none other than that defining the spin-1 weighted spherical harmonic (cf. *M.T.* p. 427, equation (270)). We shall require f to be smooth in the entire interval $0 \leqslant \theta \leqslant \pi$; we may then identify f with a spin-1 weighted spherical harmonic of integral order $l(=\lambda)$ and $m(=k_1)$:
$$S_{\pm 1} = \text{constant}\left(\frac{d}{d\theta} \mp m \operatorname{cosec}\theta\right) P_l^m. \tag{48}$$

In conformity with the standard conventions of the subject (cf. *M.T.* p. 386, equations (25) and (27) for the case $a = 0$) and with the further identifications,

$$\lambda = l > 0, k_1 = m, \quad \text{and} \quad -l \leqslant m \leqslant +l, \tag{49}$$

we may write

$$f = \text{constant}\, S_{+1}(l, m; \theta). \tag{50}$$

From the formal identity of the operators $\mathscr{D}_0^\dagger \mathscr{D}_1$ and $\mathscr{L}_0^\dagger \mathscr{L}_1$ (cf. equations (17) and (18)), we may write,

$$g = \text{constant}\, S_{+1}(l+1, \mathrm{i}k_2; \psi), \tag{51}$$

where S_{+1} with the argument $\mathrm{i}k_2$ is the analytic continuation of the standard spin-$+1$ weighted spherical harmonic into the complex m-plane along the imaginary axis. It is known (cf. Hobson 1931, p. 184) that P_l^m so extended has a cross-cut along the real μ-axis extending from the point $\mu = +1$ to $\mu = -\infty$. The implications of this cut in the definition of S_{+1} in equation (51) will be considered in §8. Meantime, we may write the required solution for P as

$$P = \text{constant}\, S_{+1}(l, m; \theta)\, S_{+1}(l+1, \mathrm{i}k_2; \psi); \tag{52}$$

and since Q satisfies the same equation,

$$Q = \text{constant}\, S_{+1}(l, m, \theta)\, S_{+1}(l+1, \mathrm{i}k_2; \psi). \tag{53}$$

In view of the relation (36) between P and Q, the constants of proportionality in the solutions (52) and (53) must be related. We find that the compatibility with equation (36) requires that the solutions for P and Q are expressible in the forms,

$$P = (l+2)\, C S_{+1}(l, m; \theta)\, S_{+1}(l+1, \mathrm{i}k_2; \psi) \tag{54}$$

and

$$Q = l C S_{+1}(l, m; \theta)\, S_{+1}(l+1, \mathrm{i}k_2; \psi), \tag{55}$$

where C is a constant which, we may without loss of generality, assume to be real. By virtue of these equations,

$$P - Q = 2C S_{+1}(l, m; \theta)\, S_{+1}(l+1, \mathrm{i}k_2; \psi). \tag{56}$$

The Weyl scalar Ψ_1 now follows from the equation (cf. equation (27))

$$\Psi_1 = \mathscr{L}_0^\dagger \mathscr{L}_1 (P - Q) - 2Q. \tag{57}$$

Inserting for Q and $(P - Q)$ from equations (55) and (56), we find

$$\Psi_1 = -2l(l+2)\, C S_{+1}(l, m; \theta)\, S_{+1}(l+1, \mathrm{i}k_2; \psi). \tag{58}$$

As required, Ψ_1 is expressed in terms of the spin-1 functions.

Considering next the Weyl scalar Ψ_0, we have by equations (24)(vi) and (56),

$$\begin{aligned} -\Psi_0 &= \mathscr{D}_{-1}^\dagger \mathscr{L}_{-1}^\dagger (P - Q) \\ &= 2C[\mathscr{L}_{-1}^\dagger S_{+1}(l, m; \theta)][\mathscr{D}_{-1}^\dagger S_{+1}(l+1, \mathrm{i}k_2; \psi)]. \end{aligned} \tag{59}$$

By making use of the relations (*M.T.* p. 243, equation (232))

$$\mathscr{D}_{-1}^\dagger S_{+1}(l+1) = -[l(l+3)]^{\frac{1}{2}} S_{+2}(l+1)$$

and

$$\mathscr{L}^\dagger_{-1} S_{+1}(l) = -[(l-1)(l+2)]^{\frac{1}{2}} S_{+2}(l), \tag{60}$$

where S_{+2} denotes the spin-2 functions, the expression for Ψ_0 becomes

$$\Psi_0 = -2[(l-1)(l+2)\, l(l+3)]^{\frac{1}{2}} C S_{+2}(l, m; \theta)\, S_{+2}(l+1, ik_2; \psi). \tag{61}$$

Consistent with the interpretation of Ψ_0 as measuring the flux of gravitational radiation, it is expressed in terms of the spin-2 weighted spherical harmonics.

Finally, from equations (24)(iii) and (56), we find

$$\begin{aligned} -\Psi_2 &= \mathscr{D}_1 \mathscr{L}_1 (P-Q) \\ &= 2C[\mathscr{L}_1 S_{+1}(l, m; \theta)][\mathscr{D}_1 S_{+1}(l+1, ik_2; \psi)]. \end{aligned} \tag{62}$$

The solution of the system of equations (21) has now been completed.

5. The equations governing ν, μ, λ, π, Ψ_2, Ψ_3, and Ψ_4 and their solution

The 'adjoint' set of already linearized equations governing the spin-coefficients ν, μ, λ and π and the Weyl scalars Ψ_2, Ψ_3 and Ψ_4 follow from the Ricci identities,

M.T. pages 46 and 47, equations (310), n, h, i, j, m and g, and the Bianchi identities,

M.T. pages 49 and 50, equations (321) and (321′), c, d, g and h.

They provide, respectively, equations (i), (iii), (ix), (vi), (x), (ii), (iv), (viii), (v) and (vii) of the following set:

$$\begin{array}{llll} \mathscr{D}_{-1}\mu + \mathscr{L}_1^\dagger \nu = 0, & \text{(i)} & \mathscr{D}_{-1}\lambda + \mathscr{L}_{-1}\nu = -\Psi_4\sqrt{2}, & \text{(vi)} \\ \mathscr{L}_{-1}\pi + \mathscr{D}_1^\dagger \lambda = 0, & \text{(ii)} & \mathscr{D}_{-1}\Psi_3 + \mathscr{L}_2^\dagger \Psi_4 = -\nu\sqrt{2}, & \text{(vii)} \\ \mathscr{D}_1^\dagger \mu + \mathscr{L}_1^\dagger \pi = +\Psi_2\sqrt{2}, & \text{(iii)} & \mathscr{L}_{-1}\Psi_3 + \mathscr{D}_2^\dagger \Psi_4 = -\lambda\sqrt{2}, & \text{(viii)} \\ \mathscr{D}_1^\dagger \Psi_3 + \mathscr{L}_0 \Psi_2 = -\pi\sqrt{2}, & \text{(iv)} & \mathscr{D}_0 \pi - \mathscr{D}_2^\dagger \nu = -\Psi_3\sqrt{2}, & \text{(ix)} \\ \mathscr{L}_1^\dagger \Psi_3 + \mathscr{D}_0 \Psi_2 = -\mu\sqrt{2}, & \text{(v)} & \mathscr{L}_0 \mu - \mathscr{L}_2^\dagger \lambda = -\Psi_3\sqrt{2}. & \text{(x)} \end{array} \tag{63}$$

The set of equations (63) can be reduced and solved by the same procedures that were followed in §4. Thus, equations (63)(i) and (ii) enable us to express μ, ν, π and λ in terms of two potentials P^+ and Q^+ in the manner:

$$\mu = -\mathscr{L}_1^\dagger P^\dagger \sqrt{2}, \quad \nu = +\mathscr{D}_{-1} P^\dagger \sqrt{2}; \tag{64}$$

and

$$\pi = +\mathscr{D}_1^\dagger Q^\dagger \sqrt{2}, \quad \lambda = -\mathscr{L}_{-1} Q^\dagger \sqrt{2}. \tag{65}$$

Next, by combining equations (63)(iii) and (iv), (iii) and (v), (vi) and (vii), and (vi) and (viii), rewritten in terms of $P^\dagger$ and $Q^\dagger$, we obtain equations that are the adjoints of (27), (28) and (29):

$$\Psi_3 + 2Q^\dagger = \mathscr{L}_0 \mathscr{L}_1^\dagger (P^\dagger - Q^\dagger), \quad \Psi_3 - 2P^\dagger = \mathscr{D}_0 \mathscr{D}_1^\dagger (P^\dagger - Q^\dagger), \tag{66}$$

$$\Psi_3 + 2P^\dagger = \mathscr{L}_2^\dagger \mathscr{L}_{-1} (P^\dagger - Q^\dagger), \quad \Psi_3 - 2Q^\dagger = \mathscr{D}_2^\dagger \mathscr{D}_{-1} (P^\dagger - Q^\dagger). \tag{67}$$

From these equations we derive the pair of equations that are analogous to equations (30) and (31); thus,

$$(\mathscr{L}_0\mathscr{L}_1^\dagger-\mathscr{D}_0\mathscr{D}_1^\dagger)(P^\dagger-Q^\dagger)=2(P^\dagger+Q^\dagger), \tag{68}$$

$$(\mathscr{L}_2^\dagger\mathscr{L}_{-1}-\mathscr{D}_2^\dagger\mathscr{D}_{-1})(P^\dagger-Q^\dagger)=2(P^\dagger+Q^\dagger). \tag{69}$$

Equations (68) and (69) are mutually consistent by virtue of the identity (19). Finally, we rewrite equation (68) in the form

$$(\mathscr{L}_0\mathscr{L}_1^\dagger-\mathscr{D}_0\mathscr{D}_1^\dagger-2)P^\dagger=(\mathscr{L}_0\mathscr{L}_1^\dagger-\mathscr{D}_0\mathscr{D}_1^\dagger+2)Q^\dagger. \tag{70}$$

Turning to Maxwell's equations, we now find that *M.T.* p. 239, equation (206) provides the relation

$$\mathscr{L}_0\mathscr{L}_1^\dagger P^\dagger=\mathscr{D}_0\mathscr{D}_1^\dagger Q^\dagger. \tag{71}$$

By combining equations (70) and (71), we obtain the equation,

$$[(\mathscr{L}_0\mathscr{L}_1^\dagger-\mathscr{D}_0\mathscr{D}_1^\dagger)^2+2(\mathscr{L}_0\mathscr{L}_1^\dagger+\mathscr{D}_0\mathscr{D}_1^\dagger)]P^\dagger=0, \tag{72}$$

which is the adjoint of equation (38); and $Q^\dagger$ satisfies the same equation.

The equations governing $P^\dagger$ and $Q^\dagger$ can be reduced and solved even as the equations governing P and Q were solved in §4. In place of the solutions (54), (55) and (56), we now have

$$P^\dagger=(l+2)C^\dagger S_{-1}(l,m;\theta)S_{-1}(l+1,ik_2;\psi), \tag{73}$$

$$Q^\dagger=lC^\dagger S_{-1}(l,m;\theta)S_{-1}(l+1,ik_2;\psi), \tag{74}$$

and

$$P^\dagger-Q^\dagger=2C^\dagger S_{-1}(l,m;\theta)S_{-1}(l+1,ik_2;\psi), \tag{75}$$

where $C^\dagger$ is a constant. In terms of the foregoing solutions, we can derive expressions for the Weyl scalars Ψ_3, Ψ_4 and Ψ_2. We find (cf. equations (58), (61) and (62)):

$$\begin{aligned}\Psi_3&=\mathscr{L}_0\mathscr{L}_1^\dagger(P^\dagger-Q^\dagger)-2Q^\dagger\\&=-2l(l+2)C^\dagger S_{-1}(l,m;\theta)S_{-1}(l+1,ik_2;\psi),\end{aligned} \tag{76}$$

$$\begin{aligned}-\Psi_4&=\mathscr{D}_{-1}\mathscr{L}_{-1}(P^\dagger-Q^\dagger)\\&=2[(l-1)(l+2)l(l+3)]^{\frac{1}{2}}C^\dagger S_{-2}(l,m;\theta)S_{-2}(l+1,ik_2;\psi),\end{aligned} \tag{77}$$

and

$$\begin{aligned}-\Psi_2&=\mathscr{D}_1^\dagger\mathscr{L}_1^\dagger(P^\dagger-Q^\dagger)\\&=2C^\dagger[\mathscr{L}_1^\dagger S_{-1}(l,m;\theta)][\mathscr{D}_1^\dagger S_{-1}(l+1,ik_2;\psi)].\end{aligned} \tag{78}$$

Finally, the requirement that Ψ_2 given by equations (62) and (78) agree will determine the ratio of the constants C and $C^\dagger$ that we have introduced and make the entire solution determinate apart from a constant of proportionality. To make the required comparison, we evaluate $\mathscr{L}_0^\dagger\mathscr{D}_0\Psi_2$ using both the expressions for Ψ_2. We find:

$$\begin{aligned}-\mathscr{L}_0^\dagger\mathscr{D}_0\Psi_2&=2C[\mathscr{L}_0^\dagger\mathscr{L}_1S_{+1}(l,m;\theta)][\mathscr{D}_0\mathscr{D}_1S_{+1}(l+1,ik_2;\psi)]\\&=-2l(l+1)CS_{+1}(l,m;\theta)[\mathscr{D}_0\mathscr{D}_1S_{+1}(l+1,ik_2;\psi)],\end{aligned} \tag{79}$$

with Ψ_2 given by equation (62), and

$$-\mathscr{L}_0^\dagger \mathscr{D}_0 \Psi_2 = 2C^\dagger[\mathscr{L}_0^\dagger \mathscr{L}_1^\dagger S_{-1}(l,m;\theta)][\mathscr{D}_0 \mathscr{D}_1^\dagger S_{-1}(l+1,ik_2;\psi)]$$
$$= -2(l+1)(l+2)C^\dagger[\mathscr{L}_0^\dagger \mathscr{L}_1^\dagger S_{-1}(l,m;\theta)]S_{-1}(l+1,ik_2;\psi), \quad (80)$$

with Ψ_2 given by equation (78). Now by making use of the 'Starobinsky' relations,

$$\mathscr{L}_0^\dagger \mathscr{L}_1^\dagger S_{-1}(l) = l(l+1)S_{+1}(l)$$

and

$$\left.\mathscr{D}_0 \mathscr{D}_1 S_{+1}(l+1) = (l+1)(l+2)S_{-1}(l+1),\right\} \quad (81)$$

provided by theorems 3, 4 and 5 in *M.T.* pages 389–390 (for the case $a=0$), we obtain, respectively, from equations (79) and (80),

$$\mathscr{L}_0^\dagger \mathscr{D}_0 \Psi_2 = 2l(l+1)^2(l+2)CS_{+1}(l,m;\theta)S_{-1}(l+1,ik_2;\psi) \quad (82)$$

and

$$\mathscr{L}_0^\dagger \mathscr{D}_0 \Psi_2 = 2l(l+1)^2(l+2)C^\dagger S_{+1}(l,m;\theta)S_{-1}(l+1,ik_2;\psi). \quad (83)$$

We conclude that

$$C = C^\dagger \quad (84)$$

assures the equality of the expressions (68) and (72) for Ψ_2.

(a) The explicit solutions for the spin-coefficients, κ, σ, λ, ν, ρ, τ, μ and π

With the solutions for P, Q, $P^\dagger$ and $Q^\dagger$ given by equations (54), (55), (73) and (74) (with $C=C^\dagger$), we find that expressions for the spin-coefficients given in equations (22), (23), (64) and (65) take the explicit forms:

$$\kappa = +(l+2)[l(l+3)]^{\frac{1}{2}}CS_{+1}(l,m;\theta)S_{+2}(l+1,ik_2;\psi)\sqrt{2},$$
$$\sigma = -l[(l-1)(l+2)]^{\frac{1}{2}}CS_{+2}(l,m;\theta)S_{+1}(l+1,ik_2;\psi)\sqrt{2},$$
$$\lambda = +l[(l-1)(l+2)]^{\frac{1}{2}}CS_{-2}(l,m;\theta)S_{-1}(l+1,ik_2;\psi)\sqrt{2},$$
$$\nu = -(l+2)[l(l+3)]^{\frac{1}{2}}CS_{-1}(l,m;\theta)S_{-2}(l+1,ik_2;\psi)\sqrt{2},$$
$$\rho = +(l+2)C[\mathscr{L}_1 S_{+1}(l,m;\theta)]S_{+1}(l+1,ik_2;\psi)\sqrt{2},$$
$$\tau = -lCS_{+1}(l,m;\theta)[\mathscr{D}_1 S_{+1}(l+1,ik_2;\psi)]\sqrt{2},$$
$$\mu = -(l+2)C[\mathscr{L}_1^\dagger S_{-1}(l,m;\theta)]S_{-1}(l+1,ik_2;\psi)\sqrt{2},$$

and

$$\pi = +lCS_{-1}(l,m;\theta)[\mathscr{D}_1^\dagger S_{-1}(l+1,ik_2;\psi)]\sqrt{2}. \quad (85)$$

6. The perturbation in the Maxwell scalar ϕ_1

As we have stated in §2, in the chosen gauge, only the Maxwell scalar ϕ_1 experiences a first-order change. To determine this change, $\phi_1^{(1)}$, we must consider the linearized version of Maxwell's equations (*M.T.* p. 52, equations (330)–(333)). In the present context they give:

$$\left.\begin{aligned} &\mathrm{D}\phi_1^{(1)} + \boldsymbol{l}^{(1)}\phi_1 = +\rho\sqrt{2}, \quad \text{(i)}; \qquad \delta^*\phi_1^{(1)} + \bar{\boldsymbol{m}}^{(1)}\phi_1 = -\pi\sqrt{2}, \quad \text{(ii)};\\ &\delta\phi_1^{(1)} + \boldsymbol{m}^{(1)}\phi_1 = +\tau\sqrt{2}, \quad \text{(iii)}; \qquad \Delta\phi_1^{(1)} + \boldsymbol{n}^{(1)}\phi_1 = -\mu\sqrt{2}, \quad \text{(iv)}, \end{aligned}\right\} \quad (86)$$

where $\boldsymbol{l}^{(1)}$, $\boldsymbol{n}^{(1)}$, $\boldsymbol{m}^{(1)}$ and $\bar{\boldsymbol{m}}^{(1)}$ denote the first-order changes in the basis vectors. Since $\phi_1(=1/\sqrt{2})$ is a constant in the background geometry, the terms in $\boldsymbol{l}^{(i)}$, etc., in the foregoing equations vanish; and we are left with

$$\left.\begin{aligned} \mathrm{D}\phi_1^{(1)} &= +\rho\sqrt{2}, \quad \text{(i)}; \qquad & \delta^*\phi_1^{(1)} &= -\pi\sqrt{2}, \quad \text{(ii)}; \\ \delta\phi_1^{(1)} &= +\tau\sqrt{2}, \quad \text{(iii)}; \qquad & \Delta\phi_1^{(1)} &= -\mu\sqrt{2}, \quad \text{(iv)}. \end{aligned}\right\} \tag{87}$$

Considering the first of these equations, and substituting for D and ρ from equations (14) and (22), we obtain

$$\mathscr{D}_0^\dagger \phi_1^{(1)} = (2\sqrt{2})\,\mathscr{L}_1 P. \tag{88}$$

With the solution for P given by equation (54), equation (88) gives

$$\begin{aligned} \mathscr{D}_0^\dagger \phi_1^{(1)} &= (2\sqrt{2})(l+2)C[\mathscr{L}_1 S_{+1}(l,m;\theta)]S_{+1}(l+1,\mathrm{i}k_2;\psi) \\ &= -\frac{2\sqrt{2}}{l+1}C[\mathscr{L}_1 S_{+1}(l,m;\theta)]\mathscr{D}_0^\dagger\mathscr{D}_1 S_{+1}(l+1,\mathrm{i}k_2;\psi), \end{aligned} \tag{89}$$

where in the second step we have made use of the equation governing S_{+1}. From this last equality it follows that

$$\phi_1^{(1)} = -\frac{2\sqrt{2}}{l+1}C[\mathscr{L}_1 S_{+1}(l,m;\theta)][\mathscr{D}_1 S_{+1}(l+1,\mathrm{i}k_2;\psi)]. \tag{90}$$

Similarly, from equation (87)(ii), we obtain successively (with the aid of equations (65), (74) and (84))

$$\begin{aligned} \mathscr{L}_0\phi_1^{(1)} &= (2\sqrt{2})\,\mathscr{D}_1^\dagger Q^\dagger \\ &= (2\sqrt{2})\,lCS_{-1}(l,m;\theta)[\mathscr{D}_1^\dagger S_{-1}(l+1,\mathrm{i}k_2;\psi)] \\ &= -\frac{2\sqrt{2}}{l+1}C[\mathscr{L}_0\mathscr{L}_1^\dagger S_{-1}(l,m;\theta)][\mathscr{D}_1^\dagger S_{-1}(l+1,\mathrm{i}k_2;\psi)]; \end{aligned} \tag{91}$$

and we obtain the alternative expression

$$\phi_1^{(1)} = -\frac{2\sqrt{2}}{l+1}C[\mathscr{L}_1^\dagger S_{-1}(l,m;\theta)][\mathscr{D}_1^\dagger S_{-1}(l+1,\mathrm{i}k_2;\psi)]. \tag{92}$$

The equality of the expressions (90) and (92) is assured by the equality of the two alternative expressions (62) and (78) for Ψ_2 (requiring in turn the equality of C and $C^\dagger$); indeed, we have

$$\phi_1^{(1)} = \frac{\sqrt{2}}{l+1}\Psi_2. \tag{93}$$

Equations (87)(iii) and (iv) yield the same expressions (91) and (93) for $\phi_1^{(1)}$.

7. The perturbations in the basis vectors

We now turn to the perturbations in the basis vectors which are directly related to the perturbations in the metric coefficients. We follow the procedure described in *M.T.* §§82–84 (pp. 443–453).

Letting

$$l^1 = l, \quad l^2 = n, \quad l^3 = m \quad \text{and} \quad l^4 = \bar{m}, \tag{94}$$

we express the perturbations, $l^{i(1)}$, in the vector l^i as a linear combination of the basis vectors, l^i, in the manner,

$$l^{i(1)} = A^i_j l^j. \tag{95}$$

The perturbations in the basis vectors are then fully described by the matrix A. The equations governing the elements of A follow from the linearized versions of the commutation relations, namely,

$$[l^i, l^j] = l^i l^j - l^j l^i = C^{ij}_k l^k, \tag{96}$$

where l^i, as tangent vectors, are to be interpreted as directional derivatives and C^{ij}_k's are the structure constants related in turn to the spin coefficients (cf. *M.T.* p. 45, equations (307)). The linearized version of equation (96) is (cf. *M.T.* 449, equation (112))

$$l^i A^j_m - l^j A^i_m = A^i_k C^{jk}_m - A^j_k C^{ik}_m + C^{ij}_k A^k_m + c^{ij}_m, \tag{97}$$

where c^{ij}_m denotes the perturbation in C^{ij}_m.

For the Bell–Szekeres space-time, the only non-vanishing structure constants are

$$C^{21}_1 = -C^{21}_2 = -\frac{\cot\psi}{\sqrt{2}} \quad \text{and} \quad C^{43}_3 = -C^{43}_4 = \frac{\cot\theta}{\sqrt{2}}. \tag{98}$$

Besides, in §§4 and 5 we have obtained the solutions for the (perturbed) spin-coefficients κ, σ, λ, ν, ρ, μ, τ and π. By considering the components of equation (97), corresponding to the structure constants and combinations of structure constants specified in *M.T.* p. 450, equations (114) and (115), we obtain the following two systems of equations which now take the places of Systems I and II considered in *M.T.* (pp. 450–452) in the context of the Kerr space-time.

System I.

$$\begin{aligned}
&\mathscr{D}_1 A^1_3 - \mathscr{D}^\dagger_1 A^2_3 = -(\tau^* + \pi)\sqrt{2} && (21, 3),\\
&\mathscr{D}_1 A^1_4 - \mathscr{D}^\dagger_1 A^2_4 = -(\tau + \pi^*)\sqrt{2} && (21, 4),\\
&\mathscr{L}^\dagger_1 A^4_1 - \mathscr{L}_1 A^3_1 = +(\mu^* - \mu)\sqrt{2} && (43, 1),\\
&\mathscr{L}_1 A^3_2 - \mathscr{L}^\dagger_1 A^4_2 = -(\rho^* - \rho)\sqrt{2} && (43, 2),\\
&\mathscr{L}^\dagger_0 (A^1_1 + A^2_2) + \mathscr{D}^\dagger_1 A^3_1 + \mathscr{D}_1 A^3_2 = -(\tau - \pi^*)\sqrt{2} && (31, 1) + (32, 2),\\
&\mathscr{L}_0 (A^1_1 + A^2_2) + \mathscr{D}^\dagger_1 A^4_1 + \mathscr{D}_1 A^4_2 = -(\tau^* - \pi)\sqrt{2} && (41, 1) + (42, 2),\\
&\mathscr{D}^\dagger_0 (A^3_3 + A^4_4) + \mathscr{L}^\dagger_1 A^1_3 + \mathscr{L}_1 A^1_4 = +(\rho + \rho^*)\sqrt{2} && (31, 3) + (41, 4),\\
&\mathscr{D}_0 (A^3_3 + A^4_4) + \mathscr{L}^\dagger_1 A^2_3 + \mathscr{L}_1 A^2_4 = -(\mu + \mu^*)\sqrt{2} && (32, 3) + (42, 4).
\end{aligned} \tag{99}$$

SYSTEM II.

$$\begin{aligned}
&\mathscr{L}_0^\dagger A_2^1+\mathscr{D}_{-1}^\dagger A_2^3 = -\kappa\sqrt{2} && (31, 2),\\
&\mathscr{L}_0 A_2^1+\mathscr{D}_{-1}^\dagger A_2^4 = -\kappa^*\sqrt{2} && (41, 2),\\
&\mathscr{D}_0^\dagger A_4^3+\mathscr{L}_{-1}^\dagger A_4^1 = +\sigma\sqrt{2} && (31, 4),\\
&\mathscr{D}_0^\dagger A_3^4+\mathscr{L}_{-1} A_3^1 = +\sigma^*\sqrt{2} && (41, 3),\\
&\mathscr{L}_0^\dagger A_1^2+\mathscr{D}_{-1} A_1^3 = +\nu^*\sqrt{2} && (32, 1),\\
&\mathscr{L}_0 A_1^2+\mathscr{D}_{-1} A_1^4 = +\nu\sqrt{2} && (42, 1),\\
&\mathscr{L}_{-1}^\dagger A_4^2+\mathscr{D}_0 A_4^3 = -\lambda^*\sqrt{2} && (32, 4),\\
&\mathscr{L}_{-1} A_3^2+\mathscr{D}_0 A_3^4 = -\lambda\sqrt{2} && (42, 3).
\end{aligned} \tag{100}$$

Substituting for κ, ρ, ν, μ, σ, τ, λ and π from equations (22), (23), (64) and (65), and introducing the abbreviations,

$$\begin{aligned}
&F_2^3 = A_2^3-2P; \quad F_3^1 = A_3^1-2Q^*; \quad F_1^3 = A_1^3-2P^{\dagger *}; \quad F_3^2 = A_3^2-2Q^\dagger;\\
&F_2^4 = A_2^4-2P^*; \quad F_4^1 = A_4^1-2Q; \quad F_1^4 = A_1^4-2P^\dagger; \quad F_4^2 = A_4^2-2Q^{\dagger *},
\end{aligned} \tag{101}$$

and

$$U = A_1^1+A_2^2, \quad V = A_3^3+A_4^4, \tag{102}$$

we find that the equations of Systems I and II can be reduced to the forms:

$$\begin{aligned}
&\mathscr{D}_1 F_3^1-\mathscr{D}_1^\dagger F_3^2 = 0, && \text{(i)}\\
&\mathscr{D}_1 F_4^1-\mathscr{D}_1^\dagger F_4^2 = 0, && \text{(ii)}\\
&\mathscr{L}_1^\dagger F_1^4-\mathscr{L}_1 F_1^3 = 0, && \text{(iii)}\\
&\mathscr{L}_1^\dagger F_2^4-\mathscr{L}_1 F_2^3 = 0, && \text{(iv)}\\
&\mathscr{L}_0^\dagger U+\mathscr{D}_1^\dagger F_1^3+\mathscr{D}_1 F_2^3 = -2\mathscr{D}_1(P-Q)-2\mathscr{D}_1^\dagger(P^\dagger-Q^\dagger)^*, && \text{(v)}\\
&\mathscr{L}_0 U+\mathscr{D}_1^\dagger F_1^4+\mathscr{D}_1 F_2^4 = -2\mathscr{D}_1(P-Q)^*-2\mathscr{D}_1^\dagger(P^\dagger-Q^\dagger), && \text{(vi)}\\
&\mathscr{D}_0^\dagger V+\mathscr{L}_1^\dagger F_3^1+\mathscr{L}_1 F_4^1 = +2\mathscr{L}_1(P-Q)+2\mathscr{L}_1^\dagger(P-Q)^*, && \text{(vii)}\\
&\mathscr{D}_0 V+\mathscr{L}_1^\dagger F_3^2+\mathscr{L}_1 F_4^2 = 2\mathscr{L}_1(P^\dagger-Q^\dagger)^*+2\mathscr{L}_1^\dagger(P^\dagger-Q^\dagger); && \text{(viii)}
\end{aligned} \tag{103}$$

and

$$\begin{aligned}
&\mathscr{D}_{-1}^\dagger F_2^3+\mathscr{L}_0^\dagger A_2^1 = 0, \quad \mathscr{D}_{-1}^\dagger F_2^4+\mathscr{L}_0 A_2^1 = 0; && \text{(i)}\\
&\mathscr{D}_0^\dagger A_4^3+\mathscr{L}_{-1}^\dagger F_4^1 = 0, \quad \mathscr{D}_0^\dagger A_3^4+\mathscr{L}_{-1} F_3^1 = 0; && \text{(ii)}\\
&\mathscr{D}_{-1} F_1^3+\mathscr{L}_0^\dagger A_1^2 = 0, \quad \mathscr{D}_{-1} F_1^4+\mathscr{L}_0 A_1^2 = 0; && \text{(iii)}\\
&\mathscr{D}_0 A_4^3+\mathscr{L}_{-1}^\dagger F_4^2 = 0, \quad \mathscr{D}_0 A_3^4+\mathscr{L}_{-1} F_3^2 = 0. && \text{(iv)}
\end{aligned} \tag{104}$$

(*a*) *The solution of equations* (103) *and* (104)

Considering first equations (104) of System II, we observe that the pairs of equations (i) and (iii) enable us to express F_2^3, F_2^4, F_1^3, F_1^4, A_2^1 and A_1^2 in terms of two real potentials H and K in the manner:

$$F_2^3 = \mathscr{L}_0^\dagger K; \quad F_2^4 = \mathscr{L}_0 K; \quad A_2^1 = -\mathscr{D}_{-1}^\dagger K; \tag{105}$$

and

$$F_1^3 = \mathscr{L}_0^\dagger H; \quad F_1^4 = \mathscr{L}_0 H; \quad A_1^2 = -\mathscr{D}_{-1} H. \tag{106}$$

(The reality of H and K is evident, as for example, from the pair of equations relating them to A_2^1 and A_1^2.) Similarly, the pair of equations (104)(ii) enables us to express F_3^1, F_4^1, A_3^4 and A_4^3 in terms of a complex potential G in the manner:

$$F_3^1 = \mathscr{D}_0^\dagger G, \quad F_4^1 = \mathscr{D}_0^\dagger G^*; \quad A_3^4 = -\mathscr{L}_{-1} G, \quad A_4^3 = -\mathscr{L}_{-1}^\dagger G^*. \tag{107}$$

Next, by substituting for A_4^3 and A_3^4 from equations (107) in the pair of equations (104)(iv) we conclude that

$$F_3^2 = \mathscr{D}_0 G, \quad F_4^2 = \mathscr{D}_0 G^*. \tag{108}$$

This completes the solution of equations (104). Turning to equations (103) of System I, we verify that equations (103)(i)–(iv) are identically satisfied by virtue of the identity,

$$\mathscr{D}_1 \mathscr{D}_0^\dagger \equiv \mathscr{D}_1^\dagger \mathscr{D}_0 \quad \text{and} \quad \mathscr{L}_1^\dagger \mathscr{L}_0 \equiv \mathscr{L}_1 \mathscr{L}_0^\dagger; \tag{109}$$

and the solutions for F_3^1, F_3^2, F_4^1, F_4^2, F_1^3, F_1^4, F_2^3 and F_2^4 that we have already found; and by substituting these known solutions in the last remaining equations (103)(v)–(viii), we find:

$$\mathscr{L}_0^\dagger(U+\mathscr{D}_1^\dagger H+\mathscr{D}_1 K) = -2\mathscr{D}_1(P-Q)-2\mathscr{D}_1^\dagger(P^\dagger-Q^\dagger)^*, \tag{110}$$

$$\mathscr{L}_0(U+\mathscr{D}_1^\dagger H+\mathscr{D}_1 K) = -2\mathscr{D}_1(P-Q)^*-2\mathscr{D}_1^\dagger(P^\dagger-Q^\dagger), \tag{111}$$

$$\mathscr{D}_0^\dagger(V+\mathscr{L}_1^\dagger G+\mathscr{L}_1 G^*) = 2\mathscr{L}_1(P-Q)+2\mathscr{L}_1^\dagger(P-Q)^*, \tag{112}$$

$$\mathscr{D}_0(V+\mathscr{L}_1^\dagger G+\mathscr{L}_1 G^*) = 2\mathscr{L}_1(P^\dagger-Q^\dagger)^*+2\mathscr{L}_1^\dagger(P^\dagger-Q^\dagger). \tag{113}$$

The difference and the sum of equations (110) and (111) and of equations (112) and (113), yield the two pairs of equations:

$$\begin{aligned} k_1(U+\mathscr{D}_1^\dagger H+\mathscr{D}_1 K)\operatorname{cosec}\theta &= \mathscr{D}_1[(P-Q)-(P-Q)^*] \\ &\quad +\mathscr{D}_1^\dagger[(P^\dagger-Q^\dagger)^*-(P^\dagger-Q^\dagger)], \end{aligned} \tag{114}$$

$$\begin{aligned} \mathrm{i}k_2(V+\mathscr{L}_1^\dagger G+\mathscr{L}_1 G^*)\operatorname{cosec}\psi &= \mathscr{L}_1[(P^\dagger-Q^\dagger)^*-(P-Q)] \\ &\quad +\mathscr{L}_1^\dagger[(P^\dagger-Q^\dagger)-(P-Q)^*]; \end{aligned} \tag{115}$$

and

$$\begin{aligned} \partial_\theta(U+\mathscr{D}_1^\dagger H+\mathscr{D}_1 K) &= -[\mathscr{D}_1(P-Q)+\mathscr{D}_1^\dagger(P^\dagger-Q^\dagger)^* \\ &\quad +\mathscr{D}_1(P-Q)^*+\mathscr{D}_1^\dagger(P^\dagger-Q^\dagger)], \end{aligned} \tag{116}$$

$$\begin{aligned} \partial_\psi(V+\mathscr{L}_1^\dagger G+\mathscr{L}_1 G^*) &= +[\mathscr{L}_1(P-Q)+\mathscr{L}_1^\dagger(P-Q)^* \\ &\quad +\mathscr{L}_1(P^\dagger-Q^\dagger)^*+\mathscr{L}_1^\dagger(P^\dagger-Q^\dagger)]. \end{aligned} \tag{117}$$

Eliminating $U+\mathscr{D}_1^\dagger H+\mathscr{D}_1 K$ from equations (114) and (116) and $V+\mathscr{L}_1^\dagger G+\mathscr{L}_1 G^*$ from equations (115) and (117), we obtain the integrability conditions:

$$\mathscr{L}_1[\mathscr{D}_1(P-Q)+\mathscr{D}_1^\dagger(P^\dagger-Q^\dagger)^*] = \mathscr{L}_1^\dagger[\mathscr{D}_1(P-Q)^*+\mathscr{D}_1^\dagger(P^\dagger-Q^\dagger)] \tag{118}$$

and

$$\mathscr{D}_1^\dagger[\mathscr{L}_1(P^\dagger-Q^\dagger)^*+\mathscr{L}_1^\dagger(P^\dagger-Q^\dagger)] = \mathscr{D}_1[\mathscr{L}_1(P-Q)+\mathscr{L}_1^\dagger(P-Q)^*]. \tag{119}$$

These conditions are identically satisfied by virtue of the equations that we have already established in proving that Ψ_2 given by equations (62) and (78) are the same. These equations and their complex conjugates ensure that

$$-\Psi_2 = \mathscr{D}_1 \mathscr{L}_1(P-Q) = \mathscr{D}_1^\dagger \mathscr{L}_1^\dagger(P^\dagger - Q^\dagger) \tag{120}$$

and

$$-\Psi_2^* = \mathscr{D}_1 \mathscr{L}_1^\dagger(P-Q)^* = \mathscr{D}_1^\dagger \mathscr{L}_1(P^\dagger - Q^\dagger)^*. \tag{121}$$

It should be noted that complex conjugation in equations (120) and (121), as in the rest of the paper, is in the context of the original conventions of the Newman–Penrose formalism in which $\boldsymbol{l}$ and $\boldsymbol{n}$ are real vectors and $\boldsymbol{m}$ and $\bar{\boldsymbol{m}}$ are complex conjugate vectors. This last fact may give rise to certain apparent contradictions which result from these directional derivatives operating on functions with an (x^1, x^2)-dependence given by (13): compare, for example, the definitions of the basis vectors in equation (2) with the expressions for the directional derivatives given in equations (14) and (15). To avoid possible misunderstandings arising on this account, we may note here, explicitly, the expressions for $(P-Q)$, $(P^\dagger - Q^\dagger)$, and their complex conjugates:

$$\begin{aligned}
(P-Q) &= 2CS_{+1}(l, m; \theta) S_{+1}(l+1, \mathrm{i}k_2; \psi), \\
(P-Q)^* &= 2CS_{-1}(l, m; \theta) S_{+1}(l+1, \mathrm{i}k_2; \psi), \\
(P^\dagger - Q^\dagger) &= 2CS_{-1}(l, m; \theta) S_{-1}(l+1, \mathrm{i}k_2; \psi), \\
(P^\dagger - Q^\dagger)^* &= 2CS_{+1}(l, m; \theta) S_{-1}(l+1, \mathrm{i}k_2; \psi).
\end{aligned} \tag{122}$$

Returning to equations (114) and (116), we readily infer that U and V will be left unspecified; and consistent with the available freedoms of gauge, we may set (cf. *M.T.* p. 485, equation (294)),

$$U = V = 0. \tag{123}$$

(For a more detailed discussion of these issues see *M.T.* §95, pp. 497–500.) In the chosen gauge, the solution of equations (114) and (115) for H, K and G may be written in the forms:

$$\begin{aligned}
H &= \frac{\sin\theta}{k_1}[(P^\dagger - Q^\dagger)^* - (P^\dagger - Q^\dagger)] + \mathscr{D}_0 X, \\
K &= \frac{\sin\theta}{k_1}[(P-Q) - (P-Q)^*] - \mathscr{D}_0^\dagger X, \\
G &= -\mathrm{i}\frac{\sin\psi}{k_2}[(P^\dagger - Q^\dagger) - (P-Q)^*] + \mathscr{L}_0 Y, \\
G^* &= -\mathrm{i}\frac{\sin\psi}{k_2}[(P^\dagger - Q^\dagger)^* - (P-Q)] - \mathscr{L}_0^\dagger Y,
\end{aligned} \tag{124}$$

where X and Y are two arbitrary real functions. Again, we may, without loss of generality, set

$$X = Y = 0. \tag{125}$$

Now substituting for $(P-Q)$, $(P^\dagger-Q^\dagger)$, and their complex conjugates from equations (122) in equations (124), we find:

$$H = \frac{2C}{k_1} S_{-1}(l+1, ik_2; \psi)[S_1(l, m; \theta)]^- \sin\theta,$$

$$K = \frac{2C}{k_1} S_{+1}(l+1, ik_2; \psi)[S_1(l, m; \theta)]^- \sin\theta,$$

$$G = \frac{2iC}{k_2} S_{-1}(l, m; \theta)[S_1(l+1, ik_2; \psi)]^- \sin\psi,$$

$$G^* = \frac{2iC}{k_2} S_{+1}(l, m; \theta)[S_1(l+1, ik_2; \psi)]^- \sin\psi. \tag{126}$$

where, in the 'bracket notation' of *M.T.* p. 112,

$$[S_n]^\pm = S_{+n} \pm S_{-n}. \tag{127}$$

Finally, with the solutions for H, K, G and G^* given in equations (126), we find that the solutions for F^i_j $(i, j = 1, 2, 3, 4; i \neq j)$ and A^i_j $(i, j = 1, 2, 3, 4; i \neq j)$ given in equations (105)–(108) become:

$$F^1_3 = \frac{2iC\sin\psi}{k_2} S_{-1}(l, m; \theta)\, \mathscr{D}^\dagger_1[S_1(l+1, ik_2; \psi)]^-,$$

$$F^1_4 = \frac{2iC\sin\psi}{k_2} S_{+1}(l, m; \theta)\, \mathscr{D}^\dagger_1[S_1(l+1, ik_2; \psi)]^-,$$

$$F^2_3 = \frac{2iC\sin\psi}{k_2} S_{-1}(l, m; \theta)\, \mathscr{D}_1[S_1(l+1, ik_2; \psi)]^-,$$

$$F^2_4 = \frac{2iC\sin\psi}{k_2} S_{+1}(l, m; \theta)\, \mathscr{D}_1[S_1(l+1, ik_2; \psi)]^-,$$

$$A^4_3 = \frac{2iC\sin\psi}{k_2} [(l-1)(l+2)]^{\frac{1}{2}} S_{-2}(l, m; \theta)[S_1(l+1, ik_2; \psi)]^-,$$

$$A^3_4 = \frac{2iC\sin\psi}{k_2} [(l-1)(l+2)]^{\frac{1}{2}} S_{+2}(l, m; \theta)[S_1(l+1, ik_2; \psi)]^-,$$

$$F^3_1 = \frac{2C\sin\theta}{k_1} S_{-1}(l+1, ik_2; \psi)\, \mathscr{L}^\dagger_1[S_1(l, m; \theta)]^-,$$

$$F^4_1 = \frac{2C\sin\theta}{k_1} S_{-1}(l+1, ik_2; \psi)\, \mathscr{L}_1[S_1(l, m; \theta)]^-,$$

$$F^3_2 = \frac{2C\sin\theta}{k_1} S_{+1}(l+1, ik_2; \psi)\, \mathscr{L}^\dagger_1[S_1(l, m; \theta)]^-,$$

$$F^4_2 = \frac{2C\sin\theta}{k_1} S_{+1}(l+1, ik_2; \psi)\, \mathscr{L}_1[S_1(l, m; \theta)]^-,$$

$$A_1^2 = \frac{2C\sin\theta}{k_1}[l(l+3)]^{\frac{1}{2}} S_{-2}(l+1, \mathrm{i}k_2; \psi)[S_1(l, m; \theta)]^-,$$

$$A_2^1 = \frac{2C\sin\theta}{k_1}[l(l+3)]^{\frac{1}{2}} S_{+2}(l+1, \mathrm{i}k_2; \psi)[S_1(l, m; \theta)]^-, \tag{128}$$

and the solution of equations (103) and (104) is completed.

(b) The solutions for the spin coefficients, α, β, γ and ϵ

With the solution for the matrix elements of $\boldsymbol{A}$ completed, the equations of System III (*M.T.* p. 452) now enable us to directly write down the solutions for the remaining spin coefficients α, β, γ and ϵ. We find:

SYSTEM III.

$$(\alpha^* + \beta)^{(1)} = \tfrac{1}{2}(\tau + \pi^*) + \frac{1}{2\sqrt{2}}(\mathscr{D}_{-1} A_2^3 - \mathscr{D}_{-1}^\dagger A_1^3 - 2\mathscr{L}_0^\dagger A_1^1) \quad (41, 1)\text{–}(42, 2),$$

$$(\epsilon - \epsilon^*)^{(1)} = \tfrac{1}{2}(\rho - \rho^*) + \frac{1}{2\sqrt{2}}(\mathscr{L}_{-1}^\dagger A_3^1 - \mathscr{L}_{-1} A_4^1 + 2\mathscr{D}_0^\dagger A_3^3) \quad (41, 4)\text{–}(31, 3),$$

$$(\gamma - \gamma^*)^{(1)} = \tfrac{1}{2}(\mu - \mu^*) + \frac{1}{2\sqrt{2}}(\mathscr{L}_{-1}^\dagger A_3^2 - \mathscr{L}_{-1} A_4^2 + 2\mathscr{D}_0 A_3^3) \quad (42, 4)\text{–}(32, 3),$$

$$(\gamma + \gamma^*)^{(1)} = \frac{1}{\sqrt{2}}(\mathscr{D}_1 A_1^1 - \mathscr{D}_1^\dagger A_1^2) \quad (21, 1),$$

$$(\epsilon + \epsilon^*)^{(1)} = \frac{1}{\sqrt{2}}(\mathscr{D}_1^\dagger A_1^1 + \mathscr{D}_1 A_2^1) \quad (21, 2),$$

$$(\alpha^* - \beta)^{(1)} = \frac{1}{\sqrt{2}}(\mathscr{L}_1^\dagger A_3^3 + \mathscr{L}_1 A_4^3) \quad (43, 3). \tag{129}$$

8. THE BOUNDEDNESS OF THE PERTURBATIONS IN REGION I

Since all the quantities describing the perturbation of the space-time in Region I are expressed in terms of the spin-weighted spherical harmonics, $S_{\pm 1}$ and $S_{\pm 2}$, with the arguments, $(l, m; \theta)$ and $(l+1, \mathrm{i}k_2; \psi)$, it is clear that the only question we need to be concerned about is the occurrence of $\mathrm{i}k_2$ among the arguments of the temporal functions. That these functions are also bounded in the domains of their definition, $(0 \leqslant \psi \leqslant \pi)$, follows from the fact that they are directly related (as we shall presently show) to the Jacobi polynomials, $P_n^{(\alpha,\beta)}(x)$, of integral order, n.

First, we observe that in considering the equation,

$$\frac{\mathrm{d}}{\mathrm{d}x}\left[(1-x^2)\frac{\mathrm{d}S}{\mathrm{d}x}\right] + \left[l(l+1) - \frac{k^2 + 2ksx + s^2}{1-x^2}\right]S = 0, \tag{130}$$

governing the spin-weighted spherical harmonics, we may, without loss of generality, restrict s to be non-negative since,

$$S_{-s}(x) \equiv S_{+s}(-x) \quad (-1 \leqslant x \leqslant +1). \tag{131}$$

It can now be directly verified that by the substitution,

$$S = (1-x)^{\alpha/2}(1+x)^{\beta/2}y(x), \tag{132}$$

where α and β are constants, unspecified at present, we can bring equation (130) to the standard form,

$$(1-x^2)\frac{d^2y}{dx^2}+[(\beta-\alpha)-(\alpha+\beta+2)x]\frac{dy}{dx}+n(n+\alpha+\beta+1)y = 0, \tag{133}$$

defining the Jacobi polynomial, $P_n^{(\alpha,\beta)}(x)$, by the choices,

$$\alpha^2 = (s+k)^2 \quad \text{and} \quad \beta^2 = (s-k)^2, \tag{134}$$

when

$$n = l-\tfrac{1}{2}(\alpha+\beta). \tag{135}$$

Therefore, when $k = ik_2$, we can let

$$\alpha = s+ik_2, \quad \beta = s-ik_2, \alpha+\beta = 2s \quad \text{and} \quad n = l-s, \quad (s \geqslant 0, l \geqslant s); \tag{136}$$

in which case,

$$S_{+s}(l, ik_2; x) = (1-x)^{ik_2/2}(1+x)^{-ik_2/2}(1-x^2)^{s/2}P_{l-s}^{(s+ik_2, s-ik_2)}(x),$$

and

$$S_{-s}(x) = S_{+s}(-x) \quad (s \geqslant 0, l \geqslant s, -1 \leqslant x \leqslant +1). \tag{137}$$

Since $P_n^{(\alpha,\beta)}(x)$ of integral order $n(\geqslant 0)$ is a polynomial of degree n (cf. Abramowitz & Stegun 1964, p. 793), we conclude that the spin-weighted harmonics are bounded for imaginary arguments, ik_2, as well. Their behaviours for $x \to \pm 1$ are the same as for the ordinary spherical harmonics (cf. Hobson 1931, p. 184): they oscillate with frequencies increasing without limit. This behaviour of the perturbations of the Bell–Szekeres space-time, as we approach the horizon in Region I, is reminiscent of the similar behaviour of the perturbations as we approach the Cauchy horizon of the Reissner–Nordström black-hole (see Chandrasekhar & Hartle 1982).

9. The Weyl and the Maxwell scalars in Region I as the null boundary at $u = 0+$ is approached

Before we address ourselves to the question as to how the perturbations in Region I, belonging to the different normal modes (or, superpositions of them) can be excited, it is convenient to assemble the expressions for the Weyl and the Maxwell scalars for $u \to 0+$, where

$$\psi \to (\pi-\theta)+0. \tag{138}$$

Making use of the relation (cf. equations (16) and (45)),

$$S_{+s}(\ldots, \ldots; \psi = \pi-\theta) \equiv S_{-s}(\ldots, \ldots; \theta), \tag{139}$$

we find from equations (58), (61), (62), (76), (77), (84), (90) and (92):

$$\begin{aligned}
\Psi_0 &= -2C[l(l-1)(l+2)(l+3)]^{\frac{1}{2}} S_{+2}(l,m;\theta)\, S_{-2}(l+1,\mathrm{i}k_2;\theta),\\
\Psi_1 &= -2Cl(l+2)\, S_{+1}(l,m;\theta)\, S_{-1}(l+1,\mathrm{i}k_2;\theta),\\
\Psi_2 &= +2C[\mathscr{L}_1 S_{+1}(l,m;\theta)][\mathscr{D}_1^{\dagger} S_{-1}(l+1,\mathrm{i}k_2;\psi)]_{\psi=\theta},\\
\Psi_3 &= -2Cl(l+2)\, S_{-1}(l,m;\theta)\, S_{+1}(l+1,\mathrm{i}k_2;\theta),\\
\Psi_4 &= -2C[l(l-1)(l+2)(l+3)]^{\frac{1}{2}} S_{-2}(l,m;\theta)\, S_{+2}(l+1,\mathrm{i}k_2;\theta);
\end{aligned} \tag{140}$$

and

$$\begin{aligned}
&\phi_0^{(1)} = 0, \quad \phi_2^{(1)} = 0,\\
&\phi_1^{(1)} = 2\sqrt{2}\frac{C}{l+1}[\mathscr{L}_1 S_{+1}(l,m;\theta)][\mathscr{D}_1^{\dagger} S_{-1}(l+1,\mathrm{i}k_2;\psi)]_{\psi=\theta}\\
&\qquad = \frac{\sqrt{2}}{l+1}\Psi_2.
\end{aligned} \tag{141}$$

The foregoing expressions for $\Psi_0, \ldots, \Psi_4$, $\phi_0^{(1)}$, $\phi_2^{(1)}$ and $\phi_1^{(1)}$ have been derived in the shear-free null tetrad-frame that we have defined in equation (2). But the background space-time was described in Chandrasekhar & Xanthopoulos (1987*a*) in the tetrad frame,

$$\begin{array}{lcccc}
 & \psi & \theta & x^1 & x^2\\
(l^i) = \frac{1}{\sqrt{2}}(& 1, & +1, & 0, & 0),\\
(n^i) = \frac{1}{\sqrt{2}}(& 1, & -1, & 0, & 0),\\
(m^i) = \frac{1}{\sqrt{2}}(& 0, & 0, & \operatorname{cosec}\theta, & -\mathrm{i}\operatorname{cosec}\psi),\\
(\bar{m}^i) = \frac{1}{\sqrt{2}}(& 0, & 0, & \operatorname{cosec}\theta, & +\mathrm{i}\operatorname{cosec}\psi),
\end{array} \tag{142}$$

adapted to the prevailing Killing fields, ∂_{x^1} and ∂_{x^2} and which further matches continuously the frame suitable for Regions II and III (cf. §§10 and 11). We readily verify that the basis vectors of the frame (142) can be obtained from those of the shear-free frame (2) by subjecting the latter, successively, to rotations of class I (with parameter $a = 1$), class II (with parameter $b = -\frac{1}{2}$), and class III (with parameter $A e^{\mathrm{i}\theta} = \frac{1}{2}e^{\mathrm{i}\pi/2}$). By this same sequence of rotations, the Weyl and the Maxwell scalars listed in equations (140) and (141) are replaced, respectively, by the combinations:

$$\begin{aligned}
\tilde{\Psi}_0 &= -\tfrac{1}{4}(\Psi_0 - 4\Psi_1 + 6\Psi_2 - 4\Psi_3 + \Psi_4),\\
\tilde{\Psi}_1 &= +\frac{\mathrm{i}}{4}(\Psi_0 - 2\Psi_1 + 2\Psi_3 - \Psi_4),\\
\tilde{\Psi}_2 &= +\tfrac{1}{4}(\Psi_0 - 2\Psi_2 + \Psi_4),
\end{aligned}$$

$$\tilde{\Psi}_3 = -\frac{\mathrm{i}}{4}(\Psi_0+2\Psi_1-2\Psi_3-\Psi_4),$$

$$\tilde{\Psi}_4 = -\tfrac{1}{4}(\Psi_0+4\Psi_1+6\Psi_2+4\Psi_3+\Psi_4); \tag{143}$$

and

$$\tilde{\phi}_0^{(1)} = -\mathrm{i}\phi_1^{(1)}, \quad \tilde{\phi}_2^{(1)} = -\mathrm{i}\phi_1^{(1)} \quad \text{and} \quad \tilde{\phi}^{(1)} = 0. \tag{144}$$

10. Linearized first-order perturbations of the space-time in Region II

We should naturally suppose that perturbations of the space-time in Region I derive from the incidence, on the null boundary at $u = 0-$ from the left (or $v = 0-$ from the right) of wave packets of gravitational and electromagnetic radiation in Region II (or III). For this reason, we shall first consider the implications of the equations governing the first-order linear perturbations of the space-time in Region II.

A general space-time appropriate to Region II, under circumstances envisaged in problems involving the collision of plane-fronted gravitational and electromagnetic waves, has been analysed in some detail in Chandrasekhar & Xanthopoulos (1985, §8; this paper will be referred to as Paper I). We specialize that analysis appropriately for the Bell–Szekeres space-time in Region II.

The metric in Region II of the Bell–Szekeres space-time is given by (Chandrasekhar & Xanthopoulos 1987*a*, equation (52))

$$\mathrm{d}s^2 = \frac{4\,\mathrm{d}u\,\mathrm{d}v}{\sqrt{(1-v^2)}} - (1-v^2)[(\mathrm{d}x^1)^2+(\mathrm{d}x^2)^2], \tag{145}$$

where, in the notation of Paper I (equation (150))

$$u = x^0+x^3 \quad \text{and} \quad v = x^0-x^3. \tag{146}$$

Also, the functions $\mathrm{e}^{2\nu}$, e^{β} and $\mathscr{E}$ introduced in Paper I (equations (161) and (162)) are, in the present context,

$$\mathrm{e}^{2\nu} = \frac{4}{\sqrt{(1-v^2)}}, \quad \mathrm{e}^{\beta} = (1-v^2) \quad \text{and} \quad \mathscr{E} = 0. \tag{147}$$

The description of the more general space-time of Paper I, in the Newman–Penrose formalism, when specialized to the Bell–Szekeres space-time, applies in the null frame,

$$\begin{array}{llllll} & & x^0 & x^3 & x^1 & x^2 \\ \boldsymbol{e}_{(1)} = +\boldsymbol{e}^{(2)} = [l_i] & = \frac{1}{\sqrt{2}}(& U, & -U, & 0, & 0), \\ \boldsymbol{e}_{(2)} = +\boldsymbol{e}^{(1)} = [n_i] & = \frac{1}{\sqrt{2}}(& U, & +U, & 0, & 0), \\ \boldsymbol{e}_{(3)} = -\boldsymbol{e}^{(4)} = [m_i] & = \frac{1}{\sqrt{2}}(& 0, & 0, & -V, & +\mathrm{i}V), \\ \boldsymbol{e}_{(4)} = -\boldsymbol{e}^{(3)} = [\bar{m}_i] & = \frac{1}{\sqrt{2}}(& 0, & 0, & -V, & -\mathrm{i}V), \end{array} \tag{148}$$

where $$U = 2(1-v^2)^{-\frac{1}{4}} \quad \text{and} \quad V = \sqrt{(1-v^2)}. \tag{149}$$

In this frame, the only non-vanishing spin-coefficients are

$$[\mu] = -\frac{1}{\sqrt{2}}\frac{v}{(1-v^2)^{\frac{3}{4}}} \quad \text{and} \quad [\gamma] = -\frac{1}{4\sqrt{2}}\frac{v}{(1-v^2)^{\frac{3}{4}}}, \tag{150}$$

and the surviving Maxwell scalar is

$$[\Phi_{22}] = \frac{1}{2\sqrt{(1-v^2)}}. \tag{151}$$

(We have enclosed μ, γ and Φ_{22} in square brackets to indicate that they have been evaluated in the frame (148).)

We shall find it advantageous to analyse the perturbations in a null frame that is obtained by subjecting the frame (148) to a rotation of class III with

$$A = -(1-v^2)^{\frac{1}{4}} \quad \text{and} \quad \theta = 0. \tag{152}$$

In the (u, v, x^1, x^2)-coordinates, the resulting frame is

$$\begin{array}{lllll} & u & v & x^1 & x^2 \\ (l^i) = -\frac{1}{\sqrt{2}} & (1, & 0, & 0, & 0), \\ (n^i) = -\frac{1}{\sqrt{2}}\sqrt{(1-v^2)} & (0, & +1, & 0, & 0), \\ (m^i) = \frac{1}{\sqrt{[2(1-v^2)]}} & (0, & 0, & 1, & -\mathrm{i}), \\ (\bar{m}^i) = \frac{1}{\sqrt{[2(1-v^2)]}} & (0, & 0, & 1, & +\mathrm{i}). \end{array} \tag{153}$$

In this frame (in contrast to the frame (148)) we have only one non-vanishing spin-coefficient:

$$\mu = \frac{v}{\sqrt{[2(1-v^2)]}}; \tag{154}$$

and besides, the surviving Maxwell scalar takes a constant value:

$$\Phi_{22} = |\phi_2|^2 = \tfrac{1}{2}. \tag{155}$$

Ignoring a physically non-significant phase factor, we may take

$$\phi_2 = \frac{1}{\sqrt{2}}. \tag{156}$$

Since the space-time we are considering has three commuting Killing vectors, namely ∂_{x^1}, ∂_{x^2} and ∂_u, we can analyse the perturbations into normal modes by supposing that all the perturbed quantities have an x^1-, x^2- and u-dependence given by

$$e^{\mathrm{i}(k_1 x^1 + k_2 x^2 + k_3 u)}, \tag{157}$$

where k_1, k_2 and k_3 are real constants. In the sequel, we shall suppress this common factor in all the quantities describing the perturbation: the symbols representing them will be their amplitudes.

For functions with an x^1-, x^2- and u-dependence given by (157), the directional derivatives D, Δ, δ and δ^* along the directions $\boldsymbol{l}$, $\boldsymbol{n}$, $\boldsymbol{m}$ and $\bar{\boldsymbol{m}}$ become

$$\mathrm{D} = -\frac{\mathrm{i}k_3}{\sqrt{2}}, \quad \Delta = -\frac{1}{\sqrt{2}}\sqrt{(1-v^2)}\frac{\mathrm{d}}{\mathrm{d}v},$$

$$\delta = \frac{k_2+\mathrm{i}k_1}{\sqrt{[2(1-v^2)]}}, \quad \delta^* = -\frac{k_2-\mathrm{i}k_1}{\sqrt{[2(1-v^2)]}}. \tag{158}$$

(*a*) *The quantities that have to be considered in the perturbation analysis*

In the background space-time, all the Weyl scalars vanish consistently with the conformal flatness of the space-time. In the perturbed space-time, we must assume that they all cease to vanish and that

$$\Psi_0, \Psi_1, \Psi_2, \Psi_3 \text{ and } \Psi_4 \tag{159}$$

are quantities of the first order of smallness (with, of course, the factor (157)!). Since ϕ_0 and ϕ_1 vanish in the background space-time, the scalars

$$\Phi_{00} = |\phi_0|^2, \quad \Phi_{11} = |\phi_1|^2 \quad \text{and} \quad \Phi_{01} = \Phi_{10}^* = \phi_0\phi_1^*, \tag{160}$$

will vanish to the second order and can be ignored. But ϕ_2 is non-vanishing and has the constant value $1/\sqrt{2}$ (see equation (156)). Accordingly, Φ_{02} and Φ_{12} will be quantities of the first order and we shall have

$$\Phi_{02} = \Phi_{20}^* = \frac{1}{\sqrt{2}}\phi_0 \quad \text{and} \quad \Phi_{12} = \Phi_{21}^* = \frac{1}{\sqrt{2}}\phi_1, \tag{161}$$

where ϕ_0 and ϕ_1 now denote their (first-order) values in the perturbed space-time. In addition, Φ_{22} will be subject to a perturbation given by

$$\Phi_{22}^{(1)} = \frac{1}{\sqrt{2}}(\phi_2^{(1)} + \phi_2^{(1)*}), \tag{162}$$

where the superscript '1' denotes the first-order change in the non-vanishing quantity.

We have already noted that in the chosen tetrad frame μ is the only non-vanishing spin-coefficient. Therefore, the remaining 11 spin-coefficients must, in the first instance, be considered as non-vanishing and of the first order in the perturbed space-time. However, by availing ourselves of the freedom to subject the perturbed basis-vectors to infinitesimal rotations, we can make the spin-coefficients λ, ρ and β vanish by suitable rotations of class I, class II and class III, respectively. We shall then be left with

$$\kappa, \sigma, \tau, \nu, \pi, \alpha, \gamma \text{ and } \epsilon; \tag{163}$$

and we must also include the first-order change $\mu^{(1)}$ in μ.

(b) The equations that are already linearized

Now, by examining the various Ricci identities (*M.T.* pp. 46 and 47, equations (310)), Maxwell's equations (*M.T.* p. 52, equations (330)–(333)) and the Bianchi identities (*M.T.* pp. 49 and 50, equations (321) and (321′)), we find that the following equations do not require allowance for the perturbations in the basis vectors: they are 'already linearized'!

Ricci Identities:

$$\kappa = 0, \quad (a); \qquad \Psi_4\sqrt{2} = -\frac{k_2 - \mathrm{i}k_1}{\sqrt{(1-v^2)}}\nu, \quad (j)$$

$$\Psi_0\sqrt{2} = -\mathrm{i}k_3\sigma, \quad (b); \qquad \Psi_1\sqrt{2} = -\frac{k_2 - \mathrm{i}k_1}{\sqrt{(1-v^2)}}\sigma, \quad (k)$$

$$\Psi_1\sqrt{2} = -\mathrm{i}k_3\tau, \quad (c); \qquad \Psi_2\sqrt{2} = -\frac{k_2 + \mathrm{i}k_1}{\sqrt{(1-v^2)}}\alpha, \quad (l)$$

$$\mathrm{i}k_3\alpha = +\frac{k_2 - \mathrm{i}k_1}{\sqrt{(1-v^2)}}\epsilon, \quad (d); \qquad \phi_1 = \frac{1}{\sqrt{(1-v^2)}}[(k_2 + \mathrm{i}k_1)\gamma - v\tau], \quad (o)$$

$$\Psi_1\sqrt{2} = -\frac{k_2 + \mathrm{i}k_1}{\sqrt{(1-v^2)}}\epsilon, \quad (e); \qquad \phi_0 = \frac{\mathrm{d}}{\mathrm{d}v}\sigma\sqrt{(1-v^2)} + \frac{k_2 + \mathrm{i}k_1}{\sqrt{(1-v^2)}}\tau, \quad (p)$$

$$\Psi_2\sqrt{2} = \sqrt{(1-v^2)}\frac{\mathrm{d}\epsilon}{\mathrm{d}v} - \mathrm{i}k_3\gamma, \quad (f); \qquad \Psi_2\sqrt{2} = -\frac{k_2 - \mathrm{i}k_1}{\sqrt{(1-v^2)}}\tau, \quad (q)$$

$$\phi_0^* = \frac{1}{\sqrt{(1-v^2)}}[(k_2 - \mathrm{i}k_1)\pi - \sigma^* v], \quad (g); \qquad \Psi_3\sqrt{2} = \frac{\mathrm{d}}{\mathrm{d}v}\alpha\sqrt{(1-v^2)} - \frac{k_2 - \mathrm{i}k_1}{\sqrt{(1-v^2)}}\gamma, \quad (r)$$

and

$$\phi_1^* + \Psi_3\sqrt{2} = \frac{\mathrm{d}}{\mathrm{d}v}\pi\sqrt{(1-v^2)} - \frac{v\tau^*}{\sqrt{(1-v^2)}} - \mathrm{i}k_3\nu. \quad (i) \qquad (164)$$

From a comparison of equations (*c*), (*e*) and (*k*) and of equations (*l*) and (*q*), we directly conclude that

$$\mathrm{i}k_3\tau = \frac{k_2 + \mathrm{i}k_1}{\sqrt{(1-v^2)}}\epsilon = \frac{k_2 - \mathrm{i}k_1}{\sqrt{(1-v^2)}}\sigma,$$

and

$$(k_2 + \mathrm{i}k_1)\alpha = (k_2 - \mathrm{i}k_1)\tau. \qquad (165)$$

Maxwell's Equations:

$$\mathrm{i}k_3\phi_1 = \frac{k_2 - \mathrm{i}k_1}{\sqrt{(1-v^2)}}\phi_0, \quad \text{(i)}$$

$$\mathrm{i}k_3\phi_2^{(1)} = \frac{k_2 - \mathrm{i}k_1}{\sqrt{(1-v^2)}}\phi_1 + 2\epsilon, \quad \text{(ii)}$$

$$\frac{\mathrm{d}}{\mathrm{d}v}[\phi_0\sqrt{(1-v^2)}] + \frac{k_2 + \mathrm{i}k_1}{\sqrt{(1-v^2)}}\phi_1 = -\sigma, \quad \text{(iii)}$$

$$\frac{\mathrm{d}}{\mathrm{d}v}[\phi_1(1-v^2)] + (k_2 + \mathrm{i}k_1)\phi_2^{(1)} = \tau\sqrt{(1-v^2)}. \quad \text{(iv)} \qquad (166)$$

Bianchi Identities:

$$ik_3 \Psi_1 = \frac{k_2 - ik_1}{\sqrt{(1-v^2)}} \Psi_0, \tag{a}$$

$$ik_3 \Psi_2 = \frac{k_2 - ik_1}{\sqrt{(1-v^2)}} \Psi_1, \tag{b}$$

$$ik_3 \Psi_3 - \frac{k_2 - ik_1}{\sqrt{(1-v^2)}} \Psi_2 = \frac{1}{\sqrt{2}}\left[ik_3 \phi_1^* + \frac{k_2 + ik_1}{\sqrt{(1-v^2)}} \phi_0^*\right], \tag{c}$$

$$ik_3 \Psi_4 - \frac{k_2 - ik_1}{\sqrt{(1-v^2)}} \Psi_3 = \frac{1}{\sqrt{2}}\left[\frac{k_2 - ik_1}{\sqrt{(1-v^2)}} \phi_1^* - \frac{d}{dv}\phi_0^* \sqrt{(1-v^2)}\right] - \frac{\sigma^*}{\sqrt{2}}, \tag{d}$$

$$\frac{k_2 + ik_1}{\sqrt{(1-v^2)}} \Psi_1 + \frac{d}{dv}[\Psi_0 \sqrt{(1-v^2)}] = -\frac{ik_3}{\sqrt{2}}\phi_0, \tag{e}$$

$$(k_2 + ik_1)\Psi_2 + \frac{d}{dv}[\Psi_1(1-v^2)] = -\frac{1}{\sqrt{2}}(k_2 - ik_1)\phi_0, \tag{f}$$

$$\frac{1}{\sqrt{(1-v^2)}}\frac{d}{dv}[(1-v^2)^{\frac{3}{2}}\Psi_2] + (k_2 + ik_1)\Psi_3 = -\frac{1}{\sqrt{2}}(k_2 + ik_1)\phi_1^*$$
$$+\frac{1}{\sqrt{2}}\sqrt{(1-v^2)}[-ik_3(\phi_2^{(1)} + \phi_2^{(1)*}) + 2(\epsilon + \epsilon^*)], \tag{g}$$

and

$$\frac{1}{(1-v^2)}\frac{d}{dv}[(1-v^2)^2\Psi_3] - \frac{1}{\sqrt{2}}\frac{d}{dv}[(1-v^2)\phi_1^*] + (k_2 + ik_1)\Psi_4$$
$$+\frac{1}{\sqrt{2}}[(k_2 - ik_1)(\phi_2^{(1)} + \phi_2^{(1)*}) + (\tau^* - 2\alpha)\sqrt{(1-v^2)}] = 0. \quad (h) \qquad (167)$$

(The labelling of the equations in (164) and (167) is the same as that in *M.T.*)

(c) *The u-independent perturbations*

The case when the perturbations are independent of u and $k_3 = 0$, as we shall presently see, is of special interest. By a systematic examination of equations (164), (166) and (167) when $k_3 = 0$, we find that

$$\kappa = \epsilon = \sigma = \alpha = \pi = \tau = \nu = \gamma = 0, \tag{168}$$

$$\Psi_0 = \Psi_1 = \Psi_2 = \Psi_3 = \Psi_4 = 0 \quad \text{and} \quad \phi_0 = \phi_1 = \phi_2^{(1)} = 0. \tag{169}$$

In other words, the space-time is left conformally flat and the prevailing electromagnetic field is unaffected. On this account, we may expect (and the expectation *is* confirmed by a detailed examination of the commutation relations) that first-order perturbations, independent of u, effect only an infinitesimal coordinate-transformation of the unperturbed space-time. We conclude that the space-time *is neutral to u-independent perturbations*. It is important to emphasize that this absence of a genuine u-independent perturbation does not arise from the imposition of any boundary condition: it is a local, not a global, result!

(d) The u-dependent perturbations

Returning to the consideration of the more general u-dependent perturbations ($k_3 \neq 0$ and $k_1^2+k_2^2 \neq 0$), we first obtain from equation (164, (p)), after expressing τ in terms of σ with the aid of the relation included in equations (165),

$$\phi_0 = \frac{\mathrm{d}}{\mathrm{d}v}\sigma\sqrt{(1-v^2)}+\frac{k_1^2+k_2^2}{\mathrm{i}k_3(1-v^2)}\sigma. \tag{170}$$

Similarly, by combining equations (166, (i)) and (166, (iii)), we obtain

$$-\sigma = \frac{\mathrm{d}}{\mathrm{d}v}\phi_0\sqrt{(1-v^2)}+\frac{k_1^2+k_2^2}{\mathrm{i}k_3(1-v^2)}\phi_0. \tag{171}$$

We can eliminate ϕ_0 or σ from these equations and obtain a second-order differential equation for either of them. Thus, by the elimination of σ, we find

$$\begin{aligned}\frac{\mathrm{d}}{\mathrm{d}v}\left[\sqrt{(1-v^2)}\frac{\mathrm{d}}{\mathrm{d}v}\phi_0\sqrt{(1-v^2)}\right]+\frac{1}{\mathrm{i}k_3}(k_1^2+k_2^2)\frac{\mathrm{d}}{\mathrm{d}v}\frac{\phi_0}{\sqrt{(1-v^2)}}\\ = -\frac{\mathrm{d}}{\mathrm{d}v}\sigma\sqrt{(1-v^2)} = -\phi_0+\frac{k_1^2+k_2^2}{\mathrm{i}k_3(1-v^2)}\sigma\\ = -\phi_0-\frac{k_1^2+k_2^2}{\mathrm{i}k_3(1-v^2)}\left[\frac{\mathrm{d}}{\mathrm{d}v}\phi_0\sqrt{(1-v^2)}+\frac{k_1^2+k_2^2}{\mathrm{i}k_3(1-v^2)}\phi_0\right]\end{aligned} \tag{172}$$

or, after simplification,

$$\frac{\mathrm{d}}{\mathrm{d}v}\left[\sqrt{(1-v^2)}\frac{\mathrm{d}}{\mathrm{d}v}\phi_0\sqrt{(1-v^2)}\right]+2\frac{k_1^2+k_2^2}{\mathrm{i}k_3\sqrt{(1-v^2)}}\frac{\mathrm{d}\phi_0}{\mathrm{d}v}+\left[1-\frac{(k_1^2+k_2^2)^2}{k_3^2(1-v^2)^2}\right]\phi_0 = 0. \tag{173}$$

Now letting,

$$v = \cos(\pi-\theta), \tag{174}$$

(consistent with the definition of θ in equations (140) and (141)), we obtain the simpler equation,

$$\frac{\mathrm{d}}{\mathrm{d}\theta}\left(\sin^2\theta\frac{\mathrm{d}\phi_0}{\mathrm{d}\theta}\right)-2\mathrm{i}q\frac{\mathrm{d}\phi_0}{\mathrm{d}\theta}-\frac{q^2}{\sin^2\theta}\phi_0 = 0, \tag{175}$$

where

$$q = (k_1^2+k_2^2)/k_3. \tag{176}$$

In view of the symmetry of equations (170) and (171) in ϕ_0 and σ, it is clear that σ will satisfy the same equation (175).

By the substitution,

$$\phi_0 = \mathrm{e}^{-\mathrm{i}q\cot\theta}y, \tag{177}$$

we find that equation (175) reduces to

$$\frac{\mathrm{d}}{\mathrm{d}\theta}\left(\sin^2\theta\frac{\mathrm{d}y}{\mathrm{d}\theta}\right) = 0. \tag{178}$$

Accordingly,

$$y = C_1\cot\theta+C_2, \tag{179}$$

where C_1 and C_2 are two constants. The general solution of equation (175) is therefore given by

$$\phi_0 = (C_1 \cot\theta + C_2)\, e^{-iq\cot\theta}. \tag{180}$$

From equation (171), we now find that

$$\sigma = (-C_2 \cot\theta + C_1)\, e^{-iq\cot\theta}, \tag{181}$$

consistent with σ satisfying the same equation as ϕ_0.

By a systematic examination of equations (164)–(167), we find that all quantities appearing in these equations can be expressed in terms of the solutions (180) and (181) for ϕ_0 and σ. We find:

$$\begin{aligned}
&\kappa = 0; \quad \epsilon = \frac{k_2 - ik_1}{k_2 + ik_1}\sigma,\\
&\tau = \frac{k_2 - ik_1}{ik_3\sqrt{(1-v^2)}}\sigma; \quad \nu = i\frac{(k_2 - ik_1)^3}{k_3^3(1-v^2)^{\frac{3}{2}}}\sigma,\\
&\alpha = \frac{(k_2 - ik_1)^2}{ik_3(k_2 + ik_1)\sqrt{(1-v^2)}}\sigma,\\
&\gamma = \frac{k_2 - ik_1}{ik_3(k_2 + ik_1)}\left[\phi_0 + \frac{v\sigma}{\sqrt{(1-v^2)}}\right],\\
&\pi = \frac{\sqrt{(1-v^2)}}{k_2 - ik_1}\left[\phi_0^* + \frac{v\sigma^*}{\sqrt{(1-v^2)}}\right],
\end{aligned} \tag{182}$$

and

$$\begin{aligned}
&\Psi_0\sqrt{2} = -ik_3\sigma,\\
&\Psi_1\sqrt{2} = -\frac{k_2 - ik_1}{\sqrt{(1-v^2)}}\sigma,\\
&\Psi_2\sqrt{2} = -\frac{(k_2 - ik_1)^2}{ik_3(1-v^2)}\sigma,\\
&\Psi_3\sqrt{2} = +\frac{(k_2 - ik_1)^3}{k_3^2(1-v^2)^{\frac{3}{2}}}\sigma,\\
&\Psi_4\sqrt{2} = -i\frac{(k_2 - ik_1)^4}{k_3^3(1-v^2)^2}\sigma,\\
&\phi_1 = \frac{k_2 - ik_1}{ik_3\sqrt{(1-v^2)}}\phi_0,\\
&\phi_2^{(1)} = -(k_2 - ik_1)\left[\frac{k_2 - ik_1}{k_3^2(1-v^2)}\phi_0 + i\frac{2\sigma}{k_3(k_2 + ik_1)}\right].
\end{aligned} \tag{183}$$

To complete the perturbation analysis, we must consider the linearized commutation relations and the remaining Ricci identities (not included in equations (164)). The analysis is straightforward and it provides the solution for the matrix $\boldsymbol{A}$, defining the perturbations in the basis vectors, and for the first-order change, $\mu^{(1)}$, in the remaining spin-coefficient μ. The solutions that we obtain are given in the Appendix.

It is manifest from the solutions listed in equations (182) and (183), that all the quantities (with the exception of κ which is zero) diverge as $v \to 1-0$. However, it can be shown that by a tetrad rotation of type II (with parameter $b = -ik_3\sqrt{(1-v^2)}/(k_2-ik_3)$) we can make all the Weyl and Maxwell scalars, *except* Ψ_4 and ϕ_2, vanish; and the divergences in Ψ_4 and ϕ_2 cannot be removed by any sequence of tetrad rotations so long as the constant C_2 in the solution for σ (equation 181) is different from zero. We conclude that there exists no u-dependent mode which is smooth on the line $v = 1$ $(u < 0)$. It would appear, then, that no superposition of the modes can represent realizable first-order perturbations in Region II. This fact, together with the neutrality of the space-time to u-independent perturbations, raises some serious problems of interpretation. We consider them (albeit tentatively) in the following concluding section.

11. Concluding remarks

The present study was undertaken with the intent of deepening our understanding of the development of horizons in the context of the collision of plane-fronted gravitational and electromagnetic waves; and of the result established in an earlier paper (Chandrasekhar & Xanthopoulos 1987*b*) that the horizons are destroyed and replaced by space-like curvature-singularities by the merest presence of other matter such as null dust or $(\epsilon = p)$-fluid. The Bell–Szekeres space-time provides the simplest example for such a study.

In carrying out a complete perturbation analysis of the Bell–Szekeres space-time, both in the region where the scattering of the waves occurs (Region I) and in the regions prior to the instant of collision (Regions II and III), we had two objectives: *first*, to find out whether a complete set of normal modes bounded in the entire Region I exists; and, *second*, to determine how these modes (or, superpositions of them) can be excited by additional incoming radiation (considered as a perturbation) accompanying the original waves forming the background space-time in Regions II and III.

The solution to the first problem is obtained in §§2–9: the entire set of equations governing the first-order perturbations in Region I is solved in full generality and a complete set of bounded normal modes, expressed in terms of the familiar spin-weighted spherical harmonics, is found. The principal novelty in the solution is the development, it predicts, of waves of ever-increasing frequency, in the x^2-direction, as the horizon at $u^2+v^2 = 1$ is approached. This high-frequency behaviour does not affect the boundedness of the solutions representing the normal modes in the entire region, $\{0 \leqslant u < 1, 0 \leqslant u^2+v^2 \leqslant 1, 0 \leqslant v < 1\}$. In particular, they are finite and smooth along the null boundaries $(u = 0, 0 \leqslant v < 1)$ and $(v = 0, 0 \leqslant u < 1)$ separating Region I from Regions II and III. Thus the solutions for the Maxwell scalar on $u = 0$, to write them out fully, are

$$\tilde{\phi}_0^{(1)} = \tilde{\phi}_2^{(1)} = -\frac{2i\sqrt{2}}{l+1} C_{l,m,k_2} [\mathscr{L}_1 S_{+1}(l, m; \theta)][\mathscr{D}_1^\dagger S_{-1}(l+1, ik_2; \theta)]\, e^{i(mx^1+k_2x^2)},$$

and

$$\tilde{\phi}_1^{(1)} = 0, \tag{184}$$

where

$$\cos(\pi-\theta) = v. \tag{185}$$

Similar expressions for the Weyl scalars can be written down in accordance with equations (140) and (143). A general wave packet of gravitational and electromagnetic radiation will be represented by a superposition of these solutions.

With respect to the second problem, our prior expectation was not confirmed. Our expectation was that a perturbation analysis of the space-time in Regions II and III, similar to the one carried out for Region I, will provide normal modes, at least for Ψ_4 and $\phi_2^{(1)}$, which can be joined continuously with suitable superpositions of the modes in Region I, along the null boundary, $(u = 0, 0 \leqslant v < 1)$. This does not appear possible. For, as we have seen, Region II does not allow any non-trivial u-independent mode. (To avoid repetition, in the discussion which follows, we shall restrict our remarks to Region II and the null boundaries, $u = 0, 0 \leqslant v < 1$ and $u < 0, v = 1$. Entirely analogous remarks will apply to Region III with appropriate interchange of the Weyl and the Maxwell scalars: e.g. Ψ_0 in place of Ψ_4 and $\phi_0^{(1)}$ in place of $\phi_2^{(1)}$.) And, u-dependent modes diverge along the *entire* line, $(u \leqslant 0, v = 1)$. Thus the normal modes for $\phi_2^{(1)}$ and Ψ_4 are:

$$\phi_2^{(1)} = -(k_2 - \mathrm{i}k_1)\exp\left\{-\frac{\mathrm{i}(k_1^2+k_2^2)}{k_3}\cot\theta + \mathrm{i}(k_1 x^1 + k_2 x^2 + k_3 u)\right\}$$
$$\times\left\{\frac{k_2-\mathrm{i}k_1}{k_3^2\sin^2\theta}(C_{1;k_1,k_2,k_3}\cot\theta + C_{2;k_1,k_2,k_3}) - \frac{2\mathrm{i}}{k_3(k_2+\mathrm{i}k_1)}(C_{2;k_1,k_2,k_3}\cot\theta - C_{1;k_1,k_2,k_3})\right\}, \tag{186}$$

and

$$\Psi_4\sqrt{2} = -\frac{\mathrm{i}(k_2-\mathrm{i}k_1)^4}{k_3^3\sin^4\theta}\exp\left\{-\frac{\mathrm{i}(k_1^2+k_2^2)}{k_3}\cot\theta + \mathrm{i}(k_1 x^1 + k_2 x^2 + k_3 u)\right\}$$
$$\times(-C_{2;k_1,k_2,k_3}\cot\theta + C_{1;k_1,k_2,k_3}), \tag{187}$$

where it may be recalled that (cf. equation (185))

$$\sin\theta = \sqrt{(1-v^2)}. \tag{188}$$

It is difficult to see how *any* superposition of these modes, *all* of which are highly divergent at $\theta = 0$ $(v = 1)$, can be matched continuously with a superposition of the modes $\tilde{\phi}_2^{(1)}$ and $\tilde{\Psi}_4$ given by equations (184), (140) and (143). It is also not clear why we are not able to assign on the null boundary, $u = 0$ and $0 \leqslant v < 1$, smooth initial data compatible with the bounded modes derived for Region I. This impasse to which we seem to be led raises several questions.

1. With respect to the analysis of the perturbations in Region I, one may ask whether one is justified in restricting oneself to the spin-weighted spherical-harmonics in the solution of equations (45) and (46). Or, should one also include in the solution 'associated' spin-weighted functions derived from the associated Legendre functions Q_l^m and $Q_l^{\mathrm{i}k_2}$ (in which case we should have logarithmic singularities for $\mu = \pm 1$ and for $\eta = 1$); or, more extremely, allow λ and k_1 in equation (47) to be non-integral? If the answers to the latter two questions are in

the affirmative, then, the 'instability' of Region I would have been established by definition! The principal reason for restricting ourselves to the bounded normal modes is that *any* wave packet of gravitational and electromagnetic radiation, with compact support, can be expanded in terms of these modes; and this, it would appear, is sufficient reason for the procedure adopted.

2. With respect to the neutrality of the space-time in Region II to u-independent perturbations: is this feature special to the Bell–Szekeres space-time? Or, is it an aspect of a more general phenomenon inherent to space-times prior to the instant of collision of plane-fronted waves? (Since this paper was written, one of us (B.C.X.) has shown that *any* diagonal metric appropriate to Regions II and III is neutral to u-independent perturbations both in the frameworks of the Einstein and the Einstein–Maxwell equations. In this respect then the Bell–Szekeres solution is not special.) (We shall presently provide an interpretation of this neutrality of the space-time in Region II to u-independent perturbations.)

3. With respect to the divergence, along the entire line, $(v = 1, u < 0)$, of the u-dependent perturbations: should we consider them as physically unacceptable? Or, is it a manifestation of the extreme 'frailty' of the space-time? (In this context, it should be noted that in the background space-time, in Region II, the Weyl scalar Ψ_0 has an impulsive character which becomes singular when $v = 1$; thus,

$$\Psi_0 = \frac{v}{2\surd(1-v^2)}\delta(u), \qquad (189).)$$

If on the other hand, one considers these divergent u-dependent modes as physically acceptable, then how are we to interpret the impossibility of matching any of them with any superposition of the normal modes of Region I?

In any event, the foregoing unanswered questions do suggest that any formulation of the problem (such as Yurtsever's) of the future time-development of pre-assigned initial values on the null boundary at $u = 0$, without an adequate assessment of the implications for the space-time in Region II of the postulated waves in this region, should be viewed with scepticism.

Finally, the following inference, from the neutrality of the space-time in Region II to u-independent perturbations is suggestive. We have shown that, to the first order, u-independent perturbations effect only a coordinate transformation on the background space-time. On this account, it is clear that after the necessary coordinate-transformation, the equations governing the *second* order u-independent perturbations, will be the same as those governing the first-order perturbations. Hence the space-time is neutral to u-independent perturbations to the second order. By induction, we may conclude that *the space-time is neutral to u-independent perturbations to all orders.*

We are indebted to Professor Roger Penrose, F.R.S., for the initial suggestion that in the perturbation analysis of the space-time in Region II, we include u-dependent perturbations. The research reported in this paper has, in part, been supported by grants from the National Science Foundation under Grant PHY-84-16691 with the University of Chicago. We are also grateful for a grant from the

Division of Physical Sciences of the University of Chicago which made the present collaboration possible.

References

Abramowitz, M. & Stegun, I. A. 1964 *Handbook of mathematical functions*. Washington, D.C.: National Bureau of Standards.
Bell, P. & Szekeres, P. 1974 *Gen. Rel. Grav.* **5**, 275–286.
Chandrasekhar, S. 1983 *The mathematical theory of black holes*. Oxford: Clarendon Press.
Chandrasekhar, S. & Hartle, J. B. 1982 *Proc. R. Soc. Lond.* A **384**, 301–315.
Chandrasekhar, S. & Xanthopoulos, B. C. 1985 *Proc. R. Soc. Lond.* A **398**, 223–259 (Paper I).
Chandrasekhar, S. & Xanthopoulos, B. C. 1986 *Proc. R. Soc. Lond.* A **408**, 175–208.
Chandrasekhar, S. & Xanthopoulos, B. C. 1987*a* *Proc. R. Soc. Lond.* A **410**, 311–336.
Chandrasekhar, S. & Xanthopoulos, B. C. 1987*b* *Proc. R. Soc. Lond.* A **414**, 1–30.
Ferrari, V. & Ibañez, J. 1987 *Gen. Rel. Grav.* **19**, 405–425.
Ferrari, V. & Ibañez, J. 1988 *Proc. R. Soc. Lond.* A **417**, 417–431.
Hobson, E. W. 1931 *Spherical and ellipsoidal harmonics*. Cambridge University Press.
Yurtsever, U. 1987 *Phys. Rev.* D **36**, 1662–1672.

Appendix. The completion of the solution for the u-dependent perturbations of Region II

In this Appendix, we complete the analysis of §10(d) by providing the solution of the linearized commutation relations for the elements of the matrix $\boldsymbol{A}$ (defining the u-dependent perturbations of the basis vectors) and the change $\mu^{(1)}$ in the spin-coefficient μ. By choosing a coordinate gauge in which

$$A_1^3 = A_1^4 = A_2^3 = A_2^4 = A_4^3 = A_3^4 = 0, \tag{A 1}$$

we find:

$$A_1^1 = \surd 2\frac{(1-v^2)^{\frac{3}{2}}}{(k_2+\mathrm{i}k_1)^2}\frac{\mathrm{d}\sigma}{\mathrm{d}v}, \qquad A_2^1 = A_3^2 = A_4^2 = 0,$$

$$A_3^1 = +\surd 2\frac{\surd(1-v^2)}{k_2-\mathrm{i}k_1}\sigma^*, \qquad A_4^1 = -\surd 2\frac{\surd(1-v^2)}{k_2+\mathrm{i}k_1}\sigma,$$

$$A_1^2 = -\surd 2\frac{(k_2-\mathrm{i}k_1)^2}{\mathrm{i}k_3^3(1-v^2)}\sigma, \qquad A_2^2 = +\surd 2\frac{(k_2-\mathrm{i}k_1)}{\mathrm{i}k_3(k_2+\mathrm{i}k_1)}\sigma,$$

$$A_3^3 = -\surd 2\frac{(k_2-\mathrm{i}k_1)}{\mathrm{i}k_3(k_2+\mathrm{i}k_1)}\sigma, \qquad A_4^4 = -\surd 2\frac{(k_2+\mathrm{i}k_1)}{\mathrm{i}k_3(k_2-\mathrm{i}k_1)}\sigma^*, \tag{A 2}$$

and

$$\mu^{(1)} = \mu^{*(1)} = -\frac{(k_2-\mathrm{i}k_1)}{\mathrm{i}k_3(k_2+\mathrm{i}k_1)}\frac{1}{\surd(1-v^2)}\frac{\mathrm{d}}{\mathrm{d}v}[(1-v^2)\,\sigma]. \tag{A 3}$$

27

On Weyl's solution for space-times with two commuting Killing fields

By Subrahmanyan Chandrasekhar, F.R.S.
The University of Chicago, Chicago, Illinois 60637, *U.S.A.*

(*Received* 7 *August* 1987)

The solution of the metric coefficients for space-times with diagonal metrics and two commuting Killing fields can be reduced to a Laplace or a wave equation in two variables and to a further pair of integrable differential equations. This reduction can be achieved in a variety of ways. The choice of a coordinate frame and the selection of the combination of metric functions that satisfies the Laplace or the wave equation depend on the physical problem that is considered. The resolution of the issues that arise is illustrated in the contexts of three physical problems; and the solution of the remaining pair of equations, of most frequent occurrence in these contexts, is obtained in explicit form.

1. Introduction

One year after Schwarzschild (1916) obtained his exact solution of Einstein's vacuum equation for spherically symmetric space-times, Weyl (1917) showed how one can, in principle, obtain the solution for the more general case of static axisymmetric space-times. Weyl's method is applicable, even more generally, to space-times with diagonal metrics and two non-null commuting Killing fields without the restriction that one of them is time-like. In essence, Weyl's method consists in showing how in these cases, the problem of solving for the metric functions can be reduced to the solution of a Laplace or a wave equation in two variables and to a further pair of integrable differential equations. However, when one seeks to apply Weyl's method for the solution of particular physical problems, one confronts three distinct problems: first, the choice of an appropriate set of coordinates in which the required physical aspects of the solutions will be plainly manifested; second, the selection of one from several combinations of metric functions (all of which satisfy the same Laplace or wave equation) which will yield solutions with the desired physical properties; and third, an algorism for solving the remaining pair of equations that will complete the solution of the problem. These problems, particularly the last one, do not seem to have been seriously addressed in the extant literature. In this paper, we shall first illustrate how in three concrete contexts the first two problems have been resolved; then show how the solution of the remaining pair of equations, of most frequent occurrence, can be obtained explicitly.

2. Three examples

Problems which require a particular choice of coordinates and combinations of metric functions which satisfy a Laplace or a wave equation (in the context of space-times with two commuting Killing fields) arise when one seeks to generalize certain known simple solutions for encompassing a wider admissible range of physical possibilities. We shall illustrate by three examples.

(*a*) *The generalization of the Schwarzschild solution*

We know that compatible with the requirement of asymptotic flatness, Schwarzschild's solution is the unique solution of Einstein's vacuum equation that represents a static black hole. Schwarzschild's solution is spherically symmetric. If we now wish to generalize Schwarschild's solution to encompass static black holes distorted by external distributions of mass, we require a family of axisymmetric solutions with event horizons, which will manifestly include Schwarzschild's solution in the limit of spherical symmetry. The manner in which such a family of solutions can be obtained is described in Chandrasekhar (1983, §112 (this book will be referred to hereafter as *M.T.*); see also Xanthopoulos 1983). The ideas underlying the method of solution described are relevant to our present purposes.

To construct static axisymmetric solutions of Einstein's vacuum equation, we start with a metric of the form

$$\mathrm{d}s^2 = \mathrm{e}^{2\nu}(\mathrm{d}t)^2 - \mathrm{e}^{2\psi}(\mathrm{d}x^1)^2 - \mathrm{e}^{2\mu_2}(\mathrm{d}x^2)^2 - \mathrm{e}^{2\mu_3}(\mathrm{d}x^3)^2, \tag{1}$$

where ν, ψ, μ_2, and μ_3 are functions of x^2 (a radial coordinate) and x^3 (an angular coordinate). As explained in detail in *M.T.* (§53, p. 278; see also Chandrasekhar 1978) when we seek solutions representing black holes, we can, by making use of the available freedom of gauge, choose a coordinate system, η and μ, in which

$$\mathrm{e}^{\mu_2-\mu_3} = \Delta(\eta) \quad \text{and} \quad \mathrm{e}^{\beta} = \mathrm{e}^{\psi+\nu} = \surd(\Delta\delta), \tag{2}$$

where

$$\Delta = \eta^2 - 1, \quad \delta = 1 - \mu^2 = \sin^2\theta, \tag{3}$$

$\eta\,(=r-1)$ is a radial coordinate and θ is the polar angle. With this choice of coordinates, the metric takes the form

$$\mathrm{d}s^2 = \surd(\Delta\delta)\left[\chi(\mathrm{d}t)^2 - \frac{1}{\chi}(\mathrm{d}\varphi)^2\right] - \mathrm{e}^{\mu_2+\mu_3}\surd\Delta\left[\frac{(\mathrm{d}\eta^2)}{\Delta} + \frac{(\mathrm{d}\mu)^2}{\delta}\right], \tag{4}$$

where

$$\chi = \mathrm{e}^{-\psi+\nu}. \tag{5}$$

It should be noted that in writing the metric in the form (4), we have already arranged for the occurrence, at $\eta = 1$, of a null surface that will eventually be identified with the event horizon of the black hole.

The metric function χ satisfies the equation (*M.T.*, equation (153), p. 584)

$$[(\eta^2-1)(\ln\chi)_{,\eta}]_{,\eta} + [(1-\mu^2)(\ln\chi)_{,\mu}]_{,\mu} = 0, \tag{6}$$

and $\mu_2+\mu_3$ is determined in terms of χ by equations which will be considered in §3.

It will be tempting at this point to assume that we can obtain the entire family of solutions describing static black holes by making immediate use of the general separable solution for $\ln \chi$, namely,

$$\ln \chi = \sum_n A_n P_n(\mu) \,[\text{or } Q_n(\mu)]\, P_n(\eta) \,[\text{or } Q_n(\eta)], \tag{7}$$

where P_n and Q_n denote the Legendre functions of the first and the second kinds. This procedure (or the somewhat different but equivalent procedure), does not yield Schwarzschild's solution in its standard form in the appropriate limit: one obtains, instead, a solution in which the curvature singularity occurs along a finite segment of the z-axis (centred at the origin) with a constant mass per unit length (cf. Wald 1984, p. 168; also Geroch & Hartle 1982).

We consider instead (see *M.T.* p. 584, equations (155)–(158))

$$\tilde{\Psi} = \chi \sqrt{(\Delta\delta)}, \tag{8}$$

which satisfies the same Laplace's equation,

$$[(\eta^2-1)(\ln \tilde{\Psi})_{,\eta}]_{,\eta} + [(1-\mu^2)(\ln \tilde{\Psi})_{,\mu}]_{,\mu} = 0. \tag{9}$$

Thus, the θ-independent solution,

$$\tilde{\Psi} = \frac{\eta-1}{\eta+1}, \tag{10}$$

and the associated solutions,

$$\chi = \left[\frac{\eta-1}{(\eta+1)^3(1-\mu^2)}\right]^{\frac{1}{2}} \quad \text{and} \quad e^{\mu_2+\mu_3} = \frac{(\eta+1)^2}{\sqrt{\Delta}}, \tag{11}$$

leads directly to the metric

$$ds^2 = \frac{\eta-1}{\eta+1}(dt)^2 - (\eta+1)^2(1-\mu^2)(d\varphi)^2 - (\eta+1)^2\left[\frac{(d\eta)^2}{\Delta} + \frac{(d\mu)^2}{\delta}\right], \tag{12}$$

which, since the luminosity radius r (in units of M) $= \eta+1$ (cf. *M.T.* p. 584, equation (151)) *is* Schwarzschild's solution in its standard form. (It should be noted that Δ vanishes for $\eta = \pm 1$, i.e. at the horizon and at the singularity; and that, more generally, Δ vanishes at the event and the Cauchy horizons when the Kerr or the Kerr–Newman metric is considered when $\eta = (r-M)\sqrt{(M^2-a^2)}$ or $\eta = (r-M)\sqrt{(M^2-a^2-Q_*^2)}$.)

From the fact that the spherically symmetric solution (10) of equation (9) leads to the Schwarzschild solution, we may conclude that the required family of solutions describing static distorted black holes will follow from the solution for $\tilde{\Psi}$ given by

$$\ln \tilde{\Psi} = \ln \frac{\eta-1}{\eta+1} + \Pi, \tag{13}$$

where Π is a suitably chosen solution of the equation

$$[(\eta^2-1)\Pi_{,\eta}]_{,\eta} + [(1-\mu^2)\Pi_{,\mu}]_{,\mu} = 0. \tag{14}$$

A further examination shows (cf. *M.T.* p. 587) that the required smoothness and flatness of the metric on the axis, $\theta = 0$, restricts the solution for Π to be of the form

$$\Pi = \sum_{n=1}^{\infty} A_n P_n(\mu) P_n(\eta), \tag{15}$$

where the coefficients A_n satisfy the condition

$$\sum_{n=1}^{\infty} A_{2n+1} P_{2n+1}(1) = 0. \tag{16}$$

The metric then reduces to the form (cf. *M.T.* p. 588, equation (182))

$$ds^2 = \frac{\eta-1}{\eta+1} e^{+\Pi}(dt)^2 - (\eta+1)^2(1-\mu^2) e^{-\Pi}(d\varphi)^2 - e^{-\Pi+f}(\eta+1)^2\left[\frac{(d\eta)^2}{\Delta}+\frac{(d\mu)^2}{\delta}\right], \tag{17}$$

where the equations governing f and their solution are postponed to §3.

(b) The generalization of the Khan–Penrose solution

When we turn to colliding plane gravitational-waves, we are concerned with space-times with two space-like commuting Killing-vectors, ∂_{x^1} and ∂_{x^2}. In this case, we choose a coordinate system in which the metric takes the form (cf. equation (5))

$$ds^2 = e^{\nu+\mu_3}\sqrt{\Delta}\left[\frac{(d\eta)^2}{\Delta}-\frac{(d\mu)^2}{\delta}\right] - \sqrt{(\Delta\delta)}\left[\chi(dx^2)^2+\frac{1}{\chi}(dx^1)^2\right], \tag{18}$$

where, now

$$\Delta = 1-\eta^2, \quad \delta = 1-\mu^2, \tag{19}$$

η measures time (in a suitable unit) from the instant of collision, μ measures the distance (also in a suitable unit), normal to the colliding fronts at the instant of collision and χ and $\nu+\mu_3$ are functions of η and μ.

In writing the metric in the form (18), we have allowed, *aposteriori*, for the fact that as a result of the collision a curvature or a coordinate singularity will develop when $\eta = 1$ or $\mu = \pm 1$.

We find that $\ln\chi$ satisfies the same equation (6), now a wave equation. In this instance, the Khan–Penrose solution follows from the solution (cf. Chandrasekhar & Ferrari 1984, equation (57) for the case $q_2 = 0$)

$$\chi = \frac{1+\eta}{1-\eta}. \tag{20}$$

Together with the associated solution (Chandrasekhar & Ferrari 1984, equation (61))

$$e^{\nu+\mu_3} = \sqrt[4]{(\Delta/\delta)}, \tag{21}$$

we are led to the metric,

$$ds^2 = \frac{\Delta^{\frac{3}{4}}}{\delta^{\frac{1}{4}}}\left[\frac{(d\eta)^2}{\Delta}-\frac{(d\mu)^2}{\delta}\right] - \sqrt{(\Delta\delta)}\left[\frac{1+\eta}{1-\eta}(dx^2)^2+\frac{1-\eta}{1+\eta}(dx^1)^2\right]. \tag{22}$$

The generalization of the metric (22) follows from the solution

$$\ln \chi = \ln \frac{1+\eta}{1-\eta} + \Pi, \tag{23}$$

where Π satisfies the same equation (14). The metric which follows has the form,

$$ds^2 = e^f \frac{\Delta^{\frac{3}{4}}}{\delta^{\frac{1}{4}}}\left[\frac{(d\eta)^2}{\Delta} - \frac{(d\mu)^2}{\delta}\right] - \sqrt{(\Delta\delta)}\left[e^{+\Pi}\frac{1+\eta}{1-\eta}(dx^2)^2 + e^{-\Pi}\frac{1-\eta}{1+\eta}(dx^1)^2\right], \tag{24}$$

where f is determined in terms of Π by equations that are considered in §3. The metric (24), which describes the space-time after the instant of collision, must be extendible in the sense described in Chandrasekhar & Xanthopoulos (1986, §3). It can be shown that the metric (24) is extendible when Π has the form (B. Xanthopoulos, personal communication)

$$\Pi = \sum A_n P_n(\mu) P_n(\eta), \tag{25}$$

where A_ns are constants. However, Ferrari & Ibanez (1987) have shown that one can obtain extendible solutions also when Π includes terms of the form $A_n P_n(\mu) Q_n(\eta)$ provided the coefficients A_n are suitably chosen. Thus, they obtain a solution when Π is given by

$$\Pi = -2Q_2(\eta) P_2(\mu). \tag{26}$$

(c) The generalization of the Bell–Szekeres solution

The Bell–Szekeres solution describes the collision of plane-shock electromagnetic waves, supported by impulsive gravitational waves in a conformally flat space-time. It arises when (cf. Chandrasekhar & Xanthopoulos 1987; equations (116)–(118))

$$\Psi = \frac{\sqrt{(\Delta\delta)}}{\chi} = 1 - H^2, \tag{27}$$

where H is a potential in terms of which the electromagnetic field is derived. Letting

$$F = \tfrac{1}{2}\ln\frac{1+H}{1-H}, \tag{28}$$

we find that F satisfies the same equation (14).

When

$$H = \eta, \quad \Psi = \Delta \quad \text{and} \quad \chi = \sqrt{(\delta/\Delta)}, \tag{29}$$

the solution for $\nu + \mu_3$ is given by

$$e^{\nu+\mu_3} = 1/\sqrt{\Delta}; \tag{30}$$

and we obtain the Bell–Szekeres solution

$$ds^2 = \left[\frac{(d\eta)^2}{\Delta} - \frac{(d\mu)^2}{\delta}\right] - \Delta(dx^1)^2 - \delta(dx^2)^2. \tag{31}$$

The generalization of the Bell–Szekeres solution follows from the solution (Chandrasekhar & Xanthopoulos 1987, equation (132))

$$F = \tfrac{1}{2}\ln\frac{1+\eta}{1-\eta}+\Pi, \tag{32}$$

where

$$\Pi = \sum A_n P_n(\mu) P_n(\eta). \tag{33}$$

We thus obtain (Chandrasekhar & Xanthopoulos, 1987, equations (137) and (140))

$$\mathrm{d}s^2 = \mathrm{e}^f\Sigma^2\left[\frac{(\mathrm{d}\eta)^2}{\Delta}-\frac{(\mathrm{d}\mu)^2}{\delta}\right]-\frac{\Delta}{\Sigma^2}(\mathrm{d}x^1)^2-\Sigma^2\delta(\mathrm{d}x^2)^2, \tag{34}$$

where

$$\Sigma = \cosh\Pi+\eta\sinh\Pi. \tag{35}$$

Again, we postpone to §3 the equations governing f and their solution.

3. The completion of the solution

In all three cases considered in §2 (and in the case considered by Xanthopoulos 1983) the metric function f was left unspecified. We now turn to the determination of f.

The equations governing f, in all four cases, are

$$\frac{\mu}{\delta}\Delta f_{,\eta}+\eta f_{,\mu} = -4\Pi_{,\mu}-4\Delta\Pi_{,\eta}\Pi_{,\mu} \tag{36}$$

and

$$\eta f_{,\eta}+\mu f_{,\mu} = -4\Pi_{,\eta}-2[\Delta(\Pi_{,\eta})^2+\delta(\Pi_{,\mu})^2], \tag{37}$$

where Π is a suitably chosen solution of equation (14) and expressible as sums of products of the Legendre functions P_n and/or Q_n.

By operating equations (36) and (37) by ∂_η and by $-\partial_\mu$, respectively, and adding, we find that the resulting equation reduces to

$$(\Delta f_{,\eta})_{,\eta}-\delta f_{,\mu,\mu} = 4\delta(\Pi_{,\mu})^2. \tag{38}$$

By adding to this equation twice equation (37), we obtain

$$\Delta f_{,\eta,\eta}-(\delta f_{,\mu})_{,\mu} = -8\Pi_{,\eta}-4\Delta(\Pi_{,\eta})^2. \tag{39}$$

In view of the linearity in f of equations (38) and (39), we can divide our consideration of these equations to the two independent pairs

$$\left.\begin{aligned}\Delta f_{,\eta,\eta}-(\delta f_{,\mu})_{,\mu} &= -8\Pi_{,\eta},\\ \delta f_{,\mu,\mu}-(\Delta f_{,\eta})_{,\eta} &= 0;\end{aligned}\right\} \tag{40}$$

and

$$\left.\begin{aligned}\Delta f_{,\eta,\eta}-(\delta f_{,\mu})_{,\mu} &= -4\Delta(\Pi_{,\eta})^2,\\ \delta f_{,\mu,\mu}-(\Delta f_{,\eta})_{,\eta} &= -4\delta(\Pi_{,\mu})^2.\end{aligned}\right\} \tag{41}$$

In considering the pairs of equations (40) and (41), we shall, in the first instance, restrict ourselves to the solution,

$$\Pi = \sum_{n=1}^{\infty} A_n P_n(\mu) P_n(\eta), \tag{42}$$

applicable to all three cases considered in §2 (cf. equations (15), (25) and (33)) and to the case considered by Xanthopoulos (1983, equations (4.11) and (4.21)). The solution for the case when Π includes terms in the Legendre functions, Q_n, of the second kind can be obtained in analogous fashion. For Π given by equation (42),

$$\Pi_{,\eta} = \sum_n A_n P'_n(\eta) P_n(\mu), \quad \Pi_{,\mu} = \sum_n A_n P'_n(\mu) P_n(\eta); \tag{43}$$

and

$$\left.\begin{aligned}(\Pi_{,\eta})^2 &= \sum_n A_n^2 [P'_n(\eta)]^2 [P_n(\mu)]^2 + 2 \sum_{n>m} A_n A_m P'_n(\eta) P'_m(\eta) P_n(\mu) P_m(\mu),\\ (\Pi_{,\mu})^2 &= \sum_n A_n^2 [P'_n(\mu)]^2 [P_n(\eta)]^2 + 2 \sum_{n>m} A_n A_m P'_n(\mu) P'_m(\mu) P_n(\eta) P_m(\eta).\end{aligned}\right\} \tag{44}$$

Again, in view of the linearity of the equations (40) and (41), it will suffice to consider the pairs,

$$\left.\begin{aligned}\Delta f_{,\eta,\eta} - (\delta f_{,\mu})_{,\mu} &= P_n(\mu) P'_n(\eta),\\ \delta f_{,\mu,\mu} - (\Delta f_{,\eta})_{,\eta} &= 0;\end{aligned}\right\} \tag{45}$$

and

$$\left.\begin{aligned}\Delta f_{,\eta,\eta} - (\delta f_{,\mu})_{,\mu} &= \Delta P'_n(\eta) P'_m(\eta) P_n(\mu) P_m(\mu),\\ \delta f_{,\mu,\mu} - (\Delta f_{,\eta})_{,\eta} &= \delta P'_n(\mu) P'_m(\mu) P_n(\eta) P_m(\eta),\end{aligned}\right\} \tag{46}$$

where, in equations (45) and (46), the factors,

$$-8A_n \quad \text{and} \quad -8A_n A_m, \tag{47}$$

have been suppressed.

(a) The solution of the pair of equations (45)

By making use of the standard recurrence relations among the Legendre functions, we can readily verify that

$$f^{(n)} = \tfrac{1}{2} P_n(\mu) P_{n-1}(\eta) \tag{48}$$

is a solution of the equation

$$\Delta f_{,\eta,\eta} - (\delta f_{,\mu})_{,\mu} = P_n(\mu) P'_n(\eta). \tag{49}$$

By writing,

$$f = f^{(n)} + f^{(n-1)}, \tag{50}$$

we find that equations (45) reduce to the pair

$$\left.\begin{aligned}\Delta f^{(n-1)}{}_{,\eta,\eta} - (\delta f^{(n-1)}{}_{,\mu})_{,\mu} &= 0,\\ \delta f^{(n-1)}{}_{,\mu,\mu} - (\Delta f^{(n-1)}{}_{,\eta})_{,\eta} &= -P_{n-1}(\eta) P'_{n-1}(\mu),\end{aligned}\right\} \tag{51}$$

i.e. apart from sign, the same pair (45) with η and μ interchanged and $n-1$ replacing n. Therefore, by writing

$$f = \tfrac{1}{2}[P_n(\mu) P_{n-1}(\eta) - P_{n-1}(\eta) P_{n-2}(\mu)] + f^{(n-2)}, \tag{52}$$

we shall obtain equations (45) with $n-2$ replacing n:

$$\left.\begin{aligned}\Delta f^{(n-2)}{}_{,\eta,\eta} - (\delta f^{(n-2)}{}_{,\mu})_{,\mu} &= P_{n-2}(\mu) P'_{n-2}(\eta),\\ \delta f^{(n-2)}{}_{,\mu,\mu} - (\Delta f^{(n-2)}{}_{,\eta})_{,\eta} &= 0.\end{aligned}\right\} \tag{53}$$

By induction, we conclude that the required solution of equations (45) is given by

$$f = \tfrac{1}{2}[P_n(\mu)P_{n-1}(\eta) - P_{n-1}(\eta)P_{n-2}(\mu) + P_{n-2}(\mu)P_{n-3}(\eta) - \cdots + P_1(\mu)P_0(\eta)] \quad (\text{if } n \text{ is odd}), \tag{54'}$$

and $$f = \tfrac{1}{2}[P_n(\mu)P_{n-1}(\eta) - P_{n-1}(\eta)P_{n-2}(\mu) + P_{n-2}(\mu)P_{n-3}(\eta) - \cdots - P_1(\eta)P_0(\mu)] \quad (\text{if } n \text{ is even}), \tag{54''}$$

where it may be recalled that a factor $-8A_n$ has been suppressed.

(b) The solution of the pair of equations (46)

In considering equations (46), it is convenient to introduce the functions,

$$Y_{nm}(\eta) = P_n(\eta)P_m(\eta) \quad \text{and} \quad Z_{nm}(\eta) = \Delta P'_n(\eta)P'_m(\eta) \tag{55}$$

first defined by F. E. Neumann (1878 p. 94, equations (1)–(6); see also Hobson 1931, §52, p. 83). They satisfy the relations,

$$[\delta Y_{nm}(\mu)_{,\mu}]_{,\mu} = -(N+M)Y_{nm}(\mu) + 2Z_{nm}(\mu) \tag{56}$$

and $$[\delta Z_{nm}(\mu)]_{,\mu,\mu} = 2NMY_{nm}(\mu) - (N+M)Z_{nm}(\mu), \tag{57}$$

where $$N = n(n+1), \quad M = m(m+1). \tag{58}$$

In terms of Y_{nm} and Z_{nm}, equations (46) take the forms

$$F = \Delta f_{,\eta,\eta} - (\delta f_{,\mu})_{,\mu} = Y_{nm}(\mu)Z_{nm}(\eta) \tag{59}$$

and $$G = \delta f_{,\mu,\mu} - (\Delta f_{,\eta})_{,\eta} = Y_{nm}(\eta)Z_{nm}(\mu). \tag{60}$$

Equations (59) and (60) are not linearly independent; for, by making use of the relations (56) and (57) we can readily verify that

$$(\Delta F)_{,\eta,\eta} - (\delta G)_{,\mu,\mu} = (\delta F_{,\mu})_{,\mu} - (\Delta G_{,\eta})_{,\eta} = -(N+M)(F-G); \tag{61}$$

and that therefore,

$$(\Delta F)_{,\eta,\eta} + (\Delta G_{,\eta})_{,\eta} = (\delta G)_{,\mu,\mu} + (\delta F_{,\mu})_{,\mu}; \tag{62}$$

a relation which is identically satisfied by virtue of the definitions of F and G.

In view of the linear dependence of equations (59) and (60) it will suffice to consider their sum, namely

$$2\eta f_{,\eta} + 2\mu f_{,\mu} = Y_{nm}(\mu)Z_{nm}(\eta) + Y_{nm}(\eta)Z_{nm}(\mu). \tag{63}$$

First, we observe that there is no difficulty in principle in obtaining the required solution of equation (63). For, by substituting for Y_{nm} and Z_{nm} the known polynomial expansions for the Legendre functions, we shall obtain an equation of the form

$$\eta f_{,\eta} + \mu f_{,\mu} = \sum_{n,m} C_{nm}(\mu^n\eta^m + \mu^m\eta^n), \tag{64}$$

where C_{nm}s are determinate constants; and the solution of this equation is clearly

$$f = \sum_{n,m} \frac{C_{nm}}{n+m}(\mu^n\eta^m + \mu^m\eta^n) + \text{const.} \tag{65}$$

Example For $n = m = 2$, equation (63) gives

$$2\eta f_{,\eta} + 2\mu f_{,\mu} = 9[P_2^2(\mu)\,\Delta\eta^2 + P_2^2(\eta)\,\delta\mu^2]. \tag{66}$$

Substituting for $P_2(\mu)$ and $P_2(\eta)$ their known expressions and expanding, we obtain

$$\eta f_{,\eta} + \mu f_{,\mu} = \tfrac{9}{8}[-18\eta^4\mu^4 + 15(\mu^2\eta^4 + \mu^4\eta^2) - (\mu^4+\eta^4) + (\mu^2+\eta^2) - 12\eta^2\mu^2]; \tag{67}$$

and the solution of this equation is

$$f = \tfrac{9}{32}[-9\eta^4\mu^4 + 10(\mu^2\eta^4 + \mu^4\eta^2) - (\mu^4+\eta^4) + 2(\mu^2+\eta^2) - 12\eta^2\mu^2] + \text{const.} \tag{68}$$

The foregoing direct method of solving equation (63) is clearly unsatisfactory for large values of n and m. We shall now describe a method of solving for f recursively. The method requires that in place of equation (63), symmetric in n and m, we consider the more general equation

$$2\eta\frac{\partial}{\partial\eta}f_{nm:jk} + 2\mu\frac{\partial}{\partial\mu}f_{nm:jk} = Y_{nm}(\mu)\,Z_{jk}(\eta) + Y_{nm}(\eta)\,Z_{jk}(\mu). \tag{69}$$

With the definitions

$$\Psi_{nm;jk} = Y_{nm}(\mu)\,Z_{jk}(\eta) + Y_{nm}(\eta)\,Z_{jk}(\mu), \tag{70}$$

and

$$\mathscr{D} = \eta\,\partial_\eta + \mu\,\partial_\mu, \tag{71}$$

equation (69) takes the form

$$\mathscr{D}f_{nm;jk} = \tfrac{1}{2}\Psi_{nm;jk}. \tag{72}$$

It should be noted that although $f_{nm;jk}$ and $\Psi_{nm;jk}$ are symmetric in n and m and in j and k, separately, they are not symmetric for the simultaneous interchange of (n, m) and (j, k).

By making use of the standard recurrence formulae for the Legendre functions we find

$$\begin{aligned}
\mu Z_{nm}(\mu)_{,\mu} &= \mu[\delta P'_n(\mu)\,P'_m(\mu)] \\
&= -2\mu^2 P'_n(\mu)\,P'_m(\mu) + \mu\delta[P''_n(\mu)\,P'_m(\mu) + P''_m(\mu)\,P'_n(\mu)] \\
&= -2\mu^2 P'_n(\mu)\,P'_m(\mu) \\
&\quad + \mu[4\mu P'_n(\mu)\,P'_m(\mu) - n(n+1)\,P_n(\mu)\,P'_m(\mu) - m(m+1)\,P_m(\mu)\,P'_n(\mu)] \\
&= [nP_n(\mu) + P'_{n-1}(\mu)]\,[P'_{m-1}(\mu) - m^2P_m(\mu)] \\
&\quad + [mP(\mu) + P'_{m-1}(\mu)]\,[P'_{n-1}(\mu) - n^2P_n(\mu)];
\end{aligned} \tag{73}$$

or, equivalently,

$$\begin{aligned}
\mu Z_{nm}(\mu)_{,\mu} &= -nm(n+m)\,Y_{nm}(\mu) + 2P'_{n-1}(\mu)\,P'_{m-1}(\mu) \\
&\quad - m(m-1)\,P_m(\mu)\,P'_{n-1}(\mu) - n(n-1)\,P_n(\mu)\,P'_{m-1}(\mu).
\end{aligned} \tag{74}$$

Combined with an analogous formula for $\eta Z_{jk}(\eta)_{,\eta}$, we obtain,

$$\begin{aligned}\mathscr{D}[Z_{nm}(\mu)\,Z_{jk}(\eta)] &= \mu Z_{nm}(\mu)_{,\mu}\,Z_{jk}(\eta)+\eta Z_{jk}(\eta)_{,\eta}\,Z_{nm}(\mu)\\ &= -nm(n+m)\,Y_{nm}(\mu)\,Z_{jk}(\eta)-jk(j+k)\,Y_{jk}(\eta)\,Z_{nm}(\mu)\\ &\quad+[2\sum_p\sum_q(2p+1)(2q+1)\,Y_{pq}(\mu)\\ &\quad-n(n-1)\sum_q(2q+1)\,Y_{nq}(\mu)-m(m-1)\sum_p(2p+1)\,Y_{mp}(\mu)]\,Z_{jk}(\eta)\\ &\quad+[2\sum_r\sum_s(2r+1)(2s+1)\,Y_{rs}(\eta)\\ &\quad-j(j-1)\sum_s(2s+1)\,Y_{js}(\eta)-k(k-1)\sum_r(2r+1)\,Y_{kr}(\eta)]\,Z_{nm}(\mu),\end{aligned}\tag{75}$$

where we have substituted for the derivatives of the Legendre functions one of Neumann's formulae (cf. Hobson 1931, p. 34, example 2),

$$\begin{aligned}P'_n &= (2n-1)\,P_{n-1}+(2n-5)\,P_{n-3}+\ldots+7P_3+3P_1, &&\text{(if } n \text{ is even)}\\ &= (2n-1)\,P_{n-1}+(2n-5)\,P_{n-3}+\ldots+5P_2+P_0, &&\text{(if } n \text{ is odd)},\end{aligned}\tag{76}$$

or,

$$P'_n = \sum_{p=n-1,\,n-3,\,\ldots}(2p+1)\,P_p.\tag{77}$$

In equation (75) (and in the sequel) the summations over p, q, r and s are over the integers,

$$\left.\begin{aligned}p &= n-2, n-4, \ldots, & q &= m-2, m-4, \ldots;\\ r &= j-2, j-4, \ldots, \quad\text{and} & s &= k-2, k-4, \ldots.\end{aligned}\right\}\tag{78}$$

Interchanging μ and η in equation (75), adding, and recalling the definition of $\Psi_{nm:jk}$ (equation (74)), we obtain,

$$\begin{aligned}&\mathscr{D}[Z_{nm}(\mu)\,Z_{jk}(\eta)+Z_{nm}(\eta)\,Z_{jk}(\mu)]\\ &\quad= -nm(n+m)\,\Psi_{nm:jk}-jk(j+k)\,\Psi_{jk:nm}\\ &\qquad+2\sum_p\sum_q(2p+1)(2q+1)\,\Psi_{pq:jk}-n(n-1)\sum_q(2q+1)\,\Psi_{nq:jk}\\ &\qquad\qquad-m(m-1)\sum_p(2p+1)\,\Psi_{mp:jk}\\ &\qquad+2\sum_r\sum_s(2r+1)(2s+1)\,\Psi_{rs:nm}-j(j-1)\sum_s(2s+1)\,\Psi_{js:nm}\\ &\qquad\qquad-k(k-1)\sum_r(2r+1)\,\Psi_{kr:nm}.\end{aligned}\tag{79}$$

Because by equation (72)

$$\mathscr{D}^{-1}\Psi_{nm:jk} = 2f_{nm:jk}\quad\text{(modulo constant)},\tag{80}$$

the '*inverse $\mathscr{D}$-transform*' of equation (79) provides the recurrence formula,

$$
\begin{aligned}
&nm(n+m)f_{nm;jk}+jk(j+k)f_{jk;nm}\\
&\quad=-\tfrac{1}{2}[Z_{nm}(\mu)\,Z_{jk}(\eta)+Z_{nm}(\eta)\,Z_{jk}(\mu)]\\
&\qquad+2\sum_p\sum_q(2p+1)(2q+1)f_{pq;jk}-n(n-1)\sum_q(2q+1)f_{nq;jk}\\
&\qquad-m(m-1)\sum_p(2p+1)f_{mp;jk}\\
&\qquad+2\sum_r\sum_s(2r+1)(2s+1)f_{rs;nm}-j(j-1)\sum_s(2s+1)f_{js;nm}\\
&\qquad\qquad-k(k-1)\sum_r(2r+1)f_{rk;nm}\quad(n,m,j\text{ and }k\neq 0).
\end{aligned}
\tag{81}
$$

We may note the following special case of the formula:

$$
\begin{aligned}
nm(n+m)f_{nm;nm}&=-\tfrac{1}{2}Z_{nm}(\mu)\,Z_{nm}(\eta)\\
&\quad-n(n-1)\sum_q(2q+1)f_{nq;nm}-m(m-1)\sum_p(2p+1)f_{mp;nm}\\
&\qquad+2\sum_p\sum_q(2p+1)(2q+1)f_{pq;nm}\quad(n,m\neq 0).
\end{aligned}
\tag{82}
$$

Because $f_{nm;jk}$ is not symmetric in the pair of indices (n,m) and (j,k), the relation (81) will not suffice to determine $f_{nm;jk}$, recursively, in terms of the functions of lower orders. A further relation is needed which we shall now obtain.

Making use of the standard recurrence relations among the Legendre functions, we obtain,

$$
\begin{aligned}
\mu Y_{nm}(\mu)_{,\mu}&=\mu P_m(\mu)\,P'_n(\mu)+\mu P_n\mu\,P'_m(\mu)\\
&=\left[-\frac{1}{m}\delta P'_m(\mu)+P_{m-1}(\mu)\right]P'_n(\mu)+\left[-\frac{1}{n}\delta P'_n(\mu)+P_{n-1}(\mu)\right]P'_m(\mu)\\
&=-\frac{n+m}{nm}Z_{nm}(\mu)+P_{m-1}(\mu)\,P'_n(\mu)+P_{n-1}(\mu)\,P'_m(\mu).
\end{aligned}
\tag{83}
$$

Combined with an analogous formula for $\eta Y_{nm}(\eta)_{,\eta}$ and substituting for $P_n(\mu)$ another of Neumann's formulae

$$
P_n(\mu)=-(1-\mu^2)\sum_{p(\neq 0)=n-1,\,n-3,\ldots}\frac{2p+1}{p(p+1)}P'_p(\mu)+\begin{cases}\mu & (\text{if } n \text{ is odd}),\\ 1 & (\text{if } n \text{ is even}).\end{cases}
\tag{84}
$$

we obtain,

$$
\begin{aligned}
&\mathscr{D}[Y_{nm}(\mu)\,Y_{jk}(\eta)]\\
&\quad=-\frac{n+m}{nm}Y_{jk}(\eta)\,Z_{nm}(\mu)-\frac{j+k}{jk}Y_{nm}(\mu)\,Z_{jk}(\eta)\\
&\qquad-Y_{jk}(\eta)\Big[\sum_p\alpha_p Z_{pm}(\mu)+\sum_q\alpha_q Z_{qn}(\mu)-\delta_n(\mu)\,P'_m(\mu)-\delta_m(\mu)\,P'_n(\mu)\Big]\\
&\qquad-Y_{nm}(\mu)\Big[\sum_r\alpha_r Z_{rk}(\eta)+\sum_s\alpha_s Z_{sj}(\eta)-\delta_j(\eta)\,P'_k(\eta)-\delta_k(\eta)\,P'_j(\eta)\Big],
\end{aligned}
\tag{85}
$$

where
$$\begin{aligned}\delta_n(\eta) &= 1 \quad \text{if} \quad n \text{ is odd}, & \delta_n(\mu) &= 1 \quad \text{if} \quad n \text{ is odd},\\ &= \eta \quad \text{if} \quad n \text{ is even}, & &= \mu \quad \text{if} \quad n \text{ is even}.\end{aligned} \tag{86}$$

$$\alpha_p = \frac{2p+1}{p(p+1)}, \tag{87}$$

and the summations over p, q, r and s are over the same integers specified in (78).

Interchanging μ and η in equation (85), adding the resulting equation to (85), and recalling the definition of $\Psi_{nm:jk}$, we obtain,

$$\begin{aligned}&\mathscr{D}[Y_{nm}(\mu)\, Y_{jk}(\eta) + Y_{nm}(\eta)\, Y_{jk}(\mu)]\\ &\quad = -\frac{n+m}{nm}\Psi_{jk;nm} - \frac{j+k}{jk}\Psi_{nm;jk}\\ &\qquad - \sum_p{}' \alpha_p \Psi_{jk:pm} - \sum_q{}' \alpha_q \Psi_{jk;qn} - \sum_r{}' \alpha_r \Psi_{nm;rk} - \sum_s{}' \alpha_s \Psi_{nm;sj}\\ &\qquad + \Phi_{jk;nm} + \Phi_{jk;mn} + \Phi_{nm;jk} + \Phi_{nm;kj},\end{aligned} \tag{88}$$

where
$$\Phi_{nm;kj} = Y_{nm}(\mu)\, \delta_k(\eta)\, P_j'(\eta) + Y_{nm}(\eta)\, \delta_k(\mu)\, P_j'(\mu), \tag{89}$$

and the prime distinguishing the summation signs indicates that the term in α_0 is to be excluded. It should be noted that while $\Phi_{nm;kj}$ is symmetric in the first pair of indices (n, m) it is not in the second pair (k, j).

We know that the inverse $\mathscr{D}$-transform of $\Psi_{nm;jk}$ is $f_{nm;jk}$. In order then to invert equation (88), we define $\phi_{nm;kj}$ by the equation

$$\mathscr{D}\phi_{nm;kj} = \tfrac{1}{2}\Phi_{nm;kj}, \tag{90}$$

and derive from equation (88) the recurrence formula,

$$\begin{aligned}&\frac{j+k}{jk} f_{nm;jk} + \frac{n+m}{nm} f_{jk;nm}\\ &\quad = -\tfrac{1}{2}[Y_{nm}(\mu)\, Y_{jk}(\eta) + Y_{nm}(\eta)\, Y_{jk}(\mu)]\\ &\qquad - \sum_r{}' \alpha_r f_{nm;rk} - \sum_s{}' \alpha_s f_{nm;sj} - \sum_p{}' \alpha_p f_{jk;pm} - \sum_q{}' \alpha_q f_{jk;qn}\\ &\qquad + \phi_{nm;jk} + \phi_{nm;kj} + \phi_{jk;nm} + \phi_{jk;mn} \quad (n, m, j \text{ and } k \neq 0).\end{aligned} \tag{91}$$

It remains to show how $\phi_{nm;kj}$ can in turn be determined recursively. By making use of the identities,

$$\begin{aligned}\mu Y_{nm,\mu} &= \mu P_n'(\mu)\, P_m(\mu) + \mu P_m'(\mu)\, P_n(\mu)\\ &= P_m(\mu)\, [nP_n(\mu) + P_{n-1}'(\mu)] + P_n(\mu)\, [\mu P_m(\mu) + P_{m-1}'(\mu)]\\ &= (n+m)\, Y_{nm}(\mu) + P_m(\mu) \sum_{p=n-2,\dots} (2p+1)\, P_p(\mu)\\ &\qquad + P_n(\mu) \sum_{q=m-2,\dots} (2q+1)\, P_q(\mu),\end{aligned} \tag{92}$$

$$\eta P_j''(\eta) = \eta \sum_{r=j-1,\dots} (2r+1)\, P_r'(\eta) = (j-1)\, P_j'(\eta) + \sum_{r=j-2,\dots} (2r+1)\, P_r'(\eta), \tag{93}$$

and
$$\eta[\eta P_j'(\eta)]' = j\eta\, P_j'(\eta) + \sum_{r=j-2,\dots} (2r+1)\, \eta P_r'(\eta). \tag{94}$$

we verify that

$$\mathscr{D}[Y_{nm}(\mu)\,\delta_k(\eta)\,P_j'(\eta)] = (n+m+j-\delta_{k,\text{odd}})\,Y_{nm}(\mu)\,\delta_k(\eta)\,P_j'(\eta)$$
$$+[\sum_q (2q+1)\,Y_{qn}(\mu)+\sum_p (2p+1)\,Y_{pm}(\mu)]\,\delta_k(\eta)\,P_j'(\eta)$$
$$+\sum_r (2r+1)\,Y_{nm}(\mu)\,\delta_k(\eta)\,P_r'(\eta), \tag{95}$$

where $$\delta_{k,\text{odd}} = 1 \quad \text{if } k \text{ is odd} \quad \text{and } 0 \text{ if } k \text{ is even.} \tag{96}$$

Interchanging η and μ in equation (95), adding the resulting equation to (95), and inverting, we obtain the formula

$$(n+m+j-\delta_{k,\text{odd}})\,\phi_{nm:kj} = \tfrac{1}{2}\Phi_{nm:kj}-\sum_r (2r+1)\,\phi_{nm:kr}$$
$$-\sum_q (2q+1)\,\phi_{nq:kj}-\sum_p (2p+1)\,\phi_{pm:kj}, \tag{97}$$

which determines $\phi_{nm:kj}$ recursively. Notice that (cf. equation (89))

$$\phi_{nm:k0} \equiv 0 \quad \text{(modulo constant) for all } n, m \text{ and } k. \tag{98}$$

The recurrence relations (81), (82) and (91) are not applicable when n, m, j or $k = 0$. The case when j or $k = 0$ does not concern us, because

$$Z_{jk} \equiv 0 \quad \text{for } j \text{ and/or } k = 0. \tag{99}$$

and $$f_{nm:j0} \equiv 0 \quad \text{(modulo constant) for all } n, m \text{ and } j. \tag{100}$$

It remains to consider when n and/or $m = 0$.

First, we observe that by virtue of the identity,

$$nP_n = (2n-1)\,\mu P_{n-1}-(n-1)\,P_{n-2}, \tag{101}$$

we have the relation,

$$nf_{n0:jk} = (2n-1)f_{n-1,1:jk}-(n-1)f_{n-2,0:jk}. \tag{102}$$

This relation enables the determination of all $f_{n0:jk}$, recursively, except $f_{10:jk}$ and $f_{00:jk}$. By a re-examination, *ab initio*, of the derivation of equation (91) for the cases $n = 1$, $m = 0$ and $n = m = 0$, we readily obtain the relations,

$$\frac{j+k}{jk}f_{10:jk} = -\tfrac{1}{2}[\eta Y_{jk}(\mu)+\mu Y_{jk}(\eta)]$$
$$-\sum_r{}' \alpha_r f_{10:rk}-\sum_s{}' \alpha_s f_{10:sj}+\phi_{10:jk}+\phi_{10:kj}+\phi_{jk:01} \tag{103}$$

and $$\frac{j+k}{jk}f_{00:jk} = -\tfrac{1}{2}[Y_{jk}(\mu)+Y_{jk}(\eta)]$$
$$-\sum_r{}' \alpha_r f_{00:rk}-\sum_s{}' \alpha_s f_{00:sj}+\phi_{00:kj}+\phi_{00:jk}. \tag{104}$$

Equations (81), (91), (97), (100), (102), (103) and (104) provide a complete set of recurrence relations for determining $f_{nm:jk}$ for all n, m, j and k.

4. The recurrence relations and their use

The completion of the solution of the different problems considered in §2 requires the determination of the functions $f_{nm:nm}(\mu,\eta)$, symmetric in the indices n and m and in the variables μ and η, for all pairs (n,m). But in deriving the relations in §3, for determining them recursively, we had to enlarge the basic definition to include the much larger class of functions $f_{nm:jk}(\mu,\eta)$, preserving, however, the symmetry in (n,m), (j,k) and (μ,η). By this enlargement of the definition, the use of the recurrence relations to determine the functions of concern to us, namely $f_{nm;nm}$, requires the prior determination of many more intermediate functions: and, therefore, is not simple.

First, we may observe that as special cases of equations (81) and (91), we have the following simple relations:

$$n(n+1)f_{1n:1n} = -\tfrac{1}{2}Z_{1n}(\mu)\,Z_{1n}(\eta), \tag{105}$$

$$\frac{1}{n}(n+1)f_{1n:1n} = -\tfrac{1}{2}Y_{1n}(\mu)\,Y_{1n}(\eta) - \sum_p{}'\alpha_p f_{1n:nn} + (\phi_{1n:n1}+\phi_{1n:1n}). \tag{106}$$

$$\begin{aligned} n(n+1)f_{1n;11}+2f_{11;1n} &= -\tfrac{1}{2}[Z_{1n}(\mu)\,Z_{11}(\eta)+Z_{1n}(\eta)\,Z_{11}(\mu)] \\ &= -\tfrac{1}{2}\Delta\delta[P_n'(\mu)+P_n'(\eta)]. \end{aligned} \tag{107}$$

The equality of the right-hand sides of equations (105) and (106), modulo a constant, provides an identity.

In the following illustrative examples, the particular recurrence relation from which the different equations follow is manifest.

Example 1

(i) $f_{11:11} = -\tfrac{1}{4}\Delta\delta$,

(ii) $2f_{11:11} = -\tfrac{1}{2}Y_{11}(\mu)\,Y_{11}(\eta)+2\phi_{11:11}$,

(iii) $\phi_{11;11} = \tfrac{1}{4}\Phi_{11:11} = \tfrac{1}{4}[Y_{11}(\eta)+Y_{11}(\mu)] = \tfrac{1}{4}(\mu^2+\eta^2)$,

(iv) $f_{11:11} = -\tfrac{1}{4}\mu^2\eta^2+\tfrac{1}{4}(\eta^2+\mu^2) \equiv -\tfrac{1}{4}\Delta\delta$ (mod constant).

Example 2

(i) $f_{12:12} = -\tfrac{1}{12}Z_{12}(\mu)\,Z_{12}(\eta) = -\tfrac{3}{4}\Delta\delta\mu\eta$,

(ii) $\tfrac{3}{2}f_{12:12} = -\tfrac{1}{2}Y_{12}(\mu)\,Y_{12}(\eta)+(\phi_{12:12}+\phi_{12:21})$;

(iii) $4\phi_{12;12} = \tfrac{1}{2}\Phi_{12:12}-\phi_{01:12}$; $4\phi_{12;21} = \tfrac{1}{2}\Phi_{12:21}-\phi_{01:21}$,

(iv) $\Phi_{12:12} = 3[\eta Y_{12}(\mu)+\mu Y_{12}(\eta)] = 3\Phi_{12:21}$;

(v) $\phi_{01:12} = \tfrac{3}{4}[\eta Y_{01}(\mu)+\mu Y_{01}(\eta)] = \tfrac{3}{2}\mu\eta = 3\phi_{01:21}$;

(vi) $4(\phi_{12:12}+\phi_{12;21}) = 2\Phi_{12;21}-4\phi_{01;21} = \mu\eta[3(\mu^2+\eta^2)-4]$,

(vii) $\tfrac{3}{2}f_{12:12} = -\tfrac{1}{4}\mu\eta[2P_2(\mu)\,P_2(\eta)-3(\mu^2+\eta^2)+4]$
$\equiv -\tfrac{9}{8}\Delta\delta\mu\eta$ (mod constant).

Example 3

(i) $$f_{11;22}+8f_{22;11} = -\tfrac{1}{4}[Z_{11}(\mu)\,Z_{22}(\eta)+Z_{11}(\eta)\,Z_{22}(\mu)]+f_{00;11}-2f_{02;11}$$
$$= -\tfrac{1}{4}[Z_{11}(\mu)\,Z_{22}(\eta)+Z_{11}(\eta)\,Z_{22}(\mu)]+2f_{00;11}-3f_{11;11},$$

(ii) $$f_{11;22}+2f_{22;11} = -\tfrac{1}{2}[Y_{11}(\mu)\,Y_{22}(\eta)+Y_{11}(\eta)\,Y_{22}(\mu)]+2(\phi_{11;22}+\phi_{22;11});$$

(iii) $$2f_{00;11} = -\tfrac{1}{2}[Y_{11}(\mu)+Y_{11}(\eta)]+2\phi_{00;11} = -\tfrac{1}{2}(\mu^2+\eta^2)+\ln\mu\eta;$$

(iv) $$4\phi_{11;22} = \tfrac{1}{2}\Phi_{11;22} = \tfrac{3}{2}[Y_{11}(\mu)\,\eta^2+Y_{22}(\eta)\,\mu^2] = 3\mu^2\eta^2,$$

(v) $$4\phi_{22;11} = \tfrac{1}{2}\Phi_{22;11}-2\phi_{02;11} = \tfrac{1}{2}[Y_{22}(\mu)+Y_{22}(\eta)]-2\phi_{02;11},$$

(vi) $$\phi_{02;11} = \tfrac{3}{8}(\mu^2+\eta^2)-\tfrac{1}{4}\ln\mu\eta;$$

(vii) $$f_{11;22}+8f_{22;11} = -\tfrac{1}{4}[Z_{11}(\mu)\,Z_{22}(\eta)+Z_{11}(\eta)\,Z_{22}(\mu)]+\tfrac{3}{4}\Delta\delta$$
$$-\tfrac{1}{2}(\mu^2+\eta^2)+\ln\mu\eta,$$

(viii) $$f_{11;22}+2f_{22;11} = -\tfrac{1}{2}[Y_{11}(\mu)\,Y_{22}(\eta)+Y_{11}(\eta)\,Y_{22}(\mu)]$$
$$+\tfrac{1}{4}[6\eta^2\mu^2+Y_{22}(\mu)+Y_{22}(\eta)]-\tfrac{3}{8}(\mu^2+\eta^2)+\tfrac{1}{4}\ln\mu\eta.$$

Example 4

(i) $$8f_{22;22} = -\tfrac{1}{4}Z_{22}(\mu)\,Z_{22}(\eta)+f_{00;22}-2f_{20;22}$$
$$= -\tfrac{1}{4}Z_{22}(\mu)\,Z_{22}(\eta)+2f_{00;22}-3f_{11;22};$$

(ii) $$f_{00;22} = -\tfrac{1}{2}[Y_{22}(\mu)+Y_{22}(\eta)]+2\phi_{00;22},$$

(iii) $$2\phi_{00;22} = \tfrac{1}{2}\Phi_{00;22} = \tfrac{1}{2}[\eta P_2'(\eta)+\mu P_2'(\mu)] = \tfrac{3}{2}(\mu^2+\eta^2);$$

(iv) $$8f_{22;22} = -\tfrac{1}{4}Z_{22}(\mu)\,Z_{22}(\eta)-[Y_{22}(\mu)+Y_{22}(\eta)]+3(\mu^2+\eta^2)-3f_{11;22};$$

or, alternatively,

(v) $$f_{22;22} = -\tfrac{1}{2}Y_{22}(\mu)\,Y_{22}(\eta)+2\phi_{22;22};$$

(vi) $$2\phi_{22;22} = \tfrac{1}{2}[\eta^2 Y_{22}(\mu)+\mu^2 Y_{22}(\eta)]-\tfrac{2}{3}\phi_{02;22},$$

(vii) $$\phi_{02;22} = \tfrac{3}{8}[\eta^2 Y_{02}(\mu)+\mu^2 Y_{02}(\eta)]-\tfrac{1}{4}\phi_{00;22};\ \phi_{00;22} = \tfrac{3}{4}(\mu^2+\eta^2);$$

(viii) $$f_{22;22} = -\tfrac{1}{2}Y_{22}(\mu)\,Y_{22}(\eta)+\tfrac{1}{2}[\eta^2 Y_{22}(\mu)+\mu^2 Y_{22}(\eta)]$$
$$-\tfrac{1}{4}[\eta^2 Y_{02}(\mu)+\mu^2 Y_{02}(\eta)]+\tfrac{1}{8}(\mu^2+\eta^2).$$

Apart from the factors specified in equation (47), equations (i, example 1), (i, example 2) and (viii, example 4), together with the 'linear terms' in equations (54′) and (54″), provide the contributions to the metric function f derived from terms in P_1 and P_2 in Π. The additional contributions derived from a term in P_3 are included in the Appendix.

5. Concluding remarks

Apart from clarifying the issues of principle that must be resolved before one routinely applies Weyl's solution to the problem of space-times with two commuting Killing fields, the principal object of this paper was to obtain the solution for the last remaining pair of equations that is of most frequent occurrence in these contexts. In §3, the solution was accomplished by obtaining a complete set of recurrence relations. But as we have seen in §4, the use of these recurrence relations requires one to evaluate many intermediate functions.

Because the solution to the basic equation (63) can be obtained very simply by a direct procedure requiring only the polynomial expansion of the function,

$$Y_{nm}(\mu)\,Z_{nm}(\eta) = P_n(\mu)\,P_m(\mu)\,(1-\eta^2)\,P'_n(\eta)\,P'_m(\eta) \tag{108}$$

– a problem which might have been considered as 'cumbersome' and looked askance at an earlier time but one which is 'child's play' with modern computers – one may reasonably ask what the purpose is for solving the problem with the aid of several recurrence relations and by a procedure that is elaborate. Perhaps there is none. But one may argue that the mere existence of the recurrence relations is a reflection of the underlying mathematical structure that is manifested, for example, by the number of identities that follow from them. Franz Neumann may have been pleased by the resurrection of his functions, Y_{nm} and Z_{nm}, in the context of Weyl's solution!

The research reported in this paper has, in part, been supported by grants from the National Science Foundation under grant PHY-84-16691 with the University of Chicago.

References

Chandrasekhar, S. 1978 *Proc. R. Soc. Lond.* A **358,** 405–420.
Chandrasekhar, S. 1983 *The mathematical theory of black holes.* Oxford: Clarendon Press.
Chandrasekhar, S. & Ferrari, V. 1984 *Proc. R. Soc. Lond.* A **396**, 55–74.
Chandrasekhar, S. & Xanthopoulos, B. C. 1986 *Proc. R. Soc. Lond.* A **408**, 175–208.
Chandrasekhar, S. & Xanthopoulos, B. C. 1987 *Proc. R. Soc. Lond.* A **410**, 311–336.
Ferrari, V. & Ibanez, J. 1987 *Gen. Rel. Grav.* **19**, 383–404.
Geroch, R. & Hartle, J. B. 1982 *J. math. Phys.* **23**, 680–692.
Hobson, E. W. 1931 *Spherical and ellipsoidal harmonics.* Cambridge University Press.
Neumann, F. E. 1878 *Beiträge zur Theorie der Kugelfunctionen* (Leipzig).
Schwarzschild, K. 1916 *Berliner Sitzungsberichte* (*Phys. Math. Klasse*) 3 Feb. 1916, p. 189.
Wald, R. M. 1984 *General relativity.* Chicago University Press.
Weyl, H. 1917 *Ann. Phys.* **54**, 117.
Xanthopoulos, B. C. 1983 *Proc. R. Soc. Lond.* A **388**, 117–131.

Appendix. Functions of order 3

In §4 the various functions of order less than or equal to two are listed. The functions of order 3 that are required for including a term in P_3 in Π are listed below.

(i) $12f_{13;13} = -\frac{1}{2}Z_{13}(\mu)\,Z_{13}(\eta)$;

(ii) $30f_{23;23} = -\frac{1}{2}Z_{23}(\mu)\,Z_{23}(\eta) - 16f_{21;23} + 10f_{01;23}$,

(iii) $\frac{2}{3}f_{33;33} = -\frac{1}{2}Y_{33}(\mu)\,Y_{33}(\eta) - 3f_{33;13} + 2\phi_{33;33}$;

(iv) $\frac{3}{2}f_{01;12} = -\frac{3}{8}[\eta Y_{12}(\mu) + \mu Y_{12}(\eta)] + \frac{15}{8}\mu\eta$,

(v) $$\tfrac{5}{6}f_{01;23} = -\tfrac{1}{2}[\eta Y_{23}(\mu)+\mu Y_{23}(\eta)]+\tfrac{5}{8}[\eta Y_{12}(\mu)+\mu Y_{12}(\eta)]+\tfrac{1}{8}\Phi_{01:23} + \tfrac{1}{12}\Phi_{23:01}-\tfrac{1}{48}\Phi_{03:01}-\tfrac{5}{8}\mu\eta;$$

(vi) $$8\phi_{33;33} = \tfrac{1}{2}\Phi_{33;33}+\tfrac{1}{4}\Phi_{33;31}-\tfrac{1}{2}\Phi_{13;33}+\tfrac{3}{4}\Phi_{13;31}+\tfrac{3}{8}\Phi_{11;33}-\tfrac{27}{16}\Phi_{11;31};$$

(vii) $$54f_{33;13}+12f_{13;33} = -\tfrac{1}{2}[Z_{33}(\mu)Z_{13}(\eta)+Z_{33}(\eta)Z_{13}(\mu)] +\tfrac{3}{2}Z_{13}(\mu)Z_{13}(\eta)+18f_{11;13},$$

(viii) $$\tfrac{4}{3}f_{33;13}+\tfrac{2}{3}f_{13;33} = -\tfrac{1}{2}[Y_{33}(\mu)Y_{13}(\eta)+Y_{33}(\eta)Y_{13}(\mu)] +\tfrac{1}{8}Z_{13}(\mu)Z_{13}(\eta)-\tfrac{3}{2}f_{33;11}+\phi_{33;31}+\phi_{33;13}+2\phi_{13;33};$$

(ix) $$8\phi_{33;13} = \tfrac{1}{2}\Phi_{33;13}-(3\phi_{33;11}+6\phi_{13;13}),$$

(x) $$3\phi_{33;11}+6\phi_{13;13} = \tfrac{1}{4}\Phi_{33;11}+\tfrac{1}{2}\Phi_{13;13}-\tfrac{3}{8}\Phi_{11;13}-\tfrac{3}{4}\Phi_{13;11}+\tfrac{27}{16}(\mu^2+\eta^2),$$

(xi) $$2f_{11;13}+12f_{13;11} = -\tfrac{1}{2}[Z_{11}(\mu)Z_{13}(\eta)+Z_{11}(\eta)Z_{13}(\mu)],$$

(xii) $$\tfrac{4}{3}f_{11;13}+2f_{13;11} = -\tfrac{1}{2}[Y_{11}(\mu)Y_{13}(\eta)+Y_{11}(\eta)Y_{13}(\mu)]+\tfrac{3}{8}\Delta\delta +\phi_{11;13}+\phi_{11;31}+2\phi_{13;11},$$

(xiii) $$\phi_{11;13}+\phi_{11;31}+2\phi_{13;11} = \tfrac{1}{8}\Phi_{11;13}+\tfrac{1}{4}\Phi_{11;31}+\tfrac{1}{4}\Phi_{13;11}-\tfrac{9}{16}(\mu^2+\eta^2);$$

(xiv) $$27f_{33;11}+f_{11;33} = -\tfrac{1}{4}[Z_{33}(\mu)Z_{11}(\eta)+Z_{33}(\eta)Z_{11}(\mu)]-\tfrac{9}{4}\Delta\delta-18f_{13;11},$$

(xv) $$2f_{33;11}+\tfrac{2}{3}f_{11;33} = -\tfrac{1}{2}[Y_{33}(\mu)Y_{11}(\eta)+Y_{33}(\eta)Y_{11}(\mu)]-3f_{11;13} +2\phi_{33;11}+2\phi_{11;33},$$

(xvi) $$2\phi_{33;11}+2\phi_{11;33} = \tfrac{1}{6}\Phi_{33;11}-\tfrac{1}{4}\Phi_{13;11}+\tfrac{1}{4}\Phi_{11;33}-\tfrac{3}{8}\Phi_{11;31}+\tfrac{3}{8}(\mu^2+\eta^2),$$

(xvii) $$6f_{21;23}+30f_{23;21} = -\tfrac{1}{2}[Z_{12}(\mu)Z_{23}(\eta)+Z_{12}(\eta)Z_{23}(\mu)]+12\Delta\delta\mu\eta+10f_{01;12},$$

(xviii) $$\tfrac{5}{6}f_{21;23}+\tfrac{3}{2}f_{23;21} = -\tfrac{1}{2}[Y_{12}(\mu)Y_{23}(\eta)+Y_{12}(\eta)Y_{23}(\mu)]+\tfrac{9}{8}\Delta\delta\mu\eta +\phi_{21;23}+\phi_{21;32}+\phi_{23;21}+\phi_{23;12};$$

(xix) $$6\phi_{21;23} = \tfrac{1}{2}\Phi_{21;23}-\tfrac{3}{8}\Phi_{21;21}-\tfrac{1}{8}\Phi_{01;23}+\tfrac{3}{8}\Phi_{01;21},$$

(xx) $$4\phi_{21;32} = \tfrac{1}{2}\Phi_{21;32}-\tfrac{1}{4}\Phi_{01;32},$$

(xxi) $$6\phi_{23;21} = \tfrac{1}{2}\Phi_{23;21}-\tfrac{3}{8}\Phi_{21;21}-\tfrac{1}{8}\Phi_{03;21}+\tfrac{3}{8}\Phi_{01;21},$$

(xxii) $$6\phi_{23;12} = \tfrac{1}{2}\Phi_{23;12}-\tfrac{3}{8}\Phi_{12;12}-\tfrac{1}{8}\Phi_{03;12}+\tfrac{3}{8}\Phi_{01;12}.$$

28

Some exact solutions of gravitational waves coupled with fluid motions

By Subrahmanyan Chandrasekhar[1], F.R.S., and Basilis C. Xanthopoulos[2]

[1] *The University of Chicago, Chicago, Illinois* 60637, *U.S.A.*
[2] *Department of Physics, University of Crete, and Research Centre of Crete, Iraklion, Greece*

(*Received* 10 *June* 1985)

Some exact solutions of Einstein's equations are found which represent the interaction of gravitational waves with a perfect fluid in which the velocity of sound equals the velocity of light. These solutions, unlike the solutions representing the collision of impulsive gravitational waves, are bounded by a space–time singularity and have some resemblance to cosmological solutions: every time-like trajectory, extended into the past, encounters the singularity. Moreover, in the generic case, matter may be considered as being created at the singularity.

1. Introduction

In an earlier paper (Chandrasekhar & Xanthopoulos 1985; this paper will be referred to hereafter as Paper I) the analysis of Chandrasekhar & Ferrari (1984), leading to the Nutku–Halil solution describing the collision of two impulsive gravitational waves, was generalized to the case when the gravitational waves are coupled, in the region of interaction, to sound waves in a perfect fluid in which the energy density, ϵ, is equal to the pressure, p. It was shown that the solution of the coupled Einstein-hydrodynamics equations can, under those circumstances, be reduced to one of supplementing *any* solution of the vacuum equations with the solution of a pair of equations (decoupled from the rest) derived from the presence of the fluid. Appropriate to the problem of collision and the subsequent development of a space–time singularity, the solution was obtained in a gauge and in a system of coordinates that limited the space–time to a finite duration and to a finite spatial extent normal to the plane wavefronts. In this paper we explore the possibility of other classes of solutions with other choices of gauge and coordinates, which do not so restrict the space–time; and we shall find solutions that are reminiscent of cosmological solutions with initial singularities.

2. The freedom in the choice of gauge and coordinates

We start with the general form of the metric,

$$ds^2 = e^{2\nu}(dx^0)^2 - e^{2\mu_3}(dx^3)^2 - e^{2\psi}(dx^1 - q_2\,dx^2)^2 - e^{2\mu_2}(dx^2)^2, \tag{1}$$

appropriate to space–times with two space-like commuting Killing vectors, where

ν, μ_3, ψ, q_2, and μ_2 are functions only of x^0 and x^3, and rewrite it in the form

$$ds^2 = \frac{e^{\mu_3+\nu}}{\sqrt{\Delta}}[(dx^0)^2 - \Delta(dx^3)^2] - e^{\beta}\left[\chi(dx^2)^2 + \frac{1}{\chi}(dx^1 - q_2\,dx^2)^2\right], \tag{2}$$

where $$\Delta = e^{2(\mu_3-\nu)}, \quad \chi = e^{-\psi+\mu_2} \quad \text{and} \quad e^{\beta} = e^{\psi+\mu_2}. \tag{3}$$

The freedom of gauge we have (derived from the freedom to impose any coordinate condition we may desire on μ_3 and ν) enables us to assume (without any loss of generality) that

$$\Delta \equiv \Delta(x^0). \tag{4}$$

Also, the freedom in the choice of coordinates we have derives from one of the field equations,

$$[\Delta^{\frac{1}{2}}(e^{\beta})_{,0}]_{,0} - [\Delta^{-\frac{1}{2}}(e^{\beta})_{,3}]_{,3} = 0, \tag{5}$$

which is valid for the vacuum and also when we have a perfect fluid-source with the equation of state $\epsilon = p$. If we assume (and it would appear that no essential restriction is involved in this assumption) that e^{β} is separable in the variables x^0 and x^3 and is, further, expressible in the form

$$e^{\beta} = f(x^3)\sqrt{\Delta}, \tag{6}$$

then equation (5) reduces to

$$\tfrac{1}{2}\Delta_{,0,0} = f_{,3,3}/f = \text{constant}. \tag{7}$$

The solution of the equations that we have chosen hitherto are

$$\left.\begin{aligned} f = \sin x^3 = \sqrt{\delta} \quad &\text{and} \quad \Delta = 1-\eta^2, \\ \text{where} \qquad \delta = 1-\mu^2 \quad &\text{and} \quad \mu = \cos x^3. \end{aligned}\right\} \tag{8}$$

The coordinate η so defined is a measure of the time in a suitable unit from a suitable origin. Similarly, μ is a measure of the distance normal to the wavefronts. In the choice of these coordinates we were guided by the fact that in the problem of the collision of impulsive gravitational waves, there is a natural origin of time ($\eta = 0$) and place ($\mu = 0$) specified by the instant and location of the collision; and the further knowledge that a space–time singularity develops at a definite later time ($\eta = 1$) and at a definite place ($\mu = 1$). But *formally* other possibilities exist: we could have chosen

$$f = \sinh x^3 \quad \text{or} \quad \cosh x^3; \tag{9}$$

in which case the corresponding solution for Δ would have been

$$\Delta = \eta^2 \pm 1. \tag{10}$$

Therefore, letting

$$\mu = \cosh x^3 \quad \text{or} \quad \sinh x^3 \tag{11}$$

we have the possibilities,

$$\Delta = \eta^2 \pm 1 \quad \text{and} \quad \delta = \mu^2 \pm 1, \tag{12}$$

where the alternative definitions of Δ and δ can be chosen independently of one another. In all four cases,

$$e^{\beta} = \sqrt{(\Delta\delta)} \tag{13}$$

We shall find that of the four possible combinations (12) we can consider, the case

$$\Delta = \eta^2+1, \quad \delta = \mu^2+1, \tag{14}$$

does not allow a simple solution within the scope of our present considerations (see §3 below). Also, it will appear that the two cases,

$$\Delta = \eta^2+1, \quad \delta = \mu^2-1 \quad \text{and} \quad \Delta = \eta^2-1, \quad \delta = \mu^2+1 \tag{15}$$

lead to solutions that are essentially the same. The two cases we are left to consider are

$$\Delta = \eta^2-1, \quad \delta = \mu^2-1 \quad \text{(case 1)}, \tag{16}$$

and

$$\Delta = \eta^2+1, \quad \delta = \mu^2-1 \quad \text{(case 2)}. \tag{17}$$

We shall find that these two cases provide solutions that have very different interpretations from the ones that have hitherto been considered.

3. The basic equations

The equations we have to consider are the same as those of Paper I, §§4, 5. However, since the present choices of coordinates are different, the signs of $\Delta_{,\eta}$ and $\delta_{,\mu}$, for the present choices, $\Delta = \eta^2 \pm 1$ and $\delta = \mu^2 \pm 1$, are the opposite of those of the earlier choice, $\Delta = 1-\eta^2$ and $\delta = 1-\mu^2$, one has to be careful about the signs in the various equations. We shall, therefore, assemble in this section the basic equations of the problem in forms that are applicable for any combinations of $\Delta = \eta^2 \pm 1$ and $\delta = \mu^2 \pm 1$.

The definitions that are always valid are

$$e^{2(\mu_3-\nu)} = \Delta, \quad e^{2\beta} = e^{2(\psi+\mu_2)} = \Delta\delta \quad \text{and} \quad \chi = e^{-\psi+\mu_2}. \tag{18}$$

The equations governing χ and q_2 are

$$\left(\frac{\Delta}{\chi}\chi_{,\eta}\right)_{,\eta} - \left(\frac{\delta}{\chi}\chi_{,\mu}\right)_{,\mu} = -\frac{1}{\chi^2}\left[\Delta(q_{2,\eta})^2 - \delta(q_{2,\mu})^2\right], \tag{19}$$

and

$$\left(\frac{\Delta}{\chi^2}q_{2,\eta}\right)_{,\eta} - \left(\frac{\delta}{\chi^2}q_{2,\mu}\right)_{,\mu} = 0. \tag{20}$$

Defining

$$Z = -\chi + i q_2, \tag{21}$$

we can combine equations (19) and (20) into the single complex equation

$$\mathscr{R}e(Z)\left[(\Delta Z_{,\eta})_{,\eta} - (\delta Z_{,\mu})_{,\mu}\right] = \Delta(Z_{,\eta})^2 - \delta(Z_{,\mu})^2. \tag{22}$$

Now letting

$$Z = \frac{1+E}{1-E}, \tag{23}$$

we obtain the Ernst equation

$$(1-|E|^2)\left[(\Delta E_{,\eta})_{,\eta} - (\delta E_{,\mu})_{,\mu}\right] = -2E^*\left[\Delta(E_{,\eta})^2 - \delta(E_{,\mu})^2\right]. \tag{24}$$

In terms of a solution of this equation, the solutions for χ and q_2 are given by

$$\chi = -\frac{1-|E|^2}{|1-E|^2} \quad \text{and} \quad \mathrm{i}q_2 = \frac{E-E^*}{|1-E|^2}. \tag{25}$$

The foregoing equations, governing χ and q_2, are the same as those that obtain for a vacuum. The coupling of the gravitational field with the source in the present context – a perfect fluid with the equation of state, $\epsilon = p$ — affects only the equations governing $\nu+\mu_3$. They are (cf. Paper I, equations (27) and (28)):

$$\frac{\eta}{\Delta}(\nu+\mu_3)_{,\mu}+\frac{\mu}{\delta}(\nu+\mu_3)_{,\eta} = \frac{1}{\chi^2}\mathscr{Re}(Z_{,\eta}Z^*_{,\mu})-\frac{8\epsilon}{\sqrt{\delta}}u_0 u_3 \mathrm{e}^{\nu+\mu_3}, \tag{26}$$

and

$$2\eta(\nu+\mu_3)_{,\eta}+2\mu(\nu+\mu_3)_{,\mu} = \frac{1}{\chi^2}[\Delta|Z_{,\eta}|^2+\delta|Z_{,\mu}|^2] +4-\frac{3\eta^2}{\Delta}-\frac{\mu^2}{\delta}-8\epsilon(u_0^2+u_3^2)\,\mathrm{e}^{\nu+\mu_3}\sqrt{\Delta}. \tag{27}$$

Turning to the equations of hydrodynamics, we find that the non-vanishing components, u_0 and u_3, of the four-velocity can be derived from a potential, ϕ, in the manner (cf. Paper I, equations (31))

$$u_0\sqrt{\epsilon} = \frac{\mathrm{e}^{-\mu_3}}{\sqrt{\Delta}}\phi_{,\mu} \quad \text{and} \quad u_3\sqrt{\epsilon} = \frac{\mathrm{e}^{-\mu_3}}{\sqrt{\delta}}\phi_{,\eta}; \tag{28}$$

and further, that ϕ is governed by the hyperbolic equation

$$\Delta\phi_{,\eta,\eta}-\delta\phi_{,\mu,\mu} = 0. \tag{29}$$

The reduction of equations (26) *and* (27)

The linearity, in $\nu+\mu_3$, of equations (26) and (27) suggests that we write (as in Paper I, equations (43) and (44))

$$\nu+\mu_3 = (\nu+\mu_3)_{\text{vac.}}+f, \tag{30}$$

where $(\nu+\mu_3)_{\text{vac.}}$ is the solution appropriate to the vacuum when $\epsilon = 0$. Since by our choice of coordinates

$$\nu-\mu_3 = (\nu-\mu_3)_{\text{vac.}}, \tag{31}$$

it follows that

$$\nu = (\nu)_{\text{vac.}}+\tfrac{1}{2}f \quad \text{and} \quad \mu_3 = (\mu_3)_{\text{vac.}}+\tfrac{1}{2}f. \tag{32}$$

We shall now dispense with the distinguishing subscript 'vac.' for ν and μ_3 and retain for them the meaning that they refer to solutions appropriate to the vacuum so that for the problem on hand,

$$\nu+\tfrac{1}{2}f \quad \text{and} \quad \mu_3+\tfrac{1}{2}f \tag{33}$$

replace what we have hitherto denoted by ν and μ_3.

From equations (26) and (27) it follows that the equations governing $\nu+\mu_3$, appropriate to the vacuum, are

$$\frac{\eta}{\Delta}(\nu+\mu_3)_{,\mu}+\frac{\mu}{\delta}(\nu+\mu_3)_{,\eta} = \frac{1}{\chi^2}\mathscr{Re}(Z_{,\eta}Z^*_{,\mu}) \tag{34}$$

and $$2\eta(\nu+\mu_3)_{,\eta}+2\mu(\nu+\mu_3)_{,\mu} = \frac{1}{\chi^2}[\Delta|Z_{,\eta}|^2+\delta|Z_{,\mu}|^2]+4-\frac{3\eta^2}{\Delta}-\frac{\mu^2}{\delta}, \tag{35}$$

while the equations governing f are

$$\frac{\eta}{\Delta}f_{,\mu}+\frac{\mu}{\delta}f_{,\eta} = -\frac{8}{\Delta\delta}\phi_{,\eta}\phi_{,\mu} \tag{36}$$

and $$\eta f_{,\eta}+\mu f_{,\mu} = -\frac{4}{\Delta\delta}[\Delta(\phi_{,\eta})^2+\delta(\phi_{,\mu})^2]. \tag{37}$$

We may also note here that in accordance with equation (28) and our present definitions (32) and (33)

$$\epsilon = \epsilon(u_0^2-u_3^2) = -\frac{e^{-2\mu_3-f}}{\Delta\delta}[\Delta(\phi_{,\eta})^2-\delta(\phi_{,\mu})^2]. \tag{38}$$

The metric to which all of the equations in this section apply is

$$ds^2 = e^{\nu+\mu_3+f}\sqrt{\Delta}\left[\frac{(d\eta)^2}{\Delta}-\frac{(d\mu)^2}{\delta}\right]-\frac{\sqrt{(\Delta\delta)}}{|E|^2-1}|(1-E)\,dx^1+i(1+E)\,dx^2|^2, \tag{39}$$

where $\Delta = \eta^2\pm 1$ and $\delta = \mu^2\pm 1$.

In our further considerations in this paper, we shall restrict ourselves to only those cases when the Ernst equation (24) allows a simple linear solution of the form

$$E = p\eta+iq\mu, \tag{40}$$

where p and q are constants. We readily verify that when

$$\Delta = \eta^2+1 \quad\text{and}\quad \delta = \mu^2+1, \tag{41}$$

equation (24) does not allow such a solution. We shall not, therefore, consider this case. Equation (24) does, however, allow a solution of the form (40) in the remaining three cases, as we shall see in detail below.

4. The case $\Delta = \eta^2-1$ and $\delta = \mu^2-1$

When $\Delta = \eta^2-1$ and $\delta = \mu^2-1$, the Ernst equation (24) allows a solution of the form

$$E = p\eta+iq\mu, \tag{42}$$

where p and q are constants and

$$p^2+q^2 = 1. \tag{43}$$

The corresponding solutions for χ and q_2 are (cf. Paper I, equations (26))

$$\chi = \frac{p^2\eta^2+q^2\mu^2-1}{(1-p\eta)^2+q^2\mu^2} \quad\text{and}\quad q_2 = \frac{2q\mu}{(1-p\eta)^2+q^2\mu^2}. \tag{44}$$

The associated solution for Z gives

$$\mathscr{Re}(Z_{,\eta}Z^*_{,\mu}) = 0 \quad\text{and}\quad \frac{1}{\chi^2}[\Delta|Z_{,\eta}|^2+\delta|Z_{,\mu}|^2] = \frac{4}{p^2\eta^2+q^2\mu^2-1}. \tag{45}$$

From equations (34) and (35) we now find

$$e^{\nu+\mu_3} = \frac{p^2\eta^2+q^2\mu^2-1}{(\eta^2-1)^{\frac{3}{4}}(\mu^2-1)^{\frac{1}{4}}}. \tag{46}$$

Combined with the first of equations (18), we have

$$e^{2\mu_3} = \frac{p^2\eta^2+q^2\mu^2-1}{(\eta^2-1)^{\frac{1}{4}}(\mu^2-1)^{\frac{1}{4}}}. \tag{47}$$

It is now convenient to introduce in place of η and μ the variables ψ and θ defined by

$$\eta = \cosh\psi \quad \text{and} \quad \mu = \cosh\theta, \tag{48}$$

when the metric takes the form (cf. Paper I, equations (98) and (99))

$$ds^2 = \frac{\Delta\, e^f}{(\sinh\psi\sinh\theta)^{\frac{1}{2}}}[(d\psi)^2-(d\theta)^2] - \frac{\sinh\psi\sinh\theta}{\Delta}\{|1+E|^2(dx^2)^2+|1-E|^2(dx^1)^2-4q(\cosh\theta)\,dx^1\,dx^2\}, \tag{49}$$

where

$$\Delta = |E|^2-1 = p^2\eta^2+q^2\mu^2-1 = p^2\sinh^2\psi+q^2\sinh^2\theta. \tag{50}$$

(a) The transformation to null coordinates

We now introduce a system of null coordinates by the definitions (cf. Paper I, equations (54) and (55))

$$u = \cosh\xi \quad \text{and} \quad v = \sinh\zeta, \tag{51}$$

where

$$\xi = \tfrac{1}{2}(\psi+\theta) \quad \text{and} \quad \zeta = \tfrac{1}{2}(\psi-\theta). \tag{52}$$

By these definitions,

$$\begin{aligned}\eta &= \cosh\psi = \cosh(\xi+\zeta) = \cosh\xi\cosh\zeta+\sinh\xi\sinh\zeta\\ &= u\sqrt{(v^2+1)}+v\sqrt{(u^2-1)},\\ \mu &= \cosh\theta = \cosh(\xi-\zeta) = \cosh\xi\cosh\zeta-\sinh\xi\sinh\zeta\\ &= u\sqrt{(v^2+1)}-v\sqrt{(u^2-1)},\end{aligned} \tag{53}$$

and

$$\begin{aligned}\sinh\psi\sinh\theta &= \sinh^2\xi\cosh^2\zeta-\cosh^2\xi\sinh^2\zeta\\ &= u^2-(v^2+1).\end{aligned} \tag{54}$$

(Other possible definitions of the null-coordinates, apart from (51), are

$$\left.\begin{aligned} &u = \cosh\xi, \quad v = \cosh\zeta;\\ &u = \sinh\xi, \quad v = \sinh\zeta;\\ \text{and} \quad &u = \sinh\xi, \quad v = \cosh\zeta.\end{aligned}\right\} \tag{51'}$$

Of these three definitions, the first two lead to coordinate singularities at $v = +u$ or $-u$ while the last, as we shall explain towards the end of this section, leads to a metric equivalent to the one we shall find with the definition (51).)

In terms of the variables ξ and ζ, equation (29) governing ϕ becomes

$$2\phi_{,\xi,\zeta}-\frac{\eta}{\sqrt{\Delta}}(\phi_{,\xi}+\phi_{,\zeta})+\frac{\mu}{\sqrt{\delta}}(\phi_{,\xi}-\phi_{,\zeta}) = 0; \tag{55}$$

or, equivalently,

$$\phi_{,\xi,\zeta}+\frac{1}{\sinh\psi\sinh\theta}[\phi_{,\xi}\sinh\zeta\cosh\zeta-\phi_{,\zeta}\sinh\xi\cosh\xi]=0. \tag{56}$$

On the other hand, by our definitions (51),

$$\phi_{,\xi}=\phi_{,u}\sinh\xi,\quad \phi_{,\zeta}=\phi_{,v}\cosh\zeta\quad\text{and}\quad \phi_{,\xi,\zeta}=\phi_{,u,v}\sinh\xi\cosh\zeta. \tag{57}$$

Making these substitutions, we find that equation (55) reduces to the form (cf. Paper I, equation (60))

$$\phi_{,u,v}+\frac{1}{u^2-(v^2+1)}(v\phi_{,u}-u\phi_{,v})=0. \tag{58}$$

We may note here, for future reference, the following relations, which one may easily verify:

$$2(u^2-v^2-1)\phi_{,\eta}\phi_{,\mu}=\tfrac{1}{2}[(u^2-1)(\phi_{,u})^2-(v^2+1)(\phi_{,v})^2], \tag{59}$$

$$\begin{aligned}\Delta(\phi_{,\eta})^2+\delta(\phi_{,\mu})^2&=\tfrac{1}{2}[(\phi_{,\xi})^2+(\phi_{,\zeta})^2]\\&=\tfrac{1}{2}[(u^2-1)(\phi_{,u})^2+(v^2+1)(\phi_{,v})^2],\end{aligned} \tag{60}$$

and

$$\Delta(\phi_{,\eta})^2-\delta(\phi_{,\mu})^2=\phi_{,\xi}\phi_{,\zeta}=(u^2-1)^{\frac{1}{2}}(v^2+1)^{\frac{1}{2}}\phi_{,u}\phi_{,v}. \tag{61}$$

Returning to equations (36) and (37) governing f, and solving for $f_{,\eta}$ and $f_{,\mu}$, we obtain

$$f_{,\eta}=+\frac{4}{\eta^2-\mu^2}\left\{\frac{\eta}{\eta^2-1}[\Delta(\phi_{,\eta})^2+\delta(\phi_{,\mu})^2]-2\mu\phi_{,\eta}\phi_{,\mu}\right\}, \tag{62}$$

and

$$f_{,\mu}=-\frac{4}{\eta^2-\mu^2}\left\{\frac{\mu}{\mu^2-1}[\Delta(\phi_{,\eta})^2+\delta(\phi_{,\mu})^2]-2\eta\phi_{,\eta}\phi_{,\mu}\right\}; \tag{63}$$

and inserting these expressions in

$$f_{,u}=(f_{,\eta}+f_{,\mu})\sqrt{(v^2+1)}+(f_{,\eta}-f_{,\mu})\frac{uv}{\sqrt{(u^2-1)}} \tag{64}$$

and

$$f_{,v}=(f_{,\eta}+f_{,\mu})\frac{uv}{\sqrt{(v^2+1)}}+(f_{,\eta}-f_{,\mu})\sqrt{(u^2-1)} \tag{65}$$

(which follow from equation (53)) and simplifying with the aid of equations (59) and (60), we find that we are left with (cf. Paper I, equations (64))

$$f_{,u}=-\frac{2(\phi_{,u})^2}{u(u^2-v^2-1)}\quad\text{and}\quad f_{,v}=+\frac{2(\phi_{,v})^2}{v(u^2-v^2-1)}. \tag{66}$$

Finally, we may note that by making use of equation (61), the expression (38) for ϵ becomes

$$\epsilon=-\frac{e^{-2\mu_3-f}}{(u^2-v^2-1)^2}(u^2-1)^{\frac{1}{2}}(v^2+1)^{\frac{1}{2}}\phi_{,u}\phi_{,v}. \tag{67}$$

(b) The solutions for ϕ and f

Letting

$$r = u^2+v^2 \quad \text{and} \quad s = u^2-(v^2+1), \tag{68}$$

we find that equation (58) reduces to

$$\phi_{,r,r}-\left(\phi_{,s,s}-\frac{1}{s}\phi_{,s}\right) = 0. \tag{69}$$

Equation (69) is clearly separable; and the separable solution, which ensures the positive-definiteness of ϵ and its boundedness for $r \to \infty$, is given by

$$\phi = C\,e^{-\alpha r}\,\alpha s I_1(\alpha s) \quad (\alpha > 0), \tag{70}$$

where α^2 is the separation constant, I_1 is the Bessel function of order 1 for a purely imaginary argument in Watson's notation (Watson 1944), and C is a constant.

For ϕ given by equation (70)

$$\phi_{,r} = -C\alpha^2\,e^{-\alpha r}\,sI_1(\alpha s) \quad \text{and} \quad \phi_{,s} = +C\alpha^2\,e^{-\alpha r}\,sI_0(\alpha s), \tag{71}$$

and the expression (67) for ϵ gives

$$\begin{aligned}\epsilon &= \frac{4}{s^2}\,uv\sqrt{(u^2-1)}\sqrt{(v^2+1)}\,\frac{e^{-f}}{\Delta}\,s^{\frac{1}{2}}[(\phi_{,s})^2-(\phi_{,r})^2]\\ &= 4uv\sqrt{(u^2-1)}\sqrt{(v^2+1)}\,\frac{e^{-f}\,s^{\frac{1}{2}}}{\Delta}\,C^2\alpha^4\,e^{-2\alpha r}\,[I_0^2(\alpha s)-I_1^2(\alpha s)],\end{aligned} \tag{72}$$

which is, indeed, positive definite since $I_0(z) > I_1(z)$.

It is clear that by definition (and by the choice of coordinates), the solution we have obtained is defined only in the domain

$$u \geqslant 1 \quad \text{and} \quad s = u^2-(v^2+1) \geqslant 0. \tag{73}$$

We shall find in § (*d*), below, that a space–time singularity occurs on the rectangular hyperbola

$$u^2 = v^2+1, \tag{74}$$

which has its vertex at $u = 1$ and $v = 0$ and whose asymptotes are along $v = \pm u$.

The general solution for ϕ can be expressed as an arbitrary linear combination (as a sum or as an integral) of the fundamental solution (70) for different Cs and αs. Thus, restricting ourselves to a sum, we can write

$$\phi = \sum_i C_i\,\alpha_i\,e^{-\alpha_i r}\,sI_1(\alpha_i s). \tag{75}$$

The corresponding solution for ϵ is

$$\begin{aligned}\epsilon = {} & 4uv\sqrt{(u^2-1)}\sqrt{(v^2+1)}\frac{e^{-f}}{\Delta}\,s^{+\frac{1}{2}}\\ & \times\Bigg\{\sum_i C_i^2\,\alpha_i^4\,e^{-2\alpha_i r}\,[I_0^2(\alpha_i s)-I_1^2(\alpha_i s)]\\ & +2\sum_{i>j} C_i\,C_j\,\alpha_i^2\,\alpha_j^2\,e^{-(\alpha_i+\alpha_j)r}\,[I_0(\alpha_i s)\,I_0(\alpha_j s)-I_1(\alpha_i s)\,I_1(\alpha_j s)]\Bigg\}.\end{aligned} \tag{76}$$

The solution for f can be found by first noting that in the r and s variables, equations (66) take the form

$$f_{,r} = -\frac{8}{s}\phi_{,r}\phi_{,s} \quad \text{and} \quad f_{,s} = -\frac{4}{s}[(\phi_{,r})^2+(\phi_{,s})^2]. \tag{77}$$

With $\phi_{,r}$ and $\phi_{,s}$ given by equations (71), we can readily verify that the corresponding solution for f is given by

$$f = -4C^2\alpha^3\,\mathrm{e}^{-2\alpha r}\,sI_1(\alpha s)\,I_0(\alpha s). \tag{78}$$

For the more general solution (75) for ϕ, the solution for f is

$$\begin{aligned} f = &-4\sum_i C_i^2\alpha_i^3\,\mathrm{e}^{-2\alpha_i r}\,sI_1(\alpha_i s)\,I_0(\alpha_j s) \\ &-8\sum_{i>j} C_iC_j\frac{\alpha_i^2\alpha_j^2}{\alpha_i+\alpha_j}\,\mathrm{e}^{-(\alpha_i+\alpha_j)r}\,s[I_1(\alpha_i s)\,I_0(\alpha_j s)+I_1(\alpha_j s)\,I_0(\alpha_i s)], \end{aligned} \tag{79}$$

(c) *The character of the distribution of ϵ in the domain $u \geqslant 1$ and $u^2-(v^2+1) \geqslant 0$*

From the known asymptotic behaviours,

$$I_0(z)\to 1 \quad \text{and} \quad I_1(z)\to \tfrac{1}{2}z \quad \text{for} \quad z\to 0 \tag{80}$$

and

$$I_0(z)\to\frac{\mathrm{e}^z}{\sqrt{(2\pi z)}}\left(1+\frac{1}{8z}+\dots\right) \quad \text{and} \quad I_1(z)\to\frac{\mathrm{e}^z}{\sqrt{(2\pi z)}}\left(1-\frac{3}{8z}+\dots\right) \quad \text{for} \quad z\to\infty, \tag{81}$$

we find that the solutions for ϕ and f given by equations (70) and (78) have the behaviours

$$\left.\begin{aligned} &\phi\to\tfrac{1}{2}C\alpha^2s^2\,\mathrm{e}^{-\alpha r} \quad (s\to 0),\\ &\phi\to C\left(\frac{\alpha s}{2\pi}\right)^{\frac{1}{2}}\mathrm{e}^{-\alpha(2v^2+1)} \quad (s\to\infty),\end{aligned}\right\} \tag{82}$$

and

$$\left.\begin{aligned} &f\to -2C^2\alpha^4s^2\,\mathrm{e}^{-2\alpha r} \quad (s\to 0),\\ &f\to -\frac{2C^2\alpha^2}{\pi}\,\mathrm{e}^{-2\alpha(2v^2+1)} = f_\infty \text{ (say)} \quad (s\to\infty).\end{aligned}\right\} \tag{83}$$

From the expression for ϵ given in equation (72) it now follows that

$$\epsilon\to 4C^2\alpha^4u^2v^2\frac{s^{\frac{1}{2}}}{\Delta}\,\mathrm{e}^{-2\alpha r} \quad (s\to 0), \tag{84}$$

since

$$f=0, \quad u^2-1=v^2 \quad \text{and} \quad v^2+1=u^2 \quad \text{when} \quad s=0. \tag{85}$$

The behaviour of Δ for $s\to 0$ is different for $p\neq 0$ and $p=0$. Thus, from the definitions (50)–(53) of the various quantities, we find

$$\Delta = u^2-v^2+1+2uv\,[uv+(p^2-q^2)\,(u^2-1)^{\frac{1}{2}}\,(v^2+1)^{\frac{1}{2}}], \tag{86}$$

so that

$$\Delta = 4p^2u^2v^2\neq 0 \quad (u^2=v^2+1;\, p\neq 0). \tag{87}$$

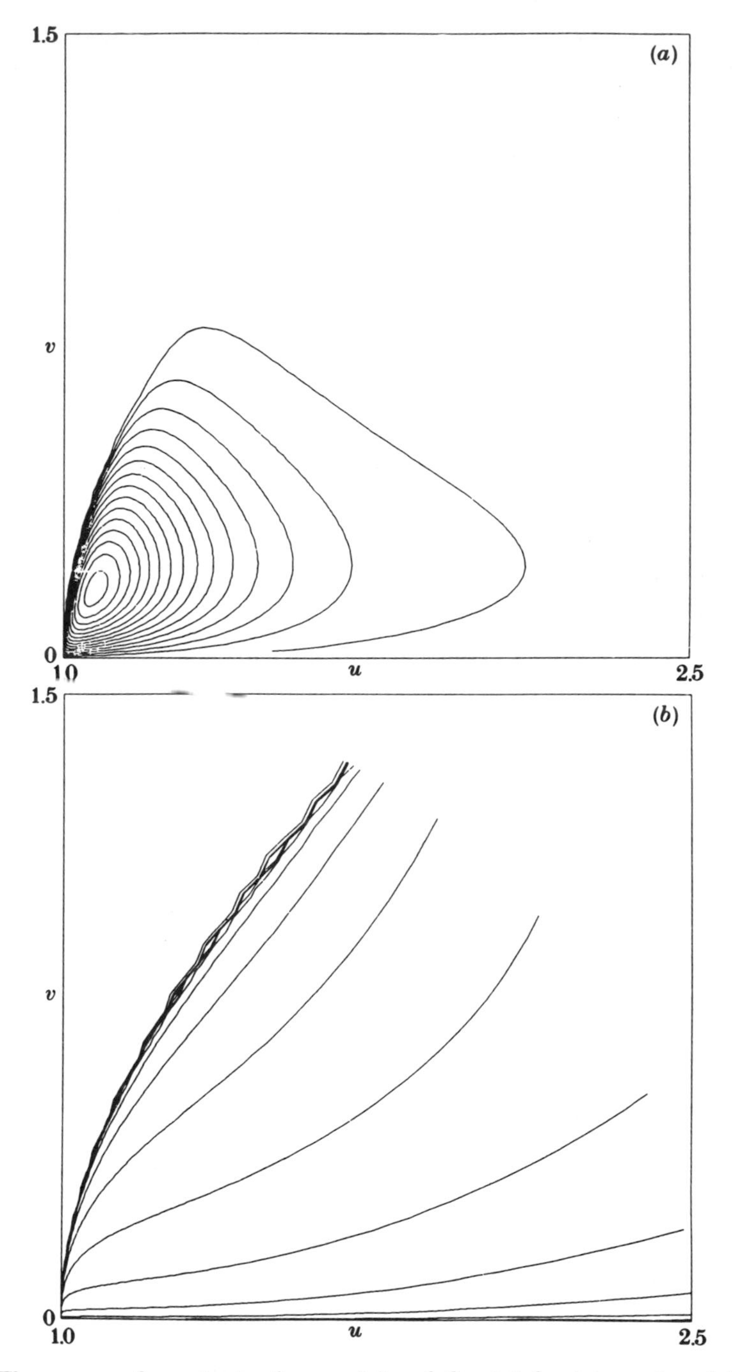

FIGURE 1. The curves of constant ϵ for $\alpha = 0.5$ and $C = 1.0$ for (a) $p = 1$ and (b) $p = 0$. The raggedness of the curves in (b), near the critical hyperbola $u^2 = v^2 + 1$, is due to errors in computer tracing.

But when $p = 0$, a more careful examination is necessary; and we find that

$$\Delta \to \frac{(u^2 - v^2 - 1)^2}{4u^2v^2} = \frac{s^2}{4u^2v^2} \quad (u^2 - v^2 - 1 \to +0). \tag{88}$$

From equations (84), (87) and (88) we now conclude that

$$\left.\begin{aligned} \epsilon &\to \frac{C^2\alpha^4}{p^2}\, s^{\frac{1}{2}}\, e^{-2\alpha r} \quad (p \neq 0), \\ &\to \frac{16C^2(\alpha uv)^4}{s^{\frac{3}{2}}}\, e^{-2\alpha r} \quad (p = 0) \end{aligned}\right\} \quad (s \to 0). \tag{89}$$

It is important to observe that ϵ generically vanishes along the hyperbola, $u^2 = v^2 + 1$; it diverges only in the exceptional case, $p = 0$.

Similarly, from the asymptotic behaviour of the various quantities for $s \to \infty$, we find that

$$\epsilon \to \frac{2C^2\alpha^2}{\pi}\, e^{-f_\infty}\, \frac{u(u^2-1)^{\frac{1}{2}}\, v(v^2+1)^{\frac{1}{2}}}{\Delta}\, s^{-\frac{3}{2}}\, e^{-2\alpha(2v^2+1)} \quad (s \to \infty). \tag{90}$$

(Note that $u\sqrt{(u^2-1)}/\Delta$ tends to a finite limit for $s \to \infty$.)

The behaviour of ϵ predicted for $s \to 0$ and $s \to \infty$ is exhibited by the curves of constant ϵ illustrated in figure 1*a* (for $q = 0$ and $p = 1$) and figure 1*b* (for $q = 1$ and $p = 0$). We are grateful to Mr S. K. Chakrabarti for obtaining computer-traced drawings for these illustrations.

(*d*) *The description of the space–time in a Newman–Penrose formalism and the occurrence of a space–time singularity on $u^2 = v^2 + 1$*

As in Paper I, §10, we can relate the spin coefficients and the Weyl scalars for the problem we are presently considering to the same quantities that pertain to the vacuum. Thus, distinguishing by tildes the quantities that obtain for $f \neq 0$ from the quantities that obtain for $f = 0$, we find that they are formally related exactly as in Paper I, equations (100)–(104), with the only difference that U must now be defined by

$$U = \frac{\sqrt{\Delta}}{(\sinh\psi \sinh\theta)^{\frac{1}{4}}} = \frac{(p^2 \sinh^2\psi + q^2 \sinh^2\theta)^{\frac{1}{4}}}{(\sinh\psi \sinh\theta)^{\frac{1}{4}}}, \tag{91}$$

instead of by (Paper I, equation (100))

$$U = \frac{\sqrt{\Delta}}{(\sin\psi \sin\theta)^{\frac{1}{4}}} = \frac{(p^2 \sin^2\psi + q^2 \sin^2\theta)^{\frac{1}{4}}}{(\sin\psi \sin\theta)^{\frac{1}{4}}}. \tag{92}$$

In particular, we have

$$\tilde{\Psi}_0 = e^{-f}(\Psi_0 - \sigma \mathbf{D} f), \quad \tilde{\Psi}_4 = e^{-f}(\Psi_4 + \lambda \mathbf{\Delta} f),$$

and

$$\tilde{\Psi}_2 = e^{-f}\, \Psi_2 + \tfrac{1}{3}\epsilon, \tag{93}$$

where

$$\mathbf{D} = \frac{1}{U\sqrt{2}}(\partial_\psi + \partial_\theta) \quad \text{and} \quad \mathbf{\Delta} = \frac{1}{U\sqrt{2}}(\partial_\psi - \partial_\theta), \tag{94}$$

with U as defined in equation (91).

The spin coefficients and the Weyl scalars for the vacuum appropriate for the metric (49) with $f = 0$ can be obtained as in Chandrasekhar & Ferrari (1984, §5). The null-tetrad basis must, however, be defined by

$$\begin{array}{llllll} & & \psi & \theta & x^1 & x^2 \\ (l_i) = +\dfrac{1}{\sqrt{2}} & (& U, & -U, & 0, & 0 \ ), \\ (n_i) = +\dfrac{1}{\sqrt{2}} & (& U, & +U, & 0, & 0 \ ), \\ (m_i) = -\dfrac{1}{\sqrt{2}} & (& 0, & 0, & V\mathcal{Q}^*_-, & +\mathrm{i}V\mathcal{Q}^*_+), \\ (\bar{m}_i) = -\dfrac{1}{\sqrt{2}} & (& 0, & 0, & V\mathcal{Q}_-, & -\mathrm{i}V\mathcal{Q}_+), \end{array} \tag{95}$$

where

$$U = \frac{\sqrt{\Delta}}{(\sinh\psi\sinh\theta)^{\frac{1}{4}}}, \quad V = (\sinh\psi\sinh\theta)^{\frac{1}{2}}, \tag{96}$$

and

$$\mathcal{Q}_\pm = \frac{E\pm 1}{\sqrt{\Delta}} \quad (E = p\cosh\psi + \mathrm{i}q\cosh\theta). \tag{97}$$

We find that the expressions for the spin coefficients and the Weyl scalars can be directly transcribed from those given by Chandrasekhar & Ferrari (1984, equations (86), (91), (94), and (95); see also the errata in Chandrasekhar & Ferrari 1985) by simply replacing the trigonometric functions by the corresponding hyperbolic functions. Thus, we now have

$$\Psi_2 = \frac{(\sinh\psi\sinh\theta)^{\frac{1}{2}}}{2\Delta^3}\left\{(p\sinh\psi - \mathrm{i}q\sinh\theta)^2 - \tfrac{1}{4}\Delta^2\frac{\sinh(\theta+\psi)\sinh(\theta-\psi)}{(\sinh\theta\sinh\psi)^2}\right\}, \tag{98}$$

$$\Psi_0 = \frac{3}{4}\,\frac{\sinh(\theta-\psi)}{(\sinh\psi\sinh\theta)^{\frac{1}{2}}(p\sinh\psi+\mathrm{i}q\sinh\theta)^3}, \tag{99}$$

$$\Psi_4 = \frac{3}{4}\,\frac{\sinh(\theta+\psi)}{(\sinh\psi\sinh\theta)^{\frac{1}{2}}(p\sinh\psi+\mathrm{i}q\sinh\theta)^3}. \tag{100}$$

The occurrence of a space–time singularity on the hyperbola, $u^2 = v^2+1$, can be deduced from the behaviour of Ψ_2 for $s\to 0$. Since $f = 0$ for $s = 0$ by equation (83), the term $\Psi_2\,\mathrm{e}^{-f}$ in $\tilde{\Psi}_2$ has the behaviour

$$\left.\begin{aligned} \mathrm{e}^{-f}\,\Psi_2 &\to \frac{\sinh^2\psi}{8\Delta s^{\frac{3}{2}}} \quad (p\neq 0), \\ &\to -\frac{3s^{\frac{5}{2}}}{8\Delta^3\sinh^2\psi} \quad (p = 0) \end{aligned}\right\} \quad (s\to 0). \tag{101}$$

Therefore, in accordance with equations (53), (87) and (88),

$$\left.\begin{aligned} \mathrm{e}^{-f}\,\Psi_2 &\to \frac{1}{8p^2 s^{\frac{3}{2}}} \quad (p\neq 0), \\ &\to -\frac{6u^4v^4}{s^{\frac{7}{2}}} \quad (p = 0) \end{aligned}\right\} \quad (s\to 0). \tag{102}$$

Comparison with the behaviour (89) of ϵ shows that $\Psi_2\,e^{-f}$ dominates ϵ when $s\to 0$. The behaviours of Ψ_2 for $s\to 0$ is therefore the same as that of $\Psi_2\,e^{-f}$. The occurrence of a space–time singularity on the hyperbola, $u^2 = v^2+1$, is manifest.

Finally, we return to the matter to which reference was made earlier in §4 (*a*), namely to the different definition, $u = \sinh\xi$ and $v = \cosh\zeta$, of the null coordinates that we could have made. It is now clear that the only difference that would have made is to interchange the roles of u and v and restrict the domain of the solution to

$$v \geqslant 1 \quad \text{and} \quad v^2 \leqslant (u^2+1), \tag{103}$$

instead of to the right of the hyperbola $u^2 = v^2+1$.

5. The case $\Delta = \eta^2+1$ and $\delta = \mu^2-1$

When $\Delta = \eta^2+1$ and $\delta = \mu^2-1$, the Ernst equation (24) allows a solution of the form

$$E = p\eta + iq\mu \tag{104}$$

where p and q are constants and

$$q^2-p^2 = 1. \tag{105}$$

(Note the difference with the condition (43) in the earlier case.) We find that the corresponding solutions for χ, q_2, $\nu+\mu_3$, and μ_3 are:

$$\chi = \frac{p^2\eta^2+q^2\mu^2-1}{(1-p\eta)^2+q^2\mu^2}, \quad q_2 = \frac{2q\mu}{(1-p\eta)^2+q^2\mu^2}, \tag{106}$$

$$e^{\nu+\mu_3} = \frac{p^2\eta^2+q^2\mu^2-1}{(\eta^2+1)^{\frac{3}{4}}(\mu^2-1)^{\frac{1}{4}}}, \quad e^{2\mu_3} = \frac{p^2\eta^2+q^2\mu^2-1}{(\eta^2+1)^{\frac{1}{4}}(\mu^2-1)^{\frac{1}{4}}}. \tag{107}$$

Now letting

$$\eta = \sinh\psi \quad \text{and} \quad \mu = \cosh\theta, \tag{108}$$

the metric (39) takes the form (cf. equation (49))

$$\begin{aligned} ds^2 = {} & \frac{\Delta\, e^f}{(\cosh\psi\,\sinh\theta)^{\frac{1}{2}}}\,[(d\psi)^2-(d\theta)^2] \\ & -\frac{\cosh\psi\,\sinh\theta}{\Delta}\,\{|1+E|^2\,(dx^2)^2+|1-E|^2\,(dx^1)^2-4q(\cosh\theta)\,dx^1\,dx^2\}, \end{aligned} \tag{109}$$

where

$$\Delta = |E|^2-1 = p^2\cosh^2\psi+q^2\sinh^2\theta. \tag{110}$$

(*a*) *The transformation to null coordinates*

We now introduce a system of null coordinates, u and v, in the same manner as in §4 (*a*), equations (51):

$$\psi = \xi+\zeta, \quad \theta = \xi-\zeta, \quad u = \cosh\xi \text{ and } v = \sinh\zeta. \tag{111}$$

By these definitions, we now have, in contrast to equations (52)–(54),

$$\eta = (u^2-1)^{\frac{1}{2}}(v^2+1)^{\frac{1}{2}}+uv = \sinh\psi, \tag{112}$$

$$\mu = u(v^2+1)^{\frac{1}{2}}-v(u^2-1)^{\frac{1}{2}} = \cosh\theta \tag{113}$$

and

$$\cosh\psi\,\sinh\theta = u(u^2-1)^{\frac{1}{2}}-v(v^2+1)^{\frac{1}{2}}. \tag{114}$$

In terms of the (ξ, ζ)-variables, equation (55) governing ϕ (which is valid in the present case, as well) takes the form

$$\phi_{,\xi,\zeta}+\frac{1}{2\cosh(\xi+\zeta)\sinh(\xi-\zeta)}[(\cosh 2\zeta)\phi_{,\xi}-(\cosh 2\xi)\phi_{,\zeta}]=0. \tag{115}$$

In the (u, v)-variables this equation becomes

$$\phi_{,u,v}+\frac{1}{2[u\sqrt{(u^2-1)}-v\sqrt{(v^2+1)}]}\left[\frac{2v^2+1}{\sqrt{(v^2+1)}}\phi_{,u}-\frac{2u^2-1}{\sqrt{(u^2-1)}}\phi_{,v}\right]=0. \tag{116}$$

With our present definitions, we find that while equations (60) and (61) are unaltered, equation (59) is replaced by

$$2\phi_{,\eta}\phi_{,\mu}=\frac{1}{2[u\sqrt{(u^2-1)}-v\sqrt{(v^2+1)}]}[(u^2-1)(\phi_{,u})^2-(v^2+1)(\phi_{,v})^2]. \tag{117}$$

Turning to equations (36) and (37) governing f and solving for $f_{,\eta}$ and $f_{,\mu}$ we now find (in contrast to equations (62) and (63)):

$$f_{,\eta}=\frac{4}{\eta^2+\mu^2}\left\{+\frac{\eta}{\eta^2+1}[\Delta(\phi_{,\eta})^2+\delta(\phi_{,\mu})^2]-2\mu\phi_{,\eta}\phi_{,\mu}\right\}, \tag{118}$$

$$f_{,\mu}=\frac{4}{\eta^2+\mu^2}\left\{-\frac{\mu}{\mu^2-1}[\Delta(\phi_{,\eta})^2+\delta(\phi_{,\mu})^2]+2\eta\phi_{,\eta}\phi_{,\mu}\right\}. \tag{119}$$

Inserting these expressions in

$$f_{,u}=(v^2+1)^{\frac{1}{2}}\left[\frac{u}{\sqrt{(u^2-1)}}f_{,\eta}+f_{,\mu}\right]+v\left[f_{,\eta}-\frac{u}{\sqrt{(u^2-1)}}f_{,\mu}\right], \tag{120}$$

$$f_{,v}=(u^2-1)^{\frac{1}{2}}\left[\frac{v}{\sqrt{(v^2+1)}}f_{,\eta}-f_{,\mu}\right]+u\left[f_{,\eta}+\frac{v}{\sqrt{(v^2+1)}}f_{,\mu}\right] \tag{121}$$

(which follow from equations (112) and (113)) and simplifying, we find

$$f_{,u}=-\frac{4\sqrt{(u^2-1)}}{(2u^2-1)[u\sqrt{(u^2-1)}-v\sqrt{(v^2+1)}]}(\phi_{,u})^2, \tag{122}$$

and

$$f_{,v}=+\frac{4\sqrt{(v^2+1)}}{(2v^2+1)[u\sqrt{(u^2-1)}-v\sqrt{(v^2+1)}]}(\phi_{,v})^2. \tag{123}$$

Finally, the expression for ϵ corresponding to equation (67) is

$$\epsilon=-\frac{e^{-2\mu_3-f}}{[u\sqrt{(u^2-1)}-v\sqrt{(v^2+1)}]^2}(u^2-1)^{\frac{1}{2}}(v^2+1)^{\frac{1}{2}}\phi_{,u}\phi_{,v}. \tag{124}$$

(b) The solutions for ϕ and f

Letting

$$r=u\sqrt{(u^2-1)}+v\sqrt{(v^2+1)}$$

and

$$s=u\sqrt{(u^2-1)}-v\sqrt{(v^2+1)}, \tag{125}$$

we find that equations (116), (122), and (123) reduce to

$$\phi_{,r,r} - \left(\phi_{,s,s} - \frac{1}{s}\phi_{,s}\right) = 0, \tag{126}$$

and

$$f_{,r} = -\frac{8}{s}\phi_{,r}\phi_{,s}, \quad f_{,s} = -\frac{4}{s}[(\phi_{,r})^2 + (\phi_{,s})^2]. \tag{127}$$

We observe that equations (126) and (127) are the same as equations (69) and (77). Therefore, the solutions for ϕ and f given in §4 (*b*) and, in particular, the fundamental solutions (70) and (78) apply equally to the present case with the difference that the variables r and s are now related differently to the null variables u and v. Because of this last difference,

$$\phi_{,u} = \frac{2u^2-1}{\sqrt{(u^2-1)}}(\phi_{,r}+\phi_{,s}) \quad \text{and} \quad \phi_{,v} = \frac{2v^2+1}{\sqrt{(v^2+1)}}(\phi_{,r}-\phi_{,s}), \tag{128}$$

and the expression (117) for ϵ gives

$$\epsilon = C^2\alpha^4(2u^2-1)(2v^2+1)\frac{e^{-f}}{\Delta}s^{\frac{1}{2}}e^{-2\alpha r}[I_0^2(\alpha s) - I_1^2(\alpha s)]. \tag{129}$$

(*c*) *The character of the distribution of* ϵ *in the domain* $u \geqslant 1$ *and* $u^2-(v^2+1) \geqslant 0$

It is evident that the solutions obtained in §§5 (*a*), (*b*) are limited to the domain,

$$u \geqslant 1, \quad \text{any} \quad v, \quad \text{and} \quad s = u\sqrt{(u^2-1)} - v\sqrt{(v^2+1)} \geqslant 0, \tag{130}$$

or since

$$s = \frac{(u^2+v^2)(u^2-v^2-1)}{u\sqrt{(u^2-1)}+v\sqrt{(v^2+1)}}, \tag{131}$$

it follows that the domain of validity of the solution is to the right of the hyperbola $u = \sqrt{(v^2+1)}$.

Since the asymptotic behaviour (80) and (81) of the Bessel functions apply equally to the present case, the behaviour of ϕ and f for $s \to 0$ and $s \to \infty$ is the same when they are expressed in terms of the variables r and s. Therefore, the behaviour as given in equations (82) and (83) is applicable, as they stand, for $s \to 0$; but for $s \to \infty$ the terms in

$$e^{-\alpha(2v^2+1)} = e^{-\alpha(r-s)}, \tag{132}$$

should be replaced by (cf. equations (125))

$$e^{-2\alpha v\sqrt{(v^2+1)}}; \tag{132'}$$

for example

$$f \to -\frac{2C^2\alpha^2}{\pi}e^{-4\alpha v\sqrt{(v^2+1)}} = f_\infty \quad \text{(say)} \quad (s \to \infty). \tag{133}$$

From the expression for ϵ given in equation (129), it now follows that

$$\epsilon \to C^2\alpha^4(u^2+v^2)^2\frac{s^{\frac{1}{2}}}{\Delta}e^{-4\alpha u\sqrt{(u^2-1)}} \quad (s \to 0). \tag{134}$$

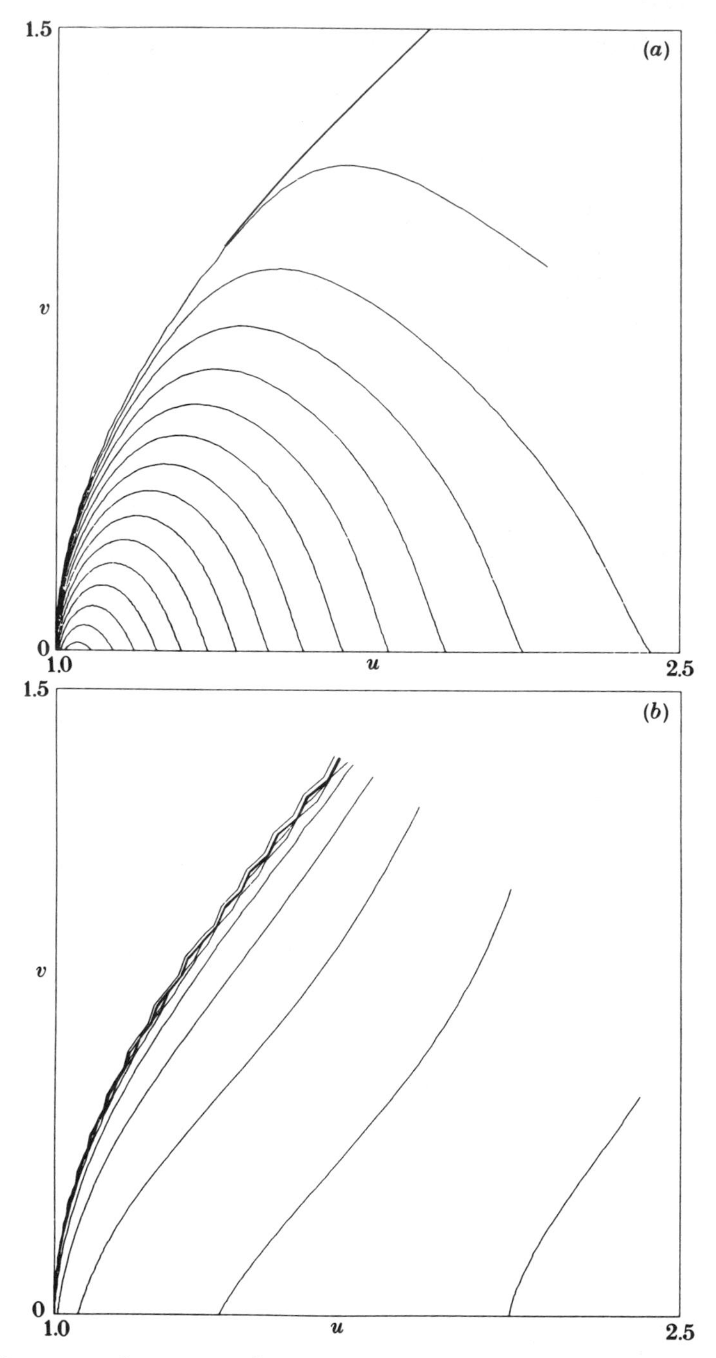

FIGURE 2. The curves of constant ϵ for $\alpha = 0.5$ and $C = 1.0$ for (*a*) $p = \sqrt{0.5}$ and (*b*) $p = 0$. The raggedness of the curves in (*b*), near the critical hyperbola $u^2 = v^2 + 1$, is due to errors in computer tracing.

Since, as may be readily verified, Δ, as now defined, has the behaviour

$$\left.\begin{aligned} \Delta &\to p^2(1+4u^2v^2) \quad (p \neq 0), \\ &\to s^2/(u^2+v^2)^2 \quad (p=0) \end{aligned}\right\} (s\to 0), \tag{135}$$

we conclude that

$$\left.\begin{aligned} \epsilon &\to C^2\alpha^4 \frac{(u^2+v^2)^2}{p^2(1+4u^2v^2)} s^{\frac{1}{2}} \,\mathrm{e}^{-4\alpha u\sqrt{(u^2-1)}} \quad (p\neq 0), \\ &\to C^2\alpha^4(u^2+v^2)^4 s^{-\frac{3}{2}} \,\mathrm{e}^{-4\alpha u\sqrt{(u^2-1)}} \quad (p=0) \end{aligned}\right\} (s\to 0). \tag{136}$$

Again, as in the case considered in §4, ϵ vanishes generically on the hyperbola, $u^2 = v^2+1$.

Similarly, we find that

$$\epsilon \to \frac{C^2\alpha^2}{2\pi}\frac{(2u^2-1)(2v^2+1)}{\Delta}\mathrm{e}^{-f_\infty}\,\mathrm{e}^{-4\alpha v\sqrt{(v^2+1)}}\, s^{-\frac{3}{2}} \quad (s\to\infty), \tag{137}$$

where f_∞ is defined in equation (133). (Note that $(2u^2-1)(2v^2+1)/\Delta$ tends to a finite limit when $s\to\infty$.)

The behaviour of ϵ predicted for $s\to 0$ and $s\to\infty$ is exhibited by the curves of constant ϵ illustrated in figure 2a ($p\neq 0$) and figure 2b ($p=0$).

(d) *The description of the space–time in a Newman–Penrose formalism and the occurrence of a space–time singularity on $u^2 = v^2+1$*

The Weyl scalars when $f\neq 0$ are related to the Weyl scalars for the vacuum (when $f=0$) by the same formulae (93) and (94). The spin coefficients and the Weyl scalars for the vacuum, appropriate for the metric (109) with $f=0$, can be obtained by defining a null-tetrad exactly as in equation (95) with U, V, Δ, and E now defined by

$$U = \frac{\sqrt{\Delta}}{(\cosh\psi\sinh\theta)^{\frac{1}{4}}}, \quad V = (\cosh\psi\sinh\theta)^{\frac{1}{4}},$$

$$E = p\sinh\psi + \mathrm{i}q\cosh\theta \quad \text{and} \quad \Delta = p^2\cosh^2\psi + q^2\sinh^2\theta. \tag{138}$$

We find that in place of equations (98)–(100) we now have

$$\Psi_2 = \frac{(\cosh\psi\sinh\theta)^{\frac{1}{2}}}{2\Delta^3}\left\{(p\cosh\psi - \mathrm{i}q\sinh\theta)^2 + \tfrac{1}{4}\Delta^2\frac{\cosh(\theta+\psi)\cosh(\theta-\psi)}{(\cosh\psi\sinh\theta)^2}\right\}, \tag{139}$$

$$\Psi_0 = -\frac{3}{4}\frac{\cosh(\theta-\psi)}{(\cosh\psi\sinh\theta)^{\frac{1}{2}}(p\cosh\psi+\mathrm{i}q\sinh\theta)^3}, \tag{140}$$

$$\Psi_4 = +\frac{3}{4}\frac{\cosh(\theta+\psi)}{(\cosh\psi\sinh\theta)^{\frac{1}{2}}(p\cosh\psi+\mathrm{i}q\sinh\theta)^3}. \tag{141}$$

(In the Appendix general expressions for the Weyl and the Ricci scalars are given, which are applicable to all cases considered in this paper.)

The occurrence of a space–time singularity on the hyperbola $u^2 = v^2+1$ (or, equivalently, $s=0$) can be deduced from the behaviour of Ψ_2 for $s\to 0$. Since $f=0$

for $s = 0$ (by equation (83)) the term $\Psi_2 \, e^{-f}$ in $\tilde{\Psi}_2$ has the behaviour (cf. equations (101))

$$\left.\begin{aligned} \Psi_2 \, e^{-f} &\to \frac{\cosh^2 \psi}{8\Delta s^{\frac{3}{2}}} \quad (p \neq 0), \\ &\to -\frac{3 s^{\frac{5}{2}}}{8\Delta^3 \cosh^2 \psi} \quad (p = 0) \end{aligned}\right\} (s \to 0); \tag{142}$$

or, by virtue of equations (112) and (135),

$$\left.\begin{aligned} \Psi_2 \, e^{-f} &\to \frac{1}{8p^2} s^{-\frac{3}{2}} \quad (p \neq 0), \\ &\to -\frac{3(u^2+v^2)^6}{8(1+4u^2v^2)} s^{-\frac{7}{2}} \quad (p = 0) \end{aligned}\right\} (s \to 0). \tag{143}$$

Comparison with the behaviour (136) of ϵ shows that $\Psi_2 \, e^{-f}$ dominates ϵ when $s \to 0$. Therefore, the behaviour of $\tilde{\Psi}_2$ for $s \to 0$ is the same as that of $\Psi_2 \, e^{-f}$, i.e. of Ψ_2. The occurrence of a space–time singularity on the hyperbola, $u^2 = v^2 + 1$, is manifest.

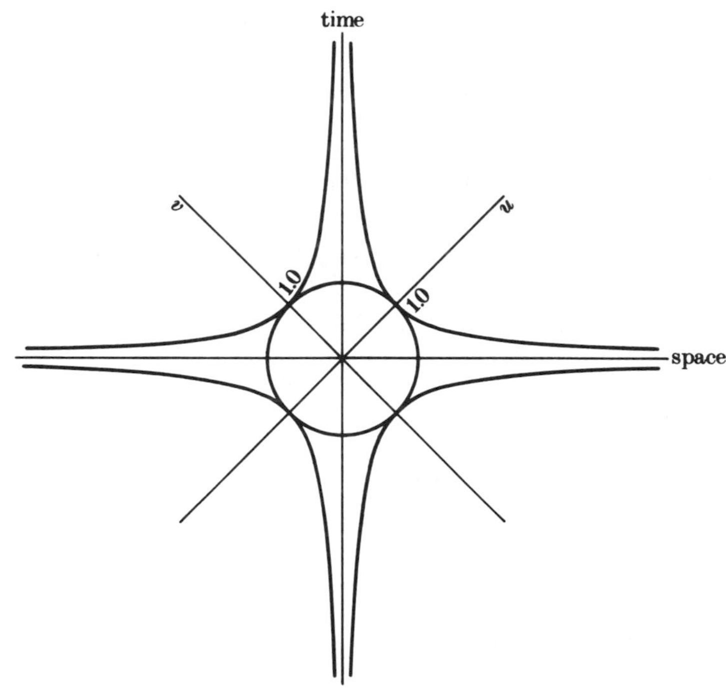

FIGURE 3. The domain in the (u, v)-plane occupied by the solutions found in the present paper and in Paper I. Solutions for impulsive waves in the region of interaction found in Paper I are represented inside the circle while the solutions found in this paper are external to the four hyperbolae.

It can be readily verified that the case $\varDelta = \eta^2 - 1$ and $\delta = \mu^2 + 1$ leads to essentially the same space–time as the case $\varDelta = \eta^2 + 1$ and $\delta = \mu^2 - 1$ we have just considered; the only difference is that the hyperbola on which the space–time singularity occurs has to be rotated to a different quadrant (see figure 3).

6. Concluding remarks

The principal aim of this paper, in seeking solutions in the alternate gauges and coordinates of Einstein's equations with a perfect-fluid source with the equation of state, $\epsilon = p$, was to supplement the solution obtained in Paper I, in a gauge and in a coordinate system appropriate for treating the collision of impulsive gravitational waves. The solutions that have been obtained enlarge the domain in the (u, v)-plane in which solutions are represented (see figure 3). Besides, the two distinct classes of solutions obtained with the choices ($\Delta = \eta^2-1$, $\delta = \mu^2-1$) and ($\Delta = \eta^2+1$, $\delta = \mu^2-1$) are inextensible since their domain of validity is bounded entirely by a space–time singularity. The solutions, in some ways, resemble the standard solutions of cosmology in that any time-like trajectory, extended into the past, encounters a singularity. Also, since ϵ vanishes in the singularity in the generic case, it would appear that a perfect fluid with $\epsilon = p$ is generated by null dust at the singularity by analogy with what happens along the null boundaries in the problem of colliding waves considered in Paper I. Finally, the distribution of ϵ that evolves from the singularity and the emergence of density maxima suggest that the solutions may have relevance for the physics of the early universe.

The research reported in this paper has, in part, been supported by grants from the National Science Foundation under grant PHY 84-16691 with the University of Chicago.

References

Chandrasekhar, S. & Ferrari, V. 1984 *Proc. R. Soc. Lond.* A **396**, 55–74.
Chandrasekhar, S. & Ferrari, V. 1985 *Proc. R. Soc. Lond.* A **398**, 429.
Chandrasekhar, S. & Xanthopoulos, B. C. 1985 *Proc. R. Soc. Lond.* A **402**, 37–65.
Watson, G. N. 1944 *A treatise on the theory of Bessel functions*. Cambridge University Press.

Appendix

The examples of space–times considered in this paper show that a metric of the form

$$ds^2 = U^2[(d\psi)^2-(d\theta)^2]-\frac{V^2}{|E|^2-1}|(1-E)\,dx^1+i(1+E)\,dx^2|^2 \tag{A 1}$$

is of frequent occurrence in the theory of space–times with two space-like Killing-vectors. For this reason, it will be convenient to have listed the spin coefficients and the Weyl and the Ricci scalars for the general case when U and V are allowed to be any two real functions and E any complex function of ψ and θ.

For a metric of the general form (A 1), we can define a null-tetrad basis exactly as in equation (95), where, besides U and V,

$$\left.\begin{aligned}\mathscr{Q}_\pm &= \frac{E\pm1}{\Delta}, \quad \Delta = |E|^2-1,\\ \text{and} \qquad \mathscr{Q}_-\mathscr{Q}_+^* &+ \mathscr{Q}_+\mathscr{Q}_-^* = 2.\end{aligned}\right\} \tag{A 2}$$

The directional derivatives $\mathbf{D}$ and $\mathbf{\Delta}$ are also defined as in equation (94).

We find that the non-vanishing spin coefficients and the Weyl and the Ricci scalars are given by:

$$\sigma = -\frac{(E^*_{,\psi}+E^*_{,\theta})}{U\Delta\sqrt{2}}, \quad \rho = -\frac{(V_{,\psi}+V_{,\theta})}{UV\sqrt{2}},$$

$$\lambda = +\frac{(E_{,\psi}-E_{,\theta})}{U\Delta\sqrt{2}}, \quad \mu = +\frac{(V_{,\psi}-V_{,\theta})}{UV\sqrt{2}},$$

$$\epsilon = +\frac{U_{,\psi}+U_{,\theta}}{2U^2\sqrt{2}}+\frac{1}{4U\Delta\sqrt{2}}[E(E^*_{,\psi}+E^*_{,\theta})-E^*(E_{,\psi}+E_{,\theta})],$$

$$\gamma = -\frac{U_{,\psi}-U_{,\theta}}{2U^2\sqrt{2}}+\frac{1}{4U\Delta\sqrt{2}}[E(E^*_{,\psi}-E^*_{,\theta})-E^*(E_{,\psi}-E_{,\theta})]; \tag{A 3}$$

$$\Psi_0 = -\frac{E^*_{,\psi}+E^*_{,\theta}}{U^2\Delta}\left\{\left(\ln\frac{V}{U}\right)_{,\psi}+\left(\ln\frac{V}{U}\right)_{,\theta}+\frac{E^*_{,\psi,\psi}+E^*_{,\theta,\theta}+2E^*_{,\psi,\theta}}{2(E^*_{,\psi}+E^*_{,\theta})}-\frac{E(E^*_{,\psi}+E^*_{,\theta})}{\Delta}\right\},$$

$$\Psi_4 = -\frac{E_{,\psi}-E_{,\theta}}{U^2\Delta}\left\{\left(\ln\frac{V}{U}\right)_{,\psi}-\left(\ln\frac{V}{U}\right)_{,\theta}+\frac{E_{,\psi,\psi}+E_{,\theta,\theta}-2E_{,\psi,\theta}}{2(E_{,\psi}-E_{,\theta})}-\frac{E^*(E_{,\psi}-E_{,\theta})}{\Delta}\right\},$$

$$\Psi_2 = \frac{1}{6U^2}\left[\left(\ln\frac{V}{U}\right)_{,\psi,\psi}-\left(\ln\frac{V}{U}\right)_{,\theta,\theta}\right]+\frac{2(E_{,\psi}E^*_{,\psi}-E_{,\theta}E^*_{,\theta})+3(E_{,\psi}E^*_{,\theta}-E_{,\theta}E^*_{,\psi})}{6U^2\Delta^2} \tag{A 4}$$

and

$$\Phi_{00} = \frac{(V_{,\psi}+V_{,\theta})(U_{,\psi}+U_{,\theta})}{U^3V}-\frac{V_{,\psi,\psi}+V_{,\theta,\theta}+2V_{,\psi,\theta}}{2U^2V}-\frac{(E_{,\psi}+E_{,\theta})(E^*_{,\psi}+E^*_{,\theta})}{2U^2\Delta^2},$$

$$\Phi_{22} = \frac{(V_{,\psi}-V_{,\theta})(U_{,\psi}-U_{,\theta})}{U^3V}-\frac{V_{,\psi,\psi}+V_{,\theta,\theta}-2V_{,\psi,\theta}}{2U^2V}-\frac{(E_{,\psi}-E_{,\theta})(E^*_{,\psi}-E^*_{,\theta})}{2U^2\Delta^2},$$

$$\Phi_{11} = -\frac{1}{4U^2}\left\{(\ln U)_{,\psi,\psi}-(\ln U)_{,\theta,\theta}-[(\ln V)_{,\psi}]^2+[(\ln V)_{,\theta}]^2\right\}-\frac{E_{,\psi}E^*_{,\psi}-E_{,\theta}E^*_{,\theta}}{4U^2\Delta^2},$$

$$\Phi_{20} = -\frac{E_{,\psi,\psi}-E_{,\theta,\theta}}{2U^2\Delta}-\frac{E^*[(E_{,\psi})^2-(E_{,\theta})^2]}{U^2\Delta^2}-\frac{E_{,\psi}V_{,\psi}-E_{,\theta}V_{,\theta}}{U^2V\Delta},$$

$$\Lambda = \frac{E_{,\psi}E^*_{,\psi}-E_{,\theta}E^*_{,\theta}}{12U^2\Delta^2}+\frac{1}{12U^2}[(\ln U)_{,\psi,\psi}-(\ln U)_{,\theta,\theta}]$$

$$+\frac{1}{9U^2V^{\frac{3}{2}}}[(V^{\frac{3}{2}})_{,\psi,\psi}-(V^{\frac{3}{2}})_{,\theta,\theta}]. \tag{A 5}$$

It may be noted that the vanishing of the Weyl scalars, Ψ_1 and Ψ_3, and of the Ricci scalars, Φ_{01} and Φ_{02}, is guaranteed by the existence of two space-like Killing vectors. Also, for the vacuum the expression for Ψ_2 can be simplified to the form

$$\Psi_2 = -\frac{1}{2U^2}\{[(\ln V)_{,\psi}]^2-[(\ln V)_{,\theta}]^2\}+\frac{(E_{,\psi}-E_{,\theta})(E^*_{,\psi}+E^*_{,\theta})}{2U^2\Delta^2}. \tag{A 6}$$

PART THREE

Cylindrical Waves

Cylindrical waves in general relativity

By Subrahmanyan Chandrasekhar, F.R.S.
The University of Chicago, Chicago, Illinois 60637, *U.S.A.*

(*Received* 1 *May* 1986)

A convenient framework is set up for constructing cylindrically symmetric solutions of the Einstein and the Einstein–Maxwell equations, and it is shown how a Hamiltonian density can be defined for space–times with cylindrical symmetry. Solutions are obtained that represent stationary monochromatic waves and satisfy all the requisite conditions of regularity. The case when the gravitational field is coupled with a perfect fluid in which the energy density is equal to the pressure is also briefly considered.

1. Introduction

Gravitational waves with cylindrical symmetry were first considered by Einstein & Rosen (1937) in a famous paper. They obtained some special solutions which have come to be called the Einstein–Rosen waves (see Appendix B for a derivation of these solutions in the framework of the present paper and for some additional historical remarks). The subject of cylindrical waves has attracted attention, periodically, since that time; and contributions have been made by Rosen (1954), Marder (1958*a*, *b*), Thorne (1965), and Stachel (1966), among others. (For a full bibliography see Kramer *et al.* (1980), pp. 220–27.) More recently, the evolution of a cylindrical pulse of gravitational radiation has been followed by numerical integrations of the relevant equations by Piran *et al.* (1985).

The object of the present paper is twofold: *first*, to present the theory of cylindrical waves in general relativity in the same frame-work in which the collision of plane-fronted waves has been studied in a series of papers published in these *Proceedings* during the past two years (Chandrasekhar & Ferrari 1984; Chandrasekhar & Xanthopoulos 1985*a*, *b*, 1986*a*, *b*); and *second*, to consider a class of separable solutions of the full set of equations which represent *stationary* (in a sense that we shall define) *monochromatic waves* and satisfy all the physical requirements. Generalizations of the solutions, when gravitation is coupled with an electromagnetic field or a perfect fluid with the equation of state ϵ (the energy density) $= p$ (the pressure), are briefly considered. The solutions obtained in this paper have an additional mathematical interest in that they are reduced to *Painlevé* transcendents (see Ince 1927, p. 345).

2. The choice of gauge and of coordinates

A space–time with two commuting space-like Killing vectors is said to have *cylindrical symmetry* about an axis if the orbits of one of them are closed while those of the other are open, in contrast to space–times that represent plane-fronted gravitational waves in which the orbits of both Killing vectors are open. (More precisely, the condition for cylindrical symmetry is that among the two commuting space-like Killing vectors there is exactly one linear combination of them, apart from a constant of proportionality, whose orbits are closed.) Letting ∂_φ and ∂_z denote the Killing vectors associated with the azimuthal angle φ and the coordinate z along the axis of symmetry, we can write the metric in the standard form,

$$ds^2 = e^{2\nu}(dt)^2 - e^{2\mu_3}(d\varpi)^2 - e^{2\psi}(dz - q_2\, d\varphi)^2 - e^{2\mu_2}(d\varphi)^2, \tag{1}$$

where ϖ denotes the radial coordinate and ν, μ_3, ψ, q_2, and μ are functions of $t\,(=x^0)$ and ϖ only. With our customary notation,

$$\Delta = e^{2(\mu_3-\nu)}, \quad \chi = e^{-\psi+\mu_2}, \quad \beta = \psi + \mu_2, \tag{2}$$

the metric takes the same form that has been considered in earlier contexts, namely,

$$ds^2 = \frac{e^{\mu_3+\nu}}{\sqrt{\Delta}}[(dt)^2 - \Delta(d\varpi)^2] - e^{\beta}\left[\chi(d\varphi)^2 + \frac{1}{\chi}(dz - q_2\, d\varphi)^2\right]. \tag{3}$$

However, in treating space–times with cylindrical symmetry, our choice of gauge and of coordinates will be different. We shall let

$$\mu_3 = \nu \quad \text{and} \quad \Delta = 1; \tag{4}$$

and we shall select

$$e^{\beta} = \varpi \tag{5}$$

as the solution of the field equation (Chandrasekhar & Ferrari 1984, equation (21); this paper will be referred to hereafter as Paper I),

$$(e^{\beta})_{,0,0} - (e^{\beta})_{,\varpi,\varpi} = 0. \tag{6}$$

With these choices, the mectric (3) becomes

$$ds^2 = e^{2\nu}[(dt)^2 - (d\varpi)^2] - \varpi[\chi(d\varphi)^2 + \frac{1}{\chi}(dz - q_2\, d\varphi)^2], \tag{7}$$

where it may be noted that

$$\mu_2 = \tfrac{1}{2}(\ln\varpi + \ln\chi) \quad \text{and} \quad \psi = \tfrac{1}{2}(\ln\varpi - \ln\chi). \tag{8}$$

If we are dealing with the vacuum equations and the solutions we seek are required to extend to $\varpi = 0$, then regularity and flatness of the metric along the symmetry axis ($\varpi = 0$) will require that the square of the norm of the Killing vector, ∂_φ, behaves like ϖ^2 for $\varpi \to 0$, i.e.

$$|\partial_\varphi|^2 = g_{\varphi\varphi} = -\varpi\left(\chi + \frac{q_2^2}{\chi}\right) = O(\varpi^2) \quad \text{for} \quad \varpi \to 0. \tag{9}$$

3. The reduction of the field equations and the 'C'-energy

The vacuum field equations have been written out in full in Paper I in a gauge and in a coordinate system suitable for colliding plane-fronted waves. They can be readily transcribed in the coordinate system of §2 suitable for space–times with cylindrical symmetry. Thus the equations governing χ and q_2 are (cf. Paper I, equations (35) and (36))

$$\left(\frac{\chi_{,0}}{\chi}\right)_{,0} - \frac{1}{\varpi}\left(\varpi\frac{\chi_{,\varpi}}{\chi}\right)_{,\varpi} = -\frac{1}{\chi^2}[(q_{2,0})^2 - (q_{2,\varpi})^2] \tag{10}$$

and

$$\left(\frac{\varpi}{\chi^2}q_{2,0}\right)_{,0} - \left(\frac{\varpi}{\chi^2}q_{2,\varpi}\right)_{,\varpi} = 0. \tag{11}$$

Alternative forms of these equations are:

$$\chi\left[\chi_{,0,0} - \frac{1}{\varpi}(\varpi\chi_{,\varpi})_{,\varpi}\right] = (\chi_{,0})^2 - (\chi_{,\varpi})^2 - (q_{2,0})^2 + (q_{2,\varpi})^2, \tag{12}$$

$$\chi\left[q_{2,0,0} - \frac{1}{\varpi}(\varpi q_{2,\varpi})_{,\varpi}\right] = 2q_{2,0}\chi_{,0} - 2q_{2,\varpi}\chi_{,\varpi}. \tag{13}$$

The remaining metric function, ν, follows by quadrature from the equations (cf. Paper I, equations (44) and (45))

$$\nu_{,0} = \frac{\varpi}{2\chi^2}(\chi_{,0}\chi_{,\varpi} + q_{2,0}q_{2,\varpi}), \tag{14}$$

$$4\nu_{,\varpi} = -\frac{1}{\varpi} + \frac{\varpi}{\chi^2}[(\chi_{,0})^2 + (\chi_{,\varpi})^2 + (q_{2,0})^2 + (q_{2,\varpi})^2]. \tag{15}$$

Equations (12) and (13) can be combined to yield an Ernst equation for $\chi + iq_2$ (as we do for $\Psi + i\Phi$ in §*a* below). However, as we shall show in Appendix A, simple solutions of the resulting Ernst equation do not provide acceptable solutions extending to $\varpi = 0$. We therefore proceed differently, as follows.

First, we define for q_2, consistently with equation (11), a potential Φ by letting,

$$\frac{\varpi}{\chi^2}q_{2,\varpi} = \Phi_{,0} \quad \text{and} \quad \frac{\varpi}{\chi^2}q_{2,0} = \Phi_{,\varpi}. \tag{16}$$

And associated with Φ, we define the function,

$$\Psi = \frac{\varpi}{\chi}. \tag{17}$$

By virtue of these definitions,

$$q_{2,\varpi} = \frac{\chi^2}{\varpi}\Phi_{,0} = \frac{\varpi}{\Psi^2}\Phi_{,0}; \quad q_{2,0} = \frac{\chi^2}{\varpi}\Phi_{,\varpi} = \frac{\varpi}{\Psi^2}\Phi_{,\varpi}, \tag{18}$$

$$\frac{\chi_{,\varpi}}{\chi} = \frac{1}{\varpi} - \frac{\Psi_{,\varpi}}{\Psi}; \quad \frac{\chi_{,0}}{\chi} = -\frac{\Psi_{,0}}{\Psi}. \tag{19}$$

With the aid of these equations, we find that equations (10) and (11) are replaced by equations of identical forms, namely,

$$\left(\frac{\Psi_{,0}}{\Psi}\right)_{,0} - \frac{1}{\varpi}\left(\varpi\frac{\Psi_{,\varpi}}{\Psi}\right)_{,\varpi} = -\frac{1}{\Psi^2}[(\Phi_{,0})^2 - (\Phi_{,\varpi})^2] \tag{20}$$

and

$$\left(\frac{\varpi}{\Psi^2}\Phi_{,0}\right)_{,0} - \left(\frac{\varpi}{\Psi^2}\Phi_{,\varpi}\right)_{,\varpi} = 0. \tag{21}$$

Alternative forms of these equations are (cf. equations (12) and (13)):

$$\Psi[\Psi_{,0,0}-\frac{1}{\varpi}(\varpi\Psi_{,\varpi})_{,\varpi}] = (\Psi_{,0})^2-(\Psi_{,\varpi})^2-(\Phi_{,0})^2+(\Phi_{,\varpi})^2, \tag{22}$$

$$\Psi[\Phi_{,0,0}-\frac{1}{\varpi}(\varpi\Phi_{,\varpi})_{,\varpi}] = 2\Phi_{,0}\Psi_{,0}-2\Phi_{,\varpi}\Psi_{,\varpi}. \tag{23}$$

In terms of Ψ and Φ, equations (14) and (15), governing ν, take the forms:

$$(\nu+\ln\sqrt{\Psi})_{,0} = \frac{\varpi}{2\Psi^2}(\Psi_{,0}\Psi_{,\varpi}+\Phi_{,0}\Phi_{,\varpi}), \tag{24}$$

$$(\nu+\ln\sqrt{\Psi})_{,\varpi} = \frac{\varpi}{4\Psi^2}[(\Psi_{,0})^2+(\Phi_{,0})^2+(\Psi_{,\varpi})^2+(\Phi_{,\varpi})^2]. \tag{25}$$

(*a*) *The reduction to an Ernst equation*

Letting

$$Z = \Psi+\mathrm{i}\Phi, \tag{26}$$

we can combine equations (22) and (23) into the single complex equation

$$\mathrm{Re}\,(Z)\left[Z_{,0,0}-\frac{1}{\varpi}(\varpi Z_{,\varpi})_{,\varpi}\right] = (Z_{,0})^2-(Z_{,\varpi})^2. \tag{27}$$

Now defining

$$E = \frac{Z-1}{Z+1} \quad \text{or} \quad Z = \frac{1+E}{1-E}, \tag{28}$$

we obtain the 'cylindrical' Ernst equation,

$$(1-|E|^2)\left[E_{,0,0}-\frac{1}{\varpi}(\varpi E_{,\varpi})_{,\varpi}\right] = -2E^*[(E_{,0})^2-(E_{,\varpi})^2]. \tag{29}$$

In terms of a solution E of this equation, the functions Ψ and Φ follow from

$$\Psi = \frac{1-|E|^2}{|1-E|^2} \quad \text{and} \quad \mathrm{i}\Phi = \frac{E-E^*}{|1-E|^2}. \tag{30}$$

And finally, we find that equations (24) and (25), in terms of E, take the very simple forms:

$$(\nu+\ln\sqrt{\Psi})_{,0} = \frac{\varpi}{(1-|E|^2)^2}(E_{,0}E^*_{,\varpi}+E^*_{,0}E_{\varpi}), \tag{31}$$

$$(\nu+\ln\sqrt{\Psi})_{,\varpi} = \frac{\varpi}{(1-|E|^2)^2}(|E_{,0}|^2+|E_{,\varpi}|^2). \tag{32}$$

(*b*) *The* 'C'-*energy*

First we observe that equations (31) and (32) can be combined to give

$$(\nu+\ln\sqrt{\Psi})_{,0}\pm(\nu+\ln\sqrt{\Psi})_{,\varpi} = \pm\frac{\varpi}{(1-|E|^2)^2}|E_{,0}\pm E_{,\varpi}|^2 \tag{33}$$

or, in terms of the radial null coordinates,

$$u = x^0 + \varpi \quad \text{and} \quad v = x^0 - \varpi, \tag{34}$$

we can write

$$\left.\begin{aligned} (\nu + \ln\sqrt{\Psi})_{,u} &= +\frac{2\varpi}{(1-|E|^2)^2}|E_{,u}|^2 \\ \text{and} \qquad (\nu + \ln\sqrt{\Psi})_{,v} &= -\frac{2\varpi}{(1-|E|^2)^2}|E_{,v}|^2. \end{aligned}\right\} \tag{35}$$

From equation (35) *it follows that the flow of* $(\nu + \ln\sqrt{\Psi})$ *is positive-definite in the outward null direction u and negative-definite in the inward null direction v. Besides* $(\nu + \ln\sqrt{\Psi})$, as we shall now show, *is expressible, in a gauge-invariant fashion, in terms of the Killing vectors,* ∂_φ *and* ∂_z.

Because

$$\left.\begin{aligned} |\partial_\varphi|^2 = g_{\varphi\varphi} = -\frac{\varpi}{\chi}(\chi^2 + q_2^2), \quad |\partial_z|^2 = g_{zz} = -\frac{\varpi}{\chi}, \\ \text{and} \qquad (\partial_\varphi \cdot \partial_z) = g_{z\varphi} = \frac{\varpi}{\chi}q_2 \end{aligned}\right\} \tag{36}$$

the area, A, of the two-dimensional surface, spanned by ∂_φ and ∂_z, is given by

$$A = [|\partial_\varphi|^2 |\partial_z|^2 - (\partial_\varphi \cdot \partial_z)^2]^{\frac{1}{2}} = \varpi. \tag{37}$$

Therefore,

$$\frac{g^{ij}A_{,i}A_{,j}}{|\partial_z|^2} = e^{-2\nu}\frac{\chi}{\varpi} = \frac{1}{\Psi}e^{-2\nu}, \tag{38}$$

and

$$\nu + \ln\sqrt{\Psi} = -\tfrac{1}{2}\ln\left(\frac{g^{ij}A_{,i}A_{,j}}{|\partial_z|^2}\right). \tag{39}$$

On essentially the foregoing grounds, Thorne (1965) considered

$$C = \nu + \ln\sqrt{\Psi} \tag{40}$$

as providing a definition of energy that may be applicable to space–times with cylindrical symmetry; and he called it the 'C'-energy. However, the arguments presented for interpreting $(\nu + \ln\sqrt{\Psi})$ as an energy are at best heuristic. A more convincing argument by O. Reula (personal communication) is the following.

We begin with the observation that the Ernst equation (29) can be derived from the Lagrangian density,

$$\mathscr{L} = \frac{\varpi}{(1-|E|^2)^2}(|E_{,0}|^2 - |E_{,\varpi}|^2). \tag{41}$$

The corresponding *Hamiltonian density* is given by

$$\mathscr{H} = E_{,0}\frac{\partial\mathscr{L}}{\partial E_{,0}} + E^*_{,0}\frac{\partial\mathscr{L}}{\partial E^*_{,0}} - \mathscr{L}, \tag{42}$$

or

$$\mathscr{H} = \frac{\varpi}{(1-|E|^2)^2}(|E_{,0}|^2 + |E_{,\varpi}|^2). \tag{43}$$

This is a conserved quantity; and it is positive-definite. We should, therefore, be justified to regard $\mathscr{H}$ as defining the *energy density* of the system. Moreover, by equation (32)

$$\mathscr{H} = (\nu + \ln\sqrt{\Psi})_{,\varpi}; \tag{44}$$

and by integrating this equation, we obtain

$$\int_{\varpi_1}^{\varpi_2} \mathscr{H}\,\mathrm{d}\varpi = (\nu + \ln\sqrt{\Psi})_{\varpi=\varpi_2} - (\nu + \ln\sqrt{\Psi})_{\varpi=\varpi_1}; \tag{45}$$

and this is positive-definite for $\varpi_2 > \varpi_1$. The quantity on the right-hand side of equation (45) is, by our interpretation of $\mathscr{H}$, the energy (per unit length in the z-direction) that is instantaneously confined in the cylindrical annulus between ϖ_2 and ϖ_1.

Reula has further pointed out that if the space–time is assumed to become asymptotically flat for $\varpi \to \infty$, then by seeking a solution of Witten's equation (Witten 1981) that tends to a constant spinor for $\varpi \to \infty$, we may be able to justify, rigorously, the *local definition* of energy density that $\mathscr{H}$ provides.

4. Separable solutions representing stationary monochromatic waves

Returning to the Ernst equation (29), we observe that it allows separable solutions of the form

$$E = \mathrm{e}^{\mathrm{i}\sigma t}\, F(\varpi), \tag{46}$$

where the radial function $F(\varpi)$, in general complex, is governed by the equation

$$(1-|F|^2)\left[\sigma^2 F + \frac{1}{\varpi}(\varpi F_{,\varpi})_{,\varpi}\right] + 2F^*[\sigma^2 F^2 + (F_{,\varpi})^2] = 0 \tag{47}$$

or, equivalently

$$F(1+|F|^2)\,\sigma^2 + (1-|F|^2)\frac{1}{\varpi}(\varpi F_{,\varpi})_{,\varpi} + 2F^*(F_{,\varpi})^2 = 0. \tag{48}$$

Equation (48) allows the integral

$$\frac{\varpi}{(1-|F|^2)^2}(FF^*_{,\varpi} - F^*F_{,\varpi}) = \text{const.} \tag{49}$$

(a) *The reduction of equation* (48) *to a real second-order equation*

Substituting

$$F = |F|\,\mathrm{e}^{\mathrm{i}\theta}, \tag{50}$$

in equation (48) and separating the real and the imaginary parts, we obtain the pair of equations,

$$F(1+F^2) + (1-F^2)\left[\frac{1}{\varpi}(\varpi F_{,\varpi})_{,\varpi} - (\theta_{,\varpi})^2 F\right] + 2F[(F_{,\varpi})^2 - (\theta_{,\varpi})^2 F^2] = 0 \tag{51}$$

and

$$F\theta_{,\varpi,\varpi} = -\theta_{,\varpi}\left[2F_{,\varpi}+\frac{F}{\varpi}+4\frac{F^2F_{,\varpi}}{1-F^2}\right], \tag{52}$$

where, for convenience of notation, we have dispensed with the vertical rules for F signifying absolute value. Also, we have taken advantage of the scale invariance of equation (47) to eliminate the appearance of σ in this equation by writing ϖ in place of $\sigma\varpi$.

Equation (52) can be rewritten in the form,

$$(\ln\theta_{,\varpi})_{,\varpi} = -\left[\ln\frac{\varpi F^2}{(1-F^2)^2}\right]_{,\varpi}. \tag{53}$$

In this form, the equation admits of immediate integration; and we have

$$\theta_{,\varpi} = k\frac{(1-F^2)^2}{\varpi F^2}, \tag{54}$$

where k is a constant of integration. Equation (54) is in fact an alternative form of the integral (49).

Now inserting the solution (54) for $\theta_{,\varpi}$ in equation (51), we obtain the following real second-order differential equation:

$$F(1+F^2)\left[1-k^2\frac{(1-F^2)^4}{\varpi^2F^4}\right]+(1-F^2)^2\frac{1}{\varpi}\frac{\mathrm{d}}{\mathrm{d}\varpi}\frac{\varpi F_{,\varpi}}{1-F^2} = 0. \tag{55}$$

With F determined by this equation, θ follows by a simple quadrature:

$$\theta = k\int^{\varpi}\frac{(1-F^2)^2}{\varpi F^2}\mathrm{d}\varpi. \tag{56}$$

(After the present investigation was completed, my attention was drawn to some papers by Morris & Dodd (1980) and by Léauté & Marcilhacy (1985, 1986) in which the foregoing reduction of the Ernst solution is also effected. But these authors do not consider the complete solution of the field equations that are obtained in the following sections.)

(*b*) *Solutions that are regular on the axis*

In this paper, we shall be concerned only with solutions that are regular on the axis and satisfy here the requirements of flatness. More general solutions will be considered in a later paper (Chandrasekhar & Persides).

From equation (49) it follows that freedom from singularity on the axis ($\varpi = 0$) requires that the constant on the right-hand of the equation be zero. Therefore, we must require:

$$FF^*_{,\varpi}-F^*F_{,\varpi} = 0. \tag{57}$$

We conclude that the real and the imaginary parts of F are in a constant ratio and that F is of the form

$$F(1+\mathrm{i}C), \tag{58}$$

where F is real and C is a real constant. The corresponding form of E is

$$E = e^{i\sigma t} F(1+iC) = e^{i(\sigma t+\alpha)} F\sqrt{(1+C^2)}, \tag{59}$$

where
$$\alpha = \tan^{-1} C. \tag{60}$$

Therefore, by displacing the origin of t and absorbing the constant factor, $\sqrt{(1+C^2)}$, in the definition of F, we can assume, without loss of generality, that E is of the form

$$E = F e^{i\sigma t} = F(\cos\sigma t + i \sin\sigma t), \tag{61}$$

where F is now real.

When F is real, equation (48) becomes

$$F(1+F^2)\sigma^2 + (1-F^2) + \frac{1}{\varpi}(\varpi F_{,\varpi})_{,\varpi} + 2F(F_{,\varpi})^2 = 0, \tag{62}$$

or, alternatively,

$$F(1+F^2)\sigma^2 + \frac{(1-F^2)^2}{\varpi}\frac{d}{d\varpi}\frac{\varpi F_{,\varpi}}{1-F^2} = 0. \tag{63}$$

Compatible with our requirements of regularity on the axis, we must seek solutions of equation (62) which satisfy the boundary conditions,

$$F = F_0\,(>0 \text{ and } <1) \quad \text{and} \quad F_{,\varpi} = 0 \quad \text{when} \quad \varpi = 0. \tag{64}$$

In §5, we shall consider the solutions of equation (62) which satisfy these boundary conditions; and we shall also ascertain their asymptotic behaviour for $\varpi \to \infty$. Meantime, we shall complete the solutions for the metric coefficients.

(c) *The solutions for the metric coefficients*

For E given by equation (61), the solutions for Ψ and Φ are, by equations (30),

$$\Psi = \frac{1-F^2}{1+F^2-2F\cos\sigma t} \quad \text{and} \quad \Phi = \frac{2F\sin\sigma t}{1+F^2-2F\cos\sigma t}. \tag{65}$$

The corresponding solution for χ is

$$\chi = \varpi\frac{1+F^2-2F\cos\sigma t}{1-F^2}. \tag{66}$$

The solutions for q_2 now follows from the equations (cf. equations (16))

$$q_{2,0} = \frac{\varpi}{\Psi^2}\Phi_{,\varpi} = \frac{2\varpi F_{,\varpi}}{1-F^2}\sin\sigma t, \tag{67}$$

and

$$q_{2,\varpi} + \frac{\varpi}{\Psi^2}\Phi_{,0} = 2\sigma\varpi\frac{F(1+F^2)}{(1-F^2)^2}\cos\sigma t - 4\varpi\frac{F^2\sigma}{(1-F^2)^2}. \tag{68}$$

We find,
$$q_2 = -\frac{2}{\sigma}\frac{\varpi F_{,\varpi}}{(1-F^2)}\cos\sigma t - 4\sigma\int_0^{\varpi}\frac{F^2\varpi}{(1-F^2)^2}d\varpi, \tag{69}$$

where we have imposed on q_2 the boundary condition that it vanishes for $\varpi = 0$ to satisfy the requirement (9) for flatness.

Finally, turning to equations (31) and (32), we find

$$(\nu + \ln \sqrt{\Psi})_{,0} = 0 \tag{70}$$

and

$$(\nu + \ln \sqrt{\Psi})_{,\varpi} = \frac{\varpi}{(1-F^2)^2}[\sigma^2 F^2 + (F_{,\varpi})^2]. \tag{71}$$

Therefore, the C-energy is independent of time. We may, on this account, describe the solution we have obtained as representing *stationary monochromatic waves.* (For definition of stationary waves, see Rayleigh 1945, vol. 1, p. 227.)

5. Solutions satisfying the required conditions of regularity on the axis and their asymptotic behaviour

To include the equation derived in the context of the Einstein–Maxwell equations in §7 below, we shall consider the more general equation (cf. equation (62)),

$$(A^2 + F^2)\, F + (A^2 - F^2)\left(F_{,\varpi,\varpi} + \frac{1}{\varpi} F_{,\varpi}\right) + 2F(F_{,\varpi})^2 = 0, \tag{72}$$

where $0 < A \leqslant 1$ is a constant and we have utilized the scale invariance of the equation to eliminate the appearance of σ in the equation by replacing $\sigma\varpi$ by ϖ.

Solutions of equation (72) satisfying the boundary conditions (64) can be obtained by inserting for F a series expansion of the form,

$$F = F_0 + F_1 \varpi^2 + F_2 \varpi^4 + \ldots, \tag{73}$$

and determining the coefficients F_1, F_2, etc., successively. Thus, we find

$$F_1 = -\frac{1}{4}\frac{A^2 + F_0^2}{A^2 - F_0^2} F_0, \quad F_2 = \frac{1}{64}\frac{(A^2 + F_0^2)(A^2 + 3F_0^2)}{(A^2 - F_0^2)^2} F_0, \text{ etc.} \tag{74}$$

In this manner, we obtain a one-parametric family of solutions labelled by F_0.

We readily verify that

$$F \to F_0 J_0(\varpi) \quad \text{when} \quad F_0 \to 0, \tag{75}$$

where J_0 is Bessel's function of order zero.

We next consider the asymptotic behaviour of the solutions of equation (72). First, we observe that with the substitution,

$$F = A\frac{e^{\frac{1}{2}f} - 1}{e^{\frac{1}{2}f} + 1}, \tag{76}$$

equation (72) reduces to the form,

$$f_{,\varpi,\varpi} + \frac{1}{\varpi} f_{,\varpi} + \sinh f = 0, \tag{77}$$

familiar in the theory of the Painlevé transcendent III (see equation (86) below and Ince 1927, p. 345; also Ablowitz & Segur 1981, p. 234, Fokas & Ablowitz 1982).

It readily follows from equation (77) that

$$f \to 0 \quad \text{as} \quad \varpi \to \infty. \tag{78}$$

Accordingly, one may expect to obtain the asymptotic behaviour of the solutions by expanding $\sinh f$ in a power series in f and truncating the series at successively higher orders. Thus, by retaining only the first term in the expansion of $\sinh f$, we obtain the equation,

$$f_{,\varpi,\varpi} + \frac{1}{\varpi} f_{,\varpi} + f = 0. \tag{79}$$

This is Bessel's equation of order zero and we may conclude that

$$f \sim \frac{1}{\sqrt{\varpi}} \cos(\varpi + b) + O(\varpi^{-\frac{3}{2}}) \quad (\varpi \to \infty), \tag{80}$$

where b is some constant. If we now consider the 'second approximation' to equation (77), namely,

$$f_{,\varpi,\varpi} + \frac{1}{\varpi} f_{,\varpi} + f + \tfrac{1}{6} f^3 = 0, \tag{81}$$

we find that the asymptotic behaviour (80) is not compatible with equation (81) even to $O(\varpi^{-\frac{1}{2}})$. A more careful analysis yields the expansion,

$$f \sim \frac{1}{\sqrt{\varpi}} \cos(\varpi + \tfrac{1}{16} \ln \varpi + b) + \frac{1}{192 \varpi^{\frac{3}{2}}} \cos 3(\varpi + \tfrac{1}{16} \ln \varpi + b) + O(\varpi^{-\frac{5}{2}}). \tag{82}$$

The validity of this expansion to the order indicated can be verified by direct substitution in equation (81). (I am indebted to Dr S. Persides for obtaining the expansion (82).) By comparison with the behaviour (80), we observe that the term of $O(\varpi^{-\frac{1}{2}})$ is different. Similarly, we find that, when we proceed to the next higher approximation of equation (77) by including the term in f^5, the term of $O(\varpi^{-\frac{3}{2}})$ in the asymptotic behaviour (82) is affected; but the term of $O(\varpi^{-\frac{1}{2}})$ is unaffected. We conclude that the required asymptotic behaviour of the solutions of equation (77), to the lowest order, is given by

$$f \sim \frac{1}{\sqrt{\varpi}} \cos(\varpi + \tfrac{1}{16} \ln \varpi + b) + O(\varpi^{-\frac{3}{2}}); \tag{83}$$

and correspondingly,

$$F \sim \frac{A}{4\sqrt{\varpi}} \cos\left(\varpi + \frac{1}{16} \ln \varpi + b\right) + O(\varpi^{-\frac{3}{2}}). \tag{84}$$

I am indebted to Mr C. Habisohn for obtaining on a computer the solutions of equation (72) for various initially assigned values of A^2 and F_0. These solutions are illustrated in figures 1, 2 and 3.

Finally, we may note that by the substitution,

$$F = A \frac{W-1}{W+1}, \tag{85}$$

equation (72) reduces to the form

$$W_{,\varpi,\varpi} = \frac{1}{W}(W_{,\varpi})^2 - \frac{1}{\varpi}W_{,\varpi} - \frac{1}{4}\left(W^3 - \frac{1}{W}\right); \tag{86}$$

and this is a special case ($\alpha = \beta = 0$ and $\nu = -\delta = -\frac{1}{4}$) of Painlevé's transcendent III (see Ince 1927, p. 345, equation (iii)).

I am indebted to Dr B. C. Xanthopoulos for pointing out that by the substitution,

$$F = -W^{-\frac{1}{2}}, \tag{87}$$

the more general equation (55) reduces to the form,

$$W_{,\varpi,\varpi} = \left(\frac{1}{2W} + \frac{1}{W-1}\right)(W_{,\varpi})^2 - \frac{1}{\varpi}W_{,\varpi} + \frac{2W(W+1)}{W-1} - \frac{2k^2}{\varpi^2}(W-1)^2\left(W - \frac{1}{W}\right), \tag{88}$$

which is a special case of Painlevé's transcendent V (see Ince 1927, p. 345, equation (v)).

6. The space–time of stationary monochromatic gravitational waves

With the solutions for the metric functions completed in §§4 and 5, the metric is given by

$$ds^2 = e^{2\nu}[(dt)^2 - (d\varpi)^2] - \frac{\varpi^2}{\Psi}(d\varphi)^2 - \Psi(dz - q_2\, d\varphi)^2, \tag{89}$$

where

$$\Psi = \frac{1-F^2}{1+F^2-2F\cos t}, \tag{90}$$

$$q_2\,\sigma = -\frac{2\varpi F_{,\varpi}}{1-F^2}\cos t - 4\int_0^{\varpi}\frac{\varpi F^2}{(1-F^2)^2}\,d\varpi, \tag{91}$$

$$(\nu + \ln\sqrt{\Psi})_{,0} = 0, \quad (\nu + \ln\sqrt{\Psi})_{,\varpi} = \frac{\varpi}{(1-F^2)^2}[F^2 + (F_{,\varpi})^2], \tag{92}$$

and F is a solution of the equation,

$$F(1+F^2) + \frac{1-F^2}{\varpi}(\varpi F_{,\varpi})_{,\varpi} + 2F(F_{,\varpi})^2 = 0, \tag{93}$$

which satisfies the boundary conditions,

$$F = F_0\,(>0 \text{ and } <1) \quad \text{and} \quad F_{,\varpi} = 0 \quad \text{for} \quad \varpi = 0, \tag{94}$$

and has the asymptotic behaviour

$$F \sim \frac{\text{const.}}{\sqrt{\varpi}}\cos\left(\varpi + \tfrac{1}{16}\ln\varpi + b\right) + O(\varpi^{-\frac{3}{2}}) \quad (\varpi \to \infty). \tag{95}$$

It should be noted that in writing the solutions in the foregoing forms, we are measuring both t and ϖ in the unit $1/\sigma$.

From the boundary conditions (94) satisfied by F and its asymptotic behaviour (95), we conclude that

$$\left.\begin{aligned} \Psi = \frac{1-F_0^2}{1+F_0^2-2F_0\cos t} = O(1), \quad q_2 = O(\varpi^2) \\ (\nu+\ln\sqrt{\Psi}) = O(\varpi^2) \quad \text{for} \quad \varpi\to 0, \end{aligned}\right\} \tag{96}$$

and

and

$$\left.\begin{aligned} \Psi\to 1, \quad q_2\to -\text{const.}\,(\sqrt{\varpi})\cos t - \text{const.}\,\varpi \\ \nu = O(\varpi) \quad \text{for} \quad \varpi\to\infty. \end{aligned}\right\} \tag{97}$$

and

The behaviours (96) of the metric functions for $\varpi\to 0$ satisfy the requirements of regularity and flatness at the poles: and the asymptotic behaviours (97) for $\varpi\to\infty$ do not also violate any of the other physical requirements. In particular, the C-energy, $(\nu+\ln\sqrt{\Psi})$, which is independent of time, increases linearly with ϖ for $\varpi\to\infty$, consistent with the interpretation of the solution as representing stationary gravitational waves.

In figures 1 and 2, we illustrate the behaviours of the solutions for F and q_2 for various initially assigned values for F_0.

Finally, it is important to observe that for no member of the one-parameter family of solutions considered is q_2 zero; and that the Einstein–Rosen waves (see Appendix B) are not included among the solutions as special cases.

(a) *The description of the space–time in Newman–Penrose formalism*

Rewriting the metric (7) in the form

$$ds^2 = e^{2\nu}[(dt)^2-(d\varpi)^2] - \frac{\varpi}{1-|\mathscr{E}|^2}|(1-\mathscr{E})\,dz + i(1+\mathscr{E})\,d\varphi|^2, \tag{98}$$

where

$$\mathscr{E} = \frac{\chi+iq_2-1}{\chi+iq_2+1}, \tag{99}$$

we can set up a null-tetrad frame as in Chandrasekhar & Xanthopoulos (1986*b*; see Appendix A) and evaluate the Weyl scalars with the aid of the expressions given in that Appendix. In the present instance

$$U = e^{\nu} \quad \text{and} \quad V = \sqrt{\varpi}, \tag{100}$$

and t and ϖ replace ψ and θ; and the expression for Ψ_2 is

$$\Psi_2 = \tfrac{1}{2}e^{-2\nu}\frac{1-E^*}{(1-E)(1-|E|^2)}\left\{\frac{1-E^*}{(1-E)(1-|E|^2)}[(E_{,0})^2-(E_{,\varpi})^2] + \frac{E_{,\varpi}}{\varpi}\right\}. \tag{101}$$

Substituting in this expression our present solution for E, namely,

$$E = e^{i\sigma t}F, \tag{102}$$

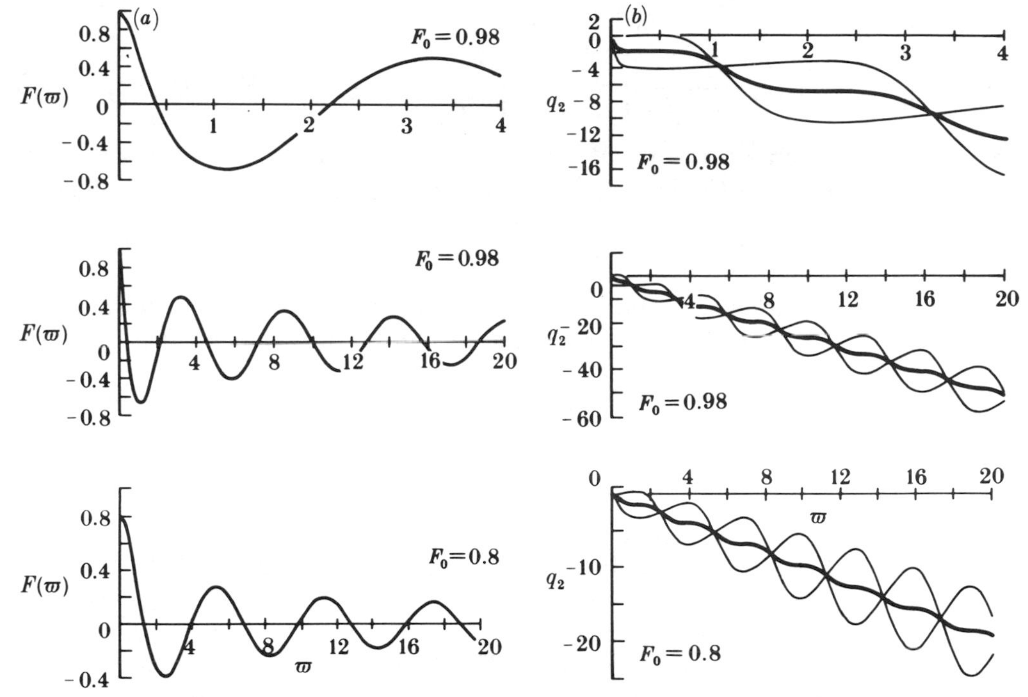

FIGURE 1. The solutions of the Einstein vacuum equations for F (the panel on the left) and q_2 (the panel on the right) for various values F_0 of F at $\varpi = 0$. The central heavily drawn curves on the right represent the time average of q_2 while the two lightly drawn curves define its limits at each ϖ as q_2 varies periodically.

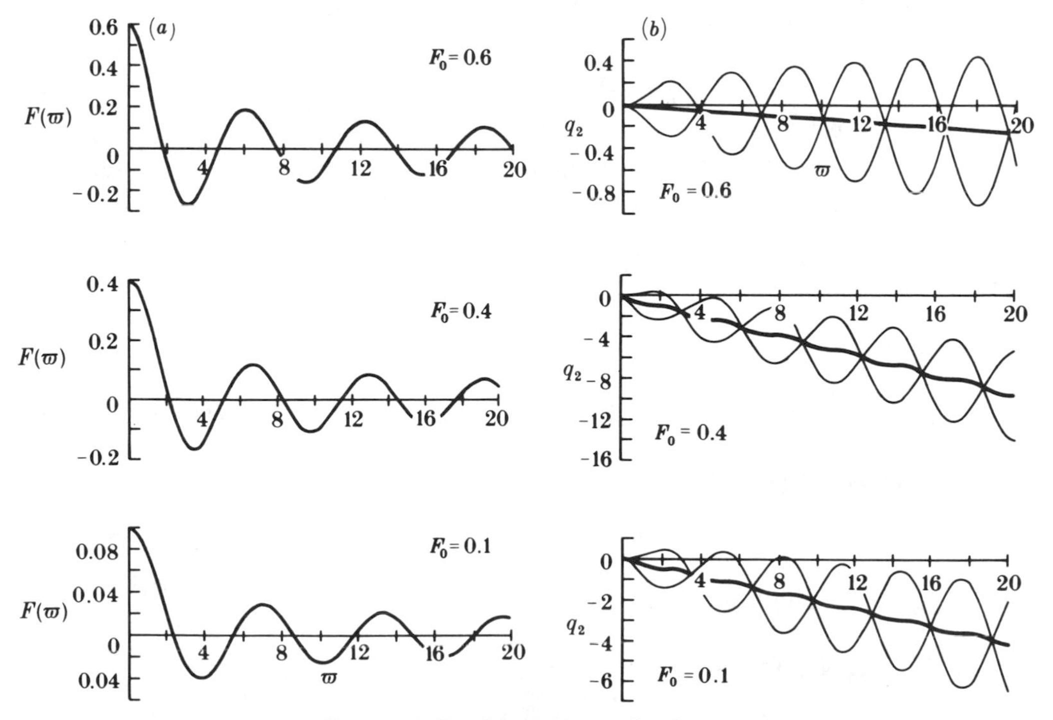

FIGURE 2. See description under figure 1.

we obtain,

$$\Psi_2 = \tfrac{1}{2}\,e^{-2\nu}\frac{1-e^{-i\sigma t}F}{(1-F^2)(1-e^{i\sigma t}F)}\left\{e^{i\sigma t}\frac{F_{,\varpi}}{\varpi}-\frac{e^{2i\sigma t}(1-e^{-i\sigma t}F)}{(1-F^2)(1-e^{i\sigma t}F)}[\sigma^2F^2+(F_{,\varpi})^2]\right\}. \quad (103)$$

From the behaviours (94) and (95) of F for $\varpi\to 0$ and $\varpi\to\infty$, it follows that Ψ_2 is bounded. And from the behaviour of ν for $\varpi\to\infty$ (cf. equation (97)) it follows that

$$\Psi_2 = O(e^{-\text{const.}\,\varpi}) \quad \text{for} \quad \varpi\to\infty. \quad (104)$$

Therefore, Ψ_2 vanishes at infinity. The remaining Weyl scalars can be similarly shown to vanish for $\varpi\to\infty$. The space–time accordingly becomes flat for $\varpi\to\infty$.

7. Stationary monochromatic waves in the Einstein–Maxwell theory

The Einstein–Maxwell equations appropriate for a space–time with two commuting space-like Killing fields are written out in full in Chandrasekhar & Xanthopoulos (1985*a*: this paper will be referred to hereafter as Paper II) in a gauge and in a coordinate-system suitable for treating colliding plane-fronted waves. The equations can be readily transcribed in the gauge and the coordinate system of §2 suitable for space–times with cylindrical symmetry. Thus, equations (11), (28), and (30) of Paper II are now replaced by

$$(\chi H_{,0})_{,0}-(\chi H_{,\varpi})_{,\varpi} = i(H_{,0}q_{2,\varpi}-H_{,\varpi}q_{2,0}), \quad (105)$$

$$\left(\varpi\frac{\chi_{,0}}{\chi}\right)_{,0}-\left(\varpi\frac{\chi_{,\varpi}}{\chi}\right)_{,\varpi} = -\frac{\varpi}{\chi^2}[(q_{2,0})^2-(q_{2,\varpi})^2],+2\chi[|H_{,0}|^2-|H_{,\varpi}|^2], \quad (106)$$

$$\left(\frac{\varpi}{\chi^2}q_{2,0}-2\,\text{Im}\,HH^*_{,\varpi}\right)_{,0}-\left(\frac{\varpi}{\chi^2}q_{2,\varpi}-2\,\text{Im}\,HH^*_{,0}\right)_{,3} = 0. \quad (107)$$

where H denotes the complex potential describing the prevailing electromagnetic field. Equation (107) allows us to define a potential Φ by the equations:

$$\Phi_{,\varpi} = \frac{\varpi}{\chi^2}q_{2,0}-2\,\text{Im}\,HH^*_{,\varpi} \quad \text{and} \quad \Phi_{,0} = \frac{\varpi}{\chi^2}q_{2,\varpi}-2\,\text{Im}\,HH^*_{,0}. \quad (108)$$

Solving these equations for $q_{2,0}$ and $q_{2,\varpi}$, we obtain

$$q_{2,0} = \frac{\varpi}{\Psi^2}(\Phi_{,\varpi}+2\,\text{Im}\,HH^*_{,\varpi}) \quad \text{and} \quad q_{2,\varpi} = \frac{\varpi}{\Psi^2}(\Phi_{,0}+2\,\text{Im}\,HH^*_{,0}), \quad (109)$$

where (cf. equation (17))

$$\Psi = \frac{\varpi}{\chi}. \quad (110)$$

The integrability condition for equations (109) is

$$\left[\frac{1}{\Psi^2}(\Phi_{,0}+2\,\text{Im}\,HH^*_{,0})\right]_{,0}-\frac{1}{\varpi}\left[\frac{\varpi}{\Psi^2}(\Phi_{,\varpi}+2\,\text{Im}\,HH^*_{,\varpi})\right]_{,\varpi} = 0. \quad (111)$$

Equations (105) and (106) written out in terms of Ψ and Φ are:

$$\Psi\left[H_{,0,0}-\frac{1}{\varpi}(\varpi H_{,\varpi})_{,\varpi}\right] = H_{,0}[\Psi_{,0}+\mathrm{i}(\Phi_{,0}+2\,\mathrm{Im}\,HH^*_{,0})] - H_{,\varpi}[\Psi_{,\varpi}+\mathrm{i}(\Phi_{,\varpi}+2\,\mathrm{Im}\,HH^*_{,\varpi})], \quad (112)$$

$$\Psi\left[\Psi_{,0,0}-\frac{1}{\varpi}(\varpi \Psi_{,\varpi})_{,\varpi}\right] = (\Psi_{,0})^2-(\Psi_{,\varpi})^2-2\Psi[|H_{,0}|^2-|H_{,\varpi}|^2] - (\Phi_{,0}+2\,\mathrm{Im}\,HH^*_{,0})^2+(\Phi_{,\varpi}+2\,\mathrm{Im}\,HH^*_{,\varpi})^2. \quad (113)$$

As in Paper II, equations (112)–(113) can be reduced to the following pair of coupled complex equations:

$$(\mathrm{Re}\,Z-|H|^2)\left[Z_{,0,0}-\frac{1}{\varpi}(\varpi Z_{,\varpi})_{,\varpi}\right] = (Z_{,0})^2-(Z_{,\varpi})^2-2H^*(Z_{,0}H_{,0}-Z_{,\varpi}H_{,\varpi}), \quad (114)$$

and

$$(\mathrm{Re}\,Z-|H|^2)\left[H_{,0,0}-\frac{1}{\varpi}(\varpi H_{,\varpi})_{,\varpi}\right] = H_{,0}Z_{,0}-H_{,\varpi}Z_{,\varpi}-2H^*[(H_{,0})^2-(H_{,\varpi})^2], \quad (115)$$

where

$$Z = \Psi+|H|^2+\mathrm{i}\Phi. \quad (116)$$

The equations governing the remaining metric function ν are:

$$\nu_{,0} = \frac{\varpi}{2\chi^2}(\chi_{,0}\chi_{,\varpi}+q_{2,0}q_{2,\varpi})+\chi(H_{,0}H^*_{,\varpi}+H^*_{,0}H_{,\varpi}), \quad (117)$$

$$4\nu_{,\varpi} = -\frac{1}{\varpi}+\frac{\varpi}{\chi^2}[(\chi_{,0})^2+(\chi_{,\varpi})^2+(q_{2,0})^2+(q_{2,\varpi})^2]+4\chi[|H_{,0}|^2+|H_{,\varpi}|^2]. \quad (118)$$

In terms of Ψ and Φ, these equations take the forms:

$$(\nu+\ln\sqrt{\Psi})_{,0} = \frac{\varpi}{2\Psi^2}[\Psi_{,0}\Psi_{,\varpi}+(\Phi_{,0}+2\,\mathrm{Im}\,HH^*_{,0})(\Phi_{,\varpi}+2\,\mathrm{Im}\,HH^*_{,\varpi})] + \frac{\varpi}{\Psi}(H_{,0}H^*_{,\varpi}+H^*_{,0}H_{,\varpi}) \quad (119)$$

and

$$(\nu+\ln\sqrt{\Psi})_{,\varpi} = \frac{\varpi}{4\Psi^2}[(\Psi_{,0})^2+(\Psi_{,\varpi})^2+(\Phi_{,0}+2\,\mathrm{Im}\,HH^*_{,0})^2 + (\Phi_{,\varpi}+2\,\mathrm{Im}\,HH^*_{,\varpi})^2]+\frac{\varpi}{\Psi}[|H_{,0}|^2+|H_{,\varpi}|^2]. \quad (120)$$

(*a*) *The reduction to an Ernst equation*

We know from previous experience (Paper II, equation (48) and Chandrasekhar 1983, §110, p. 576) that the substitution

$$H = Q(Z+1), \quad (121)$$

where Q is a constant, will reduce the pair of equations (119) and (120) to a single complex equation. In this instance, we find

$$\tfrac{1}{2}[(1-2Q^2)(Z+Z^*)-2Q^2(|Z|^2+1)]\left[Z_{,0,0}-\frac{1}{\varpi}(\varpi Z_{,\varpi})_{,\varpi}\right] = [1-2Q^2(Z^*+1)][(Z_{,0})^2-(Z_{,\varpi})^2], \quad (122)$$

where it may be noted that there is no loss of generality in assuming that Q is a real constant. By the further substitution,

$$Z = \frac{1+E}{1-E} \tag{123}$$

we obtain the Ernst equation,

$$(1-4Q^2-|E|^2)\left[E_{,0,0}-\frac{1}{\varpi}(\varpi E_{,\varpi})_{,\varpi}\right] = -2E^*[(E_{,0})^2-(E_{,\varpi})^2]. \tag{124}$$

In terms of a solution of this equation, the solutions for Ψ, Φ, and H are given by

$$\Psi = \frac{1-4Q^2-|E|^2}{|1-E|^2}, \quad i\Phi = \frac{E-E^*}{|1-E|^2}, \quad \text{and} \quad H = \frac{2Q}{1-E}. \tag{125}$$

Turning to equations (119) and (120), we can write them in terms of E by making use of the following verifiable relations:

$$\frac{\Psi_{,0}}{\Psi} = \frac{B^*E_{,0}+BE^*_{,0}}{|1-E|^2(1-4Q^2-|E|^2)}, \quad \frac{\Psi_{,\varpi}}{\Psi} = \frac{B^*E_{,\varpi}+BE^*_{,\varpi}}{|1-E|^2(1-4Q^2-|E|^2)}, \tag{126}$$

$$\frac{1}{\Psi}(i\Phi_{,0}+HH^*_{,0}-H^*H_{,0}) = \frac{B^*E_{,0}-BE^*_{,0}}{|1-E|^2(1-4Q^2-|E|^2)},$$

and

$$\frac{1}{\Psi}(i\Phi_{,\varpi}+HH^*_{,\varpi}-H^*H_{,\varpi}) = \frac{B^*E_{,\varpi}-BE^*_{,\varpi}}{|1-E|^2(1-4Q^2-|E|^2)}, \tag{127}$$

where

$$B = (1-E)(1-4Q^2-E). \tag{128}$$

We find

$$(\nu+\ln\sqrt{\Psi})_{,0} = \varpi\frac{1-4Q^2}{(1-4Q^2-|E|^2)^2}(E_{,0}E^*_{,\varpi}+E_{,\varpi}E^*_{,0}), \tag{129}$$

$$(\nu+\ln\sqrt{\Psi})_{,\varpi} = \varpi\frac{1-4Q^2}{(1-4Q^2-|E|^2)^2}(|E_{,0}|^2+|E_{,\varpi}|^2). \tag{130}$$

Therefore, the C-energy of a cylindrically symmetric Einstein–Maxwell field, under the circumstances considered, is given by the same expression, $\nu+\ln\sqrt{\Psi}$, as for a pure gravitational field; but the Hamiltonian density, $\mathscr{H}$, is now given by (cf. equation (43)),

$$\mathscr{H} = \varpi\frac{1-4Q^2}{(1-4Q^2-|E|^2)^2}(|E_{,0}|^2+|E_{,\varpi}|^2), \tag{131}$$

which is positive definite so long as $4Q^2 < 1$. *The requirement*, $4Q^2 < 1$, *must therefore be considered as a restriction on the allowed range of* Q.

(b) Separable solutions representing stationary monochromatic waves

The Ernst equation (124) allows separable solutions of the same form, as in the case of the vacuum, namely

$$E = e^{i\sigma t}F(\varpi); \tag{132}$$

and the equation governing the radial function, $F(\varpi)$, is (cf. equation (47)),

$$(1-4Q^2+|F|^2)\sigma^2 F+(1-4Q^2-|F|^2)\frac{1}{\varpi}(\varpi F_{,\varpi})_{,\varpi}+2F^*(F_{,\varpi})^2=0, \tag{133}$$

and this equation allows the integral (cf. equation (49)),

$$\frac{\varpi}{(1-4Q^2-|F|^2)^2}(FF^*_{,\varpi}-F^*F_{,\varpi})=\text{const.} \tag{134}$$

If we now restrict ourselves to solutions that are regular on the axis, we must require,

$$FF^*_{,\varpi}-F^*F_{,\varpi}=0. \tag{135}$$

And from this requirement we can conclude, as in §4(*b*), that no loss of generality is entailed by assuming that F is real and E is of the form,

$$E=F(\cos\sigma t+\mathrm{i}\sin\sigma t). \tag{136}$$

In this case, the equation governing F becomes

$$F(A^2+F^2)\sigma^2+(A^2-F^2)\frac{1}{\varpi}(\varpi F_{,\varpi})_{,\varpi}+2F(F_{,\varpi})^2=0, \tag{137}$$

where

$$A^2=1-4Q^2. \tag{138}$$

We have already considered equation (137) in §5 and shown how solutions regular at the origin can be obtained; and we have also established their asymptotic behaviour at infinity.

In terms of a solution F of equation (137), the solutions (125) for Ψ, Φ, and H are:

$$\Psi=\frac{A^2-F^2}{1+F^2-2F\cos\sigma t},\quad \Phi=\frac{2F\sin\sigma t}{1+F^2-2F\cos\sigma t},$$

and

$$H=\frac{2Q}{1-F\cos\sigma t-\mathrm{i}F\sin\sigma t}. \tag{139}$$

The solution for q_2 now follows from equations (109), (127), and the solutions for Ψ and E. We find

$$q_2=-\frac{2}{\sigma}\frac{\varpi F_{,\varpi}}{A^2-F^2}\cos\sigma t-4\sigma(1-2Q^2)\int_0^{\varpi}\frac{\varpi F^2}{(A^2-F^2)^2}\,\mathrm{d}\varpi. \tag{140}$$

And finally, equations (129) and (130) for $(\nu+\ln\sqrt{\Psi})$ give

$$(\nu+\ln\sqrt{\Psi})_{,0}=0, \tag{141}$$

$$(\nu+\ln\sqrt{\Psi})_{,\varpi}=\varpi\frac{A^2}{(A^2-F^2)^2}[\sigma^2F^2+(F_{,\varpi})^2]. \tag{142}$$

The fact that $(\nu+\ln\sqrt{\Psi})$ is independent of time confirms that we are dealing here with stationary waves as in the case of the vacuum in §6. It is also clear that the various metric functions have the same asymptotic behaviours (96) and (97) for $\varpi\to 0$ and $\varpi\to\infty$. In particular, the required conditions of regularity and flatness at the poles are satisfied.

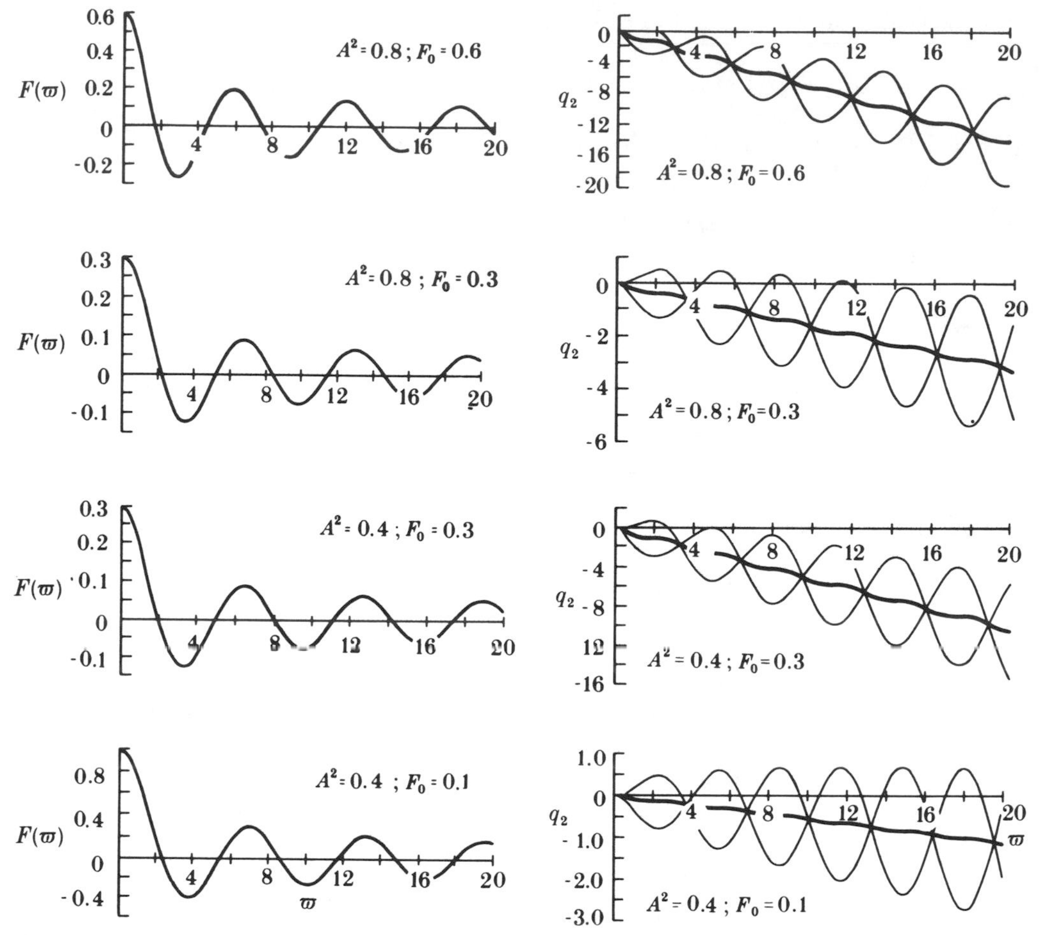

FIGURE 3. The solutions of the Einstein–Maxwell equation for F (the panel on the left) and q_2 (the panel on the right) for various values F_0 of F at $\varpi = 0$ and $A^2 (= 1-4Q^2)$. The central heavily drawn curves on the right represent the time average of q_2 while the two lightly drawn curves define its limits at each ϖ as q_2 varies periodically.

Solutions of equation (137) were obtained on a computer by Mr C. Habisohn for ($A^2 = 0.8$; $F_0 = 0.6$ and 0.4) and ($A^2 = 0.4$; $F_0 = 0.3$ and 0.1). These solutions for F and the corresponding solutions for q_2 are illustrated in figure 3.

8. CYLINDRICAL WAVES COUPLED WITH A PERFECT FLUID WITH THE EQUATION OF STATE $\epsilon = p$

With only minor modifications, the treatment in Chandrasekhar & Xanthopoulos (1985*b*), of plane-fronted gravitational waves coupled with a perfect fluid in which the energy density, ϵ, is equal to the pressure, p, can be adapted to cylindrical geometry. We shall, therefore, omit all details and state only the principal results.

First, we find that when $\epsilon = p$, the coupled Einstein-hydrodynamics equations enable us to reduce the metric (7) to the form

$$ds^2 = e^{2(\nu+f)}[(dt)^2 - (d\varpi)^2] - \varpi\left[\chi(d\varphi)^2 + \frac{1}{\chi}(dz - q_2\, d\varphi)^2\right], \tag{143}$$

where ν, χ, and q_2 represent solutions of the vacuum equations and f is the only function that derives from the presence of the fluid.

The fluid motions themselves can be derived from a potential ϕ which, in the coordinate system chosen in §2, is governed by the equation

$$\phi_{,\varpi,\varpi} - \frac{1}{\varpi}\phi_{,\varpi} - \phi_{,0,0} = 0, \tag{144}$$

while the function f is determined, in terms of ϕ, by the equations,

$$f_{,0} = -\frac{4}{\varpi}\phi_{,0}\phi_{,\varpi} \quad \text{and} \quad f_{,\varpi} = -\frac{2}{\varpi}[(\phi_{,0})^2 + (\phi_{,\varpi})^2]. \tag{145}$$

The principal hydrodynamic quantity interest, namely, ϵ, is given by

$$\epsilon = \frac{1}{\varpi^2}[(\phi_{,\varpi})^2 - (\phi_{,0})^2]\, e^{-2(\nu+f)}. \tag{146}$$

And finally, we may note that in the present context equations (31) and (32) are replaced by

$$(\nu + f + \ln\sqrt{\Psi})_{,0} = \frac{\varpi}{(1-|E|^2)^2}(E_{,0}E^*_{,\varpi} + E^*_{,0}E_{,\varpi}) - \frac{4}{\varpi}\phi_{,0}\phi_{,\varpi}, \tag{147}$$

$$(\nu + f + \ln\sqrt{\Psi})_{,\varpi} = \frac{\varpi}{(1-|E|^2)^2}(|E_{,0}|^2 + |E_{,\varpi}|^2) - \frac{2}{\varpi}[(\phi_{,0})^2 + (\phi_{,\varpi})^2]. \tag{148}$$

The fundamental solution of equation (144) which assures the positivity of ϵ is given by

$$\phi = A\, e^{\pm\lambda t}(\lambda\varpi)\, I_1(\lambda\varpi), \tag{149}$$

where A is a constant of proportionality, λ is a positive real constant and I_1 is Bessel's function of order 1 for an imaginary argument in Watson's notation (Watson 1944). The corresponding solutions for ϵ and f are:

$$\epsilon = A^2\lambda^4[I_0^2(\lambda\varpi) - I_1^2(\lambda\varpi)]\, e^{\pm 2\lambda t - 2(\nu+f)}, \tag{150}$$

and

$$f = -2\lambda^2 A^2\, e^{\pm 2\lambda t}(\lambda\varpi)\, I_0(\lambda\varpi)\, I_1(\lambda\varpi). \tag{151}$$

There is, of course, no difficulty in writing down the general solutions for ϵ and f by considering superpositions (as sums or integrals) of the fundamental solution (149). By associating these solutions with the solutions of the vacuum equations given in §§4 and 5, we can obtain solutions that represent mixtures of monochromatic gravitational and acoustic waves.

9. Concluding remarks

The present paper provides a convenient framework for constructing cylindrically symmetric solutions of the Einstein and the Einstein–Maxwell equations. Its usefulness is demonstrated by the ease and the simplicity with which the solutions representing stationary monochromatic waves have been obtained. Besides, the central role which the Painlevé transcendents play in these solutions is perhaps significant for future developments.

I am grateful to several colleagues in the preparation of this paper: to Dr B. C. Xanthopoulos for his interest and for his careful scrutiny of the entire analysis; to Dr S. Persides for his contributions to some phases of the development; to Dr O. Reula for his incisive remarks on the C-energy included in §3(*b*); to Mr C. Habisohn for his numerical integrations of the basic functions; to Professor M. J. Ablowitz for some correspondence on the Painlevé transcendents and to Professor R. Geroch for many useful discussions.

The research reported in this paper has, in part, been supported by grants from the National Science Foundation under grant PHY-84-16691 with the University of Chicago.

Appendix A. Solution derived from the Ernst equation for $\chi + iq_2$

We shall verify in this Appendix that the separable solutions derived from the Ernst equation for $\chi + iq_2$ do not satisfy the requirement of flatness at $\varpi = 0$; and as we shall show, the metric is singular as well.

It is clear that for the same solutions of the Ernst equations, χ and q_2 will be the same as Ψ and Φ. Therefore, for the solution

$$E = e^{i\sigma t}\, F(\varpi), \tag{A 1}$$

where $F(\varpi)$ is real and satisfies the same equation (62) the solutions for χ and q_2 are (cf. equations (65))

$$\chi = \frac{1-F^2}{1+F^2-2F\cos\sigma t} \quad \text{and} \quad q_2 = \frac{2F\sin\sigma t}{1+F^2-2F\cos\sigma t}. \tag{A 2}$$

On the other hand, the equations governing ν are different: they are now given by (cf. equations (70) and (71)):

$$(\nu + \tfrac{1}{4}\ln\varpi)_{,0} = 0 \tag{A 3}$$

and

$$(\nu + \tfrac{1}{4}\ln\varpi)_{,\varpi} = \frac{\varpi}{(1-F^2)^2}[\sigma^2 F^2 + (F_{,\varpi})^2]. \tag{A 4}$$

The metric representing the foregoing solutions is

$$\mathrm{d}s^2 = e^{2\nu}[(\mathrm{d}t)^2 - (\mathrm{d}\varpi)^2] - \varpi\left\{\frac{(1-F^2)^2 + 4F^2\sin^2\sigma t}{(1-F^2)(1+F^2-2F\cos\sigma t)}(\mathrm{d}\varphi)^2 \right.$$
$$\left. - \frac{4F\sin\sigma t}{1-F^2}\mathrm{d}\varphi\,\mathrm{d}z + \frac{1+F^2-2F\cos\sigma t}{1-F^2}(\mathrm{d}z)^2\right\}. \tag{A 5}$$

This metric clearly violates the requirement of flatness at $\varpi = 0$, because $g_{\varphi\varphi}$ is of $O(\varpi)$ for $\varpi \to 0$ and not $O(\varpi^2)$ as required. Besides, the equation (A 4) and the boundary conditions (94) satisfied by F,

$$(\nu + \tfrac{1}{4} \ln \varpi)_{,\varpi} = O(\varpi) \quad \text{for} \quad \varpi \to 0. \tag{A 6}$$

Therefore,

$$e^{2\nu} = O(\varpi^{-\frac{1}{2}}) \quad \text{for} \quad \varpi \to 0. \tag{A 7}$$

This singularity at $\varpi = 0$ makes the metric (A 5) unacceptable.

Appendix B. The Einstein–Rosen waves and a historical note

Einstein & Rosen (1937) consider the special case $q_2 = \Phi = 0$. The metric, then, has the form (cf. equation (89))

$$ds^2 = e^{2\nu}[(dt)^2 - (d\varpi)^2] - \frac{\varpi^2}{\Psi}(d\varphi)^2 - \Psi(dz)^2; \tag{B 1}$$

and the equations governing Ψ and ν are (cf. equations (20), (24), and (25)),

$$(\ln \Psi)_{,0,0} - \frac{1}{\varpi}[\varpi(\ln \Psi)_{,\varpi}]_{,\varpi} = 0, \tag{B 2}$$

and

$$(\nu + \ln \surd \Psi)_{,0} = \tfrac{1}{2}\varpi(\ln \Psi)_{,0}(\ln \Psi)_{,\varpi}, \tag{B 3}$$

$$(\nu + \ln \surd \Psi)_{,\varpi} = \tfrac{1}{4}\varpi\{[(\ln \Psi)_{,0}]^2 + [(\ln \Psi)_{,\varpi}]^2\}. \tag{B 4}$$

A separable solution of equation (B 1), appropriate to the problem on hand, is

$$\ln \Psi = A J_0(\sigma\varpi) \cos \sigma t, \tag{B 5}$$

where A is a constant of the proportionality and J_0 is Bessel's function of order 0. Inserting this solution in equations (B 3) and (B 4), we obtain

$$(\nu + \ln \surd \Psi)_{,0} = \tfrac{1}{4}A^2\sigma^2\varpi J_0(\sigma\varpi) J_1(\sigma\varpi) \sin 2\sigma t, \tag{B 6}$$

and

$$\begin{aligned}(\nu + \ln \surd \Psi)_{,\varpi} &= \tfrac{1}{4}A^2\sigma^2\varpi[J_0^2 \sin^2 \sigma t + J_1^2 \cos^2 \sigma t] \\ &= \tfrac{1}{8}A^2\sigma^2\varpi[J_0^2 + J_1^2 - (J_0^2 - J_1^2) \cos 2\sigma t].\end{aligned} \tag{B 7}$$

From these equations, we find

$$\nu + \ln \surd \Psi = \tfrac{1}{8}A^2[\sigma^2\varpi^2(J_0^2 + J_1^2) - 2\sigma\varpi J_0 J_1 \cos^2 \sigma t]. \tag{B 8}$$

The general solution of the equations can be obtained by considering superpositions (as sums or integrals) of the fundamental solution (B 5). Thus, by considering the 'cosine series'

$$\ln \Psi = \sum_j A_j J_0(\sigma_j \varpi) \cos \sigma_j t, \tag{B 9}$$

we obtain

$$\begin{aligned}\nu + \ln \surd \Psi = &\tfrac{1}{8}\sum_j A_j^2\{\sigma_j^2 \varpi^2[J_0^2(\sigma_j \varpi) + J_1^2(\sigma_j \varpi)] - 2\sigma_j \varpi J_0(\sigma_j \varpi) J_1(\sigma_j \varpi) \cos^2 \sigma_j t\} \\ &- \tfrac{1}{4}\varpi \sum_{j<k} A_j A_k \frac{\sigma_j \sigma_k}{\sigma_j + \sigma_k}[J_0(\sigma_j \varpi) J_1(\sigma_k \varpi) + J_0(\sigma_k \varpi) J_1(\sigma_j \varpi)] \cos(\sigma_j + \sigma_k) t \\ &- \tfrac{1}{4}\varpi \sum_{j<k} A_j A_k \frac{\sigma_j \sigma_k}{\sigma_j - \sigma_k}[J_0(\sigma_j \varpi) J_1(\sigma_k \varpi) - J_0(\sigma_k \varpi) J_1(\sigma_j \varpi)] \cos(\sigma_j - \sigma_k) t.\end{aligned} \tag{B 10}$$

We can similarly write down the solution for the corresponding 'sine series',

$$\ln \Psi = \sum_j B_j J_0(\sigma_j \varpi) \sin \sigma_j t, \tag{B 11}$$

and for the superposition of both the series (B 9) and (B 11).

It is important to observe that no member of the general solution we have obtained for the case $q_2 = \Phi = 0$ is included among the solutions representing stationary monochromatic waves considered in the text.

(a) *A historical note*

It has sometimes been stated that, as a result of his investigation with Rosen on cylindrical waves, Einstein had come to doubt the reality of gravitational waves in general relativity. The origin of these alleged doubts can be traced to a letter that Einstein wrote to Max Born in the autumn of 1936 (see *The Born–Einstein Letters* (Born 1971, p. 125)). He wrote:

> Together with a young collaborator, I arrived at the interesting result that gravitational waves do not exist, though they had been assumed a certainty to the first approximation. This shows that the non-linear general relativistic field equations can tell us more, or, rather, limit us more than we have believed up to now. If only it were not so damnably difficult to find rigorous solutions.

But one does not find any statement in the published paper which corroborates what Einstein wrote to Born. Instead, the concluding paragraph of the paper reads:

> ...the integral can be (approximately) represented by $-\frac{1}{2}a\omega^2 T$, and thus cannot vanish and always has the same sign. Progressive waves therefore produce a secular change in the metric.
>
> This is related to the fact that the waves transport energy, which is bound up with a systematic change in time of a gravitating mass localized in the axis $x = 0$.

The reference to a quantity that 'cannot vanish and always has the same sign' is, implicitly, to an expression equivalent to the Hamiltonian density $\mathscr{H}$ (defined in equation (43)) in a certain approximation. The clue to the contrary statements in Einstein's letter to Born and in the published paper is to be found in the following note added to the paper and signed by Einstein.

> The second part of this paper was considerably altered by me after the departure of Mr. Rosen for Russia since we had originally interpreted our formula results erroneously. I wish to thank my colleague Professor (H. P.) Robertson for his friendly assistance in the clarification of the original error. I thank also Mr. (B.) Hoffmann for kind assistance in translation.
>
> A. Einstein.

It is, however, curious that Einstein seems to have left his associates with an impression of ambiguity with respect to his views on the reality of gravitational waves. Thus, Infeld & Plebanski (1960, p. 201) in the concluding paragraph of their book quote Einstein as having remarked:

> We do not have any satisfactory classical theory of radiation. Ritz understood this fact. He was an intelligent man...

and they add:

> This remark seems especially apt if applied to gravitational radiation.

Appendix C. Stationary electromagnetic waves in a time-independent cylindrically symmetric space–time

(*Added* 3 *June* 1986)

It is known that the Einstein–Maxwell equations, when the space–time has two commuting Killing fields (either, both space-like or one time-like and the other space-like), can be reduced to a pair of coupled equations for two complex functions Z and H: equations (114) and (115) of this paper, equations (43) and (44) of Paper II, and equations (58) and (59) on p. 570 in *M.T.* It does not, however, seem to have been noticed that *if Z were assumed to be a constant (real or complex), then the equation governing Z is identically satisfied and the equation governing H reduces to the form of an Ernst equation for the vacuum.* The deeper implications of this fact will be explored in a subsequent paper (Chandrasekhar & Xanthopoulos 1987). Meantime, in this Appendix, we shall provide a simple illustration of how this fact can be exploited in the context of cylindrically symmetric space–times.

Consistently with the assumption of Z being a constant, no loss of generality is entailed by setting

$$Z = 1 + \mathrm{i}\alpha \quad (\alpha \text{ is a real constant}). \tag{C 1}$$

On this assumption, equation (114) is identically satisfied and equation (115), governing H, becomes

$$(1-|H|^2)\left[H_{,0,0} - \frac{1}{\varpi}(\varpi H_{,\varpi})_{,\varpi}\right] = -2H^*[(H_{,0})^2 - (H_{,\varpi})^2]; \tag{C 2}$$

and this is identical in form with the Ernst equation (29) governing the vacuum. Accordingly, in the present context, H allows a separable solution of the form (cf. equation (61)),

$$H = \mathrm{e}^{\mathrm{i}\sigma t} F(\varpi) = (\cos \sigma t + \mathrm{i} \sin \sigma t) F(\varpi), \tag{C 3}$$

where the (real) radial function F, compatible with the requirement of regularity on the axis ($\varpi = 0$), is a solution of equation (63) satisfying the boundary conditions (64). Solutions for $F(\varpi)$ are illustrated in figures 1 and 2 (in the panels on the left).

The corresponding solutions for Ψ, χ, and Φ are (cf. equations (110) and (116))

$$\Psi = 1 - |H|^2 = 1 - F^2, \quad \chi = \frac{\varpi}{1-F^2}, \quad \text{and} \quad \Phi = \alpha. \tag{C 4}$$

By equations (109) and (C 4), the equations governing q_2 are

$$q_{2,0} = 2\frac{\varpi}{\Psi^2} \operatorname{Im} HH^*_{,\varpi} \quad \text{and} \quad q_{2,\varpi} = 2\frac{\varpi}{\Psi^2} \operatorname{Im} HH^*_{,0}. \tag{C 5}$$

From these equations, it follows that

$$q_{2,0} = 0 \quad \text{and} \quad q_{2,\varpi} = -2\sigma\frac{\varpi F^2}{(1-F^2)^2}. \tag{C 6}$$

Therefore,
$$q_2 = -2\sigma \int_0^{\varpi} \frac{\varpi F^2}{(1-F^2)^2} \mathrm{d}\varpi. \tag{C 7}$$

Comparison with equation (69) shows that, as now defined, q_2 (apart from a factor $\frac{1}{2}$) has the behaviour represented by the central heavily drawn curves in figures 1 and 2 (in the panels on the right).

And, finally, from equations (119) and (120), we find that the C-energy, $\nu+\ln\sqrt{\Psi}$, is independent of time and that its dependence on ϖ is given by (cf. equation (71))

$$\nu+\ln\sqrt{(1-F^2)} = \int_0^{\varpi} \frac{\varpi}{(1-F^2)^2}[(F_{,\varpi})^2+\sigma^2F^2]\,\mathrm{d}\varpi. \tag{C 8}$$

The metric of the space–time we have derived is

$$\mathrm{d}s^2 = \mathrm{e}^{2\nu}[(\mathrm{d}t)^2-(\mathrm{d}\varpi)^2]-\frac{\varpi^2}{1-F^2}(\mathrm{d}\varphi)^2-(1-F^2)(\mathrm{d}z-q_2\,\mathrm{d}\varphi)^2, \tag{C 9}$$

where ν and q_2 are defined in equations (C 7) and (C 8). We observe that *the metric is independent of time even though the space–time supports monochromatic electromagnetic waves described by the time-dependent complex-potential H given by equation* (C 3). Also, it will be observed that the metric is regular on the axis and that the asymptotic behaviour of the various metric coefficients is the same as for the vacuum metric (89) considered in §6.

References

Ablowitz, M. J. & Segur, H. 1981 *Solitons and the inverse scattering transform.* Philadelphia: SIAM.
Born, I. (trans.) 1971 *The Born–Einstein letters.* London: Macmillan & Co.
Chandrasekhar, S. 1983 *The mathematical theory of black holes.* Oxford: Clarendon Press.
Chandrasekhar, S. & Ferrari, V. 1984 *Proc. R. Soc. Lond.* A **396**, 55 (see Errata in *Proc. R. Soc. Lond.* A **398**, 429).
Chandrasekhar, S. & Persides, S. (In preparation.)
Chandrasekhar, S. & Xanthopoulos, B. C. 1985*a* *Proc. R. Soc. Lond.* A **398**, 223.
Chandrasekhar, S. & Xanthopoulos, B. C. 1985*b* *Proc. R. Soc. Lond.* A **402**, 205.
Chandrasekhar, S. & Xanthopoulos, B. C. 1986*a* *Proc. R. Soc. Lond.* A **403**, 189.
Chandrasekhar, S. & Xanthopoulos, B. C. 1986*b* *Proc. R. Soc. Lond.* A **408**, 175.
Chandrasekhar, S. & Xanthopoulos, B. C. 1987 *Proc. R. Soc. Lond.* A. (In the press.)
Einstein, A. & Rosen, N. 1937 *J. Frank. Inst.* **223**, 43.
Fokas, A. S. & Ablowitz, M. J. 1982 *J. math. Phys.* **23**, 11.
Ince, E. L. 1927 *Ordinary differential equations.* London: Longmans, Green & Co.
Infeld, J. & Plebański, J. 1960 *Motion and relativity.* Oxford: Pergamon Press.
Kramer, D., Stephani, H., Herlt, E. & MacCallum, M. 1980 *Exact solutions of Einstein's field equations.* Cambridge University Press.
Léauté, B. & Marcilhacy, G. 1986 *J. math. Phys.* **27**, 703.
Léauté, B. & Marcilhacy, G. 1985 *J. math. Phys.* **26**, 1938.
Marder, L. 1958*a* *Proc. R. Soc. Lond.* A **244**, 524.
Marder, L. 1958*b* *Proc. R. Soc. Lond.* A **246**, 133.
Morris, H. & Dodd, R. 1980 *Phys. Lett.* A **75**, 249.
Piran, T., Safier, P. N. & Stark, R. F. 1985 *Phys. Rev.* D **32**, 3101.
Rayleigh, Baron 1945 *The theory of sound* (in two volumes). New York: Dover Publications.
Rosen, N. 1954 *Bull. Res. Council Israel* **3**, 328.
Stachel, J. J. 1966 *J. math. Phys.* **7**, 1321.
Thorne, K. 1965 *Phys. Rev.* B **138**, 251.
Watson, G. N. 1944 *A treatise on the theory of Bessel functions.* Cambridge University Press.
Witten, E. 1981 *Communs math. Phys.* **80**, 381.

On the dispersion of cylindrical impulsive gravitational waves

By Subrahmanyan Chandrasekhar[1], F.R.S.,
and Valeria Ferrari[2]

[1] *University of Chicago, Chicago, Illinois* 60637, *U.S.A.*
[2] *Dipartimento di Fisica 'G. Marconi', Universita di Roma, Rome, Italy*

(*Received* 9 *December* 1986)

An exact solution, describing the dispersion of a wave packet of gravitational radiation, having initially (at time $t = 0$) an impulsive character, is analysed. The impulsive character of the wave-packet derives from the space-time being flat, except at a radial distance $\varpi = \varpi_1$ (say) at $t = 0$, and the time-derivative of the Weyl scalars exhibiting δ-function singularities at $\varpi = \varpi_1$, when $t \to 0$. The principal feature of the dispersion is the development of a singularity of the metric function, ν, and of the Weyl scalar, Ψ_2, when the wave, after reflection at the centre, collides with the still incoming waves. The evolution of the metric functions and of the Weyl scalars, as the dispersion progresses, is illustrated graphically.

1. Introduction

It is well known that solutions of Einstein's vacuum equations exist that represent the propagation, without dispersion, of planar wave-packets of gravitational radiation (cf. Rindler 1977, §8.8). A limiting form of such wave-packets is the *impulsive wave*. It is not to be expected that such permanent impulsive waves will exist in geometries other than planar. In this paper, we shall obtain an exact solution of Einstein's equations that describes how a cylindrical wave-packet, initially impulsive, disperses and develops shock fronts and discontinuities when the wave, after reflection at the axis, encounters the still ingoing waves. The mathematical analysis is based on a recent paper by one of us (Chandrasekhar 1986; this paper will be referred to hereafter as Paper I).

2. The basic equations

We consider a space-time with cylindrical symmetry with the two commuting space-like Killing-vectors, ∂_φ and ∂_z, associated with the azimuthal angle φ and the coordinate z, along the axis of symmetry. We shall restrict ourselves to the hyper-surface orthogonal case when the metric can be written in the form (cf. Paper I, equation (B 1)):

$$ds^2 = e^{2\nu}[(dt)^2 - (d\varpi)^2] - \frac{\varpi^2}{\Psi}(d\varphi)^2 - \Psi(dz)^2, \tag{1}$$

where the metric functions ν and Ψ depend only on the time, $t\,(= x^0)$, and the radial coordinate, ϖ. The equations governing Ψ and ν are (Paper I, equations (B 2)–(B 4))

$$(\ln \Psi)_{,0,0} - \frac{1}{\varpi}[\varpi(\ln \Psi)_{,\varpi}]_{,\varpi} = 0, \tag{2}$$

$$C_{,0} = \tfrac{1}{2}\varpi(\ln \Psi)_{,0}(\ln \Psi)_{,\varpi}, \tag{3}$$

and

$$C_{,\varpi} = \tfrac{1}{4}\varpi\{[(\ln \Psi)_{,0}]^2 + [(\ln \Psi)_{,\varpi}]^2\}. \tag{4}$$

where

$$C = \nu + \ln\sqrt{\Psi}. \tag{5}$$

The quantity C denotes the 'C-energy': a concept introduced by Thorne (1965). The reasons for calling C as an 'energy' are discussed in Paper I, §3(b).

In terms of the radial null-coordinates,

$$u = x^0 + \varpi \quad \text{and} \quad v = x^0 - \varpi, \tag{6}$$

equations (3) and (4) can be combined to give

$$C_{,u} = +\tfrac{1}{2}\varpi[(\ln \Psi)_{,u}]^2 \quad \text{and} \quad C_{,v} = -\tfrac{1}{2}\varpi[(\ln \Psi)_{,v}]^2. \tag{7}$$

Alternatively, in terms of the 'Ernst function' E, defined in the present case, by (cf. I, equations (26) and (28))

$$\Psi = \frac{1+E}{1-E}, \tag{8}$$

we have

$$(\ln \Psi)_{,u} = \frac{2E_{,u}}{1-E^2}, \quad (\ln \Psi)_{,v} = \frac{2E_{,v}}{1-E^2}, \tag{9}$$

$$C_{,u} = +\frac{2\varpi}{(1-E^2)^2}(E_{,u})^2 \quad \text{and} \quad C_{,v} = -\frac{2\varpi}{(1-E^2)^2}(E_{,v})^2. \tag{10}$$

By virtue of these equations,

$$2\nu_{,u} = 2C_{,u} - (\ln \Psi)_{,u} \quad \text{and} \quad 2\nu_{,v} = 2C_{,v} - (\ln \Psi)_{,v}. \tag{11}$$

As we have noted in Paper I, Appendix B, the general solution of equation (2) is expressible as an arbitrary linear combination, as sums and/or integrals, of the fundamental solutions

$$\ln \Psi = J_0(\sigma\varpi)\cos\sigma t \quad \text{and} \quad \ln \Psi = J_0(\sigma\varpi)\sin\sigma t, \tag{12}$$

where σ is a positive real constant and J_0 is Bessel's function of order zero.

3. A solution representing a wave, impulsive at time $t = 0$ and $\varpi = \varpi_1$

We shall show that a linear superposition of the fundamental solutions (12) that represents a wave having an impulsive character at time $t = 0$ and $\varpi = \varpi_1$, is given by

$$\ln \Psi = \varpi_1 \int_0^\infty J_0(\sigma\varpi_1) J_0(\sigma\varpi)\sin\sigma t\, \mathrm{d}\sigma, \tag{13}$$

where the factor ϖ_1 has been introduced for later convenience. The foregoing integral, chosen to express $\ln \Psi$, is one of the standard discontinuous integrals which vanish for certain ranges of the arguments and which take different non-vanishing values in other ranges of the arguments (cf. Watson 1944, p. 412, equations (4) and (5)); thus

$$\ln \Psi = +\frac{\varpi_1}{2\sqrt{(\varpi\varpi_1)}} P_{-\frac{1}{2}}\left(\frac{\varpi^2+\varpi_1^2-t^2}{2\varpi\varpi_1}\right) \quad (|\varpi_1-t| < \varpi < \varpi_1+t), \tag{14a}$$

$$= -\frac{\varpi_1}{\pi\sqrt{(\varpi\varpi_1)}} Q_{-\frac{1}{2}}\left(\frac{t^2-\varpi^2-\varpi_1^2}{2\varpi\varpi_1}\right) \quad (t > \varpi_1 \quad \text{and} \quad 0 \leqslant \varpi < t-\varpi_1), \tag{14b}$$

$$= 0 \quad (\varpi > \varpi_1+t; \quad \text{and} \quad 0 \leqslant \varpi < \varpi_1 - t \quad \text{when} \quad t < \varpi_1), \tag{14c}$$

where $P_{-\frac{1}{2}}(\cos\theta)$ and $Q_{-\frac{1}{2}}(\cosh\eta)$ are the conical and the ring Legendre-functions of order $-\frac{1}{2}$ (cf. Hobson 1931, pp. 267 and 438). The functions $P_{-\frac{1}{2}}(z)$ and $Q_{-\frac{1}{2}}(z)$ are expressible in terms of the hypergeometric functions; thus,

$$P_{-\frac{1}{2}}(z) = F\left(\tfrac{1}{2}, \tfrac{1}{2}; 1; \frac{1-z}{2}\right) \quad \text{and} \quad Q_{-\frac{1}{2}}(z) = \frac{\pi}{\sqrt{(2z)}} F\left(\tfrac{3}{4}, \tfrac{1}{4}; 1; \frac{1}{z^2}\right). \tag{15}$$

These expressions are useful for the numerical evaluation of these functions.

(a) *Alternative forms of the solution in terms of the complete elliptic-integral, $K(k)$*

First, we must observe that by measuring both t and ϖ in the unit ϖ_1, the solution (14) for $\ln \Psi$ can be written more conveniently in the form

$$\ln \Psi = +\frac{1}{2\sqrt{\varpi}} P_{-\frac{1}{2}}\left(\frac{\varpi^2+1-t^2}{2\varpi}\right) \quad (|t-1| < \varpi < 1+t), \tag{16a}$$

$$= -\frac{1}{\pi\sqrt{\varpi}} Q_{-\frac{1}{2}}\left(\frac{t^2-1-\varpi^2}{2\varpi}\right) \quad (t > 1, 0 \leqslant \varpi < t-1), \tag{16b}$$

$$= 0 \quad (\varpi > 1+t \quad \text{and} \quad \varpi < 1-t \quad \text{when} \quad t < 1). \tag{16c}$$

The solutions (16*a*) and (16*b*), valid in their respective ranges, can be expressed in terms of the *complete elliptic-integral*,

$$K(k) = \int_0^{\frac{1}{2}\pi} \frac{d\theta}{\sqrt{(1-k^2\sin^2\theta)}}; \tag{17}$$

thus,
$$\ln \Psi = \frac{1}{\pi\sqrt{\varpi}} K(k) \quad (0 \leqslant k < 1), \tag{18}$$

where
$$k^2 = \frac{1}{4\varpi}[t^2-(\varpi-1)^2] \quad (|t-1| < \varpi < 1+t); \tag{19}$$

and
$$\ln \Psi = -\frac{2}{\pi\sqrt{\varpi}} F(\eta), \tag{20}$$

where
$$\eta = \operatorname{arcosh}\left[\frac{1}{2\varpi}(t^2-1-\varpi^2)\right] \quad (t > 1, 0 \leqslant \varpi < t-1) \tag{21}$$

and
$$F(\eta) = e^{-\frac{1}{2}\eta} K(e^{-\eta}) = K(k)\sqrt{k} \quad (k = e^{-\eta}). \tag{22}$$

(b) *Some useful formulae*

In obtaining the behaviours of the solution at its various fronts and discontinuities, we shall find the following formulae useful.

The functions $K(k)$ and $F(\eta)$ satisfy the differential equations,

$$k(1-k^2)\,K_{,k,k}+(1-3k^2)\,K_{,k}-kK = 0 \tag{23}$$

and

$$F_{,\eta,\eta}+F_{,\eta}\coth\eta+\tfrac{1}{4}F = 0. \tag{24}$$

The derivative of the elliptic integral $K(k)$ is given by

$$K_{,k} = \frac{1}{kk'^2}(E-k'^2K), \tag{25}$$

where

$$k' = \sqrt{(1-k^2)} \tag{26}$$

is the *complementary modulus* and

$$E = \int_0^{\frac{1}{2}\pi} d\theta\,\sqrt{(1-k^2\sin^2\theta)} \tag{27}$$

denotes the elliptic integral of the second kind. Also, by virtue of the definition (22),

$$F_{,\eta} = -\tfrac{1}{2}(K+2kK_{,k})\sqrt{k}. \tag{28}$$

Finally, we note the following behaviours of K and $K_{,k}$:

$$K = K(0)\,(1+\tfrac{1}{4}k^2+\tfrac{9}{64}k^4+\ldots), \quad K(0) = \tfrac{1}{2}\pi; \tag{29}$$

$$\lim_{k\to 0}\frac{1}{k^2}(E-k'^2K) = \tfrac{1}{4}\pi, \tag{30}$$

and

$$\lim_{k\to 1} K(k) = \ln\frac{4}{k'}. \tag{31}$$

4. The dispersion of the wave-packet represented by the solution (13)

First, we observe that for the solution specified by equation (16a, b, c),

$$\text{at}\quad t = 0, \ln\Psi \equiv 0 \quad \text{for} \quad \varpi \neq \varpi_1. \tag{32}$$

This fact, together with (3) and (4), implies that *at* $t = 0$, *the space-time is flat except at* $\varpi = 1$.

The dispersion of the wave-packet, represented by the solution (13), is, in the first instance, determined by equation (16a), according to which

$$\ln\Psi \neq 0 \quad \text{only for} \quad 1-t \leqslant \varpi \leqslant 1+t \quad (t < 1); \tag{33}$$

and, further, by equations (18), (10), and (29)

$$\ln\Psi = \frac{1}{2\sqrt{(1\pm t)}}\,(\varpi\to 1-t+0 \quad \text{and} \quad \varpi\to 1+t-0). \tag{34}$$

Therefore, $\ln\Psi$, in its dependence on t, represents a one-parameter sequence of bounded functions with compact support, whose integrals over ϖ tend asymptotically to zero linearly with t as $t\to 0$. And since, further, $\ln\Psi \equiv 0$ for $t = 0$ and $\varpi \neq 1$, we may conclude that $(\ln\Psi)_{,t}$ becomes proportional to a δ-function, centred at $\varpi = 1$, as $t \to 0$. (We are grateful to Professor R. Narasimhan for elucidating the underlying theorem.) In other words, the solution (13) for $\ln \Psi$, that we have chosen, does, indeed, represent the dispersion of an initially impulsive wave in the manner of the notes emitted by a struck string.

It follows from equations (33) and (34) that $\ln \Psi$ is non-vanishing only in the interval, allowed by causality, for an impulsive wave initially (i.e. at $t = 0$) at $\varpi = 1$, to disperse in a time $t\,(< 1)$: the fronts at $\varpi = 1 \pm t$ represent the furthest distances, in the forward and the backward directions, to which the impulsive wave can spread in a time $t\,(< 1)$.

The wavefront at $\varpi = 1 - t$, when $t < 1$, will arrive at the axis ($\varpi = 0$) at time $t = 1$; and we should expect that a singularity will develop on the axis at the instant of reflection. Indeed, it follows from equations (18) and (19):

$$\ln \Psi = \frac{1}{\pi\sqrt{\varpi}} K[\tfrac{1}{2}\sqrt{(2-\varpi)}] \quad (t = 1). \tag{35}$$

Accordingly,

$$\ln \Psi \sim \frac{K(45^0)}{\pi\sqrt{\varpi}} = \frac{1.8541}{\pi\sqrt{\varpi}} = \frac{0.5902}{\sqrt{\varpi}} \quad (t = 1, \varpi \to 0). \tag{36}$$

When $t > 1$, the wave reflected at $\varpi = 0$ will have progressed to $\varpi = t - 1$ at which point it will encounter the still incoming waves. It follows from equations (14*b*) and (16*b*), that the reflected wave, in the interval $0 \leqslant \varpi < t - 1$, will be described by the solution (20), while the wave beyond $\varpi = t - 1$, in the interval $t - 1 < \varpi < t + 1$, will be described by the solution (18).

Considering the behaviour of the solution as $\varpi = 0$ for $t > 1$, we have, in accordance with equations (21) and (22),

$$e^{\eta} \sim \frac{t^2 - 1}{\varpi} \quad \text{and} \quad e^{-\eta} = k \sim \frac{\varpi}{t^2 - 1} \quad \text{when} \quad \varpi \to 0 \quad (t \to 1). \tag{37}$$

Therefore,

$$F(\eta) \sim e^{-\frac{1}{2}\eta} K(0) = \tfrac{1}{2}\pi \left(\frac{\varpi}{t^2 - 1}\right)^{\frac{1}{2}}; \tag{38}$$

and by equation (20),

$$\ln \Psi \sim -\frac{1}{\sqrt{(t^2 - 1)}} \quad \text{when} \quad \varpi \to 0 \quad (t > 1); \tag{39}$$

$\ln \Psi$ is thus finite on the axis after reflection (even as it was before the reflection when $\ln \Psi$ was zero). *The metric* (1) *of the space-time is, therefore, regular on the axis at all times except at the instant of reflection at the axis.*

Again, we find from equations (21) and (22) that for

$$\varpi = t - 1 - \epsilon \quad (\epsilon \to +0) \tag{40}$$

$$\cosh \eta = 1 + \frac{\epsilon t}{t - 1} + O(\epsilon^2), \quad \eta^2 = \frac{2\epsilon t}{t - 1} + O(\epsilon^2), \tag{41}$$

$$k = e^{-\eta} = 1 - \eta + O(\eta^2) \quad \text{and} \quad k'^2 = 1 - k^2 = 2\eta + O(\eta^2). \tag{42}$$

From equations (20), (22) and the asymptotic behaviour (31) of $K(k)$ for $k \to 1$, we obtain

$$\ln \Psi \sim \frac{1}{2\pi\sqrt{(t-1)}} \ln \frac{\epsilon t}{32(t-1)} \quad (\varpi = t - 1 - \epsilon, \epsilon \to +0; t > 1). \tag{43}$$

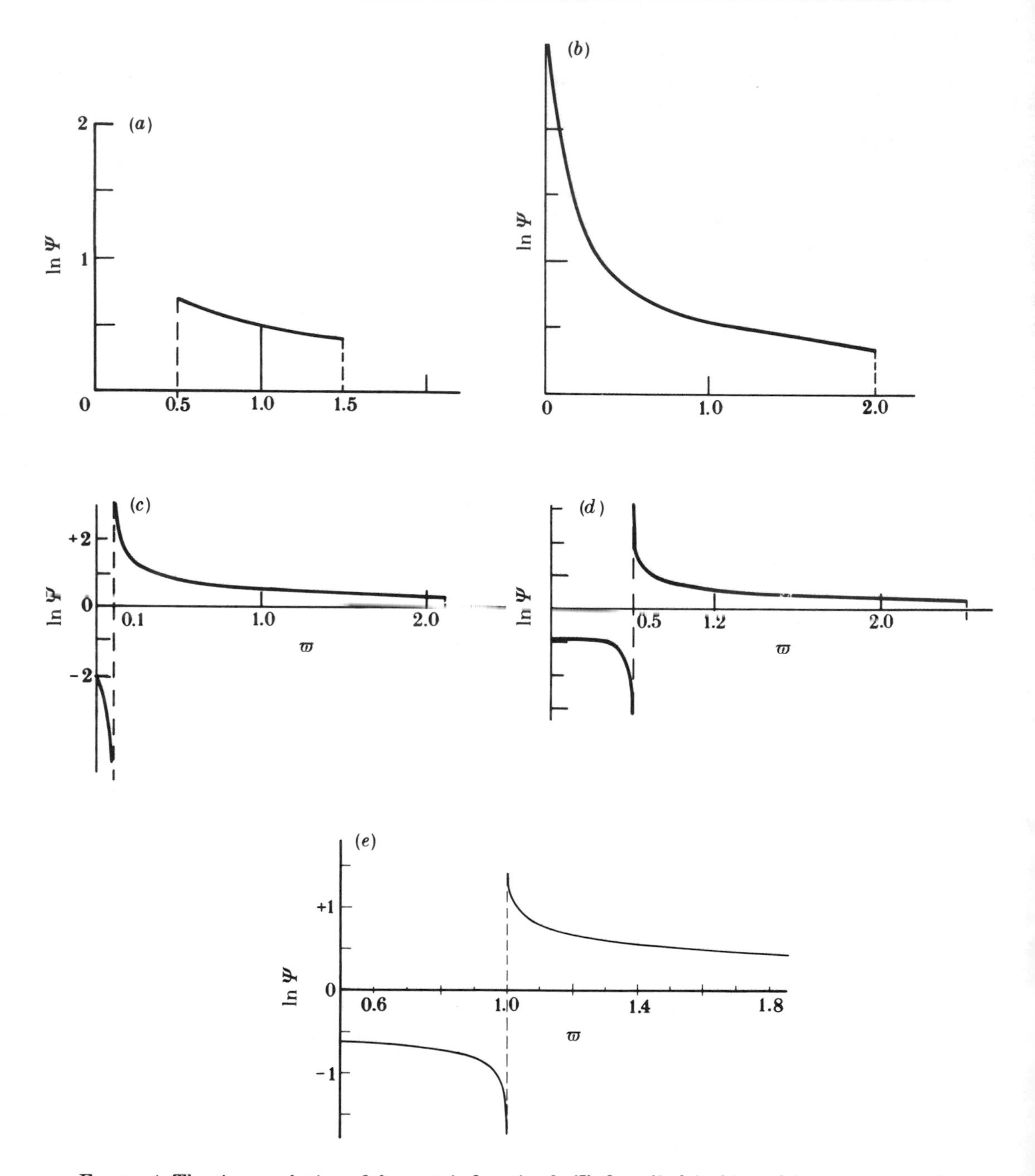

FIGURE 1. The time evolution of the metric function $\ln \Psi$ of a cylindrical impulsive gravitational wave located at $\varpi = 1.0$ at time $t = 0$. For $t < 1$, the impulsive wave disperses to an extent required by causality; namely, $1-t \leqslant \varpi \leqslant 1+t$. At $t = 1$ the front progressing inwards arrives at the axis and is reflected, forming a singularity momentarily at the instant of reflection. For $t > 1$ the reflected wave arrives at $\varpi = t-1$ and, encountering the still incoming wave, develops a singularity that propagates to infinity. (*a*) $t = 0.5$; (*b*) $t = 1.0$; (*c*) $t = 1.1$; (*d*) $t = 1.5$; (*e*) $t = 2.0$.

The corresponding behaviour of the solution for

$$\varpi = t-1+\epsilon \quad (\epsilon \to +0), \tag{44}$$

follows from equations (18) and (19). Thus,

$$k^2 = 1-\frac{\epsilon t}{2(t-1)}+O(\epsilon^2), \quad k'^2 = \frac{\epsilon t}{2(t-1)}+O(\epsilon^2) \tag{45}$$

and
$$\ln \Psi \sim -\frac{1}{2\pi\sqrt{(t-1)}} \ln \frac{\epsilon t}{32(t-1)} \quad (\varpi = t-1+\epsilon, \epsilon \to 0; t > 1). \tag{46}$$

Comparison of equations (43) and (46) shows that $\ln \Psi$ has the same behaviour on either side of $\varpi = t-1$ but in opposite senses.

Figure 1 illustrates the behaviours of the solution specified by equations (17)–(22).

5. The completion of the solution; the C-energy

We now complete the solution for the metric functions by determining ν via equations (3) and (4) governing the C-energy. Considering first equations (3) and (4), when $\ln \Psi$ is determined by equations (16a) and (18), we find, on changing the independent variables t and ϖ to t and k (defined in equation (19)), that

$$\pi^2 C_{,\varpi} = +\frac{1}{16\varpi^2}[K^2+(k'K_{,k})^2] \tag{47}$$

and
$$\pi^2 C_{,k} = -\frac{1}{4\varpi}\left[KK_{,k}+\frac{1}{2k}(2k^2-1)(K_{,k})^2\right]-\frac{1}{8k}(K_{,k})^2. \tag{48}$$

Equation (47) directly integrates to give

$$\pi^2 C = -\frac{1}{16\varpi}[K^2+(k'K_{,k})^2]+f(k), \tag{49}$$

where $f(k)$ is a function of k only. Differentiating equation (49) with respect to k, comparing the result of the differentiation with equation (48), and making use of the differential equation (23) satisfied by $K(k)$, we find

$$f_{,k} = -\frac{1}{8k}(K_{,k})^2. \tag{50}$$

The required solution for C is, therefore, given by

$$\pi^2 C = -\frac{1}{16\varpi}[K^2+(k'K_{,k})^2]-\frac{1}{8}\int\frac{\mathrm{d}k}{k}(K_{,k})^2 \quad (|t-1| < \varpi < t+1). \tag{51}$$

When the solution for $\ln \Psi$ is given by equations (16b) and (20), we find by an analogous procedure that

$$\pi^2 C = -\frac{1}{4\varpi}[F^2-4(F_{,\eta})^2]-2\int\frac{\mathrm{d}\eta}{\sinh \eta}(F_{,\eta})^2; \tag{52}$$

or, making use of equation (28) and the further relation

$$\sinh \eta = \tfrac{1}{2}k'^2/k, \tag{53}$$

we have

$$\pi^2 C = \frac{k^2}{\varpi}(K + kK_{,k})\,K_{,k} + \int \frac{k}{k'^2}(K + 2kK_{,k})^2\,dk. \tag{54}$$

The same asymptotic relations that were used in §4 to determine the behaviours of $\ln \Psi$ at its various fronts and discontinuities can now be used to establish the associated behaviours of the C-energy. We find:

$$C = -\frac{1}{64(1 \pm t)} \quad (\varpi = 1 \pm t \mp 0; t < 1), \tag{55}$$

$$C = -\frac{1}{16\pi^2\varpi}\{K^2(k) + [2E(k) - K(k)]^2\}_{k=1/\sqrt{2}} \quad (\varpi \to 0; t = 1), \tag{56}$$

$$C \sim \frac{\varpi^2 t^2}{8(t^2-1)^3} \quad (\varpi \to 0; t > 1), \tag{57}$$

and

$$C \sim \pm\frac{1}{8\pi^2\epsilon} \quad (\varpi = t - 1 \mp \epsilon, \epsilon \to +0; t > 1). \tag{58}$$

The foregoing behaviours of the C-energy are illustrated in figure 2.

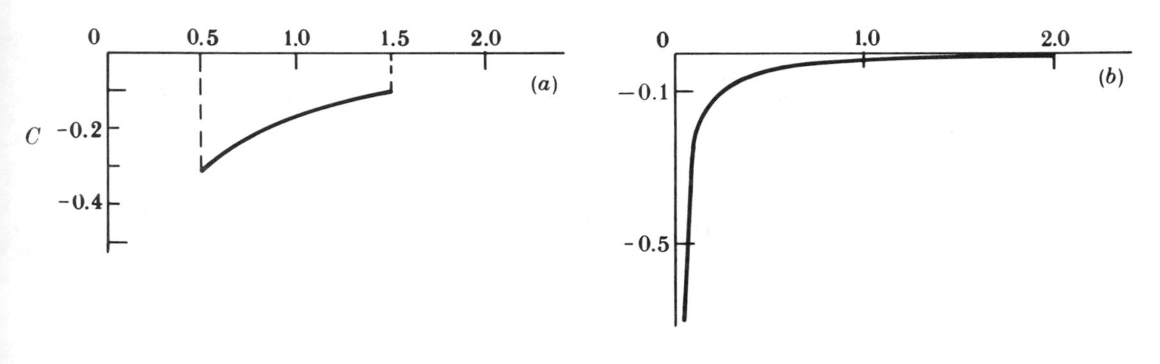

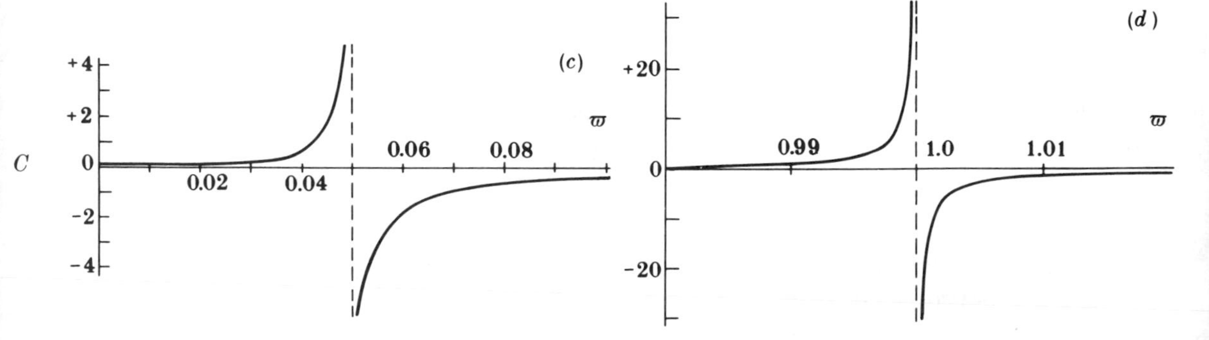

FIGURE 2. The time evolution of the C-energy of a cylindrical impulsive gravitational wave. (See also legend for figure 1.) (a) $t = 0.5$; (b) $t = 1$; (c) $t = 1.05$; (d) $t = 2.0$.

Finally, we may note that the C-energy in terms of the original solutions (14a) and (14b) is given by

$$C = -\frac{1}{64\varpi}[P^2+4(1-\mu^2)(P_{,\mu})^2]+\frac{1}{8}\int(P_{,\mu})^2\,d\mu;$$
$$P = P_{-\frac{1}{2}}(\mu) \quad\text{and}\quad \mu = (\varpi^2+1-t^2)/2\varpi, \tag{59}$$

and

$$C = -\frac{1}{16\pi^2\varpi}[Q^2-4(y^2-1)(Q_{,y})^2]-\frac{1}{2\pi^2}\int(Q_{,y})^2\,dy;$$
$$Q = Q_{-\frac{1}{2}}(y) \quad\text{and}\quad y = (t^2-1-\varpi^2)/2\varpi. \tag{60}$$

6. The flow of energy along the null directions

Two physically significant quantities are $C_{,u}$ and $C_{,v}$ which measure the flow of energy along the two null-directions. Considering first the ranges in which the solution is given by equation (18), we find, on changing the independent variables to t and k,

$$\left.\begin{aligned}\pi(\ln\Psi)_{,t} &= \frac{t}{4k\varpi^{\frac{3}{2}}}K_{,k},\\ \pi(\ln\Psi)_{,\varpi} &= -\frac{1}{2\varpi^{\frac{3}{2}}}K-\frac{1}{8\varpi^{\frac{5}{2}}k}(t^2+\varpi^2-1)K_{,k}.\end{aligned}\right\} \tag{61}$$

These expressions yield

$$\left.\begin{aligned}C_{,u} &= +\frac{1}{32\pi^2\varpi^2}\left\{K+\frac{1}{4\varpi k}[(t-\varpi)^2-1]K_{,k}\right\}^2,\\ C_{,v} &= -\frac{1}{32\pi^2\varpi^2}\left\{K+\frac{1}{4\varpi k}[(t+\varpi)^2-1]K_{,k}\right\}^2.\end{aligned}\right\} \tag{62}$$

For the ranges in which the solution (20) is valid, we have, in place of equations (61),

$$\left.\begin{aligned}\pi(\ln\Psi)_{,t} &= \frac{-2t}{\varpi^{\frac{3}{2}}\sinh\eta}F_{,\eta},\\ \pi(\ln\Psi)_{,\varpi} &= +\frac{1}{\varpi^{\frac{3}{2}}}F+\frac{1}{\varpi^{\frac{5}{2}}\sinh\eta}(t^2+\varpi^2-1)F_{,\eta};\end{aligned}\right\} \tag{63}$$

and correspondingly,

$$\left.\begin{aligned}C_{,u} &= +\frac{1}{8\pi^2\varpi^2}\left\{F+\frac{1}{\varpi\sinh\eta}[(t-\varpi)^2-1]F_{,\eta}\right\}^2,\\ C_{,v} &= -\frac{1}{8\pi^2\varpi^2}\left\{F+\frac{1}{\varpi\sinh\eta}[(t+\varpi)^2-1]F_{,\eta}\right\}^2.\end{aligned}\right\} \tag{64}$$

Rewriting these expressions in terms of the elliptic integral K (by making use of equations (28) and (53)) we obtain:

$$\left.\begin{aligned}C_{,u} &= +\frac{1}{8\pi^2\varpi^2}\left\{K\sqrt{k}-\frac{k^{\frac{3}{2}}}{\varpi k'^2}(K+2kK_{,k})[(t-\varpi)^2-1]\right\}^2,\\ C_{,v} &= -\frac{1}{8\pi^2\varpi^2}\left\{K\sqrt{k}-\frac{k^{\frac{3}{2}}}{\varpi k'^2}(K+2kK_{,k})[(t+\varpi)^2-1]\right\}^2.\end{aligned}\right\} \tag{65}$$

From the solutions (62) and (65) for $C_{,u}$ and $C_{,v}$, belonging to the solutions (18) and (20) in their respective domains of validity, we find that at the various fronts and discontinuities, they have the following behaviours:

(i) $t < 1$:

$$C_{,u}(\varpi = 1+t) = \frac{+1}{128(1+t)^2}, \quad C_{,u}(\varpi = 1-t) = \frac{+1}{128(1-t)^2}(1-\tfrac{1}{2}t)^2,$$

$$C_{,v}(\varpi = 1-t) = \frac{-1}{128(1-t)^2}, \quad C_{,v}(\varpi = 1+t) = \frac{-1}{128(1+t)^2}(1+\tfrac{1}{2}t)^2. \tag{66}$$

(ii) $t = 1, \varpi \to 0$:

$$C_{,u} = +\frac{1}{8\pi^2\varpi^2}[K(k)-E(k)]^2{}_{k=1/\sqrt{2}},$$

$$C_{,v} = -\frac{1}{8\pi^2\varpi^2}E^2(k)_{k=1/\sqrt{2}}. \tag{66'}$$

(iii) $t > 1, \varpi \to 0$:

$$C_{,u} \sim +\frac{\varpi t^2}{8(t^2-1)^3}, \quad C_{,v} \sim -\frac{\varpi t^2}{8(t^2-1)^3}. \tag{67}$$

(iv) $t > 1, \varpi = t-1-\epsilon, \epsilon \to +0$:

$$C_{,u} \sim \frac{1}{128\pi^2(t-1)^2}\left[\ln\frac{\epsilon t}{32(t-1)}\right]^2, \quad C_{,v} \sim -\frac{1}{8\pi^2\epsilon^2}. \tag{68}$$

(v) $t > 1, \varpi = t-1+\epsilon, \epsilon \to 0$:

$$C_{,u} \sim \frac{1}{128\pi^2(t-1)^2}\left[\ln\frac{\epsilon t}{32(t-1)}\right]^2, \quad C_{,v} \sim -\frac{1}{8\pi^2\epsilon^2}. \tag{69}$$

We may draw particular attention to the fact that when $t > 1$, $C_{,u}$ and $C_{,v}$ become equal in magnitude but opposite in sign as $\varpi \to 0$; and further that they have the same asymptotic behaviours on either side of $\varpi = t-1$.

7. The Weyl scalars

The Weyl scalars, for the space-time we are presently considering, can be directly read off from the expressions given in Chandrasekhar & Xanthopoulos (1986, Appendix A; the equations of this appendix will be referred to by the same numbers) by substituting

$$U = e^{\nu} \quad \text{and} \quad V = \sqrt{\varpi}, \tag{70}$$

replacing ψ and θ by t and ϖ, respectively, and remembering that E is now real. Thus, from (A 20), we now find

$$\Psi_2 = \tfrac{1}{2}e^{-2\nu}\left\{\frac{4E_{,u}E_{,v}}{(1-E^2)^2} + \frac{E_{,u}-E_{,v}}{\varpi(1-E^2)}\right\}. \tag{71}$$

We can rewrite this expression in terms of the derivatives of $\ln \Psi$ by making use of equation (9): we find:

$$\Psi_2 = \tfrac{1}{2} e^{-2\nu} \left\{ [(\ln \Psi)_{,u}][(\ln \Psi)_{,v}] + \frac{1}{2\varpi}[(\ln \Psi)_{,u} - (\ln \Psi)_{,v}] \right\}. \tag{72}$$

The derivatives of $\ln \Psi$ which occur in (72) can, in turn, be replaced in terms of the known expressions, (62) and (64), for $C_{,u}$ and $C_{,v}$, in accordance with (61) and (63):

$$(\ln \Psi)_{,u} = \mp \sqrt{\frac{2|C_{,u}|}{\varpi}} \quad \text{and} \quad \pm \sqrt{\frac{2|C_{,v}|}{\varpi}}, \tag{73}$$

where the upper signs belong to the solution (18) and the lower signs belong to the solution (20). This convention regarding the placements of the signs will be strictly adhered to in the rest of this section. With these replacements, the expression (72) for Ψ_2 becomes

$$\Psi_2 = -\frac{e^{-2\nu}}{\varpi} \left\{ \sqrt{|C_{,u} C_{,v}|} \pm \frac{1}{2\sqrt{(2\varpi)}} [\sqrt{|C_{,u}|} + \sqrt{|C_{,v}|}] \right\}. \tag{74}$$

It should be noted that in accordance with equations (62) and (65),

$$\frac{1}{\sqrt{(2\varpi)}}(\sqrt{|C_{,u}|} + \sqrt{|C_{,v}|})$$

$$= \frac{1}{4\pi\varpi^{\frac{3}{2}}} \left[K + \frac{1}{4\varpi k}(t^2 + \varpi^2 - 1) K_{,k} \right] \quad \text{(for solution (18))}, \tag{75}$$

$$= \frac{\sqrt{k}}{2\pi\varpi^{\frac{3}{2}}} \left[K - \frac{k}{\varpi k'^2}(t^2 + \varpi^2 - 1)(K + 2kK_{,k}) \right] \quad \text{(for solution (20))}; \tag{76}$$

and further, $\sqrt{|C_{,u} C_{,v}|}$ should be interpreted to stand for

$$\sqrt{|C_{,u} C_{,v}|} = \frac{1}{32\pi^2\varpi^2} \left\{ K + \frac{1}{4\varpi k}[(t-\varpi)^2 - 1] K_{,k} \right\} \left\{ K + \frac{1}{4\varpi k}[(t+\varpi)^2 - 1] K_{,k} \right\}, \tag{77}$$

for solution (18) and

$$\sqrt{|C_{,u} C_{,v}|} = \frac{k}{8\pi^2\varpi^2} \left\{ K - \frac{k}{\varpi k'^2}[(t-\varpi)^2 - 1](K + 2kK_{,k}) \right\} \times \left\{ K - \frac{k}{\varpi k'^2}[(t+\varpi)^2 - 1](K + 2kK_{,k}) \right\}, \tag{78}$$

for solution (20). In particular we find from equations (76) and (78) that $t > 1$ and $\varpi \to 0$ (cf. equation (67)).

$$\frac{1}{\sqrt{(2\varpi)}}(\sqrt{|C_{,u}|} + \sqrt{|C_{,v}|}) \sim -\varpi \frac{(2t^2+1)}{4(t^2-1)^{\frac{5}{2}}}$$

$$\text{and} \quad \sqrt{|C_{,u} C_{,v}|} \sim -\varpi \frac{t^2}{8(t^2-1)^3} \quad (t > 1, \varpi \to 0). \tag{79}$$

Accordingly,

$$e^{-2\nu}\Psi_2 \sim -\frac{1}{8(t^2-1)^3}[(2t^2+1)\sqrt{(t^2-1)} - t^2] \quad (t > 1, \varpi \to 0). \tag{80}$$

With the same substitutions (70) and replacements, we find from equation (A 23)

$$4\,e^{2\nu}\Psi_4 = \frac{2}{\varpi}(\nu_{,t}-\nu_{,\varpi})+2\frac{E_{,t}-E_{,\varpi}}{\varpi(1-E^2)}+4(\nu_{,t}-\nu_{,\varpi})\frac{E_{,t}-E_{,\varpi}}{1-E^2}$$
$$-2\left(\frac{\partial}{\partial t}-\frac{\partial}{\partial\varpi}\right)\frac{E_{,t}-E_{,\varpi}}{1-E^2}. \quad (81)$$

Simplifying this expression with the aid of equations (9) and (11), we find:

$$\Psi_4 = -e^{-2\nu}\{\tfrac{3}{2}[(\ln\Psi)_{,v}]^2+\varpi[(\ln\Psi)_{,v}]^3+(\ln\Psi)_{,v,v}\}. \quad (82)$$

Similarly, we find from equation (A 26) that

$$\Psi_0 = -e^{-2\nu}\{\tfrac{3}{2}[(\ln\Psi)_{,u}]^2-\varpi[(\ln\Psi)_{,u}]^3+(\ln\Psi)_{,u,u}\}. \quad (83)$$

The foregoing expressions for Ψ_0 and Ψ_4 can be expressed in terms of quantities which we have already evaluated (such as $C_{,u}$ and $C_{,v}$) as follows. Considering Ψ_0, for example, we have, on making use of equation (73),

$$-e^{+2\nu}\Psi_0 = \frac{3}{\varpi}|C_{,u}|\pm\varpi\left(\frac{2|C_{,u}|}{\varpi}\right)^{\frac{3}{2}}+(\ln\Psi)_{,u,u}. \quad (84)$$

We now rewrite the last term $(\ln\Psi)_{,u,u}$ on the right-hand side of equation (84) by making use of equation (2) satisfied by $\ln\Psi$:

$$(\ln\Psi)_{,u,u} = \tfrac{1}{4}[(\ln\Psi)_{,0,0}+(\ln\Psi)_{,\varpi,\varpi}+2(\ln\Psi)_{,0,\varpi}]$$
$$= \tfrac{1}{4}[2(\ln\Psi)_{,0,0}+2(\ln\Psi)_{,0,\varpi}-\frac{1}{\varpi}(\ln\Psi)_{,\varpi}]$$
$$= (\ln\Psi)_{,u,0}-\frac{1}{4\varpi}[(\ln\Psi)_{,u}-(\ln\Psi)_{,v}]. \quad (85)$$

Making use, once again, of equation (73), we have

$$-e^{2\nu}\Psi_0 = \frac{3}{\varpi}|C_{,u}|\pm\varpi\left(\frac{2|C_{,u}|}{\varpi}\right)^{\frac{3}{2}}\pm\frac{1}{4\varpi}\left(\sqrt{\frac{2|C_{,u}|}{\varpi}}+\sqrt{\frac{2|C_{,v}|}{\varpi}}\right)\mp\left(\sqrt{\frac{2|C_{,u}|}{\varpi}}\right)_{,t}. \quad (86)$$

We now reduce the last term on the right-hand side of equation (86) with the aid of equations (62) and (65). Considering first the expression for $C_{,u}$ given in equation (62), we have

$$\sqrt{\frac{2|C_{,u}|}{\varpi}} = \frac{1}{4\pi\varpi^{\frac{3}{2}}}\left\{K+\frac{1}{4\varpi k}[(t-\varpi)^2-1]\,K_{,k}\right\}. \quad (87)$$

Differentiating this equation with respect to t and remembering that (cf. equation (19))

$$k_{,t} = t/(4\varpi k), \quad (88)$$

we obtain

$$\left(\sqrt{\frac{2|C_{,u}|}{\varpi}}\right)_{,t} = \frac{1}{4\pi\varpi^{\frac{3}{2}}}\left\{\frac{3t-2\varpi}{4\varpi}\frac{K_{,k}}{k}+\frac{t[(t-\varpi)^2-1]}{16\varpi^2 k}\left(\frac{K_{,k}}{k}\right)_{,k}\right\}, \quad (89)$$

where it may be noted that $K_{,k}$ is given by equation (25) and that,

$$\left(\frac{K_{,k}}{k}\right)_{,k} = \frac{1}{k^3 k'^4}[2(2k^2-1)\,E+(2-5k^2+3k^4)\,K]. \quad (90)$$

Similarly, when $C_{,u}$ is given by equation (65),

$$\left(\sqrt{\frac{2|C_{,u}|}{\varpi}}\right)_{,t} = \frac{1}{2\pi\varpi^{\frac{3}{2}}}\left\{\frac{3t-2\varpi}{\varpi}\frac{F_{,\eta}}{\sinh\eta} + \frac{t[(t-\varpi)^2-1]}{\varpi^2\sinh\eta}\left(\frac{F_{,\eta}}{\sinh\eta}\right)_{,\eta}\right\}, \tag{91}$$

where we have made use of the relation (cf. equation (21))

$$\eta_{,t} = t/(\varpi\sinh\eta). \tag{92}$$

By virtue of the differential equation (24) satisfied by $F(\eta)$,

$$\left(\frac{F_{,\eta}}{\sinh\eta}\right)_{,\eta} = -\frac{1}{\sinh\eta}(2F_{,\eta}\coth\eta + \tfrac{1}{4}F). \tag{93}$$

Inserting this expression in equation (91) and substituting for $F_{,\eta}$ from equation (28), we obtain

$$\left(\sqrt{\frac{2|C_{,u}|}{\varpi}}\right)_{,t} = \frac{k^{\frac{3}{2}}}{2\pi k'^2\varpi^{\frac{3}{2}}}\left\{-\frac{3t-2\varpi}{\varpi}(K+2kK_{,k}) + \frac{t[(t-\varpi)^2-1]\,k}{\varpi^2k'^2}[4(K+2kK_{,k})\coth\eta - K]\right\}, \tag{94}$$

where it may be noted that

$$\coth\eta = \frac{(t^2-\varpi^2-1)}{\varpi k'^2}k. \tag{95}$$

Finally, by combining equations (86), (89), and (94) we obtain

$$-e^{2\nu}\Psi_0 = \frac{3}{\varpi}|C_{,u}| \pm \varpi\left(\frac{2|C_{,u}|}{\varpi}\right)^{\frac{3}{2}} \pm \frac{1}{4\varpi}\left(\sqrt{\frac{2|C_{,u}|}{\varpi}} + \sqrt{\frac{2|C_{,v}|}{\varpi}}\right)$$
$$+\left[\begin{array}{ll} \dfrac{-1}{4\pi\varpi^{\frac{3}{2}}}\left\{\dfrac{3t-2\varpi}{4\varpi}\dfrac{K_{,k}}{k} + \dfrac{t[(t-\varpi)^2-1]}{16\varpi^2k}\left(\dfrac{K_{,k}}{k}\right)_{,k}\right\} & (|t-1| < \varpi < t+1); \\ \dfrac{+k^{\frac{3}{2}}}{2\pi k'^2\varpi^{\frac{3}{2}}}\left\{-\dfrac{3t-2\varpi}{\varpi}(K+2kK_{,k}) + \dfrac{t[(t-\varpi)^2-1]\,k}{\varpi^2k'^2}[4(K+2kK_{,k})\coth\eta - K]\right\} & (0 \leqslant \varpi < t-1, t > 1). \end{array}\right. \tag{96}$$

Similarly, the Weyl scalar Ψ_4 is given by

$$-e^{2\nu}\Psi_4 = \frac{3}{\varpi}|C_{,v}| \pm \varpi\left(\frac{2|C_{,v}|}{\varpi}\right)^{\frac{3}{2}} \pm \frac{1}{4\varpi}\left(\sqrt{\frac{2|C_{,u}|}{\varpi}} + \sqrt{\frac{2|C_{,v}|}{\varpi}}\right)$$
$$+\left[\begin{array}{ll} \dfrac{+1}{4\pi\varpi^{\frac{3}{2}}}\left\{\dfrac{3t+2\varpi}{4\varpi}\dfrac{K_{,k}}{k} + \dfrac{t[(t+\varpi)^2-1]}{16\varpi^2k}\left(\dfrac{K_{,k}}{k}\right)_{,k}\right\} & (|t-1| < \varpi < t+1); \\ \dfrac{-k^{\frac{3}{2}}}{2\pi k'^2\varpi^{\frac{3}{2}}}\left\{-\dfrac{3t+2\varpi}{\varpi}(K+2kK_{,k})\cdot + \dfrac{t[(t+\varpi)^2-1]\,k}{\varpi^2k'^2}[4(K+2kK_{,k})\coth\eta - K]\right\} & (0 \leqslant \varpi < t-1, t > 1). \end{array}\right. \tag{97}$$

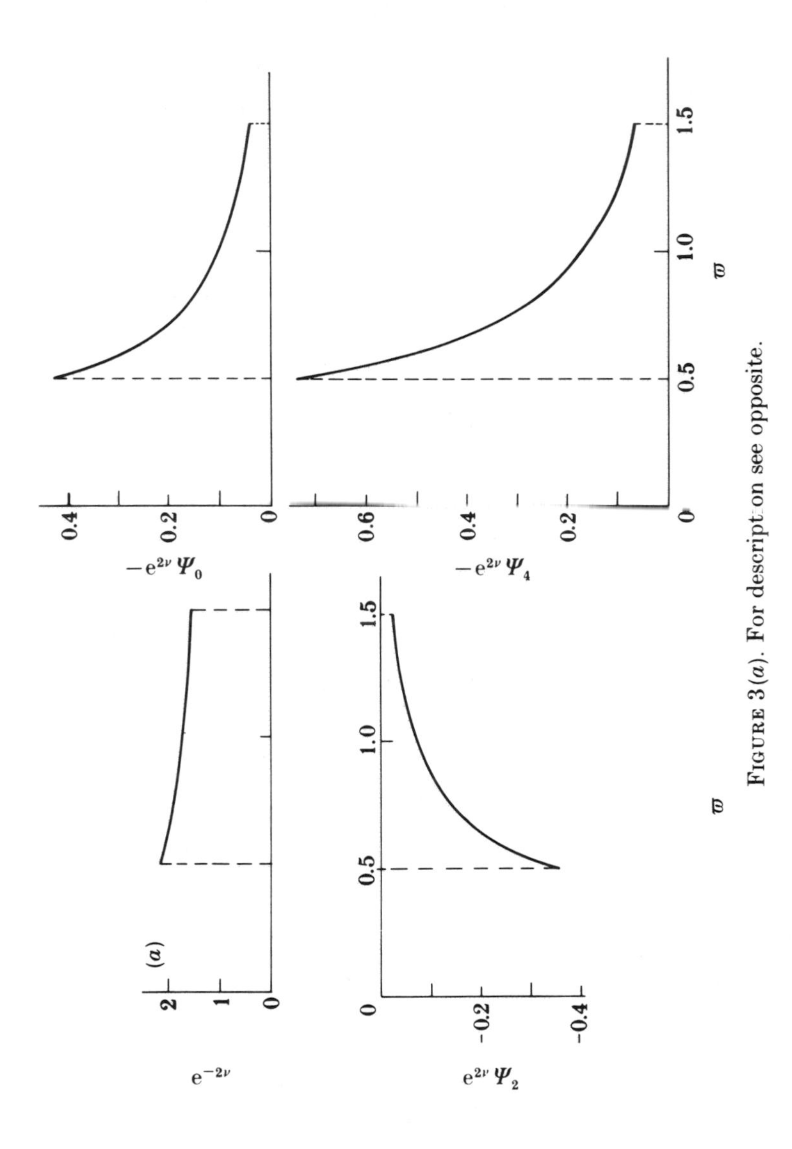

FIGURE 3(a). For description see opposite.

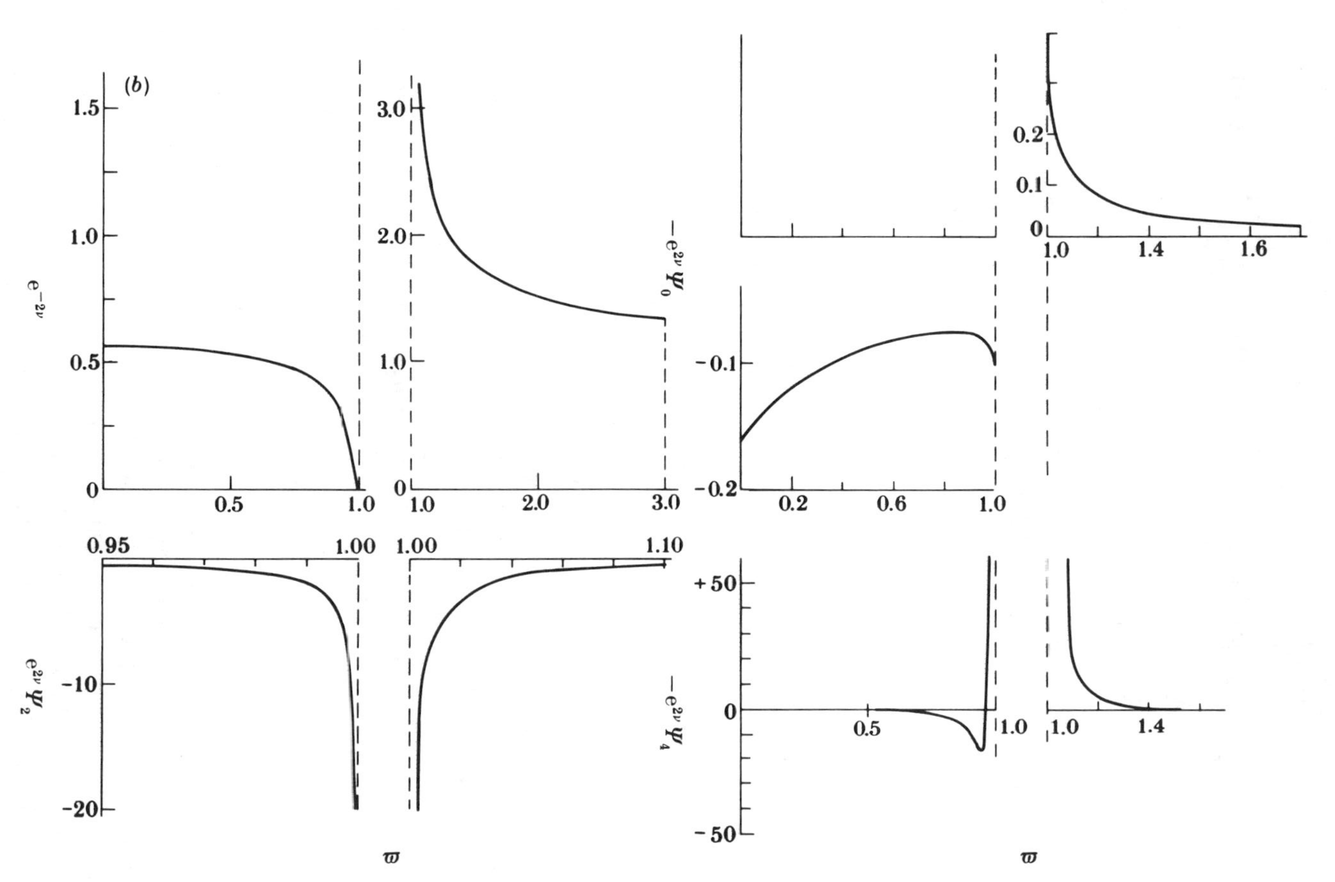

FIGURE 3. (a) The metric functions $e^{-2\nu}$ and the Weyl scalars $e^{2\nu}\Psi_2$, $-e^{2\nu}\Psi_0$ and $-e^{2\nu}\Psi_4$, at time $t = 0.5$ before reflection at the axis takes place. (b) The metric functions $e^{-2\nu}$ and the Weyl scalars $e^{2\nu}\Psi_2$, $-e^{+2\nu}\Psi_0$ and $-e^{+2\nu}\Psi_4$, at $t = 2.0$ after the instant of reflection.

Making use of equations (67), (79), and the further result,

$$\frac{k^{\frac{3}{2}}}{2\pi k'^2\varpi^{\frac{3}{2}}}\left\{-\frac{3t-2\varpi}{\varpi}(K+2kK_{,k}) + \frac{t[(t-\varpi)^2-1]\,k}{\varpi^2 k'^2}[4(K+2kK_{,k})\coth\eta - K]\right\} \sim -\frac{2t^2+1}{2(t^2-1)^{\frac{5}{2}}} \quad (\varpi\to 0), \tag{98}$$

we deduce from equation (96) that

$$e^{2\nu}\Psi_0 \sim \frac{3}{8(t^2-1)^3}[(2t^2+1)\sqrt{(t^2-1)}-t^2] \quad (t>1;\varpi\to 0). \tag{99}$$

Similarly, from equation (97), we deduce that

$$e^{2\nu}\Psi_4 \sim \frac{3}{8(t^2-1)^3}[(2t^2+1)\sqrt{(t^2-1)}-t^2] \quad (t>1;\varpi\to 0). \tag{100}$$

A comparison of equations (80), (99), and (100) now shows that

$$\Psi_0 = \Psi_4 = -3\Psi_2 \quad (\varpi = 0; t>1). \tag{101}$$

By a known theorem (cf. Chandrasekhar & Xanthopoulos 1986, §6(a)) it now follows that *the metric is of type* D *on the axis after the reflection of the dispersing impulsive waves.*

Figure 3 illustrates the behaviours of the Weyl scalars as the impulsive wave progressively disperses.

8. Concluding remarks

The cylindrical impulsive waves, whose dispersion we have analysed in this paper, differ from planar impulsive waves, that have been considered hitherto, in two important respects. *First*, it is the time derivative of the Weyl scalars themselves, that exhibit the δ-function singularity at the initial time, $t = 0$. In this respect, the comparison of the present case with the dispersion of acoustic waves along a struck string is the apposite one. And *second*, the continuation of the solution, after reflection at the centre and across the singularities that develop when the reflected wave collides with the incoming waves, is effected *analytically* and naturally without any special algorism as in the planar case. This latter feature originates in the linearity of the underlying basic equation (2), which enables us to express the discontinuous integral (13), representing the solution, explicitly as in equations (14a)–(14c). A nonlinearity of the equation at this level (as equation (77) of Paper I) would, presumably, not have made this possible. (It should be noted that at the instant of collision, $t = 1$, both C and ν tend to $-\infty$; and accordingly the Weyl scalar Ψ_2 also diverges.)

It is clear that the methods of this paper can be readily generalized to the case when the initial state consists of an arbitrary superposition of impulsive waves, of the type considered, distributed along the radial ϖ-direction: the solution for $\ln\Psi$ will then consist of a simple superposition of the solutions of the form (14) belonging to different ϖ_1s and different amplitudes. The nonlinearity of the problem will manifest itself in the solution of equations (3) and (4) for the C-energy:

the right-hand sides of these equations will include terms arising from the interference of the superposed waves (cf. Paper I, Appendix B, equation (B 10) which includes such terms); but no essential difficulty will be encountered in their solution.

While the solution for cylindrical waves, derived for the diagonal metric (1), is so easily accomplished, it is not to be expected that the solution for the general non-diagonal metric considered in Paper I (equation (7)) can be accomplished equally simply: a superposition of the monochromatic waves considered in Paper I, is not, for example, allowed. There may be other avenues of approach (not necessarily numerical!).

The research reported in this paper has, in part, been supported by grants from the National Science Foundation under grant PHY-84-16691 with the University of Chicago. We are also grateful for a grant from the Division of the Physical Sciences of the University of Chicago which enabled V.F. to visit the University of Chicago for two periods of one month, which made the present collaboration possible.

References

Chandrasekhar, S. 1986 *Proc. R. Soc. Lond.* A **408**, 209–232. (Paper I.)
Chandrasekhar, S. & Xanthopoulos, B. C. 1986 *Proc. R. Soc. Lond.* A **408**, 175–208.
Hobson, E. W. 1931 *The theory of spherical and ellipsoidal harmonics.* Cambridge University Press.
Rindler, W. 1977 *Essential relativity.* New York: Springer Verlag.
Thorne, K. 1965 *Phys. Rev.* B **138**, 251.
Watson, G. N. 1944 *A treatise on the theory of Bessel functions.* Cambridge University Press.

PART FOUR

The Two-Center Problem

The two-centre problem in general relativity: the scattering of radiation by two extreme Reissner–Nordström black-holes

By Subrahmanyan Chandrasekhar, F.R.S.
The University of Chicago, Chicago, Illinois 60637, U.S.A.

(*Received* 18 *July* 1988)

A special case of the Majumdar–Papapetrou solution of the Einstein–Maxwell equations is the static placement of two extreme Reissner–Nordström black-holes at a finite distance apart. The axi-symmetric perturbations of this special solution are analysed in prolate spheroidal coordinates. In this paper, the analysis is restricted to axial perturbations. The problem is reduced to a pair of coupled partial differential equations. The equations allow the definition of an *energy-flux* whose divergence vanishes. The asymptotic behaviours of the solutions at infinity and at the two singularities (where the horizons of the two black-holes are located) are determined. It is shown that in the vicinity of the singularities, the gravitational and the electromagnetic radiation prevails as coupled *photon-graviton* waves expressed in terms of a common amplitude and phase, for each specified angular momentum. The scattering of radiation by the two black-holes is a *four-channel* problem (in the terminology of nuclear physics). The four channels correspond to the incidence of the two sorts of radiation (gravitational and electromagnetic) from infinity and the incidence of the coupled photon-graviton waves on the horizons of the two black-holes. The flux integral, evaluated at infinity and at the two horizons, enables the formulation of the scattering problem in terms of a *scattering matrix*; and provides four identities among the scattering amplitudes as expressions of the conservation of energy in the scattering process.

In an Appendix, the equations of geodesic motion in the space-time of the two black-holes are derived; and three families of null geodesics in the meridian plane are displayed.

1. Introduction

'The most famous of the known soluble problems' in analytical dynamics, 'other than of central motion, is the *problem of two centres of gravitation* i.e., the problem of determining the motion of a free particle in a plane, attracted by two fixed Newtonian centres of force' (Whittaker 1937, p. 97). The integrability of the relevant equations of motion, in the plane containing the two centres, was discovered by Euler in 1760. The solution for the general three-dimensional motion was given by Jacobi in his 1842 lectures on dynamics ('*Vorlesungen uber Dynamik*', pp. 189–198, edited by A. Clebsch 1866) by showing the separability of the Hamilton–Jacobi equation in prolate spheroidal coordinates. The underlying reason for the integrability of this problem is that *any* ellipse, with the two centres as foci, is an exact trajectory of a particle in the field of force of either of them in the absence of the other and is, therefore, by Bonnet's theorem (cf. Whittaker 1937, §51, p. 94) an exact trajectory in the field of force of both of them.

In view of this important place that the problem of the two centres of gravitation occupies in Newtonian theory, its consideration in the frame-work of the general theory of relativity would appear worthwhile. But first, we must resolve a conceptual difficulty in considering this problem in a physically consistent framework in general relativity (and in the Newtonian theory as well): the consideration of two fixed centres of attraction, without allowing for their mutual attraction, requires (to quote E. A. Milne in a different context) the 'services of an archangel.' However, in this instance the services of an archangel can be dispensed with by simply attributing to each of the mass centres electric charges of the same sign such that the Coulomb repulsion between them exactly balances the Newtonian attraction. More generally, in the Newtonian theory we can envisage a static arrangement of any number of mass points, $M_1, M_2, \ldots, M_N$ (say), at arbitrary locations, with charges $Q_1, Q_2, \ldots, Q_N$ of the same sign such that $M_i\sqrt{G} = Q_i$ $(i = 1, \ldots, N)$ (where G denotes the constant of gravitation) and the gravitational attraction between any pair of them is balanced by their Coulomb repulsion. It is a remarkable fact that this same static arrangement is allowed in general relativity consistently with the Einstein–Maxwell equations. This is the solution of Majumdar (1947) and Papapetrou (1947). The solution was first correctly interpreted by Hartle & Hawking (1972) as representing an assemblage of extreme Reissner–Nordström black-holes. (For a derivation of the Majumdar–Papapetrou solution and the Hartle–Hawking interpretation see *The mathematical theory of black holes* (Chandrasekhar 1983, §113, pp. 588–96; this book will be referred to hereafter as *M.T.*).)

In this paper, we shall be concerned with the special case of the Majumdar–Papapetrou solution, when only two extreme Reissner–Nordström black-holes are present, with the principal object of ascertaining how two Reissner–Nordström black-holes, in proximity, will scatter and absorb incident gravitational and electromagnetic radiation differently from a single isolated Reissner–Nordström black-hole as described in *M.T.*, Chapter 5, §47.

2. The space-time of two extreme Reissner–Nordström black-holes in prolate spheroidal coordinates

The metric of the general Majumdar–Papapetrou solution of the Einstein–Maxwell equations is given by (cf. *M.T.* p. 592, equation (220))

$$\mathrm{d}s^2 = \frac{(\mathrm{d}t)^2}{U^2} - U^2[(\mathrm{d}x^1)^2 + (\mathrm{d}x^2)^2 + (\mathrm{d}x^3)^2], \tag{1}$$

where U is any solution of the three-dimensional Laplace's equation,

$$\nabla^2 U = \sum_{\alpha=1}^{3} U_{,\alpha,\alpha} = 0, \tag{2}$$

where $\alpha = 1, 2$, and 3 refers to the coordinates x^1, x^2, and x^3. The associated electrostatic potential, A, is given by

$$A = U^{-1}. \tag{3}$$

The corresponding components, $F_{0\alpha}$, of the Maxwell tensor are given by

$$F_{0\alpha} = (e^{-\psi})_{,\alpha} = A_{,\alpha}, \tag{4}$$

where
$$U = e^{\psi} \quad \text{and} \quad \sum_{\alpha=1}^{3} (e^{\psi})_{,\alpha,\alpha} = 0. \tag{5}$$

The solution for U that represents an assemblage of black-holes of masses $M_1, \ldots, M_N$ located at $\boldsymbol{r}_i$ $(i = 1, \ldots, N)$ is given by

$$U = 1 + \sum_{i=1}^{N} \frac{M_i}{|\boldsymbol{r} - \boldsymbol{r}_i|}, \tag{6}$$

where the addition of unity on the right-hand side is to ensure the asymptotic flatness of the space-time. From the form of the metric (1), when the different points $\boldsymbol{r}_i$ are widely separated, we may identify the constants M_i with the inertial masses enclosed by spheres of large enough radii surrounding the different points. Also, the charge Q_i of the mass point M_i, determined by evaluating the divergence of $F^{\alpha 0}\sqrt{(-g)}$ over a spherical volume surrounding the point $\boldsymbol{r}_i$ is found to be equal to M_i.

The metric (1), with U given by equation (6), is everywhere regular except at the points $\boldsymbol{r}_i$. The singularities at these points are only coordinate singularities. For, by evaluating the proper surface-area of a small sphere surrounding $\boldsymbol{r}_i$, we find that it does not tend to zero as the radius of the sphere (in the chart in which the metric (1) is written) tends to zero but to the finite limit $4\pi M_i^2$ that is the area of the event horizon of an extreme Reissner–Nordström black-hole of mass M_i and charge $Q_i = M_i$.

In this paper, we shall be concerned with the space-time of two extreme Reissner–Nordström black-holes located at $z = \pm 1$ (in a suitably chosen system of units) on the z-axis. Then

$$U = 1 + \frac{M_1}{\sqrt{[x^2 + y^2 + (z-1)^2]}} + \frac{M_2}{\sqrt{[x^2 + y^2 + (z+1)^2]}}, \tag{7}$$

where we have written (x, y, z) in place of (x^1, x^2, x^3).

Following Jacobi in his treatment of the two-centre problem in the Newtonian theory, we shall find it convenient to rewrite the metric in prolate spheroidal coordinates, (ψ, θ, φ) related to (x, y, z) by

$$x = \sinh\psi \sin\theta \cos\varphi, \quad y = \sinh\psi \sin\theta \sin\varphi, \quad z = \cosh\psi \cos\theta. \tag{8}$$

Then U takes the simple form

$$U = 1 + \frac{M_1}{\cosh\psi - \cos\theta} + \frac{M_2}{\cosh\psi + \cos\theta} = e^{-\nu}, \tag{9}$$

and the metric is given by

$$ds^2 = e^{2\nu}(dt)^2 - e^{-2\nu}\{(\sinh^2\psi + \sin^2\theta)[(d\psi)^2 + (d\theta)^2] + \sinh^2\psi \sin^2\theta\,(d\varphi)^2\}. \tag{10}$$

Letting $$\eta = \cosh\psi \quad \text{and} \quad \mu = \cos\theta, \tag{11}$$

we shall rewrite the metric (10) in the standard form,

$$ds^2 = e^{2\nu}(dt)^2 - e^{2\psi}(dx^1)^2 - e^{2\mu_2}(dx^2)^2 - e^{2\mu_3}(dx^3)^2, \tag{12}$$

appropriate for static axisymmetric space-times, where, in the present context

$$x^1 = \varphi, \quad x^2 = \eta, \quad \text{and} \quad x^3 = \mu. \tag{13}$$

(The use of ψ with different meanings in equations (5), (8), and (10) is unfortunate but only transitory: the meanings attached to the various symbols in equations (12) and (13) will be consistently adhered to in the sequel.)

The metric coefficients e^ν etc., are now given by

$$e^{\psi} = e^{-\nu}\sqrt{[(\eta^2-1)(1-\mu^2)]}; \quad e^{\beta} = e^{\psi+\nu} = \sqrt{[(\eta^2-1)(1-\mu^2)]}$$

$$e^{\mu_2} = e^{-\nu}\sqrt{\frac{\eta^2-\mu^2}{\eta^2-1}}, \quad e^{\mu_3} = e^{-\nu}\sqrt{\frac{\eta^2-\mu^2}{1-\mu^2}}, \tag{14}$$

where $$e^{-\nu} = 1 + \frac{M_1}{\eta-\mu} + \frac{M_2}{\eta+\mu} = U. \tag{15}$$

The only non-vanishing components of the Maxwell-tensor are F_{02} and F_{03} given by

$$F_{02} = e^{-\mu_2}\nu_{,\eta} \quad \text{and} \quad F_{03} = e^{-\mu_3}\nu_{,\mu}. \tag{16}$$

It may be noted here for future reference, that among the metric coefficients listed in (14) we have the relations:

$$e^{\mu_2-\mu_3} = \sqrt{\frac{1-\mu^2}{\eta^2-1}}, \quad e^{\psi+\nu+\mu_2-\mu_3} = 1-\mu^2 \quad \text{and} \quad e^{\psi+\nu+\mu_3-\mu_2} = \eta^2-1; \tag{17}$$

and, further, that $e^{-\nu}$ satisfies the equation,

$$[(\eta^2-1)(e^{-\nu})_{,\eta}]_{,\eta} + [(1-\mu^2)(e^{-\nu})_{,\mu}]_{,\mu} = 0, \tag{18}$$

or, alternatively,

$$[(\eta^2-1)\nu_{,\eta}]_{,\eta} + [(1-\mu^2)\nu_{,\mu}]_{,\mu} = (\eta^2-1)\nu_{,\eta}\nu_{,\eta} + (1-\mu^2)\nu_{,\mu}\nu_{,\mu}. \tag{19}$$

3. The perturbation problem and the equations governing axial perturbations

In considering the perturbation problem of the two extreme Reissner–Nordström black-holes, we shall restrict ourselves to time-dependent axisymmetric modes. Then, as explained in detail in *M.T.* (Chapter 2, §24 and Chapter 5, §42), it will suffice to consider the solution (12), with the metric coefficients specified in equations (14) and (15), as a special time-independent solution of the field-equations appropriate to the *non-stationary* axisymmetric metric,

$$ds^2 = e^{2\nu}(dt)^2 - e^{2\psi}(d\varphi - \omega\, dt - q_2\, dx^2 - q_3\, dx^3)^2 - e^{2\mu_2}(dx^2)^2 - e^{2\mu_3}(dx^3)^2, \tag{20}$$

where ν, ψ, μ_2, μ_3, ω, q_2, and q_3 are functions of $t(=x^0)$ besides x^2 $(=\eta)$ and x^3 $(=\mu)$. In other words, to obtain the equations governing the axisymmetric perturbations, we linearize the field equations for the metric (20) about the static metric (12). The resulting equations fall into two non-combining groups: the *axial* and the *polar*. The axial perturbations are characterized by non-vanishing ω, q_2, and q_3 while ν, ψ, μ_2, and μ_3 are unchanged and retain their unperturbed time-independent values. The polar perturbations, on the other hand, are characterized by first-order time-dependent changes in ν, ψ, μ_2, and μ_3 while ω, q_2, and q_3 vanish. In this paper, we shall consider only the axial perturbations; the polar perturbations will be considered in a subsequent paper (Chandrasekhar & Ferrari, in preparation).

We shall now write down the equations governing the axial perturbations. Considering first Maxwell's equations, we obtain the equations relevant to axial perturbations (as in the treatment of the isolated Reissner–Nordström black-hole in *M.T.*, §42, which we closely follow) by linearizing equations (117) on p. 226 of *M.T.* Remembering that in the present context, the only non-vanishing components of the Maxwell tensor are F_{02} and F_{03} (given in equations (16)), we obtain the equations:

$$(e^{\psi+\nu}F_{01})_{,2} = -e^{\psi+\mu_2}F_{12,0}, \tag{21}$$

$$(e^{\psi+\nu}F_{01})_{,3} = -e^{\psi+\mu_3}F_{13,0}, \tag{22}$$

$$e^{\mu_2+\mu_3}F_{01,0}+(e^{\nu+\mu_3}F_{12})_{,2}+(e^{\nu+\mu_2}F_{13})_{,3} = e^{\psi+\mu_3}F_{02}\,Q_{02}+e^{\psi+\mu_2}F_{03}\,Q_{03}, \tag{23}$$

where
$$Q_{A0} = q_{A,0}-\omega_{,A} = -Q_{0A} \quad (A = 2,3). \tag{24}$$

Taking the time-derivative of equation (23), we obtain:

$$e^{\mu_2+\mu_3}F_{01,0,0}+(e^{\nu+\mu_3}F_{12,0})_{,\eta}+(e^{\nu+\mu_2}F_{13,0})_{,\mu} = e^{\psi+\mu_3-\mu_2}\nu_{,\eta}\,Q_{02,0}+e^{\psi+\mu_2-\mu_3}\nu_{,\mu}\,Q_{03,0}, \tag{25}$$

where we have substituted for F_{02} and F_{03} from equations (16). Making use of equations (21) and (22) and simplifying with the aid of the relations (17), we rewrite the left-hand side of equation (25), successively, in the manner:

$$\begin{aligned} & e^{\mu_2+\mu_3}F_{01,0,0}-[e^{\nu+\mu_3-\psi-\mu_2}(e^{\psi+\nu}F_{01})_{,\eta}]_{,\eta} \\ & \qquad -[e^{\nu+\mu_2-\psi-\mu_3}(e^{\psi+\nu}F_{01})_{,\mu}]_{,\mu} \\ & = e^{\mu_2+\mu_3}F_{01,0,0}-\frac{1}{1-\mu^2}[e^{2\nu}(e^{\psi+\nu}F_{01})_{,\eta}]_{,\eta} \\ & \qquad -\frac{1}{\eta^2-1}[e^{2\nu}(e^{\psi+\nu}F_{01})_{,\mu}]_{,\mu}. \end{aligned} \tag{26}$$

Similarly, the right-hand side of equation (25) can be reduced to the form,

$$\begin{aligned} & e^{\psi+\nu+\mu_3-\mu_2}\,e^{-\nu}\nu_{,\eta}\,Q_{02,0}+e^{\psi+\nu+\mu_2-\mu_3}\,e^{-\nu}\nu_{,\mu}\,Q_{03,0} \\ & \qquad = (\eta^2-1)\,(e^{-\nu})_{,\eta}\,Q_{20,0}+(1-\mu^2)\,(e^{-\nu})_{,\mu}\,Q_{30,0}. \end{aligned} \tag{27}$$

We thus obtain the equation,

$$\begin{aligned} & e^{\mu_2+\mu_3}F_{01,0,0}-\frac{1}{1-\mu^2}[e^{2\nu}(e^{\psi+\nu}F_{01})_{,\eta}]_{,\eta}-\frac{1}{\eta^2-1}[e^{2\nu}(e^{\psi+\nu}F_{01})_{,\mu}]_{,\mu} \\ & \qquad = (\eta^2-1)\,(e^{-\nu})_{,\eta}\,Q_{20,0}+(1-\mu^2)\,(e^{-\nu})_{,\mu}\,Q_{30,0}. \end{aligned} \tag{28}$$

We now turn to the linearized version of the Einstein field-equations. When the source is derived from a prevailing electromagnetic field, Einstein's equations require

$$\delta R_{ab} = -2(\delta F_a^m F_{bm} + F_a^m \delta F_{bm} - \tfrac{1}{2}\eta_{ab} F^{mn}\delta F_{mn}), \tag{29}$$

where, for the case on hand,

$$F^{mn}\delta F_{mn} = -2(F_{02}\,\delta F_{02} + F_{03}\,\delta F_{03}). \tag{30}$$

From equations (29) and (30) we find

$$\delta R_{12} = -2\,e^{-\mu_2}\nu_{,\eta} F_{01} \quad \text{and} \quad \delta R_{13} = -2\,e^{-\mu_3}\nu_{,\mu} F_{01}. \tag{31}$$

On the other hand, by *M.T.*, equation (4e) on p. 141,

$$e^{-2\psi-\nu-\mu_3}[(e^{3\psi+\nu-\mu_2-\mu_3}Q_{23})_{,3} + (e^{3\psi-\nu+\mu_3-\mu_2}Q_{02})_{,0}] = +2\delta R_{12} \tag{32}$$

and

$$e^{-2\psi-\nu-\mu_2}[(e^{3\psi+\nu-\mu_2-\mu_3}Q_{23})_{,2} - (e^{3\psi-\nu+\mu_2-\mu_3}Q_{03})_{,0}] = -2\delta R_{13}, \tag{33}$$

where

$$Q_{23} = q_{2,3} - q_{3,2} = -Q_{32}. \tag{34}$$

By combining equations (31)–(33), we now obtain,

$$\begin{aligned}(e^{3\psi+\nu-\mu_2-\mu_3}Q_{23})_{,\mu} + e^{3\psi-\nu+\mu_3-\mu_2}Q_{02,0} &= -4\,e^{2\psi+\nu+\mu_3-\mu_2}\nu_{,\eta} F_{01}\\ &= +4(\eta^2-1)\,(e^{-\nu})_{,\eta}\,e^{\psi+\nu}F_{01}\end{aligned} \tag{35}$$

and

$$\begin{aligned}(e^{3\psi+\nu-\mu_2-\mu_3}Q_{23})_{,\eta} - e^{3\psi-\nu+\mu_2-\mu_3}Q_{03,0} &= +4\,e^{2\psi+\nu+\mu_2-\mu_3}\nu_{,\mu} F_{01}\\ &= -4(1-\mu^2)\,(e^{-\nu})_{,\mu}\,e^{\psi+\nu}F_{01}.\end{aligned} \tag{36}$$

By virtue of the definitions (24) and (34),

$$Q_{20,\mu} - Q_{30,\eta} = (q_{2,3} - q_{3,2})_{,0} = Q_{23,0}. \tag{37}$$

By making use of this identity, we can eliminate $Q_{02,0}$ and $Q_{03,0}$ from equation (35) and obtain:

$$\begin{aligned}Q_{23,0,0} = &\{e^{-3\psi+\nu-\mu_3+\mu_2}[(e^{3\psi+\nu-\mu_2-\mu_3}Q_{23})_{,\mu} - 4(\eta^2-1)\,(e^{-\nu})_{,\eta}\,e^{\psi+\nu}F_{01}]\}_{,\mu}\\ &+\{e^{-3\psi+\nu-\mu_2+\mu_3}[(e^{3\psi+\nu-\mu_2-\mu_3}Q_{23})_{,\eta} + 4(1-\mu^2)\,(e^{-\nu})_{,\mu}\,e^{\psi+\nu}F_{01}]\}_{,\eta}.\end{aligned} \tag{38}$$

Letting

$$X = e^{3\psi+\nu-\mu_2-\mu_3}Q_{23} = \frac{(\eta^2-1)^2\,(1-\mu^2)^2}{\eta^2-\mu^2}Q_{23} \tag{39}$$

and

$$Y = e^{\psi+\nu}F_{01} = (\eta^2-1)^{\frac{1}{2}}\,(1-\mu^2)^{\frac{1}{2}}F_{01}, \tag{40}$$

we can reduce equations (28) and (38) to the forms:

$$\begin{aligned}e^{-\psi-\nu+\mu_2+\mu_3}Y_{,0,0} = &\frac{1}{1-\mu^2}(e^{2\nu}Y_{,\eta})_{,\eta} + \frac{1}{\eta^2-1}(e^{2\nu}Y_{,\mu})_{,\mu}\\ &+(\eta^2-1)\,(e^{-\nu})_{,\eta}\,e^{-3\psi+\nu-\mu_3+\mu_2}[X_{,\mu} - 4(\eta^2-1)\,(e^{-\nu})_{,\eta}\,Y]\\ &-(1-\mu^2)\,(e^{-\nu})_{,\mu}\,e^{-3\psi+\nu-\mu_2+\mu_3}[X_{,\eta} + 4(1-\mu^2)\,(e^{-\nu})_{,\mu}\,Y].\end{aligned} \tag{41}$$

and

$$\begin{aligned}e^{-3\psi-\nu+\mu_2+\mu_3}X_{,0,0} = &\{e^{-3\psi+\nu-\mu_3+\mu_2}[X_{,\mu} - 4(\eta^2-1)\,(e^{-\nu})_{,\eta}\,Y]\}_{,\mu}\\ &+\{e^{-3\psi+\nu-\mu_2+\mu_3}[X_{,\eta} + 4(1-\mu^2)\,(e^{-\nu})_{,\mu}\,Y]\}_{,\eta}.\end{aligned} \tag{42}$$

Assuming that the perturbations have the time-dependence

$$e^{i\sigma t}, \tag{43}$$

where σ is a real positive constant, we find after some further reductions in which we make use of equations (14) and (17), that equations (41) and (42) can be brought to the forms:

$$(\eta^2-1)(U^2Y_{,\eta,\eta}-2UU_{,\eta}Y_{,\eta}-4U^2_{,\eta}Y)+U_{,\eta}X_{,\mu}-U_{,\mu}X_{,\eta}$$
$$+(1-\mu^2)(U^2Y_{,\mu,\mu}-2UU_{,\mu}Y_{,\mu}-4U^2_{,\mu}Y)+\sigma^2(\eta^2-\mu^2)U^6Y=0 \tag{44}$$

and

$$U(\eta^2-1)X_{,\eta,\eta}-4(\eta^2-1)U_{,\eta}X_{,\eta}-2\eta UX_{,\eta}$$
$$+U(1-\mu^2)X_{,\mu,\mu}-4(1-\mu^2)U_{,\mu}X_{,\mu}+2\mu UX_{,\mu}$$
$$-4U(\eta^2-1)(1-\mu^2)\left(U_{,\eta}Y_{,\mu}-U_{,\mu}Y_{,\eta}+\frac{2\mu}{1-\mu^2}YU_{,\eta}+\frac{2\eta}{\eta^2-1}YU_{,\mu}\right)$$
$$+\sigma^2(\eta^2-\mu^2)U^5X=0 \tag{45}$$

where, it may be recalled that

$$U=1+\frac{M_1}{\eta-\mu}+\frac{M_2}{\eta+\mu}. \tag{46}$$

4. The conservation theorem

It is an important and a necessary feature of equations (44) and (45) that they allow the notion of an *energy-flux* whose divergence vanishes. The *conservation theorem* which follows is essential for the formulation, in terms of a *scattering matrix* (cf. *M.T.* §§28 and 47), the problem of scattering and absorption of incident gravitational and electromagnetic radiation by the two black holes.

Multiplying equation (44) by Y^* – the complex-conjugate of Y – and similarly, the complex-conjugate of equation (44) by Y, and subtracting one from the other, we obtain,

$$(\eta^2-1)\{U^2[Y,Y^*]_{\eta,\eta}-(U^2)_{,\eta}[Y,Y^*]_\eta\}+U_{,\eta}(X_{,\mu}Y^*-X^*_{,\mu}Y)$$
$$+(1-\mu^2)\{U^2[Y,Y^*]_{\mu,\mu}-(U^2)_{,\mu}[Y,Y^*]_\mu\}-U_{,\mu}(X_{,\eta}Y^*-X^*_{,\eta}Y)=0 \tag{47}$$

where
$$[A,B]_x=A_{,x}B-B_{,x}A \tag{48}$$

is the Wronskian of A and B with respect to x. By a similar procedure, we obtain from equation (45)

$$(\eta^2-1)\{U[X,X^*]_{\eta,\eta}-4U_{,\eta}[X,X^*]_\eta\}-2\eta U[X,X^*]_\eta$$
$$+(1-\mu^2)\{U[X,X^*]_{\mu,\mu}-4U_{,\mu}[X,X^*]_\mu\}+2\mu U[X,X^*]_\mu$$
$$-4U(\eta^2-1)(1-\mu^2)\Big\{U_{,\eta}(Y_{,\mu}X^*-Y^*_{,\mu}X)-U_{,\mu}(Y_{,\eta}X^*-Y^*_{,\eta}X)$$
$$+\frac{2\mu}{1-\mu^2}U_{,\eta}(YX^*-Y^*X)+\frac{2\eta}{\eta^2-1}U_{,\mu}(YX^*-Y^*X)\Big\}=0. \tag{49}$$

Equations (47) and (49) can be rewritten in the forms:

$$(\eta^2-1)\,U^4\left\{\frac{1}{U^2}[Y,Y^*]_\eta\right\}_{,\eta}+(1-\mu^2)\,U^4\left\{\frac{1}{U^2}[Y,Y^*]_\mu\right\}_{,\mu}$$
$$+U_{,\eta}(X_{,\mu}Y^*-X^*_{,\mu}Y)-U_{,\mu}(X_{,\eta}Y^*-X^*_{,\eta}Y)=0 \quad (50)$$

and

$$(\eta^2-1)^2U^4\left\{\frac{1}{U^4(\eta^2-1)}[X,X^*]_\eta\right\}_{,\eta}+(1-\mu^2)^2U^4\left\{\frac{1}{U^4(1-\mu^2)}[X,X^*]_\mu\right\}_{,\mu}$$
$$-4(\eta^2-1)(1-\mu^2)\Big\{U_{,\eta}(Y_{,\mu}X^*-Y^*_{,\mu}X)-U_{,\mu}(Y_{,\eta}X^*-Y^*_{,\eta}X)$$
$$+\frac{2\mu}{1-\mu^2}U_{,\eta}(YX^*-Y^*X)+\frac{2\eta}{\eta^2-1}U_{,\mu}(YX^*-Y^*X)\Big\}=0. \quad (51)$$

For combining with equation (51), it is convenient to write equation (50) times $4(\eta^2-1)(1-\mu^2)$; we then have,

$$4(\eta^2-1)^2(1-\mu^2)\,U^4\left\{\frac{1}{U^2}[Y,Y^*]_\eta\right\}_{,\eta}+4(\eta^2-1)(1-\mu^2)^2U^4\left\{\frac{1}{U^2}[Y,Y^*]_\mu\right\}_{,\mu}$$
$$+4(\eta^2-1)(1-\mu^2)\{U_{,\eta}(X_{,\mu}Y^*-X^*_{,\mu}Y)-U_{,\mu}(X_{,\eta}Y^*-X^*_{,\eta}Y)\}=0. \quad (52)$$

The terms in the second and third lines of equation (51) combined with the terms in the second line of equation (52) yield

$$4(\eta^2-1)^2(1-\mu^2)^2\left\{U_{,\eta}\left[\frac{XY^*-X^*Y}{(\eta^2-1)(1-\mu^2)}\right]_{,\mu}-U_{,\mu}\left[\frac{XY^*-X^*Y}{(\eta^2-1)(1-\mu^2)}\right]_{,\eta}\right\}. \quad (53)$$

Adding the remaining terms i.e., those in the first lines of equations (51) and (52), to (53) and dividing the result by $U^4(\eta^2-1)^2(1-\mu^2)^2$, we obtain

$$\frac{1}{(1-\mu^2)^2}\left\{\frac{[X,X^*]_\eta}{U^4(\eta^2-1)}\right\}_{,\eta}+\frac{1}{(\eta^2-1)^2}\left\{\frac{[X,X^*]_\mu}{U^4(1-\mu^2)}\right\}_{,\mu}$$
$$+\frac{4}{1-\mu^2}\left\{\frac{[Y,Y^*]_\eta}{U^2}\right\}_{,\eta}+\frac{4}{\eta^2-1}\left\{\frac{[Y,Y^*]_\mu}{U^2}\right\}_{,\mu}$$
$$+\frac{4U_{,\eta}}{U^4}\left[\frac{XY^*-X^*Y}{(\eta^2-1)(1-\mu^2)}\right]_{,\mu}-\frac{4U_{,\mu}}{U^4}\left[\frac{XY^*-X^*Y}{(\eta^2-1)(1-\mu^2)}\right]_{,\eta}=0. \quad (54)$$

By expressing the terms in the last line of equation (54) in the alternative manner,

$$4\left[U_{,\eta}\frac{XY^*-X^*Y}{U^4(\eta^2-1)(1-\mu^2)}\right]_{,\mu}-4\left[U_{,\mu}\frac{XY^*-X^*Y}{U^4(\eta^2-1)(1-\mu^2)}\right]_{,\eta}, \quad (55)$$

equation (54) can be rewritten in the form,

$$\frac{\partial}{\partial\eta}\left\{\frac{[X,X^*]_\eta}{U^4(\eta^2-1)(1-\mu^2)^2}+4\frac{[Y,Y^*]_\eta}{U^2(1-\mu^2)}-4\frac{U_{,\mu}(XY^*-X^*Y)}{U^4(\eta^2-1)(1-\mu^2)}\right\}$$
$$+\frac{\partial}{\partial\mu}\left\{\frac{[X,X^*]_\mu}{U^4(\eta^2-1)^2(1-\mu^2)}+4\frac{[Y,Y^*]_\mu}{U^2(\eta^2-1)}+4\frac{U_{,\eta}(XY^*-X^*Y)}{U^4(\eta^2-1)(1-\mu^2)}\right\}=0; \quad (56)$$

or, by defining,

$$E_\eta = \frac{[X, X^*]_\eta}{U^4(\eta^2-1)(1-\mu^2)^2} + 4\frac{[Y, Y^*]_\eta}{U^2(1-\mu^2)} - 4\frac{U_{,\mu}(XY^*-X^*Y)}{U^4(\eta^2-1)(1-\mu^2)} \tag{57}$$

and

$$E_\mu = \frac{[X, X^*]_\mu}{U^4(\eta^2-1)^2(1-\mu^2)} + 4\frac{[Y, Y^*]_\mu}{U^2(\eta^2-1)} + 4\frac{U_{,\eta}(XY^*-X^*Y)}{U^4(\eta^2-1)(1-\mu^2)}, \tag{58}$$

we have

$$\frac{1}{\eta^2-\mu^2}\left(\frac{\partial E_\eta}{\partial \eta} + \frac{\partial E_\mu}{\partial \mu}\right) = 0, \tag{59}$$

where the extra factor $(\eta^2-\mu^2)^{-1}$ has been added for convenience (see equation (60) below).

Since the element of volume, in the coordinate system we are presently using, is

$$\begin{aligned} \mathrm{d}\varphi\, \mathrm{d}\eta\, \mathrm{d}\mu \sqrt{q} &= (\sinh^2\psi + \sin^2\theta)\sinh\psi \sin\theta\, \mathrm{d}\psi\, \mathrm{d}\theta\, \mathrm{d}\varphi \\ &= +(\eta^2-\mu^2)\, \mathrm{d}\eta\, \mathrm{d}\mu\, \mathrm{d}\varphi, \end{aligned} \tag{60}$$

the result of integrating equation (59) over the interior ($\mathscr{S}$) of any closed contour, C, in the (η, μ)-plane, not including the singularities at $\eta = 1$ and $\mu = \pm 1$, is, by Gauss's theorem,

$$\int_{\mathscr{S}} \left(\frac{\partial E_\eta}{\partial \eta} + \frac{\partial E_\mu}{\partial \mu}\right) \mathrm{d}\eta\, \mathrm{d}\mu = \int_{\partial\mathscr{S}=C} (E_\eta\, \mathrm{d}\mu - E_\mu\, \mathrm{d}\eta) = 0. \tag{61}$$

In the context of the scattering problem considered in §9 below, we shall apply the conservation theorem, expressed by equation (61), to the case when C consists of a circle of sufficiently large radius, η, and two small circles C_1 and C_2 surrounding the singularities at $\eta = 1$ and $\mu = \pm 1$. For this choice of C, equation (61) gives

$$\int_{\eta=\text{constant}\to\infty} E_\eta\, \mathrm{d}\mu = \int_{C_1} (E_\eta\, \mathrm{d}\mu - E_\mu\, \mathrm{d}\eta) + \int_{C_2} (E_\eta\, \mathrm{d}\mu - E_\mu\, \mathrm{d}\eta), \tag{62}$$

where all three circles are described in the positive counter-clockwise direction.

5. The asymptotic behaviour of the solutions at ∞

Considering first equation (44) governing Y, we rewrite it in the form,

$$\begin{aligned} &(\eta^2-1)\left[Y_{,\eta,\eta} - 2\frac{U_{,\eta}}{U}Y_{,\eta} - 4\left(\frac{U_{,\eta}}{U}\right)^2 Y + \sigma^2 U^4 Y\right] \\ &+ (1-\mu^2)\left[Y_{,\mu,\mu} - 2\frac{U_{,\mu}}{U}Y_{,\mu} - 4\left(\frac{U_{,\mu}}{U}\right)^2 Y + \sigma^2 U^4 Y\right] \\ &+ \frac{1}{U^2}(U_{,\eta}X_{,\mu} - U_{,\mu}X_{,\eta}) = 0. \end{aligned} \tag{63}$$

For $\eta \to \infty$, we have the behaviours,

$$U = 1 + \frac{M_1+M_2}{\eta} + \frac{M_1-M_2}{\eta^2}\mu + O(\eta^{-3}),$$

$$U^4 = 1 + 4\frac{M_1+M_2}{\eta} + 4\frac{M_1-M_2}{\eta^2}\mu + 6\frac{(M_1+M_2)^2}{\eta^2} + O(\eta^{-3}),$$

and

$$\frac{U_{,\eta}}{U} = -\frac{M_1+M_2}{\eta^2} + O(\eta^{-3}); \quad \frac{U_{,\mu}}{U} = +\frac{M_1-M_2}{\eta^2} + O(\eta^{-3}). \tag{64}$$

Inserting these behaviours in equation (63) and anticipating the result, $X = O(\eta)$ for $\eta \to \infty$ (established below), we are left with

$$(\eta^2-1)\left\{Y_{,\eta,\eta} + 2\frac{M_1+M_2}{\eta^2}Y_{,\eta} + \sigma^2\left[1 + 4\frac{M_1+M_2}{\eta} + 6\frac{(M_1+M_2)^2}{\eta^2} + 4\frac{M_1-M_2}{\eta^2}\mu\right]Y\right\}$$
$$+ (1-\mu^2)(Y_{,\mu,\mu} + \sigma^2 Y) + O(\eta^{-1}) = 0, \tag{65}$$

or, equivalently (to the same order),

$$(\eta^2-1)\left\{Y_{,\eta,\eta} + 2\frac{M_1+M_2}{\eta^2}Y_{,\eta} + \sigma^2\left[1 + 4\frac{M_1+M_2}{\eta} + 4\frac{(M_1+M_2)^2}{\eta^2}\right]Y\right\}$$
$$+ (1-\mu^2)Y_{,\mu,\mu} + \sigma^2[1-\mu^2 + 2(M_1+M_2)^2 + 4(M_1-M_2)\mu]Y + O(\eta^{-1}) = 0. \tag{66}$$

Equation (66) is separable in the variables η and μ. Thus, with the substitution

$$Y(\eta,\mu) = \mathscr{Y}(\eta)f(\mu), \tag{67}$$

where $\mathscr{Y}(\eta)$ and $f(\mu)$ are functions only of the arguments specified, we obtain the pair of equations,

$$(1-\mu^2)f_{,\mu,\mu} + \sigma^2[(1-\mu^2) + 2(M_1+M_2)^2 + 4(M_1-M_2)\mu]f = -\lambda_1 f \tag{68}$$

and

$$\mathscr{D}\mathscr{Y} = \frac{\lambda_1}{\eta^2}\mathscr{Y}, \tag{69}$$

where λ_1 is a separation constant and $\mathscr{D}$ stands for the operator,

$$\mathscr{D} = \partial_{\eta\eta} + 2\frac{M_1+M_2}{\eta^2}\partial_\eta + \sigma^2\left[1 + 4\frac{M_1+M_2}{\eta} + 4\frac{(M_1+M_2)^2}{\eta^2}\right]. \tag{70}$$

We now verify that

$$\mathscr{D}\exp\{i\sigma[\eta + 2(M_1+M_2)\lg\eta]\}$$
$$= 4i\sigma\frac{(M_1+M_2)^2}{\eta^3}\exp\{i\sigma[\eta + 2(M_1+M_2)\lg\eta]\} = O(\eta^{-3}). \tag{71}$$

Since the right-hand side of equation (71) is $O(\eta^{-3})$, in contrast to the right-hand side of equation (69) which is $O(\eta^{-1})$, we can solve equation (69) by the method of the variation of parameters (cf. Ince 1927, p. 122). Thus, we substitute

$$\mathscr{Y} = e^{i\sigma[\dots]} + V_1(\eta)e^{i\sigma[\dots]} + V_2(\eta)e^{-i\sigma[\dots]}, \tag{72}$$

and determine $V_1(\eta)$ and $V_2(\eta)$ by the equations,

$$V_1 = +\frac{\lambda_1}{2i\sigma}\int\frac{d\eta}{\eta^2[1+2(M_1+M_2)/\eta]} \sim \frac{i\lambda_1}{2\sigma\eta}+O(\eta^{-2}),$$

$$V_2 = -\frac{\lambda_1}{2i\sigma}\int\frac{e^{2i\sigma[\dots]}}{\eta^2[1+2(M_1+M_2)/\eta]}d\eta \sim \frac{\lambda_1}{4\sigma\eta^2}+O(\eta^{-3}). \tag{73}$$

Therefore to $O(1)$ the solution for $\mathscr{Y}$ is given, *independently* of λ_1, by

$$\mathscr{Y} = \exp\{\pm i\sigma[\eta+2(M_1+M_2)\lg\eta]\}+O(\eta^{-1}). \tag{74}$$

Returning to the angular equation (68) and absorbing $\sigma^2[1+2(M_1+M_2)^2]$ in the, as yet, unspecified separation constant λ_1, we shall write the equation in the form,

$$(1-\mu^2)\mathscr{C}^{-\frac{1}{2}}_{,\mu,\mu}+[\lambda_1+4\sigma^2(M_1-M_2)\mu-\sigma^2\mu^2]\mathscr{C}^{-\frac{1}{2}} = 0, \tag{75}$$

where we have replaced f by $\mathscr{C}^{-\frac{1}{2}}$ since the regular solutions of this equation (denoted by $\mathscr{C}_n^{-\frac{1}{2}}$ below), in the context of prolate spheroidal coordinates, plays the same role as the Gegenbauer polynomials, $\mathscr{C}_n^{-\frac{1}{2}}$, in the context of spherical polar coordinates. It should also be noted that equation (75) is a slightly generalized version – generalized by the inclusion of the term $4\sigma^2(M_1-M_2)\mu$ – of the standard spheroidal equation as defined by Fisher (1937).

The required regularity of the solution of equation (75) at $\mu = \pm 1$ will determine λ_1 as one of an infinite set of characteristic values – as in the case of the standard spheroidal functions (cf. Fisher 1937, p. 247); and the proper solutions, $\mathscr{C}_n^{-\frac{1}{2}}(\mu)$, belonging to the different characteristic values, will be orthogonal with the weight function $(1-\mu^2)^{-1}$); and we shall assume that they are normalized in the fashion:

$$\int_{-1}^{+1}\frac{d\mu}{1-\mu^2}\mathscr{C}_n^{-\frac{1}{2}}\mathscr{C}_m^{-\frac{1}{2}} = \delta_{nm}. \tag{76}$$

We conclude that at infinity the general solution of equation (63) can be expressed as a linear combination of the fundamental solutions,

$$Y = e^{\pm i\sigma[\eta+2(M_1+M_2)\lg\eta]}\mathscr{C}_n^{-\frac{1}{2}}(\mu) \quad (\eta\to\infty). \tag{77}$$

Turning next to equation (45), governing X, we first rewrite it in the form,

$$\begin{aligned}&(\eta^2-1)\left(X_{,\eta,\eta}-4\frac{U_{,\eta}}{U}X_{,\eta}+\sigma^2U^4X\right)-2\eta X_{,\eta}\\ &+(1-\mu^2)\left(X_{,\mu,\mu}-4\frac{U_{,\mu}}{U}X_{,\mu}+\sigma^2U^4X\right)+2\mu X_{,\mu}\\ &-4(\eta^2-1)(1-\mu^2)\left(U_{,\eta}Y_{,\mu}-U_{,\mu}Y_{,\eta}+\frac{2\mu}{1-\mu^2}U_{,\eta}Y+\frac{2\eta}{\eta^2-1}U_{,\mu}Y\right)=0.\end{aligned} \tag{78}$$

Since, as we shall presently verify, $X = O(\eta)$ for $\eta \to \infty$, we can, in view of the behaviours listed in (64), deduce from equation (78) that

$$(\eta^2-1)\left(X_{,\eta,\eta}+4\frac{M_1+M_2}{\eta^2}X_{,\eta}+\sigma^2U^4X\right)-2\eta X_{,\eta}$$

$$+(1-\mu^2)\left(X_{,\mu,\mu}-4\frac{M_1-M_2}{\eta^2}X_{,\mu}+\sigma^2U^4X\right)+2\mu X_{,\mu}+O(1)=0. \quad (79)$$

With the substitution $$Z = X/\eta, \quad (80)$$

equation (79) becomes:

$$(\eta^2-1)\left\{Z_{,\eta,\eta}+4\frac{M_1+M_2}{\eta^2}Z_{,\eta}+\sigma^2\left[1+4\frac{M_1+M_2}{\eta}+6\frac{(M_1+M_2)^2}{\eta^2}\right]Z\right\}$$

$$+(1-\mu^2)(Z_{,\mu,\mu}+\sigma^2Z)-2Z+2\mu Z_{,\mu}+4\sigma^2(M_1-M_2)\mu Z+O(\eta^{-1})=0. \quad (81)$$

This equation is clearly separable in η and μ. Thus, letting

$$Z(\eta,\mu)=\mathscr{Z}(\eta)\,g(\mu), \quad (82)$$

we obtain the pair of equations

$$\left\{\mathscr{D}+\frac{1}{\eta^2}[2\sigma^2(M_1+M_2)^2-\lambda_2]\right\}\mathscr{Z}+2\frac{M_1+M_2}{\eta^2}\mathscr{Z}_{,\eta}=0 \quad (83)$$

and

$$(1-\mu^2)(g_{,\mu,\mu}+\sigma^2g)-2g+2\mu g_{,\mu}+4\sigma^2(M_1-M_2)\mu g+\lambda_2 g=0 \quad (84)$$

where λ_2 is a separation constant and $\mathscr{D}$ is the same operator defined in equation (70).

We can now show, as in the case of equation (69), that the asymptotic behaviour of the solution of equation (83), again *independently* of λ_2, is given by

$$\mathscr{Z}(\eta)\to\exp\{\pm i\sigma[\eta+2(M_1+M_2)\lg\eta]\}+O(\eta^{-1}). \quad (85)$$

Next absorbing (σ^2-2) in equation (84) in a redefinition of λ_2, we write it in the form (cf. equation (75))

$$(1-\mu^2)\mathscr{C}^{-\frac{3}{2}}_{,\mu,\mu}+2\mu\mathscr{C}^{-\frac{3}{2}}_{,\mu}+[\lambda_2+4\sigma^2(M_1-M_2)\mu-\sigma^2\mu^2]\mathscr{C}^{-\frac{3}{2}}=0, \quad (86)$$

where we have replaced g by $\mathscr{C}^{-\frac{3}{2}}$ since, in the present context, the proper solutions, $\mathscr{C}_n^{-\frac{3}{2}}(\mu)$, of equation (86) play the same role as the Gegenbauer polynomials, $\mathscr{C}_n^{-\frac{3}{2}}$, in the context of spherical symmetry. The functions, $\mathscr{C}_n^{-\frac{3}{2}}(\mu)$, belonging to the different characteristic values of λ_2 (ensuring the regularity of the solutions at $\mu=\pm1$) are orthogonal with respect to the weight function $(1-\mu^2)^{-2}$; and we shall assume that they are normalized in the fashion,

$$\int_{-1}^{+1}\frac{d\mu}{(1-\mu^2)^2}\mathscr{C}_n^{-\frac{3}{2}}\mathscr{C}_m^{-\frac{3}{2}}=\delta_{nm}. \quad (87)$$

We conclude that at infinity, the general solution of equation (78) can be expressed as a linear combination of the fundamental solutions,

$$X=\eta\,e^{\pm i\sigma[\eta+2(M_1+M_2)\lg\eta]}\mathscr{C}_n^{-\frac{3}{2}}(\mu). \quad (88)$$

6. The flux integral at infinity

From the behaviour of the solutions for X and Y given in equations (77) and (88), it follows that in all cases the solutions at infinity must be expressible in the forms:

$$X \to \eta\, e^{\pm i\sigma[\eta + 2(M_1+M_2)\lg\eta]} \sum_n S_n \mathscr{C}_n^{-\frac{3}{2}}(\mu) \tag{89}$$

and

$$Y \to \tfrac{1}{2} e^{\pm i\sigma[\eta + 2(M_1+M_2)\lg\eta]} \sum_n R_n \mathscr{C}_n^{-\frac{1}{2}}(\mu) \quad (\eta \to \infty), \tag{90}$$

where S_n and R_n are complex constants. For incident gravitational and electromagnetic waves, when the exponential factors appear with $+i\sigma$, the complex constants S_n and R_n can be arbitrarily specified as boundary conditions on the required solutions.

From the behaviours,

$$U \to 1 \quad \text{and} \quad U_{,\eta} = -U_{,\mu} = O(\eta^{-2}) \quad \text{for} \quad \eta \to \infty, \tag{91}$$

it follows from equations (57) and (58) that

$$E_\eta \to \frac{[X, X^*]_\eta}{\eta^2(1-\mu^2)^2} + 4\frac{[Y, Y^*]_\eta}{(1-\mu^2)} \quad (\eta \to \infty) \tag{92}$$

and

$$E_\mu = O(\eta^{-2}). \tag{93}$$

Accordingly, the term in $E_\mu \, d\eta$ in the flux integral does not contribute for $\eta \to \infty$, while

$$\int_{-1}^{+1} E_{\eta\to\infty}\, d\mu \to \pm 2i\sigma\left[\int_{-1}^{+1} \frac{d\mu}{(1-\mu^2)^2} \Big|\sum_n S_n \mathscr{C}_n^{-\frac{3}{2}}(\mu)\Big|^2 + \int_{-1}^{+1} \frac{d\mu}{1-\mu^2} \Big|\sum_n R_n \mathscr{C}_n^{-\frac{1}{2}}(\mu)\Big|^2\right]. \tag{94}$$

Making use of the orthogonality properties, (76) and (87), of the functions $\mathscr{C}_n^{-\frac{1}{2}}(\mu)$ and $\mathscr{C}_n^{-\frac{3}{2}}(\mu)$, we find:

$$\int_{-1}^{+1} E_{\eta\to\infty}\, d\mu = \pm 2i\sigma \left(\sum_n |S_n|^2 + \sum_n |R_n|^2\right). \tag{95}$$

7. The asymptotic behaviours of the solutions at the singular points ($\eta = 1$, $\mu = \pm 1$)

We now turn to the asymptotic behaviours of the solutions near the singularities at $\eta = 1$ and $\mu = \pm 1$. It will clearly suffice to consider the behaviours at $\eta = 1$ and $\mu = 1$; the behaviours at $\eta = 1$ and $\mu = -1$ will follow by simply replacing M_1 by M_2.

Let

$$\eta = 1 + \xi \quad \text{and} \quad \mu = 1 - \zeta. \tag{96}$$

At the singularity both

$$\xi \quad \text{and} \quad \zeta \to 0 \tag{97}$$

and

$$\eta^2 - 1 \simeq 2\xi, \quad 1-\mu^2 \simeq 2\zeta \quad \text{and} \quad \eta^2 - \mu^2 \simeq 2(\xi + \zeta). \tag{98}$$

From the definition of U in equation (15) it follows that

$$U \sim \frac{M_1}{\xi+\zeta}, \quad U_{,\eta} = -U_{,\mu} \sim -\frac{M_1}{(\xi+\zeta)^2}. \tag{99}$$

Inserting the behaviours (98) and (99) in equations (44) and (45), we obtain:

$$\xi\left[Y_{,\xi,\xi}+\frac{2}{\xi+\zeta}Y_{,\xi}-\frac{4}{(\xi+\zeta)^2}Y\right]-\frac{1}{2M_1}(X_{,\xi}-X_{,\zeta})$$
$$+\zeta\left[Y_{,\zeta,\zeta}+\frac{2}{\xi+\zeta}Y_{,\zeta}-\frac{4}{(\xi+\zeta)^2}Y\right]+\sigma^2\frac{M_1^4}{(\xi+\zeta)^3}Y = 0 \tag{100}$$

and

$$\xi\left(X_{,\xi,\xi}+\frac{4}{\xi+\zeta}X_{,\xi}\right)+\zeta\left(X_{,\zeta,\zeta}+\frac{4}{\xi+\zeta}X_{,\zeta}\right)-(X_{,\xi}+X_{,\zeta})$$
$$+8\frac{M_1}{(\xi+\zeta)^2}[(\xi-\zeta)\,Y+\xi\zeta(Y_{,\xi}-Y_{,\zeta})]+\sigma^2\frac{M_1^4}{(\xi+\zeta)^3}X = 0. \tag{101}$$

In terms of the variables,

$$x = \tfrac{1}{2}(\xi+\zeta) \quad\text{and}\quad y = \tfrac{1}{2}(\xi-\zeta), \tag{102}$$

(where it will be noted that x is one-half the radial distance to the singularity), equations (100) and (101) become

$$(xY_{,x,x}+2Y_{,x}+2yY_{,x,y}+xY_{,y,y}+2\frac{y}{x}Y_{,y})-\frac{4}{x}Y+\sigma^2\frac{M_1^4}{4x^3}Y-\frac{1}{M_1}X_{,y} = 0 \tag{103}$$

and

$$(xX_{,x,x}+2X_{,x}+2yX_{,x,y}+xX_{,y,y}+2\frac{y}{x}X_{,y})$$
$$+2\frac{y}{x}X_{,y}+\sigma^2\frac{M_1^4}{4x^3}X+8\frac{M_1}{x^2}[yY+\tfrac{1}{2}(x^2-y^2)\,Y_{,y}] = 0. \tag{104}$$

For our present purposes, it is important to transform equations (103) and (104) to the variables x and ν where,

$$\nu = \frac{(\eta+\mu)^2-(\eta-\mu)^2-4}{4(\eta-\mu)} = \frac{\eta\mu-1}{\eta-\mu} \simeq \frac{\xi-\zeta}{\xi+\zeta} = \frac{y}{x}, \tag{105}$$

is the cosine of the angle between the radius vector, $\eta-\mu$, at the singularity and the direction $\mu = +1$. For this transformation of variables, $(x, y) \to (x, \nu)$,

$$\frac{\partial}{\partial x} = \frac{\partial}{\partial x}-\frac{\nu}{x}\frac{\partial}{\partial \nu}, \quad \frac{\partial}{\partial y} = \frac{1}{x}\frac{\partial}{\partial \nu}, \quad \frac{\partial^2}{\partial y^2} = \frac{1}{x^2}\frac{\partial^2}{\partial^2\nu^2},$$

$$\frac{\partial^2}{\partial x\,\partial y} = \frac{1}{x}\frac{\partial^2}{\partial\nu\,\partial x}-\frac{\nu}{x^2}\frac{\partial^2}{\partial\nu^2}-\frac{1}{x^2}\frac{\partial}{\partial\nu},$$

$$\frac{\partial^2}{\partial x^2} = \frac{\partial^2}{\partial x^2}+\frac{2\nu}{x^2}\frac{\partial}{\partial\nu}-\frac{2\nu}{x}\frac{\partial^2}{\partial x\,\partial\nu}+\frac{\nu^2}{x^2}\frac{\partial^2}{\partial\nu^2}. \tag{106}$$

By virtue of the relations,

$$x\frac{\partial^2}{\partial x^2}+2\frac{\partial}{\partial x}+2y\frac{\partial^2}{\partial x\,\partial y}+x\frac{\partial^2}{\partial y^2}+2\frac{y}{x}\frac{\partial}{\partial y} = x\frac{\partial^2}{\partial x^2}+2\frac{\partial}{\partial x}+\frac{(1-\nu^2)}{x}\frac{\partial^2}{\partial \nu^2}, \tag{107}$$

and equations (106), equations (103) and (104) reduce to

$$xY_{,x,x}+2Y_{,x}+\sigma^2\frac{M_1^4}{4x^3}Y+\frac{1}{x}\left[(1-\nu^2)\,Y_{,\nu,\nu}-4Y-\frac{1}{M_1}X_{,\nu}\right] = 0, \tag{108}$$

$$xX_{,x,x}+2X_{,x}+\sigma^2\frac{M_1^4}{4x^3}X+\frac{1}{x}\{(1-\nu^2)\,X_{,\nu,\nu}+2\nu X_{,\nu}$$
$$+4M_1[(1-\nu^2)\,Y_{,\nu}+2\nu Y]\} = 0. \tag{109}$$

For the treatment of equations (108) and (109) the following lemma is essential:

LEMMA. *The solution of the equation,*

$$x\frac{\mathrm{d}^2F}{\mathrm{d}x^2}+2\frac{\mathrm{d}F}{\mathrm{d}x}+\sigma^2\frac{M_1^4}{4x^3}F-\lambda\frac{F}{x} = 0, \tag{110}$$

where λ is an arbitrary constant, is, for $x\to 0$, given, independently of λ, by

$$F\to \text{constant}\,\mathrm{e}^{\pm \mathrm{i}\sigma M_1^2/(2x)}+O(x) \quad (x\to 0). \tag{111}$$

Proof. Rewriting equation (110) in the form,

$$\frac{1}{x}\frac{\mathrm{d}}{\mathrm{d}x}\left(x^2\frac{\mathrm{d}F}{\mathrm{d}x}\right)+\sigma^2\frac{M_1^4}{4x^3}F-\lambda\frac{F}{x} = 0, \tag{112}$$

and letting

$$x = y^{-1}, \tag{113}$$

we obtain the equation,

$$\frac{\mathrm{d}^2F}{\mathrm{d}y^2}+\tfrac{1}{4}\sigma^2M_1^4F-\frac{\lambda}{y^2}F = 0. \tag{114}$$

From this equation we readily deduce that

$$F\to(\text{constant})\,\mathrm{e}^{\pm \mathrm{i}\sigma M_1^2 y/2}+O(y^{-1}) \quad (y\to\infty); \tag{115}$$

and the lemma stated follows.

In view of the lemma we have just established, equations (108) and (109), for $x\to 0$, can be separated by the substitutions,

$$Y = \mathrm{e}^{\pm \mathrm{i}\sigma M_1^2/(2x)}f(\nu) \quad \text{and} \quad X = \mathrm{e}^{\pm \mathrm{i}\sigma M_1^2/(2x)}g(\nu). \tag{116}$$

We thus obtain the pair of equations:

$$(1-\nu^2)f_{,\nu,\nu}-4f-\frac{1}{M_1}g_{,\nu} = -(\lambda_1+4)f, \tag{117}$$

$$(1-\nu^2)\,g_{,\nu,\nu}+2\nu g_{,\nu}+4M_1[(1-\nu^2)f_{,\nu}+2\nu f] = -\lambda_2 g, \tag{118}$$

where λ_1 and λ_2 are two separation constants. Differentiating equation (118) with respect to ν and letting

$$g_{,\nu} = G, \tag{119}$$

we obtain the equation,

$$(1-\nu^2)\,G_{,\nu,\nu}+4M_1(1-\nu^2)f_{,\nu,\nu} = -(\lambda_2+2)\,G-8M_1 f; \tag{120}$$

while equation (117) becomes

$$(1-\nu^2)f_{,\nu,\nu} = -\lambda_1 f+\frac{1}{M_1}G. \tag{121}$$

Now eliminating $f_{,\nu,\nu}$ from equation (120) we obtain

$$(1-\nu^2)\,G_{,\nu,\nu} = -(\lambda_2+6)\,G+4M_1(\lambda_1-2)f. \tag{122}$$

Replacing λ_2+6 by λ_2 and writing

$$2M_1 f = F, \tag{123}$$

we obtain the pair of equations:

$$(1-\nu^2)\,F_{,\nu,\nu} = -\lambda_1 F+2G, \tag{124}$$

$$(1-\nu^2)\,G_{,\nu,\nu} = 2(\lambda_1-2)\,F-\lambda_2 G. \tag{125}$$

By the linear substitution,

$$H = AF+BG, \tag{126}$$

where A and B are constants, unspecified at present, we can bring equations (124) and (125) to the diagonal form

$$(1-\nu^2)\,H_{,\nu,\nu} = -\kappa_\pm H, \tag{127}$$

where κ_+ and κ_- are the characteristic roots of the matrix

$$\begin{vmatrix} \kappa-\lambda_1 & 2(\lambda_1-2) \\ 2 & \kappa-\lambda_2 \end{vmatrix} = 0; \tag{128}$$

and the roots are:

$$\kappa_\pm = \tfrac{1}{2}(\lambda_1+\lambda_2)\pm\tfrac{1}{2}\sqrt{[(\lambda_1-\lambda_2)^2+16(\lambda_1-2)]}. \tag{129}$$

The corresponding characteristic vectors are determined by

$$A_\pm = -\tfrac{1}{4}\{(\lambda_1-\lambda_2)\pm\sqrt{[(\lambda_1-\lambda_2)^2+16(\lambda_1-2)]}\}\,B_\pm. \tag{130}$$

In order that H governed by equation (127) reduces to the Gegenbauer polynomials, $C_{l+1}^{-\frac{1}{2}}(\nu)$, (as required by the regularity of the solutions at $\nu = \pm 1$), it is necessary (but *not* sufficient) that $\kappa_\pm$ be of the form

$$\kappa_\pm = m\pm n, \tag{131}$$

where m and n are both positive integers. From equation (129) it now follows that

$$(\lambda_1+\lambda_2)^2 = 4m^2 \quad \text{and} \quad (\lambda_1-\lambda_2)^2+16(\lambda_1-2) = 4n^2. \tag{132}$$

On solving these equations, we find:

$$\lambda_1 = m-2-\sqrt{(n^2-4m+12)} \quad \text{and} \quad \lambda_2 = m+2+\sqrt{(n^2-4m+12)}. \tag{133}$$

Now by setting

$$m = l^2 \quad \text{and} \quad n = l, \tag{134}$$

when $$\kappa_{\pm} = l(l \pm 1), \tag{135}$$

the solutions for H_+ and H_- of equation (127) become proportional to the Gegenbauer polynomials $C_{l+1}^{-\frac{1}{2}}(\nu)$ and $C_l^{-\frac{1}{2}}(\nu)$. Thus, we may write

$$H_+ = A_{l+1} C_{l+1}^{-\frac{1}{2}}(\nu) \quad \text{and} \quad H_- = B_l C_l^{-\frac{1}{2}}(\nu), \tag{136}$$

where A_{l+1} and B_l are arbitrary constants (which can be complex) and $C_{l+1}^{-\frac{1}{2}}(\nu)$ and $C_l^{-\frac{1}{2}}(\nu)$ now denote the Gegenbauer polynomials *in their standard normalizations* (see equation (172) below). The constants λ_1 and λ_2, for the choice (134), are

$$\lambda_1 = (l^2-2) - i\surd(3l^2-12); \quad \lambda_2 = (l^2+2) + i\surd(3l^2-12), \tag{137}$$

while equation (130) gives,

$$A_{\pm} = \mp\tfrac{1}{2}[(l \mp 2) \mp i\surd(3l^2-12)] B_{\pm}. \tag{138}$$

From equations (126), (136), and (138) it now follows that the equations determining F and G are:

$$-\tfrac{1}{2}[(l-2) - i\surd(3l^2-12)]F + G = H_+ = A_{l+1} C_{l+1}^{-\frac{1}{2}}(\nu) \tag{139}$$

and
$$+\tfrac{1}{2}[(l+2) + i\surd(3l^2-12)]F + G = H_- = B_l C_l^{-\frac{1}{2}}(\nu). \tag{140}$$

Solving for F and G, we obtain (cf. equations (119) and (123)):

$$F = 2M_1 f = -\frac{1}{l}\left[A_{l+1} C_{l+1}^{-\frac{1}{2}}(\nu) - B_l C_l^{-\frac{1}{2}}(\nu)\right] \tag{141}$$

and

$$G = g_{,\nu} = \frac{1}{2l}\left\{[l+2+i\surd(3l^2-12)] A_{l+1} C_{l+1}^{-\frac{1}{2}}(\nu) + [l-2-i\surd(3l^2-12)] B_l C_l^{-\frac{1}{2}}(\nu)\right\}. \tag{142}$$

Making use of the relation

$$\frac{d}{d\nu} C_{l+2}^{-\frac{3}{2}}(\nu) = -3 C_{l+1}^{-\frac{1}{2}}(\nu), \tag{143}$$

we can integrate equation (142) to obtain

$$g = -\frac{1}{6l}\left\{[l+2+i\surd(3l^2-12)] A_{l+1} C_{l+2}^{-\frac{3}{2}}(\nu) + [(l-2) - i\surd(3l^2-12)] B_l C_{l+1}^{-\frac{3}{2}}(\nu)\right\}. \tag{144}$$

The general solutions for f and g are now expressible as linear combinations of the fundamental solutions (141) and (144). Thus, by absorbing the common factor $-l^{-1}$ in the constants A_{l+1} and B_l, we can write

$$2M_1 f = \sum_{l \geqslant 2} (A_{l+1} C_{l+1}^{-\frac{1}{2}} - B_l C_l^{-\frac{1}{2}}) = \sum_{l \geqslant 2} (A_{l+1} - B_{l+1}) C_{l+1}^{-\frac{1}{2}} \tag{145}$$

and

$$g = \sum_{l \geqslant 2} (\alpha_{l+2} A_{l+1} C_{l+2}^{-\frac{3}{2}} + \beta_{l+2} B_l C_{l+1}^{-\frac{3}{2}}) = \sum_{l \geqslant 2} (\alpha_{l+2} A_{l+1} + \beta_{l+3} B_{l+1}) C_{l+2}^{-\frac{3}{2}}, \tag{146}$$

where
$$\alpha_{l+2} = \tfrac{1}{6}[l+2+i\surd(3l^2-12)], \quad \beta_{l+2} = \tfrac{1}{6}[l-2-i\surd(3l^2-12)]. \tag{147}$$

(It may be noted parenthetically here that

$$C_2^{-\frac{1}{2}} \equiv C_3^{-\frac{3}{2}} \equiv 0 \quad \text{and that} \quad \beta_4 = 0.) \tag{148}$$

With the definitions,

$$S_{l+1} = \surd(A_{l+1}^2 + B_{l+1}^2) \quad \text{and} \quad \cot\delta_{l+1} = A_{l+1}/B_{l+1} \tag{149}$$

the solutions (145) and (146) can be written more conveniently in the forms:

$$2M_1 f = \sum_{l\geqslant 2} S_{l+1}(\cos\delta_{l+1} - \sin\delta_{l+1})\, C_{l+1}^{-\frac{1}{2}}(\nu) \tag{150}$$

and

$$g = \sum_{l\geqslant 2} S_{l+1}(\alpha_{l+2}\cos\delta_{l+1} + \beta_{l+3}\sin\delta_{l+1})\, C_{l+2}^{-\frac{3}{2}}(\nu). \tag{151}$$

With the further definitions,

$$S(1\,; l+1) = S_{l+1}(\cos\delta_{l+1} - \sin\delta_{l+1}) \tag{152}$$

and

$$S(2\,; l+2) = S_{l+1}(\alpha_{l+2}\cos\delta_{l+1} + \beta_{l+3}\sin\delta_{l+1}), \tag{153}$$

we have, more compactly,

$$2M_1 f = \sum_{l\geqslant 2} S(1, l+1)\, C_{l+1}^{-\frac{1}{2}}(\nu) \tag{154}$$

and

$$g = \sum_{l\geqslant 2} S(2, l+2)\, C_{l+2}^{-\frac{3}{2}}(\nu). \tag{155}$$

It should be noted that the 'phases', δ_{l+1}, in the solutions (150) and (151), for each $l(\geqslant 2)$ can be arbitrarily specified (even as the coefficients S_{l+1}).

It is important to observe that the solutions (150) and (151) allow for the incidence of a pure gravitational or a pure electromagnetic wave. Thus, with the choices,

$$S_{l+1} = 0 \quad \text{for all } l \text{ except } l = n \text{ and} \quad \tan\delta_{n+1} = 1, \tag{156}$$

we have the solutions,

$$f = 0 \quad \text{and} \quad g = \frac{1}{\surd 2}(\alpha_{n+2} + \beta_{n+3})\, S_{n+1}\, C_{n+2}^{-\frac{3}{2}}(\nu), \tag{157}$$

which represents a pure gravitational wave with angular momentum corresponding to n. Similarly, with the choices

$$S_{l+1} = 0 \quad \text{for all } l \text{ except } l = n \text{ and} \quad \tan\delta_{n+1} = -\alpha_{n+2}/\beta_{n+3}, \tag{158}$$

we have the solutions,

$$g = 0 \quad \text{and} \quad 2M_1 f = +\frac{\alpha_{n+2} + \beta_{n+3}}{\surd(\alpha_{n+2}^2 + \beta_{n+3}^2)} S_{n+1}\, C_{n+1}^{-\frac{1}{2}}(\nu), \tag{159}$$

which represents a pure electromagnetic wave with an angular momentum n.

More generally, we can arrange for the incidence of a superposition of gravitational (or electromagnetic) waves belonging to different ns without any admixture of electromagnetic (or gravitational) waves.

8. The flux integral at $\eta = 1$ and $\mu = 1$

We have seen that the solutions for X and Y, near the singularity at $\eta = 1$ and $\mu = 1$, have the behaviours,

$$X = e^{\pm i\sigma M_1^2/(2x)} g(\nu) \quad \text{and} \quad Y = e^{\pm i\sigma M_1^2/(2x)} f(\nu), \tag{160}$$

where $g(\nu)$ and $f(\nu)$ have the forms given in equations (150) and (151) (where, it should be noted, the coefficients, S_{l+1}, and the phases, δ_{l+1}, will in general be complex). For gravitational and electromagnetic waves that are absorbed by the black hole at the singularity, the exponential factor occurs with $-i\sigma M_1^2/(2x)$.

We shall now evaluate the flux integral (see equation (62)):

$$\int_{C_1} (E_\eta \, d\mu - E_\mu \, d\eta), \tag{161}$$

where C_1 is a circle with its centre at $\eta = 1$ and $\mu = 1$ and of radius $x \to 0$. By changing the variables of integrations, successively, from (η, μ) to (ξ, ζ) to (x, y), in accordance with equations (96) and (102), we have

$$\int_{C_1} (E_\eta \, d\mu - E_\mu \, d\eta) = -\int_{C_1} (E_\xi \, d\zeta - E_\zeta \, d\xi)$$

$$= -\int_{C_1} [(E_\xi - E_\zeta) \, dx - (E_\xi + E_\zeta) \, dy]; \tag{162}$$

or, since x is constant along C_1,

$$\int_{C_1} (E_\xi + E_\zeta) \, dy = +x \int_{-1}^{+1} (E_\xi + E_\zeta) \, d\nu, \tag{163}$$

where the integration over y has been changed to an integration over ν in accordance with the transformations (105) and (106). It should be noted that because the directions $\nu = 1$ and $\mu = 1$ coincide, contours described in the sense of increasing ν and μ are both counter-clockwise.

The expressions for E_ξ and E_ζ that have to be used in evaluating the integral over ν in equation (163) follow from the definitions (57) and (58) and the respective transformation equations. We find:

$$E_\xi = \frac{[X, X^*]_\xi}{8U^4 \xi \zeta^2} + 4\frac{[Y, Y^*]_\xi}{2U^2 \zeta} - 4\frac{U_{,\mu}(XY^* - X^*Y)}{4U^4 \xi \zeta}$$

$$= \frac{x}{M_1^4 (1-\nu)(1-\nu^2)} \{[X, X^*]_x + [X, X^*]_y\}$$

$$+4\frac{x}{M_1^2(1-\nu)} \{[Y, Y^*]_x + [Y, Y^*]_y\} - \frac{4}{M_1^3(1-\nu^2)}(XY^* - X^*Y); \tag{164}$$

and, similarly,

$$E_\zeta = \frac{x}{M_1^4(1+\nu)(1-\nu^2)}\{[X,X^*]_x - [X,X^*]_y\}$$

$$+4\frac{x}{M_1^2(1+\nu)}\{[Y,Y^*]_x - [Y,Y^*]_y\} + \frac{4}{M_1^3(1-\nu^2)}(XY^* - X^*Y). \quad (165)$$

We now evaluate the foregoing expressions for X and Y given by equations (160) for the positive exponent, $+\mathrm{i}\sigma M_1^2/(2x)$. We find (cf. equation (106))

$$[X,X^*]_x + [X,X^*]_y = X^*\left(\frac{\partial}{\partial x} + \frac{1-\nu}{x}\frac{\partial}{\partial \nu}\right)X - \text{complex conjugate}$$

$$= -\frac{\mathrm{i}\sigma M_1^2}{2x^2}|g(\nu)|^2 + \frac{1-\nu}{x}g^*(\nu)\,g'(\nu) - \text{complex conjugate}$$

$$= -\frac{\mathrm{i}\sigma M_1^2}{x^2}|g(\nu)|^2 + \frac{1-\nu}{x}(g^*g' - g^{*\prime}g); \quad (166)$$

and, similarly,

$$[X,X^*]_x - [X,X^*]_y = -\frac{\mathrm{i}\sigma M_1^2}{x^2}|g(\nu)|^2 - \frac{1+\nu}{x}(g^*g' - g^{*\prime}g). \quad (167)$$

The corresponding expressions for the sum and difference of the Wronskians involving Y are given by simply replacing g by f in equations (166) and (167). We thus find:

$$E_\xi = \frac{x}{M_1^4(1-\nu)(1-\nu^2)}\left\{-\frac{\mathrm{i}\sigma M_1^2}{x^2}|g(\nu)|^2 + \frac{1-\nu}{x}(g^*g' - g^{*\prime}g)\right\}$$

$$+\frac{4x}{M_1^2(1-\nu)}\left\{-\frac{\mathrm{i}\sigma M_1^2}{x^2}|f(\nu)|^2 + \frac{1-\nu}{x}(f^*f' - f^{*\prime}f)\right\} - \frac{4}{M_1^3(1-\nu^2)}(gf^* - g^*f) \quad (168)$$

and

$$E_\zeta = \frac{x}{M_1^4(1+\nu)(1-\nu^2)}\left\{-\frac{\mathrm{i}\sigma M_1^2}{x^2}|g(\nu)|^2 - \frac{1+\nu}{x}(g^*g' - g^{*\prime}g)\right\}$$

$$+\frac{4x}{M_1^2(1+\nu)}\left\{-\frac{\mathrm{i}\sigma M_1^2}{x^2}|f(\nu)|^2 - \frac{1+\nu}{x}(f^*f' - f^{*\prime}f)\right\} + \frac{4}{M_1^3(1-\nu^2)}(gf^* - g^*f). \quad (169)$$

From these equations, it follows that

$$x(E_\xi + E_\zeta) = -\frac{2\mathrm{i}\sigma}{M_1^2}\frac{|g(\nu)|^2}{(1-\nu^2)^2} - 8\mathrm{i}\sigma\frac{|f(\nu)|^2}{1-\nu^2}, \quad (170)$$

Accordingly,

$$\lim_{x\to 0}\int_{-1}^{+1} x(E_\xi + E_\zeta)\,\mathrm{d}\nu = -\frac{2\mathrm{i}\sigma}{M_1^2}\left[\int_{-1}^{+1}\frac{\mathrm{d}\nu}{(1-\nu^2)^2}|g(\nu)|^2 + 4M_1^2\int_{-1}^{+1}\frac{\mathrm{d}\nu}{(1-\nu^2)}|f(\nu)|^2\right] \quad (171)$$

For the particular case, when f and g are given by equations (154) and (155), the integrals over $|f(\nu)|^2$ and $|g(\nu)|^2$ can be evaluated explicitly by making use of the orthogonality relations:

$$\int_{-1}^{+1} C_n^\gamma C_m^\gamma \frac{d\nu}{(1-\nu^2)^{-\gamma+\frac{1}{2}}} = N_n^\gamma \delta_{nm}, \tag{172}$$

where

$$N_n^\gamma = \frac{\pi\Gamma(2\gamma+n)}{2^{2\gamma-1}(\gamma+n)\,n!\,[\Gamma(\gamma)]^2}. \tag{173}$$

We find:

$$\frac{2i\sigma}{M_1^2}\left\{\sum_{l\geqslant 2} |S(2,l+2)|^2 N_{l+2}^{-\frac{3}{2}} + \sum_{l\geqslant 2} |S(1,l+1)|^2 N_{l+1}^{-\frac{1}{2}}\right\}. \tag{174}$$

With the further definitions,

$$\left.\begin{aligned} g(\delta_{l+1}) &= \alpha_{l+2}\cos\delta_{l+1} + \beta_{l+3}\sin\delta_{l+1}, \\ f(\delta_{l+1}) &= \cos\delta_{l+1} - \sin\delta_{l+1}, \end{aligned}\right\} \tag{175}$$

the expressions for $S(2,l+2)$ and $S(1,l+1)$, explicitly, are (cf. equations (154) and (155))

$$S(2,l+2) = S_{l+1}\,g(\delta_{l+1}); \quad S(1,l+1) = S_{l+1}f(\delta_{l+1}). \tag{176}$$

Inserting these expressions in (174), we obtain,

$$-\frac{2i\sigma}{M_1^2}\sum_{l\geqslant 2} |S_{l+1}|^2 \{|g(\delta_{l+1})|^2 N_{l+2}^{-\frac{3}{2}} + |f(\delta_{l+1})|^2 N_{l+1}^{-\frac{1}{2}}\}, \tag{177}$$

where, it will be noted that, the quantity in curly brackets is also positive definite. Accordingly, we define

$$Q^2(\delta_{l+1}) = |g(\delta_{l+1})|^2 N_{l+2}^{-\frac{3}{2}} + |f(\delta_{l+1})|^2 N_{l+1}^{-\frac{1}{2}}; \tag{178}$$

and the expression (177) takes the simple form

$$-\frac{2i\sigma}{M_1^2}\sum_{l\geqslant 2} |S_{l+1}\,Q(\delta_{l+1})|^2. \tag{179}$$

(a) A vector representation of the radiation field near the singularities

The expression of the radiation field near the singularities as a superposition of the elementary solution pair,

$$X = e^{\pm i\sigma M_i^2/(2x)} S_{l+1}\,g(\delta_{l+1})\,C_{l+2}^{-\frac{3}{2}}(\nu), \tag{180}$$

$$Y = e^{\pm i\sigma M_i^2/(2x)} S_{l+1} f(\delta_{l+1})\,C_{l+1}^{-\frac{1}{2}}(\nu), \tag{181}$$

(where i $(=1, 2)$ distinguishes the singularities at $\eta = 1$ and $\mu = \pm 1$) suggests that we consider $(X, 2M_i Y)$ as the components of a vector,

$$\boldsymbol{I} = (X, 2M_i\,Y), \tag{182}$$

and write the general solution for the radiation field as a single vector-equation in the form:

$$\boldsymbol{I} = e^{\pm i\sigma M_i^2/(2x)} \sum_{l\geqslant 2} \boldsymbol{S}(l+1, \delta_{l+1}), \tag{183}$$

where
$$\boldsymbol{S}(l+1, \delta_{l+1}) = S_{l+1}[g(\delta_{l+1})\, C_{l+2}^{-\frac{3}{2}}(\nu), f(\delta_{l+1})\, C_{l+1}^{-\frac{1}{2}}(\nu)]. \tag{184}$$

For the solution (183) the flux integral at the singularities is given by

$$\pm \frac{2i\sigma}{M_i^2} \sum_{l \geqslant 2} |S_{l+1}\, Q(\delta_{l+1})|^2. \tag{185}$$

The fact that in the vicinity of the singularities, the prevailing gravitational and electromagnetic fields are best represented by a vector with two components and the elementary solutions (180) and (181) are expressed in terms of the same amplitude (S_{l+1}) and phase (δ_{l+1}), is clearly a result of the strong curvature of the space-time near the horizons of the black holes. In view of this intimate association between the electromagnetic and the gravitational fields, we may properly describe the radiation field in these regions as that of *coupled photon–graviton waves* (in the manner of Siamese twins).

This association of electromagnetic and gravitational waves belonging to spins 1 and 2 suggests that the scattering of electromagnetic waves by a pair of magnetic monopoles may result in a similar association of scalar waves (associated with the Higgs boson) and electromagnetic waves belonging to spins 0 and 1. (I am indebted to Professor Y. Nambu for this suggestion.)

9. The scattering of radiation by the two black-holes: the scattering matrix

With the flux integrals evaluated, at infinity and at the singularities, the problem of the scattering of incident gravitational and electromagnetic radiation can be treated (at least at a formal level) in the same manner in which the scattering by an isolated Reissner–Nordström black-hole has been treated (*M.T.* Chapter 5, §47).

In the case of the isolated Reissner–Nordström black-hole, the problem of scattering is, in the terminology of nuclear physics, a four-channel problem (cf. Blatt & Weisskopf 1952, pp. 521–525; I am greatly indebted to Professor Roland Winston for this reference and for numerous discussions in which he clarified for me the basic concepts and notions). The four channels correspond to the incidence of radiation of the two sorts from $+\infty$ and from $-\infty$ (the horizons); and this is consistent with a scattering matrix of order four that is needed for a complete description of the phenomenon (see *M.T.* p. 260, equations (354) and (355) for the explicit form of the matrix).

When two black-holes are present, it is clear that we have a six-channel problem (see figure 1 which should be compared with Fig. 2.1 in Blatt & Weisskopf 1952, p. 522). The six channels correspond to the incidence of radiation of the two sorts from infinity and from the two singularities (i.e. the horizons of the two black-holes). Since however the radiation field in the vicinity of the singularities prevails as coupled photon–graviton waves and is described in terms of a two-component vector, we have in effect only four channels. Besides distinguishing the four principal channels, we must further distinguish in each channel the different states of angular momentum.

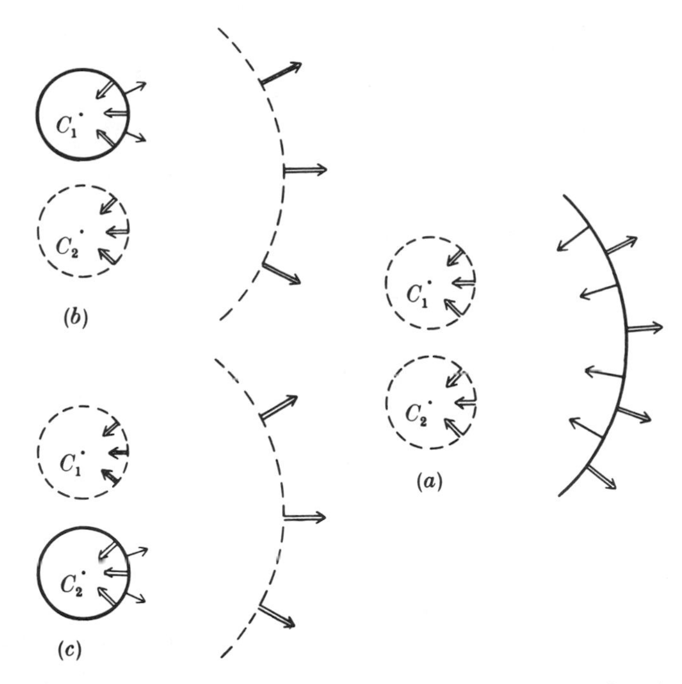

FIGURE 1. The different channels to be considered in the scattering by the two extreme Reissner–Nordström black-holes. (*a*) Gravitational *or* electromagnetic radiation is incident from infinity, parts of which are reflected as gravitational *and* electromagnetic radiation to infinity while other parts, as coupled photon-graviton waves, are absorbed by the two black-holes whose horizons are located at C_1 and C_2; (*b*) and (*c*) radiation as coupled photon-graviton waves is incident on the horizon at C_1 (or C_2), parts of which emerge at infinity as gravitational and electromagnetic radiation while other parts are absorbed by the black holes at C_1 and C_2.

The circumstances envisaged in the contexts of the four channels are:

I.

$$\text{At } \infty: \qquad X = e^{+i\sigma[\dots]}\mathscr{C}_n^{-\frac{3}{2}}(\mu) + e^{-i\sigma[\dots]}\sum_l S(2,n;\infty\,|\,2,l;\infty)\,\mathscr{C}_l^{-\frac{3}{2}}(\mu),$$

$$\text{At } \infty: \qquad 2Y = \qquad + e^{-i\sigma[\dots]}\sum_l S(2,n;\infty\,|\,1,l;\infty)\,\mathscr{C}_l^{-\frac{1}{2}}(\mu),$$

$$\text{At } C_i(i=1,2): \quad I = \qquad e^{-i\sigma M_i^2/(2x)}\sum_{l\geqslant 2} \boldsymbol{S}(2,n;\infty\,|\,l+1,\delta_{l+1};C_i); \tag{186}$$

II.

$$\text{At } \infty: \qquad X = \qquad + e^{-i\sigma[\dots]}\sum_l S(1,n;\infty\,|\,2,l;\infty)\,\mathscr{C}_l^{-\frac{3}{2}}(\mu),$$

$$\text{At } \infty: \qquad 2Y = e^{+i\sigma[\dots]}\mathscr{C}_n^{-\frac{1}{2}}(\mu) + e^{-i\sigma[\dots]}\sum_l S(1,n;\infty\,|\,1,l;\infty)\,\mathscr{C}_l^{-\frac{1}{2}}(\mu),$$

$$\text{At } C_i(i=1,2): \quad I = \qquad e^{-i\sigma M_i^2/(2x)}\sum_{l\geqslant 2} \boldsymbol{S}(1,n;\infty\,|\,l+1,\delta_{l+1};C_i); \tag{187}$$

III & IV:

At ∞: $$X = e^{-i\sigma[\dots]}\sum_l S(n+1, \delta_{n+1}; C_i \,|\, 2, l; \infty)\,\mathscr{C}_l^{-\frac{3}{2}}(\mu),$$

At ∞: $$2Y = e^{-i\sigma[\dots]}\sum_l S(n+1, \delta_{n+1}; C_i \,|\, 1, l; \infty)\,\mathscr{C}_l^{-\frac{1}{2}}(\mu),$$

At $C_i(i = 1, 2)$: $$\boldsymbol{I} = e^{+i\sigma M_i^2/(2x)}\boldsymbol{S}(n+1, \delta_{n+1}; C_i) + e^{-i\sigma M_i^2/(2x)}\sum_{l\geqslant 2}\boldsymbol{S}(n+1, \delta_{n+1}; C_i \,|\, l+1, \delta_{l+1}; C_i),$$

At $C_j(j \neq i)$: $$\boldsymbol{I} = e^{-i\sigma M_j^2/(2x)}\sum_{l\geqslant 2}\boldsymbol{S}(n+1, \delta_{n+1}; C_i \,|\, l+1, \delta_{l+1}; C_j). \tag{188}$$

In the foregoing, the notation in which the various 'matrix-elements' (such as $S(2, n; \infty \,|\, 1, l; \infty)$) are written is manifest: the symbols to the left of the vertical rule specify the channel in which the radiation (gravitational, electromagnetic, or coupled photon–graviton, wave) is incident and the symbols to the right, the radiation that is 'induced' in the particular channel. Thus, $S(2, n; \infty \,|\, 1, l; \infty)$ is the amplitude of the radiation of spin 1 and angular momentum l that is reflected at ∞ as a result of the incidence of radiation of spin 2 and angular momentum n from ∞.

Now, making use of the expressions for the flux integrals, at ∞ and at the singularities, given in equations (95) and (185), we find that, for the circumstances specified in equations (186)–(188), the divergence theorem as expressed in equation (62) yields the following identities:

$$1 = \sum_l |S(2, n; \infty \,|\, 2, l; \infty)|^2 + \sum_l |S(2, n; \infty \,|\, 1, l; \infty)|^2 + \sum_{i=1,2}\frac{1}{M_i^2}\left\{\sum_{l\geqslant 2}|S(2, n; \infty \,|\, l+1, \delta_{l+1}; C_i)\,Q(\delta_{l+1})|^2\right\}, \tag{189}$$

$$1 = \sum_l |S(1, n; \infty \,|\, 2, l; \infty)|^2 + \sum_l |S(1, n; \infty \,|\, 1, l; \infty)|^2 + \sum_{i=1,2}\frac{1}{M_i^2}\left\{\sum_{l\geqslant 2}|S(1, n; \infty \,|\, l+1, \delta_{l+1}; C_i)\,Q(\delta_{l+1})|^2\right\}, \tag{190}$$

and

$$\begin{aligned}\frac{1}{M_i^2}|S_{n+1}\,Q(\delta_{n+1})|^2 = &\sum_l |S(n+1, \delta_{n+1}; C_i \,|\, 2, l; \infty)|^2 \\ &+\sum_l |S(n+1, \delta_{n+1}; C_i \,|\, l, l; \infty)|^2 \\ &+\frac{1}{M_i^2}\sum_{l\geqslant 2}|S(n+1, \delta_{n+1}; C_i \,|\, l+1, \delta_{l+1}; C_i)\,Q(\delta_{l+1})|^2 \\ &+\frac{1}{M_j^2}\sum_{l\geqslant 2}|S(n+1, \delta_{n+1}; C_i \,|\, l+1, \delta_{l+1}; C_j)\,Q(\delta_{l+1})|^2 \\ &\qquad (i = 1, 2; j \neq i).\end{aligned} \tag{191}$$

In terms of the *scattering matrix*, $\boldsymbol{S}(\alpha, n \mid \beta, l)$, defined in the manner,

$$\boldsymbol{S}(\alpha, n \mid \beta, l) = \begin{bmatrix} S(2, n; \infty \mid 2, l; \infty) & S(2, n; \infty \mid 1, l; \infty) & \dfrac{1}{M_1} S(2, n; \infty \mid l+1, \delta_{l+1}; C_1)\, Q(\delta_{l+1}) & \dfrac{1}{M_2} S(2, n; \infty \mid l+1, \delta_{l+1}; C_2)\, Q(\delta_{l+1}) \\ S(1, n; \infty \mid 2, l; \infty) & S(1, n; \infty \mid 1, l; \infty) & \dfrac{1}{M_1} S(1, n; \infty \mid l+1, \delta_{l+1}; C_1)\, Q(\delta_{l+1}) & \dfrac{1}{M_2} S(1, n; \infty \mid l+1, \delta_{l+1}; C_2)\, Q(\delta_{l+1}) \\ \dfrac{M_1 S(n+1, \delta_{n+1}; C_1 \mid 2, l; \infty)}{S_{n+1}\, Q(\delta_{n+1})} & \dfrac{M_1 S(n+1, \delta_{n+1}; C_1 \mid 1, l; \infty)}{S_{n+1}\, Q(\delta_{n+1})} & \dfrac{S(n+1, \delta_{n+1}; C_1 \mid l+1, \delta_{l+1}; C_1)\, Q(\delta_{l+1})}{S_{n+1}\, Q(\delta_{n+1})} & \dfrac{M_1 S(n+1, \delta_{n+1}; C_1 \mid l+1, \delta_{l+1}; C_2)\, Q(\delta_{l+1})}{M_2 S_{n+1}\, Q(\delta_{n+1})} \\ \dfrac{M_2 S(n+1, \delta_{n+1}; C_2 \mid 2, l; \infty)}{S_{n+1}\, Q(\delta_{n+1})} & \dfrac{M_2 S(n+1, \delta_{n+1}; C_2 \mid 1, l; \infty)}{S_{n+1}\, Q(\delta_{n+1})} & \dfrac{M_2 S(n+1, \delta_{n+1}, C_2 \mid l+1, \delta_{l+1}; C_1)\, Q(\delta_{l+1})}{M_1 S_{n+1}\, Q(\delta_{n+1})} & \dfrac{S(n+1, \delta_{n+1}; C_2 \mid l+1, \delta_{l+1}; C_2)\, Q(\delta_{l+1})}{S_{n+1}\, Q(\delta_{n+1})} \end{bmatrix} \tag{192}$$

equations (189)–(191) express the equality,

$$\sum_{\beta, l} | S(\alpha, n; \beta, l)|^2 = 1. \tag{193}$$

This equation is no more than an expression of the conservation of energy in the scattering process.

It follows from entirely general considerations that any non-consumptive scattering process must conform to the *principle of reciprocity* which states that 'the probability for a transition (i.e. a process) proceeding one way is equal to the probability for the same transition (i.e. process) but with the sense of time reversed' (Blatt & Weisskopf 1952, p. 529). When the scattering process is described by a scattering matrix such as (192), the reciprocity theorem requires:

$$S(\alpha, n; \beta, l) = S(\beta, l; \alpha, n). \tag{194}$$

In particular, we must have

$$S(1, n; \infty \mid 2, l; \infty) = S(2, n; \infty \mid 1, l; \infty),$$

$$\frac{M_i S(n+1, \delta_{n+1}; C_i \mid 2, l; \infty)}{S_{n+1}\, Q(\delta_{n+1})} = \frac{1}{M_i} S(2, n; \infty \mid l+1, \delta_{l+1}; C_i)\, Q(\delta_{l+1}) \quad (i = 1, 2),$$

$$\frac{M_i S(n+1, \delta_{n+1}; C_i \mid 1, l; \infty)}{S_{n+1}\, Q(\delta_{n+1})} = \frac{1}{M_i} S(1, n; \infty \mid l+1, \delta_{l+1}; C_i)\, Q(\delta_{l+1}) \quad (i = 1, 2),$$

$$M_1 S(n+1, \delta_{n+1}; C_1 \mid l+1, \delta_{l+1}; C_2) = M_2 S \mid n+1, \delta_{n+1}; C_2 \mid l+1, \delta_{l+1}; C_1). \tag{195}$$

By virtue of the relation (193), it follows from equation (194) that

$$\sum_{\alpha, n} |S(\alpha, n \mid \beta, l)|^2 = 1; \tag{196}$$

in other words the *scattering matrix* $\boldsymbol{S}(\alpha, n \mid \beta, l)$ is *unitary*.

10. Concluding remarks

The scattering of radiation by the two extreme Reissner–Nordström black-holes provides a larger scope, than the scattering by an isolated Reissner–Nordström black-hole, for exploring the implications of the general theory of relativity for the reversible transformation of gravitational into electromagnetic energy. Even though the underlying problem has been analysed in the simplest context of axisymmetric axial perturbations, several novel features, both on the analytical and on the physical side, have emerged.

On the analytical side, the coupled pair of partial differential equations that govern the problem allow the definition of an energy flux whose divergence vanishes; and the divergence theorem provides a generalization of the constancy of the Wronskian of two independent (generally, the complex-conjugate) solutions of the one-dimensional Schrödinger equation that underlies the treatment of isolated black-holes. Also, besides the somewhat unconventional manner of separating the variables, in determining the asymptotic behaviours of the solutions at infinity (in the treatment of equations (69) and (83) in §5) and at the singularities (in the use of the lemma (equations (110) and (111) in the treatment of equations (108) and (109) in §7), the separations, themselves, are essential for the application of the divergence theorem towards the definition of the scattering matrix (equation (192)). Among the minor features, reference may be made to the generalization of the standard spheroidal equation, in equations (75) and (82), and the definition of the functions which play the same role in spheroidal coordinates as the familiar Gegenbauer polynomials, $C_n^{-\frac{1}{2}}(\mu)$ and $C_n^{-\frac{3}{2}}(\mu)$, in polar coordinates.

On the physical side, the most important feature is that gravitational and electromagnetic radiation in the vicinity of the horizons of the black hole prevails as coupled photon–graviton waves represented by a two-component vector (§8*a*). This last fact reduces the scattering problem to one, formally, in four channels rather than one in six channels as might have been expected. The principal result of the entire analysis is, of course, the expression of the conservation of energy in the scattering process in terms of the scattering amplitudes as in equation (193).

Finally, as to the directions in which the analysis of the present paper can be extended, the consideration of the polar perturbations is the first and obvious one. Dr Valeria Ferrari and I are at present investigating this extension. At a more concrete level, it will be useful to study in greater depth (perhaps, also numerically) the basic equations (44) and (45) and the generalized spheroidal functions defined by equations (75) and (86).

I am grateful to Dr Valeria Ferrari, Dr Robert Wald, and Dr Basilis Xanthopoulos for their careful scrutiny of the entire analysis and for many useful discussions; and to Dr Roland Winston for his clarification of the many issues that arise in the context of multi-channel scattering. The research reported in this paper has, in part, been supported by grants from the National Science Foundation under Grant PHY-84-16691 with the University of Chicago.

APPENDIX. A DISPLAY OF NULL GEODESICS IN THE MERIDIAN PLANE

There exist exhaustive studies of geodesics in the space-times of isolated black-holes (see, for example, *M.T.* Chapter 3 (§§19 and 20), Chapter 5 (§44), and Chapter 7 (§§61–64) for the classification, enumeration, and display of the null and the time-like geodesics in the space-times of the Schwarzschild, the Reissner–Nordström, and the Kerr black-holes). But so far, there exists no study of geodesics in the field of more than one central mass. On this account, a brief consideration of the geodesics in the space-time of two extreme Reissner–Nordström black-holes may be of some interest.

The equations governing geodesic motion in the space-time with the metric (10), can be derived from the Lagrangian,

$$2\mathscr{L} = \frac{(\dot{t})^2}{U^2} - U^2\{(\sinh^2\psi+\sin^2\theta)[(\dot{\psi})^2+(\dot{\theta})^2]+(\sinh^2\psi\sin^2\theta)(\dot{\varphi})^2\}, \qquad \text{(A 1)}$$

where dots denote differentiation with respect to a suitably chosen affine parameter. The constancy of the Lagrangian leads to the integral

$$\frac{(\dot{t})^2}{U^2} - U^2\{(\sinh^2\psi+\sin^2\theta)[(\dot{\psi})^2+(\dot{\theta})^2]+(\sinh^2\psi\sin^2\theta)(\dot{\varphi})^2\} = \delta_1, \qquad \text{(A 2)}$$

where

$$\left.\begin{aligned}\delta_1 &= 1 \quad \text{for time-like geodesics}\\ &= 0 \quad \text{for null geodesics.}\end{aligned}\right\} \qquad \text{(A 3)}$$

In addition, we have the integrals,

$$p_t = \frac{\dot{t}}{U^2} = E = \text{constant} \qquad \text{(A 4)}$$

and

$$p_\varphi = U^2(\sinh^2\psi\sin^2\theta)\dot{\varphi} = L_z = \text{constant}, \qquad \text{(A 5)}$$

which follow from the static axisymmetric character of the space-time. In equation (A 5), L_z denotes the resolved component of the angular momentum about the axis of symmetry. Eliminating $\dot{t}$ and $\dot{\varphi}$ from equation (A 2), by making use of equations (A 4) and (A 5), we have the more useful form of the energy integral:

$$(\dot{\psi})^2+(\dot{\theta})^2 = \frac{1}{(\sinh^2\psi+\sin^2\theta)}\left\{\left(E^2-\frac{\delta_1}{U^2}\right)-\frac{L_z^2}{U^4\sinh^2\psi\sin^2\theta}\right\}. \qquad \text{(A 6)}$$

Finally, the momenta p_ψ and p_θ, conjugate to ψ and θ, are given by

$$p_\psi = U^2(\sinh^2\psi+\sin^2\theta)\,\dot{\psi} \quad \text{and} \quad p_\theta = U^2(\sinh^2\psi+\sin^2\theta)\,\dot{\theta}. \qquad \text{(A 7)}$$

The equations governing geodesic motion follow from the remaining Lagrangian equations. Thus,

$$\begin{aligned}\dot{p}_\psi = -\frac{\partial\mathscr{L}}{\partial\psi} = \frac{(\dot{t})^2}{U^3}\frac{\partial U}{\partial\psi} &\\ +U\frac{\partial U}{\partial\psi}\{[(\dot{\psi})^2+(\dot{\theta})^2](\sinh^2\psi+\sin^2\theta)+(\dot{\varphi})^2\sinh^2\psi\sin^2\theta\} &\\ +U^2\{[(\dot{\psi})^2+(\dot{\theta})^2]\sinh\psi\cosh\psi+(\dot{\varphi})^2\sinh\psi\cosh\psi\sin^2\theta\}; &\qquad \text{(A 8)}\end{aligned}$$

or, eliminating $(\dot{\psi}^2+\dot{\theta}^2)$ and $\dot{\varphi}^2$ with the aid of equations (A 5) and (A 6), we have:

$$\dot{p}_\psi = -\delta_1\left(\frac{1}{U}\frac{\partial U}{\partial\psi}+\frac{\sinh\psi\cosh\psi}{\sinh^2\psi+\sin^2\theta}\right)+L_z^2\frac{\cosh\psi}{U^2(\sinh^2\psi+\sin^2\theta)\sinh^3\psi}$$
$$+E^2\left(\frac{\partial U^2}{\partial\psi}+U^2\frac{\sinh\psi\cosh\psi}{\sinh^2\psi+\sin^2\theta}\right). \quad \text{(A 9)}$$

Similarly, we find:

$$\dot{p}_\theta = -\delta_1\left(\frac{1}{U}\frac{\partial U}{\partial\theta}+\frac{\sin\theta\cos\theta}{\sinh^2\psi+\sin^2\theta}\right)+L_z^2\frac{\cos\theta}{U^2(\sinh^2\psi+\sin^2\theta)\sin^3\theta}$$
$$+E^2\left(\frac{\partial U^2}{\partial\theta}+U^2\frac{\sin\theta\cos\theta}{\sinh^2\psi+\sin^2\theta}\right). \quad \text{(A 10)}$$

Equation (A 9) and (A 10) must be combined with

$$\dot{p}_\psi = U^2(\sinh^2\psi+\sin^2\theta)\,\ddot{\psi}+[U^2(\sinh^2\psi+\sin^2\theta)]^{\cdot}\,\dot{\psi} \quad \text{(A 11)}$$

and

$$\dot{p}_\theta = U^2(\sinh^2\psi+\sin^2\theta)\,\ddot{\theta}+[U^2(\sinh^2\psi+\sin^2\theta)]^{\cdot}\,\dot{\theta}. \quad \text{(A 12)}$$

A useful relation which follows from the foregoing equations is

$$\dot{p}_\psi\dot{\theta}-\dot{p}_\theta\dot{\psi} = U^2(\sinh^2\psi+\sin^2\theta)\,(\ddot{\psi}\dot{\theta}-\ddot{\theta}\dot{\psi}). \quad \text{(A 13)}$$

One can derive from equations (A 9)–(A 13) the following equation which determines the orbit in the rotating (ψ,θ)-plane:

$$\frac{1}{1+(\mathrm{d}\psi/\mathrm{d}\theta)^2}\frac{\mathrm{d}^2\psi}{\mathrm{d}\theta^2} = \frac{1}{Q^2}\left\{\left[(\lg F)_{,\psi}+\frac{L_z^2\cosh\psi}{F^2(E^2-\delta_1/U^2)\sinh^3\psi}\right]\right.$$
$$\left.-\left[(\lg F)_{,\theta}+\frac{L_z^2\cos\theta}{F^2(E^2-\delta_1/U^2)\sin^3\theta}\right]\frac{\mathrm{d}\psi}{\mathrm{d}\theta}\right\}$$
$$+\frac{\delta_1}{Q^2(E^2U^2-\delta_1)}\left[(\lg U)_{,\psi}-(\lg U)_{,\theta}\frac{\mathrm{d}\psi}{\mathrm{d}\theta}\right], \quad \text{(A 14)}$$

where

$$Q^2 = 1-\frac{L_z^2}{E^2U^4\sinh^2\psi\sin^2\theta} \quad \text{and} \quad F = U^2\sqrt{(\sinh^2\psi+\sin^2\theta)}. \quad \text{(A 15)}$$

Since $\mathrm{d}\psi/\mathrm{d}\theta$ becomes infinite at several points along a given orbit, equation (A 14) is not useful for a direct numerical integration; the derivation of the equation has been omitted on this account. For numerical integration, equations (A 9)–(A 13) appear to be more useful. This is particularly the case for null geodesics $(\delta_1 = 0)$ in the meridian plane $(L_z = 0)$. For this case, equations (A 6), (A 9), (A 10), and (A 13) simplify considerably and we are left with

$$\dot{\psi}^2+\dot{\theta}^2 = \frac{E^2}{\sinh^2\psi+\sin^2\theta}, \quad \text{(A 16)}$$

$$\dot{p}_\psi = E^2\left(\frac{\partial U^2}{\partial\psi}+U^2\frac{\sinh\psi\cosh\psi}{\sinh^2\psi+\sin^2\theta}\right), \quad \text{(A 17)}$$

$$\dot{p}_\theta = E^2\left(\frac{\partial U^2}{\partial\theta}+U^2\frac{\sin\theta\cos\theta}{\sinh^2\psi+\sin^2\theta}\right), \quad \text{(A 18)}$$

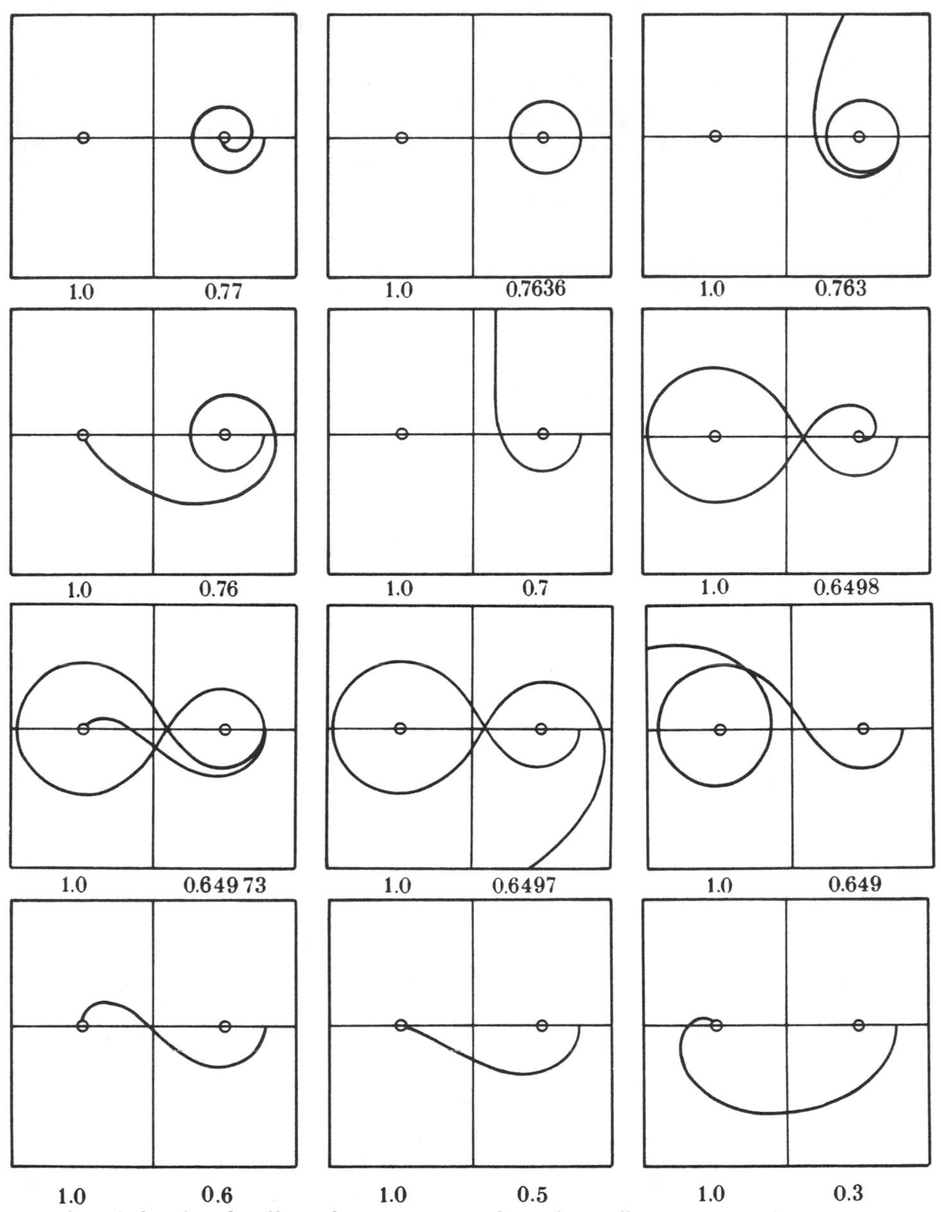

FIGURE A 1. A family of null geodesics in a meridian plane all starting at the same point $z = 1.5431$ ($\psi = 1, \theta = 0$) on the line joining the two mass points. The mass of the black hole to the left is 1 in all cases, while it varies along the sequence from 0.77 to 0.3. The mass of the black hole on the right is indicated at the bottom of each panel.

and

$$\dot{\theta}\ddot{\psi} - \dot{\psi}\ddot{\theta} = \frac{E^2}{(\sinh^2\psi + \sin^2\theta)^2}(\dot{\theta}\sinh\psi\cosh\psi - \dot{\psi}\sin\theta\cos\theta) + \frac{E^2}{U^2(\sinh^2\psi + \sin^2\theta)}\left(\dot{\theta}\frac{\partial U^2}{\partial\psi} - \dot{\psi}\frac{\partial U^2}{\partial\theta}\right). \quad \text{(A 19)}$$

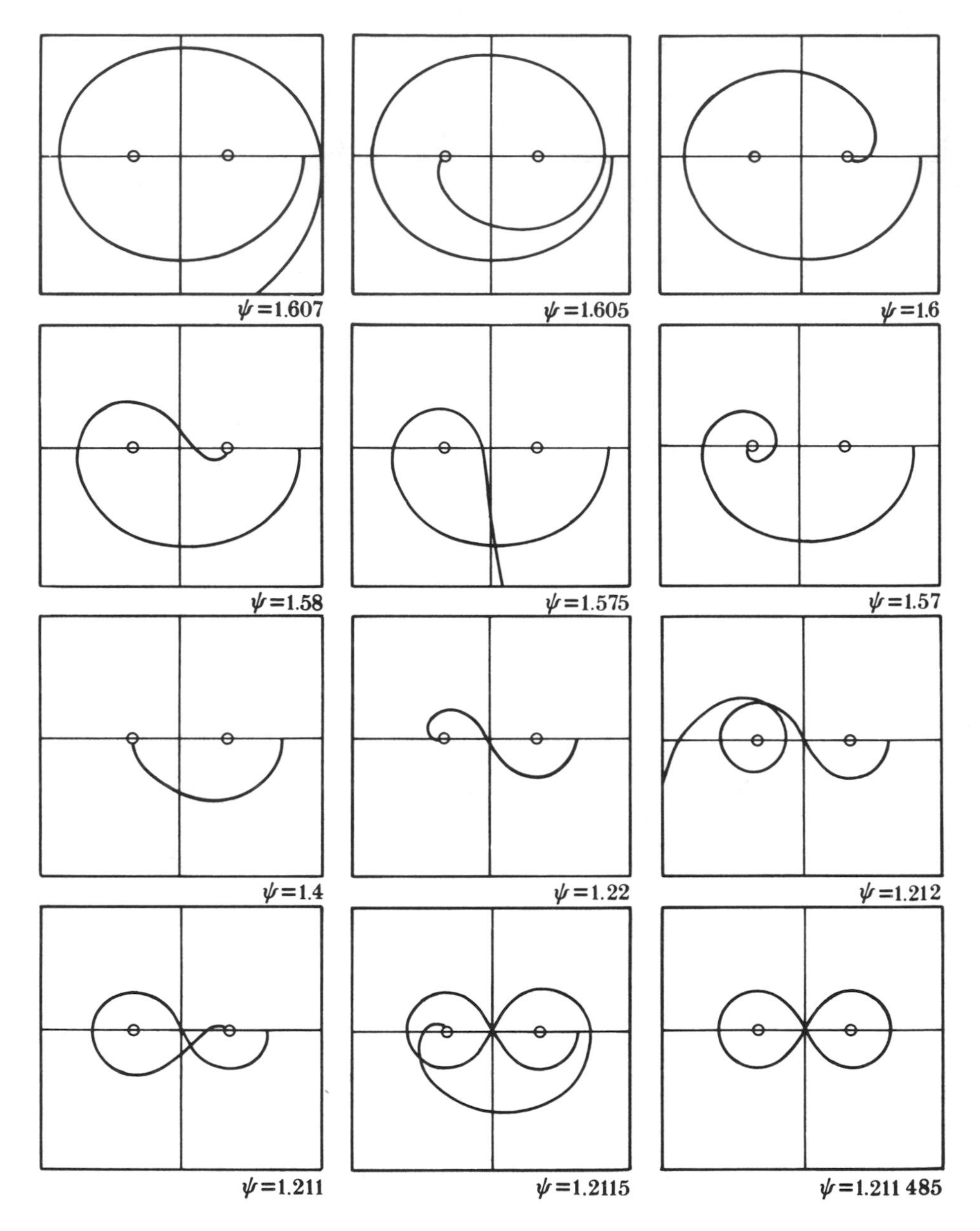

FIGURE A 2. A family of null geodesics on the meridian plane when the masses of the two black-holes are both equal to 1: the trajectories start on the z-axis at various distances varying from ($z = 2.5942$; $\psi = 1.607$, $\theta = 0$) to ($z = 1.8281$; $\psi = 1.211485$, $\theta = 0$). The values ψ at which the various trajectories start on the z-axis are indicated at the bottom of each panel.

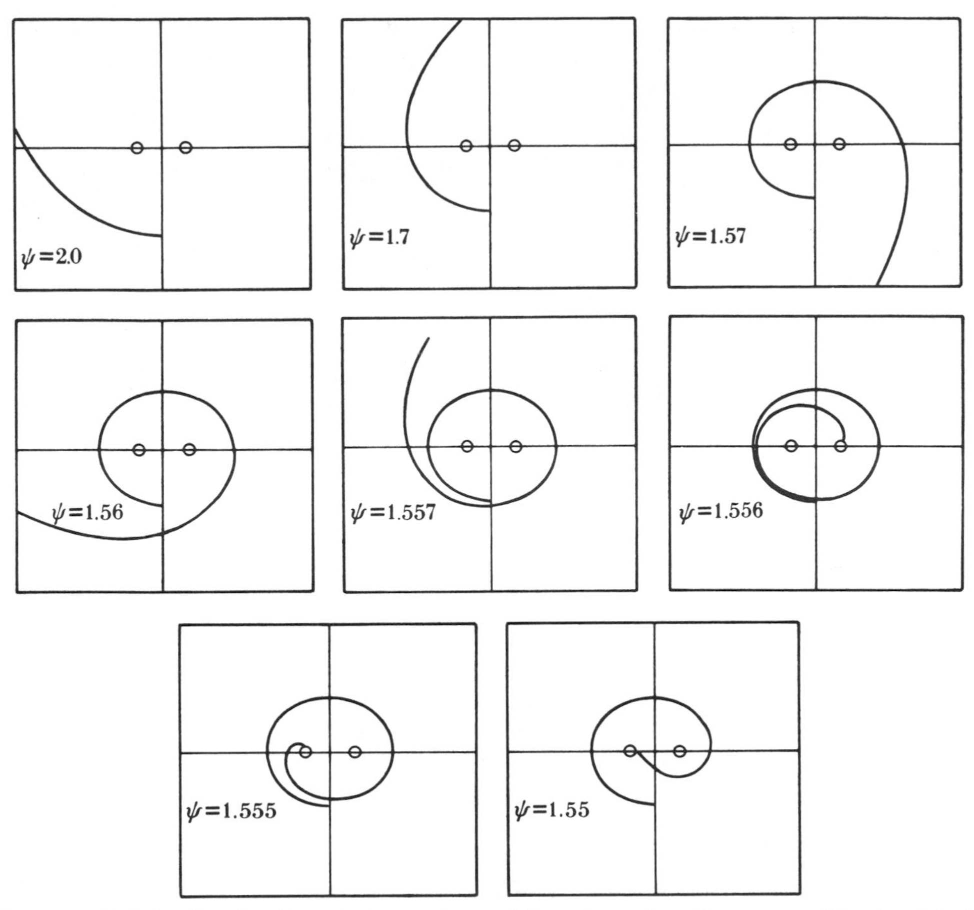

FIGURE A 3. A family of null geodesics on the meridian plane when the masses of the two black-holes are both equal to 1: the trajectories start on the x-axis (perpendicular to the z-axis) at various distances varying from ($x = 3.6269$, $\psi = 2.0$; $\theta = \pi/2$) to ($x = 2.1293$, $\psi = 1.5$; $\theta = \pi/2$). The values ψ at which the various trajectories start on the x-axis are indicated at the left of each panel.

We supplement these equations by the dot derivative of equation (A 16):

$$\dot{\psi}\ddot{\psi} + \dot{\theta}\ddot{\theta} = -\frac{E^2}{(\sinh^2\psi + \sin^2\theta)^2}(\dot{\psi}\sinh\psi\cosh\psi + \dot{\theta}\sin\theta\cos\theta). \qquad \text{(A 20)}$$

By suitably combining equations (A 19) and (A 20), we obtain the pair of equations:

$$\ddot{\psi} = \frac{1}{\sinh^2\psi + \sin^2\theta}[(\dot{\theta}^2 - \dot{\psi}^2)\sinh\psi\cosh\psi - 2\dot{\psi}\dot{\theta}\sin\theta\cos\theta] + \frac{1}{U^2}\left(\frac{\partial U^2}{\partial\psi}\dot{\theta} - \frac{\partial U^2}{\partial\theta}\dot{\psi}\right)\dot{\theta} \qquad \text{(A 21)}$$

and

$$\ddot{\theta} = \frac{1}{\sinh^2\psi + \sin^2\theta}[(\dot{\psi}^2 - \dot{\theta}^2)\sin\theta\cos\theta - 2\dot{\psi}\dot{\theta}\sinh\psi\cosh\psi] - \frac{1}{U^2}\left(\frac{\partial U^2}{\partial\psi}\dot{\theta} - \frac{\partial U^2}{\partial\theta}\dot{\psi}\right)\dot{\psi}. \qquad \text{(A 22)}$$

Mr C. Habisohn has directly integrated equations (A 7), (A 17) and (A 18) and generously provided the displays exhibited in figures A 1, A 2 and A 3.

It may be worth noting that the Hamilton–Jacobi equation,

$$2\frac{\partial S}{\partial \tau} = U^2\left(\frac{\partial S}{\partial t}\right)^2 - \frac{1}{U^2(\sinh^2\psi+\sin^2\theta)}\left[\left(\frac{\partial S}{\partial \psi}\right)^2+\left(\frac{\partial S}{\partial \theta}\right)^2\right] - \frac{1}{U^2\sinh^2\psi\sin^2\theta}\left(\frac{\partial S}{\partial \varphi}\right)^2, \tag{A 23}$$

with the substitution,

$$S = \tfrac{1}{2}\delta_1\tau - Et + L_z\varphi + S_1(\psi,\theta), \tag{A 24}$$

reduces to

$$\left(\frac{\partial S_1}{\partial \psi}\right)^2 + \left(\frac{\partial S_1}{\partial \theta}\right)^2 = \left(E^2U^4 - \delta_1 U^2 - \frac{L_z^2}{\sinh^2\psi\sin^2\theta}\right)(\sinh^2\psi+\sin^2\theta). \tag{A 25}$$

It does not appear that even for null geodesics ($\delta_1 = 0$) in the meridian plane ($L_z = 0$) the equation can be separated.

References

Blatt, J. M. & Weisskopf, V. F. 1952 *Theoretical nuclear physics*. New York: John Wiley & Sons.
Chandrasekhar, S. 1983 *The mathematical theory of black holes*. Oxford: Clarendon Press.
Fisher, E. 1937 *Phil. Mag.* **7** (24), 245–256.
Ince, E. L. 1927 *Ordinary differential equations*. London: Longmans, Green & Co.
Jacobi, C. G. J. 1866 *Vorlesungen uber Dynamik* (ed. A. Clebsch). Berlin: Druck & Verlag von George Reimer.
Hartle, J. B. & Hawking, S. W. 1972 *Communs math. Phys.* **26**, 87–101.
Majumdar, S. D. 1947 *Phys. Rev.* **72**, 390–398.
Papapetrou, A. 1947 *Proc. R. Irish. Acad.* **51**, 191–205.
Whittaker, E. T. 1937 *Analytical Dynamics*. Cambridge University Press.

32

A one-to-one correspondence between the static Einstein–Maxwell and stationary Einstein-vacuum space-times

By Subrahmanyan Chandrasekhar, F.R.S.
The University of Chicago, Chicago, Illinois 60637, *U.S.A.*

(*Received* 23 *December* 1988)

A one-to-one correspondence is established between the static solutions of the Einstein–Maxwell equations and the stationary solutions of the Einstein-vacuum equations, that enables one to directly write down a solution for the one from a known solution of the other, and conversely, by a simple transcription. The directness of the correspondence is achieved by writing the metric for static Einstein–Maxwell space-times in a coordinate system and a gauge adapted to the two-centre problem and the metric for stationary Einstein-vacuum space-times in a coordinate system and a gauge adapted to black holes with event horizons.

1. Introduction

A special case of the Majumdar–Papapetrou solution, representing the static placement of two extreme Reissner–Nordström black holes at a finite distance apart, is an axisymmetric solution of the Einstein–Maxwell equations. At the base of this solution is a three-dimensional Laplace's equation in prolate spheroidal coordinates (cf. Chandrasekhar 1989, equations (18) and (19); this paper will be referred to hereafter as Paper I). In exploring the relation of this three-dimensional Laplace's equation to the two-dimensional Laplace's equation, the Ernst equations, and the X- and Y-equations that govern static and stationary axisymmetric space-times (for an account of these matters see *The mathematical theory of black holes*, Chandrasekhar 1983, §§53 and 112, hereafter referred to as *M.T.*), a one-to-one correspondence was found which enables one to pass freely from a static axisymmetric solution of the Einstein–Maxwell equations to a stationary axisymmetric solution of the Einstein-vacuum equations, and *conversely*. I was later informed by Dr B. Xanthopoulos that a method of effecting a transformation of a *real* solution of the Einstein–Maxwell equations to a *real* stationary solution of the Einstein-vacuum equations had been discovered by Bonnor (1961; see also Perjés 1968 and Misra *et al.* 1973). But Bonnor's method is based on the assumption that 'it is possible to perform a complex transformation of the parameters of the solution' (Kramer *et al.* 1980, p. 333). In contrast, the method that will be described in this paper is a very direct one: as will be shown, to pass from a static axisymmetric solution of the Einstein–Maxwell equations to a stationary solution of the Einstein-vacuum equations (and conversely) is no more than a simple matter of transcription.

2. The equations governing static Einstein–Maxwell space-times in prolate spheroidal coordinates

We start with a metric of the form

$$ds^2 = e^{\psi+\nu}\left[\chi(dt)^2 - \frac{1}{\chi}(d\varphi)^2\right] - e^{\mu_2+\mu_3}\left[\frac{(dx^2)^2}{\sqrt{\Delta}} + (dx^3)^2\sqrt{\Delta}\right], \tag{1}$$

appropriate for static axisymmetric space-times, where

$$\psi, \mu_2, \mu_3, \chi = e^{-\psi+\nu} \quad \text{and} \quad \sqrt{\Delta} = e^{\mu_3-\mu_2} \tag{2}$$

are functions only of the spatial coordinates x^2 and x^3. We choose prolate spheroidal coordinates

$$x^2 = \eta\,(=\cosh\psi) \quad \text{and} \quad x^3 = \mu\,(=\cos\theta); \tag{3}$$

and compatible with this choice, we adopt the solution (Paper I, equations (14) and (15))

$$e^\beta = e^{\psi+\nu} = \sqrt{[(\eta^2-1)(1-\mu^2)]} \quad \text{and} \quad \Delta = (\eta^2-1)/(1-\mu^2), \tag{4}$$

of the field equation (*M.T.* equation (23), p. 565)

$$[e^{\mu_3-\mu_2}(e^\beta)_{,2}]_{,2} + [e^{\mu_2-\mu_3}(e^\beta)_{,3}]_{,3} = 0. \tag{5}$$

[Note that for the adopted solution

$$[e^{\mu_3-\mu_2}(e^\beta)_{,2}]_{,2} - [e^{\mu_2-\mu_3}(e^\beta)_{,3}]_{,3} = 2. \tag{6}$$]

In view of equation (4), the metric becomes

$$ds^2 = \left[\chi(dt)^2 - \frac{1}{\chi}(d\varphi)^2\right]\sqrt{[(\eta^2-1)(1-\mu^2)]} - e^{\mu_2+\mu_3}\left[(d\eta)^2\sqrt{\frac{1-\mu^2}{\eta^2-1}} + (d\mu)^2\sqrt{\frac{\eta^2-1}{1-\mu^2}}\right], \tag{7}$$

or, alternatively,

$$ds^2 = \left\{\chi(dt)^2 - \frac{1}{\chi}(d\varphi)^2 - e^{\mu_2+\mu_3}\left[\frac{(d\eta)^2}{\eta^2-1} + \frac{(d\mu)^2}{1-\mu^2}\right]\right\}\sqrt{[(\eta^2-1)(1-\mu^2)]}, \tag{8}$$

where, it may be noted that

$$e^{2\nu} = \chi\sqrt{[(\eta^2-1)(1-\mu^2)]}. \tag{9}$$

(a) *Maxwell's equations*

For static axisymmetric space-times, the metric function ω, defining the 'dragging' of the inertial frame in stationary rotating systems, is zero; and consistently with *M.T.* equations (2)–(6) in §109 (p. 564), there is no loss of generality in assuming that the only non-vanishing tetrad components of the Maxwell tensor are F_{02} and F_{03} derived from an electrostatic potential B in the manner:

$$e^{\psi+\mu_2}F_{03} = -B_{,2} \quad \text{and} \quad e^{\psi+\mu_3}F_{02} = +B_{,3}. \tag{10}$$

And B is governed by the equation

$$(e^{-\psi+\nu-\mu_2+\mu_3}B_{,2})_{,2}+(e^{-\psi+\nu+\mu_2-\mu_3}B_{,3})_{,3}=0; \tag{11}$$

or, by virtue of the definitions (2) and (4),

$$\left(B_{,2}\chi\sqrt{\frac{\eta^2-1}{1-\mu^2}}\right)_{,2}+\left(B_{,3}\chi\sqrt{\frac{1-\mu^2}{\eta^2-1}}\right)_{,3}=0. \tag{12}$$

Letting (as usual) $\quad \Delta=\eta^2-1, \quad \delta=1-\mu^2, \quad$ (13)

and $\quad \Psi=\sqrt{(\Delta\delta)}/\chi, \quad$ (14)

we can rewrite equation (11), alternatively, in the form

$$\left(\frac{\Delta}{\Psi}B_{,2}\right)_{,2}+\left(\frac{\delta}{\Psi}B_{,3}\right)_{,3}=0. \tag{15}$$

(b) The remaining field equations

The remaining field equations are provided by the appropriate specializations of *M.T.*, §109, equations (19), (24), and (25). They are:

$$[e^{\beta+\mu_3-\mu_2}(\psi-\nu)_{,2}]_{,2}+[e^{\beta+\mu_2-\mu_3}(\psi-\nu)_{,3}]_{,3}$$
$$=-2\,e^{-\psi+\nu}[e^{\mu_3-\mu_2}(B_{,2})^2+e^{\mu_2-\mu_3}(B_{,3})^2], \tag{16}$$

$$2\,e^{\mu_3-\mu_2}(\beta_{,2}\mu_{3,2}+\psi_{,2}\nu_{,2})-2\,e^{\mu_2-\mu_3}(\beta_{,3}\mu_{2,3}+\psi_{,3}\nu_{,3})$$
$$=e^{-\beta}\{[e^{\mu_3-\mu_2}(e^{\beta})_{,2}]_{,2}-[e^{\mu_2-\mu_3}(e^{\beta})_{,3}]_{,3}\}$$
$$+2\,e^{-2\psi}[e^{\mu_3-\mu_2}(B_{,2})^2-e^{\mu_2-\mu_3}(B_{,3})^2], \tag{17}$$

$$(\psi+\nu)_{,2,3}-(\psi+\nu)_{,2}\mu_{2,3}-(\psi+\nu)_{,3}\mu_{3,2}+\psi_{,2}\psi_{,3}+\nu_{,2}\nu_{,3}=-2\,e^{-2\psi}B_{,2}B_{,3}. \tag{18}$$

Substituing for χ, β, and $\mu_3-\mu_2$ from equations (2) and (4), we find that equation (16) becomes

$$[\Delta(\lg\chi)_{,2}]_{,2}+[\delta(\lg\chi)_{,3}]_{,3}=2\frac{\chi}{\sqrt{(\Delta\delta)}}[\Delta(B_{,2})^2+\delta(B_{,3})^2], \tag{19}$$

or, in terms of the variable Ψ,

$$\Psi[(\Delta\Psi_{,2})_{,2}+(\delta\Psi_{,3})_{,3}]=\Delta(\Psi_{,2})^2+\delta(\Psi_{,3})^2-2\Psi[\Delta(B_{,2})^2+\delta(B_{,3})^2]. \tag{20}$$

Equation (15) takes a similar form:

$$\Psi[(\Delta B_{,2})_{,2}+(\delta B_{,3})_{,3}]=\Delta B_{,2}\Psi_{,2}+\delta B_{,3}\Psi_{,3}. \tag{21}$$

After some further reductions (in which we make use of equation (6)), we find that equations (17) and (18) can be reduced to the forms,

$$\eta(\mu_2+\mu_3)_{,2}+\mu(\mu_2+\mu_3)_{,3}+\frac{3}{2}\left(\frac{\eta^2}{\Delta}-\frac{\mu^2}{\delta}\right)-2$$
$$=\frac{1}{2}\left(\Delta\frac{\chi_{,2}^2}{\chi^2}-\delta\frac{\chi_{,3}^2}{\chi^2}\right)+2\,e^{-2\psi}[\Delta(B_{,2})^2-\delta(B_{,3})^2] \tag{22}$$

and

$$-\frac{\mu}{\delta}(\mu_2+\mu_3)_{,2}+\frac{\eta}{\Delta}(\mu_2+\mu_3)_{,3}-\frac{\eta\mu}{\Delta\delta} = \frac{\chi_{,2}\chi_{,3}}{\chi^2}+4\,e^{-2\psi}B_{,2}B_{,3}. \tag{23}$$

3. The transformation to the conjugate variables and the reduction to the X- and Y-equations

A transformation of variables, defined as 'conjugation', has been described in *M.T.*, §52(*a*), p. 276 and §109 (*e*), p. 572. In the present context, it corresponds to introducing the variables, $\tilde{\chi}$, $\tilde{\Psi}$, and $\tilde{B}$ in place of χ, Ψ and B and related to them in the manner:

$$\tilde{\chi} = -1/\chi; \quad \tilde{\Psi} = \sqrt{(\Delta\delta)}/\tilde{\chi} = -\chi\sqrt{(\Delta\delta)} = e^{-2\nu}; \tag{24}$$

$$B_{,2} = +\tilde{\chi}\tilde{B}_{,3}\sqrt{\frac{\delta}{\Delta}} = -\frac{1}{\chi}\tilde{B}_{,3}\sqrt{\frac{\delta}{\Delta}} = +\frac{\delta}{\tilde{\Psi}}\tilde{B}_{,3}, \tag{25}$$

$$B_{,3} = -\tilde{\chi}\tilde{B}_{,2}\sqrt{\frac{\Delta}{\delta}} = +\frac{1}{\chi}\tilde{B}_{,2}\sqrt{\frac{\Delta}{\delta}} = -\frac{\Delta}{\tilde{\Psi}}\tilde{B}_{,2}. \tag{26}$$

Expressing $\tilde{B}_{,2}$ and $\tilde{B}_{,3}$ in terms of $B_{,2}$ and $B_{,3}$ and the tetrad components F_{02} and F_{03} of the Maxwell tensor, we find:

$$\tilde{B}_{,2} = +\chi B_{,3}\sqrt{\frac{\delta}{\Delta}} = +\chi\,e^{\mu_2-\mu_3}e^{\psi+\mu_3}F_{02} = e^{\nu+\mu_2}F_{02}, \tag{27}$$

and

$$\tilde{B}_{,3} = -\chi B_{,2}\sqrt{\frac{\Delta}{\delta}} = +\chi\,e^{\mu_3-\mu_2}e^{\psi+\mu_2}F_{03} = e^{\nu+\mu_3}F_{03}, \tag{28}$$

(It should be noted that $\tilde{B}_{,2}$ and $\tilde{B}_{,3}$ as given by these equations define the *tensor* components of the Maxwell tensor.)

There are a number of useful relations which follow from the foregoing definitions and which are needed in the further reductions. We enumerate them in the following lemma.

Lemma.

(i)
$$e^{-2\psi}B_{,2}B_{,3} = -e^{2\nu}\frac{\tilde{B}_{,2}\tilde{B}_{,3}}{\tilde{\Psi}^2}; \tag{29}$$

(ii)
$$e^{-2\psi}\Delta(B_{,2})^2 = +e^{2\nu}\delta\frac{\tilde{B}^2_{,3}}{\tilde{\Psi}^2}; \quad e^{-2\psi}\delta(B_{,3})^2 = +e^{2\nu}\Delta\frac{\tilde{B}^2_{,2}}{\tilde{\Psi}^2}; \tag{30}$$

(iii)
$$\frac{\chi_{,2}}{\chi} = \frac{\tilde{\Psi}_{,2}}{\tilde{\Psi}}-\frac{\eta}{\Delta}; \quad \frac{\chi_{,3}}{\chi} = \frac{\tilde{\Psi}_{,3}}{\tilde{\Psi}}+\frac{\mu}{\delta}; \tag{31}$$

(iv)
$$\eta\left[\lg\frac{\eta^2-\mu^2}{\sqrt{(\Delta\delta)}}\right]_{,2}+\mu\left[\lg\frac{\eta^2-\mu^2}{\sqrt{(\Delta\delta)}}\right]_{,3} = -\frac{1}{\Delta}+\frac{1}{\delta}, \tag{32}$$

(v)
$$-\frac{\mu}{\delta}\left[\lg\frac{\eta^2-\mu^2}{\sqrt{(\Delta\delta)}}\right]_{,2}+\frac{\eta}{\Delta}\left[\lg\frac{\eta^2-\mu^2}{\sqrt{(\Delta\delta)}}\right]_{,3} = 0. \tag{33}$$

Making use of the relations (i), (ii) and (iii) of the lemma, we find that equations (22) and (23) can be brought to the forms:

$$\eta[(\mu_2+\mu_3)+\lg\tilde{\Psi}]_{,2}+\mu[(\mu_2+\mu_3)+\lg\tilde{\Psi}]_{,3}+\frac{1}{\Delta}-\frac{1}{\delta}$$
$$=\frac{1}{2\tilde{\Psi}^2}\{\Delta(\tilde{\Psi}_{,2})^2-\delta(\tilde{\Psi}_{,3})^2-4\,e^{2\nu}[\Delta(\tilde{B}_{,2})^2-\delta(\tilde{B}_{,3})^2]\}, \quad (34)$$

$$-\frac{\mu}{\delta}[(\mu_2+\mu_3)+\lg\tilde{\Psi}]_{,2}+\frac{\eta}{\Delta}[(\mu_2+\mu_3)+\lg\tilde{\Psi}]_{,3}=\frac{1}{\tilde{\Psi}^2}(\tilde{\Psi}_{,2}\tilde{\Psi}_{,3}-4\,e^{2\nu}\tilde{B}_{,2}\tilde{B}_{,3}). \quad (35)$$

Now letting
$$f=\mu_2+\mu_3+2\nu-\lg\frac{\eta^2-\mu^2}{\sqrt{(\Delta\delta)}}, \quad (36)$$

we find that, with the aid of the relations (iv) and (v) of the lemma, equations (34) and (35) reduce to:

$$\eta f_{,2}+\mu f_{,3}=2\{\Delta[(\nu_{,2})^2-e^{-2\nu}(\tilde{B}_{,2})^2]-\delta[(\nu_{,3})^2-e^{-2\nu}(\tilde{B}_{,3})^2], \quad (37)$$

$$-\frac{\mu}{\delta}f_{,2}+\frac{\eta}{\Delta}f_{,3}=4(\nu_{,2}\nu_{,3}-e^{-2\nu}\tilde{B}_{,2}\tilde{B}_{,3}). \quad (38)$$

Similarly, we find that in place of equations (20) and (21) we now have the following equations in the conjugate variables:

$$(\Delta\nu_{,2})_{,2}+(\delta\nu_{,3})_{,3}=e^{-2\nu}[\Delta(\tilde{B}_{,2})^2+\delta(\tilde{B}_{,3})^2], \quad (39)$$

$$(\Delta\tilde{B}_{,2})_{,2}+(\delta\tilde{B}_{,3})_{,3}=2(\Delta\tilde{B}_{,2}\nu_{,2}+\delta\tilde{B}_{,3}\nu_{,3}). \quad (40)$$

Finally, we may note that by virtue of equation (9) and the definition (36) of f, we can rewrite the metric (8) in the form

$$ds^2=e^{2\nu}(dt)^2-e^{-2\nu}\left\{e^{f}(\eta^2-\mu^2)\left[\frac{(d\eta)^2}{\eta^2-1}+\frac{(d\mu)^2}{1-\mu^2}\right]+(\eta^2-1)(1-\mu^2)(d\varphi)^2\right\}. \quad (41)$$

(*a*) *The space-time of two extreme Reissner–Nordström black-holes*

We observe that equations (37) and (38) *trivially* allow the solution:

$$f=0, \quad \tilde{B}_{,2}=\pm e^{\nu}\nu_{,2}, \quad \text{and} \quad \tilde{B}_{,3}=\pm e^{\nu}\nu_{,3}, \quad (42)$$

or,
$$f=0 \quad \text{and} \quad \tilde{B}=\pm e^{\nu}. \quad (43)$$

For $\tilde{B}=\pm e^{\nu}$, both equations (39) and (40) reduce to the same equation,

$$(\Delta\nu_{,2})_{,2}+(\delta\nu_{,3})_{,3}=\Delta(\nu_{,2})^2+\delta(\nu_{,3})^2; \quad (44)$$

or, as can be readily verified, $e^{-\nu}$ is a solution of Laplace's equation (cf. Paper I, equations (18) and (19)),

$$[\Delta(e^{-\nu})_{,2}]_{,2}+[\delta(e^{-\nu})_{,3}]_{,3}=0, \quad (45)$$

which allows the solution

$$e^{-\nu}=1+\frac{M_1}{\eta-\mu}+\frac{M_2}{\eta+\mu}. \quad (46)$$

With $e^{-\nu}$ given by equation (46) we recover the metric,

$$ds^2 = e^{2\nu}(dt)^2 - e^{-2\nu}\left\{(\eta^2-\mu^2)\left[\frac{(d\eta)^2}{\eta^2-1}+\frac{(d\mu)^2}{1-\mu^2}\right]+(\eta^2-1)(1-\mu^2)(d\varphi)^2\right\}, \tag{47}$$

that describes the space-time of two extreme Reissner–Nordström black-holes at a finite distance apart on the axis.

(b) The reduction to the X- and Y-equations

In view of the special solution,

$$\tilde{B} = \pm e^{\nu}, \tag{48}$$

which equations (37)–(40) allow, the substitutions,

$$X = e^{\nu}+\tilde{B} \quad \text{and} \quad Y = e^{\nu}-\tilde{B}, \tag{49}$$

would appear to be apposite. By these substitutions, we find that equations (37) and (38) become:

$$\eta f_{,2}+\mu f_{,3} = \frac{8}{(X+Y)^2}(\Delta X_{,2}Y_{,2}-\delta X_{,3}Y_{,3}), \tag{50}$$

$$-\frac{\mu}{\delta}f_{,2}+\frac{\eta}{\Delta}f_{,3} = \frac{8}{(X+Y)^2}(X_{,2}Y_{,3}+X_{,3}Y_{,2}); \tag{51}$$

while by rewriting equations (39) and (40) in the alternative forms

$$[\Delta(e^{\nu})_{,2}]_{,2}+[\delta(e^{\nu})_{,3}]_{,3} = e^{-\nu}\{\Delta[(e^{\nu})_{,2}]^2+\delta[(e^{\nu})_{,3}]^2+\Delta(\tilde{B}_{,2})^2+\delta(\tilde{B}_{,3})^2\}, \tag{52}$$

$$(\Delta\tilde{B}_{,2})_{,2}+(\delta\tilde{B}_{,3})_{,3} = 2e^{-\nu}[\Delta(e^{\nu})_{,2}\tilde{B}_{,2}+\delta(e^{\nu})_{,3}\tilde{B}_{,3}], \tag{53}$$

we obtain by addition and subtraction the pair of equations:

$$\tfrac{1}{2}(X+Y)[(\Delta X_{,2})_{,2}+(\delta X_{,3})_{,3}] = \Delta(X_{,2})^2+\delta(X_{,3})^2, \tag{54}$$

$$\tfrac{1}{2}(X+Y)[(\Delta Y_{,2})_{,2}+(\delta Y_{,3})_{,3}] = \Delta(Y_{,2})^2+\delta(Y_{,3})^2. \tag{55}$$

Equations (54) and (55) are the same equations that were first encountered in the reduction of the equations governing stationary axisymmetric vacuum space-times (Chandrasekhar 1978, equations (41) and (42); see also *M.T.* §53, equations (57) and (58)). The implications of this reduction of the equations governing static Einstein–Maxwell and stationary Einstein-vacuum space-times to the same pair of X- and Y-equations are considered in the following section. Meantime, as a footnote, it is worth noticing that if, for example, $Y = 0$, equations (54) and (55) reduce to the single equation,

$$\tfrac{1}{2}X[(\Delta X_{,2})_{,2}+(\delta X_{,3})_{,3}] = \Delta(X_{,2})^2+\delta(X_{,3})^2; \tag{56}$$

and this equation, with the substitution $X = e^{\nu}$, becomes identical with equation (44). In other words, the Laplace's equation that underlies the solution representing the two extreme Reissner–Nordström black holes is none other than what the X- and Y-equations become when one or the other is zero!

4. The one-to-one correspondence

By writing the metric appropriate to stationary axisymmetric space-times in the form (*M.T.* equation (1), p. 273)

$$ds^2 = e^{2\nu}(dt)^2 - e^{2\psi}(d\varphi - \omega\, dt)^2 - e^{2\mu_2}(dx^2)^2 - e^{2\mu_3}(dx^3)^2. \tag{57}$$

with the choices,

$$e^\beta = e^{\psi+\nu} = \sqrt{[(\eta^2-1)(1-\mu^2)]} \quad \text{and} \quad e^{\mu_3-\mu_2} = \sqrt{(\eta^2-1)} \tag{58}$$

compatible with equation (15), where $\eta\ (= x^2)$ is the radial coordinate and $\mu\ (= x^3)$ is the cosine of a polar angle, we obtain,

$$ds^2 = \left[\chi(dt)^2 - \frac{1}{\chi}(d\varphi - \omega\, dt)^2\right]\sqrt{[(\eta^2-1)(1-\mu^2)]} - e^{\mu_2+\mu_3}\left[\frac{(d\eta)^2}{\eta^2-1} + \frac{(d\mu)^2}{1-\mu^2}\right]\sqrt{(\eta^2-1)}, \tag{59}$$

where

$$\chi = e^{-\psi+\nu}. \tag{60}$$

It has been shown (Chandrasekhar 1978, equations (40)–(42); see also *M.T.* §53, p. 280) that with the definitions

$$X = \chi + \omega \quad \text{and} \quad Y = \chi - \omega, \tag{61}$$

X and Y satisfy the same equations (54) and (55); and further that $\mu_2 + \mu_3$ is determined by the equations (*M.T.* p. 281, equations (64) and (65)):

$$\eta(\mu_2+\mu_3)_{,2} + \mu(\mu_2+\mu_3)_{,3} = \frac{2}{(X+Y)^2}(\Delta X_{,2} Y_{,2} - \delta X_{,3} Y_{,3}) - \frac{3}{2\Delta} + \frac{1}{2\delta}, \tag{62}$$

$$-\frac{\mu}{\delta}(\mu_2+\mu_3)_{,2} + \frac{\eta}{\Delta}(\mu_2+\mu_3)_{,3} = \frac{2}{(X+Y)^2}(X_{,2} Y_{,3} + X_{,3} Y_{,2}). \tag{63}$$

Now letting,

$$f = \mu_2 + \mu_3 - \lg\frac{\eta^2-\mu^2}{\Delta^{\frac{3}{4}}\delta^{\frac{1}{4}}}, \tag{64}$$

we find, with the aid of equations (32) and (33), that f is determined by the equations,

$$\eta f_{,2} + \mu f_{,3} = \frac{2}{(X+Y)^2}(\Delta X_{,2} Y_{,2} - \delta X_{,3} Y_{,3}), \tag{65}$$

$$-\frac{\mu}{\delta} f_{,2} + \frac{\eta}{\Delta} f_{,3} = \frac{2}{(X+Y)^2}(X_{,2} Y_{,3} + X_{,3} Y_{,2}). \tag{66}$$

A comparison with equations (50) and (51) now shows that

$$f_{\text{E.M.}} = 4 f_{\text{E}}, \tag{67}$$

where we have now distinguished the fs belonging to the Einstein–Maxwell and the Einstein-vacuum space-times by the subscripts 'E.M.' and 'E'. By virtue of the definitions (64), the relation (67) is equivalent to

$$e^{f_{\text{E.M.}}} = e^{4(\mu_2+\mu_3)_{\text{E}}}\Delta^3\delta/(\eta^2-\mu^2)^4. \tag{68}$$

We conclude that *the correspondence,*

$$e^{\nu} \rightleftarrows \chi, \tilde{B} \rightleftarrows \omega \quad \text{and} \quad e^{f} \rightleftarrows e^{4(\mu_2+\mu_3)}\Delta^3\delta/(\eta^2-\mu^2)^4, \tag{69}$$

enables one to pass freely from a metric of the form (41) *and an electrostatic potential* $\tilde{B}$, *appropriate as a solution of the static Einstein–Maxwell equations to a metric of the form* (59) *appropriate as a solution of the stationary Einstein-vacuum equations, and conversely.*

As an example we may ask for the stationary vacuum solution that corresponds to the static placement of two extreme Reissner–Nordström black holes. Since the latter solution is derived from

$$f = 0, \tilde{B} = \pm e^{\nu} \quad \text{and} \quad e^{-\nu} \quad \text{a solution of Laplace's equation,} \tag{70}$$

the corresponding stationary vacuum solution follows from

$$\chi = \pm\omega, \quad \chi^{-1} \quad \text{a solution of Laplace's equation,}$$

and

$$e^{\mu_2+\mu_3} = (\eta^2-\mu^2)/\Delta^{\frac{3}{4}}\delta^{\frac{1}{4}}. \tag{71}$$

The associated metric is given by

$$ds^2 = [\pm 2\, d\varphi\, dt - U(d\varphi)^2]\sqrt{[(\eta^2-1)(1-\mu^2)]} - \frac{\eta^2-\mu^2}{(\Delta\delta)^{\frac{1}{4}}}\left[\frac{(d\eta)^2}{\eta^2-1} + \frac{(d\mu)^2}{1-\mu^2}\right], \tag{72}$$

where we may choose for $U(=e^{-\nu})$ the same solution (46). It does not appear that the space-time described by the metric (72) has much physical significance. However, in the paper following this one (Chandrasekhar & Xanthopoulos 1989) it will be shown how the correspondence enunciated does enable us to derive a solution of physical interest.

I am grateful to Dr Basilis Xanthopoulos for useful discussions and a scrutiny of the entire analysis. The research reported in this paper has, in part, been supported by grants from the National Science Foundation under grant PHY-84-16691 with the University of Chicago.

References

Bonnor, W. B. 1961 *Z. Phys.* **161**, 439.
Chandrasekhar, S. 1978 *Proc. R. Soc. Lond.* A **358**, 405–420.
Chandrasekhar, S. 1983 *The mathematical theory of black holes.* Oxford: Clarendon Press.
Chandrasekhar, S. 1989 *Proc. R. Soc. Lond.* A **421**, 227–258.
Chandrasekhar, S. & Xanthopoulos, B. C. 1989 *Proc. R. Soc. Lond.* A **423**, 387–400. (Following paper.)
Kramer, D., Stephani, H., Herlt, E. & MacCallum, M. 1980 In *Exact solutions of Einstein's field equations*, p. 333. Cambridge University Press.
Misra, R. M., Pandey, D. B., Srivastava, D. C. & Tripathi, S. N. 1973 *Phys. Rev.* D **7**, 1587–1589.
Perjés, Z. 1968 *Nuovo Cim.* **55** B, 600–603.

33

Two black holes attached to strings

By Subrahmanyan Chandrasekhar[1], F.R.S.,
and B. C. Xanthopoulos[2]

[1] *The University of Chicago, Chicago, Illinois* 60637, *U.S.A.*
[2] *Department of Physics, University of Crete, and Research Centre of Crete, Iraklion, Greece*

(*Received* 23 *December* 1988)

An axisymmetric solution of the Einstein–Maxwell equations is found which represents the static placement of two charged black holes of equal mass (M) and opposite charge ($\pm Q$) with $|Q| > M$. The space-time, external to the event horizons of the two black holes, is asymptotically flat and entirely smooth except for the occurrence, on the axis, of a simple conical singularity with deficit. In other words, a 'string' stretches along the axis of symmetry and provides support for the black holes. In the extended space-time, interior to the horizons of the black holes, time-like curvature singularities, with two spatial dimensions, do occur. And, finally, the surface gravity that prevails on the horizons of the two black holes vanishes. This transgression of the two theorems, excluding the existence of multiple black holes except those of the extreme Reissner–Nordström type and requiring $|Q| \leqslant M$ for isolated black holes, is made possible by relaxing the strict requirements of smoothness to the extent of allowing conical singularities.

The solution obtained in this paper is, at the classical level, analogous to the equilibrium solution one has found, at the quantal level, for Dirac magnetic monopoles connected by strings.

1. Introduction

In this paper, we shall obtain a static axisymmetric solution of the Einstein–Maxwell equations that follows from the simplest solution of the X- and Y-equations (see §3(*b*) of the preceding paper, Chandrasekhar (1989*b*), referred to hereafter as Paper I; familiarity with the notations and definitions of this paper will be assumed). The solution represents the static placement, on the axis of symmetry, of two charged black holes of equal mass (M) and opposite charge ($\pm Q$) with $|Q| > M$. Besides, the two black holes are attached to 'strings'. A string for the purposes of this paper is a line in space-time along which a *conical singularity* (with no associated curvature singularity) occurs characterized by a *deficit* defined by the difference between 2π and the limiting ratio of the circumference to the proper radius of a small circle described normal to the line. For the solution that is obtained the string (in general) stretches along the entire axis with the deficit along the part of the axis joining the black holes being less than that along this axis extending to $+\infty$ and $-\infty$ from the north pole of the one and the south pole of the other. Except for the occurrence of the simple conical singularity along the axis, the space-time external to the horizons of the two black holes is smooth and asymptotically flat. But interior to the horizons, in the extended space-time, three-dimensional time-like curvature singularities do occur.

It is a noteworthy fact that by allowing for the presence of strings, we can transgress both the rule excluding multiple black holes (except extreme Reissner–Nordström black holes as allowed by the Majumdar–Papapetrou solution (cf. Rubak 1988; see also Bunting & Masood-ul-Alam 1987)) and the rule excluding $|Q| > M$, for isolated charged black holes (to avoid naked singularities).

In this paper an 'event horizon' (or simply a 'horizon') is defined as a two-surface having the following properties: (1) it is a smooth null-surface spanned by a null vector which is also a Killing vector; (2) no time-like or null trajectory can cross the surface from the 'interior' to the 'exterior'; it acts as a one-way membrane; (3) the surface gravity on the null surface is a constant; (4) there are no naked singularities external to the horizon; and (5) the space-time is asymptotically flat. These properties are shared by the event horizons of black holes, which in addition satisfy a global condition of Hawking & Ellis (1973, p. 312). Our definition is thus somewhat restrictive; but is of greater applicability as for example to the horizons formed by colliding waves.

2. The derivation of the metric

First, we may recall that by the substitutions,

$$X = \frac{1+F}{1-F} \quad \text{and} \quad Y = \frac{1+G}{1-G}, \tag{1}$$

the X- and Y-equations become (see Chandrasekhar 1983, *The mathematical theory of black holes* (hereafter referred to as *M.T.*)):

$$(1-FG)\,[(\Delta F_{,2})_{,2}+(\delta F_{,3})_{,3}] = -2G[\Delta(F_{,2})^2+\delta(F_{,3})^2] \tag{2}$$

$$(1-FG)\,[(\Delta G_{,2})_{,2}+(\delta G_{,3})_{,3}] = -2F[\Delta(G_{,2})^2+\delta(G_{,3})^2]. \tag{3}$$

The simplest solution these equations allow is

$$F = -p\eta - q\mu \quad \text{and} \quad G = -p\eta + q\mu, \tag{4}$$

where p and q are two real positive constants subject to the condition

$$p^2-q^2 = 1. \tag{5}$$

The solution for the stationary axisymmetric space-time that follows from the solution (4) was found in Chandrasekhar (1978, §8, equations (103) and (106)); and the solutions for the functions χ, ω, and $\mu_2+\mu_3$ that occur in the metric written in the form of Paper I, equation (59), are:

$$\chi = -\frac{p^2\Delta+q^2\delta}{(1+p\eta)^2-q^2\mu^2}, \quad \omega = -\frac{2q\mu}{(1+p\eta)^2-q^2\mu^2} \tag{6}$$

and

$$e^{\mu_2+\mu_3} = \text{const.}\,\frac{p^2\Delta+q^2\delta}{(\eta^2-1)^{\frac{3}{4}}(1-\mu^2)^{\frac{1}{4}}}. \tag{7}$$

By the correspondence established in Paper I §4, equation (69), the solutions for e^{ν}, the electrostatic potential $\tilde{B}$, and the function f that occur in the description of Einstein–Maxwell space-time with the metric written in the form of Paper I, equation (41), are:

$$e^{\nu} = -\frac{p^2\Delta + q^2\delta}{(1+p\eta)^2 - q^2\mu^2}, \quad \tilde{B} = -\frac{2q\mu}{(1+p\eta)^2 - q^2\mu^2}, \quad e^{f} = \frac{(p^2\Delta + q^2\delta)^4}{\alpha^8(\eta^2-\mu^2)^4}, \tag{8}$$

where α is a real positive constant at our disposal. The explicit form of the metric that follows is

$$ds^2 = \frac{(p^2\Delta + q^2\delta)^2}{[(1+p\eta)^2 - q^2\mu^2]^2}(dt)^2 - \frac{[(1+p\eta)^2 - q^2\mu^2]^2}{(p^2\Delta + q^2\delta)^2} \times \left\{\frac{(p^2\Delta + q^2\delta)^4}{\alpha^8(\eta^2-\mu^2)^3}\left[\frac{(d\eta)^2}{\eta^2-1} + \frac{(d\mu)^2}{1-\mu^2}\right] + (\eta^2-1)(1-\mu^2)(d\varphi)^2\right\}. \tag{9}$$

For future reference, we may note that

$$\tilde{B}_{,2} = \frac{4pq\mu(1+p\eta)}{[(1+p\eta)^2 - q^2\mu^2]^2} \quad \text{and} \quad \tilde{B}_{,3} = -\frac{2q[(1+p\eta)^2 + q^2\mu^2]}{[(1+p\eta)^2 - q^2\mu^2]^2} \tag{10}$$

and that the non-vanishing *tetrad* components of the Maxwell tensor, F_{0j} $(j = 2, 3)$, are given by (cf. I, equations (27) and (28))

$$F_{0j} = e^{-\nu-\mu_j}\tilde{B}_{,j} \quad (j = 2, 3;\ \text{tetrad components}), \tag{11}$$

while the corresponding *tensor* components are given by

$$F_{0j} = \tilde{B}_{,j} \quad (j = 2, 3;\ \text{tensor components}). \tag{12}$$

Finally, we may note that a metric Misra *et al.* (1973) wrote down, in an attempt to find a static solution of the Einstein–Maxwell equations by a complex transformation of the Kerr black-hole solution, can be obtained from equation (9) by replacing q by iq and $p^2 - q^2 = 1$ by $p^2 + q^2 = 1$. The space-time so obtained, while asymptotically flat, is unphysical in that it has naked singularities.

3. The space-time as representing two charged black holes with a conical singularity along the axis

There are two principal features of the space-time that require to be examined: the conical singularity along the axis of symmetry and the (coordinate) singularity at $\eta = 1$ and $\mu = \pm 1$. (It should be noted that *even though we shall continue to refer to the points,* $(\eta = 1, \mu = \pm 1)$, *as 'singularities', they are in fact only coordinate singularities,* as will be established in § (*c*) below.)

(*a*) *The conical singularity along the axis*

Normally, one requires, as in discussions relating to distorted black-holes (cf. Geroch & Hartle 1982, *M.T.*, §112*a*, p. 586, and Xanthopoulos 1983) that space-time be locally flat along the axis. This condition of local flatness requires that the limiting ratio of the circumference to the proper radius of a small circle, described orthogonal to the axis, is 2π. One has a conical singularity if this condition is not satisfied; and it is one of *deficit* (or *excess*) if the ratio in question is less than (or greater than) 2π. Current ideas concerning strings require that the conical singularity along their length is *always* one of deficit in accordance with its interpretation as the mass per unit length. (For a general review of the subject, see Vilenkin (1985).)

For the space-time described by the metric (9), the condition for local flatness must be examined, separately, for

$$|z| > 1 \quad \text{and} \quad |z| < 1, \quad \text{where} \quad z = \sqrt{[(\eta^2-1)(1-\mu^2)]}, \tag{13}$$

as required by the definition of the prolate-spheroidal coordinates we are presently using. For $|z| > |1$, the ratio to be considered is (cf. *M.T.*, p. 586, equation (172))

$$\underset{\mu^2\to 1;\,\theta\to 0 \text{ or } \pi}{\text{limit}} \left[\frac{\alpha^8(\eta^2-\mu^2)^3}{(p^2\Delta+q^2\delta)^4}\frac{(\eta^2-1)\sin^2\theta}{(\mathrm{d}\theta)^2}\right] = \frac{\alpha^8(\eta^2-1)^4}{p^8(\eta^2-1)^4} = \frac{\alpha^8}{p^8}, \tag{14}$$

where it will be noted that the numerator (apart from a factor) is the square of an infinitesimal arc along the circumference of an infinitesimal circle of angular radius $\mathrm{d}\theta$ described about the z-axis, at height z, and the denominator (apart from the same factor) is the square of the same arc determined by the square of the proper radius. Accordingly, the deficit (or excess) is given by

$$2\pi\,\delta_{|z|>1} = 2\pi(1-\alpha^4/p^4). \tag{15}$$

Similarly, for $|z| < 1$, the ratio to be considered is

$$\underset{\eta^2\to 1,\,\psi\to 0}{\text{limit}} \left[\frac{\alpha^8(\eta^2-\mu^2)^3}{(p^2\Delta+q^2\delta)^4}\frac{(1-\mu^2)\sinh^2\psi}{(\mathrm{d}\psi)^2}\right] = \frac{\alpha^8(1-\mu^2)^4}{q^8(1-\mu^2)^4} = \frac{\alpha^8}{q^8}; \tag{16}$$

and the corresponding deficit (or excess) is given by

$$2\pi\,\delta_{|z|<1} = 2\pi(1-\alpha^4/q^4). \tag{17}$$

If one stipulates, as a physical requirement, that the conical singularity be always one of deficit, then it will suffice to require that

$$\alpha \leqslant q, \tag{18}$$

because $p > q$. If $\alpha = q$, there is no string connecting the two (coordinate) singularities at $\mu = \pm 1$; but strings extend to $\pm\infty$ along $z > +1$ and $z < -1$.

(*b*) *The asymptotic behaviour at infinity and the mass at* ($\eta = 1, \mu = \pm 1$)

For $\eta \to \infty$, the metric (9) behaves like

$$\mathrm{d}s^2 = \left(1-\frac{4}{p\eta}\right)(\mathrm{d}t)^2 - \frac{p^8}{\alpha^8}[(\mathrm{d}\eta)^2+\eta^2(\mathrm{d}\theta)^2] - \eta^2(\sin^2\theta)(\mathrm{d}\varphi)^2. \tag{19}$$

From the behaviour of the coefficient of $(\mathrm{d}t)^2$, it follows that the combined mass present at the two singularities is $2/p$. Since the space-time is symmetric about the equatorial plane ($\theta = \frac{1}{2}\pi$ and $\mu = 0$), we conclude that the mass, M, at each of the two singularities is given by

$$M = 1/p. \tag{20}$$

We also observe that the space-time at infinity is characterized by the same conical singularity that prevails for $|z| > 1$ and that otherwise, the space-time is asymptotically flat.

(*c*) *The smoothness of the manifold as* $\eta \to 1+0$ *and* $\mu \to 1-0$

To ascertain the behaviour of the metric as the (coordinate) singularity at $\eta = 1+0$ and $\mu = 1-0$ is approached – entirely similar considerations will apply to the behaviour as the other singularity at $\eta = 1+0$ and $\mu = -1+0$ is approached – it is convenient to introduce the pair of variables (ξ, ζ) and (x, y), defined as in Chandrasekhar (1989*a*, equations (96) and (102); this paper will be referred to hereafter as Paper II):

$$(\eta = 1+\xi, \mu = 1-\zeta) \quad \text{and} \quad [x = \tfrac{1}{2}(\xi+\zeta), y = \tfrac{1}{2}(\xi-\zeta)]. \tag{21}$$

Besides x (equal to one-half of the radial coordinate-distance $\eta-\mu$) and y (equal to the coordinate distance measured perpendicular to the z-axis), we define the polar angle $\arccos \nu$ (valid only to the first order in ξ and ζ) by

$$y = x\nu, \quad \xi = x(1+\nu), \quad \text{and} \quad \zeta = x(1-\nu). \tag{22}$$

To investigate the behaviour of the metric (9) for $x \to 0$, we first rewrite it in the form:

$$\mathrm{d}s^2 = \frac{(p^2\Delta+q^2\delta)^2}{[(1+p\eta)^2-q^2\mu^2]^2}\left\{(\mathrm{d}t)^2 - \frac{[(1+p\eta)^2-q^2\mu^2]^4}{\alpha^8(\eta^2-\mu^2)^3}\left[\frac{(\mathrm{d}\eta)^2}{\eta^2-1}+\frac{(\mathrm{d}\mu)^2}{1-\mu^2}\right]\right\}$$
$$-\frac{[(1+p\eta)^2-q^2\mu^2]^2}{(p^2\Delta+q^2\delta)^2}(\eta^2-1)(1-\mu^2)(\mathrm{d}\varphi)^2, \tag{23}$$

and substitute the following directly verifiable relations, valid to the orders specified:

$$p^2\Delta+q^2\delta = 2x(p^2+q^2+\nu)+O(x^2), \tag{24}$$

$$(1+p\eta)^2-q^2\mu^2 = 2(1+p)[1+p\xi+(p-1)\zeta]+O(x^2), \tag{25}$$

$$\eta^2-\mu^2 = 4x[1+\tfrac{1}{2}(\xi-\zeta)]+O(x^3), \tag{26}$$

$$(\eta^2-1)(1-\mu^2) = 4x^2(1-\nu^2)[1+\tfrac{1}{2}(\xi-\zeta)]+O(x^4), \tag{27}$$

and

$$\frac{(\mathrm{d}\eta)^2}{\eta^2-1}+\frac{(\mathrm{d}\mu)^2}{1-\mu^2} \simeq \frac{1}{2\xi\zeta[1+\frac{1}{2}(\xi-\zeta)]}[\zeta(1-\tfrac{1}{2}\zeta)(\mathrm{d}\xi)^2+\xi(1+\tfrac{1}{2}\xi)(\mathrm{d}\zeta)^2]$$
$$= \frac{1}{x[1+\frac{1}{2}(\xi-\zeta)]}\left[(\mathrm{d}x)^2+x^2\frac{(\mathrm{d}\nu)^2}{1-\nu^2}-x^2\,\mathrm{d}x\,\mathrm{d}\nu+O(x^3)\right]. \tag{28}$$

We find

$$\mathrm{d}s^2 = \frac{(p^2+q^2+\nu)^2}{(1+p)^2}\left\{x^2(\mathrm{d}t)^2-\frac{(1+p)^4}{4\alpha^8x^2}[1+4(2p-1)x]\left[(\mathrm{d}x)^2+x^2\frac{(\mathrm{d}\nu)^2}{1-\nu^2}-x^2\,\mathrm{d}x\,\mathrm{d}\nu\right]\right\}$$
$$-4\frac{(1+p)^2}{(p^2+q^2+\nu)^2}(1-\nu^2)(\mathrm{d}\varphi)^2, \tag{29}$$

or, equivalently,

$$ds^2 = \frac{(p^2+q^2+\nu)^2}{(1+p)^2}\left\{x^2(dt)^2 - \frac{(1+p)^4}{4\alpha^8}\left[\frac{1}{x^2}+\frac{4(2p-1)}{x}\right](dx)^2\right\}$$
$$-\frac{(p^2+q^2+\nu)^2(1+p)^2}{4\alpha^8}\left[\frac{(d\nu)^2}{1-\nu^2} - dx\,d\nu\right] - \frac{4(1+p)^2}{(p^2+q^2+\nu)^2}(1-\nu^2)(d\varphi)^2. \quad (30)$$

The singularity in the coefficient of $(dx)^2$ in the metric (30) can be eliminated by the substitution,

$$t = \tau + \frac{(1+p)^2}{\alpha^4}\left[(2p-1)\lg x - \frac{1}{2x}\right]. \quad (31)$$

We thus obtain

$$ds^2 = \frac{(p^2+q^2+\nu)^2}{(1+p)^2}\left\{x^2(d\tau)^2 + \frac{(1+p)^2}{\alpha^4}dx\,d\tau + O(1)(dx)^2\right\}$$
$$-\frac{(1+p)^2(p^2+q^2+\nu)^2}{4\alpha^8}\left[\frac{(d\nu)^2}{1-\nu^2} - dx\,d\nu\right] - \frac{4(1+p)^2}{(p^2+q^2+\nu)^2}(1-\nu^2)(d\varphi)^2, \quad (32)$$

where the coefficient of $(dx)^2$, while not determined at this level of approximation, is of necessity $O(1)$; and this fact suffices to *establish the smoothness of the manifold at* $\eta = 1+0$ *and* $\mu = 1-0$ *and the coordinate nature of the singularity at this point* (*and, similarly, also at* $\eta = 1+0$ *and* $\mu = -1+0$).

(*d*) *The coordinate singularities at* $\eta = 1$ *and* $\mu = \pm 1$ *represent regular null surfaces*

By equation (32), the metric on the two-surface $x = \text{const.}$ for $x \to 0$ is given by

$$d\sigma^2 = \frac{(1+p)^2(p^2+q^2+\nu)^2}{4\alpha^8(1-\nu^2)}(d\nu)^2 + \frac{4(1+p)^2(1-\nu^2)}{(p^2+q^2+\nu)^2}(d\varphi)^2. \quad (33)$$

The determinant of this metric $(=(1+p)^4/\alpha^8)$ being a constant, the surface area, $4\pi S$, tends to the finite value

$$4\pi S = 4\pi\frac{(1+p)^2}{\alpha^4} \quad \text{as} \quad x \to 0. \quad (34)$$

As this surface, $x = 0$, is also null (cf. equation (32)), it follows that *the coordinate singularity at* $\eta = 1+0$ *and* $\mu = 1-0$ *represents the event horizon of a black hole*. Clearly, the same is true of the other coordinate singularity at $\eta = 1+0$ and $\mu = -1+0$.

It also follows from the metric (33) directly that conical singularities occur at both the north and the south poles of the horizon with the deficits;

$$2\pi\left[1 - \frac{4\alpha^4}{(p^2+q^2+\nu)^2}\right] \quad \text{for} \quad \nu = \pm 1, \quad (35)$$

or, $$2\pi(1-\alpha^4/p^4) \quad \text{for} \quad \nu = +1 \quad \text{and} \quad 2\pi(1-\alpha^4/q^4) \quad \text{for} \quad \nu = -1. \quad (36)$$

These deficits are the same as those found for $|z| > 1$ and $|z| < 1$ in § (a) (see equations (15) and (17)). We may therefore conclude that *the strings are attached to the horizons of the two black holes.*

(e) The absence of curvature singularities in the space external to the horizons

The curvature invariant $R^{ab}R_{ab}$ was evaluated with the aid of the symbolic manipulation language MACSYMA with the result,

$$R^{ab}R_{ab} = \frac{64\alpha^{16}q^4(\eta^2-\mu^2)^6 K^2}{(p^2\Delta+q^2\delta)^8[(1+p\eta)^2-q^2\mu^2]^8}, \tag{37}$$

where

$$K = -\delta^3 - 4p\eta[\delta^2+p^2(\mu^2+1)(\eta^2-\mu^2)] \\ -2p^2[\delta^2(\eta^2+\mu^2)+2(\eta^2-\mu^2)] - p^4[\mu^4\delta+\eta^2(\eta^2-\mu^2)+\eta^2\mu^2(3\Delta+2\delta)]. \tag{38}$$

In particular,

$$K \sim -16p^2(1+p)^2 x \quad \text{and} \quad R^{ab}R_{ab} \to \frac{2^{10}\alpha^{16}p^4(p-1)^2}{(1+p)^2(p^2+q^2+\nu)^8} \quad (x \to 0). \tag{39}$$

Since, for the allowed ranges of η and μ, namely $\eta > 1$ and $-1 \leqslant \mu \leqslant +1$,

$$p^2\Delta+q^2\delta > 0 \quad \text{and} \quad (1+p\eta)^2-q^2\mu^2 > 0 \quad (p^2 = q^2+1), \tag{40}$$

it follows from equations (37) and (39) that $R^{ab}R_{ab}$ is positive and bounded in the entire three-dimensional space, (η, μ, φ), including the coordinate singularities at $\eta = 1$ and $\mu = \pm 1$. *The absence of naked curvature singularities follows.*

(f) The space-time interior to the horizons

As the surface $x = 0$ is smooth, it is clear that we get an extension of the space-time into the interior of the horizons (as in the analogous considerations of Hartle & Hawking (1972) and *M.T.*, p. 595 in the context of the Majumdar–Papapetrou solution) by simply letting x to be negative or by equation (26), $\eta^2-\mu^2 < 0$. Accordingly, in the extended space-time, the allowed ranges of η and μ are

$$-1 < \eta \leqslant 1 \quad \text{and} \quad \mu \geqslant 1. \tag{41}$$

With these reductions, the metric of the space-time interior to the horizons is given by

$$ds^2 = \frac{(p^2\Delta+q^2\delta)^2}{[(1+p\eta)^2-q^2\mu^2]^2}(dt)^2 - \frac{[(1+p\eta)^2-q^2\mu^2]^2}{(p^2\Delta+q^2\delta)^2} \\ \times \left\{\frac{(p^2\Delta+q^2\delta)^4}{\alpha^8(\mu^2-\eta^2)^3}\left[\frac{(d\mu)^2}{\mu^2-1}+\frac{(d\eta)^2}{1-\eta^2}\right]+(\mu^2-1)(1-\eta^2)(d\varphi)^2\right\}. \tag{42}$$

The curvature scalar, $R^{ab}R_{ab}$, continues to be given by equations (37) and (38) for the presently allowed ranges (41) for η and μ. A curvature singularity is now possible for the admissible roots of

$$(1+p\eta)^2-q^2\mu^2 = 0, \tag{43}$$

or
$$\mu = \pm(1+p\eta)/q. \tag{44}$$
For the allowed range, $(-1, +1)$, of η, a curvature singularity can occur for μ in the range
$$\pm(p-1)/q = \pm q/(p+1) \quad \text{and} \quad \pm(1+p)/q. \tag{45}$$
Because, however, μ is required to be positive and greater than one, the admissible range of μ is
$$1 < \mu \leqslant (1+p)/q; \tag{46}$$
and the corresponding range of η is
$$-(1-q)/p < \eta \leqslant 1. \tag{47}$$
Therefore, *in the space-time interior to the horizons, time-like singularities of two spatial dimensions occur.*

(g) *The charges enclosed at $\eta = 1$ and $\mu = \pm 1$*

The net charge, Q_V, enclosed in a given volume V, is obtained by integrating over V, Maxwell's equation,
$$(F^{\alpha 0}\sqrt{-g})_{,\alpha} = J^0\sqrt{-g}, \tag{48}$$
where $F^{\alpha 0}$ denotes the *tensor* components of the Maxwell tensor and α refers to a spatial coordinate. In view of the axisymmetry of the space-time, it will suffice to consider volumes V which are also axisymmetric. Then, with the definitions,
$$E_\eta = F^{20}\sqrt{-g} \quad \text{and} \quad E_\mu = F^{30}\sqrt{-g}, \tag{49}$$
the integration over V gives
$$4\pi\, Q_V = \int_V \left(\frac{\partial E_\eta}{\partial \eta} + \frac{\partial E_\mu}{\partial \mu}\right) \mathrm{d}\eta\, \mathrm{d}\mu\, \mathrm{d}\varphi, \tag{50}$$
or, by Gauss's theorem,
$$2Q_V = \int_{\partial\mathscr{S}} (E_\eta\, \mathrm{d}\mu - E_\mu\, \mathrm{d}\eta), \tag{51}$$
where $\partial\mathscr{S}$ denotes the boundary of V in a meridianal plane.

By substituting for the tensor components F_{0j} from equation (12), we obtain,
$$F^{j0}\sqrt{-g} = -F_{0j}\,g^{00}g^{jj}\sqrt{-g} = -\tilde{B}_{,j}\,g^{00}g^{jj}\sqrt{-g} \quad (j = 2, 3;\ \text{no summation over } j), \tag{52}$$
or, explicitly,
$$E_\eta = F^{20}\sqrt{-g} = \frac{[(1+p\eta)^2 - q^2\mu^2]^2}{(p^2\Delta + q^2\delta)^2}(\eta^2 - 1)\,\tilde{B}_{,2} \tag{53}$$
and
$$E_\mu = F^{30}\sqrt{-g} = \frac{[(1+p\eta)^2 - q^2\mu^2]^2}{(p^2\Delta + q^2\delta)^2}(1 - \mu^2)\,\tilde{B}_{,3} \tag{54}$$

(i) *The total net charge*

The total net charge, Q, of the system can be obtained by evaluating the integral (51) over a bounding curve $\partial\mathscr{S}$ tending to infinity.

Now by substituting for $\tilde{B}_{,2}$ and $\tilde{B}_{,3}$ in equations (53) and (54) from equations (10), we obtain

$$E_\eta = 4pq\frac{\mu(1+p\eta)}{(p^2\Delta+q^2\delta)^2}(\eta^2-1) \tag{55}$$

and

$$E_\mu = -2q\frac{(1+p\eta)^2+q^2\mu^2}{(p^2\Delta+q^2\delta)^2}(1-\mu^2). \tag{56}$$

From these equations, it follows that

$$E_\eta \to 4q\mu/p^2\eta \quad \text{and} \quad E_\mu \to -2q(1-\mu^2)/p^2\eta^2 \quad (\eta\to\infty). \tag{57}$$

Accordingly,

$$2Q = \int_{\eta\to\infty}(E_\eta\,\mathrm{d}\mu - E_\mu\,\mathrm{d}\eta) = 0. \tag{58}$$

We conclude that *the charges at $\eta = 1$ and $\mu = \pm 1$ are of opposite signs.*

(ii) *The charge at $(\eta = 1, \mu = 1)$*

The charge Q_{+1} located at $(\eta = 1, \mu = 1)$ can be determined by evaluating the integral (50) over a spherical volume of radius $x\,(=\tfrac{1}{2}(\eta-\mu))$ and letting $x\to 0$. By a procedure entirely analogous to that adopted in the reduction of a similar integral in Paper II (§8, equations (161)–(163)), we find from equations (53) and (54) and making use of equations (22), (24), and (25) that

$$2Q_{+1} = x\int_{-1}^{+1}(E_\xi+E_\zeta)\,\mathrm{d}\nu, \tag{59}$$

where

$$E_\xi = 2\frac{(p+1)^2}{x(p^2+q^2+\nu)^2}(1+\nu)\,\tilde{B}_{,\xi} \tag{60}$$

and

$$E_\zeta = 2\frac{(p+1)^2}{x(p^2+q^2+\nu)^2}(1-\nu)\,\tilde{B}_{,\zeta}. \tag{61}$$

and $\tilde{B}_{,\xi}$ and $\tilde{B}_{,\zeta}$ are to be evaluated directly from a suitable expansion for $\tilde{B}$, given by equation (9), for $x\to 0$. With the aid of relations (21) and (25), we readily find that

$$\tilde{B} = -\frac{q}{1+p}(1-2px)+O(x^2). \tag{62}$$

Since, in addition (cf. Paper II, equations (106))

$$\tilde{B}_{,\xi} = \frac{1}{2}\left(\frac{\partial}{\partial x}+\frac{1-\nu}{x}\frac{\partial}{\partial\nu}\right)\tilde{B} \quad \text{and} \quad \tilde{B}_{,\zeta} = \frac{1}{2}\left(\frac{\partial}{\partial x}-\frac{1+\nu}{x}\frac{\partial}{\partial\nu}\right)\tilde{B}, \tag{63}$$

we find, by combining equations (60)–(63), that

$$E_\xi+E_\zeta = 2\frac{(p+1)^2}{x(p^2+q^2+\nu)^2}\tilde{B}_{,x} = \frac{4pq(p+1)}{x(p^2+q^2+\nu)^2}. \tag{64}$$

Inserting this expression for $E_\xi + E_\zeta$ in equation (59), we obtain

$$Q_{+1} = 2pq(p+1)\int_{-1}^{+1} \frac{d\nu}{(p^2+q^2+\nu)^2} = \frac{p+1}{pq}. \tag{65}$$

By a similar calculation, we find

$$Q_{-1} = -\frac{p+1}{pq} = -Q_{+1}, \tag{66}$$

consistent with the result (58).

(h) The vacuum space-time ($q = 0, p = 1$)

From the divergence of the expression (66) for the charges of the two black holes for $q \to 0$, it is clear that the vacuum space-time (which we obtain for $\tilde{B} = 0$, $q = 0$, and $p = 1$) must be considered *ab initio*,

Setting $q = 0$ and $p = \alpha = 1$ in equation (9), we obtain the metric,

$$ds^2 = \left(\frac{\eta-1}{\eta+1}\right)^2 (dt)^2 - \frac{(\eta+1)^6(\eta-1)^2}{(\eta^2-\mu^2)^3}\left[\frac{(d\eta)^2}{\eta^2-1} + \frac{(d\mu)^2}{1-\mu^2}\right] - \frac{(\eta+1)^3(1-\mu^2)}{\eta-1}(d\varphi)^2. \tag{67}$$

This metric is seen to be the same as the '$\delta = 2$ Voorhees–Zipoy solution' (cf. Kramer *et al.* 1980, p. 201); but it is not written in the form (67) and neither is a meaningful analysis of the space-time given.

For $\eta \to \infty$, the metric (67) tends to

$$ds^2 = (1-4/\eta)(dt)^2 - (d\eta)^2 - \eta^2(d\theta)^2 - \eta^2(\sin^2\theta)(d\varphi)^2. \tag{68}$$

The space-time is, therefore, asymptotically flat, appropriately, as is to be expected, for a mass, $M = 1$, at each of the two singularities at ($\eta = 1, \mu = \pm 1$); and this is consistent with equation (20) for $p = 1$.

By a procedure exactly analogous to that followed in §§(*c*) and (*d*), it can be shown that the metric of the two-surface, $x =$ const., (centred at $\eta = 1, \mu = 1$) for $x \to 0$, continues to be given by equation (35) for $p = \alpha = 1$ and $q = 0$, i.e. by

$$d\sigma^2 = \frac{(1+\nu)^2}{1-\nu^2}(d\nu)^2 + 16\frac{1-\nu^2}{(1+\nu)^2}(d\varphi)^2. \tag{69}$$

The 'point' ($\eta = 1, \mu = 1$) is, therefore, in reality a surface with the area,

$$4\pi S = 16\pi. \tag{70}$$

The coordinate singularities at ($\eta = 1, \mu = \pm 1$), therefore, represent horizons (*not* smooth, as we shall presently verify).

The conical singularity at $\nu = \pm 1$ is determined by (cf. equation (35))

$$1 - 4/(1+\nu)^2 \quad \text{for} \quad \nu = \pm 1. \tag{71}$$

The surface is locally flat at $\nu = +1$; but is characterized by a divergent *excess* at $\nu = -1$, a fact which portends a curvature singularity.

To determine whether curvature singularities occur in the space-time described by the metric (67), the completely contracted Riemann-tensor, $R^{abcd}R_{abcd}$, was evaluated, again, with the aid of the symbolic manipulation language MACSYMA with the result,

$$R^{abcd}R_{abcd} = -192\frac{(\eta^2-\mu^2)^5}{(\eta-1)^6(\eta+1)^{14}}(4\eta^2\mu^2-10\eta\mu^2+7\mu^2-\eta^4+4\eta^3-7\eta^2+6\eta-3). \tag{72}$$

It follows that the space-time exhibits a curvature singularity for $\eta = 1$, i.e. on the part of the z-axis joining $\mu = +1$ and $\mu = -1$. And, since moreover,

$$R^{abcd}R_{abcd} \to \frac{48}{(1+\nu)^6} \quad \text{for} \quad \eta = 1+0 \quad \text{and} \quad \mu = 1-0 \quad (\text{i.e.}, x \to 0), \tag{73}$$

the singularity extends to $\nu = -1$. In other words, on the 'horizons' of the two black holes, curvature singularities occur at the south pole of the one (at $\eta = 1$, $\mu = +1$) and the north pole of the other (at $\eta = 1, \mu = -1$).

We conclude that *the vacuum space-time* ($p = 1, q = 0$), exhibits naked singularities, and is, therefore, physically unacceptable.

4. Physical aspects of the space-time of the two black holes

The analysis of §3 has established that the static axisymmetric solution of the Einstein–Maxwell equations, derived from the simplest solution of the X- and Y-equations, represents two charged black holes located on the axis at $z = +1$ and $z = -1$, and a string (in general) stretched along the entire z-axis. The two black holes are of equal mass (M) and opposite charge ($\pm Q$) given by

$$M = 1/p \quad \text{and} \quad Q = \pm(p+1)/pq, \tag{74}$$

where p and q are two positive real constants related by

$$p^2-q^2 = 1 \quad (p > q > 0). \tag{75}$$

The conical singularity along the axis is characterized by the deficits (in units of 2π),

$$\delta_{|z|<1} = 1-\alpha^4/q^4 \quad \text{and} \quad \delta_{|z|>1} = 1-\alpha^4/p^4, \tag{76}$$

where, without loss of generality, we may suppose that α is a real positive constant and that

$$\alpha \leqslant q, \tag{77}$$

so that $\delta_{|z|<1}$ and $\delta_{|z|>1}$ are both deficits (not excesses). And, finally, the surface area, S, of the horizons of the two black holes (in units of 4π) is given by

$$S = (1+p)^2/\alpha^4. \tag{78}$$

Except for the conical singularity on the z-axis, the space, external to the horizons, is smooth. In the manifold extended into the interior of the horizons, time-like singularities with two spatial dimensions occur. But the space-time exhibits no naked singularities.

An important feature of these charged black holes is provided by the inequality,

$$|Q|-M = (1+p-q)/pq > 0. \tag{79}$$

They, therefore, transgress the rule, $|Q| \leqslant M$ for isolated black holes, to avoid the occurrence of naked singularities.

(*a*) *The surface gravity on the horizons*

An important feature of event horizons is the constant surface-gravity, κ, that prevails on them. Apart from a constant of proportionality, κ is determined by the equation

$$(A^i A_i)_{,j} = -2\kappa A_j, \tag{80}$$

where A^i is the Killing vector that becomes null on the horizon. (For static and stationary black holes A^i is defined as the Killing vector that is asymptotically a time-translation; then κ is determined uniquely.) For the case on hand, we shall carry out the evaluation of κ in the coordinate system, (τ, x, ν, φ), which is smooth on the horizon. From the metric (32), written in these coordinates, for $x \to 0$, it is clear that the Killing vector that becomes null on the horizon (i.e. $x = 0$) is

$$\partial_\tau = (A^i) = (1, 0, 0, 0). \tag{81}$$

The non-vanishing covariant components of (A^i) are, for $x \to 0$,

$$A_\tau = g_{\tau\tau} = \frac{(p^2+q^2+\nu)^2}{(1+p)^2} x^2 \quad \text{and} \quad A_x = g_{x\tau} = \frac{(p^2+q^2+\nu)^2}{2\alpha^4} \tag{82}$$

Besides,

$$A^i A_i = g_{\tau\tau} = \frac{(p^2+q^2+\nu)^2}{(1+p)^2} x^2 \quad (x \to 0). \tag{83}$$

Accordingly,

$$(A^i A_i)_{,x} = 2x \frac{(p^2+q^2+\nu)^2}{(1+p)^2} = -2\kappa A_x = -\kappa \frac{(p^2+q^2+\nu)^2}{\alpha^4} \quad (x \to 0). \tag{84}$$

Hence,

$$\kappa = 0. \tag{85}$$

The surface gravity on the horizons of the two black holes vanishes identically. This fact suggests that these two black holes are in some sense generically related to the extreme Reissner–Nordström black holes, on the horizons of which the surface gravity also vanishes (cf. Wald 1984, p. 331, equation 12.5.4). The relation is considered in §(*b*) below.

(*b*) *The relation to the extreme Reissner–Nordström black holes*

To exhibit the relationship of the black holes attached to strings with the extreme Reissner–Nordström black holes, we shall consider the special case,

$$\alpha = q, \tag{86}$$

when

$$\delta_{|z|<1} = 0, \tag{87}$$

and

$$\delta_{|z|>1} = 1 - \frac{q^4}{p^4} = \frac{1}{p}(p-q)\left(1+\frac{q}{p}\right)\left(1+\frac{q^2}{p^2}\right). \tag{88}$$

In this case, the condition for local flatness is met for $|z| < 1$, i.e. for the part of the z-axis joining the two black holes; but strings stretch to $\pm\infty$ from the north pole of the one and the south pole of the other.

From the relation (75), it follows that

$$p-q = 1/(p+q). \tag{89}$$

Therefore, $p \to q$ when both p and q tend to infinity. In this limit, equations (79) and (88) give,

$$\frac{1}{M}(|Q|-M) = \frac{1}{q}(1+p-q) \to \frac{1}{p} \tag{90}$$

and

$$\delta_{|z|>1} \to 2/p^2 \quad (p \to \infty). \tag{91}$$

Thus, in this limit, *with the evanescence of the strings, we recover the extreme Reissner–Nordström black holes.* It should, however, be noted that in this same limit, M, $|Q|$, and S also vanish and the space-time becomes Minkowskian. This is consistent with the fact that the two black holes are of opposite charges (unlike the extreme Reissner–Nordström black holes of the Majumdar–Papapetrou solution, all of which have charges of the same sign).

(c) *A comparison with the analogous equilibrium configurations of magnetic monopoles*

The multiple black-hole solutions that have been found so far are of two kinds: first, we have the Majumdar–Papapetrou solution which allows a static assemblage of extreme Reissner–Nordström black holes, with charges of the same sign, at arbitrary locations, in which (speaking in Newtonian terms) the gravitational attraction between any pair of them is exactly balanced by the Coulomb repulsion between them; secondly, we have the solution found in this paper of two black holes of equal mass (M) and opposite charge ($\pm Q$) with $|Q| > M$ with strings attached to them. A special case of this second class of solutions is when there is no string connecting the two black holes, but strings stretching to $+\infty$ and $-\infty$ from the north pole of the one and the south pole of the other.

The two classes of black-hole solutions that we have described are analogous to static configurations of magnetic monopoles that have been found. First, we have the Bogomil'nyi–Prasad–Sommerfield monopoles (with twice the charge of the Dirac monopole) of which we can contemplate a static assemblage (with charges of the same sign) in which the magnetic (Coulomb) repulsion is balanced by the attraction derived from a scalar field. Second, we have the possibility of two Dirac-monopoles, of opposite charge, held in place by a connecting string. These two classes of solutions are manifestly similar to the black-hole solutions we have described. It is a remarkable fact that possibilities, contemplated only in recent years at the quantal level, are inherent, already at the classical level, in general relativity.

We are grateful to Professor Y. Nambu for discussions relating to what is known about magnetic monopoles and, more particularly, for the remarks made in the preceding paragraph.

5. Concluding observations

The fact that two of the standard theorems of the subject, the non-existence of solutions describing assemblages of black holes except those of the extreme

Reissner–Nordström type and the rule $|Q| \leqslant M$ for isolated charged black holes – can be transgressed with a minimal violation of the smoothness requirements on the space-time manifold by allowing only simple conical singularities – suggests the fragility of the theorems of general relativity to the strict enforcement of smoothness conditions. Does it follow then that minimal and physically permissible departures from those conditions may enlarge the physical domain of general relativity? And, could one conclude from the very natural way in which strings emerge in the binary black-hole solution of this paper, that strings are indeed *predicted* by the general theory of relativity?

A final observation bearing on the attitude to physical problems of general relativity that we have maintained in our studies of the mathematical theory of black holes and of colliding waves: in the theory of space-times with two Killing fields – one time-like and one space-like or both space-like – the basic governing equations are one or more Laplace's equations, the Ernst equations, and the X- and Y-equations. The simplest solutions of the Laplace and the Ernst equations have provided almost all of the fundamental solutions describing black holes and colliding waves. (For a documentation of this statement see Chandrasekhar (1987), pp. 164–166.) But the simplest solution of the X- and Y-equations has so far been left out. At long last, it has found its place: it provides the first non-trivial binary black-hole solution with supporting strings!

We are grateful to Professor Y. Nambu for discussions relating to magnetic monopoles. The research reported in this paper has, in part, been supported by grants from the National Science Foundation under grant PHY-84-16691 with the University of Chicago. We are also grateful for a grant from the Division of Physical Sciences of the University of Chicago which has enabled our continued collaboration by making possible periodic visits by B. C. X. to the University of Chicago.

References

Bunting, G. L. & Masood-ul-Alam, A. K. M. 1987 *Gen. Rel. Grav.* **19**, 147–154.
Chandrasekhar, S. 1978 *Proc. R. Soc. Lond.* A **358**, 405–420.
Chandrasekhar, S. 1983 *The mathematical theory of black holes.* Oxford: Clarendon Press.
Chandrasekhar, S. 1987 *Truth and beauty: aesthetics and motivations in science.* University of Chicago Press.
Chandrasekhar, S. 1989*a* *Proc. R. Soc. Lond.* A **421**, 227–258. (Paper II.)
Chandrasekhar, S. 1989*b* *Proc. R. Soc. Lond.* A **423**, 379–386. (Paper I.)
Geroch, R. & Hartle, J. B. 1982 *J. math. Phys.* **23**, 680–692.
Hartle, J. B. & Hawking, S. W. 1972 *Commun. math. Phys.* **26**, 87–101.
Hawking, S. W. & Ellis, G. F. R. 1973 *The large-scale structure of space-time.* Cambridge University Press.
Kramer, D., Stephani, H., Herlt, E. & MacCallum, M. 1980 *Exact solutions of Einstein's field equations.* Cambridge University Press.
Misra, R. M., Pandey, D. B., Srivastava, D. C. & Tripathi, S. N. 1973 *Phys. Rev.* D **7**, 1587–1589.
Rubak, P. 1988 *Class. Quantum Grav.* **5**, L155–L159.
Vilenkin, A. 1985 *Phys. Rep.* **121**, 263–315.
Wald, R. 1984 *General relativity.* University of Chicago Press.
Xanthopoulos, B. C. 1983 *Proc. R. Soc. Lond.* A **388**, 117–131.

PART FIVE

Additional Papers

34

The flux integral for axisymmetric perturbations of static space-times

By Subrahmanyan Chandrasekhar[1], F.R.S., and V. Ferrari[2]

[1] *University of Chicago, Chicago, Illinois* 60637, *U.S.A.*

[2] *ICRA* (*International Centre for Relativistic Astrophysics*), *Dipartimento di Fisica 'G. Marconi', Universita di Roma, Rome, Italy*

(*Received* 21 *August* 1989)

The axisymmetric perturbations of static space-times with prevailing sources (a Maxwell field or a perfect fluid) are considered; and it is shown how a flux integral can be derived directly from the relevant linearized equations. The flux integral ensures the conservation of energy in the attendant scattering of radiation and the sometimes accompanying transformation of one kind of radiation into another. The flux integral derived for perturbed Einstein–Maxwell space-times will be particularly useful in this latter context (as in the scattering of radiation by two extreme Reissner–Nordström black-holes) and in the setting up of a scattering matrix. And the flux integral derived for a space-time with a perfect-fluid source will be directly applicable to the problem of the non-radial oscillations of a star with accompanying emission of gravitational radiation and enable its reformulation as a problem in scattering theory.

1. Introduction

It could perhaps be stated that the principal object in studying the perturbations of a static or a stationary space-time is to understand the attendant scattering of radiation and the sometimes accompanying transformation of one kind of radiation into another in terms of the unitarity and the time-reversibility of a suitably defined scattering matrix. The scattering matrix essentially expresses the conservation of energy in the scattering process that takes place.

In the case of the reflexion and absorption of radiation by the Schwarzschild, the Reissner–Nordström, and the Kerr black-holes, the understanding derives from the existence of one or more Wronskians whose constancy assures the sum of the reflexion and the absorption coefficients to be unity and thus the conservation of energy. In the examples of the Schwarzschild and the Reissner–Nordström black-holes, the constancy of the Wronskians expresses no more than the elementary fact that the Wronskian of a solution and its complex conjugate of a one-dimensional Schrödinger equation with a real potential, is a constant. But the deeper significance of this simple result is that it is a very special case of a far more general fact, that for perturbations of static or stationary space-times, there exists a three-dimensional vector, $\boldsymbol{E}$, whose divergence vanishes; and that, therefore, by Gauss's theorem, the existence of a flux integral follows (cf. equations (52) and (53) below). Thus, in the problem recently studied

(Chandrasekhar 1989) of the scattering and the accompanying transformation of gravitational radiation into electromagnetic radiation (and conversely) by the axisymmetric axial modes of perturbation of two extreme Reissner–Nordström black-holes, it was possible to set up a scattering matrix only by making use of the flux-integral (Chandrasekhar 1989, equation (56)) involving certain Wronskians. (In the Appendix, we provide a derivation of this flux integral, more generally than in the context of the particular problem considered.)

For treating the scattering of radiation by the same two Reissner–Nordström black-holes (or, more generally, by any static, axisymmetric Einstein–Maxwell space-time) by the alternate polar modes of perturbation, an analogous flux integral of the then applicable linearized equations is a necessary prerequisite. Since the number of coupled equations governing the polar perturbations is eight, in contrast to two for axial perturbations, the derivation of the necessary flux integral is neither simple nor straightforward: it requires, in particular, the writing of the linearized equations in a form that manifests the inner relationships among them. The existence and the exhibition of these relationships is one of the prime motivations of this paper.

With respect to the flux integral that equations, governing the first-order perturbations of static and stationary space-times of general relativity, allow, some of those whom we consulted (e.g. A. Ashtekar, J. Friedman, R. Sorkin and R. Wald) were aware that the existence of such an integral can indeed be inferred on general grounds (cf. J. Friedman 1978). More recently, Lee & Wald (1990) have established that the existence of such a conserved flux (or, a *simplectic current* in their terminology) can always be inferred for any field theory derived from a suitably defined Lagrangian action. Also, Burnett & Wald (1990) have shown how an explicit expression for this conserved simplectic-current can be derived for a perturbed space-time of general relativity. The expressions they have derived do indeed reduce to the ones derived in this paper when simplified for the contexts that are considered.

If one can thus obtain the flux integrals derived in this paper from a general theory, one may well question the usefulness of obtaining them *ab initio* from the equations applicable to the separate cases. Although the answer to this question depends on one's point of view, the following facts may be relevant. The reduction of Burnett & Wald's general expression to the forms derived in this paper (equations (50)–(52), (63)–(65), and (132)–(134)) requires the explicit use of what we have called the initial-value equations (equations (6) and (7) and their special forms). These equations, while they are not included in the general theory in a natural way, they are essential ingredients in the *ab initio* derivations. Besides, it will appear that the inclusion in the flux integral, of terms that derive from different sources (e.g. a Maxwell-field or a perfect fluid), is simple and straightforward, once the expression for the vacuum has been obtained in its simplest form. Apart from these factors, for us the overriding motivation derives from the insight one gains in the intimate relationships that exist among the equations themselves, a motivation that may not have a general appeal.

The plan of the paper is the following. In §2 the mathematical background of the problems to which the paper is addressed is described; and the basic equations are assembled in forms that we shall need. The paper then divides into two parts. In Part I, the equations are specialized appropriately for Einstein–Maxwell space-times; and the flux integrals for the vacuum and for the Einstein–Maxwell space-times are derived (equations (50)–(52) and (63)–(65)). In Part II, the case when the prevailing source is a perfect fluid is considered. The analysis in this part is directly applicable to the problem of non-radial oscillations of a static spherical star; and the flux integral that is derived (equations (132)–(134)) reduces the associated problem – the emission of gravitational radiation and the attendant damping of the oscillations – to one in scattering theory that can be described by a scattering matrix.

2. The formulation of the problem and the basic equations

We consider a static axisymmetric space-time. The metric of such a space-time can be written in the form,

$$ds^2 = e^{2\nu}(dt)^2 - e^{2\psi}(d\varphi)^2 - e^{2\mu_2}(dx^2)^2 - e^{2\mu_3}(dx^3)^2, \tag{1}$$

where ν, ψ, μ_2, and μ_3 are functions of the spatial coordinates, x^2 and x^3, only. (In accordance with the attitude expressed in Chandrasekhar (1988), in writing the metric in the form (1) we have not made use of the available gauge freedom to 'simplify' the form of the metric as is customary.)

We consider axisymmetric perturbations of the space-time described by the metric (1) with sources which we leave unspecified for the present. As has been explained in *The mathematical theory of black holes* (Chandrasekhar 1983, §§24 and 42; this book will be referred to hereafter as *M.T.*); and in Chandrasekhar (1989, §3) the axisymmetric perturbations of a static space-time fall into two non-combining groups: the *axial* and the *polar*. The metric of the axially perturbed space-time is of the form,

$$ds^2 = e^{2\nu}(dt)^2 - e^{2\psi}(d\varphi - \omega\, dt - q_2\, dx^2 - q_3\, dx^3)^2$$
$$- e^{2\mu_2}(dx^2)^2 - e^{2\mu_3}(dx^3)^2, \tag{2}$$

where ν, ψ, μ_2, and μ_3 retain the values they have in the static space-time, while ω, q_2, and q_3, describing the perturbation, are functions of x^2, x^3, and t. Such axial perturbations of an Einstein–Maxwell space-time, in the context of the two-centre problem, have been considered in Chandrasekhar (1989); and the flux integral appropriate to that problem was derived (Chandrasekhar 1989, equation (56)). The derivation of the flux integral, without the specialization to the two-centre problem, follows along the same lines and is given in the Appendix. In the main text of the paper, we shall be concerned only with the more difficult problem of the polar perturbations. Under such perturbations, the metric retains the same form (1) but the functions ν, ψ, μ_2, and μ_3 suffer infinitesimal increments, $\delta\nu$, $\delta\psi$, $\delta\mu_2$, and $\delta\mu_3$, respectively. We shall further suppose – no loss of generality is implied by this supposition – that these perturbations (and others) have a common time dependence given by

$$e^{i\sigma t}, \tag{3}$$

where σ is a constant. In view of the applications contemplated in Part II, σ is allowed to be complex though we shall normally restrict it to be real (unless stated otherwise).

The equations governing the perturbations, $\delta\nu$, $\delta\psi$, $\delta\mu_2$, and $\delta\mu_3$ can be readily written down by appropriately linearizing the equations listed in *M.T.* (pages 141 and 142, equations (4) and (5)). In writing down the various linearized equations, we shall adopt the following definitions:

$$\left.\begin{aligned}
\alpha_0 &= \psi+\mu_2+\mu_3-\nu;\\
\alpha_2 &= \psi+\nu+\mu_3-\mu_2, \quad \beta_2 = \psi+\mu_3+\nu;\\
\alpha_3 &= \psi+\nu+\mu_2-\mu_3, \quad \beta_3 = \psi+\mu_2+\nu;\\
A_2 &= \psi_{,2}\mu_{3,2}+\mu_{3,2}\nu_{,2}+\nu_{,2}\psi_{,2}, \quad X_2 = e^{\alpha_2}A_2;\\
A_3 &= \psi_{,3}\mu_{2,3}+\mu_{2,3}\nu_{,3}+\nu_{,3}\psi_{,3}, \quad X_3 = e^{\alpha_3}A_3;\\
\delta A_2 &= \psi_{,2}\,\delta(\mu_3+\nu)_{,2}+\mu_{3,2}\,\delta(\nu+\psi)_{,2}+\nu_{,2}\,\delta(\psi+\mu_3)_{,2};\\
\delta A_3 &= \psi_{,3}\,\delta(\mu_2+\nu)_{,3}+\mu_{2,3}\,\delta(\nu+\psi)_{,3}+\nu_{,3}\,\delta(\psi+\mu_2)_{,3};\\
\delta X_2 &= e^{\alpha_2}(A_2\,\delta\alpha_2+\delta A_2), \quad \delta X_3 = e^{\alpha_3}(A_3\,\delta\alpha_3+\delta A_3);\\
\delta f_2 &= \psi_{,2}\,\delta(\nu+\mu_3)+\mu_{3,2}\,\delta(\psi+\nu)+\nu_{,2}\,\delta(\psi+\mu_3);\\
\delta f_3 &= \psi_{,3}\,\delta(\nu+\mu_2)+\mu_{2,3}\,\delta(\psi+\nu)+\nu_{,3}\,\delta(\psi+\mu_2).
\end{aligned}\right\} \tag{4}$$

With these definitions, the equations governing the perturbations are:

$$\begin{aligned}
&[e^{\alpha_2}(\delta\psi_{,2}+\psi_{,2}\,\delta\alpha_2)]_{,2}+[e^{\alpha_3}(\delta\psi_{,3}+\psi_{,3}\,\delta\alpha_3)]_{,3}\\
&\quad+\sigma^2 e^{\alpha_0}\delta\psi = -[\delta R_{11}+\delta(\psi+\nu+\mu_2+\mu_3)R_{11}]\sqrt{-g};
\end{aligned} \tag{5i}$$

$$\begin{aligned}
&[e^{\alpha_2}(\delta\nu_{,2}+\nu_{,2}\,\delta\alpha_2)]_{,2}+[e^{\alpha_3}(\delta\nu_{,3}+\nu_{,3}\,\delta\alpha_3)]_{,3}\\
&\quad+\sigma^2 e^{\alpha_0}\delta(\psi+\mu_2+\mu_3) = +[\delta R_{00}+\delta(\psi+\nu+\mu_2+\mu_3)R_{00}]\sqrt{-g};
\end{aligned} \tag{5ii}$$

$$\begin{aligned}
&[e^{\alpha_3}(\delta\mu_{2,3}+\mu_{2,3}\,\delta\alpha_3)]_{,3}+[e^{\alpha_2}(\delta\beta_{2,2}+\beta_{2,2}\,\delta\alpha_2)]_{,2}-2\delta X_2\\
&\quad+\sigma^2 e^{\alpha_0}\delta\mu_2 = -[\delta R_{22}+\delta(\psi+\nu+\mu_2+\mu_3)R_{22}]\sqrt{-g};
\end{aligned} \tag{5iii}$$

$$\begin{aligned}
&[e^{\alpha_2}(\delta\mu_{3,2}+\mu_{3,2}\,\delta\alpha_2)]_{,2}+[e^{\alpha_3}(\delta\beta_{3,3}+\beta_{3,3}\,\delta\alpha_3)]_{,3}-2\delta X_3\\
&\quad+\sigma^2 e^{\alpha_0}\delta\mu_3 = -[\delta R_{33}+\delta(\psi+\nu+\mu_2+\mu_3)R_{33}]\sqrt{-g};
\end{aligned} \tag{5iv}$$

$$\begin{aligned}
&\{e^{\alpha_2}[\delta(\psi+\mu_3)_{,2}+(\psi+\mu_3)_{,2}\,\delta\alpha_2]\}_{,2}-\delta X_2\\
&\quad+\{e^{\alpha_3}[\delta(\psi+\mu_2)_{,3}+(\psi+\mu_2)_{,3}\,\delta\alpha_3]\}_{,3}-\delta X_3\\
&\qquad= -[\delta G_{00}+\delta(\psi+\nu+\mu_2+\mu_3)G_{00}]\sqrt{-g}.
\end{aligned} \tag{5v}$$

In addition, we have the *initial-value* equations,

$$\begin{aligned}
&e^{\alpha_2}[\delta(\psi+\mu_3)_{,2}+\psi_{,2}\,\delta(\psi-\mu_2)+\mu_{3,2}\,\delta(\mu_3-\mu_2)-\nu_{,2}\,\delta(\psi+\mu_3)]_{,0}\\
&\qquad= -2\,e^{\psi+2\nu+\mu_3}\delta R_{02},
\end{aligned} \tag{6}$$

$$\begin{aligned}
&e^{\alpha_3}[\delta(\psi+\mu_2)_{,3}+\psi_{,3}\,\delta(\psi-\mu_3)+\mu_{2,3}\,\delta(\mu_2-\mu_3)-\nu_{,3}\,\delta(\psi+\mu_2)]_{,0}\\
&\qquad= -2\,e^{\psi+2\nu+\mu_2}\delta R_{03}.
\end{aligned} \tag{7}$$

We call these the initial-value equations since, as we shall find in §§4 and 11, δR_{02} and δR_{03} are in turn time derivatives of functions describing the perturbations of the prevailing sources. The equations accordingly allow immediate integration with respect to time.

Alternative forms of equations (6) and (7) which we shall find useful are:

$$e^{\alpha_2}[\delta(\psi+\mu_3)_{,2}+(\psi+\mu_3)_{,2}\,\delta\alpha_2]_{,0}-(e^{\alpha_2}\delta f_2)_{,0} = -2\,e^{\psi+2\nu+\mu_3}\delta R_{02}, \tag{8}$$

$$e^{\alpha_3}[\delta(\psi+\mu_2)_{,3}+(\psi+\mu_2)_{,3}\,\delta\alpha_3]_{,0}-(e^{\alpha_3}\delta f_3)_{,0} = -2\,e^{\psi+2\nu+\mu_2}\delta R_{03}. \tag{9}$$

PART I. THE FLUX INTEGRAL FOR POLAR PERTURBATIONS OF STATIC EINSTEIN–VACUUM AND EINSTEIN–MAXWELL SPACE-TIMES

3. Linearized Maxwell's equations

In static Einstein–Maxwell space-times, the only non-vanishing components of the Maxwell tensor can be F_{02} and F_{03} related by the equation,

$$(e^{\nu+\mu_2}F_{02})_{,3}-(e^{\nu+\mu_3}F_{03})_{,2} = 0. \tag{10}$$

For the polar perturbations of the space-time, we are presently considering, besides the first-order changes, δF_{02} and δF_{03} in F_{02} and F_{03}, we must also allow for a first-order δF_{23} induced by the perturbations. The equations governing these perturbations follow directly from *M.T.*, p. 220, equations (118). Thus, letting

$$e^{\psi+\nu}\delta F_{23} = i\sigma Y, \tag{11}$$

in accordance with our general assumption (3), that all quantities describing the perturbation have the common time-dependent factor $e^{i\sigma t}$ (which is suppressed) we find:

$$\delta F_{03} = -F_{03}\,\delta(\psi+\mu_2)+e^{-\psi-\mu_2}Y_{,2}, \tag{12}$$

$$\delta F_{02} = -F_{02}\,\delta(\psi+\mu_3)-e^{-\psi-\mu_3}Y_{,3}, \tag{13}$$

and

$$\begin{aligned}&\{e^{\nu+\mu_3}[e^{-\psi-\mu_2}Y_{,2}+F_{03}\,\delta(\nu+\mu_3-\psi-\mu_2)]\}_{,2}\\ &+\{e^{\nu+\mu_2}[e^{-\psi-\mu_3}Y_{,3}-F_{02}\,\delta(\nu+\mu_2-\psi-\mu_3)]\}_{,3} = -\sigma^2\,e^{-\psi-\nu+\mu_2+\mu_3}Y.\end{aligned} \tag{14}$$

4. Linearized Einstein's equations

For the static Einstein–Maxwell space-time considered, the non-vanishing components of the Ricci tensor (by Einstein's equations) are:

$$R_{00} = R_{11} = F_{02}^2+F_{03}^2; \quad R_{22} = -R_{33} = F_{03}^2-F_{02}^2; \tag{15}$$

and for the polar perturbations of the space-time:

$$\left.\begin{aligned}\delta R_{00} &= +\delta R_{11} = 2F_{03}\,\delta F_{03}+2F_{02}\,\delta F_{02};\\ \delta R_{22} &= -\delta R_{33} = 2F_{03}\,\delta F_{03}-2F_{02}\,\delta F_{02};\end{aligned}\right\} \tag{16}$$

$$\delta R_{02} = +2F_{03}\,\delta F_{23} \quad \text{and} \quad \delta R_{03} = -2F_{02}\,\delta F_{23}. \tag{17}$$

The linearized Einstein's equations for the problem on hand now follow from equations (5i–v) by substituting from (16) for the terms on the right-hand sides of these equations. With the further definitions,

$$\left.\begin{aligned} \Phi_2 &= F_{02}^2\,\delta(\nu+\mu_2-\psi-\mu_3)\sqrt{-g}, \\ \Phi_3 &= F_{03}^2\,\delta(\nu+\mu_3-\psi-\mu_2)\sqrt{-g}, \\ \Psi_2 &= 2\,e^{\nu+\mu_2}F_{02}\,Y_{,3}, \quad \Psi_3 = 2\,e^{\nu+\mu_3}F_{03}\,Y_{,2}, \end{aligned}\right\} \tag{18}$$

the equations take the forms:

$$[e^{\alpha_2}(\delta\psi_{,2}+\psi_{,2}\,\delta\alpha_2)]_{,2}+[e^{\alpha_3}(\delta\psi_{,3}+\psi_{,3}\,\delta\alpha_3)]_{,3}$$
$$= -(\Phi_2+\Phi_3)+(\Psi_2-\Psi_3)-\sigma^2\,e^{\alpha_0}\delta\psi; \tag{19i}$$

$$[e^{\alpha_2}(\delta\nu_{,2}+\nu_{,2}\,\delta\alpha_2)]_{,2}+[e^{\alpha_3}(\delta\nu_{,3}+\nu_{,3}\,\delta\alpha_3)]_{,3}$$
$$= +(\Phi_2+\Phi_3)-(\Psi_2-\Psi_3)-\sigma^2\,e^{\alpha_0}\delta(\psi+\mu_2+\mu_3); \tag{19ii}$$

$$[e^{\alpha_3}(\delta\mu_{2,3}+\mu_{2,3}\,\delta\alpha_3)]_{,3}+[e^{\alpha_2}(\delta\beta_{2,2}+\beta_{2,2}\,\delta\alpha_2)]_{,2}-2\delta X_2$$
$$= +(\Phi_2-\Phi_3)-(\Psi_2+\Psi_3)-\sigma^2\,e^{\alpha_0}\delta\mu_2; \tag{19iii}$$

$$[e^{\alpha_2}(\delta\mu_{3,2}+\mu_{3,2}\,\delta\alpha_2)]_{,2}+[e^{\alpha_3}(\delta\beta_{3,3}+\beta_{3,3}\,\delta\alpha_3)]_{,3}-2\delta X_3$$
$$= -(\Phi_2-\Phi_3)+(\Psi_2+\Psi_3)-\sigma^2\,e^{\alpha_0}\delta\mu_3; \tag{19iv}$$

$$\{e^{\alpha_2}[\delta(\psi+\mu_3)_{,2}+(\psi+\mu_3)_{,2}\,\delta\alpha_2]\}_{,2}+\{e^{\alpha_3}[\delta(\psi+\mu_2)_{,3}+(\psi+\mu_2)_{,3}\,\delta\alpha_3]\}_{,3}$$
$$= (\delta X_2+\delta X_3)-(\Phi_2+\Phi_3)+(\Psi_2-\Psi_3). \tag{19v}$$

Since, in accordance with equations (11) and (17),

$$\delta R_{02} = 2i\sigma\,e^{-\psi-\nu}F_{03}\,Y, \quad \delta R_{03} = -2i\sigma\,e^{-\psi-\nu}F_{02}\,Y, \tag{20}$$

the initial-value equations (8) and (9) now give

$$e^{\alpha_2}[\delta(\psi+\mu_3)_{,2}+(\psi+\mu_3)_{,2}\,\delta\alpha_2]-e^{\alpha_2}\delta f_2 = -2\,e^{\nu+\mu_3}F_{03}\,Y, \tag{21}$$

$$e^{\alpha_3}[\delta(\psi+\mu_2)_{,3}+(\psi+\mu_2)_{,3}\,\delta\alpha_3]-e^{\alpha_3}\delta f_3 = +2\,e^{\nu+\mu_2}F_{02}\,Y. \tag{22}$$

From these equations, it follows that

$$\{e^{\alpha_2}[\delta(\psi+\mu_3)_{,2}+(\psi+\mu_3)_{,2}\,\delta\alpha_2]\}_{,2}-(e^{\alpha_2}\delta f_2)_{,2}$$
$$+\{e^{\alpha_3}[\delta(\psi+\mu_2)_{,3}+(\psi+\mu_2)_{,3}\,\delta\alpha_3]\}_{,3}-(e^{\alpha_3}\delta f_3)_{,3}$$
$$= -2(e^{\nu+\mu_3}F_{03}\,Y)_{,2}+2(e^{\nu+\mu_2}F_{02}\,Y)_{,3} = \Psi_2-\Psi_3. \tag{23}$$

Together with equation (19v), we obtain the useful identity:

$$(e^{\alpha_2}\delta f_2)_{,2}+(e^{\alpha_3}\delta f_3)_{,3} = \delta X_2+\delta X_3-(\Phi_2+\Phi_3). \tag{24}$$

Alternative forms of equations (19iii, iv)

Returning to equation (19iii) and remembering the definition of β_2 (in equations (4)) we can write,

$$[e^{\alpha_3}(\delta\mu_{2,3}+\mu_{2,3}\,\delta\alpha_3)]_{,3}+\{e^{\alpha_2}(\delta(\psi+\mu_3)_{,2}+(\psi+\mu_3)_{,2}\,\delta\alpha_2]\}_{,2}$$
$$+[e^{\alpha_2}(\delta\nu_{,2}+\nu_{,2}\,\delta\alpha_2)]_{,2} = 2\delta X_2+(\Phi_2-\Phi_3)-(\Psi_2+\Psi_3)-\sigma^2\,e^{\alpha_0}\delta\mu_2. \tag{25}$$

Replacing the second term, $\{\ \}_{,2}$, on the left-hand side of equation (25) by the term, $\{e^{\alpha_3}[2\to 3]\}_{,3}$, with the aid of equation (19v), we obtain:

$$[e^{\alpha_3}(\delta\mu_{2,3}+\mu_{2,3}\,\delta\alpha_3)]_{,3}-\{e^{\alpha_3}[\delta(\psi+\mu_2)_{,3}+(\psi+\mu_2)_{,3}\,\delta\alpha_3]\}_{,3}$$
$$+[e^{\alpha_2}(\delta\nu_{,2}+\nu_{,2}\,\delta\alpha_2)]_{,2}=\delta X_2-\delta X_3+2\Phi_2-2\Psi_2-\sigma^2\,e^{\alpha_0}\delta\mu_2. \tag{26}$$

On the other hand, by the initial-value equation (22),

$$-\{e^{\alpha_3}[\delta(\psi+\mu_2)_{,3}+(\psi+\mu_2)_{,3}\,\delta\alpha_3]\}_{,3}=-(e^{\alpha_3}\delta f_3+2\,e^{\nu+\mu_2}F_{02}\,Y)_{,3}$$
$$=-(e^{\alpha_3}\delta f_3)_{,3}-2(e^{\nu+\mu_2}F_{02})_{,3}\,Y-\Psi_2. \tag{27}$$

Inserting this last expression for $\{\ \}_{,3}$ in equation (26), we obtain the required form of equation (19iii):

$$[e^{\alpha_3}(\delta\mu_{2,3}+\mu_{2,3}\,\delta\alpha_3)]_{,3}-(e^{\alpha_3}\delta f_3)_{,3}+[e^{\alpha_2}(\delta\nu_{,2}+\nu_{,2}\,\delta\alpha_2)]_{,2}$$
$$=\delta X_2-\delta X_3+2\Phi_2-\Psi_2+2(e^{\nu+\mu_2}F_{02})_{,3}\,Y-\sigma^2\,e^{\alpha_0}\delta\mu_2. \tag{28}$$

By an analogous sequence of transformations, we obtain from equation (19iv):

$$[e^{\alpha_2}(\delta\mu_{3,2}+\mu_{3,2}\,\delta\alpha_2)]_{,2}-(e^{\alpha_2}\delta f_2)_{,2}+[e^{\alpha_3}(\delta\nu_{,3}+\nu_{,3}\,\delta\alpha_3)]_{,3}$$
$$=\delta X_3-\delta X_2+2\Phi_3+\Psi_3-2(e^{\nu+\mu_3}F_{03})_{,2}\,Y-\sigma^2\,e^{\alpha_0}\delta\mu_3. \tag{29}$$

5. The evaluation of certain Wronskians

In preparation for deriving the flux integral of the linearized Einstein's and Maxwell's equations, we shall obtain equations for the Wronskians,

$$[A,A^*]_i=A_{,i}A^*-A^*_{,i}A \quad (i=2,3) \tag{30}$$

of the metric functions, $\delta\psi$, $\delta\mu_2$, $\delta\mu_3$, and $\delta\nu$ describing the perturbations. The procedure, as we shall apply it to the different equations, will always be the same. Thus, considering equation (19i), we multiply the equation by $\delta\psi^*$ and subtract from it the complex conjugate of the equation multiplied by $\delta\psi$. By this procedure, we obtain

$$\{e^{\alpha_2}[\delta\psi,\delta\psi^*]_2\}_{,2}+\{e^{\alpha_3}[\delta\psi,\delta\psi^*]_3\}_{,3}$$
$$+\delta\psi^*[(e^{\alpha_2}\psi_{,2}\,\delta\alpha_2)_{,2}+(e^{\alpha_3}\psi_{,3}\,\delta\alpha_3)_{,3}]-\text{c.c.}=-\delta\psi^*[(\Phi_2+\Phi_3)-(\Psi_2-\Psi_3)]+\text{c.c.}, \tag{31}$$

where c.c. denotes the complex conjugate of the expression immediately preceding. An alternative form of equation (31) which we shall find useful is

$$\{e^{\alpha_2}[\delta\psi,\delta\psi^*]_2\}_{,2}+[e^{\alpha_2}\psi_{,2}\,\delta\psi^*\delta(\nu+\mu_3-\mu_2)-\text{c.c.}]_{,2}$$
$$+\{e^{\alpha_3}[\delta\psi,\delta\psi^*]_3\}_{,3}+[e^{\alpha_3}\psi_{,3}\,\delta\psi^*\delta(\nu+\mu_2-\mu_3)-\text{c.c.}]_{,3}$$
$$-(e^{\alpha_2}\psi_{,2}\,\delta\psi^*_{,2}\,\delta\alpha_2+e^{\alpha_3}\psi_{,3}\,\delta\psi^*_{,3}\,\delta\alpha_3)+\text{c.c.}$$
$$=-\delta\psi^*[(\Phi_2+\Phi_3)-(\Psi_2-\Psi_3)]+\text{c.c.} \tag{32}$$

In simplifying equations such as the foregoing, we shall make repeated use of the following elementary relations:

$$\left.\begin{aligned} A(A+B+C+\ldots)^*-\text{c.c.} &= A(B+C+\ldots)^*-\text{c.c.} \\ \text{and} \qquad AB^*-\text{c.c.} &= -A^*B+\text{c.c.} \end{aligned}\right\} \tag{33}$$

Next, by combining two of the terms in the first line of equation (29) with the aid of the relation,

$$\mu_{3,2}\,\delta\alpha_2-\delta f_2 = \mu_{3,2}\,\delta(\mu_3-\mu_2)-\psi_{,2}\,\delta(\nu+\mu_3)-\nu_{,2}\,\delta(\psi+\mu_3), \tag{34}$$

and applying to the resulting equation the same procedure that we have described in the context of equation (19i), we obtain:

$$\begin{aligned} &\{e^{\alpha_2}[\delta\mu_3,\delta\mu_3^*]_2\}_{,2}+\delta\mu_3^*\{e^{\alpha_2}[\mu_{3,2}\,\delta(\mu_3-\mu_2)-\psi_{,2}\,\delta(\nu+\mu_3)-\nu_{,2}\,\delta(\psi+\mu_3)]\}_{,2}-\text{c.c.} \\ &+\delta\mu_3^*[e^{\alpha_3}(\delta\nu_{,3}+\nu_{,3}\,\delta\alpha_3)]_{,3}-\text{c.c.} = \delta\mu_3^*[2\Phi_3+\Psi_3-2(e^{\nu+\mu_3}F_{03})_{,2}\,Y]-\text{c.c.} \\ &\qquad\qquad +\delta\mu_3^*(\delta X_3-\delta X_2)-\text{c.c.} \end{aligned} \tag{35}$$

We rewrite the second term in the first line of equation (35) in the manner:

$$\begin{aligned} &\{e^{\alpha_2}\delta\mu_3^*[\mu_{3,2}\,\delta(\mu_3-\mu_2)-\psi_{,2}\,\delta(\nu+\mu_3)-\nu_{,2}\,\delta(\psi+\mu_3)]=-\text{c.c.}\}_{,2} \\ &-e^{\alpha_2}\delta\mu_{3,2}^*[\mu_{3,2}\,\delta(\mu_3-\mu_2)-\psi_{,2}\,\delta(\nu+\mu_3)-\nu_{,2}\,\delta(\psi+\mu_3)]+\text{c.c.} \\ =&-\{e^{\alpha_2}\delta\mu_3^*(\mu_{3,2}\,\delta\mu_2+\psi_{,2}\,\delta\nu+\nu_{,2}\,\delta\psi)-\text{c.c.}\}_{,2} \\ &-e^{\alpha_2}\delta\mu_{3,2}^*[\mu_{3,2}\,\delta(\mu_3-\mu_2)-\psi_{,2}\,\delta(\nu+\mu_3)-\nu_{,2}\,\delta(\psi+\mu_3)]+\text{c.c.} \end{aligned} \tag{36}$$

We thus obtain:

$$\begin{aligned} &\{e^{\alpha_2}[\delta\mu_3,\delta\mu_3^*]_2\}_{,2}-[e^{\alpha_2}\delta\mu_3^*(\mu_{3,2}\,\delta\mu_2+\psi_{,2}\,\delta\nu+\nu_{,2}\,\delta\psi)-\text{c.c.}]_{,2} \\ &-e^{\alpha_2}\delta\mu_{3,2}^*[\mu_{3,2}\,\delta(\mu_3-\mu_2)-\psi_{,2}\,\delta(\nu+\mu_3)-\nu_{,2}\,\delta(\psi+\mu_3)]+\text{c.c.} \\ &+\delta\mu_3^*[e^{\alpha_3}(\delta\nu_{,3}+\nu_{,3}\,\delta\alpha_3)]_{,3}-\text{c.c.} \\ =&\ \delta\mu_3^*(\delta X_3-\delta X_2)+\delta\mu_3^*[2\Phi_3+\Psi_3-2(e^{\nu+\mu_3}F_{03})_{,2}\,Y]-\text{c.c.} \end{aligned} \tag{37}$$

By adding to equation (37) the equation which follows from it by the transformations,

$$\{{}_2\rightleftharpoons{}_3, \quad Y\to -Y, \quad \text{and} \quad \Psi\to-\Psi\}, \tag{38}$$

we obtain the equation,

$$\begin{aligned} &\{e^{\alpha_2}[\delta\mu_3,\delta\mu_3^*]_2\}_{,2}+\{e^{\alpha_3}[\delta\mu_2,\delta\mu_2^*]_3\}_{,3} \\ &+\delta\mu_3^*[e^{\alpha_3}(\delta\nu_{,3}+\nu_{,3}\,\delta\alpha_3)]_{,3}+\delta\mu_2^*[e^{\alpha_2}(\delta\nu_{,2}+\nu_{,2}\,\delta\alpha_2)]_{,2}-\text{c.c.} \\ &-[e^{\alpha_2}\delta\mu_3^*(\mu_{3,2}\,\delta\mu_2+\psi_{,2}\,\delta\nu+\nu_{,2}\,\delta\psi)-\text{c.c.}]_{,2} \\ &-[e^{\alpha_3}\delta\mu_2^*(\mu_{2,3}\,\delta\mu_3+\psi_{,3}\,\delta\nu+\nu_{,3}\,\delta\psi)-\text{c.c.}]_{,3} \\ &-e^{\alpha_2}\delta\mu_{3,2}^*[\mu_{3,2}\,\delta(\mu_3-\mu_2)-\psi_{,2}\,\delta(\nu+\mu_3)-\nu_{,2}\,\delta(\psi+\mu_3)]+\text{c.c.} \\ &-e^{\alpha_3}\delta\mu_{2,3}^*[\mu_{2,3}\,\delta(\mu_2-\mu_3)-\psi_{,3}\,\delta(\nu+\mu_2)-\nu_{,3}\,\delta(\psi+\mu_2)]+\text{c.c.} \\ =&\ \delta\mu_3^*[2\Phi_3+\Psi_3-2(e^{\nu+\mu_3}F_{03})_{,2}\,Y]+\delta\mu_2^*[2\Phi_2-\Psi_2+2(e^{\nu+\mu_2}F_{02})_{,3}\,Y]-\text{c.c.} \\ &+(\delta\mu_3-\delta\mu_2)^*(\delta X_2-\delta X_2)-\text{c.c.} \end{aligned} \tag{39}$$

We simplify equation (39) further by making use of the following identity which follows from applying our standard procedure to equation (19v):

$$\delta(\psi+\mu_2+\mu_3)^*\{[e^{\alpha_2}(\delta\nu_{,2}+\nu_{,2}\,\delta\alpha_2)]_{,2}+[e^{\alpha_3}(\delta\nu_{,3}+\nu_{,3}\,\delta\alpha_3)]_{,3}\}-\text{c.c.}$$
$$=\delta(\psi+\mu_2+\mu_3)^*[(\Phi_2+\Phi_3)-(\Psi_2-\Psi_3)]-\text{c.c.} \qquad (40)$$

Using this identity, we can rewrite the terms in the second line of equation (39), successively, in the manner:

$$\begin{aligned}
&\delta\mu_2^*[e^{\alpha_2}(\delta\nu_{,2}+\nu_{,2}\,\delta\alpha_2)]_{,2}+\delta\mu_3^*[e^{\alpha_3}(\delta\nu_{,3}+\nu_{,3}\,\delta\alpha_3)]_{,3}-\text{c.c.}\\
&=-\{\delta(\psi+\mu_3)^*[e^{\alpha_2}(\delta\nu_{,2}+\nu_{,2}\,\delta\alpha_2)]_{,2}+\delta(\psi+\mu_2)^*[e^{\alpha_3}(\delta\nu_{,3}+\nu_{,3}\,\delta\alpha_3)]_{,3}-\text{c.c.}\}\\
&\quad+\delta(\psi+\mu_2+\mu_3)^*[(\Phi_2+\Phi_3)-(\Psi_2-\Psi_3)]-\text{c.c.}\\
&=-\{e^{\alpha_2}\delta(\psi+\mu_3)^*[\delta\nu_{,2}+\nu_{,2}\,\delta(\nu-\mu_2)]-\text{c.c.}\}_{,2}\\
&\quad-\{e^{\alpha_3}\delta(\psi+\mu_2)^*[\delta\nu_{,3}+\nu_{,3}\,\delta(\nu-\mu_3)]-\text{c.c.}\}_{,3}\\
&\quad+\delta(\psi+\mu_3)^*_{,2}\,e^{\alpha_2}(\delta\nu_{,2}+\nu_{,2}\,\delta\alpha_2)+\delta(\psi+\mu_2)^*_{,3}\,e^{\alpha_3}(\delta\nu_{,3}+\nu_{,3}\,\delta\alpha_3)-\text{c.c.}\\
&\quad+\delta(\psi+\mu_2+\mu_3)^*[(\Phi_2+\Phi_3)-(\Psi_2-\Psi_3)]-\text{c.c.}
\end{aligned} \qquad (41)$$

Replacing now the terms in the second line of equation (39) by the second equality in equation (41) and adding the resulting equation to equation (32), we obtain our principal equation:

$$\begin{aligned}
&\underline{[\![e^{\alpha_2}\{[\delta\mu_3,\delta\mu_3^*]_2+[\delta\psi,\delta\psi^*]_3\}]\!]}_{,2}+\underline{[\![e^{\alpha_3}\{[\delta\mu_2,\delta\mu_2^*]_3+[\delta\psi,\delta\psi^*]_3\}]\!]}_{,3}\\
&-\{e^{\alpha_2}\delta(\psi+\mu_3)^*[\underline{\delta\nu_{,2}}+\nu_{,2}\,\delta(\nu-\mu_2)]_{,2}-\text{c.c.}\}_{,2}+\delta(\psi+\mu_3)^*_{,2}\,e^{\alpha_2}(\delta\nu_{,2}+\nu_{,2}\,\delta\alpha_2)-\text{c.c.}\\
&-\{e^{\alpha_3}\delta(\psi+\mu_2)^*[\underline{\delta\nu_{,3}}+\nu_{,3}\,\delta(\nu-\mu_3)]_{,3}-\text{c.c.}\}_{,3}+\delta(\psi+\mu_2)^*_{,3}\,e^{\alpha_3}(\delta\nu_{,3}+\nu_{,3}\,\delta\alpha_3)-\text{c.c.}\\
&+[e^{\alpha_2}\psi_{,2}\,\delta\psi^*\delta(\nu+\mu_3-\mu_2)-\text{c.c.}]_{,2}-e^{\alpha_2}\psi_{,2}\,\delta\psi^*_{,2}\,\delta\alpha_2+\text{c.c.}\\
&+[e^{\alpha_3}\psi_{,3}\,\delta\psi^*\delta(\nu+\mu_2-\mu_3)-\text{c.c.}]_{,3}-e^{\alpha_3}\psi_{,3}\,\delta\psi^*_{,3}\,\delta\alpha_3+\text{c.c.}\\
&-[e^{\alpha_2}\delta\mu_3^*(\mu_{3,2}\,\delta\mu_2+\psi_{,2}\,\delta\nu+\nu_{,2}\,\delta\psi)-\text{c.c.}]_{,2}\\
&-[e^{\alpha_3}\delta\mu_2^*(\mu_{2,3}\,\delta\mu_3+\psi_{,3}\,\delta\nu+\nu_{,3}\,\delta\psi)-\text{c.c.}]_{,3}\\
&-e^{\alpha_2}\delta\mu^*_{3,2}[\mu_{3,2}\,\delta(\mu_3-\mu_2)-\psi_{,2}\,\delta(\nu+\mu_3)-\nu_{,2}\,\delta(\psi+\mu_3)]+\text{c.c.}\\
&-e^{\alpha_3}\delta\mu^*_{2,3}[\mu_{2,3}\,\delta(\mu_2-\mu_3)-\psi_{,3}\,\delta(\nu+\mu_2)-\nu_{,3}\,\delta(\psi+\mu_2)]+\text{c.c.}\\
&+\delta(\mu_3-\mu_2)^*\{e^{\alpha_2}[A_2\,\delta(\psi+\nu)+\delta A_2]-e^{\alpha_3}[A_3\,\delta(\psi+\nu)+\delta A_3]\}-\text{c.c.}\\
&=-2\delta\psi^*[(\Phi_2+\Phi_3-(\Psi_2-\Psi_3)]+\text{c.c.}\\
&\quad+\delta\mu_3^*[-(\Phi_2-\Phi_3)+\Psi_2-2(e^{\nu+\mu_3}F_{03})_{,2}\,Y]-\text{c.c.}\\
&\quad+\delta\mu_2^*[+(\Phi_2-\Phi_3)-\Psi_3+2(e^{\nu+\mu_2}F_{02})_{,3}\,Y]-\text{c.c.}
\end{aligned} \qquad (42)$$

6. The flux integral for the vacuum

It is convenient at this stage to consider equation (42) for the vacuum when there will be no terms on the right-hand side of this equation. The terms in the resulting 'homogeneous' equation are of two kinds: the 'inside' terms that are enclosed in the brackets, $\{\ \}_{,i}$ and $[\ ]_{,i}$ $(i = 2, 3)$, and require to be differentiated and the 'outside' terms which are not so enclosed and do not require to be differentiated. Leaving aside the terms in the Wronskians and the terms underlined in the second and the third lines of equation (42) and picking out of the inside terms those which occur with the factor e^{α_2}, we have

$$[-\delta(\psi+\mu_3)^*\nu_{,2}\,\delta(\nu-\mu_2)+\delta\psi^*\psi_{,2}\,\delta(\nu+\mu_3-\mu_2) -\delta\mu_3^*(\mu_{3,2}\,\delta\mu_2+\psi_{,2}\,\delta\nu+\nu_{,2}\,\delta\psi)]-\text{c.c.}, \tag{43}$$

or, after some rearrangements,

$$=\{-\delta\mu_2[\psi_{,2}\,\delta\psi^*+\mu_{3,2}\,\delta\mu_3^*-\nu_{,2}\,\delta(\psi+\mu_3)^*] -(\psi+\nu)_{,2}\,\delta(\psi+\nu)\,\delta\mu_3^*-(\psi-\nu)_{,2}\,\delta\psi\,\delta\nu^*\}-\text{c.c.} \tag{44}$$

A similar reduction applies to the terms with the factor e^{α_3}. Thus the inside terms in equation (42) reduce to

$$\begin{aligned}&[\![e^{\alpha_2}\{[\delta\mu_3,\delta\mu_3^*]_2+[\delta\psi,\delta\psi^*]_2-[\delta\nu_{,2}\,\delta(\psi+\mu_3)^*-\text{c.c.}]\\ &\quad-\delta\mu_2[\psi_{,2}\,\delta\psi^*+\mu_{3,2}\,\delta\mu_3^*-\nu_{,2}\,\delta(\psi+\mu_3)^*]+\text{c.c.}\}]\!]_{,2}\\ &-\{e^{\alpha_2}[(\psi+\nu)_{,2}\,\delta(\psi+\nu)\,\delta\mu_3^*+(\psi-\nu)_{,2}\,\delta\psi\,\delta\nu^*]-\text{c.c.}\}_{,2}\\ &+\text{terms }(2\rightleftharpoons 3).\end{aligned} \tag{45}$$

We keep the terms in $[\![\ ]\!]_{,i}$ $(i = 2, 3)$ as they are, but expand by differentiation the terms $\{\ \}_{,i}$ $(i = 2, 3)$ by making use of the equations,

$$\left.\begin{aligned}&[e^{\alpha_2}(\psi-\nu)_{,2}]_{,2}+[e^{\alpha_3}(\psi-\nu)_{,3}]_{,3}=0,\\ &[e^{\alpha_2}(\psi+\nu)_{,2}]_{,2}=e^{\alpha_2}A_2-e^{\alpha_3}A_3,\\ &[e^{\alpha_3}(\psi+\nu)_{,3}]_{,3}=e^{\alpha_3}A_3-e^{\alpha_2}A_2,\end{aligned}\right\} \tag{46}$$

valid for the background space-time. We find:

$$\begin{aligned}&\{e^{\alpha_2}(\psi+\nu)_{,2}\,\delta(\psi+\nu)\,\delta\mu_3^*\}_{,2}+\{e^{\alpha_3}(\psi+\nu)_{,3}\,\delta(\psi+\nu)\,\delta\mu_2^*\}_{,3}\\ &=e^{\alpha_2}(\psi+\nu)_2[\delta(\psi+\nu)\,\delta\mu_3^*]_{,2}+e^{\alpha_3}(\psi+\nu)_{,3}[\delta(\psi+\nu)\,\delta\mu_2^*]_{,3}\\ &\quad+(e^{\alpha_2}A_2-e^{\alpha_3}A_3)\,\delta(\mu_3-\mu_2)^*\delta(\psi+\nu)\end{aligned} \tag{47}$$

and

$$\{e^{\alpha_2}(\psi-\nu)_{,2}\,\delta\psi\,\delta\nu^*\}_{,2}+\{e^{\alpha_3}(\psi-\nu)_{,3}\,\delta\psi\,\delta\nu^*\}_{,3} =e^{\alpha_2}(\psi-\nu)_{,2}(\delta\psi\,\delta\nu^*)_{,2}+e^{\alpha_3}(\psi-\nu)_{,3}(\delta\psi\,\delta\nu^*)_{,3}. \tag{48}$$

Including these terms along with the outside terms already present in equation

(42), we find, after some obvious cancellations, that the terms, with the factor e^{α_2} that we are left with, are:

$$\begin{aligned}
&-(\psi+\nu)_{,2}[\delta(\psi+\nu)_{,2}\,\delta\mu_3^*+\delta(\psi+\nu)\,\delta\mu_{3,2}^*]+\text{c.c.}\\
&-(\psi-\nu)_{,2}(\delta\psi\,\delta\nu_{,2}^*+\delta\psi_{,2}\,\delta\nu^*)+\text{c.c.}\\
&+\delta(\psi+\mu_3)_{,2}^*\,\delta\nu_{,2}+\delta(\psi+\mu_3)_{,2}^*\,\nu_{,2}\,\delta(\psi+\nu)-\text{c.c.}\\
&+\delta(\mu_3-\mu_2)^*[\psi_{,2}\,\delta(\mu_3+\nu)_{,2}+\mu_{3,2}\,\delta(\nu+\psi)_{,2}]-\text{c.c.}\\
&-\delta\mu_{3,2}^*[\mu_{3,2}\,\delta(\mu_3-\mu_2)-\psi_{,2}\,\delta(\nu+\mu_3)-\nu_{,2}\,\delta(\psi+\mu_3)]+\text{c.c.}\\
&-\psi_{,2}\,\delta\psi_{,2}^*\,\delta(\psi+\nu+\mu_3-\mu_2)+\text{c.c.}
\end{aligned}\tag{49}$$

On simplifying these terms, appealing several times to the initial-value equation (6), with the right hand side not equal to zero, we find that they vanish identically! Therefore, after all these reductions, we are left only with the terms included in $[\![\;]\!]_{,i}$ $(i=2,3)$ in equation (45); and the term with the factor $-\delta\mu_2$ in the second line of this equation, again by virtue of the initial-value equation, becomes,

$$+\delta\mu_2\,\delta(\psi+\mu_3)_{,2}^*-\text{c.c.}$$

Thus, with the definitions,

$$\begin{aligned}
E_2 = e^{\alpha_2}\{&[\delta\mu_3,\delta\mu_3^*]_2+[\delta\psi,\delta\psi^*]_2-[\delta\nu_{,2}\,\delta(\psi+\mu_3)^*-\text{c.c.}]\\
&+[\delta\mu_2\,\delta(\psi+\mu_3)_{,2}^*-\text{c.c.}\},
\end{aligned}\tag{50}$$

$$\begin{aligned}
E_3 = e^{\alpha_3}\{&[\delta\mu_2,\delta\mu_2^*]_3+[\delta\psi,\delta\psi^*]_3-[\delta\nu_{,3}\,\delta(\psi+\mu_2)^*-\text{c.c.}]\\
&+[\delta\mu_3\,\delta(\psi+\mu_2)_{,2}^*-\text{c.c.}]\}
\end{aligned}\tag{51}$$

we obtain the flux integral,

$$E_{2,2}+E_{3,3}=0.\tag{52}$$

We call this the '*flux integral*' because, by Gauss's theorem

$$\int_{C_1}(E_2\,dx^3-E_3\,dx^2)=\int_{C_2}(E_2\,dx^3-E_3\,dx^2),\tag{53}$$

where C_1 and C_2 are any two closed contours, one inside the other, in the (x^2, x^3)-plane, provided no singularity of $\boldsymbol{E}$ (if any) occurs inside the area included between C_1 and C_2; and, therefore, *the outward normal fluxes of* $\boldsymbol{E}$ *across* C_1 *and* C_2 *are equal.*

7. The flux integral for the Einstein–Maxwell space-time

We find that the flux integral must now include the Wronskians, $[Y, Y^*]_i$ $(i = 2,3)$, besides those already included in equation (42). The required equation follows from equation (14) by applying to it our standard procedure, namely, by multiplying it by Y^* and subtracting from it the complex conjugate of the equation multiplied by Y. We find:

$$
\begin{aligned}
&\{e^{\nu+\mu_3-\psi-\mu_2}[Y, Y^*]_2\}_{,2}+\{e^{\nu+\mu_2-\psi-\mu_3}[Y, Y^*]_3\}_{,3} \\
&+Y^*[e^{\nu+\mu_3}F_{03}\,\delta(\nu+\mu_3-\psi-\mu_2)]_{,2}-\text{c.c.} \\
&-Y^*[e^{\nu+\mu_2}F_{02}\,\delta(\nu+\mu_2-\psi-\mu_3)]_{,3}-\text{c.c.} = 0,
\end{aligned} \tag{54}
$$

or, alternatively,

$$
\begin{aligned}
&2[\![e^{\alpha_2}\{e^{-2\psi}[Y, Y^*]_2-e^{-\psi+\mu_2}F_{03}[Y\,\delta(\nu+\mu_3-\psi-\mu_2)^*-\text{c.c.}]\}]\!]_{,2} \\
&+2[\![e^{\alpha_3}\{e^{-2\psi}[Y, Y^*]_3+e^{-\psi+\mu_3}F_{02}[Y\,\delta(\nu+\mu_2-\psi-\mu_3)^*-\text{c.c.}]\}]\!]_{,3} \\
&\qquad = [\Psi_2\,\delta(\nu+\mu_2-\psi-\mu_3)^*-\Psi_3\,\delta(\nu+\mu_3-\psi-\mu_2)^*]-\text{c.c.}
\end{aligned} \tag{55}
$$

We add this equation to equation (42). The resulting equation will have on its right-hand side the terms,

$$
\begin{aligned}
&-2\delta\psi^*[+(\Phi_2+\Phi_3)-(\Psi_2-\Psi_3)]+\text{c.c.} \\
&+\delta\mu_3^*[-(\Phi_2-\Phi_3)+\Psi_2-2(e^{\nu+\mu_3}F_{03})_{,2}\,Y]-\text{c.c.} \\
&+\delta\mu_2^*[+(\Phi_2-\Phi_3)-\Psi_3+2(e^{\nu+\mu_2}F_{02})_{,3}\,Y]-\text{c.c.} \\
&+[\Psi_2\,\delta(\nu+\mu_2-\psi-\mu_3)^*-\Psi_3\,\delta(\nu+\mu_3-\psi-\mu_2)^*]-\text{c.c.}
\end{aligned} \tag{56}
$$

The reduction of the inside terms of the combined equations (42) and (55) will proceed as before and in place of (45), we will now have:

$$
\begin{aligned}
&[\![e^{\alpha_2}\{[\delta\mu_3, \delta\mu_3^*]_2+[\delta\psi, \delta\psi^*]_2+2\,e^{-2\psi}[Y, Y^*]_2 \\
&-[\delta\nu_{,2}\,\delta(\psi+\mu_3)^*-\text{c.c.}]-\delta\mu_2[\psi_{,2}\,\delta\psi^*+\mu_{3,2}\,\delta\mu_3^*-\nu_{,2}\,\delta(\psi+\mu_3)^*]+\text{c.c.} \\
&-2\,e^{-\psi+\mu_2}F_{03}[Y\,\delta(\nu+\mu_3-\psi-\mu_2)^*-\text{c.c.}]\}]\!]_{,2} \\
&-\{e^{\alpha_2}[(\psi+\nu)_{,2}\,\delta(\psi+\nu)\,\delta\mu_3^*-(\psi-\nu)_{,2}\,\delta\psi\,\delta\nu^*]-\text{c.c.}\}_{,2} \\
&+\text{terms }(2\rightleftharpoons 3, Y\to -Y).
\end{aligned} \tag{57}
$$

In expanding the terms $\{\ \}_{,i}$, $(i = 2, 3)$, by differentiation, we must now use the equations

$$
\left.
\begin{aligned}
&[e^{\alpha_2}(\psi-\nu)_{,2}]_{,2}+[e^{\alpha_3}(\psi-\nu)_{,3}]_{,3} = -2(F_{02}^2+F_{03}^2)\sqrt{-g}, \\
&[e^{\alpha_2}(\psi+\nu)_{,2}]_{,2} = e^{\alpha_2}A_2-e^{\alpha_3}A_3+(F_{02}^2-F_{03}^2)\sqrt{-g}, \\
&[e^{\alpha_3}(\psi+\nu)_{,3}]_{,3} = e^{\alpha_3}A_3-e^{\alpha_2}A_2+(F_{03}^2-F_{02}^2)\sqrt{-g},
\end{aligned}
\right\} \tag{58}
$$

in place of equations (46) valid for the vacuum. The use of equations (58) will lead to additional terms in the reduction of the outside terms. Additional terms will also arise from the use of the inhomogeneous initial-value equations (21) and (22) instead of the homogeneous equations (appropriate for the vacuum). A careful scrutiny of the reduction of the outside terms, in the case of the vacuum, shows that in the present context, the reduction, besides all the terms which vanish (as in the case of the vacuum), contributes to the right-hand side of the combined equations (42) and (55) the following terms, in addition to the terms (56) we already have:

$$+[\delta(\psi+\nu)\,\delta\mu_3^* - \delta(\psi+\nu)\,\delta\mu_2^*](F_{02}^2 - F_{03}^2)\sqrt{-g} - \text{c.c.}$$
$$-2\,\delta\psi\,\delta\nu^*(F_{02}^2 + F_{03}^2)\sqrt{-g} + \text{c.c.}$$
$$-2[\delta\mu_{3,2}^*\,e^{\nu+\mu_3}F_{03}\,Y - \delta\mu_{2,3}^*\,e^{\nu+\mu_2}F_{02}\,Y] + \text{c.c.}$$
$$+2[\delta(\psi+\nu)_{,2}\,e^{\nu+\mu_3}F_{03}\,Y^* - \delta(\psi+\nu)_{,3}\,e^{\nu+\mu_2}F_{02}\,Y^*] - \text{c.c.} \tag{59}$$

By a fairly straightforward reduction, we find that the terms (56) and (59) combine, surprisingly, to give:

$$-[e^{\nu+\mu_3}F_{03}\,\delta(\psi+\nu+\mu_3)^*Y - \text{c.c.}]_{,2}$$
$$+[e^{\nu+\mu_2}F_{02}\,\delta(\psi+\nu+\mu_2)^*Y - \text{c.c.}]_{,3}. \tag{60}$$

Therefore, the result of the reduction of the combined equations (42) and (55) is:

$$[\![e^{\alpha_2}\{[\delta\mu_3, \delta\mu_3^*]_2 + [\delta\psi, \delta\psi^*]_2 + 2\,e^{-2\psi}[Y, Y^*]_2$$
$$-[\delta\nu_{,2}\,\delta(\psi+\mu_3)^* - \text{c.c.}] - \{\delta\mu_2[\psi_{,2}\,\delta\psi^* + \mu_{3,2}\,\delta\mu_3^* - \nu_{,2}\,\delta(\psi+\mu_3)^*] - \text{c.c.}\}$$
$$+2\,e^{-\psi+\mu_2}F_{03}[\delta(\psi+\nu+\mu_3)^*Y - \text{c.c.}]$$
$$-2\,e^{-\psi+\mu_2}F_{03}[\delta(\nu+\mu_3-\psi-\mu_2)^*Y - \text{c.c.}]\}]\!]_{,2}$$
$$+[\![2 \rightleftharpoons 3, Y \to -Y]\!]_{,3} = 0. \tag{61}$$

The term in the second line of equation (61), which occurs with the factor $\delta\mu_2$, can be simplified by appealing once again to the initial-value equation. The term becomes

$$+\{\delta\mu_2\,\delta(\psi+\mu_3)_{,2}^* - 2\,e^{\nu+\mu_3}F_{03}\,Y\,\delta\mu_2^*\} - \text{c.c.} \tag{62}$$

Inserting the expression in (61) and simplifying, we obtain the flux integral,

$$E_{2,2} + E_{3,3} = 0, \tag{63}$$

where now,

$$E_2 = e^{\alpha_2}\{[\delta\mu_3, \delta\mu_3^*]_2 + [\delta\psi, \delta\psi^*]_2 + 2\,e^{-2\psi}[Y, Y^*]_2$$
$$-[\delta\nu_{,2}\,\delta(\psi+\mu_3)^* - \delta\nu_{,2}^*\,\delta(\psi+\mu_3)] + [\delta\mu_2\,\delta(\psi+\mu_3)_{,2}^* - \delta\mu_2^*\,\delta(\psi+\mu_3)_{,2}]$$
$$+4F_{03}\,e^{-\psi+\mu_2}[Y\,\delta\psi^* - Y^*\,\delta\psi]\}, \tag{64}$$

$$E_3 = e^{\alpha_3}\{[\delta\mu_2, \delta\mu_2^*]_3 + [\delta\psi, \delta\psi^*]_3 + 2\,e^{-2\psi}[Y, Y^*]_3$$
$$-[\delta\nu_{,3}\,\delta(\psi+\mu_2)^* - \delta\nu_{,3}^*\,\delta(\psi+\mu_2)] + [\delta\mu_3\,\delta(\psi+\mu_2)_{,3}^* - \delta\mu_3^*\,\delta(\psi+\mu_2)_{,3}]$$
$$-4F_{02}\,e^{-\psi+\mu_3}[Y\,\delta\psi^* - Y^*\,\delta\psi]\}. \tag{65}$$

The remarkable simplicity of these final expressions for E_2 and E_3 is noteworthy; in particular, the parallel patterns in which the terms representing the vacuum and the Maxwell field occur.

Application to the Reissner–Nordström space-time

The application of the flux integral (65) to the polar perturbations of the Reissner–Nordström space-time has greater interest than as a mere exercise. The interest derives from the fact that, in this instance, we are presented with two independent one-dimensional potential-scattering problems. There are thus two functions, Z_1 and Z_2 which satisfy one-dimensional Schrödinger equations with real potentials and, correspondingly, two Wronskians, $[Z_1, Z_1^*]_{r_*}$ and $[Z_2, Z_2^*]_{r_*}$ which are constants (for details see *M.T.*, §42 (*d*), pp. 230–235). The question that arises is how two independent flux integrals can result from the single integrand (53). To answer this question, we must first clarify the circumstance which gives rise to this situation.

The two functions, Z_1 and Z_2, are related to the amplitudes, H_1 and H_2, of the two kinds of radiation – electromagnetic and gravitational – that prevail. Thus (cf. *M.T.*, p. 259, equation (347))

$$\left.\begin{aligned} Z_1 &= H_1 \cos\Psi + H_2 \sin\Psi \\ \text{and} \qquad Z_2 &= H_2 \cos\Psi - H_1 \sin\Psi, \end{aligned}\right\} \tag{66}$$

where

$$\tan\Psi = \frac{2Q_* \sqrt{(l-1)(l+2)}}{3M + \sqrt{[9M^2 + 4Q_*^2(l-1)(l+2)]}}. \tag{67}$$

Since Z_1 and Z_2 satisfy independent equations, we can set either,

$$Z_1 = 0, \quad H_1 = -H_2 \tan\Psi, \quad Z_2 = H_2 \sec\Psi, \quad \text{and} \quad [Z_2, Z_2^*]_{r_*} = \text{const.}, \tag{68}$$

or

$$Z_2 = 0, \quad H_2 = +H_1 \tan\Psi, \quad Z_1 = H_1 \sec\Psi, \quad \text{and} \quad [Z_1, Z_1^*]_{r_*} = \text{const.} \tag{69}$$

A consequence of the two cases, (68) and (69), that we have distinguished (surprisingly not explicitly noted before) is that the radiation field that prevails (when, for example, an arbitrary superposition of electromagnetic and gravitational waves is incident on the black hole) is a mixture of the two kinds of radiation – photons and gravitons – with correlated phases and amplitudes, as specified in equations (68) and (69). This character of the prevailing radiation field was noted in the context of the scattering of radiation by two extreme Reissner–Nordström black-holes (Chandrasekhar, 1989, §8); but it is present quite generally.

We now turn to the application of the flux integral (63). Since the Reissner–Nordström space-time is spherically symmetric, we can identify x^2 with the radial coordinate r and x^3 with $\mu = \cos\theta$ (where θ denotes the polar angle) and apply equation (53) when C_1 and C_2 are circles of radii r_1 and r_2 both exceeding the radius r_+ of the event horizon. We infer that

$$\int_{-1}^{+1} E_r \, d\mu = \text{const.} \quad (r > r_+). \tag{70}$$

The metric coefficients of the static Reissner–Nordström space-time in the chosen coordinate system $(r, \mu = \cos\theta)$ are

$$e^{2\nu} = e^{-2\mu_2} = \Delta/r^2, \quad e^{2\mu_3} = r^2/(1-\mu^2), \quad e^{2\psi} = r^2(1-\mu^2), \tag{71}$$

$$\Delta = r^2 - 2Mr + Q_*^2, \tag{72}$$

where Q_* denotes the charge of the black hole. And the only non-vanishing component of the Maxwell tensor is

$$F_{02} = -Q_*/r^2. \tag{73}$$

It will be noticed that ν, μ_2, and $\psi+\mu_3$ are independent of μ and that $\psi_{,r} = \mu_{3,r}$. The associated separated solutions for the perturbations, $\delta\nu$, $\delta\psi$, $\delta\mu_2$, $\delta\mu_3$, and $Y(= e^{\psi+\nu}\delta F_{23}/i\sigma)$ are given by (*M.T.*, p. 231, equations (158)–(160)):

$$\left.\begin{aligned} \delta\nu = N(r)\,P_l(\mu); \quad \delta\psi &= T(r)\,P_l(\mu) - V(r)\,\mu P_{l,\mu}, \\ \delta\mu_2 = L(r)\,P_l(\mu); \quad \delta\mu_3 &= [T(r) - \kappa V(r)]\,P_l(\mu) + V(r)\,\mu P_{l,\mu}, \\ \delta\psi + \delta\mu_3 &= [2T(r) - \kappa V(r)]\,P_l(\mu), \quad \kappa = l(l+1), \end{aligned}\right\} \tag{74}$$

and

$$Y = -(r\,e^{\psi}/2Q_*)\,B_{23}(r)\,P_{l,\mu}\sqrt{(1-\mu^2)}. \tag{75}$$

We also have the relations,

$$\left.\begin{aligned} T = V - L + B_{23}, \quad X &= \tfrac{1}{2}(l-1)(l+2)\,V = nV, \\ \text{and} \qquad \delta(\psi+\mu_3) &= -2(L+X-B_{23})\,P_l. \end{aligned}\right\} \tag{76}$$

With the solutions given in equations (74)–(78), we find:

$$\begin{aligned} &[\delta\mu_3, \delta\mu_3^*]_2 + [\delta\psi, \delta\psi^*]_2 \\ &= \{2[T, T^*]_2 + \kappa^2[V, V^*]_2 - \kappa[T, V^*]_2 + \kappa[T^*, V]_2\}\,P_l^2 \\ &\quad + 2[V, V^*]_2[\mu P_l'(\mu)]^2 - 2\kappa[V, V^*]_2\,\mu P_l P_l', \end{aligned} \tag{77}$$

$$2\,e^{-2\psi}[Y, Y^*]_2 = (r^2/2Q_*^2)\,[B_{23}, B_{23}^*]_2(1-\mu^2)\,(P_l')^2, \tag{78}$$

and

$$\begin{aligned} &[-\delta\nu_{,2}\,\delta(\psi+\mu_3)^* + \delta\mu_2\,\delta(\psi+\mu_3)^*_{,2}] - \text{c.c.} \\ &\qquad = 2[N_{,2}(L+X-B_{23})^* - L(L+X-B_{23})^*_{,2}]\,P_l^2 - \text{c.c.} \end{aligned} \tag{79}$$

Remembering that $F_{03} = 0$, we find in accordance with equations (77)–(79):

$$\begin{aligned} &\int_{-1}^{+1} E_r\,d\mu \\ &= 2\frac{\Delta r^2}{2l+1}\left\{\frac{2\kappa}{n^2}[X, X^*]_2 + 2[L+X-B_{23}, L^*+X^*-B_{23}^*]_2 + \frac{\kappa r^2}{2Q_*^2}[B_{23}, B_{23}^*]_2\right. \\ &\quad + 2[N_{,r}(L+X-B_{23})^* - N_{,r}^*(L+X-B_{23})] \\ &\quad \left. - 2[L(L+X-B_{23})^*_{,2} - L^*(L+X-B_{23})_{,2}]\right\}. \end{aligned} \tag{80}$$

The further reduction of this expression requires a discriminating use of the solutions for L, N, X, and B_{23} (due to B. Xanthopoulos) given in *M.T.*, p. 235, equations (190)–(196) as well as equations (173)–(179) given on p. 233. The result of the reduction (after 'miraculous' cancellations) is

$$\frac{4\Delta\kappa}{(2l+1)\,r^2}[Z_i, Z_i^*]_r = \frac{4\kappa}{2l+1}[Z_i, Z_i^*]_{r_*} \quad (i = 1, 2), \tag{81}$$

for the two cases (68) and (69). The flux integral thus gives the same two conserved quantities.

PART II. THE FLUX INTEGRAL FOR POLAR PERTURBATIONS OF A STATIC SPACE-TIME WITH A PREVAILING FLUID SOURCE

8. Introduction

The polar perturbations of a static spherical star leading to damped non-radial oscillations with accompanying emission of gravitational radiation is one of the central problems of relativistic astrophysics. Its study was initiated by Thorne & Campolattaro (1967) and continued by Thorne and others, notably, by Detweiler & Ipser (1973) and Lindblom & Detweiler (1983, 1985; the first of these papers gives a fairly complete bibliography). But it does not seem to have been noticed that there is a flux integral governing these oscillations which, besides providing a useful additional constraint, recasts the problem of the non-radial oscillations of a star as a problem in scattering theory.

In this Part II, we shall be concerned only with the derivation of the flux integral. We postpone to a later paper the reformulation of the general problem of the non-radial oscillations of a star in the light of the flux integral. We do not also consider the problem of the axial perturbations: it is known that the scattering of gravitational waves, belonging to the axial modes (described by the Regge–Wheeler equation outside the star) does not affect and, in turn, is not affected by the axial modes of perturbation of the fluid (characterized by motions purely in the φ-direction).

9. The equations governing equilibrium*

Our restriction to axisymmetric space-times with a perfect-fluid source, effectively requires us to consider only a spherically symmetric distribution of matter described in terms of an isotropic pressure (p) and an energy-density (ϵ), i.e. a star. The metric of the space-time, both in the interior of the star and in the vacuum outside the star, is well-known and can be found in any text book. We shall simply quote the principal equations in a notation that we shall use.

The metric is of the standard form (1), where in the present context,

$$e^{2\psi} = r^2(1-\mu^2), \quad e^{2\mu_3} = r^2/(1-\mu^2), \tag{82}$$

$r = x^2$ is a radial coordinate and $\mu = \cos\theta$. Inside the star, with its centre at $r = 0$, the metric functions ν and μ_2 are given by

$$\nu_{,r} = -p_{,r}/(\epsilon+p), \quad e^{-2\mu_2} = 1-2M(r)/r, \tag{83}$$

where

$$M(r) = \int_0^r \epsilon r^2 \,dr. \tag{84}$$

The equation of hydrostatic equilibrium is

$$[1-2M(r)/r]\,p_{,r} = -(\epsilon+p)[M(r)/r^2+rp]. \tag{85}$$

* See note on page 667.

Three additional relations we shall find useful are

$$2\epsilon = r^{-2}[+1-e^{-2\mu_2}(1-2r\mu_{2,r})], \tag{86}$$

$$2p = r^{-2}[-1+e^{-2\mu_2}(1+2r\nu_{,r})], \tag{87}$$

and

$$\epsilon+p = r^{-1}e^{-2\mu_2}(\nu+\mu_2)_{,r}. \tag{88}$$

Outside the star (where $\epsilon = p = 0$) it follows from equations (83) and (88) that

$$\nu = -\mu_2 \quad \text{and} \quad e^{-2\mu_2} = 1-2M/r, \tag{89}$$

where

$$M = \int_0^R \epsilon r^2\, dr \tag{90}$$

is the mass of the star and R its radius. The metric is of course that of Schwarzschild.

10. The equations of hydrodynamics and their linearization

The equations of hydrodynamics, in the tetrad frame that is adopted in this paper, follow from the intrinsic divergence of the energy–momentum tensor,

$$T_{(a)(b)} = (\epsilon+p)\,u_{(a)}\,u_{(b)}-p\eta_{(a)(b)}, \tag{91}$$

where we have enclosed the indices in parentheses (temporarily!) to denote that they are tetrad indices. These equations are:

$$(\epsilon+p)\,u^{(b)}u_{(a)|(b)} = p_{,(a)}-u_{(a)}\,u^{(b)}p_{,(b)} \tag{92}$$

and

$$\epsilon_{,(a)}\,u^{(a)}+(\epsilon+p)\,u^{(a)}{}_{|(a)} = 0, \tag{93}$$

where the vertical rule in $u_{(a)|(b)}$ denotes intrinsic differentiation as defined in *M.T.*, p. 37. Making use of the Ricci rotation coefficients listed in *M.T.* (p. 82, equation (91)), we find that the explicit form of equations (92) and (93) (when the only non-vanishing components of the four-velocity $u_{(a)}$, are $u_{(0)}$, $u_{(2)}$, and $u_{(3)}$) are:

$$\begin{aligned}
&(\epsilon+p)^{-1}\{-e^{-\nu}[u_{(2)}^2+u_{(3)}^2]\,p_{,0}+e^{-\mu_2}u_{(0)}\,u_{(2)}\,p_{,2}+e^{-\mu_3}u_{(0)}\,u_{(3)}\,p_{,3}\}\\
&\quad = +e^{-\nu}[u_{(0)}\,u_{(0),0}+\mu_{2,0}\,u_{(2)}^2+\mu_{3,0}\,u_{(3)}^2]\\
&\qquad -e^{-\mu_2}u_{(2)}[\nu_{,2}\,u_{(0)}+u_{(0),2}]-e^{-\mu_3}u_{(3)}[\nu_{,3}\,u_{(0)}+u_{(0),3}];\\
&(\epsilon+p)^{-1}\{+e^{-\mu_2}[u_{(0)}^2-u_{(3)}^2]\,p_{,2}-e^{-\nu}u_{(2)}\,u_{(0)}\,p_{,0}+e^{-\mu_3}u_{(2)}\,u_{(3)}\,p_{,3}\}\\
&\quad = -e^{-\mu_2}[u_{(2)}\,u_{(2),2}+\nu_{,2}\,u_{(0)}^2-\mu_{3,2}\,u_{(3)}^2]\\
&\qquad +e^{-\nu}u_{(0)}[\mu_{2,0}\,u_{(2)}+u_{(2),0}]-e^{-\mu_3}u_{(3)}[\mu_{2,3}\,u_{(2)}+u_{(2),3}];\\
&(\epsilon+p)^{-1}\{+e^{-\mu_3}[u_{(0)}^2-u_{(2)}^2]\,p_{,3}-e^{-\nu}u_{(3)}\,u_{(0)}\,p_{,0}+e^{-\mu_2}u_{(3)}\,u_{(2)}\,p_{,2}\}\\
&\quad = -e^{-\mu_3}[u_{(3)}\,u_{(3),3}+\nu_{,3}\,u_{(0)}^2-\mu_{2,3}\,u_{(2)}^2]\\
&\qquad +e^{-\nu}u_{(0)}[\mu_{3,0}\,u_{(3)}+u_{(3),0}]-e^{-\mu_2}u_{(2)}[\mu_{3,2}\,u_{(3)}+u_{(3),2}],
\end{aligned} \tag{94}$$

and

$$\begin{aligned}
&e^{-\nu}\{\epsilon_{,0}\,u_{(0)}+(\epsilon+p)[u_{(0),0}+(\psi+\mu_2+\mu_3)_{,0}\,u_{(0)}]\}\\
&-e^{-\mu_2}\{\epsilon_{,2}\,u_{(2)}+(\epsilon+p)[u_{(2),2}+(\psi+\nu+\mu_3)_{,2}\,u_{(2)}]\}\\
&-e^{-\mu_3}\{\epsilon_{,3}\,u_{(3)}+(\epsilon+p)[u_{(3),3}+(\psi+\nu+\mu_2)_{,3}\,u_{(3)}]\}=0. \qquad (95)
\end{aligned}$$

(a) *The linearized equations*

In the static spherically symmetric background, the only non-vanishing component of the four-velocity is $u_{(0)}(=1)$; and besides,

$$p_{,3}=\nu_{,3}=\mu_{2,3}=0. \qquad (96)$$

In the perturbed space-time $u_{(0)}$ continues to be 1 (i.e. it differs from unity only in the second order) and $u_{(2)}$ and $u_{(3)}$ are quantities of the first order. We shall write,

$$u_{(2)}=\delta u_2=\xi_{2,0} \quad\text{and}\quad u_{(3)}=\delta u_3=\xi_{3,0}, \qquad (97)$$

where ξ_2 and ξ_3 are the associated Lagrangian displacements in the radial and in the θ-directions. From equations (94) we deduce that

$$e^{\nu-\mu_2}\delta[(\epsilon+p)^{-1}p_{,2}+\nu_{,2}]=\delta u_{2,0}=\xi_{2,0,0} \qquad (98)$$

and

$$e^{\nu-\mu_3}\delta[(\epsilon+p)^{-1}p_{,3}+\nu_{,3}]=\delta u_{3,0}=\xi_{3,0,0}, \qquad (99)$$

or, explicitly,

$$\delta p_{,2}+\delta(\epsilon+p)\,\nu_{,2}+(\epsilon+p)\,\delta\nu_{,2}=-\sigma^2(\epsilon+p)\,e^{-\nu+\mu_2}\xi_2, \qquad (100)$$

and

$$\delta p_{,3}+(\epsilon+p)\,\delta\nu_{,3}=-\sigma^2(\epsilon+p)\,e^{-\nu+\mu_3}\xi_3, \qquad (101)$$

where we have assumed, as hitherto, that all the perturbed quantities have the common time-dependent factor, $e^{i\sigma t}$.

Turning next to equation (95), we find:

$$\begin{aligned}
\delta\epsilon+(\epsilon+p)\,\delta(\psi+\mu_2+\mu_3)&=e^{\nu-\mu_2}\{\epsilon_{,2}\,\xi_2+(\epsilon+p)[\xi_{2,2}+(\psi+\nu+\mu_3)_{,2}\,\xi_2]\}\\
&\quad+e^{\nu-\mu_3}(\epsilon+p)(\xi_{3,3}+\psi_{,3}\,\xi_3). \qquad (102)
\end{aligned}$$

(b) *The equation that follows from the conservation of baryon number*

To complete the system of hydrodynamic equations, we need an additional equation. It is provided by the conservation of the baryon number N which is assumed to be some known function of ϵ and p. In other words,

$$N=N(\epsilon,p) \qquad (103)$$

and

$$[Nu^{(a)}]_{|(a)}=N_{,(a)}\,u^{(a)}+Nu^{(a)}{}_{|(a)}=0. \qquad (104)$$

From equations (103) and (104) it readily follows that (cf. equation (102))

$$\begin{aligned}
\delta p+\gamma p\,\delta(\psi+\mu_2+\mu_3)&=e^{\nu-\mu_2}\{p_{,2}\,\xi_2+\gamma p[\xi_{2,2}+(\psi+\nu+\mu_3)_{,2}\,\xi_2]\}\\
&\quad+e^{\nu-\mu_3}\gamma p(\xi_{3,3}+\psi_{,3}\,\xi_3), \qquad (105)
\end{aligned}$$

where

$$\gamma=\frac{1}{p\,\partial N/\partial p}\left[N-(\epsilon+p)\frac{\partial N}{\partial\epsilon}\right]. \qquad (106)$$

The physical content of equation (105) is no more than that the changes in the pressure and in the energy density of a fluid element, as it moves, take place adiabatically.

With the further definition,

$$Q = (\epsilon + p)/\gamma p, \tag{107}$$

equation (105) takes the alternative form,

$$Q\,\delta p + (\epsilon+p)\,\delta(\psi+\mu_2+\mu_3) = e^{\nu-\mu_2}\{Qp_{,2}\,\xi_2 + (\epsilon+p)[\xi_{2,2}+(\psi+\nu+\mu_3)_{,2}\,\xi_2]\} + e^{\nu-\mu_3}(\epsilon+p)(\xi_{3,3}+\psi_{,3}\,\xi_3). \tag{108}$$

An eliminant of equations (102) and (108) is

$$\delta\epsilon = Q\,\delta p + e^{\nu-\mu_2}(\epsilon_{,2} - Qp_{,2})\,\xi_2. \tag{109}$$

From this equation, we obtain the physically meaningful relation.

$$(\delta\epsilon\xi_2^* - \delta\epsilon^*\xi_2) = Q(\delta p\xi_2^* - \delta p^*\xi_2). \tag{110}$$

11. The linearized Einstein's equations

For a perfect-fluid source, the components of the Ricci tensor (by Einstein's equations) are:

$$\left.\begin{aligned}
&-\tfrac{1}{2}R_{00} = (\epsilon+p)\,u_0^2 - \tfrac{1}{2}(\epsilon-p); \quad -\tfrac{1}{2}R_{11} = \tfrac{1}{2}(\epsilon-p),\\
&-\tfrac{1}{2}R_{22} = (\epsilon+p)\,u_2^2 + \tfrac{1}{2}(\epsilon-p); \quad -\tfrac{1}{2}R_{33} = (\epsilon+p)\,u_3^2 + \tfrac{1}{2}(\epsilon-p),\\
&-\tfrac{1}{2}R_{02} = (\epsilon+p)\,u_0\,u_2; \quad -\tfrac{1}{2}R_{03} = (\epsilon+p)\,u_0\,u_3,\\
&\qquad\qquad -\tfrac{1}{2}R_{23} = (\epsilon+p)\,u_2\,u_3.
\end{aligned}\right\} \tag{111}$$

For a static spherically symmetric space-time subject to polar perturbations, the components of the Ricci tensor, inclusive of terms of the first order, are

$$\left.\begin{aligned}
&R_{00} = -(\epsilon+3p); \quad G_{00} = -2\epsilon; \quad R_{11} = R_{22} = R_{33} = -(\epsilon-p),\\
&R_{02} = -2(\epsilon+p)\,u_2 = -2(\epsilon+p)\,\xi_{2,0},\\
&R_{03} = -2(\epsilon+p)\,u_3 = -2(\epsilon+p)\,\xi_{3,0}.
\end{aligned}\right\} \tag{112}$$

We can now write down the linearized Einstein's equations as we have in §4, equations (19i–v), for the Einstein–Maxwell space-time. The left-hand sides of these equations remain unchanged while, on the right-hand sides, the terms in $(\Phi_2 \pm \Phi_3)$ and $(\Psi_2 \pm \Psi_3)$ are replaced, respectively, by

$$+[\delta(\epsilon-p) + (\epsilon-p)\,\delta(\psi+\nu+\mu_2+\mu_3)]\sqrt{-g}, \tag{113i}$$

$$-[\delta(\epsilon+3p) + (\epsilon+3p)\,\delta(\psi+\nu+\mu_2+\mu_3)]\sqrt{-g}, \tag{113ii}$$

$$+[\delta(\epsilon-p) + (\epsilon-p)\,\delta(\psi+\nu+\mu_2+\mu_3)]\sqrt{-g}, \tag{113iii}$$

$$+[\delta(\epsilon-p) + (\epsilon-p)\,\delta(\psi+\nu+\mu_2+\mu_3)]\sqrt{-g}, \tag{113iv}$$

$$+2[\delta\epsilon + \epsilon\,\delta(\psi+\nu+\mu_2+\mu_3)]\sqrt{-g}. \tag{113v}$$

It is convenient to have the initial-value equations (6) and (7) written out explicitly. We have:

$$e^{\alpha_2}[\delta(\psi+\mu_3)_{,2}+\psi_{,2}\,\delta(\psi-\mu_2)+\mu_{3,2}\,\delta(\mu_3-\mu_2)-\nu_{,2}\,\delta(\psi+\mu_3)] = 2\,e^{\psi+2\nu+\mu_3}(\epsilon+p)\,\xi_2, \quad (114)$$

$$e^{\alpha_3}[\delta(\psi+\mu_2)_{,3}+\psi_{,3}\,\delta(\psi-\mu_3)+\mu_{2,3}\,\delta(\mu_2-\mu_3)-\nu_{,3}\,\delta(\psi+\mu_2)] = 2\,e^{\psi+2\nu+\mu_2}(\epsilon+p)\,\xi_3. \quad (115)$$

Again, we need to rewrite equations (113iii) and (113iv) as in §4. By following the same procedure, we find that in place of equations (28) and (29) we now have:

$$[e^{\alpha_3}(\delta\mu_{2,3}+\mu_{2,3}\,\delta\alpha_3)]_{,3}-(e^{\alpha_3}\delta f_3)_{,3}+[e^{\alpha_2}(\delta\nu_{,2}+\nu_{,2}\,\delta\alpha_2)]_{,2} -2[e^{\psi+2\nu+\mu_2}(\epsilon+p)\,\xi_3]_{,3} = \delta X_2-\delta X_3-\sigma^2 e^{\alpha_0}\delta\mu_2 -[\delta(\epsilon+p)+(\epsilon+p)\,\delta(\psi+\nu+\mu_2+\mu_3)]\sqrt{-g} \quad (116)$$

and

$$[e^{\alpha_2}(\delta\mu_{3,2}+\mu_{3,2}\,\delta\alpha_2)]_{,2}-(e^{\alpha_2}\delta f_2)_{,2}+[e^{\alpha_3}(\delta\nu_{,3}+\nu_{,3}\,\delta\alpha_3)]_{,3} -2[e^{\psi+2\nu+\mu_3}(\epsilon+p)\,\xi_2]_{,2} = \delta X_3-\delta X_2-\sigma^2 e^{\alpha_0}\delta\mu_3 -[\delta(\epsilon+p)+(\epsilon+p)\,\delta(\psi+\nu+\mu_2+\mu_3)]\sqrt{-g}. \quad (117)$$

And finally in place of equation (41), we have,

$$\delta\mu_2^*[e^{\alpha_2}(\delta\nu_{,2}+\nu_{,2}\,\delta\alpha_2)]_{,2}+\delta\mu_3^*[e^{\alpha_3}(\delta\nu_{,3}+\nu_{,3}\,\delta\alpha_3)]_{,3} = -\delta(\psi+\mu_3)^*[e^{\alpha_2}(\delta\nu_{,2}+\nu_{,2}\,\delta\alpha_2)]_{,2}-\delta(\psi+\mu_2)^*[e^{\alpha_3}(\delta\nu_{,3}+\nu_{,3}\,\delta\alpha_3)]_{,3} -[\delta(\epsilon+3p)\,\delta(\psi+\mu_2+\mu_3)^*+(\epsilon+3p)\,\delta(\psi+\mu_2+\mu_3)^*\delta\nu]\sqrt{-g}. \quad (118)$$

12. The flux integral

We consider the same combination of Wronskians as in equation (42) but evaluated with the aid of the linearized Einstein's equations of §11 (instead of those of §4). It is clear that the left-hand side of the equation will be unaffected; and we find that on the right-hand side (the last three lines of equation (42)) are replaced by

$$2\delta\mu_3^*[e^{\psi+2\nu+\mu_3}(\epsilon+p)\,\xi_2]_{,2}+2\delta\mu_2^*[e^{\psi+2\nu+\mu_2}(\epsilon+p)\,\xi_3]_{,3}-\text{c.c.} +2\{[\delta(\psi+\mu_2+\mu_3)^*(\delta p+p\delta\nu)+\delta\psi^*[\delta\epsilon+\epsilon\delta(\nu+\mu_2+\mu_3)]\}\sqrt{-g}-\text{c.c.} \quad (119)$$

The reduction of the 'inside' terms (on the left-hand side of equation (42)) will result in the same equation (45). But the expansion (by differentiation) of the terms $\{\ \}_{,i}$ should now be effected by making use of the equations,

$$\left.\begin{aligned} &[e^{\alpha_2}(\psi-\nu)_{,2}]_{,2}+[e^{\alpha_3}(\psi-\nu)_{,3}]_{,3} = 2(\epsilon+p)\sqrt{-g},\\ &[e^{\alpha_2}(\psi+\nu)_{,2}]_{,2} = e^{\alpha_2}A_2-e^{\alpha_3}A_3-2p\sqrt{-g},\\ &[e^{\alpha_2}(\psi+\nu)_{,3}]_{,3} = e^{\alpha_3}A_3-e^{\alpha_2}A_2-2p\sqrt{-g},\end{aligned}\right\} \quad (120)$$

instead of equations (46). The use of equations (120) will lead to additional terms in the further reduction of the outside terms. Additional terms will also arise from the use of the initial-value equations, (113) and (114), instead of the homogeneous equations appropriate to the vacuum. A careful scrutiny of the reduction of the outside terms, in the case of the vacuum, shows that in the present context, the reduction, besides all the terms which vanish, contributes to the right-hand side of equation (42) only the following terms, in addition to the terms (119) we already have:

$$\begin{aligned}&2(\epsilon+p)(\delta\psi\,\delta\nu^*-\delta\psi^*\,\delta\nu)\sqrt{-g}\\&-2p[\delta(\psi+\nu)\,\delta(\mu_2+\mu_3)^*-\delta(\psi+\nu)^*\delta(\mu_2+\mu_3)]\sqrt{-g}\\&+2[e^{\psi+2\nu+\mu_3}(\epsilon+p)\,\xi_2\,\delta\mu^*_{3,2}+e^{\psi+2\nu+\mu_2}(\epsilon+p)\,\xi_3\,\delta\mu^*_{2,3}]-\text{c.c.}\\&+2[e^{\psi+2\nu+\mu_3}(\epsilon+p)\,\delta(\psi+\nu)^*_{,2}\,\xi_2+e^{\psi+2\nu+\mu_2}(\epsilon+p)\,\delta(\psi+\nu)^*_{,3}\,\xi_3]-\text{c.c.}\end{aligned}\tag{121}$$

Combining the terms (119) and (121), we have

$$\begin{aligned}&2[e^{\psi+2\nu+\mu_3}(\epsilon+p)\,\xi_2\,\delta\mu_3^*]_{,2}+2[e^{\psi+2\nu+\mu_2}(\epsilon+p)\,\xi_3\,\delta\mu_2^*]_{,3}-\text{c.c.}\\&+2\{\delta(\psi+\mu_2+\mu_3)^*(\delta p+p\delta\nu)+\delta\psi^*[\delta\epsilon+\epsilon\delta(\nu+\mu_2+\mu_3)]\\&\quad-(\epsilon+p)\,\delta\psi^*\delta\nu-p\delta(\psi+\nu)\,\delta(\mu_2+\mu_3)^*]\}\sqrt{-g}-\text{c.c.}\\&+2[e^{\psi+2\nu+\mu_3}(\epsilon+p)\,\xi_2\,\delta(\psi+\nu)^*_{,2}+e^{\psi+2\nu+\mu_2}(\epsilon+p)\,\xi_3\,\delta(\psi+\nu)^*_{,3}]-\text{c.c.},\end{aligned}\tag{122}$$

or, after some rearrangements,

$$\begin{aligned}&2[e^{\psi+2\nu+\mu_3}(\epsilon+p)\,\xi_2\,\delta\mu_3^*]_{,2}+2[e^{\psi+2\nu+\mu_2}(\epsilon+p)\,\xi_3\,\delta\mu_2^*]_{,3}-\text{c.c.}\\&+2[e^{\psi+2\nu+\mu_3}(\epsilon+p)\,\xi_2\,\delta(\psi+\nu)^*_{,2}+e^{\psi+2\nu+\mu_2}(\epsilon+p)\,\xi_3\,\delta(\psi+\nu)^*_{,3}]-\text{c.c.}\\&+2[\delta\psi^*\underline{[\delta\epsilon+(\epsilon+p)\,\delta(\mu_2+\mu_3)]}+\delta p\,\delta(\psi+\mu_2+\mu_3)^*\}\sqrt{-g}-\text{c.c.}\end{aligned}\tag{123}$$

By equation (102), the underlined term in the foregoing expression is

$$\begin{aligned}&2[\![e^{\psi+2\nu+\mu_3}\{\epsilon_{,2}\,\xi_2+(\epsilon+p)\,[\xi_{2,2}+(\psi+\nu+\mu_3)_{,2}\,\xi_2]\}\\&\qquad+e^{\psi+2\nu+\mu_2}(\epsilon+p)\,(\xi_{3,3}+\psi_{,3}\,\xi_3)]\!]\,\delta\psi^*-\text{c.c.}\end{aligned}\tag{124}$$

Inserting this expression in (123), we find after some lengthy but straightforward reductions, that we are left with

$$\begin{aligned}&2[e^{\psi+2\nu+\mu_3}(\epsilon+p)\,\xi_2\,\delta(\psi+\mu_3)^*]_{,2}+2[e^{\psi+2\nu+\mu_2}(\epsilon+p)\,\xi_3\,\delta(\psi+\mu_2)^*]_{,3}-\text{c.c.}\\&+2\,e^{\psi+2\nu+\mu_3}(\epsilon+p)\,\xi_2\,\delta\nu^*_{,2}+2\,e^{\psi+2\nu+\mu_2}(\epsilon+p)\,\xi_3\,\delta\nu^*_{,3}-\text{c.c.}\\&+2\delta p\,\delta(\psi+\mu_2+\mu_3)^*\sqrt{-g}-\text{c.c.}\end{aligned}\tag{125}$$

The terms in the second line of equation (125) can be simplified by making use of the equations,

$$(\epsilon+p)\,\xi_2^*\,\delta\nu_{,2}-\text{c.c.}=-\xi_2^*[\delta p_{,2}+\delta(\epsilon+p)\,\nu_{,2}]+\text{c.c.}$$

and

$$(\epsilon+p)\,\xi_3^*\,\delta\nu_{,3}-\text{c.c.}=-\xi_3^*\,\delta p_{,3}+\text{c.c.},\tag{126}$$

which follow directly from equations (100) and (101); and we find, successively,

$$\begin{aligned}
&-2\,e^{\psi+2\nu+\mu_3}\xi_2[\delta p^*_{,2}+\delta(\epsilon+p)^*\nu_{,2}]-2\,e^{\psi+2\nu+\mu_2}\xi_3\,\delta p^*_{,3}+\text{c.c.}\\
&=-2(e^{\psi+2\nu+\mu_3}\xi_2\,\delta p^*)_{,2}-2(e^{\psi+2\nu+\mu_2}\xi_3\,\delta p^*)_{,3}+\text{c.c.}\\
&\quad+2[(e^{\psi+2\nu+\mu_3}\xi_2)_{,2}+(e^{\psi+2\nu+\mu_2}\xi_3)_{,3}]\,\delta p^*-\text{c.c.}\\
&\quad-2\,e^{\psi+2\nu+\mu_3}\xi_2\,\delta(\epsilon+p)^*\nu_{,2}+\text{c.c.}\\
&=-2(e^{\psi+2\nu+\mu_3}\xi_2\,\delta p^*)_{,2}-2(e^{\psi+2\nu+\mu_2}\xi_3\,\delta p^*)_{,3}+\text{c.c.}\\
&\quad+2\{e^{\psi+2\nu+\mu_3}[\xi_{2,2}+(\psi+\nu+\mu_3)_{,2}\,\xi_2]+e^{\psi+2\nu+\mu_2}(\xi_{3,3}+\psi_{,3}\,\xi_3)\}\,\delta p^*-\text{c.c.}\\
&\quad-2\,e^{\psi+2\nu+\mu_3}\xi_2\,\delta\epsilon^*\nu_{,2}+\text{c.c.},
\end{aligned}\tag{127}$$

where, in passing from the first to the second equality, we have made use of the equilibrium equation, $p_{,2}=-\nu_{,2}/(\epsilon+p)$. The terms in the last but one line of equation (127) can be simplified by making use of equation (108) 'contracted' with δp^*. In this manner, the terms in the second line of (125) reduce to

$$\begin{aligned}
&-2(e^{\psi+2\nu+\mu_3}\xi_2\,\delta p^*)_{,2}-2(e^{\psi+2\nu+\mu_2}\xi_3\,\delta p^*)_{,3}+\text{c.c.}\\
&-2\,e^{\psi+2\nu+\mu_3}\xi_2[\delta\epsilon^*\nu_{,2}+Qp_{,2}(\epsilon-p)^{-1}\delta p^*]+\text{c.c.}\\
&+2\delta p^*\delta(\psi+\mu_2+\mu_3)\sqrt{-g}-\text{c.c.},
\end{aligned}\tag{128}$$

where the terms in the second line vanish by virtue of the relation $p_{,2}=-\nu_{,2}(\epsilon+p)$ and equation (110). Also the terms in the last lines of (125) and (128) cancel each other. Therefore the reduction of the entire equation (42), with the terms (119) on the right-hand side, is to provide the same terms (cf. equation (45)),

$$\begin{aligned}
&[\![e^{\alpha_2}\{[\delta\mu_3,\delta\mu_3^*]_2+[\delta\psi,\delta\psi^*]_2-[\delta\nu_{,2}\,\delta(\psi+\mu_3)^*-\text{c.c.}]\\
&-\delta\mu_2\{\psi_{,2}\,\delta\psi^*+\mu_{3,2}\,\delta\mu_3^*-\nu_{,2}\,\delta(\psi+\mu_3)^*\}+\text{c.c.}]\!]_{,2}\\
&+[\![2\rightleftharpoons 3]\!]_{,3},
\end{aligned}\tag{129}$$

on the left-hand side of the equation and the terms,

$$\begin{aligned}
&2[e^{\psi+2\nu+\mu_3}(\epsilon+p)\,\xi_2\,\delta(\psi+\mu_3)^*]_{,2}+2[e^{\psi+2\nu+\mu_2}(\epsilon+p)\,\xi_3\,\delta(\psi+\mu_2)^*]_{,3}-\text{c.c.}\\
&-2(e^{\psi+2\nu+\mu_3}\xi_2\,\delta p^*)_{,2}-2(e^{\psi+2\nu+\mu_2}\xi_3\,\delta p^*)_{,3}+\text{c.c.},
\end{aligned}\tag{130}$$

on the right-hand side of the equation. And finally the term which occurs with the factor $-\delta\mu_2$ in (129) can be replaced by

$$+\delta\mu_2[\delta(\psi+\mu_3)^*_{,2}-2\,e^{\nu+\mu_2}(\epsilon+p)\,\xi_2^*]-\text{c.c.},\tag{131}$$

by making use of the initial-value equation (114). With this replacement, we obtain the flux integral, once again, in the form

$$E_{2,2}+E_{3,3}=0,\tag{132}$$

where

$$E_2 = e^{\alpha_2}[\![[\delta\mu_3, \delta\mu_3^*]_2 + [\delta\psi, \delta\psi^*]_2 - [\delta\nu_{,2}\,\delta(\psi+\mu_3)^* - \delta^*\nu_{,2}\,\delta(\psi+\mu_3)]$$
$$+ [\delta\mu_2\,\delta(\psi+\mu_3)^*_{,2} - \delta\mu_2^*\,\delta(\psi+\mu_3)_{,2}]$$
$$-2\,e^{\nu+\mu_2}\{[(\epsilon+p)\,\delta(\psi+\mu_3-\mu_2)^* - \delta p^*]\,\xi_2 - \text{c.c.}\}]\!] \quad (133)$$

and

$$E_3 = e^{\alpha_3}[\![[\delta\mu_2, \delta\mu_2^*]_3 + [\delta\psi, \delta\psi^*]_3 - [\delta\nu_{,3}\,\delta(\psi+\mu_2)^* - \delta^*\nu_{,3}\,\delta(\psi+\mu_2)]$$
$$+ [\delta\mu_3\,\delta(\psi+\mu_2)^*_{,3} - \delta\mu_3^*\,\delta(\psi+\mu_2)_{,3}]$$
$$-2\,e^{\nu+\mu_3}\{[(c+p)\,\delta(\psi+\mu_2-\mu_3)^* - \delta p^*]\,\xi_3 - \text{c.c.}\}]\!]. \quad (134)$$

13. Concluding remarks

The derivation of the flux integral, for the axisymmetric perturbations of a static space-time, directly from the appropriate linearized equations demonstrates vividly that the 'magic' of the exact theory extends to its linearized versions. It is specially noteworthy that allowance for different prevailing sources is accomplished parallelly with such ease. It is equally remarkable how every single equation governing the source plays its own equal role, a feature particularly evident in the reductions of §12 for a perfect-fluid source.

As we have stated in the introductory section, the present investigation arose when we wished to extend our earlier study (Chandrasekhar 1989) of the scattering of radiation by two extreme Reissner–Nordström black-holes by axial modes to include polar modes. The flux integral of §7 now provides the necessary means for this extension. Similarly, as we have indicated, the flux-integral derived in §12 will enable a reformulation of the problem of the non-radial oscillations of a star. We are presently studying these matters.

We are grateful to Greg Burnett, John Friedman, Rafael Sorkin, and Robert Wald for helpful and critical discussions of the various questions – both concrete and general – that arise in the context of the problem considered. The research reported in this paper has, in part, been supported by grants from the National Science Foundation under Grant PHY-84-16691 with the University of Chicago. We are also grateful for a grant from the Division of Physical Sciences of the University of Chicago which has enabled our continued collaboration by making possible periodic visits by Valeria Ferrari to the University of Chicago.

Appendix. The flux integral for the axial perturbations of a static Einstein–Maxwell space-time

For axial perturbations, the metric functions, ν, ψ, μ_2, and μ_3 and the components F_{02} and F_{03} of the Maxwell-tensor retain their values in the static space-time while the perturbation is represented by the first-order quantities, ω, q_2, and q_3 in the metric (2) and the component F_{01} of the Maxwell-tensor. In Chandrasekhar (1989) the relevant perturbation equations have been written down in the particular context of the two-centre problem. Without this specialization, one readily finds that equations (41) and (42) of Chandrasekhar (1989) are replaced by

$$e^{-3\psi-\nu+\mu_2+\mu_3}X_{,0,0} = [e^{-3\psi+\nu-\mu_2+\mu_3}(X_{,2}-4\,e^{\psi+\mu_2}F_{03}\,Y)]_{,2}$$
$$+[e^{-3\psi+\nu-\mu_3+\mu_2}(X_{,3}+4\,e^{\psi+\mu_3}F_{02}\,Y)]_{,3}, \quad \text{(A 1)}$$

$$e^{-3\psi-\nu+\mu_2+\mu_3}Y_{,0,0} = (e^{\nu+\mu_3-\psi-\mu_2}Y_{,2})_{,2}+(e^{\nu+\mu_2-\psi-\mu_3}Y_{,3})_{,3}$$
$$-e^{-2\psi+\nu+\mu_2}F_{02}(X_{,3}+4\,e^{\psi+\mu_3}F_{02}\,Y)+e^{-2\psi+\nu+\mu_3}F_{03}(X_{,2}-4\,e^{\psi+\mu_2}F_{03}\,Y), \quad \text{(A 2)}$$

where

$$X = e^{3\psi+\nu-\mu_2-\mu_3}Q_{23} \quad \text{and} \quad Y = e^{\psi+\nu}F_{01}. \quad \text{(A 3)}$$

Applying to equations (A 1) and (A 2) our standard procedure, we obtain

$$\{e^{-3\psi+\nu-\mu_2+\mu_3}[X,X^*]_2\}_{,2}+\{e^{-3\psi+\nu-\mu_3+\mu_2}[X,X^*]_3\}_{,3}$$
$$+4\{X^*(e^{-2\psi+\nu+\mu_2}F_{02}\,Y)_{,3}-X(e^{-2\psi+\nu+\mu_2}F_{02}\,Y^*)_{,3}$$
$$-X^*(e^{-2\psi+\nu+\mu_3}F_{03}\,Y)_{,2}+X(e^{-2\psi+\nu+\mu_3}F_{03}\,Y^*)_{,2}\} = 0 \quad \text{(A 4)}$$

and

$$\{e^{\nu+\mu_3-\psi-\mu_2}[Y,Y^*]_2\}_{,2}+\{e^{\nu+\mu_2-\psi-\mu_3}[Y,Y^*]_3\}_{,3}$$
$$-4\,e^{-2\psi+\nu+\mu_2}F_{02}(Y^*X_{,3}-YX^*_{,3})+4\,e^{-2\psi+\nu+\mu_3}F_{03}(Y^*X_{,2}-YX^*_{,2}) = 0. \quad \text{(A 5)}$$

By adding equation (A 4) to four times equation (A 5), we obtain,

$$\{e^{-3\psi+\nu-\mu_2+\mu_3}[X,X^*]_2+4\,e^{-\psi+\nu+\mu_3-\mu_2}[Y,Y^*]_2-4\,e^{-2\psi+\nu+\mu_3}F_{03}(YX^*-Y^*X)\}_{,2}$$
$$+\{e^{-3\psi+\nu-\mu_3+\mu_2}[X,X^*]_3+4\,e^{-\psi+\nu+\mu_2-\mu_3}[Y,Y^*]_3+4\,e^{-2\psi+\nu+\mu_2}F_{02}(YX^*-Y^*X)\}_{,3} = 0. \quad \text{(A 6)}$$

Equation (A 6) represents the flux integral for the problem on hand.

Application to the Reissner–Nordström space-time

In accordance with the solution for the axial perturbations of the Reissner–Nordström space-time given in *M.T.* (equations (131), (132), (137), (138), and (143) on pp. 228 and 229) we have:

$$X = e^{3\psi+\nu-\mu_2-\mu_3}Q_{23} = rH_2\,C_{l+2}^{-\frac{3}{2}}(\mu), \quad \text{(A 7)}$$

$$Y = e^{\psi+\nu}F_{01} = -\frac{3}{\sqrt{[(l-1)(l+2)]}}H_1\,C_{l+1}^{-\frac{1}{2}}(\mu), \quad \text{(A 8)}$$

where $C_{l+2}^{-\frac{3}{2}}$ and $C_{l+1}^{-\frac{1}{2}}$ are the Gegenbauer functions in their standard normalizations. From equations (A 7) and (A 8) we find:

$$e^{-3\psi+\nu-\mu_2+\mu_3}[X,X^*]_2 = \frac{\Delta}{r^2}[H_2,H_2^*]_r\frac{[C_{l+2}^{-\frac{3}{2}}(\mu)]^2}{(1-\mu^2)^2} \quad \text{(A 9)}$$

and

$$4\,e^{-\psi+\nu+\mu_3-\mu_2}[Y,Y^*]_2 = \frac{9}{(l-1)(l+2)}\frac{\Delta}{r^2}[H_1,H_1^*]_r\frac{[C_{l+1}^{-\frac{1}{2}}(\mu)]^2}{(1-\mu^2)}, \quad \text{(A 10)}$$

or, after integration over μ,

$$\int_{-1}^{+1} \{e^{-3\psi+\nu-\mu_2+\mu_3}[X, X^*]_2 + 4\,e^{-\psi+\nu+\mu_3-\mu_2}[Y, Y^*]_2\}\,d\mu$$

$$= \frac{9}{(l+\frac{1}{2})(l+2)(l+1)\,l(l-1)}\{[H_2, H_2^*]_{r_*} + [H_1, H_1^*]_{r_*}\} = \text{const.} \qquad \text{(A 11)}$$

Since $F_{03} = 0$ for the Reissner–Nordström space-time, it is manifest that equation (A 11) is consistent with the required constancy of the Wronskians for the two cases (68) and (69).

References

Burnett, G. & Wald, R. 1990 *Proc. R. Soc. Lond.* A **430**. (In the press.)
Chandrasekhar, S. 1983 *The mathematical theory of black holes.* Oxford: Clarendon Press.
Chandrasekhar, S. 1988 *Proc. R. Soc. Lond.* A **415**, 329–345.
Chandrasekhar, S. 1989 *Proc. R. Soc. Lond.* A **421**, 227–258.
Detweiler, S. L. & Ipser, J. R. 1978 *Astrophys. J.* **185**, 685–207.
Detweiler, S. L. & Lindblom, L. 1985 *Astrophys. J.* **292**, 12–15.
Friedman, J. L. 1978 *Communs math. Phys.* **62**, 247–278.
Lee, J. & Wald, R. 1990 (In the press.)
Lindblom, L. & Detweiler, S. L. 1983 *Astrophys. J. Suppl.* **53**, 73–92.
Thorne, K. S. & Campolattaro, A. 1967 *Astrophys. J.* **149**, 591–611.

Author's Note

Unfortunately the sign conventions used in §9 and in §§10–12 are "mixed." The field equation used in §9 is based on the convention $G_{(a)(b)} = +2T_{(a)(b)}$ (with $T_{(a)(b)}$ as defined in equation (91)); but the equation is used with the opposite sign in §§10–12. However, the analysis in §§10–12 has been carried through consistently since the only equilibrium equation that is used is $\nu_{,r} = -p_{,r}/(\varepsilon + p)$: and this equation is the same on both conventions.

35

The Teukolsky–Starobinsky constant for arbitrary spin

By Subrahmanyan Chandrasekhar

University of Chicago, The Enrico Fermi Institute, 933 East 56th Street, Chicago, Illinois 60637, U.S.A.

A determinantal expression for the Teukolsky–Starobinsky constant for arbitrary integral and half-odd integral spins is obtained.

1. Introduction

One of the many remarkable properties of the separated radial and angular functions (in terms of which the propagation, in Kerr geometry, of electromagnetic and gravitational waves, of spins $s = 1$ and $s = 2$, is described) is the relation between the functions belonging to opposite spins, $+s$ and $-s$. The relation is expressed by the '*Teukolsky–Starobinsky identities*' and the associated '*Teukolsky–Starobinsky constant*' (cf. *Mathematical theory of black holes* (Chandrasekhar 1983) §§70 and 81, pp. 387–391 and 436–441, Theorems I–V). Although the identities themselves are fairly straightforward to derive from the governing equations, a direct evaluation of the Teukolsky–Starobinsky constant (as on pp. 438–441 in *M.T.* for the case $s = 2$) is 'shrouded in complexity till the very last step when the veil is lifted.' But an alternative method of evaluation (Chandrasekhar 1984; this paper will be referred to hereafter as Paper I; for an earlier statement concerning this method, see Wald (1973)) based on the notion of *algebraically special solutions*, requires no more than an evaluation of a simple determinant of order 4.

More recently, Kalnins *et al.* (1989) have proved identities relating the solutions for opposite spins of equations that are the *formal*, generalizations of Teukolsky's equations for $s = \pm 1$ and $s = \pm 2$. Though the consideration of these equations, for a physically consistent description of particles or fields with spins greater than two, in curved space-times, is problematical, they, nevertheless, have a mathematical interest of their own.

In the Appendix to their paper, Kalnins *et al.* list the Teukolsky–Starobinsky constant, $|\mathscr{C}_s|^2$, for integral or half odd-integral spins $s \leqslant 4$. They do not state whether they were evaluated directly (as for the case $s = 2$) with or without the aid of computers.

In this short paper, we show how the notion of algebraically special solutions extended to Teukolsky's equations for $s > 2$, enables us to express the Teukolsky–Starobinsky constant explicitly as a determinant of order $2s$ whose evaluation even for $s = 3$ requires only a modest effort.

2. A general determinantal expression for $|\mathscr{C}_s|^2$

The formal generalizations, for arbitrary integral or half odd-integral spins $s > 0$, of Teukolsky's original equations for $s = \pm 1$ and ± 2, are (cf. *M.T.* p. 397, equations (96) and (97)):

$$\{\Delta \mathscr{D}_{1-s} \mathscr{D}_0^\dagger - [2\mathrm{i}\sigma(2s-1)\, r + \lambda]\} P_{+s} = 0, \tag{1}$$

$$\{\Delta \mathscr{D}_{1-s}^\dagger \mathscr{D}_0 + [2\mathrm{i}\sigma(2s-1)\, r - \lambda]\} P_{-s} = 0, \tag{2}$$

where (in the notation of *M.T.* adopted in Paper I),

$$\mathscr{D}_n^\dagger = (\mathscr{D}_n)^* = \frac{\mathrm{d}}{\mathrm{d}r} - \mathrm{i}\sigma \frac{\varpi^2}{\Delta} + 2n \frac{r-M}{\Delta}, \tag{3}$$

$$\varpi^2 = r^2 + \alpha^2, \quad \alpha^2 = a^2 + am/\sigma, \quad \text{and} \quad \Delta = r^2 - 2Mr + a^2. \tag{4}$$

Considering equations (1) and (2), Kalnins *et al.* have shown that the functions P_{+s} and P_{-s}, belonging to $+s$ and $-s (s > 0)$ are related by the identities,

$$\Delta^s \mathscr{D}_0^{\dagger 2s} P_{+s} = \mathscr{C}_1 P_{-s} \tag{5}$$

and

$$\Delta^s \mathscr{D}_0^{2s} P_{-s} = \mathscr{C}_2 P_{+s}, \tag{6}$$

where $\mathscr{C}_1$ and $\mathscr{C}_2$ are (complex) constants. It follows from these equations that

$$\Delta^s \mathscr{D}_0^{\dagger 2s} \Delta^s \mathscr{D}_0^{2s} P_{-s} = \mathscr{C}_1 \mathscr{C}_2 P_{-s}. \tag{7}$$

Therefore,

$$\Delta^s \mathscr{D}_0^{\dagger 2s} \Delta^s \mathscr{D}_0^{2s} \equiv \mathscr{C}_1 \mathscr{C}_2 = |\mathscr{C}_s|^2 \bmod [\Delta \mathscr{D}_{1-s}^\dagger \mathscr{D}_0 + 2\mathrm{i}\sigma(2s-1)\, r - \lambda = 0]. \tag{8}$$

An obvious way of determining $|\mathscr{C}_n|^2$ is to evaluate the right-hand side of equation (8) directly as on p. 438 of *M.T.* for $s = 2$. A much simpler way is to seek *an algebraically special solution for* P_{+s} *or* P_{-s} *when the one or the other vanishes identically.* By equation (5), *a non-vanishing solution for* P_{+s}, *associated with an identically vanishing* P_{-s}, *is possible, only if* $\mathscr{C}_2 = 0$ *and*

$$\mathscr{D}_0^{\dagger 2s} P_{+s} = 0. \tag{9}$$

Manifestly, algebraically special solutions describe waves propagated only in the forward or in the backward direction.

A non-trivial solution of equation (9) can be found by the same procedure as in Paper I for the case $s = 2$.

Since

$$\mathscr{D}_0^\dagger = \frac{\varpi^2}{\Delta}\left(\frac{\mathrm{d}}{\mathrm{d}r_*} - \mathrm{i}\sigma\right) \tag{10}$$

where r_* is a new radial coordinate defined by the equation (cf. *M.T.*, p. 398, equations (100) and (101)),

$$\frac{\mathrm{d}r_*}{\mathrm{d}r} = \frac{\varpi^2}{\Delta}, \tag{11}$$

it is manifest that

$$\mathscr{D}_0^\dagger (r^n \, \mathrm{e}^{\mathrm{i}\sigma r_*}) = n r^{n-1} \, \mathrm{e}^{\mathrm{i}\sigma r_*}. \tag{12}$$

Accordingly, the general solution of equation (9) is

$$P_{+s} = \mathrm{e}^{\mathrm{i}\sigma r_*} \sum_{j=0}^{2s-1} A_j r^j, \tag{13}$$

where $A_{2s-1}, \ldots, A_1, A_0$ are $2s$ constants of integration. The requirement, now, that P_{+s} given by equation (13) satisfies equation (1) leads to the condition

$$\Delta\left[\frac{\mathrm{d}}{\mathrm{d}r}+\mathrm{i}\sigma\frac{\varpi^2}{\Delta}-2(s-1)\frac{(r-M)}{\Delta}\right]\mathrm{e}^{\mathrm{i}\sigma r_*}\sum_{j=1}^{2s-1} A_j j r^{j-1}-[2\mathrm{i}\sigma(2s-1)\,r+\lambda]\,\mathrm{e}^{\mathrm{i}\sigma r_*}\sum_{j=0}^{2s-1} A_j r^j = 0, \tag{14}$$

or, more explicitly,

$$\{2\mathrm{i}\sigma r^2-2(s-1)\,r+2[(s-1)M+\mathrm{i}\sigma\alpha^2]\}\sum_{j=1}^{2s-1} A_j j r^{j-1}$$
$$+(r^2-2Mr+a^2)\sum_{j=2}^{2s-1} A_j j(j-1)\,r^{j-2}-[2\mathrm{i}\sigma(2s-1)\,r+\lambda]\sum_{j=0}^{2s-1} A_j r^j = 0. \tag{15}$$

The coefficients of the $2s$ different powers of r in equation (15) must vanish separately. For a non-trivial solution for the coefficients $A_j (j = 2s-1, \ldots, 1, 0)$, the determinant of the system of equations which follow from equation (15) must vanish; and we find:

$$\begin{vmatrix}
\lambda & 2i\sigma & 0 & 0 & 0 & 0 \\
2(2s-1)[(s-1)M-i\sigma\alpha^2] & \lambda+(2s-2) & 4i\sigma & 0 & 0 & 0 \\
-(2s-1)(2s-2)a^2 & 2(2s-2)[(s-2)M-i\sigma\alpha^2] & \lambda+2(2s-3) & 6i\sigma & 0 & 0 \\
0 & -(2s-2)(2s-3)a^2 & 2(2s-3)[(s-3)M-i\sigma\alpha^2] & \lambda+3(2s-4) & 8i\sigma & 0 \\
0 & 0 & -(2s-3)(2s-4)a^2 & 2(2s-4)[(s-4)M-i\sigma\alpha^2] & \lambda+4(2s-5) & 10i\sigma \\
0 & 0 & 0 & -(2s-4)(2s-5)a^2 & 2(2s-5)[(s-5)M-i\sigma\alpha^2] & \lambda+5(2s-6) \\
0 & 0 & 0 & 0 & -(2s-5)(2s-6)a^2 & 2(2s-6)[(s-6)M-i\sigma\alpha^2] \\
0 & 0 & 0 & 0 & 0 & -(2s-6)(2s-7)a^2
\end{vmatrix} = 0. \quad (16)$$

A simpler form of the determinant, which exhibits the underlying symmetries of the equations, is obtained by dividing the successive columns by $2s-1, 2s-2, \ldots, ., 1$. (Note the repetition of '1' for the last two columns.) We thus obtain

$$|\mathscr{C}_s|^2 = \begin{bmatrix}
\lambda/(2s-1) & 2i\sigma/(2s-2) & 0 & 0 & 0 & 0 \\
2(\bar{M}-M) & 1+\lambda/(2s-2) & 4i\sigma/(2s-3) & 0 & 0 & 0 \\
-(2s-2)a^2 & 2(\bar{M}-2M) & 2+\lambda/(2s-3) & 6i\sigma/(2s-4) & 0 & 0 \\
0 & -(2s-3)a^2 & 2(\bar{M}-3M) & 3+\lambda/(2s-4) & 8i\sigma/(2s-5) & 0 \\
0 & 0 & -(2s-4)a^2 & 2(\bar{M}-4M) & 4+\lambda/(2s-5) & 10i\sigma/(2s-6) \\
0 & 0 & 0 & -(2s-5)a^2 & 2(\bar{M}-5M) & 5+\lambda/(2s-6) \\
0 & 0 & 0 & 0 & -(2s-6)a^2 & 2(\bar{M}-6M) \\
0 & 0 & 0 & 0 & 0 & -(2s-7)a^2
\end{bmatrix}, \quad (17)$$

where
$$\overline{M} = sM - i\sigma\alpha^2. \tag{18}$$

It should be noted that the determinants (16) and (17) are to be extended (in an obvious way) beyond the rows and columns shown and truncated appropriately for a specified s.

We shall now consider special cases of equation (17).

$s = \frac{3}{2}$: $\overline{M} = \frac{3}{2}M - i\sigma\alpha^2$; $2(\overline{M} - M) = M - 2i\sigma\alpha^2 = -2(\overline{M} - 2M)^*$

$$|\mathscr{C}_{\frac{3}{2}}|^2 = \begin{vmatrix} \frac{1}{2}\lambda & 2i\sigma & 0 \\ 2(\overline{M}-M) & 1+\lambda & 4i\sigma \\ -a^2 & 2(\overline{M}-2M) & \lambda \end{vmatrix}$$

$$= \tfrac{1}{2}\lambda^2(\lambda+1) + 8\sigma^2(a^2 - \lambda\alpha^2). \tag{19}$$

$s = 2$: $A = \frac{1}{3}\lambda$, $B = 1 + \frac{1}{2}\lambda$:

$$\overline{M} - M = M - i\sigma\alpha^2 = -(\overline{M} - 3M)^* = \tfrac{1}{2}\beta; \quad \overline{M} - 2M = -i\sigma\alpha^2.$$

$$|\mathscr{C}_2|^2 = \left|\begin{array}{cccc} A & i\sigma & 0 & 0 \\ \beta & B & 4i\sigma & 0 \\ \hdashline -2a^2 & -2i\sigma\alpha^2 & 2B & 6i\sigma \\ 0 & -a^2 & -\beta^* & 3A \end{array}\right|$$

$$= 6|AB - i\sigma\beta|^2 + 24A\sigma^2(2a^2 - A\alpha^2). \tag{20}$$

$s = \frac{5}{2}$: $A = \frac{1}{4}\lambda$, $B = \frac{1}{3}\lambda + 1$, $C = \frac{1}{2}\lambda + 2$;

$$\beta = 2(\overline{M} - M) = 3M - 2i\sigma\alpha^2 = -2(\overline{M} - 4M)^*;$$
$$\gamma = 2(\overline{M} - 2M) = M - 2i\sigma\alpha^2 = -2(\overline{M} - 3M)^*.$$

$$|\mathscr{C}_{\frac{5}{2}}|^2 = \left|\begin{array}{ccccc} A & \frac{2}{3}i\sigma & 0 & 0 & 0 \\ \beta & B & 2i\sigma & 0 & 0 \\ -3a^2 & \gamma & C & 6i\sigma & 0 \\ \hdashline 0 & -2a^2 & -\gamma^* & 3B & 8i\sigma \\ 0 & 0 & -a^2 & -\beta^* & 4A \end{array}\right|$$

$$= 12C|AB - \tfrac{2}{3}i\sigma\beta|^2 + 96A^2a^2\sigma^2$$
$$+ 48\,\mathrm{Im}\, i\sigma(AB - \tfrac{2}{3}i\sigma\beta)(A\gamma^* - 2i\sigma a^2). \tag{21}$$

$s = 3$: $A = \frac{1}{5}\lambda$, $B = 1 + \frac{1}{4}\lambda$, $C = 2 + \frac{1}{3}\lambda$;

$$\tfrac{1}{2}\beta = \overline{M} - M = 2M - i\sigma\alpha^2 = -(\overline{\overline{M}} - 5M)^*;$$
$$\tfrac{1}{2}\gamma = \overline{M} - 2M = M - i\sigma\alpha^2 = -(\overline{M} - 4M)^*; \quad \overline{M} - 3M = -i\sigma\alpha^2.$$

$$
|\mathscr{C}_3|^2 = \left|\begin{array}{cccccc}
A & \frac{1}{2}\mathrm{i}\sigma & 0 & 0 & 0 & 0 \\
\beta & B & \frac{4}{3}\mathrm{i}\sigma & 0 & 0 & 0 \\
-4a^2 & \gamma & C & 3\mathrm{i}\sigma & 0 & 0 \\
0 & -3a^2 & -2\mathrm{i}\sigma\alpha^2 & \frac{3}{2}C & 8\mathrm{i}\sigma & 0 \\
\hdashline
0 & 0 & -2a^2 & -\gamma^* & 4B & 10\mathrm{i}\sigma \\
0 & 0 & 0 & -a^2 & -\beta^* & 5A
\end{array}\right|
$$

$$
\begin{aligned}
&= (30C^2 - 120\sigma^2\alpha^2)\,|AB - \tfrac{1}{2}\mathrm{i}\beta\sigma|^2 + \tfrac{160}{3}\sigma^2|A\gamma + 2\mathrm{i}\sigma a^2|^2 \\
&\quad + 480Aa^2\sigma^2\,\mathrm{Re}\,(AB + \tfrac{1}{2}\mathrm{i}\sigma\beta^*) \\
&\quad + 80C\,\mathrm{Im}\,\mathrm{i}\sigma(AB - \tfrac{1}{2}\mathrm{i}\sigma\beta)\,(A\gamma^* - 2\mathrm{i}\sigma\alpha^2).
\end{aligned} \tag{22}
$$

It may be noted here that in all cases, the determinants are most easily evaluated by expanding them with respect to the minors of the last two rows.

The expressions (21) and (22) for $|\mathscr{C}_2|^2$ and $|\mathscr{C}_3|^2$, agree with those given by Kalnins *et al.*

3. Concluding remark

The structure of the determinantal expression (17) for $|\mathscr{C}_n|^2$ suggests that there may perhaps lie concealed some recurrence relations among them.

The research reported in this paper has, in part, been supported by grants from the National Science Foundation under Grant PHY-89-18388 with the University of Chicago.

References

Chandrasekhar, S. 1983 *The mathematical theory of black holes.* Oxford: Clarendon Press.

Chandrasekhar, S. 1984 *Proc. R. Soc. Lond.* A **392**, 1.

Kalkins, E. G., Miller, W. Jr & Williams, G. C. 1989 *J. math. Phys.* **30**, 2925.

Wald, R. M. 1973 *J. math. Phys.* **14**, 1453.

Received 9 April 1990; accepted 5 June 1990

36

ON STARS, THEIR EVOLUTION AND THEIR STABILITY

Nobel lecture, 8 December, 1983

by

SUBRAHMANYAN CHANDRASEKHAR

The University of Chicago, Chicago, Illinois 60637, USA

1. *Introduction*

When we think of atoms, we have a clear picture in our minds: a central nucleus and a swarm of electrons surrounding it. We conceive them as small objects of sizes measured in Ängstroms ($\sim 10^{-8}$ cm); and we know that some hundred different species of them exist. This picture is, of course, quantified and made precise in modern quantum theory. And the success of the entire theory may be traced to two basic facts: *first*, the Bohr radius of the ground state of the hydrogen atom, namely,

$$\frac{h^2}{4\pi^2 me^2} \simeq 0.5 \times 10^{-8} \text{ cm}, \tag{1}$$

where h is Planck's constant, m is the mass of the electron and e is its charge, provides a correct measure of atomic dimensions; and *second*, the reciprocal of *Sommerfeld's fine-structure constant,*

$$\frac{hc}{2\pi e^2} \sim 137, \tag{2}$$

gives the maximum positive charge of the central nucleus that will allow a stable electron-orbit around it. This maximum charge for the central nucleus arises from the effects of special relativity on the motions of the orbiting electrons.

We now ask: can we understand the basic facts concerning stars as simply as we understand atoms in terms of the two combinations of natural constants (1) and (2). In this lecture, I shall attempt to show that in a limited sense we can.

The most important fact concerning a star is its mass. It is measured in units of the mass of the sun, $\odot$, which is 2×10^{33} gm: stars with masses very much less than, or very much more than, the mass of the sun are relatively infrequent. The current theories of stellar structure and stellar evolution derive their successes largely from the fact that the following combination of the dimensions of a mass provides a correct measure of stellar masses:

$$\left(\frac{hc}{G}\right)^{3/2} \frac{1}{H^2} \simeq 29.2 \odot, \tag{3}$$

where G is the constant of gravitation and H is the mass of the hydrogen atom. In the first half of the lecture, I shall essentially be concerned with the question: how does this come about?

2. *The role of radiation pressure*

A central fact concerning normal stars is the role which radiation pressure plays as a factor in their hydrostatic equilibrium. Precisely the equation governing the hydrostatic equilibrium of a star is

$$\frac{dP}{dr} = -\frac{GM(r)}{r^2}\rho, \tag{4}$$

where P denotes the total pressure, ρ the density, and $M(r)$ is the mass interior to a sphere of radius r. There are two contributions to the total pressure P : that due to the material and that due to the radiation. On the assumption that the matter is in the state of a perfect gas in the classical Maxwellian sense, the material or the gas pressure is given by

$$p_{\text{gas}} = \frac{k}{\mu H}\rho T, \tag{5}$$

where T is the absolute temperature, k is the Boltzmann constant, and μ is the mean molecular weight (which under normal stellar conditions is ~ 1.0). The pressure due to radiation is given by

$$p_{\text{rad}} = \frac{1}{3}aT^4, \tag{6}$$

where a denotes Stefan's radiation-constant. Consequently, if radiation contributes a fraction $(1-\beta)$ to the total pressure, we may write

$$P = \frac{1}{1-\beta}\frac{1}{3}aT^4 = \frac{1}{\beta}\frac{k}{\mu H}\rho T. \tag{7}$$

To bring out explicitly the role of the radiation pressure in the equilibrium of a star, we may eliminate the temperature, T, from the foregoing equations and express P in terms of ρ and β instead of in terms of ρ and T. We find:

$$T = \left(\frac{k}{\mu H}\frac{3}{\text{a}}\frac{1-\beta}{\beta}\right)^{1/3}\rho^{1/3} \tag{8}$$

and

$$P = \left[\left(\frac{k}{\mu H}\right)^4 \frac{3}{a}\frac{1-\beta}{\beta^4}\right]^{1/3}\rho^{4/3} = C(\beta)\rho^{4/3}\ \text{(say)}. \tag{9}$$

The importance of this ratio, $(1-\beta)$, for the theory of stellar structure was first emphasized by Eddington. Indeed, he related it, in a famous passage in his book on *The Internal Constitution of the Stars,* to the 'happening of the stars'.[1] A more rational version of Eddington's argument which, at the same time, isolates the combination (3) of the natural constants is the following:

There is a general theorem[2] which states that the pressure, P_c, at the centre of a star of a mass M in hydrostatic equilibrium in which the density, $\rho(r)$, at a point at a radial distance, r, from the centre does not exceed the mean density, $\bar{\rho}(r)$, interior to the same point r, must satisfy the inequality,

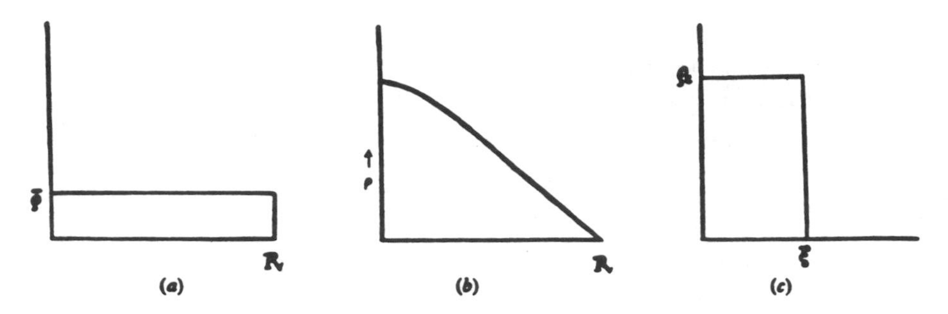

Fig. 1. A comparison of an inhomogeneous distribution of density in a star (*b*) with the two homogeneous configurations with the constant density equal to the mean density (*a*) and equal to the density at the centre (*c*).

$$\frac{1}{2}G\left(\frac{4}{3}\pi\right)^{1/3}\bar{\rho}^{4/3}M^{2/3}\leqslant P_c\leqslant\frac{1}{2}G\left(\frac{4}{3}\pi\right)^{1/3}\rho_c^{4/3}M^{2/3}, \tag{10}$$

where $\bar{\rho}$ denotes the mean density of the star and ρ_c its density at the centre. The content of the theorem is no more than the assertion that the actual pressure at the centre of a star must be intermediate between those at the centres of the two configurations of uniform density, one at a density equal to the mean density of the star, and the other at a density equal to the density ρ_c at the centre (see Fig. 1). If the inequality (10) should be violated then there must, in general, be some regions in which adverse density gradients must prevail; and this implies instability. In other words, we may consider conformity with the inequality (10) as equivalent to the condition for the stable existence of stars.

The right-hand side of the inequality (10) together with P given by equation (9), yields, for the stable existence of stars, the condition,

$$\left[\left(\frac{k}{\mu H}\right)^4\frac{3}{a}\frac{1-\beta_c}{\beta_c^4}\right]^{1/3}\leqslant\left(\frac{\pi}{6}\right)^{1/3}GM^{2/3}, \tag{11}$$

or, equivalently,

$$M\geqslant\left(\frac{6}{\pi}\right)^{1/2}\left[\left(\frac{k}{\mu H}\right)^4\frac{3}{a}\frac{1-\beta_c}{\beta_c^4}\right]^{1/2}\frac{1}{G^{3/2}}, \tag{12}$$

where in the foregoing inequalities, β_c is a value of β at the centre of the star. Now Stefan's constant, a, by virtue of Planck's law, has the value

$$a=\frac{8\pi^5k^4}{15h^3c^3}. \tag{13}$$

Inserting this value a in the inequality (12) we obtain

$$\mu^2M\left(\frac{\beta_c^4}{1-\beta_c}\right)^{1/2}\geqslant\frac{(135)^{1/2}}{2\pi^3}\left(\frac{hc}{G}\right)^{3/2}\frac{1}{H^2}=0.1873\left(\frac{hc}{G}\right)^{3/2}\frac{1}{H^2}. \tag{14}$$

We observe that the inequality (14) has isolated the combination (3) of natural constants of the dimensions of a mass; by inserting its numerical value given in equation (3), we obtain the inequality,

$$\mu^2 M \left(\frac{\beta_c^4}{1-\beta_c} \right)^{1/2} \geqslant 5.48 \odot . \tag{15}$$

This inequality provides an upper limit to $(1-\beta_c)$ for a star of a given mass. Thus,

$$1-\beta_c \leqslant 1-\beta_*, \tag{16}$$

where $(1-\beta_*)$ is uniquely determined by the mass M of the star and the mean molecular weight, μ, by the quartic equation,

$$\mu^2 M = 5.48 \left(\frac{1-\beta_*}{\beta_*^4} \right)^{1/2} \odot . \tag{17}$$

In Table 1, we list the values of $1-\beta_*$ for several values of $\mu^2 M$. From this table it follows in particular, that for a star of solar mass with a mean molecular weight equal to 1, the radiation pressure at the centre cannot exceed 3 percent of the total pressure.

Table 1
The maximum radiation pressure, $(1-\beta_*)$, at the centre of a star of a given mass, M.

$1-\beta_*$	$M\mu^2/\odot$	$1-\beta_*$	$M\mu^2/\odot$
0.01	0.56	0.50	15.49
.03	1.01	.60	26.52
.10	2.14	.70	50.92
.20	3.83	.80	122.5
.30	6.12	.85	224.4
0.40	9.62	0.90	519.6

What do we conclude from the foregoing calculation? We conclude that to the extent equation (17) is at the base of the equilibrium of actual stars, to that extent the combination of natural constants (3), providing a mass of proper magnitude for the measurement of stellar masses, is at the base of a physical theory of stellar structure.

3. *Do stars have enough energy to cool?*

The same combination of natural constants (3) emerged soon afterward in a much more fundamental context of resolving a paradox Eddington had formulated in the form of an aphorism: 'a star will need energy to cool.' The paradox arose while considering the ultimate fate of a gaseous star in the light of the then new knowledge that white-dwarf stars, such as the companion of Sirius, exist, which have mean densities in the range 10^5-10^7 gm cm^{-3}. As Eddington stated[3]

> I do not see how a star which has once got into this compressed state is ever going to get out of it.... It would seem that the star will be in an awkward predicament when its supply of subatomic energy fails.

The paradox posed by Eddington was reformulated in clearer physical terms by R. H. Fowler.[4] His formulation was the following:

> The stellar material, in the white-dwarf state, will have radiated so much energy that it has less energy than the same matter in normal atoms expanded at the absolute zero of temperature. If part of it were removed from the star and the pressure taken off, what could it do?

Quantitatively, Fowler's question arises in this way.

An estimate of the electrostatic energy, E_V, per unit volume of an assembly of atoms, of atomic number Z, ionized down to bare nuclei, is given by

$$E_V = 1.32\times10^{11} Z^2 \rho^{4/3}, \tag{18}$$

while the kinetic energy of thermal motions, E_{kin}, per unit volume of free particles in the form of a perfect gas of density, ρ, and temperature, T, is given by

$$E_{\text{kin}} = \frac{3}{2}\frac{k}{\mu H}\rho\, T = \frac{1.24\times10^8}{\mu}\rho\, T. \tag{19}$$

Now if such matter were released of the pressure to which it is subject, it can resume a state of ordinary normal atoms only if

$$E_{\text{kin}} > E_V, \tag{20}$$

or, according to equations (18) and (19), only if

$$\rho < \left(0.94\times10^{-3}\frac{T}{\mu Z^2}\right)^3. \tag{21}$$

This inequality will be clearly violated if the density is sufficiently high. This is the essence of Eddington's paradox as formulated by Fowler. And Fowler resolved this paradox in 1926 in a paper[4] entitled 'Dense Matter' — one of the great landmark papers in the realm of stellar structure: in it the notions of Fermi statistics and of electron degeneracy are introduced for the first time.

4. *Fowler's resolution of Eddington's paradox; the degeneracy of the electrons in white-dwarf stars*

In a completely degenerate electron gas all available parts of the phase space, with momenta less than a certain 'threshold' value p_0 — the Fermi 'threshold' — are occupied consistently with the Pauli exclusion-principle i.e., with two electrons per 'cell' of volume h^3 of the six-dimensional phase space. Therefore, if $n(p)\,dp$ denotes the number of electrons, per unit volume, between p and $p+dp$, then the assumption of complete degeneracy is equivalent to the assertion,

$$\left.\begin{aligned} n(p) &= \frac{8\pi}{h^3}p^2 && (p \leq p_0), \\ &= 0 && (p > p_0). \end{aligned}\right\} \tag{22}$$

The value of the threshold momentum p_0, is determined by the normalization condition

$$n = \int_0^{p_0} n(p)dp = \frac{8\pi}{3h^3}p_0^3, \tag{23}$$

where n denotes the total number of electrons per unit volume.

For the distribution given by (22), the pressure p and the kinetic energy E_{kin} of the electrons (per unit volume), are given by

$$P = \frac{8\pi}{3h^3}\int_0^{p_0} p^3 v_p \, dp \tag{24}$$

and

$$E_{\text{kin}} = \frac{8\pi}{h^3}\int_0^{p_0} p^2 \, T_p \, dp, \tag{25}$$

where v_p and T_p are the velocity and the kinetic energy of an electron having a momentum p.

If we set

$$v_p = p/m \text{ and } T_p = p^2/2m, \tag{26}$$

appropriate for non-relativistic mechanics, in equations (24) and (25), we find

$$P = \frac{8\pi}{15h^3 m}p_0^5 = \frac{1}{20}\left(\frac{3}{\pi}\right)^{2/3}\frac{h^2}{m}n^{5/3} \tag{27}$$

and

$$E_{\text{kin}} = \frac{8\pi}{10h^3 m}p_0^5 = \frac{3}{40}\left(\frac{3}{\pi}\right)^{2/3}\frac{h^2}{m}n^{5/3}. \tag{28}$$

Fowler's resolution of Eddington's paradox consists in this: at the temperatures and densities that may be expected to prevail in the interiors of the white-dwarf stars, the electrons will be highly degenerate and E_{kin} must be evaluated in accordance with equation (28) and *not* in accordance with equation (19); and equation (28) gives,

$$E_{\text{kin}} = 1.39 \times 10^{13} \, (\rho/\mu)^{5/3}. \tag{29}$$

Comparing now the two estimates (18) and (29), we see that, for matter of the density occurring in the white dwarfs, namely $\rho \approx 10^5$ gm cm^{-3}, the total kinetic energy is about two to four times the negative potential-energy; and Eddington's paradox does not arise. Fowler concluded his paper with the following highly perceptive statement:

> The black-dwarf material is best likened to a single gigantic molecule in its lowest quantum state. On the Fermi-Dirac statistics, its high density can be achieved in one and only one way, in virtue of a correspondingly great energy content. But this energy can no more be expended in radiation than the energy of a normal atom or molecule. The only difference between black-dwarf matter and a normal molecule is that the molecule can exist in a free state while the black-dwarf matter can only so exist under very high external pressure.

5. *The theory of the white-dwarf stars; the limiting mass*

The internal energy (= $3P/2$) of a degenerate electron gas that is associated with a pressure P is *zero-point energy;* and the essential content of Fowler's paper is that this zero-point energy is so great that we may expect a star to eventually settle down to a state in which all of its energy is of this kind. Fowler's argument can be more explicitly formulated in the following manner.[5]

According to the expression for the pressure given by equation (27), we have the relation,

$$P = K_1\rho^{5/3} \text{ where } K_1 = \frac{1}{20}\left(\frac{3}{\pi}\right)^{2/3}\frac{h^2}{m(\mu_e H)^{5/3}}, \tag{30}$$

where μ_e is the mean molecular weight per electron. An equilibrium configuration in which the pressure, P, and the density ρ, are related in the manner,

$$P = K\rho^{1+1/n}, \tag{31}$$

is an *Emden polytrope* of index n. The degenerate configurations built on the equation of state (30) are therefore polytropes of index 3/2; and the theory of polytropes immediately provides the relation,

$$K_1 = 0.4242\,(GM^{1/3}R) \tag{32}$$

or, numerically, for K_1 given by equation (30),

$$\log_{10}(R/R_\odot) = -\frac{1}{3}\log_{10}(M/\odot) - \frac{5}{3}\log_{10}\mu_e - 1.397. \tag{33}$$

For a mass equal to the solar mass and $\mu_e = 2$, the relation (33) predicts $R = 1.26\times 10^{-2}R_\odot$ and a mean density of 7.0×10^5 gm/cm^3. These values are precisely of the order of the radii and mean densities encountered in white-dwarf stars. Moreover, according to equations (32) and (33), the radius of the white-dwarf configuration is inversely proportional to the cube root of the mass. On this account, finite equilibrium configurations are predicted for all masses. And it came to be accepted that the white-dwarfs represent the last stages in the evolution of all stars.

But it soon became clear that the foregoing simple theory based on Fowler's premises required modifications. For, the electrons at their threshold energies, at the centres of the degenerate stars, begin to have velocities comparable to that of light as the mass increases. Thus, already for a degenerate star of solar mass (with $\mu_e = 2$) the central density (which is about six times the mean density) is 4.19×10^6 gm/cm^3; and this density corresponds to a threshold momentum $p_0 = 1.29\, mc$ and a velocity which is $0.63\, c$. Consequently, the equation of state must be modified to take into account the effects of special relativity. And this is easily done by inserting in equations (24) and (25) the relations,

$$v_p = \frac{p}{m\,(1+p^2/m^2c^2)^{1/2}} \text{ and } T_p = mc^2\left[(1+p^2/m^2c^2)^{1/2}-1\right], \tag{34}$$

in place of the non-relativistic relations (26). We find that the resulting equation of state can be expressed, parametrically, in the form

$$P = A f(x) \text{ and } \rho = Bx^3, \tag{35}$$

where

$$A = \frac{\pi m^4 c^5}{3h^3},\ B = \frac{8\pi m^3 c^3 \mu_e H}{3h^3} \tag{36}$$

and

$$f(x) = x(x^2+1)^{1/2}(2x^2-3)+3\sinh^{-1} x. \tag{37}$$

And similarly

$$E_{\text{kin}} = Ag(x), \tag{38}$$

where

$$g(x) = 8x^3\left[(x^2+1)^{1/2}-1\right]-f(x). \tag{39}$$

According to equations (35) and (36), the pressure approximates the relation (30) for low enough electron concentrations ($x \ll 1$); but for increasing electron concentrations ($x \gg 1$), the pressure tends to[6]

$$P = \frac{1}{8}\left(\frac{3}{\pi}\right)^{1/3} hc\, n^{4/3}. \tag{40}$$

This limiting form of relation can be obtained very simply by setting $v_p = c$ in equation (24); then

$$P = \frac{8\pi c}{3h^3}\int_0^{p_0} p^3\, dp = \frac{2\pi c}{3h^3}p_0^4; \tag{41}$$

and the elimination of p_0 with the aid of equation (23) directly leads to equation (40).

While the modification of the equation of state required by the special theory of relativity appears harmless enough, it has, as we shall presently show, a dramatic effect on the predicted mass-radius relation for degenerate configurations.

The relation between P and ρ corresponding to the limiting form (41) is

$$P = K_2 \rho^{4/3} \text{ where } K_2 = \frac{1}{8}\left(\frac{3}{\pi}\right)^{1/3} \frac{hc}{(\mu_e H)^{4/3}}. \tag{42}$$

In this limit, the configuration is an Emden polytrope of index 3. And it is well known that when the polytropic index is 3, the mass of the resulting equilibrium configuration is uniquely determined by the constant of proportionality, K_2, in the pressure-density relation. We have accordingly,

$$M_{\text{limit}} = 4\pi \left(\frac{K_2}{\pi G}\right)^{3/2} (2.018) = 0.197 \left(\frac{hc}{G}\right)^{3/2} \frac{1}{\mu_e H)^2} = 5.76 \mu_e^{-2} \odot. \tag{43}$$

(In equation (43), 2.018 is a numerical constant derived from the explicit solution of the Lane-Emden equation for $n = 3$.)

It is clear from general considerations[7] that *the exact mass-radius relation for the degenerate configurations must provide an upper limit to the mass of such configurations given by equation* (43); *and further, that the mean density of the configuration must tend to infinity,* while the radius tends to zero, and $M \rightarrow M_{\text{limit}}$. These conditions, straightforward as they are, can be established directly by considering the equilibrium of configurations built on the exact equation of state given by equations (35)–(37). It is found that the equation governing the equilibrium of such configurations can be reduced to the form[8,9]

$$\frac{1}{\eta^2} \frac{d}{d\eta}\left(\eta^2 \frac{d\phi}{d\eta}\right) = -\left(\phi^2 - \frac{1}{y_0^2}\right)^{3/2}, \tag{44}$$

where

$$y_0^2 = x_0^2 + 1, \tag{45}$$

and mcx_0 denotes the threshold momentum of the electrons at the centre of the configuration and η measures the radial distance in the unit

$$\left(\frac{2A}{\pi G}\right)^{1/2} \frac{1}{By_0} = l_1 y_0^{-1} \text{ (say)}. \tag{46}$$

By integrating equation (44), with suitable boundary conditions and for various initially prescribed values of y_0, we can derive the exact mass-radius relation, as well as the other equilibrium properties, of the degenerate configurations. The principal results of such calculations are illustrated in Figures 2 and 3.

The important conclusions which follow from the foregoing considerations are: *first,* there is an upper limit, M_{limit}, to the mass of stars which can become degenerate configurations, as the last stage in their evolution; and *second,* that stars with $M > M_{\text{limit}}$ must have end states which cannot be predicted from the considerations we have presented so far. And finally, we observe that the

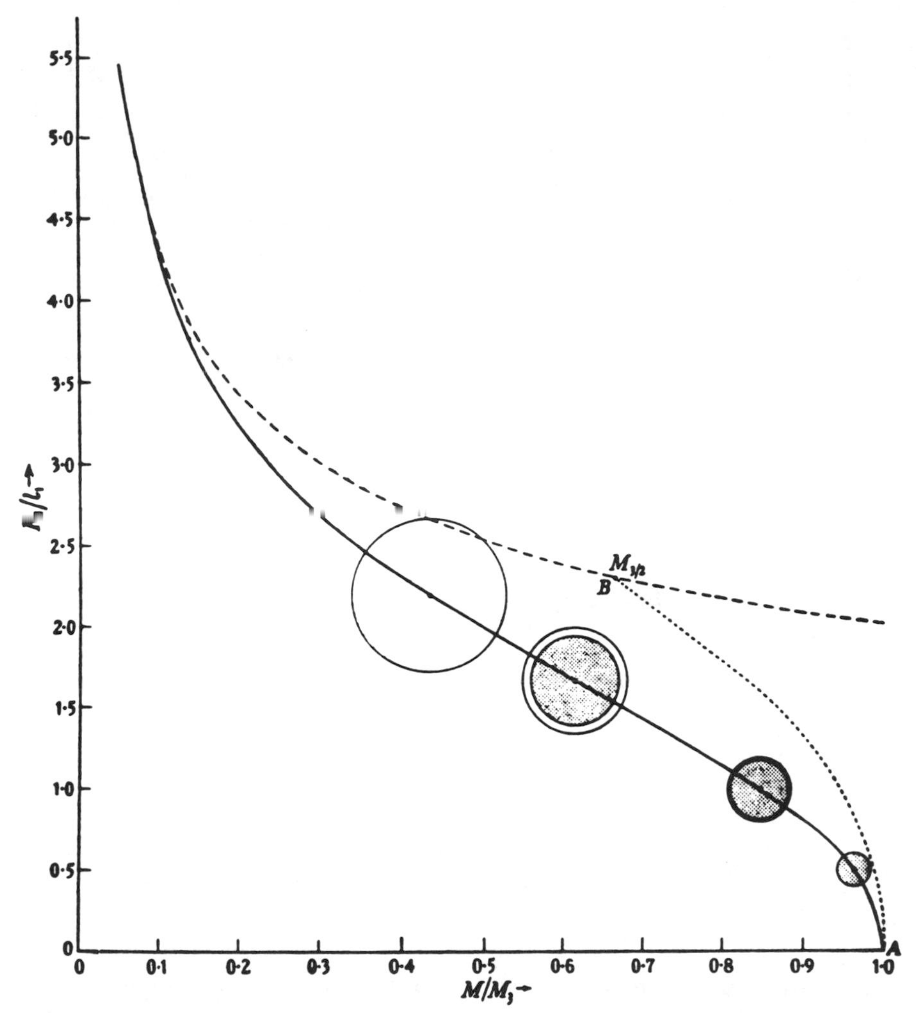

Fig. 2. The full-line curve represents the exact (mass-radius)-relation (l_1 is defined in equation (46) and M_3 denotes the limiting mass). This curve tends asymptotically to the ---- curve appropriate to the low-mass degenerate configurations, approximated by polytropes of index 3/2. The regions of the configurations which may be considered as relativistic ($\varrho > (K_1/K_2)^3$) are shown shaded. (From Chandrasekhar, S., *Mon. Not. Roy. Astr. Soc.*, 95, 207 (1935).)

combination of the natural constant (3) now emerges in the fundamental context of M_{limit} given by equation (43): its significance for the theory of stellar structure and stellar evolution can no longer be doubted.

6. *Under what conditions can normal stars develop degenerate cores?*

Once the upper limit to the mass of completely degenerate configurations had been established, the question that required to be resolved was how to relate its existence to the evolution of stars from their gaseous state. If a star has a mass less than M_{limit}, the assumption that it will eventually evolve towards the completely degenerate state appears reasonable. But what if its mass is greater than M_{limit}?

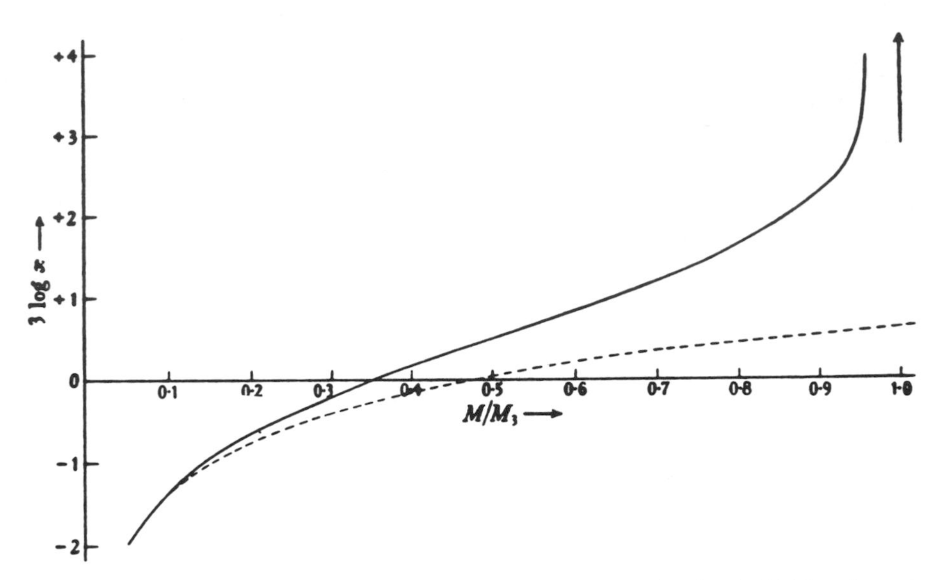

Fig. 3. The full-line curve represents the exact (mass-density)-relation for the highly collapsed configurations. This curve tends asymptotically to the dotted curve as $M \to 0$. (From Chandrasekhar, S., *Mon. Not. Roy. Astr. Soc.*, 95, 207 (1935).)

Clues as to what might ensue were sought in terms of the equations and inequalities of §§2 and 3.[10, 11]

The first question that had to be resolved concerns the circumstances under which a star, initially gaseous, will develop degenerate cores. From the physical side, the question, when departures from the perfect-gas equation of state (5) will set in and the effects of electron degeneracy will be manifested, can be readily answered.

Suppose, for example, that we continually and steadily increase the density, at constant temperature, of an assembly of free electrons and atomic nuclei, in a highly ionized state and initially in the form of a perfect gas governed by the equation of state (5). At first the electron pressure will increase linearly with ρ; but soon departures will set in and eventually the density will increase in accordance with the equation of state that describes the fully degenerate electron-gas (see Fig. 4). The remarkable fact is that this limiting form of the equation of state is independent of temperature.

However, to examine the circumstances when, during the course of evolution, a star will develop degenerate cores, it is more convenient to express the electron pressure (as given by the classical perfect-gas equation of state) in terms of ρ and β_e defined in the manner (cf. equation (7)).

$$p_e = \frac{k}{\mu_e H}\rho T = \frac{\beta_e}{1-\beta_e}\frac{1}{3}a\,T^4, \tag{47}$$

where p_e now denotes the electron pressure. Then, analogous to equation (9), we can write

$$p_e = \left[\left(\frac{k}{\mu_e H}\right)^4 \frac{3}{a}\frac{1-\beta_e}{\beta_e}\right]^{1/3} \rho^{4/3}. \tag{48}$$

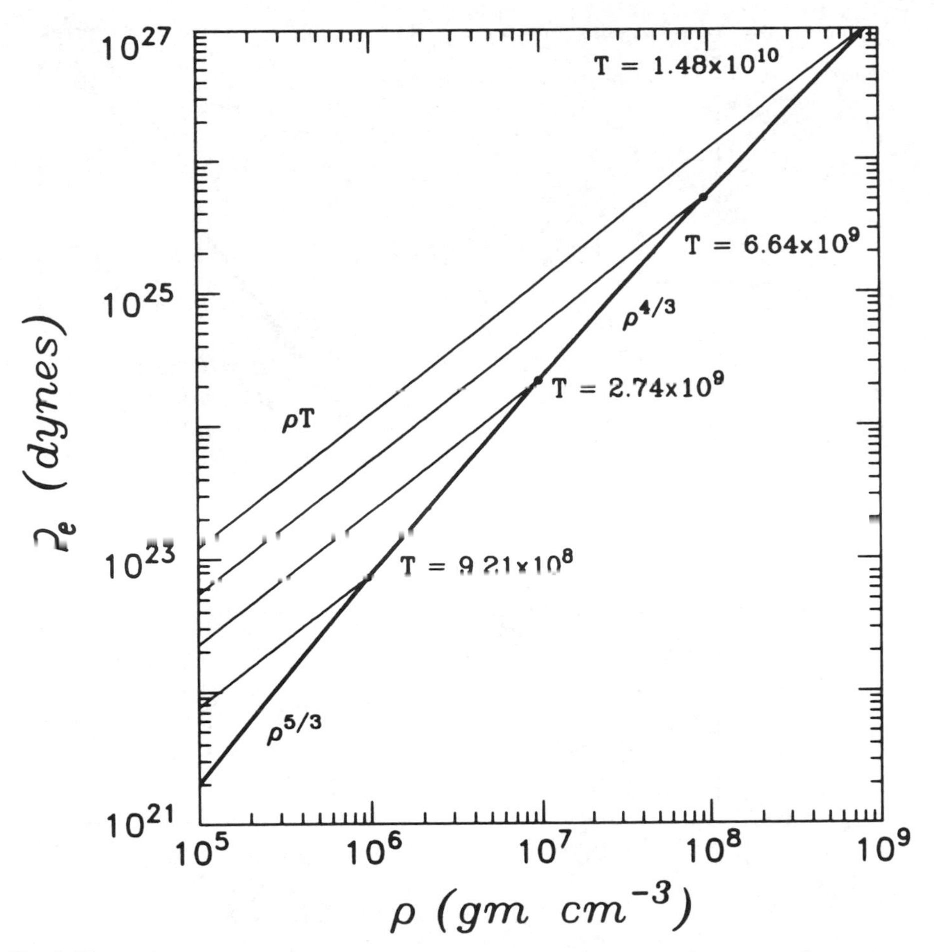

Fig. 4. Illustrating how by increasing the density at constant temperature degeneracy always sets in.

Comparing this with equation (42), we conclude that if

$$\left[\left(\frac{k}{\mu_e H}\right)^4 \frac{3}{a}\frac{1-\beta_e}{\beta_e}\right]^{1/3} > K_2 = \frac{1}{8}\left(\frac{3}{\pi}\right)^{1/3}\frac{hc}{(\mu_e H)^{4/3}}, \tag{49}$$

the pressure p_e given by the classical perfect-gas equation of state will be greater than that given by the equation if degeneracy were to prevail, not only for the prescribed ρ and T, but for *all* p and T *having the same* β_e.

Inserting for a its value given in equation (13), we find that the inequality (49) reduces to

$$\frac{960}{\pi^4}\frac{1-\beta_e}{\beta_e} > 1, \tag{50}$$

or equivalently,

$$1-\beta_e > 0.0921 = 1-\beta_\omega \text{ (say)}. \tag{51}$$

(See Fig. 5)

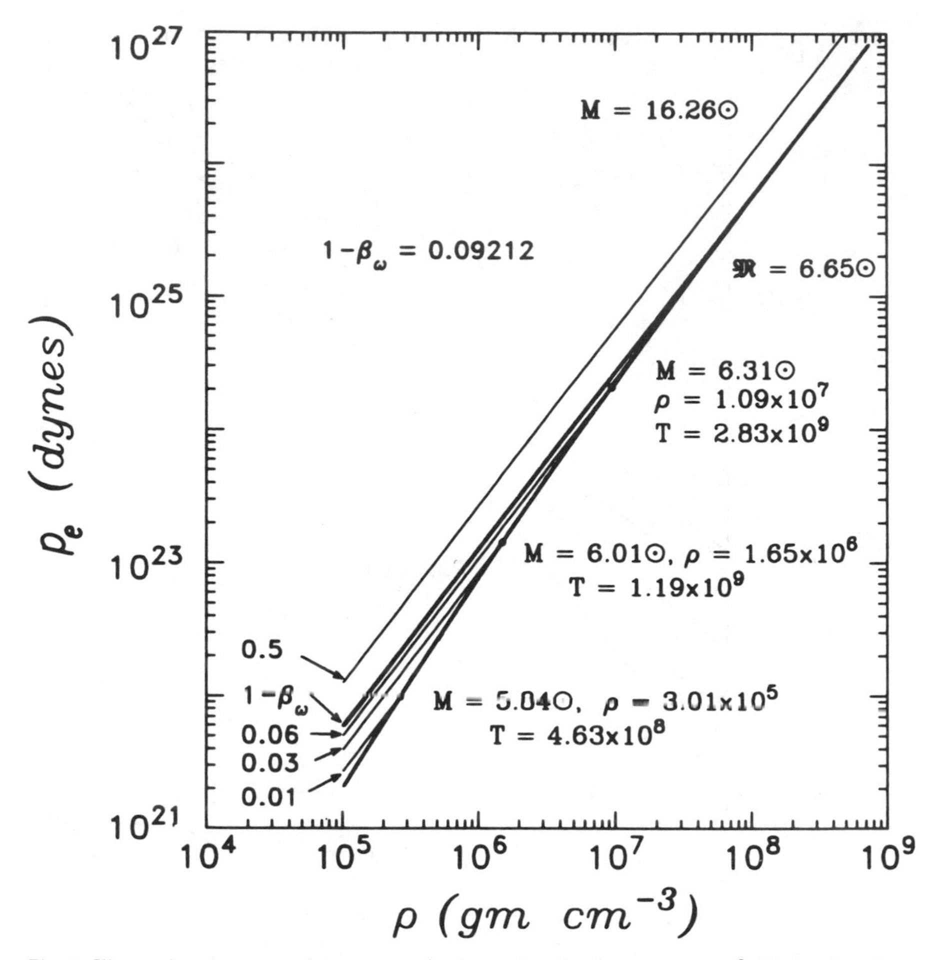

Fig. 5. Illustrating the onset of degeneracy for increasing density at constant β. Notice that there are no intersections for β> 0.09212. In the figure, 1-β is converted into mass of a star built on the standard model.

For our present purposes, the principal content of the inequality (51) is the criterion that for a star to develop degeneracy, it is necessary that the radiation pressure be less than 9.2 percent of (p_e+p_{rad}). This last inference is so central to all current schemes of stellar evolution that the directness and the simplicity of the early arguments are worth repeating.

The two principal elements of the early arguments were these: *first,* that radiation pressure becomes increasingly dominant as the mass of the star increases; and *second,* that the degeneracy of electrons is possible only so long as the radiation pressure is not a significant fraction of the total pressure — indeed, as we have seen, it must not exceed 9.2 percent of (p_e+p_{rad}). The second of these elements in the arguments is a direct and an elementary consequence of the physics of degeneracy; but the first requires some amplification.

That radiation pressure must play an increasingly dominant role as the mass of the star increases is one of the earliest results in the study of stellar structure that was established by Eddington. A quantitative expression for this fact is given by Eddington's *standard model* which lay at the base of early studies summarized in his *The Internal Constitution of the Stars.*

On the standard model, the fraction β (= gas pressure/total pressure) is a constant through a star. On this assumption, the star is a polytrope of index 3 as is apparent from equation (9); and, in consequence, we have the relation (cf. equation (43))

$$M = 4\pi \left[\frac{C(\beta)}{\pi G}\right]^{3/2} (2.018) \tag{52}$$

where $C(\beta)$ is defined in equation (9). Equation (52) provides a quartic equation for β analogous to equation (17) for β^*. Equation (52) for $\beta = \beta_\omega$ gives

$$M = 0.197\, \beta_\omega^{-3/2} \left(\frac{hc}{G}\right)^{3/2} \frac{1}{(\mu H)^2} = 6.65\, \mu^{-2} \odot = \boldsymbol{M} \text{ (say)}. \tag{53}$$

On the standard model, then stars with masses exceeding $\boldsymbol{M}$ will have radiation pressures which exceed 9.2 percent of the total pressure. Consequently stars with $M > \boldsymbol{M}$ cannot, at any stage during the course of their evolution, develop degeneracy in their interiors. Therefore, for such stars an eventual white-dwarf state is not possible unless they are able to eject a substantial fraction of their mass.

The standard model is, of course, only a model. Nevertheless, except under special circumstances, briefly noted below, experience has confirmed the standard model, namely that the evolution of stars of masses exceeding 7–8 $\odot$ must proceed along lines very different from those of less massive stars. These conclusions, which were arrived at some fifty years ago, appeared then so convincing that assertions such as these were made with confidence:

> Given an enclosure containing electrons and atomic nuclei (total charge zero) what happens if we go on compressing the material indefinitely? (1932)[10]
>
> The life history of a star of small mass must be essentially different from the life history of a star of large mass. For a star of small mass the natural white-dwarf stage is an initial step towards complete extinction. A star of large mass cannot pass into the white-dwarf stage and one is left speculating on other possibilities. (1934)[8]

And these statements have retained their validity.

While the evolution of the massive stars was thus left uncertain, there was no such uncertainty regarding the final states of stars of sufficiently low mass.[11] The reason is that by virtue, again, of the inequality (10), the maximum central pressure attainable in a star must be less than that provided by the degenerate equation of state, so long as

$$\frac{1}{2} G \left(\frac{4}{3}\pi\right)^{1/3} M^{2/3} < K_2 = \frac{1}{8}\left(\frac{3}{\pi}\right)^{1/3} \frac{hc}{\mu_e H)^{4/3}} \tag{54}$$

or, equivalently

$$M < \frac{3}{16\pi}\left(\frac{hc}{G}\right)^{3/2} \frac{1}{(\mu_e H)^2} = 1.74\, \mu_e^{-2} \odot. \tag{55}$$

We conclude that there can be no surprises in the evolution of stars of mass less than 0.43 ⊙ (if $\mu_e = 2$). The end stage in the evolution of such stars can only be that of the white dwarfs. (Parenthetically, we may note here that the inequality (55) implies that the so-called 'mini' black-holes of mass $\sim 10^{15}$ gm cannot naturally be formed in the present astronomical universe.)

7. Some brief remarks on recent progress in the evolution of massive stars and the onset of gravitational collapse

It became clear, already from the early considerations, that the inability of the massive stars to became white dwarfs must result in the development of much more extreme conditions in their interiors and, eventually, in the onset of gravitational collapse attended by the super-nova phenomenon. But the precise manner in which all this will happen has been difficult to ascertain in spite of great effort by several competent groups of investigators. The facts which must be taken into account appear to be the following.*

In the first instance, the density and the temperature will steadily increase without the inhibiting effect of degeneracy since for the massive stars considered $1-\beta_e > 1-\beta_\omega$. On this account, 'nuclear ignition' of carbon, say, will take place which will be attended by the emission of neutrinos. This emission of neutrinos will effect a cooling and a lowering of $(1-\beta_e)$; but it will still be in excess of $1-\beta_\omega$. The important point here is that the emission of neutrinos acts selectively in the central regions and is the cause of the lowering of $(1-\beta_e)$ in these regions. The density and the temperature will continue to increase till the next ignition of neon takes place followed by further emission of neutrinos and a further lowering of $(1-\beta_e)$. This succession of nuclear ignitions and lowering of $(1-\beta_e)$ will continue till $1-\beta_e < 1-\beta_\omega$ and a relativistically degenerate core with a mass approximately that of the limiting mass (=1.4 ⊙ for $\mu_e = 2$) forms at the centre. By this stage, or soon afterwards, instability of some sort is expected to set in (see following §8) followed by gravitational collapse and the phenomenon of the super-nova (of type II). In some instances, what was originally the highly relativistic degenerate core of approximately 1.4 ⊙, will be left behind as a neutron star. That this happens sometimes is confirmed by the fact that in those cases for which reliable estimates of the masses of pulsars exist, they are consistently close to 1.4 ⊙. However, in other instances — perhaps, in the majority of the instances — what is left behind, after all 'the dust has settled', will have masses in excess of that allowed for stable neutron stars; and in these instances black holes will form.

In the case of less massive stars ($M \sim 6-8$ ⊙) the degenerate cores, which are initially formed, are not highly relativistic. But the mass of core increases with the further burning of the nuclear fuel at the interface of the core and the mantle; and when the core reaches the limiting mass, an explosion occurs following instability; and it is believed that this is the cause underlying super-nova phenomenon of type I.

* I am grateful to Professor D. Arnett for guiding me through the recent literature and giving me advice in the writing of this section.

From the foregoing brief description of what may happen during the late stages in the evolution of massive stars, it is clear that the problems one encounters are of exceptional complexity, in which a great variety of physical factors compete. This is clearly not the occasion for me to enter into a detailed discussion of these various questions. Besides, Professor Fowler may address himself to some of these matters in his lecture that is to follow.

8. *Instabilities of relativistic origin: (1) The vibrational instability of spherical stars*

I now turn to the consideration of certain types of stellar instabilities which are derived from the effects of general relativity and which have no counterparts in the Newtonian framework. It will appear that these new types of instabilities of relativistic origin may have essential roles to play in discussions pertaining to gravitational collapse and the late stages in the evolution of massive stars.

We shall consider first the stability of spherical stars for purely radial perturbations. The criterion for such stability follows directly from the linearized equations governing the spherically symmetric radial oscillations of stars. In the framework of the Newtonian theory of gravitation, the stability for radial perturbations depends only on an average value of the adiabatic exponent, Γ_1, which is the ratio of the fractional Lagrangian changes in the pressure and in the density experienced by a fluid element following the motion; thus,

$$\Delta P/P = \Gamma_1 \, \Delta\rho/\rho. \tag{56}$$

And the Newtonian criterion for stability is

$$\overline{\Gamma}_1 = \int_0^M \Gamma_1(r)\, P(r)\, d\,M(r) \div \int_0^M P(r)\, d\,M(r) > \frac{4}{3}. \tag{57}$$

If $\overline{\Gamma}_1 < 4/3$, *dynamical instability* of a global character will ensue with an *e*-folding time measured by the time taken by a sound wave to travel from the centre to the surface.

When one examines the same problem in the framework of the general theory of relativity, one finds [12] that, again, the stability depends on an average value of Γ_1; but contrary to the Newtonian result, the stability now depends on the radius of the star as well. Thus, one finds that no matter how high $\overline{\Gamma}_1$ may be, instability will set in provided the radius is less than a certain determinate multiple of the *Schwarzschild radius*,

$$R_s = 2\, GM/c^2. \tag{58}$$

Thus, if for the sake of simplicity, we assume that Γ_1 is a constant through the star and equal to 5/3, then the star will become dynamically unstable for radial perturbations, if $R_1 < 2.4\, R_s$. And further, if $\Gamma_1 \rightarrow \infty$, instability will set in for all $R < (9/8)\, R_s$. *The radius* $(9/8)\, R_s$ *defines, in fact, the minimum radius which any gravitating mass, in hydrostatic equilibrium, can have in the framework of general relativity.* This important result is implicit in a fundamental paper by Karl Schwarzschild published in 1916. (Schwarzschild actually proved that for a star in which the energy density is uniform, $R > (9/8) R_s$.

In one sense, the most important consequence of this instability of relativistic origin is that if Γ_1 (again assumed to be a constant for the sake of simplicity) differs from and greater than 4/3 only by a small positive constant, then the instability will set in for a radius R which is a large multiple of R_s; and, therefore, under circumstances when the effects of general relativity, on the structure of the equilibrium configuation itself, are hardly relevant. Indeed, it follows[13] from the equations governing radial oscillations of a star, in a first post-Newtonian approximation to the general theory of relativity, that instability for radial perturbations will set in for all

$$R < \frac{K}{\Gamma_1 - 4/3} \frac{2GM}{c^2}, \tag{59}$$

where K is a constant which depends on the *entire** march of density and pressure in the equilibrium configuration in the Newtonian frame-work. Thus, for a polytrope of index n, the value of the constant is given by

$$K = \frac{5-n}{18}\left[\frac{2(11-n)}{(n+1)\,\xi_1^4\,1\theta_1'\,1^3}\int_0^{\xi_1}\theta\left(\frac{d\theta}{d\xi}\right)^2 \xi^2\,d\xi + 1\right], \tag{60}$$

where θ is the Lane-Emden function in its standard normalization ($\theta = 1$ at $\xi = 0$), ξ is the dimensionless radial coordinate, ξ_1 defines the boundary of the polytrope (where $\theta = 0$) and θ_1' is the derivative of θ at ξ_1.

Table 2
Values of the constant K in the inequality (59)
for various polytropic indices, n.

n	K	n	K
0	0.452381	3.25	1.28503
1.0	.565382	3.5	1.49953
1.5	.645063	4.0	2.25338
2.0	.751296	4.5	4.5303
2.5	.900302	4.9	22.906
3.0	1.12447	4.95	45.94

In Table 2, we list the values of K for different polytropic indices. It should be particularly noted that K increases without limit for $n \to 5$ and the configuration becomes increasingly centrally condensed.** Thus, already for $n = 4.95$ (for which polytropic index $\rho_c = 8.09 \times 10^6\,\bar{\rho}$), $K \sim 46$. In other words, for the highly centrally condensed massive stars (for which Γ_1 may differ from 4/3 by as little

* It is for this reason that we describe the instability as *global*.

** Since this was written, it has been possible to show (Chandrasekhar and Lebovitz 13*a*) that for $n \to 5$, the asymptotic behaviour of K is given by

$$K \to 2.3056/(5-n).$$

and, further, that along the polytropic sequence, the criterion for instability (59) can be expressed alternatively in the form

$$R < 0.2264\,\frac{2GM}{c^2}\frac{1}{\Gamma_1 - 4/3}\left(\frac{\rho_c}{\bar{\rho}}\right)^{1/3} \qquad (\rho_c > 10^6\,\bar{\rho})$$

as 0.01),* the instability of relativistic origin will set in, already, when its radius falls below $5 \times 10^3\ R_s$. Clearly this relativistic instability must be considered in the contexts of these problems.

A further application of the result described in the preceding paragraph is to degenerate configurations near the limiting mass[14]. Since the electrons in these highly relativistic configurations have velocities close to the velocity of light, the effective value of Γ_1 will be very close to 4/3 and the post-Newtonian relativistic instability will set in for a mass slightly less than that of the limiting mass. On account of the instability for radial oscillations setting in for a mass less than M_{limit}, the period of oscillation, along the sequence of the degenerate configurations, must have a minimum. This minimum can be estimated to be about two seconds (see Fig. 6). Since pulsars, when they were discovered, were known to have periods much less than this minimum value, the possibility of their being degenerate configurations near the limiting mass was ruled out; and this was one of the deciding factors in favour of the pulsars being neutron stars. (But by a strange irony, for reasons we have briefly explained in § 7, pulsars which have resulted from super-nova explosions have masses close to 1.4 ⊙!)

Finally, we may note that the radial instability of relativistic origin is the underlying cause for the *existence* of a maximum mass for stability: it is a direct consequence of the equations governing hydrostatic equilibrium in general relativity. (For a complete investigation on the periods of radial oscillation of neutron stars for various admissible equations of state, see a recent paper by Detweiler and Lindblom[15].)

9. Instabilities of relativistic origin: (2) The secular instability of rotating stars derived from the emission of gravitational radiation by non-axisymmetric modes of oscillation

I now turn to a different type of instability which the general theory of relativity predicts for rotating configurations. This new type of instability[16] has its origin in the fact that the general theory of relativity builds into rotating masses a dissipative mechanism derived from the possibility of the emission of gravitational radiation by non-axisymmetric modes of oscillation. It appears that this instability limits the periods of rotation of pulsars. But first, I shall explain the nature and the origin of this type of instability.

It is well known that a possible sequence of equilibrium figures of rotating homogeneous masses is the Maclaurin sequence of oblate spheroids[17]. When one examines the second harmonic oscillations of the Maclaurin spheroid, in a frame of reference rotating with its angular velocity, one finds that for two of these modes, whose dependence on the azimuthal angle is given by $e^{2i\varphi}$, the characteristic frequencies of oscillation, σ, depend on the eccentricity e in the manner illustrated in Figure 7. It will be observed that one of these modes becomes neutral (i.e., $\sigma = 0$) when $e = 0.813$ and that the two modes coalesce when $e = 0.953$ and become complex conjugates of one another beyond this

* By reason of the dominance of the radiation pressure in these massive stars and of β being very close to zero.

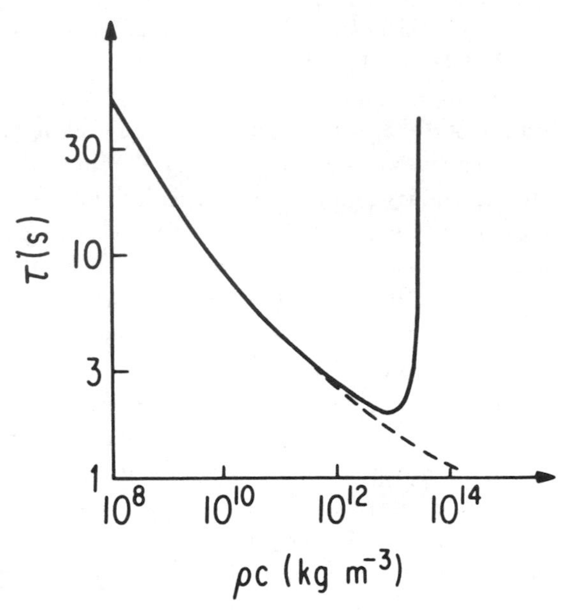

Fig. 6. The variation of the period of radial oscillation along the completely degenerate configurations. Notice that the period tends to infinity for a mass close to the limiting mass. There is consequently a minimum period of oscillation along these configurations; and the minimum period approximately 2 seconds. (From J. Skilling, *Pulsating Stars* (Plenum Press, New York, 1968), p. 59.)

point. Accordingly, the Maclaurin spheroid becomes *dynamically unstable* at the latter point (first isolated by Riemann). On the other hand, the origin of the neutral mode at $e = 0.813$ is that at this point a new equilibrium sequence of triaxial ellipsoids—the ellipsoids of Jacobi—bifurcate. On this latter account, Lord Kelvin conjectured in 1883 that

> if there be any viscosity, however slight ... the equilibrium beyond $e = 0.81$ cannot be secularly stable.

Kelvin's reasoning was this: viscosity dissipates energy but not angular momentum. And since for equal angular momenta, the Jacobi ellipsoid has a lower energy content than the Maclaurin spheroid, one may expect that the action of viscosity will be to dissipate the excess energy of the Maclaurin spheroid and transform it into the Jacobi ellipsoid with the lower energy. A detailed calculation[18] of the effect of viscous dissipation on the two modes of oscillation, illustrated in Figure 7, does confirm Lord Kelvin's conjecture. It is found that viscous dissipation makes the mode, which becomes neutral at $e = 0.813$, unstable beyond this point with an e-folding time which depends inversely on the magnitude of the kinematic viscosity and which further decreases monotonically to zero at the point, $e = 0.953$ where the dynamical instability sets in.

Since the emission of gravitational radiation dissipates *both* energy and angular momentum, it does *not* induce instability in the Jacobi mode; instead it

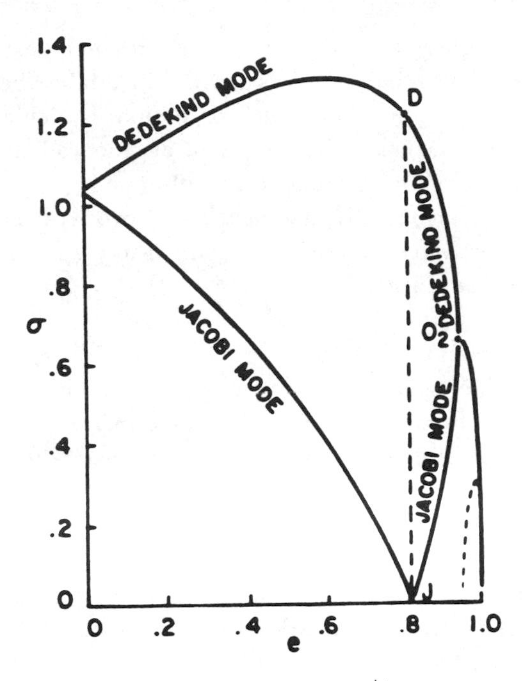

Fig. 7. The characteristic frequencies (in the unit $(\pi G \varrho)^{1/2}$) of the two even modes of second-harmonic oscillation of the Maclaurin spheriod. The Jacobi sequence bifurcates from the Maclaurin sequence by the mode that is neutral ($\sigma = 0$) at $e = 0.813$; and the Dedekind sequence bifurcates by the alternative mode at D. At O_2 ($e = 0.9529$) the Maclaurin spheroid becomes dynamically unstable. The real and the imaginary parts of the frequency, beyond O_2, are shown by the full line and the dashed curves, respectively. Viscous dissipation induces instability in the branch of the Jacobi mode; and radiation-reaction induces instability in the branch DO_2 of the Dedekind mode.

induces instability in the *alternative* mode at the same eccentricity. In the first instance this may appear surprising; but the situation we encounter here clarifies some important issues.

If instead of analyzing the normal modes in the rotating frame, we had analyzed them in the inertial frame, we should have found that the mode which becomes unstable by radiation reaction at $e = 0.813$, is in fact neutral at this point. And the neutrality of *this* mode in the inertial frame corresponds to the fact that the neutral deformation at this point is associated with the bifurcation (at this point) of a new triaxial sequence—the sequence of the Dedekind ellipsoids. These Dedekind ellipsoids, while they are congruent to the Jacobi ellipsoids, they differ from them in that they are at rest in the inertial frame and owe their triaxial figures to internal vortical motions. An important conclusion that would appear to follow from these facts is that in the framework of general relativity we can expect secular instability, derived from radiation reaction, to arise from a Dedekind mode of deformation (which is quasi-stationary in the inertial frame) rather than the Jacobi mode (which is quasi-stationary in the rotating frame).

A further fact concerning the secular instability induced by radiation reaction, discovered subsequently by Friedman[19] and by Comins[20], is that the modes belonging to higher values of m ($= 3, 4, , ,$) become unstable at smaller eccentricities though the e-folding times for the instability becomes rapidly longer. Nevertheless it appears from some preliminary calculations of Friedman[21] that it is the secular instability derived from modes belonging to $m = 3$ (or 4) that limit the periods of rotation of the pulsars.

It is clear from the foregoing discussions that the two types of instabilities of relativistic origin we have considered are destined to play significant roles in the contexts we have considered.

10. *The mathematical theory of black holes*

So far, I have considered only the restrictions on the last stages of stellar evolution that follow from the existence of an upper limit to the mass of completely degenerate configurations and from the instabilities of relativistic origin. From these and related considerations, the conclusion is inescapable that black holes will form as one of the natural end products of stellar evolution of massive stars; and further that they must exist in large numbers in the present astronomical universe. In this last section I want to consider very briefly what the general theory of relativity has to say about them. But first, I must define precisely what a black hole is.

A black hole partitions the three-dimensional space into two regions: an inner region which is bounded by a smooth two-dimensional surface called the *event horizon;* and an outer region, external to the event horizon, which is asymptotically flat; and it is required (as a part of the definition) that no point in the inner region can communicate with any point of the outer region. This incommunicability is guaranteed by the impossibility of any light signal, originating in the inner region, crossing the event horizon. The requirement of asymptotic flatness of the outer region is equivalent to the requirement that the black hole is isolated in space and that far from the event horizon the space-time approaches the customary space-time of terrestrial physics.

In the general theory of relativity, we must seek solutions of Einstein's vacuum equations compatible with the two requirements I have stated. It is a startling fact that compatible with these very simple and necessary requirements, the general theory of relativity allows for stationary (i.e., time-independent) black-holes exactly a single, unique, two-parameter family of solutions. This is the Kerr family, in which the two parameters are the mass of the black hole and the angular momentum of the black hole. What is even more remarkable, the metric describing these solutions is simple and can be explicitly written down.

I do not know if the full import of what I have said is clear. Let me explain.

Black holes are macroscopic objects with masses varying from a few solar masses to millions of solar masses. To the extent they may be considered as stationary and isolated, to that extent, they are all, every single one of them, described *exactly* by the Kerr solution. This is the only instance we have of an exact description of a macroscopic object. Macroscopic objects, as we see them all around us, are governed by a variety of forces, derived from a variety of approximations to a variety of physical theories. In contrast, the only elements in the construction of black holes are our basic concepts of space and time. They are, thus, almost by definition, the most perfect macroscopic objects there are in the universe. And since the general theory of relativity provides a single unique two-parameter family of solutions for their descriptions, they are the simplest objects as well.

Turning to the physical properties of the black holes, we can study them best by examining their reaction to external perturbations such as the incidence of waves of different sorts. Such studies reveal an analytic richness of the Kerr space-time which one could hardly have expected. This is not the occasion to elaborate on these technical matters[22]. Let it suffice to say that contrary to every prior expectation, all the standard equations of mathematical physics can be solved exactly in the Kerr space-time. And the solutions predict a variety and range of physical phenomena which black holes must exhibit in their interaction with the world outside.

The mathematical theory of black holes is a subject of immense complexity; but its study has convinced me of the basic truth of the ancient mottoes,

The simple is the seal of the true

and

Beauty is the splendour of truth.

REFERENCES

1. Eddington, A. S., *The Internal Constitution of the Stars* (Cambridge University Press, England, 1926), p. 16.
2. Chandrasekhar, S., *Mon. Not. Roy. Astr. Soc.*, 96, 644 (1936).
3. Eddington, A. S., *The Internal Constitution of the Stars* (Cambridge University Press, England, 1926), p. 172.
4. Fowler, R. H., *Mon. Not. Roy. Astr. Soc.*, 87, 114 (1926).
5. Chandrasekhar, S., *Phil. Mag.*, 11, 592 (1931).
6. Chandrasekhar, S., *Astrophys. J.*, 74, 81 (1931).
7. Chandrasekhar, S., *Mon. Not. Roy. Astr. Soc.*, 91, 456 (1931).
8. Chandrasekhar, S., *Observatory*, 57, 373 (1934).
9. Chandrasekhar, S., *Mon. Not. Roy. Astr. Soc.*, 95, 207 (1935).
10. Chandrasekhar, S., *Z. f. Astrophysik*, 5, 321 (1932).
11. Chandrasekhar, S., *Observatory*, 57, 93 (1934).
12. Chandrasekhar, S., *Astrophys. J.*, 140, 417 (1964); see also *Phys. Rev. Lett.*, 12, 114 and 437 (1964).
13. Chandrasekhar, S., *Astrophys. J.*, 142, 1519 (1965).

13 a. Chandrasekhar, S. and Lebovitz, N. R., *Mon. Not. Roy. Astr. Soc.*, 207, 13 *P* (1984).

14. Chandrasekhar, S. and Tooper, R. F., *Astrophys. J.*, 139, 1396 (1964).
15. Detweiler, S. and Lindblom, L., *Astrophys. J. Supp.*, 53, 93 (1983).
16. Chandrasekhar, S., *Astrophys. J.*, 161, 561 (1970); see also *Phys. Rev. Lett.*, 24, 611 and 762 (1970).
17. For an account of these matters pertaining to the classical ellipsoids see Chandrasekhar, S., *Ellipsoidal Figures of Equilibrium* (Yale University Press, New Haven, 1968).
18. Chandrasekhar, S., *Ellipsoidal Figures of Equilibrium* (Yale University Press, New Haven, 1968), Chap. 5, § 37.
19. Friedman, J. L., *Comm. Math. Phys.*, 62, 247 (1978); see also Friedman, J. L. and Schutz, B. F., *Astrophys. J.*, 222, 281 (1977).
20. Comins, N., *Mon. Not. Roy. Astr. Soc.*, 189, 233 and 255 (1979).
21. Friedman, J. L., *Phys. Rev. Lett.*, 51, 11 (1983).
22. The author's investigations on the mathematical theory of black holes, continued over the years 1974–1983, are summarized in his last book *The Mathematical Theory of Black Holes* (Clarendon Press, Oxford, 1983).

The reader may wish to consult the following additional references:

1. Chandrasekhar, S., 'Edward Arthur Milne: his part in the development of modern astrophysics,' *Quart. J. Roy. Astr. Soc.*, 21, 93–107 (1980).
2. Chandrasekhar, S., *Eddington: The Most Distinguished Astrophysicist of His Time* (Cambridge University Press, 1983).

How One May Explore the Physical Content of the General Theory of Relativity

S. CHANDRASEKHAR

1. Introduction. The general theory of relativity, in its exact nonlinear form, must predict phenomena which have no counterparts in the weak Newtonian limit and are qualitatively different. As is well known, in the weak limit, the predictions relating to the deflection of light when traversing a gravitational field and to the consequent time delay, to the precession of the Kepler orbit as manifested by the orbit of Mercury, and to the changing period of a binary star in an eccentric orbit due to the emission of gravitational radiation, have been quantitatively confirmed. But all these effects relate to departures from the Newtonian theory by at most a few parts in a million and of no more than three or four parameters in a post-Newtonian expansion of the exact equations of the theory. My concern is different: it relates to the exact theory.

An example will clarify my meaning. Consider Dirac's relativistic theory of the electron. In the first instance, one naturally asked if his equation will lead to the correct formula of Sommerfeld for the fine-structure separation of the spectral lines of one-electron atoms. It did. But the novel content of this theory is the prediction of the positron and how electron-positron pairs will be created by γ-rays of sufficient energy interacting with matter. It is this prediction of anti-matter that is the central feature of the Dirac theory. In the same way, the general theory of relativity must predict phenomena that have no parallels in the Newtonian theory. And we ask what they may be. The very formulation of this equation requires, perhaps, some explanation. One did not, for example, ask how one might explore the physical content of quantum electrodynamics or any similar innovative theory in physics. Why is the general theory of relativity different? The reason is this: it has generally been the case, that when fundamentally new ideas are formulated in a physical context, the predictions following from those ideas are on the border of experimental possibilities, and often are well inside the scope of one's physical intuition. But this is not the case with general relativity. The possibility of observing the strong effects of gravity are far beyond the scope of even one's imagination; they are a million or more times what we can experience or contemplate. Besides, we have no base for physical intuition to play its part since one cannot even guess the sort of circumstances one should envisage. For these reasons, an entirely different manner of exploration is called for.

2. The singularity theorem. The essential feature of the general theory of relativity that distinguishes it from the Newtonian theory is provided by the singularity theorems of Penrose and Hawking. Very roughly stated, what the theorems assert is the following: these exists a large class of physically realizable initial conditions which, when evolved according to the equations of general relativity, will necessarily lead to the development of singularities in space-time. Stated differently, unlike in the Newtonian theory, the development of singularities is generic to general relativity. This is a qualitative difference. It is illustrated by the following example.

Consider an initial distribution of matter, devoid of any internal pressure (i.e., dust), that is perfectly spherically symmetric (see Figure 1). In the Newtonian theory, such an initial distribution of matter will gravitationally collapse to the center in a finite time—the time of free fall—and a singularity of infinite density will form at the center. But this development of a singularity in the Newtonian theory is not generic to the theory in the sense that the slightest departure from spherical symmetry or the slightest amount of internal pressure or rotation will prevent its occurrence. In other words, the minutest departure from the initial conditions stated will efface the singularity. But this is not the case with general relativity. While the initial conditions stated will lead to a singularity, departures from spherical symmetry and allowance for internal pressure or rotation, even of finite amounts, will not prevent the development of the singularity if, during the process of gravitational collapse, a point of no return is transgressed. (At the point of no return, a *trapped surface*—eventually to become the event horizon of a black hole—forms.) It is this consequence of the singularity theorems that assures us that black holes will form in nature as the result of the gravitational collapse of a star of mass greater than say five or six solar masses during the last stages of their evolution when they have exhausted their source of energy. There remains of course the question, whether during the gravitational collapse of a star, the point of no return will be transgressed. That is an astrophysical question with which I am not presently concerned; it suffices to say that an instability of general relativistic origin is ultimately the cause.

The necessary formation of black holes, starting from a wide class of initial conditions, is an example of a phenomenon predicted by general relativity that has no parallel in the Newtonian theory. The question arises: are there other examples? Before I consider this question, it is necessary that I say a few things about the mathematical theory of black holes.

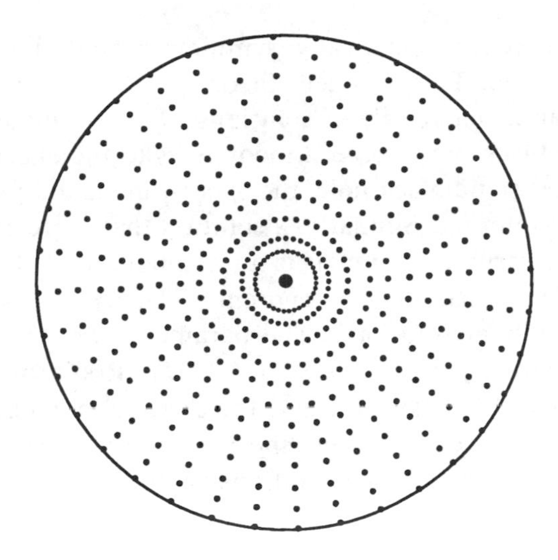

FIGURE 1. A perfectly spherically symmetric distribution of matter, devoid of internal pressure, will gravitationally collapse to produce a singularity at the center in the frameworks both of Newtonian theory and of general relativity.

3. On the black-hole solutions of general relativity. I shall restrict myself in the first instance to isolated black holes. A black hole isolated in space represents a space-time with the following properties:

1. It is an exact solution of the Einstein-vacuum or the Einstein–Maxwell equations.

(I shall explain later why it is important to include the Einstein–Maxwell equations along with the Einstein-vacuum equations.)

2. The three-dimensional space is divided into two non-overlapping regions—an internal region and an external region—separated by a smooth convex two-dimensional null surface.

3. The null surface is an *event horizon*, meaning that *the internal region is incommunicable to the external region.* That is, no signal originating in the internal region can cross the boundary (the surface of separation) and emerge into the external region; and conversely, no time-like or null trajectory which originates in the external region and enters the internal region by crossing the boundary separating them can ever recross the boundary to emerge into the external region: it is lost forever.

4. The space-time is asymptotically flat.

It is a remarkable fact that, consistent with the foregoing requirements, there are exactly four black-hole solutions. The black-hole solutions are of two kinds: static and stationary. The distinction is one of whether the black hole is nonrotating or rotating.

Consider first the vacuum solutions. A solution that represents a static black hole is necessarily spherically symmetric; and it is given by the Schwarzschild solution. This is Israel's theorem. The Schwarzschild solution provides a one-parameter family of solutions. The parameter is the inertial mass, M, of the black hole. Israel's theorem is to the effect that given the mass of an isolated static black hole, the space-time is uniquely specified by the Schwarzschild solution. Similarly, a solution that represents a stationary black hole is necessarily axisymmetric; it is unique and it is given by the Kerr solution. This is Robinson's theorem. The Kerr solution provides a two-parameter family of solutions. The parameters are the mass, M, and the angular momentum per unit mass $a(= J/M)$. Robinson's theorem is to the effect that given the mass, M, and the angular momentum, J, of the black hole, the space-time is uniquely specified by the Kerr solution.

Turning next to the black-hole solutions of the Einstein–Maxwell equations, we have simple generalizations of the Schwarzschild and the Kerr solutions: the Reissner–Nordström and the Kerr–Newman solutions.

The Reissner–Nordström solution describes a spherically symmetric, charged, static black hole. It provides a two-parameter family of solutions; the parameters are the mass, M, and the charge, $Q(|Q| \leq M)$ of the black hole. (The mass and the charge are measured in the units $c = G = 1$.) The Reissner–Nordström black hole has an event horizon exactly as the Schwarzschild black hole. But unlike the Schwarzschild black hole, the Reissner–Nordström black hole has a second horizon inside the event horizon. It is called the *Cauchy horizon* since a time-like or a null trajectory can cross this horizon and escape into a world that is outside the domain of dependence of the external world. An observer who crosses the Cauchy horizon emancipates himself from his past. The two horizons coalesce when $|Q| = M$. And for $|Q| > M$, the space-time has no horizon but the singularity at the center remains. Solutions with $|Q| > M$ are unphysical since the space-times exhibit naked singularities; and these are forbidden by Penrose's cosmic censorship hypothesis. It may be noted here that the Newtonian attraction and the Coulomb repulsion between two extreme Reissner–Nordström black holes with charges of the same sign exactly balance.

The Kerr–Newman solution represents a rotating charged black hole. It provides a three-parameter family; the parameters are the mass, M, the charge, Q, and the angular momentum per unit mass, a. Absence of naked singularities require that $M^2 > Q^2 + a^2$.

The mathematical theory of the Schwarzschild, the Reissner–Nordström, and the Kerr black holes is exceptionally complete and exceptionally rich. Thus, contrary to every prior expectation, all the standard equations of mathematical physics can be separated and solved in Kerr geometry. The equations include: Hamilton–Jacobi equations governing the geodesic motion of particles and polarized photons, Maxwell's equations, Dirac's relativistic equation of the electron, and the (linearized) equations governing the propagation and scattering of gravitational waves. The manner of separation of Dirac's equation in particular has led to a fruitful re-examination of the century-old problem of the separability of the partial differential equations of mathematical physics.

Returning to the black-hole solutions of general relativity, one can ask whether there are other solutions besides those we have described if we relax the requirement that they are isolated. For example, are there other static or stationary solutions that describe assemblages of the black holes? It has recently been shown by Rubak that, consistent with the other requirements of smoothness and asymptotic flatness, the only multiple black-hole solution that is allowed is that of Majumdar and Papapetrou that describes assemblages of extreme Reissner–Nordström black holes. I shall return to a more detailed consideration of this solution in §5.

4. The theory of black holes and the theory of colliding plane-waves patterned after it. In the general theory of relativity, one can construct plane-fronted gravitational waves confined between two parallel planes with a finite energy per unit area; and therefore, one can, in the limit, construct *impulsive* gravitational waves with a δ-function energy profile. Parenthetically, it may be noted that one cannot, similarly, construct such impulsive waves in electromagnetic theory. For a δ-function profile of the energy will imply a square root of a δ-function profile for the field variables; and the square root of a δ-function is simply not permissible—it is not an allowed mathematical or physical concept.

In 1971, Khan and Penrose considered the collision of two impulsive plane-fronted gravitational waves with parallel polarizations. And they showed that the result of the collision is the development of a space-like singularity not unlike the singularity in the interior of black holes. This phenomenon, illustrated in Figures 2 and 3, is not manifested in any linearized version of the theory: the occurrence of the singularity by a focussing of the colliding waves, in no way depends on their amplitudes. Clearly in this context (and in others, quite generally) nothing short of an exact solution of the problem will suffice to disclose a new phenomenon predicted by general relativity. In any event, the occurrence of a singularity in this example suggested to Penrose that here is a new realm in the physics of general relativity for exploration. However, there was no substantial progress in this area before one realized that the mathematical theory of black holes is, structurally, very closely related to the mathematical theory of colliding waves. This fact is, in itself, a matter for some surprise; one should not have thought that two theories dealing with such disparate circumstances will be as closely related as they are. Indeed, by developing the mathematical theory of colliding waves with a view to constructing a mathematical structure architecturally similar to the mathematical theory of black holes, one discovered a variety of new implications of the theory that simply could not have been anticipated.

To describe in detail how the mathematical theory of colliding waves was

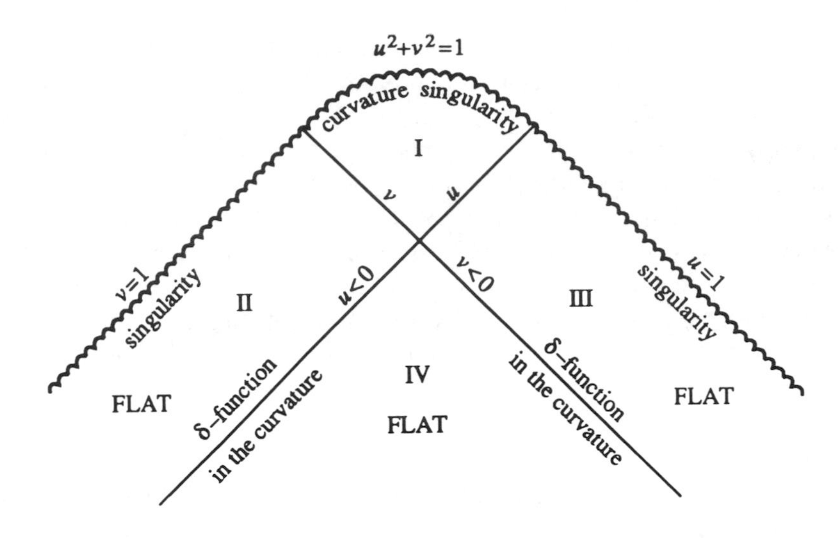

FIGURE 2. To illustrate the space-time resulting from the collision of impulsive gravitational waves.

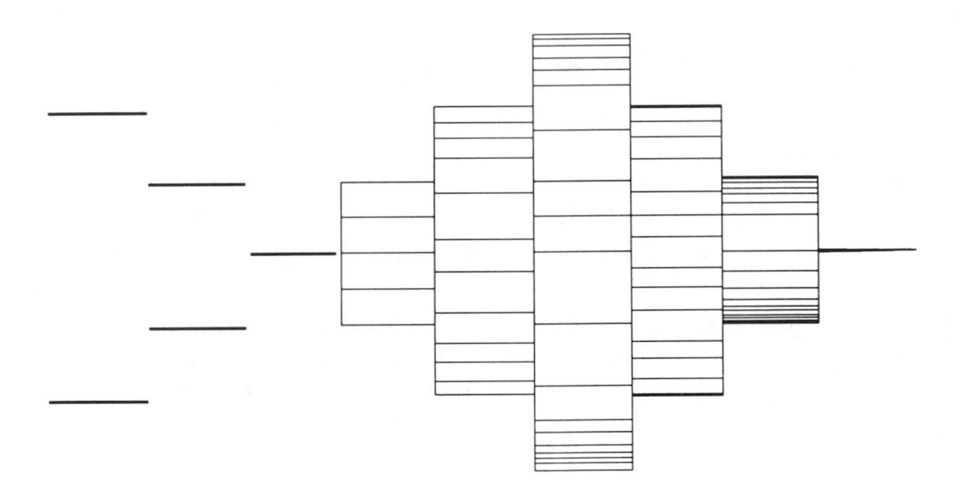

FIGURE 3. Illustrating the development of a singularity when two plane impulsive gravitational waves collide head on: following the collision the two waves scatter off each other and focus to form a singularity.

fashioned in the manner described will require a descent into technical matters that is not appropriate to this occasion. I must restrict myself to an impressionistic description that is still precise enough not to obscure the solid mathematical base.

First, consider the base of the mathematical theory of black holes. The entire set of field equations that govern stationary space-times (and static space-times as special cases) can be reduced to the complex equations,

$$(1 - |E|^2)\{[(1-\eta^2)E,_\eta],_\eta - [(1-\mu^2)E,_\mu],_\mu\}$$
$$= -2E^*[(1-\eta^2)(E,_\eta)^2 - (1-\mu^2)(E,_\mu)^2], \tag{1}$$

for the vacuum and for a special class of Einstein–Maxwell space-times, and

$$(1 - 4Q^2 - |E|^2)\{[(1-\eta^2)E,_\eta],_\eta - [(1-\mu^2)E,_\mu],_\mu\}$$
$$= -2E^*[(1-\eta^2)(E,_\eta)^2 - (1-\mu^2)(E,_\mu)^2], \tag{2}$$

valid only for a particular class of Einstein–Maxwell space-times. These are the Ernst equations. (I postpone for the present, the consideration of an equivalent pair of X- and Y-equations.)

In equations (1) and (2), E is a complex function of the two variables η and μ. In a gauge and a coordinate system adapted to black-hole solutions (but not restricted to them) η is a radial coordinate and $\mu = \cos\theta$, where θ is a polar angle. Besides, $\eta = 1$ defines the event horizon. In equation (2), Q is a real constant and

$$|Q| < \frac{1}{2}. \tag{3}$$

It can be shown that if $E_{\text{vac.}}$ is a solution of equation (1) then

$$E_{\text{E.M.}} = E_{\text{vac.}}\sqrt{(1-4Q^2)} \tag{4}$$

is a solution of equation (2).

Finally, it is important to observe that more than one combination of the metric functions and the associated electromagnetic potentials will satisfy the *same* Ernst equation. A particular context may require one choice while another context may require a different choice. And further, once the choice of the combination of the metric functions (and the associated electromagnetic potentials) and of a particular solution of the Ernst equation have been made, the complete specification of the space-time is relatively straightforward and requires no more than elementary quadratures.

It is a remarkable fact that all the known black-hole solutions follow from the simplest solutions of equations (1) and (2), namely

$$E = p\eta + iq\mu, \tag{5}$$

of equation (1) and

$$E = (p\eta + iq\mu)\sqrt{(1-4Q^2)} \tag{6}$$

of equation (2) where p and q are two real constants subject to the condition

$$p^2 + q^2 = 1. \tag{7}$$

The Schwarzschild and the Reissner–Nordström solutions follow, for example, when

$$E = \eta \quad \text{and} \quad E = \eta\sqrt{(1-4Q^2)} \qquad (q = 0). \tag{8}$$

The charge of the black hole for the Reissner–Nordström solution is $2Q$. Similarly the Kerr and the Kerr–Newman solutions follow from the general solutions (5) and (6). It is of course necessary that in deriving the black-hole solutions, we make the "right" choice of the combination of the metric functions and electromagnetic potentials that satisfies the Ernst equation.

I now turn to the mathematical theory of colliding waves. Before I proceed further, I should make it clear that most of what I have to say from now on derives from work done in close collaboration with Valeria Ferrari of Rome and Basilis Xanthopoulos of Crete. I have been exceptionally fortunate in my collaboration with them.

Space-times representing the collision of two plane-fronted waves are characterized by two space-like Killing vectors (∂_{x^1} and ∂_{x^2}, that span the wave fronts), in contrast to stationary axisymmetric space-times which are characterized by one space-like (∂_φ) and one time-like (∂_t) Killing vector. Nevertheless the equations governing the space-times of colliding gravitational and electromagnetic waves can be reduced to the same Ernst equations (1) and (2). But the coordinates η and μ have different meanings: $0 \leq \eta \leq 1$ is time-like and measures time (in an appropriate unit) from the instant of collision at $\eta = 0$, and $-1 \leq \mu \leq +1$ is space-like and measures distance (again in an appropriate unit) normal to the wave fronts. It should be noted that the choice of the coordinates allows, a posteriori, for the fact that as a result of the collision, a curvature or a coordinate singularity will develop when $\eta = 1$ and $\mu = \pm 1$.

The origin of the structural similarity of the mathematical theory of black holes and of colliding waves stems from the circumstance that in both cases the Einstein and the Einstein–Maxwell equations are reducible to the same basic equations and, as we shall see, even to the same solution. The richness and the diversity of the physical situations that are described, in spite of this identity, results from the different combinations of the metric functions and the electromagnetic potentials which can be associated with the same solution of the Ernst equation.

Thus, the fundamental solution of Khan and Penrose which describes the collision of two impulsive gravitational waves with parallel polarizations, is, in a well-defined mathematical sense, the analogue of the Schwarzschild solution: The Khan–Penrose and the Schwarzschild solutions follow from the solution $E = \eta$ of the Ernst equation but, of course, for different combinations of the metric functions. Similarly, the solution $E = p\eta + iq\mu$ leads to the Nutku–Halil solution which describes the more general case when the colliding impulsive waves have nonparallel polarizations (Chandrasekhar and Ferrari 1984). Thus, the Khan–Penrose and the Nutku–Halil solutions play the same role in the theory of colliding waves as the Schwarzschild and the Kerr solutions play in the theory of black holes.

Turning next to the collision of plane impulsive gravitational waves together with electromagnetic waves in the framework of the Einstein–Maxwell equations, one had to confront some conceptual difficulties. Penrose had raised the question: would an impulsive gravitational wave with its associated δ-function profile in the Weyl-tensor, imply a similar δ-function profile for the energy-momentum tensor of the Maxwell field? If that should happen, then the expression for the Maxwell-tensor would involve the square root of the δ-function; and "one would be at a loss to know how to interpret such a function." Besides, there was a formidable problem of satisfying the many conditions at the various null boundaries. On these accounts, all efforts to obtain a solution compatible with carefully formulated initial conditions failed. However, when it was realized that the Khan–Penrose and the Nutku-Halil solutions follow from the simplest solution of the Ernst equation it was natural to seek a solution of the Einstein–Maxwell equations which will reduce to the Nutku–Halil solution when the Maxwell field is switched off (the analogue of the Kerr–Newman solution). The problem is not a straightforward one since in the framework of the Einstein–Maxwell equations, we do not have an Ernst equation which reduces to the Ernst equation for the particular combination of the metric functions that is appropriate for the Nutku-Halil vacuum solution. The technical problems that are presented can be successfully overcome and a solution can be obtained which satisfies all the necessary boundary conditions and physical requirements (Chandrasekhar and Xanthopoulos, 1985; see Figure 4). That we can obtain a physically consistent solution by this "inverted procedure" is a manifestation of the firm aesthetic base of the general theory of relativity.

With the derivation of solutions for colliding waves that are the analogues of the known black-hole solutions, one might have thought that the theory of colliding waves, patterned after the theory of black holes, was completed. But that is not the case. By asking for solutions that follow from the same simplest solution of the Ernst equation (for space-times with two space-like Killing fields) for other allowed combinations of the metric functions and electromagnetic potentials, one was led to solutions describing space-times with entirely unexpected features. Thus, solutions for both the Einstein-vacuum and the Einstein–Maxwell equations were found (Chandrasekhar and Xanthopoulos, 1987) which violated the then commonly held belief that colliding waves must invariably lead to the development of curvature singularities. One found instead that event horizons formed; and further that the space-time, (extended beyond the horizons) included a domain which is a mirror image of the one that was left behind, and a further domain which included hyperbolic arc-like curvature singularities reminiscent of the Kerr and the Kerr–Newman black holes (see Figure 5). It is remarkable that a space-time resulting from the collision of waves should bear such a close

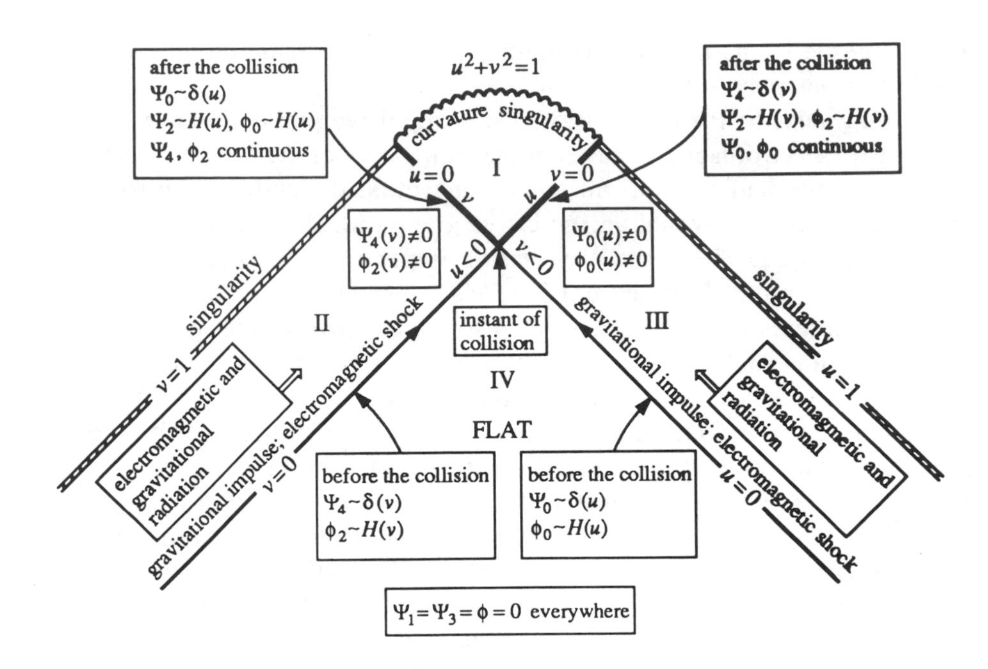

FIGURE 4. The space-time diagram for two colliding plane impulsive gravitational waves. The colliding waves are propagated along the null directions, u and v . The time coordinate is along the vertical, and the spatial direction of propagation (in space) is along the horizontal. The plane of the wavefronts (on which the geometry is invariant) is orthogonal to the plane of the diagram. The instant of the collision is at the origin of the (u, v)-coordinates.

The flat portion of the space-time, prior to the arrival of either wave, is region IV. These waves produce a spray of gravitational and electromagnetic radiation which fills regions II and III; and the region of the space-time in which the waves scatter off each other and focus is region I. The result of the collision is the development of a curvature space-time singularity on $u^2 + v^2 = 1$; and $v = 1$ or $u = 1$ are space-time singularities for observers who do *not* observe the collision. Singular behaviours in the Weyl scalars, Ψ_4 , Ψ_2 , and Ψ_0 (as Dirac δ-functions or Heaviside (H) step-functions, or both, expressing the shock-wave character of the colliding electromagnetic waves) occur, as indicated, along the null boundaries separating the different regions.

resemblance to Alice's anticipations with respect to the world *Through the Looking Glass.* "It [the passage in the Looking-Glass House] is very like our

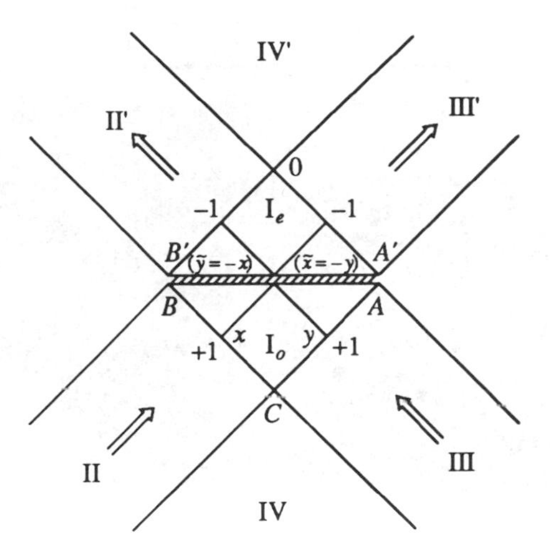

FIGURE 5. An alternative representation of the space-time diagram, which achieves manifest symmetry between the region of the space-time before and after the formation of the null-surface, is provided by the transformation to the coordinates $x = 1 - 2u^2$ and $y = 1 - 2v^2$. The regions marked I_0, II, III, IV in this figure and in Figure 4 correspond. A permissible extension beyond the domain of dependence, is by simply attaching to region I_e, regions II$'$, III$'$, and IV$'$ isometric to regions II, III, and IV. By this C^0 extension the two worlds joined together are mirror images of one another.

passage as far as you can see, only you know it may be quite different on beyond" (see Figure 6, p. 238).

In Table 1 (see pp. 239–241), we describe more fully the various solutions that have been derived for black holes and for colliding waves. The visual pattern manifested by this table is a reflection of the structural unity of the subject.

The inner relationships between the theory of the black holes and the theory of colliding waves is equally manifest in the simpler context when the metric is diagonal. In this case the basic equations reduce to the two-dimensional Laplacean equation,

$$[(\eta^2 - 1)(lg\Psi)_{,\eta}]_{,\eta} + [(1 - \mu^2)(lg\Psi)_{,\mu}]_{,\mu} = 0. \tag{9}$$

This equation can of course be solved exactly; and the solutions that are relevant in the two theories are listed in Table 2 (see p. 242).

In developing the theory of colliding waves in parallel with the theory of black holes, we have in effect systematically examined the consequences of

FIGURE 6. Alice crossing the 'event horizon' in *Through the Looking Glass.*

TABLE 1.

Killing vectors	Field equations	Solution for Ernst equation for			Solution	Description
		E	$E^\dagger$	$\tilde{E}^\dagger$		
$\partial_t, \partial_\phi$	Einstein-vacuum	does not exist		η	Schwarzschild	Black hole; static; spherically symmetric event horizon; space-like singularity at centre; type D; parameter: mass
$\partial_t, \partial_\phi$	Einstein-vacuum	does not exist		$p\eta + iq\mu$; $p^2 + q^2 = 1$	Kerr	Black hole; stationary, axisymmetric event and Cauchy horizons; ergosphere; time-like ring singularity in equatorial plane; type D; parameters: mass and angular momentum
$\partial_t, \partial_\phi$	Einstein–Maxwell	does not exist		$\eta\sqrt{1 - 4Q^2}$	Reissner–Nordström	Charged black hole; static; spherically symmetric event and Cauchy horizons; time-like singularity at centre; type D; parameters: mass and charge
$\partial_t, \partial_\phi$	Einstein–Maxwell	does not exist		$p\eta + iq\mu$; $p^2 + q^2 = 1 - 4Q^2$	Kerr–Newman	Charged black hole; stationary, axisymmetric event and Cauchy horizons; ergosphere; time-like ring singularity in equatorial plane; type D; parameters: mass, charge and angular momentum
$\partial_{x^1}, \partial_{x^2}$	Einstein-vacuum	η		interchanges x^1 and x^2	Khan and Penrose	Collision of impulsive gravitational waves; parallel polarizations; develops space-like curvature singularity

Continued on next page

Table 1. (continued)

Killing vectors	Field equations	Solution for Ernst equation for E	$E^{\dagger}$	$\tilde{E}^{\dagger}$	Solution	Description
$\partial_{x^1}, \partial_{x^2}$	Einstein-vacuum	$p\eta + iq\mu$; $p^2 + q^2 = 1$		interchanges x^1 and x^2	Nutku–Halil	Collision of impulsive gravitational waves; nonparallel polarizations; develops space-like curvature singularity (weaker than Khan–Penrose)
$\partial_{x^1}, \partial_{x^2}$	Einstein–Maxwell	does not exist	$E^{\dagger} \leftrightarrows E_{vac} = (p\eta + iq\mu) \times\sqrt{1 - 4Q^2}$	interchanges x^1 and x^2	Chandrasekhar and Xanthopoulos	Collision of impulsive gravitational waves and accompanying gravitational and electromagnetic shock waves; nonparallel polarizations; develops space-like curvature singularity
$\partial_{x^1}, \partial_{x^2}$	Einstein-vacuum		η	interchanges x^1 and x^2	Chandrasekhar and Xanthopoulos	Collision of impulsive gravitational waves and accompanying gravitational shock waves; parallel polarizations; develops very strong space-like curvature singularity; type D
$\partial_{x^1}, \partial_{x^2}$	Einstein-vacuum		$p\eta + iq\mu$; $p^2 + q^2 = 1$	interchanges x^1 and x^2	Chandrasekhar and Xanthopoulos	Collision of impulsive gravitational waves and accompanying gravitational shock waves; nonparallel polarizations; develops a horizon and subsequent time-like arc singularities; type D
$\partial_{x^1}, \partial_{x^2}$	Einstein–Maxwell	does not exist	$\eta\sqrt{(1 - 4Q^2)}$	interchanges x^1 and x^2	Chandrasekhar and Xanthopoulos	Collision of impulsive gravitational waves and accompanying gravitational and electromagnetic shock waves; parallel polarizations; develops a horizon and subsequent three-dimensional time-like singularities; type D

Continued on next page

TABLE 1. (continued)

Killing vectors	Field equations	Solution for Ernst equation for E	$E^{\dagger}$	$\tilde{E}^{\dagger}$	Solution	Description
$\partial_{x^1}, \partial_{x^2}$	Einstein–Maxwell	does not exist	$p\eta + iq\mu$; $p^2 + q^2 = 1 - 4Q^2$	interchanges x^1 and x^2	Chandrasekhar and Xanthopoulos	Collision of impulsive gravitational waves and accompanying gravitational and electromagnetic shock waves; nonparallel polarizations; develops a horizon and subsequent time-like arc singularities; type D
$\partial_{x^1}, \partial_{x^2}$	Einstein–Maxwell ($H = E_{\text{vac.}}$)	does not exist	η	interchanges x^1 and x^2	Bell–Szekeres	Collision of impulsive gravitational waves and accompanying electromagnetic shock waves; parallel polarizations; space-time conformally flat; develops a horizon; permits extension with no subsequent singularities
$\partial_{x^1}, \partial_{x^2}$	Einstein–Maxwell ($H = E_{\text{vac.}}$)	does not exist	$p\eta + iq\mu$; $p^2 + q^2 = 1$	interchanges x^1 and x^2	Bell–Szekeres	same as above
$\partial_{x^1}, \partial_{x^2}$	Einstein–Maxwell ($H = E_{\text{vac.}}$)	does not exist	Ehlers transform of $p\eta + iq\mu$	interchanges x^1 and x^2	Chandrasekhar and Xanthopoulos	Collision of impulsive gravitational waves and accompanying gravitational and electromagnetic shock waves; develops a horizon and subsequent time-like arc singularities; type D
$\partial_{x^1}, \partial_{x^2}$	Einstein-hyrodynamics ($\varepsilon = p$)	$p\eta + iq\mu$		interchanges x^1 and x^2	Chandrasekhar and Xanthopoulos	Collision of impulsive gravitational and accompanying gravitational shock waves and null dust ($R_{ij} = Ck_ik_j$); nonparallel polarizations; develops weakened space-like singularity; transforms null dust into a perfect fluid with $\varepsilon = p$

TABLE 2. Basic equation: $[(1-\eta^2)(\lg\Psi)_{,\eta}]_{,\eta} - [(1-\mu^2)(\lg\Psi)_{,\mu}]_{,\mu} = 0$

Killing vectors	Field equations	Solution	Remarks
$\partial_t, \partial_\phi$	Einstein vacuum	$\lg\widetilde{\Psi}^\dagger = \lg\frac{\eta-1}{\eta+1}$	Schwarzschild solution: spherically symmetric static black hole
$\partial_t, \partial_\phi$	Einstein vacuum	$\lg\widetilde{\Psi}^\dagger = \lg\frac{\eta-1}{\eta+1} + \Sigma_n A_n P_n(\mu)P_n(\eta)$	Distorted black holes (when $\Sigma A_{2n+1}P_n(1) = 0$) (Weyl's solutions)
$\partial_{x^1}\partial_{x^2}$	Einstein vacuum	$\lg\chi = \lg\frac{1+\eta}{1-\eta}$	Khan–Penrose solution for colliding impulsive waves with parallel polarizations
$\partial_{x^1}, \partial_{x^2}$	Einstein vacuum	$\lg\chi = \lg\frac{1+\eta}{1-\eta} + \Sigma_n A_n Q_n(\mu)Q_n(\eta)$	Collision of impulsive gravitational waves with accompanying gravitational shock waves; parallel polarizations
$\partial_{x^1}, \partial_{x^2}$	Einstein–Maxwell ($H = E_{\text{vac.}}$)	$\frac{1}{2}\lg\frac{1+H}{1-H} = \lg\frac{1+\eta}{1-\eta}$	Collision of impulsive gravitational waves with accompanying electromagnetic shock waves in conformally flat space-time; parallel polarizations; Bell–Szekeres solution
$\partial_{x^1}, \partial_{x^2}$	Einstein–Maxwell ($H = E_{\text{vac.}}$)	$\frac{1}{2}\lg\frac{1+H}{1-H} = \lg\frac{1+\eta}{1-\eta} + \Sigma_n A_n P_n(\mu)P_n(\eta)$	Collision of impulsive gravitational waves with accompanying gravitational and electromagnetic shock waves; parallel polarizations

adopting for the Ernst equation, in its various contexts, its simplest solution. While this approach may appear as an exceedingly formal one, it has nevertheless disclosed possibilities that one could not have, in any way, foreseen: the development of horizons and subsequent time-like singularities or the transformation of null dust into a perfect fluid.

5. Binary black-hole solutions. We have seen how the simplest solution of the complex Ernst equation provides almost all the fundamental solutions describing black holes and colliding plane impulsive gravitational waves. But there exists an alternative pair of real equations—the X- and Y-equations—to which also the stationary axisymmetric vacuum solutions can be reduced. These equations are

$$\frac{1}{2}(X+Y)[(\Delta X_{,2})_{,2}+(\delta X_{,3})_{,3}]=\Delta(X_{,2})^2+\delta(X_{,3})^2 \tag{10}$$

and

$$\frac{1}{2}(X+Y)[(\Delta Y_{,2})_{,2}+(\delta Y_{,3})_{,3}]=\Delta(Y_{,2})^2+\delta(Y_{,3})^2, \tag{11}$$

where

$$\Delta=\eta^2-1 \quad \text{and} \quad \delta=1-\mu^2. \tag{12}$$

By the substitutions,

$$X=\frac{1+F}{1-F} \text{ and } Y=\frac{1+G}{1-G}, \tag{13}$$

equations (10) and (11) become

$$(1-FG)[(\Delta F_{,2})_{,2}+(\delta F_{,3})_{,3}]=-2G[\Delta(F_{,2})^2+\delta(F_{,3})^2] \tag{14}$$

and

$$(1-FG)[(\Delta G_{,2})_{,2}+(\delta G_{,3})_{,3}]=-2F[\Delta(G_{,2})^2+\delta(G_{,3})^2]. \tag{15}$$

These equations are real counterparts of the complex Ernst equation (1). And like the Ernst equation, equations (14) and (15) allow the simple solution,

$$F=-p\eta-q\mu \quad \text{and} \quad G=-p\eta+q\mu, \tag{16}$$

where p and q are two real positive constants subject to the condition,

$$p^2-q^2=1. \tag{17}$$

It has always been my belief that the solutions for X and Y derived from the simplest solution (16) of equations (14) and (15) must provide, in a suitable context, a space-time of some real physical significance. But the X- and Y-equations have remained, like Cinderella, the ignored and neglected stepsister of the Ernst equation (see Figure 7). And like Cinderella, the X- and Y-equations have been rescued and restored. I should like to tell this tale

FIGURE 7. *X*- and *Y*-functions as Cinderella being rescued.

since it illustrates how one may explore the physical content of the general theory of relativity.

Like all tales that are told, I must start in a seemingly remote territory—in this instance the two-center problem.

As E. T. Whittaker has written: "the most famous of the known soluble problems" in analytical dynamics, "other than of central motion, is *the problem of two centres of gravitation* i.e., the problem of determining the motion of a free particle in a plane, attracted by two fixed centres of Newtonian centres of force" (*Analytical dynamics*, Cambridge Univ. Press, 1937, p. 97). The integrability of the relevant equations of motion, in a plane containing the two centers, was discovered by Euler in 1760. The solution for the general three-dimensional motion was given by Jacobi in his 1842 lectures on dynamics by showing the separability of the Hamilton–Jacobi equation in prolate spheroidal coordinates. The underlying reason for the integrability of this problem is that any ellipse, with the two centers as foci, is an exact trajectory of a particle in the field of force of either of them in the absence of the other and is therefore by Bonnet's theorem—a theorem that could well have been known to Newton—an exact trajectory in the field of force of both of them.

It has generally been true that problems that are central to the Newtonian theory play an equally central role in general relativity. On this account, it would appear worthwhile to consider the two-center problem in general relativity. But first we must resolve a conceptual difficulty in considering the problem consistently in general relativity and, strictly, in the Newtonian theory as well: the consideration of two fixed centers of attraction, without allowing for their mutual attraction, requires (to quote E. A. Milne in a different context) the "services of an archangel." However, in this instance, the services of an archangel can be dispensed with by attributing to each of the mass centers electric charges of the same sign so that the Coulomb repulsion between them exactly balances the Newtonian attraction. More generally, in the Newtonian theory we can envisage a static arrangement of any number of mass points $M_1, M_2, \dots, M_N$ (say), at arbitrary locations, with charges $Q_1, Q_2, \dots, Q_N$ of the same sign such that $M_i\sqrt{G} = Q_i \quad (i = 1, \dots, N)$ (where G denotes the constant of gravitation) and the gravitational attraction between any pair of them is balanced by their Coulomb repulsion. It is a remarkable fact that this same static arrangement is allowed in general relativity in conformity with the Einstein–Maxwell equations. This is the solution of Majumdar and Papapetrou. The solution was correctly interpreted by Hartle and Hawking as representing an assemblage of extreme Reissner–Nordström black holes.

As we have stated earlier, the Majumdar–Papapetrou solution is the sole static (or stationary) multiple black-hole solution that is compatible with smoothness of the space-time external to the event horizons and asymptotic flatness; there is none other.

Since we are presently interested in the two-center problem, we shall consider the special case of the Majumdar–Papapetrou solution when only two extreme Reissner–Nordström black holes are present and the space-time is axisymmetric about the line joining the centers of the two black holes.

With the two-center problem in view, we shall, following Jacobi, write the metric in prolate spheroidal coordinates:

$$x = \sinh\psi \sin\theta \cos\varphi\,, \quad y = \sinh\psi \sin\theta \sin\varphi\,, \quad z = \cosh\psi \cos\theta\,, \tag{18}$$

or, more conveniently, in terms of

$$x^2 = \eta = \cosh\psi\,, \quad x^3 = \mu = \cos\theta\,, \quad x^1 = \varphi. \tag{19}$$

In static Einstein–Maxwell space-times, the only nonvanishing (tensor) components of the Maxwell tensor are F_{02} and F_{03}; they can be derived from a scalar potential $\widetilde{B}$:

$$F_{02} = \widetilde{B}_{,2} \quad \text{and} \quad F_{03} = \widetilde{B}_{,3}. \tag{20}$$

And it can be shown that the metric of a static Einstein–Maxwell space-time in the coordinates chosen can be written in the form

$$ds^2 = e^{2\nu}(dt)^2 - e^{-2\nu}\left\{e^f(\eta^2-\mu^2)\left[\frac{(d\eta)^2}{\eta^2-1} + \frac{(d\mu)^2}{1-\mu^2}\right] + (\eta^2-1)(1-\mu^2)(d\varphi)^2\right\}, \tag{21}$$

where f and e^ν are the sole metric functions to be determined.

A specially simple class of solutions is obtained when

$$f = 0\,, \quad \widetilde{B} = \pm e^\nu\,, \tag{22}$$

and $e^{-\nu}$ is a solution of the three-dimensional Laplace's equation,

$$[(\eta^2-1)(e^{-\nu})_{,2}]_{,2} + [(1-\mu^2)(e^{-\nu})_{,3}]_{,3} = 0. \tag{23}$$

The Majumdar–Papapetrou solution of two extreme Reissner–Nordström black holes located on the z-axis follows from the solution

$$e^{-\nu} = 1 + \frac{M_1}{\eta-\mu} + \frac{M_2}{\eta+\mu}\,, \tag{24}$$

where M_1 and M_2 denote the masses of the two black holes. Their charges, both of the same sign, are $Q_1 = \pm M_1$ and $Q_2 = \pm M_2$. The two black holes are located at the coordinate singularities $\eta = 0$ and $\mu = \pm 1$. These are not points. They represent smooth two-dimensional surfaces with areas $4\pi M_1^2$

and $4\pi M_2^2$ consistently with the fact that $\eta = 0$ and $\mu = \pm 1$ represent the event horizons of the two extreme Reissner–Nordström black holes of masses M_1 and M_2 and charges Q_1 and Q_2 (of the same sign) equal to M_1 and M_2 respectively. The fact that, for the basic solution, representing the static placement of two extreme Reissner–Nordström black holes on the axis of symmetry, $f = 0$ confirms the fundamental correctness of writing the metric of static axisymmetric Einstein–Maxwell space-times in the form (21). And in view of the special solutions (21) which the Einstein–Maxwell equations allow, the substitutions,

$$X = e^{\nu} + \tilde{B} \quad \text{and} \quad Y = e^{\nu} - \tilde{B}, \tag{25}$$

would appear apposite. By these substitutions, we find that X and Y, in fact, satisfy equations (10) and (11). In other words, *the three-dimensional Laplacean equation*, (23), *which underlies the solution representing the static placement of two extreme Reissner–Nordström black holes, is none other than what the X- and Y-equations become when one or the other is zero.*

As we have stated, the X- and Y-equations were first derived in the context of stationary axisymmetric vacuum space-times. The metric of such space-times, in a gauge and in a coordinate system adapted to black hole solutions (but not restricted to them) can be written in the form,

$$\begin{aligned} ds^2 = {} & \left[\chi(dt)^2 - \frac{1}{\chi}(d\varphi - \omega\, dt)^2\right]\sqrt{[(\eta^2-1)(1-\mu^2)]} \\ & - e^{\mu_2+\mu_3}\left[\frac{(d\eta)^2}{\eta^2-1} + \frac{(d\mu)^2}{1-\mu^2}\right]\sqrt{(\eta^2-1)}, \end{aligned} \tag{26}$$

where χ, ω, and $\mu_2 + \mu_3$ are functions η and μ. With the definitions

$$X = \chi + \omega \quad \text{and} \quad Y = \chi - \omega, \tag{27}$$

we find that X and Y satisfy the same equations (10) and (11)! By virtue of this relation between static Einstein–Maxwell and stationary Einstein vacuum space-times, one establishes the following one-to-one correspondence between them.

The correspondence

$$e^{\nu} \rightleftarrows \chi, \; \tilde{B} \rightleftarrows \omega, \quad \text{and} \quad e^{f} = e^{4(\mu_2+\mu_3)}\Delta^3\delta/(\eta^2-\mu^2)^4, \tag{28}$$

enables one to pass freely from a metric of the form (21) *and an electrostatic potential $\tilde{B}$, appropriate as a solution of the Einstein–Maxwell equations to a metric of the form* (26) *appropriate as a solution of the stationary Einstein vacuum equations, and conversely.*

This one-to-one correspondence is a manifestation of a natural and a harmonious blending of Einstein's relativity and Maxwell's electrodynamics in a single unified structure. It reminds one of the magnificent hall built for the five princes in the Indian epic, the Mahabharatta. Here is van Buitenen's translation of the description of the great hall in the Mahabharatta:

> The hall, which had solid golden pillars, great king, measured ten thousand cubits in circumference. Radiant and divine, it had a superb color like the fire, or the sun, or the moon. Made with the best materials, garlanded with gem-encrusted walls, gilded with precious stones and treasures, it was build well ... and possessed the matchless beauty that Maya imparted to it.
>
> ... Inside the hall Maya built a peerless lotus pond, covered with beryl leaves and lotuses with gem-studded stalks, filled with lilies and water plants and inhabited by many flocks of fowl. Blossoming lotuses embellished it, and turtles and fishes adorned it. Steps descended gently into it; the water was not muddy and it was plentiful in all seasons; and the pearl-drop flowers that covered it were stirred by a breeze ... and it was thick with precious stones and gems.
>
> ... One time the princely Dhārtarāstra came, in the middle hall, upon a crystal slab, and thinking it was water, the flustered prince raised his robe; ... Again, seeing a pond with crystalline water adorned with crystalline lotuses, he thought it was land and fell into the water .... He once tried a door which appeared to be open, and hurt his forehead; another time, thinking the door was closed, he shrank from the doorway.

And as the prince in the Mahabharatta, one realizes, when wandering through the great hall of general relativity, that what one had believed to be Einstein's hall, is in fact a corridor leading to Maxwell's hall; and when one is certain that one is examining the gems in Maxwell's hall, one has inadvertently slipped into Einstein's hall. "So matchless is the beauty" that Einstein has "imparted to it."

To return to the X- and Y-equations. Since X or Y equal to zero provides the unique binary black-hole solution with a static space-time that is entirely smooth exterior to the horizons and asymptotically flat, it behooves us to consider the space-time that follows from the (next!) simplest solution of the X- and Y-equations, namely, that derived from the solutions of F and G given in equation (16). We find that the metric of the space-time is given by (Chandrasekhar and Xanthopoulos, 1989)

(29)

$$ds^2 = \frac{(p^2\Delta + q^2\delta)^2}{[(1+p\eta)^2 - q^2\mu^2]^2} \times \left\{ (dt)^2 - \frac{[(1+p\eta)^2 - q^2\mu^2]^4}{\alpha^8(\eta^2-\mu^2)^3} \left[\frac{(d\eta)^2}{\eta^2 - 1} + \frac{(d\mu)^2}{(1-\mu^2)} \right] \right\} - \frac{[(1+p\eta)^2 - q^2\mu^2]^2}{(p^2\Delta + q^2\delta)^2}(\eta^2 - 1)(1-\mu^2)(d\varphi)^2,$$

where

(30) $$\Delta = \eta^2 - 1, \qquad \delta = 1 - \mu^2,$$

p and q are real positive constants subject to the condition,

(31) $$p^2 - q^2 = 1 \quad \text{and} \quad \alpha \le q$$

is a real positive constant. The electrostatic potential is given by

$$\widetilde{B} = -\frac{2q\mu}{(1+p\eta)^2 = q^2\mu^2}. \tag{32}$$

An examination of the metric (29) shows that it represents two charged black holes located on the z-axis at $z = +1$ and $z = -1$ and a *string* (in general) stretched along the entire z-axis. A string is a line in space-time along which a *conical singularity* (with no associated curvature singularity) occurs characterized by a *deficit* defined by the difference between 2π and the limiting ratio of the circumference to the proper radius of a small circle described normal to the line.

The two black holes are of equal mass,

$$M = 1/p, \tag{33}$$

and opposite charge,

$$Q = \pm(p+1)/pq. \tag{34}$$

The conical singularity along the axis is characterized by the deficits (in units of 2π)

$$\delta_{|z|<1} = 1 - \alpha^4/q^4 \quad \text{and} \quad \delta_{|z|>1} = 1 - \alpha^4/p^4. \tag{35}$$

where, by the assumption, $\alpha \leq q$, both are deficits (not excesses). We observe that when $\alpha = q$,

$$\delta_{|z|<1} = 0; \tag{36}$$

and the condition for local flatness is met for $|z| < 1$, i.e., for the part of the z-axis joining the two black holes; but strings stretch to $\pm\infty$ from the north pole of the one and the south pole of the other.

The surface area, S, of the horizons of the two black holes (in units of 4π) is given by

$$S = (1+p)^2/\alpha^4. \tag{37}$$

Except for the conical singularity on the axis, the space external to the horizons is smooth. In the manifold extended into the interior of the horizons, time-like singularities with two spatial dimensions occur. But the space-time exhibits no naked singularities and is asymptotically flat.

An important feature of these black holes is provided by the inequality,

$$|Q| - M = (1+p-q)/pq > 0. \tag{38}$$

The restriction $|Q| < M$ for the Reissner–Nordström black holes is not applicable to these charged black holes.

And finally, it can be shown that the surface gravity on the horizons of the two black holes vanishes identically. This fact suggests that these two black holes are in some sense generically related to the extreme Reissner–Nordström black holes on the horizons of which the surface gravity also vanishes.

A comparison with the analogous equilibrium configurations of magnetic monopoles. The multiple black-hole solutions that have been found so far are of two kinds: first, we have the Majumdar–Papapetrou solution which allows a static assemblage of extreme Reissner–Nordström black holes, with charges of the same sign, at arbitrary locations, in which (speaking in Newtonian terms) the gravitational attraction between any pair of them is exactly balanced by the Coulomb repulsion between them; and second, we have the solution, that we have described, of two black holes of equal mass (M) and opposite charge ($\pm Q$, $|Q| > M$) with strings attached to them. A special case of this second class of solution is when there is no string connecting the two black holes but strings stretching to $+\infty$ and $-\infty$ from the north pole of the one and the south pole of the other.

The two classes of black-hole solutions that we have described are analogous to static configurations of magnetic monopoles that have been found. First, we have the Bogomil'nyi–Prasad–Sommerfeld monopoles (with twice the charge of the Dirac monopole) of which we can contemplate a static assemblage (with charges of the same sign) in which the magnetic (Coulomb) repulsion is balanced by the attraction derived from a scalar field. Second, we have the possibility of two Dirac monopoles, of opposite charge, held in place by a connecting string. These two classes of solutions are manifestly similar to the black-hole solutions we have described. It is a remarkable fact that possibilities, contemplated only in recent years at the quantal level, are inherent, already at the classical level, in general relativity. A question that occurs is: Could one conclude from the very natural way in which strings emerge in the binary black-hole solution, that strings are indeed *predicted* by the general theory of relativity?

Concluding remark. The physical insights that we seem to have achieved by developing the mathematical theory of colliding waves deliberately patterned after the mathematical theory of black holes, and by seeking ad hoc the simplest solutions of the equations which simultaneously describe the static Einstein–Maxwell and the stationary Einstein-vacuum equations, suggest that one of the ways in which one may explore the physical content of the general theory of relativity is to allow one's sensibility to its aesthetic base guide in the formulation of problems with conviction in the harmonious coherence of its mathematical structure.

References

The nature of the lecture precludes giving a list of references in the conventional style. In the context of the different sections the reader may wish to consult the following references:

§3 For the theory of black holes:
S. Chandrasekhar, *The mathematical theory of black holes*, Clarendon Press, Oxford, 1983.

§4 The principal references for this section are:

K. Khan and R. Penrose, Nature, London **229** (1971), 185.
Y. Nutku and M. Halil, Phys. Rev. Lett. **39** (1977), 1379.
S. Chandrasekhar and V. Ferrari, Proc. Roy. Soc. London Ser. A **396** (1984), 55–74.
S. Chandrasekhar and B. C. Xanthopoulos, Proc. Roy. Soc. London Ser. A **398** (1985), 223–259.
——, Proc. Roy. Soc. London Ser. A **408** (1986), 175–208.

For a more complete discussion along the same lines, see:

S. Chandrasekhar, *Truth and beauty : æsthetic motivations in science,* University of Chicago Press, Chicago, 1987, pp. 144–169.

§5 The discussion in this section is based on:

S. Chandrasekhar, Proc. Roy. Soc. London Ser. A **358** (1978), 405–420.
——, Proc. Roy. Soc. London Ser. A **423** (1989), 379–386.
S. Chandrasekhar and B. C. Xanthopoulos, Proc. Roy. Soc. London Ser. A **423** (1989), 387–400.

The University of Chicago
The Enrico Fermi Institute
Chicago, Illinois 60637

Publications by S. Chandrasekhar

PAPERS

1928

Thermodynamics of the Compton Effect with Reference to the Interior of the Stars. *Indian Journal of Physics* 3, 241–50

1929

[1] The Compton Scattering and the New Statistics. *Proceedings of the Royal Society*, A, 125, 231–37

1930

[1] The Ionization-Formula and the New Statistics. *Philosophical Magazine* 9, 292–99.

On the Probability Method in the New Statistics. *Philosophical Magazine* 9, 621–24

1931

[1] The Dissociation Formula according to the Relativistic Statistics. *Monthly Notices of the Royal Astronomical Society* 91, 446–55

[1] The Highly Collapsed Configurations of a Stellar Mass. *Monthly Notices of the Royal Astronomical Society* 91, 456–66

[1] The Maximum Mass of Ideal White Dwarfs. *The Astrophysical Journal*, 74, 81–82

[1] The Stellar Coefficients of Absorption and Opacity. *Proceedings of the Royal Society*, A, 133, 241–54

[1] The Density of White Dwarf Stars. *Philosophical Magazine* 11 (suppl.), 592–96

0_2 Eridani B. *Zeitschrift für Astrophysik* 3, 302–5

1932

Ionization in Stellar Atmospheres, III (with E. A. Milne). *Monthly Notices of the Royal Astronomical Society* 92, 150–86

[1] Model Stellar Photospheres. *Monthly Notices of the Royal Astronomical Society* 92, 186–95

The Stellar Coefficients of Absorption and Opacity, II. *Proceedings of the Royal Society*, A, 135, 472–90

[1] Some Remarks on the State of Matter in the Interior of Stars. *Zeitschrift für Astrophysik* 5, 321–27

1933

[1] The Equilibrium of Distorted Polytropes. I, The Rotational Problem. *Monthly Notices of the Royal Astronomical Society* 93, 390–405

[1] The Equilibrium of Distorted Polytropes. II, The Tidal Problem. *Monthly Notices of the Royal Astronomical Society*, 93, 449–61

Superior numbers identify titles included in these *Selected Papers* by volume.

[1]The Equilibrium of Distorted Polytropes. III, The Double-Star Problem. *Monthly Notices of the Royal Astronomical Society* 93, 462–71

[1]The Equilibrium of Distorted Polytropes. IV, The Rotational and the Tidal Distortions as Functions of the Density Distribution. *Monthly Notices of the Royal Astronomical Society* 93, 539–74

[1]The Solar Chromosphere. *Monthly Notices of the Royal Astronomical Society* 94, 14–35

1934

[1]The Radiative Equilibrium of Extended Stellar Atmospheres. *Monthly Notices of the Royal Astronomical Society* 94, 444–58

On the Hypothesis of the Radial Ejection of High-Speed Atoms for the Wolf-Rayet Stars and the Novae. *Monthly Notices of the Royal Astronomical Society* 94, 522–38

The Solar Chromosphere (Second Paper). *Monthly Notices of the Royal Astronomical Society* 94, 726–37

The Relation between the Chromosphere and the Prominences. *The Observatory* 57, 65–68

[1]The Physical State of Matter in the Interior of Stars. *The Observatory* 57, 93–99

The Radiative Equilibrium of Extended Stellar Atmospheres. *The Observatory* 57, 225–27

[1]Stellar Configurations with Degenerate Cores. *The Observatory* 57, 373–77

An Analysis of the Problems of the Stellar Atmospheres. *Astronomical Journal of the Soviet Union* 11, 550–96

The Stellar Coefficient of Absorption. *Zeitschrift für Astrophysik* 8, 167

1935

[1]The Highly Collapsed Configurations of a Stellar Mass (Second Paper). *Monthly Notices of the Royal Astronomical Society* 95, 207–25

[1]Stellar Configurations with Degenerate Cores. *Monthly Notices of the Royal Astronomical Society* 95, 226–60

[1]Relativistic Degeneracy (with C. Møller). *Monthly Notices of the Royal Astronomical Society* 95, 673–76

[1]Stellar Configurations with Degenerate Cores (Second Paper). *Monthly Notices of the Royal Astronomical Society* 95, 676–93

[1]The Radiative Equilibrium of the Outer Layers of a Star with Special Reference to the Blanketing Effect of the Reversing Layer. *Monthly Notices of the Royal Astronomical Society* 96, 21–42

[1]The Radiative Equilibrium of a Planetary Nebula. *Zeitschrift für Astrophysik* 9, 266–89

[1]The Nebulium Emission in Planetary Nebulae. *Zeitschrift für Astrophysik* 10, 36–39

Étude des Atmosphères Stellaires. *Memoirs de la Société Royale des Sciences de Liège* 20, 3–89

Stjernernes Struktur. *Nordisk Astronomisk Tidsskrift* 16, 37–44

[1]Production of Electron Pairs and the Theory of Stellar Structure (with L. Rosenfeld). *Nature* 135, 999

[1]On the Effective Temperatures of Extended Photospheres. *Proceedings of the Cambridge Philosophical Society* 31, 390–93

1936

[1]The Pressure in the Interior of a Star. *Monthly Notices of the Royal Astronomical Society* 96, 644–47

The Equilibrium of Stellar Envelopes and the Central Condensations of Stars. *Monthly Notices of the Royal Astronomical Society* 96, 647–60

The Profile of the Absorption Lines in Rotating Stars, Taking into Account the Variation of Ionization Due to Centrifugal Force (with P. Swings). *Monthly Notices of the Royal Astronomical Society* 96, 883–89

On the Distribution of the Absorbing Atoms in the Reversing Layers of Stars and the Formation of Blended Absorption Lines (with P. Swings). *Monthly Notices of the Royal Astronomical Society* 97, 24–37

[1]On the Maximum Possible Central Radiation Pressure in a Star of a Given Mass. *The Observatory* 59, 47–48

On Trumpler's Stars (with A. Beer). *The Observatory* 59, 168–70

1937

[1]The Pressure in the Interior of a Star. *The Astrophysical Journal* 85, 372–79

[1]The Opacity in the Interior of a Star. *The Astrophysical Journal* 86, 78–83

[1]Partially Degenerate Stellar Configurations. *The Astrophysical Journal* 86, 623–25

On a Class of Stellar Models. *Zeitschrift für Astrophysik* 14, 164–88

[1]The Cosmological Constants. *Nature* 139, 757–58

1938

A Method of Deriving the Constants of the Velocity Ellipsoid from the Observed Radial Speeds of the Stars (with W. M. Smart). *Monthly Notices of the Royal Astronomical Society* 98, 658–63

On a Generalization of Lindblad's Theory of Star-Streaming. *Monthly Notices of the Royal Astronomical Society* 98, 710–26

[1]Ionization and Recombination in the Theory of Stellar Absorption Lines and Nebular Luminosity. *The Astrophysical Journal* 87, 476–95

[1]An Integral Theorem on the Equilibrium of a Star. *The Astrophysical Journal* 87, 535–52

1939

[1]The Minimum Central Temperature of a Gaseous Star. *Monthly Notices of the Royal Astronomical Society* 99, 673–85

The Lane-Emden Function $\theta_{3.25}$. *The Astrophysical Journal* 89, 116–18

Review of *Stellar Dynamics*, by W. M. Smart. *The Astrophysical Journal* 89, 679–87

The Dynamics of Stellar Systems, I–VIII. *The Astrophysical Journal* 90, 1–154

[1]The Internal Constitution of the Stars. *Proceedings of the American Philosophical Society* 81, 153–87

1940

The Dynamics of Stellar Systems, IX–XIV. *The Astrophysical Journal* 92, 441–642

1941

The Time of Relaxation of Stellar Systems, I. *The Astrophysical Journal* 93, 285–304

The Time of Relaxation of Stellar Systems, II (with R. E. Williamson). *The Astrophysical Journal* 93, 305–22

The Time of Relaxation of Stellar Systems, III. *The Astrophysical Journal* 93, 323–36

[1]The White Dwarfs and Their Importance for Theories of Stellar Evolution. *Conférences du Collège de France, Colloque International d'Astrophysique,* III, White Dwarfs. 17–23 Juillet 1939 (Paris: Hermann & Company), 41–50

[3]A Statistical Theory of Stellar Encounters. *The Astrophysical Journal* 94, 511–25

[1]Stellar Models with Isothermal Cores (with L. R. Henrich). *The Astrophysical Journal* 94, 525–36

1942

[1]An Attempt to Interpret the Relative Abundances of the Elements and Their Isotopes (with L. R. Henrich. *The Astrophysical Journal* 95, 288–98

[3] The Statistics of the Gravitational Field Arising from a Random Distribution of Stars. I, The Speed of Fluctuations (with J. von Neumann). *The Astrophysical Journal* 95, 489–531

[1] A Note on the Perturbation Theory for Distorted Stellar Configurations (with W. Krogdahl). *The Astrophysical Journal* 96, 151–54

[1] On the Evolution of the Main-Sequence Stars (with M. Schönberg). *The Astrophysical Journal* 96, 161–72

1943

[3] The Statistics of the Gravitational Field Arising from a Random Distribution of Stars. II, The Speed of Fluctuations; Dynamical Friction; Spatial Co-Relations (with J. von Neumann). *The Astrophysical Journal* 97, 1–27

[3] Dynamical Friction. I, General Considerations: The Coefficient of Dynamical Friction. *The Astrophysical Journal* 97, 255–62

[3] Dynamical Friction. II, The Rate of Escape of Stars from Clusters and the Evidence for the Operation of Dynamical Friction. *The Astrophysical Journal* 97, 263–73

[3] Dynamical Friction. III, A More Exact Theory of the Rate of Escape of Stars from Clusters. *The Astrophysical Journal* 98, 54–80

On the Negative Hydrogen Ion and Its Absorption Coefficient (with M. K. Krogdahl). *The Astrophysical Journal* 98, 205–8

[3] Stochastic Problems in Physics and Astronomy. *Reviews of Modern Physics* 15, 1–89

[3] New Methods in Stellar Dynamics. *Annals of the New York Academy of Sciences* 45, 131–62

1944

[3] The Statistics of the Gravitational Field Arising from a Random Distribution of Stars. III, The Correlations in the Forces Acting at Two Points Separated by a Finite Distance. *The Astrophysical Journal* 99, 25–46

[3] The Statistics of the Gravitational Field Arising from a Random Distribution of Stars. IV, The Stochastic Variation of the Force Acting on a Star. *The Astrophysical Journal* 99, 47–53

[3] On the Stability of Binary Systems. *The Astrophysical Journal* 99, 54–58

On the Radiative Equilibrium of a Stellar Atmosphere. *The Astrophysical Journal* 99, 180–90

[3] Galactic Evidences for the Time-Scale of the Universe. *Science* 99, 133–36

[2] On the Radiative Equilibrium of a Stellar Atmosphere, II. *The Astrophysical Journal* 100, 76–86

On the Absorption Continuum of the Negative Oxygen Ion (with R. Wildt). *The Astrophysical Journal* 100, 87–93

On the Radiative Equilibrium of a Stellar Atmosphere, III. *The Astrophysical Journal* 100, 117–27

[2] Some Remarks on the Negative Hydrogen Ion and Its Absorption Coefficient. *The Astrophysical Journal* 100, 176–80

On the Radiative Equilibrium of a Stellar Atmosphere, IV (with C. U. Cesco and J. Sahade). *The Astrophysical Journal* 100, 355–59

The Negative Ions of Hydrogen and Oxygen in Stellar Atmospheres. *Reviews of Modern Physics* 16, 301–6

1945

[1] Ralph Howard Fowler, 1889–1944. *The Astrophysical Journal* 101, 1–5

[2] On the Radiative Equilibrium of a Stellar Atmosphere, V. *The Astrophysical Journal* 101, 95–107

On the Radiative Equilibrium of a Stellar Atmosphere, VI (with C.U. Cesco and J. Sahade). *The Astrophysical Journal* 101, 320–27

[2]On the Radiative Equilibrium of a Stellar Atmosphere, VII. *The Astrophysical Journal* 101, 328–47

[2]On the Radiative Equilibrium of a Stellar Atmosphere, VIII. *The Astrophysical Journal* 101, 348–55

Photographs of the Corona Taken during the Total Eclipse of the Sun on July 9, 1945, at Pine River, Manitoba, Canada (with W.A. Hiltner). *The Astrophysical Journal* 102, 135–36

[2]On the Continuous Absorption Coefficient of the Negative Hydrogen Ion. *The Astrophysical Journal* 102, 223–31

[2]On the Continuous Absorption Coefficient of the Negative Hydrogen Ion, II. *The Astrophysical Journal* 102, 395–401

[2]The Radiative Equilibrium of an Expanding Planetary Nebula. I, Radiation Pressure in Lyman$_{-\alpha}$. *The Astrophysical Journal* 102, 402–28

[2]The Formation of Absorption Lines in a Moving Atmosphere. *Reviews of Modern Physics* 17, 138–56

Reports on the Progress of Astronomy: Stellar Dynamics. *Monthly Notices of the Royal Astronomical Society* 105, 124–34

1946

The Motion of an Electron in the Hartree Field of a Hydrogen Atom (with F.H. Breen). *The Astrophysical Journal* 103, 41–70

On the Radiative Equilibrium of a Stellar Atmosphere, IX. *The Astrophysical Journal* 103, 165–92

[2]On the Radiative Equilibrium of a Stellar Atmosphere, X. *The Astrophysical Journal* 103, 351–70

[2]On the Radiative Equilibrium of a Stellar Atmosphere, XI. *The Astrophysical Journal* 104, 110–32

On a New Theory of Wëizsacker on the Origin of the Solar System. *Reviews of Modern Physics* 18, 94–102

[2]On the Radiative Equilibrium of a Stellar Atmosphere, XII. *The Astrophysical Journal* 104, 191–202

[2]On the Continuous Absorption Coefficient of the Negative Hydrogen Ion, III (with F.H. Breen). *The Astrophysical Journal* 104, 430–45

[2]The Continuous Spectrum of the Sun and the Stars (with G. Münch). *The Astrophysical Journal* 104, 446–57

[2]A New Type of Boundary-Value Problem in Hyperbolic Equations. *Proceedings of the Cambridge Philosophical Society* 42, 250–60

1947

[2]On the Radiative Equilibrium of a Stellar Atmosphere, XIII. *The Astrophysical Journal* 105, 151–63

[2]On the Radiative Equilibrium of a Stellar Atmosphere, XIV. *The Astrophysical Journal* 105, 164–203

[2]On the Radiative Equilibrium of a Stellar Atmosphere, XV. *The Astrophysical Journal* 105, 424–34

On the Radiative Equilibrium of a Stellar Atmosphere, XVI (with F.H. Breen). *The Astrophysical Journal* 105, 435–40

[2]On the Radiative Equilibrium of a Stellar Atmosphere, XVII. *The Astrophysical Journal* 105, 441–60

On the Radiative Equilibrium of a Stellar Atmosphere, XVIII (with F. H. Breen). *The Astrophysical Journal* 105, 461–70

On the Radiative Equilibrium of a Stellar Atmosphere, XIX (with F. H. Breen). *The Astrophysical Journal* 106, 143–44

[2]On the Radiative Equilibrium of a Stellar Atmosphere, XX. *The Astrophysical Journal* 106, 145–51

[2]On the Radiative Equilibrium of a Stellar Atmosphere, XXI. *The Astrophysical Journal* 106, 152–216

The Scientist. Pp. 159–79 in *The Works of the Mind,* ed. Robert B. Heywood. Chicago: University of Chicago Press

[2]The Transfer of Radiation in Stellar Atmospheres. *Bulletin of the American Mathematical Society* 53, 641–711

[2]The Story of Two Atoms. *The Scientific Monthly* 64, 313–21

James Hopwood Jeans. *Science* 105, 224–26

Solar Research and Theoretical Astrophysics. *Science* 106, 213–14

1948

[2]On the Radiative Equilibrium of a Stellar Atmosphere, XXII. *The Astrophysical Journal* 107, 48–72

[2]On the Radiative Equilibrium of a Stellar Atmosphere, XXII (concluded). *The Astrophysical Journal* 107, 188–215

On the Radiative Equilibrium of a Stellar Atmosphere, XXIII (with F. H. Breen), *The Astrophysical Journal* 107, 216–19

On the Radiative Equilibrium of a Stellar Atmosphere, XXIV. *The Astrophysical Journal* 108, 92–111

[2]The Softening of Radiation by Multiple Compton Scattering. *Proceedings of the Royal Society,* A, 192, 508–18

[3]On a Class of Probability Distributions. *Proceedings of the Cambridge Philosophical Society* 45, 219–24

1949

The Theory of Statistical and Isotropic Turbulence. *Physical Review* 75, 896–97

On the Decay of Isotropic Turbulence. *Physical Review* 75, 1454–55; erratum, *Physical Review,* 76, 158

[3]Brownian Motion, Dynamical Friction and Stellar Dynamics. *Reviews of Modern Physics* 21, 383–88

The Isothermal Function (with G. W. Wares). *The Astrophysical Journal* 109, 551–54

The Functions $G_{n,m}(\tau)$ and $G'_{n,m}(\tau)$ of Order 6 ($m = 6$ and $m \geqslant n$). *The Astrophysical Journal* 109, 555

[3]On Heisenberg's Elementary Theory of Turbulence. *Proceedings of the Royal Society,* A, 200, 20–33

[3]Turbulence—A Physical Theory of Astrophysical Interest (Henry Norris Russell Lecture). *The Astrophysical Journal* 110, 329–39

1950

[3]On the Integral Equation Governing the Distribution of the True and the Apparent Rotational Velocities of Stars (with G. Münch). *The Astrophysical Journal* 111, 142–56

[3]The Theory of Axisymmetric Turbulence. *Philosophical Transactions of the Royal Society* 242, 557–77

[3]The Decay of Axisymmetric Turbulence. *Proceedings of the Royal Society,* A, 203, 358–64

The Invariant Theory of Isotropic Turbulence in Magneto-Hydrodynamics. *Proceedings of the Royal Society,* A, 204, 435–49

[3]The Theory of the Fluctuations in Brightness of the Milky Way, I (with G. Münch). *The Astrophysical Journal* 112, 380–92

[3]The Theory of Fluctuations in Brightness of the Milky Way, II (with G. Münch). *The Astrophysical Journal* 112, 393–98

1951

[1]The Angular Distribution of the Radiation at the Interface of Two Adjoining Media. *Canadian Journal of Physics* 29, 14–20

Polarization of the Sunlit Sky (with D. D. Elbert). *Nature* 167, 51–55

[3]On Stellar Statistics (with G. Münch). *The Astrophysical Journal* 113, 150–65

[3]The Invariant Theory of Isotropic Turbulence in Magneto-Hydrodynamics. *Proceedings of the Royal Society*, A, 204, 435–39

[3]The Invariant Theory of Isotropic Turbulence in Magneto-Hydrodynamics, II. *Proceedings of the Royal Society*, A, 207, 301–6

[3]The Fluctuations of Density in Isotropic Turbulence. *Proceedings of the Royal Society*, A, 210, 18–25

[3]The Gravitational Instability of an Infinite Homogeneous Turbulent Medium. *Proceedings of the Royal Society*, A, 210, 26–29

[3]The Theory of the Fluctuations in Brightness of the Milky Way, III (with G. Münch). *The Astrophysical Journal* 114, 110–22

The Structure, the Composition, and the Source of Energy of the Stars. Pp. 598–674 in *Astrophysics—A Topical Symposium*. New York: McGraw-Hill.

[3]Some Aspects of the Statistical Theory of Turbulence. *Proceedings of the 4th Symposium of the American Mathematical Society: Fluid Dynamics*, 1–17. University of Maryland.

1952

[3]The Theory of the Fluctuations in Brightness of the Milky Way, IV (with G. Münch). *The Astrophysical Journal* 115, 94–102

[3]The Theory of the Fluctuations in Brightness of the Milky Way, V (with G. Münch). *The Astrophysical Journal* 115, 103–23

The X- and Y-Functions for Isotropic Scattering, I (with D. D. Elbert and A. Franklin). *The Astrophysical Journal* 115, 244–68

[3]On Turbulence Caused by Thermal Instability. *Philosophical Transactions of the Royal Society* 244, 357–84

On the Inhibition of Convection by a Magnetic Field. *Philosophical Magazine* 43, 501–32

Convection under Terrestrial and Astrophysical Conditions. *Publications of the Astronomical Society of the Pacific* 64, 98–104

The Thermal Instability of a Fluid Sphere Heated Within. *Philosophical Magazine* 43, 1317–29

[3]A Statistical Basis for the Theory of Stellar Scintillation. *Monthly Notices of the Royal Astronomical Society* 112, 475–83

1953

The Onset of Convection by Thermal Instability in Spherical Shells. *Philosophical Magazine* 44, 233–41; correction, *Philosophical Magazine* 44, 1129–30

The Stability of Viscous Flow between Rotating Cylinders in the Presence of a Magnetic Field. *Proceedings of the Royal Society*, A, 216, 293–309

The Instability of a Layer of Fluid Heated Below and Subject to Coriolis Forces. *Proceedings of the Royal Society*, A, 217, 306–27

Some Aspects of the Statistical Theory of Turbulence. *Proceedings of the Symposium on Applied Mathematics*, The American Mathematical Society, 4, 1–17

[3] Magnetic Fields in Spiral Arms (with E. Fermi). *The Astrophysical Journal* 118, 113–15

[3] Problems of Gravitational Stability in the Presence of a Magnetic Field (with E. Fermi). *The Astrophysical Journal* 118, 116–41

The Roots of $J_{-(1+1/2)}(\lambda\eta)J_{1+1/2}(\lambda) - J_{1+1/2}(\lambda\eta)J_{-(1+1/2)}(\lambda) = 0$ (with D. Elbert). *Proceedings of the Cambridge Philosophical Society* 49, 446–48

[2] Shift of the 1^1S State of Helium (with D. Elbert and G. Herzberg). *The Physical Review* 91, 1172–73

Problems of Stability in Hydrodynamics and Hydromagnetics (George Darwin Lecture). Monthly Notices of the Royal Astronomical Society, 113, 667–78

1954

The Gravitational Instability of an Infinite Homogeneous Medium when Coriolis Force is Acting and a Magnetic Field is Present. *The Astrophysical Journal* 119, 7–9

[3] On the Pulsation of a Star in which there is a Prevalent Magnetic Field (with D. N. Limber). *The Astrophysical Journal* 119, 10–13

[4] The Stability of Viscous Flow between Rotating Cylinders in the Presence of a Radial Temperature Gradient. *Journal of Rational Mechanics and Analysis* 3, 181–207

The Roots of $Y_n(\lambda\eta)J_n(\lambda) - J_n(\lambda\eta)Y_n(\lambda) = 0$ (with D. Elbert). *Proceedings of the Cambridge Philosophical Society* 50, 266–68

The Instability of a Layer of Fluid Heated Below and Subject to the Simultaneous Action of a Magnetic Field and Rotation. *Proceedings of the Royal Society,* A, 225, 173–84

[4] On the Characteristic Value Problems in High Order Differential Equations which Arise in Studies on Hydrodynamic and Hydromagnetic Stability. *American Mathematical Monthly* 61 (Suppl.), 32–45

The Stability of Viscous Flow between Rotating Cylinders. *Mathematika* 1, 5–13

On the Inhibition of Convection by a Magnetic Field, II. *Philosophical Magazine* 45, 1177–91

Examples of the Instability of Fluid Motion in the Presence of a Magnetic Field. *Proceedings of the 5th Symposium on Applied Mathematics,* 19–27

[2] The Illumination and Polarization of the Sunlit Sky on Rayleigh Scattering (with D. Elbert). *Transactions of the American Philosophical Society* 44, 643–728

1955

The Character of the Equilibrium of an Incompressible Heavy Viscous Fluid of Variable Density. *Proceedings of the Cambridge Philosophical Society* 51, 162–78

[4] The Character of the Equilibrium of an Incompressible Fluid Sphere of Variable Density and Viscosity Subject to Radial Acceleration. *Quarterly Journal of Mechanics and Applied Mathematics* 8, 1–21

The Instability of a Layer of Fluid Heated Below and Subject to Coriolis Forces, II (with D. Elbert). *Proceedings of the Royal Society,* A, 231, 198–210

[2] Energies of the Ground States of He, Li^+, and O^{6+} (with G. Herzberg). *Physical Review* 98, 1050–54

The Gravitational Instability of an Infinite Homogeneous Medium when a Coriolis Acceleration is Acting. Pp. 344–47 in *Vistas in Astronomy,* vol. I. New York: Pergamon Press.

1956

[3] On Force-Free Magnetic Fields. *Proceedings of the National Academy of Sciences* 42, 1–5

[3] The Equilibrium of Magnetic Stars (with K. H. Prendergast). *Proceedings of the National Academy of Sciences* 42, 5–9

[3] On Cowling's Theorem on the Impossibility of Self-Maintained Axisymmetric Homoge-

nous Dynamos (with G. E. Backus). *Proceedings of the National Academy of Sciences* 42, 105–09

On the Stability of the Simplest Solution of the Equations of Hydromagnetics. *Proceedings of the National Academy of Sciences* 42, 273–76

[3]Axisymmetric Magnetic Fields and Fluid Motions. *The Astrophysical Journal* 124, 232–43

[3]Effect of Internal Motions on the Decay of a Magnetic Field in a Fluid Conductor. *The Astrophysical Journal* 124, 244–65

The Instability of a Layer of Fluid Heated Below and Subject to the Simultaneous Action of a Magnetic Field and Rotation, II. *Proceedings of the Royal Society,* A, 237, 476–84

[3]Hydromagnetic Oscillations of a Fluid Sphere with Internal Motions. *The Astrophysical Journal* 124, 571–79

1957

[3]On Cosmic Magnetic Fields. *Proceedings of the National Academy of Sciences* 43, 24–27

[4]Thermal Convection (Rumford Medal Lecture). *Proceedings of the Academy of Arts and Sciences* 86, 323–39

On the Expansion of Functions which Satisfy Four Boundary Conditions (with W. H. Reid). *Proceedings of the National Academy of Sciences* 43, 521–27

[3]On Force-Free Magnetic Fields (with P. C. Kendall). *The Astrophysical Journal* 126, 457–60

[4]Properties of an Ionized Gas of Low Density in a Magnetic Field, III (with A. N. Kaufman and K. M. Watson). *Annals of Physics* 2, 435–70

On the Expansion of Functions Satisfying Four Boundary Conditions. *Mathematika* 4, 140–45

The Thermal Instability of a Rotating Fluid Sphere Heated Within. *Philosophical Magazine* 2, 845–58

The Thermal Instability of a Rotating Fluid Sphere Heated Within, II. *Philosophical Magazine* 2, 1282–84

1958

[3]On Force-Free Magnetic Fields (with L. Woltjer). *Proceedings of the National Academy of Sciences* 44, 285–89

[4]The Stability of the Pinch (with A. N. Kaufman and K. M. Watson. *Proceedings of the Royal Society,* A, 245, 435–55

The Stability of Viscous Flow Between Rotating Cylinders. *Proceedings of the Royal Society,* A, 246, 301–11

[2]On the Continuous Absorption Coefficient of the Negative Hydrogen Ion, IV. *The Astrophysical Journal* 128, 114–23

[3]On the Equilibrium Configurations of an Incompressible Fluid with Axisymmetric Motions and Magnetic Fields. *Proceedings of the National Academy of Sciences* 44, 842–47

[2]On the Diffuse Reflection of a Pencil of Radiation by a Plane-Parallel Atmosphere. *Proceedings of the National Academy of Sciences* 44, 933–40

[4]Adiabatic Invariants in the Motions of Charged Particles. Pp. 3–22 in *The Plasma in a Magnetic Field: A Symposium on Magnetohydrodyanmics.* Stanford: Stanford University Press.

[4]Properties of an Ionized Gas of Low Density in a Magnetic Field, IV (with A. N. Kaufman and K. M. Watson). *Annals of Physics* 5, 1–25

On Orthogonal Functions which Satisfy Four Boundary Conditions. III, Tables for Use in Fourier-Bessel-Type Expansions (with D. Elbert). *The Astrophysical Journal Supplement* 3, 453–58

[4]Variational Methods in Hydrodynamics. *Proceedings of the Eighth Symposium in Applied Mathematics* 8, 139–41

[2]On the Continuous Absorption Coefficient of the Negative Hydrogen Ion, V (with D. Elbert). *The Astrophysical Journal* 128, 633–35

1959

The Oscillations of a Viscous Liquid Globe. *Proceedings of the London Mathematical Society* 9, 141–49

[4]The Thermodynamics of Thermal Instability in Liquids. Pp. 103–14 in *Max-Planck-Festschrift 1958*. Berlin: Veb Deutscher Verlag der Wissenschaften.

1960

The Hydrodynamic Stability of Inviscid Flow between Coaxial Cylinders. *Proceedings of the National Academy of Sciences* 46, 137–41

The Hydrodynamic Stability of Viscid Flow between Coaxial Cylinders. *Proceedings of the National Academy of Sciences* 46, 141–43

The Stability of Non-Dissipative Couette Flow in Hydromagnetics. *Proceedings of the National Academy of Sciences* 46, 253–57

[4]The Virial Theorem in Hydromagnetics. *Journal of Mathematical Analysis and Applications* 1, 240–52

The Stability of Inviscid Flow between Rotating Cylinders. *Journal of the Indian Mathematical Society* 24, 211–21

1961

[5]The Geodesics in Gödel's Universe (with J. P. Wright). *Proceedings of the National Academy of Sciences* 48, 341–47

[2]Diffuse Reflection by a Semi-Infinite Atmosphere (with H. G. Horak). *The Astrophysical Journal* 134, 45–56

[4]Adjoint Differential Systems in the Theory of Hydrodynamic Stability. *Journal of Mathematics and Mechanics* 10, 683–90

The Stability of Viscous Flow between Rotating Cylinders in the Presence of a Magnetic Field, II (with D. Elbert). *Proceedings of the Royal Society,* A, 262, 443–54

The Stability of Viscous Flow in a Curved Channel in the Presence of a Magnetic Field (with D. Elbert and N. Lebovitz). *Proceedings of the Royal Society,* A, 264, 155–64

[4]A Theorem on Rotating Polytropes. *The Astrophysical Journal* 134, 662–64

1962

[4]On Super-Potentials in the Theory of Newtonian Gravitation (with N. Lebovitz). *The Astrophysical Journal* 135, 238–47

[4]On the Oscillations and the Stability of Rotating Gaseous Masses (with N. Lebovitz). *The Astrophysical Journal* 135, 248–60

An Interpretation of Double Periods in β Canis Majoris Stars (with N. Lebovitz). *The Astrophysical Journal* 135, 305–6

The Stability of Spiral Flow between Rotating Cylinders. *Proceedings of the Royal Society,* A, 265, 188–95. Appendix (with L. Lee), 196–97

[4]The Stability of Viscous Flow between Rotating Cylinders, II (with D. Elbert). *Proceedings of the Royal Society,* A, 268, 145–52

[4]On Superpotentials in the Theory of Newtonian Gravitation. II, Tensors of Higher Rank (with N. Lebovitz). *The Astrophysical Journal* 136, 1032–36

[4]The Potentials and the Superpotentials of Homogeneous Ellipsoids (with N. Lebovitz). *The Astrophysical Journal* 136, 1037–47

On the Point of Bifurcation along the Sequence of the Jacobi Ellipsoids. *The Astrophysical Journal* 136, 1048–68

[4]On the Oscillations and the Stability of Rotating Gaseous Masses. II, The Homogeneous, Compressible Model (with N. Lebovitz). *The Astrophysical Journal* 136, 1069–81

[4]On the Oscillations and the Stability of Rotating Gaseous Masses. III, The Distorted Polytropes (with N. Lebovitz). *The Astrophysical Journal* 136, 1082–1104

[4]On the Occurence of Multiple Frequencies and Beats in the β Canis Majoris Stars (with N. Lebovitz). *The Astrophysical Journal* 136, 1105–7

An Approach to the Theory of Equilibrium and the Stability of Rotating Masses via the Virial Theorem and its Extensions. Pp. 9–14 in *Proceedings of the Fourth U. S. National Congress on Applied Mechanics.*

1963

On the Stability of the Jacobi Ellipsoids (with N. Lebovitz). *The Astrophysical Journal* 137, 1142–61

On the Oscillations of the Maclaurin Spheroid Belonging to the Third Harmonics (with N. Lebovitz). *The Astrophysical Journal* 137, 1162–71

The Equilibrium and the Stability of the Jeans Spheroids (with N. Lebovitz). *The Astrophysical Journal* 137, 1172–84

[4]The Points of Bifurcation along the Maclaurin, the Jacobi, and the Jeans Sequences. *The Astrophysical Journal* 137, 1185–1202

[4]Non-Radial Oscillations and Convective Instability of Gaseous Masses (with N. Lebovitz). *The Astrophysical Journal* 138, 185–99

[5]The Virial Theorem in General Relativity in the Post-Newtonian Approximation (with G. Contopoulos). *Proceedings of the National Academy of Sciences* 49, 608–13

[4]The Ellipticity of a Slowly Rotating Configuration (with P. H. Roberts). *The Astrophysical Journal* 138, 801–8

A General Variational Principle Governing the Radial and the Non-Radial Oscillations of Gaseous Masses. *The Astrophysical Journal* 138, 896–97

[4]The Equilibrium and the Stability of the Roche Ellipsoids. *The Astrophysical Journal* 138, 1182–1213

1964

[4]A General Variational Principle Governing the Radial and the Non-Radial Oscillations of Gaseous Masses. *The Astrophysical Journal* 139, 664–74

Otto Struve, 1897–1963. *The Astrophysical Journal* 139, 423

The Case for Astronomy. *Proceedings of the American Philosophical Society* 108, 1–6

[5]Dynamical Instability of Gaseous Masses Approaching the Schwarzschild Limit in General Relativity. *Physical Review Letters* 12, 114–16; erratum, *Physical Review Letters* 12, 437–8

[5]The Dynamical Instability of the White-Dwarf Configurations Approaching the Limiting Mass (with R. F. Tooper). *The Astrophysical Journal* 139, 1396–98

[5]The Dynamical Instability of Gaseous Masses Approaching the Schwarzschild Limit in General Relativity. *The Astrophysical Journal* 140, 417–33

The Equilibrium and the Stability of the Darwin Ellipsoids. *The Astrophysical Journal* 140, 599–620

[4]Non-Radial Oscillations of Gaseous Masses (with N. Lebovitz). *The Astrophysical Journal* 140, 1517–28

[4]The Virial Equations of the Various Orders (chapter 1 in *The Higher Order Virial Equations and Their Applications to the Equilibrium and Stability of Rotating Configurations*). Lectures in Theoretical Physics, vol. 6. Boulder: University of Colorado Press

On the Ellipsoidal Figures of Equilibrium of Homogeneous Masses (with N. Lebovitz). *Astrophysica Norvegica* 9, 323–32

1965

The Equilibrium and the Stability of the Dedekind Ellipsoids. *The Astrophysical Journal* 141, 1043–55

Post-Newtonian Equations of Hydrodynamics and the Stability of Gaseous Masses in General Relativity. *Physical Review Letters* 14, 241–44

[4]The Stability of a Rotating Liquid Drop. *Proceedings of the Royal Society,* A, 286, 1–26

[4]The Equilibrium and the Stability of the Riemann Ellipsoids, I. *The Astrophysical Journal* 142, 890–921

[5]The Post-Newtonian Equations of Hydrodynamics in General Relativity. *The Astrophysical Journal* 142, 1488–1512

[5]The Post-Newtonian Effects of General Relativity on the Equilibrium of Uniformly Rotating Bodies. I, The Maclaurin Spheroids and the Virial Theorem. *The Astrophysical Journal* 142, 1513–18

[5]The Stability of Gaseous Masses for Radial and Non-Radial Oscillations in the Post-Newtonian Approximation of General Relativity. *The Astrophysical Journal* 142, 1519–40

1966

[4]The Equilibrium and the Stability of the Riemann Ellipsoids. II. *The Astrophysical Journal* 145, 842–77

1967

[5]The Post-Newtonian Effects of General Relativity on the Equilibrium of Uniformly Rotating Bodies. II, The Deformed Figures of the Maclaurin Spheroids. *The Astrophysical Journal* 147, 334–52

[5]Virial Relations for Uniformly Rotating Fluid Masses in General Relativity. *The Astrophysical Journal* 147, 383–84

[5]On a Post-Galilean Transformation Appropriate to the Post-Newtonian Theory of Einstein, Infeld, and Hoffman (with G. Contopoulos). *Proceedings of the Royal Society,* A, 298, 123–41

[5]The Post-Newtonian Effects of General Relativity on the Equilibrium of Uniformly Rotating Bodies. III, The Deformed Figures of the Jacobi Ellipsoids. *The Astrophysical Journal* 148, 621–44

[5]The Post-Newtonian Effects of General Relativity on the Equilibrium of Uniformly Rotating Bodies. IV, The Roche Model. *The Astrophysical Journal* 148, 645–49

Ellipsoidal Figures of Equilibrium—An Historical Account. *Communications on Pure and Applied Mathematics* 20, 251–65

1968

[4]The Pulsations and the Dynamical Stability of Gaseous Masses in Uniform Rotation (with N. R. Lebovitz). *The Astrophysical Journal* 152, 267–91

[4]The Virial Equations of the Fourth Order. *The Astrophysical Journal* 152, 293–304

[4]A Tensor Virial-Equation for Stellar Dynamics (with E. P. Lee). *Monthly Notices of the Royal Astronomical Society* 139, 135–39

Astronomy in Science and Human Culture (Jawaharlal Nehru Memorial Lecture). New Delhi: Indraprastha Press, 22 pages

1969

The Richtmyer Memorial Lecture—Some Historical Notes. *American Journal of Physics* 37, 577–84

[5]Conservation Laws in General Relativity and in the Post-Newtonian Approximations. *The Astrophysical Journal* 158, 45–54

[5]The Second Post-Newtonian Equations of Hydrodynamics in General Relativity (with Y. Nutku). *The Astrophysical Journal* 158, 55–79

The Effect of Viscous Dissipation on the Stability of the Roche Ellipsoids. *Publications of the Ramanujan Institute* 1, 213–22

1970

[5]Solutions of Two Problems in the Theory of Gravitational Radiation. *Physical Review Letters* 24, 611–15; erratum, *Physical Review Letters* 24, 762

[5]The $2\frac{1}{2}$ Post-Newtonian Equations of Hydrodynamics and Radiation Reaction in General Relativity (with F. P. Esposito). *The Astrophysical Journal* 160, 153–79

[5]Post-Newtonian Methods and Conservation Laws. Pp. 81–108 in *Relativity,* ed. M. Carmeli, S. I. Fickler, and L. Witten. New York: Plenum Press

The Oscillations of a Rotating Gaseous Mass in the Post-Newtonian Approximation to General Relativity. Pp. 182–95 in *Quanta,* ed. P. G. O. Freund, C. J. Goebel, and Y. Nambu. Chicago: University of Chicago Press

The Instability of Congruent Darwin Ellipsoids, II. *The Astrophysical Journal* 160, 1043–48

[5]The Effect of Gravitational Radiation on the Secular Stability of the Maclaurin Spheroid. *The Astrophysical Journal* 161, 561–69

The Evolution of the Jacobi Ellipsoid by Gravitational Radiation. *The Astrophysical Journal* 161, 571–78

1971

[5]Criterion for the Instability of a Uniformly Rotating Configuration in General Relativity (with J. L. Friedman). *Physical Review Letters* 26, 1047–50

[5]The Post-Newtonian Effects of General Relativity on the Equilibrium of Uniformly Rotating Bodies. V, the Deformed Figures of the Maclaurin Spheroids (Continued). *The Astrophysical Journal* 167, 447–53

[5]The Post-Newtonian Effects of General Relativity on the Equilibrium of Uniformly Rotating Bodies. VI, The Deformed Figures of the Jacobi Ellipsoids (Continued). *The Astrophysical Journal* 167, 455–63

[4]Some Elementary Applications of the Virial Theorem to Stellar Dynamics (with D. Elbert). *Monthly Notices of the Royal Astronomical Society* 155, 435–47

1972

[5]On the "Derivation" of Einstein's Field Equations. *American Journal of Physics* 40, 224–34

[5]A Limiting Case of Relativistic Equilibrium (in honor of J. L. Synge). Pp. 185–99 in *General Relativity,* ed. L. O'Raifeartaigh. Oxford: Clarendon Press

[5]On the Stability of Axisymmetric Systems to Axisymmetric Perturbations in General Relativity. I, The Equations Governing Nonstationary, Stationary, and Perturbed Systems (with J. L. Friedman). *The Astrophysical Journal* 175, 379–405

[5]On the Stability of Axisymmetric Systems to Axisymmetric Perturbations in General Relativity. II, A Criterion for the Onset of Instability in Uniformly Rotating Configurations and the Frequency of the Fundamental Mode in Case of Slow Rotation (with J. L. Friedman). *The Astrophysical Journal* 176, 745–68

[5]On the Stability of Axisymmetric Systems to Axisymmetric Perturbations in General Relativity. III, Vacuum Metrics and Carter's Theorem (with J. L. Friedman). *The Astrophysical Journal* 177, 745–56

[5][Stability of Stellar Configurations in General Relativity] Proceedings at Meeting of the Royal Astronomical Society. *The Observatory* 92, 116–20

[5]The Increasing Role of General Relativity in Astronomy (Halley Lecture). *The Observatory* 92, 160–74

1973

[5]On the Stability of Axisymmetric Systems to Axisymmetric Perturbations in General Relativity. IV, Allowance for Gravitational Radiation in an Odd-Parity Mode (with J. L. Friedman). *The Astrophysical Journal* 181, 481–95

[5]On a Criterion for the Occurrence of a Dedekind-like Point of Bifurcation along a Sequence of Axisymmetric Systems. I, Relativistic Theory of Uniformly Rotating Configurations (with J. L. Friedman). *The Astrophysical Journal* 185, 1–18

[5]On a Criterion for the Occurrence of a Dedekind-like Point of Bifurcation along a Sequence of Axisymmetric Systems. II, Newtonian Theory for Differentially Rotating Configurations (with N. R. Lebovitz). *The Astrophysical Journal* 183, 19–30

P. A. M. Dirac on His Seventieth Birthday. *Contemporary Physics* 13, 389–94

A Chapter in the Astrophysicist's View of the Universe. Pp. 34–44 in *The Physicist's Conception of Nature*, ed. J. Mehra. Dordrecht, Holland: D. Reidel

1974

[5]On a Criterion for the Onset of Dynamical Instability by a Non-Axisymmetric Mode of Oscillation along a Sequence of Differentially Rotating Configurations. *The Astrophysical Journal* 187, 169–74

The Black Hole in Astrophysics: The Origin of the Concept and Its Role. *Contemporary Physics* 14, 1–24

[5]On the Slowly Rotating Homogeneous Masses in General Relativity (with J.C. Miller). Monthly Notices of the Royal Astronomical Society 167, 63–79

[5]The Stability of Relativistic Systems. Pp. 63–81 in *Gravitational Radiation and Gravitational Collapse*, ed. C. DeWitt-Morette. Dordrecht, Holland: D. Reidel

[5]The Stability of Stellar Masses in General Relativity. Pp. 162–65 in *Proceedings of the First European Astronomical Meeting, Athens*, vol. 3. New York: Springer-Verlag

[5]The Deformed Figures of the Dedekind Ellipsoids in the Post-Newtonian Approximation to General Relativity (with D. Elbert). *The Astrophysical Journal* 192, 731–46. Corrections and Amplifications to this paper in *The Astrophysical Journal* 220 (1978): 303–11, are incorporated in the present version

Development of General Relativity. *Nature* 252, 15–17

1975

[6]On the Equations Governing the Perturbations of the Schwarzschild Black Hole. *Proceedings of the Royal Society*, A, 343, 289–98

[6]The Quasi-Normal Modes of the Schwarzschild Black Hole (with S. Detweiler). *Proceedings of the Royal Society*, A, 344, 441–52

[6]On the Equations Governing the Axisymmetric Perturbations of the Kerr Black Hole (with S. Detweiler). *Proceedings of the Royal Society*, A, 345, 145–67

On Coupled Second-Harmonic Oscillations of the Congruent Darwin Ellipsoids. *The Astrophysical Journal* 202, 809–14

1976

Verifying the Theory of Relativity. *Notes and Records of the Royal Society* 30, 249–60

[6]On a Transformation of Teukolsky's Equation and the Electromagnetic Perturbations of the Kerr Black Hole. *Proceedings of the Royal Society*, A, 348, 39–55

[6]The Solution of Maxwell's Equations in Kerr Geometry. *Proceedings of the Royal Society*, A, 349, 1–8

[6]The Solution of Dirac's Equation in Kerr Geometry. *Proceedings of the Royal Society*, A, 349, 571–75

On the Equations Governing the Gravitational Perturbations of the Kerr Black Hole (with S. Detweiler). *Proceedings of the Royal Society*, A, 350, 165–74

1977

On the Reflexion and Transmission of Neutrino Waves by a Kerr Black Hole (with S. Detweiler). *Proceedings of the Royal Society*, A, 352, 325–38

1978

[6]The Kerr Metric and Stationary Axisymmetric Gravitational Fields. *Proceedings of the Royal Society*, A, 358, 405–20

[6]The Gravitational Perturbations of the Kerr Black Hole. I, The Perturbations in the Quantities which Vanish in the Stationary State. *Proceedings of the Royal Society*, A, 358, 421–39

[6]The Gravitational Perturbations of the Kerr Black Hole. II, The Perturbations in the Quantities which are Finite in the Stationary State. *Proceedings of the Royal Society*, A, 358, 441–65

[5]The Deformed Figures of the Dedekind Ellipsoids in the Post-Newtonian Approximation to General Relativity: Corrections and Amplifications (with D. Elbert). *The Astrophysical Journal* 220, 303–13

Why Are the Stars As They Are? Pp. 1–14 in *Physics and Astrophysics of Neutron Stars and Black Holes*, ed. R. Giacconi and R. Ruffini. Holland: North-Holland Publishing Co.

On the Linear Perturbations of the Schwarzschild and the Kerr Geometries. Pp. 528–38 in *Physics and Astrophysics of Neutron Stars and Black Holes*, ed. R. Giacconi and R. Ruffini. Holland: North-Holland Publishing Co.

1979

[6]The Gravitational Perturbations of the Kerr Black Hole. III, Further Amplifications. *Proceedings of the Royal Society*, A, 365, 425–51

[6]On the Equations Governing the Perturbations of the Reissner-Nordström Black Hole. *Proceedings of the Royal Society*, A, 365, 453–65

Einstein and General Relativity: Historical Perspectives (1978 Oppenheimer Memorial Lecture). *American Journal of Physics* 47, 212–17

[6]On the Metric Perturbations of the Reissner-Nordström Black Hole (with B. C. Xanthopoulos). *Proceedings of the Royal Society*, A, 367, 1–14

An Introduction to the Theory of the Kerr Metric and Its Perturbations. Pp. 370–453 in *General Relativity—An Einstein Centenary Survey*, ed. W. Israel and S. Hawking. Cambridge: Cambridge University Press

Beauty and the Quest for Beauty in Science. *Physics Today* 32, 25–30

[5]Einstein's General Theory of Relativity and Cosmology. In *The Great Ideas Today*, 90–138. Encyclopaedia Britannica

C. T. Rajagopal (1903–78) (with A. Weil). *Nature* 279, 358

1980

[6]One One-Dimensional Potential Barriers Having Equal Reflexion and Transmission Coefficients. *Proceedings of the Royal Society*, A, 369, 425–33

Edward Arthur Milne: His Part in the Development of Modern Astrophysics (The 1979 Milne Lecture). *The Quarterly Journal of the Royal Astronomical Society* 21, 93–107

The Role of General Relativity in Astronomy: Retrospect and Prospect. Pp. 45–61 in *Highlights of Astronomy*, vol. 5, ed. P. A. Wayman. Dordrecht, Holland: D. Reidel

[5]The General Theory of Relativity: The First Thirty Years. *Contemporary Physics* 21, 429–49

[6]The Gravitational Perturbations of the Kerr Black Hole. IV, The Completion of the Solution. *Proceedings of the Royal Society,* A, 372, 475–84

1981

Review: *Oort and the Universe. Science* 211, 273

1982

[6]On Crossing the Cauchy Horizon of a Reissner-Nordström Black Hole (with J. B. Hartle). *Proceedings of the Royal Society,* A, 384, 301–15

[6]On the Potential Barriers Surrounding the Schwarzschild Black-Hole. Pp. 120–46 in *Spacetime and Geometry: The Alfred Schild Lectures,* ed. R.A. Matzner and L.C. Shepley. Austin: University of Texas Press

1984

[5]On the Onset of Relativistic Instability in Highly Centrally Condensed Stars (with N. R. Lebovitz). *Monthly Notices of the Royal Astronomical Society* 207, 13P–16P

[6]On Algebraically Special Perturbations of Black Holes. *Proceedings of the Royal Society,* A, 392, 1–13

[6]On Stars, Their Evolution and Their Stability (Nobel Lecture). Stockholm: The Nobel Foundation, 1984, 58–80

The General Theory of Relativity: Why "It is Probably the Most Beautiful of all Existing Theories." *Journal of Astrophysics and Astronomy* 5, 3–11

[6]On the Nutku-Halil Solution for Colliding Impulsive Gravitational Waves (with V. Ferrari), *Proceedings of the Royal Society,* A, 396, 55–74

1985

[6]On Colliding Waves in the Einstein-Maxwell Theory (with B. C. Xanthopoulos). *Proceedings of the Royal Society,* A, 398, 223–59

[6]On the Collision of Impulsive Gravitational Waves when Coupled with Fluid Motions (with B. C. Xanthopoulos). *Proceedings of the Royal Society,* A, 402, 37–65

The Pursuit of Science: Its Motivations. *Current Science* (India) 54, 161–69

[6]Some Exact Solutions of Gravitational Waves when Coupled with Fluid Motions (with B. C. Xanthopoulos). *Proceedings of the Royal Society,* A, 402, 205–24

1986

[6]On the Collision of Impulsive Gravitational Waves when Coupled with Null Dust (with B. C. Xanthopoulos). *Proceedings of the Royal Society,* A, 403, 189–98

[6]A New Type of Singularity Created by Colliding Gravitational Waves (with B. C. Xanthopoulos). *Proceedings of the Royal Society,* A, 408, 175–208

[6]Cylindrical Waves in General Relativity. *Proceedings of the Royal Society,* A, 408, 209–32

The Aesthetic Base of the General Theory of Relativity (Karl Schwarzschild Lecture). *Mitteilungen der Astronomischen Gesellschaft* 67, Hamburg, 19–49

[3]Marian Smoluchowski as the Founder of the Physics of Stochastic Phenomena. Pp. 21–28 in *Polish Men of Science: Marian Smoluchowski: His Life and Scientific Work,* ed. R. S. Ingarden. Warsaw: Polish Scientific Publishers

1987

[6]On Colliding Waves that Develop Time-like Singularities: A New Class of Solutions of the Einstein-Maxwell Equations (with B.C. Xanthopoulos). *Proceedings of the Royal Society,* A, 410, 311–36

[6]On the Dispersion of Cylindrical Impulsive Gravitational Waves (with V. Ferrari). *Proceedings of the Royal Society,* A, 412, 75–91

[6]The Effect of Sources on Horizons that May Develop when Plane Gravitational Waves Collide (with B. C. Xanthopoulos). *Proceedings of the Royal Society,* A, 414, 1–30

1988

[6]On Weyl's Solution for Space-Times with Two Commuting Killing Fields. *Proceedings of the Royal Society,* A, 415, 329–45

[6]A Perturbation Analysis of the Bell-Szekeres Space-Time (with B. C. Xanthopoulos). *Proceedings of the Royal Society,* A, 420, 93–123

Massless Particles from a Perfect Fluid. *Nature* 333, 506

[5]A Commentary on Dirac's Views on "The Excellence of General Relativity." Pp. 49–56 in *Festi-Val: Festschrift for Val Telegdi,* ed. K. Winter. North-Holland: Elsevier

To Victor Ambartsumian on His 80th Birthday. *Astrofisika* 29 (Russian: in English), 7–8

1989

[6]The Two-Centre Problem in General Relativity: The Scattering of Radiation by Two Extreme Reissner-Nordström Black-Holes. *Proceedings of the Royal Society,* A, 421, 227–58

[6]A one-to-one Correspondence between the Static Einstein-Maxwell and Stationary Einstein-Vacuum Space-Times. *Proceedings of the Royal Society,* A, 423, 379–86

[6]Two Black Holes Attached to Strings (with B. C. Xanthopoulos). *Proceedings of the Royal Society,* A, 423, 387–400

The Perception of Beauty and the Pursuit of Science. *Bulletin of the American Academy of Arts and Sciences* 43, 14–29

1990

[6]The Flux Integral for Axisymmetric Perturbations of Static Space-Times (with V. Ferrari). *Proceedings of the Royal Society,* A, 428, 325–49

Science and Scientific Attitudes. *Nature* 344, 285–86

[6]The Teukolsky-Starobinsky Constant for Arbitrary Spin. *Proceedings of the Royal Society,* A, 430, 433–38

[6]How One May Explore the Physical Content of the General Theory of Relativity. Proceedings of the Yale Symposium in Honor of the 150th anniversary of the birth of J. Willard Gibbs, The American Mathematical Society and the American Physical Society, 227–51

BOOKS

An Introduction to the Study of Stellar Structure. Chicago: University of Chicago Press, 1939. Repr. New York: Dover, 1958, 1967. Translations in Japanese (Tokyo: Kodansha Press, Ltd., 1972) and Russian have appeared

Principles of Stellar Dynamics. Chicago: University of Chicago Press, 1942. Repr. New York: Dover, 1960

Radiative Transfer. Oxford: Clarendon Press, 1950. Repr. New York: Dover, 1960. A Russian translation exists

Plasma Physics: Notes Compiled by S. K. Trehan from a Course Given by S. Chandrasekhar at the University of Chicago. Chicago: The University of Chicago Press, 1960. Repr. 1962, 1975

Hydrodynamic and Hydromagnetic Stability. Oxford: Clarendon Press, 1961. Repr. New York: Dover, 1970, 1981. A Russian translation exists

Ellipsoidal Figures of Equilibrium. New Haven: Yale University Press, 1969. Repr. New York: Dover, 1987. A Russian translation exists

The Mathematical Theory of Black Holes. Oxford: Clarendon Press, 1983. Repr. in Russian (Moscow: Mer Press), 1986

Eddington: The Most Distinguished Astrophysicist of His Time. Cambridge: Cambridge University Press, 1983

Truth and Beauty: Aesthetics and Motivations in Science. Chicago: University of Chicago Press, 1987

Selected Papers (six volumes). Chicago: University of Chicago Press, 1989–91

Acknowledgments

The author and publisher are grateful to the following societies and publishers for their permission to reprint in this volume papers that originally appeared in print under their auspicies.

American Mathematical Society
The American Physical Society
The Royal Society
The University of Texas Press

Corrigenda

Because of a renumbering of the later papers in volume 5 after Professor Thorne had written his foreword, the following references to paper numbers have to be changed as shown below:

Page	*Paragraph*	*Line*		
xvi	3	4	*Replace* 17–22	*by* 18–24
xviii	3	8	17, 22	*by* 18, 24
	4	3	31, 32	*by* 33, 34
		6	23, 24, 25, 27	*by* 25, 26, 27, 29
		9	24	*by* 27
		8 (from bottom)	26	*by* 38
		5 (from bottom)	28	*by* 30
xix		2	29, 30	*by* 31, 32
	2	3	33	*by* 35
		4	34	*by* 36
		6	36, 37	*by* 38, 39
	3	5	35, 38	*by* 37, 40

In volume 1, page xi, in paragraph three, instead of Robert Mullikin, read T. W. Mullikin.